ABOUT THE AUTHORS

CHARLOTTE PRATT received a B.S. in biology from the University of Notre Dame and a Ph.D. in biochemistry from Duke University. She is a protein chemist who has conducted research in blood coagulation and inflammation and was a member of the Center for Thrombosis and Hemostasis at the University of North Carolina at Chapel Hill. She is currently affiliated with Seattle Pacific University, where she has taught biochemistry. Her interests include molecular evolution, enzyme action, and the relationship between metabolic processes and disease. She has written numerous research and review articles, has worked as a textbook editor, and is a co-author, with Donald Voet and Judith G. Voet, of *Fundamentals of Biochemistry.*

KATHLEEN CORNELY holds a B.S. in chemistry from Bowling Green (Ohio) State University, an M.S. in biochemistry from Indiana University, and a Ph.D. in nutritional biochemistry from Cornell University. Her experimental research has included a wide range of studies of protein purification and chemical modification. She currently serves as a Professor of Chemistry and Biochemistry at Providence College where she has taught courses in biochemistry, organic chemistry, and general chemistry. Her recent research efforts have focused on issues related to chemical education, particularly the use of case studies in biochemistry education. Several of these case studies have been published in the chemical education literature and others are collected in her book *Cases in Biochemistry.* She also serves on the editorial board of *Biochemistry and Molecular Biology Education* and is a member of the Educational and Professional Development Committee of the American Society for Biochemistry and Molecular Biology.

PREFACE

WE SET OUT TO WRITE AN INTRODUCTORY BIOCHEMISTRY TEXTBOOK because we believed that there was a need for a new approach to the subject and that students could clearly benefit from our efforts. The result of our inspiration is *Essential Biochemistry,* a textbook that focuses on the chemistry of biochemistry and places it in biological context. Our experience with students as well as advances in cognitive theory prompted us to combine relatively short text chapters with extensive problems sets in order to maximize opportunities for students to learn through problem-solving. We also recognized the advantages of providing well-integrated multimedia exercises to reinforce and extend principles introduced in the text. For this reason, the text and media components were developed in tandem. We believe that this book provides an ideal balance of elements that will ease the efforts of instructors and facilitate learning by students.

A Modern Approach

Writing a new biochemistry textbook provides an opportunity not just to convey the latest research findings but also to approach the subject with a fresh perspective. Biochemistry is an enormous field, and it has always been a challenge to give students a solid foundation in the subject, particularly in a one-semester course. We saw the need for a modern textbook that would introduce students to a large and ever-growing subject, that would provide broad coverage without being overwhelming, that would explore important topics in detail, that would not minimize chemical rigor, and that would provide students with knowledge and tools that they could apply to other areas of chemistry and biology.

Essential Biochemistry differs somewhat from other textbooks in its organization and in the way material is presented. For example, there are no separate chapters devoted to carbohydrate and lipid structure. We believe that students can learn about molecular structures and terminology as they study molecules in the context of their biological functions and their metabolic transformations. We also chose to focus on aspects of biochemistry that tend to receive little coverage in other courses or present a challenge to many students. Thus, we include discrete chapters on motor proteins, enzyme mechanisms, enzyme kinetics, oxidative phosphorylation, photosynthesis, and DNA repair.

By departing somewhat from the traditional table of contents, we hope to provide students with a solid introduction to modern biochemistry. We believe that depth of coverage is often more important than breadth. Wherever possible, we describe the physiological context of biochemical processes. Finally, we present examples of chemical phenomena to illuminate general themes of biochemistry, not necessarily to illustrate all the details of a biochemical process.

In a short textbook, every example must count. For this reason, a single topic explored in depth not only tells an interesting story, it can provide insights into a number of biochemical principles. For example, myoglobin appears repeatedly in Chapter 4 to explain various aspects of protein structure and function. In Chapter 6, chymotrypsin highlights various features of enzyme action. Regulation of fuel metabolism by insulin and glucagon introduces the principles of signal transduction in Chapter 16.

In a similar vein, different topics within a chapter are linked by placing them in the broader context of a biological "story," often a disease. In Chapter 3, the genetic nature of cystic fibrosis provides a backdrop for topics ranging from DNA sequencing to protein expression. The generation of a nerve impulse ties together information about membrane permeability, transport, and fusion in Chapter 8. In Chapter 14, lipid metabolism is linked to atherosclerosis; in Chapter 18, cancer is the framework for a discussion of DNA repair.

In our experience, students sometimes miss the forest for the trees. To counteract this tendency, we have intentionally left out some details, particularly in the chapters on metabolic pathways. This allows us to focus on some general themes, including the stepwise nature of pathways (Chapter 10) and their evolution (Chapter 11).

It is our hope that by approaching biochemistry as a guidebook rather than as a catalog, this textbook will allow students to master the subject at several levels while minimizing the need for rote memorization.

Problem-Based Learning

Developments in cognitive learning theory as well as the results of classroom research indicate that students learn more when they can construct their own knowledge, for example, by answering questions and solving problems. Students who are actively engaged with the material are more likely to retain information. In fact, we designed *Essential Biochemistry* so that students can take an active role in their education. For example, each chapter begins with a list of Learning Objectives, and brief questions periodically prompt the students to review particular objectives. A checklist at the end of each chapter helps students organize their study efforts.

Most notably, each chapter includes an extensive problem set. The 20 chapters of *Essential Biochemistry* are intentionally succinct so that students can extend their learning through active problem-solving. Virtually all of the problems require analysis rather than simple recitation of facts. Many problems are case studies based on data from research publications and clinical reports. Not only do these problems provide a glimpse of the "real world" of science and medicine, they present students with novel situations and raw data that must be interpreted and analyzed. Complete solutions to all problems are placed in an appendix so that students can receive immediate feedback.

Of course, problem-solving is not the only route to understanding, and productive learning must incorporate both student-centered and instructor-centered approaches. By providing a generous selection of problem-solving opportunities, *Essential Biochemistry* can accommodate courses with varying emphasis on problem-based learning.

Multimedia Components

From the outset, we intended the media components to fully integrate with and complement the text. Although the book can stand alone, a full appreciation of the structural and dynamic aspects of biochemistry requires a medium more versatile than the printed page. The media package that accompanies *Essential*

Biochemistry in the form of a CD and Web site consists of exercises (including review exercises), an online glossary-based quiz, resources for instructors, relevant Web sites, updated references, and structure files corresponding to molecular images from the textbook.

The most important of these media elements are the exercises available on the CD and Web site. Four of these exercises—covering thermodynamics, acid–base chemistry, kinetics, and redox chemistry—are designed to review and reinforce concepts that tend to challenge poorly prepared students. The other 27 exercises, which focus on biochemical topics, use a variety of formats to actively engage students. Many of these exercises incorporate Chime-based molecular images that can be rotated, zoomed, and altered in format (e.g., spacefilling, ball-and-stick, wireframe) in order to highlight various aspects of structure. Students learn to use the simple Chime command menu to manipulate molecules and are thus equipped to explore the vast amount of structural information available in the biochemistry community at large, much of which is also accessible through Chime and similar programs.

In addition, exercises based on computer animations foster visual-based learning. This is an advantage for processes that involve motion or changes over time and that are difficult to "see" on paper, such as oxidative phosphorylation, fatty acid metabolism, and DNA replication. In all cases, students are invited to participate in learning by manipulating structures and answering questions.

Although the media exercises use the textbook as a starting point, they do not simply echo material in the book, but present it in a complementary format and, in some cases, extend principles introduced in the text by applying them to novel cases. At the appropriate place in each chapter, a note in the margin calls out the corresponding exercise. Additional reminders appear throughout the chapter to alert readers to instances where a particular point is illustrated in a media exercise. Each exercise is different because no single format can adequately address all aspects of a biochemical structure or function. Consequently, some of the exercises are brief enough to include during a lecture, whereas others require more time for student exploration.

For the Instructor: How to Use This Book

Although we believe our approach to biochemistry is innovative, the overall order of chapters in *Essential Biochemistry* follows a traditional syllabus. The 20 chapters are organized into three parts that span the major themes of biochemistry, including structure-function relationships, the transformation of matter and energy, and how genetic information is stored and made accessible. Students learn best when they can see the "big picture" and incorporate new knowledge into an existing paradigm. Accordingly, this book includes an early chapter on the genetic basis of macromolecular structure and function (Chapter 3) and introduces metabolic processes by providing an overview of fuel acquisition, storage, and mobilization (Chapter 9).

To provide as much context as possible for students who have taken other courses in chemistry and biology, who are likely to be planning careers in the health sciences, and who are continually exposed to news about science and medicine, *Essential Biochemistry* includes material on practical applications of biochemistry, health and disease, and other topics familiar to students. These items are scattered throughout the book in the form of chapter openers, items in the text, boxes, and problems. For example, there is coverage of controversial topics such as gene therapy, cloning, and GM foods. Health-related concerns include the action of the anthrax toxin, the link between emphysema and smoking, alcohol metabolism, diet and heart disease, antibiotic resistance, and obesity and diabetes. There are even problems based on examples of "junk" science such as homeopathy and "vitamin O."

Notes on Organization

Chapter 1 of *Essential Biochemistry* provides a preview of biological molecules and thermodynamics. Relevant cell biology topics, such as descriptions of organelles, are placed in the appropriate chapters (e.g., mitochondria in Chapter 12, chloroplasts in Chapter 13, and peroxisomes in Chapter 14).

Chapter 3, which introduces DNA structure and function, is placed early in the book to provide a brief summary of the central dogma, which should be very familiar to students by this point in their education. DNA sequencing is also covered in Chapter 3 (rather than in the chapter on DNA replication) because of its importance for understanding protein sequences, which are discussed in Chapter 4. The outlines of cloning and PCR are included, again because these techniques underlie much of the research on proteins. This material is not meant to be comprehensive, because we assume that recombinant DNA technology is covered in other courses. Genomics is introduced in Chapter 3 (and not reserved for a higher-level discussion at the end of the book) in order to make it easier for students to understand the variation and evolution of genes and organisms before they encounter the numerous protein structures and enzyme functions in later chapters.

There is no discrete chapter on amino acid chemistry. This subject is more effectively conveyed in the context of protein structure (Chapters 4 and 5) and function (Chapters 6 and 7). In Chapter 4, the 20 standard amino acids are grouped according to widely used hydropathy scales, rather than by side chain structure per se, which facilitates subsequent understanding of protein folding and membrane protein structure.

Chapter 5 describes the major structural proteins (actin, tubulin, collagen, and keratin) and their associated motor proteins (myosin and kinesin). This material reinforces aspects of protein structure introduced in Chapter 4 and provides the biochemical underpinnings of topics typically covered in other biology courses.

An explanation of how enzymes work (enzyme mechanisms, Chapter 6) precedes a discussion of enzyme kinetics (Chapter 7). This allows students to grasp the importance of enzymes and to focus on the chemistry of enzyme-catalyzed reactions before delving into the more quantitative aspects of enzyme kinetics (which often intimidate students with poor mathematical preparation). The discussion of enzyme inhibition in Chapter 7 focuses exclusively on the physiologically relevant competitive inhibition and mixed inhibition (of which noncompetitive inhibition is a special case). In Chapter 7, phosphofructokinase—rather than the traditional asparate transcarbamoylase—serves as an example of allosteric enzyme regulation because the enzyme conformational changes are dramatic and easily illustrated. For a more traditional approach to enzyme chemistry, an instructor can begin with the introductory section of Chapter 6 (Section 6-1), follow this with material on enzyme kinetics and inhibition (Sections 7-1 through 7-3), and then return to catalytic mechanisms and specific features of chymotrypsin (Sections 6-2 through 6-4).

Chapter 8 emphasizes the molecular properties of fatty acids and other lipids as a prelude to a discussion of membrane structure. This arrangement allows the characteristics of different lipids to be presented in the appropriate context as membrane components (in Chapter 8) or as metabolic fuels (Chapters 9 and 14).

Chapter 9, an overview of mammalian fuel metabolism, comes before rather than after individual chapters on metabolism. This makes it easier for students to keep track of key principles (e.g., the citric acid cycle receives acetyl-CoA groups not just from carbohydrate degradation). The chapter begins with digestion as a source of raw materials for human metabolism, which allows a discussion of metabolic fuels and their storage, mobilization, and catabolism, so that subsequent chapters can be more easily placed in context. To aid the student, an overview figure introduced in Chapter 9 is repeated in each subsequent chapter, with the portion relevant to that chapter highlighted.

Chapter 10 introduces glucose structure as a prelude to glycolysis and other pathways of carbohydrate metabolism. The structures and metabolism of some other sugars are included in Chapter 10, but because there is no chapter solely dedicated to carbohydrate structure, students are spared the need to memorize structures and nomenclature that appear nowhere else in the book.

In order to emphasize biochemical principles and to minimize the number of structures and enzymes presented to students, some steps of some metabolic pathways are not shown. For example, the reactions in the rearrangement phase of the pentose phosphate pathway (Chapter 10) and the Calvin cycle (Chapter 13) are not shown explicitly. Only the major pathways of lipid metabolism are shown in Chapter 14, and much of the bulk of traditional amino acid and nucleotide metabolism has also been excluded from Chapter 15.

A short chapter on photosynthesis (Chapter 13) follows the chapter on oxidative phosphorylation (Chapter 12) so that students can more easily discern the similarities between these processes.

Chapter 15 covers amino acid and nucleotide metabolism by focusing on reactions involving nitrogen. This allows a complete overview of the "biological" nitrogen cycle by following nitrogen fixation, assimilation, synthesis and degradation of amino acids and nucleotides, and nitrogen disposal via the urea cycle. By covering nitrogen metabolism in this manner, students are exposed to all the relevant pathways and can focus on important reactions without getting bogged down in a comprehensive recounting of all the reactions of all these pathways. Thus, there is no separate chapter on nucleotides (their structures are first presented in Chapter 3 in the context of DNA structure, and their metabolism is described in the context of amino acid–derived biomolecules in Chapter 15).

Signal transduction is covered in the context of regulation of mammalian fuel metabolism (Chapter 16), which creates an opportunity to summarize the major features of the metabolic pathways described in Chapters 10–15.

In order to focus on principles rather than details, the discussions of DNA replication (Chapter 17) and transcription (Chapter 19) do not make a sharp distinction between prokaryotic and eukaryotic systems. The overall processes are presented using examples from both types of systems, and brief notes explain how they differ.

A chapter on cancer and DNA repair (Chapter 18) provides an opportunity to tie DNA metabolism to various aspects of cell biology (e.g., cell cycle control and apoptosis) and reinforces understanding of DNA structure and function (first presented in Chapter 3) by showing how DNA is damaged and how it can be repaired.

Because replication, transcription, and translation are typically also covered in other courses, Chapters 17–20 focus primarily on some of the biochemical details of these processes, such as topoisomerase action, nucleosome structure, mechanisms of polymerases and other enzymes, structures of accessory proteins, mechanisms for proofreading during polymerization and aminoacylation, and chaperone-assisted protein folding.

Pedagogical Features

Each chapter of *Essential Biochemistry* is designed to be self-contained so that it can be covered at any point in the syllabus.

- Each chapter begins with a paragraph **(This Chapter in Context)** to help orient the reader to the main topics of the chapter and how they relate to surrounding chapters.

- A short example of a **biochemical application** opens each chapter.
- A list of **Learning Objectives** precedes the text of each chapter. Students are periodically prompted to review the objectives and to answer **Study Questions** that reinforce each objective.
- Reminders to explore the **Media Exercises** appear at appropriate places in the chapter. The text includes additional cross-references to specific topics in the exercises. The exercises animate complex processes and show detailed molecular structures using a Chime-based interactive format. Four Review Exercises provide supplemental background material. Most of the media exercises are designed so that students can proceed at their own pace, viewing animations, manipulating molecular structures, and answering questions.
- Sentences summarizing **key points** are in italics. **Key terms** are in boldfaced. Their definitions are also included in the glossary and form the basis of the online quiz. **Key equations** are boxed for emphasis.
- **Sample Calculations** illustrate the use of important equations in thermodynamics (Chapter 1), acid–base chemistry (Chapter 2), binding phenomena (Chapter 4), enzyme kinetics (Chapter 7), transport processes (Chapter 8), equilibria (Chapter 9), and redox chemistry (Chapter 12).
- Some material that is of a higher level or that is thematically distinct from the bulk of the chapter is set off in **boxes** so as not to distract the reader from the main thread of the discussion.
- Illustrations include **photos from research publications** and **computer-generated molecular models** designed specifically for *Essential Biochemistry*. Many small figures are incorporated directly into the text. An **overview figure** illustrating all the major metabolic pathways is introduced in Chapter 9 and revisited in subsequent chapters on metabolism. Chapters 10 and 14, which focus heavily on pathways, include an additional summary figure as a study aid.
- Each chapter ends with a large selection of **Problems**, including some multistep case-type problems based on the recent literature. The problems require students to apply information rather than simply recall memorized details. **Complete Solutions** to all problems are provided in the appendix.
- An **annotated list of Selected Readings** following each chapter includes recent short papers, mostly reviews, that students are likely to find useful as sources of additional information. Some **Relevant Web Links** are included on the accompanying Web site (www.wiley.com/college/pratt).
- Each chapter includes a **Summary** of the main points. A **Checklist** at the end of the chapter reminds students to review the Learning Objectives, solve the problems, complete the relevant media exercises, take the glossary-based quiz, explore the Web sites listed online, and consult the list of selected readings for further information.
- Four **Instructor's Resources**, part of the media package, include lecture-ready slides and student activities on the Human Genome Project, bacterial drug resistance, proteomics, and phylogenetic trees.

Acknowledgments

We would like to thank everyone who helped develop *Essential Biochemistry,* including Senior Editor Patrick Fitzgerald; Developmental Editor Ellen Ford; Media Editor Linda Muriello; the Production Management Services of

Ingrao Associates; Production Editor Sandra Dumas; Senior Designer Kevin Murphy; Photo Editor Hilary Newman; Illustration Coordinator Anna Melhorn; Media Project Manager Holly Rioux; and Editorial Assistants Justin Bow and Dana Kasowitz. Special thanks goes to the media team at Science Technologies, James Caras, Paige Caras and Barrie Kitto, for their superb work on the media package.

We also thank all the reviewers who provided essential feedback on manuscript and media, corrected errors, and made valuable suggestions for improvements throughout the writing and development process. They include:

Paul Azari, Colorado State University
Allan Bieber, Arizona State University
Jeffrey Brodsky, University of Pittsburgh
Carolyn S. Brown, Clemson University
Kim Colvert, Ferris State University
Charles Crittell, East Central University
David W. Eldridge, Baylor University
Jeffrey Evans, University of Southern Mississippi
Wilson Francisco, Arizona State University
Edward Funkhouser, Texas A&M University
Don Heck, Iowa State University
James R. Heitz, Mississippi State University
Todd Hrubey, Butler University
Christine Hrycyna, Purdue University
Barrie Kitto, University of Texas, Austin
S. Madhavan, University of Nebraska
Marilee Parsons, University of Michigan
Scott Pattison, Ball State University
Richard Posner, Northern Arizona State University
Russell Rasmussen, Wayne State College
Melvin Schnindler, Michigan State University
Tammy Stobb, St. Cloud State University
Michael Sypes, Pennsylvania State University
Linette Watkins, SW Texas State University
Lisa Wen, Western Illinois University
Beulah Woodfin, University of New Mexico

Many of the molecular graphics that illustrate this book were created using publicly available coordinates from the Protein Data Bank (www.rcsb.org). The figures were rendered with the Swiss-Pdb Viewer [Guex, N., Peitsch, M.C., SWISS-MODEL and the Swiss-Pdb Viewer: an environment for comparative protein modeling, *Electrophoresis* **18,** 2714–2723 (1997); program available at www.expasy.ch/spdbv] and Pov-Ray (available at www.povray.org).

BRIEF CONTENTS

CONTENTS

PART I
MOLECULAR STRUCTURE AND FUNCTION

PART II
METABOLIC REACTIONS

PART III
MANAGEMENT OF GENETIC INFORMATION

CHAPTER 18
Cancer and DNA Repair 556

CHAPTER 19
Transcription and RNA 580

CHAPTER 20
Protein Synthesis 612

FOR THE STUDENT

HOW TO USE THIS BOOK

Welcome to Biochemistry! Your success in this course will depend to a great extent on your willingness to take an active role in your education. Learning biochemistry requires more than simply reading the textbook, although we recommend that as a first step! *Essential Biochemistry* has been designed and written with you in mind, and we urge you to take advantage of all it has to offer.

Biochemical knowledge is cumulative; it is not something that can be learned all at once. We advise you to keep up with your reading and other assignments so that you have plenty of time to reflect, ask questions, and, if necessary, seek help from your instructor. As you read each chapter of the textbook, make sure you understand how it fits into the course syllabus. Use the study aids provided in the textbook: First, note the list of learning objectives at the start of the chapter. At the appropriate points, you will be asked to review each objective and answer one or more study questions. Be sure to view the media exercises that expand on material covered in the textbook. These exercises include animations of dynamic biochemical processes and interactive molecular graphics. You can enrich your understanding of biochemistry by exploring the exercises and answering the questions they pose. Consult the review exercises if your chemistry background is weak.

As you study, note the key sentences that are highlighted in italics. Be able to define terms in boldface, and test your knowledge by taking the online quiz. Most importantly, solve the problems at the end of each chapter. You should make every effort to complete all the problems without looking at the answers in the appendix. Developing problem-solving skills will facilitate your understanding of biochemistry and will help pave the way to success in any future academic or career endeavor.

Finally, use the summary and checklist at the end of every chapter as a guide to help you study. Take advantage of the additional resources available—such as the list of selected readings and Internet sites—if you need help, are curious about biochemistry, or need up-to-date information as a starting point for a class project.

In writing *Essential Biochemistry,* we endeavored to select topics that would provide a solid introduction to modern biochemistry, which is a vast and ever-changing field. We realize that most students who use this book will not become biochemists. Nevertheless, it is our hope that you will come to understand the major themes of biochemistry and see how they relate to current and future developments in science and medicine.

Charlotte W. Pratt
Kathleen Cornely

PART I

MOLECULAR STRUCTURE AND FUNCTION

CHAPTER 1

The Chemical Basis of Life

A look at a microfossil such as this one, about 3400 million years old, drives both speculation and formal scientific research. Are these the marks of an early life form or just the result of a random geochemical process? Biochemistry provides leads for analyzing such specimens. The process whereby a living cell takes up matter and energy and uses them leaves a characteristic signature. For example, cells prefer to take up lighter isotopes of carbon and iron, leaving behind localized variations in the naturally occurring isotope ratios. Even more telling is the identification of certain hydrocarbon molecules—synthesized only by living organisms—that apparently persist for billions of years. Adding molecular evidence to physical analysis of microfossils makes it easier to classify such fossils as the remnants of living organisms.

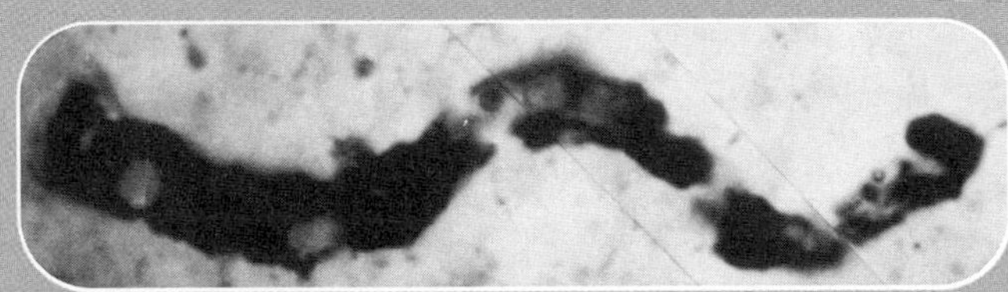

[Courtesy J. William Schopf, UCLA.]

LEARNING OBJECTIVES

1. Understand that biological molecules are composed of a subset of all possible elements and functional groups.
2. Understand the basic structures of the four major types of small biomolecules.
3. Understand the general structures and functions of the three major kinds of biological polymers.
4. Understand the meanings of free energy, enthalpy, and entropy and how they are related.
5. Understand how living organisms obey the laws of thermodynamics.
6. Understand the requirements for the evolution of modern cells.
7. Understand the differences between prokaryotes and eukaryotes.

THIS CHAPTER IN CONTEXT

This first chapter offers a preview of the study of biochemistry, broken down into three sections that reflect how topics in this book are organized. First come brief descriptions of the four major types of small biological molecules and their polymeric forms. Next is a summary of the thermodynamics that apply to metabolic reactions. Finally, there is a discussion of the origin of self-replicating life forms and their evolution into modern cells. These short discussions will introduce you to some of the key players and major themes of biochemistry so that you can begin to fit new knowledge of biochemistry within the framework of biology and chemistry that is already familiar.

1 WHAT IS BIOCHEMISTRY?

Biochemistry is the scientific discipline that seeks to explain life at the molecular level. It uses the tools and terminology of chemistry to describe the various attributes of living organisms. Biochemistry offers answers to such fundamental questions as "What are we made of?" and "How do we work?" Biochemistry is also a practical science: It generates powerful techniques that underlie advances in other fields such as genetics, cell biology, and immunology; it offers insights into the treatment of diseases such as cancer and diabetes; and it improves the efficiency of industries such as wastewater treatment and the synthesis of pesticides and drugs.

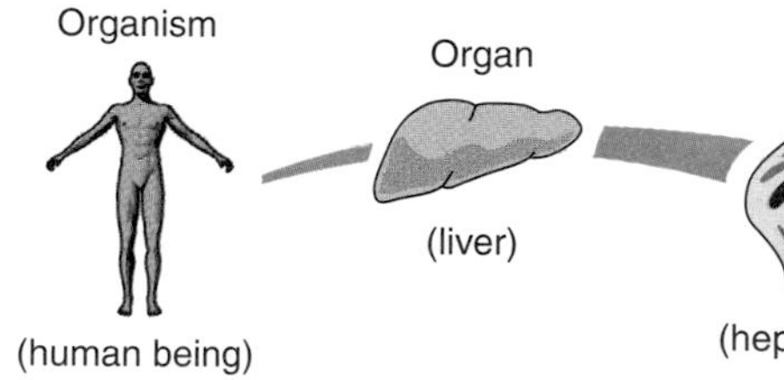

FIGURE 1-1

Levels of organization in a living organism.
Biochemistry focuses on the structures and functions of molecules. Interactions between molecules give rise to higher-order structures (for example, organelles), which may themselves be components of larger entities, leading ultimately to the entire organism.

Biochemistry has traditionally been a science of reductionism; that is, it attempts to explain the whole by breaking it into smaller parts and examining each part separately. For biochemists, this means isolating and characterizing an organism's component molecules (Fig. 1-1). *A thorough understanding of each molecule's physical structure and chemical reactivity should, in theory, lead to an understanding of how molecules cooperate and combine to form larger functional units and, ultimately, the intact organism.*

Nevertheless, a holistic approach is indispensable for unraveling the secrets of nature. Just as a watch completely disassembled no longer resembles a watch, information about a multitude of biological molecules does not necessarily reveal how an organism lives. Some molecular interactions are too complex to be teased apart in the laboratory, so it may be necessary to examine a cultured organism to see how it fares when a particular molecule is modified or absent. In addition, so much is known about so many molecules that the sheer volume of data—much of it accessible only by computer—makes it difficult to see how it all fits together. Thus, a biochemist's laboratory is as likely to hold racks of test tubes as flasks of bacteria or a computerized database. The material included in this book reflects both traditional and modern approaches to biochemical understanding.

The chapters in this book are divided into three groups that roughly correspond to three major themes of biochemistry:

1. *What are living organisms made of?* Some molecules are responsible for the physical shapes of cells. Others carry out various activities in

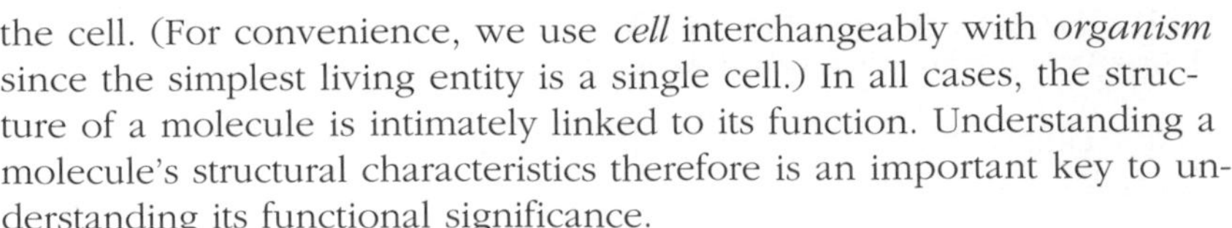

the cell. (For convenience, we use *cell* interchangeably with *organism* since the simplest living entity is a single cell.) In all cases, the structure of a molecule is intimately linked to its function. Understanding a molecule's structural characteristics therefore is an important key to understanding its functional significance.

2. *How do organisms acquire and use energy?* The ability of a cell to carry out metabolic reactions—to synthesize its constituents and to move, grow, and reproduce—requires the input of energy. A cell must extract this energy from the environment and spend it or store it in a manageable form.

3. *How does an organism maintain its identity across generations?* Modern human beings look much as they did 100,000 years ago. Certain bacteria have persisted for millions, if not billions, of years. In all organisms, information specifying a cell's structural composition and functional capacity must be faithfully maintained and transmitted each time the cell divides.

Even within its own lifetime, a cell may dramatically alter its shape or metabolic activities, but it does so within certain limits. Throughout this book we will examine how regulatory processes define an organism's ability to respond to changing internal and external conditions. In addition, we will examine the diseases that can result from a molecular defect in any of the cell's attributes—its structural components, its metabolism, or its ability to follow genetic instructions.

2 BIOLOGICAL MOLECULES

A staggering number of different molecules are found in even the simplest organisms, yet this number represents only an infinitesimal portion of all the molecules that are chemically possible. For one thing, *only a small subset of the known elements occur in biomolecules* (Fig. 1-2). The most abundant of these are C, N, O, and H, followed by Ca, P, K, S, Cl, Na, and Mg. Certain **trace elements** are also present in very small quantities.

Virtually all the molecules in a living organism contain carbon. Keep in mind that even though the term *organic chemistry* is used to describe all carbon-containing molecules, not all of these are biological in origin. Most molecules in a living cell belong to one of a few structural classes, which are described below. Similarly, *the chemical reactivity of biomolecules is limited relative to the reactivity of all organic compounds.* A few of the functional groups and intramolecular linkages that are common in biochemistry are listed in Table 1-1. It may be helpful to review them, because we will refer to them throughout this book.

1 H												5 B	6 C	7 N	8 O	9 F	
												13 Al	14 Si	15 P	16 S	17 Cl	
11 Na	12 Mg			23 V	24 Cr	25 Mn	26 Fe	27 Co	28 Ni	29 Cu	30 Zn			33 As	34 Se	35 Br	
19 K	20 Ca				42 Mo						48 Cd					53 I	
					74 W												

FIGURE 1-2

Elements found in biological molecules. The most abundant elements are most darkly shaded; trace elements are most lightly shaded. Not every organism contains every trace element.

TABLE 1-1 Common functional groups and linkages in biochemistry

Compound name	Structure[a]	Functional group
Amine[b]	$\mathrm{RNH_2}$ or $\mathrm{RNH_3^+}$ $\mathrm{R_2NH}$ or $\mathrm{R_2NH_2^+}$ $\mathrm{R_3N}$ or $\mathrm{R_3NH^+}$	$\mathrm{{-}N{<}}$ or $\mathrm{{-}\overset{+}{N}{-}}$ (with two further bonds) (amino group)
Alcohol	ROH	$\mathrm{{-}OH}$ (hydroxyl group)
Thiol	RSH	$\mathrm{{-}SH}$ (sulfhydryl group)
Ether	ROR	$\mathrm{{-}O{-}}$ (ether linkage)
Aldehyde	$\mathrm{R{-}C({=}O){-}H}$	$\mathrm{{-}C({=}O){-}}$ (carbonyl group), $\mathrm{R{-}C({=}O){-}}$ (acyl group)
Ketone	$\mathrm{R{-}C({=}O){-}R}$	$\mathrm{{-}C({=}O){-}}$ (carbonyl group), $\mathrm{R{-}C({=}O){-}}$ (acyl group)
Carboxylic acid[b] (Carboxylate)	$\mathrm{R{-}C({=}O){-}OH}$ or $\mathrm{R{-}C({=}O){-}O^-}$	$\mathrm{{-}C({=}O){-}OH}$ (carboxyl group) or $\mathrm{{-}C({=}O){-}O^-}$ (carboxylate group)
Ester	$\mathrm{R{-}C({=}O){-}OR}$	$\mathrm{{-}C({=}O){-}O{-}}$ (ester linkage)
Amide	$\mathrm{R{-}C({=}O){-}NH_2}$ $\mathrm{R{-}C({=}O){-}NHR}$ $\mathrm{R{-}C({=}O){-}NR_2}$	$\mathrm{{-}C({=}O){-}N{<}}$ (amido group)
Imine[b]	$\mathrm{R{=}NH}$ or $\mathrm{R{=}NH_2^+}$ $\mathrm{R{=}NR}$ or $\mathrm{R{=}NHR^+}$	$\mathrm{{>}C{=}N{-}}$ or $\mathrm{{>}C{=}\overset{+}{N}{<}^{H}}$ (imino group)
Phosphoric acid ester[b]	$\mathrm{R{-}O{-}P({=}O)(OH){-}OH}$ or $\mathrm{R{-}O{-}P({=}O)(O^-){-}O^-}$	$\mathrm{{-}O{-}P({=}O)(OH){-}O{-}}$ (phosphoester linkage) $\mathrm{{-}P({=}O)(OH){-}OH}$ or $\mathrm{{-}P({=}O)(O^-){-}O^-}$ (phosphoryl group, P_i)
Diphosphoric acid ester[b] (Phosphodiester)	$\mathrm{R{-}O{-}P({=}O)(OH){-}O{-}P({=}O)(OH){-}OH}$ or $\mathrm{R{-}O{-}P({=}O)(O^-){-}O{-}P({=}O)(O^-){-}O^-}$	$\mathrm{{-}O{-}P({=}O)(OH){-}O{-}P({=}O)(OH){-}O{-}}$ (phosphodiester linkage) $\mathrm{{-}P({=}O)(OH){-}O{-}P({=}O)(OH){-}OH}$ or $\mathrm{{-}P({=}O)(O^-){-}O{-}P({=}O)(O^-){-}O^-}$ (diphosphoryl group, pyrophosphoryl group, PP_i)

[a]R represents any carbon-containing group. In a molecule with more than one R group, the groups may be the same or different.
[b]Under physiological conditions, these groups are ionized and hence bear a positive or negative charge.

Go to **Exercise 1/Biological Molecules** to explore different ways of visualizing the four classes of small biological molecules. The structures presented in Exercise 1 and the questions based on them address Learning Objective 2.

REVIEW LEARNING OBJECTIVE 1

- What are the common functional groups and linkages in biological molecules?

Cells contain four major types of biomolecules

Most of the cell's small molecules can be divided into four groups. Although each group contains many members, *they are united under a single structural or functional definition.* Identifying a particular molecule's group may help predict its chemical properties and perhaps its role in the cell.

1. Amino acids

The simplest compounds are the **amino acids,** so named because they contain an amino group (—NH_2) and a carboxylic acid group (—COOH). Under physiological conditions, these groups are actually ionized to —NH_3^+ and —COO^-. The common amino acid alanine—like other small molecules—can be depicted in different ways, for example, by a structural formula, a ball-and-stick model, or a space-filling model (Fig. 1-3). Other amino acids resemble alanine in basic structure but instead of a methyl group (—CH_3), they have another group that may also contain N, O, or S; for example,

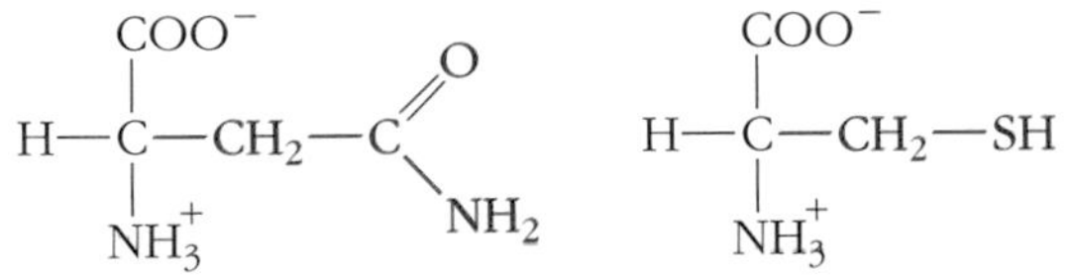

COO^- / H—C—CH_2—C(=O)—NH_2 / NH_3^+

Asparagine

COO^- / H—C—CH_2—SH / NH_3^+

Cysteine

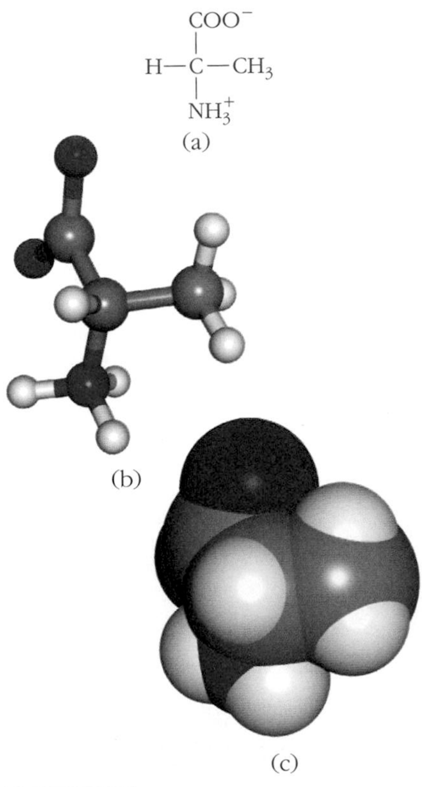

FIGURE 1-3

Representations of alanine.
The structural formula (a) indicates all the atoms and the major bonds. Minor bonds such as the C—O and N—H bonds are implied. Because the central carbon atom has tetrahedral geometry, its four bonds do not lie flat in the plane of the paper: The horizontal bonds actually extend slightly above the plane of the page, and the vertical bonds extend slightly behind it. This tetrahedral arrangement is more accurately depicted in the ball-and-stick model (b). Here, the atoms are color-coded by convention: C is gray, N is blue, O is red, and H is white. This ball-and-stick representation reveals the identities of the atoms and their positions in space but does not indicate their relative size or electrical charge. In a space-filling model (c), each atom is presented as a sphere whose radius corresponds to the distance of closest approach by another atom. This model most accurately depicts the actual size of the molecule, but it obscures some of its atoms and linkages.

2. Carbohydrates

Simple **carbohydrates** (also called **monosaccharides** or just sugars) have the formula $(CH_2O)_n$. Glucose, a monosaccharide with six carbon atoms, has the formula $C_6H_{12}O_6$. It is sometimes convenient to draw it as a ladder-like chain:

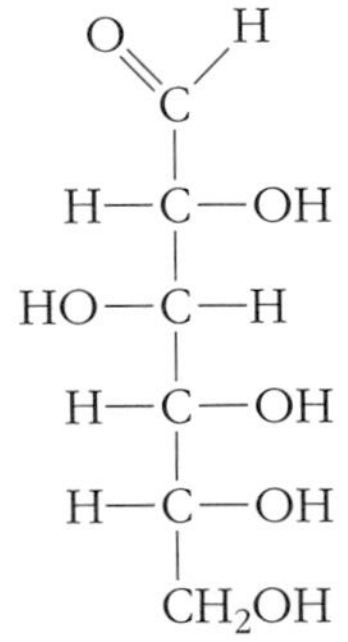

O=C—H
H—C—OH
HO—C—H
H—C—OH
H—C—OH
CH_2OH

Glucose

However, glucose forms a cyclic structure in solution:

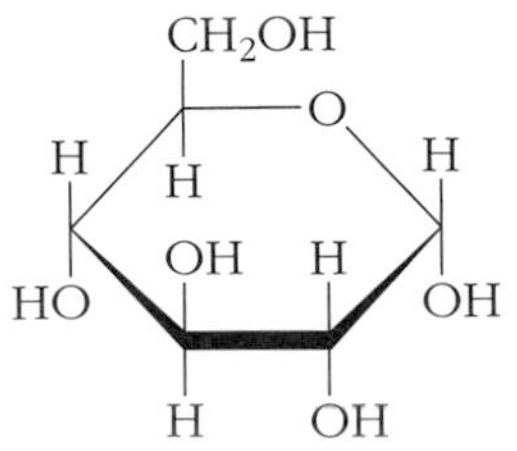

CH_2OH; ring O; H, H, H; OH, H; HO, OH; H, OH

Glucose

In this representation, the darker bonds project in front of the page and the lighter bonds project behind it. In many monosaccharides, one or more hydroxyl groups are replaced by other groups, but the ring structure and multiple —OH groups of these molecules allow them to be easily recognized as carbohydrates.

3. Nucleotides

A five-carbon sugar, a nitrogen-containing ring, and one or more phosphate groups are the components of **nucleotides.** For example, adenosine triphosphate (ATP) contains the nitrogenous group adenine linked to the monosaccharide ribose to which a triphosphate group is also attached:

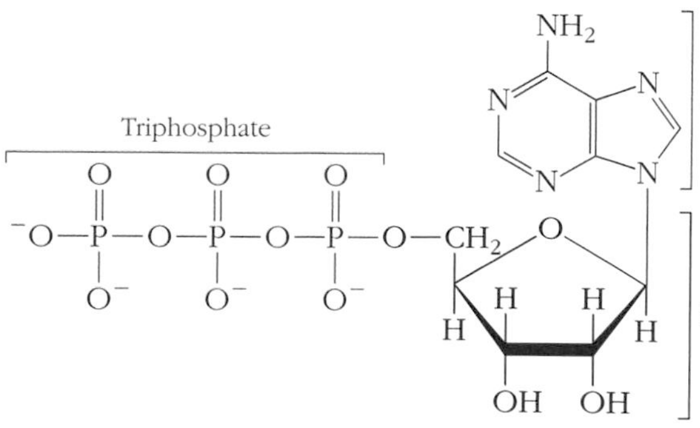

Triphosphate

Adenine

Ribose

Adenosine triphosphate (ATP)

4. Lipids

The fourth major group of biomolecules consists of the **lipids.** These compounds cannot be described by a single structural formula since they are a diverse collection of molecules. However, they all have in common a tendency to be poorly soluble in water because of their hydrocarbon-like nature. For example, palmitic acid consists of a chain of 15 carbons attached to a carboxylic acid group, which is ionized under physiological conditions. The anionic lipid is therefore called palmitate.

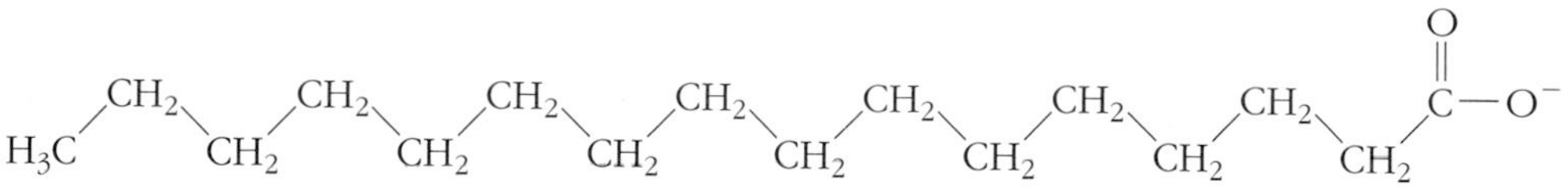

Palmitate

Cholesterol, although it differs significantly in structure from palmitate, is also poorly soluble in water because of its hydrocarbon-like composition.

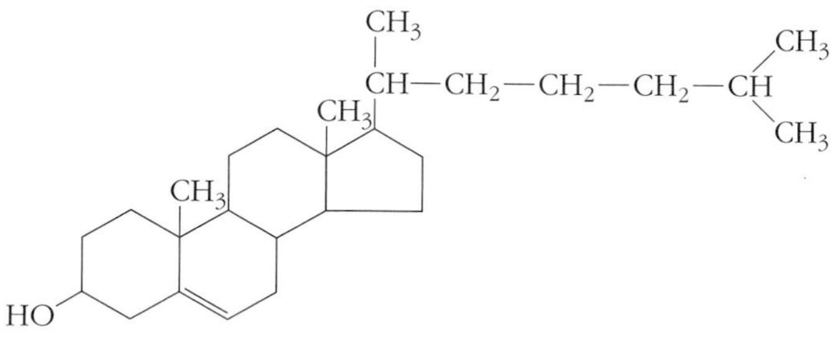

Cholesterol

Cells also contain a few other small molecules that cannot be easily classified into the groups above or that consist of more than one type of molecule.

REVIEW LEARNING OBJECTIVE 2

- Give the structural or functional definitions for amino acids, monosaccharides, nucleotides, and lipids.

There are three major kinds of biological polymers

In addition to small molecules consisting of relatively few atoms, organisms contain macromolecules that may consist of thousands of atoms. Such huge molecules are not synthesized in one piece but are built from smaller units. This is a universal feature of nature: *A few kinds of building blocks can be combined in different ways to produce a wide variety of larger structures.* This is advantageous for a cell, which can get by with a limited array of raw materials. In addition, the very act of chemically linking individual units **(monomers)** into longer strings **(polymers)** is a way of encoding information (the sequence of the monomeric units) in a stable form. Biochemists use certain units of measure to describe both large and small molecules (Box 1-A).

Amino acids, monosaccharides, and nucleotides each form polymeric structures with widely varying properties. In most cases, the individual monomers become covalently linked in head-to-tail fashion:

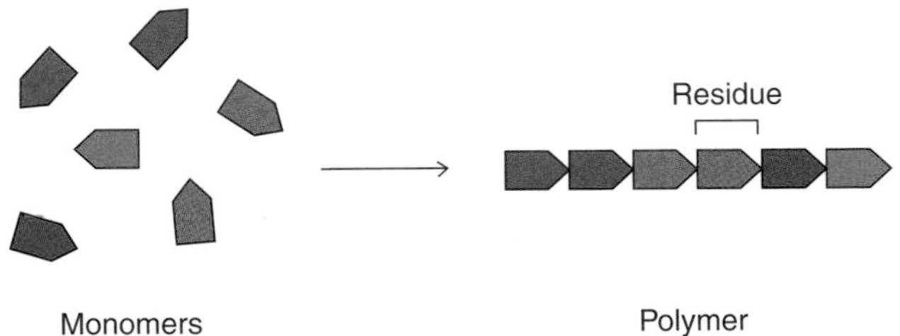

The monomers are called **residues** after they have been incorporated into the polymer. Strictly speaking, lipids do not form polymers, although they do

A CLOSER LOOK

BOX 1-A Quantitative biochemistry

Custom rules the biochemist's choice of terms to quantify objects on a molecular scale. For example, the mass of a molecule can be expressed in atomic mass units; however, the masses of biological molecules—especially very large ones—are typically given without units. Here it is understood that the mass is expressed relative to one-twelfth the mass of an atom of the common carbon isotope ^{12}C (12.011 atomic mass units). Occasionally, units of Daltons (D) are used, often with the prefix kilo, k (kD). This is useful for macromolecules such as proteins, many of which have masses in the range from 20,000 (20 kD) to over 1,000,000 (1000 kD).

The standard metric prefixes are also necessary for expressing the minute concentrations of biomolecules in living cells. Concentrations are usually given as moles per liter ($mol \cdot L^{-1}$ or M), with the appropriate prefix such as m, μ, or n:

mega	(M)	10^6
kilo	(k)	10^3
milli	(m)	10^{-3}
micro	(μ)	10^{-6}
nano	(n)	10^{-9}
pico	(p)	10^{-12}
femto	(f)	10^{-15}

For example, the concentration of the sugar glucose in human blood is about 5 mM, but many intracellular molecules are present at concentrations of μM or less.

Distances are customarily expressed in angstroms, Å ($1\ Å = 10^{-10}$ m) or, less frequently, in nanometers, nm ($1\ nm = 10^{-9}$ m). For example, the distance between the centers of carbon atoms in a C—C bond is about 1.5 Å, and the diameter of a DNA molecule is about 20 Å.

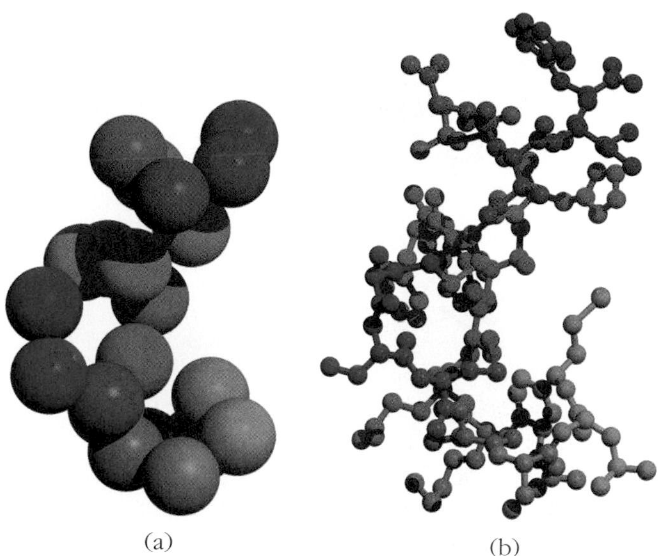

FIGURE 1-4

Structure of human endothelin.
The 21 amino acid residues of this polypeptide, shaded from blue to red, form a compact structure. In (a), each amino acid residue is represented by a sphere. The ball-and-stick model (b) shows all the atoms. [*Structure (pdb 1EDN) determined by B.A. Wallace and R.W. Jones.*]

tend to aggregate to form larger structures. Exercise 1 presents the general structures and functions of the major types of biological polymers.

1. Proteins

Polymers of amino acids are called **polypeptides** or **proteins.** Twenty different amino acids serve as building blocks for proteins, which may contain many hundreds of amino acid residues. Because the functional groups of the 20 amino acids have different sizes, shapes, and chemical properties, the exact **conformation** (three-dimensional shape) of the polypeptide chain depends on its amino acid composition. For example, the small polypeptide endothelin, with 21 residues, assumes a compact shape in which the polymer bends and folds to accommodate the functional groups of its amino acid residues (Fig. 1-4).

The 20 different amino acids can be combined in almost any order and in almost any proportion to produce myriad polypeptides, all of which have unique three-dimensional shapes. This property makes proteins as a class the most structurally variable and therefore the most functionally versatile of all the biopolymers. Accordingly, *proteins perform a wide variety of tasks in the cell, such as mediating chemical reactions and providing structural support.*

2. Nucleic acids

Polymers of nucleotides are termed **polynucleotides** or **nucleic acids,** better known as DNA and RNA. Unlike polypeptides, with 20 different amino acids available for polymerization, each nucleic acid is made from just four different nucleotides. For example, the residues in RNA contain the nitrogenous ring compounds adenine, cytosine, guanine, and uracil, which are abbreviated A, C, G, and U, respectively (Fig. 1-5). In part because nucleotides are much less variable in structure and chemical properties than amino acids, nucleic acids tend to have more regular structures than proteins. *This is in keeping with their primary role as carriers of genetic information, which is contained in their sequence of nucleotide residues rather than in their three-dimensional shape.* Nevertheless, some nucleic acids do bend and fold into compact globular shapes, as proteins do.

CGUACG

(a)

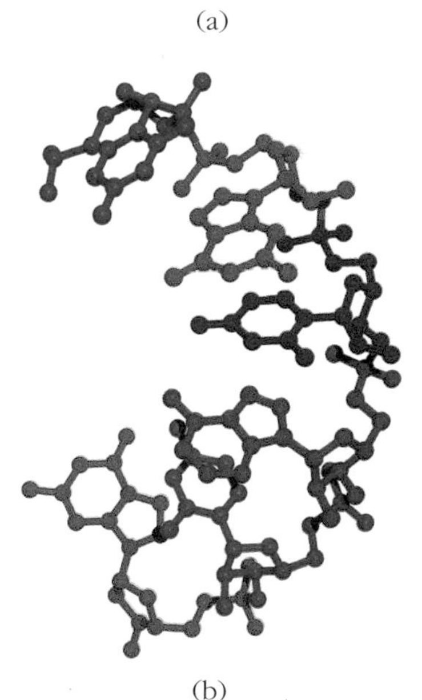

FIGURE 1-5

Structure of a nucleic acid.
(a) Sequence of nucleotide residues, using one-letter abbreviations. (b) Ball-and-stick model of the polynucleotide RNA. [*Structure (pdb ARF0108) determined by R. Biswas, S. N. Mitra, and M. Sundaralingam.*]

3. Polysaccharides

Polysaccharides usually contain only one or a few different types of monosaccharide residues, so even though a cell may synthesize dozens of different

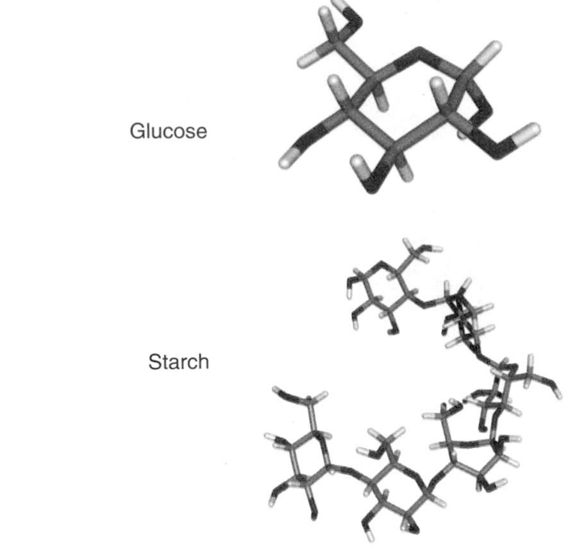

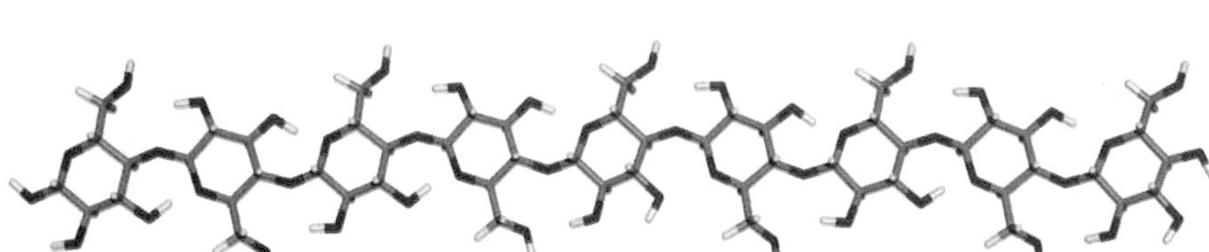

FIGURE 1-6

Glucose and its polymers. Both starch and cellulose are polysaccharides containing glucose residues. They differ in the type of chemical linkage between the monosaccharide units. Starch molecules have a loose helical conformation, whereas cellulose molecules are extended and relatively stiff.

kinds of monosaccharides, most of its polysaccharides are homogeneous polymers. This tends to limit their potential for carrying genetic information in the sequence of their residues (as nucleic acids do) or for adopting a large variety of shapes and metabolic functions (as proteins do). On the other hand, *polysaccharides perform essential cell functions by serving as fuel storage molecules and by providing structural support.* For example, plants link the monosaccharide glucose, which is a fuel for virtually all cells, into the polysaccharide starch for long-term storage. Glucose monomers are also the building blocks for cellulose, the extended polymer that helps make plant cell walls rigid (Fig. 1-6).

The brief descriptions of biological polymers given above are generalizations, meant to convey some appreciation for the possible structures and functions of these macromolecules. *Exceptions to the generalizations abound.* For example, some small polysaccharides encode information that allows cells bearing the molecules on their surfaces to recognize each other. Likewise, some nucleic acids perform structural roles, for example, by serving as scaffolding in ribosomes, the small organelles where protein synthesis takes place. Under certain conditions, proteins are called on as fuel-storage molecules. A summary of the major and minor functions of proteins, polysaccharides, and nucleic acids is presented in Table 1-2.

REVIEW LEARNING OBJECTIVE 3

- Why is it efficient for macromolecules to be polymers?
- What is the relationship between a monomer and a residue?
- Give the structural definitions and major functions of proteins, polysaccharides, and nucleic acids.

TABLE 1-2 Functions of biopolymers

Biopolymer	Encode information	Carry out metabolic reactions	Store energy	Support cellular structures
Proteins	—	✔ (major)	✔ (minor)	✔ (major)
Nucleic acids	✔ (major)	✔ (minor)	—	✔ (minor)
Polysaccharides	✔ (minor)	—	✔ (major)	✔ (major)

✔ major function
✔ minor function

3 ENERGY AND METABOLISM

Assembling small molecules into polymeric macromolecules requires energy. And unless the monomeric units are readily available, a cell must synthesize the monomers, which also requires energy. In fact, *cells require energy for all the functions of living, growing, and reproducing.*

For a review of the theory and equations of thermodynamics that are relevant to the study of biochemistry, go to **Review Exercise 1/Thermodynamics.** The material presented there addresses Learning Objectives 4 and 5.

It is useful to describe the energy in biological systems using the terminology of thermodynamics (the study of heat and power). An organism, like any chemical system, is subject to the laws of thermodynamics. According to the first law, energy cannot be created or destroyed. However, it can be transformed. For example, the energy of a river flowing over a dam can be harnessed as electricity, which can then be used to produce heat or perform mechanical work. Cells can be considered to be very small machines that use chemical energy to drive metabolic reactions, which may also produce heat or carry out mechanical work.

Free energy, enthalpy, and entropy

The energy relevant to biochemical systems is called the Gibbs free energy (after the scientist who defined it) or just **free energy.** It is abbreviated G and has units of joules per mol ($J \cdot mol^{-1}$). Free energy has two components: enthalpy and entropy. ***Enthalpy,*** *abbreviated* ***H,*** *is taken to be equivalent to the heat content of the system.* ***Entropy,*** *abbreviated* ***S,*** *is a measure of a system's disorder or randomness.* The more ways a system's components can be arranged, the greater its entropy. For example, consider a pool table at the start of a game when all 15 balls are arranged in a neat triangle (a state of high order or low entropy). After play has begun, the balls are scattered across the table, which is now in a state of disorder and high entropy (Fig. 1-7).

(a)

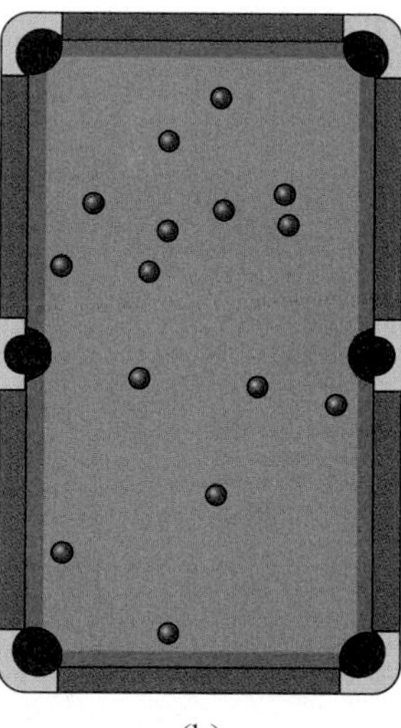
(b)

FIGURE 1-7
Illustration of entropy. Entropy is a measure of a system's randomness or disorder. (a) Entropy is low when all the balls are arranged in a single area of the pool table. (b) Entropy is high after the balls have been scattered, because there are now a large number of different possible arrangements of the balls on the table.

Free energy, enthalpy, and entropy are related by the equation

$$G = H - TS \qquad [1\text{-}1]$$

where T represents temperature in Kelvin (equivalent to degrees Celsius plus 273). Temperature is a coefficient of the entropy term because the entropy of a system varies with temperature; typically, a system becomes more disordered as the temperature rises. The enthalpy of a chemical system can be measured, although with some difficulty, but it is next to impossible to measure a system's entropy because this would require counting all the possible arrangements of its components. Therefore, it is more practical to deal with changes in these quantities (change is indicated by the Greek letter Delta, Δ) so that

$$\Delta G = \Delta H - T\Delta S \qquad [1\text{-}2]$$

Biochemists can measure how the free energy, enthalpy, and entropy of a system differ before and after a chemical reaction. For example, some chemical reactions are accompanied by the release of heat to the surroundings ($\Delta H < 0$), whereas others absorb heat from the surroundings ($\Delta H > 0$). Similarly, the entropy change can be positive or negative. See Review Exercise 1 for additional explanations of free energy, enthalpy, and entropy.

What makes a process spontaneous?

A china cup dropped from a great height will break, but the pieces will never reassemble themselves to restore the cup. The thermodynamic explanation is that the broken pieces have less free energy than the intact cup. *In order for a process to occur, the overall change in free energy* **(ΔG)** *must be negative.* For a chemical reaction, this means that the free energy of the products must be less than the free energy of the reactants:

$$\Delta G = G_{\text{products}} - G_{\text{reactants}} < 0 \qquad [1\text{-}3]$$

When ΔG is less than zero, the reaction is said to be **spontaneous** or **exergonic.** A **nonspontaneous** or **endergonic** reaction has a free energy change greater than zero, and in this case, the reverse reaction is spontaneous.

$A \rightarrow B$	$B \rightarrow A$
$\Delta G > 0$	$\Delta G < 0$
Nonspontaneous	Spontaneous

Note that thermodynamic spontaneity does not indicate how *fast* a reaction occurs, only whether it will occur as written. (The rate of a reaction depends on other factors, such as the concentrations of the reacting molecules, the temperature, and the presence of catalysts.) When a reaction, such as $A \rightarrow B$, is at equilibrium, the rate of the forward reaction is equal to the rate of the reverse reaction, so there is no net change in the system. In this situation, $\Delta G = 0$.

A quick examination of Equation 1-2 reveals that *a reaction that occurs with a decrease in enthalpy and an increase in entropy is spontaneous at all temperatures because* ΔG *is always less than zero.* These results are consistent with everyday experience. For example, heat moves spontaneously from a hot object to a cool object, and items that are neatly arranged tend to become disordered, never the other way around. (This is a manifestation of the second law of thermodynamics, which holds that the disorder of a system plus its surroundings always increases.) Accordingly, reactions in which the enthalpy increases and entropy decreases do not occur. If enthalpy and entropy both increase or both decrease during a reaction, the value of ΔG then depends on the temperature, which governs whether the $T\Delta S$ term of Equation 1-2 is greater than or less than the ΔH term. This means that a large increase in entropy can offset an unfavorable (positive) change in enthalpy. Conversely, the release of a large amount of heat ($\Delta H < 0$) during a reaction can offset an unfavorable decrease in entropy (see Sample Calculation 1-1).

 Review Exercise 1 includes additional calculations related to free energy changes.

REVIEW LEARNING OBJECTIVE 4

- What do enthalpy and entropy mean and how are they related to free energy?
- What does the value of ΔG reveal about a biochemical process?

Sample Calculation 1-1

Problem

Use the information below to calculate the free energy change for the reaction A → B at 25°C. Is the reaction spontaneous?

	Enthalpy ($kJ \cdot mol^{-1}$)	Entropy ($J \cdot K^{-1} \cdot mol^{-1}$)
A	60	22
B	75	97

Solution

First, calculate ΔH and ΔS:

$$\begin{aligned}\Delta H &= H_B - H_A \\ &= 75\ kJ \cdot mol^{-1} - 60\ kJ \cdot mol^{-1} \\ &= 15\ kJ \cdot mol^{-1} \\ &= 15{,}000\ J \cdot mol^{-1}\end{aligned}$$

$$\begin{aligned}\Delta S &= S_B - S_A \\ &= 97\ J \cdot K^{-1} \cdot mol^{-1} - 22\ J \cdot K^{-1} \cdot mol^{-1} \\ &= 75\ J \cdot K^{-1} \cdot mol^{-1}\end{aligned}$$

Next, substitute these values into Equation 1-2. To express the temperature in Kelvin, add 273 to the temperature in degrees Celcius: 273 + 25 = 298K.

$$\begin{aligned}\Delta G &= \Delta H - T\Delta S \\ &= 15{,}000\ J \cdot mol^{-1} - 298K\ (75\ J \cdot K^{-1} \cdot mol^{-1}) \\ &= 15{,}000 - 22{,}350\ J \cdot mol^{-1} \\ &= -7350\ J \cdot mol^{-1} \\ &= -7.35\ kJ \cdot mol^{-1}\end{aligned}$$

Because ΔG is less than zero, the reaction is spontaneous. Even though the change in enthalpy is unfavorable, the large increase in entropy makes ΔG favorable.

Why is life thermodynamically possible?

In order to exist, life must be thermodynamically spontaneous. Does this hold at the molecular level? Many of a cell's metabolic reactions have free energy changes that are less than zero, but some reactions, when analyzed in a test tube (***in vitro,*** literally "in glass"), do not. Nevertheless, these reactions are able to proceed ***in vivo*** (in a living organism) because they occur in concert with other reactions that are thermodynamically favorable. Consider two reactions *in vitro,* one spontaneous ($\Delta G < 0$) and one nonspontaneous ($\Delta G > 0$):

$$\text{A} \rightarrow \text{B} \qquad \Delta G = -20\ \text{kJ} \cdot \text{mol}^{-1}\ \text{(spontaneous)}$$

$$\text{B} \rightarrow \text{C} \qquad \Delta G = +15\ \text{kJ} \cdot \text{mol}^{-1}\ \text{(nonspontaneous)}$$

When the reactions are combined, their ΔG values are added, so the overall process has a negative change in free energy:

$$\text{A} + \text{B} \rightarrow \text{B} + \text{C} \qquad \Delta G = -20\ \text{kJ} \cdot \text{mol}^{-1} + 15\ \text{kJ} \cdot \text{mol}^{-1}$$

$$\text{A} \rightarrow \text{C} \qquad \Delta G = -5\ \text{kJ} \cdot \text{mol}^{-1}$$

Cells couple unfavorable metabolic processes with favorable ones so that the net change in free energy is negative.

Most macroscopic life on earth today is sustained by the energy of the sun (this was not always the case, nor is it true of all organisms). In photosynthetic organisms, such as green plants, light energy excites certain molecules so that their subsequent chemical reactions occur with a net negative change in free energy. These thermodynamically favorable (spontaneous) reactions are coupled to the unfavorable synthesis of monosaccharides from atmospheric CO_2 (Fig. 1-8). In this process, the carbon is **reduced.** Reduction is equivalent to the addition of hydrogen or the removal of oxygen (the oxidation states of carbon are reviewed in Table 1-3). The plant—or an animal that eats the plant—can then break down the monosaccharide to use it as a fuel to power other metabolic activities. In the process, the carbon is **oxidized** (oxygen is added to it or hydrogen is removed), ultimately becoming CO_2. The oxidation of carbon is thermodynamically favorable, so it can be coupled to energy-requiring processes such as the synthesis of building blocks and their polymerization to form macromolecules.

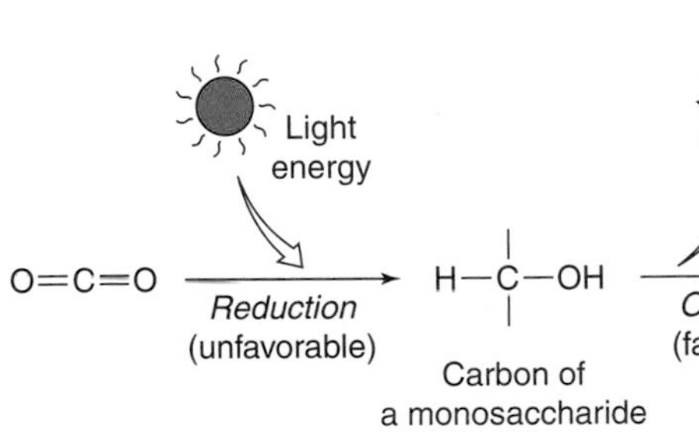

FIGURE 1-8

Reduction and reoxidation of carbon compounds.
The sun provides the free energy to convert CO_2 to reduced compounds such as monosaccharides. The reoxidation of these compounds to CO_2 is thermodynamically spontaneous, so that free energy can be made available for other metabolic processes. Note that free energy is not actually a substance that is physically released from a molecule.

A living organism—with its high level of organization of atoms, molecules, and larger structures—represents a state of low entropy relative to its surroundings. Yet the organism can maintain this thermodynamically unfavorable state as long as it continuously obtains free energy from its food. When the organism ceases to obtain a source of free energy from its surroundings or exhausts its stored food, it reaches equilibrium ($\Delta G = 0$), which results in death. Review Exercise 1 includes an overview of the thermodynamics of living systems.

REVIEW LEARNING OBJECTIVE 5

- How can thermodynamically unfavorable reactions proceed *in vivo*?
- Why must an organism have a steady supply of food?
- Describe the cycle of carbon reduction and oxidation in photosynthesis and breakdown of a compound such as a monosaccharide.

4 THE ORIGIN AND EVOLUTION OF LIFE

Every living cell originates from the division of a parental cell. Thus, the ability to **replicate** (make a replica or copy of itself) is one of the universal

TABLE 1-3 Oxidation states of carbon

Compound*	Formula	
Carbon dioxide	$O{=}C{=}O$	most oxidized (least reduced)
Acetic acid	$H_3C{-}C({=}O)OH$	
Carbon monoxide	$C{\equiv}O$	
Formic acid	$H{-}C({=}O)OH$	
Acetone	$H_3C{-}C({=}O){-}CH_3$	
Acetaldehyde	$H_3C{-}C({=}O)H$	
Formaldehyde	$H_2C{=}O$	
Acetylene	$H{-}C{\equiv}C{-}H$	
Ethanol	$H_3C{-}CH_2{-}OH$	
Ethene	$H_2C{=}CH_2$	
Ethane	$H_3C{-}CH_3$	
Methane	CH_4	least oxidized (most reduced)

*Compounds are listed in order of decreasing oxidation state of the red carbon atom.

characteristics of living organisms. *In order to leave descendants that closely resemble itself, a cell must contain a set of instructions—and the means for carrying them out—that can be transmitted from generation to generation.* Over time, the instructions change gradually, so that species also change, or evolve (Box 1-B). By carefully examining an organism's genetic information and the cellular machinery that supports it, biochemists can draw some conclusions about the organism's development from more ancient life forms. The history of evolution is therefore contained not just within the fossil record, but also in the molecular makeup of all living cells. For example, nucleic

A CLOSER LOOK

BOX 1-B How does evolution work?

Observing evolutionary changes is relatively straightforward, but the mechanisms whereby evolution occurs are prone to misunderstanding. Populations change over time and new species arise as a result of natural selection. Selection operates on individuals, but its effects can be seen in a population only over a period of time. Most populations are collections of individuals who share an overall genetic makeup but also exhibit small variations due to random changes (mutations) in their genetic material as it is passed from parent to offspring. In general, the survival of an individual depends on how well suited it is to the particular conditions under which it lives.

Individuals whose genetic makeup grants them the greatest rate of survival have more chances to leave offspring with the same genetic makeup. Consequently, their characteristics become widespread in a population, and, over time, the population appears to adapt to its environment. A species that is well-suited to its environment tends to persist; a poorly adapted species fails to reproduce and therefore dies out. For this reason, one sign of evolutionary "success" is to be called a living fossil. In the case of the coelocanth, the living fish closely resembles its fossil forms, which are 400 million years old.

Because evolution is the result of random variations and changing probabilities for successful reproduction, it is inherently random and unpredictable. Furthermore, natural selection acts on the raw materials at hand. It cannot create something out of nothing but must operate in increments. For example, the insect wing did not suddenly appear in the offspring of a wingless parent but most likely developed bit by bit, over many generations, by modification of a gill or heat-exchange appendage. Each step of the wing's development would have been subject to natural selection, so that an individual that bore the appendage must have been more likely to survive, perhaps by being able to first glide and then actually fly in pursuit of food or to evade predators.

Although we tend to think of evolution as an imperceptibly slow process, occurring on a geological time scale, it is ongoing and accessible to observation and quantification in the laboratory. For example, under optimal conditions, one complete generation in *E. coli* requires only about 20 minutes. In the laboratory, a culture of *E. coli* cells can progress through about 2500 generations in a year (in contrast, 2500 human generations would require about 60,000 years). Hence, it is possible to subject a population of cultured bacterial cells to some "artificial" selection, for example, by making an essential nutrient scarce, and observe how the genetic composition of the population changes over time as it adapts to the new conditions.

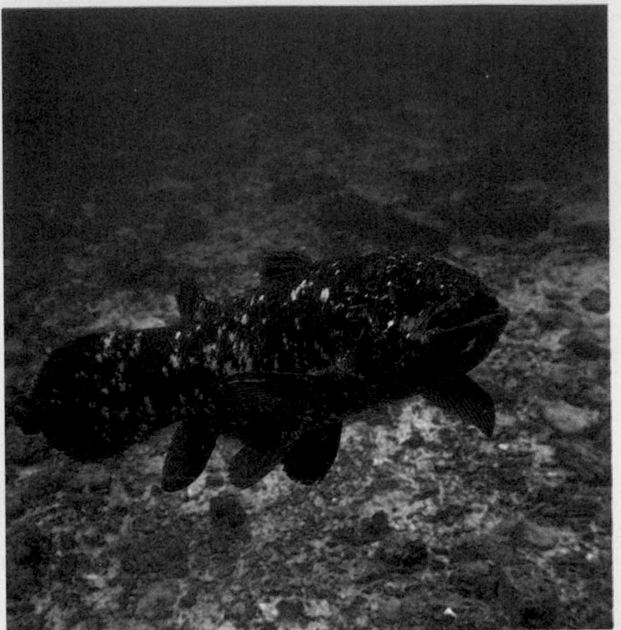

[Peter Scoones/Science Photo Library/Photo Researchers.]

acids participate in the storage and transmission of genetic information in all organisms, and the oxidation of glucose is an almost universal means for generating metabolic free energy. Consequently, DNA, RNA, and glucose must have been present in the ancestor of all cells.

The prebiotic world

A combination of theory and experimental data leads to several plausible scenarios for the emergence of life from nonbiological (prebiotic) materials on the early earth. In one scenario, inorganic compounds such as H_2, H_2O, NH_3, and CH_4—which may have been present in the early atmosphere—could have given rise to simple biomolecules, such as amino acids, when sparked by lightning. Laboratory experiments with the same raw materials and electrical discharges to simulate lightning do in fact yield these molecules (Fig. 1-9). Other experiments suggest that hydrogen cyanide (HCN), formaldehyde (HCOH), and phosphate could have been converted to nucleotides with a similarly modest input of energy.

Over time, simple molecular building blocks could have accumulated and formed larger structures, particularly in shallow waters where evaporation would have had a concentrating effect. Eventually, conditions would have been ripe for the assembly of functional, living cells. Charles Darwin proposed that life might have arisen in some "warm little pond"; however, the early earth was probably a much more violent place, with frequent meteorite impacts and volcanic activity.

An alternative scenario, supported by studies of the metabolism of some modern bacteria, has the first cells developing at deep-sea hydrothermal vents, some of which are characterized by temperatures as high as 350°C and clouds of gaseous H_2S and metal sulfides (giving them the name "black smokers"; Fig. 1-10). In the laboratory, incubating a few small molecules in the presence of iron sulfide and nickel sulfide at 100°C yields acetic acid, an organic compound with a newly formed C—C bond:

$$\underset{\text{Methyl thiol}}{CH_3SH} + \underset{\text{Carbon monoxide}}{CO} + H_2O \xrightarrow{\text{FeS, NiS}} \underset{\text{Acetic acid}}{CH_3COOH} + H_2S$$

Under similar conditions, amino acids spontaneously form short polypeptides. Although the high temperatures that are necessary for their synthesis also tend to break them down, these compounds would have been stable in the cooler water next to the hydrothermal vent.

Regardless of how they formed, *the first biological building blocks would have had to polymerize.* This process might have been stimulated when the organic molecules—often bearing anionic (negatively charged) groups—aligned themselves on a cationic (positively charged) mineral surface.

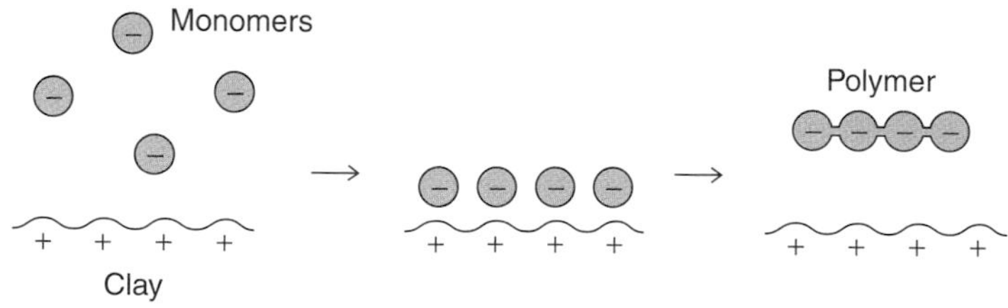

In fact, in the laboratory, common clay promotes the polymerization of nucleotides into RNA.

Primitive polymers would have had to gain the capacity for self-replication. Otherwise, no matter how stable or chemically versatile, such molecules would never have given rise to anything larger or more complicated: The probability of assembling a fully functional cell from a solution of thousands of separate small molecules is practically nil. Because RNA in modern cells represents a form of genetic information and participates in all aspects of expressing that information, it may be similar to the first self-replicating biopolymer. It might have made a copy of itself by first making a **complement,** a sort of mirror

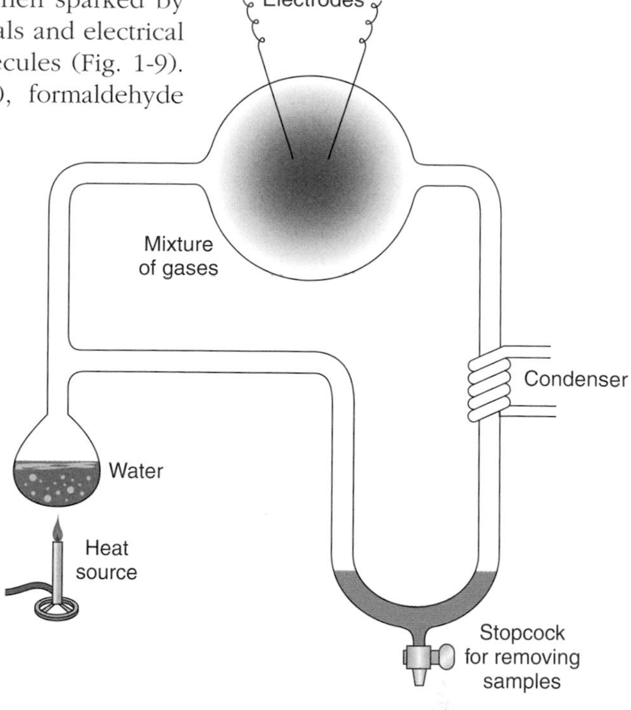

FIGURE 1-9

Laboratory synthesis of biological molecules.
A mixture of gases—H_2, H_2O, NH_3, and CH_4—is subject to an electrical discharge. Newly formed compounds, such as amino acids, accumulate in the aqueous phase as water vapor condenses. Samples of the reaction products can be removed via the stopcock.

FIGURE 1-10

A hydrothermal vent.
Life may have originated at one of these "black smokers," where high temperatures, H_2S, and metal sulfides might have stimulated the formation of biological molecules. [*B. Murton/ Southhampton Oceanography Centre/Science Photo Library/Photo Researchers.*]

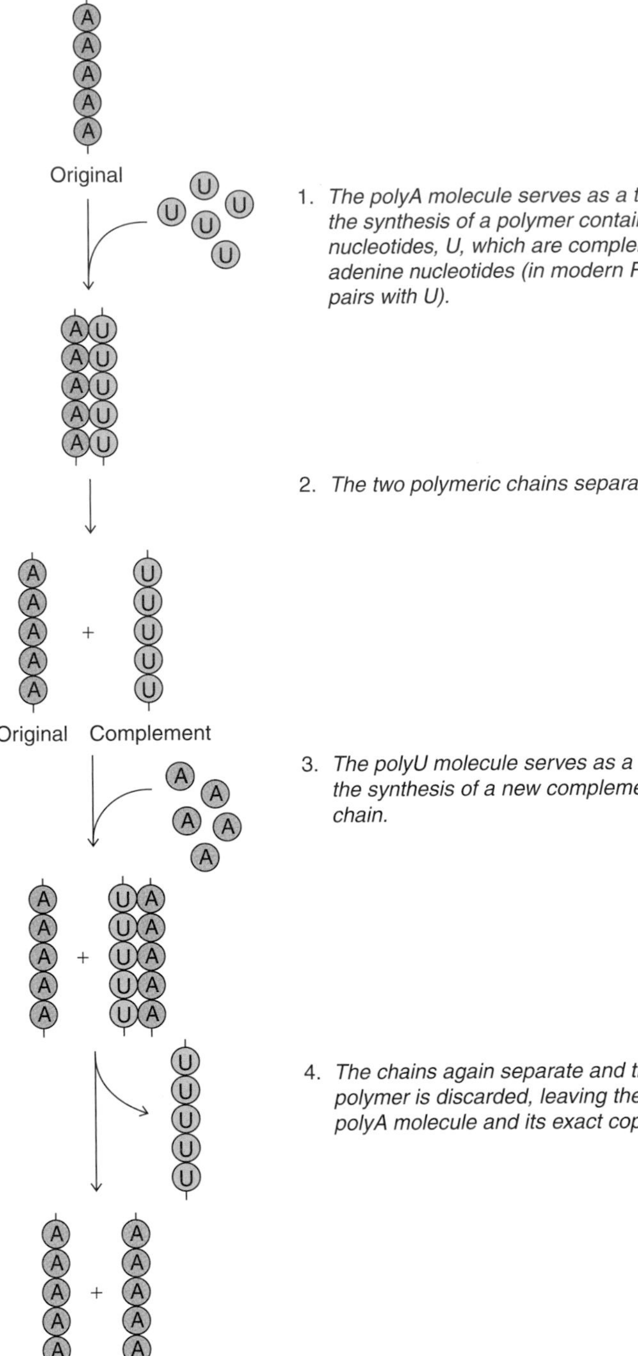

Original
1. The polyA molecule serves as a template for the synthesis of a polymer containing uracil nucleotides, U, which are complementary to adenine nucleotides (in modern RNA, A pairs with U).
2. The two polymeric chains separate.
Original
Complement
3. The polyU molecule serves as a template for the synthesis of a new complementary polyA chain.
4. The chains again separate and the polyU polymer is discarded, leaving the original polyA molecule and its exact copy.
Original
Copy

FIGURE 1-11

Possible mechanism for the self-replication of a primitive RNA molecule. For simplicity, the RNA molecule is shown as a polymer of adenine nucleotides, A.

image, that could then make a complement of *itself,* which would be identical to the original molecule (Fig. 1-11).

REVIEW LEARNING OBJECTIVE 6

- Explain how simple prebiotic compounds could give rise to biological monomers and polymers.
- Why is replication a requirement for life?

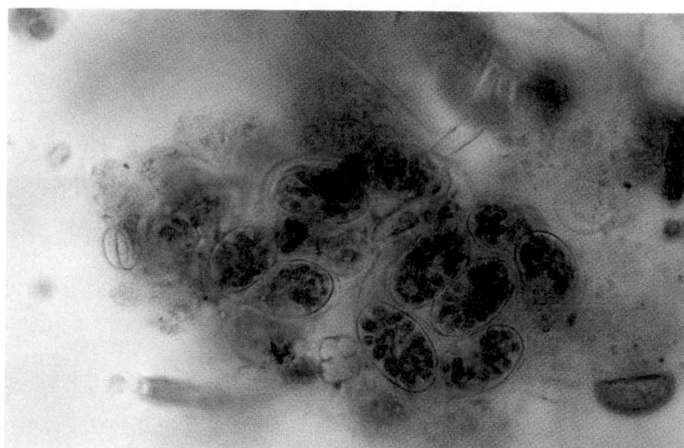

FIGURE 1-12

Cyanobacteria.
The first photosynthetic organisms were probably similar to these bacterial cells, which use the sun's energy to convert CO_2 to reduced organic compounds. [*Sinclair Stammers/Science Photo Library/Photo Researchers.*]

Origins of modern cells

A replicating molecule's chances of increasing in number depend on many factors. **Natural selection**—the phenomenon whereby the entities best suited to the prevailing conditions are the likeliest to survive and multiply—would have favored a replicator that was chemically stable and had a ready supply of building blocks and free energy for synthesizing copies of itself. Accordingly, it would have been advantageous to become enclosed in some sort of membrane that could prevent valuable small molecules from diffusing away. Natural selection would also have favored replicating systems that developed the means for synthesizing their own building blocks and for more efficiently harnessing sources of free energy.

The first cells were probably able to "fix" CO_2—that is, convert it to reduced organic compounds—using the free energy released in the oxidation of readily available inorganic compounds such as H_2S or Fe^{2+}. Vestiges of these processes can be seen in modern metabolic reactions that involve sulfur and iron.

Later, photosynthetic organisms similar to present-day cyanobacteria (also called blue-green algae; Fig. 1-12) used the sun's energy to fix CO_2:

$$CO_2 + H_2O \rightarrow (CH_2O) + O_2$$

The concomitant oxidation of H_2O to O_2 dramatically increased the concentration of atmospheric O_2 and made it possible for **aerobic** (oxygen-using) organisms to avail themselves of this powerful oxidizing agent. The **anaerobic** origins of life are still visible in the most basic metabolic reactions of modern organisms; these reactions proceed in the absence of oxygen. Now that the earth's atmosphere contains about 18% oxygen, anaerobic organisms have not disappeared, but they have been restricted to microenvironments where O_2 is scarce, such as the digestive systems of animals or underwater sediments.

The earth's present-day life forms are of two types, which are distinguished by their cellular architecture:

1. ***Prokaryotes*** *are small unicellular organisms that lack a discrete nucleus and internal membrane system.* This group comprises two subgroups that are remarkably different metabolically although they are similar in appearance: the **Eubacteria** (usually just called **Bacteria**), exemplified by *Escherichia coli;* and the **Archaea,** best known as organisms that inhabit extreme environments, although they are actually found almost everywhere (Fig. 1-13).

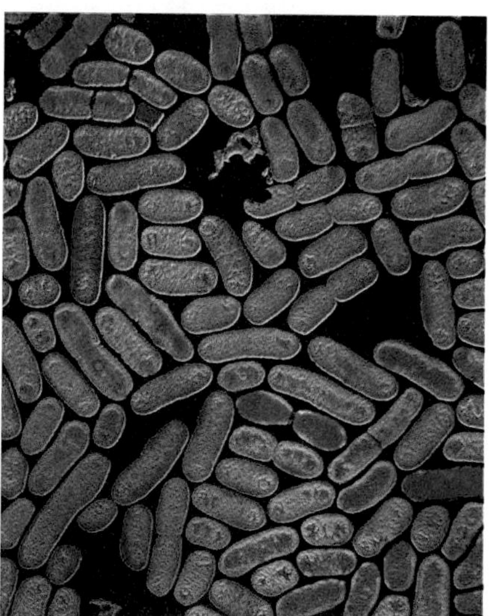

FIGURE 1-13

Prokaryotic cells.
These single-celled *Escherichia coli* bacteria lack internal membrane systems. [*E. Gray/Science Photo Library/Photo Researchers.*]

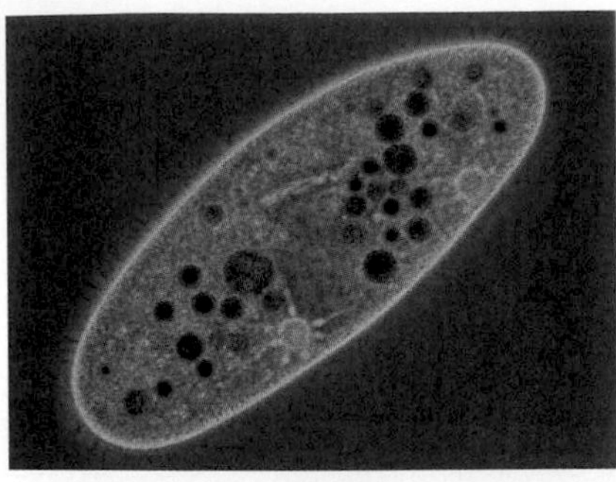

FIGURE 1-14

A eukaryotic cell.
The paramecium, a one-celled organism, contains a nucleus and other membrane-bounded compartments. Compare this cell to the bacterial cells shown in Fig. 1-13. [*Dr. David Patterson/Science Photo Library/Photo Researchers.*]

2. ***Eukaryotic*** *cells are usually larger than prokaryotic cells and contain a nucleus and other membrane-bounded cellular compartments* (such as mitochondria, chloroplasts, and endoplasmic reticulum). Eukaryotes may be unicellular or multicellular. This group (also called the **Eukarya**) includes microscopic organisms as well as familiar macroscopic plants and animals (Fig. 1-14).

By analyzing the sequences of nucleotides in certain genes that are present in all species, it is possible to construct a diagram that indicates how the Bacteria, Archaea, and Eukarya might be related. In theory, *the number of sequence differences between two groups of organisms indicates how long ago they diverged from a common ancestor:* Species with similar sequences have a longer shared evolutionary history than species with dissimilar sequences. This sort of analysis can generate an evolutionary tree such as the one shown in Fig. 1-15. Unfortunately, the branches of these trees tend to shift, depending on which genes are selected for sequence analysis.

To further complicate the elucidation of their evolutionary history, eukaryotic cells exhibit characteristics of both Bacteria and Archaea. Eukaryotic cells also contain organelles that were almost certainly once free-living cells. Specifically, the chloroplasts of plant cells, which carry out photosynthesis, closely resemble the photosynthetic cyanobacteria. The mitochondria of plant and animal cells, which are the site of much of the eukaryotic cell's aerobic metabolism, resemble certain bacteria. In fact, both chloroplasts and mitochondria contain their own genetic material and protein-synthesizing machinery.

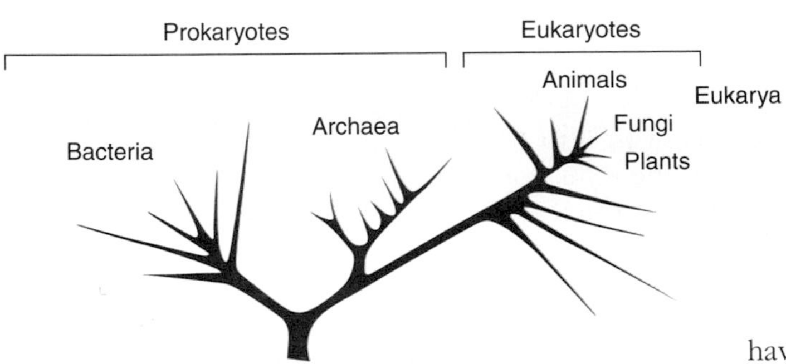

FIGURE 1-15

Evolutionary tree based on nucleotide sequences.
In this example, the Bacteria separated before the Archaea and Eukarya diverged. Note that the closely spaced fungi, plants, and animals are actually more similar to each other than are many groups of prokaryotes. [*After Wheelis, M.L., Kandler, O., and Woese, C.R.,* Proc. Natl. Acad. Sci. USA ***89,*** *2930–2934 (1992).*]

It is likely that an early eukaryotic "host" cell and its prokaryotic "dependents" developed gradually from a mixed population of archaeal and bacterial cells. Over many generations of living in close proximity and sharing each other's metabolic products, some of these cells co-evolved a stable arrangement that eventually produced the mosaic-like modern eukaryote (Fig. 1-16).

At some point, cells in dense populations might have traded their individual existence for a colonial lifestyle. This would have allowed for a division of labor as cells became specialized and would have eventually produced multicellular organisms.

The earth currently sustains about 14 million different species (although estimates vary widely). Perhaps some 500 million species have appeared and vanished over the course of evolutionary history. It is unlikely that the earth harbors more than a few mammals that have yet to be discovered, but new microbial species are routinely described. And although the number of known prokaryotes (about 5000) is much less than the number of known eukaryotes (for example, there are about 750,000 known species of insects), prokaryotic metabolic strategies are amazingly varied. Nevertheless, by documenting char-

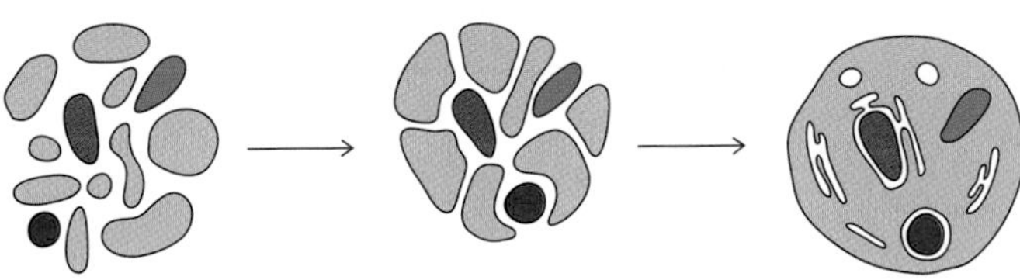

FIGURE 1-16

Possible origin of eukaryotic cells.
The close association of different kinds of free-living cells gradually led to the modern eukaryotic cell, which appears to be a mosaic of bacterial and archaeal features and contains organelles that resemble whole bacterial cells.

acteristics that are common to all species, we can derive far-reaching conclusions about what life is made of, what sustains it, and how it has developed over the eons.

REVIEW LEARNING OBJECTIVE 7

- Why were the first organisms anaerobic, and why are they now relatively scarce?
- Describe the differences between prokaryotes and eukaryotes.
- Why do eukaryotic cells appear to be mosaics?

SUMMARY

1. The most abundant elements in biological molecules are C, N, O, and H, but a variety of other elements are also present in living systems.
2. The major classes of small molecules in cells are amino acids, monosaccharides, nucleotides, and lipids. The major types of biological polymers are proteins, nucleic acids, and polysaccharides.
3. Free energy has two components: enthalpy (heat content) and entropy (disorder). Free energy decreases in a spontaneous process.
4. Life is thermodynamically possible because unfavorable endergonic processes are coupled to favorable exergonic processes.
5. The earliest cells may have evolved in concentrated solutions of molecules or near hydrothermal vents.
6. Eukaryotic cells contain membrane-bounded organelles. Prokaryotic cells, which are smaller and simpler, include the Bacteria and the Archaea.

CHECKLIST

1. Review the Learning Objectives listed on page 2.
2. Complete Exercise 1/Biological Molecules, which explores the general structures of the major biomolecules and their polymeric forms in different viewing formats.
3. Complete Review Exercise 1/Thermodynamics, which provides definitions and calculations as well as applications to biochemical systems.
4. Apply your knowledge by solving the problems at the end of this chapter. Check your results in the Solutions appendix.
5. Be able to define the boldfaced terms (consult the glossary at the end of the book). Test your understanding by taking the Chapter 1 quiz (www.wiley.com/college/pratt).
6. Explore the websites listed at www.wiley.com/college/pratt.
7. Consult the list of Selected Readings for background information or additional details on biomolecules, thermodynamics, and the origin of life.

GLOSSARY TERMS

trace element
amino acid
carbohydrate
monosaccharide
nucleotide
lipid
monomer
polymer
residue
polypeptide
protein
conformation
polynucleotide
nucleic acid
polysaccharide
free energy (G)
enthalpy (H)
entropy (S)
ΔG
spontaneous process
endergonic reaction
nonspontaneous process
exergonic reaction
in vitro
in vivo
reduction
oxidation
replication
complement
natural selection
aerobic
anaerobic
prokaryote
Bacteria
Archaea
eukaryote
Eukarya

PROBLEMS

1. Identify the functional groups in the following molecules:

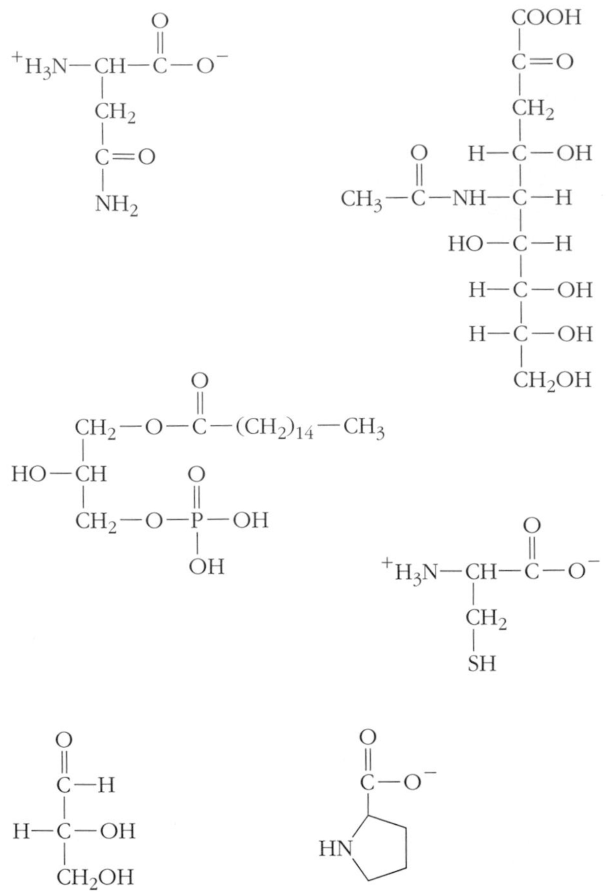

$$^{+}H_3N-CH-\overset{\overset{O}{\|}}{C}-O^-$$
CH_2
$C=O$
NH_2

$COOH$
$C=O$
CH_2
$H-C-OH$
$CH_3-\overset{\overset{O}{\|}}{C}-NH-C-H$
$HO-C-H$
$H-C-OH$
$H-C-OH$
CH_2OH

$CH_2-O-\overset{\overset{O}{\|}}{C}-(CH_2)_{14}-CH_3$
$HO-CH$
$CH_2-O-\overset{\overset{O}{\|}}{P}-OH$
OH

$$^{+}H_3N-CH-\overset{\overset{O}{\|}}{C}-O^-$$
CH_2
SH

$\overset{\overset{O}{\|}}{C}-H$
$H-C-OH$
CH_2OH

$\overset{\overset{O}{\|}}{C}-O^-$
HN

2. Name the four types of biological small molecules. Which three are capable of forming polymeric structures? What are the names of the polymeric structures that are formed?

3. To which of the four classes of biomolecules do the following compounds belong?

(a) CH_2OH, O, H, H, OH, OH, H, HO, H, H, $NH-\overset{O}{\overset{\|}{C}}-CH_3$

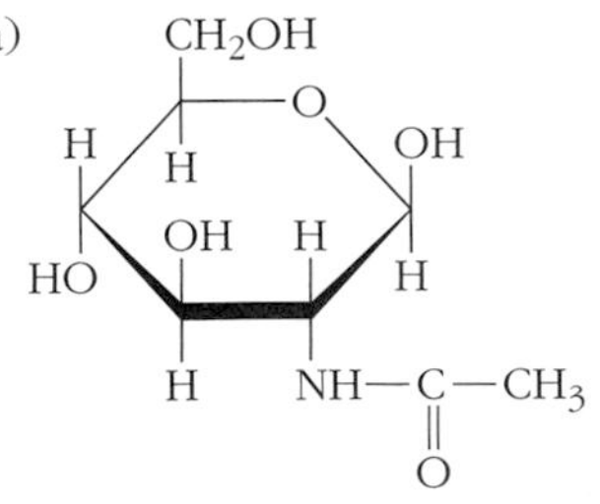

(b) $R-\overset{\overset{O}{\|}}{C}-O$

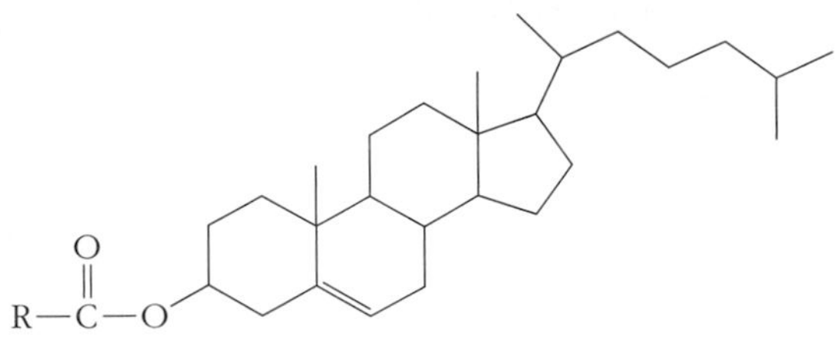

(c) $HS-CH_2-CH_2-\underset{NH_3^+}{\underset{|}{CH}}-COO^-$

(d) $^{-}O-\overset{\overset{O}{\|}}{\underset{O^-}{P}}-O-CH_2$, O, H, H, H, H, OH, OH, NH_2, N, O, N

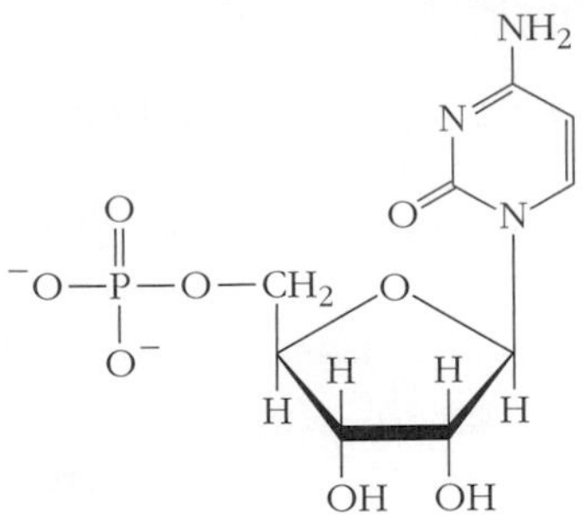

4. There are 20 different amino acids in proteins (see Fig. 4-2). Each has the same basic structure with the exception of the R group that is unique to each amino acid.

(a) Identify the functional groups present in all amino acids.

(b) Draw the structure of the amino acid alanine. What is special about the central carbon atom of alanine?

5. Draw the "straight chain" structure of glucose. What functional groups are present in the glucose molecule?

6. What are the structural components of biological molecules called nucleotides?

7. Compare the solubilities in water of alanine, palmitate, and cholesterol, and explain your reasoning.

8. What polymeric molecule forms a more regular structure, RNA or protein? Explain this observation in terms of the cellular roles of the two different molecules.

9. What are the two major biological roles of polysaccharides?

10. What is the sign of the entropy change for each of the following processes?

(a) Water freezes.

(b) Water evaporates.

(c) Dry ice sublimes.

(d) Sodium chloride dissolves in water.

(e) Several different types of lipid molecules assemble to form a membrane.

(f) A cell combusts glucose in order to obtain energy for cellular work.

11. Which has the greater entropy, a polymeric molecule or a mixture of its constituent monomers?

12. Consider the conversion of glucose to glucose-6-phosphate:

$$\text{Glucose} + \text{phosphate} \rightleftharpoons \text{glucose-6-phosphate} + H_2O \qquad \Delta G^{\circ\prime} = 13.8 \text{ kJ} \cdot \text{mol}^{-1}$$

(a) Is the reaction favorable? Explain.

(b) Suppose the synthesis of glucose-6-phosphate is coupled with the hydrolysis of ATP.

$$\text{ATP} + H_2O \rightleftharpoons \text{ADP} + \text{phosphate} \qquad \Delta G^{\circ\prime} = -30.5 \text{ kJ} \cdot \text{mol}^{-1}$$

Write the overall equation for the coupled process and calculate the $\Delta G^{\circ\prime}$ of the coupled reaction. Is the production of glucose-6-phosphate favorable? Explain.

13. For the reaction in which reactant A is converted to product B, tell whether this process is favorable at (a) 4°C and (b) 37°C.

	H (kJ · mol^{-1})	S (J · K^{-1} · mol^{-1})
A	54	22
B	60	43

14. For a given reaction, the value of ΔH is 15 kJ · mol^{-1} and the value of ΔS is 51 J · K^{-1} · mol^{-1}. Above what temperature will this reaction be spontaneous?

15. Does entropy increase or decrease in the following reactions in aqueous solution?

(a) 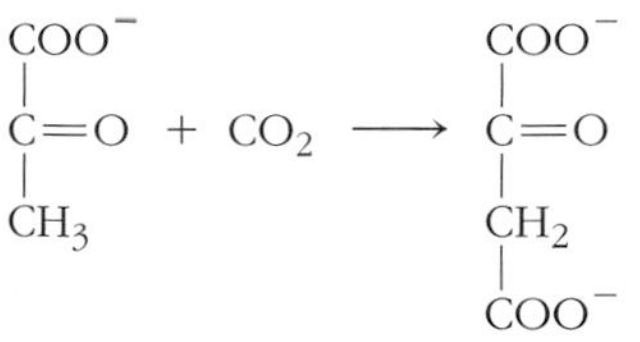 $^-OOC{-}C({=}O){-}CH_3 + CO_2 \longrightarrow {^-OOC}{-}C({=}O){-}CH_2{-}COO^-$

(b) 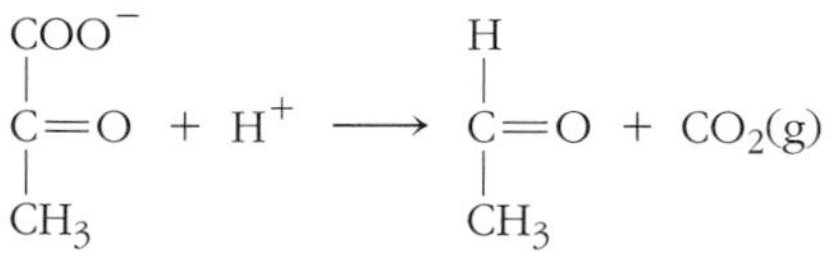 $^-OOC{-}C({=}O){-}CH_3 + H^+ \longrightarrow H{-}C({=}O){-}CH_3 + CO_2(g)$

16. Which of the following processes are never spontaneous?

(a) A reaction that occurs with any size decrease in enthalpy and any size increase in entropy.
(b) A reaction that occurs with a small increase in enthalpy and a large increase in entropy.
(c) A reaction that occurs with a large decrease in enthalpy and a small decrease in entropy.
(d) A reaction that occurs with any size increase in enthalpy and any size decrease in entropy.

17. The ΔG value at 37°C for the arginine kinase reaction was recently determined. The ΔH value for the reaction is −8.19 kJ · mol^{-1}and the value of ΔS is 2.2 J · K^{-1} · mol^{-1}.

(a) Is the reaction exothermic or endothermic?
(b) What is the value of ΔG for the reaction? Is the reaction spontaneous?
(c) Which component makes a greater contribution to the free energy value: the ΔH or ΔS value? Comment on the significance of this observation.

18. Indicate whether the process described in each statement is an oxidation or reduction process.

(a) A plant synthesizes monosaccharides from carbon dioxide during photosynthesis.
(b) An animal eats the plant and breaks down the monosaccharide in order to obtain energy for cellular processes.

19. For each of the following reactions, tell whether the reactant is being oxidized or reduced. Reactions may not be balanced.

(a) 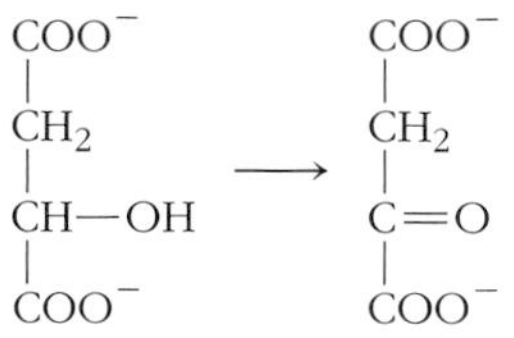 $^-OOC{-}CH_2{-}CH(OH){-}COO^- \longrightarrow {^-OOC}{-}CH_2{-}C({=}O){-}COO^-$

(b) $CH_3{-}(CH_2)_{14}{-}C({=}O){-}O^- \longrightarrow 8\ CH_3{-}C({=}O){-}O^-$

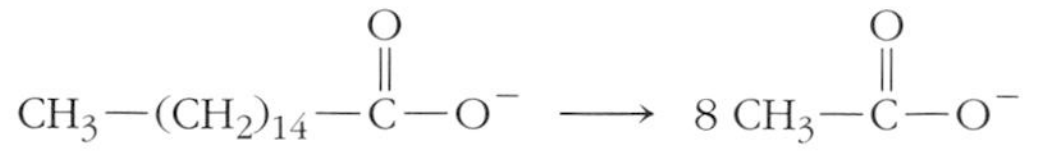

(c) $^-OOC{-}CH{=}CH{-}COO^- \longrightarrow {^-OOC}{-}CH_2{-}CH(OH){-}COO^-$

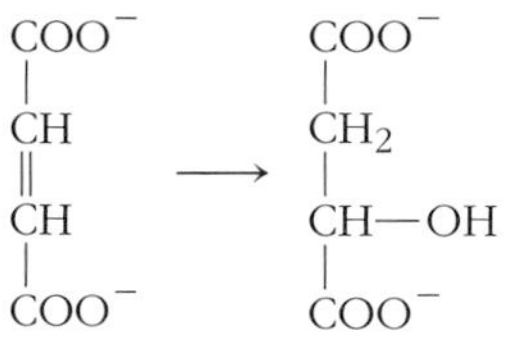

(d) 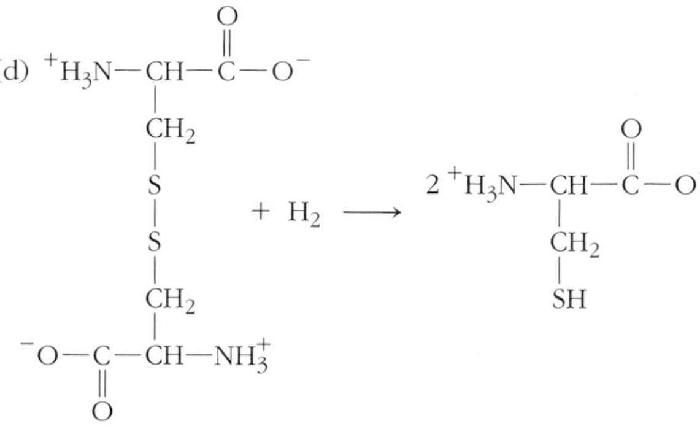 $^+H_3N{-}CH(COO^-){-}CH_2{-}S{-}S{-}CH_2{-}CH(NH_3^+){-}COO^- + H_2 \longrightarrow 2\ {^+H_3N}{-}CH(CH_2{-}SH){-}C({=}O){-}O^-$

20. In order to give rise to more highly complex structures, what capabilities did the first biological molecules have to have?

21. In some cells, lipids such as palmitate (shown on page 7), rather than monosaccharides, serve as the primary metabolic fuel.

(a) Consider the oxidation state of palmitate's carbon atoms and explain how it fits into a scheme such as the one shown in Fig. 1-8.
(b) On a per-carbon basis, which would make more free energy available for metabolic reactions: palmitate or glucose?

22. Why is molecular information so important for classifying and tracing the evolutionary relatedness of bacterial species but less important for vertebrate species?

23. The first theories to explain the similarities between bacteria and mitochondria or chloroplasts suggested that an early eukaryotic cell actually engulfed but failed to fully digest a free-living prokaryotic cell. Why is such an event unlikely to account for the origin of mitochondria or chloroplasts?

SELECTED READINGS

Nisbet, E.G. and Sleep, N.H., The habitat and nature of early life, *Nature* **409,** 1083–1091 (2001). [Explains some of the hypotheses regarding the early earth and the origin of life, including the possibility that life originated at hydrothermal vents.]

Pace, N.R., A molecular view of microbial diversity and the biosphere, *Science* **276,** 734–740 (1997). [Describes the three domains of life and their general metabolic strategies.]

Tinoco, I., Jr., Sauer, K., and Wang, J.C., *Physical Chemistry: Principles and Applications in Biological Sciences* (3rd ed.), Chapters 2–5, Prentice Hall (1996). [This and other physical chemistry textbooks present the basic equations of thermodynamics.]

CHAPTER 2

Aqueous Chemistry

Geochemical evidence suggests that several times during its history, the earth was entirely frozen, looking much like present-day Antarctica. According to the "snowball earth" theory, the planet's oceans were topped with ice up to a kilometer thick and glaciers covered the continents. Each 5- to 50-million-year-long global freeze probably ended when volcanoes spewing CO_2 triggered a greenhouse effect that melted the ice. During the frozen periods, the last ending about 570 million years ago, only simple organisms such as bacteria and algae existed, and these were largely confined to the lightless oceans beneath the ice crust. The oceans remained liquid because water's molecular structure makes ice less dense than liquid water. The life forms that survived under the ice later gave rise to the multicellular organisms that now populate a more hospitable planet. A similar phenomenon occurs every winter, when ice forms on a lake, leaving in the liquid water beneath the fish, frogs, and insects that emerge when the ice melts the following spring.

[British Antarctic Survey/Science Photo Library/Photo Researchers.]

LEARNING OBJECTIVES

1. Understand how the polar water molecule forms hydrogen bonds.
2. Understand the different types of noncovalent forces acting on biological molecules.
3. Understand why water dissolves ionic and polar substances.
4. Understand why the hydrophobic effect is driven by entropy.
5. Understand why amphiphiles form micelles or bilayers.
6. Understand the nature and relationship of the H^+ and OH^- components of water.
7. Understand that a p*K* value describes an acid's tendency to ionize.
8. Understand the relationship between pH and the p*K* of an acid.
9. Understand how buffers work.

THIS CHAPTER IN CONTEXT

This chapter, like the first, serves as an introduction to a variety of topics in the remainder of this book. First, water is the medium of life. It surrounds biological molecules and participates in their chemical transformations. The interaction of water with other substances—either through electrostatic or hydrophobic effects—is a key determinant of molecular structure in proteins (discussed in Chapters 4 and 5) and in membranes (Chapter 8). Similarly, water's reactive components—its H^+ and OH^- ions—influence the structures of biological molecules as well as their chemical reactivity. As we will see, acid–base chemistry in aqueous solution is central to the mechanisms of enzymes (Chapters 6 and 7) and the metabolic transformations of matter and energy in cells (Chapters 9–15).

LIVING ORGANISMS CAN BE FOUND IN VIRTUALLY EVERY PORTION of the earth that contains liquid water. In the polar ice caps, prokaryotes and small eukaryotes survive in the spaces between ice crystals (Fig. 2-1). The hot waters near hydrothermal vents host the prokaryote ***Pyrolobus fumarii***, which grows best at a temperature of 105°C and can tolerate 113°C. Living organisms have even been discovered several kilometers below the earth's surface, but only where water is present.

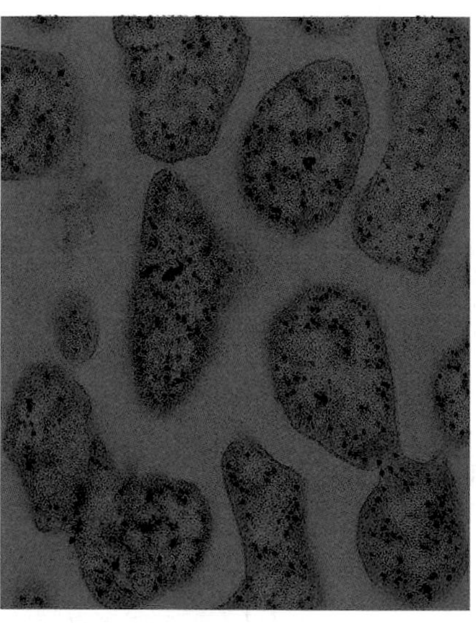

FIGURE 2-1

Methanococcoides burtonii. These bacterial cells from an Antarctic lake survive at temperatures as low as −2.5°C. [*Dr. M. Rohde, GBF/Science Photo Library/Photo Researchers.*]

Water is a fundamental requirement for life, so it is important to understand the structural and chemical properties of water. Not only are most biological molecules surrounded by water, but their molecular structure is in part governed by how their component groups interact with water. And water plays a role in how these molecules assemble to form larger structures or undergo chemical transformation. In fact, water itself—or its H^+ and OH^- constituents—participates directly in many biochemical processes. Therefore, an examination of water is a logical prelude to exploring the structures and functions of biomolecules in the following chapters.

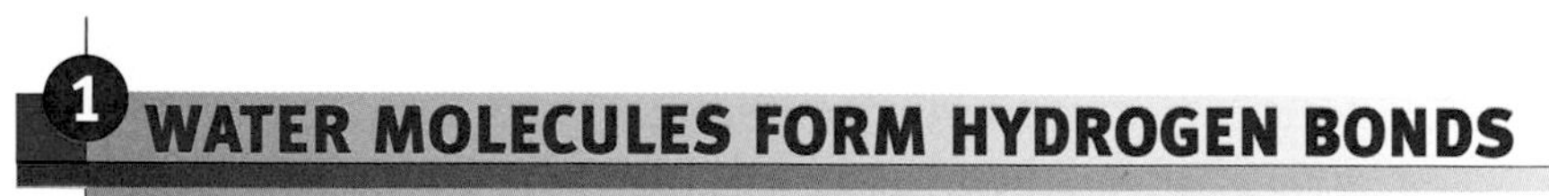

1 WATER MOLECULES FORM HYDROGEN BONDS

What is the nature of the substance that accounts for about 70% of the mass of most organisms? The human body, for example, is about 60% by weight water, most of it in the interstitial fluid (the fluid surrounding cells) and inside cells:

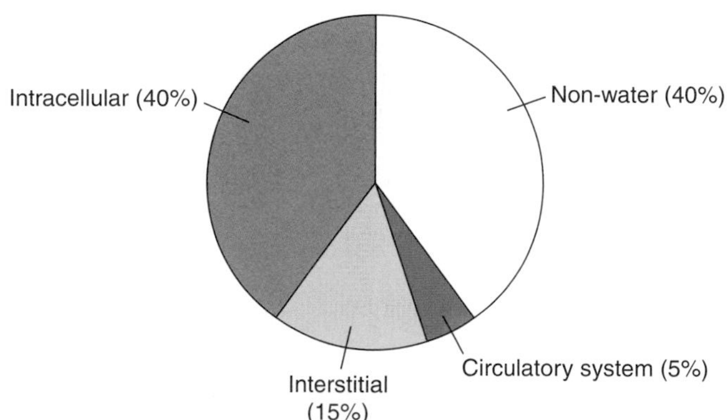

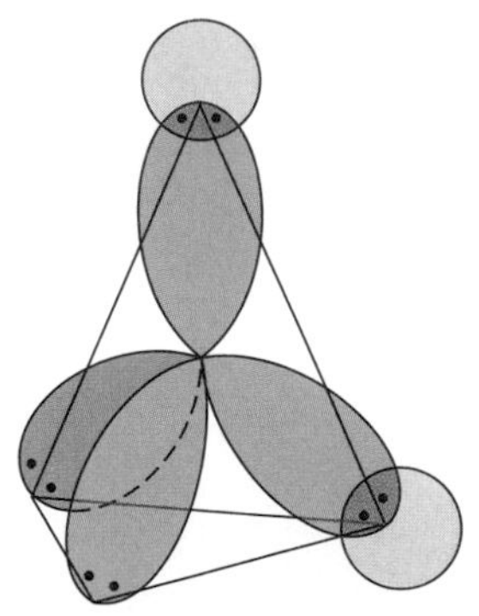

FIGURE 2-2

Electronic structure of the water molecule. Four electron orbitals, in an approximately tetrahedral arrangement, surround the central oxygen. Two orbitals participate in bonding to hydrogen (gray), and two contain unshared electron pairs.

In an individual H_2O molecule, the central oxygen atom forms covalent bonds with two hydrogen atoms, leaving two unshared pairs of electrons. The molecule therefore has approximately tetrahedral geometry, with the oxygen atom at the center of the tetrahedron, the hydrogen atoms at two of the four corners, and electrons at the other two corners (Fig. 2-2).

As a result of this electronic arrangement, *the water molecule is* ***polar;*** that is, it has an uneven distribution of charge. The oxygen atom bears a partial negative charge (indicated by the symbol δ^-), and each hydrogen atom bears a partial positive charge (indicated by the symbol δ^+):

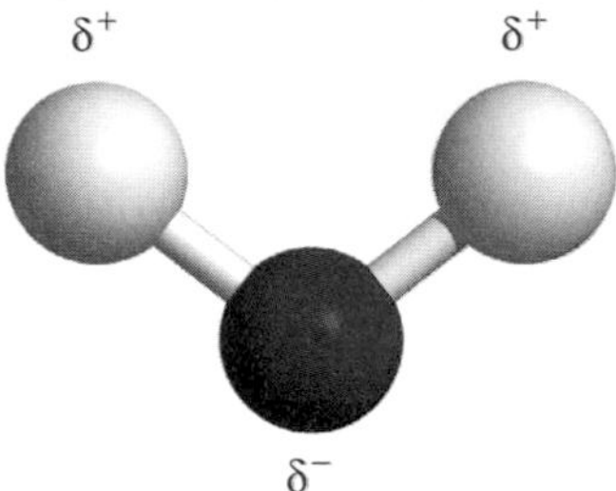

This polarity is the key to many of water's unique physical properties.

Neighboring water molecules tend to orient themselves so that each partially positive hydrogen is aligned with a partially negative oxygen:

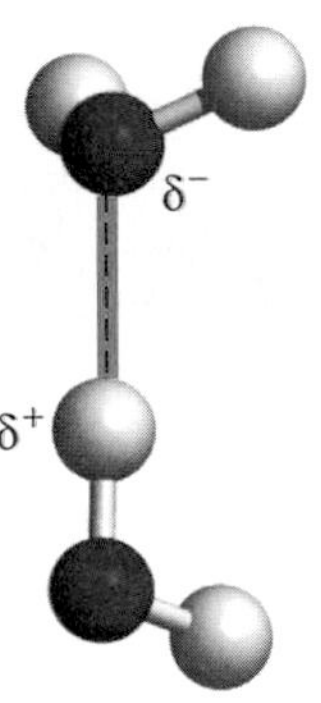

This interaction, shaded yellow here, is known as a **hydrogen bond.** Traditionally shown as a simple electrostatic attraction between oppositely charged particles, the hydrogen bond is now known to have some covalent character. In other words, the electrons of the water molecules spend some time in the hydrogen bond as well as in the covalent O—H bonds.

Each water molecule can potentially participate in four hydrogen bonds, since it has two hydrogen atoms to "donate" to a hydrogen bond and two pairs of unshared electrons that can "accept" a hydrogen bond. In ice, a crystalline form of water, each water molecule does indeed form hydrogen bonds with four other water molecules (Fig. 2-3). This regular, lattice-like structure breaks down to some extent when the ice melts.

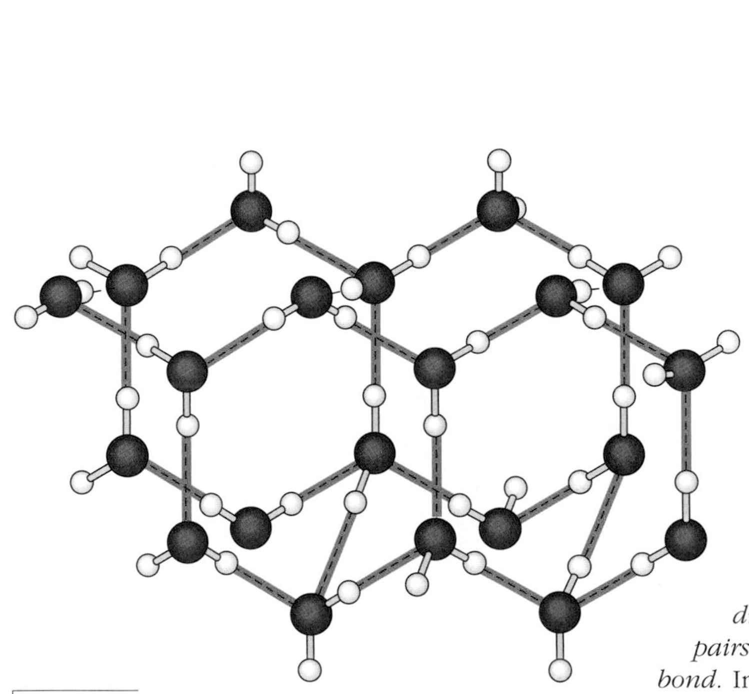

FIGURE 2-3

Structure of ice. Each water molecule acts as a donor for two hydrogen bonds and an acceptor for two hydrogen bonds, thereby interacting with four other water molecules in the crystal. (Only two layers of water molecules are shown here.)

In liquid water, each molecule still forms hydrogen bonds with up to four other water molecules, but each bond has a lifetime of only about 10^{-12} s. As a result, *the structure of water is continually flickering as water molecules rotate, bend, and reorient themselves.* Theoretical calculations and spectroscopic data suggest that water molecules spontaneously form small hydrogen-bonded clusters such as the 6-membered ring shown here:

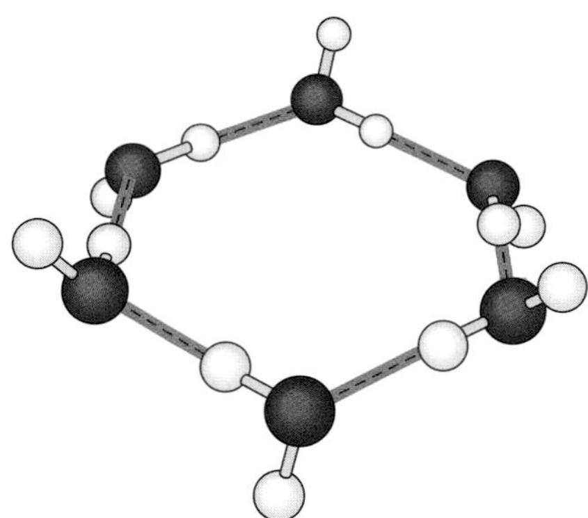

Although transient, such structures maximize the number of hydrogen bonds formed while minimizing strain.

Because of its ability to form hydrogen bonds, water is highly cohesive. This accounts for its high surface tension, which allows certain insects to walk on water (Fig. 2-4). The cohesiveness of water molecules also explains why water remains a liquid whereas similar molecules such as CH_4 and H_2S are gases at room temperature (25°C).

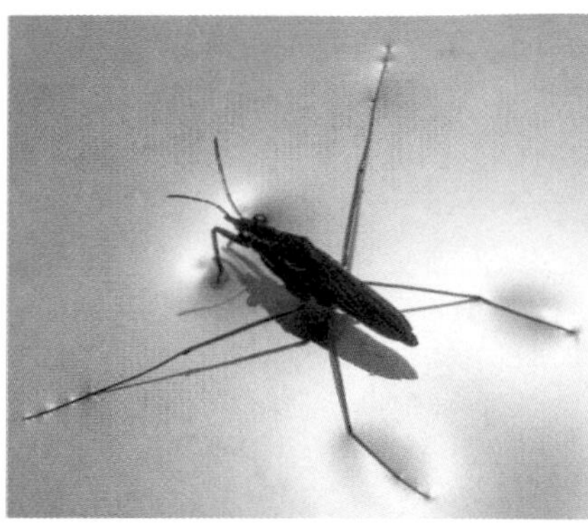

FIGURE 2-4

A water strider supported by the surface tension of water.
[© *E.R. Degginger/Bruce Coleman, Inc.*]

REVIEW LEARNING OBJECTIVE 1

- Why is a water molecule polar?
- What is a hydrogen bond and why does it form between water molecules?
- Describe the structure of liquid water.

Hydrogen bonds are one type of electrostatic force

Biochemists are mainly concerned not with the powerful covalent bonds that define basic molecular constitutions, but with the much weaker noncovalent bonds, including hydrogen bonds, that govern the final three-dimensional shapes of molecules and how they interact with each other. For example, about 460 kJ · mol^{-1} of energy is required to break a covalent O—H bond. But a hydrogen bond in water has a strength of only about 20 kJ · mol^{-1}. Other noncovalent interactions are weaker still.

Among the noncovalent interactions that occur in biological molecules are electrostatic interactions between charged groups such as carboxylate ($—COO^-$) and amino ($—NH_3^+$) groups. These **ionic interactions** are intermediate in strength to covalent bonds and hydrogen bonds (Fig. 2-5).

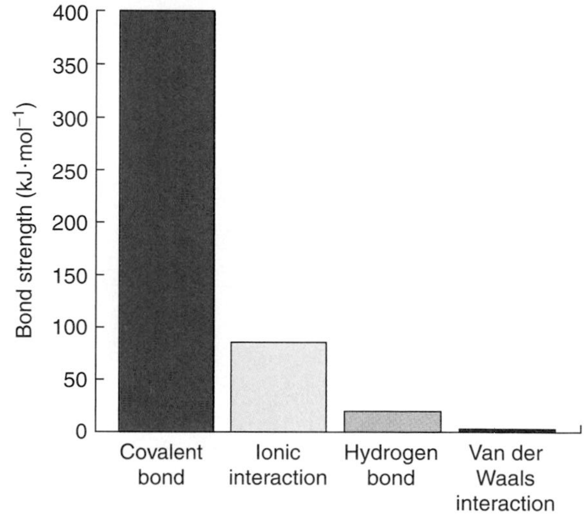

FIGURE 2-5

Relative strengths of bonds in biological molecules.

Hydrogen bonds, despite their partial covalent nature, are classified as a type of electrostatic interaction. At about 1.8 Å, they are longer and hence weaker than a covalent O—H bond (which is about 1 Å long). However, a completely noninteracting O and H would approach no closer than about 2.7Å, which is the sum of their **van der Waals radii** (the van der

Waals radius of an isolated atom is the distance from its nucleus to its effective electronic surface).

O —1 Å— H Covalent bond

O —1.8 Å— H Hydrogen bond

O —1.5 Å—|—1.2 Å— H Nonbonding

Hydrogen bonds usually involve N—H, O—H, and S—H groups as hydrogen donors, and the electronegative N, O, or S atoms as hydrogen acceptors (**electronegativity** is a measure of an atom's affinity for electrons). *Water, therefore, can form hydrogen bonds not just with other water molecules but with a wide variety of other compounds that bear N-, O-, or S-containing functional groups.*

H—O—H······O(H)(R) Water–alcohol

H₂O······H—N(R)(R) Water–amine

Likewise, these functional groups can form hydrogen bonds among themselves. For example, the complementarity of bases in DNA and RNA is determined by their ability to form hydrogen bonds with each other:

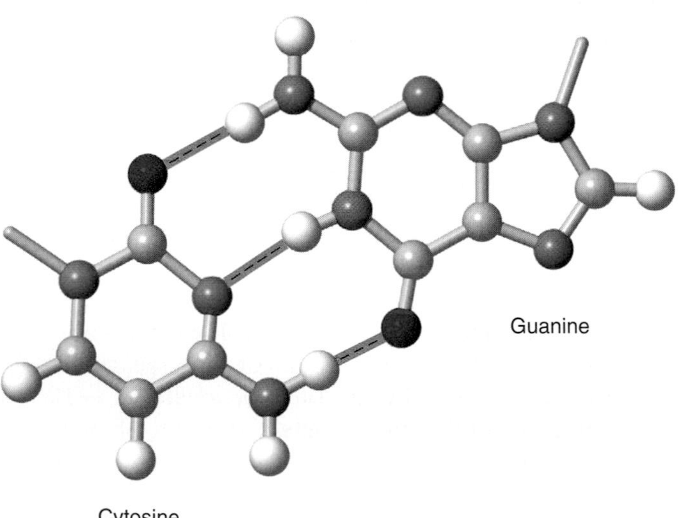

Other electrostatic interactions occur between particles that are polar but not actually charged, for example, two carbonyl groups:

$$>C=O^{\delta^-}\cdots\cdots {}^{\delta^+}C=O$$

These forces, called **van der Waals interactions,** are usually weaker than hydrogen bonds. The interaction shown above, between two strongly polar groups, is known as a **dipole–dipole interaction** and has a strength of about 9 kJ · mol^{-1}. Very weak van der Waals interactions, called **London dispersion forces,** occur between nonpolar molecules as a result of small fluctuations in their distribution of electrons that create a temporary separation of charge. Nonpolar groups such as methyl groups can therefore experience a small attractive force, in this case about 0.3 kJ · mol^{-1}:

$$\text{—}\overset{\delta^-}{\mathrm{C}}\text{—}\overset{\delta^+}{\mathrm{H}}\cdots\cdots\overset{\delta^-}{\mathrm{H}}\text{—}\overset{\delta^+}{\mathrm{C}}\text{—}$$

FIGURE 2-6

The cumulative effect of small forces. Just as the fictional giant Gulliver was restrained by many small tethers at the hands of the tiny Lilliputians, the structures of macromolecules are constrained by the effects of many weak noncovalent interactions. [*Hulton Archive/ Getty Images.*]

Not surprisingly, these forces act only when the groups are very close, and their strength quickly falls off as the groups draw apart. If the groups approach too closely, however, their van der Waals radii collide and a strong repulsive force overcomes the attractive force.

Although hydrogen bonds and van der Waals interactions are individually weak, biological molecules usually contain multiple groups capable of participating in these intermolecular interactions, so that their cumulative effect can be significant (Fig. 2-6).

REVIEW LEARNING OBJECTIVE 2

- Describe the nature and relative strength of ionic interactions, hydrogen bonds, and van der Waals interactions.

Water dissolves many compounds

The ability of the water molecule to form hydrogen bonds and participate in other electrostatic interactions explains why water is an effective solvent for a wide variety of compounds. The polar nature of the water molecule allows it to associate with ions (for example, the Na^+ and Cl^- ions from the salt NaCl) by aligning its partial charges accordingly. Because the interactions between the polar water molecules and the ions are stronger than the attractive forces between the Na^+ and Cl^- ions, the salt dissolves (the dissolved particle is called a **solute**). Each solute ion surrounded by water molecules is said to be **solvated** (or **hydrated,** to indicate that the solvent is water).

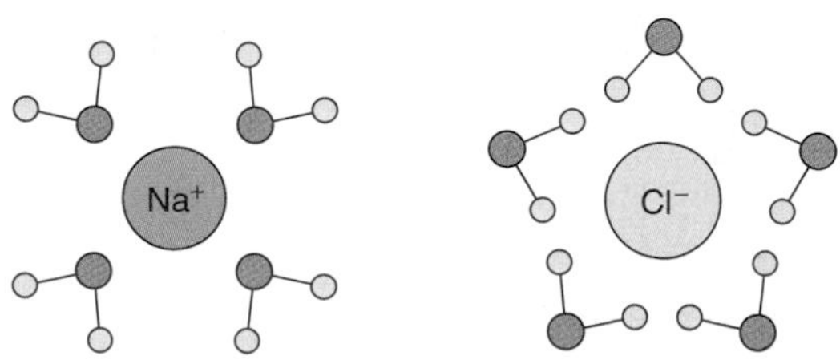

Biological molecules that bear polar or ionic groups are also readily solubilized, in this case because their groups can form hydrogen bonds with the

solvent water molecules. Glucose, for example, with its six hydrogen-bonding oxygens, is highly soluble in water:

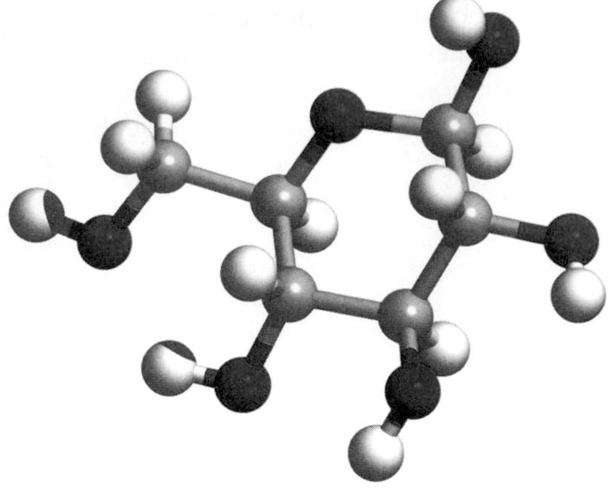

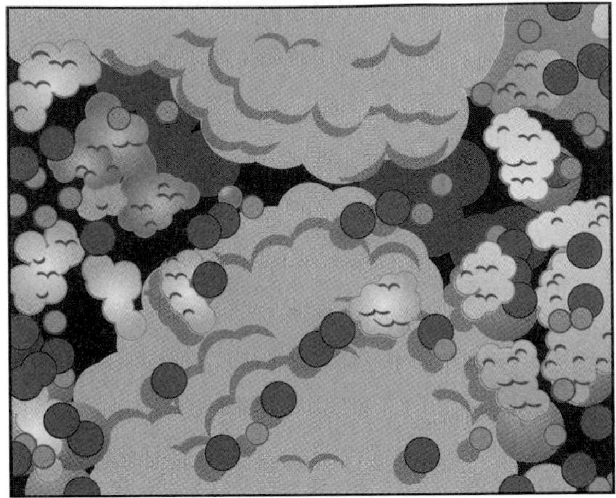

FIGURE 2-7

A solution of molecules inside a cell. Large numbers of ions and molecules of all sizes form a dense solution. Water molecules (not shown here) fill the spaces between these molecules. [*After Goodsell, D.S., Am. Scientist* ***80,*** *458 (1992).*]

The concentration of glucose in human blood is about 5 mM. In a solution of 5 mM glucose in water, there are about 10,000 water molecules for every glucose molecule (the water molecules are present at a concentration of about 55.5 M). However, biological molecules are never found alone in such dilute conditions *in vivo,* because a large number of ions, small molecules, and large polymers collectively form a solution that is more like a hearty stew than a thin watery soup (Fig. 2-7).

Inside a cell, the spaces between molecules may be only a few Å wide, enough room for only two water molecules to fit. This allows solute molecules, each with a coating of properly oriented water molecules, to slide past each other. This thin coating of water may be enough to keep molecules from coming into van der Waals contact (van der Waals interactions are weak but attractive), thereby helping maintain the cell's contents in a fluid state.

REVIEW LEARNING OBJECTIVE 3

- What happens when an ionic substance dissolves?

2 THE HYDROPHOBIC EFFECT

Glucose and other readily hydrated substances are said to be **hydrophilic** (water-loving). In contrast, a compound such as dodecane (a C_{12} alkane),

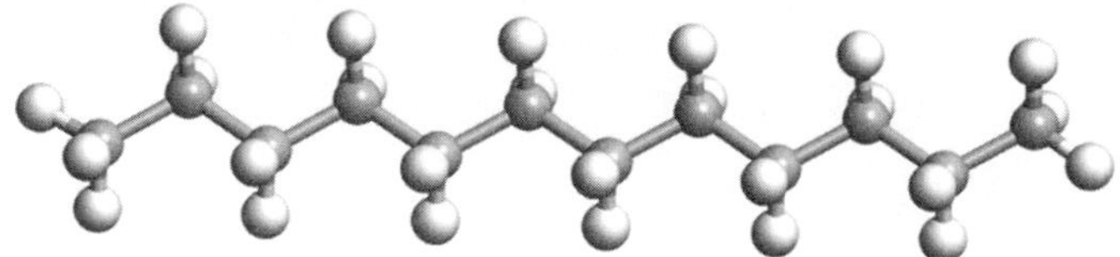

which lacks polar groups, is relatively insoluble in water and is said to be **hydrophobic** (water-fearing). Although pure hydrocarbons are rare in biological systems, many biological molecules contain hydrocarbon-like portions that are insoluble in water.

When a nonpolar substance such as vegetable oil (which consists of hydrocarbon-like molecules) is added to water, it does not dissolve but forms

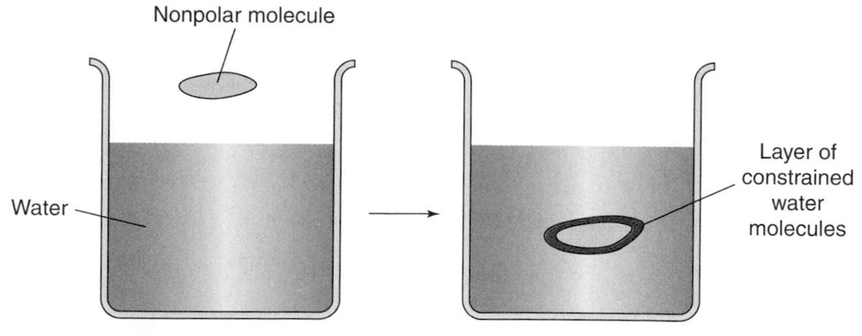

FIGURE 2-8

Hydration of a nonpolar molecule.
When a nonpolar molecule (green) is added to water, the system loses entropy because the water molecules surrounding the nonpolar solute (orange) lose their freedom to form hydrogen bonds. The loss of entropy is a property of the entire system, not just the water molecules nearest the solute, because these molecules are continually changing places with water molecules from the rest of the solution. The loss of entropy presents a thermodynamic barrier to the hydration of a nonpolar solute.

a separate phase. In order for the water and oil to mix, free energy must be added to the system (for example, by vigorous stirring or the application of heat). Why is it thermodynamically unfavorable to dissolve a hydrophobic substance in water? One possibility is that enthalpy is required to break the hydrogen bonds among solvent water molecules in order to create a "hole" into which a nonpolar molecule can fit. Experimental measurements, however, show that the free energy barrier (ΔG) to the solvation process depends much more on the entropy term (ΔS) than on the enthalpy term (ΔH; recall from Chapter 1 that $\Delta G = \Delta H - T\Delta S$; Equation 1-2). This is because when a hydrophobic molecule is hydrated, it becomes surrounded by a layer of water molecules that cannot participate in normal hydrogen bonding but instead must align themselves so that their polar ends are not oriented toward the nonpolar solute. *This constraint on the structure of water represents a loss of entropy in the system,* because now the highly mobile water molecules have lost some of their freedom to rapidly form, break, and re-form hydrogen bonds with other water molecules (Fig. 2-8). The loss of entropy is not due to the formation of a frozen "cage" of water molecules around the nonpolar solute, as commonly pictured, because in liquid water, the solvent molecules are in constant motion.

When a large number of nonpolar molecules are introduced into a sample of water, they do not disperse and become individually hydrated, each surrounded by a layer of water molecules. Instead, the nonpolar molecules tend to clump together, removing themselves from contact with water molecules. (This explains why small oil droplets coalesce into one large oily phase.) Although the entropy of the nonpolar molecules is thereby reduced, this thermodynamically unfavorable event is more than offset by the increase in the entropy of the water molecules, which regain their ability to interact freely with other water molecules (Fig. 2-9).

The exclusion of nonpolar substances from an aqueous solution is known as the ***hydrophobic effect.*** It is a powerful force in biochemical systems, even though it is not a bond or attractive interaction in the conventional sense. The nonpolar molecules do not experience any additional attractive force among themselves; they aggregate only because they are driven out of the aqueous phase by the unfavorable entropy cost of individually hydrating them.

The hydrophobic effect governs the structures and functions of many biological molecules. For example, the polypeptide chain of a protein folds into

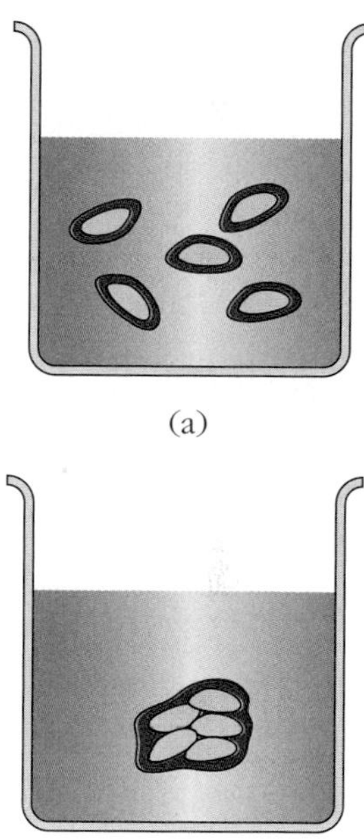

FIGURE 2-9

Aggregation of nonpolar molecules in water.
(a) The individual hydration of dispersed nonpolar molecules (green) decreases the entropy of the system because the hydrating water molecules (orange) are not as free to form hydrogen bonds. (b) Aggregation of the nonpolar molecules increases the entropy of the system, since the number of water molecules required to hydrate the aggregated solutes is less than the number of water molecules required to hydrate the dispersed solute molecules. This increase in entropy accounts for the spontaneous aggregation of nonpolar substances in water.

a globular mass so that its hydrophobic groups are in the interior, away from the solvent, and its polar groups are on the exterior, where they can interact with water. Similarly, the structure of the lipid membrane that surrounds all cells is maintained by the hydrophobic effect acting on the lipids.

REVIEW LEARNING OBJECTIVE 4

- Why do polar molecules dissolve more easily than nonpolar substances in water?
- What does entropy have to do with the solubility of nonpolar substances in water?

Amphiphilic molecules experience both hydrophilic interactions and the hydrophobic effect

Consider a molecule such as the fatty acid palmitate:

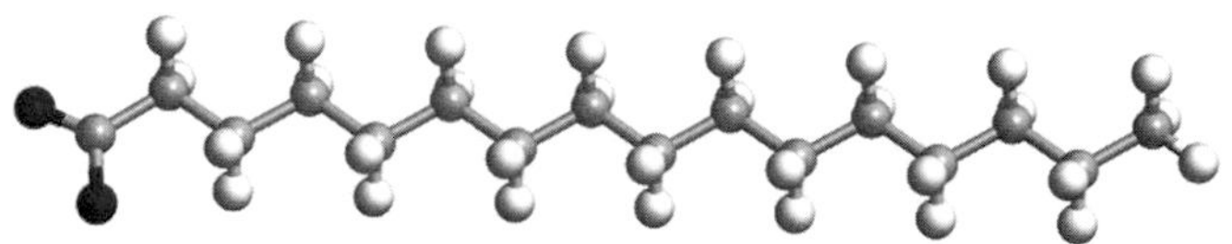

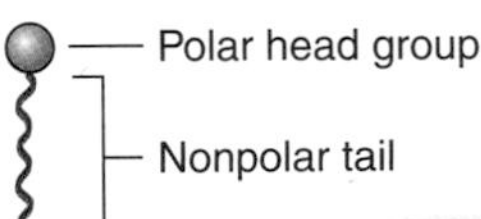

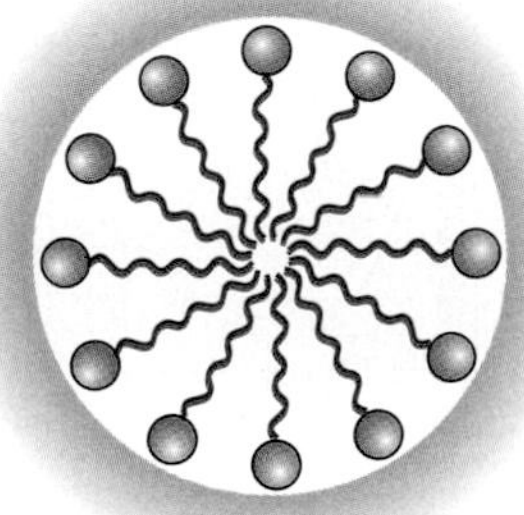

FIGURE 2-10

A micelle formed by amphiphilic molecules.
The hydrophobic tails of the molecules aggregate, out of contact with water, due to the hydrophobic effect. The polar head groups are exposed to and can interact with the solvent water molecules.

The hydrocarbon "tail" of the molecule (on the left) is nonpolar while its carboxylate "head" (on the right) is strongly polar. Molecules such as this one, which have both hydrophobic and hydrophilic portions, are said to be **amphiphilic** or **amphipathic.** What happens when amphiphilic molecules are added to water? In general, *the polar groups of amphiphiles orient themselves toward the solvent molecules and are therefore hydrated, while the nonpolar groups tend to aggregate due to the hydrophobic effect.* As a result, the amphiphiles may form a spherical **micelle,** a particle with a solvated surface and a hydrophobic core (Fig. 2-10).

Depending in part on the relative sizes of the hydrophilic and hydrophobic portions of the amphiphiles, the molecules may form a sheet rather than a spherical micelle. The amphiphilic lipids that provide the structural basis of biological membranes form two-layered sheets called **bilayers,** in which a hydrophobic layer is sandwiched between hydrated polar surfaces (Fig. 2-11). The structures of biological membranes are discussed in more detail in Chapter 8. The formation of micelles or bilayers is thermodynamically favored because the hydrogen-bonding capacity of the polar groups is satisfied through

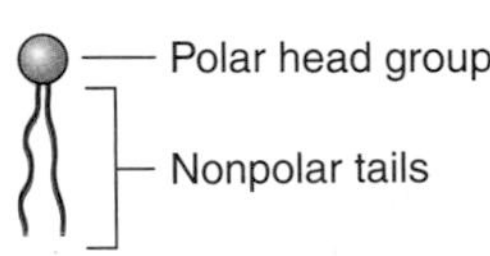

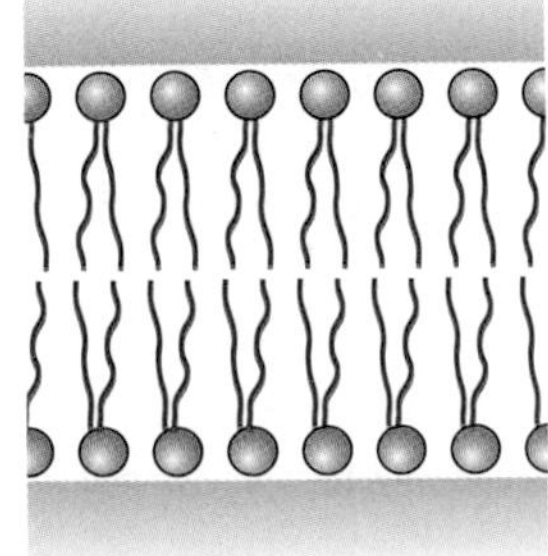

FIGURE 2–11

A lipid bilayer.
The amphiphilic lipid molecules form two layers so that their polar head groups are exposed to the solvent while their hydrophobic tails are sequestered in the interior of the bilayer, away from water. The likelihood of amphiphilic molecules forming a bilayer rather than a micelle depends in part on the sizes and nature of the hydrophobic and hydrophilic groups. One-tailed lipids tend to form micelles (see Fig. 2–10), and two-tailed lipids tend to form bilayers.

interactions with solvent water molecules, and the nonpolar groups are sequestered from the solvent.

The hydrophobic core of a lipid bilayer is a barrier to diffusion

To eliminate its solvent-exposed edges, a lipid bilayer tends to close up to form a **vesicle,** shown cut in half:

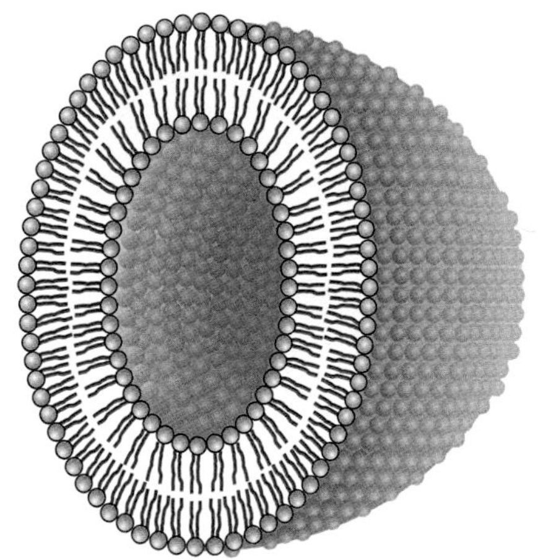

Many of the subcellular compartments (organelles) in eukaryotic cells are really just more elaborate versions of the lipid vesicle shown here.

When the vesicle forms, it traps a volume of the aqueous solution. Polar solutes in the enclosed compartment tend to remain there because they cannot easily pass through the hydrophobic interior of the bilayer. The energetic cost of transferring a hydrated polar group through the nonpolar lipid tails is too great. (In contrast, small nonpolar molecules such as O_2 can pass through the bilayer relatively easily.)

Normally, substances that are present at high concentrations tend to diffuse to regions of lower concentration (this movement "down" a concentration gradient is a spontaneous process driven by the increase in entropy of the solute molecules). *A barrier such as a bilayer can prevent this diffusion* (Fig. 2-12). This helps explain why cells, which are universally enclosed by some sort of membrane, can maintain their specific concentrations of ions, small molecules, and biopolymers, even when the external concentrations of these substances are quite different (Fig. 2-13). The solute composition of

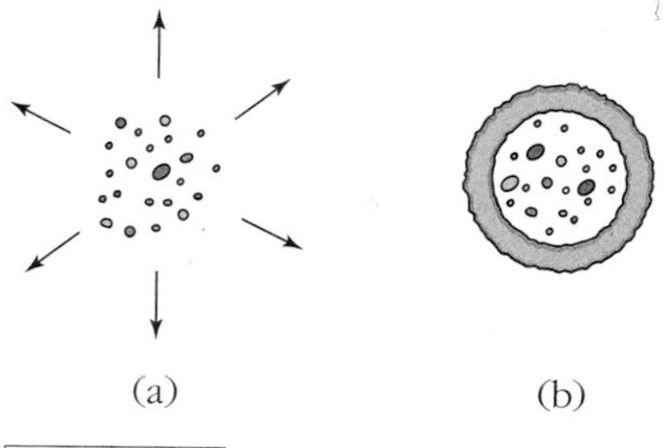

FIGURE 2-12

A bilayer prevents the diffusion of polar substances.
(a) Solutes spontaneously diffuse from a region of high concentration to a region of low concentration. (b) A lipid barrier, which presents a thermodynamic barrier to the passage of polar substances, prevents the diffusion of polar substances out of the inner compartment (it also prevents the inward diffusion of polar substances from the external solution).

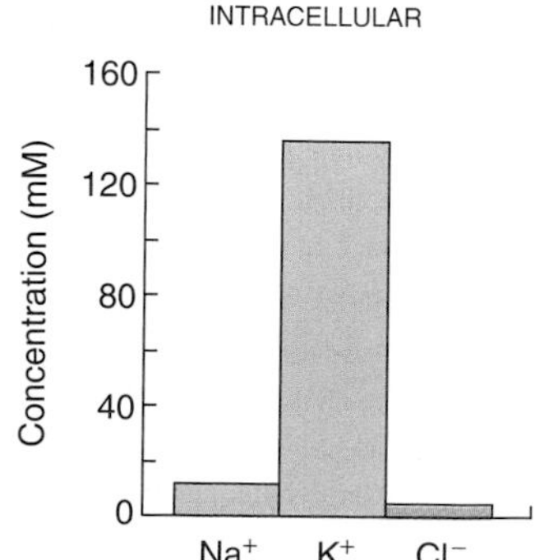

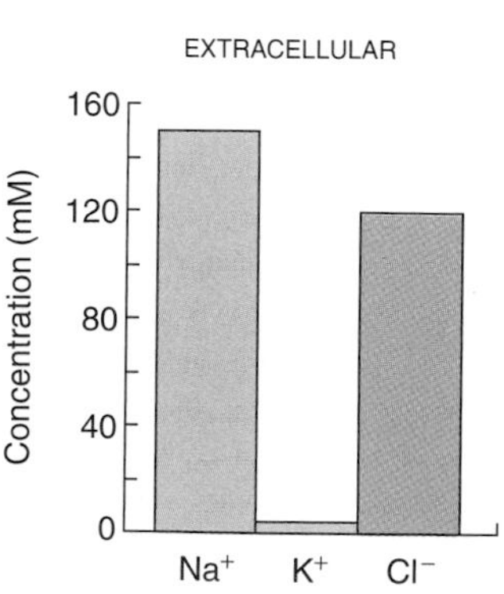

FIGURE 2-13

Ionic composition of intracellular and extracellular fluid.
Human cells contain much higher concentrations of potassium than either sodium or chloride; the opposite is true of the fluid outside the cell. The cell membrane helps maintain the concentration differences.

intracellular compartments and other biological fluids is carefully regulated. Not surprisingly, organisms spend a considerable amount of metabolic energy to maintain the proper concentrations of water and salts, and losses of one or the other must be compensated (Box 2-A).

REVIEW LEARNING OBJECTIVE 5

- Can a molecule be both hydrophilic and hydrophobic?
- Why is a lipid bilayer a barrier to the diffusion of polar molecules?

Go to **Review Exercise 2/Acids, Bases, and pH** for an overview of the meaning and importance of pH in biochemical systems. The information and questions presented in Review Exercise 2 address Learning Objectives 6, 7, 8, and 9.

3 ACID–BASE CHEMISTRY

Water is not merely an inert medium for biochemical processes; it is an active participant. Its chemical reactivity in biological systems is in part a result of its ability to ionize:

$$H_2O \rightleftharpoons H^+ + OH^-$$

A CLOSER LOOK

BOX 2-A Sweat and exercise

Animals, including humans, generate heat, even at rest, due to their metabolic activity. Some of this heat is lost to the environment by radiation, convection, and conduction, and—in terrestrial animals—by the vaporization of water. Evaporation has a significant cooling effect because about 2.5 kJ of heat is given up for every gram (mL) of water lost. In humans and certain other animals, an increase in skin temperature triggers the activity of sweat glands, which secrete a solution containing (in humans) about 50 mM Na^+, 5 mM K^+, and 45 mM Cl^-—about three times as dilute as the extracellular fluid (see Fig. 2-13). The body is cooled as the sweat evaporates from its surface.

The evaporation of water accounts for a small portion of a resting body's heat loss, but sweating is the main mechanism for dissipating heat generated when the body is highly active. During vigorous exercise or exertion at high ambient temperatures, the body may experience a fluid loss of up to 2 L per hour. Athletic training not only improves the performance of the muscles and cardiopulmonary system, it also increases the capacity for sweating so that the athlete begins to sweat at a lower skin temperature and loses less salt in the secretions of the sweat glands. But regardless of training, a fluid loss representing more than 2% of the body's weight may impair cardiovascular function. In fact, "heat exhaustion" in humans is usually due to dehydration rather than an actual increase in body temperature.

Numerous studies have concluded that athletes seldom drink enough before or during exercise. Ideally, fluid intake should match the losses due to sweat, and the rate of intake should keep pace with the rate of sweating. So what should the conscientious athlete drink? For activities lasting less than about 90 minutes, especially when periods of high intensity alternate with brief periods of rest, water alone is sufficient. Commercial sports drinks containing carbohydrates can replace the water lost as sweat and also provide a source of energy. However, this carbohydrate boost may be an advantage only during prolonged sustained activity, such as during a marathon, when the body's own carbohydrate stores are depleted. A marathon runner or a manual laborer in the hot sun might benefit from the salt found in sports drinks, but most athletes don't need the supplemental salt (although it does make the carbohydrate solution more palatable). A normal diet usually contains enough Na^+ and Cl^- to offset the losses in sweat.

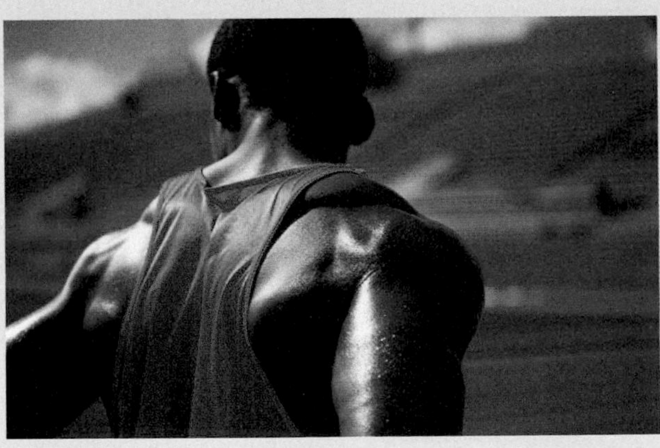

[White/Packert/Photographers' Choice/Getty Images.]

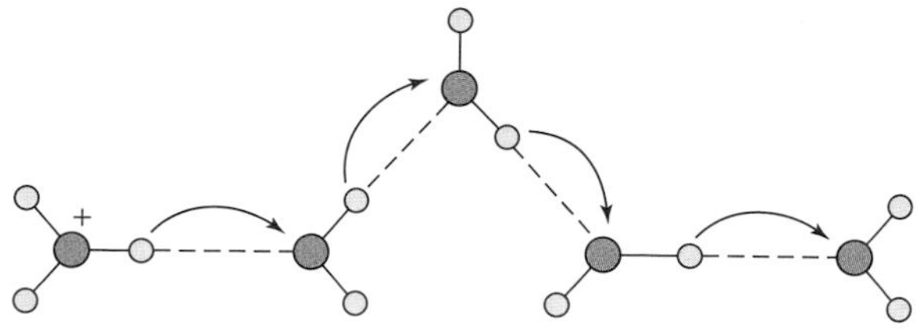

FIGURE 2-14

Proton jumping. A proton associated with one water molecule (as a hydronium ion) appears to jump rapidly through a network of hydrogen-bonded water molecules.

The products of water's dissociation are a hydrogen ion or proton (H^+) and a hydroxide ion (OH^-).

Aqueous solutions do not actually contain lone protons. Instead, the H^+ can be visualized as combining with a water molecule to produce a **hydronium ion (H_3O^+):**

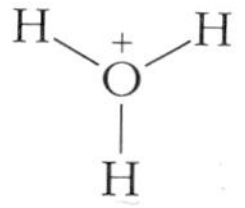

$$H_3O^+$$

However, the H^+ is somewhat delocalized so that it probably exists as part of a larger, fleeting structure such as

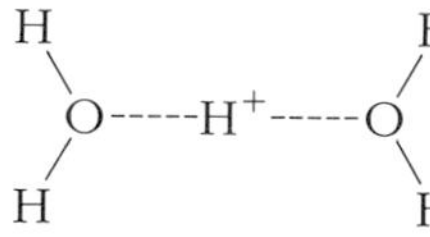

$$H_2O\text{-----}H^+\text{-----}OH_2 \quad \text{or} \quad H_9O_4^+$$

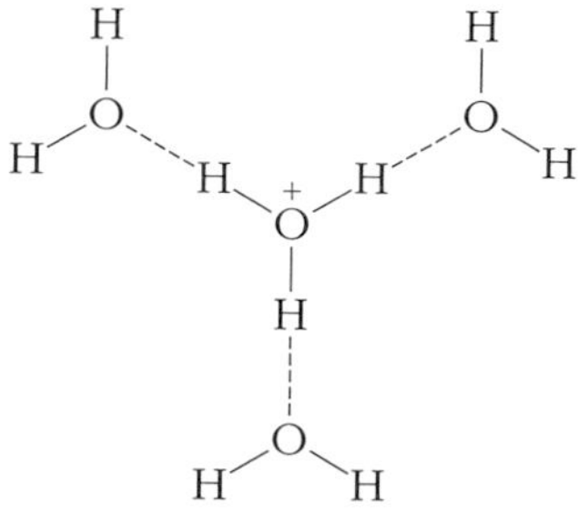

Because a proton does not remain associated with a single water molecule, it appears to be relayed through a hydrogen-bonded network of water molecules (Fig. 2-14). This rapid **proton jumping** means that the effective mobility of H^+ in water is much greater than the mobility of other ions that must physically diffuse among water molecules. Consequently, acid–base reactions are among the fastest biochemical reactions.

[H⁺] and [OH⁻] are inversely related

Pure water exhibits only a slight tendency to ionize, so the resulting concentrations of H^+ and OH^- are actually quite small. The ionization of water can be described by a dissociation constant, K, which is equivalent to the concentrations of the reaction products divided by the concentration of unionized water:

$$K = \frac{[H^+][OH^-]}{[H_2O]} \qquad [2\text{-}1]$$

The square brackets indicate the molar concentrations of the indicated species.

Because the concentration of H_2O (55.5 M) is so much greater than $[H^+]$ or $[OH^-]$, it is considered to be constant, and K is redefined as K_w:

$$K_w = K[H_2O] = [H^+][OH^-] \qquad [2\text{-}2]$$

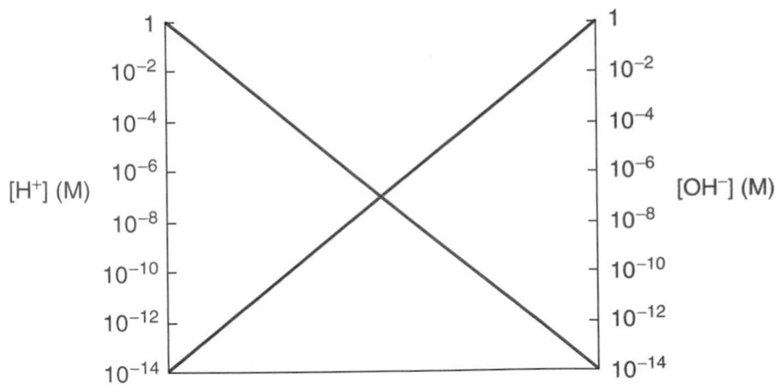

FIGURE 2-15

Relationship between $[H^+]$ and $[OH^-]$. The product of $[H^+]$ and $[OH^-]$ is K_w, which is equal to 10^{-14}. Consequently, when $[H^+]$ is greater than 10^{-7} M, $[OH^-]$ is less than 10^{-7} M, and vice versa.

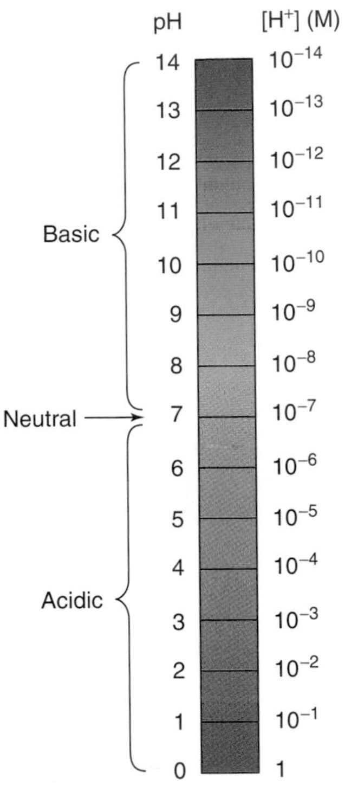

FIGURE 2-16

Relationship between pH and $[H^+]$. Because pH is equal to $-\log [H^+]$, the greater the $[H^+]$, the lower the pH. A solution with a pH of 7 is neutral, a solution with a pH < 7 is acidic, and a solution with a pH > 7 is basic.

$\boldsymbol{K_w}$, the **ionization constant of water,** is 10^{14} at 25°C. In a sample of pure water, $[H^+] = [OH^-]$, so $[H^+]$ and $[OH^-]$ must both be equal to 10^{-7} M:

$$K_w = 10^{-14} = [H^+][OH^-] = (10^{-7}\ M)(10^{-7}\ M) \qquad [2\text{-}3]$$

Since *the product of* $[H^+]$ *and* $[OH^-]$ *in any solution must be equal to* 10^{-14}, a hydrogen ion concentration greater than 10^{-7} M is balanced by a hydroxide ion concentration less than 10^{-7} M (Fig. 2-15).

A solution in which $[H^+] = [OH^-] = 10^{-7}$ M is said to be **neutral;** a solution with $[H^+] > 10^{-7}$M ($[OH^-] < 10^{-7}$ M) is **acidic;** and a solution with $[H^+] < 10^{-7}$ M ($[OH^-] > 10^{-7}$ M) is **basic.** To more easily describe such solutions, the hydrogen ion concentration is expressed as a **pH:**

$$\text{pH} = -\log [H^+] \qquad [2\text{-}4]$$

Accordingly, a neutral solution has a pH of 7, an acidic solution has a pH < 7, and a basic solution has a pH > 7 (Fig. 2-16). Note that because the pH scale is logarithmic, *a difference of one pH unit is equivalent to a tenfold difference in* $[H^+]$. The so-called "physiological" pH, the normal pH of human blood, is a near-neutral 7.4 (see Sample Calculation 2-1). The pH values of some other body fluids are listed in Table 2-1.

REVIEW LEARNING OBJECTIVE 6

- What is pH and how is it related to $[H^+]$ and $[OH^-]$?

Sample Calculation 2-1

Problem

What is the hydrogen ion concentration of human blood plasma, whose pH is 7.4?

Solution

Since pH $= -\log [H^+]$,

$$\begin{aligned}[H^+] &= 10^{-\text{pH}} \\ &= 10^{-7.4} \\ &= 4.0 \times 10^{-8}\ M\end{aligned}$$

TABLE 2-1 pH values of some biological fluids

Fluid	pH
Pancreatic juice	8.0
Blood	7.4
Saliva	6.6
Urine	4.5
Gastric juice	1.5

The pH of a solution can be altered

The pH of a sample of water can be changed by adding a substance that affects the existing balance between $[H^+]$ and $[OH^-]$. Adding an acid increases the concentration of $[H^+]$ and decreases the pH; adding a base has the opposite effect. *Biochemists define an* ***acid*** *as a substance that can donate a proton, and a* ***base*** *as a substance that can accept a proton.* For example, adding hydrochloric acid (HCl) to a sample of water increases the hydrogen ion concentration ($[H^+]$ or $[H_3O^+]$) because the HCl donates a proton to water:

$$HCl + H_2O \rightarrow H_3O^+ + Cl^-$$

Note that in this reaction, H_2O acts as a base that accepts a proton from the added acid.

Similarly, adding the base sodium hydroxide (NaOH) increases the pH (decreases $[H^+]$) by introducing hydroxide ions that can recombine with existing hydrogen ions:

$$NaOH + H_3O^+ \rightarrow Na^+ + 2\,H_2O$$

In this reaction, H_3O^+ is the acid that donates a proton to the added base. The final pH of the solution depends on how much H^+ (for example, from HCl) has been introduced or how much H^+ has been removed from the solution by its reaction with a base (for example, the OH^- ion of NaOH). For substances such as HCl and NaOH, which ionize completely, the calculation of pH is straightforward (see Sample Calculation 2-2). Review Exercise 2 includes definitions and examples of pH, acids, and bases.

Sample Calculation 2-2

Problem

Calculate the pH of 1 L of water to which is added (a) 10 mL of 5 M HCl or (b) 10 mL of 5 M NaOH.

Solution

(a) The final concentration of HCl is $\frac{(0.01\text{ L})(5\text{ M})}{1.01\text{ L}} \approx 0.05\text{ M}$

Since HCl dissociates completely, the added $[H^+]$ is equal to [HCl], or 0.05 M (the existing hydrogen ion concentration, 10^{-7} M, can be ignored because it is much smaller).

$$\begin{aligned} pH &= -\log[H^+] \\ &= -\log 0.05 \\ &= 1.30 \end{aligned}$$

(b) The final concentration of NaOH is 0.05 M. Since NaOH dissociates completely, the added $[OH^-]$ is 0.05 M. Use Equation 2-2 to calculate $[H^+]$.

$$K_w = 10^{-14} = [H^+][OH^-]$$

$$[H^+] = 10^{-14}/[OH^-]$$

$$= 10^{-14}/(0.05\ \text{M})$$

$$= 2 \times 10^{-13}\ \text{M}$$

$$\text{pH} = -\log[H^+]$$

$$= -\log(2 \times 10^{-13})$$

$$= 12.7$$

p*K* values describe an acid's tendency to ionize

Most biologically relevant acids and bases, unlike HCl and NaOH, do not dissociate completely when added to water. In other words, proton transfer to or from water is not complete. Therefore, the final concentrations of the acidic and basic species (including water itself) must be expressed in terms of an equilibrium. For example, acetic acid partially ionizes, or donates only some of its protons to water:

$$CH_3COOH + H_2O \rightleftharpoons CH_3COO^- + H_3O^-$$

The equilibrium constant for this reaction takes the form

$$K = \frac{[CH_3COO^-][H_3O^+]}{[CH_3COOH][H_2O]} \qquad [2\text{-}5]$$

Because the concentration of H_2O is much higher than the other concentrations, it is considered constant and is incorporated into the value of K, which is then formally known as $\boldsymbol{K_a}$, the **acid dissociation constant:**

$$K_a = K[H_2O] = \frac{[CH_3COO^-][H^+]}{[CH_3COOH]} \qquad [2\text{-}6]$$

The acid dissociation constant for acetic acid is 1.74×10^{-5}. The larger the value of K_a, the more likely the acid is to ionize, that is, the greater its tendency to donate a proton to water. The smaller the value of K_a, the less likely the compound is to donate a proton.

Acid dissociation constants, like hydrogen ion concentrations, often represent very small quantities. Therefore, it is convenient to transform the K_a to a **p*K*** value as follows:

$$pK = -\log K_a \qquad [2\text{-}7]$$

For acetic acid,

$$pK = -\log(1.74 \times 10^{-5}) = 4.76 \qquad [2\text{-}8]$$

The larger an acid's K_a, the smaller its pK and the greater its "strength" as an acid.

Consider an acid such as the ammonium ion, NH_4^+:

$$NH_4^+ \rightleftharpoons NH_3 + H^+$$

Its K_a is 5.62×10^{-10}, which corresponds to a pK of 9.25. This indicates that the ammonium ion is a relatively weak acid—a compound that tends not to ionize. On the other hand, ammonia (NH_3), which is the **conjugate base** of the acid NH_4^+, is a strong base, a compound that readily accepts a proton.

REVIEW LEARNING OBJECTIVE 7

- How does the addition of an acid or base affect the pH of a solution?
- What is the relationship between pK and a molecule's tendency to lose a proton?

The pH of a solution of acid is related to the p*K*

When an acid (represented as the proton donor HA) is added to water, *the final hydrogen ion concentration of the solution depends on the acid's tendency to ionize:*

$$HA \rightleftharpoons A^- + H^+$$

In other words, the final pH depends on the equilibrium between HA and A^-,

$$K_a = \frac{[A^-][H^+]}{[HA]} \qquad [2\text{-}9]$$

so that

$$[H^+] = K_a \frac{[HA]}{[A^-]} \qquad [2\text{-}10]$$

We can express $[H^+]$ as a pH, and K_a as a pK, which yields

$$-\log[H^+] = -\log K_a - \log \frac{[HA]}{[A^-]} \qquad [2\text{-}11]$$

or

$$pH = pK + \log \frac{[A^-]}{[HA]} \qquad [2\text{-}12]$$

Equation 2-12 is known as the **Henderson–Hasselbalch equation.** It relates the pH of a solution to the pK of an acid and the concentration of the acid (HA) and its conjugate base (A^-). This equation makes it possible to perform practical calculations such as predicting the pH of solution containing known amounts of an acid and its conjugate base (which is usually in the form of a salt). The pK values of some compounds are listed in Table 2-2 (see Sample Calculations 2-3 and 2-4). Consult Review Exercise 2 for more discussion of the meaning of pK and applications of the Henderson–Hasselbalch equation.

Sample Calculation 2-3

Problem

Calculate the pH of a 1 L solution to which has been added 6 mL of 1.5 M acetic acid and 5 mL of 0.4 M sodium acetate.

TABLE 2-2 pK values of some acids

Name	Formula[a]	pK
Trifluoroacetic acid	CF_3COOH	0.18
Hydrofluoric acid	HF	3.20
Acetic acid	CH_3COOH	4.76
Cacodylic acid	$(CH_3)_2As(O)OH$	6.27
N-(2-acetamido)-2-amino-ethanesulfonic acid (ACES)	$H_2NCOCH_2\overset{+}{N}H_2CH_2CH_2SO_3^-$	6.90
Imidazole	(imidazolium ring: $\overset{+}{N}H$, N–H)	7.00
N-2-hydroxyethylpiperazine-N′-2-ethanesulfonic acid (HEPES)	$HOCH_2CH_2\overset{+}{H}N$ (piperazine ring) $NCH_2CH_2SO_3^-$	7.55
Glycinamide	$^{+}H_3NCH_2CONH_2$	8.20
Tris(hydroxymethyl)-aminomethane (Tris)	$(HOCH_2)_3C\overset{+}{N}H_3$	8.30
Boric acid	H_3BO_3	9.24
Ammonium ion	NH_4^+	9.25
Phenol	C_6H_5OH	9.90
Methylammonium ion	$CH_3NH_3^+$	10.6

[a]The acidic hydrogen is highlighted in red.

Solution

First, calculate the final concentrations of acetic acid (HA) and acetate (A^-). The final volume of the solution is 1 L + 6 mL + 5 mL = 1.011 L.

$$[HA] = \frac{(0.006\ \text{L})(1.5\ \text{M})}{1.011\ \text{L}} \approx 0.009\ \text{M}$$

$$[A^-] = \frac{(0.005\ \text{L})(0.4\ \text{M})}{1.011\ \text{L}} \approx 0.002\ \text{M}$$

Next, substitute these values into the Henderson–Hasselbalch equation using the pK for acetic acid given in Table 2-1:

$$\text{pH} = \text{p}K + \log\frac{[A^-]}{[HA]}$$

$$\text{pH} = 4.76 + \log\frac{0.002}{0.009}$$

$$= 4.76 - 0.65$$

$$= 4.11$$

Sample Calculation 2-4

Problem

How many mL of a 2 M solution of boric acid must be added to 600 mL of a solution of 0.01 M sodium borate in order for the pH to be 9.45?

Solution

Rearrange the Henderson–Hasselbalch equation to isolate the [HA] term (the required concentration of boric acid):

$$\mathrm{pH} = \mathrm{p}K + \log \frac{[\mathrm{A}^-]}{[\mathrm{HA}]}$$

$$\log \frac{[\mathrm{A}^-]}{[\mathrm{HA}]} = \mathrm{pH} - \mathrm{p}K$$

$$\log[\mathrm{A}^-] - \log[\mathrm{HA}] = \mathrm{pH} - \mathrm{p}K$$

$$\log[\mathrm{HA}] = \log[\mathrm{A}^-] - \mathrm{pH} + \mathrm{p}K$$

Substitute the known quantities: the concentration of borate ($[\mathrm{A}^-]$), the desired pH, and the pK of boric acid (from Table 2-1).

$$\log[\mathrm{HA}] = \log 0.01 - 9.45 + 9.24 = -2.21$$

$$[\mathrm{HA}] = 10^{-2.21} = 0.0062\ \mathrm{M}$$

Next, calculate the volume (x) of 2 M boric acid to add to the 600 mL solution:

$$\frac{x(2\ \mathrm{M})}{0.6\ \mathrm{L} + x} = 0.0062\ \mathrm{M}$$

$$2x = 0.00372 + 0.0062x$$

$$x = \frac{0.00372}{(2 - 0.0062)}$$

$$x = 0.0019\ \mathrm{L\ or\ } 1.9\ \mathrm{mL}$$

Many of the functional groups, including carboxylate and amino groups, on biological molecules act as acids and bases. Their ionization states depend on their respective pK values and on the pH ($[\mathrm{H}^+]$) of the solution. For example, at physiological pH, a polypeptide bears multiple ionic charges because its carboxylic acid ($-\mathrm{COOH}$) groups are ionized to carboxylate ($-\mathrm{COO}^-$) groups and its amino ($-\mathrm{NH_2}$) groups are protonated ($-\mathrm{NH_3^+}$). This is because the pK values for the carboxylic acid groups are about 2–4, and the pK values for the amino groups are about 9–11. Consequently, below pH 2, both the carboxylic acid and amino groups would be mostly protonated; above pH 11, both groups would be mostly deprotonated.

$-\mathrm{COOH}$	$-\mathrm{COO}^-$	$-\mathrm{COO}^-$
$-\mathrm{NH_3^+}$	$-\mathrm{NH_3^+}$	$-\mathrm{NH_2}$
$\mathrm{pH} < 2$	$4 < \mathrm{pH} < 9$	$\mathrm{pH} > 11$

REVIEW LEARNING OBJECTIVE 8

- What is the relationship between pH and the pK of an acid?

Buffers

When a strong acid such as HCl is added to pure water, all the added acid contributes directly to a decrease in pH. But when HCl is added to a solution containing a weak acid in equilibrium with its conjugate base (A^-), the pH does not change so dramatically, because some of the added protons combine with the conjugate base to re-form the acid and therefore do not contribute to an increase in $[H^+]$.

$$HCl \rightarrow H^+ + Cl^- \quad \text{large increase in } [H^+]$$

$$HCl + A^- \rightarrow HA + Cl^- \quad \text{small increase in } [H^+]$$

Conversely, when a strong base (such as NaOH) is added to the solution, some of the added hydroxide ions accept protons from the acid to form H_2O and therefore do not contribute to a decrease in $[H^+]$.

$$NaOH \rightarrow Na^+ + OH^- \quad \text{large decrease in } [H^+]$$

$$NaOH + HA \rightarrow Na^+ + A^- + H_2O \quad \text{small decrease in } [H^+]$$

The weak acid/conjugate base system (HA/A⁻) acts as a ***buffer*** *against the added acid or base by preventing the dramatic changes in pH that would otherwise occur.* Review Exercise 2 illustrates how buffers minimize pH changes upon the addition of strong acid or base.

The buffering activity of a weak acid, such as acetic acid, can be traced by titrating the acid with a strong base (Fig. 2-17). At the start of the titration, all the acid is present in its protonated (HA) form. As base (for example, NaOH) is added, protons begin to dissociate from the acid, producing A^-. The continued addition of base eventually causes all the protons to dis-

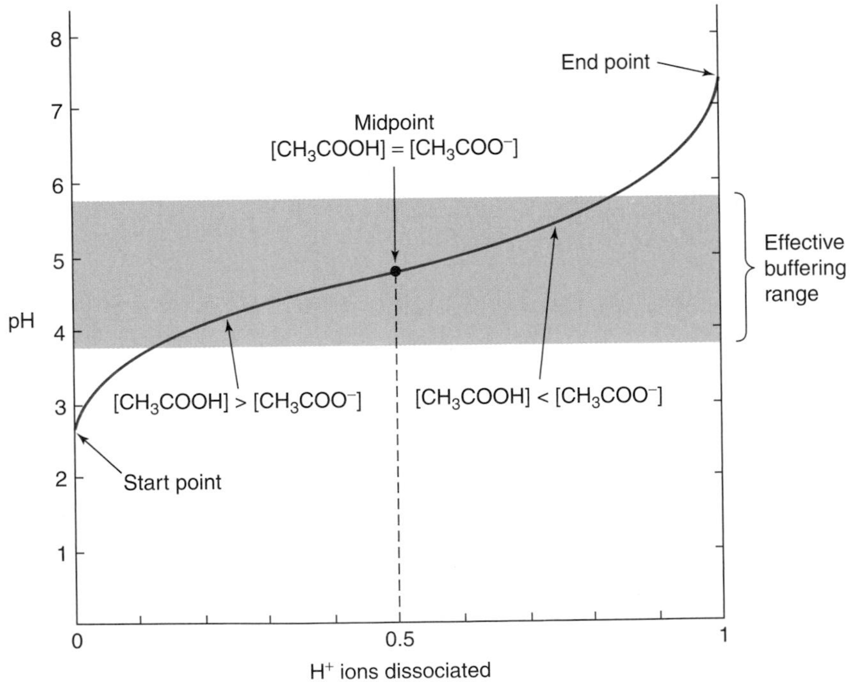

FIGURE 2-17

Titration of acetic acid.
At the start point (before base is added), the acid is present in its CH_3COOH form. As small amounts of base are added, protons dissociate until, at the midpoint of the titration (where pH = p*K*), $[CH_3COOH] = [CH_3COO^-]$. The addition of more base causes more protons to dissociate until all the acid is in the CH_3COO^- form (the end point). The shaded area indicates the effective buffering range of acetic acid. Within one pH unit of the p*K*, additions of acid or base do not greatly perturb the pH of the solution.

sociate, leaving all the acid in its conjugate base (A^-) form. When exactly half the protons have dissociated, $[HA] = [A^-]$. The pH at which this occurs is the midpoint of the titration. In fact, when $[HA] = [A^-]$, the log ($[A^-]/[HA]$) term of the Henderson–Hasselbalch equation (Equation 2-12) becomes zero (log 1 = 0), and pH = pK. In other words, *the pK of an acid is the pH at which it is half-dissociated and the concentrations of the acid and its conjugate base are equal.* For a substance with more than one acidic proton (for example, phosphoric acid, H_3PO_4, with three hydrogens that can dissociate), there are multiple pK values (see Sample Calculation 2-5). Review Exercise 2 includes additional calculations involving buffers.

Sample Calculation 2-5

Problem

For phosphoric acid (with pK values of 2.15, 6.82, and 12.38), determine which molecular species predominates at pH values of (a) 1.5, (b) 4, (c) 9, and (d) 13.

Solution

From the pK values, we know that:
Below pH 2.15, the fully protonated H_3PO_4 species predominates.
At pH 2.15, $[H_3PO_4] = [H_2PO_4^-]$.
Between pH 2.15 and 6.82, the $H_2PO_4^-$ species predominates.
At pH 6.82, $[H_2PO_4^-] = [HPO_4^{2-}]$.
Between pH 6.82 and 12.38, the HPO_4^{2-} species predominates.
At pH 12.38, $[HPO_4^{2-}] = [PO_4^{3-}]$.
Above pH 12.38, the fully deprotonated PO_4^{3-} species predominates.
Therefore, the predominant species at the indicated pH values are (a) H_3PO_4, (b) $H_2PO_4^-$, (c) HPO_4^{2-}, and (d) PO_4^{3-}.

The broad, flat shape of the titration curve shown in Figure 2-17 indicates that the pH does not change drastically with added acid or base when the pH is near the pK. *The effective buffering capacity of an acid is generally taken to be within one pH unit of its pK.* For acetic acid (pK = 4.76), this would be pH 3.76–5.76.

In the laboratory, buffered solutions are used to maintain a constant pH. Without buffering, fluctuations in pH would alter the ionization states of purified macromolecules, which might then undergo changes in their structure or function. In the human body, a variety of different molecules act as buffers to maintain a constant pH (Box 2-B).

REVIEW LEARNING OBJECTIVE 9

- How does a solution of a weak acid and its conjugate base minimize changes in pH when a strong acid or base is added to the solution?

A CLOSER LOOK

BOX 2-B Acid–base balance in humans

The cells of the human body typically maintain an internal pH of 6.9–7.4. The body does not normally have to defend itself against strong inorganic acids, but many metabolic processes generate H^+, which must be buffered so that it does not cause the pH of blood to drop below 7.4. The functional groups of proteins and phosphate groups can serve as biological buffers; however, the most important buffering system involves CO_2 (itself a product of metabolism) in the blood plasma (plasma is the fluid component of blood).

Like all gases, CO_2 is soluble in water; however, CO_2 also combines with water to form carbonic acid, H_2CO_3:

$$CO_2 + H_2O \rightleftharpoons H_2CO_3$$

This freely reversible reaction is accelerated *in vivo* by the enzyme carbonic anhydrase, which is present in most tissues and is particularly abundant in red blood cells. Carbonic acid ionizes to bicarbonate, HCO_3^-,

$$H_2CO_3 \rightleftharpoons H^+ + HCO_3^-$$

so that the overall reaction is

$$CO_2 + H_2O \rightleftharpoons H^+ + HCO_3^-$$

The p*K* for this process is 6.35 (the ionization of HCO_3^- to CO_3^{2-} occurs with a p*K* of 10.33 and is therefore not significant at physiological pH).

Although a p*K* of 6.35 appears to be just outside the range of a useful physiological buffer (which would be within one pH unit of 7.4), the effectiveness of the bicarbonate buffer system is augmented by the fact that excess hydrogen ions not only can be buffered but also can be eliminated from the body. This is possible because after the H^+ combines with HCO_3^- to re-form H_2CO_3, which rapidly equilibrates with $CO_2 + H_2O$, some of the CO_2 can be given off as a gas in the lungs. If it becomes necessary to retain more H^+ to maintain a constant pH, breathing can be adjusted so that less gaseous CO_2 is lost during exhalation.

Normal conditions

$$H^+ + HCO_3^- \rightleftharpoons H_2CO_3 \rightleftharpoons H_2O + CO_2$$

Excess acid

$$\mathbf{H^+} + HCO_3^- \Rightarrow H_2CO_3 \rightleftharpoons H_2O + CO_2$$

$$H^+ + HCO_3^- \rightleftharpoons \mathbf{H_2CO_3} \Rightarrow H_2O + CO_2$$

$$H^+ + HCO_3^- \rightleftharpoons H_2CO_3 \rightleftharpoons H_2O + \mathbf{CO_2}$$

Insufficient acid

$$H^+ + HCO_3^- \rightleftharpoons H_2CO_3 \Leftarrow H_2O + \mathbf{CO_2}$$

$$H^+ + HCO_3^- \Leftarrow \mathbf{H_2CO_3} \rightleftharpoons H_2O + CO_2$$

$$\mathbf{H^+} + HCO_3^- \rightleftharpoons H_2CO_3 \rightleftharpoons H_2O + CO_2$$

Certain illnesses and genetic diseases lead to overproduction or retention of hydrogen ions and a blood pH less than 7.35; this is known as **metabolic acidosis** (the opposite effect, metabolic alkalosis, a pH $>$ 7.45, is not as common). For example, severe diarrhea can contribute to the loss of ions, including HCO_3^-, and renal failure may prevent the normal excretion of metabolic acids. The resulting drop in pH can lead to diminished cardiac function and neurological impairment, which may be evident as hypotension and seizures.

Individuals with metabolic acidosis often exhibit very rapid and deep breathing. The increased respiration helps compensate for the acidosis by "blowing off" more acid in the form of CO_2 derived from H_2CO_3. However, this mechanism also impairs O_2 uptake by the lungs. Metabolic acidosis can be treated by administering bicarbonate. In chronic metabolic acidosis, the mineral component of bone serves as a buffer, which leads to the loss of calcium, magnesium, and phosphate and ultimately to osteoporosis and fractures.

Impaired pulmonary function can contribute to **respiratory acidosis,** characterized by the failure to eliminate sufficient CO_2 through the lungs. In this case, the kidneys can partially compensate for the increased acid load by excreting more H^+ in the form of NH_4^+. Respiratory acidosis is best treated by curing the underlying pulmonary disorder, since O_2 absorption is likely to be affected also.

SUMMARY

1. Water molecules are polar and form hydrogen bonds with each other and with other polar molecules bearing hydrogen bond donor or acceptor groups.
2. The electrostatic forces acting on biological molecules also include ionic interactions and van der Waals interactions.

3. Water dissolves polar and ionic substances.
4. Nonpolar (hydrophobic) substances tend to aggregate rather than disperse in water in order to minimize the decrease in entropy that would be required for water molecules to surround each nonpolar molecule. This is the hydrophobic effect.
5. Amphiphilic molecules, which contain both polar and nonpolar groups, may aggregate to form micelles or bilayers.
6. The dissociation of water produces hydroxide ions (OH^-) and protons (H^+) whose concentration can be expressed as a pH value. The pH of a solution can be altered by adding an acid (which donates protons) or a base (which accepts protons).
7. The tendency for an acid to ionize, or produce protons, is expressed as its pK value.
8. The Henderson–Hasselbalch equation relates the pH of a solution of a weak acid and its conjugate base to the pK and the concentrations of the acid and base. Such a solution is said to be buffered because the equilibrium between the acid and its conjugate base resists changes in pH when more acid or base is added.

CHECKLIST

1. Review the Learning Objectives listed on page 24.
2. Complete Review Exercise 2/Acids, Bases, and pH to learn more about pH, acids, bases, and buffers, and to solve problems using the Henderson–Hasselbalch equation.
3. Apply your knowledge by solving the problems at the end of this chapter. Check your results in the Solutions appendix.
4. Be able to define the boldfaced terms (consult the glossary at the end of the book). Test your understanding by taking the Chapter 2 quiz (www.wiley.com/college/pratt).
5. Explore the websites listed at www.wiley.com/college/pratt.
6. Consult the list of Selected Readings for background information or additional details on water structure, the hydrophobic effect, and acid–base chemistry.

GLOSSARY TERMS

polarity
hydrogen bond
ionic interaction
van der Waals radius
electronegativity
van der Waals interaction
dipole–dipole interaction
London dispersion forces
solvation
hydration
hydrophilic
hydrophobic
hydrophobic effect
amphiphilic
amphipathic
micelle
bilayer
vesicle
hydronium ion
proton jumping
ionization constant of water (K_w)
neutral solution
acidic solution
basic solution
pH
acid
base
acid dissociation constant (K_a)
pK
conjugate base
Henderson–Hasselbalch equation
buffer
metabolic acidosis
respiratory acidosis

PROBLEMS

1. The H—C—H bond angle in the perfectly tetrahedral CH_4 molecule is 109°. Explain why the H—O—H bond angle in water is only about 104.5°.
2. Each C═O bond in CO_2 is polar, yet the whole molecule is nonpolar. Explain.
3. Is ammonia a polar molecule?
4. Consider the following molecules and their melting points listed below. How can you account for the differences in melting points among these molecules of similar size?

	Molecular weight ($g \cdot mol^{-1}$)	Melting point (°C)
Water, H_2O	18.0	0
Ammonia, NH_3	17.0	−77
Methane, CH_4	16.0	−182

5. Identify the hydrogen bond acceptor and donor groups in the following molecules. Use an arrow to point toward each acceptor and away from each donor.

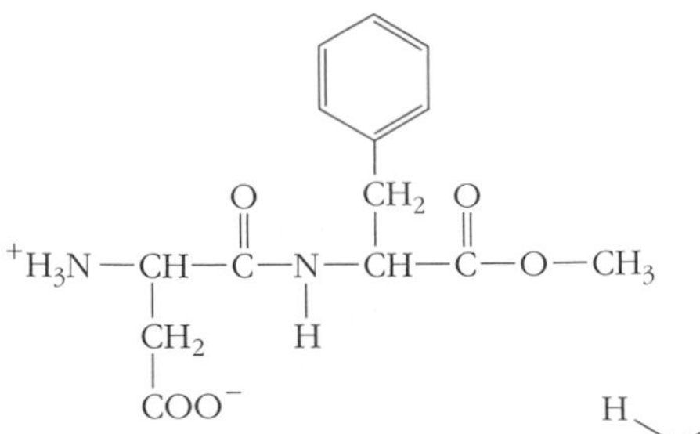

Aspartame: $^{+}H_3N{-}CH(CH_2COO^-){-}C({=}O){-}N(H){-}CH(CH_2C_6H_5){-}C({=}O){-}O{-}CH_3$

Sulfanilamide: $H_2N{-}C_6H_4{-}S(O)_2{-}NH_2$

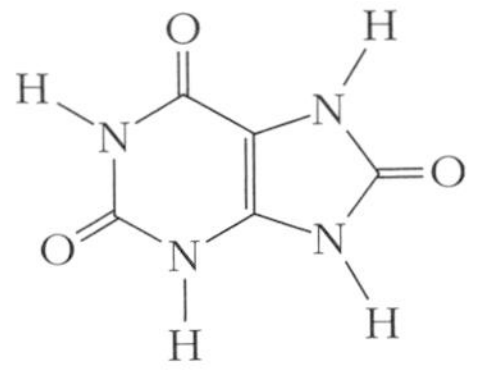

Uric acid

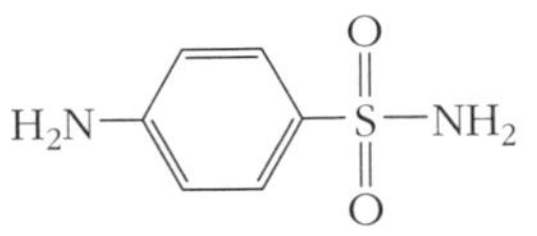

6. Do intermolecular hydrogen bonds form in the compounds below? Draw the hydrogen bonds where appropriate.

(A) $H_3C{-}C({=}O){-}H$

(B) 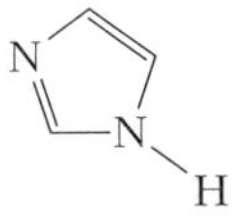 imidazole (ring with N and N–H)

(C) $H_3C{-}CH_2{-}OH$

(D) $H_3C{-}CH_2{-}Cl$ and $H{-}O{-}H$

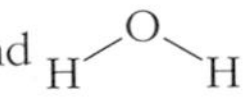

(E) $H_3C{-}CH_2{-}O{-}CH_2{-}CH_3$ and $H{-}N(H){-}H$ (with N)

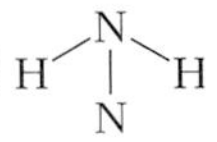

7. What are the most important intermolecular interactions in the following molecules?

(a) $H_3C{-}C({=}O){-}CH_3$

(b) $H_3C{-}C({=}O){-}OH$

(c) $H_3C{-}CH_2{-}CH_3$

(d) CsCl

8. Water is unusual in that its solid form is less dense than its liquid form. This means that when a pond freezes in the winter, ice is found as a layer on top of the pond, not the bottom. What are the biological advantages of this?

9. Rank the water solubility of the following compounds.

(A) $H_3C{-}CH_2{-}O{-}CH_3$

(B) $H_3C{-}C({=}O){-}NH_2$

(C) $H_2N{-}C({=}O){-}NH_2$

(D) $H_3C{-}CH_2{-}CH_3$

(E) $H_3C{-}CH_2{-}CH({=}O)$

10. Ammonium sulfate, $(NH_4)_2SO_4$, is a water-soluble salt. Draw the structures of the hydrated ions that form when ammonium sulfate dissolves in water.

11. The amino acid glycine is sometimes drawn as Structure A below. However, the structure of glycine is more accurately represented by Structure B. Glycine has the following properties: white crystalline solid, high melting point, and high water solubility. Why does Structure B more accurately represent the structure of glycine than Structure A?

$H_2N{-}CH_2{-}COOH$ — Structure A

$^{+}H_3N{-}CH_2{-}COO^-$ — Structure B

12. A typical spherical bacterial cell with a diameter of 1 μm contains 1000 molecules of a certain protein.

(a) What is the protein's concentration in units of mM? (Hint: The volume of a sphere is $4\pi r^3/3$.)

(b) How many molecules of glucose does the cell contain if the internal glucose concentration is 5 mM?

13. Practitioners of homeopathic medicine believe that good health can be restored by administering a small amount of a harmful substance that stimulates the body's natural healing processes. The remedy is typically prepared by serially diluting an animal or plant extract. In a so-called 30X dilution, the active substance is diluted tenfold, 30 times in succession.

(a) If the substance is initially present at a concentration of 1M, what is its concentration after a 30X dilution?

(b) A typical homeopathic dose is a few drops. How many molecules of the active substance would be present in 1 mL?

(c) When confronted with the results of an exercise such as the one in part b, proponents of homeopathy claim that a molecule can leave an imprint or memory of itself on the surrounding water molecules. What information presented in this chapter might be used to support this claim? To refute it?

14. Considper the structures of the molecules below. Are these molecules polar, nonpolar, or amphiphilic? Which are capable of forming micelles? Which are capable of forming bilayers?

(A) 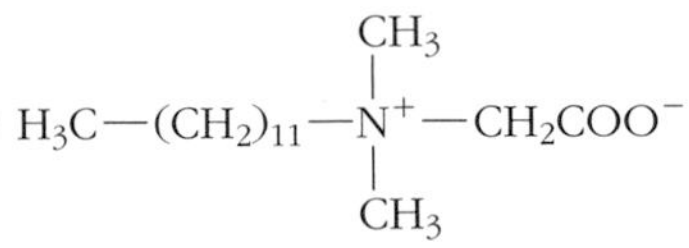 $H_3C{-}(CH_2)_{11}{-}N^+(CH_3)_2{-}CH_2COO^-$

(B) $H_3C{-}(CH_2)_{11}{-}CH_3$

(C) 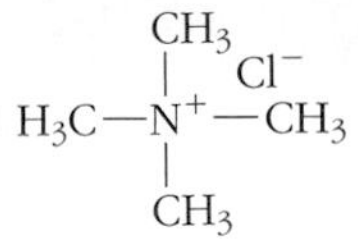 $H_3C{-}N^+(CH_3)_2{-}CH_3$ Cl^-

(D) 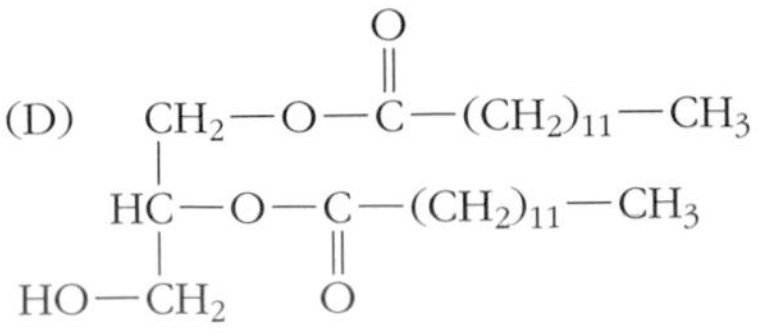 $CH_2{-}O{-}C({=}O){-}(CH_2)_{11}{-}CH_3$
$HC{-}O{-}C({=}O){-}(CH_2)_{11}{-}CH_3$
$HO{-}CH_2$

(E) $H_3C{-}CH(OH){-}COO^-$

15. The compound bis-(2-ethylhexyl)sulfosuccinate (abbreviated AOT) is capable of forming "reverse" micelles in the hydrocarbon solvent isooctane (its IUPAC name is 2,2,4-trimethylpentane). Scientists have investigated the use of reverse micelles for the extraction of water-soluble proteins. A two-phase system is formed: the hydrocarbon phase containing the reverse micelles and the water phase containing the protein. After a certain period of time, the protein is transferred to the reverse micelle.

$H_3C{-}(CH_2)_3{-}CH(CH_2CH_3){-}CH_2{-}O{-}C({=}O){-}CH_2$
$H_3C{-}(CH_2)_3{-}CH(CH_2CH_3){-}CH_2{-}O{-}C({=}O){-}CH{-}S(O_2){-}O^-$

Bis-(2-ethylhexyl)sulfosuccinate (AOT)

(a) Draw the structure of the reverse micelle that AOT would form in isooctane.
(b) Where would the protein be located in the reverse micelle?

16. Many household soaps are amphiphilic substances, often the salts of long-chain fatty acids, that form water-soluble micelles. An example is sodium dodecyl sulfate (SDS).

$H_3C{-}(CH_2)_{11}{-}O{-}S(O_2){-}O^-Na^+$

Sodium dodecyl sulfate (SDS)

(a) Identify the polar and nonpolar portions of the SDS molecule.
(b) Draw the structure of the micelle formed by SDS.
(c) Explain how the SDS micelles "wash away" water-insoluble substances such as cooking grease.

17. A device sometimes called a laundry ball is advertised as a replacement for environmentally harmful detergent. The laundry ball is a rubber-coated, baseball-sized plastic sphere containing fluid that is said to serve as an organizing template for water molecules.
(a) Evaluate this claim based on your knowledge of water structure.
(b) Magnets inside the laundry ball are said to rip apart clusters of water molecules to make it easier for individual water molecules to access and remove dirt molecules. If true, would this action help wash away dirt?
(c) The instructions recommend that a laundry ball, rather than the usual detergent, be added to a load of laundry. Why do the instructions recommend washing with hot rather than cold water?

18. Explain why water forms nearly spherical droplets on the surface of a freshly waxed car. Why doesn't water bead on a clean windshield?

19. Just as a dissolved substance tends to move spontaneously down its concentration gradient, water also tends to move from an area of high concentration (low solute concentration) to an area of low concentration (high solute concentration), a process known as osmosis.
(a) Explain why a lipid bilayer would be a barrier to osmosis.
(b) Why are isolated human cells placed in a solution that typically contains about 150 mM NaCl? What would happen if the cells were placed in pure water?

20. A specialized protein pump in the red blood cell membrane exports Na^+ ions and imports K^+ ions in order to maintain the sodium and potassium ion concentrations shown in Figure 2–13. Does the movement of these ions occur spontaneously, or is this an energy-requiring process? Explain.

21. Which of the following substances might be able to cross a bilayer? Which substances could not? Explain your answer.
(a) CO_2
(b) Glucose (Fischer projection: CHO; H–C–OH; HO–C–H; H–C–OH; H–C–OH; CH_2OH)
(c) 2,4-Dinitrophenol (DNP) (benzene ring with OH, NO_2 at position 2, NO_2 at position 4)
(d) Ca^{2+}

22. Estimate the amount of Na^+ lost in sweat during 15 minutes of vigorous exercise. What is the mass of potato chips (200 mg Na^+ per ounce) you would have to consume in order to replace the lost sodium?

23. The bacterium *E. coli* has the ability to change both its water content and its cytoplasmic K^+ concentration to adapt to changes in the solute concentration of its growth medium. (The solute concentration is referred to as *osmolarity*. A solution with a high concentration of solute has a high osmolarity and a solution with a low concentration of solute has a low osmolarity.) The bacterial cell consists of a cytoplasmic compartment bounded by a membrane, which is in turn surrounded by a cell wall. The cell wall is porous and can accommodate the passage of water and ions. (The membrane is not as porous, but does allow passage of water and ions.) In addition, the cell wall is also fairly elastic and can stretch to accommodate increases in the cytoplasmic compartment volume.

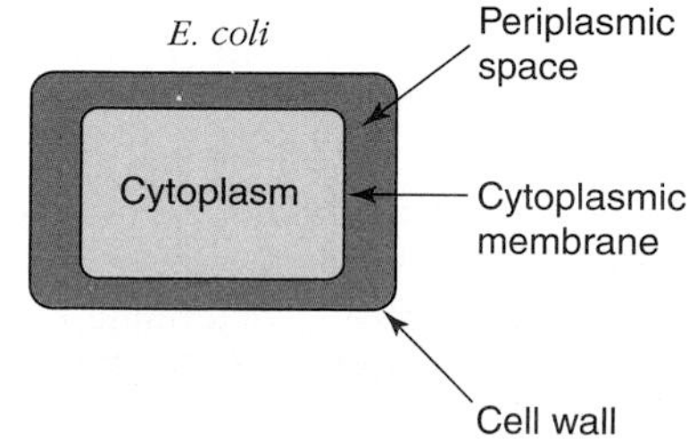

(a) Under nongrowing conditions, *E. coli* responds to changes in osmolarity by regulating its cytoplasmic water content. What would happen to the cytoplasmic volume if *E. coli* were grown in a high osmolarity medium? In a low osmolarity medium?

(b) Under growth conditions, *E. coli* also regulates cytoplasmic K^+ content (in addition to water) in response to growth medium osmolarity. This prevents the large changes in cell volume that would occur in response to changes in growth medium osmolarity if only water content were regulated. How might *E. coli* regulate both the cytoplasmic concentrations of K^+ and water when grown in low-osmolarity medium? What happens when *E. coli* is grown in high-osmolarity medium?

[From Record, M.T., et al., *Trends Biochem. Sci.* **23,** 143–148 (1998).]

24. Compare the concentrations of H_2O and H^+ in a sample of pure water at pH 7.0 at 25°C.

25. Explain why H_3O^+ is the strongest acid that can exist in a biological system.

26. Calculate $[H^+]$ for saliva and urine (see Table 2–1).

27. Fill in the blanks of the following table:

	Acid, base, or neutral?	**pH**	**$[H^+]$ (M)**	**$[OH^-]$ (M)**
A		5.60		
B				4.5×10^{-7}
C	neutral			
D			2.1×10^{-3}	

28. Calculate the pH of 500 mL of water to which (a) 20 mL of 1.0 M HNO_3 or (b) 15 mL of 1.0 M KOH has been added.

29. What is the pH of 1.0 L of water to which (a) 1.5 mL of 3.0 M HCl or (b) 1.5 mL of 3.0 M NaOH has been added?

30. What is the pH of a solution of 1.0×10^{-9} M HCl?

31. Give the conjugate base of the following acids:

(a) $HC_2O_4^-$
(b) HSO_3^-
(c) $H_2PO_4^-$
(d) HCO_3^-
(e) $HAsO_4^{2-}$
(f) HPO_4^{2-}
(g) HO_2^-

32. Give the conjugate acid of the species listed in the above problem.

33. The weak acid phosphoric acid, H_3PO_4, has three ionizable protons. The titration curve for phosphoric acid is shown.

(a) Write the equation for the dissociation of each of the three protons.

(b) Use the titration curve to determine the p*K* of each dissociable proton. Calculate the K_a for each dissociable proton.

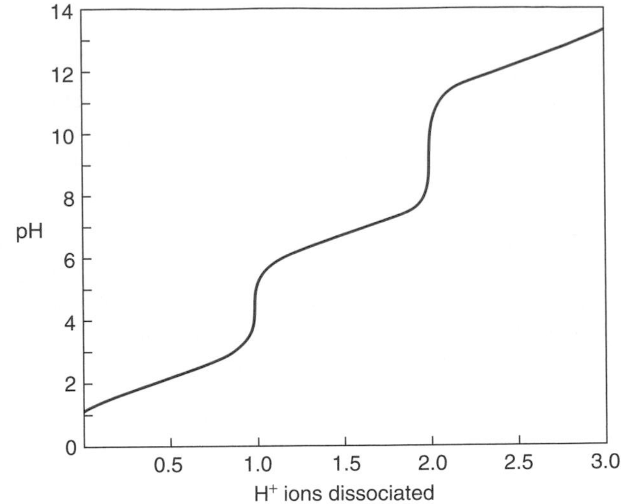

(c) Label the titration curve at the points that correspond to the following concentrations:

(i) high $[H_3PO_4]$
(ii) $[H_3PO_4] = [H_2PO_4^-]$
(iii) high $[H_2PO_4^-]$
(iv) $[H_2PO_4^-] = [HPO_4^{2-}]$
(v) high $[HPO_4^{2-}]$
(vi) $[HPO_4^{2-}] = [PO_4^{3-}]$
(vii) high $[PO_4^{3-}]$

(d) Phosphate buffers are often used in biochemistry because of their ability to buffer solutions of approximately pH 7, which is near physiological pH. Which two phosphate species would be used to buffer a solution of pH 7?

34. Identify the acidic hydrogen in the following compounds.

$$\mathrm{HO{-}C(CH_2{-}COOH)_2{-}COOH}$$

Citric acid

Piperidine ($N{H_2}^+$ in ring)

$$\mathrm{{}^+H_3N{-}CH(CH_2CH_2CH_2CH_2NH_3^+){-}C(=O){-}OH}$$

Lysine

$$\mathrm{HO{-}C(=O){-}C(=O){-}OH}$$

Oxalic acid

Barbiturate

$$\mathrm{O(C_4H_8)NH^+{-}CH_2{-}CH_2{-}S(=O)_2{-}O^-}$$

4-Morphine ethanesulfonic acid (MES)

35. Rank the following according to their strength as acids:

	Acid	K_a	pK
A	Citrate		4.76
B	Succinic acid	6.17×10^{-5}	
C	Succinate	2.29×10^{-6}	
D	Formic acid	1.78×10^{-4}	
E	Citric acid		3.13

36. The amino acid glycine ($H_2N—CH_2—COOH$) has pK values of 2.35 and 9.78. Indicate the structure and net charge of the molecular species that predominates at pH 2, 7, and 10.

37. Several hours after a meal, partially digested food leaves the stomach and enters the small intestine, where pancreatic juice is added. How does the pH of the partially digested mixture change as it passes from the stomach into the intestine (see Table 2-1)?

38. A solution is made by mixing 50 mL of a stock solution of 2.0 M K_2HPO_4 and 25 mL of a stock solution of 2.0 M KH_2PO_4. The solution is diluted to a final volume of 200 mL. What is the final pH of the solution? (See Sample Calculation 2-5 for the appropriate pK values).

39. What is the volume (in mL) of 6 M acetic acid that would have to be added to 500 mL of a solution of 0.20 M sodium acetate in order to achieve a pH of 5.0?

40. What is the volume (in mL) of 6 M NaOH that would have to be added to 500 mL of a solution of 0.20 M acetic acid in order to achieve a pH of 5.0?

41. Which would be a more effective buffer at pH 8.0 (see Table 2-2)?
(a) 10 mM HEPES buffer or 10 mM glycinamide buffer
(b) 10 mM Tris buffer or 20 mM Tris buffer
(c) 10 mM boric acid or 10 mM sodium borate

42. An experiment requires the buffer HEPES, pH = 8.0 (see Table 2–2).
(a) Write an equation for the dissociation of HEPES in water. Identify the weak acid and the conjugate base.
(b) What is the effective buffer range for HEPES?
(c) The buffer will be prepared by making 1.0 L of a 0.10 M solution of HEPES. Sodium hydroxide will be added until the desired pH is achieved. Describe how you will make 1.0 L of 0.10 M HEPES. (HEPES is supplied by the chemical company as a sodium salt with a molecular weight of 260.3 $g \cdot mol^{-1}$.)
(d) What is the volume (in mL) of a stock solution of 6.0 M HCl that must be added to the 0.1 M HEPES to achieve the desired pH of 8.0? Describe how you will make the buffer.

43. One liter of a 0.1 M Tris buffer (see Table 2-2) is prepared and adjusted to a pH of 8.2.
(a) Write the equation for the dissociation of Tris in water. The structure of the conjugate base form of Tris is shown. Identify the weak acid and the conjugate base.
(b) What is the effective buffering range of Tris?
(c) What are the concentrations of the conjugate acid and weak base at pH 8.2?
(d) What is the ratio of conjugate base to weak acid if 1.5 mL of 3.0 M HCl is added to 1.0 L of the buffer? What is the new pH? Has the buffer functioned effectively? Compare the pH change to that of Problem 29a in which the same amount of acid was added to the same volume of pure water.
(e) What is the ratio of conjugate base to weak acid if 1.5 mL of 3.0 M NaOH is added to 1.0 L of the buffer? What is the new pH? Has the buffer functioned effectively? Compare the pH change to that of Problem 29b in which the same amount of base was added to the same volume of pure water.

44. One liter of a 0.1 M Tris buffer (see Table 2-2) is prepared and adjusted to a pH of 2.0.
(a) What is the ratio of conjugate base to weak acid at this pH?
(b) What is the ratio of conjugate base to weak acid if 1.5 mL of 3.0 M HCl is added to 1.0 L of the buffer? What is the pH? Has the buffer functioned effectively? Explain.
(c) What is the ratio of conjugate base to weak acid if 1.5 mL of 3.0 M NaOH is added to 1.0 L of the buffer? What is the pH? Has the buffer functioned effectively?

45. Explain why metabolically generated CO_2 does not accumulate in tissues but is quickly converted to carbonic acid by the action of carbonic anhydrase in red blood cells.

46. As stated in Box 2-B, impaired pulmonary function can contribute to respiratory acidosis. Using the appropriate equations, explain how the failure to eliminate sufficient CO_2 through the lungs leads to acidosis.

47. A hen's egg loses carbon dioxide through its shell as it ages. What is the corresponding change in pH of the egg contents? Explain, using appropriate equations.

48. Metabolic acidosis often occurs in patients with impaired circulation from cardiac arrest. Treatment of acidosis using mechanical hyperventilation (a standard treatment for acidosis) cannot be used with these patients because they often have acute lung injury (ALI). Recently, a group of physicians at San Francisco General Hospital advocated using tris(hydroxymethyl)aminomethane (Tris) to treat the metabolic acidosis of these patients.

$$\begin{array}{c} CH_2OH \\ | \\ HOH_2C—C—NH_2 \\ | \\ CH_2OH \end{array}$$

Tris(hydroxymethyl)aminomethane

(a) How would mechanical hyperventilation help alleviate acidosis in patients who do not have ALI?
(b) Acidosis is sometimes treated by administering sodium bicarbonate ($NaHCO_3$) to the patients. Why would this treatment be effective in treating metabolic acidosis? Why would this treatment also be unacceptable for patients with ALI?
(c) Explain how Tris works to treat metabolic acidosis. Why is this treatment acceptable for patients with ALI?
[From Kallet, R.H., et al., *Am. J. Respir. Crit. Care Med.* **161,** 1149–1153 (2000).]

49. You are an emergency-room physician and you have just admitted a patient, Susan M., a 22-year-old female, who was brought to the emergency room around 9 P.M. by her roommate Anne S. Anne tells you that she had come home and found that Susan was disoriented and had trouble speaking. Anne brought Susan to the emergency room when Susan began to suffer from nausea and vomiting. She is also hyperventilating. Anne reveals that Susan had been depressed lately, and shows you an empty aspirin bottle she had found. The aspirin bottle, when full, would have contained 250 tablets. Susan admits that she took the tablets around 7 P.M. that evening. You draw blood from Susan and the laboratory performs the analyses shown in the table.

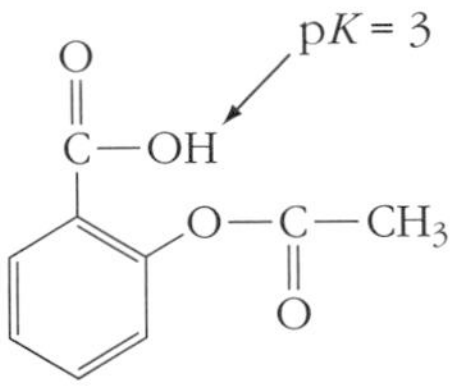

$pK = 3$

Acetylsalicylic acid (aspirin)

In the emergency room, Susan is given a stomach lavage with saline and two doses of activated charcoal to adsorb the aspirin. Eight hours later, she was still experiencing nausea and vomiting, and her respiratory rate increased, and further treatment was required. You carry out a gastric lavage with a sodium bicarbonate solution, pH = 8.5. Susan's blood pH begins to drop around 24 hours after the aspirin ingestion and finally returns to normal at 60 hours after the ingestion.

	Susan M, 2 h after aspirin ingestion	**Susan M, 10 h after aspirin ingestion**	**Normal values**
pCO_2 (mm Hg)	26	19	35–45
HCO_3^- (mM)	18	21	22–26
pO_2 (mm Hg)	113	143	75–100
pH	7.44	7.55	7.35–7.45
Blood aspirin concentration (mg/dL)	57	117	

(a) Aspirin, or acetylsalicylic acid, is converted in the stomach in the presence of aqueous acid and stomach esterases (which act as catalysts) to salicylic acid and acetic acid. Write a balanced equation for this chemical transformation.

(b) Since Anne has brought Susan into the emergency room only two hours after the overdose, you suspect that Susan's stomach might contain undissolved aspirin that is continuing to be absorbed. The fact that Susan is still experiencing nausea and vomiting 10 hours after the ingestion confirms your suspicion and you decide to use a gastric lavage at pH 8.5 to effectively remove the undissolved aspirin. This treatment solubilizes the aspirin so that it can easily be removed from the stomach. Calculate the percentage of protonated and unprotonated forms of salicylic acid at the pH of the stomach (usually around 2.0).

(c) Calculate the percentage of protonated and unprotonated forms of salicylic acid at the pH of the gastric lavage. Why does the lavage result in increased solubility of the drug? (Assume that the p*K* values for the carboxylate group in salicylic acid and acetylsalicylic acid are the same.)

(d) It has been shown that salicylates act *directly* on the nervous system to stimulate respiration. Thus, the patient is hyperventilating due to her salicylate overdose. Explain how the salicylate-induced hyperventilation leads to the values of pO_2 and pCO_2 seen in the patient.

(e) Explain how the salicylate-induced hyperventilation leads to the values of pH and $[HCO_3^-]$ seen in the patient. Illustrate your answer with the appropriate equations.

(f) Determine the ratio of HCO_3^- to H_2CO_3 in the patient's blood 10 hours after aspirin ingestion. How does this compare to the ratio of HCO_3^- to H_2CO_3 in normal blood? Can the H_2CO_3/HCO_3^- system serve as an effective buffer in this patient? Explain.

(g) Sixty hours after aspirin ingestion, the patient's blood pH has returned to normal (pH = 7.4). Describe how the carbonic/bicarbonate buffering system responded to bring the patient's blood pH back to normal, using appropriate equations.

[From Krause, D.S., Wolf, B.A., and Shaw, L.M., *Therapeutic Drug Monitoring* **14,** 441–451 (1992).]

SELECTED READINGS

Ellis, R.J., Macromolecular crowding: Obvious but underappreciated, *Trends Biochem. Sci.* **26,** 597–603 (2001). [Discusses how the large number of macromolecules in cells could affect reaction equilibria and reaction rates.]

Gerstein, M. and Levitt, M., Simulating water and the molecules of life, *Sci. Am.* **279(11),** 101–105 (1998). [Describes the structure of water and how water interacts with other molecules.]

Halperin, M.L. and Goldstein, M.B., *Fluid, Electrolyte, and Acid–Base Physiology: A Problem-Based Approach* (3rd ed.), W.B. Saunders (1999). [Includes extensive problem sets with explanations of basic science as well as clinical effects of acid–base disorders.]

Kropman, M.F. and Bakker, H.J., Dynamics of water molecules, *Science* **291,** 2118–2120 (2001). [Describes how water molecules in solvation shells move more slowly than bulk water molecules.]

CHAPTER 3

From Genes to Proteins

Were the Neanderthals predecessors of modern humans or an evolutionary dead end? Studies of bone morphology and cultural artifacts of the Neanderthals, who inhabited Europe and western Asia until about 30,000 years ago, yield conflicting theories. Enter the molecular biologist, armed with powerful techniques for recovering, amplifying, and analyzing traces of ancient DNA. Bones with typical Neanderthal features (large, muscular, and having a protruding brow) yielded stretches of mitochondrial DNA whose sequences fall far outside the range of variation seen in the mitochondrial DNA of contemporary humans. The results suggest that the Neanderthal lineage diverged some 555,000 to 690,000 years ago from the lineage that subsequently led to modern humans about 150,000 years ago. Thus, modern humans do not appear to be descendants of the heavyset Neanderthals, although they surely coexisted with them for many thousands of years.

[John Reader/Science Photo Library/Photo Researchers.]

LEARNING OBJECTIVES

1. Understand the relationships between purines, pyrimidines, nucleosides, nucleotides, and nucleic acids.
2. Understand the molecular structure of DNA.
3. Understand the structural differences between DNA and RNA.
4. Understand how DNA denatures and renatures.
5. Understand the flow of biological information from DNA to RNA to protein.
6. Understand how DNA is sequenced.
7. Understand why restriction enzymes are useful.
8. Understand how recombinant DNA molecules are constructed.
9. Understand the polymerase chain reaction.
10. Understand what can be learned by examining an organism's genome.

THIS CHAPTER IN CONTEXT

All the structural components of cells and the machinery that carries out the cell's activities are ultimately specified by the cell's genetic material—DNA. Therefore, before examining all the various types of biological molecules and their metabolic transformations, we must consider the nature of DNA: What is its chemical structure and how is its biological information stored and expressed as protein? We will also examine methods for determining the sequences of nucleotides in DNA and for manipulating these sequences in the laboratory. Finally, we will see how information about an organism's DNA can be used to explore questions about the proteins encoded by the DNA. In the following four chapters, we will consider the structures and functions of those proteins.

IT'S NO SECRET THAT CHILDREN PHYSICALLY RESEMBLE THEIR parents because they inherit their parents' genetic material, DNA. But DNA does more than just specify eye color and hair color: It contains a complete set of instructions for the development of an embryo into a fully independent organism that can eat, breathe, grow, and reproduce. The DNA may also determine the likelihood that the individual will develop a particular disease.

Some diseases appear to run in families because defective DNA is passed from parent to offspring. For example, about 1 in 3000 babies in the United States is born with **cystic fibrosis (CF),** the most common inherited disease in individuals of northern European extraction. The most serious symptom of CF is the obstruction of the airways by thick, sticky mucus, which also creates an ideal environment for bacterial growth (Fig. 3-1). Historically, individuals with CF died in childhood, but the use of antibiotics to prevent lung infections has now extended survival well into adulthood. Cystic fibrosis is one of many genetic diseases in which a protein is absent or abnormal because the segment of DNA—the gene—that encodes the protein is defective. The molecular defect that causes CF was not known until the "CF gene" was identified.

Understanding the structure and function of DNA is essential for understanding the nature of genetic information, whether it is linked to a specific disease or not. By dissecting, examining, and cataloguing segments of DNA, it becomes possible to see how this molecule's biochemical properties can be translated into the physical attributes and metabolic activities of an entire organism.

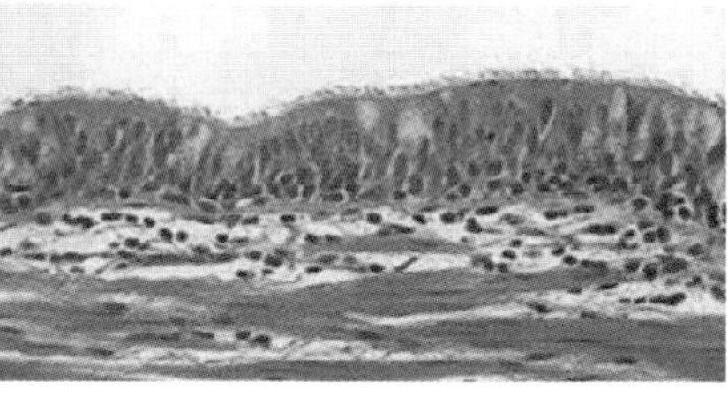

(a)

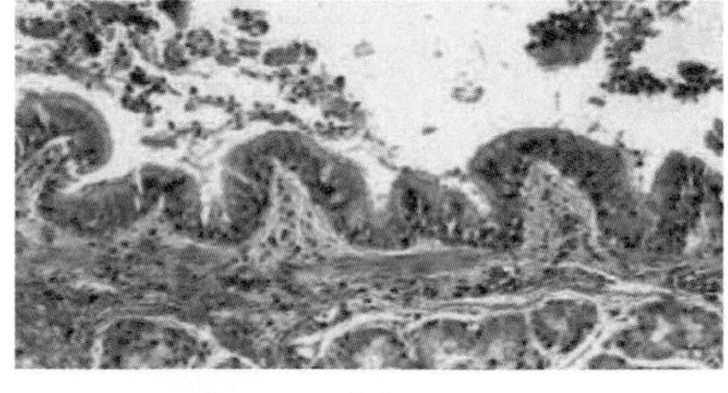

(b)

FIGURE 3-1

Micrographs of normal and cystic fibrosis lung tissue.

(a) In a lung from a healthy individual, the space above the tissue surface is clear. (b) In a lung from an individual with cystic fibrosis, the extracellular space contains bits of dehydrated secretions. [*Photos courtesy James M. Wilson, University of Pennsylvania. From* Journal of Clinical Investigation, ***vol. 103,*** *no. 3, February 1999, p. 303. Reproduced with permission of Copyright Clearance Center.*]

DNA IS THE GENETIC MATERIAL

Gregor Mendel was certainly not the first to notice that an organism's characteristics (for example, flower color or seed shape in pea plants) were passed to its progeny, but in 1865 he was the first to describe their predictable patterns of inheritance. By 1903, Mendel's inherited factors (now called **genes**)

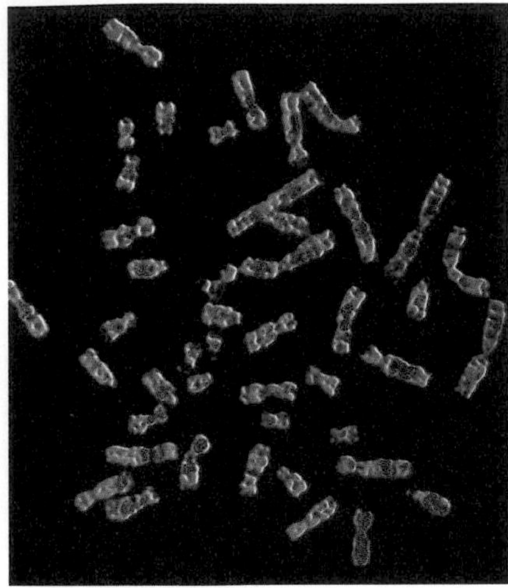

FIGURE 3-2

Human chromosomes from amniocentesis. [*CNRI/Photo Researchers.*]

were recognized as belonging to **chromosomes** (a word that means "colored bodies"), which are visible by light microscopy (Fig. 3-2).

Nucleic acids had been discovered in 1869 by Friedrich Miescher, who isolated this material from the white blood cells in pus on surgical bandages. However, when it became clear that chromosomes were composed of both proteins and nucleic acids, nucleic acids were dismissed as possible carriers of genetic information due to their lack of complexity: They contained only four different types of structural units called **nucleotides.** In contrast, proteins contained 20 different types of amino acids and exhibited great diversity in composition, size, and shape—attributes that seemed more appropriate for carriers of genetic information.

Years later, microbiologists showed that a substance from a dead pathogenic (disease-causing) strain of the bacterium *Streptococcus pneumoniae* could transform cells from a normal strain to the pathogenic type. In 1944, Oswald Avery, Colin MacLeod, and Maclyn McCarty showed that this transforming substance was **deoxyribonucleic acid (DNA),** but their results did not garner much attention. Another seven years went by, until Alfred Hershey and Martha Chase demonstrated that in **bacteriophages** (viruses that infect bacterial cells and that consist only of protein and DNA), the DNA but not the protein was the infectious agent (Fig. 3-3).

By this time, DNA was known to contain chains of polymerized nucleotides—abbreviated A, C, G, and T—but these were thought to occur as simple repeating tetranucleotides, for example,

—ACGT-ACGT-ACGT-ACGT—

When Erwin Chargaff showed in 1950 that the nucleotides in DNA were not all present in equal numbers and that the nucleotide composition varied among species, it became apparent that DNA might be complex enough to be the genetic material after all, and the race was on to decipher its molecular structure.

The DNA structure ultimately elucidated by James Watson and Francis Crick in 1953 incorporated Chargaff's observations. Specifically, Chargaff noted that the amount of A is equal to the amount of T; the amount of C is equal to the amount of G; and the total amount of A + G is equal to the total amount of C + T. *Chargaff's "rules" could be satisfied by a molecule with two* ***polynucleotide*** *strands (polymers of nucleotides) in which A and C in one strand pair with T and G in the other.*

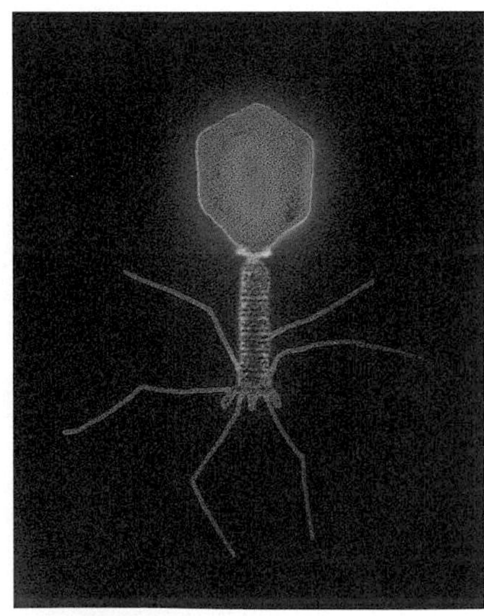

FIGURE 3-3

A T-type bacteriophage.
The phage consists mostly of a protein coat surrounding a molecule of DNA. Alfred Hershey and Martha Chase identified the DNA as the infectious agent. [*Dept. of Microbiology, Biozentrum/Science Photo Library/Photo Researchers.*]

The nucleotide components of nucleic acids

Each nucleotide of DNA includes a nitrogen-containing **base.** The bases adenine (A) and guanine (G) are known as **purines** because they resemble the organic group purine:

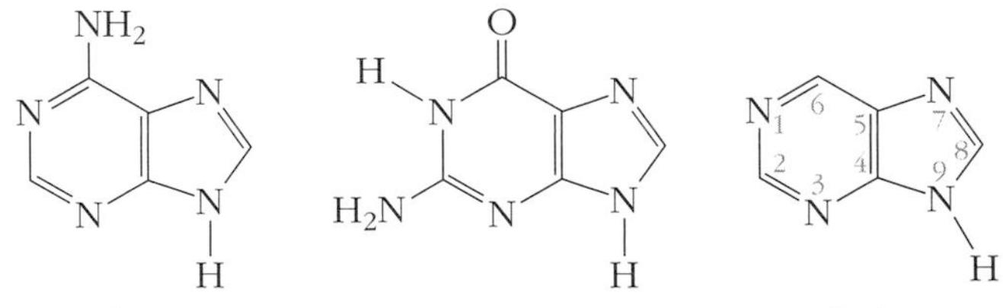

Adenine Guanine Purine

The bases cytosine (C) and thymine (T) are known as **pyrimidines** because they resemble the organic compound pyrimidine:

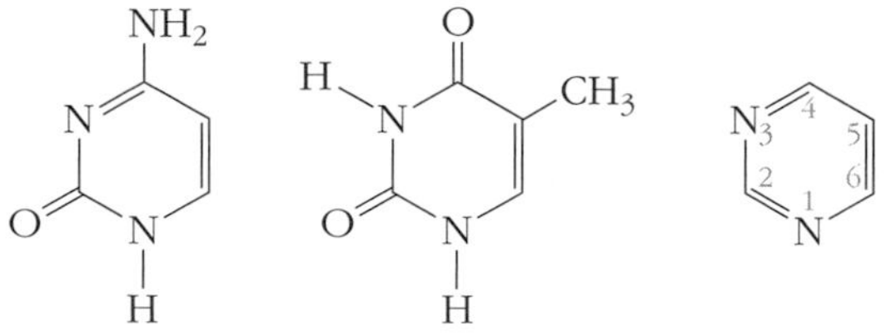

Cytosine Thymine Pyrimidine

In **ribonucleic acid (RNA),** the pyrimidine uracil (U) takes the place of thymine:

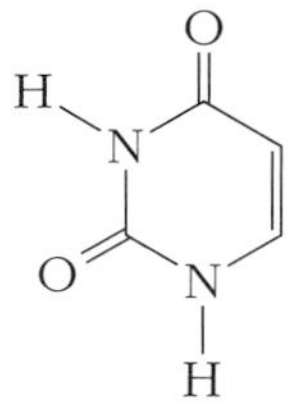

Uracil

so that DNA contains the bases A, C, G, and T, and RNA contains A, C, G, and U.

N9 of purines and N1 of pyrimidines are linked to a five-carbon sugar to form a **nucleoside.** *In DNA, the sugar is 2′ deoxyribose; in RNA, the sugar is ribose:*

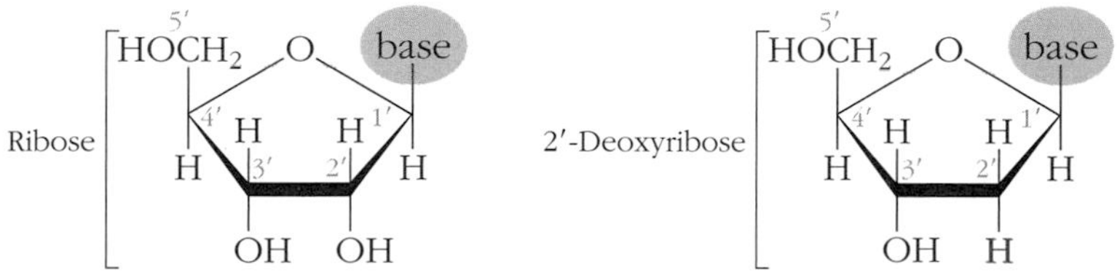

Ribose 2′-Deoxyribose

(the sugar atoms are numbered with primes to distinguish them from the atoms of the attached bases).

A nucleotide is a nucleoside to which one or more phosphate groups are linked, usually at C5′ of the sugar. Depending on whether there are 1, 2, or 3 phosphate groups, the nucleotide is known as a nucleoside monophosphate, nucleoside diphosphate, or nucleoside triphosphate and is represented by a three-letter abbreviation, for example,

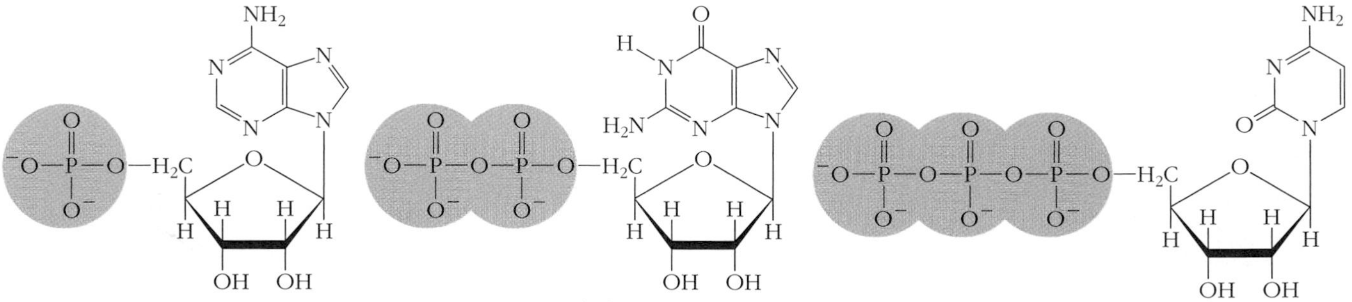

Adenosine monophosphate (AMP) Guanosine diphosphate (GDP) Cytidine triphosphate (CTP)

Deoxynucleotides are named in a similar fashion, and their abbreviations are preceded by "d." The deoxy counterparts of the compounds shown above would therefore be deoxyadenosine monophosphate (dAMP), deoxyguanosine diphosphate (dGDP), and deoxycytidine triphosphate (dCTP). The names and abbreviations of the bases, nucleosides, and nucleotides are summarized in Table 3-1.

REVIEW LEARNING OBJECTIVE 1

- What are the structural components of nucleic acids?

The structure of DNA

Go to **Exercise 2/Nucleic Acid Structure** for a closer look at the chemical structure and conformation of DNA and RNA. The structures presented in Exercise 2 and the questions based on those structures address Learning Objectives 2 and 3.

In a nucleic acid, the linkage between nucleotides is called a **phosphodiester bond,** because a single phosphate group forms ester bonds to both C5′ and C3′. (When a nucleoside triphosphate is added to the polynucleotide chain, two of its phosphate groups are eliminated.) Once incorporated into a polynucleotide, the nucleotide is formally known as a nucleotide **residue.** Nucleotides consecutively linked by phosphodiester bonds form a polymer in which the bases project out from a backbone of repeating sugar–phosphate groups.

5′ end
Phosphodiester bond
Adenine
Guanine
Thymine
Cytosine
3′ end

TABLE 3-1 Nucleic acid bases, nucleosides, and nucleotides

Base	Nucleoside[a]	Nucleotides[a]
Adenine (A)	Adenosine	Adenylate; adenosine monophosphate (AMP) adenosine diphosphate (ADP) adenosine triphosphate (ATP)
Guanine (G)	Guanosine	Guanylate; guanosine monophosphate (GMP) guanosine diphosphate (GDP) guanosine triphosphate (GTP)
Cytosine (C)	Cytidine	Cytidylate; cytidine monophosphate (CMP) cytidine diphosphate (CDP) cytidine triphosphate (CTP)
Thymine (T)[b]	Thymidine	Thymidylate; thymidine monophosphate (TMP) thymidine diphosphate (TDP) thymidine triphosphate (TTP)
Uracil (U)[c]	Uridine	Uridylate; uridine monophosphate (UMP) uridine diphosphate (UDP) uridine triphosphate (UTP)

[a]Nucleosides and nucleotides containing 2′-deoxyribose rather than ribose may be called deoxynucleosides and deoxynucleotides. The nucleotide abbreviation is then preceded by "d."

[b]Thymine is found in DNA but not in RNA.

[c]Uracil is found in RNA but not in DNA.

The end of the polymer that bears a phosphate group attached to C5′ is known as the **5′ end,** and the end that bears a free OH group at C3′ is the **3′ end.** By convention, the base sequence in a polynucleotide is read from the 5′ end (on the left) to the 3′ end (on the right).

DNA contains two polynucleotide strands whose bases pair through hydrogen bonding (hydrogen bonds are discussed in Section 2-1). Two hydrogen bonds link adenine and thymine, and three hydrogen bonds link guanine and cytosine:

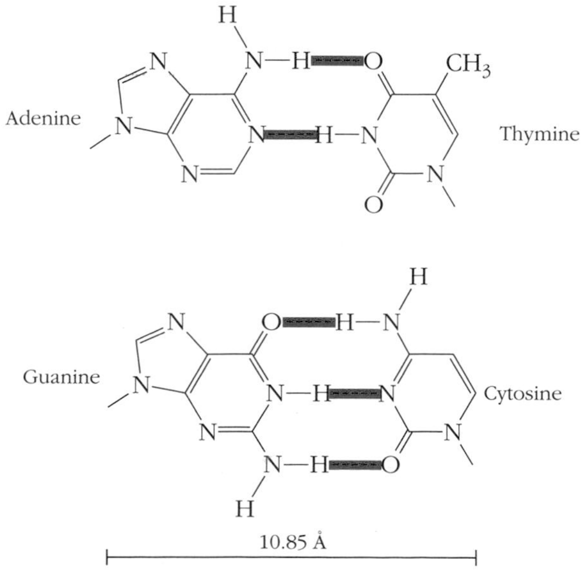

Because all the ***base pairs*** *contain a purine and a pyrimidine, they have the same molecular dimensions* (about 11 Å wide). Consequently, the

sugar–phosphate backbones of the two strands of DNA are separated by a constant distance, regardless of whether the base pair is A:T, G:C, T:A, or C:G.

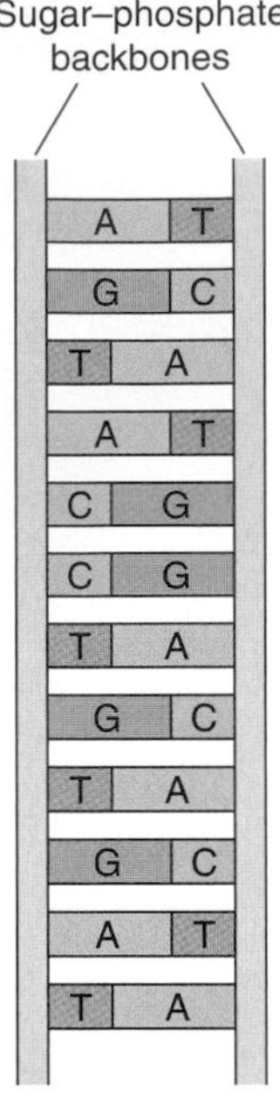

Although the DNA is shown here as a ladder-like structure—with the two sugar–phosphate backbones as the vertical supports and the base pairs as the rungs—the two strands of DNA twist around each other to generate the familiar double helix. This conformation allows successive base pairs, which are essentially planar, to stack on top of each other with a center-to-center distance of only 3.4 Å.

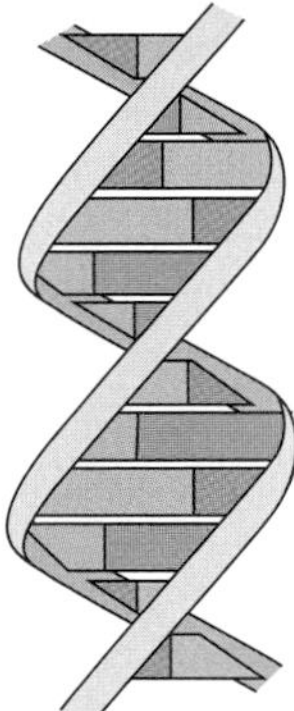

In fact, Watson and Crick derived this model for DNA not just from Chargaff's rules, but also from Rosalind Franklin's and Maurice Wilkins' studies of the diffraction (scattering) of an X-ray beam by a DNA fiber, which suggested a helix with a repeating spacing of 3.4 Å.

The major features of the DNA molecule (Fig. 3-4) include the following:

1. The two polynucleotide strands are **antiparallel;** that is, their phosphodiester bonds run in opposite directions. One strand has a $5' \rightarrow 3'$ orientation, and the other has a $3' \rightarrow 5'$ orientation.
2. The DNA "ladder" is twisted in a right-handed fashion. (If you climbed the DNA helix as if it were a spiral staircase, you would hold the outer railing—the sugar–phosphate backbone—with your right hand.)
3. The diameter of the helix is about 20 Å, and it completes a turn about every 10 base pairs, which corresponds to an axial distance of about 34 Å.

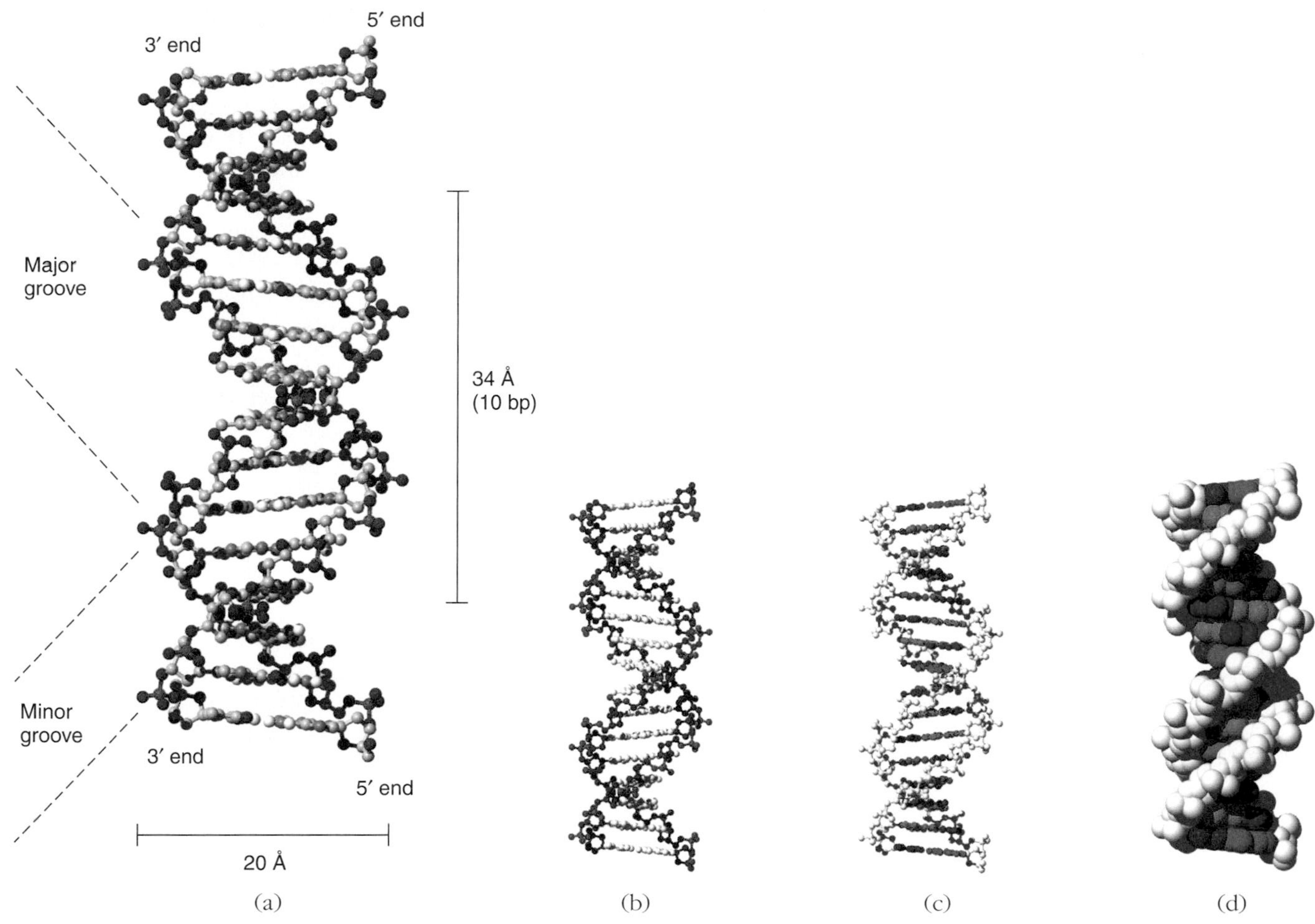

FIGURE 3-4

Model of DNA.
(a) Ball-and-stick model showing all atoms: C gray, O red, N blue, and P gold (H atoms are not shown). (b) Model highlighting the sugar–phosphate backbone, with deoxyribose groups in red and phosphate groups in gold. (c) Model with bases highlighted: A green, G yellow, C blue, and T red. (d) Space-filling model with bases color-coded as in part (c).

4. The twisting of the DNA "ladder" into a helix creates two grooves of unequal width, the **major** and **minor grooves.**

5. The sugar–phosphate backbones define the exterior of the helix and are exposed to the solvent (see Fig. 3-4b). The negatively charged phosphate groups bind Mg^{2+} cations *in vivo,* which helps minimize electrostatic repulsion between these groups.

6. The base pairs are located in the center of the helix, approximately perpendicular to the helix axis (see Fig. 3-4c).

7. The base pairs stack on top of each other, so the core of the helix is solid (see Fig. 3-4d). Although the planar faces of the base pairs are not accessible to the solvent, their edges are exposed in the major and minor grooves (this allows certain DNA-binding proteins to recognize specific bases). The structural features of the DNA helix, including base pairing and stacking, are highlighted in Exercise 2.

The size of a DNA segment is expressed in units of base pairs **(bp)** or kilobase pairs (1000 bp, abbreviated **kb**). Most naturally occurring DNA molecules comprise thousands to millions of base pairs. A short single-stranded polymer of nucleotides is usually called an **oligonucleotide** (*oligo* is Greek for "few").

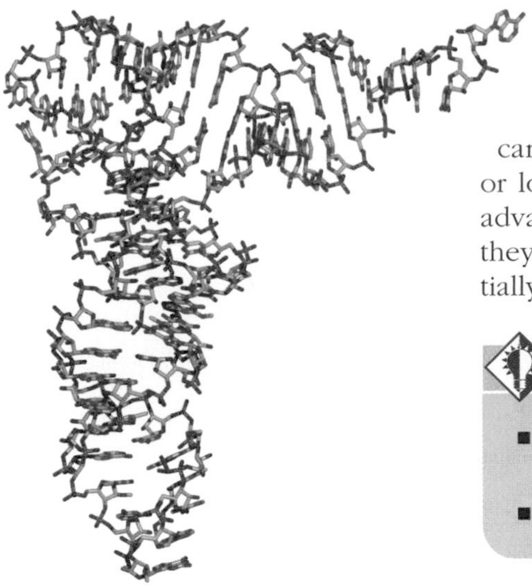

In nature, DNA seldom assumes a perfectly regular conformation because of small sequence-dependent irregularities. For example, base pairs can roll or twist like propeller blades, and the helix may wind more tightly or loosely at certain nucleotide sequences. DNA-binding proteins may take advantage of these small variations to locate their specific binding sites, and they in turn may further distort the DNA helix by causing it to bend or partially unwind.

REVIEW LEARNING OBJECTIVE 2

- Describe the arrangement of the base pairs and sugar–phosphate backbones in DNA.
- What did Chargaff's rules reveal about the structure of DNA?

FIGURE 3-5

A transfer RNA molecule.
This 76-nucleotide single-stranded RNA molecule folds back on itself so that base pairs form between complementary segments. [*Structure (pdb 4TRA) determined by E. Westhoff, P. Dumas, and D. Moras.*]

The structure of RNA

RNA, which is a single-stranded polynucleotide, has greater conformational freedom than DNA, whose structure is constrained by the requirements of regular base-pairing between its two strands. *An RNA strand can fold back on itself so that base pairs form between different segments of the same strand.* Consequently, RNA molecules tend to assume intricate "knotted" shapes (Fig. 3-5). Unlike DNA, whose regular structure is suited for the long-term storage of genetic information, RNA can assume more active roles in expressing that information. For example, the molecule shown in Figure 3-5, which carries the amino acid phenylalanine, must interact with a number of proteins and other RNA molecules during protein synthesis.

The bases of RNA are also capable of base-pairing with a complementary single strand of DNA to produce an RNA–DNA hybrid double helix. A double helix involving RNA is wider and flatter than the standard DNA helix (its diameter is about 26 Å and it makes one helical turn every 11 residues). In addition, its base pairs are inclined to the helix axis by about 20° (Fig. 3-6). These structural differences relative to the standard DNA helix primarily reflect the presence of the 2′ OH groups in RNA.

A double-stranded DNA helix can adopt this same helical conformation; it is known as **A-DNA.** The standard DNA helix shown in Figure 3-4 is known as **B-DNA.** Other conformations of DNA have been described, and there is some evidence that they may exist *in vivo,* at least for certain nucleotide sequences, but their functional significance is not completely understood.  Exercise 2 explores the conformations of RNA and A-DNA.

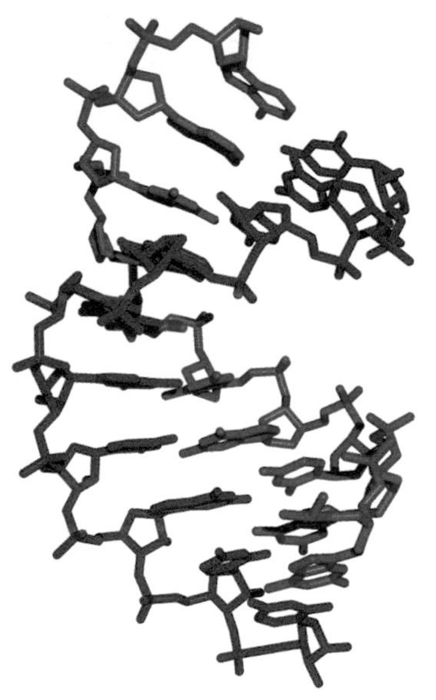

FIGURE 3-6

An RNA–DNA hybrid helix.
In a double helix formed by one strand of RNA (red) and one strand of DNA (blue), the planar base pairs are tilted and the helix does not wind as steeply as in a standard DNA double helix (compare with Fig. 3-4). [*Structure (pdb 1FIX) determined by N. C. Horton and B. C. Finzel.*]

REVIEW LEARNING OBJECTIVE 3

- How do DNA and RNA differ?

DNA denaturation and renaturation

The pairing of polynucleotide strands in a double-stranded nucleic acid is possible because bases in each strand form hydrogen bonds with complementary bases in the other strand: A is the complement of T (or U), and G is the complement of C. However, the structural stability of the DNA helix does not depend significantly on hydrogen bonding between complementary bases. (If the strands were separated, the bases could still satisfy their hydrogen-bonding requirements by forming hydrogen bonds with solvent water molecules.) Instead, *stability depends mostly on* ***stacking interactions,*** *which are a form of van der Waals interaction, between adjacent base pairs.* A view down the helix axis shows that stacked base pairs do not overlap exactly, due to the winding of the helix (Fig. 3-7). Although individual stacking interactions are weak, they are additive along the length of a DNA molecule.

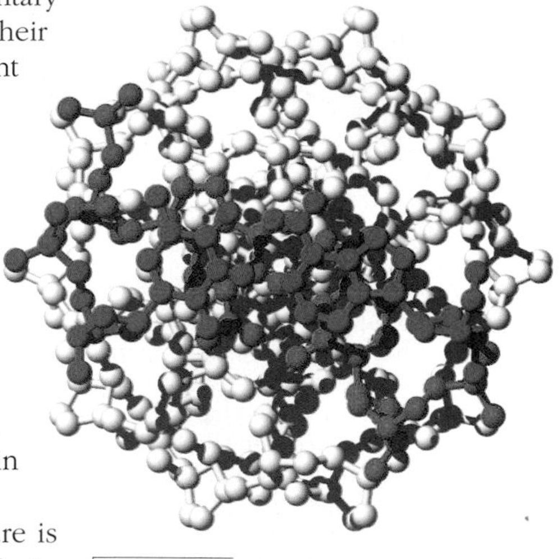

FIGURE 3-7

Axial view of DNA base pairs. A view down the central axis of the DNA helix shows the overlap of neighboring base pairs (only the first two nucleotides are highlighted).

The stacking interactions between neighboring G:C base pairs are stronger than those of A:T base pairs (this is not related to the fact that G:C base pairs have one more hydrogen bond than A:T base pairs). Consequently, *DNA rich in G and C is harder to disrupt than DNA with a high proportion of A and T.* These differences can be quantified in the **melting temperature (T_m)** of the DNA.

To determine the melting point of a sample of DNA, the temperature is slowly increased. At a sufficiently high temperature, the base pairs begin to unstack, hydrogen bonds break, and the two strands begin to separate. This process continues as the temperature rises, until the two strands come completely apart. The melting or **denaturation** of the DNA can be recorded as a melting curve (Fig. 3-8) by monitoring an increase in the absorbance of ultraviolet (260 nm) light (the aromatic bases absorb more light when unstacked). The midpoint of the melting curve (that is, the temperature at which half the DNA has separated into single strands) is the T_m. Table 3-2 lists the GC content and the melting point of the DNA from different species. Since manipulating DNA in the laboratory frequently requires the thermal separation of paired DNA strands, it is sometimes helpful to know the DNA's GC content.

When the temperature is lowered slowly, denatured DNA can **renature;** that is, *the separated strands can re-form a double helix by reestablishing hydrogen bonds between the complementary strands and by restacking the base pairs.* The maximum rate of renaturation occurs at about 20–25°C below the melting temperature. If the DNA is cooled too rapidly, it may not fully renature because base pairs may form randomly between short complementary segments. At low temperatures, the improperly paired segments are frozen in place since they do not have enough thermal energy to melt

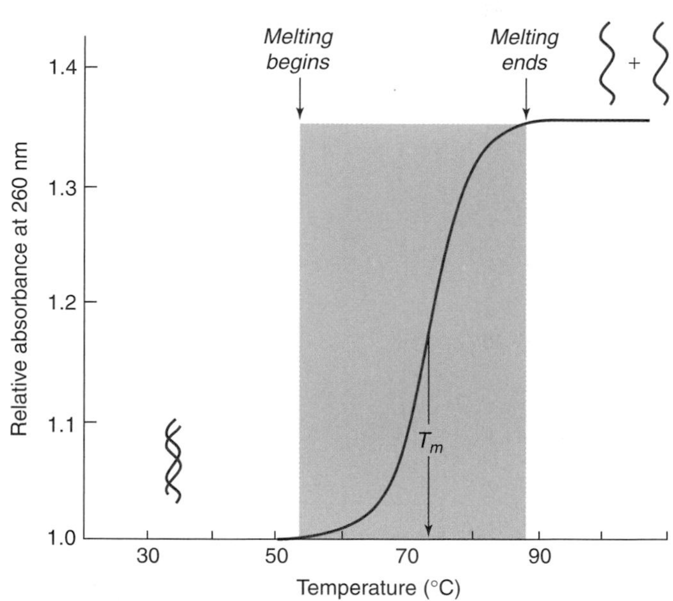

FIGURE 3-8

A DNA melting curve. Denaturation (melting or strand separation) in DNA results in an increase in ultraviolet absorbance relative to the absorbance at 25°C. The melting point T_m of the DNA sample is defined as the midpoint of the melting curve.

TABLE 3-2 GC content and melting points of DNA

Source of DNA	GC content (%)	T_m (°C)
Dictyostelium discoideum	23.0	79.5
Clostridium butyricum	37.4	82.1
Homo sapiens	40.3	86.5
Streptomyces albus	72.3	100.5

[Data from *Molecular Biology LabFax,* vol. I., Brown, T.A. (ed.), Academic Press (1998), pp. 233–237.]

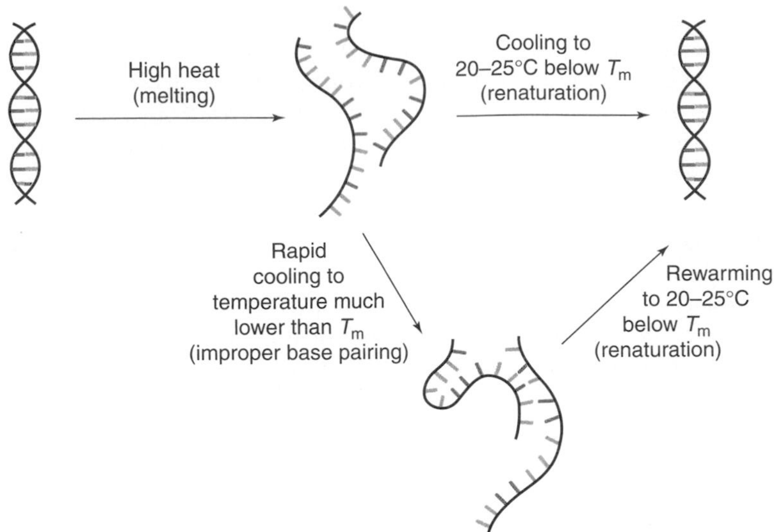

FIGURE 3-9

Renaturation of DNA.
DNA strands that have been melted apart can renature at a temperature 20–25°C less than T_m. At much lower temperatures, base pairs may form between short complementary segments within and between the single strands. Correct renaturation is possible only if the sample is rewarmed so that the improperly paired strands can separate and reanneal.

apart and find their correct complements (Fig. 3-9). The absolute rate of renaturation of denatured DNA depends on the length of the double-stranded molecule: Short segments come together **(anneal)** faster than longer segments, because the bases in each strand must locate their partners along the length of the complementary strand.

Short single-stranded nucleic acids (either RNA or DNA) can hybridize with much longer single-stranded molecules, provided they find a complementary region. This is the principle behind the use of natural or synthetic oligonucleotide **probes** to identify a specific sequence in a much longer nucleic acid or a mixture of nucleic acids. In one procedure, the single-stranded nucleic acid to be tested is immobilized on an inert support such as a nitrocellulose membrane and then incubated with the probe, which has been tagged with a radioactive or fluorescent group. The probe hybridizes with its complementary sequence, the unbound probe is washed away, and the bound probe (and therefore the target nucleic acid sequence) is revealed by the presence of radioactivity or fluorescence (Fig. 3-10).

REVIEW LEARNING OBJECTIVE 4

- Describe the molecular events in DNA denaturation and renaturation.
- How can a single-stranded oligonucleotide be used to reveal the presence of a particular segment of DNA?

2 THE CENTRAL DOGMA

The complementarity of the two strands of DNA is essential for its function as the storehouse of genetic information, since this information must be **replicated** (copied) for each new generation. As first suggested by Watson and Crick, the separated strands of DNA direct the synthesis of complementary strands, thereby generating two identical double-stranded molecules (Fig. 3-11). The parental strands are said to act as templates for the assembly of the new strands because their sequence of nucleotides determines the sequence

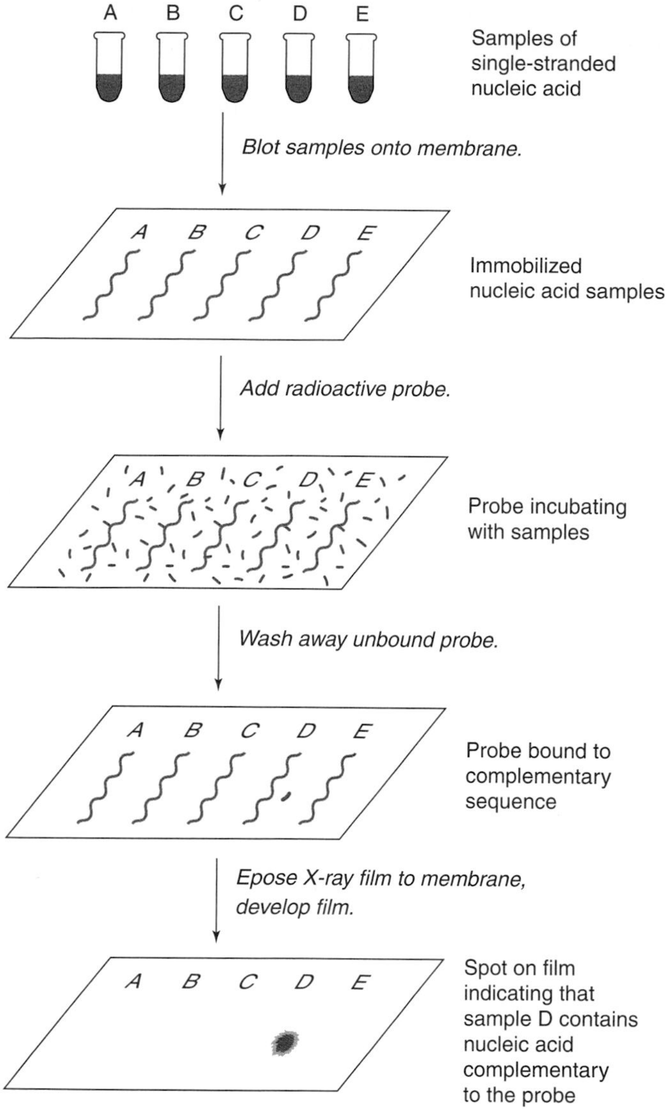

FIGURE 3-10

Procedure for detecting a specific nucleic acid sequence through hybridization of a radioactive probe.

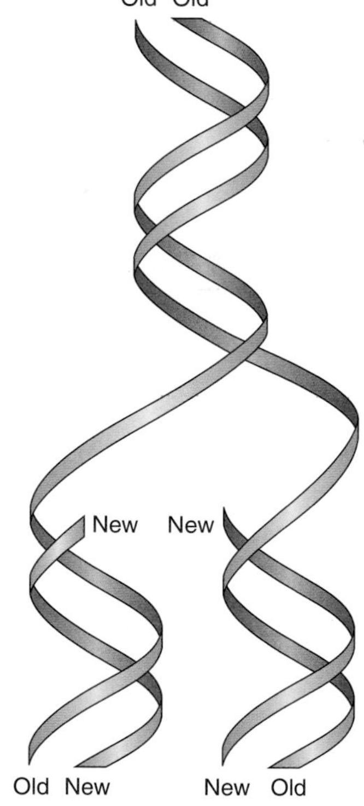

FIGURE 3-11

DNA replication.
The double helix unwinds so that each parental strand can serve as a template for the synthesis of a new complementary strand. The result is two identical DNA molecules.

of nucleotides in the new strands. Thus, genetic information—in the form of a sequence of nucleotide residues—is transmitted each time a cell divides.

A similar phenomenon is responsible for the **expression** of that genetic information, a process in which the information is used to direct the synthesis of proteins that carry out the cell's activities. First, *a portion of the DNA is* ***transcribed*** *to produce a complementary strand of RNA, then the RNA is* ***translated*** *into protein.* This paradigm is known as the Central Dogma of Molecular Biology, formulated by Francis Crick, and can be shown schematically:

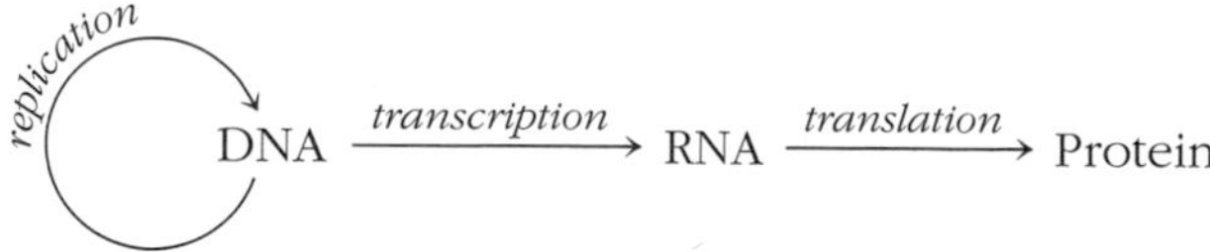

Even in the simplest organisms, DNA is an enormous molecule, and many organisms contain several different DNA molecules (for example, the chromosomes of eukaryotes). An organism's complete set of genetic information is called its **genome.** A genome may comprise several hundred to perhaps 50,000 genes.

To transcribe a gene, one of the two strands of DNA serves as a template for the synthesis of a complementary strand of RNA. The RNA therefore has the same sequence (except for the substitution of U for T) and the same $5' \rightarrow 3'$ orientation as the nontemplate strand of DNA. This strand of DNA is often called the **coding strand** (the template strand is called the **noncoding strand**).

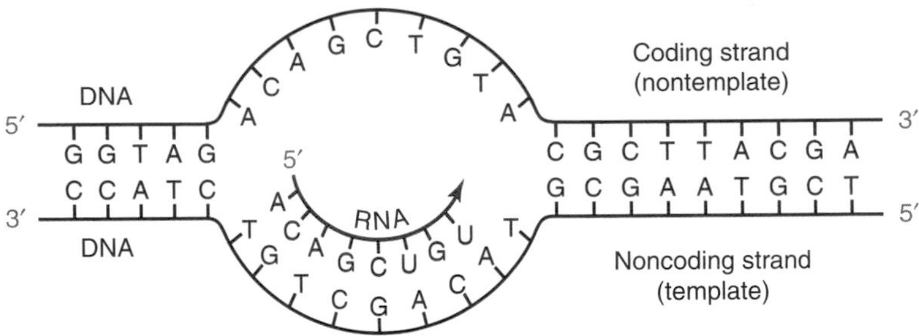

The transcribed RNA is known as **messenger RNA (mRNA)** because it carries the same genetic message as the gene.

The mRNA is translated in the **ribosome,** a cellular particle consisting of protein and **ribosomal RNA (rRNA).** At the ribosome, small molecules called **transfer RNA (tRNA),** which carry amino acids, recognize sequential sets of three bases (known as **codons**) in the mRNA through complementary base-pairing (a tRNA molecule is shown in Fig. 3-5). The ribosome covalently links the amino acids carried by successive tRNAs to form a protein. *The protein's amino acid sequence therefore ultimately depends on the nucleotide sequence of the DNA.*

DNA 5′—C–T–C–A–G–T–G–C–C— 3′
3′—G–A–G–T–C–A–C–G–G— 5′

mRNA 5′—C–U–C–A–G–U–G–C–C— 3′

tRNAs G–A–G–U–C–A–C–G–G
Leucine Serine Alanine

Protein – Leucine – Serine – Alanine –

The correspondence between amino acids and mRNA codons is known as the **genetic code.** There are a total of 64 codons: 3 of these are "stop" signals that terminate translation, and the remaining 61 represent, with some redundancy, the 20 standard amino acids found in proteins. Table 3-3 shows which codons specify which amino acids. In theory, knowing a gene's nucleotide sequence should be equivalent to knowing the amino acid sequence of the protein encoded by the gene. However, as we shall see, genetic information is often "processed" at several points before the protein reaches its mature form. Keep in mind that the rRNA and tRNA required for protein synthesis are also encoded by genes. The "products" of these genes are the result of transcription without translation.

TABLE 3-3 The standard genetic code[a]

First position (5′ end)	Second position U	Second position C	Second position A	Second position G	Third position (3′ end)
U	UUU Phe	UCU Ser	UAU Tyr	UGU Cys	U
	UUC Phe	UCC Ser	UAC Tyr	UGC Cys	C
	UUA Leu	UCA Ser	UAA Stop	UGA Stop	A
	UUG Leu	UCG Ser	UAG Stop	UGG Trp	G
C	CUU Leu	CCU Pro	CAU His	CGU Arg	U
	CUC Leu	CCC Pro	CAC His	CGC Arg	C
	CUA Leu	CCA Pro	CAA Gln	CGA Arg	A
	CUG Leu	CCG Pro	CAG Gln	CGG Arg	G
A	AUU Ile	ACU Thr	AAU Asn	AGU Ser	U
	AUC Ile	ACC Thr	AAC Asn	AGC Ser	C
	AUA Ile	ACA Thr	AAA Lys	AGA Arg	A
	AUG Met	ACG Thr	AAG Lys	AGG Arg	G
G	GUU Val	GCU Ala	GAU Asp	GGU Gly	U
	GUC Val	GCC Ala	GAC Asp	GGC Gly	C
	GUA Val	GCA Ala	GAA Glu	GGA Gly	A
	GUG Val	GCG Ala	GAG Glu	GGG Gly	G

[a]The 20 amino acids are abbreviated: Ala, alanine; Arg, arginine; Asn, asparagine; Asp, aspartate; Cys, cysteine; Gly, glycine; Gln, glutamine; Glu, glutamate; His, histidine; Ile, isoleucine; Leu, leucine; Lys, lysine; Met, methionine; Phe, phenylalanine; Pro, proline; Ser, serine; Thr, Threonine;Trp, tryptophan; Tyr, tyrosine; and Val, valine.

REVIEW LEARNING OBJECTIVE 5

- How does DNA encode genetic information and how is this information expressed?
- What is the relationship between the nucleotide sequence in a gene and the amino acid sequence of a protein?

3 SEQUENCING

Because an organism's genetic material influences the organism's entire repertoire of activities, it is vitally important to unravel the sequence of nucleotides in that organism's DNA, even by examining one gene at a time.

Go to **Exercise 3/DNA Sequencing** for an overview of the most commonly used technique for sequencing DNA. The material covered in Exercise 3 addresses Learning Objective 6.

Tens of thousands of genes have been catalogued through genome-sequencing projects (discussed below in Section 3-5). Other genes have been identified through studies of the genes' protein products. If a defective protein is associated with a particular disease, identifying the gene may be a necessary part of understanding the development and possibly the treatment for the disease. In the case of the cystic fibrosis gene, however, the gene product had never been identified. One of the diagnostic signs of CF is high chloride concentrations in sweat (according to medieval folklore, a baby who tasted salty when kissed was predicted to soon die). But neither this characteristic nor other symptoms, such as the thick mucus in the airways, could unequivocally implicate a defect in any known protein. In order to identify the molecular

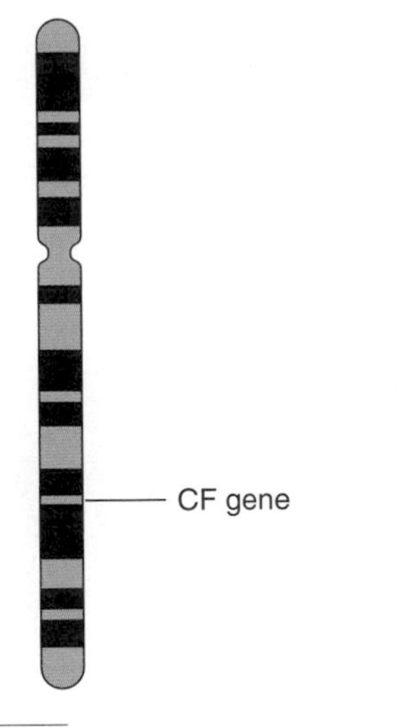

FIGURE 3-12

Cytogenetic map of human chromosome 7. The approximate location of the CF gene on the long arm of chromosome 7 was inferred from the locations of two DNA markers that are inherited along with a defective CF gene.

basis of the disease, researchers had to identify the gene which, when **mutated** (altered), produces the characteristic CF symptoms.

Teams of researchers analyzed the DNA of affected individuals (who had two copies of the defective CF gene) and of family members who were asymptomatic carriers (who had one normal and one defective copy of the gene). Individuals with one or two copies of the defective CF gene shared two other genetic features that can be detected in a laboratory test. These two **DNA markers** were known to be located on the long arm of chromosome 7 (there are 22 pairs of human chromosomes, plus the X and Y chromosomes, for a total of 46). The cystic fibrosis gene was therefore also likely to be located on the long arm of chromosome 7 (Fig. 3-12), since genes that are inherited together are usually physically close.

Using some of the laboratory techniques discussed in Section 3-4, a segment of normal human chromosome 7 was isolated and used for the final search for the CF gene. The researchers gradually closed in on a DNA segment that appeared to be present in a number of mammalian species, which provided an important clue that the segment contained an essential gene (about 98% of mammalian DNA does not encode any proteins). The nucleotide sequence of the cystic fibrosis gene was finally deduced in 1989.

Dideoxy DNA sequencing

The most widely used technique for determining the sequence of nucleotides in a segment of DNA was developed by Frederick Sanger in 1975. The Sanger sequencing technique is also known as **dideoxy sequencing** because it uses dideoxy nucleotides, that is, nucleotides lacking both 2′ and 3′ hydroxyl groups:

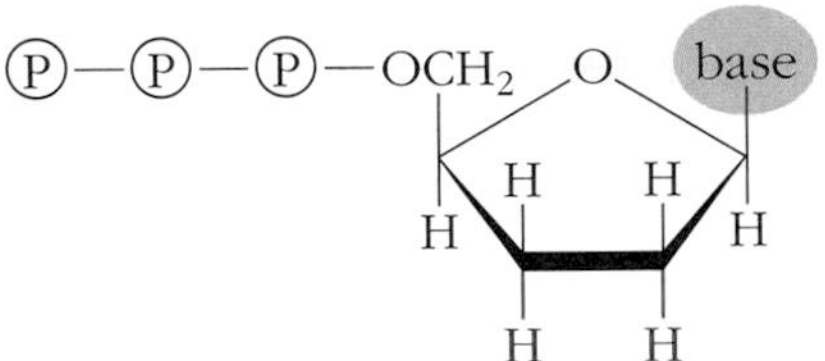

2′,3′-Dideoxynucleoside triphosphate (ddNTP)

A dideoxynucleoside triphosphate can be abbreviated **ddNTP,** where N represents any of the four bases. We will describe the protocol used in automated dideoxy sequencing (Fig. 3-13; the procedure for manual sequencing is conceptually similar). First, the DNA is denatured to separate its two strands. The DNA is then incubated with a mixture of all four deoxynucleoside triphosphates (dATP, dCTP, dGTP, and dTTP) and a bacterial **DNA polymerase,** an enzyme that catalyzes the polymerization of the nucleotides in the order determined by their base-pairing with a single-stranded DNA template. Because the DNA polymerase cannot begin a new nucleotide strand but can only extend a preexisting chain, a short single-stranded **primer** that base pairs with the template strand is added to the mixture. Keep in mind that the reaction mixture actually contains millions of molecules of the template, the primer, and the polymerase.

The reaction mixture also includes small amounts of four dideoxynucleotides (ddATP, ddCTP, ddGTP, and ddTTP), each of which is tagged with a different fluorescent dye. *As the DNA polymerase proceeds to synthesize a new DNA chain, it occasionally adds one of the ddNTPs in place of the corresponding dNTP.* This halts further extension of the DNA chain because the

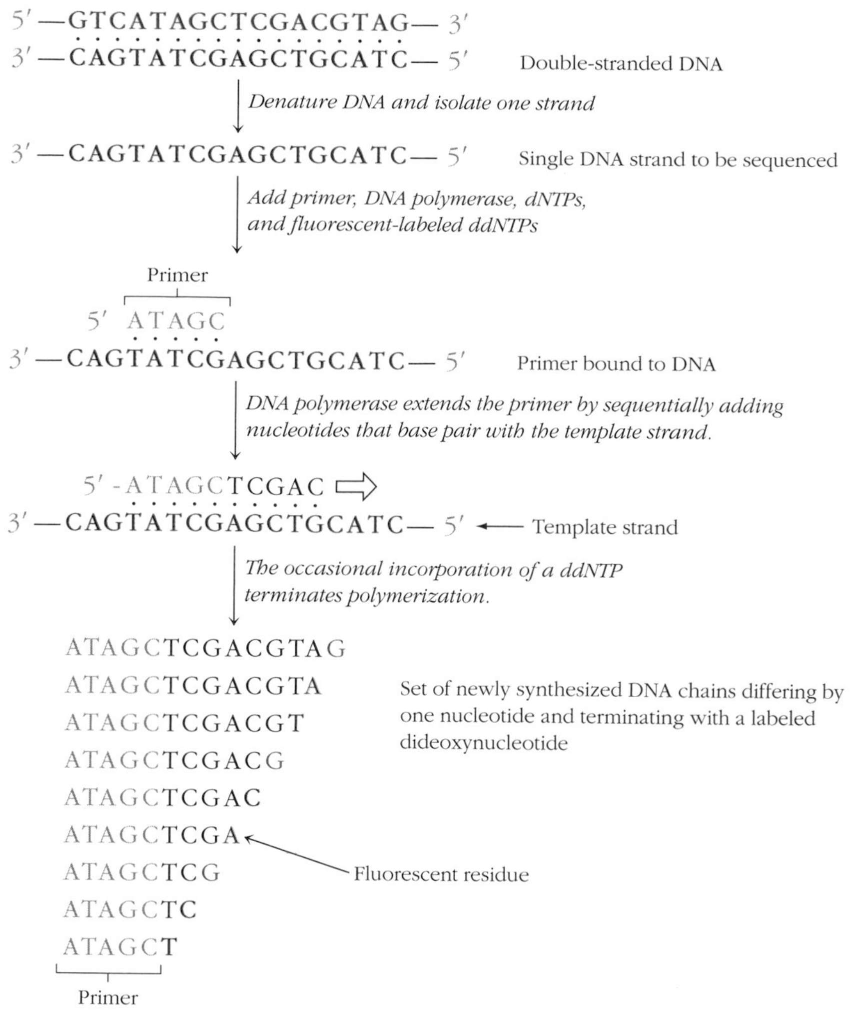

5′—GTCATAGCTCGACGTAG— 3′
3′—CAGTATCGAGCTGCATC— 5′ Double-stranded DNA

Denature DNA and isolate one strand

3′—CAGTATCGAGCTGCATC— 5′ Single DNA strand to be sequenced

Add primer, DNA polymerase, dNTPs, and fluorescent-labeled ddNTPs

Primer
5′ ATAGC
3′—CAGTATCGAGCTGCATC— 5′ Primer bound to DNA

DNA polymerase extends the primer by sequentially adding nucleotides that base pair with the template strand.

5′-ATAGCTCGAC ⇨
3′—CAGTATCGAGCTGCATC— 5′ ← Template strand

The occasional incorporation of a ddNTP terminates polymerization.

ATAGCTCGACGTAG
ATAGCTCGACGTA
ATAGCTCGACGT
ATAGCTCGACG
ATAGCTCGAC
ATAGCTCGA
ATAGCTCG
ATAGCTC
ATAGCT
Primer

Set of newly synthesized DNA chains differing by one nucleotide and terminating with a labeled dideoxynucleotide

Fluorescent residue

FIGURE 3-13

Dideoxy DNA sequencing procedure.

newly incorporated ddNTP, which lacks a 3′ OH group, cannot form the 3′ portion of a 3′–5′ phosphodiester bond to the next nucleotide (see page 56).

The concentrations of the ddNTPs in the reaction mixture are lower than the concentrations of their corresponding **dNTPs,** so the DNA polymerase can assemble new chains of varying length before the random incorporation of a ddNTP halts further polymerization. The result is a population of truncated DNA strands, each capped by a fluorescent ddNTP residue. *The fragments differ in length by a single nucleotide since they all started from identical primers.*

The reaction products are subjected to **electrophoresis,** a procedure in which the molecules move through a gel-like matrix under the influence of an electric field. Because all the DNA segments have a uniform charge density, they are separated on the basis of their size (the smallest molecules move fastest). The separated molecules are excited by a laser so that the fluorescent dye attached to each dideoxynucleotide residue emits its characteristic color (Fig. 3-14). The order of appearance of the colors corresponds to the order of bases in the newly synthesized DNA. Keep in mind that this sequence is complementary to the DNA strand that was used as a template in the sequencing reaction. With automation, a single reaction can yield a sequence

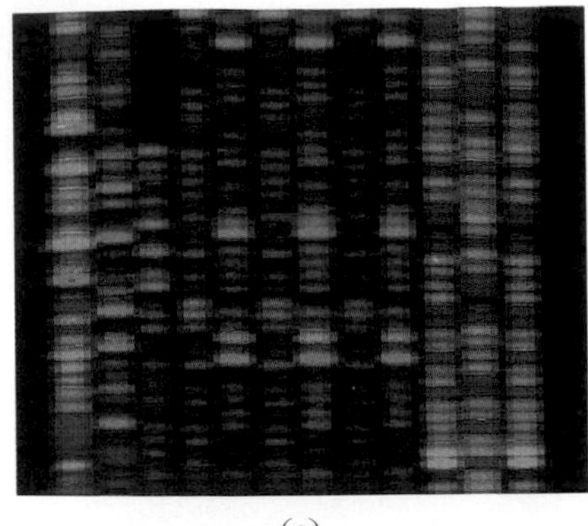

(a)

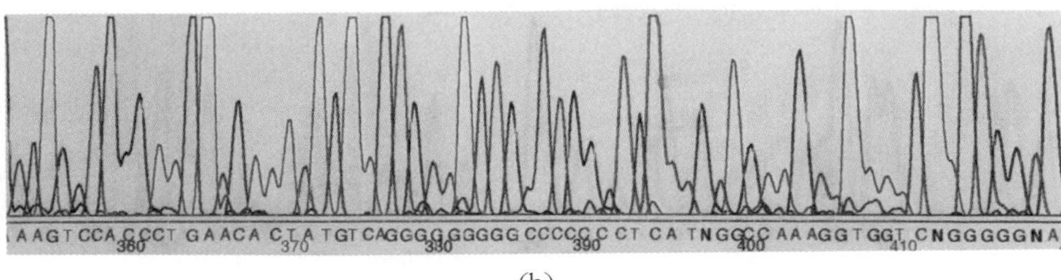

(b)

FIGURE 3-14

Results of dideoxy sequencing.
(a) An electrophoretogram of fluorescent DNA fragments. The DNA fragments generated by dideoxy sequencing are subjected to electrophoresis. Each lane of the electrophoretogram contains a series of DNA fragments, with the largest (slowest migrating) at the top and the smallest (fastest migrating) at the bottom. The color of each fragment identifies the dideoxynucleotide at its end (A green, G yellow, C blue, and T red). (b) A scan of a sequencing electrophoretogram. Each of the four colored curves represents a dideoxynucleotide. The peaks indicate the position of that nucleotide in the sequence. [*Photo (a) Yoav Levy/Phototake. Photo (b) Klaus Guldbrandsen/Science Photo Library/Photo Researchers.*]

of 400–1000 nucleotides.  Exercise 3 illustrates the major steps of DNA sequencing, from DNA purification to computer analysis of sequence data.

REVIEW LEARNING OBJECTIVE 6

- Summarize the steps of dideoxy DNA sequencing.

Many genes are sequenced in pieces

Many genes are too long to be sequenced in a single pass. Therefore, a number of overlapping DNA segments are prepared and individually sequenced. The sequence of the entire gene can then be reconstructed by aligning the overlapping sequences:

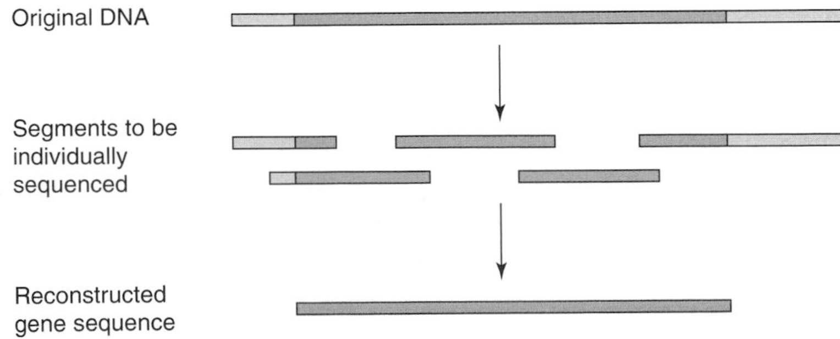

This sort of approach was required for sequencing the CF gene, which spans about 250,000 bp of DNA.

As is the case for most mammalian genes, only certain portions of the CF gene directly correspond to a protein product, because segments of the mRNA molecule transcribed from the gene are excised (an event called **splicing**) before the message is translated into protein (splicing is discussed further in Section 19-3). In addition, sequences at each end of the mRNA are not translated. After splicing, the mRNA is only 6129 nucleotides long. Of this molecule, 4440 nucleotides (or 4440 ÷ 3 = 1480 codons) specify the 1480 amino acid residues of the protein product.

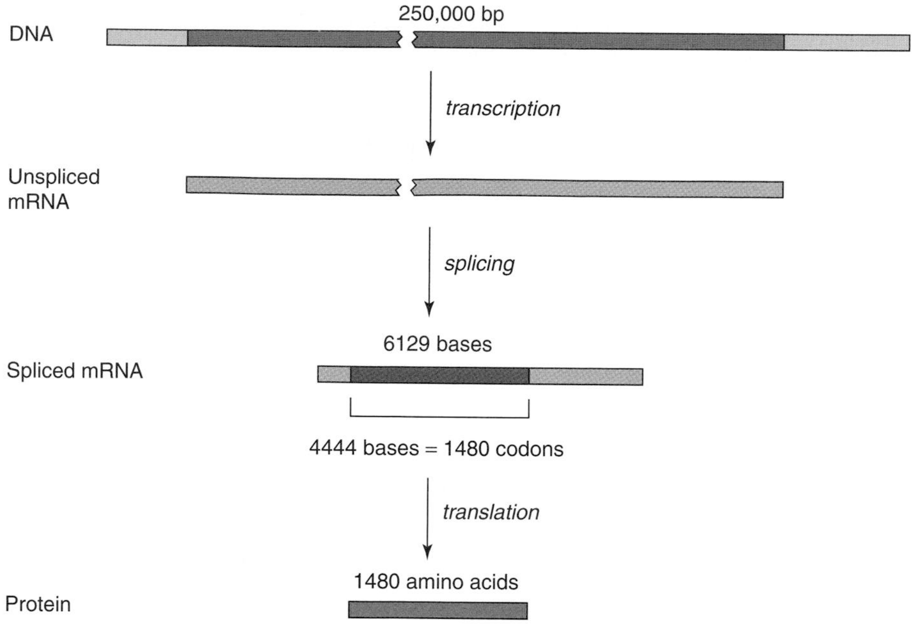

Matching every three bases in the derived mRNA sequence with the appropriate amino acid (see Table 3-3) yielded the amino acid sequence of the protein.

Because this protein sequence was derived from normal human DNA, other experiments were required to prove that it corresponded to the protein that was missing or abnormal in individuals with CF. A second round of sequencing experiments showed that in about 70% of CF patients, the gene is missing three nucleotides. This results in the deletion of a single phenylalanine (Phe) residue at position 508 (the 508th amino acid residue in the encoded protein):

Normal gene		504	505	506	507	508	509	510	511	512	
mRNA	· · ·	GAA	AAT	ATC	ATC	TTT	GGT	GTT	TCC	TAT	· · ·
Protein	· · ·	Glu	Asn	Ile	Ile	Phe	Gly	Val	Ser	Tyr	· · ·

Mutated gene		504	505	506	507	508	509	510	511	512	
mRNA	· · ·	GAA	AAT	ATC	AT–	––T	GGT	GTT	TCC	TAT	· · ·
Protein	· · ·	Glu	Asn	Ile	Ile		Gly	Val	Ser	Tyr	· · ·

Note that although the deletion affects both codons 507 and 508, the redundancy of the genetic code means that the isoleucine (Ile) at position 507 is not affected (because codons ATC and ATT both specify Ile). The protein lacking Phe 508 is abnormally processed by the cell so that very little is available.

More than 200 other mutations scattered throughout the CF gene have been identified, accounting for most of the remaining 30% of CF cases. Some of these mutations cause milder forms of the disease that are not detected until adulthood. Genes linked to several hundred other human diseases have been identified and sequenced using some of the same approaches used to track down the CF gene.

Once a gene's sequence has been determined, this information (that is, the sequence of the coding strand) is customarily deposited in a public database such as GenBank. The information in a sequence database can be accessed electronically so that it is possible to compare a given sequence to sequences in other genes. Such comparisons are vital for assigning functions to newly discovered genes, since *genes with similar functions in different species tend to have similar sequences. In addition, sequence similarities often indicate a common origin and shared evolutionary history for the species harboring these sequences.* The division of prokaryotes into two groups (Archaea and Bacteria; see Section 1-4) rests on the results of sequencing studies.

4 MANIPULATING DNA IN THE LABORATORY

Molecular biologists have devised clever procedures for isolating DNA in sufficient quantities to sequence or manipulate in some other way. Many of the techniques take advantage of naturally occurring enzymes and the ability of many cells to take up foreign DNA molecules.

Restriction enzymes: scissors that cut DNA at specific sequences

Bacteria contain DNA-cleaving enzymes known as **restriction endonucleases** (or **restriction enzymes**) that catalyze the breakage of phosphodiester bonds at or near specific nucleotide sequences. These enzymes can thereby destroy foreign DNA that enters the cell (for example, phage DNA). In this way, the bacterial cell "restricts" the growth of the phage. The bacterial cell protects its own DNA from endonucleolytic digestion by methylating it (adding a $—CH_3$ group) at the same sites recognized by its restriction endonucleases. In the laboratory, the most useful restriction enzymes are those that cleave at the recognition site. Hundreds of these enzymes have been examined; some are listed in Table 3-4 along with their recognition sequences and cleavage sites.

Restriction enzymes typically recognize a 4- to 8-base sequence that is identical, when read in the same $5' \rightarrow 3'$ direction, on both strands. DNA with this form of symmetry is said to be **palindromic** (words such as "madam" and "noon" are palindromes). One restriction enzyme isolated from *E. coli* is known as *Eco*RI (the first three letters are derived from the genus and species names). Its recognition sequence is

5′—G↓AATTC—3′
3′—CTTAA↑G—5′

The arrows indicate the phosphodiester bonds that are cleaved. Note that the sequence reads the same on both strands.

Because the *Eco*RI cleavage sites are symmetrical but staggered, the enzyme generates DNA fragments with single-stranded extensions known as **sticky ends:**

—G AATTC—
—CTTAA G—

TABLE 3-4 Recognition and cleavage sites of some restriction endonucleases

Enzyme	Recognition/cleavage site[a]
*Alu*I	AG I CT
*Msp*I	C I CGG
*Asu*I[b]	G I GNCC
*Eco*RI	G I AATTC
*Eco*RV	GAT I ATC
*Pst*I	CTGCA I G
*Sau*I	CC I TNAGG
*Not*I	GC I GGCCGC

[a]The sequence of one of the two DNA strands is shown. The vertical bar indicates the cleavage site.

[b]N represents any nucleotide.

[An exhaustive source of information on restriction enzymes is available through the Restriction Enzyme Database: rebase.neb.com/rebase/rebase.html.]

In contrast, the *E. coli* restriction enzyme known as *Eco*RV cleaves both strands of DNA at the center of its 6 bp recognition sequence so that the resulting DNA fragments have **blunt ends:**

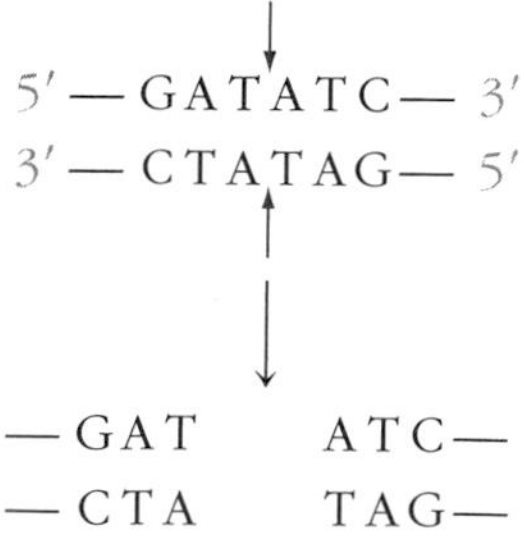

Restriction enzymes have many uses in the laboratory. For example, they are indispensible for reproducibly breaking large pieces of DNA into smaller pieces of manageable size. **Restriction digests** of well-characterized DNA molecules, such as the 48,502 bp *E. coli* bacteriophage λ, yield **restriction fragments** of predictable size (Fig. 3-15). DNA can also be broken by shaking or stirring, since it is a such a long molecule, but it tends to break at random points, not at specific sites. The reproducibility of restriction enzymes also makes them useful tools in the forensics laboratory (Box 3-A).

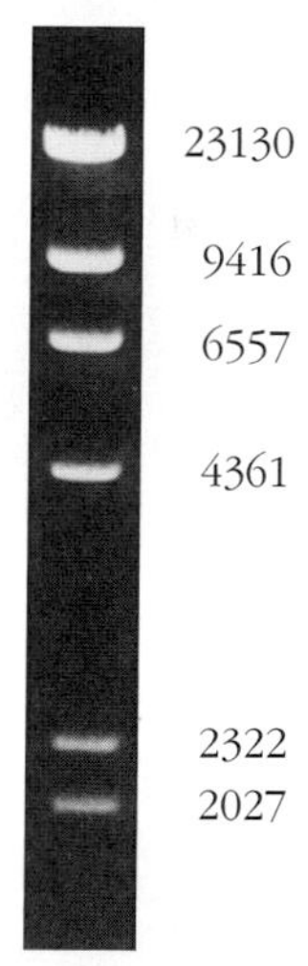

FIGURE 3-15

Digestion of bacteriophage λ DNA by the restriction enzyme *Hind*III.
The restriction enzyme cleaves the DNA to produce eight fragments of defined size, six of which are large enough to be separated by electrophoresis in an agarose gel. The numbers indicate the number of base pairs in each fragment. [*Copyright 2002/2003 New England Biolabs catalog. Reprinted with permission.*]

Recombinant DNA

When different samples of DNA are digested with the same sticky end–generating restriction endonuclease, all the fragments have identical sticky ends. If the fragments are mixed together, the sticky ends may find their complements and re-form base pairs. The discontinuities in the sugar–phosphate backbone can then be mended by a **DNA ligase** (an enzyme that forms new phosphodiester bonds between adjacent nucleotides). *These cutting-and-pasting reactions are essential tools for constructing* ***recombinant DNA*** *molecules.* For example, a segment of mammalian DNA can be excised from a chro-

A CLOSER LOOK

BOX 3-A DNA in the courtroom

Most of the human genome is identical in all individuals, but there are sites where variations or polymorphisms (meaning "many forms") occur. Many genetic polymorphisms result from minor sequence differences and have no functional consequences because they occur in regions of the DNA that do not encode genes. However, the substitution of one base pair for another may create or eliminate a recognition sequence for a particular restriction endonuclease. The resulting **restriction fragment length polymorphism (RFLP)** can be used to characterize the DNA from different individuals. The procedure for detecting RFLPs is shown below.

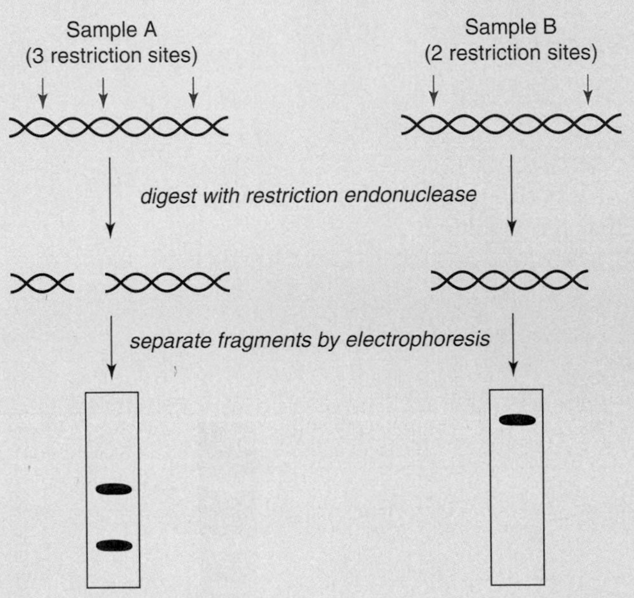

Since human DNA differs in sequence every thousand base pairs on average, RFLP analysis is a powerful tool for identifying criminals: It can link a suspect to a small sample of DNA-containing biological material from the scene of a crime. In this classic RFLP analysis, Suspect 1 matches the biological specimen.

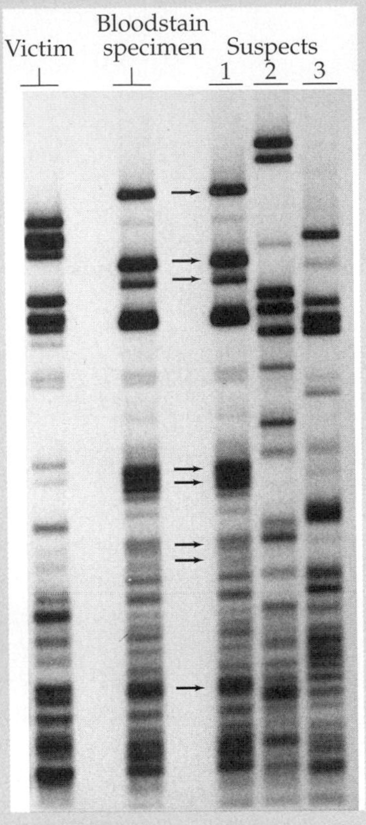

Because the probability of two individuals yielding the same pattern of restriction fragment length polymorphisms is about one in 10^6, the courts have used RFLP analysis both to convict and to exonerate, with a high degree of confidence. Current DNA-profiling techniques that examine polymorphisms at several sites are even more accurate.

[Courtesy Cellmark Diagnostics, U.K.]

mosome and inserted into another piece of DNA, such as a circular **plasmid,** that has been cut by the same restriction enzyme and that therefore has the same sticky ends. DNA ligase seals the breaks in the nucleotide strands, leaving an unbroken double-stranded recombinant DNA molecule consisting of the plasmid with a foreign DNA insert (Fig. 3-16).

Plasmids are small, circular DNA molecules present in many bacterial cells. A single cell may contain multiple copies of a plasmid, which replicates independently of the bacterial chromosome and usually does not contain genes essential for the host's normal activities. However, plasmids often do carry genes for specialized functions, such as resistance to certain antibiotics (these genes usually encode proteins that inactivate the antibiotics). An antibiotic resistance gene allows the **selection** of cells that harbor the plasmid:

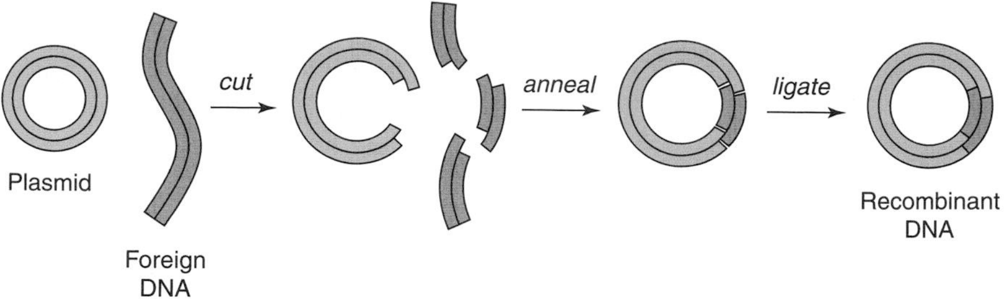

FIGURE 3-16

Production of a recombinant DNA molecule.
A small circular plasmid and a sample of DNA are cut with the same restriction enzyme, generating complementary sticky ends, so that the fragment of foreign DNA can be ligated into the plasmid.

Only cells that contain the plasmid can survive in the presence of the antibiotic. Growing large quantities of plasmid-laden cells is one way to produce large amounts of the foreign DNA insert. (It can be removed later by treating the plasmid with the same restriction enzyme used to insert the foreign DNA.) A piece of DNA that is amplified in this way is said to be **cloned** (which means "copied"; Box 3-B). The plasmid that contains the foreign DNA is called a **cloning vector.**

A CLOSER LOOK

BOX 3-B What is a clone?

Strictly speaking, a **clone** is an identical copy of an original, an organism whose genetic makeup is identical to that of another. For example, a colony of bacterial cells growing on a culture plate consists of clones of the single bacterial cell that was used to inoculate the plate. A clone can be produced only through the reproduction of a parental organism.

Clones are not unique to the laboratory. They abound in nature, particularly in species that rarely undergo sexual reproduction (in which genes from two parents are more or less randomly allocated to the offspring). Plants propagated by cuttings are also clones, since their genes are identical to those of the original stock. Human clones appear with a frequency of about 1%—the birth rate for identical twins.

To scientists, the act of cloning also refers to the production of multiple identical copies of a gene through recombinant DNA technology. Hence the "cloned" gene is the one that is amplified or grown as part of a plasmid or other construct in a colony of cells in a culture plate or flask.

For several decades, developmental biologists have been cloning organisms with easily manipulated eggs, such as amphibians. More recently, it has become possible to clone mammals by fusing a nonreproductive cell (either from an adult animal or from cells grown in a flask) with an egg whose chromosomes have been removed. Sheep, cows, goats, mice, pigs, and cats have been cloned in this way, although the technique has an extremely low success rate and has few commercial applications.

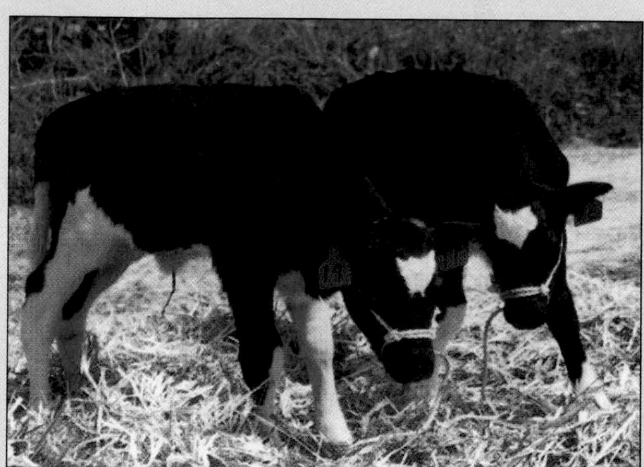

Cloned calves.

Nevertheless, these studies have proved that cloning a human is at least technically feasible, thereby igniting widespread debate about the ethics of ever carrying out such a procedure.

[© AP/Wide World Photos.]

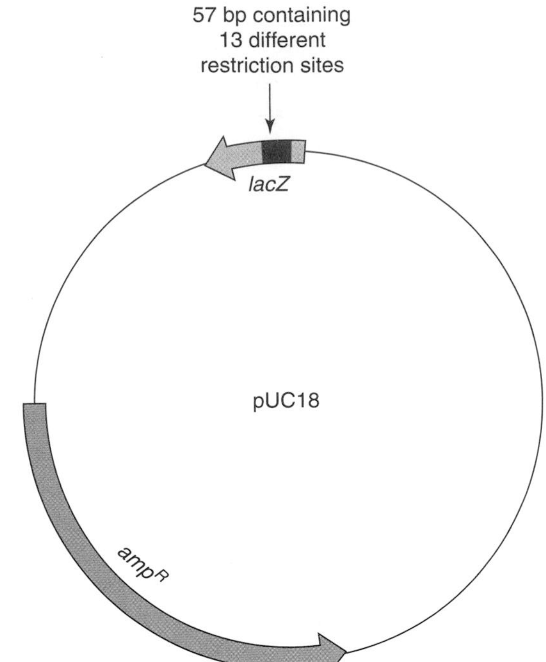

FIGURE 3-17

Map of the cloning vector pUC18. This circular DNA molecule has a gene for resistance to ampicillin so that bacterial cells containing the plasmid can be selected by their ability to grow in the presence of the antibiotic. The plasmid also has a site comprising recognition sequences for 13 different restriction enzymes. Insertion of a foreign DNA at this site interrupts the *lacZ* gene, which encodes the enzyme β-galactosidase.

An example of a cloning vector is the *E. coli* plasmid pUC18 (Fig. 3-17). This double-stranded 2686-bp molecule contains a gene (called *amp*^{R}) for resistance to ampicillin and a gene (called *lacZ*) encoding the enzyme β-galactosidase, which catalyzes the hydrolysis of certain galactose derivatives. The *lacZ* gene has been engineered to contain several restriction sites, any one of which can be used as an insertion point for a piece of foreign DNA with compatible sticky ends. Interrupting the *lacZ* gene with foreign DNA prevents the synthesis of the β-galactosidase protein.

Colonies of bacterial cells harboring the intact plasmid can be detected when their β-galactosidase cleaves a galactose derivative that generates a blue dye:

Galactose — 5-Bromo-4-chloro-3-indole

Colonies of bacterial cells in which a foreign DNA insert has interrupted the *lacZ* gene are unable to cleave the galactose derivative and therefore do not turn blue (Fig. 3-18). This method for identifying (screening) bacterial cells containing plasmids with the insert is known as **blue-white screening.** A single white colony can then be removed from the culture plate and grown. Bacterial cells that lack the plasmid entirely would also be white. However, these cells are eliminated by including ampicillin in the culture medium, which kills cells that don't contain the *amp*^{R} gene.

pUC18 is just one of a huge number of cloning vectors that have been developed to accommodate different sizes of DNA inserts (Table 3-5). Vectors

TABLE 3-5 Types of cloning vectors

Vector	Size of DNA insert (kb)
Plasmid	<20
Cosmid	40–45
Bacterial artificial chromosome	40–400
Yeast artificial chromosome	400–1000

also differ by the cell type in which the vector can grow (for example, bacterial, fungal, insect, or mammalian cells) and in the strategy used to select cells containing the foreign DNA.

If a gene that has been isolated and cloned in a host cell is also expressed (transcribed and translated into protein), it may affect the metabolism of that cell. The functions of some gene products have been assessed in this way. Sometimes a specific combination of vector and host cell are chosen so that large quantities of the gene product can be isolated from the cultured cells or from the medium in which they grow. This is a far more economical way to purify certain proteins, such as the human hormone insulin, which are difficult to obtain directly from human tissues. Introducing a foreign gene into a single host cell via a cloning vector alters the genetic makeup of that cell and all its descendants. Producing a **transgenic organism**—that is, a multicellular organism containing a foreign gene—is more difficult (Box 3-C).

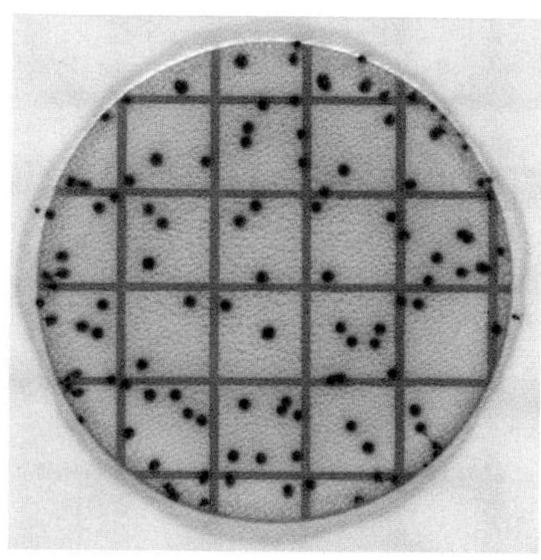

FIGURE 3-18

Culture dish used in blue-white screening. Blue colonies arise from cells whose plasmids have an intact β-galactosidase gene. White colonies arise from cells whose plasmids contain an insert that interrupts the β-galactosidase gene. [© *2001 Becton, Dickinson and Company (Clontech).*]

REVIEW LEARNING OBJECTIVES 7 AND 8

- Why are restriction endonucleases useful for constructing recombinant DNA?
- How can cells containing recombinant DNA molecules be identified by selection or screening?

The polymerase chain reaction

At one time, scientists requiring large amounts of a particular DNA sequence had to laboriously isolate the target DNA and clone it using an appropriate vector and host cell. That changed in 1985, with the invention of the **polymerase chain reaction (PCR)** by Kary Mullis. Although PCR cannot entirely replace cloning as a research tool, it provides a relatively easy and rapid way to amplify a segment of DNA. *One of the advantages of PCR over traditional cloning techniques is that the starting material need not be pure* (this makes the technique ideal for analyzing complex mixtures such as tissues or biological fluids).

The steps of PCR are outlined in Fig. 3-19. The reaction mixture contains the DNA sample, a DNA polymerase, all four deoxynucleotides, and two oligonucleotide primers that are complementary to the 3′ ends of the two strands of the target sequence.

First, the sample is heated to 90–95°C to separate the DNA strands. Next, the temperature is lowered to about 55°C, cool enough for the primers to hybridize with the DNA strands. The temperature is then increased to about 75°C, and the DNA polymerase synthesizes new DNA strands by extending the primers. The three steps—strand separation, primer binding, and primer

A CLOSER LOOK

BOX 3-C Transgenic organisms and gene therapy

To produce a transgenic mammal, cloned DNA is injected into fertilized eggs, which are implanted in a foster mother. Some of the resulting embryos' cells (possibly including their reproductive cells) will contain the foreign gene. When the animals mature, they must be bred in order to yield offspring whose cells all contain the foreign gene. The transgenic mouse on the left is larger than normal because its cells contain copies of the rat growth hormone gene.

It is also possible to produce animals that lack a particular gene. These "gene knockouts" can serve as animal models for human diseases. For example, mice lacking the cystic fibrosis gene can be used to study the development of the disease and to test potential treatments.

Transgenic plants are produced by introducing recombinant DNA into a few cells, which can often develop into an entire plant whose cells all contain the foreign DNA. Desirable traits such as resistance to insect pests have been introduced into several important crop species. For example, about 20% of the U.S. corn (maize) crop is genetically modified to produce a protein that is toxic to plant-eating insects. So-called Bt corn plants have been engineered to express an insecticidal toxin from the bacterium *Bacillus thuringiensis*.

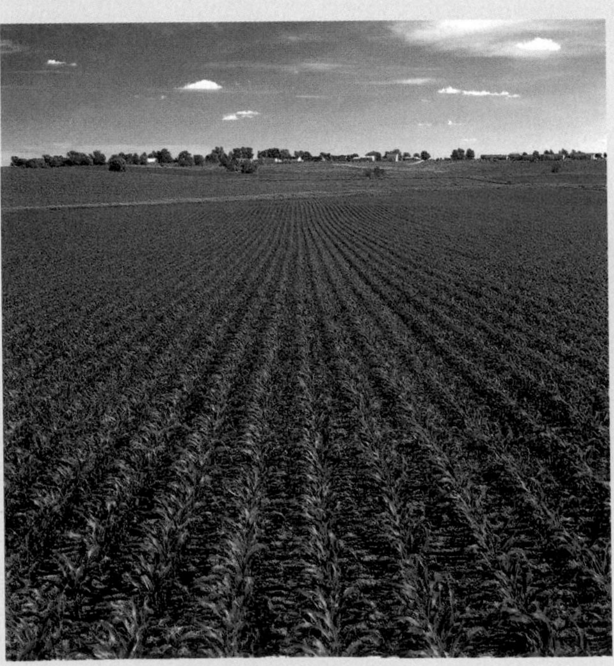

However, the huge acreage devoted to the transgenic corn (also known as a genetically modified or "GM" food) raises some concerns: It increases the selective pressure on insects to evolve resistance to the toxin, and it increases the likelihood that the toxin gene will be transferred to other plant species, with disastrous effects on the insects that feed on them.

Less controversial are transgenic plants that have been engineered for better nutrition. For example, researchers are developing a strain of rice with foreign genes that encode enzymes necessary to synthesize β-carotene (an orange pigment that is the precursor of vitamin A) and a gene for the iron-storage protein ferritin. The white grains on the right represent the wild-type. The grains on the left have been engineered to store up to three times more iron. They also synthesize β-carotene (the precursor of vitamin A), which gives them their yellow color. The genetically modified rice is intended to help alleviate vitamin A deficiencies (which afflict some 400 million people) and iron deficiencies (an estimated 30% of the world's population suffers from iron deficiency).

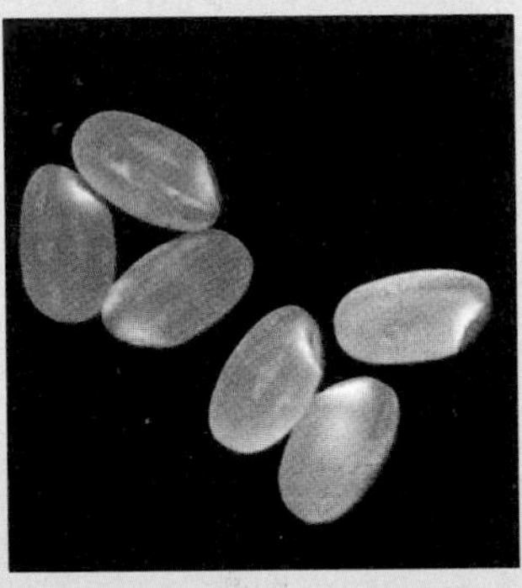

The aim of ***gene therapy*** *is to introduce a functional gene into an individual in order to compensate for an existing malfunctioning gene*. For example, to correct the genetic defect in cystic fibrosis, a normal cystic fibrosis gene would have to be introduced into the appropriate cells (mainly in the lungs) and the gene would have to be expressed at a level sufficient to yield enough of the protein product to eliminate the life-threatening symptoms of the disease. Sadly, gene therapy has had few successes due to numerous practical obstacles. For one thing, the human immune system tends to destroy the viruses that are often used as gene vectors, and even when the gene is delivered and taken up by the target cells, it may be expressed only transiently. Gene therapy *in utero* (in which a gene is introduced to a fetus) may ultimately be more successful since fetal cells are immunologically naive and tend to take up foreign DNA more efficiently.

[Mouse photo courtesy Ralph Brinster, School of Veterinary Medicine, University of Pennsylvania; Bt corn photo from Arthur G. Smith III/Grant Heilman Photography; rice photo courtesy Ingo Potrykus, Swiss Federal Institute of Technology.]

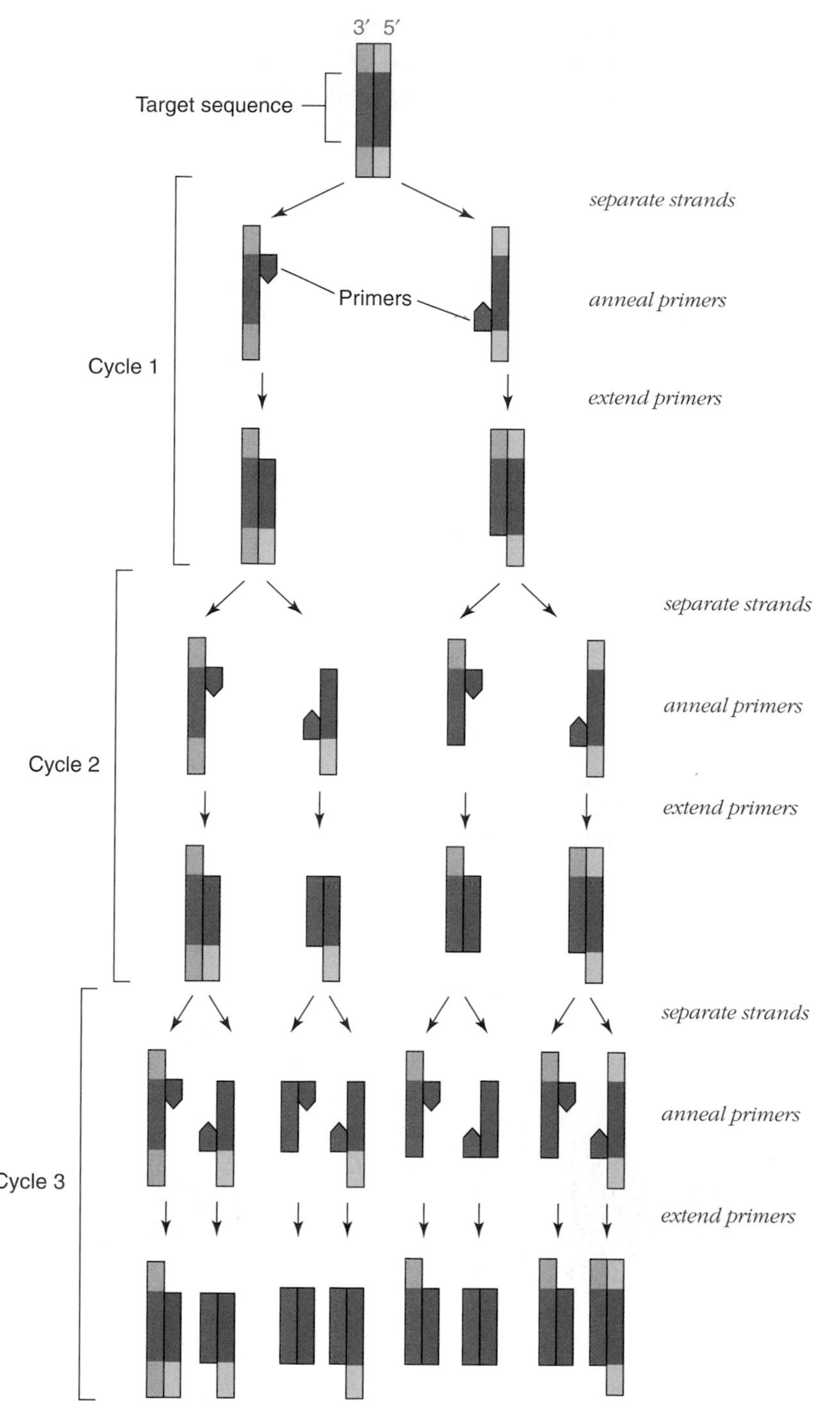

FIGURE 3-19

The polymerase chain reaction. Each cycle consists of separation of DNA strands, binding of primers to the 3′ ends of the target sequence, and extension of the primers by DNA polymerase. The target DNA doubles in concentration with each cycle.

extension—are repeated as many as 40 times. Because the primers represent the two ends of the target DNA, this sequence is preferentially amplified so that *it doubles in concentration with each reaction cycle.* For example, 20 cycles of PCR can theoretically yield $2^{20} = 1{,}048{,}576$ copies of the target sequence in a matter of hours. The DNA can then be cloned, sequenced, or used for another purpose, such as RFLP analysis (see Box 3-A).

One of the keys to the success of PCR is the use of bacterial DNA polymerases that can withstand the high temperatures required for strand separation (these temperatures inactivate most enzymes). Commercial PCR kits usually contain DNA polymerase from *Thermus aquaticus* (which lives in hot springs) or *Pyrococcus furiosus* (which inhabits geothermally heated marine sediments) since their enzymes perform optimally at high temperatures.

If the primer used to amplify the target DNA contains an error, that is, one or more bases that cannot base pair with the target sequence, then all the resulting PCR products will contain that same error. The deliberate use of a mismatched primer is the essence of **site-directed mutagenesis** (also called ***in vitro*** **mutagenesis**), which is a technique for generating DNA molecules in which specific nucleotides have been intentionally substituted by others (Fig. 3-20). If the nucleotides are part of the protein-coding portion of the gene, the sequence of the resulting protein may be altered. The mutagenized protein can then be purified and studied.

In PCR, no new DNA is synthesized unless the primers first find their complementary sequences in the original DNA sample, so *PCR can be used to verify the presence of that sequence.* In fact, PCR is the most efficient way to detect certain pathogenic organisms such as *Borrelia burgdorferi* and *Hel-*

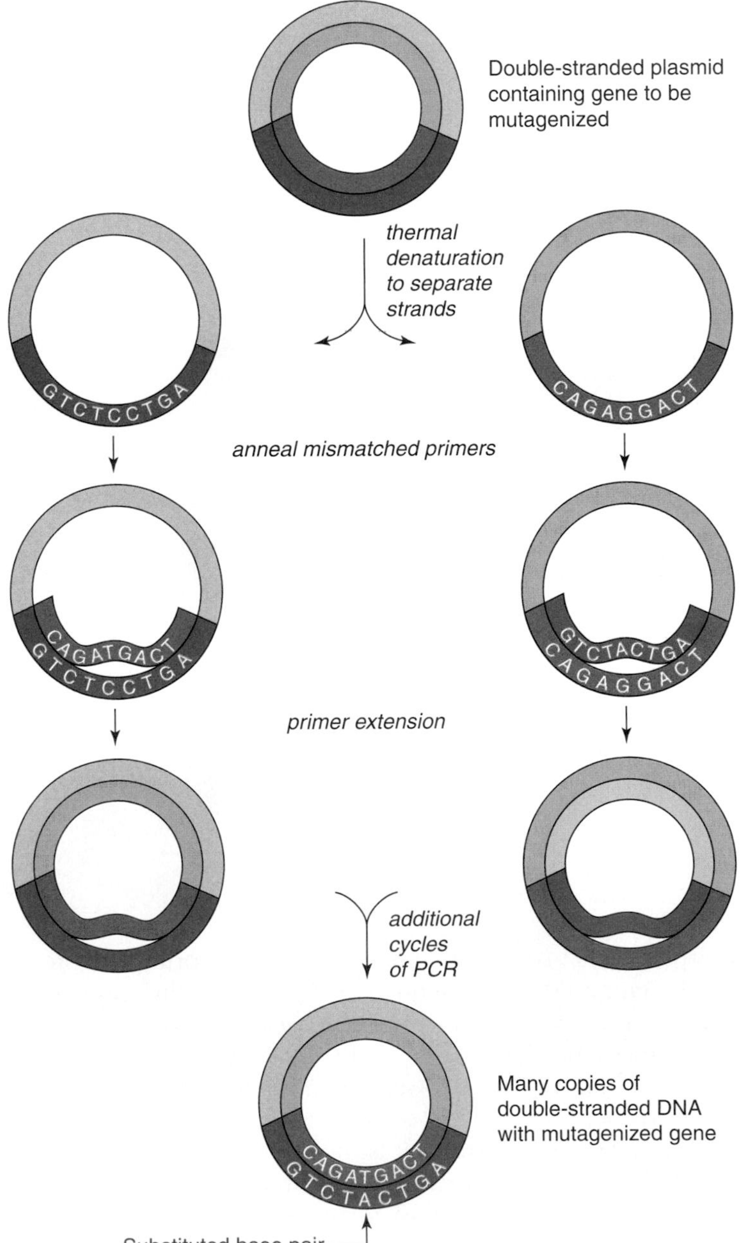

FIGURE 3-20
Site-directed mutagenesis.

FIGURE 3-21

Pathogens that are identified through PCR.
(a) *Borrelia burgdorferi,* which causes Lyme disease. (b) *Helicobacter pylori,* which causes gastric ulcers. [*Science Photo Library/Photo Researchers.*]

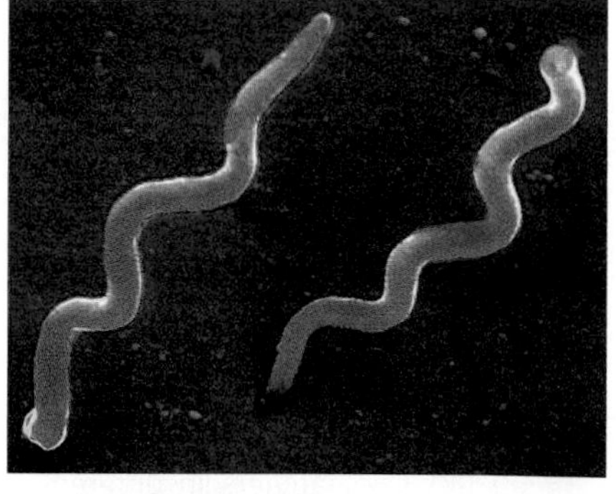

(a)

icobacter pylori (Fig. 3-21) since the diseases resulting from infection by these organisms (Lyme disease and stomach ulcers, respectively) do not always exhibit diagnostically useful symptoms. For obvious reasons, PCR is also preferred over culturing as a way to detect the presence of highly contagious bacteria and viruses, including the human immunodeficiency virus (HIV).

REVIEW LEARNING OBJECTIVE 9

- Summarize the steps of PCR.
- Summarize the steps required to construct a gene containing altered base pairs.

5 GENOMICS

As soon as DNA sequencing became routine, large amounts of sequence data made it possible to compare genes within and between species. This led to efforts to sequence entire genomes, beginning with the small DNA molecules of bacteria and progressing to the enormous chromosomes of humans and other mammals. Some of the organisms whose genomes have been sequenced are listed in Table 3-6. This list includes species that were already widely used as model organisms for different types of biochemical studies (Box 3-D).

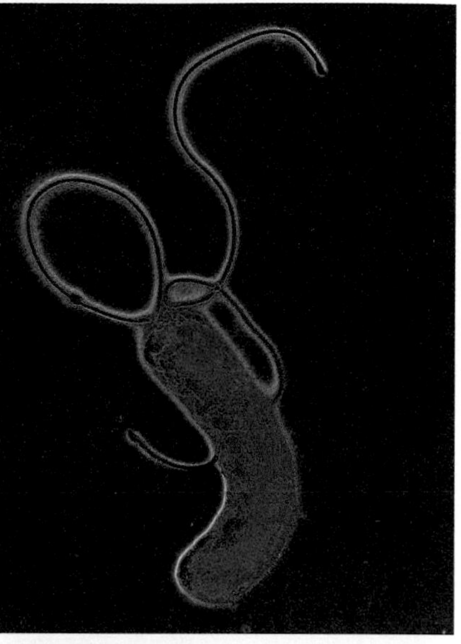

(b)

TABLE 3-6 Genome size and gene number of some organisms

Organism	Genome size (kb)	Number of genes
Bacteria		
Mycoplasma genitalium	580	482
Haemophilus influenzae	1830	1740
Synechocystis PCC6803	3573	3168
Mycobacterium tuberculosis	4412	3924
Escherichia coli	4639	4288
Archaea		
Methanococcus jannaschii	1740	1738
Archaeoglobus fulgidus	2178	2436
Fungi		
Saccharomyces cerevisiae (yeast)	12,069	6034
Plants		
Arabidopsis thaliana	125,000	25,500
Oryza sativa (rice)	450,000	~35,000
Zea mays (corn)	~5,000,000	~25,000
Animals		
Caenorhabditis elegans (nematode)	97,000	19,099
Drosophila melanogaster (fruitfly)	180,000	13,061
Homo sapiens	3,200,000	~35,000

A CLOSER LOOK

BOX 3-D Some model organisms for genome studies

A few of the world's estimated 14 million nonhuman species are studied in much greater detail than the others. These model organisms, often chosen initially because of their ready availability or ease of culture, make it possible for researchers in different laboratories to directly compare experimental results and build a coherent body of knowledge from studies of the organisms' genetics, metabolism, and behavior.

A detailed picture of these organisms' genomes can therefore borrow from and contribute to an existing database of biochemical information. Model organisms representing some major life forms are the bacterium *Escherichia coli*, the yeast *Saccharomyces cerevisiae*, the nematode (roundworm) *Caenorhabditis elegans*, and the plant *Arabidopsis thaliana*.

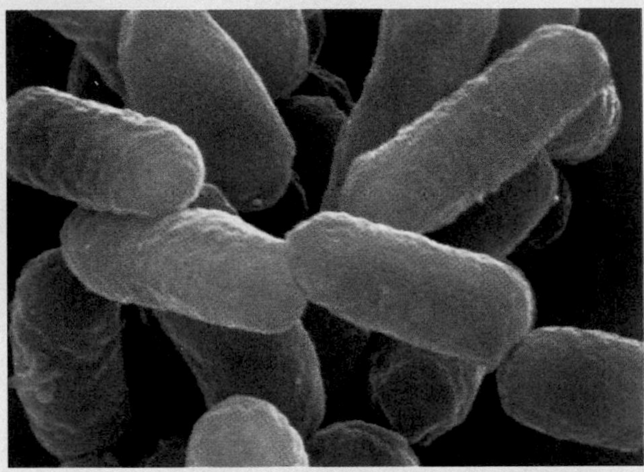

E. coli, a normal inhabitant of the mammalian digestive tract, has served as a model organism for decades. It is a metabolically versatile bacterium, using a variety of compounds as sources of carbon and nitrogen, and it tolerates both aerobic and anaerobic conditions. Because of its simple nutritional requirements, it is easy to grow in the laboratory. In fact, it grows rapidly, with a doubling time of 20 minutes under optimal conditions. Even before its genome was sequenced, many of *E. coli*'s genes had been identified and mapped to its single circular chromosome.

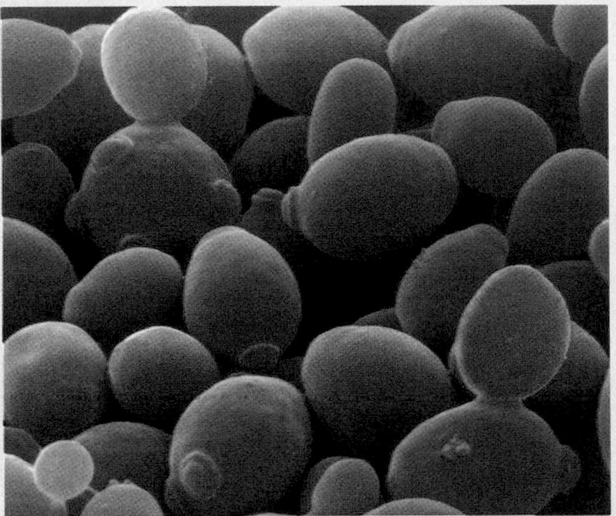

Baker's yeast, *S. cerevisiae*, is one of the simplest eukaryotic organisms. Its genes are distributed among 16 chromosomes. Although not as metabolically versatile as *E. coli*, *S. cerevisiae* can grow in defined media so that its chemical and physical environment can be varied for different experimental purposes. Extensive knowledge of its physiology makes it easier to assign functions to newly discovered genes.

(continued on next page)

Complexity and gene number

Not surprisingly, *organisms with the simplest lifestyles tend to have the least amount of DNA and the fewest genes.* For example, *M. genitalium* and *H. influenzae* (see Table 3-6) are human parasites that depend on their host to provide nutrients; these organisms do not contain as many genes as free-living bacteria such as *Synechocystis* (a photosynthetic bacterium). Multicellular organisms generally have even more DNA and more genes, presumably to support the activities of their many specialized cell types. Interestingly, humans contain only about two or three times as many genes as nematodes or fruit flies, suggesting that organismal complexity results not just from the raw number of genes but from how the genes are expressed.

In prokaryotic genomes, all but a few percent of the DNA represents genes for proteins and RNA. The proportion of noncoding DNA generally increases with the complexity of the organism. For example, about 30% of

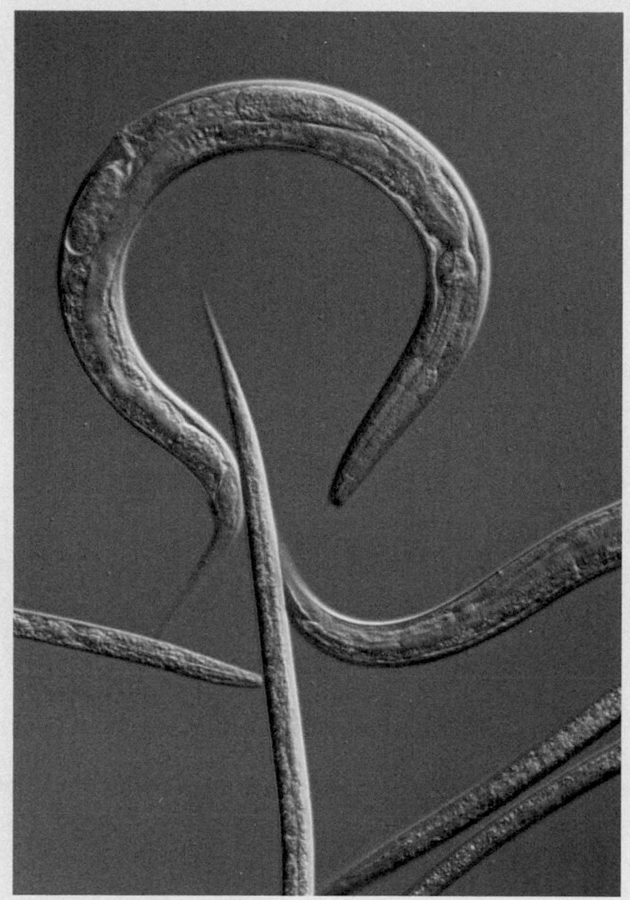

C. elegans is a small (1 mm) and transparent roundworm. As a multicellular organism, it bears genes not found in unicellular organisms, such as genes encoding hormones and hormone receptors that are involved in coordinating the activities of its 959 cells. In fact, about 300 of these cells are part of the worm's nervous system. Remarkably, the origins and developmental fates of all the cells have been documented. Consequently, *C. elegans* is invaluable for developmental and neurological studies.

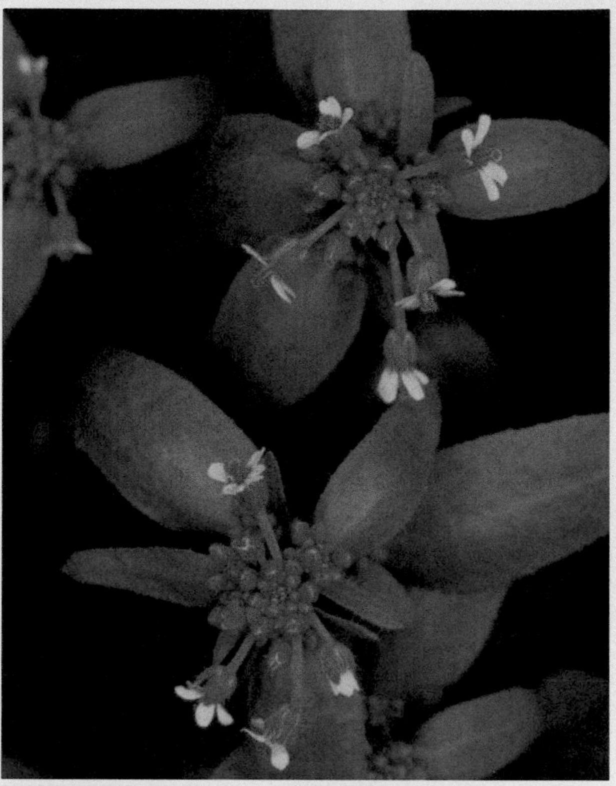

The plant kingdom is represented by *A. thaliana*, a small member of the mustard family. Although not an economically important crop species, *A. thaliana* has a short generation time and readily takes up foreign DNA. This makes it an ideal plant for laboratory study. Because its genome is compact and contains relatively little repetitive DNA, *A. thaliana* was the first plant species to have its genome sequenced.

[Fig. 1: Dr. Kari Lounatmaa/Science Photo Library/Photo Researchers; Fig. 2: Andrew Syred/Science Photo Library/Photo Researchers; Fig. 3: Sinclair Stammers/Science Photo Library/Photo Researchers; Fig.4: Dr. Jeremy Burgess/Science Photo Library/Photo Researchers.]

the yeast genome, about half of the *Arabidopsis* genome, and over 98% of the human genome is noncoding DNA. Although up to 24% of the human genome may be transcribed, the protein-coding segments account for only about 1.1% of the total.

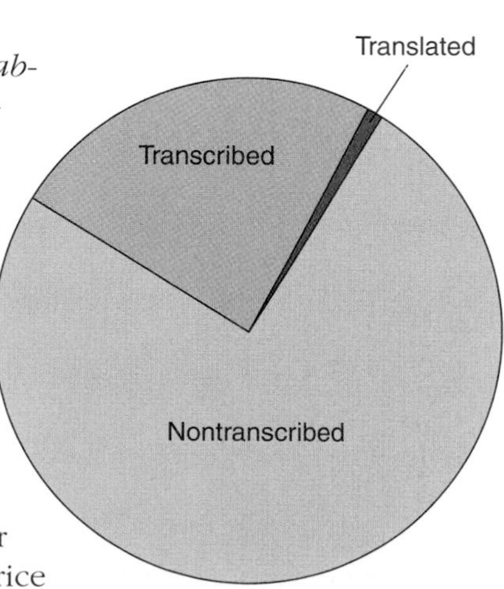

Much of the noncoding DNA consists of repeating sequences with no known function. The presence of **repetitive DNA** helps explain why certain very large genomes actually include only a modest number of genes. For example, compare the maize (corn) and rice genomes in Table 3-6: Both have about the same number of genes, but the maize genome is 10 times larger than the rice genome. About half of the maize genome appears to be composed of **transposable elements,** short segments of DNA that are copied many times and inserted randomly into the chromosomes.

The human genome contains several types of repetitive DNA, including several hundred thousand Alu sequences. These elements are about 300 bp long but vary somewhat in nucleotide sequence. Their name comes from the fact that they usually contain a recognition site for the restriction enzyme *Alu*I (see Table 3-4). Alu sequences may represent the remnants of a transposable element. The human genome also harbors DNA sequences ranging from a few bases to several hundred bases that are repeated in tandem (end-to-end) up to thousands of times. The number of repeats may vary between individuals, even in the same family.

Genome sequencing

Breaking a genome into smaller pieces for sequencing is conceptually simple; the challenge lies in ensuring that the sequenced fragments overlap enough so that they can be ordered and the complete sequence reassembled. Stretches of reassembled sequence, called **contigs,** are often separated by gaps that include regions of highly repetitive DNA, where the large number of tandem repeats makes it impossible to accurately align overlapping fragments.

The construction of a **genome map** combines the information in contigs with a physical map, which is based on the locations of restriction sites and other landmarks, such as the chromosomal positions of known genes. The genome map is not just a catalog of nucleotide sequences but provides a framework for classifying genes.

Several criteria can be used to spot genes in "raw" DNA sequence data:

1. *Identification of* ***open reading frames (ORFs).*** An ORF is a stretch of nucleotides that can be potentially translated. It begins with a "start" codon: ATG in the coding strand of DNA, which corresponds to AUG in RNA. This codon specifies methionine, the initial residue of all newly synthesized proteins. The ORF ends with one of the "stop" codons (see Table 3-3): DNA coding sequences of TAA, TAG, or TGA, which correspond to the mRNA stop codons UAA, UAG, and UGA.

2. *Comparison to a library of* ***expressed sequence tags (ESTs).*** To construct an EST, a cell's mRNA is isolated and used as a template for the *in vitro* synthesis of **complementary DNA (cDNA)** molecules. This is accomplished using an enzyme called reverse transcriptase, which can synthesize a DNA strand that is complementary to an RNA molecule. The segments of cDNA correspond to portions of the genome that are expressed (transcribed into RNA), and therefore can represent or "tag" protein-coding genes.

3. *Sequence similarities.* Genes with similar functions in different species often resemble each other. An inexact match might still indicate a protein's functional category, such as enzyme or hormone receptor, even if its exact role in the cell cannot be determined.

A genome map, such as the one shown in Fig. 3-22, indicates the placement and orientation of all genes. Arrows pointing in opposite directions indicate genes encoded by different strands of the double-stranded chromosome. Wherever possible, genes are color-coded by category. A map of the human genome has the same general format. However, human genes are typ-

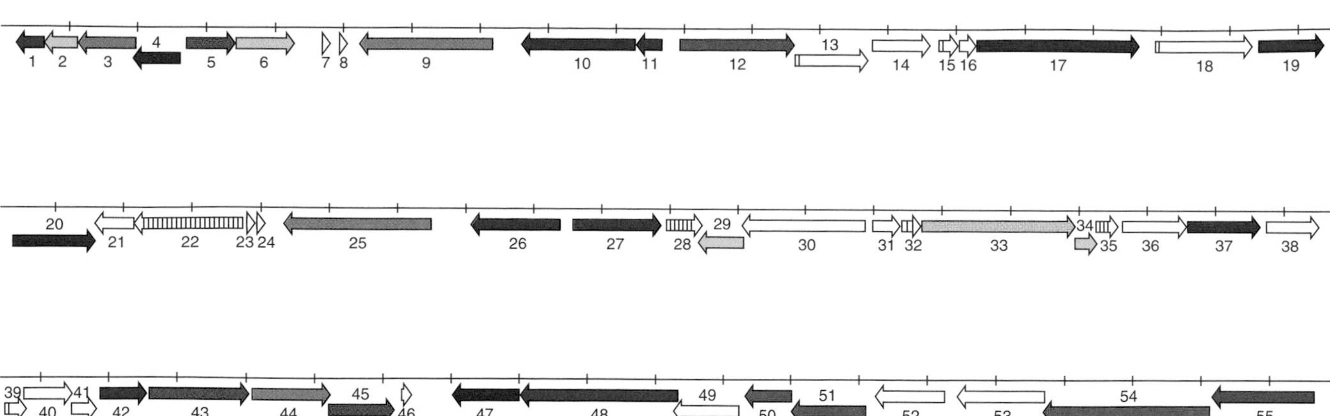

FIGURE 3-22

A portion of a genome map.
Genes located in the first 58 kb of the 1668 kb genome of *Helicobacter pylori* are shown color-coded by category. For example, brown represents genes involved in intermediary metabolism, yellow represents genes involved in DNA replication, and orange represents genes involved in nucleotide synthesis. White arrows represent ORFs of unknown function. The arrowheads indicate the direction of transcription, and the tick marks correspond to stretches of 1 kb. [*After Tomb, J.-F., et al., Nature* ***388,*** *539–547 (1997).*]

ically much longer (27 kb on average), since they contain sequences that are spliced out of the transcript before translation. In addition, the spaces between genes are much larger in the human genome.

In all sequenced genomes, including the human genome, *between 30% and 50% of the ORFs are* ***orphan genes*** *with no assigned functions* (these are the genes without color in Fig. 3-22). Some of these sequences probably represent completely novel genes whose protein products have not yet been discovered. The others may simply be counterparts of known genes but are too different in sequence to be recognized as such.

How accurate are the current genome maps? The known human DNA sequences account for about 95% of the total nuclear DNA. The unsequenced gaps most likely correspond to highly repetitive segments that contain few if any genes. Within the sequenced regions, the error rate is estimated to be less than one base in 10,000. Still, there is no single sequence that represents the human genome, because scattered throughout the 3.2 billion bp are about 3 million **single-nucleotide polymorphisms (SNPs),** locations where the nucleotide sequence varies. On average, the genomes of two individuals differ about every 1000 bp. However, less than 1% of these SNPs result in protein variants, and probably fewer still have functional consequences. Nevertheless, SNPs are the basis of genetic diversity and may ultimately provide clues to an individual's susceptibility to different diseases.

What do genome data tell us?

The number of genes and their putative functions provide a rough snapshot of the metabolic capabilities of a given organism. E. coli, for example, devotes at least 21% of its genes to synthesizing and breaking down biological molecules (the number of these genes might actually be higher, since 38% of *E. coli*'s genes are orphans; Fig. 3-23). An unusual number of genes belonging to one category might indicate some unusual metabolic property in an organism. This sort of knowledge could be useful for developing drugs to inhibit the growth of a pathogenic organism according to its unique metabolism.

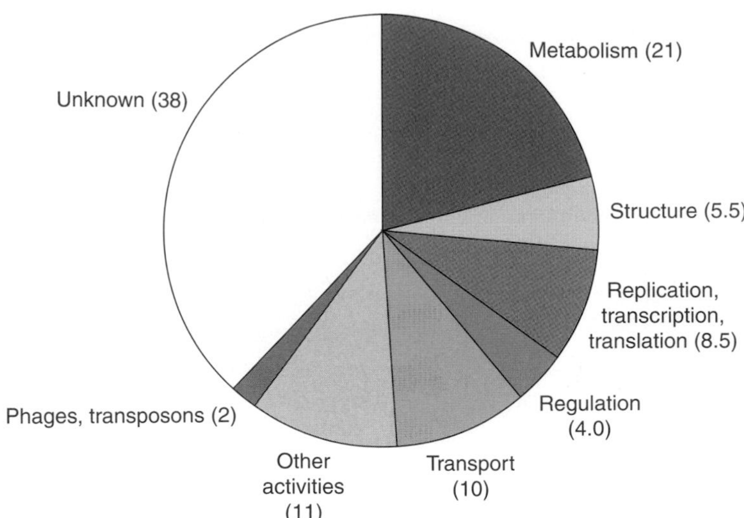

FIGURE 3-23

Functions of *E. coli* genes.
The numbers indicate the percentage of a total of 4288 genes. [*Data from Blattner, F.R., et al., Science* ***277****, 1458 (1997)*.]

Of course, compiling a catalog of an organism's genes is not as useful as knowing which genes are actually expressed, and when (Box 3-E).

One practical outcome of **genomics,** the study of genomes, is based on the insight that *many human genes have a* ***homolog*** *in a model organism such as yeast, nematode, or mouse (homologous genes are derived from a common ancestor and have similar sequences and functions in different species).* This makes it worthwhile to alter or delete that gene in the model organism and observe whether any function is lost. It is also possible to systematically mutate every gene in *S. cerevisiae* or *C. elegans* in order to assign a function to each gene. These findings can then be extrapolated to the human genome, which may contain a family of similar genes.

The putative function of the cystic fibrosis gene product was identified through its sequence similarity to a large family of proteins involved in the transport of substances across cell membranes (recall from Section 2-2 that only hydrophobic substances can spontaneously traverse a lipid bilayer; all other substances require a protein transporter). Each member of this protein family has one or two segments that position the protein in the membrane. The CF protein also contains an additional domain thought to play a regulatory role. Accordingly, the protein was named the cystic fibrosis transmembrane conductance regulator (CFTR).

When the cloned CFTR gene was expressed in different cell types, its function could be studied. The CFTR protein is, in fact, a membrane protein that acts as a channel to allow Cl^- to exit the cell (Fig. 3-24). It also appears to regulate Na^+ uptake by the cell. Consequently, a defective or absent CFTR protein disrupts the normal distribution of Na^+ and Cl^-. In the CF lung, the concentrations of the ions are low in the extracellular space. As a result, the

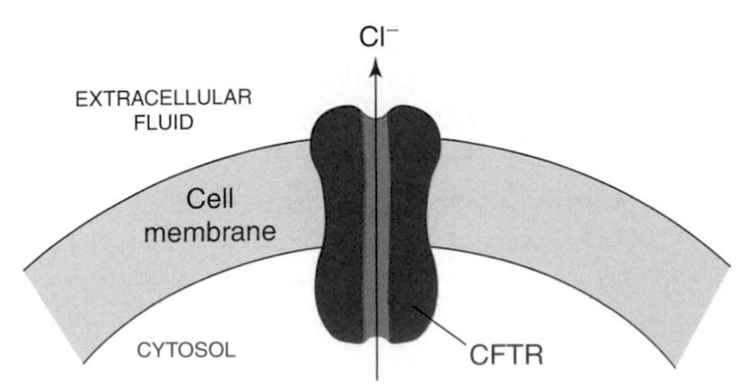

FIGURE 3-24

The cystic fibrosis transmembrane conductance regulator.
In this schematic view, the CFTR protein is positioned in the cell membrane so that it can provide a channel for Cl^- to exit the cell.

A CLOSER LOOK

BOX 3-E Transcriptomics and proteomics

At virtually every point in its development, a cell contains a population of mRNA molecules that represent the genes that are turned on, or transcribed, at that time. The study of these mRNAs is known as **transcriptomics.** Identifying and quantifying all the mRNA transcripts (the **transcriptome**) from a single cell type yields a profile of active genes. This can be done by assembling short strands of DNA of known sequences on a solid support, then allowing them to hybridize with fluorescent-labeled mRNAs from a cell preparation. The strength of fluorescence indicates how much mRNA binds to a particular complementary DNA sequence. The collection of DNA sequences is called a **microarray** or **DNA chip** because thousands of sequences fit in a few square centimeters. The microarray may represent an entire genome or just a few selected genes.

Cancer researchers can use DNA chips to profile the pattern of gene expression in tumor cells since different types of tumors synthesize different proteins. This information may be useful in deciding how to treat the cancer.

Unfortunately, the correlation between the amount of a particular mRNA and the amount of its protein is not perfect; some mRNAs are rapidly degraded whereas others are translated many times, yielding large quantities of the corresponding protein. Hence, the most reliable way to assess gene expression would be through **proteomics**—by examining a cell's **proteome,** the complete set of proteins that are synthesized over the cell's lifetime or at a particular point. However, this approach is limited by the technical problems of detecting minute quantities of thousands of different proteins. Nucleic acids can be amplified by PCR, but there is no comparable procedure for amplifying proteins.

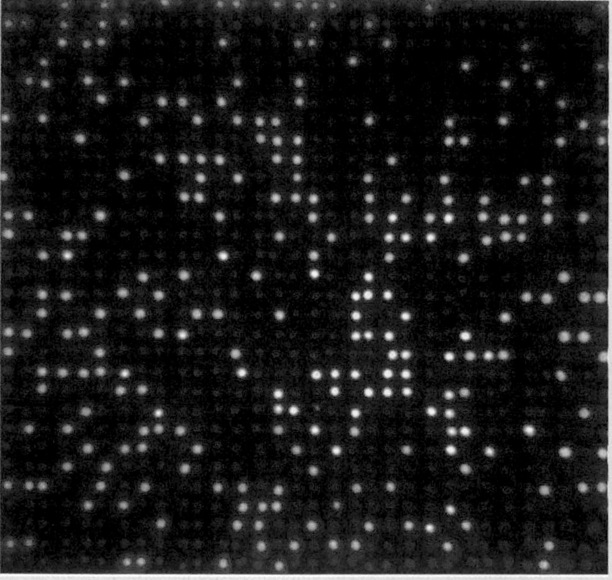

[Voker Steger/Science Photo Library/Photo Researchers.]

water that would normally be drawn by high concentrations of these ions is absent. In a normal lung, the extracellular fluid is thin and watery, but in the CF lung, the fluid is thick and viscous. In the sweat gland, a defective CFTR alters the transport of Na^+ and Cl^-, causing the salty sweat that is diagnostic of CF.

Monogenetic diseases such as CF are relatively rare. Most diseases result from interactions among multiple genes, many of which are not yet discovered, and from environmental factors. But by looking at the genome as a whole, researchers can link genetic diseases, as well other traits such as susceptibility to infection, to certain patterns of genetic variation and, ultimately, derive a better understanding of how genetic information defines an organism's growth and development from birth until death.

REVIEW LEARNING OBJECTIVE 10

- Describe the rough correlation between gene number and organismal lifestyle.
- How are genes identified?
- Why is it useful to identify homologs of human genes?

SUMMARY

1. The genetic material in virtually all organisms consists of DNA, a polymer of nucleotides. A nucleotide contains a purine or pyrimidine base linked to a ribose group (in RNA) or a deoxyribose group (in DNA) that also bears one or more phosphate groups.
2. DNA contains two antiparallel helical strands of nucleotides linked by phosphodiester bonds. Each base pairs with a complementary base in the opposite strand: A with T and G with C. The structure of RNA, which is single-stranded and contains U rather than T, is more variable.
3. Nucleic acid structures are stabilized primarily by stacking interactions between bases. The separated strands of DNA can reanneal.
4. The central dogma summarizes how the sequence of nucleotides in DNA is transcribed into RNA, which is then translated into protein according to the genetic code.
5. The sequence of nucleotides in a segment of DNA is commonly determined by the dideoxy method, in which labeled complementary copies of a DNA strand are synthesized. The presence of dideoxynucleotides, which cannot support further synthesis, generates a set of fragments of different lengths that are separated and analyzed to deduce the sequence of the original template strand. Gene sequencing may reveal mutations that cause disease.
6. DNA molecules can be experimentally manipulated. Restriction enzymes cleave DNA at specific sequences, and the resulting fragments can be used to generate recombinant DNA molecules that are then grown (cloned) in host cells.
7. Alternatively, large amounts of a particular DNA segment can be obtained by the polymerase chain reaction, in which a DNA polymerase makes complementary copies of a selected segment of DNA. PCR can also be used to introduce mutations into the DNA.
8. The complete genomic sequences from different species, including humans, reveal that the simplest organisms contain the fewest genes. The proportion of noncoding DNA increases with the complexity of the organism. Much of the human genome consists of repetitive DNA.
9. Genes can be identified by the presence of an open reading frame, by comparison with a library of expressed sequence tags that represent transcribed genes, and by comparison with other genes of known sequence. Although around half of all identified genes have no assigned function, some of the metabolic capabilities of an organism can be inferred from its complement of genes.

CHECKLIST

1. Review the Learning Objectives listed on page 52.
2. Complete Exercise 2/Nucleic Acid Structure, which explores the chemical and conformational features of nucleic acids.
3. Complete Exercise 3/DNA Sequencing, which describes how DNA is sequenced.
4. Apply your knowledge by solving the problems at the end of this chapter. Check your results in the Solutions appendix.
5. Be able to define the boldfaced terms (consult the glossary at the end of the book). Test your understanding by taking the Chapter 3 quiz (www.wiley.com/college/pratt).
6. Explore the websites listed at www.wiley.com/college/pratt.
7. Consult the list of Selected Readings for background information or additional details on nucleic acid structure, sequencing, recombinant DNA technology, and genomics.

GLOSSARY TERMS

cystic fibrosis (CF)
gene
chromosome
nucleic acid
bacteriophage
polynucleotide
base
purine
pyrimidine
DNA
RNA
nucleoside
nucleotide
deoxynucleotide
phosphodiester bond
residue
5′ end
3′ end
base pair
sugar–phosphate backbone
antiparallel
major groove
minor groove
bp
kb
oligonucleotide
A-DNA
B-DNA
stacking interactions
melting temperature (T_m)
denaturation
renaturation
anneal
probe
replication
gene expression
transcription
translation
genome
coding strand
noncoding strand
messenger RNA (mRNA)
ribosome
ribosomal RNA (rRNA)
transfer RNA (tRNA)
codon
genetic code
mutation
DNA marker
dideoxy DNA sequencing
dNTP
ddNTP
DNA polymerase
primer
electrophoresis
splicing
restriction endonuclease
palindrome
sticky ends
blunt ends

restriction digest
restriction fragment
restriction fragment length polymorphism (RFLP)
DNA ligase
recombinant DNA
plasmid
selection
clone
cloning vector
blue-white screening
transgenic organism
gene therapy
polymerase chain reaction (PCR)
site-directed mutagenesis
repetitive DNA
transposable element
contig
genome map
open reading frame (ORF)
expressed sequence tag (EST)
complementary DNA (cDNA)
orphan gene
single-nucleotide polymorphism (SNP)
genomics
homologous genes
transcriptomics
transcriptome
microarray (DNA chip)
proteomics
proteome

PROBLEMS

1. The identification of DNA as the genetic material began with Griffith's "transformation" experiment conducted in 1928. Griffith worked with *Pneumococcus,* an encapsulated bacterium that causes pneumonia. Wild-type *Pneumococcus* forms smooth colonies when plated on agar and causes death when injected into mice. A mutant *Pneumococcus* lacking the enzymes needed to synthesize the polysaccharide capsule (required for virulence) forms rough colonies when plated on agar and does not cause death when injected into mice. Griffith found that heat-treated wild-type *Pneumococcus* did not cause death when injected into the mice because the heat treatment destroyed the polysaccharide capsule. However, if Griffith mixed heat-treated wild-type *Pneumococcus* and the mutant unencapsulated *Pnemococcus* together and injected this mixture, the mice died. Even more surprisingly, upon autopsy, Griffith found live, encapsulated *Pneumococcus* bacteria in the mouse tissue. Griffith concluded that the mutant *Pneumococcus* had been "transformed" into disease-causing *Pneumococcus,* but he could not explain how this occurred. Using your current knowledge of how DNA works, explain how the mutant *Pneumococcus* became "transformed."

2. In 1944, Avery, MacLeod, and McCarty set out to identify the "transforming factor," the chemical agent capable of transforming mutant unencapsulated *Pneumococcus* to the deadly encapsulated form. They isolated a viscous substance with the chemical and physical properties of DNA and showed that this substance was capable of "transformation." Transformation could still occur if proteases (enzymes that degrade proteins) or ribonucleases (enzymes that degrade RNA) were added prior to the experiment. What did these treatments tell the investigators about the molecular identity of the "transforming factor"?

3. In 1952, Alfred Hershey and Martha Chase carried out experiments using bacteriophages, which are virus-like particles that infect bacteria. Bacteriophages consist of nucleic acid enclosed by a protein capsid (coat). Hershey and Chase first labeled the bacteriophages with the radioactive isotopes ^{35}S and ^{32}P. Because proteins contain sulfur but not phosphorus, and DNA contains phosphorus but not sulfur, each type of molecule was separately labeled. The radiolabeled bacteriophages were allowed to infect the bacteria, then the preparation was placed in a blender to shear the bacteriophage "ghosts" from the bacterial cells. Next, the ghosts were separated from the bacteria by centrifugation. The ghosts were found to contain most of the ^{35}S label, whereas 30% of the ^{32}P was found in progeny bacteriophages.

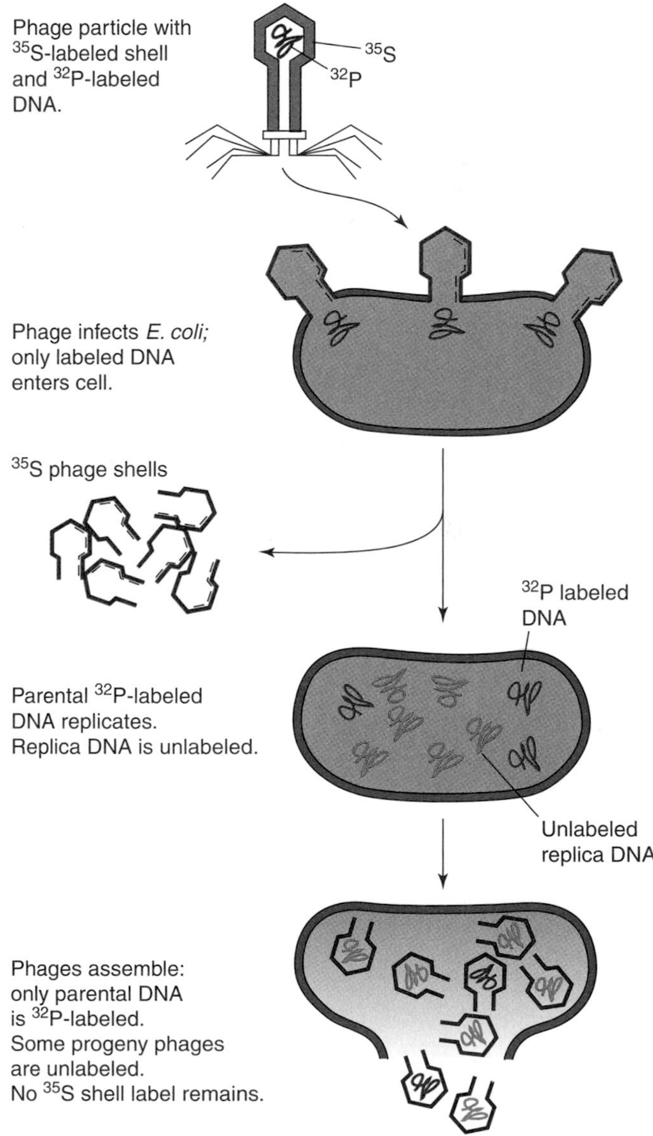
Phage particle with
^{35}S-labeled shell and ^{32}P-labeled DNA.
^{35}S
^{32}P
Phage infects *E. coli;* only labeled DNA enters cell.
^{35}S phage shells
^{32}P labeled DNA
Parental ^{32}P-labeled DNA replicates.
Replica DNA is unlabeled.
Unlabeled replica DNA
Phages assemble: only parental DNA is ^{32}P-labeled.
Some progeny phages are unlabeled.
No ^{35}S shell label remains.

What can you conclude from the results of this experiment?

4. Explain the structural basis for Chargaff's observation that the total amount of A + G in DNA is equal to the total amount of C + T. Does this rule hold for RNA?

5. A diploid organism with a 30,000 kb haploid genome contains 19% T residues. Calculate the number of A, C, G, and T residues in the DNA of each cell in this organism.

6. Identify the base pair highlighted in green in Figure 3-7.

7. The adenine derivative inosine can base pair with cytosine, adenine, and uracil. Show the structures of these base pairs.

Inosine

8. Explain why DNA denatures more easily at pH > 11.

9. Explain whether the following statement is true or false: Because a G:C base pair is stabilized by three hydrogen bonds, whereas an A:T base pair is stabilized by only two hydrogen bonds, GC-rich DNA is harder to melt than AT-rich DNA.

10. Explain why hydrogen bonding does not make a significantly large contribution to the overall stability of the DNA molecule.

11. Draw melting curves that would be obtained from the DNA of *Dictyostelium discoideum* and *Streptomyces albus* (see Table 3-2).

12. What might you find in comparing the GC content of DNA from *Thermus aquaticus* or *Pyrococcus furiosus* and DNA from bacteria in a typical backyard pond?

13. Explain why the melting temperature of a sample of double-helical DNA increases when the Na^+ concentration increases.

14. You have a short piece of synthetic RNA that you want to use as a probe to identify a gene in a sample of DNA. The RNA probe has a tendency to hybridize with sequences that are only weakly complementary. Should you increase or decrease the temperature to improve your chances of tagging the correct sequence?

15. Discuss the shortcomings of the following definitions for "gene":
   (a) A gene is the information that determines an inherited characteristic such as flower color.
   (b) A gene is a segment of DNA that encodes a protein.
   (c) A gene is a segment of DNA that is transcribed in all cells.

16. The semiconservative nature of DNA replication (as shown in Fig. 3-11) was proposed by Watson and Crick in 1953, but it wasn't experimentally verified until 1958. Meselson and Stahl grew bacteria on a "heavy nitrogen" source (ammonium chloride containing the isotope ^{15}N) for many generations so that virtually every nitrogen atom in the bacterial DNA was the ^{15}N isotope. This resulted in DNA that was denser than normal. The food source was abruptly switched to one containing only ^{14}N. Bacteria were harvested and the DNA isolated by density gradient centrifugation.
   (a) What is the density of the DNA of the first-generation daughter DNA molecules? Explain.
   (b) What is the density of the DNA isolated after two generations? Explain.

17. The primer used in sequencing a cloned DNA segment often includes the recognition sequence for a restriction endonuclease. Explain.

18. Heat-stable DNA polymerases are sometimes used for dideoxy sequencing at high temperatures, especially when the template DNA has a high content of G + C. Explain.

19. How many different amino acids could theoretically be encoded by nucleic acids containing four different nucleotides if (a) each nucleotide coded for one amino acid; (b) consecutive sequences of two nucleotides coded for one amino acid; (c) consecutive sequences of three nucleotides coded for one amino acid; (d) consecutive sequences of four nucleotides coded for one amino acid?

20. Examine the following nucleotide sequence from the coding strand of DNA. What is the amino acid sequence of the encoded protein?

GTAATTCAAAATGCCTTACGC
CCCTGGAGACGAAAAGAAGGGT
GCTATTACGTATTTGAAGAAG
GCCACCTCTGAGTAAATGTGAC

21. A mutation occurs when there is a base change in the DNA sequence. Some base changes do not lead to changes in the amino acid sequence of the resulting protein. Explain why.

22. Which restriction enzymes in Table 3-4 generate sticky ends? Blunt ends?

23. Which is more likely to be called a "rare cutter": a restriction enzyme with a four-base recognition sequence or a restriction enzyme with an eight-base recognition sequence?

24. Restriction enzymes are used to construct restriction maps of DNA. These are diagrams of specific DNA molecules that show the sites where the restriction enzymes cleave the DNA. To construct a restriction map, purified samples of the DNA are treated with restriction enzymes, either alone or in combination, and then the reaction products are separated by agarose gel electrophoresis, a technique that separates the DNA fragments based on size. (The smaller fragments travel more quickly and are found near the bottom of the gel, whereas the larger fragments are found near the top of the gel.)

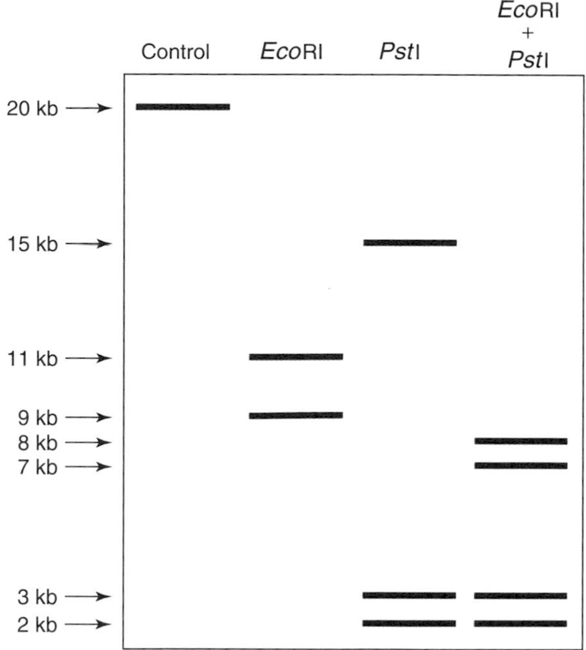

Use the results of the agarose gel electrophoresis separation (above) to construct a restriction map for the sample of DNA.

25. Could you perform PCR with an ordinary DNA polymerase, that is, one that is destroyed by high temperatures? What modifications would you make in the PCR protocol?

26. The search for a gene sometimes starts with a DNA library, which is a set of cloned DNA fragments representing all the sequences in an organism's genome. If you were to construct a human DNA library to search for novel genes, would you choose to clone the DNA fragments in plasmids or in yeast artificial chromosomes? Explain.

27. A researcher trying to identify the gene for a known protein might begin by looking closely at the protein's sequence in order to design a single-stranded oligonucleotide probe that will hybridize with the DNA of the gene. Why would the researcher focus on a segment of the protein containing a methionine (Met) or tryptophan (Trp) residue (see Table 3-3)?

28. Examine the sequence of the protein in Solution 3-20. Assume that the corresponding DNA sequence is *not* known. Using the amino acid sequence as a guide, design a pair of nine-base deoxynucleotide primers that could be used for PCR amplification of the protein-coding portion of the gene. (Hint: DNA polymerase can extend a primer only from its 3′ end.) How many different pairs of primers could you choose from?

29. Is it possible for the same segment of DNA to encode two different proteins? Explain.

30. The number of short interspersed repeats, such as Alu sequences, in the human genome varies among individuals. Explain why individuals with different numbers of Alu sequences give rise to restriction fragment length polymorphisms.

SELECTED READINGS

Collins, F.S., Cystic fibrosis: Molecular biology and therapeutic implications, *Science* **256,** 774–779 (1992). [Describes the discovery of the CFTR gene and its protein product.]

Conn, G.I. and Draper, D.E., RNA structure, *Curr. Opin. Struct. Biol.* **8,** 278–285 (1998). [Briefly reviews the structure and folding of some well-studied RNA molecules.]

Dickerson, R.E., DNA structure from A to Z, *Methods Enzymol.* **211,** 67–111 (1992). [Describes the various crystallographic forms of DNA.]

International Human Genome Sequencing Consortium, Initial sequencing and analysis of the human genome, *Nature* **409,** 860–921 (2001) and Venter, J.C., et al., The sequence of the human genome, *Science* **291,** 1304–1351 (2001). [These and other papers in the same issues of *Nature* and *Science* describe the data that constitute the draft sequence of the human genome and discuss how this information can be used in understanding biological function, evolution, and human health.]

Jurka, J., Repeats in genomic DNA: Mining and meaning, *Curr. Opin. Struct. Biol.* **8,** 333–337 (1998). [Reviews the evolutionary history and function of repetitive DNA sequences.]

Pingoud, A. and Jeltsch, A., Recognition and cleavage of DNA by type-II restriction endonucleases, *Eur. J. Biochem.* **246,** 1–22 (1997). [Includes an overview of different types of restriction enzymes.]

Quinton, P.M., Physiological basis of cystic fibrosis: A historical perspective, *Physiol. Rev.* **79 (Suppl. 1),** S3–S22 (1999). [Summarizes research on the cause of cystic fibrosis.]

Thieffry, D., Forty years under the central dogma, *Trends Biochem. Sci.* **23,** 312–316 (1998). [Traces the origins, acceptance, and shortcomings of the idea that nucleic acids contain biological information.]

Wu, R., Development of the primer-extension approach: A key role in DNA sequencing, *Trends Biochem. Sci.* **19,** 429–433 (1994). [Recounts the development of the dideoxy method for sequencing DNA.]

CHAPTER 4

Myoglobin and Hemoglobin: A Study of Protein Structure and Function

Among the proteins produced in the human body is the 165-amino-acid hormone erythropoietin (EPO), which signals the bone marrow to produce more red blood cells. EPO is synthesized primarily by the kidneys, so individuals with kidney disease are often deficient in EPO and develop anemia (lack of red blood cells). EPO can be isolated with difficulty from urine, but its gene has been cloned, so recombinant EPO is widely available. Treating individuals with recombinant EPO boosts red blood cell production. This fact has not been lost on some athletes, particularly in endurance sports such as cycling; they have found that EPO also increases their red cell counts, thereby increasing their ability to deliver oxygen to their muscles. The illegal use of EPO to enhance athletic performance is difficult to detect. For one thing, EPO is already present in the body, and an increase in red blood cells by itself is not proof of EPO doping. A clever abuser may even take iron supplements to mask the telltale drop in stored iron that accompanies the increased production of red cells. Successfully detecting the presence of recombinant EPO takes advantage of subtle differences between the natural and recombinant proteins: Although both have the same amino acid sequence, recombinant EPO produced by cultured cells is not derivatized with carbohydrate in the same manner as EPO produced naturally by the kidneys.

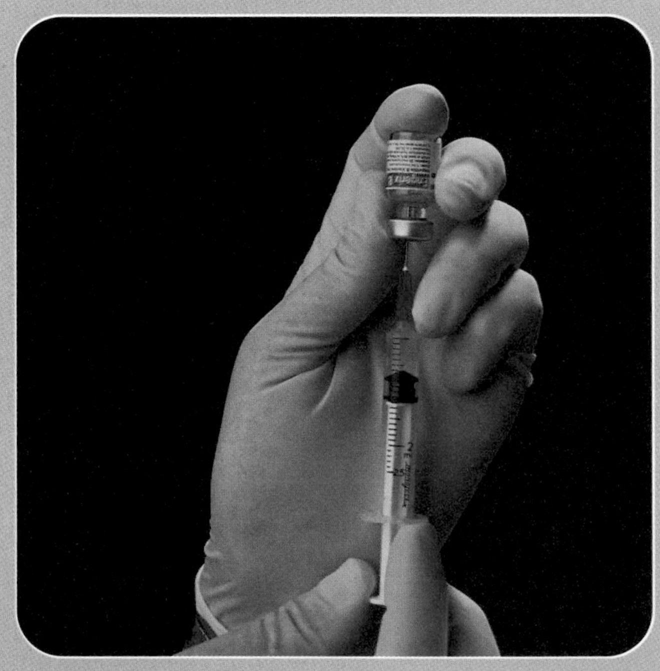

[Saturn Stills/Science Photo Library/Photo Researchers.]

LEARNING OBJECTIVES

1. Understand the chemical characteristics of the R groups that distinguish the 20 standard amino acids.
2. Understand how a protein's amino acid sequence is determined.
3. Understand the four levels of protein structure.

4. Understand the constraints on polypeptide conformation.
5. Understand the structural characteristics of the α helix and the β sheet and how these differ from irregular secondary structure.
6. Understand that proteins have a hydrophilic surface and a hydrophobic core.
7. Understand the forces that stabilize proteins and guide protein folding.
8. Understand how O_2 binds to myoglobin.
9. Understand the advantages of quaternary structure.
10. Understand how hemoglobin is related to myoglobin in structure, function, and evolution.
11. Understand how O_2 binds cooperatively to hemoglobin.
12. Understand how the Bohr effect and BPG modulate hemoglobin function *in vivo*.

THIS CHAPTER IN CONTEXT

Most of an organism's genetic information directs the synthesis of proteins, which are the molecules that carry out virtually all the metabolic work of the organism. In this chapter, we take a close look at the oxygen-binding proteins myoglobin and hemoglobin. Myoglobin is an intracellular protein that gives vertebrate muscles their red color, and hemoglobin is the major protein of red blood cells. These two proteins provide a wealth of information on how molecular structure is related to biological function. The chapter first looks at the chemical properties of the amino acid components of proteins, along with methods for determining the sequence of amino acids in a protein. Next comes a discussion of how the protein backbone and side chains fold into a unique three-dimensional shape stabilized by noncovalent forces. Detailed structural information about proteins such as myoglobin and hemoglobin leads to an understanding of how these proteins bind oxygen and how their physiological function is modulated by other factors. A comparison of the molecular structures of these protein also sheds light on how proteins evolve.

PROTEINS ARE THE WORKHORSES OF THE CELL. THEY PROVIDE STRUCTURAL stability and motors for movement; they form the molecular machinery for harvesting free energy and using it to carry out other metabolic activities; they participate in the expression of genetic information; and they mediate communication between the cell and its environment. In coming chapters we will describe in more detail these protein-driven phenomena, but for now we will focus on protein structure and how it relates to protein function.

Proteins come in many shapes and sizes (Fig. 4-1). The essence of their biological function is their interaction with other molecules, including other proteins. Some proteins act independently, whereas others form large aggregates. In every case, it is important to understand the protein's structure. What information dictates its three-dimensional shape? What forces stabilize that

Myoglobin (sperm whale)
Facilitates oxygen diffusion in muscle

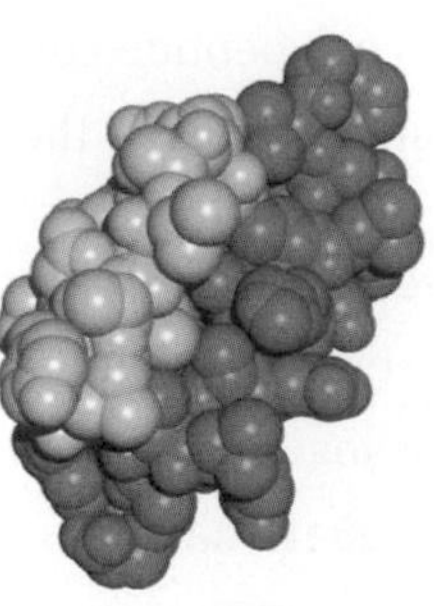

Insulin (pig)
Released from the pancreas to signal the availability of the metabolic fuel glucose
(more in Section 16-2)

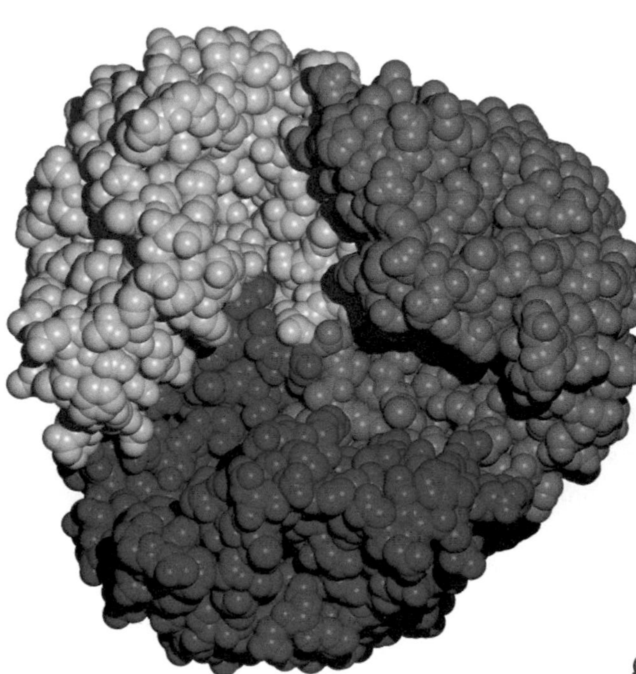

Maltoporin (*E. coli*)
Permits sugars to cross the bacterial cell membrane
(more in Section 8-3)

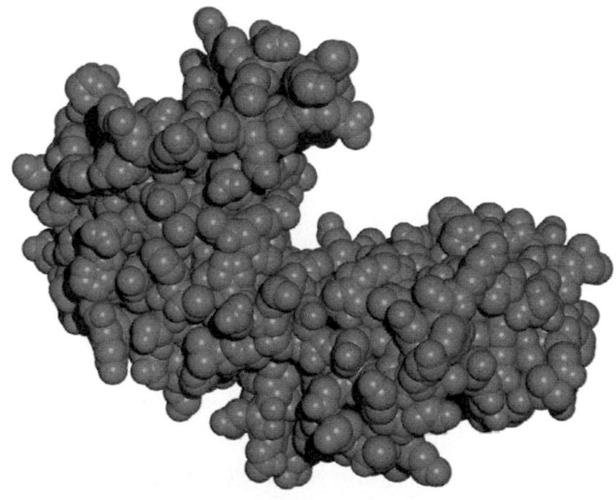

Phosphoglycerate kinase (yeast)
Catalyzes one of the central reactions in metabolism
(more in Section 10-2)

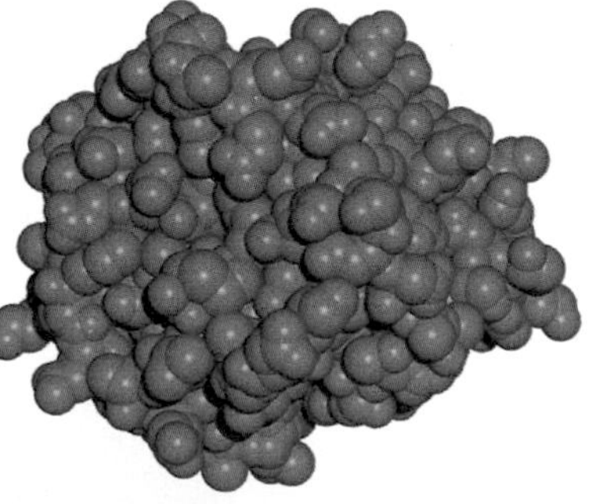

Chymotrypsin (cow)
Degrades dietary proteins in the small intestine
(more in Chapter 6)

FIGURE 4-1

A gallery of protein structure and function.
These space-filling models are all shown at approximately the same scale. In proteins that consist of more than one chain of amino acids, the chains are shaded differently. [*Structure of myoglobin (pdb 1MBD) determined by S.E.V. Phillips; structure of insulin (pdb 1ZNI) determined by M.G.W. Turkenburg, J.L. Whittingham, G.G. Dodson, E.J. Dodson, B. Xiao, and G.A. Bentley; structure of maltoporin (pdb 1MPM) determined by R. Dutzler, and T. Schirmer; structure of phosphoglycerate kinase (pdb 3PGK) determined by P.J. Shaw, N.P. Walker, and H.C. Watson; structure of chymotrypsin (pdb 4CHA) determined by H. Tsukada and D.M. Blow; structure of DNA polymerase (pdb 1KFS) determined by C.A. Brautigan and T.A. Steitz; structure of collagen (pdb 1BBE) determined by G. Nemethy, K.D. Gibson, K.A. Palmer, C.N. Yoon, G. Paterlini, A. Zagari, S. Rumsey, and H.A. Scheraga; and structure of plastocyanin (pdb 1PND) determined by B.A. Fields, J.M. Guss, and H.C. Freeman.*]

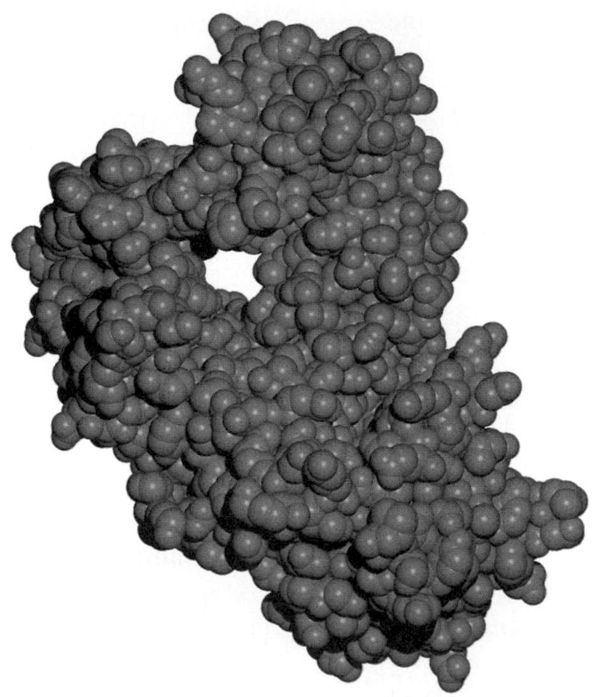

DNA polymerase (*E. coli* Klenow fragment)
Synthesizes a new DNA chain using an existing DNA strand as a template
(more in Section 17-2)

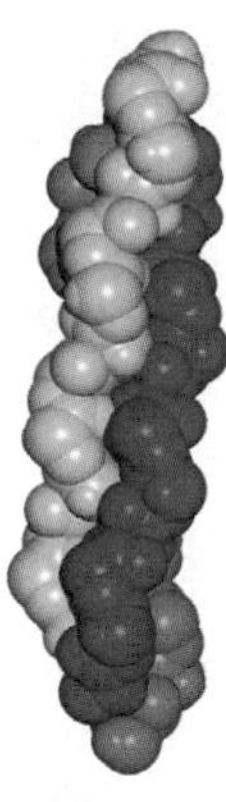

Fragment of collagen (theoretical model)
Provides strength and rigidity in bones and connective tissue
(more in Section 5-6)

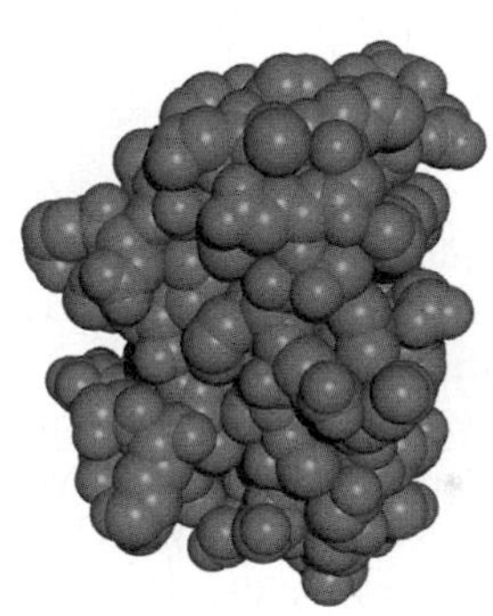

Plastocyanin (poplar)
Shuttles electrons as part of the apparatus for converting light energy to chemical energy
(more in Section 13-2)

shape? Which of its functional groups mediate its interactions with other substances? To answer some of these questions, we will look at two oxygen-binding proteins, myoglobin and hemoglobin, whose structure, function, and evolution have been studied for many decades.

1 PROTEINS ARE CHAINS OF AMINO ACIDS

A **protein** is a biological molecule that consists of one or more **polypeptides,** which are chains of polymerized amino acids. A cell may contain dozens of different amino acids, but only 20 of these—called the "standard" amino acids—are commonly found in proteins. As introduced in Section 1-2, an **amino acid** is a small molecule containing an amino group ($—NH_3^+$) and a carboxylate group ($—COO^-$) as well as a side chain of variable structure, called an **R group:**

$$\begin{array}{c} COO^- \\ | \\ H—C—R \\ | \\ NH_3^+ \end{array}$$

Go to **Exercise 4/Amino Acids** for a detailed look at the three-dimensional shapes and chemical properties of the 20 standard amino acids. The structures presented in Exercise 4 and the questions based on them address Learning Objective 1.

Note that at physiological pH, the carboxyl group is unprotonated and the amino group is protonated, so that an isolated amino acid bears both a negative and a positive charge.

The 20 amino acids have different chemical properties

The identities of the R groups distinguish the 20 standard amino acids. The R groups can be classified by their overall chemical characteristics as hydrophobic, polar, or charged, as shown in Figure 4-2, which also includes the one- and three-letter codes for each amino acid. These compounds are formally called **α-amino acids** because the amino and carboxylate (acid) groups are both attached to a central carbon atom known as the **α carbon** (abbreviated **Cα**). There are two possible arrangements for the substituents of Cα, but one predominates in biological systems (Box 4-A).

It is advisable to become familiar with the structures of the standard amino acids, since their side chains ultimately help determine the three-dimensional shape of the protein as well as its chemical reactivity.

The hydrophobic amino acids

As their name implies, *the hydrophobic amino acids have essentially nonpolar side chains that interact very weakly or not at all with water.* The aliphatic (hydrocarbon-like) side chains of **alanine (Ala), valine, (Val), leucine (Leu), isoleucine (Ile),** and **phenylalanine (Phe)** obviously fit into this group. Although **methionine** (**Met,** with an S atom) and **tryptophan** (**Trp,** with an NH group) include atoms that can form hydrogen bonds, the bulk of their side chains is nonpolar. **Proline (Pro)** is unique among the amino acids because its aliphatic side chain is also covalently linked to its amino group.

In proteins, the hydrophobic amino acids are almost always located in the interior of the molecule, among other hydrophobic groups, where they do not interact with water. And because they lack reactive functional groups, the hydrophobic side chains seldom participate in mediating chemical reactions.

The polar amino acids

The side chains of the polar amino acids can interact with water because they contain hydrogen-bonding groups. **Serine (Ser), threonine (Thr),** and **tyrosine (Tyr)** have hydroxyl groups; **cysteine (Cys)** has a thiol group; and **asparagine (Asn)** and **glutamine (Gln)** have amide groups. All these amino acids, along with **histidine** (**His,** which bears a polar imidazole ring), can be found on the solvent-exposed surface of a protein, although they also

Hydrophobic amino acids

Alanine (Ala, A)

Leucine (Leu, L)

Polar amino acids

Serine (Ser, S)

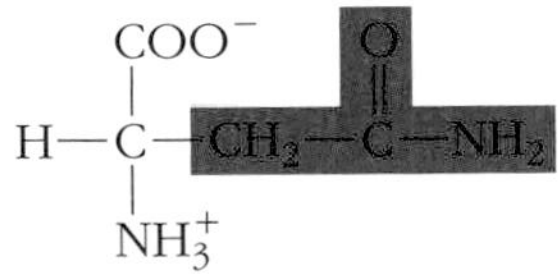

Asparagine (Asn, N)

Charged amino acids

Aspartate (Asp, D)

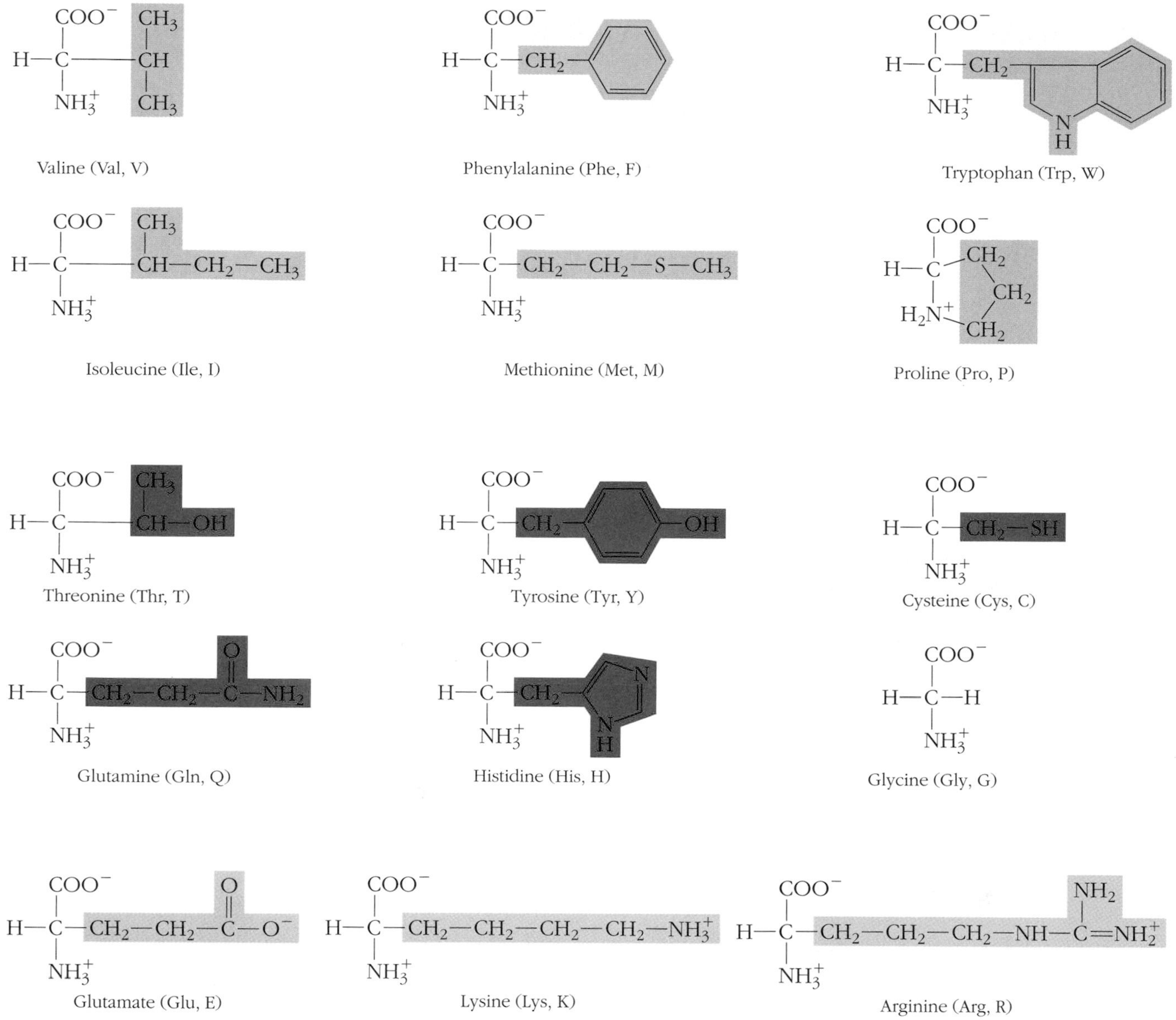

Valine (Val, V)

Phenylalanine (Phe, F)

Tryptophan (Trp, W)

Isoleucine (Ile, I)

Methionine (Met, M)

Proline (Pro, P)

Threonine (Thr, T)

Tyrosine (Tyr, Y)

Cysteine (Cys, C)

Glutamine (Gln, Q)

Histidine (His, H)

Glycine (Gly, G)

Glutamate (Glu, E)

Lysine (Lys, K)

Arginine (Arg, R)

FIGURE 4-2

Structures and abbreviations of the 15 standard amino acids.
The amino acids can be classified according to the chemical properties of their R groups as hydrophobic, polar, or charged. The side chain (R group) of each amino acid is shaded. The three-letter code is usually the first three letters of the amino acid's name. The one-letter code is derived as follows: If only one amino acid begins with a particular letter, that letter is used: C = cysteine, H = histidine, I = isoleucine, M = methionine, S = serine, and V = valine. If more than one amino acid begins with a particular letter, the letter is assigned to the most abundant amino acid: A = alanine, G = glycine, L = leucine, P = proline, and T = threonine. Most of the others are phonetically suggestive: D = aspartate ("asparDate"), F = phenylalanine ("Fenylalanine"), N = asparagine ("asparagiNe"), R = arginine ("aRginine"), W = tryptophan ("tWyptophan"), and Y = tyrosine ("tYrosine"). The rest are assigned as follows: E = glutamate (near D, aspartate), K = lysine, and Q = glutamine (near N, asparagine). The carbon atoms of amino acids are sometimes assigned Greek letters, beginning with Cα, the carbon to which the R group is attached. Thus, glutamate has a γ-carboxylate group, and lysine has an ε-amino group.

A CLOSER LOOK

BOX 4-A Chirality in nature

Nineteen of the 20 standard amino acids are asymmetric or chiral molecules. Their **chirality** or handedness (from the Greek *cheir,* hand) results from the asymmetry of the alpha carbon. The four different substituents of Cα can be arranged in two ways. For alanine, a small amino acid with a methyl R group, the possibilities are

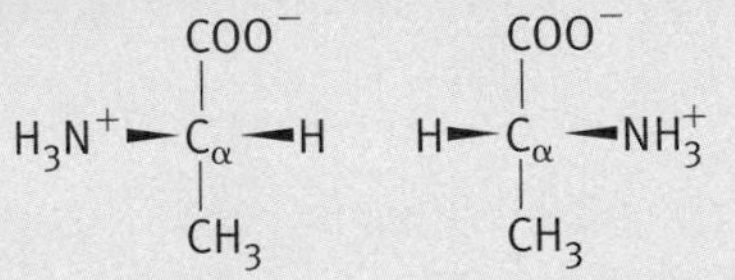

$$H_3N^+—C_\alpha(COO^-)(CH_3)—H \qquad H—C_\alpha(COO^-)(CH_3)—NH_3^+$$

You can use a simple model-building kit to satisfy yourself that the two structures are not identical. They are nonsuperimposable mirror images, like right and left hands.

The amino acids found in proteins all have the form on the left. For historical reasons, these are designated L-amino acids (from the Greek *levo,* left). Their mirror images, which are rarely if ever found in proteins, are the D-amino acids (from *dextro,* right). The L and D designations were originally used to indicate that model compounds with similar configurations rotated polarized light to the left and right, respectively. However, for other compounds, an L or D label does not indicate which direction the light is rotated.

Molecules related by mirror symmetry, such as the two forms of alanine shown above, are physically indistinguishable and are usually present in equal amounts in synthetic preparations. However, the two forms behave differently in biological systems, which are inherently chiral. This lesson was brought home in the 1960s when pregnant women with morning sickness were given the sedative thalidomide, which was a mixture of right- and left-handed forms. The active form of the drug has the structure shown below.

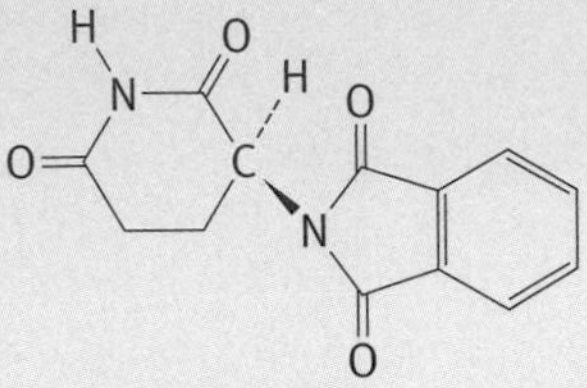

Thalidomide

Tragically, its mirror image caused severe birth defects.

An organism's ability to distinguish chiral molecules results from the handedness of its molecular constituents—proteins that contain all L-amino acids, polynucleotides that coil in a right-handed helix (see Fig. 3-4), and so on. But if pairs of chiral molecules are chemically indistinguishable, how did handedness originate in biological systems?

Some studies suggest that the conditions on the early earth might have had a slight stabilizing effect on L-amino acids relative to D-amino acids. But no purely physical phenomenon can account for the exclusive polymerization of L-amino acids in proteins. A more probable explanation is that by chance, natural selection favored a self-replicating polymer of L-amino acids. This could have occurred if the polymer served as a chiral template that allowed the assembly of a new polymer containing L- but not D-amino acids. Over time, all-L polymers would predominate. In experimental systems with self-replicating synthetic molecules, chiral polymers arise and perpetuate themselves even in an environment where both D and L monomers are present in equal amounts.

occur in the protein interior, provided that their hydrogen-bonding requirements are satisfied by their proximity to other hydrogen bond donor or acceptor groups. **Glycine (Gly),** which lacks a hydrogen-bonding side chain (it has only an H atom), is classified as a polar amino acid by default, because it is neither hydrophobic nor charged.

Depending on the presence of nearby groups that increase their polarity, *some of the polar side chains can ionize at physiological pH values.* For example, the neutral (basic) form of His can accept a proton to form an imidazolium ion (an acid):

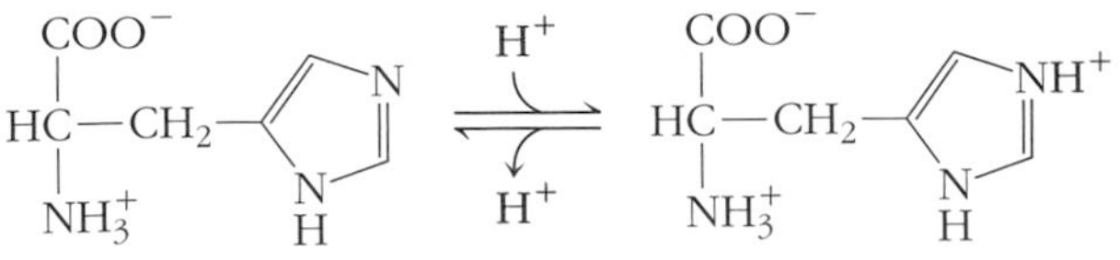

$$\text{Base} \underset{H^+}{\overset{H^+}{\rightleftharpoons}} \text{Acid}$$

As we shall see, the ability of His to act as an acid or a base gives it great versatility in catalyzing chemical reactions.

Similarly, the thiol group of Cys can be deprotonated, yielding a thiolate anion:

$$\begin{array}{c} COO^- \\ | \\ HC{-}CH_2{-}SH \\ | \\ NH_3^+ \end{array} \underset{H^+}{\overset{H^+}{\rightleftharpoons}} \begin{array}{c} COO^- \\ | \\ HC{-}CH_2{-}S^- \\ | \\ NH_3^+ \end{array}$$

More commonly, however, cysteine's thiol group undergoes oxidation with another thiol group, such as another Cys side chain, to form a **disulfide bond:**

$$\begin{array}{c} COO^- \\ | \\ HC{-}CH_2{-}S{-}S{-}CH_2{-}CH \\ | \\ NH_3^+ \end{array} \quad \begin{array}{c} COO^- \\ | \\ \\ | \\ NH_3^+ \end{array}$$

Disulfide bond

In rare cases, the very weakly acidic hydroxyl groups of Ser, Thr, and Tyr ionize to yield hydroxide groups that can act as strong bases in chemical reactions.

The charged amino acids

Four amino acids have side chains that are virtually always charged under physiological conditions. **Aspartate (Asp)** and **glutamate (Glu),** which bear carboxylate groups, are negatively charged. **Lysine (Lys)** and **arginine (Arg)** are positively charged. These side chains are usually located on the protein's surface, where their charged groups can be surrounded by water molecules or interact with other ionic species.

REVIEW LEARNING OBJECTIVE 1

- Know the structures and the one- and three-letter abbreviations for the 20 standard amino acids.
- Divide the 20 amino acids into groups that are hydrophobic, polar, and charged.
- Which polar amino acids are sometimes charged?

Peptide bonds link amino acids in proteins

The polymerization of amino acids to form a polypeptide chain involves the condensation of the carboxylate group of one amino acid with the amino group of another (a **condensation reaction** is one in which a water molecule is eliminated):

$$\begin{array}{c} R_1 \\ | \\ H_3\overset{+}{N}{-}C{-}COO^- \\ | \\ H \end{array} + \begin{array}{c} H \quad R_2 \\ | \quad\; | \\ H{-}\overset{+}{N}{-}C{-}COO^- \\ | \quad\; | \\ H \quad H \end{array}$$

$$\downarrow \rightarrow H_2O$$

$$\begin{array}{c} R_1 \;\; O \qquad R_2 \\ | \quad \| \qquad\; | \\ H_3\overset{+}{N}{-}C{-}C{-}N{-}C{-}COO^- \\ | \qquad\quad | \quad | \\ H \qquad\;\; H \quad H \end{array}$$

Peptide bond

The resulting amide bond linking the two amino acids is called a **peptide bond.** The remaining portions of the amino acids are called amino acid **residues.**

By convention, a chain of amino acid residues linked by peptide bonds is written or drawn so that the residue with a free amino group is on the left (this end of the polypeptide is called the **N-terminus**), and the residue with a free carboxylate group is on the right (this end is called the **C-terminus**):

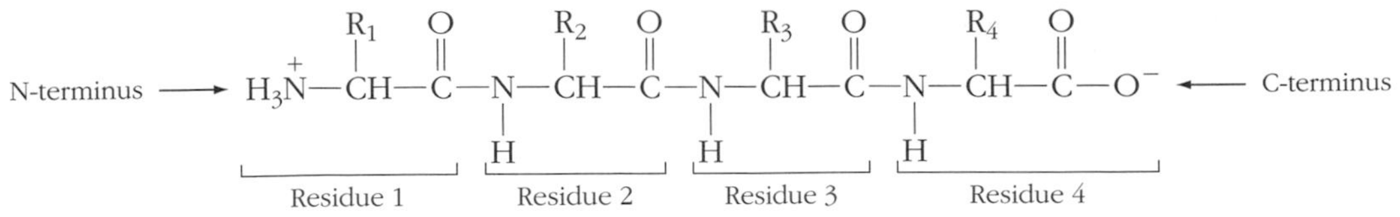

$$\text{N-terminus} \longrightarrow \underbrace{H_3\overset{+}{N}-\overset{R_1}{\overset{|}{CH}}-\overset{O}{\overset{\|}{C}}}_{\text{Residue 1}}-\underbrace{\underset{\underset{H}{|}}{N}-\overset{R_2}{\overset{|}{CH}}-\overset{O}{\overset{\|}{C}}}_{\text{Residue 2}}-\underbrace{\underset{\underset{H}{|}}{N}-\overset{R_3}{\overset{|}{CH}}-\overset{O}{\overset{\|}{C}}}_{\text{Residue 3}}-\underbrace{\underset{\underset{H}{|}}{N}-\overset{R_4}{\overset{|}{CH}}-\overset{O}{\overset{\|}{C}}-O^-}_{\text{Residue 4}} \longleftarrow \text{C-terminus}$$

Note that, except for the two terminal groups, the charged amino and carboxylate groups of each amino acid are eliminated in forming peptide bonds. *The electrostatic properties of the polypeptide therefore depend primarily on the identities of the side chains (R groups) that project out from the polypeptide* ***backbone.*** Exercise 4 shows how a peptide bond forms between two amino acids.

The p*K* values of all the charged and ionizable groups in amino acids are given in Table 4-1 (recall from Section 2-3 that a p*K* value is a measure of a group's tendency to ionize). Thus, it is possible to calculate the net charge of a protein at a given pH (see Sample Calculation 4-1). At best, this value is only an estimate, since the side chains of polymerized amino acids do not behave as they do in free amino acids. This is because of the electronic effects of the peptide bond and other functional groups that may be brought

TABLE 4-1 p*K* values of ioinizable groups in polypeptides

Group[a]		p*K*
C-terminus	$-COOH$	3.5
Asp	$-CH_2-\overset{O}{\overset{\|}{C}}-OH$	3.9
Glu	$-CH_2-CH_2-\overset{O}{\overset{\|}{C}}-OH$	4.1
His	$-CH_2-$(imidazole ring with NH^+ and N–H)	6.0
Cys	$-CH_2-SH$	8.4
N-terminus	$-NH_3^+$	9.0
Tyr	$-CH_2-C_6H_4-OH$	10.5
Lys	$-CH_2-CH_2-CH_2-CH_2-NH_3^+$	10.5
Arg	$-CH_2-CH_2-CH_2-NH-\overset{NH_2}{\overset{\mid}{C}}=NH_2^+$	12.5

[a]The ionizable proton is indicated in red.

into proximity when the polypeptide chain folds into a three-dimensional shape. The chemical properties of a side chain's immediate neighbors, its **microenvironment,** may alter its polarity, thereby altering its tendency to lose or accept a proton.

Sample Calculation 4-1

Problem
Estimate the net charge of the polypeptide chain below at physiological pH (7.4) and at pH 5.0.

Ala–Arg–Val–His–Asp–Gln

Solution
The polypeptide contains the following ionizable groups, whose pK values are listed in Table 4-1: the N-terminus (pK = 9.0), Arg (pK = 12.5), His (pK = 6.0), Asp (pK = 3.9), and the C-terminus (pK = 3.5).

At pH 7.4, the groups whose pK values are less than 7.4 will be mostly deprotonated, and the groups with pK values greater than 7.4 will be mostly protonated. The polypeptide therefore has a net charge of 0:

Group	Charge
N-terminus	+1
Arg	+1
His	0
Asp	−1
C-terminus	−1
net charge	0

At pH 5.0, His is likely to be protonated, giving the polypeptide a net charge of +1:

Group	Charge
N-terminus	+1
Arg	+1
His	+1
Asp	−1
C-terminus	−1
net charge	+1

Nevertheless, *the chemical and physical properties of proteins depend on their constituent amino acids,* so proteins exhibit different behaviors under given laboratory conditions. In fact, a protein's net charge at a particular pH can be exploited as a means to purify it from a mixture containing many other proteins (such as a preparation of cellular material). The desired protein may bind strongly to a negatively or positively charged surface while contaminating proteins fail to bind and can therefore be washed away (Fig. 4-3). The desired protein is then recovered by altering the pH or salt concentration. This technique is known as **ion exchange.**

Amino acid sequencing

Proteins vary widely in their amino acid composition. Most polypeptides contain between 100 and 1000 amino acid residues (myoglobin, for example, has 153 residues), although some contain thousands of amino acids (Table 4-2).

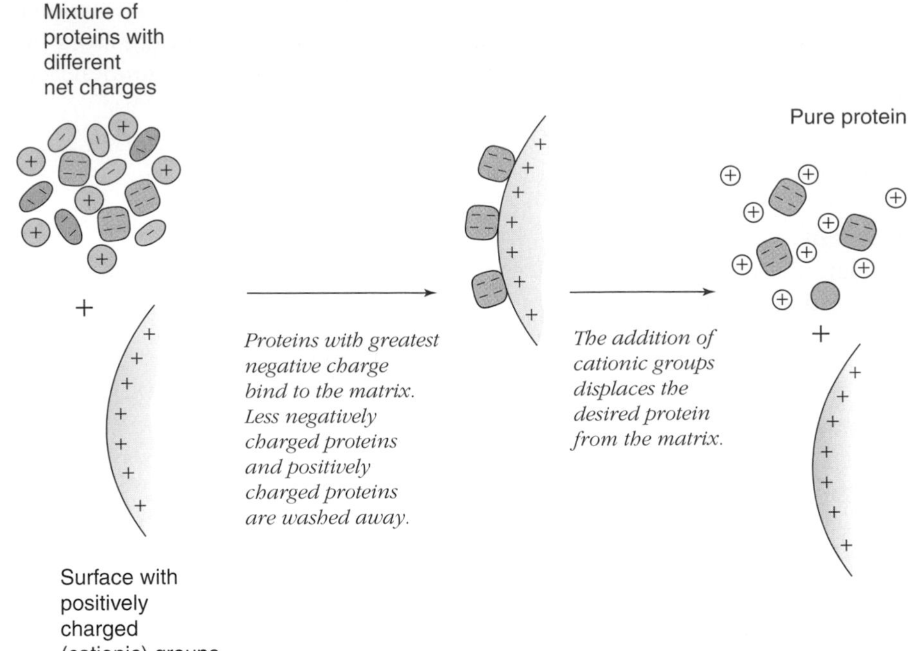

FIGURE 4-3
Protein purification by ion exchange.

Polypeptides smaller than about 40 residues are often called **oligopeptides** (*oligo* is Greek for "few") or just **peptides.** Since there are 20 different amino acids that can be polymerized to form polypeptides, even peptides of similar size can differ dramatically from each other, depending on their complement of amino acids. For example, myoglobin has 20 Lys residues but only 1 Cys residue. The similarly sized protein ribonuclease A has 8 Lys and 8 Cys residues.

TABLE 4-2 Composition of some proteins

Protein	Number of amino acid residues	Number of polypeptide chains	Molecular mass (D)
Insulin (bovine)	51	2	5733
Rubredoxin (*Pyrococcus*)	53	1	5878
Myoglobin (human)	153	1	17,053
Phosphorylase kinase (yeast)	416	1	44,552
Hemoglobin (human)	574	4	61,972
Reverse transcriptase (HIV)	986	2	114,097
Nitrite reductase (*Alcaligenes*)	1029	3	111,027
C-reactive protein (human)	1030	5	115,160
Pyruvate decarboxylase (yeast)	1112	2	121,600
Immunoglobulin (mouse)	1316	4	145,228
Ribulose bisphosphate carboxylase (spinach)	5048	16	567,960
Glutamine synthetase (*Salmonella*)	5628	12	621,600
Carbamoyl phosphate synthetase (*E. coli*)	5820	8	637,020

The potential for sequence variation is enormous. For a modest-sized polypeptide of 100 residues, there are 20^{100} or 1.27×10^{130} possible amino acid sequences. This number is clearly unattainable in nature, since there are only about 10^{79} atoms in the universe, but it illustrates the tremendous structural variability available to proteins.

Unraveling the amino acid sequence of a protein may be relatively straightforward if its gene has been sequenced (see Section 3-3). In this case, it is just a matter of reading successive sets of three nucleotides in the DNA as a sequence of amino acids in the protein. However, this exercise may not be accurate if the gene's mRNA is spliced before being translated, or if the protein is hydrolyzed or otherwise covalently modified immediately after it is synthesized. And of course, nucleic acid sequencing is of no use if the protein's gene has not been identified. The alternative is to use chemical methods for directly determining the protein's amino acid sequence.

The procedure for sequencing a protein has several steps:

1. First, *the sample of protein to be sequenced is purified* so that it is free of other proteins.

2. If the protein contains more than one kind of polypeptide chain, *the chains are separated* so that each can be individually sequenced. If two polypeptides contain Cys residues that are linked by a disulfide bond, the bond must be broken (reduced) to separate the chains, and the free sulfhydryl groups must be blocked by alkylation so that they cannot re-form disulfide bonds (Fig. 4-4).

3. Techniques for sequencing polypeptides are reliable for only about 40 to 100 residues. Therefore, *most large polypeptides must be broken down into smaller pieces that can be individually sequenced.* Cleavage can be accomplished chemically, for example, by treating the polypep-

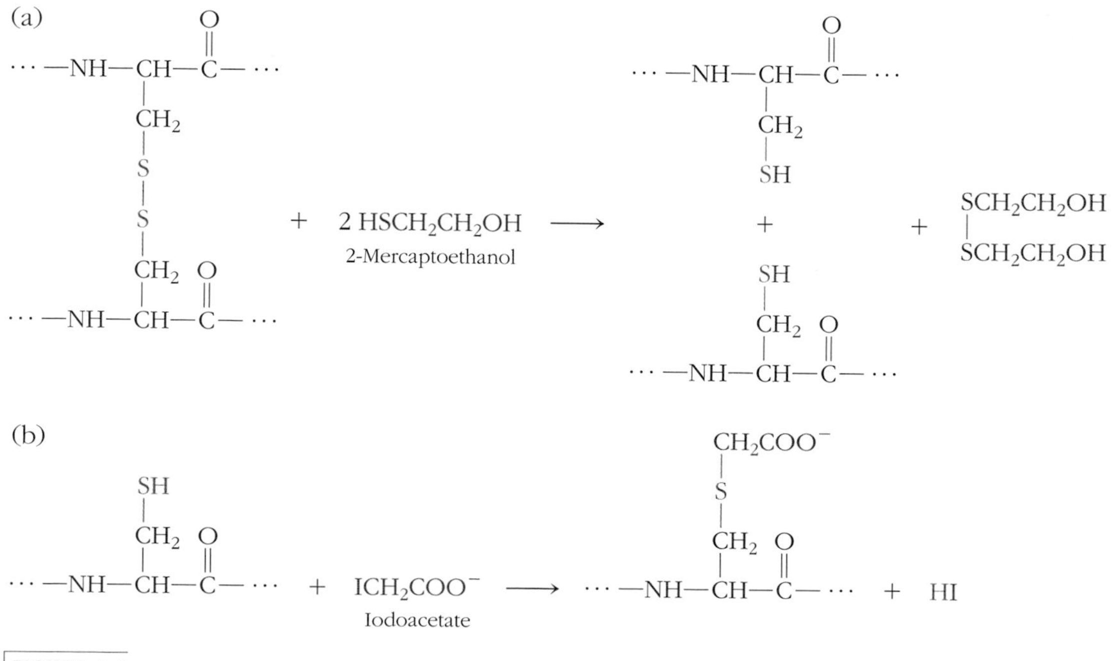

(a)

$$\cdots\text{—NH—CH(CH}_2\text{—S—S—CH}_2\text{)—C(=O)—}\cdots \text{ (two chains linked by disulfide)} + 2\,HSCH_2CH_2OH \longrightarrow 2\ \cdots\text{—NH—CH(CH}_2\text{SH)—C(=O)—}\cdots + HOCH_2CH_2S\text{—}SCH_2CH_2OH$$

2-Mercaptoethanol

(b)

$$\cdots\text{—NH—CH(CH}_2\text{SH)—C(=O)—}\cdots + ICH_2COO^- \longrightarrow \cdots\text{—NH—CH(CH}_2\text{—S—CH}_2COO^-\text{)—C(=O)—}\cdots + HI$$

Iodoacetate

FIGURE 4-4

Reduction and alkylation of a protein containing a disulfide bond. (a) A reducing agent such as 2-mercaptoethanol is used to break the disulfide bonds linking two polypeptide chains. (b) Each of the resulting free Cys residues is then "blocked" by treatment with an alkylating agent such as iodoacetate.

TABLE 4-3 Specificities of some proteases

Protease	Residue preceding cleaved peptide bond[a]
Chymotrypsin	Phe, Trp, Tyr
Elastase	Ala, Gly, Ser, Val
Thermolysin	Ile, Met, Phe, Trp, Tyr, Val
Trypsin	Arg, Lys

[a]Cleavage does not occur if the following residue is Pro.

tide with cyanogen bromide (CNBr), which cleaves the peptide bond on the C-terminal side of Met residues. Cleavage can also be accomplished with **proteases,** enzymes that hydrolyze specific peptide bonds. For example, the protease trypsin cleaves the peptide bond on the C-terminal side of the positively charged residues Arg and Lys:

$$\cdots\text{—NH—}\underset{\substack{|\\ CH_2CH_2CH_2CH_2NH_3^+}}{\text{CH}}\text{—}\overset{O}{\overset{\|}{\text{C}}}\text{—NH—}\overset{R}{\text{CH}}\text{—}\overset{O}{\overset{\|}{\text{C}}}\text{—}\cdots \xrightarrow[\text{trypsin}]{H_2O} \cdots\text{—NH—}\underset{\substack{|\\ CH_2CH_2CH_2CH_2NH_3^+}}{\text{CH}}\text{—}\overset{O}{\overset{\|}{\text{C}}}\text{—}O^- + H_3\overset{+}{N}\text{—}\overset{R}{\text{CH}}\text{—}\overset{O}{\overset{\|}{\text{C}}}\text{—}\cdots$$

Lys (or Arg)

Some commonly used proteases and their preferred cleavage sites are listed in Table 4-3.

4. Each peptide is sequenced using a procedure devised by Pehr Edman. *In the* ***Edman degradation*** *procedure, the N-terminal residue of a peptide is derivatized and cleaved off, leaving a peptide one residue shorter, which is then subject to another round of derivatization and cleavage* (Fig. 4-5). As each N-terminal residue is released, it is identified, and the sequence of the peptide is deduced from the order of appearance of the derivatized amino acids.

5. To reconstruct the sequence of the intact polypeptide, *a different set of fragments that overlaps the first is generated,* so that the two sets of sequenced fragments can be aligned.

Set 1 (cleaved with trypsin): Val—Leu—Lys ↓ Ser—Phe—Gly—Arg ↓ Tyr—Ala—Gln—Thr

Set 2 (cleaved with chymotrypsin): Val—Leu—Lys—Ser—Phe ↑ Gly—Arg—Tyr ↑ Ala—Gln—Thr

REVIEW LEARNING OBJECTIVE 2

- Describe the steps required to sequence a protein.

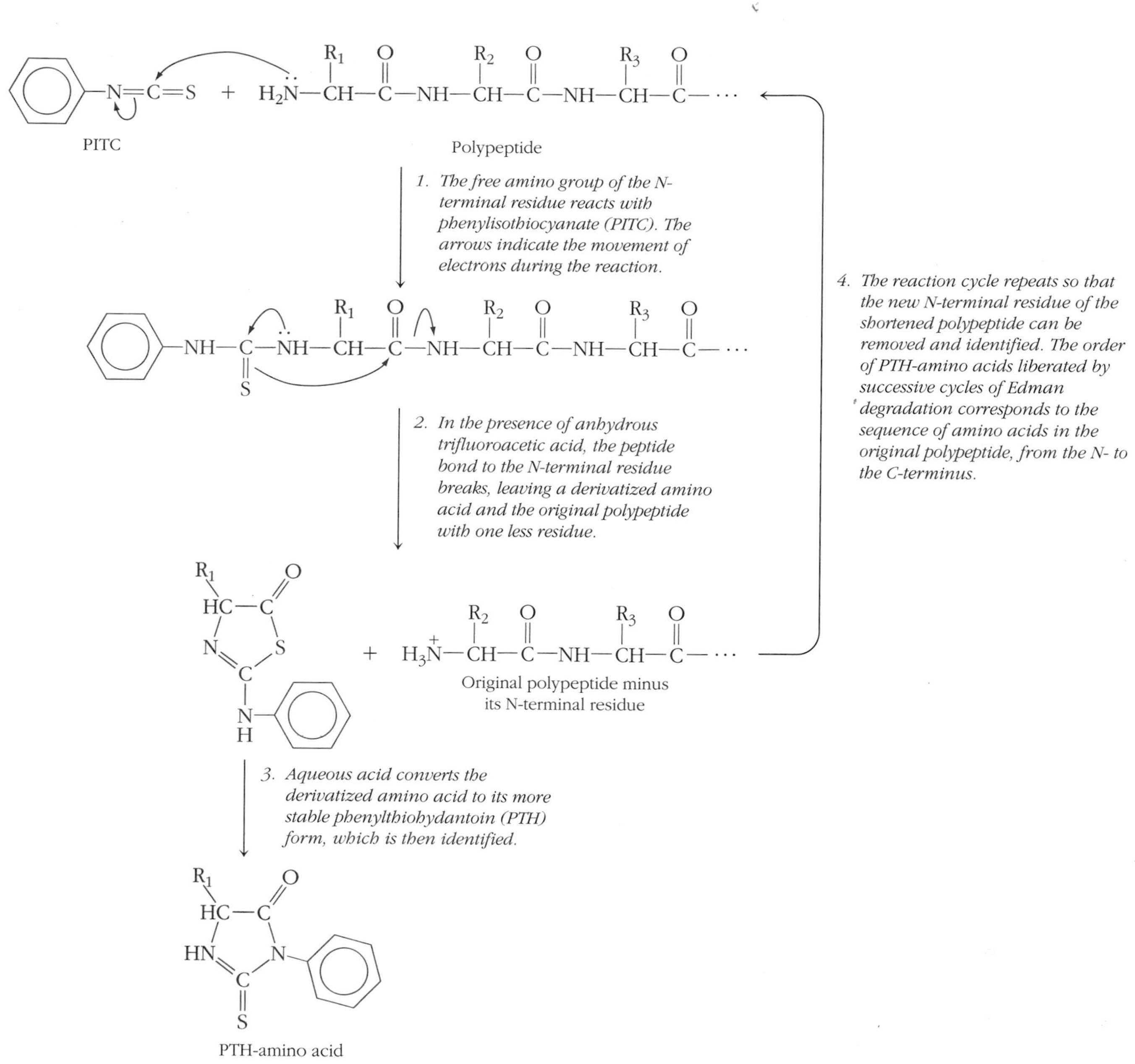

PITC

Polypeptide

1. The free amino group of the N-terminal residue reacts with phenylisothiocyanate (PITC). The arrows indicate the movement of electrons during the reaction.

2. In the presence of anhydrous trifluoroacetic acid, the peptide bond to the N-terminal residue breaks, leaving a derivatized amino acid and the original polypeptide with one less residue.

Original polypeptide minus its N-terminal residue

3. Aqueous acid converts the derivatized amino acid to its more stable phenylthiohydantoin (PTH) form, which is then identified.

PTH-amino acid

4. The reaction cycle repeats so that the new N-terminal residue of the shortened polypeptide can be removed and identified. The order of PTH-amino acids liberated by successive cycles of Edman degradation corresponds to the sequence of amino acids in the original polypeptide, from the N- to the C-terminus.

FIGURE 4-5

Protein sequencing by Edman degradation.

The amino acid sequence is the first level of protein structure

The sequence of amino acids in a polypeptide is called the protein's **primary structure.** There are as many as four levels of structure in a protein such as hemoglobin (Fig. 4-6). Under physiological conditions, a polypeptide very seldom assumes a linear extended conformation but instead folds up to form a more compact shape, usually consisting of several layers. The local folding arrangement of the polypeptide backbone (exclusive of the side chains) is known as **secondary structure.** The complete three-dimensional conformation of the polypeptide, including its backbone atoms and all its side chains, is the polypeptide's **tertiary structure.** In a protein such as hemoglobin, which consists of more than one polypeptide chain, the **quaternary structure** refers to the spatial arrangement of all the chains. In the following sections

Primary structure
The sequence of amino acid residues

Secondary structure
The localized conformation of the polypeptide backbone

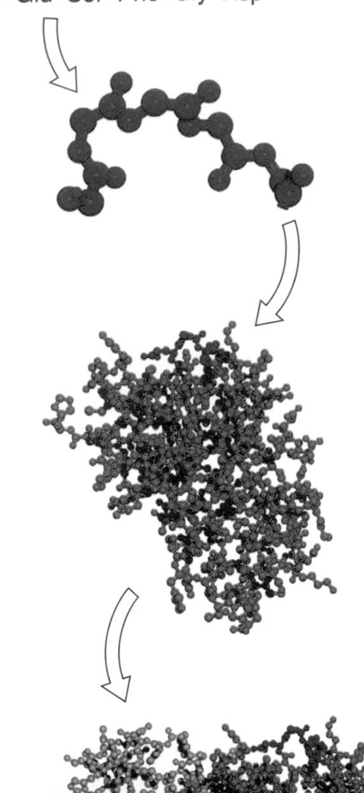

Tertiary structure
The three-dimensional structure of an entire polypeptide, including all its side chains

Quaternary structure
The spatial arrangement of polypeptide chains in a protein with multiple subunits

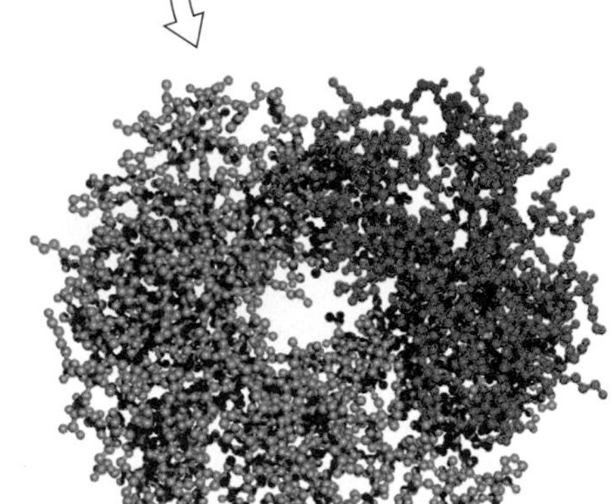

FIGURE 4-6
Levels of protein structure in hemoglobin. [*Structure of human hemoglobin (pdb 2HHB) determined by G. Fermi and M.F. Perutz.*]

we will consider the second, third, and fourth levels of protein structure.  Exercise 6 also illustrates the different levels of protein structure using hemoglobin as an example.

REVIEW LEARNING OBJECTIVE 3

- Describe the four levels of protein structure.

 Go to **Exercise 5/Protein Secondary Structure** for a detailed examination of the α helix and β sheet and their combinations. The structures shown in Exercise 5 and the questions based on them illuminate Learning Objective 5.

2 SECONDARY STRUCTURE: THE CONFORMATION OF THE PEPTIDE GROUP

In the peptide bond that links successive amino acids in a polypeptide chain, the electrons are somewhat delocalized so that the peptide bond has two resonance forms:

Due to this partial (about 40%) double-bond character, there is no rotation around the C—N bond. In a polypeptide backbone, the repeating N—Cα—C units of the amino acid residues therefore can be considered to be a series of planar peptide groups (where each plane contains the atoms involved in the peptide bond):

Here the H atom and R group attached to Cα are not shown.

The polypeptide backbone can still rotate around the N—Cα and Cα—C bonds, although rotation is somewhat limited. For example, a sharp bend at Cα could bring the carbonyl oxygens of neighboring residues too close:

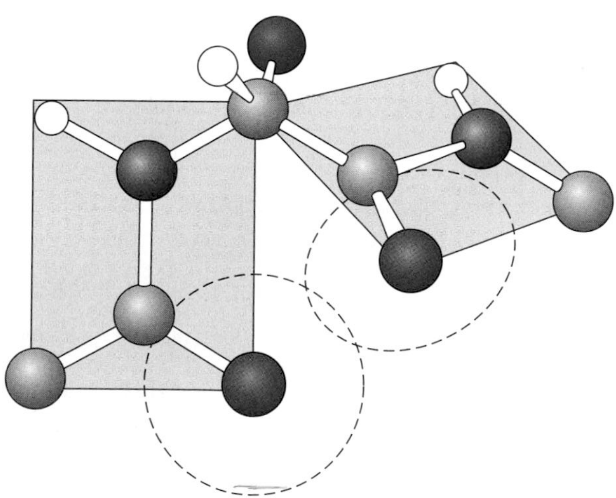

Here the atoms are color-coded: C gray, O red, N blue, and H white.

As the resonance structures above indicate, the atoms involved in the peptide bond are strongly polar, with a tendency to form hydrogen bonds. The backbone amide groups are hydrogen bond donors, and the carbonyl oxygens are hydrogen bond acceptors. *Under physiological conditions, the polypeptide chain folds so that it can satisfy as many of these hydrogen-bonding requirements as possible.* At the same time, *the polypeptide backbone must adopt a conformation (a secondary structure) that minimizes steric strain.* Two kinds of secondary structure commonly found in proteins meet these criteria: the α helix and the β sheet.

REVIEW LEARNING OBJECTIVE 4

- Which amino acid atoms constitute the polypeptide backbone?
- What limits the rotation of the backbone?

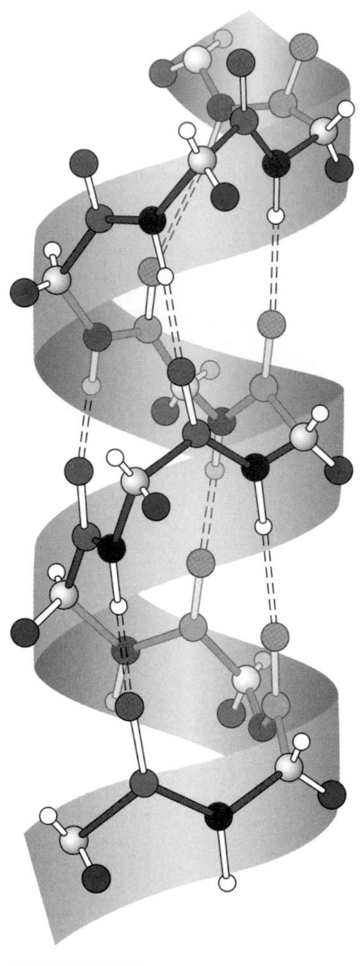

FIGURE 4-7

The α helix.
In this conformation, the polypeptide backbone twists in a right-handed fashion so that hydrogen bonds (dashed lines) form between C=O and N—H groups four residues farther along. Atoms are color-coded: Cα light gray, carbonyl C dark gray, O red, N blue, side chain purple, H white. [*Based on a drawing by Irving Geis.*]

The α helix

The **α helix** was first identified through model-building studies carried out by Linus Pauling. In this type of secondary structure, the polypeptide backbone twists in a right-handed helix (the DNA helix is also right-handed; see Section 3-1 for an explanation). There are 3.6 residues per turn of the helix, and for every turn, the helix rises 5.4 Å along its axis. In the α helix, the carbonyl oxygen of each residue forms a hydrogen bond with the backbone NH group four residues ahead. The backbone hydrogen-bonding tendencies are thereby met, except for the four residues at each end of the helix (Fig. 4-7). Most α helices are about 12 residues long.

Like the DNA helix, whose side chains fill the helix interior (see Fig. 3-4d), the α helix is solid—the atoms of the polypeptide backbone are in van der Waals contact. However, in the α helix, the side chains extend outward from the helix (Fig. 4-8). Exercise 5 shows the backbone conformation, hydrogen-bonding pattern, and side-chain dispositions of a segment of α helix.

The β sheet

Pauling, along with Robert Corey, also built models of the **β sheet.** This type of secondary structure consists of aligned strands of polypeptide whose hydrogen-bonding requirements are met by bonding between neighboring strands. The strands of a β sheet can be arranged in two ways (Fig. 4-9): In a **parallel β sheet,** neighboring chains run in the same direction; in an **antiparallel β sheet,** neighboring chains run in opposite directions. Each residue forms two hydrogen bonds with a neighboring strand, so all hydrogen-bonding requirements are met, except in the first and last strands of the sheet.

A single β sheet may contain from 2 to more than 12 polypeptide strands, with an average of 6 strands, and each strand has an average length of 6 residues. In a β sheet, the amino acid side chains extend from both faces (Fig. 4-10). Exercise 5 shows the backbone conformations, hydrogen-bonding patterns, and side-chain dispositions of parallel and antiparallel β sheets.

Proteins also contain irregular secondary structure

α Helices and β sheets are classified as **regular secondary structure,** because *their component residues exhibit backbone conformations that are the same*

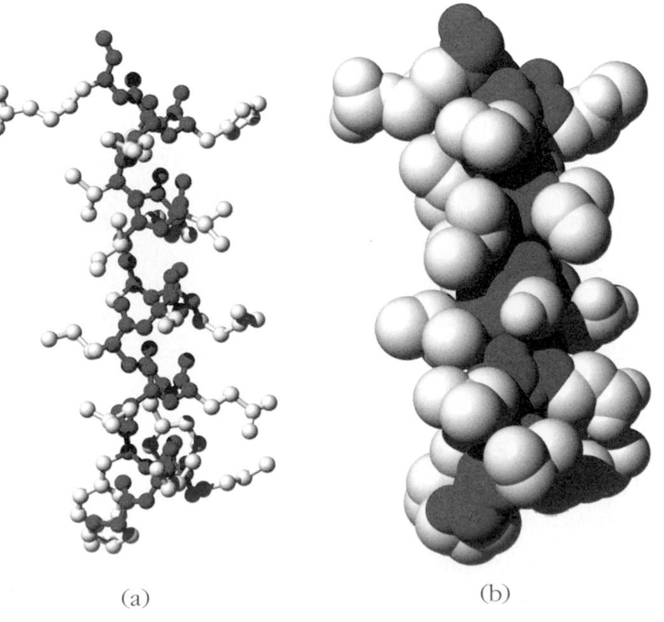

FIGURE 4-8

An α helix from myoglobin.
In (a) a ball-and-stick model and (b) a space-filling model of residues 100–118 of myoglobin, the backbone atoms are green and the side chains are gray. [*Structure of sperm whale myoglobin (pdb 1MBD) determined by S.E.V. Phillips.*]

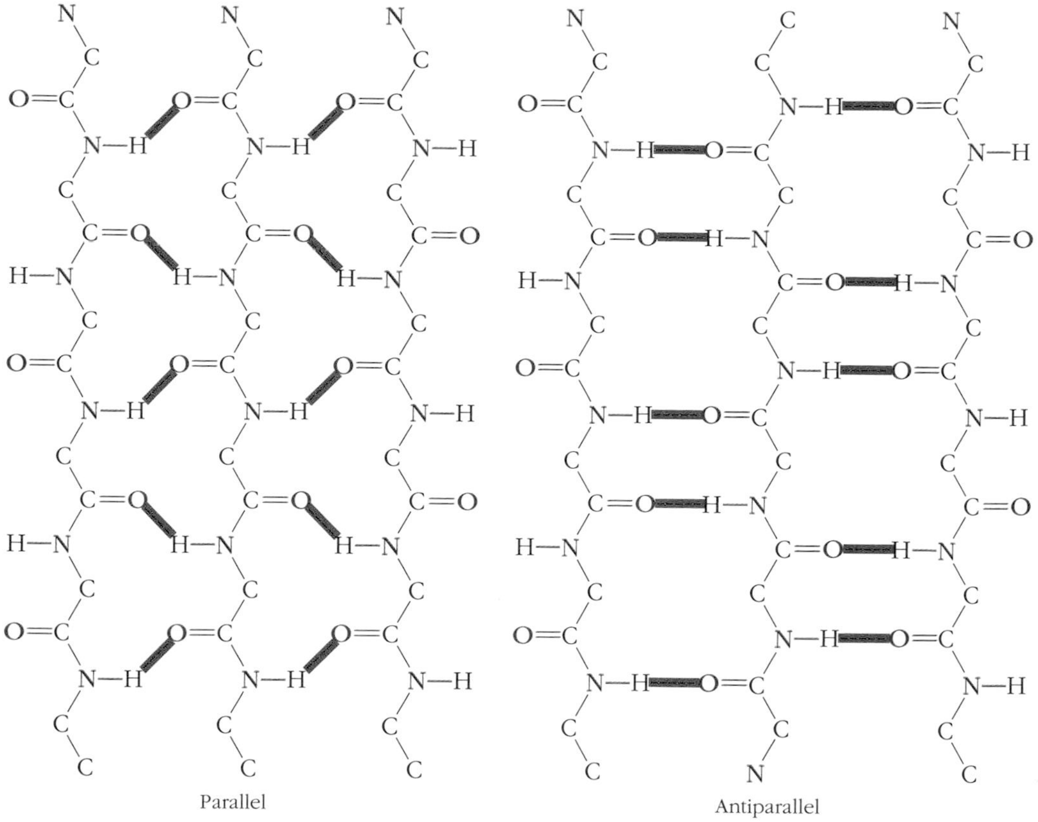

Parallel

Antiparallel

FIGURE 4-9

β Sheets.
In a parallel β sheet and an antiparallel β sheet, the polypeptide backbone is extended. In both types of β sheet, hydrogen bonds form between the amide and carbonyl groups of adjacent strands. The H and R attached to Cα are not shown.

from one residue to the next. In fact, these elements of secondary structure are easily recognized in the three-dimensional structures of a huge variety of proteins, regardless of their amino acid composition. Of course, depending on the identities of the side chains and other groups that might be present, α helices and β sheets may be slightly distorted from their ideal conformations. For example, the final turn of some α helices becomes "stretched out" (longer and thinner) than the rest of the helix.

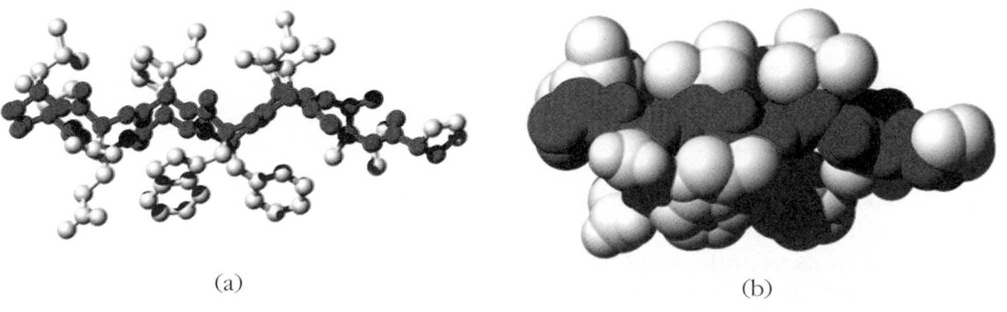

FIGURE 4-10

Side view of two parallel strands of a β sheet from carboxypeptidase A.
In (a) a ball-and-stick model and (b) a space-filling model, the backbone atoms are green. The amino acid side chains (gray) point alternately to each side of the β sheet. [*Structure of carboxypeptidase (pdb 3CPA) determined by W.N. Lipscomb.*]

In every protein, elements of secondary structure (individual α helices or strands in a β sheet) are linked together by polypeptide ***loops*** *of various sizes.* A loop may be a relatively simple hairpin turn, as in the connection of two antiparallel β strands (which are shown here as flat arrows):

Or it may be quite long, especially if it joins successive strands in a parallel β sheet:

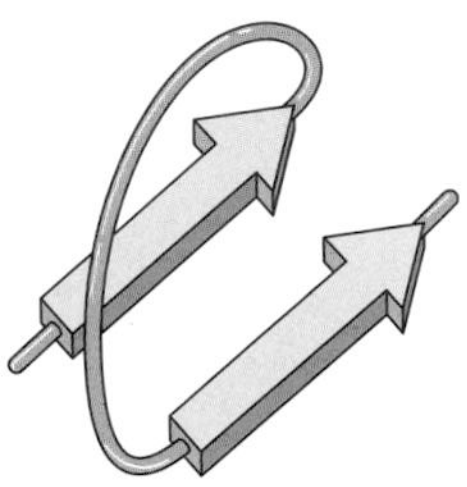

Usually, the loops that link β strands or α helices consist of residues with **irregular secondary structure;** that is, the polypeptide does not adopt a defined secondary structure in which successive residues have the same backbone conformation. Note that "irregular" does not mean "disordered": The peptide backbone almost always adopts a single, unique conformation.

Most proteins contain a combination of regular and irregular secondary structure. On average, 31% of residues are in α helices, and 23% are in β sheets, and much of the remainder is in irregular loops of different sizes.  Exercise 5 includes examples of proteins that contain a combination of α helices and β sheets.

REVIEW LEARNING OBJECTIVE 5

- Summarize the structural features of α helices and β sheets.
- Distinguish regular and irregular secondary structure.
- Why does virtually every protein contain loops?

3 TERTIARY STRUCTURE AND PROTEIN STABILITY

 Go to **Exercise 6/Myoglobin and Hemoglobin** for a detailed look at the tertiary structure of myoglobin. The structures presented in Exercise 6 and the questions based on those structures address Learning Objectives 6 and 8.

The three-dimensional shape of a protein, known as its tertiary structure, includes its regular and irregular secondary structure (that is, the overall folding of its peptide backbone) as well as the spatial arrangement of all its side chains. In a fully folded protein under physiological conditions, virtually every atom has a designated place.

One of the most powerful techniques for probing the atomic structures of macromolecules, including proteins, is **X-ray crystallography** (Box 4-B). The structure of myoglobin, the first protein structure to be determined by X-ray crystallography, came to light in 1958 through the efforts of John

A CLOSER LOOK

BOX 4-B X-Ray crystallography

Most proteins are too small to be directly visualized, even by electron microscopy, but their atomic structures are accessible to high-energy probes in the form of X-rays. X-Ray crystallography is performed on samples of protein that have been induced to form crystals. A protein preparation must be exceptionally pure in order to crystallize without imperfections. A protein crystal, often no more than 0.5 mm in diameter, usually contains 40–70% water by volume and is therefore more gel-like than solid.

A crystal of streptavidin.

When bombarded with a narrow beam of X-rays, the electrons of the atoms in the crystal scatter the X-rays, which reinforce and cancel each other to produce a ***diffraction pattern*** of light and dark spots. The pattern can be captured on a piece of X-ray film or recorded by computer (as shown here).

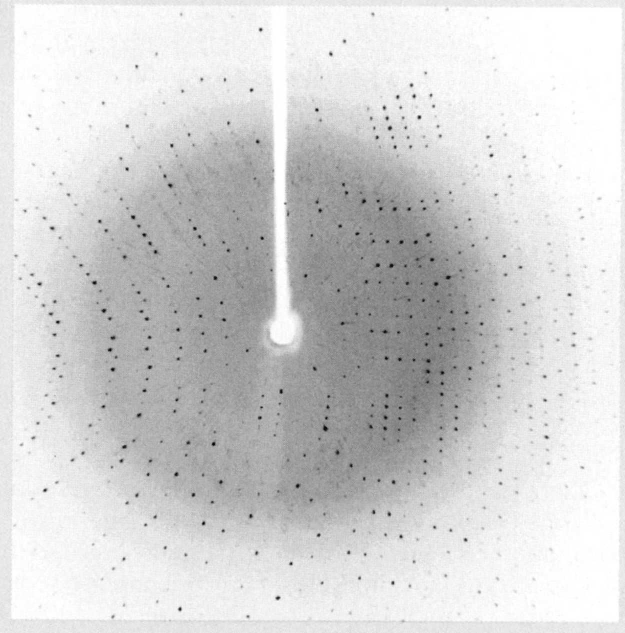

Mathematical analysis of the intensities of the diffracted X-rays yields a three-dimensional map of electron density in the crystallized molecule. The level of detail of the image depends in part on the quality of the crystal. Slight conformational variations among the crystallized protein molecules often limit resolution to about 2 Å. However, this is usually sufficient to trace the polypeptide backbone and discern the general shapes of the side chains.

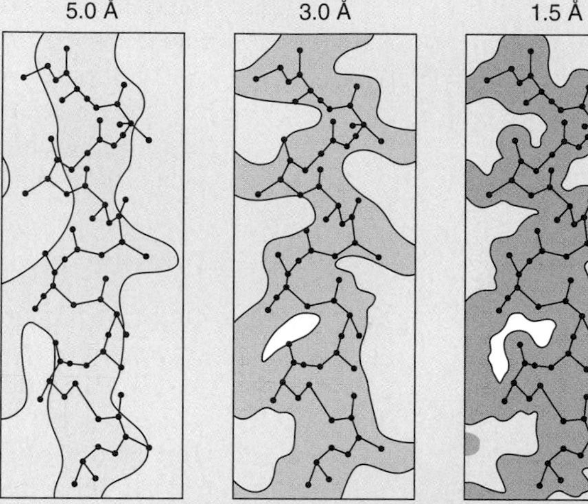

In the example shown here, the gray areas represent electron density, and the protein structure is superimposed in red. As the resolution improves, it becomes easier to trace the pattern of the protein backbone (a portion of an α helix).

Are X-ray structures accurate? One might expect a crystallized molecule to be utterly immobile, quite different from a protein in solution, which undergoes bending and stretching movements among its many bonds. But in fact, crystallized proteins retain some of their ability to move. They can sometimes bind small molecules that diffuse into the crystal (which is about half water) and can sometimes mediate chemical reactions. These activities would not be possible if the structure of the crystallized protein did not closely approximate the structure of the protein in solution. Finally, nuclear magnetic resonance (NMR) methods for determining the structures of small proteins in solution appear to yield results consistent with X-ray crystallographic data.

[Photos courtesy Isolde Le Trong, David Teller, and Ronald Stenkamp, University of Washington. Drawing based on one by Wayne Hendrickson.]

(a)

(b)

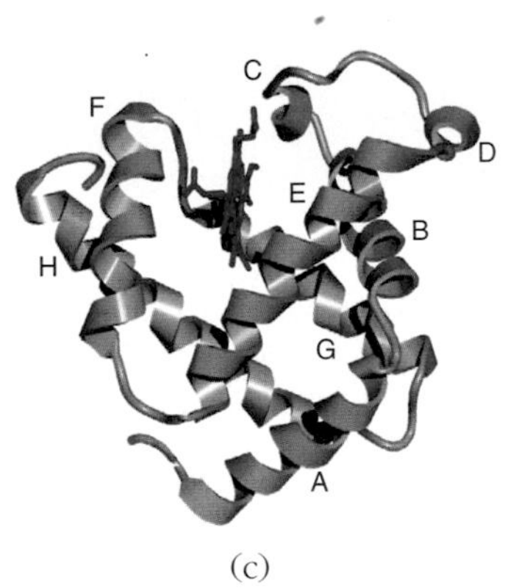

(c)

FIGURE 4-11

Models of myoglobin structure. (a) Space-filling model. All atoms (except H) are shown (C gray, O red, N blue). The heme group, where oxygen binds, is purple. (b) Polypeptide backbone. The trace connects the α carbons of successive amino acid residues. (c) Ribbon diagram. The ribbon represents the overall conformation of the backbone. The eight α helices are labeled A–H.

Kendrew, who painstakingly determined the conformation of every backbone and side-chain group. At that time, the amino acid sequence of myoglobin was not known. Today, protein crystallographers usually take advantage of amino acid sequence information to simplify the task of elucidating protein structures. Kendrew's results—coming just a few years after Watson and Crick had published their elegant model of DNA—were a bit of a disappointment. The myoglobin structure lacked the simplicity and symmetry of a molecule such as DNA and was more irregular and complex than expected.

The structure Kendrew described is a compact and rounded mass about $44 \times 44 \times 25$ Å (Fig. 4-11a). Myoglobin is said to be a **globular** protein (**fibrous** proteins, in contrast, are usually highly elongated). Partially buried in the protein mass is an organic group known as **heme;** this is where oxygen binds. Myoglobin's tertiary structure can be simplified by showing just the peptide backbone (Fig. 4-11b). Alternatively, the structure can be represented by a ribbon that passes through Cα of each residue (Fig. 4-11c). This makes it easy to identify elements of secondary structure. Myoglobin is unusual among proteins in that it lacks β sheets and is almost entirely α helical (compare with the proteins shown in Exercise 5). All but 32 of its 153 residues are part of 8 helices, which range in length from 7 to 26 amino acid residues and are labeled A through H. See Exercise 6 for additional views of myoglobin's secondary and tertiary structure.

Proteins have hydrophobic cores

Globular proteins typically contain at least two layers of secondary structure (myoglobin, for example, is two or three helices thick; see Fig. 4-11c). This means that a protein has definite surface and core regions. On the protein's surface, some backbone and side-chain groups are exposed to the solvent, and in the core, these groups are sequestered from the solvent. In other words, *the protein comprises a hydrophilic surface and a hydrophobic core.*

A polypeptide segment that has folded into a single structural unit with a hydrophobic core is often called a **domain.** Some small proteins, such as myoglobin, consist of a single domain. Larger proteins may contain several domains, which may be structurally similar or dissimilar (Fig. 4-12).

The core of a small hydrophobic protein (or a domain) is typically rich in regular secondary structure. This is because the formation of α helices and β sheets, which are internally hydrogen-bonded, minimizes the hydrophobicity of the polar backbone groups. Irregular secondary structure (loops) are more often found on the surface of the protein (or domain), where the polar backbone groups can form hydrogen bonds to water molecules.

The requirement for a hydrophobic core and a hydrophilic surface also places constraints on amino acid sequence. The location of a particular side chain in a protein's tertiary structure is related to its hydrophobicity: *The greater a residue's hydrophobicity, the more likely it is to be located in the protein interior.* In the protein interior, side chains pack together, leaving very little empty space or space that could be occupied by a water molecule.

Table 4-4 lists two scales for assessing the hydrophobicity of amino acid side chains. For example, highly hydrophobic residues such as Phe and Met are almost always buried. Polar side chains, like hydrogen-bonding backbone groups, can participate in hydrogen bonding in the protein interior, which helps "neutralize" their polarity and allows them to be buried in a nonpolar environment. When a charged residue occurs in the protein interior, it is almost always located near another residue with the opposite charge, so that the two groups can interact electrostatically to form an **ion pair.** By color-coding the amino acid residues of myoglobin according to their hydrophobicity, it is easy to see that hydrophobic side chains cluster in the interior

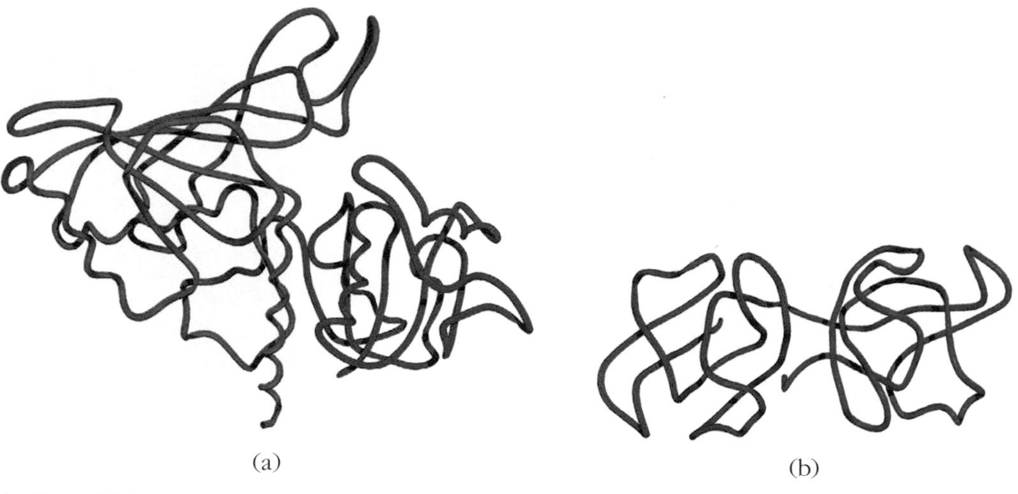

FIGURE 4-12

Examples of two-domain proteins.
(a) Glyceraldehyde-3-phosphate dehydrogenase. The small domain is red, and the large domain is green. (b) B crystallin. The two similar domains are colored red and gold. [*Structure of the dehydrogenase (pdb 1GPD) determined by D. Moras, K.W. Olsen, M.N. Sabesan, M. Buehner, G.C. Ford, and M.G. Rossmann; structure of B crystallin (pdb 4GCR) determined by C. Slingsby, S. Najmudin, V. Nalini, H.P.C. Driessen, T.L. Blundell, D.S. Moss, and P. Lindley.*]

TABLE 4-4 Hydrophobicity scales

Residue	Scale A[a]	Scale B[b]
Phe	2.8	3.7
Met	1.9	3.4
Ile	4.5	3.1
Leu	3.8	2.8
Val	4.2	2.6
Cys	2.5	2.0
Trp	−0.9	1.9
Ala	1.8	1.6
Thr	−0.7	1.2
Gly	−0.4	1.0
Ser	−0.8	0.6
Pro	−1.6	−0.2
Tyr	−1.3	−0.7
His	−3.2	−3.0
Gln	−3.5	−4.1
Asn	−3.5	−4.8
Glu	−3.5	−8.2
Lys	−3.9	−8.8
Asp	−3.5	−9.2
Arg	−4.5	−12.3

[a]Scale A is from Kyte, J. and Doolittle, R.F., *J. Mol. Biol.* **157,** 105–132 (1982).
[b]Scale B is from Engelman, D.M., Steitz, T.A., and Goldman, A., *Annu. Rev. Biophys. Biophys. Chem.* **15,** 321–353 (1986).

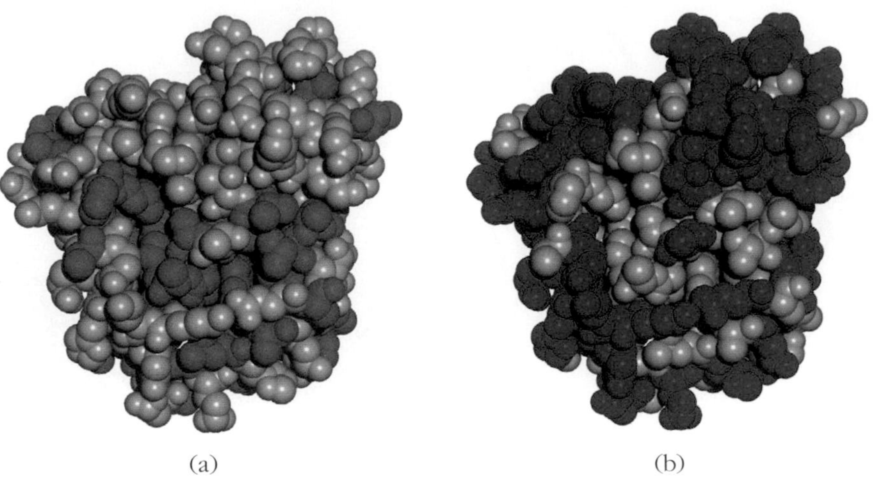

FIGURE 4-13

Hydrophobic and hydrophilic residues in myoglobin. (a) All hydrophobic side chains (Ala, Ile, Leu, Met, Phe, Pro, Trp, and Val) are colored green. These residues are located mostly in the protein interior. (b) All polar and charged side chains are colored red. These are located primarily on the protein surface.

while hydrophilic side chains predominate on the surface (Fig. 4-13).  Exercise 6 also illustrates the distribution of side chains in myoglobin.

REVIEW LEARNING OBJECTIVE 6

- How do the surface and core of a protein differ?
- Which residues are likely to be found in each region?

How are proteins stabilized?

Surprisingly, the fully folded conformation of a protein is only marginally more stable than its unfolded form. The difference in thermodynamic stability amounts to about 0.4 kJ · mol^{-1} per amino acid, or about 40 kJ · mol^{-1} for a 100-reside polypeptide. This is equivalent to the amount of free energy required to break two hydrogen bonds (about 20 kJ · mol^{-1} each). This quantity seems incredibly small, considering the number of potential interactions among all of a protein's backbone and side-chain atoms. Nevertheless, each protein does fold into a unique and stable three-dimensional arrangement of atoms.

The largest force governing protein structure is the hydrophobic effect (introduced in Section 2-2), which causes nonpolar groups to aggregate in order to minimize their contact with water. The hydrophobic effect is driven by the increase in entropy of the solvent water molecules, which would otherwise have to order themselves around each hydrophobic group. As we have seen, hydrophobic side chains are located predominantly in the interior of a protein. This arrangement stabilizes the folded polypeptide backbone, since unfolding it or extending it would expose the hydrophobic side chains to the solvent (Fig. 4-14).

(a) Folded

(b) Unfolded

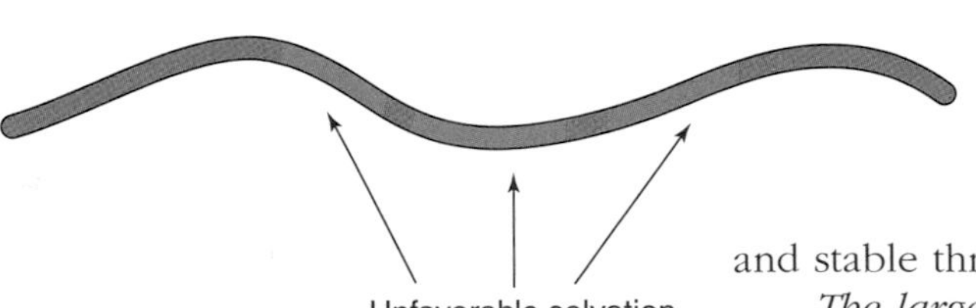

FIGURE 4-14

The hydrophobic effect in protein folding. (a) In a folded protein, hydrophobic regions (represented by green segments of the polypeptide chain) are sequestered in the protein interior. (b) Unfolding the protein exposes these segments to water. This is energetically unfavorable, because the presence of the hydrophobic groups interrupts the hydrogen-bonded network of water molecules.

Hydrogen bonding by itself is not a major determinant of protein stability, because in an unfolded protein, polar groups could just as easily form energetically equivalent hydrogen bonds with water molecules. Instead, hydrogen bonding may help the protein fine-tune a folded conformation that is already largely stabilized by the hydrophobic effect.

Do cross-links stabilize proteins?

Many folded polypeptides appear to be held in place by cross-links of various kinds, the most common being ion pairs, disulfide bonds, and inorganic ions such as zinc. Do these cross-links help stabilize the protein?

An ion pair can form between oppositely charged side chains or the N- and C-terminal groups (Fig. 4-15). Although the resulting electrostatic interaction is strong, it does not contribute much to protein stability. This is because the favorable free energy of the electrostatic interaction is offset by the loss of entropy when the side chains become fixed in the ion pair. For a buried ion pair, there is the additional energetic cost of desolvating the charged groups in order for them to enter the hydrophobic core.

Disulfide bonds (shown on page 97) can form within and between polypeptide chains. Experiments show that when the Cys residues of certain proteins are blocked with alkyl groups (see Fig. 4-4), the proteins may still fold and function normally. This suggests that disulfide bonds are not essential for stabilizing these proteins. In fact, disulfides are rare in intracellular proteins, since the cytoplasm is a reducing environment. They are more plentiful in proteins that are secreted to an extracellular (oxidizing) environment (Fig. 4-16). Here, the bonds may help prevent protein unfolding under relatively harsh extracellular conditions.

Domains containing cross-links called **zinc fingers** are common in DNA-binding proteins. These structures consist of 20–60 residues with one or two Zn^{2+} ions. The Zn^{2+} ions are tetrahedrally coordinated by the side chains of Cys and/or His and sometimes Asp or Glu (Fig. 4-17). Protein domains this

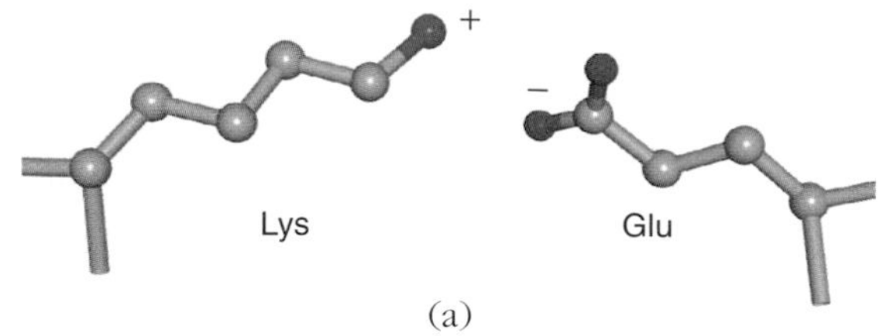

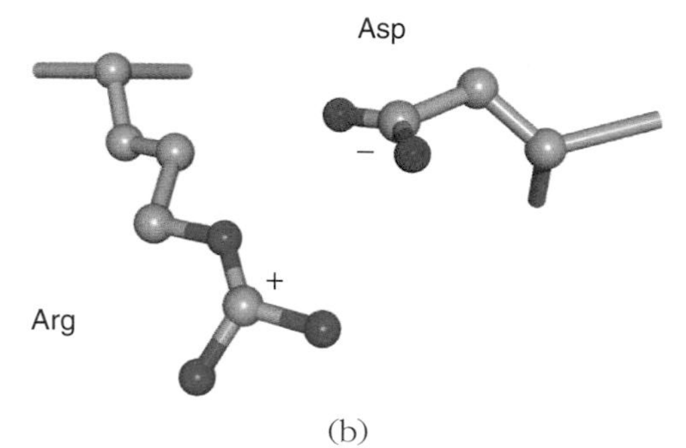

FIGURE 4-15

Examples of ion pairs in myoglobin. (a) The ε-amino group of Lys 77 interacts with the carboxylate group of Glu 18. (b) The carboxylate group of Asp 60 interacts with Arg 45. The atoms are color-coded: C gray, N blue, and O red.

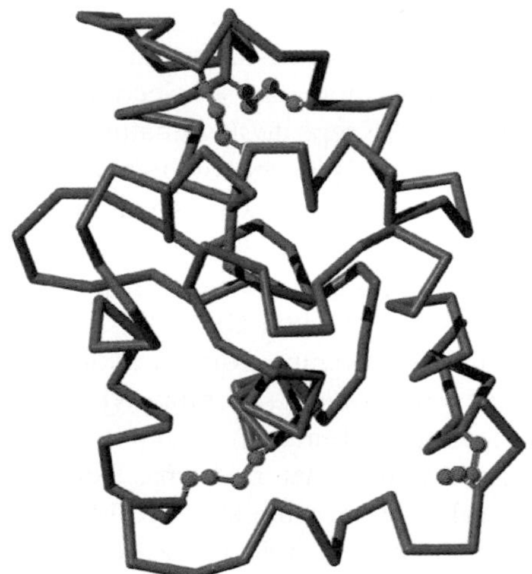

FIGURE 4-16

Disulfide bonds in lysozyme, an extracellular protein. This 129-residue enzyme from hen egg white contains 8 Cys residues (yellow), which form 4 disulfide bonds that link different sites on the polypeptide backbone. [*Structure (pdb 1E8L) determined by H. Schwalbe, S.B. Grimshaw, A. Spencer, M. Buck, J. Boyd, C.M. Dobson, C. Redfield, and L.J. Smith.*]

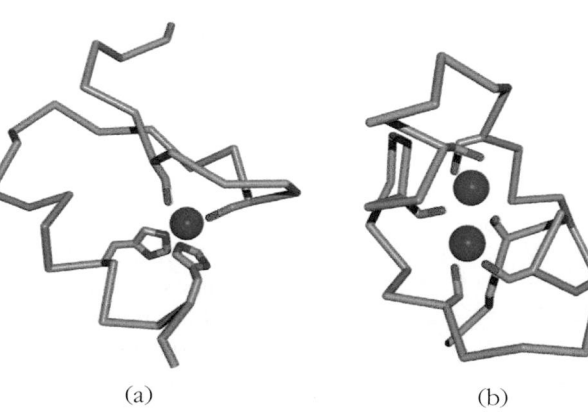

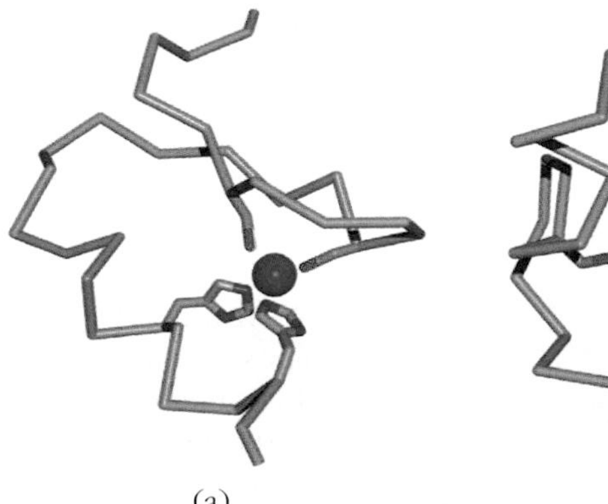

FIGURE 4-17

Zinc fingers. (a) A zinc finger with one Zn^{2+} (purple sphere) coordinated by 2 Cys and 2 His residues, from *Xenopus* transcription factor IIIA. (b) A zinc finger with two Zn^{2+} coordinated by 6 Cys residues, from the yeast transcription factor GAL4. [*Structure of transcription factor IIIA (pdb 1TF6) determined by R.T. Nolte, R.M. Conlin, S.C. Harrison, and R.S. Brown; structure of GAL4 (pdb 1D66) determined by R. Marmorstein and S. Harrison.*]

size are too small to assume a stable tertiary structure without a metal ion cross-link. Zinc is an ideal ion for stabilizing proteins: It can interact with ligands (S, N, or O) provided by several amino acids, and it has only one oxidation state (unlike Cu or Fe ions, which readily undergo oxidation–reduction reactions under cellular conditions).

Protein folding

In the cell, a newly synthesized polypeptide begins to fold as soon as it emerges from the ribosome, so part of the polypeptide may adopt its mature tertiary structure before the entire chain has been synthesized. It is difficult to monitor this process in the cell, so studies of protein folding *in vitro* usually use full-length polypeptides that have been chemically unfolded (**denatured**) and then allowed to refold (**renature**). In the laboratory, proteins can be denatured by adding highly soluble substances such as salts or urea (NH_2—CO—NH_2). Large amounts of these solutes interfere with the structure of the solvent water, thereby attenuating the hydrophobic effect and causing the protein to unfold. When the solutes are removed, the proteins renature.

Protein renaturation experiments demonstrate that *protein folding is not a random process.* That is, the protein does not just happen upon its most stable tertiary structure (its **native structure**) by trial and error, but approaches this conformation through one or a few alternative pathways. During this process, small elements of secondary structure form first, then these coalesce under the influence of the hydrophobic effect to produce a mass with a hydrophobic core. Finally, small rearrangements yield the native tertiary structure (Fig. 4-18). A protein's native structure is not necessarily rigid and inflexible. In fact, some minor movement, primarily the result of bending and stretching of individual bonds, is required for most proteins to carry out their biological functions.

All the information required for a protein to fold is contained in its amino acid sequence. Unfortunately, there are no completely reliable methods for predicting how a polypeptide chain will fold. In fact, it is difficult to determine whether a given amino acid sequence will form an α helix, β sheet, or irregular secondary structure. This presents a formidable obstacle to assigning three-dimensional structures—and possible functions—to the burgeoning number of proteins identified through genome sequencing (see Section 3-5).

In the laboratory, certain small proteins can be repeatedly denatured and renatured, but in the cell, protein folding is more complicated and may require the assistance of other proteins. Some of these are known as **molecular chaperones** and are described in more detail in Section 20-4. Proteins that do not fold properly—often due to genetic mutations that substitute one amino acid for another—can lead to a variety of diseases (Box 4-C).

For some proteins, the pathway to full functionality requires additional steps beyond polypeptide folding. Some proteins contain several polypeptide chains, which must fold individually before assembling. In addition, many proteins undergo **processing** before reaching their mature forms. Depending on the protein, this might mean removal of some amino acid residues or the covalent attachment of another group such as a lipid or carbohydrate

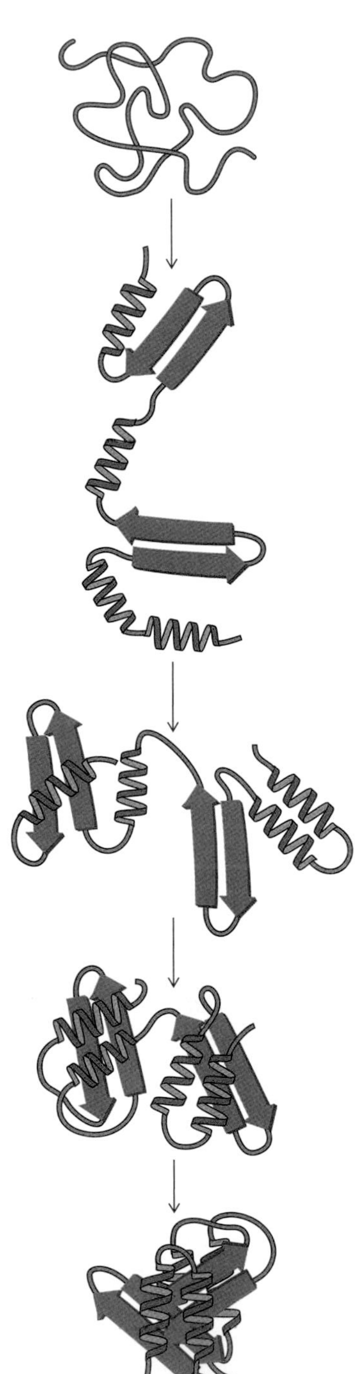

FIGURE 4-18

Model of protein folding.
In this hypothetical example, the polypeptide first forms elements of secondary structure (α helices and β sheets). These coalesce into a single globular mass, and small adjustments generate the final stable tertiary structure. [*After Goldberg, M.E., Trends Biochem. Sci.* ***10,*** *389 (1985).*]

A CLOSER LOOK

BOX 4-C Protein misfolding and disease

What happens to proteins that fold improperly? The chaperones that help new proteins to fold can also help misfolded proteins to refold. If the protein cannot be salvaged in this way, it is usually degraded to its component amino acids.

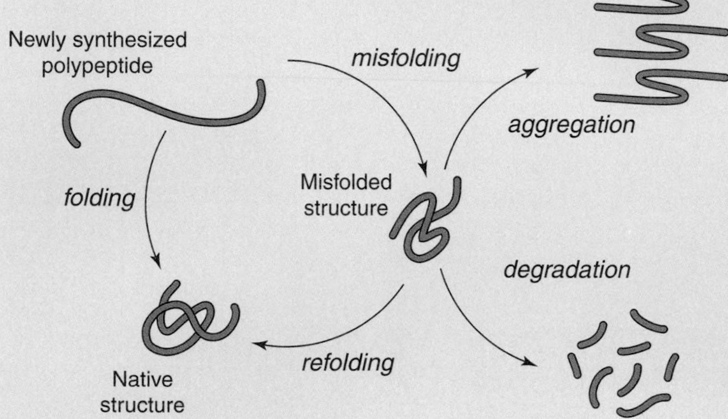

The operation of this quality control system explains what happens in the most common form of cystic fibrosis (see Section 3-3): A mutated form of the protein folds incorrectly and therefore never reaches its intended cellular destination.

Other diseases may result when misfolded proteins are not immediately refolded or degraded but instead aggregate, often as long insoluble fibers (the misfolded proteins tend to aggregate because their hydrophobic regions are not properly tucked away in the protein core). Among the human diseases characterized by such fibrous deposits (also called **amyloid deposits**) are Alzheimer's disease, Parkinson's disease, and spongiform encephalopathies (including "mad cow disease"). These disorders are not necessarily caused by a mutated protein that is prone to misfold and aggregate; in some cases, a protein that is normally stable as a solitary unit can be induced to unfold and refold so as to aggregate.

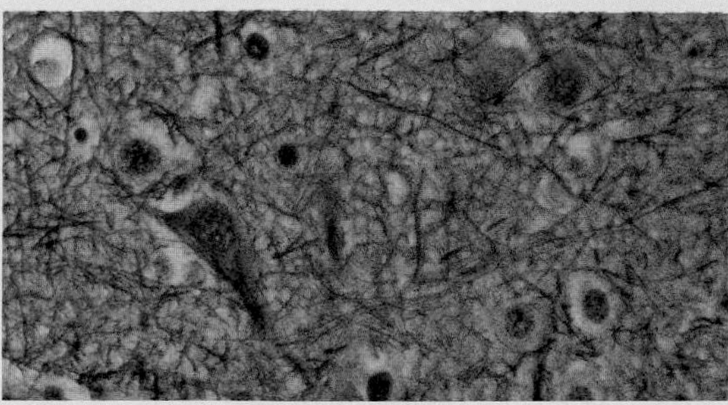

Fibrous deposits in the brain of an individual with Alzheimer's disease.

Remarkably, fibers of insoluble proteins are often structurally similar, with a preponderance of β structure. Even myoglobin, which consists almost entirely of α helices in its native state, can be induced to adopt a β-rich fibrillar form. In such fibers, the polypeptide chains form β sheets that align side-to-side so that the β strands are perpendicular to the long axis of the fiber. Experimental evidence suggests that *virtually any polypeptide chain can be induced to form an amyloid fiber.* Once formed, the fibers resist degradation by cellular proteases, possibly due to the large number of intermolecular hydrogen bonds among the β sheets. It remains a mystery why a misfolded protein aggregates with others of its kind, and not with any of the thousands of other proteins in the cell.

Although amyloid fibers can occur throughout the body, they appear to be deadliest when they occur in the brain. The deposition of intracellular or extracellular amyloid fibers is strongly correlated with the development of a neurodegenerative disease. However, there is some evidence that the loss of neural cell function begins even before the fibers appear. In fact, experiments with cultured cells suggest that misfolded proteins are most toxic at an early stage of aggregation, whereas the mature fibers are relatively harmless.

[Science Photo Library/Photo Researchers.]

(Fig. 4-19). The attached groups usually have a discrete biological function and may also help stabilize the folded conformation of the protein. Finally, some proteins, such as myoglobin, become functional only after associating with a specific organic molecule.

REVIEW LEARNING OBJECTIVE 7

- What forces stabilize protein structure?
- How does an unfolded polypeptide reach its native conformation?

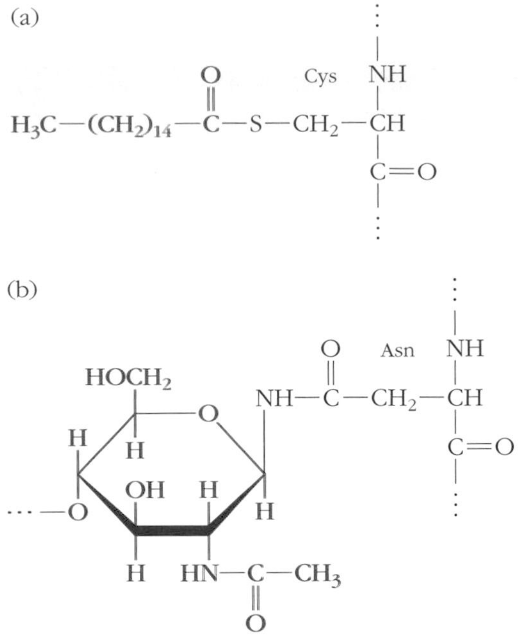

(a)

(b)

FIGURE 4-19

Some covalent modifications of proteins. (a) A 16-carbon fatty acid (palmitate, in red) is linked by a thioester bond to a Cys residue. (b) A chain of several carbohydrate units (here only one sugar residue is shown, in red) is linked to the amide N of an Asn side chain.

4 MYOGLOBIN, AN OXYGEN-BINDING PROTEIN

The fully functional myoglobin molecule contains a polypeptide chain plus the iron-containing porphyrin derivative known as heme:

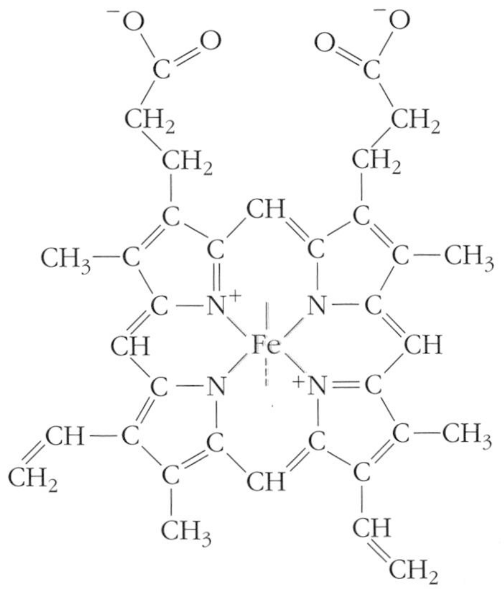

The heme is a type of **prosthetic group,** an organic compound that allows a protein to carry out some function that the polypeptide alone cannot perform—in this case, binding oxygen.

The planar heme is tightly wedged into a hydrophobic pocket between helices E and F of myoglobin (see Fig. 4-11a). It is oriented so that its two nonpolar vinyl ($—CH{=}CH_2$) groups are buried and its two polar propionate ($—CH_2—CH_2—COO^-$) groups are exposed to the solvent. The central iron atom, with six possible coordination bonds, is liganded by four N atoms of the porphyrin ring system. A fifth ligand is provided by a His residue of myoglobin known as His F8 (the eighth residue of helix F). Molecular oxygen (O_2) can bind reversibly to form the sixth coordination bond. (This is what allows certain heme-containing proteins, such as myoglobin and hemoglobin, to function physiologically as oxygen carriers.) Residue His E7 (the seventh residue of helix E) forms a hydrogen bond to the O_2 molecule (Fig. 4-20). By itself, heme is not an effective oxygen carrier because the central Fe(II) (or Fe^{2+}) atom is easily oxidized to the ferric Fe(III) (or Fe^{3+}) state, which cannot bind O_2. Oxidation does not readily take place when the heme is part of a protein such as myoglobin or hemoglobin. Exercise 6 shows the position of the heme group in myoglobin.

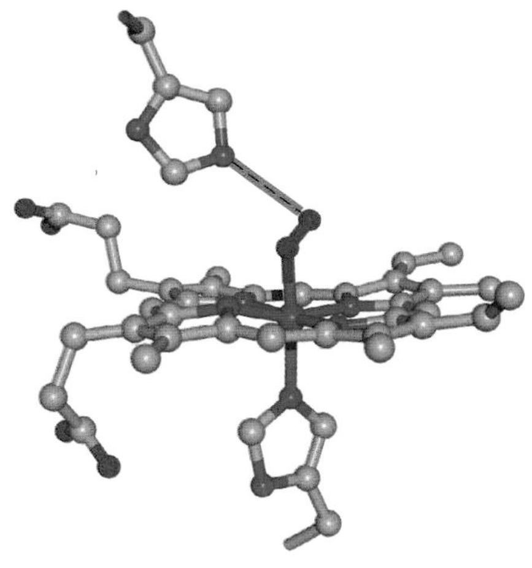

FIGURE 4-20

Oxygen binding to the heme group of myoglobin.
The central Fe(II) atom of the heme group (purple) is liganded to four porphyrin N atoms and to the N of His F8 below the porphyrin plane. O_2 (red) binds reversibly to the sixth coordination site, above the porphyrin plane. Residue His E7 forms a hydrogen bond to O_2. [*Structure of myoglobin (pdb 1MBO) determined by S.E.V. Phillips.*]

Oxygen binding depends on the oxygen concentration

The muscles of diving mammals are especially rich in myoglobin (John Kendrew used sperm whale as his source of the protein). At one time, myoglobin was believed to be an oxygen-storage protein—which would be advantageous during a long dive—but it most likely just facilitates oxygen diffusion through muscle cells.

Myoglobin's O_2-binding behavior can be quantified. To begin, the reversible binding of O_2 to myoglobin (Mb) is described by a simple equilibrium:

$$\text{Mb} + \text{O}_2 \rightleftharpoons \text{MbO}_2$$

with a dissociation constant, K:

$$K = \frac{[\text{Mb}][\text{O}_2]}{[\text{MbO}_2]} \quad [4\text{-}1]$$

where the square brackets indicate molar concentrations. The proportion of the total myoglobin molecules that have bound O_2 is a called the **fractional saturation** and is abbreviated ***Y*:**

$$Y = \frac{[\text{MbO}_2]}{[\text{Mb}] + [\text{MbO}_2]} \quad [4\text{-}2]$$

Since $[\text{MbO}_2]$ is equal to $[\text{Mb}][\text{O}_2]/K$ (Equation 4-1, rearranged),

$$Y = \frac{[\text{O}_2]}{K + [\text{O}_2]} \quad [4\text{-}3]$$

O_2 is a gas, so its concentration is expressed as $\boldsymbol{pO_2}$, the **partial pressure of oxygen,** in units of torr (where 760 torr = 1 atm). Thus,

$$Y = \frac{pO_2}{K + pO_2} \quad [4\text{-}4]$$

In other words, *the amount of O_2 bound to myoglobin (Y) is a function of the oxygen concentration (pO_2) and the affinity of myoglobin for O_2 (K).*

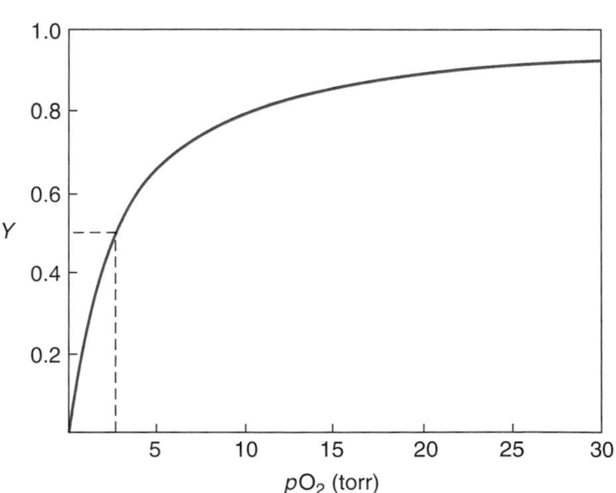

FIGURE 4-21

Myoglobin oxygen-binding curve.
The relationship between the fractional saturation of myoglobin (Y) and the oxygen concentration (pO_2) is hyperbolic. When $pO_2 = K = 2.8$ torr, myoglobin is half-saturated ($Y = 0.5$).

A plot of fractional saturation (Y) versus pO_2 yields a hyperbola (Fig. 4-21). As the O_2 concentration increases, more and more O_2 molecules bind to the heme groups of myoglobin molecules, until at very high O_2 concentrations, virtually all the myoglobin molecules have bound O_2. Myoglobin is then said to be **saturated** with oxygen. The oxygen concentration at which myoglobin is half-saturated—that is, the concentration of O_2 at which Y is half-maximal—is equivalent to K. For convenience, K is usually called $\boldsymbol{p_{50}}$, the oxygen pressure at 50% saturation. For human myoglobin, p_{50} is 2.8 torr (see Sample Calculation 4-2). We will see the physiological significance of myoglobin's O_2-binding affinity in the following section.

Sample Calculation 4-2

Problem

Calculate the fractional saturation of myoglobin when pO_2 = 1 torr, 10 torr, and 100 torr.

Solution

Use Equation 4-4 and let K = 2.8 torr.

When $pO_2 = 1$ torr, $Y = \dfrac{1}{2.8 + 1} = 0.26$

When $pO_2 = 10$ torr, $Y = \dfrac{10}{2.8 + 10} = 0.78$

When $pO_2 = 100$ torr, $Y = \dfrac{100}{2.8 + 100} = 0.97$

REVIEW LEARNING OBJECTIVE 8

- Why does myoglobin require a prosthetic group?
- What is the relationship between myoglobin's fractional saturation and the oxygen concentration?

5 HEMOGLOBIN, A PROTEIN WITH QUATERNARY STRUCTURE

Go to **Exercise 6/Myoglobin and Hemoglobin** for a detailed look at the structure of hemoglobin, including its allosteric mechanism. The structures presented in Exercise 6 and the questions based on them address Learning Objectives 10 and 11.

Most proteins, especially those with molecular masses greater than 100 kD, consist of more than one polypeptide chain. The individual chains, called **subunits,** may all be identical, in which case the protein is known as a **homodimer, homotetramer,** and so on. If the chains are not all identical, the prefix ***hetero-*** is used. Hemoglobin, for example, is a heterotetramer containing two α chains and two β chains. The spatial arrangement of these four polypeptides is known as hemoglobin's quaternary structure.

The forces that hold subunits together are similar to those that determine the tertiary structures of the individual polypeptides. The interface (the area of contact) between two subunits is often hydrophobic, with closely packed side chains. Hydrogen bonds, ion pairs, and disulfide bonds dictate the exact geometry of the interacting subunits.

Among the most common quaternary structures in proteins are symmetrical arrangements of two or more identical subunits (Fig. 4-22). Even in proteins with nonidentical subunits, the symmetry is based on groups of subunits. Hemoglobin, with two αβ units, therefore can be considered to be a dimer of dimers (see Fig. 4-22c).

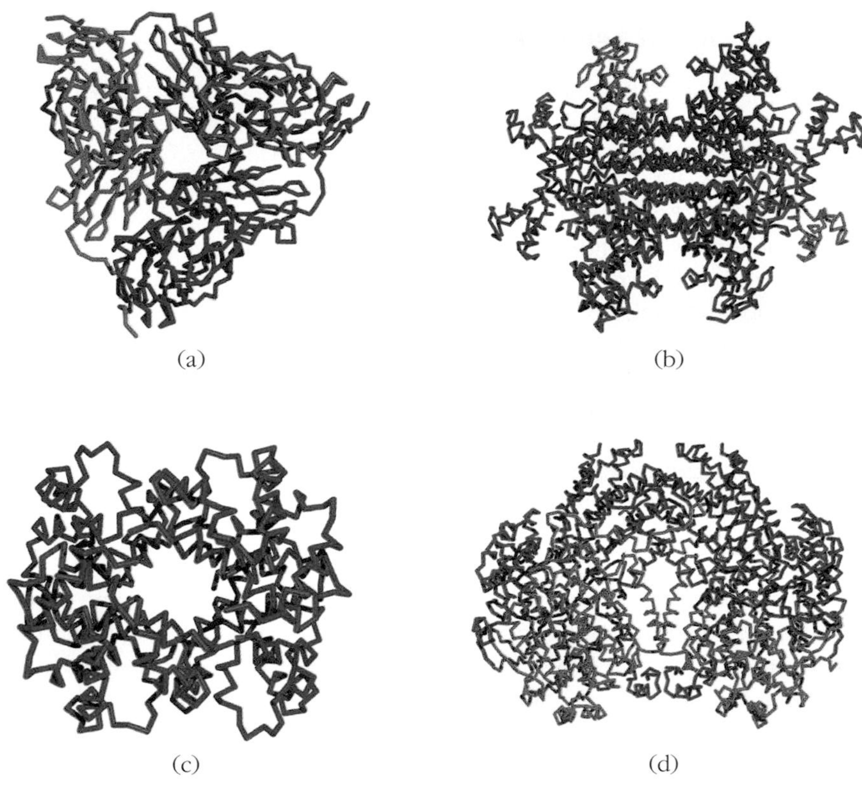

(a) (b) (c) (d)

FIGURE 4-22

Some proteins with quaternary structure.

The alpha carbon backbone of each polypeptide is shown. (a) Nitrite reductase, an enzyme with three identical subunits, from *Alcaligenes*. [*Structure (pdb 1AS8) determined by M.E.P Murphy, E.T. Adams, and S. Turley.*] (b) *E. coli* fumarase, a homotetrameric enzyme. [*Structure (pdb 1FUQ) determined by T. Weaver and L. Banaszak.*] (c) Human hemoglobin, a heterotetramer with two α subunits (blue) and 2 β subunits (red). [*Structure (pdb 2HHB) determined by G. Fermi and M.F. Perutz.*] (d) Bacterial methane hydroxylase, whose two halves (*right* and *left* in this image) each contain three kinds of subunits. [*Structure (pdb 1MMO) determined by A.C. Rosenzweig, C.A. Frederick, S.J. Lippard, and P. Nordlund.*]

Advantages of quaternary structure

The advantages of multisubunit protein structure are many. For starters, the cell can construct extremely large proteins through the incremental addition of small building blocks (we will see examples of this in the next chapter). This is clearly an asset for certain structural proteins which—due to their enormous size—cannot be synthesized all in one piece or must be assembled outside the cell. Moreover, the impact of the inevitable errors in transcription and translation can be minimized if the affected polypeptide is small and readily replaced.

Finally, the interaction between subunits in a multisubunit protein affords an opportunity for the subunits to influence each other's behavior. The result is a way of regulating function that is not possible in single-subunit proteins or in multisubunit proteins whose subunits each operate independently. Hemoglobin, which has four interacting oxygen-binding sites, provides an excellent example of how cooperative behavior serves a biological purpose.

REVIEW LEARNING OBJECTIVE 9

- What are the advantages of quaternary structure?

FIGURE 4-23

Tertiary structures of myoglobin and the α and β chains of hemoglobin. Backbone traces of α globin (blue) and β globin (red) are aligned with myoglobin (green) to show their structural similarity. The heme group of myoglobin is shown in gray.

The evolution of hemoglobin

Each of the four polypeptide subunits of hemoglobin, called a **globin,** looks a lot like myoglobin. The hemoglobin α chains, the hemoglobin β chains, and myoglobin have remarkably similar tertiary structures (Fig. 4-23). All have a heme group in a hydrophobic pocket, a His F8 that ligands the Fe(II) atom, and a His E7 that forms a hydrogen bond to O_2. Exercise 6 includes a comparison of the tertiary structures of myoglobin and the hemoglobin α and β chains.

Somewhat surprisingly, the amino acid sequences of the three globin polypeptides are only 18% identical. Figure 4-24 shows the aligned sequences, with the necessary gaps (for example, the hemoglobin α chain has no D helix). The lack of striking sequence similarities among these proteins highlights an important principle of protein three-dimensional structure: *Certain tertiary structures—for example, the backbone folding pattern of a globin polypeptide—can accommodate a variety of amino acid sequences. In fact, many proteins with completely unrelated sequences adopt similar structures.*

Clearly, the globins are **homologous proteins** that have evolved from a common ancestor through genetic mutation (see Section 3-5). The α and β chains of human hemoglobin share a number of residues; some of these are also identical in human myoglobin. A few residues are found in all vertebrate hemoglobin and myoglobin chains. The **invariant residues,** those that are identical in all the globins, are essential for the structure and/or function of the proteins and cannot be replaced by other residues. Some positions are under less selective pressure to maintain a particular amino acid match and can be **conservatively substituted** by a similar amino acid (for example, Ile for Leu or Ser for Thr). Still other positions are **variable,** meaning that they can accommodate a variety of residues, none of which is critical for the protein's structure or function. *By looking at the similarities and differences in sequences among evolutionarily related proteins such as the*

```
Helix        A                B           C           D         E
Mb     G-LSDGEWQLVLNVWGKVEADIPGHGQEVLIRLFKGHPETLEKFDKFKHLKSEDEMKASEDLKKHGATVLTALG
Hb α   V-LSPADKTNVKAAWGKVGAHAGEYGAEALERMFLSFPTTKTYFPHF-DLSH-----GSAQVKGHGKKVADALT
Hb β   VHLTPEEKSAVTALWGKV--NVDEVGGEALGRLLVVYPWTQRFFESFGDLSTPDAVMGNPKVKAHGKKVLGAFS

Helix              F               G                        H
Mb     GILKKKGHHEAEIKPLAQSHATKHKIPVKYLEFISECIIQVLQSKHPGDFGADAQGAMNKALELFRKDMASNYKELGFQG
Hb α   NAVAHVDDMPNALSALSDLHAHKLRVDPVNFKLLSHCLLVTLAAHLPAEFTPAVHASLDKFLASVSTVLTSKYR
Hb β   DGLAHLDNLKGTFATLSELHCDKLHVDPENFRLLGNVLVCVLAHHFGKEFTPPVQAAYQKVVAGVANALAHKYH
```

FIGURE 4-24

The amino acid sequences of myoglobin and the hemoglobin α and β chains. The sequence of human myoglobin (Mb) and the human hemoglobin (Hb) chains are written so that their helical segments (bars labeled A through H) are aligned. Residues that are identical in the α and β globins are shaded yellow; residues identical in myoglobin and the α and β globins are shaded green; and residues that are invariant in all vertebrate myoglobin and hemoglobin chains are shaded blue. The one-letter abbreviations for amino acids are given in Figure 4-2. [*After Dickerson, R.E. and Geis, I.,* Hemoglobin, *Benjamin/Cummings (1983), pp. 68–69.*]

globins, it is possible to deduce considerable information about elements of protein structure that are central to protein function.

Sequence analysis also provides a window on the course of globin evolution, since *the number of sequence differences roughly corresponds to the time since the genes diverged.* An estimated 1.1 billion years ago, a single globin gene was duplicated, possibly by aberrant genetic recombination, leaving two globin genes that then could evolve independently (Fig. 4-25). Over time, the gene sequences diverged by mutation. One gene became the myoglobin gene. The other coded for a monomeric hemoglobin, which is still found in some primitive vertebrates such as the lamprey (an organism that originated about 425 million years ago). Subsequent duplication of the hemoglobin gene and additional sequence changes yielded the α and β globins, which made possible the evolution of a tetrameric hemoglobin (whose structure is abbreviated $\alpha_2\beta_2$). Additional gene duplications and mutations produced the ζ chain (from the α chain) and the γ and ε chains (from the β chain). In fetal mammals, hemoglobin has the composition $\alpha_2\gamma_2$, and early human embryos synthesize a $\zeta_2\varepsilon_2$ hemoglobin. In primates, a recent duplication of the β chain gene has yielded the δ chain. An $\alpha_2\delta_2$ hemoglobin occurs as a minor component (about 1%) of adult human hemoglobin. At present, the δ chain appears to have no unique biological function, but it may eventually evolve one.

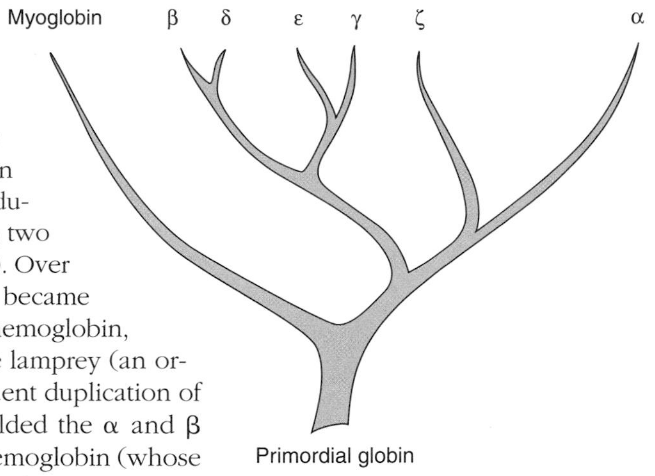

Myoglobin β δ ε γ ζ α

Primordial globin

FIGURE 4-25

Evolution of the globins.
Duplication of a primordial globin gene allowed the separate evolution of myoglobin and a monomeric hemoglobin. Additional duplications among the hemoglobin genes gave rise to six different globin chains that combine to form tetrameric hemoglobin variants at various times during development.

REVIEW LEARNING OBJECTIVE 10

- How can the sequences of homologous proteins provide information about residues that are essential or nonessential for a protein's function?

Oxygen binding to hemoglobin

A milliliter of human blood contains about 5 billion red blood cells, each of which is packed with about 300 million hemoglobin molecules. Consequently, blood can carry far more oxygen than a comparable volume of pure water. Like myoglobin, hemoglobin binds O_2 reversibly, but it does not exhibit the simple behavior of myoglobin. A plot of fractional saturation (Y) versus pO_2 for hemoglobin is sigmoidal (S-shaped) rather than hyperbolic (Fig. 4-26).

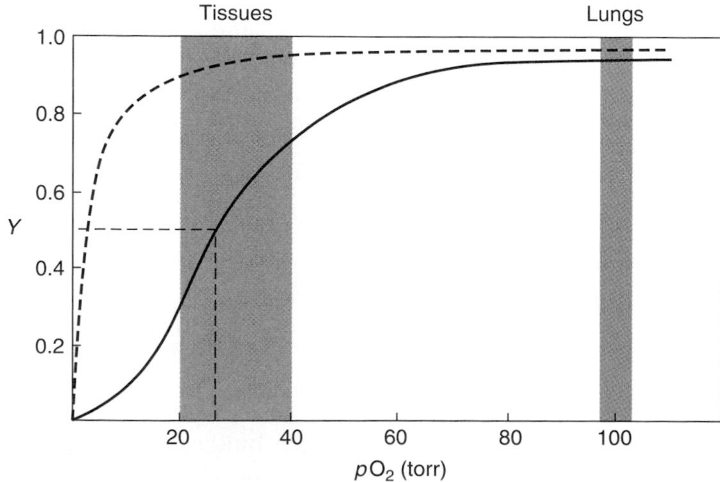

FIGURE 4-26

Oxygen binding to hemoglobin.
The relationship between fractional saturation (Y) and oxygen concentration (pO_2) is sigmoidal. The pO_2 at which hemoglobin is half-saturated (p_{50}) is 26 torr. For comparison, myoglobin's O_2-binding curve is indicated by the dashed line. The difference in oxygen affinity between hemoglobin and myoglobin ensures that O_2 bound to hemoglobin in the lungs is released to myoglobin in the muscles. This oxygen delivery system is efficient because the tissue pO_2 corresponds to the part of the hemoglobin-binding curve where the O_2 affinity falls off most sharply.

Furthermore, hemoglobin's overall oxygen affinity is lower than that of myoglobin: Hemoglobin is half-saturated at an oxygen pressure of 26 torr (p_{50} = 26 torr), whereas myoglobin is half-saturated at 2.8 torr.

Why is hemoglobin's binding curve sigmoidal? At low O_2 concentrations, hemoglobin appears to be reluctant to bind the first O_2, but as the pO_2 increases, O_2 binding increases sharply, until hemoglobin is almost fully saturated. A look at the binding curve in reverse shows that at high O_2 concentrations, oxygenated hemoglobin is reluctant to give up its first O_2, but as the pO_2 decreases, all the O_2 molecules are suddenly given up. This behavior suggests that the binding of the first O_2 increases the affinity of the remaining O_2-binding sites. Apparently, *hemoglobin's four heme groups are not independent but communicate with each other in order to work in a unified fashion.* This is known as **cooperative binding** behavior. In fact, the fourth O_2 taken up by hemoglobin binds with about 100 times greater affinity than the first.

Hemoglobin's relatively low oxygen affinity and its cooperative binding behavior are the keys to its physiological function (see Fig. 4-26). In the lungs, where the pO_2 is about 100 torr, hemoglobin is about 95% saturated with O_2. In the tissues, where the pO_2 is only about 20 to 40 torr, hemoglobin's oxygen affinity drops off rapidly, so that it is only about 55% saturated when the pO_2 is 30 torr. Under these conditions, *the O_2 released from hemoglobin is readily taken up by myoglobin in muscle cells, since myoglobin's affinity for O_2 is much higher,* even at the lower oxygen pressure. Myoglobin can therefore relay O_2 from red blood cells in the capillaries to the cell's mitochondria, where it is consumed in the oxidative reactions that sustain muscle activity.

What is the structural basis for cooperative behavior?

The four heme groups of hemoglobin must be able to sense each other's oxygen-binding status so that they can bind or release their O_2 in concert. But the four heme groups are 25 to 37 Å apart, too far for them to communicate via an electronic signal. Therefore, the signal must be mechanical. In a mechanism worked out by Max Perutz, who also elucidated the crystal structure of hemoglobin, the four globin subunits undergo conformational changes when they bind O_2.

In **deoxyhemoglobin** (hemoglobin without any bound O_2), the heme Fe atom has five ligands, so the porphyrin ring is somewhat dome-shaped and the Fe lies about 0.6 Å out of the plane of the porphyrin ring. As a result, the heme group is bowed slightly toward His F8 (Fig. 4-27). When O_2 binds to produce **oxyhemoglobin,** the Fe—now with six ligands—moves into the center of the porphyrin plane. This movement of the Fe atom pulls His F8 further toward the heme group, and this in turn drags the entire F helix so that it moves by up to 1 Å. The F helix cannot move in this manner unless the entire protein alters its conformation, culminating in the rotation of one αβ unit relative to the other. Consequently, *hemoglobin has two quaternary structures, corresponding to the oxy and deoxy states.*

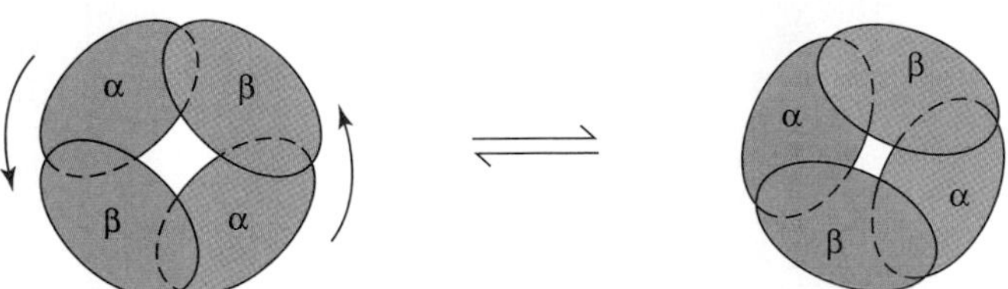

The shift in conformation between the oxy and deoxy states primarily involves rotation of one αβ unit relative to the other. Oxygen binding decreases the size of the central cavity between the four subunits and alters some of the contacts between subunits. The two conformational states of hemoglobin are formally known as **T** (for "tense") and **R** (for "relaxed"). The T state corresponds to deoxyhemoglobin, and the R state corresponds to oxyhemoglobin. Exercise 6 explores the conformational differences between oxy- and deoxyhemoglobin.

Deoxyhemoglobin is reluctant to bind the first O_2 molecule because the protein is in the deoxy (T) conformation, which is unfavorable for O_2 binding (the Fe atom lies out of the heme plane). However, once O_2 has bound, probably to the α chain in each αβ pair, the entire tetramer snaps into the oxy (R) conformation as the Fe atom and the F helix move. An intermediate conformation is not possible, because the contacts between the αβ units do not allow it (Fig. 4-28).

Subsequent O_2 molecules bind with higher affinity because the protein is already in the oxy (R) conformation, which is favorable for O_2 binding. Similarly, oxyhemoglobin tends to retain its bound O_2 molecules until the oxygen pressure drops significantly. Then, some O_2 is released, triggering the change to the deoxy (T) conformation. This decreases the affinity of the remaining bound O_2 molecules, making it easier for hemoglobin to unload its bound oxygen.

Hemoglobin and many other proteins with multiple binding sites are known as **allosteric proteins** (from the Greek *allos,* meaning "other," and *stereos,* meaning "space"). In these proteins, *the binding of a small molecule (called a* ***ligand****) to one site alters the ligand-binding affinity of the other sites.* In principle, the ligands need not be identical, and their binding may either increase or decrease the binding activity of the other sites. In hemoglobin, the ligands are all oxygen molecules, and O_2 binding to one part of the protein increases the O_2 affinity of the rest of the protein.

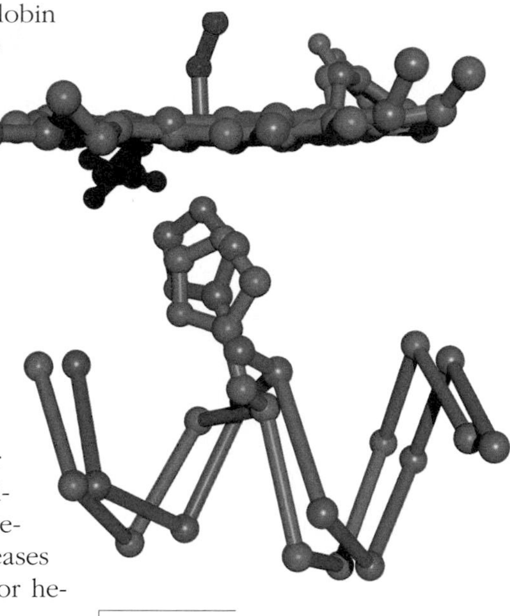

FIGURE 4-27

Conformational changes in hemoglobin upon O_2 binding.
In deoxyhemoglobin (blue), the porphyrin ring is slightly bowed toward His F8 (shown in ball-and-stick form). The remainder of the F helix is represented by its alpha carbon atoms. In oxyhemoglobin (purple), the heme group becomes planar, pulling His F8 and its attached F helix. The bound O_2 is shown in red.

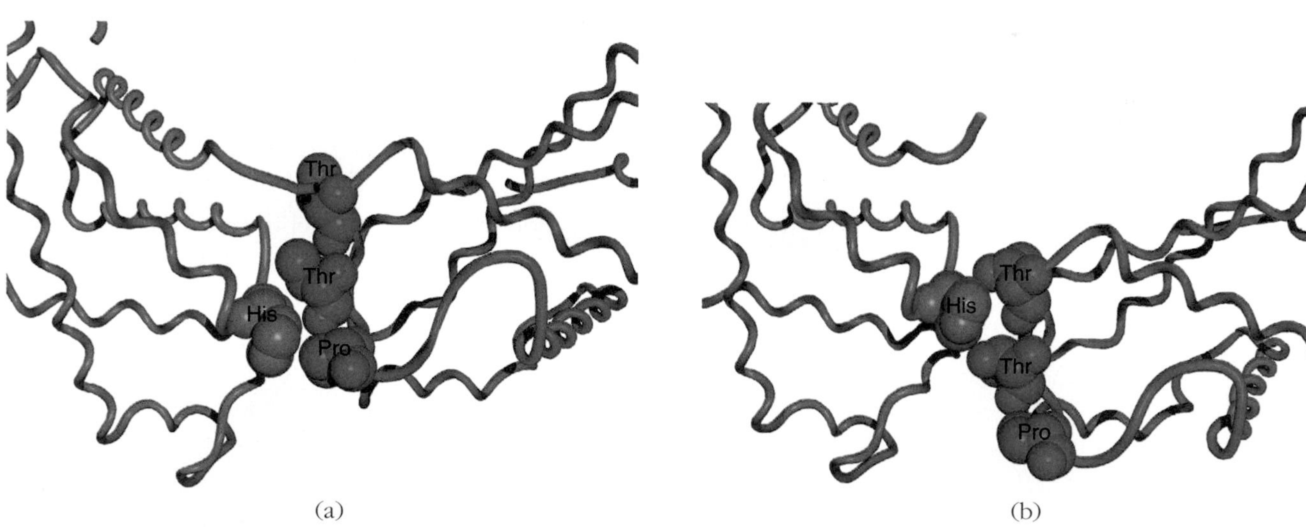

FIGURE 4-28

Some of the subunit interactions in hemoglobin.
The interactions between the αβ units of hemoglobin include contacts between side chains. The relevant residues are shown in space-filling form. (a) In deoxyhemoglobin, a His residue on the β chain (blue, *left*) fits between a Pro and a Thr residue on the α chain (green, *right*). (b) Upon oxygenation, the His residue moves between two Thr residues. An intermediate conformation (between the deoxy and oxy conformations) is disallowed in part because the highlighted side chains would experience strain. [*Structure of human deoxyhemoglobin (pdb 2HHB) determined by G. Fermi and M.F. Perutz; structure of human oxyhemoglobin (pdb 1HHO) determined by B. Shaanan.*]

REVIEW LEARNING OBJECTIVE 10

- Why are the different O_2 affinities of myoglobin and hemoglobin physiologically important?

REVIEW LEARNING OBJECTIVE 11

- How does hemoglobin bind O_2 cooperatively?

Additional factors regulate oxygen binding *in vivo*

Decades of study have revealed the detailed chemistry behind hemoglobin's activity (and have also discovered how molecular defects can lead to disease; Box 4-D). The conformational change that transforms deoxyhemoglobin to oxyhemoglobin alters the microenvironments of several ionizable groups in the protein, including the two N-terminal amino groups of the α subunits and the two His residues near the C-terminus of the β subunits. As a result, these groups become more acidic and release H^+ when O_2 binds to the protein:

$$\mathrm{Hb \cdot H^+ + O_2 \rightleftharpoons Hb \cdot O_2 + H^+}$$

Therefore, increasing the pH of a solution of hemoglobin (decreasing $[H^+]$) favors O_2 binding by "pushing" the reaction to the right as written above. Decreasing the pH (increasing $[H^+]$) favors O_2 dissociation by "pushing" the reaction to the left. *The reduction of hemoglobin's oxygen-binding affinity when the pH decreases is known as the* ***Bohr effect*****.**

The Bohr effect plays an important role in O_2 transport *in vivo*. Tissues release CO_2 as they consume O_2 in respiration. The dissolved CO_2 enters red blood cells, where it is rapidly converted to bicarbonate (HCO_3^-) by the action of the enzyme carbonic anhydrase (see Box 2-B):

$$\mathrm{CO_2 + H_2O \rightleftharpoons HCO_3^- + H^+}$$

The H^+ released in this reaction induces hemoglobin to unload its O_2 (Fig. 4-29). In the lungs, the high concentration of oxygen promotes O_2 binding

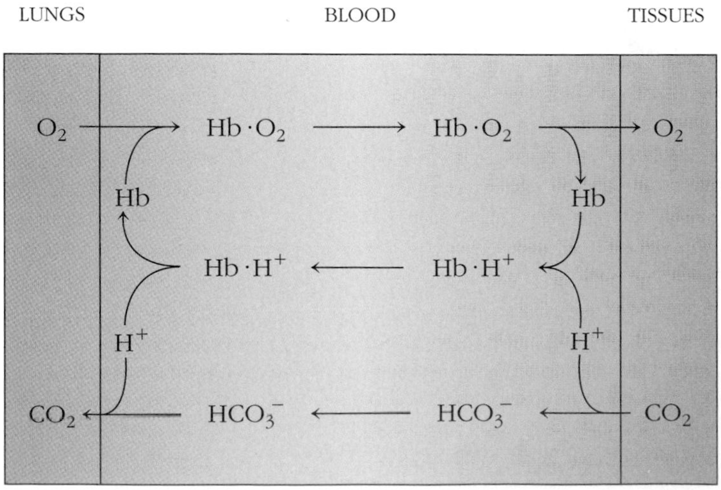

FIGURE 4-29

Oxygen transport and the Bohr effect. Hemoglobin picks up O_2 in the lungs. In the tissues, H^+ derived from the metabolic production of CO_2 decreases hemoglobin's affinity for O_2, thereby promoting O_2 release to the tissues. Back in the lungs, hemoglobin binds more O_2, releasing the protons, which recombine with bicarbonate to re-form CO_2.

A CLOSER LOOK

BOX 4-D Sickle-cell anemia, the first "molecular" disease

About 5% of world's population carries an inherited disorder of hemoglobin. Hundreds of hemoglobin variants are known, comprising amino acid substitutions and mutations that reduce production of a globin polypeptide. One of the better-known hemoglobin variants is sickle-cell hemoglobin (known as hemoglobin S). Individuals with two copies of the defective gene develop sickle-cell anemia, a debilitating disease that predominantly affects populations of African descent.

The discovery of the molecular defect that causes sickle-cell anemia was a groundbreaking event in biochemistry. The disease was first described in 1910, but until about midcentury there was no direct evidence that sickle-cell anemia or any genetic disease was the result of an alteration in the molecular structure of a protein. Then in 1949, Linus Pauling (who was already on his way to discovering the α helix) showed that hemoglobin from patients with sickle-cell anemia had a different electrical charge than hemoglobin from healthy individuals. Eight years later, in 1957, Vernon Ingram identified a single amino acid difference: Glu at position A3 in the β chain is replaced by Val in sickle-cell hemoglobin. This was the first evidence that an alteration in a gene caused an alteration in the amino acid sequence of the corresponding polypeptide.

In normal hemoglobin, the switch from the oxy to the deoxy conformation exposes a hydrophobic patch on the protein surface between the E and F helices. The hydrophobic Val residues on hemoglobin S are optimally positioned to bind to this patch. The resulting hydrophobic interactions lead to the rapid aggregation of hemoglobin S molecules to form long rigid fibers.

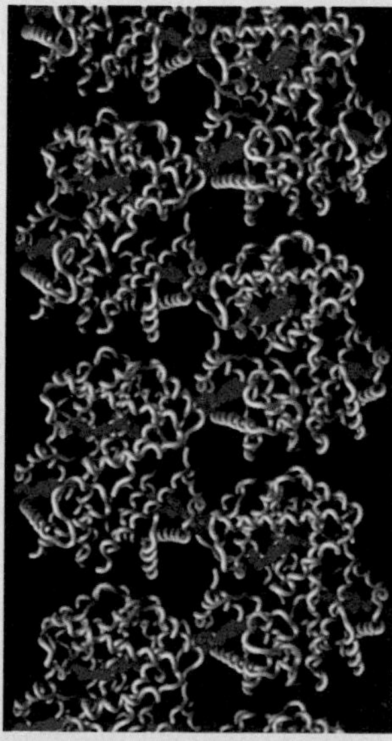

In this model of polymerized hemoglobin S molecules, the heme groups are red and the mutant Val residues are blue.

These fibers physically distort the red blood cell, producing the familiar sickle shape. Because hemoglobin S aggregation occurs only among deoxy hemoglobin S molecules, sickling tends to occur when the red blood cells pass through oxygen-poor capillaries. The misshapen cells can obstruct blood flow and rupture, leading to the intense pain, organ damage, and loss of red blood cells that characterize the disease.

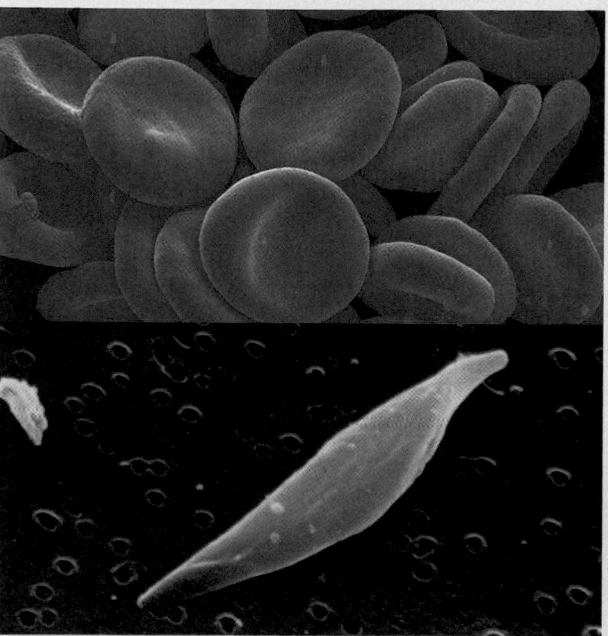

The high frequency of the gene for sickle-cell anemia (that is, the mutated β globin gene) was at first puzzling: Genes that lead to disabling diseases tend to be rare because individuals with two copies of the gene usually die before they can pass the gene to their offspring. However, carriers of the sickle-cell variant appear to have a selective advantage. They are protected against malaria, a parasitic disease that afflicts about 300 million people and kills about 3 million each year, mostly children. In fact, the sickle-cell hemoglobin variant is common in regions of the world where malaria is endemic. In heterozygotes (individuals with one normal and one defective β globin gene), only about 2% of red blood cells undergo sickling. This has a significant antimalarial effect, however, since red blood cells harboring *Plasmodium falciparum,* the protozoan that causes malaria, are especially prone to sickling. The spleen removes the sickled cells from the circulation, thereby eliminating the parasite as well.

[Hemoglobin S photo courtesy W. Royer and D. Harrington, *J. Mol. Biol.* **272**, 398–407 (1992). Red blood cells (top) Andrew Skred/Science Photo Library/Photo Researchers. Sickled cells (bottom) Jackie Lewin, Royal Free Hospital/Science Photo Library/Photo Researchers.]

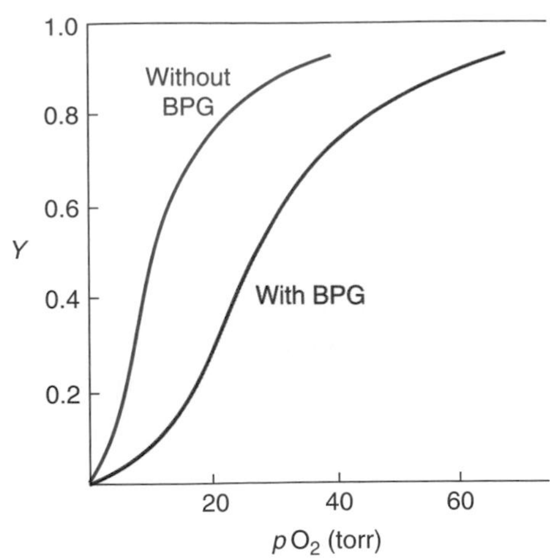

FIGURE 4-30

Effect of BPG on hemoglobin. BPG binds to deoxyhemoglobin but not to oxyhemoglobin. It therefore reduces hemoglobin's O_2 affinity by stabilizing the deoxy conformation.

to hemoglobin. This causes the release of protons that can then combine with bicarbonate to re-form CO_2, which is breathed out.

Red blood cells use one additional mechanism to fine-tune hemoglobin function. These cells contain a three-carbon compound, 2,3-bisphosphoglycerate (BPG):

$$
\begin{array}{c}
{}^{-}O \diagdown \;\; \diagup\!\!\!\!= O \\
C \\
| \\
H-C-OPO_3^{2-} \\
| \\
H-C-OPO_3^{2-} \\
| \\
H
\end{array}
$$

2,3-Bisphosphoglycerate (BPG)

BPG binds in the central cavity of hemoglobin, but only in the T (deoxy) state. The five negative charges in BPG interact with positively charged groups in deoxyhemoglobin; in oxyhemoglobin, these cationic groups have moved away and the central cavity is too narrow to accommodate BPG. Thus, *the presence of BPG stabilizes the deoxy conformation of hemoglobin.* Without BPG, hemoglobin would bind O_2 too tightly to release it to cells. In fact, hemoglobin stripped of its BPG *in vitro* exhibits very strong O_2 affinity, even at low pO_2 (Fig. 4-30).  Exercise 6 shows how the central cavity of deoxyhemoglobin can accommodate BPG.

The fetus takes advantage of this chemistry to obtain O_2 from its mother's hemoglobin. Fetal hemoglobin has the subunit composition $\alpha_2\gamma_2$. In the γ chains, position H21 is not His (as it is in the mother's β chain) but Ser. His H21 bears one of the positive charges important for binding BPG in adult hemoglobin. The absence of this interaction in fetal hemoglobin reduces BPG binding. Consequently, *hemoglobin in fetal red blood cells has a higher O_2 affinity than adult hemoglobin, which helps transfer O_2 from the maternal circulation to the fetus.*

REVIEW LEARNING OBJECTIVE 12

- How do the Bohr effect and BPG regulate O_2 transport *in vivo*?

SUMMARY

1. The 20 amino acid constituents of proteins are differentiated by the chemical properties of their side chains, which can be roughly classified as hydrophobic, polar, or charged.

2. Amino acids are linked by peptide bonds to form a polypeptide. The sequence of amino acids (a protein's primary structure) is determined by cleaving the polypeptide into smaller segments that can be individually sequenced by Edman degradation, and then aligning the sequences of overlapping peptides.

3. Protein secondary structure includes the α helix and β sheet, in which hydrogen bonds form between backbone carbonyl and amide groups. Irregular secondary structure has no regularly repeating conformation.

4. The three-dimensional shape (tertiary structure) of a protein includes its backbone and all side chains. A globular protein such as myoglobin has a hydrophobic core and is stabilized primarily by the hydrophobic effect. Ion pairing, disulfide bonds, and other cross-links may also help stabilize proteins.

5. A denatured protein may refold to achieve its native structure. In a cell, chaperones assist protein folding.

6. Myoglobin contains a heme prosthetic group that reversibly binds oxygen. The amount of O_2 bound depends on the O_2 concentration and on myoglobin's affinity for oxygen.

7. Proteins with quaternary structure, such as hemoglobin, have multiple subunits. Hemoglobin's α and β chains are

homologous to myoglobin, indicating a common evolutionary origin.

8. O_2 binds cooperatively to hemoglobin with low affinity so that hemoglobin can efficiently bind O_2 in the lungs and deliver it to myoglobin in the tissues.

9. Hemoglobin is an allosteric protein whose four subunits alternate between the T (deoxy) and R (oxy) conformations in response to O_2 binding to the heme groups. The deoxy conformation is favored by low pH (the Bohr effect) and by the presence of BPG.

CHECKLIST

1. Review the Learning Objectives listed on pages 90–91.

2. Complete Exercise 4/Amino Acids, which explores the chemical characteristics of the 20 standard amino acids.

3. Complete Exercise 5/Protein Secondary Structure, which illustrates the conformations of the α helix and β sheet.

4. Complete Exercise 6/Myoglobin and Hemoglobin, which shows the structures of these proteins and explains how hemoglobin binds oxygen cooperatively.

5. Apply your knowledge by solving the problems at the end of this chapter. Check your results in the Solutions appendix.

6. Be able to define the boldfaced terms (consult the glossary at the end of the book). Test your understanding by taking the Chapter 4 quiz (www.wiley.com/college/pratt).

7. Explore the websites listed at www.wiley.com/college/pratt.

8. Consult the list of Selected Readings for background information or additional details on amino acid structure, protein sequencing, protein three-dimensional structure, and myoglobin and hemoglobin function.

GLOSSARY TERMS

protein
polypeptide
α-amino acid
R group
Cα
disulfide bond
chirality
condensation reaction
peptide bond
residue
N-terminus
C-terminus
backbone
microenvironment
ion exchange
oligopeptide
peptide
protease
Edman degradation
primary structure
secondary structure
tertiary structure
quaternary structure
α helix
parallel β sheet
antiparallel β sheet
regular secondary structure
loop
irregular secondary structure
X-ray crystallography
diffraction pattern
heme
domain
ion pair
zinc finger
denaturation
renaturation
native structure
molecular chaperone
amyloid deposits
processing
prosthetic group
Y
pO_2
saturation
p_{50}
subunit
homo-
hetero-
globin
homologous proteins
invariant residue
conservative substitution
variable residue
cooperative binding
deoxyhemoglobin
oxyhemoglobin
T state
R state
allosteric protein
ligand
Bohr effect

PROBLEMS

1. At what pH would an amino acid bear both a COOH and an NH_2 group?

2. Which of the 20 standard amino acids are

(a) cyclic
(b) aromatic
(c) sometimes charged at physiological pH
(d) technically not hydrophobic, polar, or charged
(e) basic
(f) acidic
(g) sulfur-containing

3. Histones are basic proteins that bind to DNA. What amino acids are found in abundance in histones and why? What important intermolecular interactions form between histones and DNA?

4. A sample of the amino acid tyrosine is barely soluble in water. Would a polypeptide containing only Tyr residues, poly(Tyr), be more or less soluble, assuming the total number of Tyr remains constant?

5. The pK values of the amino and carboxyl groups in free amino acids differ from the pK values of the N- and C-termini of polypeptides. Explain.

6. Amino acids (except for glycine) are chiral; that is, they contain a chiral carbon and thus can form mirror-image isomers, or enantiomers. Biochemists use "D" and "L" to distinguish enantiomers instead of the *RS* system used by chemists.

(a) The structure of L-alanine is shown. Label the chiral carbon and draw the structure of D-alanine.

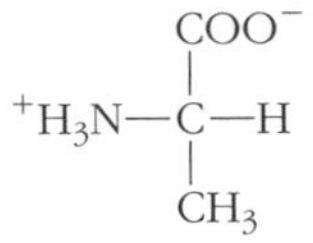

COO^-
|
$^+H_3N—C—H$
|
CH_3

L-Alanine

(b) The vast majority of proteins in living systems consist of L-amino acids. However, D-amino acids are found in some short bacterial peptides that make up cell walls. Why might this be advantageous to the bacterium?

7. Rank the solubility of the following amino acids in water at pH 6.5: His, Phe, Ser, Ile, Asp, Trp.

8. Biochemists sometimes link a recombinant protein to a protein known as green fluorescent protein (GFP), which was first purified from bioluminescent jellyfish. The fluorophore in GFP (shown below) is a derivative of three consecutive amino acids that undergo cyclization of the polypeptide chain and an oxidation. Identify (a) the three residues, and the bonds that result from the (b) cyclization and (c) oxidation reactions

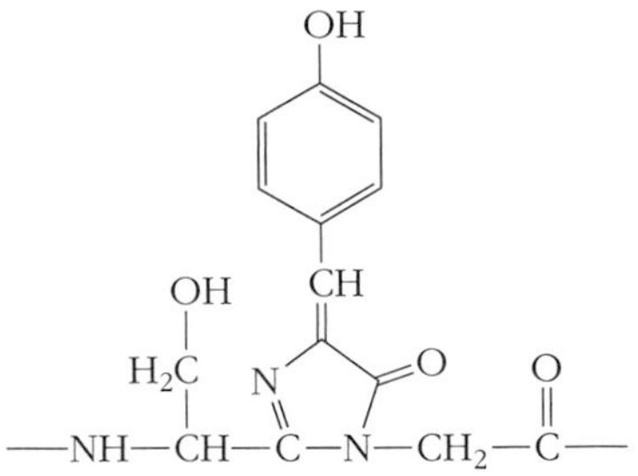

OH

OH CH

H_2C N O O

$—NH—CH—C—N—CH_2—C—$

9. Why are the terms *protein* and *polypeptide* not interchangeable?

10. Draw the structure of the dipeptide Lys–Glu at pH 7.0. Label the following: (a) peptide bond, (b) N-terminus, (c) C-terminus, (d) an α-amino group and an ε-amino group, (e) an α-carboxylate group and a γ-carboxylate group.

11. The artificial nonnutritive sweetener Aspartame is a dipeptide with the sequence Asp–Phe. The carboxyl terminus is methylated. Draw the structure of Aspartame at pH 7.0.

12. In 1975, two pentapeptides with opiate-like activity were isolated from the brain. The pentapeptides, called Met-enkephalin and Leu-enkephalin, have the sequence shown below. Draw the structures of the two enkephalins at pH 7.0.

Met-enkephalin Tyr–Gly–Gly–Phe–Met
Leu-enkephalin Tyr–Gly–Gly–Phe–Leu

13. Glutathione is a Cys-containing tripeptide found in red blood cells, which functions to remove hydrogen peroxide and organic peroxides that can irreversibly damage hemoglobin and cell membranes.

(a) The sequence of glutathione is γ-Glu–Cys–Gly. The γ-Glu indicates that the first peptide bond forms between the γ-carboxyl group of the Glu side chain and the α-amino group of Cys. Draw the structure of glutathione.

(b) Glutathione is sometimes abbreviated as "GSH" to show the importance of the Cys side chain in reactions. For example, two molecules of GSH react with an organic peroxide as shown below. The glutathione is oxidized to "GSSG" and the organic peroxide is reduced to a less harmful alcohol. Draw the structure of GSSG.

$$2\ GSH + R—O—O—H \xrightarrow{\text{glutathione peroxidase}} GSSG + ROH + H_2O$$

14. In your laboratory, you can perform ion exchange using different media that consist of tiny plastic beads whose surfaces are covered with ionic groups such as diethylaminoethyl (DEAE) or carboxymethyl (CM):

DEAE: $—CH_2—CH_2—N^+H(CH_2CH_3)_2$
CM: $—CH_2—COO^-$

These media are used to separate polypeptides or proteins with different overall charges. Conditions are chosen such that the desired polypeptide or protein binds to the ion exchange media whereas undesired polypeptides or proteins do not. Would you choose media containing DEAE or CM to try to separate the peptide shown below (using one-letter codes) from a mixture of different peptides at pH 7.0?

Peptide: GLEKSLVRLGDVQPSLGKESRAKKFQRQ

15. The sequence of crinia-angiotensin, an angiotensin II-like undecapeptide from the skin of the Australian frog, is determined. A single round of Edman degradation releases DNP-Ala. A second sample of the peptide is then treated with chymotrypsin. Two fragments are released with the following amino acid compositions: Fragment I (His, Pro, Phe, Val) and Fragment II (Ala, Asp, Arg, Gly, Pro, Ile, Tyr). Next, a third sample of peptide is treated with trypsin, which results in two fragments with the following amino acid compositions: Fragment III (Ala, Asp, Arg, Gly, Pro) and Fragment IV (His, Ile, Pro, Phe, Tyr, Val). Treatment of another sample with elastase yields three fragments, two of which are sequenced: Fragment V (His–Pro–Phe) and Fragment VI (Ala–Pro–Gly). What is the sequence of the undecapeptide?

16. The peptide hormone glucagon from the Nile tilapia was treated with chymotrypsin and the resulting fragments were sequenced. A second sample of the polypeptide was treated with trypsin and the fragments were sequenced. What is the sequence of the polypeptide?

Chymotrypsin fragments	**Trypsin fragments**
LMNNKRSGAAE	AQDFVRWLMNNK
SNDY	HSEGTFSNDYSK
HSEGTF	RSGAAE
LEDRKAQDF	YLEDRK
VRW	
SKY	

17. Before sequencing, a protein whose two identical polypeptide chains are linked by a disulfide bond must be reduced and alkylated. Why should reduction and alkylation also be performed for a single polypeptide chain that includes an intramolecular disulfide bond?

18. You must cleave the following peptide into smaller fragments. Which of the proteases listed in Table 4-3 would theoretically yield the most fragments? The fewest?

NMTQGRCKPVNTFVHEPLVDVQNVCFKE

19. Seedlings use seed storage proteins as an important nitrogen source during germination. A new seed storage protein was discovered in *Brassica nigra* and named BN. The protein was purified and an attempt was made to sequence the protein by Edman degradation. However, this was unsuccessful because the amino terminus was blocked. Based on comparisons with other proteins in the same family whose sequences are known, the investigators hypothesized that the amino terminal residue was *N*-acetyl serine.

(a) Draw the structure of *N*-acetyl-serine.

(b) If *N*-acetyl-serine is indeed the N-terminal residue, why would Edman degradation be unsuccessful?

(c) The amino acid sequence of BN was finally determined in the following manner: The BN protein was first treated with 2-mercaptoethanol to reduce any disulfide bridges. This treatment revealed that BN consists of two chains, a light chain and a heavy chain. Next, the two chains were separated and three separate samples of each chain were treated with different proteases. The fragments obtained were individually sequenced by Edman degradation. The sequence is shown below (the first five residues of the light chain are missing due to a blocked amino terminus). Why was it necessary to carry out a minimum of two different proteolytic cleavages of the protein using different proteases?

Light chain
R I P K C R K E F Q Q A Q H L R A
C Q Q W L H K Q A N Q S G G G P S

Heavy chain
P Q G P Q Q R P P L L Q Q C C N E
K H Q E E P L C V C P T L K G A S
K A V R Q Q I R Q Q G Q Q Q G Q Q
G Q Q L Q R E I S R I Y Q T A T H
L P R V C N I P R V S I C P F Q K
T M P G P

(d) One of the enzymes used by the investigators was trypsin. Indicate the sequences of the fragments that would result from trypsin digestion.

(e) Choose a second protease to cleave both the light and heavy chains into smaller fragments. What protease did you choose, and why? Indicate the sequences of the fragments that would result from the digestion of the protease you chose.

20. Proline is known as a helix disrupter; it sometimes appears at the end of an α helix but never in the middle. Explain.

21. In site-directed mutagenesis experiments, Gly is often successfully substituted for Val, but Val cannot substitute for Gly. Explain.

22. Choose the amino acid in the following pairs that would be more likely to appear on the solvent-exposed surface of a protein.

(a) His or Pro
(b) Ser or Ala
(c) Phe or Tyr
(d) Met or Cys
(e) Asn or Ile

23. You are performing site-directed mutagenesis to test predictions about which residues are essential for a protein's function. Which of each pair of amino acid substitutions listed below would you expect to disrupt protein function the most?

(a) Val replaced by Ala or Phe.
(b) Lys replaced by Asp or Arg.
(c) Gln replaced by Glu or Asn.
(d) Pro replaced by His or Gly.

24. Draw two amino acid side chains that can interact with one another via the following intermolecular interactions:

(a) ion pair
(b) hydrogen bond
(c) van der Waals interaction (London dispersion forces)

25. A type of muscular dystrophy, called severe childhood autosomal recessive muscular dystrophy (SCARMD), results from a mutation in the gene for a 50 kD muscle protein. The defective protein leads to muscle necrosis. Detailed studies of this protein have revealed that an arginine residue at position 98 has been mutated to a histidine. Why might replacing an arginine with histidine result in a defective protein?

26. Laboratory techniques for randomly linking together amino acids typically generate an insoluble polypeptide, yet a naturally occurring polypeptide of the same length is usually soluble. Explain.

27. Proteins can be unfolded, or denatured, by agents that alter the balance of weak noncovalent forces that maintain the native conformation. How would the following agents cause a protein to denature? Be specific about the type of intermolecular forces that would be affected.

(a) heat
(b) pH
(c) amphiphilic detergents
(d) reducing agents such as 2-mercaptoethanol

28. In the early 1970s, Christian Anfinsen carried out a denaturation experiment with ribonuclease. Ribonuclease, a pancreatic enzyme that catalyzes the digestion of RNA, consists of a single chain of 124 amino acids cross-linked by four disulfide bonds. Urea and 2-mercaptoethanol were added to a solution of ribonuclease, which caused it to unfold, or denature. The loss of tertiary structure resulted in a loss of biological activity. When the denaturing agent (urea) and the reducing agent (mercaptoethanol) were simultaneously removed, the ribonuclease spontaneously folded back up to its native conformation and regained full enzymatic activity in a process called renaturation. What is the significance of this experiment?

29. Insulin consists of two chains, a shorter A chain and a longer B chain. The two chains are held together with disulfide bonds. *In vivo,* insulin is processed from proinsulin, a single polypeptide chain. The C chain is removed from proinsulin to form insulin.

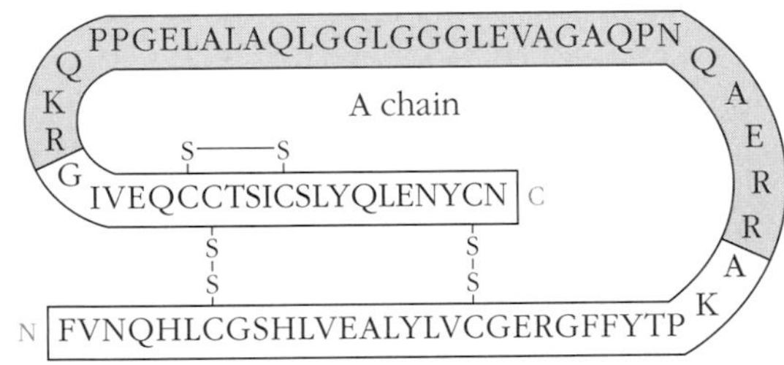

A denaturation/renaturation experiment similar to the one carried out by Anfinsen with ribonuclease (see Problem 28) was carried out using insulin. However, in contrast to Anfinsen's results, less than 10% of the activity of insulin was recovered when the urea and 2-mercaptoethanol were removed by dialysis (this is the level of activity to be expected if the disulfide bridges formed randomly). However, when the experiment was repeated with proinsulin, full activity was restored upon renaturation. Explain these observations.

30. In the mid-1980s, scientists noted that if cells were incubated at 42°C instead of the normal 37°C, the synthesis of a group of proteins dramatically increased. For lack of a better name, the scientists called these "heat-shock proteins." It was later determined that the heat-shock proteins were chaperones. Why do you think that the cell would increase its synthesis of chaperones when the temperature is increased?

31. X-Ray crystallographic analysis of a protein crystal sometimes fails to reveal the positions of the first few residues of a polypeptide chain. Explain.

32. During evolution, why do insertions, deletions, and substitutions of amino acids occur more often in loops than in elements of regular secondary structure?

33. The restriction endonucleases *Eco*RI and *Eco*RV are dimeric (two-subunit) enzymes (see Section 3-4). Based on how these proteins interact with DNA, do you expect them to be homo- or heterodimeric?

34. A protein with two identical subunits can often be rotated 180° (halfway) around its axis so as to generate an identical structure; such a protein is said to have rotational symmetry. Why is it not possible for a protein to have mirror symmetry (that is, its halves would be related as if reflected in a mirror)?

35. Below is a list of the first 10 residues of the B helix in myoglobin from different organisms. Based on this information, which positions (a) cannot tolerate substitution, (b) can tolerate conservative substitution, and (c) are highly variable?

Position	**1**	**2**	**3**	**4**	**5**	**6**	**7**	**8**	**9**	**10**
Human	D	I	P	G	H	G	Q	E	V	L
Chicken	D	I	A	G	H	G	H	E	V	L
Alligator	K	L	P	E	H	G	H	E	V	I
Turtle	D	L	S	A	H	G	Q	E	V	I
Tuna	D	Y	T	T	M	G	G	L	V	L
Carp	D	F	E	G	T	G	G	E	V	L

36. Insulin consists of two polypeptide chains, called A and B, that are joined by disulfide bonds. The smaller A chain consists of 21 amino acids in humans. The longer B chain has 30 amino acids in humans. Insulin from various animals is similar to but not identical to human insulin, as illustrated in the table below, which shows the variations at positions A8, A9, A10, B1, B2, B27, and B30. All other amino acids are the same.

Position	A8	A9	A10	B1	B2	B27	B30
Human	Thr	Ser	Ile	Phe	Val	Thr	Thr
Cow	Ala	Ser	Val	Phe	Val	Thr	Ala
Pig	Thr	Ser	Ile	Phe	Val	Thr	Ala
Horse	Thr	Gly	Ile	Phe	Val	Thr	Ala
Rabbit	Thr	Ser	Ile	Phe	Val	Thr	Ser
Dog	Thr	Ser	Ile	Phe	Val	Thr	Ala
Chicken	His	Asn	Thr	Ala	Ala	Ser	Ala
Duck	Glu	Asn	Pro	Ala	Ala	Ser	Thr

(a) What animals would serve as the best source of insulin to be used for treating diabetics? Explain.

(b) Would the p*I* values of the animal insulins be the same as, greater than, or less than the p*I* of human insulin?

(c) Some people develop allergies to animal insulin because their immune systems recognize the proteins as foreign. Explain why the immune system would be able to distinguish animal insulin from human insulin.

37. In prokaryotes, the error rate in protein synthesis may be as high as 5×10^{-4} per codon. How many polypeptides containing (a) 500 residues or (b) 2000 residues would you expect to contain an amino acid substitution?

38. Hexacoordinate Fe(II) in heme is bright red. Pentacoordinate Fe(II) is blue. Explain how these electronic changes account for the different colors of arterial (scarlet) and venous (purple) blood.

39. Explain why globin alone or heme alone is not effective as an oxygen carrier.

40. Invariant residues are those that must be essential for the structure and/or function of the protein and cannot be replaced by other residues. Name two amino acids residues in the globin chains that are the most important invariant residues and explain the reason for your answer.

41. Hemoglobin is 50% saturated with oxygen when $pO_2 = 26$ torr. If hemoglobin exhibited hyperbolic binding (as myoglobin does) with 50% saturation at 26 torr, what would be the fractional saturation when $pO_2 = 30$ torr and 100 torr? What does this tell you about the physiological importance of hemoglobin's sigmoidal oxygen-binding curve?

42. Highly active muscle generates lactic acid by respiration so fast that the blood passing through the muscle actually experiences a drop in pH from about 7.4 to about 7.2. Under these conditions, hemoglobin releases about 10% more O_2 than it does at pH 7.4. Explain.

43. The side chain of Asp 94 on the β chain of hemoglobin is in close proximity to the imidazole ring of His 146 in the deoxy form of hemoglobin but not the oxy form.

(a) What kind of interaction occurs between Asp 94 and His 146 in deoxyhemoglobin?

(b) The proximity of Asp 94 alters the pK value of the imidazole ring of His. In what way?

44. Would you expect hemoglobin from the llama, a species native to the Andes of South America, to have a higher or lower oxygen affinity than human hemoglobin?

45. The developing fetus has a different kind of hemoglobin than most normal adults. Fetal hemoglobin (hemoglobin F) consists of two α chains and two γ chains, whereas adult hemoglobin (hemoglobin A) consists of two α chains and two β chains. Fetal hemoglobin is synthesized beginning at the third month of gestation and continues up through birth. After birth, hemoglobin F synthesis declines and synthesis of hemoglobin A increases (synthesis of the γ chain declines and synthesis of the β chain increases). By the time the baby is six months old, 98% of its hemoglobin is hemoglobin A.

(a) Hemoglobin F has a higher oxygen affinity than hemoglobin A. Why do you think this would be advantageous to the developing fetus?

(b) In the graph below, which curve represents fetal hemoglobin?

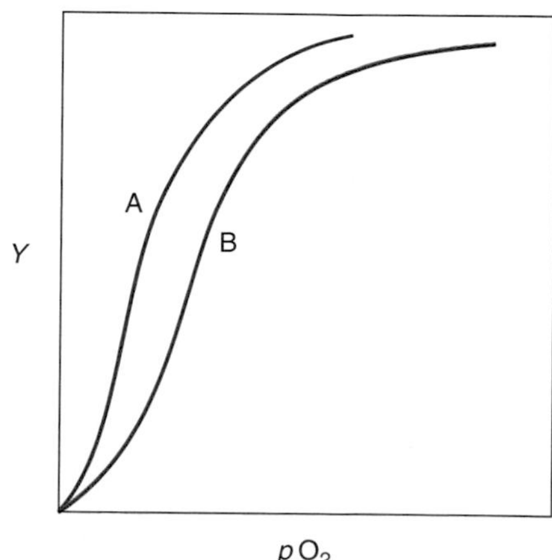

(c) The γ chain differs slightly in structure from the β chain and has fewer Lys and His residues than the β chain. This means that hemoglobin F has fewer positive charges in its central cavity than does hemoglobin A. Given this information, explain why hemoglobin F has a higher affinity for oxygen than does hemoglobin A.

46. Oxygen dissociation curves are shown for hypothetical oxygen carriers A and B.

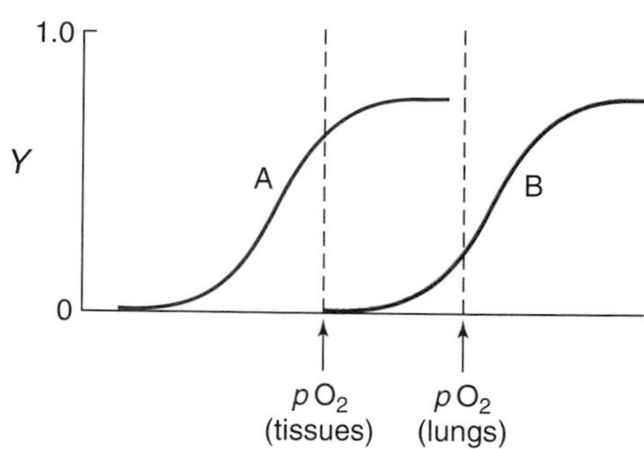

(a) What would be the disadvantage of A as an oxygen carrier?

(b) What would be the disadvantage of B as an oxygen carrier?

(c) Draw a curve for an effective oxygen carrier and explain why you drew the curve the way you did.

47. The oxygen-binding curves for normal hemoglobin (Hb A) and a mutant hemoglobin (Hb Great Lakes) are shown in the figure.

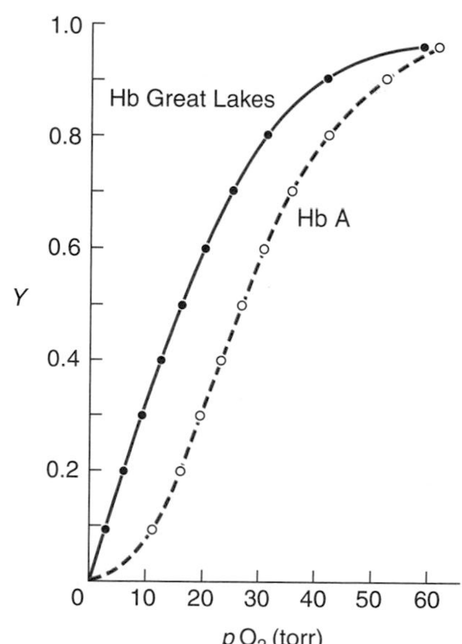

(a) Compare the shapes of the curves for the two hemoglobins and comment on their significance.

(b) Which hemoglobin has a higher affinity for oxygen when $pO_2 = 20$ torr?

(c) Which hemoglobin has a higher affinity for oxygen when $pO_2 = 75$ torr?

(d) Which hemoglobin is most efficient at delivering oxygen from arterial blood (pO_2 = 75 torr) to active muscle (pO_2 = 20 torr)?

48. Drinking a few drops of a preparation called "vitamin O," which consists of oxygen and sodium chloride dissolved in water, purportedly increases the concentration of oxygen in the body. (a) Use your knowledge of oxygen transport to evaluate this claim. (b) Would vitamin O be more or less effective if it were infused directly into the bloodstream?

49. *Plasmodium falciparum,* the protozoan that causes malaria, slightly decreases the pH of the red blood cells it infects. Invoke the Bohr effect to explain why *Plasmodium*-infected cells are more likely to undergo sickling in individuals with the Hb S variant.

50. The drug hydroxyurea can be used to treat sickle-cell anemia, although it is not often used because of undesirable side effects. Hydroxyurea is thought to function by stimulating the afflicted person's synthesis of fetal hemoglobin. In a clinical study, patients who took hydroxyurea showed a 50% reduction in frequency of hospital admissions for severe pain, and there was also a decrease in the frequency of fever and abnormal chest X-rays. Why would this drug alleviate the symptoms of sickle-cell anemia?

Mutant hemoglobins

Changes in the DNA sequence for the genes that encode the α and β chains of hemoglobin have resulted in mutated hemoglobins with altered amino acid sequences. In some cases, the mutation is benign and the hemoglobin molecules function more or less normally. But in other cases, the mutation results in serious physical complications for the individual as the ability of the mutant hemoglobin to deliver oxygen to cells is compromised. Mutations that occur in amino acid residues known to be important for hemoglobin function are particularly detrimental. Mutant hemoglobins are often unstable, which affects oxygen delivery and may also result in destruction of the red blood cells. The role of certain critical amino acids in hemoglobin is summarized in the table below. Use this information to answer Problems 51–55.

51. In the mutant hemoglobin Hb Ohio (β142Ala → Asp), the substitution of Asp for Ala results in the displacement of the G helix relative to the H helix in the β chain. This decreases the stability of the β146His–β94 Asp ion pair. Draw an oxygen-binding curve that compares the relative p_{50} values of Hb A and Hb Ohio. What is the effect of the decreased stability of the His–Asp ion pair on Hb Ohio? [Adapted from Moo-Penn, W.F. et al., *Blood* **56,** 246–250 (1980).]

52. Investigators of mutant human hemoglobins often subject the proteins to cellulose acetate electrophoresis at pH 8.5, a pH at which most hemoglobins are negatively charged. The proteins are applied at the negative pole and migrate toward the positive pole when the current is turned on. The more negatively charged the hemoglobin, the faster it migrates to the positive pole. Results for normal (Hb A) and sickle-cell hemoglobin (Hb S) are shown. Draw a diagram that shows the results of electrophoresis of Hb A and Hb Ohio (β142Ala → Asp).

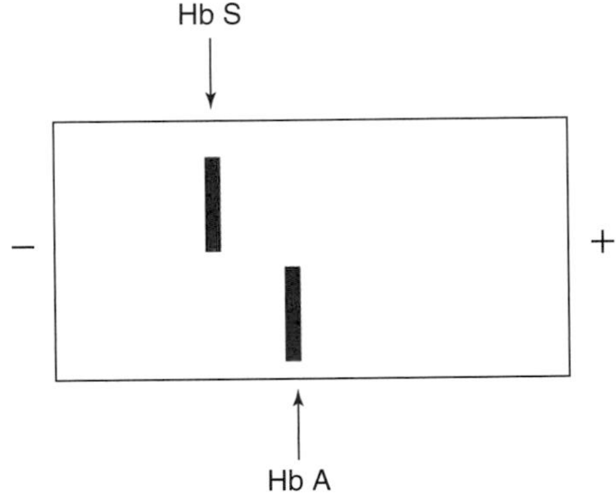

53. Hb Milledgeville (α44Pro → Leu) results in a mutated hemoglobin with altered oxygen affinity. Explain how the oxygen affinity is altered.

54. Hb Providence (β44Lys → Asn), isolated in the mid-1970s, results from a single point mutation of the DNA.

(a) What is the change in the DNA that occurred to produce the mutant hemoglobin?

(b) Compare the oxygen affinity of Hb Providence with that of Hb A.

Chain	Position	Amino acid	Role	Significance
α	44	Pro	Participates in the formation of the $\alpha_1\beta_2$ interface in the deoxy form but not the oxy form.	Stabilizes the deoxy form.
α	141 (C terminus)	Arg	Its COO^- forms an ion pair with Lys 127 and its side chain forms an ion pair with Asp 126 in the deoxy form.	Stabilizes the deoxy form.
β	82	Lys	Forms an ion pair to BPG in the central cavity.	Stabilizes the deoxy form.
β	146	His	The side chain imidazole ring forms an ion pair with Asp 94. It also forms an ion pair with BPG in the central cavity.	Stabilizes the deoxy form.

(c) There are actually two forms of Hb Providence in affected individuals. Hb Providence in which the β44Lys has been replaced with Asn is referred to as "Hb Providence Asn." This mutant hemoglobin can undergo deamidation to produce "Hb Providence Asp." Draw the reaction that converts Hb Providence Asn to Hb Providence Asp.

(d) Draw a diagram that shows the cellulose acetate electrophoresis results for Hb A, Hb Providence Asn, and Hb Providence Asp.

(e) Compare the oxygen affinity of Hb Providence Asp with Hb Providence Asn.

55. Hb Syracuse (β146His → Pro) is a mutant hemoglobin with altered oxygen affinity.

(a) Draw a diagram showing the cellulose acetate electrophoresis results for Hb Syracuse and Hb A.

(b) Evaluate the ability of Hb Syracuse to respond to normal allosteric effectors of hemoglobin. How is the oxygen affinity affected as a result?

56. Rank the stability of the following mutant hemoglobins:

Hb Hammersmith	β42Phe → Ser
Hb Bucuresti	β42Phe → Leu
Hb Sendagi	β42Phe → Val
Hb Bruxelles	β42Phe → 0 (Phe is deleted)

57. There is an abundant protein found in the plasma that is referred to as histidine-proline-rich glycoprotein (HPRG) in part because of its high histidine content (13 mol %). The human HPRG contains a pentapeptide GHHPH sequence repeated in tandem 12 times. Investigators hypothesized that its high histidine content might allow HPRG to play a role in regulating local pH in the blood. The local pH in blood may drop half a pH unit during lactic acidosis or even a full pH unit in hypoxia or ischemia. The binding of HPRG to glycosaminoglycans was investigated. Glycosaminoglycans are anionic polysaccharides that are the major component of the ground substance that forms the matrix of the extracellular spaces of the connective tissue in blood vessel walls. The binding of HPRG to the glycosaminoglycan heparin was measured (the repeating disaccharide unit of heparin is shown below). Based on their results, the investigators proposed a model that describes how binding of HPRG to glycosaminoglycans may allow HPRG to regulate local blood pH.

Heparin

(a) How does the binding of HPRG to heparin depend on pH? Give structural reasons for the pH dependence of binding.

(b) The same binding studies were carried out after HPRG had been reacted with diethylpyrocarbonate (DEPC), a compound that specifically reacts with histidine residues. The reaction is shown below. How does this affect heparin binding?

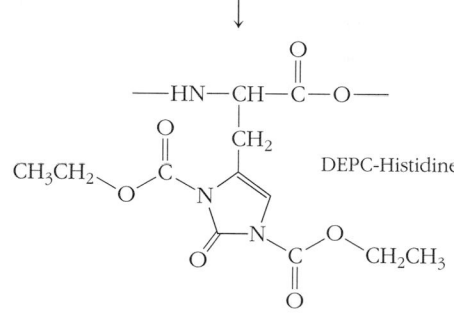

—HN—CH—C(=O)—O— (histidine side chain CH_2–imidazole)

+

$CH_3CH_2—O—C(=O)—O—C(=O)—O—CH_2CH_3$

Diethylpyrocarbonate (DEPC)

↓

DEPC-Histidine

(c) The pH-dependence of HPRG binding to heparin is shown.

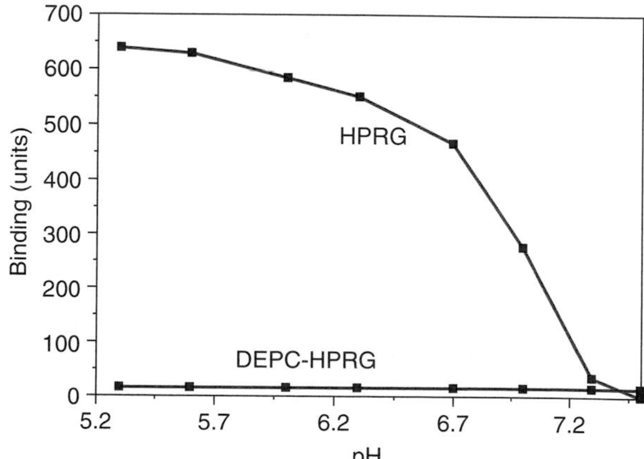

Next, the investigators measured the ability of different concentrations of Cu^{2+} and Zn^{2+} to promote HPRG binding to heparin at pH 7.3. The results are shown on the next page. In addition, the binding of HPRG to heparin in the presence of 5.2 nM Zn^{2+} was measured at pH 6.0 and at pH 7.4. What is your interpretation of these results?

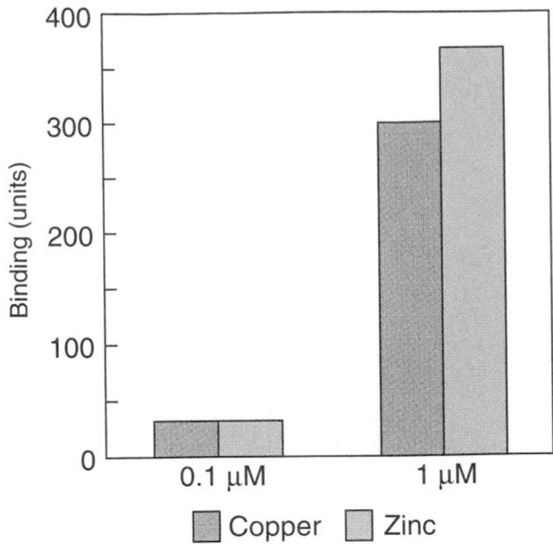

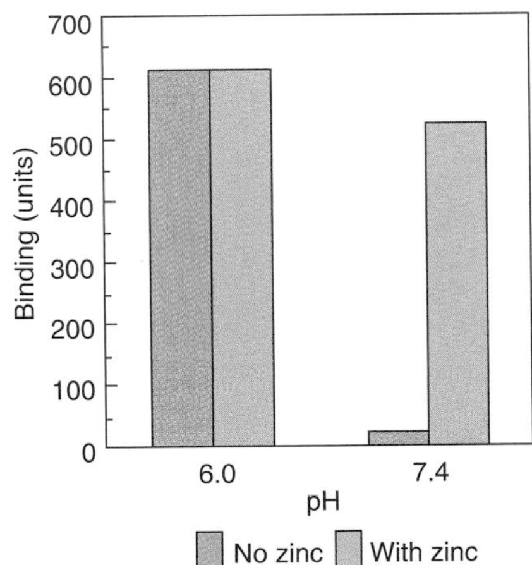

(d) Local cellular pH can decrease from one-half to one pH unit, depending on a variety of circumstances, including ischemia, hypoxia, and inflammation due to lactic acidosis. In addition, metabolic acidosis is often one of the symptoms in complications following surgery. The investigators have proposed that HPRG acts to relieve the acidosis in these conditions. Propose a model that explains the mechanism of pH regulation by HPRG.

(e) Other plasma proteins have been studied for their ability to bind to glycosaminoglycans. One such protein is kininogen, which is a lysine-rich protein. Like HPRG, kininogen is able to bind to glycosaminoglycans, but this binding is far less sensitive to small fluctuations in physiological pH. Why does kininogen easily bind to glycosaminoglycans?

(f) Why is the binding of kininogen less sensitive to physiological pH changes? [Adapted from Borza, D.-B. and Morgan, W.T., *J. Biol. Chem.* **273,** 5493–5499 (1998).]

58. While most human beings are able to hold their breath for only a minute or two, other species are able to stay under water for much longer periods of time. Various physiological adaptations allow some organisms to deliver oxygen to tissues while submerged. For example, deep sea–diving mammals, such as whales and seals, are able to stay under water for long periods of time because their muscles contain many-fold higher concentrations of myoglobin than humans. Crocodiles are also able to stay submerged for periods of time exceeding one hour. This adaptation allows the crocodile to kill small animals by drowning them. However, the crocodile does not have large amounts of myoglobin in its muscle, as the deep sea–diving mammals do, so its physiological adaptation must be different. In 1995, Nagai and colleagues described a possible mechanism that allows crocodile hemoglobin to deliver a large fraction of its bound oxygen to the tissues. They suggested that bicarbonate (HCO_3^-) binds to hemoglobin to promote the dissociation of oxygen in a manner similar to BPG in humans. Their findings are important because information gathered in experiments like those described here may allow scientists to design effective blood replacements.

(a) Explain why having a higher concentration of myoglobin allows whales and seals to stay submerged for a long period of time.

(b) Consider the hypothesis that bicarbonate serves as an allosteric modulator of oxygen binding to hemoglobin in crocodiles. What is the source of HCO_3^- in crocodile tissues?

(c) Draw oxygen-binding curves for crocodile hemoglobin in the presence and absence of bicarbonate. Which conditions give rise to a greater p_{50} value for crocodile hemoglobin? What does this tell you about the oxygen-binding affinity of hemoglobin under those conditions?

(d) The investigators examined the bicarbonate-binding site on crocodile hemoglobin by constructing human–crocodile chimeric hemoglobins in which amino acids in human hemoglobin were replaced with the amino acids found at the same positions in crocodile hemoglobin. The intention was to make a synthetic human hemoglobin that resembled crocodile hemoglobin in its ability to bind bicarbonate. The investigators found that the bicarbonate-binding site was located at the $\alpha_1\beta_2$ subunit interface, where the two subunits slide with respect to each other during the oxy $\leftrightarrow$ deoxy transition. Based on their results, the researchers modeled a *stereochemically plausible* binding site that included the phenolate anion of Tyr 41β, the ε-amino group of Lys 38β, and the phenolate anion of Tyr42α. What kinds of interactions do you think occur between these groups and the bicarbonate anion?

(e) In order to create an engineered human hemoglobin molecule that had the same bicarbonate binding properties as crocodile hemoglobin, 12 amino acids were changed. Not all of these residues directly interact with bicarbonate—perhaps only three of them do, as described in part (d). What might be the role of the other nine amino acids?

(f) Other animals have similarly adapted to using small molecules as allosteric effectors to encourage hemoglobin to release its oxygen. Whereas humans use BPG and crocodiles use HCO_3^-, birds use *myo*-inositol pentaphosphate and fish use ATP or GTP. What structural characteristics do all of these molecules have in common and how would the molecules bind to hemoglobin? [Adapted from Komiyama, N.H., Miyazaki, G., Tame, J., and Nagai, K., *Nature* **373,** 244–246 (1995).]

SELECTED READINGS

Branden, C. and Tooze, J., *Introduction to Protein Structure* (2nd ed.), Garland Publishing (1999). [A well-illustrated book with chapters introducing amino acids and protein structure, plus chapters on specific proteins categorized by their structure and function.]

Bucciantini, M., Giannoni, E., Chiti, F., Baroni, F., Formigli, L., Zurdo, J., Taddei, N., Ramponi, G., Dobson, C.M., and Stefani, M., Inherent toxicity of aggregates implies a common mechanism for protein misfolding diseases, *Nature* **416,** 507–511 (2002). [Provides evidence that a variety of misfolded proteins can form fibrous aggregates that can potentially damage cells.]

Eaton, W.A., Henry, E.R., Hofrichter, J., and Mozzarelli, A., Is cooperative oxygen binding by hemoglobin really understood? *Nature Struct. Biol.* **6,** 351–357 (1999). [Reviews the physical evidence for the quaternary structural changes underlying hemoglobin's allosteric behavior.]

Hsia, C.C.W., Respiratory function of hemoglobin, *New Engl. J. Med.* **338,** 239–247 (1998). [A short review of hemoglobin's physiological role.]

Laity, J.H., Lee, B.M., and Wright, P.E., Zinc finger proteins: New insights into structural and functional diversity, *Curr. Opin. Struct. Biol.* **11,** 39–46 (2001). [Describes well-known and novel zinc finger structures in DNA-binding and other proteins.]

Richardson, J.S., Richardson, D.C., Tweedy, N.B., Gernert, K.M., Quinn, T.P., Hecht, M.H., Erickson, B.W., Yan, Y., McClain, R.D., Donlan, M.E., and Surles, M.C., Looking at proteins: Representations, folding, packing, and design, *Biophys. J.* **63,** 1186–1209 (1992). [A highly readable review showing a variety of ways of visualizing protein folding patterns.]

CHAPTER 5

Cytoskeletal and Motor Proteins

Bioengineers on the lookout for naturally occurring substances with commercially desirable properties have begun to decipher the secrets of spider silk. In particular, dragline silk (which serves as a safety line and frame for the web) is a proteinaceous fiber whose toughness is exceeded only by Kevlar, the key component of bulletproof vests. Spider silk, like the more familiar silk produced by the larvae of the silkworm moth, is made of a protein called fibroin, which consists of Ala- and Gly-rich β sheets interspersed with more amorphous segments of polypeptide. Spider silk contains less crystal-like β sheet than silkworm silk and is therefore less rigid and more extensible than silkworm silk. Ideally, spider fibroin could be woven into durable fabrics or used for making ropes, parachutes, and bandages. The commercial production of this silk could occur at ambient temperatures and in aqueous solution, in contrast to the high temperatures and organic solvents used to synthesize fibers such as Kevlar and nylon. Furthermore, silk is not a petroleum-based product, and, because it is a protein, it is 100% biodegradable.

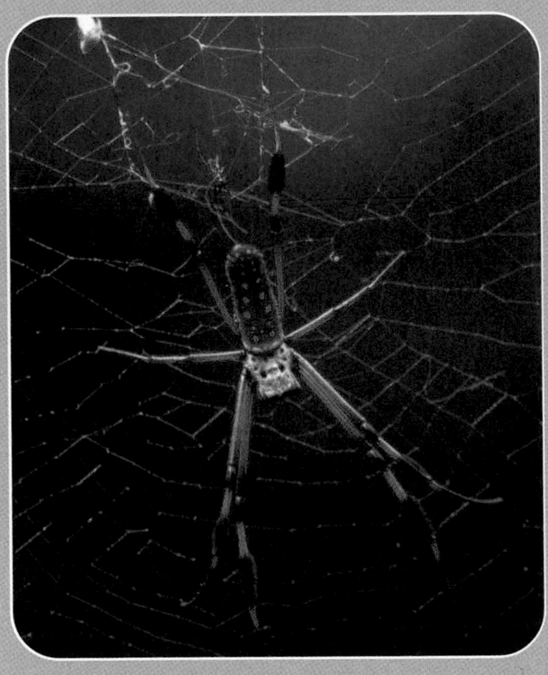

[William Ervin/Science Photo Library/Photo Researchers.]

LEARNING OBJECTIVES

1. Understand how the physical properties of microfilaments, microtubules, intermediate filaments, and collagen fibers depend on their polypeptide composition.
2. Understand how fibrous proteins assemble and, In some cases, disassemble.
3. Understand how the motor proteins myosin and kinesin couple the steps of ATP hydrolysis to physical movement.

THIS CHAPTER IN CONTEXT

Most proteins, such as myoglobin and hemoglobin (described in the preceding chapter), are globular proteins. They are responsible for carrying out the bulk of the metabolic reactions of cells: harvesting and storing free energy, transforming biological compounds into others, and decoding genetic information. But many of the most abundant proteins are fibrous proteins that are elongated and often insoluble. These proteins determine the shape and other physical attributes of cells and organisms. In this second chapter on proteins, we look at some of the proteins that form the filaments and fibers of the cytoskeleton in order to further explore the relationship between molecular structure and function. We also examine how cytoskeletal and motor proteins mediate cellular movements.

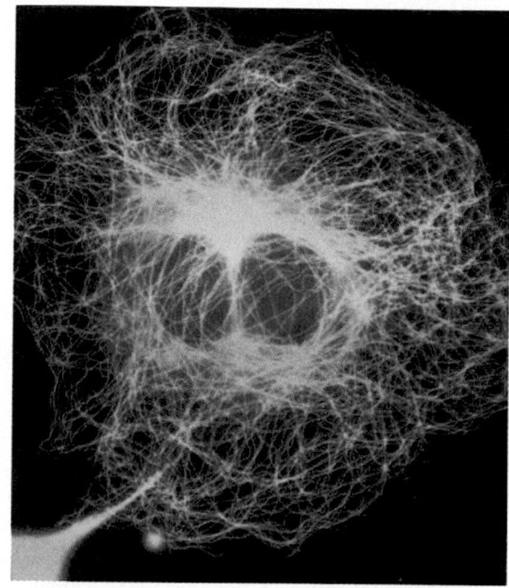

Microfilaments

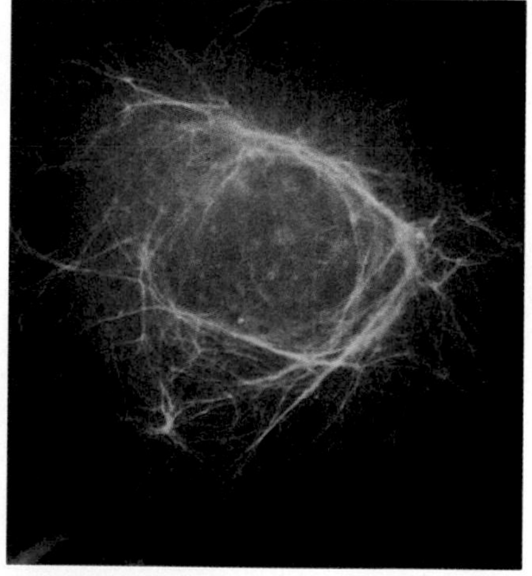

Intermediate filaments

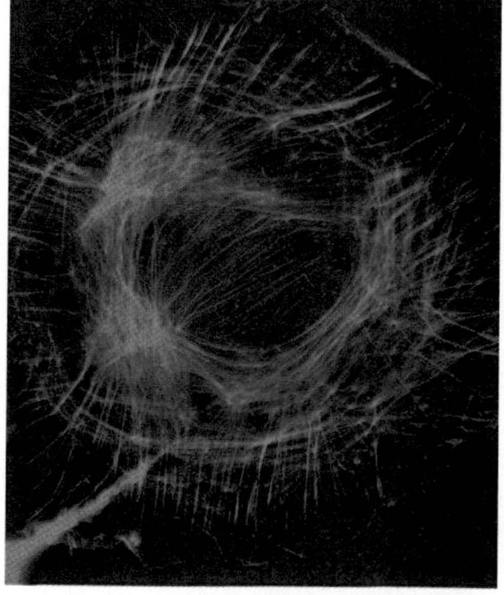

Microtubules

NATURE ABOUNDS IN FILAMENTS AND FIBERS, AS WELL AS CABLES, rods, and tubes. From an engineering perspective, the bodies of living organisms consist of systems of simple structural elements linked in ways that allow them to act on each other as ratchets, pulleys, springs, and levers. Essentially all of the fundamental components of a piece of machinery (for example, a bicycle) have counterparts in the biological world. This principle applies also to individual cells, in which it is possible to identify the rods, tubes, and cables that determine the cell's shape. It is also possible to describe the molecular machinery that bends, pulls, and twists these structures so as to cause the cell to reorganize its constituents, change its shape, and even crawl or swim.

The cell's structural components (collectively known as the **cytoskeleton**) represent some of the most abundant proteins in nature. A typical eukaryotic cell contains three types of cytoskeletal proteins that form fibers extending throughout the cell (Fig. 5-1): These are microfilaments (with a diameter of about 70 Å), intermediate filaments (with a diameter of about 100 Å), and microtubules (with a diameter of about 240 Å). In large multicellular organisms, fibers of the protein collagen provide structural support extracellularly. Bacterial cells also contain proteins that form structures similar to microfilaments and microtubules.

Other than forming fibrous structures that are visible under a microscope, the cytoskeletal proteins, along with collagen, have little in common. *Their conformations are markedly different, and they associate in characteristic ways to produce fibers with distinct physical properties.* In the following sections, note how the structure of each protein influences the overall structure and flexibility of the fiber as well as the fiber's ability to disassemble and reassemble.

FIGURE 5-1

Distribution of cytoskeletal fibers in a single cell.
To make these micrographs, each type of fiber was labeled with a fluorescent probe that binds specifically to one type of cytoskeletal protein. Note how the distribution of microfilaments differs somewhat from that of intermediate filaments and microtubules. [*Courtesy John Victor Small, Austrian Academy of Sciences, Salzburg.*]

Whereas the supportive role of fibrous proteins in cellular architecture may seem obvious, it turns out that many of the dynamic functions of cells are also intricately tied to the cytoskeleton. The movements of cells and the movements of organelles within cells reflect the action of **motor proteins** that operate along tracks provided by cytoskeletal fibers. Motor proteins provide some additional lessons in protein structure and function due to the mechanism whereby they convert metabolic free energy (for example, the energy of ATP) into mechanical work, that is, into molecular movement.

1 MICROFILAMENTS ARE MADE OF ACTIN

Go to **Exercise 7/Actin and Myosin** for a detailed look at the structures of actin and myosin and their dynamics. The material presented in Exercise 7 illuminates Learning Objectives 1, 2, and 3.

A major portion of the eukaryotic cytoskeleton consists of **microfilaments,** or polymers of actin. In many cells, a network of microfilaments supports the plasma membrane and therefore determines cell shape (see Fig. 5-1). Certain proteins cross-link individual actin polymers to help form bundles of microfilaments, thereby increasing their strength. Exercise 7 includes an overview of the different types of cytoskeletal elements.

Actin is a polymer of globular subunits

Monomeric actin is a globular protein (that is, its shape is compact and rounded), and it contains about 375 amino acids (Fig. 5-2). On its surface is a cleft in which adenosine triphosphate (ATP) binds. The adenosine group slips into a pocket on the protein, and the ribose hydroxyl groups and the phosphate groups form hydrogen bonds with the protein.

Polymerized actin is sometimes referred to as **F-actin** (for filamentous actin, to distinguish it from **G-actin,** the globular monomeric form). *The actin polymer is actually a double chain of subunits in which each subunit contacts four neighboring subunits* (Fig. 5-3). Each actin subunit has the same orientation (for example, all the nucleotide-binding sites point up in Fig. 5-3), so the assembled fiber has a distinct polarity. The end with the ATP site is known as the **(−) end,** and the opposite end is the **(+) end.**

Initially, polymerization of actin monomers is slow, because actin dimers and trimers are unstable. But once a longer polymer has formed, subunits

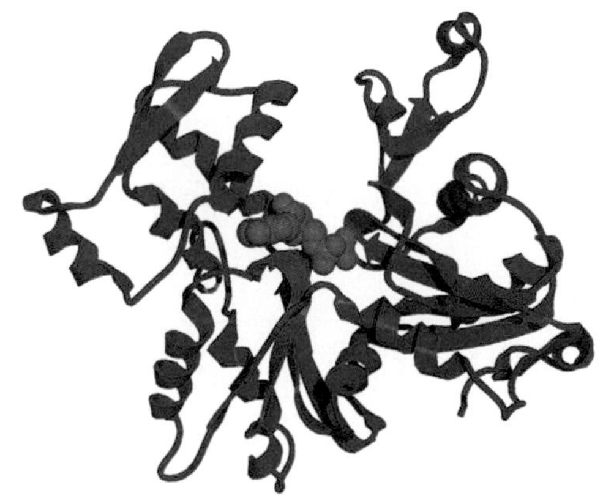

FIGURE 5-2

Actin monomer.
This 372-residue protein assumes a globular shape with a cleft where ATP (green) binds. [*Structure of rabbit actin (pdb 1J6Z) determined by L.R. Otterbein, P. Graceffa, and R. Dominguez.*]

add to both ends. Addition is usually much more rapid at the (+) end (hence its name) than at the (−) end (Fig. 5-4).

The presence of ATP in the actin cleft is not essential for polymerization to occur, but afterward, ATP is hydrolyzed (split by the addition of water) to ADP + inorganic phosphate (P_i):

$$\text{Adenosine triphosphate (ATP)} + H_2O \longrightarrow \text{Adenosine diphosphate (ADP)} + \text{Inorganic phosphate } (P_i)$$

Consequently, *most of the actin subunits in a microfilament contain bound ADP.* Only the most recently added subunits still contain ATP. Because ATP-actin and ADP-actin probably assume slightly different conformations, proteins that interact with microfilaments may be able to distinguish rapidly polymerizing (ATP-rich) microfilaments from longer-established (ADP-rich) microfilaments.

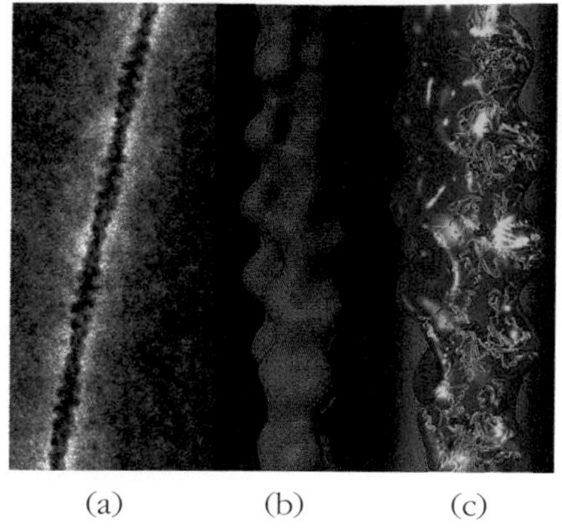

FIGURE 5-3

Actin polymer.
(a) Electron micrograph of an actin filament. (b) X-Ray structure of an actin filament. (c) Theoretical model based on the X-ray structure of actin monomers. [*Photo courtesy of Daniel Stoffler (Scripps Research Institute, La Jolla, CA), Michel O. Steinmetz (Paul Scherrer Institute, Villigen, Switzerland), Andreas Hoenger (European Molecular Biology Laboratory, Heidelberg, Germany) and Ueli Aebi (M.E. Muller Institute for Structural Biology, Biozentrum, Basel, Switzerland), Actin: from cell biology to atomic detail,* J. Struct. Biol. ***119,*** *295–320 (1997), Fig. 1b, 8a, and 8b.*]

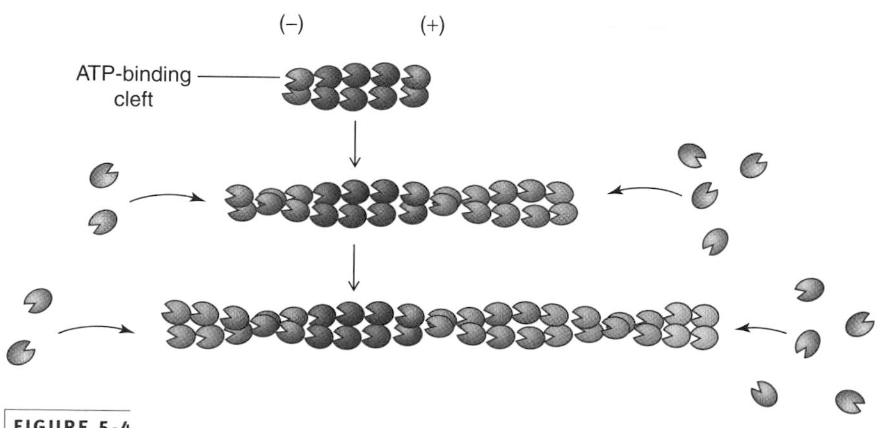

FIGURE 5-4

Microfilament assembly.
A microfilament grows as subunits add to its ends. Subunits usually add more rapidly to the (+) end, which therefore grows faster than the (–) end. Actual microfilaments are much longer than depicted here.

REVIEW LEARNING OBJECTIVE 1

- What is the basis of microfilament polarity?

Microfilaments continuously extend and retract

Microfilaments are dynamic structures. *Polymerization of actin monomers is a reversible process, so the polymer undergoes constant shrinking and growing as subunits add to and dissociate from one or both ends of the microfilament* (see Fig. 5-4). When the net rate of addition of subunits to one end of a microfilament matches the net rate of removal of subunits at the other end, the polymer is said to be **treadmilling** (Fig. 5-5).

Calculations suggest that under cellular conditions, the equilibrium between monomeric actin and polymeric actin favors the polymer. However, the growth of microfilaments *in vivo* is limited by **capping proteins** that bind to and block further polymerization at the (+) or (−) ends. A process that removes a microfilament cap will target growth to the uncapped end.

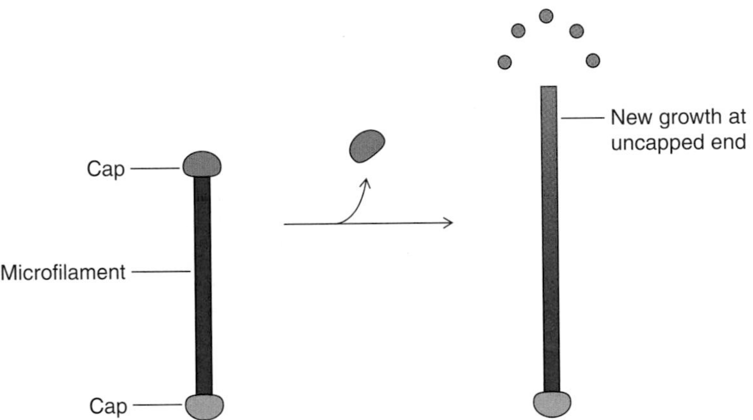

New microfilament growth can also occur as branches form along existing microfilaments. The variations in microfilament structure, including treadmilling, branching, and capping, are animated in Exercise 7.

A supply of actin monomers to support microfilament growth in one area must come at the expense of microfilament disassembly elsewhere. In a cell, certain proteins sever microfilaments by binding to a polymerized actin subunit and inducing a small structural change that weakens actin–actin interactions and thereby increases the likelihood that the microfilament will break at that point. Actin subunits can then dissociate from the newly exposed ends, unless they are subsequently capped.

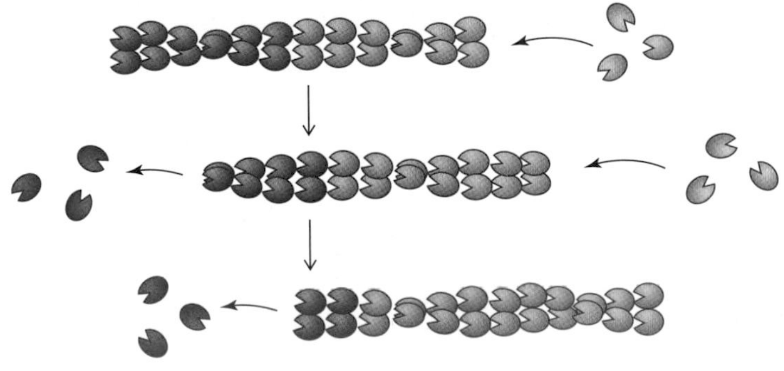

FIGURE 5-5

Microfilament treadmilling.
Net assembly at one end balances net dissociation at the other end.

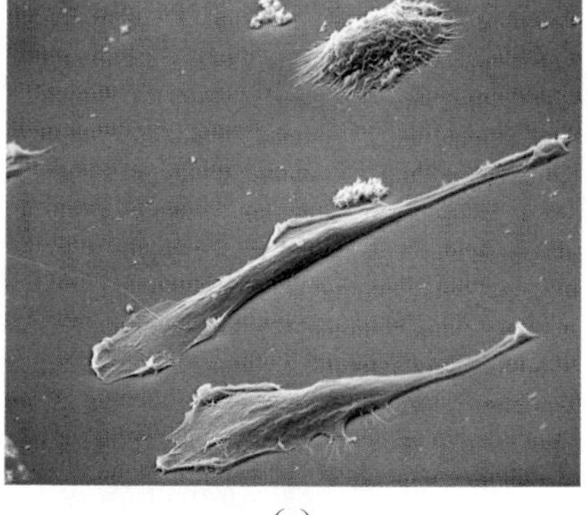
(a)

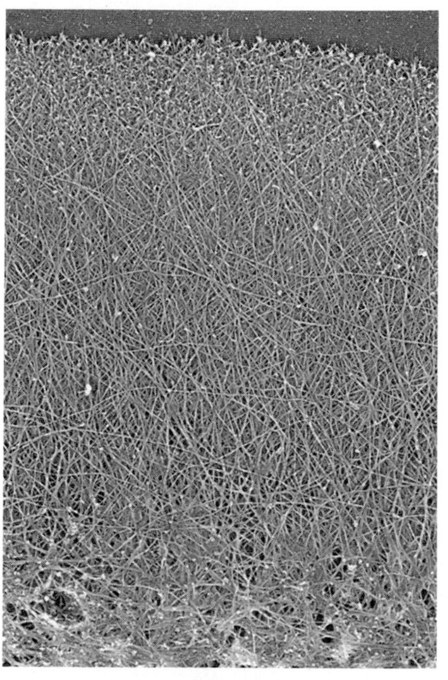
(b)

FIGURE 5-6

Microfilament dynamics in cell crawling.
(a) Scanning electron micrograph of crawling cells. The leading edge of the cell is ruffled where it has become detached from the surface and is in the process of extending. The trailing edge or tail of the cell, still attached to the surface, is gradually pulled toward the leading edge. The rate of actin polymerization is greatest at the leading edge.
(b) Organization of actin filaments in a fish epithelial cell. At the leading edge of the cell (top), the microfilaments form a dense and highly branched network. Deeper within the cell (bottom), the microfilaments are sparser. [*Photo (a) courtesy Guenter Albrecht-Buehler. Photo (b) courtesy Tatyana Svitkina, Northwestern University Medical School. From* J. Cell Biol., ***vol 139,*** *no. 2, October 20, 1997, 397–415.*]

Capping, branching, and severing proteins, along with other proteins whose activity is sensitive to extracellular signals, regulate the assembly and disassembly of microfilaments. A cell containing a network of microfilaments can therefore change its shape as microfilaments lengthen in one area and regress in another. Certain cells use this system to move. When a cell crawls along a surface, actin polymerization extends its "leading" edge while depolymerization helps retract its "trailing" edge (Fig. 5-6a). The high density of growing microfilament ends at the leading edge of the cell (Fig. 5-6b) illustrates how the rapid formation and outward extension of microfilaments can modulate cell shape and drive cell locomotion. Cellular locomotion is normal and necessary in many cells such as those that mediate wound healing and fight infection, but in tumor cells, actin-driven cell movement helps promote **metastasis,** the spread of cancer to another part of the body.

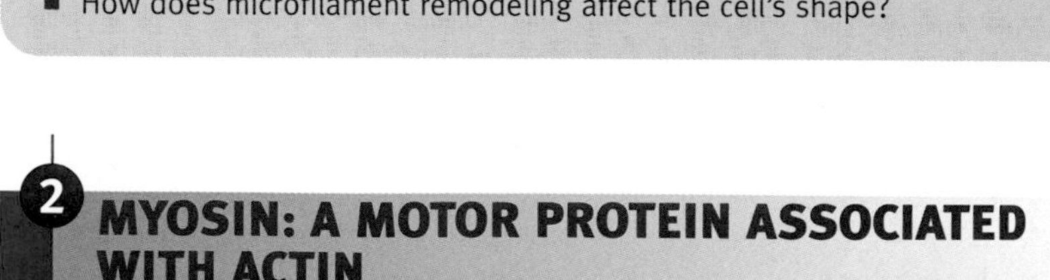

REVIEW LEARNING OBJECTIVE 2

- How does microfilament remodeling affect the cell's shape?

2 MYOSIN: A MOTOR PROTEIN ASSOCIATED WITH ACTIN

Not only do microfilaments provide structural support and generate cell movement by assembly and disassembly, they also participate in generating tensile force. This system is well-developed in muscle cells, where actin filaments are an essential part of the contractile apparatus.

Myosin is a motor protein that works with actin to produce movement by transducing chemical energy (in the form of ATP) to mechanical work. Like actin, myosin is present in nearly all eukaryotic cells. At least 15 different types of myosin have been described, but we will focus on the myosin that is involved in the contraction of skeletal muscle cells. Exercise 7 illustrates the actin and myosin assemblies in skeletal muscle.

Myosin has two heads and a long tail

Muscle myosin, with a total molecular mass of about 540 kD, consists mostly of two large polypeptides that form two globular heads attached to a long tail (Fig. 5-7). Each head includes a binding site for actin and a binding site for an adenine nucleotide. In the tail region, the two polypeptides twist around each other to form a single rod-like **coiled coil** (we will look more closely at the

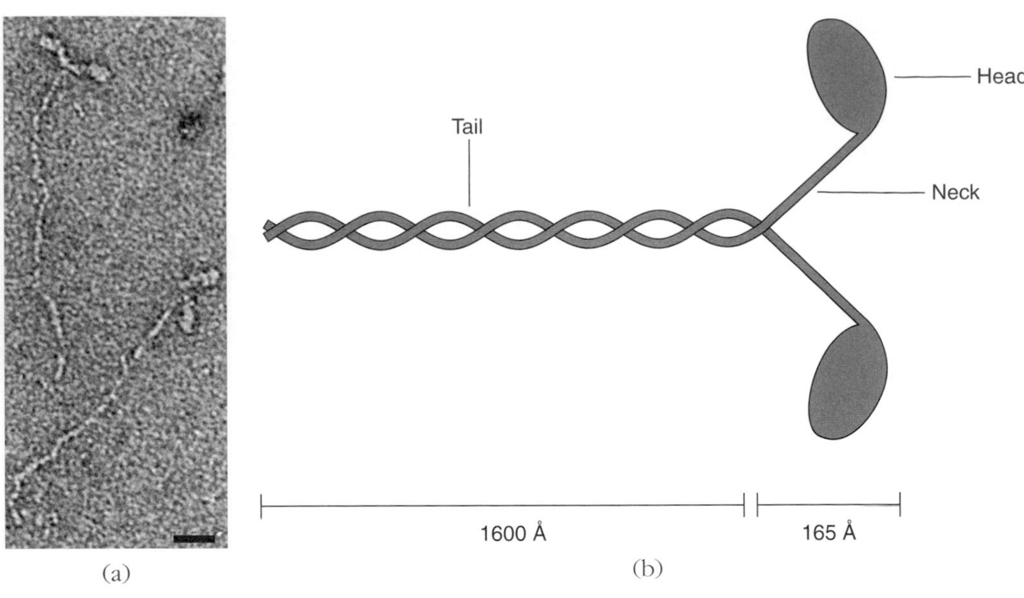

FIGURE 5-7

Overall structure of muscle myosin.
(a) Electron micrograph. (b) Drawing of a myosin molecule. Myosin's two globular heads are connected via necks to myosin's tail, where the polypeptide chains coil around each other to form a long fibrous structure. [*Photo courtesy John Trinick. From Burgess, S.A., Walker, M.L., White, H.D., and Trinick, J.,* J. Cell Biol., ***139***, *675–681 (1997), Fig. 1.*]

structure of coiled coils in Section 5-5). The "neck" that joins each myosin head to the tail region consists of an α helix about 100 Å long, around which are wrapped two small polypeptides (Fig. 5-8). These so-called light chains help stiffen the neck helix so that it can act as a lever. Exercise 7 includes an overview of myosin structure.

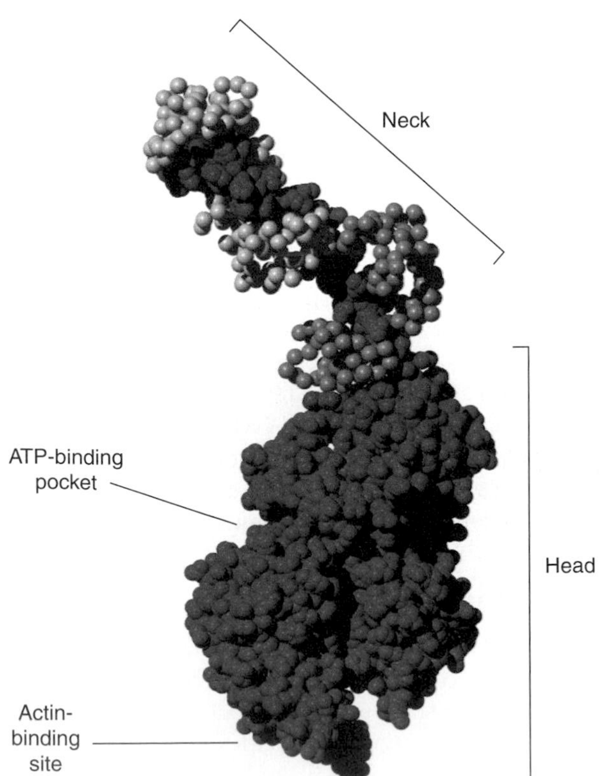

FIGURE 5-8

Myosin head and neck region.
The myosin neck forms a molecular lever between the head domain and the tail. Two light chains (light blue and dark blue) help stabilize the α-helical neck. The actin-binding site is at the far end of the myosin head. ATP binds in a cleft near the middle of the head. Only the alpha carbons of the light chains are visible in this model. [*Structure of chicken myosin (pdb 2MYS) determined by I. Rayment and H.M. Holden.*]

Each myosin head can interact noncovalently with a subunit in an actin filament, but the two heads act independently, and only one head binds to the actin filament at a given time. *In a series of steps that include protein conformational changes and the hydrolysis of ATP, the myosin head releases its bound actin subunit and rebinds another subunit closer to the (+) end of the actin filament.* Repetition of this reaction cycle allows myosin to progressively walk along the length of the actin filament.

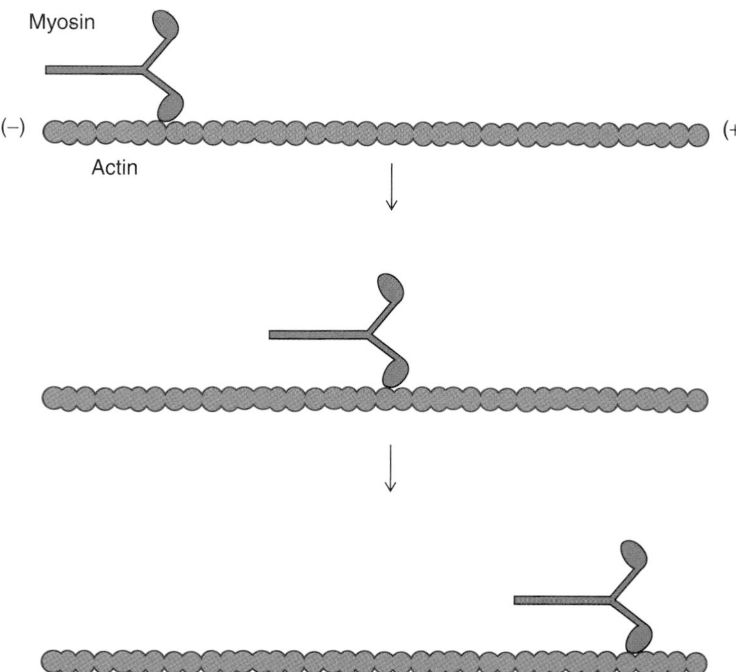

In a muscle cell, hundreds of myosin tails associate to form a **thick filament** with the head domains sticking out (Fig. 5-9). These heads act as cross-bridges to **thin filaments,** which each consist of an actin filament and actin-binding proteins that regulate the accessibility of the actin subunits to myosin heads. When a muscle contracts, the multitude of myosin heads individually bind and release actin, like rowers working asynchronously, which causes the thin and thick filaments to slide past each other (Fig. 5-10). Because of the arrangement of filaments in the muscle cell, the action of

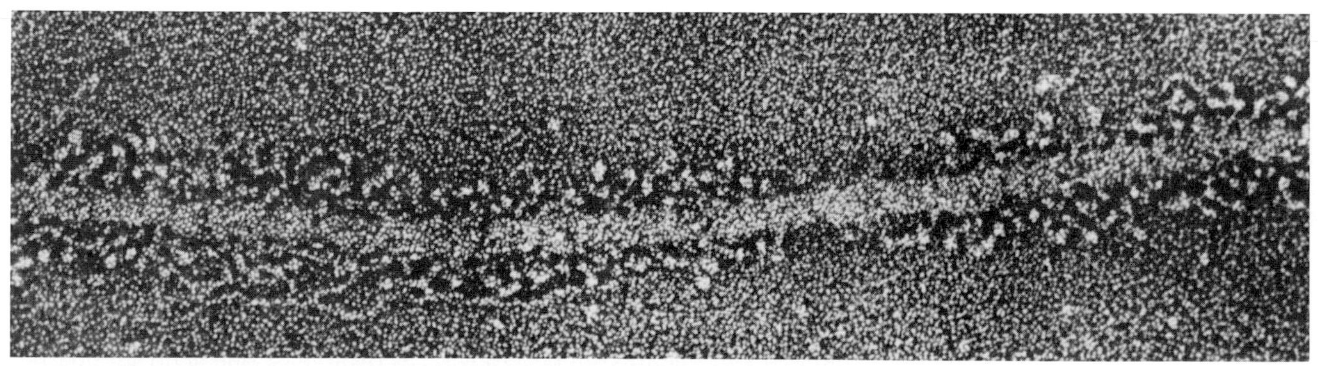

FIGURE 5-9

Electron micrograph of a thick filament.
The heads of many myosin molecules project laterally from the thick rod formed by the aligned myosin tails. [*From Trinick, J. and Elliott, A.,* J. Mol. Biol., ***131,*** *135 (1977).*]

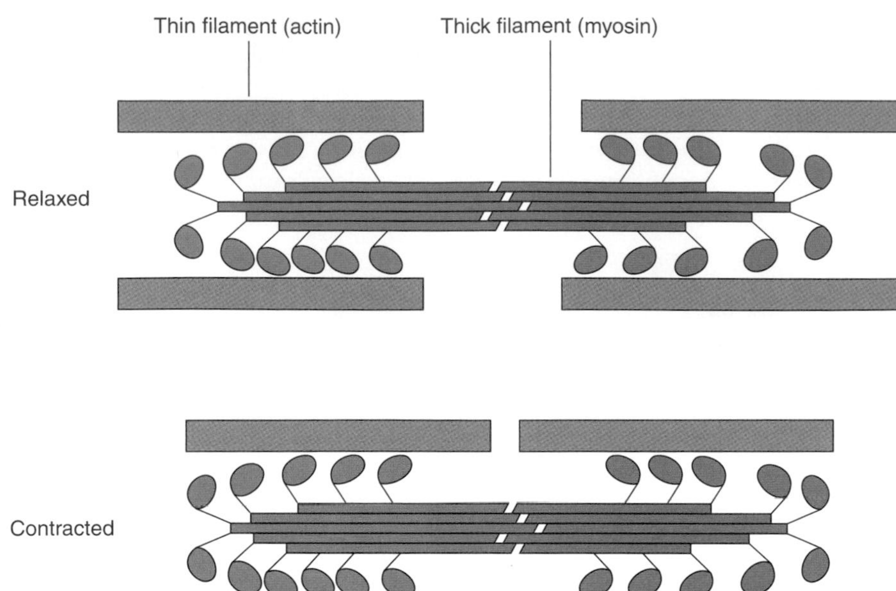

FIGURE 5-10

Movement of thin and thick filaments during muscle contraction.

myosin on actin results in an overall shortening of the muscle. This phenomenon is commonly called contraction, but the muscle does not undergo any compression and its volume remains constant—it actually becomes thicker around the middle. A shortening on the order of 20% of a muscle's length is typical; 40% is extreme.

Myosin operates through a lever mechanism

How does myosin work? The key to its mechanism is the hydrolysis of ATP that is bound to the myosin head. Although the ATP-binding site is about 35 Å away from the actin-binding site, *the conversion of ATP to ADP + P_i elicits conformational changes in the myosin head that are communicated to the actin-binding site as well as to the lever (the neck region). The chemical reaction of ATP hydrolysis thereby drives the physical movement of myosin along an actin filament.* In other words, the free energy released by hydrolysis of ATP is transformed into mechanical work.

The four steps of the myosin–actin reaction sequence are shown in Fig. 5-11. Note how each event at the nucleotide-binding site correlates with a conformational change related to either actin binding or bending of the lever. An α helix makes an ideal lever because it can be quite long. It is also relatively incompressible, so it can pull the coiled-coil myosin tail along with it. Altogether, the lever swings by about 70° relative to the myosin head. The return of the lever to its original conformation (Step 4 of the reaction cycle) is the force-generating step. When corrected for the difference in mass, the myosin–actin system has a power output comparable to a typical automobile. The ATP-dependent conformational changes in the myosin–actin reaction cycle are animated in Exercise 7.

Each cycle of ATP hydrolysis moves the myosin head by an estimated 50–100 Å. Since individual actin subunits are spaced about 55 Å apart along the thin filament, the myosin head advances by at least one actin subunit per reaction cycle. Because the reaction cycle involves several steps, some of which are essentially irreversible, the entire cycle is unidirectional.

Myosin proteins occur in many cells, not just in muscle. For example, myosin works with actin during **cytokinesis** (the splitting of the cell into two

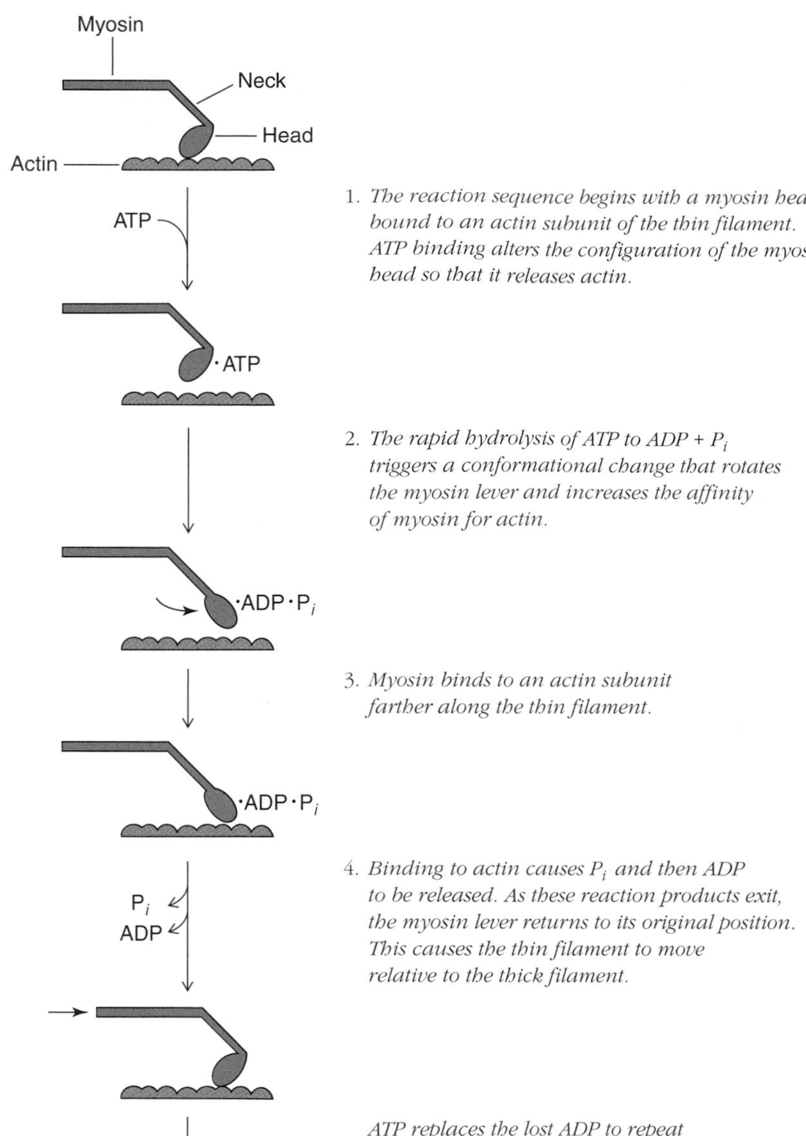

Myosin
Neck
Head
Actin
ATP
·ATP
·ADP·P_i
·ADP·P_i
P_i
ADP
1. The reaction sequence begins with a myosin head bound to an actin subunit of the thin filament. ATP binding alters the configuration of the myosin head so that it releases actin.
2. The rapid hydrolysis of ATP to ADP + P_i triggers a conformational change that rotates the myosin lever and increases the affinity of myosin for actin.
3. Myosin binds to an actin subunit farther along the thin filament.
4. Binding to actin causes P_i and then ADP to be released. As these reaction products exit, the myosin lever returns to its original position. This causes the thin filament to move relative to the thick filament.
ATP replaces the lost ADP to repeat the reaction cycle.

FIGURE 5-11

The myosin–actin reaction cycle.
For simplicity, only one myosin head is shown.

halves following mitosis), and some myosin proteins use their motor activity to transport certain cell components along microfilament tracks. Myosin molecules may also act as tension rods to cross-link the microfilaments of the cytoskeleton. This is one reason why mutations in myosin in the sensory cells of the ear cause deafness and other abnormalities (Box 5-A).

REVIEW LEARNING OBJECTIVE 3

- Describe the steps of the myosin–actin reaction cycle.
- Which events at the nucleotide-binding site correlate with protein conformational changes?

A CLOSER LOOK

BOX 5-A Myosin mutations and deafness

Inside the cochlea, the spiral-shaped organ of the inner ear, are thousands of hair cells, each of which is topped with a bundle of bristles known as **stereocilia.** Each stereocilium contains several hundred cross-linked actin filaments and is therefore extremely rigid, except at its base, where there are fewer actin filaments. Sound waves deflect the stereocilia at the base, initiating an electrical signal that is transmitted to the brain.

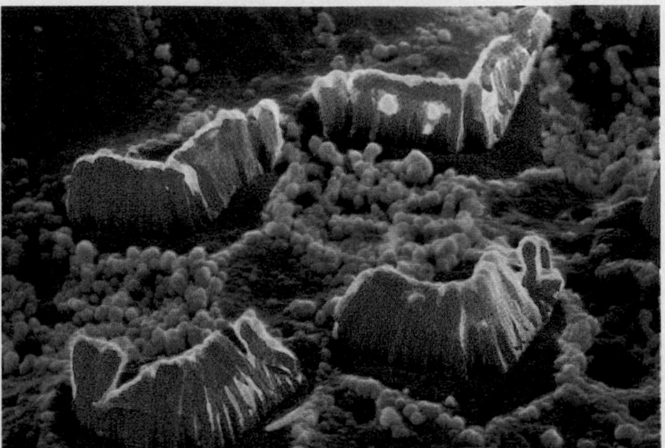

Electron micrograph of stereocilia of a hair cell.

Myosin molecules probably help control the tension inside each stereocilium, so the ratcheting activity of the myosin motors along the actin filaments may adjust the sensitivity of the hair cells to different degrees of stimulus. Other myosin molecules whose tails bind certain cell constituents may use their motor activity to redistribute these substances along the length of the actin filaments. Abnormalities in any of these proteins could interfere with normal hearing.

About half of all cases of deafness have a genetic basis, and over 100 different genes have been linked to deafness. Some of these genes have been identified through mapping studies (similar to those carried out for the cystic fibrosis gene; see Section 3-3). Geneticists working on these genes are often hampered by the fact that deaf people tend to marry other deaf people, so that their offspring, who usually have normal hearing, inherit not one but two different and defective genes.

Among the cytoskeletal proteins whose genes have been linked to deafness is myosin type VIIa, which is considered to be an "unconventional" myosin because it differs somewhat from the "conventional" myosin of skeletal muscle (also known as myosin type II). Gene-sequencing studies indicate that myosin VIIa has 2215 residues and forms a dimer with two heads and a long tail. Its head domains probably operate by the same mechanism as muscle myosin, converting the chemical energy of ATP into mechanical energy for movement relative to an actin filament. Several dozen mutations in the myosin VIIa gene have been identified, including premature stop codons, amino acid substitutions, and deletions—all of which compromise the protein's function. Such mutations are responsible for many cases of **Usher syndrome,** the most common form of deaf-blindness in the United State. Usher syndrome is characterized by profound hearing loss, retinitis pigmentosa (which leads to blindness), and sometimes vestibular (balance) problems.

The congenital deafness of Usher syndrome results from the failure of the cochlear hair cells to develop properly. The unresponsiveness of the stereocilia to sound waves probably also accounts for their inability to respond normally to the movement of fluid in the inner ear, which is necessary for maintaining balance. Abnormal myosin also plays a role in the blindness that often develops in individuals with Usher syndrome, usually by the second or third decade. The intracellular transport function of myosin VIIa is responsible for distributing bundles of pigment in the retina. In retinitis pigmentosa, retinal neurons gradually lose their ability to transmit signals in response to light; in advanced stages of the disease, pigment actually becomes clumped on the retina.

Unfortunately, understanding the molecular defect that underlies these cases of Usher syndrome does not offer an immediate cure for its symptoms. Nor does it explain the types of Usher syndrome not caused by mutations in the myosin type VIIa gene, or the many other types of congenital hearing disorders.

[P. Motta, Dept. of Anatomy, University La Sapienza, Rome/Science Photo Library/Photo Researchers.]

3 TUBULIN FORMS HOLLOW TUBES

Like microfilaments, **microtubules** are cytoskeletal fibers built from small globular protein subunits. Consequently, they share with microfilaments the ability to assemble and disassemble on a time scale that allows the cell to rapidly change shape in response to external or internal stimuli. Compared to a microtubule, however, a microfilament is a thin and flexible rod. *A microtubule is about three times thicker and much more rigid because it is constructed as a hollow tube.* Consider the following analogy: A rod of metal with the dimensions of a pencil is easily bent. The same quantity of metal, fashioned into a hollow tube with a larger diameter but the same length, is much more resistant to bending.

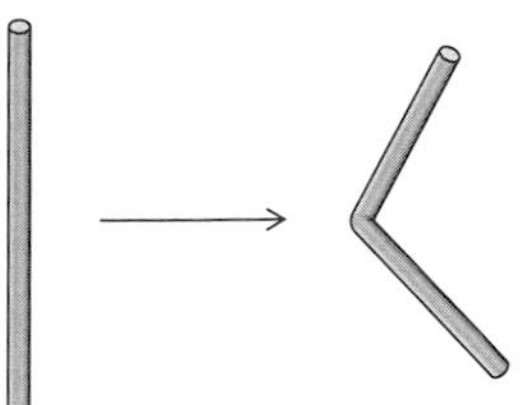

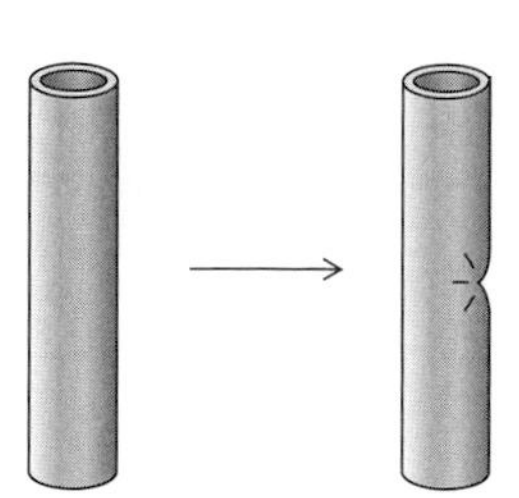

Bicycle frames, plant stems, and bones are built on this same principle. Cells use hollow microtubules to reinforce other elements of the cytoskeleton (see Fig. 5-1), to construct cilia and flagella, and to align and separate pairs of chromosomes during mitosis. The central rod among the bundle of stereocilia in the inner ear (see Box 5-A) is actually built from microtubules rather than actin filaments.

FIGURE 5-12
Structure of β-tubulin.
The strands of the two β sheets are shown in blue, and the 12 α helices that surround them are green. [*Structure of pig tubulin (pdb 1TUB) determined by E. Nogales and K.H. Downing.*]

The tubulin dimer

The basic structural unit of a microtubule is the protein tubulin. Two monomers, known as α-tubulin and β-tubulin, form a dimer, and a microtubule grows by the addition of tubulin dimers. Each tubulin monomer contains about 450 amino acids, 40% of them identical in α- and β-tubulin. The core of tubulin consists of a four-stranded and a 6-stranded β sheet surrounded by 12 α helices (Fig. 5-12).

The structure of the tubulin dimer was determined by **electron crystallography,** because tubulin does not form crystals suitable for X-ray crystallography (see Box 4-B). Electron crystallography measurements were made by recording the diffraction of electron beams by a two-dimensional sheet of tubulin (rather than a three-dimensional crystal). One advantage of this method is that it mapped the electron density of tubulin in a dense array and therefore revealed some of the interactions that occur in fully assembled microtubules. X-Ray crystallography would have yielded structural information for only a discrete unit—the tubulin dimer.

Each tubulin subunit has a nucleotide-binding site. Unlike actin and myosin, tubulin binds a guanine nucleotide, either guanosine triphosphate (GTP) or its hydrolysis product, guanosine diphosphate (GDP). When the dimer forms, the α-tubulin GTP-binding site becomes buried in the interface between the monomers. The nucleotide-binding site in β-tubulin remains exposed to the solvent (Fig. 5-13). After the tubulin dimer is incorporated into a microtubule and another dimer binds on top of it, the β-tubulin nucleotide-binding site is also sequestered from solvent. The GTP is then hydrolyzed, but the resulting GDP remains bound to β-tubulin because it cannot diffuse away (the GTP in the α-tubulin subunit remains where it is and is not hydrolyzed).

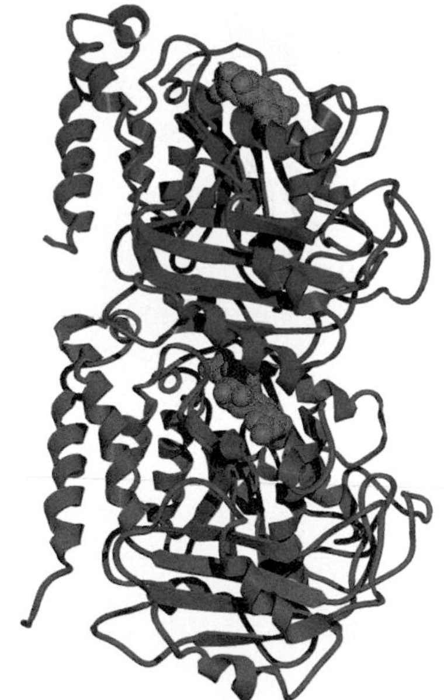

FIGURE 5-13
The tubulin dimer.
The guanine nucleotide (gold) in the α-tubulin subunit (bottom) is inaccessible in the dimer, whereas the nucleotide in the β-tubulin subunit (top) is more exposed to the solvent. [*Structure of the tubulin dimer (pdb 1TUB) determined by E. Nogales and K.H. Downing.*]

Microtublule dynamics

Assembly of a microtubule begins with the end-to-end association of tubulin dimers to form a short linear **protofilament** (Fig. 5-14). Protofilaments then align side-to-side in a curved sheet, which wraps around on itself to form a hollow tube of 13 protofilaments. The microtubule extends as tubulin dimers

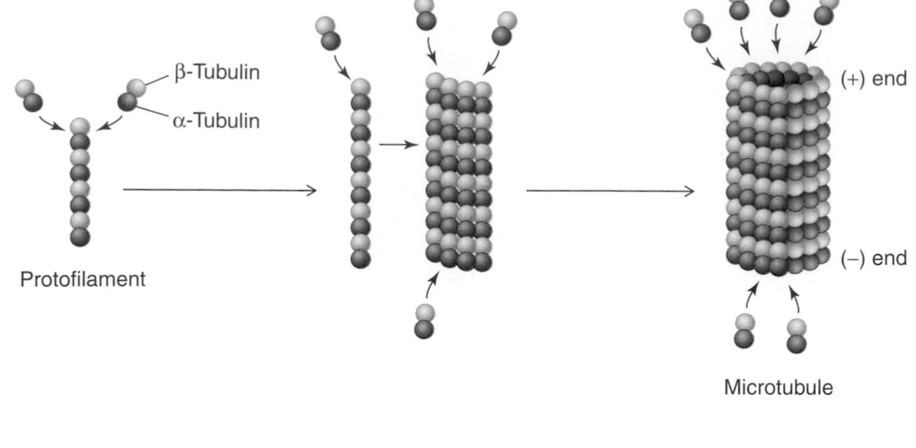

FIGURE 5-14

Assembly of a microtubule.
αβ Dimers of tubulin initially form a linear protofilament. Protofilaments associate side by side, ultimately forming a tube. Tubulin dimers can add to either end of the microtubule, but growth is about twice as fast at the (+) end.

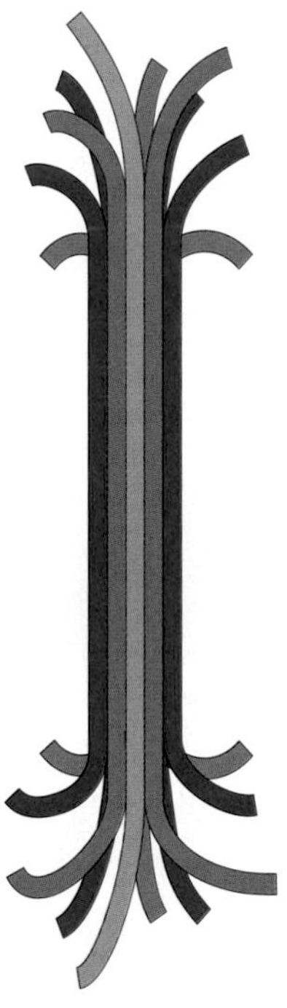

FIGURE 5-15

A depolymerizing microtubule.
Protofilaments apparently separate before tubulin dimers dissociate from the microtubule.

add to both ends. Like a microfilament, the microtubule is polar and one end grows more rapidly. *The (+) end, terminating in β-tubulin, grows about twice as fast as the (−) or α-tubulin end, because tubulin dimers bind preferentially to the (+) end.*

Disassembly of a microtubule also takes place at both ends but occurs more rapidly at the (+) end. Under conditions that favor depolymerization, the ends of the microtubule appear to fray (Fig. 5-15). This suggests that tubulin dimers do not simply dissociate individually from the microtubule ends, but that the interactions between protofilaments weaken before the tubulin dimers come loose.

Under certain conditions, microtubule treadmilling can occur, when tubulin subunits add to the (+) end as fast as they leave the (−) end. *In vivo*, the (−) ends are often anchored to some sort of organizing center in the cell. This means that most microtubule growth and regression occurs at the (+) end, nearer the cell periphery. Microtubule dynamics are regulated by proteins that cross-link microtubules and promote or prevent depolymerization. Gravity may also play a role in coordinating microtubule growth. Experiments in space suggest that although gravity has no effect on individual molecules (forces stemming from random thermal motion would be much greater), in a dynamic assembly/disassembly system such as a cell-wide microtubule network, gravitational force helps organize microtubules.

REVIEW LEARNING OBJECTIVE 1

- How does a microtubule differ structurally from a microfilament?

Some drugs affect microtubules

Compounds that interfere with microtubule dynamics can have drastic physiological effects. One reason is that during mitosis, paired chromosomes separate along a spindle made of microtubules (Fig. 5-16). The drug colchicine, a product of the meadow saffron plant, causes microtubules to depolymerize, thereby blocking cell division. Colchicine binds at the interface between α- and β-tubulin in a dimer, facing the inside of the microtubule cylinder. The bound drug may induce a slight conformational change that weakens the lateral contacts between protofilaments. If enough colchicine is present, microtubules shorten and eventually disappear. Colchicine was first used over 2000 years ago to treat **gout** (an inflammation stemming from the precipitation of uric acid in the joints) because it inhibits the action of the white blood cells that mediate inflammation.

Colchicine

Taxol® binds to β-tubulin subunits in a microtubule, but not to free tubulin, so it stabilizes the microtubule, preventing its depolymerization. The taxol–tubulin interaction appears to include close contacts between taxol's phenyl groups and hydrophobic residues such as Phe, Val, and Leu. Taxol was originally extracted from the slow-growing and endangered Pacific yew tree, but it can also be purified from more renewable sources or chemically synthesized. *Taxol is used as an anticancer agent because it blocks cell division and is therefore toxic to rapidly dividing cells such as tumor cells.*

Taxol

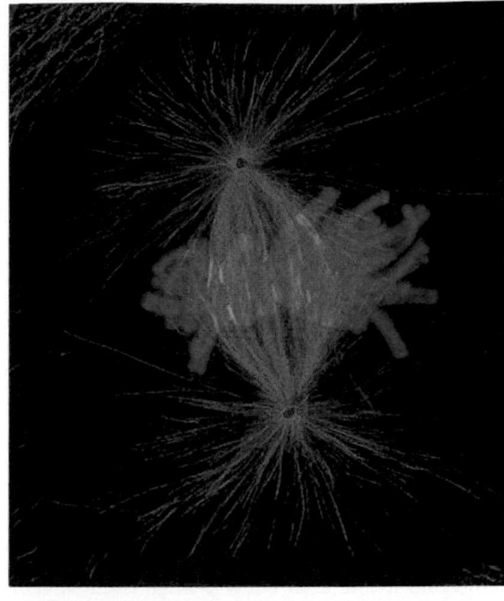

FIGURE 5-16

Microtubules in a dividing cell. During mitosis, a microtubule-based spindle (green fluorescence) links paired chromosomes (blue fluorescence) to two points at opposite ends of the cell. Retraction of microtubules toward these poles draws the chromosomes apart before the cell splits in half. [*Photo by Dr. Alexey Khodjakov and Dr. Conly L. Rieder, Division of Molecular Medicine, Wadsworth Center, Albany, NY.*]

REVIEW LEARNING OBJECTIVE 2

- Why does normal physiological function require regulation of the assembly and disassembly of cytoskeletal elements?

4 KINESIN: A MICROTUBULE MOTOR PROTEIN

Just as many cells contain molecular motors that move along microfilament tracks (for example, myosin VIIa; see Box 5-A), they contain motors that move along microtubule tracks. Kinesin is one of these motor proteins. Although it was discovered only in 1985, it is nearly as well understood as myosin, with which it shares some important structural and functional characteristics. There are several different types of kinesins; we will describe the prototypical one, also known as conventional kinesin.

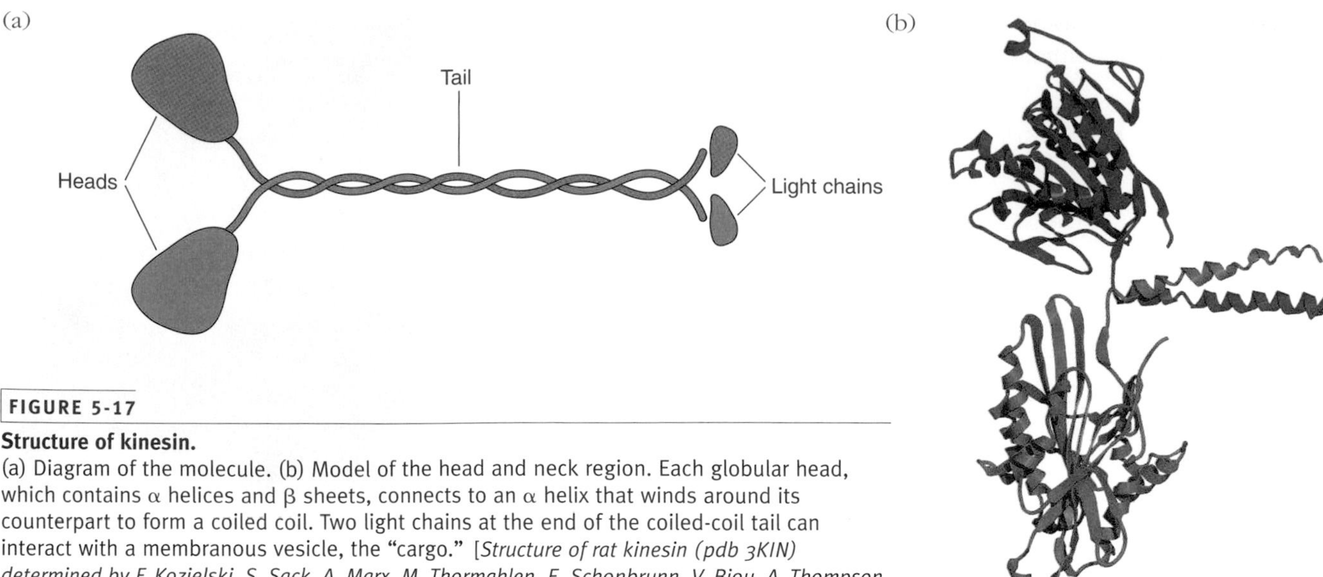

FIGURE 5-17

Structure of kinesin.
(a) Diagram of the molecule. (b) Model of the head and neck region. Each globular head, which contains α helices and β sheets, connects to an α helix that winds around its counterpart to form a coiled coil. Two light chains at the end of the coiled-coil tail can interact with a membranous vesicle, the "cargo." [*Structure of rat kinesin (pdb 3KIN) determined by F. Kozielski, S. Sack, A. Marx, M. Thormahlen, E. Schonbrunn, V. Biou, A. Thompson, E.-M. Mandelkow, and E. Mandelkow.*]

Kinesin, like myosin, is a relatively large protein (with a molecular mass of 380 kD) and has two large globular heads and a coiled-coil tail domain (Fig. 5-17). Each 100-Å-long head consists of an 8-stranded β sheet flanked by 3 α helices on each side and includes a tubulin-binding site and a nucleotide-binding site. The light chains, situated at the opposite end of the protein, bind to proteins in the membrane shell of a **vesicle.** The vesicle and its contents become kinesin's cargo.

Kinesin moves its cargo toward the (+) end of a microtubule by stepping along the length of a single protofilament.

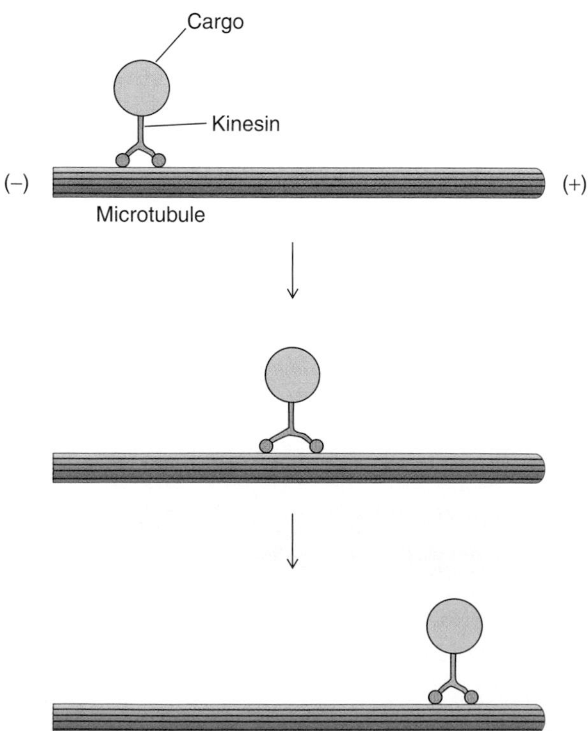

Other microtubule-associated motor proteins appear to use a similar mechanism but move toward the (−) end of the microtubule.

Kinesin's two heads work together

The motor activity of kinesin requires the chemical energy of ATP hydrolysis. However, *kinesin cannot follow the myosin lever mechanism because its head domains are not rigidly fixed to its neck regions.* In kinesin, a relatively flexible polypeptide segment joins each head to an α helix that eventually becomes part of the coiled-coil tail (see Fig. 5-17b). (Recall that in myosin, the lever is a long α helix that extends from the head to the coiled-coil region and is stiffened by the two light chains; see Fig. 5-8). Nevertheless, the relative flexibility of kinesin's neck is critical for its function.

Kinesin's two heads are not independent but work in a coordinated fashion. According to one model for kinesin action, the two heads alternately bind to successive β-tubulin subunits along a protofilament. Conformational changes elicited by ATP binding and hydrolysis are relayed to other regions of the molecule (Fig. 5-18). *This transforms the free energy of ATP into the mechanical movement of kinesin.* Each ATP-binding event yanks the trailing head forward

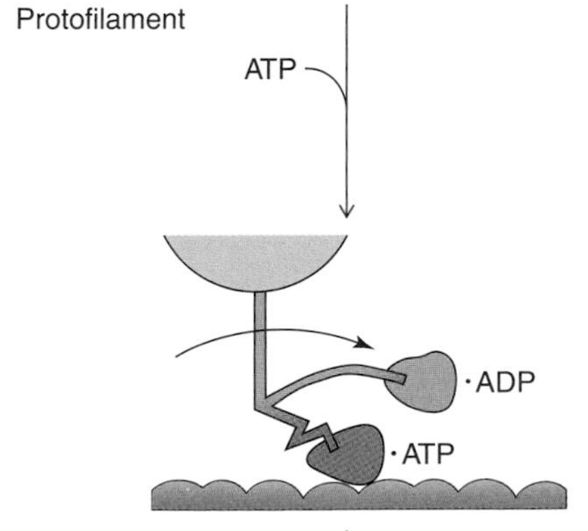

1. *ATP binding to the leading head (orange) induces a conformational change in which the neck docks against the head. This movement swings the trailing head (yellow) forward by 180° toward the (+) end of the microtubule. This is the force-generating step.*

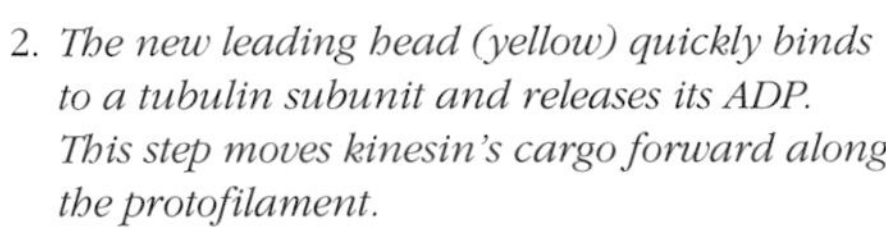

2. *The new leading head (yellow) quickly binds to a tubulin subunit and releases its ADP. This step moves kinesin's cargo forward along the protofilament.*

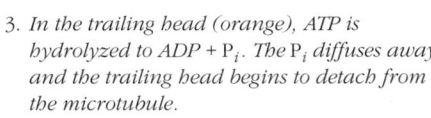

3. *In the trailing head (orange), ATP is hydrolyzed to ADP + P_i. The P_i diffuses away, and the trailing head begins to detach from the microtubule.*

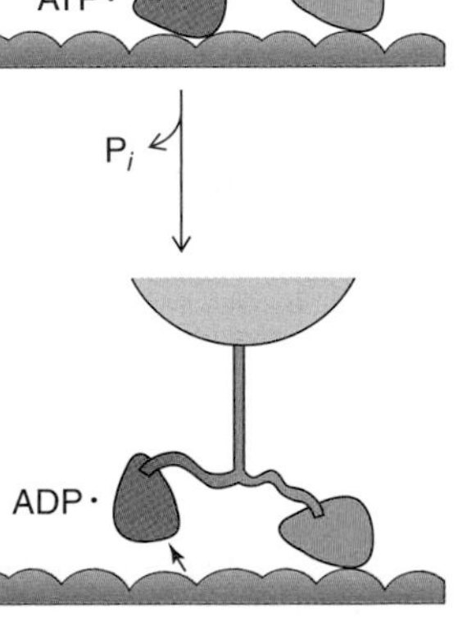

ATP binds to the leading head to repeat the reaction cycle.

FIGURE 5-18

The hand-over-hand model for kinesin action.
This scheme begins with one kinesin head bound to a tubulin subunit in a protofilament of a microtubule. The trailing head has ADP in its nucleotide-binding site. For clarity, the relative size of the neck region is exaggerated. In the alternative inchworm model, one head hydrolyzes ATP and moves forward, while the other, or trailing, head, catches up with the advancing head.

by about 160 Å, so that the net movement of the attached cargo is about 80 Å, or the length of a tubulin dimer. Experiments with immobilized kinesin also provide evidence that kinesin may function more like an inchworm. In this case, only one head advances, dragging the other one behind it.

REVIEW LEARNING OBJECTIVE 3

- Compare and contrast the myosin–actin and kinesin–tubulin reaction cycles.

Kinesin is a processive motor

As in the myosin–actin system (see Fig. 5-1), the kinesin–tubulin reaction cycle proceeds in only one direction. Although most molecular movement is associated with ATP binding, ATP hydrolysis is a necessary part of the reaction cycle. The slowest step of the reaction cycle shown in Figure 5-18 is the dissociation of the trailing kinesin head from the microtubule. ATP binding to the leading head may help promote trailing-head release as part of the forward-swing step. Because the free head quickly rebinds tubulin, the kinesin heads spend most of their time bound to the microtubule track.

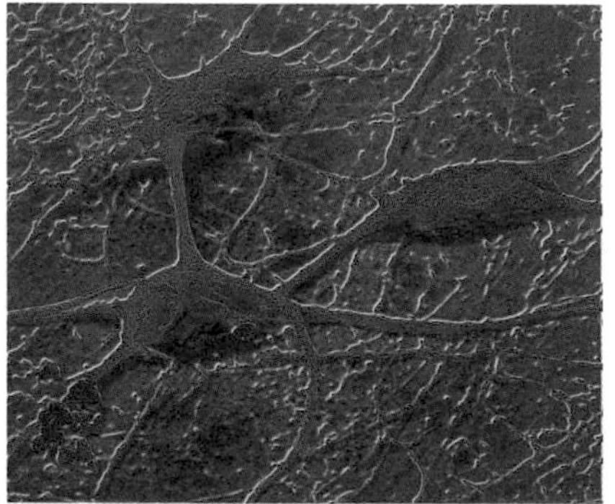

FIGURE 5-19

Electron micrograph of neurons. Microtubule-associated motor proteins move cargo between the cell body and the ends of the axon and other cell processes. [*CNRI/Science Photo Library/Photo Researchers.*]

One consequence of kinesin's almost constant hold on a microtubule is that many—perhaps 100 or more—cycles of ATP hydrolysis and kinesin advancement can occur before the motor dissociates from its microtubule track. *Kinesin is therefore said to have high* ***processivity.*** A motor protein such as myosin, which dissociates from an actin filament after a single stroke, is not processive.

In a muscle cell, low processivity is permitted because the many myosin–actin interactions occur more or less simultaneously to cause the thin and thick filaments to slide past each other (see Fig. 5-10). *High processivity is advantageous for a transport engine such as kinesin, because its cargo (which is relatively large and bulky) can be moved long distances without being lost.* Consider the need for efficient transport in a neuron. Neurotransmitters and membrane components are synthesized in the cell body, where ribosomes are located, but must be moved to the end of the axon, which may be several meters long in extreme cases (Fig. 5-19).

REVIEW LEARNING OBJECTIVE 3

- Why is kinesin a processive motor?

5 KERATIN IS AN INTERMEDIATE FILAMENT

In addition to microfilaments and microtubules, eukaryotic cells—particularly those in multicellular organisms—contain **intermediate filaments.** With a diameter of about 100 Å, these fibers are intermediate in thickness to microfilaments and microtubules. *Intermediate filaments are exclusively structural proteins.* They play no part in cell motility, and they have no associated motor proteins. However, they do interact with microfilaments and microtubules via cross-linking proteins.

Intermediate filament proteins as a group are much more heterogeneous than the highly conserved actin and tubulin. For example, humans contain over 50 intermediate filament genes. In many cells, intermediate filaments are much more abundant than microfilaments or microtubules and are most prominent in the dead remnants of epidermal cells—that is, in the hard outer layers of the skin—where they may account for 85% of the total protein (Fig. 5-20). The best-known intermediate filament proteins are the keratins, a large group of proteins that include the "soft" keratins, which help define internal body structures, and the "hard" keratins of skin, hair, and claws.

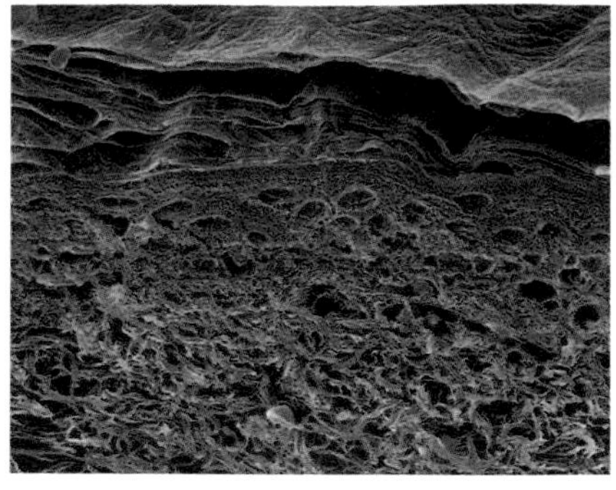

FIGURE 5-20

Scanning electron micrograph of sectioned human skin.
The layers of dead epidermal cells at the top consist mostly of keratin. [*Science Photo Library/Photo Researchers.*]

Keratin forms coiled coils

The basic structural unit of an intermediate filament is a dimer of α helices that wind around each other—that is, a coiled coil (the same structural motif found in myosin and kinesin tails). The amino acid sequence in such a structure consists of seven-residue repeating units where the first and fourth residues are predominantly nonpolar. In an α helix, these nonpolar residues line up along one side (Fig. 5-21). Because a nonpolar group appears on average every 3.5 residues but there are 3.6 residues per α-helical turn, the strip of nonpolar residues actually winds slightly around the surface of the helix. Two helices whose nonpolar strips contact each other therefore adopt a coiled structure with a left-handed twist (Fig. 5-22). Additional views of a coiled-coil structure, including the positions of its nonpolar residues, are shown in Exercise 8.

Go to **Exercise 8/Keratin and Collagen** for a detailed look at the coiled-coil structure of keratin and the triple-helical structure of collagen. The molecular structures presented in Exercise 8 and the questions based on those structures address Learning Objective 1.

Each intermediate filament subunit contains a stretch of α helix flanked by nonhelical regions at the N- and C-termini. Two of these polypeptides interact in register (parallel and with ends aligned) to form a coiled coil. The dimers then associate in a staggered antiparallel arrangement to form higher-order fibrous structures (Fig. 5-23). The fully assembled intermediate filament may consist of 16 to 32 polypeptides in cross-section. Note that no nucleotides are required for intermediate filament assembly. The N- and C-terminal domains may help align subunits during polymerization and interact with proteins that cross-link intermediate filaments to other cell components. Keratin fibers themselves are cross-linked through disulfide bonds between Cys residues on adjacent chains.

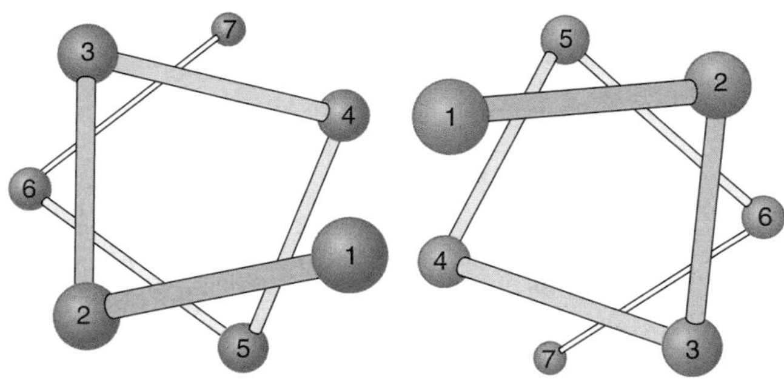

FIGURE 5-21

Arrangement of residues in a coiled coil.
This view down the axis of two seven-residue α helices shows that amino acids at positions 1 and 4 line up on one side of each helix. Nonpolar residues occupying these positions form a hydrophobic strip along the sides of the helices.

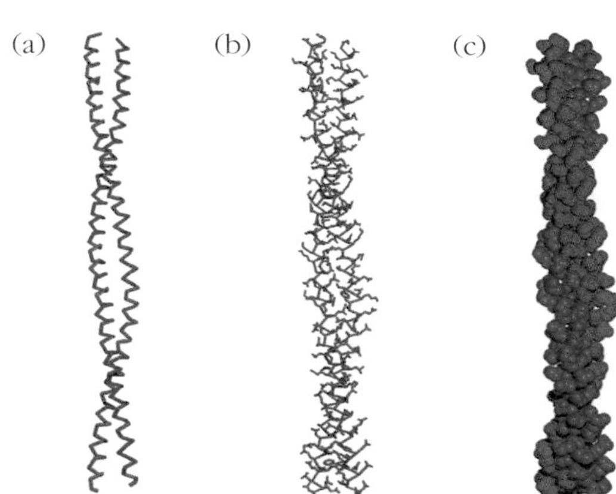

FIGURE 5-22

Three views of a coiled coil.
These models show a segment of the coiled coil from the protein tropomyosin. (a) Backbone model. (b) Stick model. (c) Space-filling model. Each α-helical chain contains 100 residues. The nonpolar strips along each helix contact each other, so that the two helices wind around each other in a gentle left-handed coil. [*Structure of tropomyosin (pdb 1C1G) determined by F.G. Whitby and G.N. Phillips, Jr.*]

that support layers of cells in tissues. Often, several types of collagen are found together. Not surprisingly, defects in collagen affect a variety of organ systems (Box 5-C).

REVIEW LEARNING OBJECTIVE 1

- How are collagen fibers stabilized and cross-linked?

A CLOSER LOOK

BOX 5-C Genetic collagen diseases

Hundreds of mutations in collagen genes or in the genes for proteins that process collagen molecules have been identified. Because most tissues contain more than one type of collagen, the physiological manifestations of collagen mutations are highly variable.

Defects in collagen type I (the major form in bones and tendons) cause the congenital disease **osteogenesis imperfecta.** The primary symptoms of the disease include bone fragility leading to easy fracture, long-bone deformation, and abnormalities of the skin and teeth.

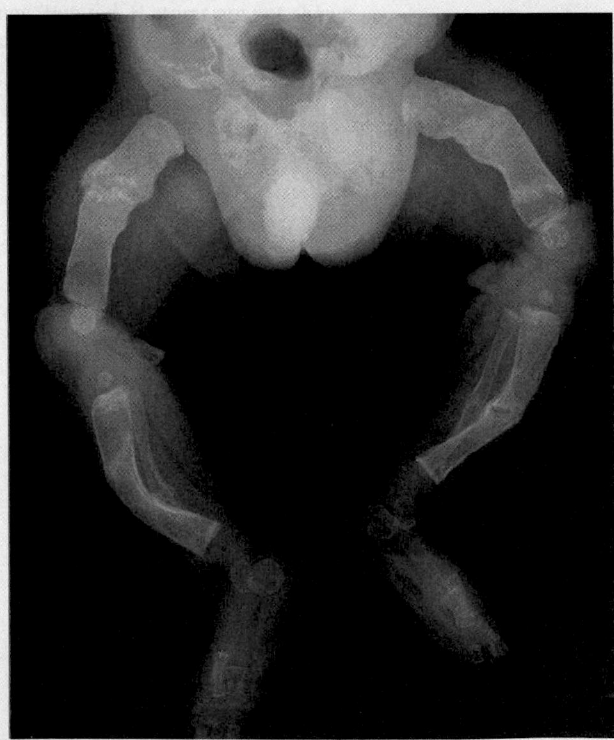

This X-ray of the lower legs of a child with a moderately severe case of osteogenesis imperfecta shows that the long bones are bowed and abnormally thin.

Collagen type I, a trimeric molecule, contains two different types of polypeptide chains. Therefore, the severity of the disease depends in part on whether one or two chains in a collagen molecule are affected. Furthermore, *the location and nature of the mutation determine whether the abnormal collagen retains some normal function.* For example, in one severe form of osteogenesis imperfecta, a 599-base deletion in a collagen gene represents the loss of a large portion of triple helix. The resulting protein is unstable and is degraded intracellularly. Milder cases of osteogenesis imperfecta result from amino acid substitutions, for example, the replacement of Gly by a bulkier residue. Other amino acid changes may slow intracellular processing and excretion of collagen polypeptides, which affects the assembly of collagen fibers. Osteogenesis imperfecta affects about one in 10,000 people, with most cases arising from new mutations. Since individuals with severe cases of the disease do not survive to reproductive age, their particular genetic defect is seldom passed on.

Mutations in collagen type II, a form found in cartilage, lead to osteoarthritis. This genetic disease, which becomes apparent in childhood, is distinct from the osteoarthritis that can develop later in life, often after years of wear and tear on the joints. Defects in the proteins that process collagen extracellularly and help assemble collagen fibers lead to disorders such as dermatosparaxis, which is characterized by extreme skin fragility.

Ehlers-Danlos syndrome results from abnormalities in collagen type III, a molecule that is abundant in most tissues but is scarce is skin and bone. Symptoms of this phenotypically variable disorder include easy bruising, thin or elastic skin, and joint hyperextensibility. In one form of the disease, which is accompanied by a high risk for arterial rupture, the molecular defect is a mutation in a collagen type III gene. In another form of the disease, in which individuals often suffer from scoliosis (curvature of the spine), the collagen genes are normal. In these cases, the disease results from a deficiency of lysine hydroxylase, the enzyme that modifies Lys residues so that they can participate in collagen cross-links (see page 157). Ehlers-Danlos syndrome is both rarer and less severe than osteogenesis imperfecta, with many affected individuals surviving to adulthood. About half the individuals with the syndrome have a parent with the same defect; the other cases arise from new mutations. Very mild forms of the disorder often go unrecognized.

[ISM/Phototake.]

SUMMARY

1. The microfilament elements of a cell's cytoskeleton are built from ATP-binding globular actin subunits that polymerize as a double chain. Polymerization is reversible, so microfilaments undergo growth and regression. Their dynamics may be modified by proteins that mediate microfilament capping, branching, and severing.
2. Myosin, a protein with a coiled-coil tail and two globular heads, interacts with actin filaments to perform mechanical work. ATP-driven conformational changes allow the myosin head to bind, release, and rebind actin. This mechanism, in which myosin acts as a lever, is the basis of muscle contraction.
3. GTP-biding tubulin dimers polymerize to form a hollow microtubule. Polymerization is more rapid at one end, and the microtubule can disassemble rapidly by fraying. Drugs that affect microtubule dynamics interfere with cell division.
4. The motor protein kinesin has two globular heads connected by flexible necks to a coiled-coil tail. Kinesin transports vesicular cargo along the length of a microtubule by a processive stepping mechanism that is driven by conformational changes triggered by ATP binding and hydrolysis.
5. The intermediate filament keratin contains two long α helical chains that coil around each other so that their hydrophobic residues are in contact. Keratin filaments associate and are cross-linked to form semipermanent structures.
6. Collagen polypeptides contain a large amount of proline and hydroxyproline and include a Gly residue at every third position. Each chain forms a narrow left-handed helix, and three chains coil around each other to form a right-handed triple helix with Gly residues at its center. Covalent cross-links strengthen collagen fibers.

CHECKLIST

1. Review the Learning Objectives listed on page 136.
2. Complete Exercise 7/Actin and Myosin, which shows the reaction cycle that drives the movement of myosin relative to actin.
3. Complete Exercise 8/Keratin and Collagen, which explores fibrous protein structure, including the keratin coiled coil and the collagen triple helix.
4. Apply your knowledge by solving the problems at the end of this chapter. Check your results in the Solutions appendix.
5. Be able to define the boldfaced terms (consult the glossary at the end of the book). Test your understanding by taking the Chapter 5 quiz (www.wiley.com/college/pratt).
6. Explore the websites listed at www.wiley.com/college/pratt.
7. Consult the list of Selected Readings for background information or additional details on actin, myosin, tubulin, kinesin, keratin, and collagen.

GLOSSARY TERMS

cytoskeleton
motor protein
microfilament
F-actin
G-actin
(−) end
(+) end
treadmilling
capping protein
metastasis
coiled coil
thick filament
thin filament
stereocilia
Usher syndrome
cytokinesis
microtubule
electron crystallography
protofilament
gout
vesicle
processivity
intermediate filament
collagen
triple helix
osteogenesis imperfecta
Ehlers–Danlos syndrome

PROBLEMS

1. Compare and contrast globular and fibrous proteins.
2. Of the six proteins highlighted in this chapter (actin, myosin, tubulin, kinesin, keratin, and collagen), (a) which are exclusively structural (not involved in cell shape changes); (b) which are considered to be motor proteins; (c) which are not motor proteins but can undergo structural changes; and (d) which contain nucleotide-binding sites?
3. Explain why microfilaments and microtubules are polar whereas intermediate filaments are not.

4. In order to obtain crystals suitable for X-ray crystallography, G-actin was first mixed with another protein.
   (a) Why was this step necessary?
   (b) What additional information was required to solve the structure of G-actin?

5. Is a myosin a fibrous protein or a globular protein? Explain.

6. Myosin type V is a two-headed myosin that operates as a transport motor to move its attached cargo along actin filaments. Its mechanism is similar to that of muscle myosin, but it acts processively, like kinesin. The reaction cycle diagrammed below begins with both myosin V heads bound to an actin filament. ADP is bound to the leading head, and the nucleotide-binding site in the trailing head is empty.

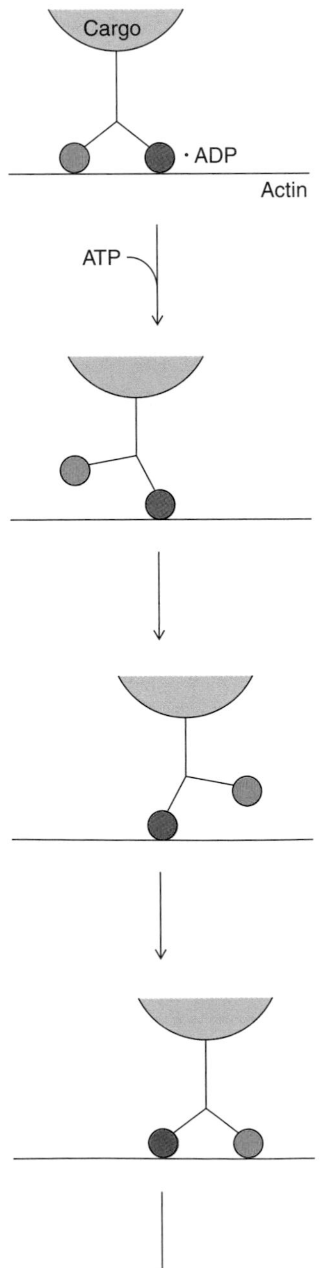

Based on your knowledge of the muscle myosin reaction cycle (Fig. 5-11), propose a reaction mechanism for myosin V, starting with the entry of ATP. How does each step of the ATP hydrolysis reaction correspond to a conformational change in myosin V? [From Walker, M.L., Burgess, S.A., Sellers, J.R., Wang, F., Hammer, J.A. III, Trinick, J., and Knight, P.J., Two-headed binding of a processive myosin to F-actin, *Nature* **405,** 804–807 (2000).]

7. How could a microtubule-binding protein distinguish a rapidly growing microtubule from one that was growing more slowly?

8. Why would fraying be a faster mechanism for microtubule disassembly than dissociation of tubulin subunits as dimers?

9. Microtubule ends with GTP bound tend to be straight whereas microtubule ends with GDP bound are curved and tend to fray. An experiment was carried out in which monomers of β-tubulin were allowed to polymerize in the presence of a nonhydrolyzable analog of GTP called guanylyl-(αβ)-methylenediphosphonate (GMP·CPP). Compare the stability of this polymer with one polymerized in the presence of GTP.

GMP · CPP

10. Explain why colchicine (which promotes microtubule depolymerization) and taxol (which prevents depolymerization) both prevent cell division.

11. A recent review listed literally dozens of compounds which are being investigated as possible therapeutic agents for treating cancer. Tubulin was the target for all of the compounds. Why is tubulin a good target for anticancer drugs?

12. The three-dimensional structure of paclitaxel (Taxol®) binding to the tubulin dimer has recently been solved. Paclitaxel binds to a pocket in β-tubulin on the inner surface of the microtubule and causes a conformational change that counteracts the effects of GTP hydrolysis. How does the binding of paclitaxel affect the stability of the microtubule? [From Amos, L.A., and Löwe, J., *Chem. Biol.* **6,** R65–R69 (1999).]

13. Gout is a disease characterized by the deposition of sodium urate crystals in the synovial fluid. The crystals are ingested by polymorphonuclear leukocytes (PMNs), which are white blood cells that circulate in the bloodstream before they crawl through tissues to reach sites of injury. Upon ingestion of the crystals, a series of biochemical reactions leads to a release of mediators from the PMNs that cause inflammation, which causes pain in

the joints. Colchicine is used as a drug to treat gout because colchicine has the ability to inhibit the mobility of the PMNs. How does colchicine accomplish this?

14. Microtubule structures involved in cell division are less stable than microtubules found in axon extensions of nerve cells. Why is this the case?

15. Early cell biologists, examining living cells under the microscope, observed that the movement of certain cell constituents was rapid, linear, and targeted (that is, directed toward a particular point).

(a) Why are these qualities inconsistent with diffusion as a mechanism for redistributing cell components?

(b) List the minimum requirements for an intracellular transport system that is rapid, linear, and targeted.

16. Explain why the movement of myosin along an actin filament is "hopping," whereas the movement of kinesin along a microtubule is "walking."

17. Hydrophobic residues usually appear at the first and fourth positions in the seven-residue repeats of polypeptides that form coiled coils.

(a) Why do polar or charged residues usually appear in the remaining five positions?

(b) Why is the sequence Ile–Gln–Glu–Val–Glu–Arg–Asp more likely than the sequence Trp–Gln–Glu–Tyr–Glu–Arg–Asp to appear in a coiled coil?

18. Globular proteins are typically constructed from several layers of secondary structure, with a hydrophobic core and a hydrophilic surface. Is this true for a fibrous protein such as keratin?

19. The digestive tract of the larvae of clothes moths is a strongly reducing environment. Why is this beneficial to the larvae?

20. Describe the primary, secondary, tertiary, and quaternary structures of (a) actin and (b) collagen. Why is it difficult to assign the four structural categories to these proteins?

21. The highly pathogenic anaerobic bacterium *Clostridium perfringens* is responsible for gas gangrene, a condition in which animal tissue structure is destroyed. This bacterium secretes an enzyme that efficiently catalyzes the hydrolysis of the peptide bond indicated below. X and Y can be any amino acid. How does the secretion of this enzyme contribute to the invasiveness of this bacterium in human tissues? Why does this enzyme not affect the bacterium itself?

—X—Gly—Pro—Y—

↓ Bacterial enzyme

$—X—COO^- + NH_3^+—Gly—Pro—Y—$

22. An enzyme found in tadpoles cleaves the bond indicated in the figure. What kind of enzyme is this, and what is its role in the developing tadpole? [From Stolow, M.A., Bauzon, D.D., Li, J., Sedgwick, T., Liang, V.C., Sang, Q.A., and Shi, Y.B., *Mol. Biol. Cell* **7,** 1471–1483 (1996).]

—Gly—Pro—Gln—Gly—↓—Ile—Ala

23. The term "collagen" refers to a family of diverse protein molecules. Members of the collagen family from the same species, or collagen molecules from different species, are often characterized by their melting temperature (T_m). The collagen is subjected to increasing temperature and monitored for structural changes. At high temperatures, the intermolecular forces holding the three chains together in the triple helix are disrupted and the chains come apart, resulting in denatured collagen, or gelatin (a substance that is liquid at high temperatures but solidifies upon cooling). Because the intermolecular forces holding the collagen strands together are cooperative, the collagen triple helix tends to come apart all at once. The T_m is defined as that temperature at which the collagen is half-denatured and is used as a criterion for collagen stability.

T_m measurements for collagen from three different species are presented in the table below. One is from an Antarctic ice fish, one is from shark skin, and one is from chick skin. (a) Assign each collagen to the proper organism, and (b) explain the correlation between imino acid content and T_m.

Collagen	Imino acid content per 1000 residues	T_m (°C)
A	212	41
B	143	6
C	191	29

24. A group of investigators synthesized a series of collagen-like peptides, each containing 30 amino acid residues, in order to study the important interactions among the three chains in the triple helix. Three peptides were synthesized: Peptide 1, which is $(Pro–Hyp–Gly)_{10}$, Peptide 2, which is $(Pro–Hyp–Gly)_4$–Glu–Lys–Gly–$(Pro–Hyp–Gly)_5$ and Peptide 3, which is Gly–Lys–Hyp–Gly–Glu–Hyp–Gly–Pro–Lys–Gly–Asp–Ala–$(Gly–Ala–Hyp)_2$–$(Gly–Pro–Hyp)_4$. Their properties are summarized in the table.

Peptide	Forms trimers?	Imino acid content	T_m(°C)	
1	yes	67%	pH = 1 pH = 7 pH = 11	61 58 60
2	yes	60%	pH = 1 pH = 7 pH = 13	44 46 49
3	yes	30%	pH = 1 pH = 7 pH = 13	18 26.5 19

(a) Rank the stability of the three collagen-like peptides. What is the reason for the observed stability?
(b) Compare the T_m values of Peptide 3 at the various pH values. Why does Peptide 3 have a maximum T_m value at pH = 7? What interactions are primarily responsible?
(c) Compare the T_m values of Peptide 1 at the various pH values. Why is there less of a difference of T_m values at different pH values for Peptide1 than for Peptide 3?
(d) Consider the answers to the above questions. Which factor is more important in stabilizing the structure of the collagen-like peptides: ion pairing or imino acid content?

[From Venugopal, M.G., Ramshaw, J.A.M., Braswell, E., Zhu, D., and Brodsky, B., *Biochemistry* **33,** 7948–7956 (1994).]

25. The deep-sea hydrothermal vent worm *Riftia pachyptila* resides under extreme conditions of high temperature, low oxygen content, and drastic temperature changes. The worm has a thick collagen-containing cuticle that protects it from its harsh environment. Recently the structure of this collagen was investigated. Sequence analyses indicated that the collagen had the customary –Gly–X–Y– triplet but that hydroxyproline occurred only in the X position and that Y was often a glycosylated threonine (a galactose sugar residue was covalently attached via a condensation reaction between a hydroxyl group on the galactose and the hydroxyl group of the threonine side chain). Experiments were carried out using synthetic peptides in order to evaluate the stability of each. Results are shown in the table.

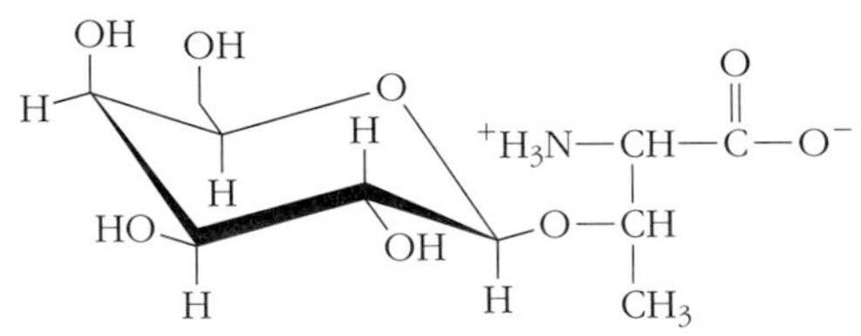

Gal-Thr

Synthetic peptide	Forms a triple helix?	T_m (°C)
(Pro–Pro–Gly)$_{10}$	yes	41
(Pro–Hyp–Gly)$_{10}$	yes	60
(Gly–Pro–Thr)$_{10}$	no	N/A
(Gly–Pro–Thr(Gal))$_{10}$	yes	41

(a) Compare the melting temperatures of (Pro–Pro–Gly)$_{10}$ and (Pro–Hyp–Gly)$_{10}$. What is the structural basis for the difference?
(b) Compare the melting temperature of (Pro–Pro–Gly)$_{10}$ and (Gly–Pro–Thr(Gal))$_{10}$ and provide an explanation.
(c) Why was (Gly–Pro–Thr)$_{10}$ included by the investigators?

[From Bann, J.G., Peyton, D.H., and Bächinger, H.P., *FEBS Lett.* **473,** 237–240 (2000).]

26. Several investigators have sought to explain why hydroxyproline-containing collagen molecules have increased stability. Some suggested that the hydroxyl group on the pyrrolidine ring of Hyp participates in a network of hydrogen bonds with "bridging" water molecules. However, this has been called into question by others who have noted that triple helices made of (Pro–Pro–Gly)$_{10}$ and (Pro–Hyp–Gly)$_{10}$ are stable in methanol. To investigate the question of stability further, one group of investigators synthesized a synthetic collagen that contains 4-fluoroproline (Flp), which resembles hydroxyproline except that fluorine substitutes for the hydroxyl group. The melting points of three synthetic collagen molecules were measured and are shown in the table.

Synthetic peptide	Forms a triple helix?	T_m (°C)
(Pro–Pro–Gly)$_{10}$	yes	41
(Pro–Hyp–Gly)$_{10}$	yes	60
(Pro–Flp–Gly)$_{10}$	yes	91

(a) Compare the structure of hydroxyproline and fluoroproline. Why do you suppose the investigators chose fluorine?
(b) Compare the melting points of the three synthetic collagens. What factor contributes the most to the stability of the molecule?

[From Holmgren, S., Taylor, K., Bretscher, L.E., and Raines, R., *Nature* **392,** 666–667 (1998).]

27. Explain why gelatin, which is mostly collagen, is nutritionally inferior to other types of protein.

Biosynthesis of Collagen

The pathway for the biosynthesis of Type I collagen consists of a number of steps, as shown in the figure.

(1) Two α1(I) chains and one α2 chain are transcribed from the appropriate gene and translated into protein.
(2) Prolyl hydroxylase and lysyl hydroxylase catalyze the hydroxylation of selected Pro and Lys residues on the polypeptide chains. Ascorbate (more commonly known as Vitamin C) is an essential cofactor for this reaction. Hydroxylation of these residues contributes to the stability of the collagen fiber.
(3) Glycosyltransferase enzymes catalyze the addition of sugar residues to selected amino acid side chains, including Lys, on the polypeptide chains. Glycosylation contributes to the stability of the collagen fiber.
(4) The two α(I) chains and one α2 chain form a triple helix called procollagen. The N- and C-terminal regions of the chains assist in proper alignment. The N- and C-terminal ends are cleaved by procollagen peptidases after the procollagen is formed, resulting in tropocollagen.
(5) Tropocollagen units assemble to form the type I collagen fiber.
(6) The collagen fiber is strengthened by the formation of covalent cross-links. Connective tissue disorders re-

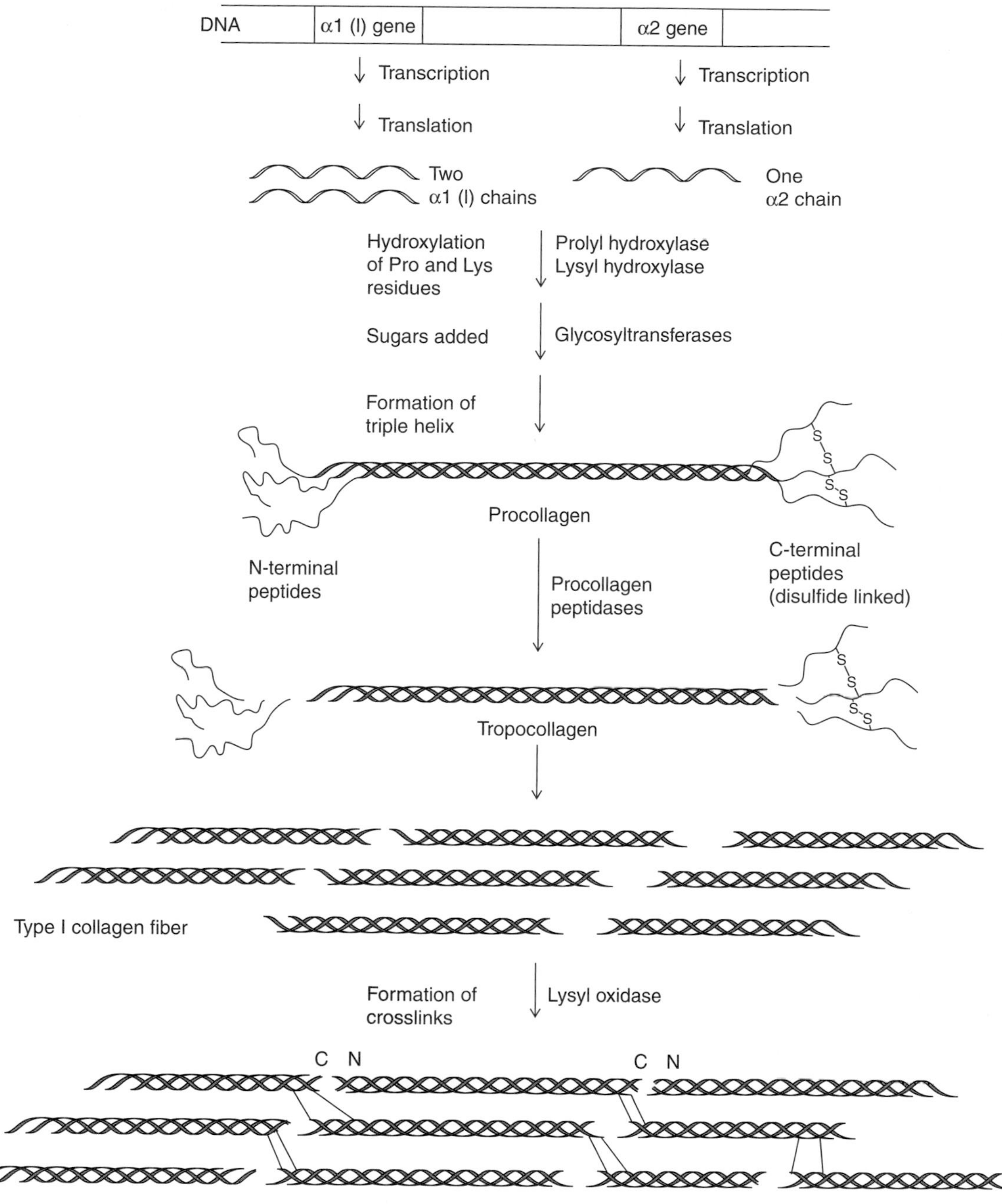

Biosynthesis of Type I collagen
Steps up through the formation of procollagen occur intracellularly in fibroblasts; procollagen is then secreted from the cell and subsequent steps occur extracellularly.

sult if collagen cannot be synthesized properly. Any step that is not properly carried out can result in defective collagen. Use the information provided in the figure to answer Problems 28–34.

28. You are a physician with a 45-year-old male patient with swollen gums, petechiae (tiny red spots which result from red blood cells leaking from capillaries into the skin) and thick and scaly skin on the face and fingers. The patient reports occasional gum bleeding and joint pain. You suspect that the patient has scurvy, a disease caused by the deficiency of vitamin C (ascorbic acid). He tells you that he smokes 10 cigarettes a day, is trying to lose weight by eating only high-protein foods such as eggs, cottage cheese, and meat, and is under a great deal of stress.

(a) Which of the patient's behaviors are likely to cause scurvy? Explain.

(b) Why would a deficiency of ascorbic acid cause his particular symptoms?

(c) What advice would you give your patient?

[From Hodges, R.E., Hood, J., Canham, J.E., Sauberlich, H.E., and Baker, E.M., *Am. J. Clin. Nutr.* **24,** 432–443 (1971).]

29. Radioactively labeled [^{14}C]-proline is incorporated into collagen in cultured fibroblasts. The radioactivity is detected in the collagen protein. But collagen synthesized in the presence of [^{14}C]-hydroxyproline is not radiolabeled. Explain why.

30. The effect of the drug minoxidil was measured in cultured human skin fibroblasts and the data are shown in the table below. Fibroblasts were incubated with [^{3}H]-proline or [^{3}H]-lysine to allow collagen to incorporate the radioactively labeled amino acids into the polypeptide chains. Following treatment with minoxidil (controls were untreated), the cells were harvested and the subjected to enzyme treatment, which hydrolyzed all cellular protein. The amount of [^{3}H]-proline or [^{3}H]-lysine incorporation into collagen was then measured in terms of $^{3}H_2O$ release.

Culture conditions	Prolyl hydroxylase activity (cpm, $^{3}H_2O$/mg protein)	Lysyl hydroxylase activity (cpm, $^{3}H_2O$/mg protein)
Control	13,056	12,402
Minoxidil	13,242	1,936

(a) What is the effect of minoxidil on the cultured fibroblasts?
(b) Why would this drug be effective in treating fibrosis (a skin condition associated with an accumulation of collagen)?
(c) What are the potential dangers in the long-term use of minoxidil to treat hair loss?

[From Murad, S., Walker, L.C., Tajima, S., and Pinnell, Sr. R., *Arch. Biochem. Biophys.* **308,** 42–47 (1994) and Price, V.H., *New Engl. J. Med.* **341,** 964–973 (1999).]

31. The disease lathyrism results in animals from the ingestion of seeds from the sweet pea *Lathyrus odoratus*. The plant contains a compound that inhibits lysyl oxidase. The disease is characterized by abnormalities of the bones, joints, and blood vessels. Give a biochemical explanation for the symptoms of this disease.

32. A portion of the gene (the sequence of the nontemplate, or coding strand of the DNA) for the α2(I) chain of type I procollagen is shown below (the entire gene contains 4859 base pairs). Also shown is the gene isolated from the skin fibroblasts of a patient with a mutant α2(I) gene. The patient was a near-term fetus delivered by cesarean section. The neonate developed severe respiratory distress and died shortly after birth. An autopsy showed foreshortening of all extremities, skeletal abnormalities including bowed ribs, and little calcified bone in the skull.

Normal α2(I) procollagen gene

. . .CTGGTGCTGTTGGCCCAAGAGGTCCTAGTGGCCCAC. . .

Mutant α2(I) procollagen gene

. . .CTGGTGCTGTTGGCCCAAGAGATCCTAGTGGCCCAC. . .

(a) What is the sequence of the protein encoded by both the normal and mutant genes? (Note that you must first determine the correct open reading frame.) What is the change in the protein sequence?
(b) How would the T_m values for normal collagen and mutant collagen compare?
(c) What is the patient's disease and why did the patient die?

[From Baldwin, C.T., Constantinou, C.D., Dumars, K.W., and Prockop, D.J., *J. Biol. Chem.* **264,** 3002–3006 (1989).]

33. The papaya plant contains the enzyme papain, which is a collagenase.

(a) Why is papain used in the kitchen as a meat tenderizer?
(b) Fresh papaya should not be added to gelatin desserts (see Problem 27), but cooked papaya can be used. Explain.

34. Because collagen molecules are difficult to isolate from the connective tissue of animals, studies of collagen structure often use synthetic peptides. Would it be practical to try to purify collagen molecules from cultures of bacterial cells that have been engineered to express collagen genes?

SELECTED READINGS

Beck, K. and Brodsky, B., Supercoiled protein motifs: The collagen triple-helix and the α-helical coiled coil, *J. Struct. Biol.* **122,** 17–29 (1998). [Includes some of the history of the study of collagens and coiled-coil proteins such as myosin and keratin.]

Burkhard, P., Stetefeld, J., Strelkov, S.V., Coiled coils: A highly versatile protein folding motif, *Trends Cell Biol.* **11,** 82–88 (2001). [Describes different classes of coiled-coil proteins and how they function in cytoskeletal, motor, and other proteins.]

Cooper, J.A. and Schafer, D.A., Control of actin assembly and disassembly at filament ends, *Curr. Opin. Cell Biol.* **12,** 97–103 (2000). [Provides an overview of principles of microfilament dynamics and some of the key protein players.]

Mandelkow, E. and Johnson, K.A., The structural and mechanochemical cycle of kinesin, *Trends Biochem. Sci.* **23,** 429–433 (1998). [Summarizes the motor mechanism of kinesin.]

Nogales, E., Structural insights into microtubule function, *Annu. Rev. Biochem.* **69,** 277–302 (2000). [Discusses tubulin and microtubule structure along with other proteins and drugs that affect microtubule structure.]

Spudich, J.A., The myosin swinging cross-bridge model, *Nature Rev. Mol. Cell Biol.* **2,** 387–391 (2001). [Summarizes the history and current state of models for myosin action.]

Vale, R.D. and Milligan, R.A., The way things move: Looking under the hood of molecular motor proteins, *Science* **288,** 88–95 (2000). [Compares and contrasts the molecular mechanisms of myosin and kinesin motors.]

CHAPTER 6

Chymotrypsin: A Model Enzyme

On occasion, biochemistry meets the needs of fashion, as in the production of "stonewashed" blue jeans. At one time, the desired wear and fading were achieved by tumbling the garments with pumice stones in enormous rock tumblers. Currently, the abrading effect is achieved not with actual stones but with an enzyme solution. The cotton fabric of the jeans consists almost entirely of cellulose, a polymer of glucose residues. The enzyme cellulase, usually purified from fungi or bacteria, partially digests the outer surface of the cellulose fibers, softening and fading the garment. Enzymatic stonewashing is more energy efficient than the traditional method, and eliminates the possibility of a stray stone finding its way inside the pocket of a brand-new pair of jeans.

[Chris Hackett/The Image Bank/Getty Images.]

LEARNING OBJECTIVES

1. Understand how enzymes differ from nonbiological catalysts.
2. Understand that the height of the activation energy barrier determines the rate of a reaction.
3. Understand how an enzyme lowers the activation energy for a reaction.
4. Understand acid–base catalysis, covalent catalysis, and metal ion catalysis.
5. Understand the role of the catalytic triad in peptide bond hydrolysis catalyzed by chymotrypsin.
6. Understand the mechanisms of transition state stabilization in chymotrypsin.
7. Understand the effects of proximity, orientation, and desolvation in enzyme catalysis.
8. Understand the chemical basis for substrate specificity among serine proteases.
9. Understand how zymogens such as chymotrypsin are activated.

THIS CHAPTER IN CONTEXT

We have seen how some proteins perform relatively passive roles (for example, the cytoskeletal protein keratin, described in Section 5-5). We have also seen how certain proteins carry out their physiological functions by reversibly binding small molecules (for example, myoglobin and hemoglobin, described in Chapter 4). Still other proteins, known as enzymes, are even more active, directly participating in the chemical reactions by which matter and energy are transformed inside living cells. In this chapter, we examine the fundamental features of enzymes, including the thermodynamic underpinnings of their activity. We describe various mechanisms by which enzymes accelerate chemical reactions, focusing primarily on the digestive enzyme chymotrypsin to illustrate how different structural features influence catalytic activity. The following chapter continues the discussion of enzymes by describing how enzymatic activity is quantified and how it can be regulated.

1 WHAT IS AN ENZYME?

In vitro, under physiological conditions, the peptide bond that links amino acids in peptides and proteins has a half-life of about 20 years. That is, after 20 years, about half of the peptide bonds in a given sample of a peptide will have been broken down through **hydrolysis** (cleavage by water):

$$\cdots\!-\!\underset{\displaystyle\mathrm{H}}{\mathrm{N}}\!-\!\overset{\displaystyle\mathrm{R}}{\mathrm{CH}}\!-\!\overset{\displaystyle\mathrm{O}}{\overset{\|}{\mathrm{C}}}\!-\!\underset{\displaystyle\mathrm{H}}{\mathrm{N}}\!-\!\overset{\displaystyle\mathrm{R}}{\mathrm{CH}}\!-\!\overset{\displaystyle\mathrm{O}}{\overset{\|}{\mathrm{C}}}\!-\!\cdots \;+\; \mathrm{H_2O}$$

$$\downarrow$$

$$\cdots\!-\!\underset{\displaystyle\mathrm{H}}{\mathrm{N}}\!-\!\overset{\displaystyle\mathrm{R}}{\mathrm{CH}}\!-\!\mathrm{C}\!\begin{smallmatrix}\nearrow\!\!\mathrm{O}\\ \searrow\!\mathrm{O^-}\end{smallmatrix} \;+\; \mathrm{H}\!-\!\underset{\displaystyle\mathrm{H}}{\overset{\displaystyle\mathrm{H}}{\overset{+}{\mathrm{N}}}}\!-\!\overset{\displaystyle\mathrm{R}}{\mathrm{CH}}\!-\!\overset{\displaystyle\mathrm{O}}{\overset{\|}{\mathrm{C}}}\!-\!\cdots$$

Obviously, the long half-life of the peptide bond is advantageous for living organisms, since many of their structural and functional characteristics depend on the integrity of proteins. On the other hand, many proteins—some hormones, for example—must be broken down very rapidly, so that their biological effects can be limited. Clearly, an organism must be able to accelerate the rate of peptide bond hydrolysis.

In general, there are two ways to increase the rate of a chemical reaction, including hydrolysis:

1. *Increasing the temperature (adding energy in the form of heat) accelerates the reaction.* Unfortunately, this is not very practical, since the vast majority of organisms cannot regulate their internal temperature and thrive only within relatively narrow temperature ranges. Furthermore, an increase in temperature increases the rates of all chemical reactions, not just the desired reaction.

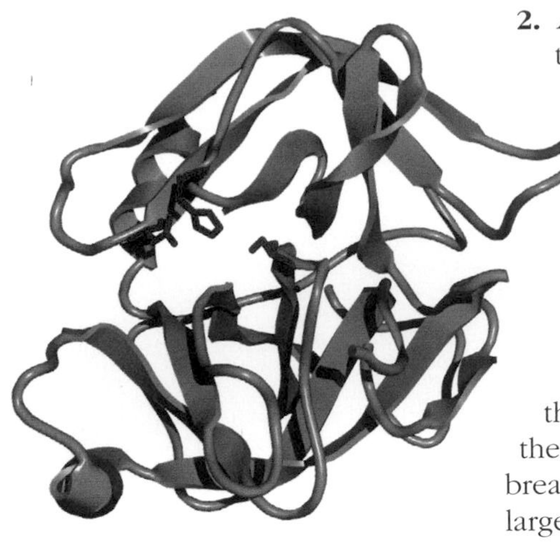

FIGURE 6-1

X-Ray structure of chymotrypsin. The polypeptide chain (gray) folds into two domains. Three residues essential for the enzyme's activity are shown in red. [*Structure (pdb 4CHA) determined by H. Tsukada and D.M. Blow.*]

2. *Adding a **catalyst** accelerates the reaction.* A catalyst is a substance that participates in the reaction yet emerges at the end in its original form. A huge variety of chemical catalysts are known. For example, the catalytic converter in an automobile engine contains a mixture of platinum and palladium that accelerates the conversion of carbon monoxide and unburned hydrocarbons to the relatively harmless carbon dioxide. *Living systems use catalysts called **enzymes** to increase the rates of chemical reactions.*

Most enzymes are proteins, but a few are made of RNA (these are called **ribozymes** and are described more fully in Section 19-4). One of the best-studied enzymes is chymotrypsin, a digestive protein that is synthesized in the pancreas and secreted into the small intestine, where it helps break down dietary proteins. Perhaps because it can be purified in relatively large quantities from the pancreas of cows, chymotrypsin was one of the first enzymes to be crystallized (it is also widely used in the laboratory; for example, see Table 4-3). Chymotrypsin's 241 amino acid residues form a compact two-domain structure (Fig. 6-1). Hydrolysis of polypeptide substrates takes place in a cleft between the two domains, near the side chains of three residues (His 57, Asp 102, and Ser 195). This area of the enzyme is known as the **active site.** The active sites of nearly all known enzymes are located in similar crevices on the enzyme surface.

Chymotrypsin catalyzes the hydrolysis of peptide bonds at a rate of about 190 per second, which is about 1.7×10^{11} times faster than in the absence of a catalyst. This is also orders of magnitude faster than would be possible with a simple chemical catalyst. In addition, *chymotrypsin and other enzymes act under mild conditions* (atmospheric pressure and physiological temperature), whereas many chemical catalysts require extremely high temperatures and pressures for optimal performance.

Chymotrypsin's catalytic power is not unusual: *Rate enhancements of 10^8 to 10^{12} are typical of enzymes* (Table 6-1 gives the rates of some enzyme-catalyzed reactions). Of course, the slower the rate of the uncatalyzed reac-

TABLE 6-1 Rate enhancements of enzymes

Enzyme	Half-time (uncatalyzed)[a]	Uncatalyzed rate (s^{-1})	Catalyzed rate (s^{-1})	Rate enhancement (catalyzed rate/ uncatalyzed rate)
Orotidine-5′-monophosphate decarboxylase	78,000,000 years	2.8×10^{-16}	39	1.4×10^{17}
Staphylococcal nuclease	130,000 years	1.7×10^{-13}	95	5.6×10^{14}
Adenosine deaminase	120 years	1.8×10^{-10}	370	2.1×10^{12}
Chymotrypsin	20 years	1.0×10^{-9}	190	1.7×10^{11}
Triose phosphate isomerase	1.9 days	4.3×10^{-6}	4,300	1.0×10^{9}
Chorismate mutase	7.4 hours	2.6×10^{-5}	50	1.9×10^{6}
Carbonic anhydrase	5 s	1.3×10^{-1}	1,000,000	7.7×10^{6}

[a]The half-times of very slow reactions were estimated by extrapolating from measurements made at very high temperatures.

[Data mostly from Radzicka, R. and Wolfenden, R., *Science* **267,** 90–93 (1995).]

tion, the greater the opportunity for rate enhancement by an enzyme (see, for example, orotidine-5′-monophosphate decarboxylase in Table 6-1). Interestingly, even relatively fast reactions are subject to enzymatic catalysis in biological systems. For example, the conversion of CO_2 to carbonic acid in water

$$CO_2 + H_2O \rightleftharpoons H_2CO_3$$

has a half-time of 5 seconds (half the molecules will have reacted within 5 seconds). This reaction is accelerated over a millionfold by the enzyme carbonic anhydrase (see Table 6-1).

Another feature that sets enzymes apart from nonbiological catalysts is their **reaction specificity.** Most enzymes are highly specific for their reactants (called **substrates**) and products. The functional groups in the active site of an enzyme are so carefully arranged that the enzyme can select its substrates from among many others that are similar in size and shape, and can then mediate a single chemical reaction involving those substrates. This reaction specificity stands in marked contrast to the permissiveness of most organic catalysts, which can act on many different kinds of substrates and, for a given substrate, often yield more than one product.

Chymotrypsin and some other digestive enzymes are somewhat unusual in acting on a relatively broad range of substrates and, at least in the laboratory, catalyzing several types of reactions. For instance, chymotrypsin catalyzes the hydrolysis of the peptide bond following almost any large nonpolar residue such as Phe, Trp, or Tyr. It can also catalyze the hydrolysis of other amide bonds and ester bonds. This behavior has proved to be convenient for quantifying the activity of purified chymotrypsin. An artificial substrate such as *p*-nitrophenylacetate (an ester) is readily hydrolyzed by the action of chymotrypsin (the name of the enzyme appears next to the reaction arrow to indicate that it participates as a catalyst):

$$H_3C-\overset{O}{\overset{\|}{C}}-O-C_6H_4-NO_2$$

p-Nitrophenylacetate (colorless)

H_2O → chymotrypsin → $2\,H^+$

$$H_3C-\overset{O}{\overset{\|}{C}}-O^- \quad + \quad {}^-O-C_6H_4-NO_2$$

Acetate — *p*-Nitrophenolate (yellow)

The *p*-nitrophenolate reaction product is bright yellow, so the progress of the reaction can be easily monitored in a spectrophotometer at 410 nm.

Finally, enzymes differ from nonbiological catalysts in that *the activities of many enzymes are regulated so that the organism can respond to changing conditions or follow genetically determined developmental programs.* For this reason, biochemists seek to understand *how* enzymes work as well as *when* and *why.* These aspects of enzyme behavior are fairly well understood for chymotrypsin, which makes it an ideal subject to showcase the fundamentals of enzyme activity.

REVIEW LEARNING OBJECTIVE 1

- How do enzymes differ from nonbiological catalysts?

A note on nomenclature

The enzymes that catalyze biochemical reactions have been classified into six subgroups according to the type of reaction carried out (Table 6-2). Basically, *all biochemical reactions involve either the addition of some substance to another or its removal, or the rearrangement of that substance.* Keep in mind that although the substrates of many biochemical reactions appear to be quite large (for example, proteins or nucleic acids), the action really involves just a few chemical bonds and a few small groups (sometimes H_2O or even just an electron).

The name of an enzyme frequently provides a clue to its function. In some cases, an enzyme is named by incorporating the suffix *-ase* into the name of its substrate. For example, fumarase is an enzyme that acts on fumarate (Reaction 7 in the citric acid cycle; see Section 11-2). Chymotrypsin can similarly be called a proteinase, a protease, or a peptidase. Most enzyme names contain more descriptive words (also ending in *-ase*) to indicate the nature of the reaction catalyzed by that enzyme. For example, pyruvate decarboxylase catalyzes the removal of a CO_2 group from pyruvate:

$$\underset{\text{Pyruvate}}{CH_3{-}C({=}O){-}COO^-} \xrightarrow[\text{pyruvate decarboxylase}]{H^+} \underset{\text{Acetaldehyde}}{CH_3{-}CHO} + CO_2$$

Alanine aminotransferase catalyzes the transfer of an amino group from alanine to an α-keto acid:

$$\underset{\text{Alanine}}{H_3C{-}CH(\overset{+}{N}H_3){-}COO^-} + \underset{\alpha\text{-Ketoglutarate}}{{}^-OOC{-}C({=}O){-}CH_2{-}CH_2{-}COO^-}$$

$$\rightleftharpoons \text{ alanine aminotransferase}$$

$$\underset{\text{Pyruvate}}{H_3C{-}C({=}O){-}COO^-} + \underset{\text{Glutamate}}{{}^-OOC{-}CH(\overset{+}{N}H_3){-}CH_2{-}CH_2{-}COO^-}$$

Such a descriptive naming system tends to break down in the face of the many thousands of known enzyme-catalyzed reactions, but it is adequate for the small number of well-known reactions that are included in this book. A more precise classification scheme systematically groups enzymes in a four-level hierarchy and assigns each enzyme a unique number. For example, chymotrypsin is known as EC 3.4.21.1 (*EC* stands for Enzyme Commission, part of the nomenclature committee of the International Union of Biochemistry and Molecular Biology).

Keep in mind that even within an organism, more than one protein may catalyze a given chemical reaction. Multiple enzymes catalyzing the same re-

TABLE 6-2 Enzyme classification

Class of enzyme	Type of reaction catalyzed
1. Oxidoreductases	Oxidation–reduction reactions
2. Transferases	Transfer of functional groups
3. Hydrolases	Hydrolysis reactions
4. Lyases	Group elimination to form double bonds
5. Isomerases	Isomerization reactions
6. Ligases	Bond formation coupled with ATP hydrolysis

action are called **isozymes.** Although they usually share a common evolutionary origin, isozymes differ in their catalytic properties. Consequently, the various isozymes that are expressed in different tissues or at different developmental stages can carry out slightly different metabolic functions.

2 HOW DO ENZYMES WORK?

In a biochemical reaction, the reacting species must come together and undergo electronic rearrangements that result in the formation of products. Let us consider an idealized transfer reaction in which compound A—B reacts with compound C to form two new compounds, A and B—C:

Go to **Review Exercise 3/Elementary Kinetics** for an overview of the thermodynamics of chemical reactions and the effects of catalysts.

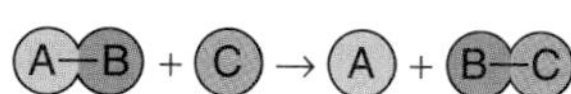

In order for the first two compounds to react, they must approach closely enough for their constituent atoms to interact. Normally, atoms that approach too closely repel each other. But if the groups have sufficient free energy, they can pass this point and react with each other to form products. The progress of the reaction can be depicted on a diagram (Fig. 6-2) in which the horizontal axis represents the progress of the reaction **(the reaction coordinate)** and the vertical axis represents the free energy **(*G*)** of the system. The energy-requiring step of the reaction is shown as an energy barrier, called the **free energy of activation** or **activation energy** and symbolized $\Delta G^{\ddagger}$. The point of highest energy is known as the **transition state** and can be considered a sort of intermediate between the reactants and products.

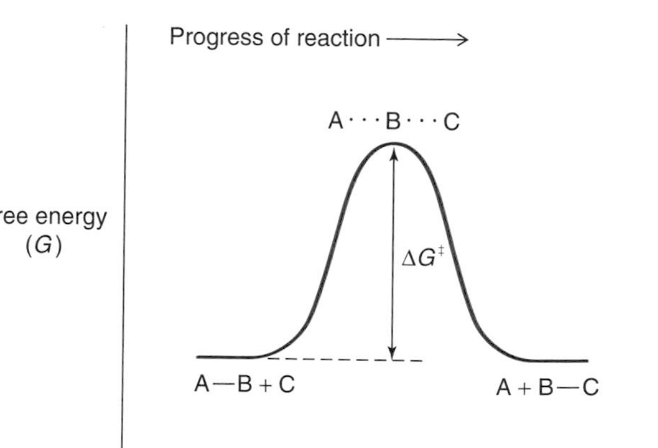

FIGURE 6-2

Reaction coordinate diagram for the reaction A—B + C → A + B—C. The progress of the reaction is shown on the horizontal axis, and the free energy is shown on the vertical axis. The transition state of the reaction, represented as A · · · B · · · C, is the point of highest free energy. The free energy difference between the reactants and the transition state is the free energy of activation ($\Delta G^{\ddagger}$).

The lifetimes of transition states are extremely short, on the order of 10^{-14} to 10^{-13} seconds. Because they are too shortlived to be accessible to most analytical techniques, the transition states of most reactions cannot be identified with absolute certainty. However, it is useful to visualize the transition state as a molecular species in the process of breaking and forming bonds. For the reaction above, we can represent this as A · · · B · · · C. The reactants require free energy ($\Delta G^{\ddagger}$) to reach this point, so the analogy of going uphill in order to undergo reaction is appropriate.

The height of the activation energy barrier determines the rate of a reaction (the amount of product formed per unit time). The higher the activation

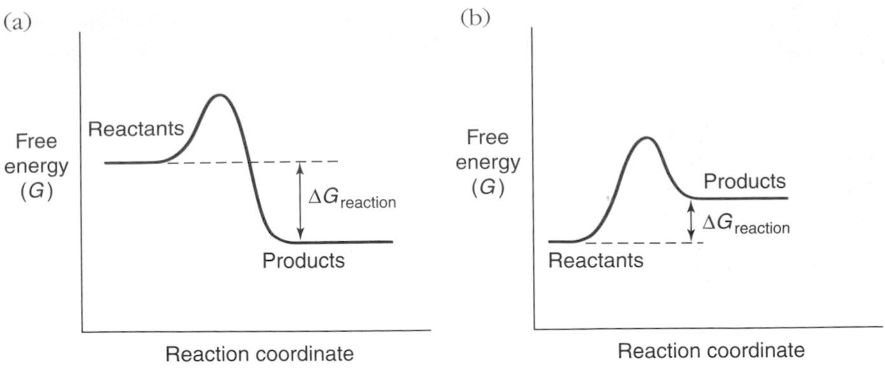

FIGURE 6-3

Reaction coordinate diagram for a reaction in which reactants and products have different free energies.
The free energy change for the reaction (ΔG) is equivalent to $G_{products} - G_{reactants}$. (a) When the free energy of the reactants is greater than that of the products, the free energy change for the reaction is negative, so the reaction proceeds spontaneously. (b) When the free energy of the products is greater than that of the reactants, the free energy change for the reaction is positive, so the reaction does not proceed spontaneously (however, the reverse reaction does proceed).

energy barrier, the less likely the reaction is to occur (the slower it is), because very few reactant molecules have enough free energy to reach the transition state per unit time. The lower the energy barrier, the more likely the reaction is to occur (the faster it is), because more reactant molecules have enough free energy to achieve the transition state per unit time. Note that the transition state, at the peak, can potentially roll down either side of the free energy hill. Therefore, not all the reactants that get together to form a transition state actually proceed all the way to products; they may return to their original state. Similarly, the products (A and B—C in this case) can react, pass through the same transition state (A · · · B · · · C), and yield the original reactants (A—B and C).

In nature, the free energies of the reactants and products of a chemical reaction are seldom identical, so it would be more accurate to draw a reaction coordinate diagram as shown in Fig. 6-3. If the products have a lower free energy than the reactants (Fig. 6-3a), then the overall free energy change of the reaction (**$\Delta G_{reaction}$**, or $G_{products} - G_{reactants}$) is less than zero. A negative free energy change indicates that a reaction proceeds spontaneously as written. Note that "spontaneously" does not mean "quickly." A reaction with a negative free energy change is thermodynamically favorable, but the height of the activation energy barrier ($\Delta G^{\ddagger}$) determines how fast the reaction actually occurs. If the products have greater free energy than the reactants (Fig. 6-3b), then the overall free energy change for the reaction ($\Delta G_{reaction}$) is greater than zero. The reaction does not proceed as written (unless free energy is added), but it does proceed in the reverse direction.

REVIEW LEARNING OBJECTIVE 2

- Why does a free energy barrier separate reactants and products in a chemical reaction?
- What is the relationship between the height of the activation energy barrier and the rate of the reaction?

A catalyst lowers the activation energy barrier

A catalyst, whether inorganic or enzymatic, decreases the activation energy barrier ($\Delta G^{\ddagger}$) of a reaction (Fig. 6-4). *It does so by interacting with the reacting molecules such that they are more likely to assume the transition state.* A catalyst speeds up the reaction because as more reacting molecules achieve the transition state per unit time, more molecules of product can form per unit time. (An increase in temperature increases the rate of a reaction for a similar reason: The input of thermal energy allows more molecules to achieve the transition state per unit time.) Thermodynamic calculations indicate that lowering $\Delta G^{\ddagger}$ by about 5.7 $kJ \cdot mol^{-1}$ accelerates the reaction tenfold. A rate increase of 10^6 requires lowering $\Delta G^{\ddagger}$ by six times this amount, or about 34 $kJ \cdot mol^{-1}$.

An enzyme catalyst does not alter the net free energy change for a reaction; it merely provides a pathway from reactants to products that passes through a transition state that has lower free energy than the transition state of the uncatalyzed reaction. *An enzyme, therefore, lowers the height of the activation energy barrier ($\Delta G^{\ddagger}$) by lowering the energy of the transition state.* The hydrolysis of a peptide bond is always thermodynamically favorable, but the reaction occurs quickly only when a catalyst (such as the protease chymotrypsin) is available to lower the free energy of the transition state.

REVIEW LEARNING OBJECTIVE 3

- How does an enzyme lower the activation energy of a reaction?

Enzymes use chemical catalytic mechanisms

The idea that living organisms contain agents that can promote the change of one substance into another has been around at least since the early nineteenth century, when scientists began to analyze the chemical transformations carried out by organisms such as yeast. However, it took some time to appreciate that these catalytic agents were not part of some "vital force" present only in intact, living organisms. At one point, in 1878, the word "enzyme"

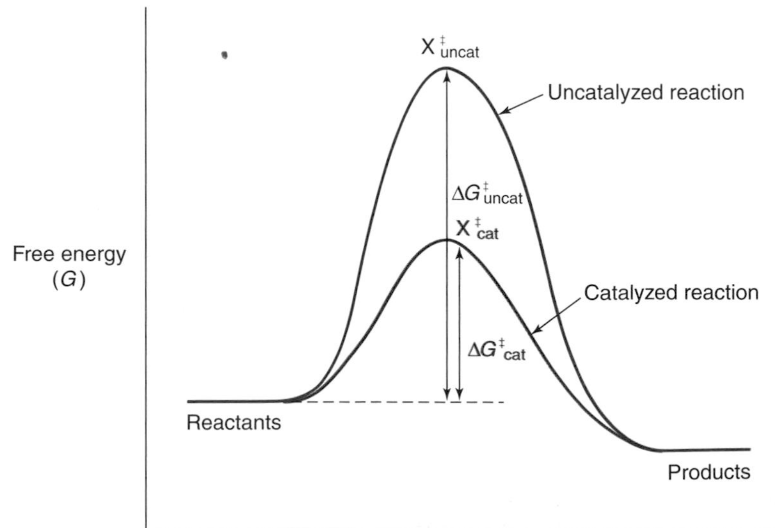

FIGURE 6-4

Effect of a catalyst on a chemical reaction.
Here, the reactants proceed through a transition state symbolized by $X^{\ddagger}$ during their conversion to products. A catalyst lowers the free energy of activation ($\Delta G^{\ddagger}$) for the reaction so that $\Delta G^{\ddagger}_{cat} < \Delta G^{\ddagger}_{uncat}$. Lowering the free energy of the transition state ($X^{\ddagger}$) accelerates the reaction because more reactants are able to achieve the transition state per unit time.

was coined to indicate that there was something *in* yeast (Greek *en* = in, *zyme* = yeast), rather than the yeast itself, that was responsible for breaking down (fermenting) sugar. In fact, the action of enzymes can be explained in purely chemical terms. What we currently know about enzyme mechanisms rests solidly on a foundation of knowledge about simple chemical catalysts.

In an enzyme, certain functional groups contributed by amino acid side chains at the enzyme's active site perform the same catalytic function as small chemical catalysts. In some cases, the protein portion of an enzyme cannot provide the required catalytic groups, so a tightly bound **cofactor** actually serves as the catalyst. For example, many oxidation–reduction reactions require a metal ion cofactor, since a metal ion can exist in multiple oxidation states, unlike an amino acid side chain. Some enzyme cofactors are organic molecules known as **coenzymes,** which may be derived from vitamins. Enzymatic activity still requires the protein, which helps position the cofactor and reactants for the reaction (this situation in reminiscent of myoglobin and hemoglobin, where the globin and heme group together function to bind oxygen; see Section 4-4).

There are three basic kinds of chemical catalytic mechanisms used by enzymes: acid–base catalysis, covalent catalysis, and metal ion catalysis. We will examine each of these, using model reactions to illustrate some of their fundamental features.

1. Acid–base catalysis

Nearly all enzyme mechanisms include **acid–base catalysis,** in which a proton is transferred between the enzyme and the substrate. This mechanism of catalysis can be further divided into **acid catalysis** and **base catalysis.** Some enzymes use one or the other; many use both. Consider the following model reaction, the tautomerization of a ketone to an enol (**tautomers** are interconvertible isomers that differ in the placement of a hydrogen and a double bond):

$$\begin{array}{c} \text{R} \\ | \\ \text{C=O} \\ | \\ \text{CH}_2 \\ | \\ \text{H} \end{array} \rightleftharpoons \left[\begin{array}{c} \text{R} \\ | \\ \text{C}\cdots\text{O}^- \\ \vdots \\ \text{CH}_2 \\ \vdots \\ \text{H}^+ \end{array}\right] \rightleftharpoons \begin{array}{c} \text{R} \\ | \\ \text{C—O—H} \\ \| \\ \text{CH}_2 \end{array}$$

Ketone — Transition state — Enol

Here the transition state is shown in square brackets to indicate that it is an unstable, transient species. The dotted lines represent bonds in the process of breaking or forming. The uncatalyzed reaction occurs slowly because formation of the carbanion-like transition state has a high activation energy barrier (a **carbanion** is a compound in which the carbon atom bears a negative charge).

If a catalyst (symbolized H—A) donates a proton to the ketone's oxygen atom, it reduces the unfavorable carbanion character of the transition state, thereby lowering its energy and hence lowering the activation energy barrier for the reaction:

$$\begin{array}{c} \text{R} \\ | \\ \text{C=O} \\ | \\ \text{CH}_2 \\ | \\ \text{H} \end{array} + \text{H—A} \rightleftharpoons \left[\begin{array}{c} \text{R} \\ | \\ \text{C}\cdots\text{O}^-\cdots\text{H}^+\cdots\text{A}^- \\ \vdots \\ \text{CH}_2 \\ \vdots \\ \text{H}^+ \end{array}\right] \rightleftharpoons \begin{array}{c} \text{R} \\ | \\ \text{C—O—H} \\ \| \\ \text{CH}_2 \end{array} + \text{H—A}$$

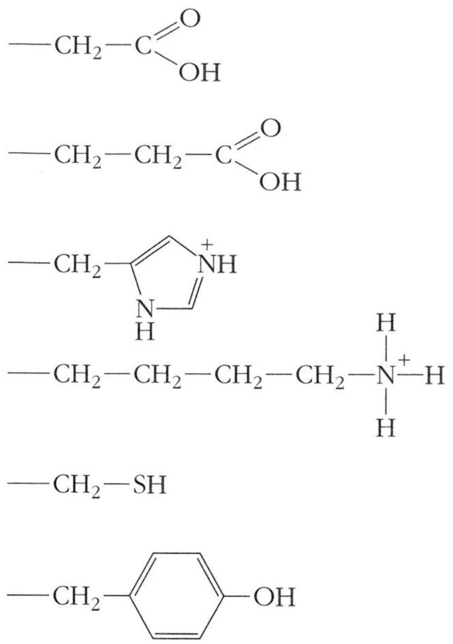

Asp	$—CH_2—C(=O)OH$
Glu	$—CH_2—CH_2—C(=O)OH$
His	$—CH_2$—(imidazole ring, $\overset{+}{N}H$, N—H)
Lys	$—CH_2—CH_2—CH_2—CH_2—\overset{+}{N}H_3$
Cys	$—CH_2—SH$
Tyr	$—CH_2$—(benzene ring)—OH

FIGURE 6-5

Amino acid side chains that can act as acid–base catalysts.
These groups can act as acid or base catalysts, depending on their state of protonation in the enzyme's active site. The side chains are shown in their protonated forms, with the acidic proton highlighted.

This is an example of acid catalysis, since the catalyst acts as an acid by donating a proton. Note that *the catalyst is returned to its original form at the end of the reaction.*

The same keto–enol tautomerization reaction shown above can be accelerated by a catalyst that can accept a proton, that is, by a base catalyst. Here, the catalyst is shown as :B, where the dots represent unpaired electrons:

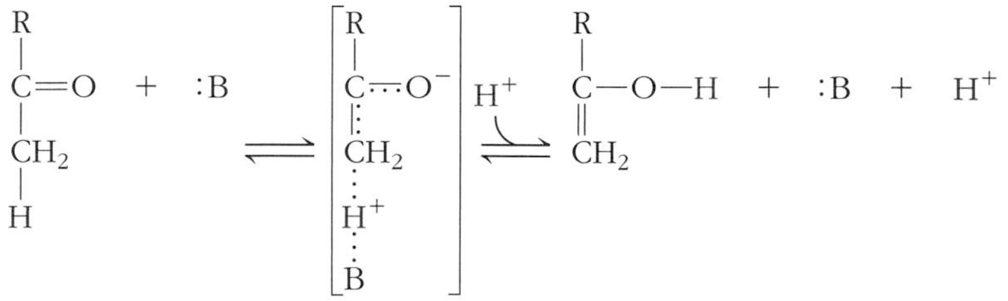

$$R—C(=O)—CH_2—H + :B \rightleftharpoons \left[R—C{\cdots}O^- \vdots CH_2 \cdots H^+ \cdots B\right] H^+ \rightleftharpoons R—C(—O—H)=CH_2 + :B + H^+$$

Base catalysis lowers the energy of the transition state and thereby accelerates the reaction.

In enzyme active sites, several amino acid side chains can potentially act as acid or base catalysts. These are the groups whose pK values are in or near the physiological pH range. The residues most commonly identified as acid–base catalysts are shown in Fig. 6-5. Because the catalytic functions of these residues depend on their state of protonation or deprotonation, the catalytic activity of the enzyme may be sensitive to changes in pH.

REVIEW LEARNING OBJECTIVE 4

- How can an acid or base accelerate a reaction?

2. Covalent catalysis

In **covalent catalysis,** the second major chemical reaction mechanism used by enzymes, a covalent bond forms between the catalyst and the substrate during formation of the transition state. Consider as a model reaction the decarboxylation of acetoacetate. In this reaction, the movement of electron pairs among atoms is indicated by red curved arrows that emanate from the electrons (either an unshared pair or the electrons of a covalent bond) and point to an electron-deficient center.

$$H_3C{-}\overset{\overset{\displaystyle O}{\|}}{C}{-}CH_2{-}C\overset{\displaystyle O}{\underset{\displaystyle O^-}{\lessgtr}} \xrightarrow{CO_2} \left[H_3C{-}\overset{\overset{\displaystyle O^-}{|}}{C}{=}CH_2\right] \xrightarrow{H^+} H_3C{-}\overset{\overset{\displaystyle O}{\|}}{C}{-}CH_3$$

Acetoacetate — Enolate — Acetone

The transition state, an enolate, has a high free energy of activation. This reaction can be catalyzed by a primary amine (RNH_2), which reacts with the carbonyl group of acetoacetate to form an **imine,** a compound containing a C=N bond (this adduct is known as a **Schiff base**):

$$H_3C{-}\overset{\overset{\displaystyle O}{\|}}{C}{-}CH_2{-}C\overset{\displaystyle O}{\underset{\displaystyle O^-}{\lessgtr}} + RNH_2 \longrightarrow H_3C{-}\overset{\overset{\displaystyle R\diagdown\overset{+}{N}\diagup H}{\|}}{C}{-}CH_2{-}C\overset{\displaystyle O}{\underset{\displaystyle O^-}{\lessgtr}} + OH^-$$

Acetoacetate — Schiff base (imine)

In this covalent intermediate, the protonated nitrogen atom acts as an electron sink to reduce the enolate character of the transition state in the decarboxylation reaction:

$$H_3C{-}\overset{\overset{\displaystyle R\diagdown\overset{+}{N}\diagup H}{\|}}{C}{-}CH_2{-}C\overset{\displaystyle O}{\underset{\displaystyle O^-}{\lessgtr}} \xrightarrow{CO_2} H_3C{-}\overset{\overset{\displaystyle R\diagdown\ddot{N}\diagup H}{|}}{C}{=}CH_2 \xrightarrow{H^+} H_3C{-}\overset{\overset{\displaystyle R\diagdown\overset{+}{N}\diagup H}{\|}}{C}{-}CH_3$$

Finally, the Schiff base decomposes, which regenerates the amine catalyst and releases the product, acetone:

$$H_3C{-}\overset{\overset{\displaystyle R\diagdown\overset{+}{N}\diagup H}{\|}}{C}{-}CH_3 + OH^- \longrightarrow H_3C{-}\overset{\overset{\displaystyle O}{\|}}{C}{-}CH_3 + RNH_2$$

Schiff base — Acetone

In enzymes that use covalent catalysis, an electron-rich group in the enzyme active site forms a covalent adduct with a substrate. This covalent complex can sometimes be isolated; it is much more stable than a transition state. *Enzymes that use covalent catalysis undergo a two-part reaction process so that the reaction coordinate diagram contains two energy barriers with the reaction intermediate between them* (Fig. 6-6).

Many of the same groups that make good acid–base catalysts (Fig. 6-5) also make good covalent catalysts because they contain unshared electron pairs (Fig. 6-7). Covalent catalysis is often called **nucleophilic catalysis** because the catalyst is a **nucleophile,** that is, an electron-rich group in search of an electron-poor center (a compound with an electron deficiency is known as an **electrophile**).

REVIEW LEARNING OBJECTIVE 4

- How does a covalent catalyst accelerate a reaction?
- Why must a reaction that proceeds via covalent catalysis have two transition states?

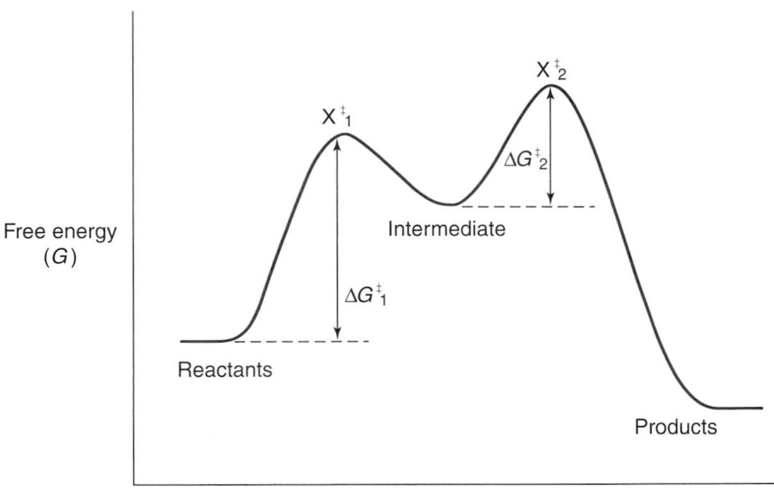

FIGURE 6-6

Reaction coordinate diagram for a reaction accelerated by covalent catalysis. Two transition states flank the covalent intermediate. The relative heights of the activation energy barriers (the energies of the two transition states, $X^{\ddagger}_1$ and $X^{\ddagger}_2$) vary depending on the reaction.

3. Metal ion catalysis

Metal ions may participate in enzymatic reactions by mediating oxidation–reduction reactions, as mentioned above, or by promoting the reactivity of other groups in the enzyme's active site through electrostatic effects. A protein-bound metal ion can also interact directly with the reacting substrate. For example, during the conversion of acetaldehyde to ethanol as catalyzed by the liver enzyme alcohol dehydrogenase, a zinc ion stabilizes the developing negative charge on the oxygen atom during formation of the transition state:

$$\underset{\text{Acetaldehyde}}{H_3C{-}C(=O)H} \rightleftharpoons \left[H_3C{-}C(\cdots O^{-}\cdots Zn^{2+})H\right] \overset{2\,H}{\rightleftharpoons} \underset{\text{Ethanol}}{H_3C{-}C(OH)(H){-}H}$$

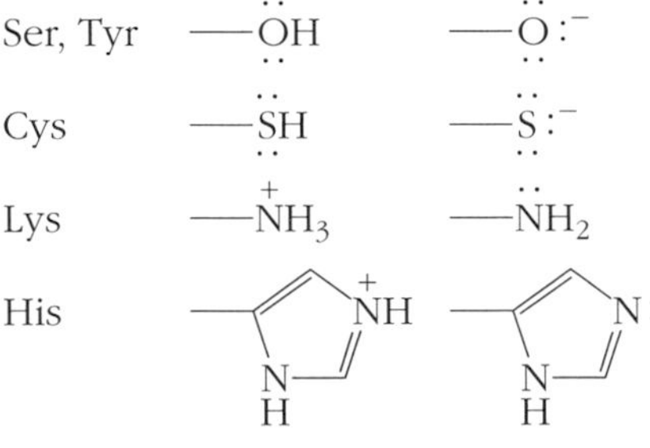

FIGURE 6-7

Protein groups that can act as covalent catalysts. In their deprotonated forms, these groups act as nucleophiles. They attack electron-deficient centers to form covalent intermediates.

REVIEW LEARNING OBJECTIVE 4

- How do metal ions accelerate reactions?

The catalytic triad of chymotrypsin promotes peptide bond hydrolysis

Chymotrypsin uses both acid–base catalysis and covalent catalysis to accelerate peptide bond hydrolysis. These activities depend on three active site residues whose identities and catalytic importance have been the object of intense study—still ongoing—since the 1960s. Two of chymotrypsin's catalytic residues, His and Ser, were identified using a technique called **chemical labeling.** When chymotrypsin is incubated with the compound diisopropylphosphofluoridate (DIPF), one of its 27 Ser residues (Ser 195) becomes

Go to **Exercise 9/Serine Proteases** for a detailed look at the structure and mechanism of chymotrypsin and related enzymes. The molecular structures presented in Exercise 9 and the questions based on those structures address Learning Objectives 5, 6, and 8.

covalently tagged with the diisopropylphospho (DIP) group, and the enzyme loses activity.

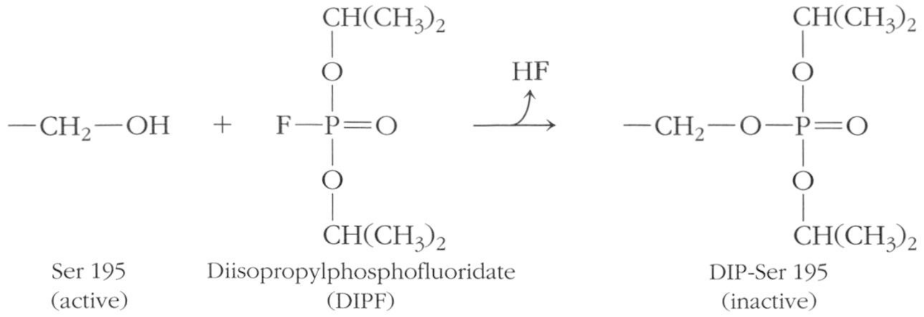

$$-CH_2-OH + F-P(=O)(-O-CH(CH_3)_2)_2 \xrightarrow{HF} -CH_2-O-P(=O)(-O-CH(CH_3)_2)_2$$

Ser 195 (active) — Diisopropylphosphofluoridate (DIPF) — DIP-Ser 195 (inactive)

This observation provided strong evidence that Ser 195 is essential for catalysis. Chymotrypsin is therefore known as a **serine protease.** It is one of a large family of enzymes that use the same Ser-dependent catalytic mechanism.

A catalytically important His residue—His 57—was identified using a specific labeling reagent that mimics the enzyme's natural substrate. In this case, tosyl-phenylalanine chloromethylketone (TPCK), whose Phe residue is one of chymotrypsin's preferred substrate residues, binds to the active site of chymotrypsin, where the chloromethylketone group (a powerful alkylating agent) reacts with a nearby His residue:

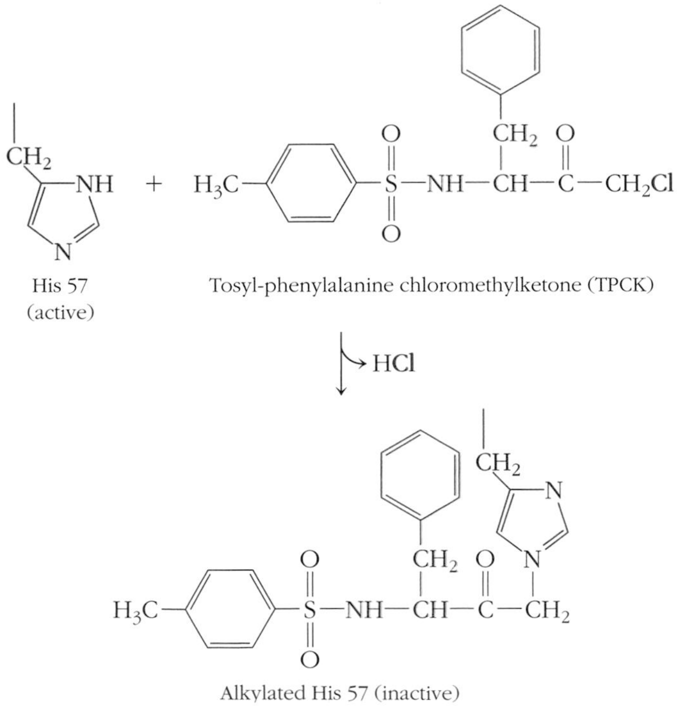

His 57 (active) + $H_3C-C_6H_4-SO_2-NH-CH(CH_2C_6H_5)-C(=O)-CH_2Cl$

Tosyl-phenylalanine chloromethylketone (TPCK)

↓ HCl

$H_3C-C_6H_4-SO_2-NH-CH(CH_2C_6H_5)-C(=O)-CH_2-$(His 57 imidazole N)

Alkylated His 57 (inactive)

Reagents that bind to an enzyme's active site by mimicking a substrate and then "mark" catalytic groups by forming covalent bonds with them are called **affinity labels.**

The third residue involved in catalysis by chymotrypsin—Asp 102—was identified only after the fine structure of chymotrypsin was visualized through X-ray crystallography. The Asp side chain is mostly buried in the active site, but it is close enough to His 57 to form a hydrogen bond with the imidazole ring. The importance of Asp 102 is supported by the observation that this residue appears in other proteases that have active site His and Ser residues.

The arrangement of the Asp, His, and Ser residues of chymotrypsin and other serine proteases is called the **catalytic triad** (Fig. 6-8). The substrate's **scissile bond** (the bond to be cleaved by hydrolysis) is positioned near Ser

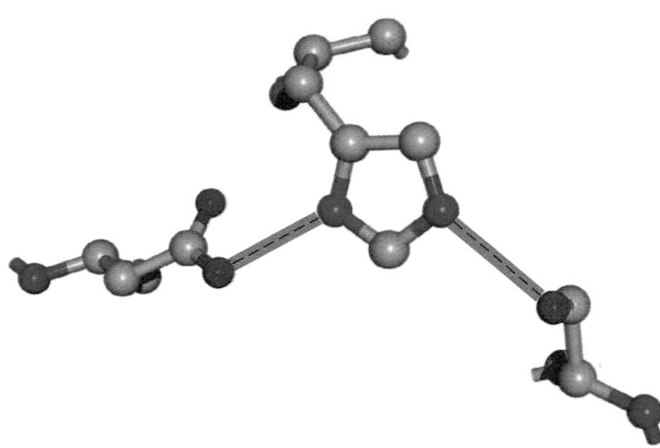

FIGURE 6-8

The catalytic triad of chymotrypsin.
Asp 102 (left), His 57 (center), and Ser 195 (right) are arrayed in a hydrogen-bonded network. Atoms are color coded (C gray, N blue, O red) and the hydrogen bonds are shaded yellow.

195 when the substrate binds to the enzyme. The side chain of Ser is not normally a strong enough nucleophile to attack an amide bond. However, His 57, acting as a base catalyst, abstracts a proton from Ser 195 so that the oxygen can act as a covalent catalyst. Asp 102 promotes catalysis by stabilizing the resulting positively charged imidazole group of His 57. **Exercise 9 shows three-dimensional views of chymotrypsin's catalytic triad.**

Chymotrypsin-catalyzed peptide bond hydrolysis actually occurs in two phases that correspond to the formation and breakdown of a covalent reaction intermediate. The steps of catalysis are detailed in Figure 6-9. Nucleophilic attack of Ser 195 on the substrate's carbonyl carbon leads to a transition state in which the carbonyl carbon assumes tetrahedral geometry. This collapses to an intermediate in which the N-terminal portion of the substrate remains covalently attached to the enzyme. The second part of the reaction, during which the oxygen of a water molecule attacks the carbonyl carbon, also includes a tetrahedral transition state. Although the enzyme-catalyzed reaction requires multiple steps, the net reaction is the same as the uncatalyzed reaction shown on page 167. **The chymotrypsin reaction is animated in Exercise 9.**

The roles of Asp, His, and Ser in peptide bond hydrolysis, as catalyzed by chymotrypsin and other members of the serine protease family, have been tested though site-directed mutagenesis (see Section 3-4). Replacing the catalytic Asp with another residue decreases the rate of substrate hydrolysis about 5000-fold. Adding a methyl group to His by chemical labeling (so that it can't accept or donate a proton) has a similar effect. Replacing the catalytic Ser with another residue decreases enzyme activity about a millionfold. Surprisingly, replacing all three catalytic residues—Asp, His, and Ser—through site-directed mutagenesis does not completely abolish protease activity: The modified enzyme still catalyzes peptide bond hydrolysis at a rate about 50,000 times greater than the rate of the uncatalyzed reaction. Clearly, chymotrypsin and its relatives rely on the acid–base catalysis and covalent catalysis carried out by the Asp–His–Ser catalytic triad, but these enzymes must have additional catalytic mechanisms that allow them to achieve reaction rates 10^{11} times greater than the rate of the uncatalyzed reaction.

REVIEW LEARNING OBJECTIVE 5

- How can chemical labeling identify an enzyme's catalytic residues?
- Which residues constitute chymotrypsin's catalytic triad?
- How do these residues participate in peptide bond hydrolysis as catalyzed by chymotrypsin?

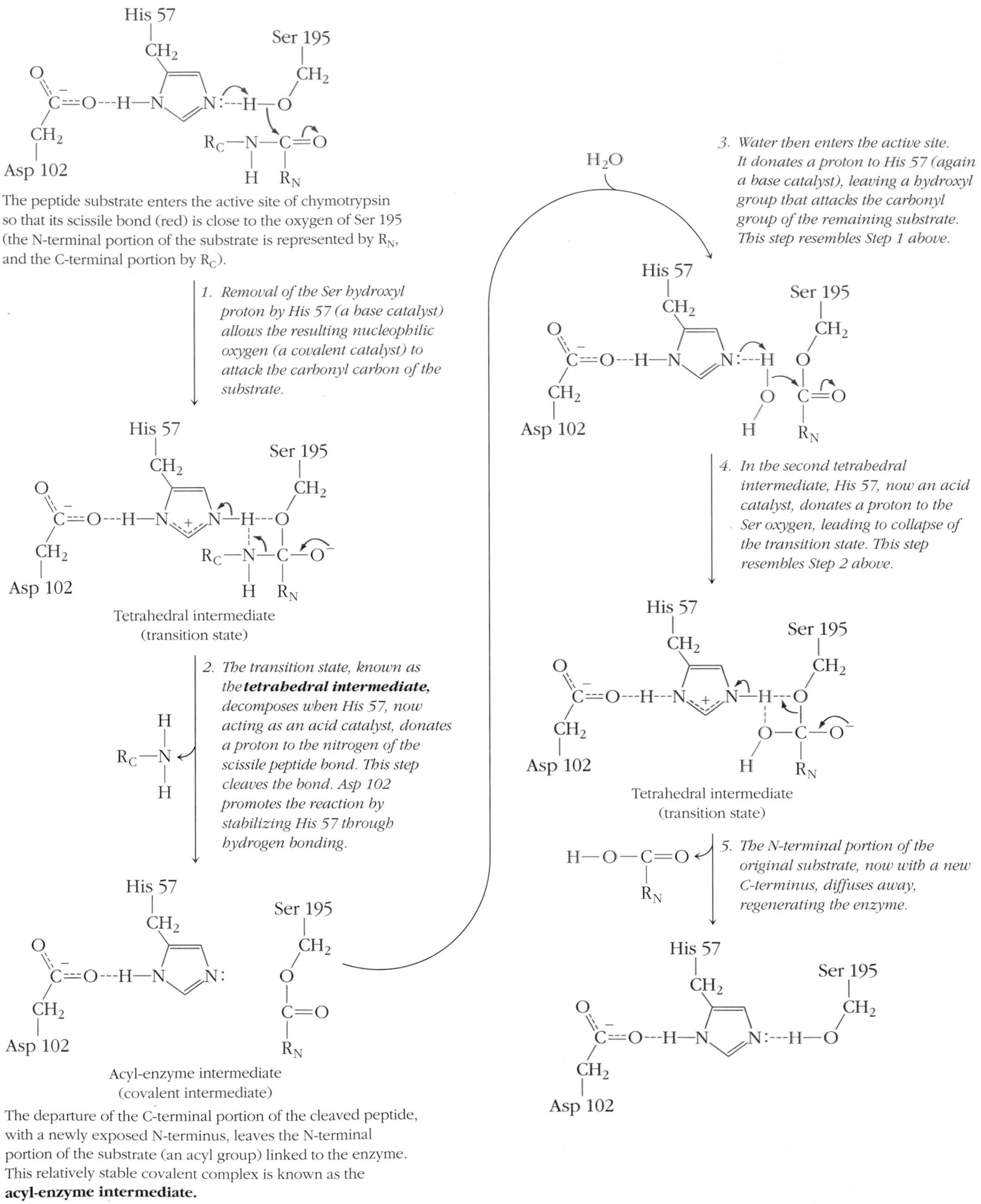

The peptide substrate enters the active site of chymotrypsin so that its scissile bond (red) is close to the oxygen of Ser 195 (the N-terminal portion of the substrate is represented by R_N, and the C-terminal portion by R_C).

1. Removal of the Ser hydroxyl proton by His 57 (a base catalyst) allows the resulting nucleophilic oxygen (a covalent catalyst) to attack the carbonyl carbon of the substrate.

Tetrahedral intermediate
(transition state)

2. The transition state, known as the ***tetrahedral intermediate,*** *decomposes when His 57, now acting as an acid catalyst, donates a proton to the nitrogen of the scissile peptide bond. This step cleaves the bond. Asp 102 promotes the reaction by stabilizing His 57 through hydrogen bonding.*

Acyl-enzyme intermediate
(covalent intermediate)

The departure of the C-terminal portion of the cleaved peptide, with a newly exposed N-terminus, leaves the N-terminal portion of the substrate (an acyl group) linked to the enzyme. This relatively stable covalent complex is known as the **acyl-enzyme intermediate.**

H_2O

3. Water then enters the active site. It donates a proton to His 57 (again a base catalyst), leaving a hydroxyl group that attacks the carbonyl group of the remaining substrate. This step resembles Step 1 above.

4. In the second tetrahedral intermediate, His 57, now an acid catalyst, donates a proton to the Ser oxygen, leading to collapse of the transition state. This step resembles Step 2 above.

Tetrahedral intermediate
(transition state)

5. The N-terminal portion of the original substrate, now with a new C-terminus, diffuses away, regenerating the enzyme.

FIGURE 6-9

The catalytic mechanism of chymotrypsin and other serine proteases.

3 THE UNIQUE PROPERTIES OF ENZYME CATALYSTS

If only a few residues in an enzyme directly participate in catalysis (for example, Asp, His, and Ser in chymotrypsin), why are enzymes so large? One obvious answer is that the catalytic residues must be precisely aligned in the active site, so a certain amount of surrounding structure is required to hold them in place. In 1894, long before the first enzyme structure had been determined (and several decades before it had been shown that enzymes are proteins), Emil Fischer noted the exquisite substrate specificity of enzymes and proposed that the substrate fit the enzyme like a key in a lock (Fig. 6-10).

The **lock-and-key model** has some validity in explaining how enzymes increase reaction rates by bringing substrates into contact with the reactive groups at an enzyme's active site. Unfortunately, it does not fully describe enzyme action. For one thing, it does not explain how an active site that perfectly accommodates substrates could also accommodate products before they are released from the enzyme. Furthermore, *very tight binding of substrates would be counterproductive, since the substrates would then have an even greater energetic barrier to surpass in order to achieve the transition state* (Fig. 6-11).

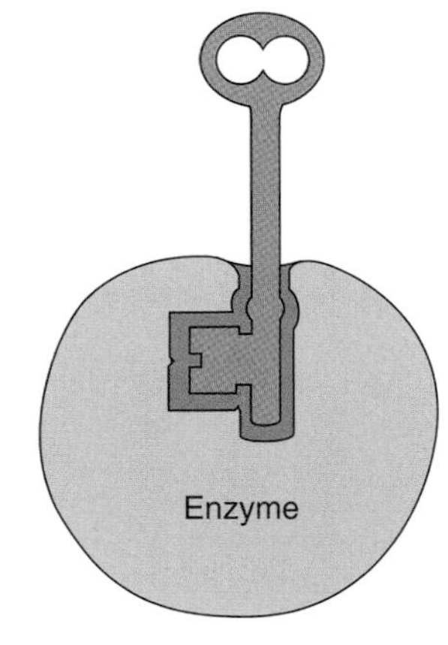

FIGURE 6-10

Lock-and-key model of enzyme action. In this early model, the substrate (key) was believed to bind tightly to a site on the enzyme (lock) that was exactly complementary to it in size, shape, and charge.

Transition state stabilization

The lock-and-key model does contain a grain of truth, a principle first formulated by Linus Pauling in 1946. He proposed that *an enzyme increases the reaction rate not by binding tightly to the substrates, but by binding tightly to the reaction's transition state* (that is, substrates that have been strained toward the structures of the products). In other words, the tightly bound key of the lock-and-key model is the transition state, not the substrate. Pauling's theory of enzyme action is consistent with the chemical catalytic mechanisms outlined in the preceding section, in which the catalyst stabilizes the transition state of a reaction. In an enzyme, tight binding (stabilization) of the transition state occurs in addition to acid–base, covalent, or metal ion catalysis.

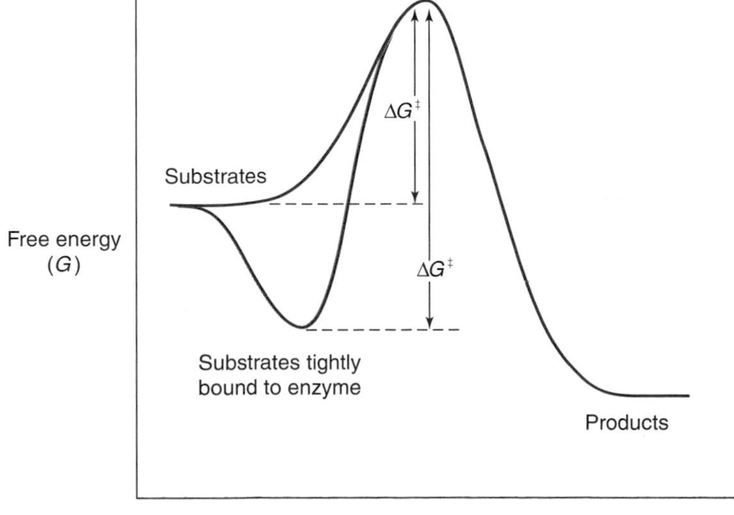

FIGURE 6-11

Effect of very tight substrate binding on enzyme catalysis.
The red line traces the free energy changes as the reactants pass through a transition state to products. The blue line traces the free energy changes for the same reaction when the reactants (substrates) are very tightly bound to the enzyme (that is, they are at a very low free energy level) before they react. The activation energy barrier ($\Delta G^{\ddagger}$) is therefore increased because the tight binding must be overcome before the substrates can reach the transition state.

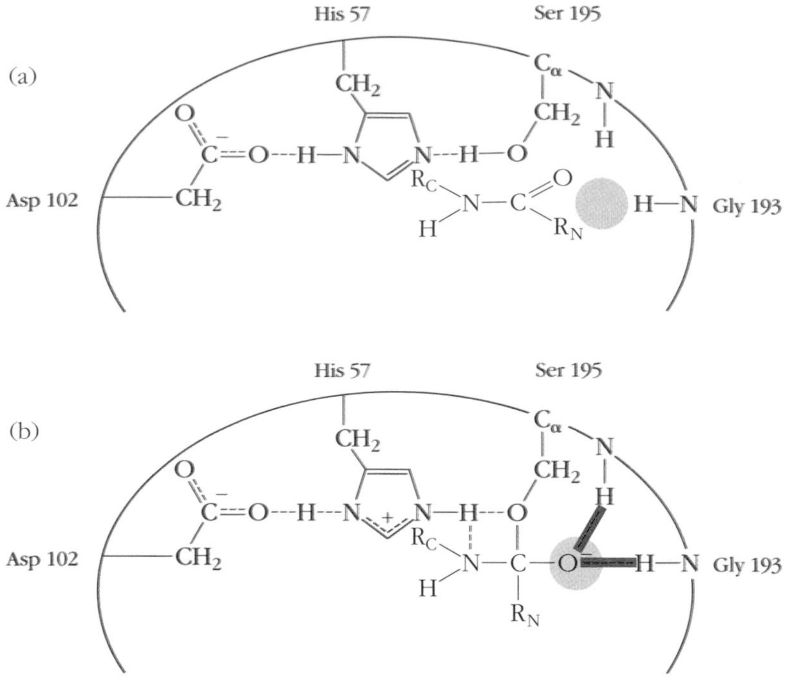

FIGURE 6-12

Transition state stabilization in the oxyanion hole.
(a) The chymotrypsin active site is shown with the oxyanion hole shaded in color. The carbonyl carbon of the peptide substrate has trigonal geometry, so the carbonyl oxygen cannot occupy the oxyanion hole. (b) Nucleophilic attack by the oxygen of Ser 195 on the substrate carbonyl group leads to formation of the transition state, in which the carbonyl carbon assumes tetrahedral geometry. At this point, the substrate's anionic oxygen (the oxyanion) can move into the oxyanion hole, where it forms hydrogen bonds (shaded yellow) with two enzyme backbone groups.

Transition state stabilization appears to be an important part of the chymotrypsin reaction. In this case, the two tetrahedral transition states of the reaction (as shown in Fig. 6-9) are stabilized through interactions that do not occur at any other point in the reaction. Rate acceleration is believed to result from an increase in both the number and strength of the bonds that form between active site groups and the substrate in the transition state.

1. During formation of the tetrahedral transition states, the planar peptide group of the substrate changes its geometry, and the carbonyl oxygen, now an anion, moves into a previously unoccupied cavity near the Ser 195 side chain. In this cavity, called the **oxyanion hole,** the substrate oxygen can form two new hydrogen bonds with the backbone NH groups of Ser 195 and Gly 193 (Fig. 6-12). The backbone NH group of the substrate residue preceding the scissile bond forms another hydrogen bond to Gly 193 (not shown in Fig. 6-12). Thus, the transition state is stabilized (its energy lowered) by three hydrogen bonds that cannot form when the enzyme first binds its substrate. The stabilizing effect of these three new hydrogen bonds could account for a significant portion of chymotrypsin's catalytic power, since the energy of a standard hydrogen bond is about 20 $kJ \cdot mol^{-1}$ and the reaction rate increases tenfold for every decrease in $\Delta G^{\ddagger}$ of 5.7 $kJ \cdot mol^{-1}$.
Go to Exercise 9 to see chymotrypsin's oxyanion hole

2. NMR studies, which can identify individual hydrogen-bonding interactions, suggest that the hydrogen bond between Asp 102 and His 57 becomes shorter during formation of the two transition states (Fig. 6-13). Such a bond is called a **low-barrier hydrogen bond** because the hydrogen is shared equally between the original donor and acceptor atoms (in a standard hydrogen bond, the proton still "belongs" to the donor atom and there is an energy barrier for its full transfer to the acceptor atom). A decrease in bond length from ~2.8 Å to ~2.5 Å in forming the low-barrier hydrogen bond is accompanied by a three- to fourfold increase in bond strength. The low-barrier hy-

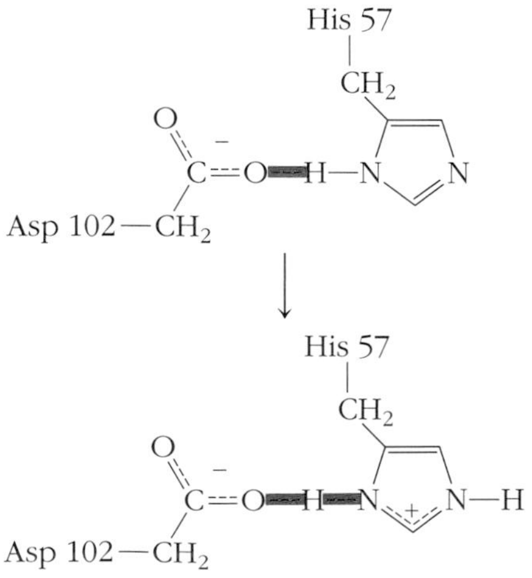

FIGURE 6-13

Formation of a low-barrier hydrogen bond during catalysis in chymotrypsin.
The Asp 102—His 57 hydrogen bond becomes shorter and stronger, so that the imidazole proton comes to be shared equally between the O of Asp and the N of histidine in a low-barrier hydrogen bond.

drogen bond that forms during catalysis in chymotrypsin may help stabilize the transition state and thereby accelerate the reaction.

REVIEW LEARNING OBJECTIVE 6

- What are the deficiencies and merits of the lock-and-key model for enzyme action?
- What is the role of the oxyanion hole in chymotrypsin?
- How does a low-barrier hydrogen bond promote catalysis?

Proximity, orientation, and electrostatic catalysis

There is considerable debate over the absolute contribution of transition state stabilization to the overall catalytic power of enzymes. Such issues are difficult to address experimentally, even for well-studied enzymes such as chymotrypsin. What other factors might account for the catalytic power of enzymes, especially those whose mechanisms do not involve oxyanion holes or low-barrier hydrogen bonds?

It seems obvious that enzymes increase reaction rates by bringing reacting groups into close proximity so as to increase the frequency of collisions that can lead to a reaction. Furthermore, when substrates bind to an enzyme, their translational and rotational motions are frozen out so that they can be oriented properly for reaction (Fig. 6-14). Nevertheless, calculations suggest that these **proximity and orientation effects** can account for no more than about a thousand-fold rate enhancement. *An enzyme must therefore be more than a template for assembling and aligning reacting groups.* Again, the structure of the enzyme itself offers a clue to its catalytic abilities.

In nearly all cases, an enzyme's active site is somewhat removed from the solvent, with its catalytic residues in some sort of cleft or pocket on the enzyme surface. In fact, upon binding substrates, some enzymes undergo a pronounced conformational change so that they almost fully enclose the substrates. This phenomenon is known as **induced fit.** A classic example is hexokinase, which catalyzes the phosphorylation of glucose by ATP (Reaction 1 of glycolysis; Section 10-2):

$$\text{Glucose (HO–CH}_2\text{)} + \text{ATP} \xrightarrow{\text{hexokinase}} \text{Glucose-6-phosphate } ({}^{-2}O_3P\text{–O–CH}_2) + \text{ADP}$$

The enzyme consists of two hinged lobes with the active site located between them (Fig. 6-15a). When glucose binds to hexokinase, the lobes swing together, engulfing the sugar (Fig. 6-15b). The result of hinge bending is that the substrate glucose is positioned near the substrate ATP such that a phosphoryl group can be easily transferred from the ATP to a hydroxyl group of the sugar. Not even a water molecule can enter the closed active site. This is beneficial, since water in the active site could lead to wasteful hydrolysis of ATP:

$$\text{ATP} + H_2O \rightarrow \text{ADP} + P_i$$

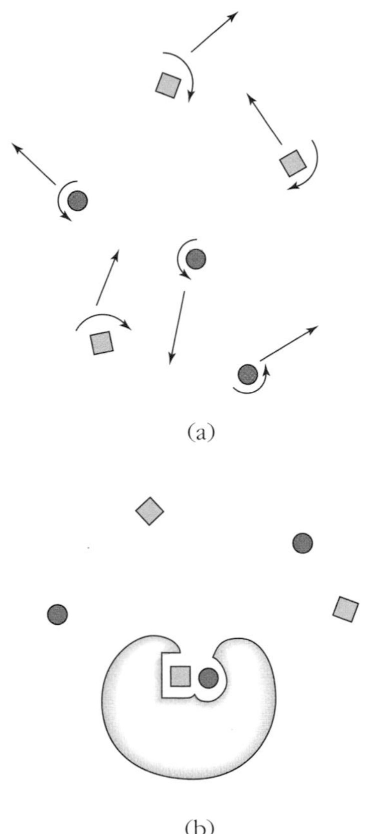

FIGURE 6-14

Proximity and orientation effects in catalysis. In order to react, two groups must come together and collide with the correct orientation. (a) Reactants in solution are separated in space and have translational and rotational motions that must be overcome. (b) When the reactants bind to an enzyme, their motion is limited, and they are held in close proximity and with the correct alignment for a productive reaction.

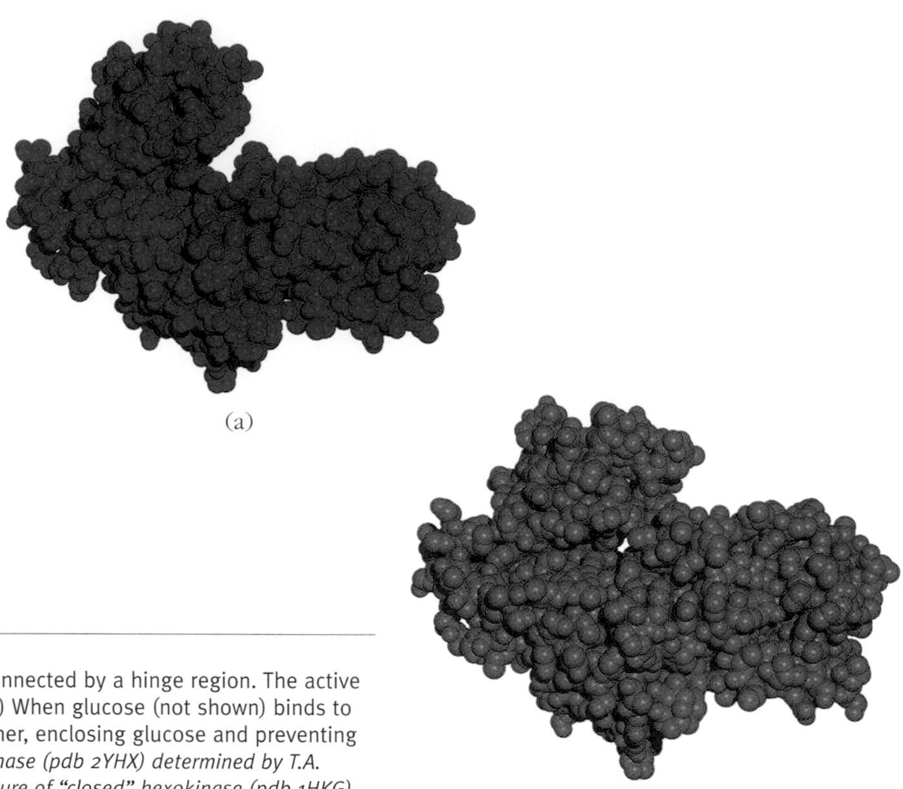

FIGURE 6-15

Conformational changes in hexokinase.
(a) The yeast enzyme consists of two lobes connected by a hinge region. The active site is located in a cleft between the lobes. (b) When glucose (not shown) binds to the active site, the enzyme lobes swing together, enclosing glucose and preventing the entry of water. [*Structure of "open" hexokinase (pdb 2YHX) determined by T.A. Steitz, C.M. Anderson, and R.E. Stenkamp; structure of "closed" hexokinase (pdb 1HKG) determined by W.S. Bennett, Jr. and T.A. Steitz.*]

A glucose molecule in solution is surrounded by water molecules in a hydration shell (see Section 2-1). These ordered water molecules must be disrupted in order for glucose to react with another molecule. These water molecules must also be shed in order for glucose to fit into the active site of an enzyme, such as hexokinase. However, once the desolvated substrate is in the enzyme active site, the reaction can proceed quickly, because there are no solvent molecules to interfere. In solution, rearranging the hydrogen bonds of surrounding water molecules as the reactants approach each other and pass through the transition state is energetically costly. *By sequestering substrates in the active site, an enzyme can eliminate the energy barrier imposed by the ordered solvent molecules, thereby accelerating the reaction.*

This phenomenon is sometimes described as **electrostatic catalysis** since the nonaqueous active site allows more powerful electrostatic interactions between the enzyme and substrate than could occur in aqueous solution (for example, a low-barrier hydrogen bond can form in an active site but not in the presence of solvent molecules that would form ordinary hydrogen bonds). This desolvation or electrostatic effect could explain why mutating an enzyme's catalytic residues does not completely obliterate catalytic activity, provided that the substrates can still bind in the active site. Keep in mind that the active site cavity is not necessarily nonpolar, just nonaqueous. The surface of the active site must include polar and ionized groups that can mediate chemical catalysis or stabilize the transition state.

REVIEW LEARNING OBJECTIVE 7

- What are the roles of proximity and orientation effects in enzyme catalysis?
- Why does the exclusion of water from an active site promote catalysis?

4 WHAT ELSE CAN CHYMOTRYPSIN TELL US?

Chymotrypsin serves as a model for the structures and functions of a large family of evolutionarily related serine proteases. The simplest organisms often contain just a single nonspecific chymotrypsin-like protease whose function is to break down proteins. The most complex organisms contain many serine proteases, some with exquisite substrate specificity, whose various physiological functions are likewise highly specialized. This section describes some additional aspects of serine protease function, focusing on the digestive enzymes as well as on more specialized enzymes with similar chemistry.

FIGURE 6-16

X-Ray structures of chymotrypsin, trypsin, and elastase. The superimposed backbone traces of bovine chymotrypsin (blue), bovine trypsin (green), and porcine elastase (red) are shown along with the side chains of the active-site Asp, His, and Ser residues. [*Chymotrypsin structure (pdb 4CHA) determined by H. Tsukada and D.M. Blow; trypsin structure (pdb 3PTN) determined by J. Walker, W. Steigemann, T.P. Singh, H. Bartunik, W. Bode, and R. Huber; elastase structure (3EST) determined by E.F. Meyer, G. Cole, R. Radhakrishnan, and O. Epp.*]

Evolution of serine proteases

The first three proteases to be examined in detail were the digestive enzymes chymotrypsin, trypsin, and elastase, which have strikingly similar three-dimensional structures (Fig. 6-16). This was not expected on the basis of their limited sequence similarity (Table 6-3). However, careful examination revealed that most of the sequence variation is on the enzyme surface, and the positions of the catalytic residues in the three actives sites are virtually identical. It is believed that *these proteins diverged from a common ancestor and have retained their overall structure and catalytic mechanism.*

Some bacterial proteases with a catalytically essential Ser are structurally related to the mammalian digestive serine proteases. However, the bacterial serine protease subtilisin (Fig. 6-17) shows no sequence similarity to chymotrypsin and no overall structural similarity, although it has the same Asp–His–Ser catalytic triad and an oxyanion hole in its active site. Subtilisin provides a classic example of **convergent evolution,** the phenomenon whereby unrelated proteins evolve similar characteristics.

As many as five groups of serine proteases, each with a different overall backbone conformation, have undergone convergent evolution to arrive at the same Asp, His, and Ser catalytic groups. In some other hydrolases, the substrate is attacked by a nucleophilic Ser or Thr residue that is located in a catalytic triad such as His–His–Ser or Asp–Lys–Thr. It would appear that natural selection favors this sort of arrangement of catalytic residues.

TABLE 6-3 Percent sequence identity among three serine proteases

Bovine trypsin	100%
Bovine chymotrypsin	53%
Porcine elastase	48%

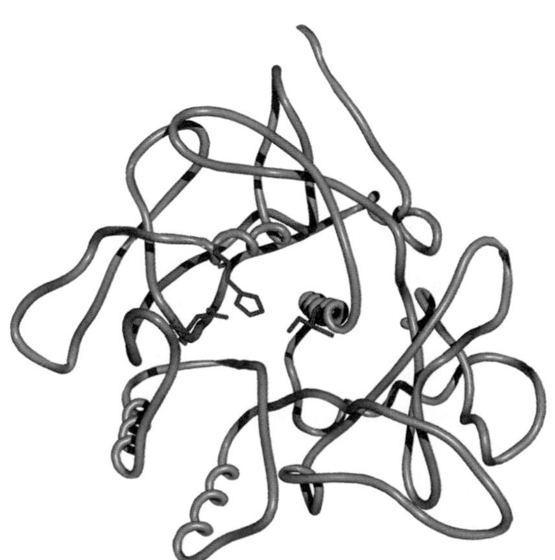

FIGURE 6-17

Structure of subtilisin from *Bacillus amyloliquefaciens*. The residues of the catalytic triad are highlighted in red. Compare the structure of this enzyme with those of the three serine proteases shown in Figure 6-16. [*Subtilisin structure (pdb 1CSE) determined by W. Bode.*]

What determines substrate specificity?

Chymotrypsin, trypsin, and elastase differ significantly from each other in their substrate specificity. Chymotrypsin preferentially cleaves peptide bonds following large hydrophobic residues. Trypsin prefers the basic residues Arg and Lys, and elastase cleaves the peptide bonds following small hydrophobic residues such as Ala, Gly, and Val (these residues predominate in elastin, an animal protein responsible for the elasticity of some tissues). *The varying specificities of these enzymes are largely explained by the chemical character of the so-called* ***specificity pocket,*** *a hole on the enzyme surface at the active site that accommodates the residue on the N-terminal side of the scissile peptide bond* (Fig. 6-18). In chymotrypsin, the specificity pocket is about 10 Å deep and 5 Å wide, which offers a snug fit for an aromatic ring (whose dimensions are 6 Å × 3.5 Å). The specificity pocket in trypsin is similarly sized but has an Asp residue rather than Ser at the bottom. Consequently, the trypsin specificity pocket readily binds the side chain of Arg or Lys, which has a diameter of about 4 Å and a basic group at the end. In elastase, the specificity pocket is only a small depression due to the replacement of two Gly residues on the walls of the specificity pocket (residues 216 and 226 in chymotrypsin) with the bulkier Val and Thr. Elastase therefore preferentially binds small nonpolar side chains. Although these same side chains could easily enter the chymotrypsin and trypsin specificity pockets, they do not fit well enough to immobilize the substrate at the active site, as required for efficient catalysis.

REVIEW LEARNING OBJECTIVE 8

- How are chymotrypsin, trypsin, and elastase similar?
- What is the chemical basis for their different substrate specificity?

How is chymotrypsin activated?

Relatively nonspecific proteases could do considerable damage to the cells where they are synthesized, unless their activity is carefully controlled. In many organisms, *the activity of proteases is limited by the action of protease inhibitors* (some of which are discussed further below) *and by synthesizing the proteases as inactive precursors (called* ***zymogens****) that are later activated when and where they are needed.*

The inactive precursor of chymotrypsin is called chymotrypsinogen, and it is synthesized by the pancreas along with the zymogens of trypsin (trypsinogen), elastase (proelastase), and other hydrolytic enzymes. All these

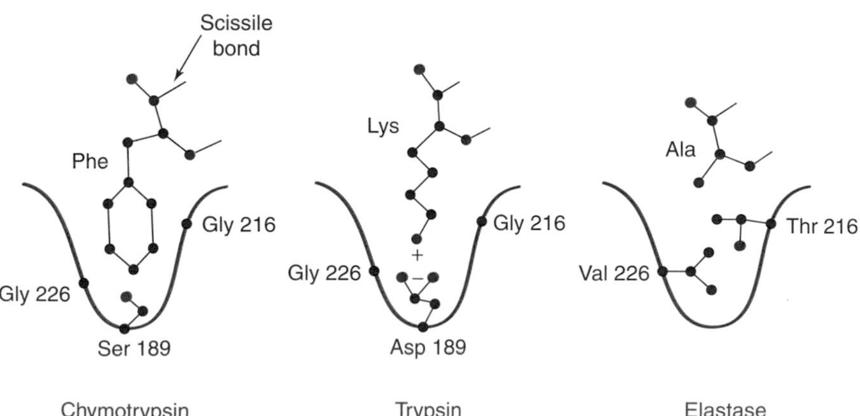

FIGURE 6-18

Specificity pockets of three serine proteases. The side chains of key residues that determine the size and nature of the specificity pocket are shown along with a representative substrate for each enzyme. Chymotrypsin prefers large hydrophobic side chains; trypsin prefers Lys or Arg; and elastase prefers Ala, Gly, or Val. For convenience, the residues of all three enzymes are numbered to correspond to the sequence of residues in chymotrypsin.

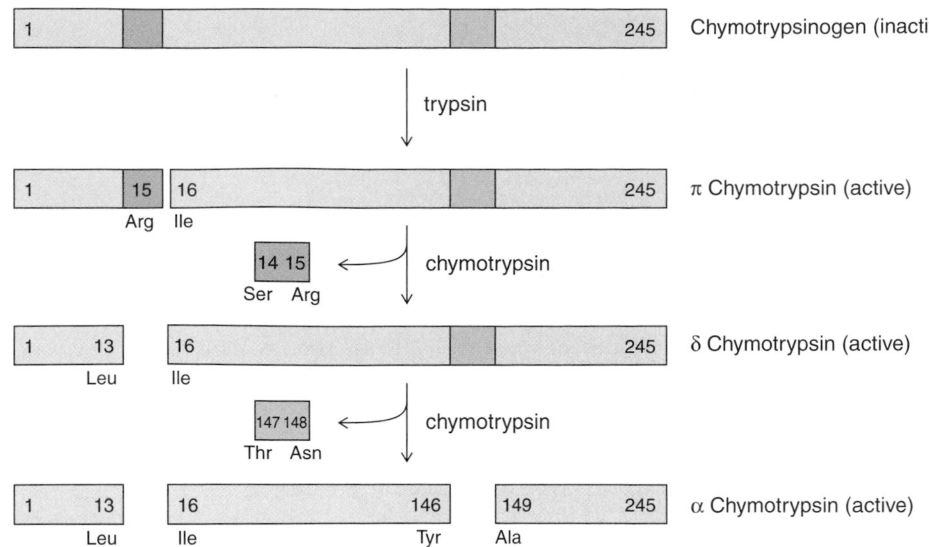

FIGURE 6-19

Activation of chymotrypsinogen. Trypsin activates chymotrypsinogen by catalyzing hydrolysis of the Arg 15—Ile 16 bond of the zymogen. The resulting active chymotrypsin then excises the Ser 14–Arg 15 dipeptide (by cleaving the Leu 13—Ser 14 bond) and the Thr 147–Asn 148 dipeptide (by cleaving the Tyr 146—Thr 147 and Asn 148—Ala 149 bonds). All three species of chymotrypsin (π, δ, and α) have proteolytic activity.

zymogens are activated by proteolysis after they are secreted into the small intestine. An intestinal protease called enteropeptidase activates trypsinogen by catalyzing the hydrolysis of its Lys 6—Ile 7 bond. Enteropeptidase catalyzes a highly specific reaction; it appears to recognize a string of Asp residues near the N-terminus of its substrate:

$$\mathrm{H_3\overset{+}{N}{-}Val{-}Asp{-}Asp{-}Asp{-}Asp{-}Lys{-}Ile{-}\cdots}$$

$$\xrightarrow[\text{enteropeptidase}]{H_2O}$$

$$\mathrm{H_3\overset{+}{N}{-}Val{-}Asp{-}Asp{-}Asp{-}Asp{-}Lys{-}COO^-} \; + \; \mathrm{H_3\overset{+}{N}{-}Ile{-}\cdots}$$

Trypsin, now itself active, cleaves the N-terminal peptide of the other pancreatic zymogens, including trypsinogen. The activation of trypsinogen by trypsin is an example of **autoactivation.**

The Arg 15—Ile 16 bond of chymotrypsinogen is susceptible to trypsin-catalyzed hydrolysis. Cleavage of this bond generates a species of active chymotrypsin (called π chymotrypsin), which then undergoes two autoactivation steps to generate fully active chymotrypsin (also called α chymotrypsin; Fig. 6-19). A similar process, in which zymogens are sequentially activated through proteolysis, occurs during blood coagulation (Box 6-A).

The two dipeptides that are excised during chymotrypsinogen activation are far removed from the active site (Fig. 6-20). How does their removal boost catalytic activity? A comparison of the X-ray structures of chymotrypsin and chymotrypsinogen reveals that the conformations of their active site Asp, His, and Ser residues are virtually identical (in fact, the zymogen can catalyze hydrolysis extremely slowly). However, the substrate specificity pocket and the oxyanion hole are incompletely formed in the zymogen. Proteolysis of the zymogen elic-

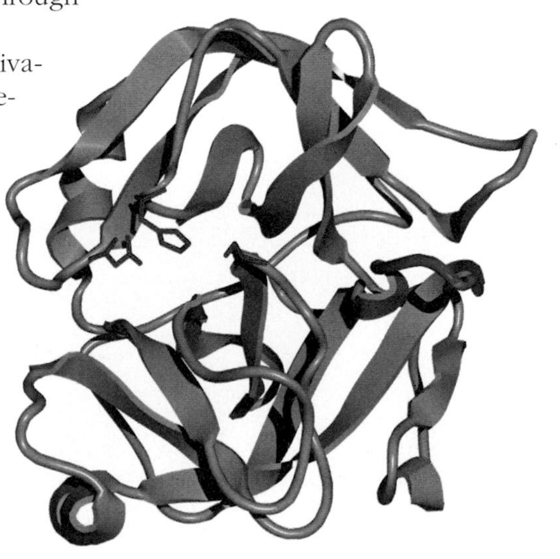

FIGURE 6-20

Location of the dipeptides removed during the activation of chymotrypsinogen. The Ser 14—Arg 15 dipeptide (lower right, green) and Thr 147—Asn 148 dipeptide (right, blue) are located at some distance from the active site residues (red) in chymotrypsinogen. [*Chymotrypsinogen structure (pdb 2CGA) determined by D. Wang, W. Bode, and R. Huber.*]

A CLOSER LOOK

BOX 6-A Blood coagulation requires a cascade of proteases

When a blood vessel is injured by mechanical force, infection, or some other pathological process, red and white blood cells and the plasma (fluid) that surrounds them can leak out. Except in the most severe trauma, the loss of blood can be halted through formation of a clot at the site of injury. The clot consists of aggregated platelets (tiny enucleated cells that rapidly adhere to the damaged vessel wall and to each other) and a mesh of the protein fibrin, which reinforces the platelet plug and traps larger particles such as red blood cells.

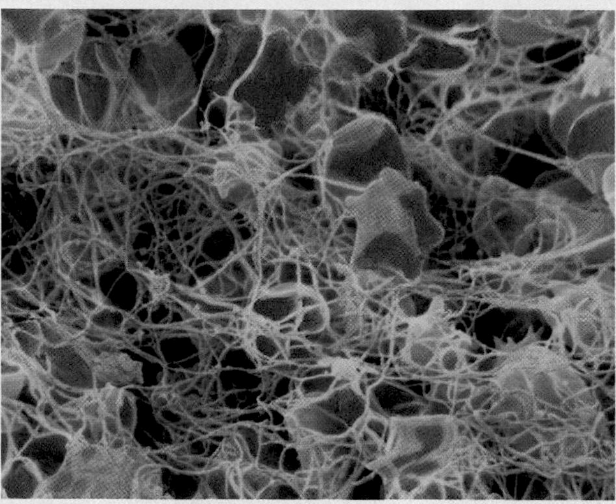

Fibrin polymers can form rapidly because they are generated at the site of injury from the soluble protein fibrinogen, which circulates in the blood plasma. Fibrinogen is an elongated molecule with a molecular mass of 340,000 and consists of three pairs of polypeptide chains. The proteolytic removal of short (14 or 16 residue) peptides from the N-termini of four of the six chains causes the protein to polymerize in end-to-end and side-to-side fashion to produce a thick fiber.

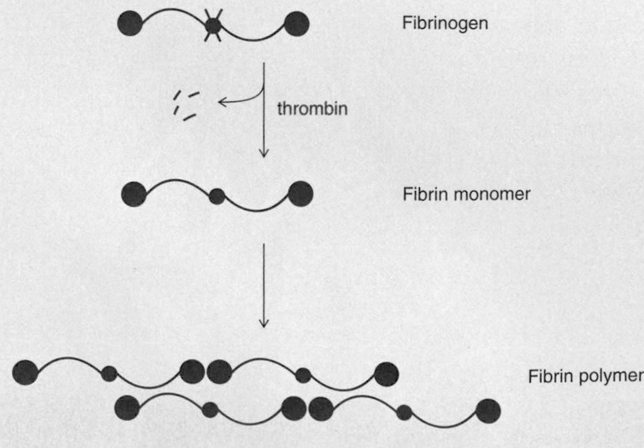

The conversion of fibrinogen to fibrin is the final step of **coagulation,** a series of proteolytic reactions involving a number of proteins and additional factors from platelets and damaged tissue. The enzyme responsible for cleaving fibrinogen to fibrin is known as thrombin. It is similar to trypsin in its sequence (38% of their residues are identical) and in its structure and catalytic mechanism.

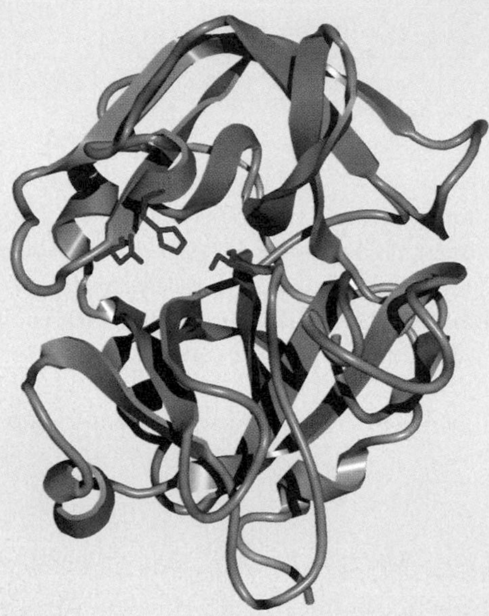

Compare the structure of thrombin shown here with that of chymotrypsin (Fig. 6-1). The catalytic residues (Asp, His, and Ser) are highlighted in red.

Like trypsin, thrombin cleaves peptide bonds following Arg residues, but it is highly specific for the two cleavage sites in the fibrinogen sequence. Thrombin, like fibrinogen, circulates as an inactive precursor. Its zymogen, called prothrombin, contains a serine protease domain along with several other structural motifs. These elements interact with other coagulation factors to help ensure that thrombin—and therefore fibrin—is produced only when needed.

A serine protease known as factor Xa catalyzes the specific hydrolysis of prothrombin to generate thrombin. Factor Xa (the *a* stands for *active*) is the protease form of the zymogen factor X. In the initial stages of coagulation, factor X is activated by a protease known as factor VIIa, working in association with an accessory protein called tissue factor. During the later stages of coagulation, factor Xa is generated by the activity of factor IXa, which is generated from its zymogen by the activity of factor XIa or factor VIIa–tissue factor. Factor XIa is in turn generated by the proteolysis of its zymogen by trace amounts of thrombin produced earlier in

(continued)

BOX 6-A *(continued)*

the coagulation process. This cascade of activation reactions is depicted below.

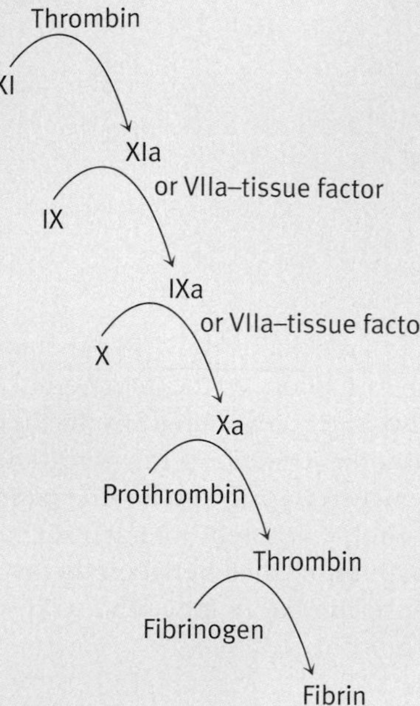

Note that the proteins are named according to their order of discovery, not their order of action. Many of these enzymatic reactions require accessory factors that are not shown in the simplified diagram.

The coagulation reactions have an amplifying effect, because each protease is a catalyst for the activation of another catalyst. Thus, a very small amount of factor IXa can activate a larger amount of Factor Xa, which can then activate an even larger amount of thrombin. This amplification effect is reflected in the plasma concentrations of the coagulation factors.

Plasma concentrations of some human coagulation factors

Factor	Concentration (μM)[a]
XI	0.06
IX	0.09
VII	0.01
X	0.18
Prothrombin	1.39
Fibrinogen	8.82

[Concentrations calculated from data in High, K.A. and Roberts, H.R., eds., *Molecular Basis of Thrombosis and Hemostasis,* Marcel Dekker, 1995.]

All the coagulation proteases appear to have evolved from a trypsin-like enzyme but have acquired narrow substrate specificity and a correspondingly narrow range of physiological activities.

[Photo: P. Motta/Dept. of Anatomy, University La Sapienza, Rome/Science Photo Library/Photo Researchers; Thrombin structure (pdb 1PPB) determined by W. Bode.]

its small conformational changes that open up the substrate specificity pocket and oxyanion hole. *The enzyme becomes maximally active only when it can bind its substrates specifically (via the specificity pocket) and can decrease the energy of the transition state (through binding interactions in the oxyanion hole).*

REVIEW LEARNING OBJECTIVE 9

- How is chymotrypsin activated and why is this step necessary?

Protease inhibitors limit protease activity

The pancreas, in addition to synthesizing the zymogens of digestive proteases, synthesizes small proteins that act as **protease inhibitors.** The liver also produces a variety of protease inhibitors that circulate in the bloodstream. If the pancreatic enzymes were prematurely activated or escaped from the pancreas through trauma, they would be rapidly inactivated by protease inhibitors. *The inhibitors pose as protease substrates but are not completely hydrolyzed.* For

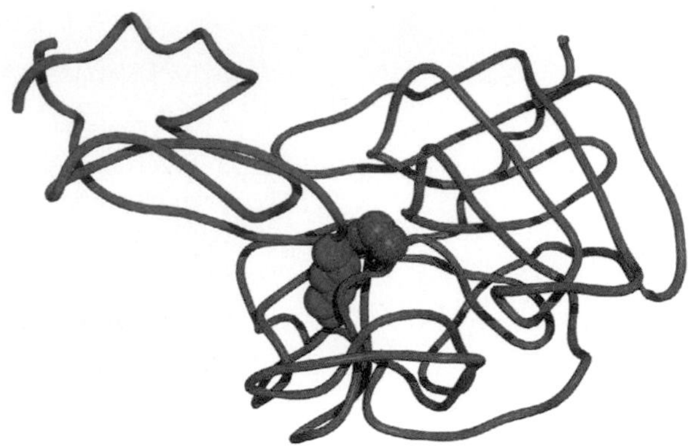

FIGURE 6-21

The complex of trypsin with bovine pancreatic trypsin inhibitor. Ser 195 of trypsin (gold) attacks the peptide bond of Lys 15 (green) in the inhibitor, but the reaction is arrested on the way to the tetrahedral intermediate stage. [*Complex structure (pdb 2PTC) determined by R. Huber and J. Deisenhofer.*]

example, when trypsin attacks a Lys residue of bovine pancreatic trypsin inhibitor, the reaction halts during formation of the tetrahedral intermediate. The inhibitor remains in the active site, preventing any further catalytic activity (Fig. 6-21; enzyme inhibitors are described in more detail in Section 7-3). The noncovalent complex between trypsin and bovine pancreatic trypsin inhibitor is one of the strongest protein–protein interactions known, with a dissociation constant of 10^{-14} M. An imbalance between the activities of proteases and the activities of protease inhibitors may lead to diseases, for example, emphysema (Box 6-B).

SUMMARY

1. Enzymes accelerate chemical reactions with high specificity under mild conditions.
2. A reaction coordinate diagram illustrates the change in free energy between the reactants and products as well as the activation energy required to reach the transition state. The higher the activation energy, the fewer the reactant molecules that can reach the transition state, and the slower the reaction.
3. An enzyme provides a route from reactants (substrates) to products that has a lower activation energy than the uncatalyzed reaction. Enzymes, sometimes with the assistance of a cofactor, use chemical catalytic mechanisms such as acid–base catalysis, covalent catalysis, and metal ion catalysis.
4. In chymotrypsin, an Asp–His–Ser triad catalyzes peptide bond hydrolysis through acid–base and covalent catalysis and by stabilizing the transition state via the oxyanion hole and low-barrier hydrogen bonds.
5. In addition to transition state stabilization, enzymes use proximity and orientation effects and electrostatic catalysis to facilitate reactions.
6. Serine proteases that have evolved from a common ancestor share their overall structure and catalytic mechanism but differ in their substrate specificity.
7. The activities of some proteases are limited by their synthesis as zymogens that are later activated and by their interaction with protease inhibitors.

CHECKLIST

1. Review the Learning Objectives listed on page 166.
2. Complete Review Exercise 3/Elementary Kinetics for an overview of chemical reactions and catalysis.
3. Complete Exercise 9/Serine Proteases, which explores the structure and mechanism of chymotrypsin and related enzymes.
4. Apply your knowledge by solving the problems at the end of this chapter. Check your results in the Solutions appendix.
5. Be able to define the boldfaced terms (consult the glossary at the end of the book). Test your understanding by taking the Chapter 6 quiz (www.wiley.com/college/pratt).
6. Explore the websites listed at www.wiley.com/college/pratt.
7. Consult the list of Selected Readings for background information or additional details on catalysis, enzyme mechanisms, and specific serine proteases.

A CLOSER LOOK

BOX 6-B Emphysema, elastase, and smoking

In the chronic lung disease **emphysema,** the alveoli (small air sacs where gas exchange takes place) degenerate, forming larger air spaces, and the lungs lose some of their elastic recoil. Breathing requires greater effort as the ability of the lung tissue to expand and contract is lost. In these micrographs of lung tissue, note the increased size of the alveolar spaces in lung tissue from an individual with emphysema (right) relative to lung tissue from a normal individual.

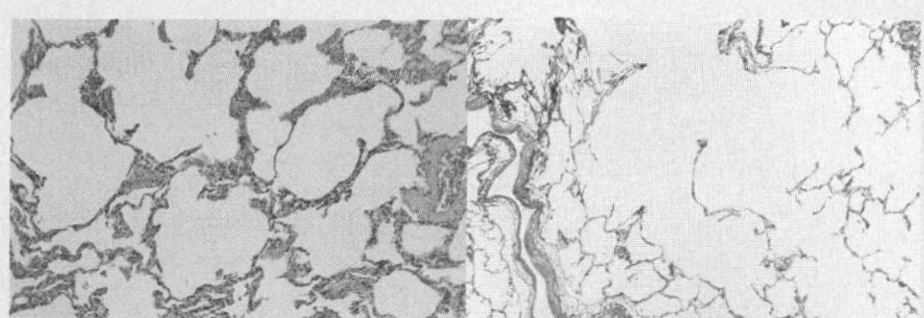

Individuals with a mutation in the gene for α1-proteinase inhibitor, which is the most abundant of the circulating protease inhibitors, are more likely to develop emphysema. When first discovered, the link between α1-proteinase inhibitor deficiency and lung tissue destruction was puzzling, particularly because α1-proteinase inhibitor was known at that time as α1-antitrypsin. α1-Proteinase inhibitor does indeed inhibit trypsin, but this activity is unlikely to be physiologically important in the lung. The true target of α1-proteinase inhibitor appears to be an elastase secreted by neutrophils, a type of white blood cell. Neutrophil elastase is a serine protease that is evolutionarily related to pancreatic elastase, and its physiological substrate appears to be the protein elastin, which provides lungs with their elastic properties. The destructive power of neutrophil elastase is required for a normal response to tissue injury or infection, but it is confined to small areas because of the high concentrations of circulating α1-proteinase inhibitor. When α1-proteinase inhibitor is deficient as a result of a mutation, neutrophil elastase catalyzes the hydrolysis of elastin to a greater extent. The increased rate of elastin breakdown in lung tissue is believed to be the cause of emphysema.

Smokers are also more likely to develop emphysema than nonsmokers. This does not seem surprising, since tobacco smoke contains a multitude of irritants that lead to chronic inflammation of the lung tissue and the release of elastase from neutrophils. However, smokers who develop emphysema have normal concentrations of α1-proteinase inhibitor, which should help limit the extent of elastase-induced lung damage. The solution to this apparent paradox came from the discovery that reactive oxygen species—either in tobacco smoke itself or released by white blood cells in response to smoke—can oxidize a Met residue in α1-proteinase inhibitor.

$$-NH-\underset{\displaystyle CH_2-CH_2-S-CH_3}{CH}-\overset{O}{\overset{\|}{C}}- \qquad -NH-\underset{\displaystyle CH_2-CH_2-S(=O)-CH_3}{CH}-\overset{O}{\overset{\|}{C}}-$$

Methionine — Methionine sulfoxide

This residue is normally the "bait" for neutrophil elastase, the inhibitor residue that binds in the specificity pocket of elastase.

The Met residue (highlighted in purple) is located in a flexible loop on the inhibitor surface and blocks catalysis by the elastase active site. α1-Proteinase inhibitor with an oxidized Met residue is about a thousand times less effective as an elastase inhibitor, accounting for the runaway elastase activity and loss of elastin that leads to emphysema.

[Left photo, Biodisc/Visuals Unlimited; right photo, Science Vu/Visuals Unlimited; structure (pdb 1PSI) determined by J.P. Abrahams, P.R. Elliott, D.A. Lomas, and R.W. Carrell.]

GLOSSARY TERMS

hydrolysis
catalyst
enzyme
ribozyme
active site
reaction specificity
substrate
isozymes
reaction coordinate
$\Delta G^{\ddagger}$
transition state
$\Delta G_{\text{reaction}}$
cofactor
coenzyme
acid–base catalysis
acid catalysis
base catalysis
tautomer
carbanion
covalent catalysis
imine
Schiff base
nucleophile
electrophile
metal ion catalysis
chemical labeling
serine protease
affinity label
catalytic triad
scissile bond
lock-and-key model
oxyanion hole
low-barrier hydrogen bond
proximity and orientation effects
induced fit
electrostatic catalysis
convergent evolution
specificity pocket
zymogen
autoactivation
protease inhibitor
coagulation
emphysema

PROBLEMS

1. Why are most proteins globular rather than fibrous proteins?

2. Explain why the motor proteins myosin and kinesin (described in Chapter 5) are enzymes and write the reaction that each catalyzes.

3. The reactions catalyzed by the enzymes listed below are presented in this chapter. To which class does each enzyme belong? Explain your answers.

(a) Pyruvate decarboxylase
(b) Alanine aminotransferase
(c) Alcohol dehydrogenase
(d) Hexokinase
(e) Chymotrypsin

4. To which class do the enzymes that catalyze the following reactions belong?

(a) O=C—H, CH—OH, $CH_2—O—PO_3^{2-}$ ⟶ CH_2OH, C=O, $CH_2—O—PO_3^{2-}$

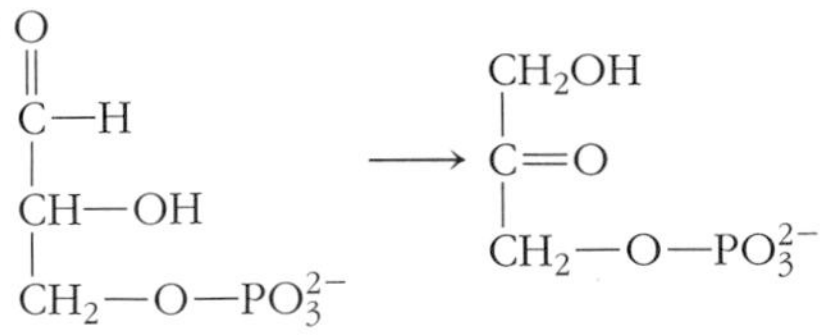

(b) COO^-, $CH—O—PO_3^{2-}$, CH_2OH ⟶ COO^-, $C—O—PO_3^{2-}$, $\|CH_2$

(c) COO^-, C=O, CH_3 ⟶ COO^-, HO—C—H, CH_3

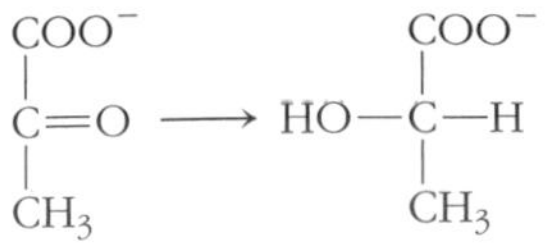

(d) CH_2OH, O, H, H, OH, H, HO, $O—PO_3^{2-}$, H, OH $\xrightarrow{HPO_4^{2-}}$ CH_2OH, O, H, H, OH, H, HO, OH, H, OH

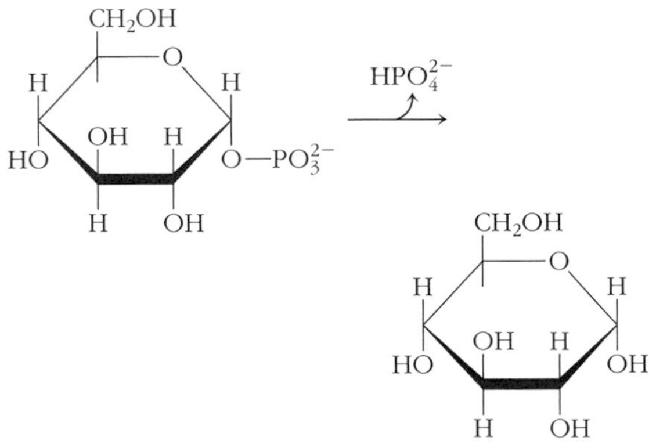

5. Draw the oxidized product of the reaction catalyzed by succinate dehydrogenase. To which class of enzymes does succinate dehydrogenase belong?

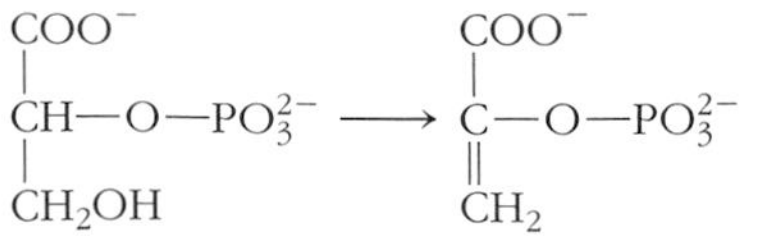

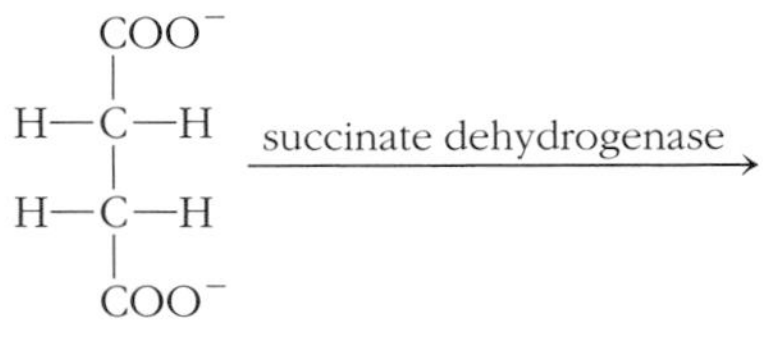

COO^-, H—C—H, H—C—H, COO^- (Succinate) $\xrightarrow{\text{succinate dehydrogenase}}$

6. Malate dehydrogenase catalyzes a reaction in which C2 of malate is oxidized. Draw the structure of the product. To which class of enzymes does malate dehydrogenase belong?

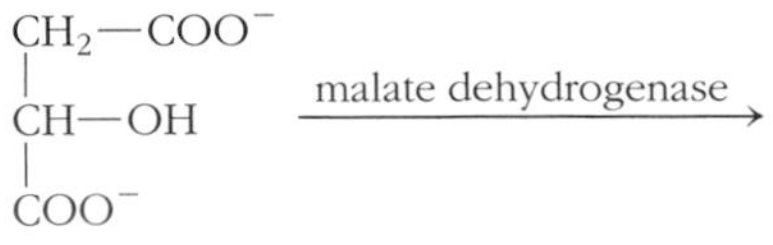

$CH_2—COO^-$, CH—OH, COO^- (Malate) $\xrightarrow{\text{malate dehydrogenase}}$

7. Examine the reaction catalyzed by hexokinase on page 183. Draw the product of the reaction catalyzed by creatine kinase, which acts on creatine in a similar manner. What would you predict to be the usual function of a kinase?

NH_2—C(=NH_2^+)—N(CH_3)—CH_2COO^-

Creatine

8. Propose a name for the enzymes that catalyze the following reactions (reactions may not be balanced):

(a) Glucose-6-phosphate → 6-Phosphoglucono-δ-lactone

(b) Isocitrate (CH_2—COO^- / CH—COO^- / HO—CH—COO^-) → Succinate (CH_2—COO^- / CH_2—COO^-) + Glyoxylate (H—C(=O)—COO^-)

(c) 1,3-Bisphosphoglycerate (O=C—OPO_3^{2-} / HC—OH / $CH_2OPO_3^{2-}$) + ADP → ATP + 3-Phosphoglycerate (O=C—O^- / HC—OH / $CH_2OPO_3^{2-}$)

(d) Pyruvate (CH_3—C(=O)—COO^-) + CO_2 → Oxaloacetate (CH_2—COO^- / C=O / COO^-)

9. What is the relationship between the rate of an enzyme-catalyzed reaction and the rate of the corresponding uncatalyzed reaction? Do enzymes enhance the rates of slow uncatalyzed reactions as much as they enhance the rates of fast uncatalyzed reactions?

10. Approximately how much does staphylococcal nuclease (Table 6-1) decrease the activation energy ($\Delta G^{\ddagger}$) of its reaction (the hydrolysis of a phosphodiester bond)?

11. The rate of an enzyme-catalyzed reaction is measured at several temperatures, generating the curve shown here. Explain why the enzyme activity increases with temperature and then drops off sharply.

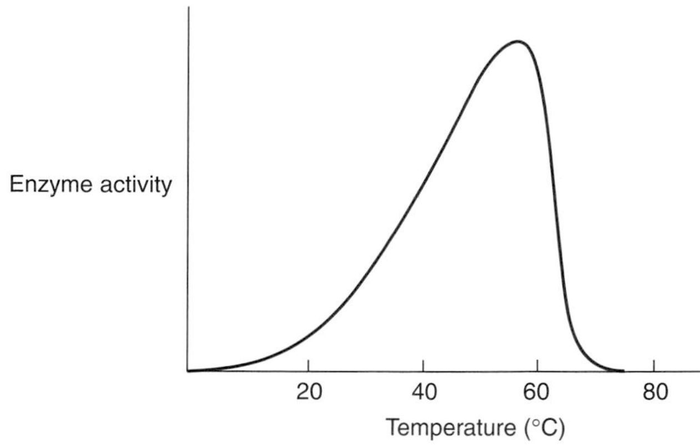

12. Draw pairs of free energy diagrams that show the differences between the following reactions:

(a) A fast reaction versus a slow reaction
(b) A one-step reaction versus a two-step reaction
(c) An endergonic reaction versus an exergonic reaction

13. When substrates bind to an enzyme, their free energy may be lowered. Why doesn't this binding defeat catalysis? (Refer to Fig. 6-11.)

14. Substrates and reactive groups in an enzyme's active site must be precisely aligned in order for a productive reaction to occur. Why, then, is some conformational flexibility also a requirement for catalysis?

15. Under certain conditions, peptide bond formation is more thermodynamically favorable than peptide bond hydrolysis. Would you expect chymotrypsin to catalyze peptide bond formation?

16. What is the relationship between the nucleophilicity and the acidity of an amino acid side chain?

17. The enzyme adenosine deaminase catalyzes the conversion of adenosine to inosine.

Adenosine + H_2O —(adenosine deaminase)→ Inosine + NH_3

1,6-Dihydropurine nucleoside

The compound 1,6-dihydropurine ribonucleoside binds to the enzyme with a much greater affinity than the adenosine substrate. What does this reveal about the mechanism of adenosine deaminase?

18. Amino acids such as Gly, Ala, and Val are not known to participate directly in acid–base or covalent catalysis.
(a) Explain why this is the case.
(b) Mutating a Gly, Ala, or Val residue in an enzyme's active site can still have dramatic effects on catalysis. Why?

19. Many genetic mutations prevent the synthesis of a protein or give rise to an enzyme with diminished catalytic activity. Is it possible for a mutation to increase the catalytic activity of an enzyme?

20. Ribozymes are RNA molecules that catalyze chemical reactions.
(a) What features of a nucleic acid would be important for it to act as an enzyme?
(b) Why can RNA but not DNA act as a catalyst?

21. Using what you know about the structures of the amino acid side chains and the mechanisms presented in this chapter, assign roles to the following amino acid side chains in enzymatic mechanisms.

	Cation binding	**Anion binding**	**Proton transfer**
Asp and Glu			
His			
Lys and Arg			
Cys and Ser			
Tyr			

	Forms covalent bonds with acyl groups	**Possible hydrogen bond donor**	**Possible hydrogen bond acceptor**
Asp and Glu			
His			
Lys and Arg			
Cys and Ser			
Tyr			

22. Design chloromethylketone affinity labels for trypsin and elastase.

23. Draw the mechanism of the reaction of TPCK with chymotrypsin.

24. Use what you know about the mechanism of chymotrypsin to explain why TPCK and DIPF inactivate the enzyme.

25. Sulfhydryl groups can react with the alkylating reagent *N*-ethylmaleimide (NEM). When NEM is added to a solution of creatine kinase, Cys 278 is alkylated, but no other Cys residues in the protein are modified. What can you infer about the role of Cys 278 based on this information?

26. Sarin is an organophosphorous compound similar to DIPF. It was used in 1995 by terrorists who released it on a Japanese subway. Injuries resulted from the reaction of sarin with a serine esterase involved in nerve transmission called acetylcholinesterase. Draw the structure of the enzyme's catalytic residue modified by sarin.

$$(H_3C)_2CH{-}O{-}\overset{\overset{\displaystyle O}{\|}}{\underset{\underset{\displaystyle CH_3}{|}}{P}}{-}F$$

Sarin

27. Enzymes with Cys residues in their active sites can often be inactivated by iodoacetate (ICH_2COO^-). Show the chemical reaction for the modification of the Cys side chain by iodoacetate. Why would such a modification inactivate an enzyme that required a Cys side chain for activity?

28. Angiogenin, an enzyme involved in the formation of blood vessels, was treated with bromoacetate ($BrCH_2COO^-$). An essential histidine was modified, and the activity of the enzyme was reduced by 95%. Show the chemical reaction for the modification of the histidine side chain with bromoacetate. Why would such a modification inactivate an enzyme that required a His side chain for activity? [From Shapiro, R., Weremowicz, S., Riordan, J.F., and Vallee, B., *Proc. Natl. Acad. Sci.* **84,** 8783–8787 (1987).]

29. In a chemical labeling study of an enzyme, a reagent that covalently modifies Lys residues also abolishes enzyme activity. Why doesn't this observation prove that the active site includes a Lys residue?

30. Diagram the hydrogen bonding interactions of the catalytic triad His–Lys–Ser during catalysis in a hypothetical hydrolytic enzyme.

31. Chymotrypsin hydrolyzes the artificial substrate *p*-nitrophenylacetate (see page 169) using a mechanism similar to that of peptide bond hydrolysis. The product *p*-nitrophenolate is bright yellow and absorbs light at 410 nm; thus the reaction can be monitored spectrophotometrically. When *p*-nitrophenylacetate is mixed with chymotrypsin, *p*-nitrophenolate initially forms extremely rapidly. This is followed by a "steady-state" phase in which the product forms at a uniform rate.

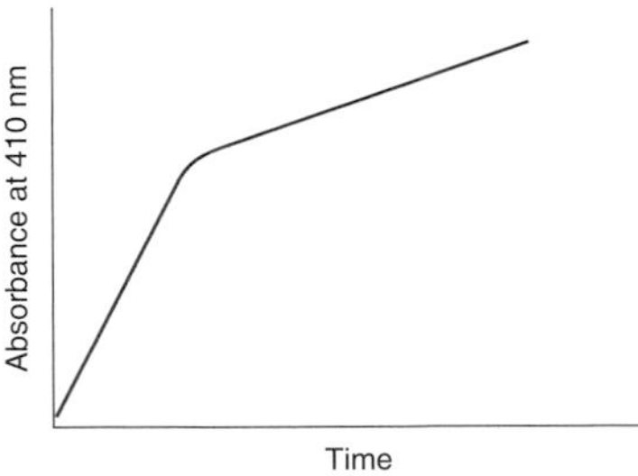

(a) Using what you know about the mechanism of chymotrypsin, explain these results.
(b) Draw a reaction coordinate diagram for the reaction.
(c) Would *p*-nitrophenylacetate be useful for monitoring the activity of trypsin *in vitro*?

32. Solutions of chymotrypsin prepared for experimental purposes are stored in a dilute solution of hydrochloric acid as a "preservative." Explain.

33. If the active site of chymotrypsin excludes water when its substrate binds, how is it possible for a water molecule to participate in the second stage of catalysis (as shown in Fig. 6-9)?

34. The protease from the human immunodeficiency virus I (HIV I), the target of the protease inhibitor "cocktail" of drugs now on the market, possesses a catalytic dyad of two aspartate residues (one residue in each of two identical subunits: Asp 25 and Asp 25′). The protease has a mechanism similar to that of chymotrypsin, except that the tetrahedral intermediate is *not* covalently bound to the enzyme. Using the diagram provided as a starting point, draw a mechanism for HIV I protease.

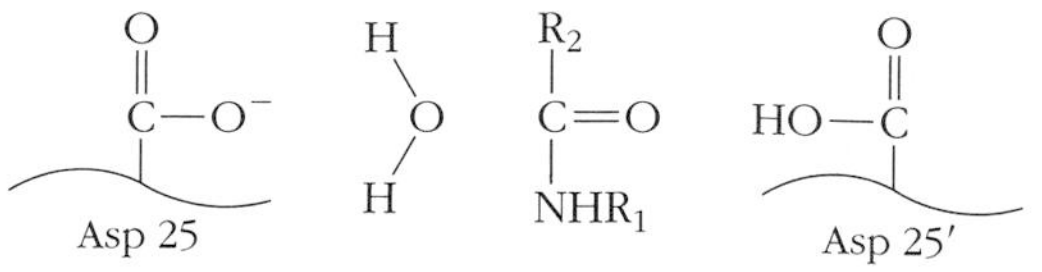

Asp 25 Asp 25′ NHR_1 R_2

35. Chymotrypsin, trypsin, and elastase are members of a family of enzymes referred to as serine proteases. The enzyme papain (found in papaya) is also a protease but is a member of a family of enzymes referred to as cysteine proteases. The amino acids in papain's active site that are critical for peptide bond hydrolysis are a protonated histidine and a deprotonated cysteine.

(a) Using what you know about chymotrypsin's mechanism, draw a reaction mechanism for the hydrolysis of a peptide bond by papain.

(b) Does the mechanism you have drawn employ acid–base catalysis, covalent catalysis, or both?

(c) Draw a reaction coordinate diagram that is consistent with the mechanism of papain.

(d) Natives of tropical countries who cooked meat wrapped in papaya leaves noted that the meat was more tender as a result. Provide a biochemical explanation for this observation.

(e) The pK values for the active-site Cys and His in papain are 4.2 and 8.2. The pH optimum for the reaction is 6.0. Assign the pK values to the appropriate amino acid side chains and explain your reasoning.

36. RNase A is a digestive enzyme secreted by the pancreas into the small intestine, where it hydrolyzes RNA into its component nucleotides. The optimum pH for RNase A is about 6, and the pK values of the two histidines that serve as catalytic residues are 5.4 and 6.4. The first step of the mechanism is shown.

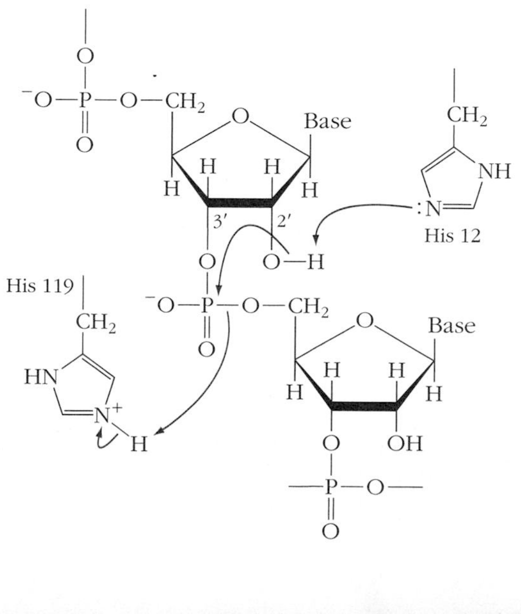

His 119 His 12 Base Base

(a) Assign the appropriate pK values to His 12 and His 119.

(b) Explain why the pH optimum of ribonuclease is pH 6.

(c) Ribonuclease catalyzes the hydrolysis of RNA but not DNA. Explain why.

37. *E. coli* ribonuclease HI is an enzyme that catalyzes the hydrolysis of phosphodiester bonds in RNA. Its proposed mechanism involves a "carboxylate relay."

(a) Using the structures below as a guide, draw arrows that indicate how the hydrolysis reaction might be initiated.

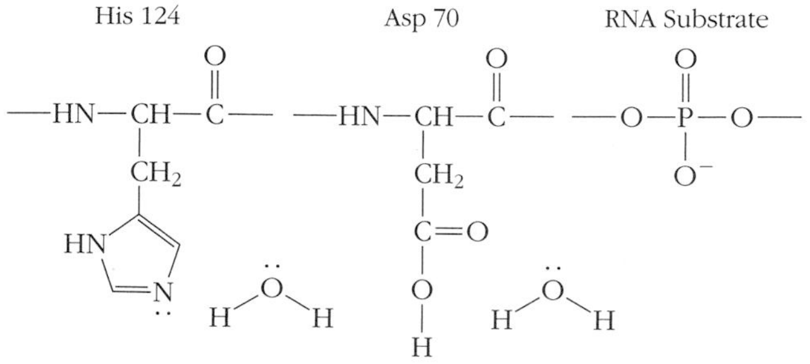

His 124 Asp 70 RNA Substrate

(b) The pK values for all of the histidines in ribonuclease HI were determined and include values of 7.1, 5.5, and <5.0. Which value is most likely to correspond to His 124? Explain.

(c) Substituting an alanine residue for His 124 resulted in a dramatic decrease in enzyme activity. Explain.

[From Oda, Y., Yoshida, M., and Kanaya, S., *J. Biol. Chem.* **268,** 88–92 (1993).]

38. The enzyme carbonic anhydrase is one of the fastest enzymes known. It catalyzes the hydration of carbon dioxide to form bicarbonate and a hydrogen ion:

$$CO_2 + H_2O \rightleftharpoons H^+ + HCO_3^-$$

The enzyme's active site contains a zinc ion that is coordinated to the imidazole rings (abbreviated "Im" in the figure) of three histidine residues. (A fourth histidine residue is located nearby and participates in catalysis.) A fourth coordination position is occupied by a water molecule. Draw the reaction mechanism (the first step is shown) of the hydration of carbon dioxide and show the regeneration of the enzyme.

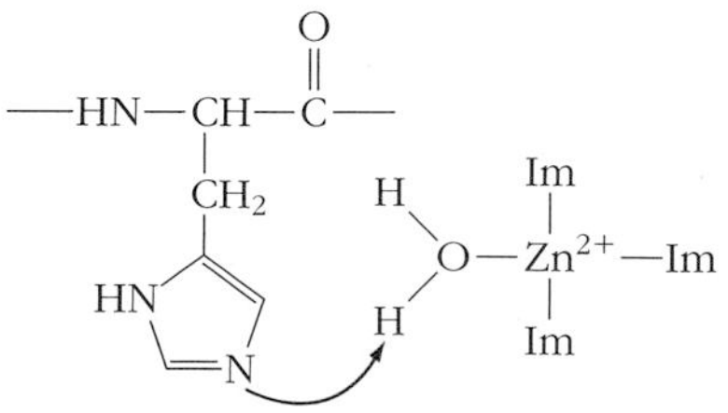

39. The amino acid Asp 189 lies at the base of the substrate specificity pocket in the enzyme trypsin.

(a) How is this related to trypsin's substrate specificity? What kinds of interactions take place between the Asp 189 and the amino acid side chain on the carboxyl side of the scissile bond?

(b) In site-directed mutagenesis studies, Asp 189 was replaced with lysine. How do you think this would affect substrate specificity?

(c) The investigators who carried out the experiment described in part (b) analyzed the three-dimensional structure of the mutant enzyme and found that Lys 189 is actually not located in the substrate specificity pocket. Instead, the Lys side chain reaches out of the base of the pocket, rendering the specificity pocket nonpolar. With this additional information, how would the substrate specificity differ in the Lys 189 mutant enzyme?

[From Graf, L., Craik, C.S., Patthy, A., Roczniak, S., Fletterick, R.J., and Rutter, W.J., *Biochemistry* **26,** 2616–2623 (1987).]

40. An enzyme involved in the regulation of blood pressure was recently purified and characterized. It is an aspartyl aminopeptidase, which hydrolyzes peptide bonds on the carboxyl side of aspartate residues. The investigators who purified the enzyme were interested in learning about the substrate binding site, so they synthesized a series of artificial peptides and tested the ability of the enzyme to hydrolyze them. The peptides and their k_{cat} values are listed in the table below. The residues are labeled P1-P1′-P2′-P3′, and the hydrolyzed bond is between P1 and P1′. The k_{cat} value is a turnover number, which is the number of catalytic events per second.

(a) It is likely that both the P1′ residue and the P2′ residue fit in adjacent "pockets" on the aspartyl aminopeptidase enzyme. Describe the characteristics of these pockets on the enzyme, using the kinetic data in the table.

(b) The investigators have proposed that the "natural" or endogenous substrate for aspartyl aminopeptidase is angiotensin II, which is involved in the regulation of blood pressure. However, the authors also mention that a potential exogenous substrate might be the artificial sweetener aspartame (Nutrasweet®), which is the methylated dipeptide aspartylphenylalanine (Asp–Phe–OCH_3). Based on the results presented here, do you think that aspartame would be a good substrate for the enzyme?

(c) Draw the structure of aspartame and the products of its hydrolysis by aspartyl aminopeptidase.

(d) Under nondenaturing conditions, the enzyme appears to be the same size as the protein ferritin, which has a molecular mass of 440 kD. Under denaturing conditions, the purified enzyme behaves like a single polypeptide of 55 kD. What can you determine about the enzyme's structure from these data?

Peptide (P1-P1′-P2′-P3′)	k_{cat} (s^{-1})
Asp–Ala–Ala–Leu	5.3
Asp–Phe–Ala–Leu	9.9
Asp–Lys–Ala–Leu	2.8
Asp–Asp–Ala–Leu	9.8
Asp–Ala–Phe–Leu	17.2
Asp–Ala–Lys–Leu	5.0
Asp–Ala–Asp–Leu	2.3

[From Wilk, S., Wilk, E., and Magnusson, R.P., *J. Biol. Chem.* **273,** 15961–15970 (1998).]

41. Asn and Gln residues in proteins are sometimes nonenzymatically hydrolytically deamidated to Asp and Glu residues, respectively. It has been suggested that deamidation might function as a molecular timer for protein turnover, since deamidation of some proteins increases their susceptibility to degradation by cellular proteases.

(a) Write the balanced chemical equations for the hydrolytic deamidation of Asn and Gln to Asp and Glu.

(b) Why might deamidation increase the susceptibility of some proteins to degradation by proteases?

(c) An exhaustive study of proteins known to undergo deamidation has revealed that the rate of deamidation of Asn and Gln is influenced by the amino acids preceding and following the labile amino acids. The results for Asn are shown below. What statements can you make concerning the role of the protein's primary sequence in the deamidation process?

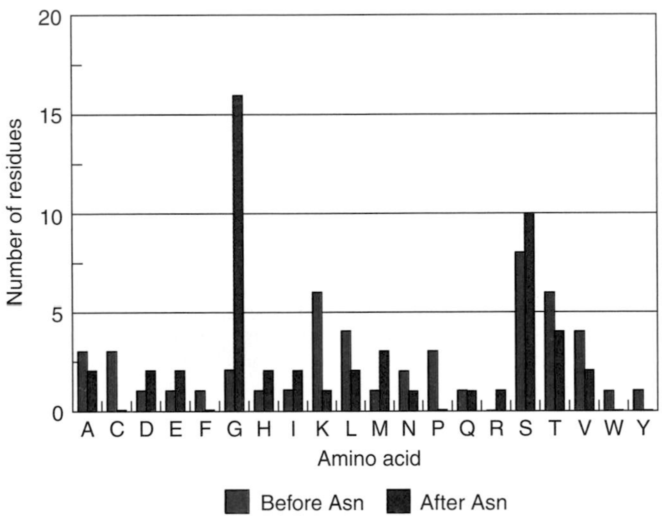

Frequency with which each of the 20 amino acids occurs before and after the labile Asn residues in a set of proteins known to undergo nonenzymatic deamidation

(d) The mechanism of deamidation of the amide side chain involves an acid catalyst (HA). The first step of the reaction is shown below. Draw the rest of the reaction mechanism, including the intermediates. What amino acid side chains might stabilize the reaction intermediates?

CH_2
|
CH_2
|
C=O
|
NH_2 H–A

(e) Since the deamidation of Asn and Gln residues is known to be nonenzymatic, the HA acid catalyst cannot be provided by an enzyme. Instead, the *catalytic* groups are believed to be provided by neighboring amino acids in the protein undergoing deamidation. Refer to your answer to (d) and examine the mechanism you have written. Describe how amino acid side chains that either precede or follow the labile Asn could serve as catalytic groups in the deamidation process.

(f) Amino terminal Gln residues undergo deamidation much more rapidly than internal Gln residues. Write a mechanism for this deamidation process, which includes the formation of a five-membered pyrrolidone ring. Amino terminal Asn residues do not undergo deamidation. Explain why.

(g) It has been observed that Asn and Gln residues in the interior of a protein are deamidated at a much slower rate than Asn and Gln residues on the surface of the protein. Explain why.

[From Wright, H.T., *Crit. Rev. Biochem. Mol. Biol.* **26,** 1–52 (1991).]

42. Imidazole reacts with *p*-nitrophenylacetate to produce the *N*-acetylimidazolium and *p*-nitrophenolate ions. The mechanism is shown below.

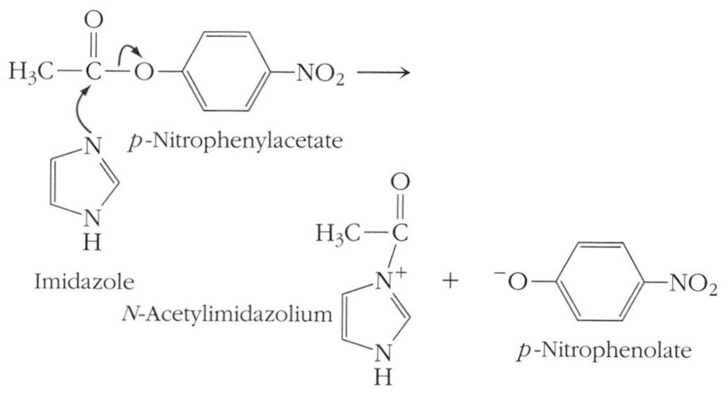

$H_3C-C(=O)-O-C_6H_4-NO_2$ ⟶

p-Nitrophenylacetate

Imidazole

N-Acetylimidazolium + $^-O-C_6H_4-NO_2$

p-Nitrophenolate

(a) The compound shown also reacts with *p*-nitrophenylacetate to form *p*-nitrophenolate. Draw the reaction mechanism for this reaction.

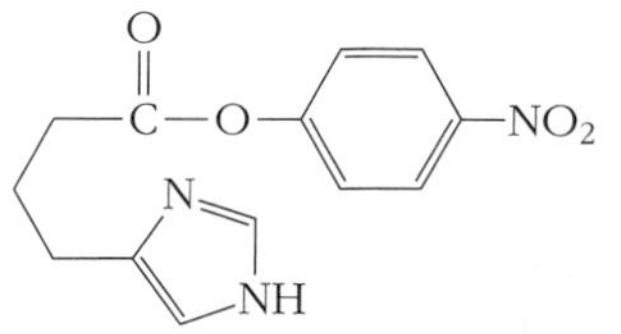

(b) The compound described in (a) reacts 24 times faster than imidazole. Explain why.

(c) What can the results of this experiment tell us about the way enzymes speed up reaction rates?

43. Chymotrypsin is usually described as an enzyme that catalyzes hydrolysis of peptide bonds following Phe, Trp, or Tyr residues. Is this information consistent with the description of the bonds cleaved during chymotrypsin activation (Fig. 6-18)? What does this tell you about chymotrypsin's substrate specificity?

44. Describe how the activation of the digestive enzyme chymotrypsin follows a cascade mechanism. Draw a diagram of your cascade.

45. A genetic defect in coagulation factor IX causes hemophilia b, a disease characterized by a tendency to bleed profusely after very minor trauma. However, a genetic defect in coagulation factor XI has no clinical symptoms. Explain this discrepancy in terms of the mechanism for activation of coagulation proteases shown in Box 6-A.

46. Why is the broad substrate specificity of chymotrypsin advantageous? Why would this be a disadvantage for some other proteases?

47. You are a physician and you have a patient who is suffering from premature pancreatic zymogen activation, which means that the pancreatic enzymes are being activated in the pancreas (rather than in the small intestine) and pancreatic tissue is being damaged as a result. What would be most effective: a chymotrypsin inhibitor, a trypsin inhibitor, or an elastase inhibitor? Explain your answer.

48. A trypsin inhibitor is located in the pancreas of mammals. What is the reason for its presence?

49. Some plants contain compounds that inhibit serine proteases. It has been hypothesized that these compounds protect the plant from proteolytic enzymes of insects and microorganisms that would damage the plant. Tofu, or bean curd, possesses these compounds. Manufacturers of the product treat the tofu to eliminate serine protease inhibitors. Why is this treatment necessary?

50. Not all proteases are synthesized as zymogens or have inhibitors that block their activity. What limits the potentially destructive power of these proteases?

SELECTED READINGS

Cannon, W.R. and Benkovic, S.J., Solvation, reorganization energy, and biological catalysis, *J. Biol. Chem.* **273,** 26257–26260 (1998). [Summarizes some general features of enzyme function.]

Evans, M.D. and Pryor, W.A., Cigarettes, smoking, emphysema, and damage to α1-proteinase inhibitor, *Am. J. Physiol.* **266,** L593–611 (1994). [Discusses how an absent or damaged inhibitor contributes to lung disease.]

Fersht, A., *Structure and Mechanism in Protein Science: A Guide to Enzyme Catalysis and Protein Folding,* W.H. Freeman (1999). [Includes detailed reaction mechanisms for chymotrypsin and other enzymes.]

Perona, J.J. and Craik, C.S., Evolutionary divergence of substrate specificity within the chymotrypsin-like protease fold, *J. Biol. Chem.* **272,** 29987–29990 (1997). [Summarizes research identifying the structural basis of substrate specificity in chymotrypsin and related enzymes.]

Wilmouth, R.C., Edman, K., Neutze, R., Wright, P.A., Clifton, I.J., Schneider, T.R., Schofield, C.J., and Hajdu, J., X-ray snapshots of serine protease catalysis reveal a tetrahedral intermediate, *Nature Struct. Biol.* **8,** 689–694 (2001). [Reports the first structural evidence for a tetrahedral intermediate in the hydrolysis reaction catalyzed by a serine protease.]

CHAPTER 7

Enzyme Kinetics and Inhibition

Ethylene glycol, the active ingredient in automotive antifreeze, annually poisons hundreds of people and pets who find its sweet taste irresistible. Ethylene glycol itself is not toxic, but its metabolic by-products can produce potentially lethal damage to the central nervous system, heart, and kidneys. The first step in the breakdown of ethylene glycol is catalyzed by the enzyme alcohol dehydrogenase; blocking this step is effective treatment for ethylene glycol poisoning. Traditionally, individuals who have ingested ethylene glycol are given ethanol (the "alcohol" of beer and wine) since alcohol dehydrogenase preferentially acts on ethanol rather than ethylene glycol. This gives the body time to eliminate in the urine the unreacted ethylene glycol. Ethanol treatment has problems of its own, primarily its depressive effect on the central nervous system. However, ethanol is inexpensive and usually readily available. A safer—but far costlier—alternative is to inhibit the activity of alcohol dehydrogenase with the drug 4-methylpyrazole, which prevents the breakdown of ethylene glycol and has no significant side effects.

[Larry Stepanowicz/Fundamental Photographs.]

LEARNING OBJECTIVES

1. Understand the relationship between reaction velocity and substrate concentration for an enzyme-catalyzed reaction.
2. Understand how a rate equation and rate constant characterize a simple chemical reaction.
3. Understand the meaning of K_M and V_{max} in the Michaelis–Menten equation.
4. Understand the meaning of k_{cat} and k_{cat}/K_M.
5. Understand how kinetic parameters are experimentally determined.
6. Understand that not all enzymes follow the Michaelis–Menten model.
7. Understand the difference between irreversible and reversible enzyme inhibition.
8. Understand the mechanism of competitive inhibition and its effect on K_M and V_{max}.
9. Understand the mechanism of mixed inhibition and its effect on K_M and V_{max}.
10. Understand how phosphoenolpyruvate acts as an allosteric inhibitor of phosphofructokinase.

THIS CHAPTER IN CONTEXT

In the preceding chapter, we examined the basic features of enzyme catalytic activity, primarily by exploring the various catalytic mechanisms used by chymotrypsin. This chapter extends the discussion of enzymes by introducing enzyme kinetics, the mathematical analysis of enzyme activity. Here we describe how an enzyme's reaction speed and specificity can be quantified and how this information can be used to evaluate the enzyme's physiological function. We also look at the regulation of enzyme activity by inhibitors that bind to the enzyme and alter its activity. The discussion focuses on the most common modes of inhibition—competitive enzyme inhibition, mixed inhibition, and allosteric regulation—since these are the regulatory mechanisms that we will see again in the chapters describing enzyme-catalyzed metabolic reactions.

1 INTRODUCTION TO ENZYME KINETICS

Understanding the structure and chemical mechanism of an enzyme (for example, chymotrypsin, as discussed in Chapter 6) often reveals a great deal about how that enzyme functions *in vivo.* However, structural information alone does not provide a full accounting of an enzyme's physiological role. For example, one might need to know exactly how fast the enzyme catalyzes a reaction or how well it recognizes different substrates or how its activity is affected by other substances. These questions are not trivial—consider that a single cell contains thousands of different enzymes, all operating simultaneously and in the presence of each others' substrates and products. To fully describe enzyme activity, enzymologists apply mathematical tools to quantify an enzyme's catalytic power and its substrate affinity as well as its response to inhibitors. This analysis is part of the area of study known as enzyme kinetics (from the Greek *kinetos,* which means "moving").

Go to **Review Exercise 3/Elementary Kinetics** for a summary of the general features of reaction rates and the effect of catalysts. The material in this exercise addresses Learning Objective 2.

Long before the first enzymes were purified, scientists were examining the enzymatic reactions of crude preparations of yeast cells and other organisms. Mathematical analysis of these impure enzymes was possible because researchers could measure the concentrations of substrates and reaction products and observe how these quantities changed over time. For example, consider the simple reaction catalyzed by triose phosphate isomerase, which interconverts two three-carbon sugars (trioses):

O=C(H)–CH(OH)–$CH_2OPO_3^{2-}$ → (triose phosphate isomerase) → $HOCH_2$–C(=O)–$CH_2OPO_3^{2-}$

Glyceraldehyde-3-phosphate → Dihydroxyacetone phosphate

Over the course of the reaction, the concentration of the substrate glyceraldehyde-3-phosphate falls as the concentration of the product dihy-

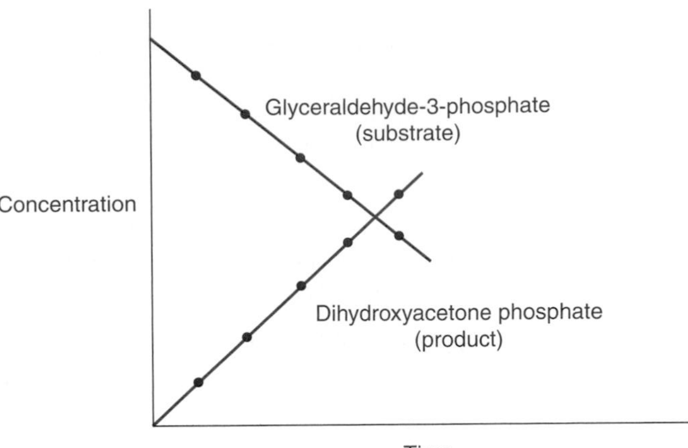

FIGURE 7-1

Progress of the triose phosphate isomerase reaction.
Over time, the concentration of the substrate glyceraldehyde-3-phosphate decreases and the concentration of the product dihydroxyacetone phosphate increases.

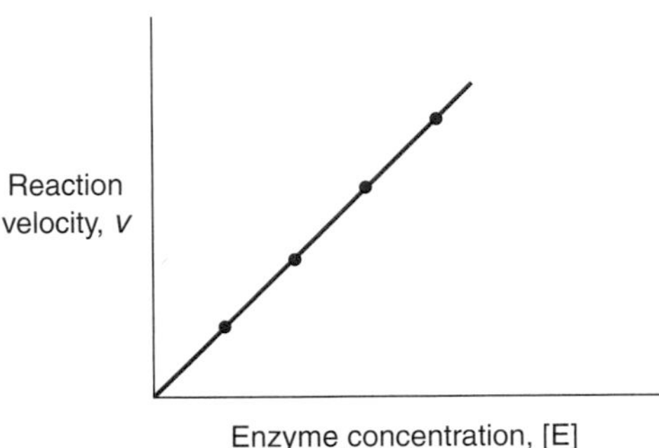

FIGURE 7-2

Relationship between enzyme concentration and reaction velocity.
The more enzyme present, the faster the reaction.

droxyacetone phosphate rises (Fig. 7-1). *The progress of this or any reaction can be expressed as a* ***velocity (v),*** *either the rate of disappearance of the substrate (S) or the rate of appearance of the product (P):*

$$v = -\frac{d[S]}{dt} = \frac{d[P]}{dt} \qquad [7\text{-}1]$$

where [S] and [P] represent the concentrations of the substrate and product, respectively. Not surprisingly, *the more catalyst (enzyme) present, the faster the reaction* (Fig. 7-2).

When the enzyme concentration is held constant, the reaction velocity varies with the substrate concentration, but in a nonlinear fashion (Fig. 7-3). The shape of this velocity versus substrate curve is an important key to understanding how enzymes interact with their substrates. *The hyperbolic, rather than linear, shape of the curve suggests that an enzyme physically combines with its substrate to form an* ***enzyme–substrate (ES) complex.*** Therefore, the enzyme-catalyzed conversion of S to P

$$S \xrightarrow{E} P$$

can be more accurately written as

$$E + S \rightarrow ES \rightarrow E + P$$

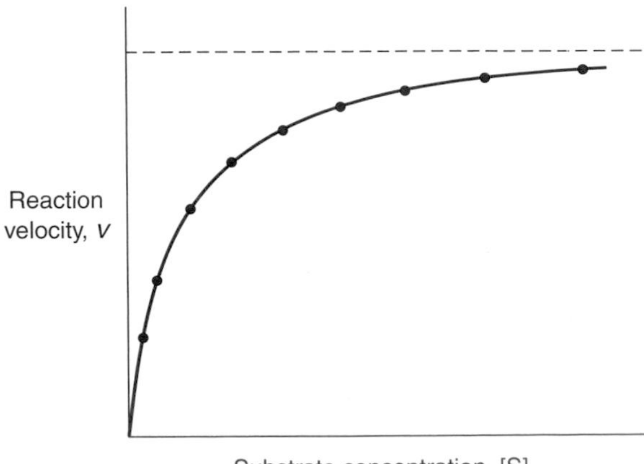

FIGURE 7-3

A plot of reaction velocity versus substrate concentration.
Varying amounts of substrate are added to a fixed amount of enzyme. The reaction velocity is measured for each substrate concentration and plotted. The resulting curve takes the form of a hyperbola, a mathematical function in which the values first increase steeply but eventually approach a maximum.

As small amounts of substrate are added to the enzyme preparation, enzyme activity (measured as the reaction velocity) appears to increase almost linearly. However, the enzyme's activity increases less dramatically as more substrate is added. At very high substrate concentrations, enzyme activity appears to level off as it approaches a maximum value. This behavior shows that at low substrate concentrations, the enzyme quickly converts all the substrate to product, but as more substrate is added, the enzyme becomes **saturated** with substrate—that is, there are many more substrate molecules than enzyme molecules, so not all the substrate can be converted to product in a given time. These so-called saturation kinetics are a feature of many binding phenomena, including the binding of O_2 to myoglobin (see Fig. 4-21 in Section 4-4).

The curve shown in Fig. 7-3 reveals considerable information about a given enzyme and substrate under a chosen set of reaction conditions. All simple enzyme-catalyzed reactions yield a hyperbolic velocity versus substrate curve, but the exact shape of the curve depends on the enzyme, its concentration, the concentrations of enzyme inhibitors, the pH, the temperature, and so on. By analyzing such curves, it is possible to address some basic questions, for example,

1. How fast does the enzyme operate?
2. How efficiently does the enzyme convert different substrates to products?
3. How susceptible is the enzyme to various inhibitors and how do these inhibitors affect enzyme activity?

The answers to these questions, in turn, may reveal

1. Whether an enzyme is likely to catalyze a particular reaction *in vivo*.
2. What substances are likely to serve as physiological regulators of the enzyme's activity.
3. Which enzyme inhibitors might be effective drugs.

REVIEW LEARNING OBJECTIVE 1

- Describe two ways to express the velocity of an enzymatic reaction.
- Why does a plot of velocity versus substrate concentration yield a curve?

2 DERIVATION AND MEANING OF THE MICHAELIS–MENTEN EQUATION

Go to **Exercise 10/ Enzyme Kinetics** for a summary of the basic principles of enzyme kinetics and how parameters such as K_M and V_{max} are determined. The material presented in this exercise addresses Learning Objectives 3, 4, and 5.

The mathematical analysis of enzyme behavior centers on the equation that describes the hyperbolic shape of the velocity versus substrate plot (Fig. 7-3). We can analyze an enzyme-catalyzed reaction, such as the one catalyzed by triose phosphate isomerase, by conceptually breaking it down into smaller steps and using the terms that apply to simple chemical processes.

The equations of chemical kinetics

Consider a **unimolecular reaction** (one that involves a single reactant):

$$A \rightarrow B$$

The progress of this reaction can be mathematically described by a **rate equation** in which the reaction rate (the velocity) is expressed in terms of a constant (the **rate constant**) and the reactant concentration [A]:

$$v = -\frac{d[\text{A}]}{dt} = k[\text{A}] \qquad [7\text{-}2]$$

Here, $\boldsymbol{k}$ is the rate constant and has units of reciprocal seconds (s^{-1}). This equation shows that the reaction velocity is directly proportional to the concentration of reactant A. Such a reaction is said to be **first order** because its rate depends on the concentration of one substance.

A **bimolecular** or **second-order reaction** can be written

$$\text{A} + \text{B} \rightarrow \text{C}$$

Its rate equation is

$$v = -\frac{d[\text{A}]}{dt} = -\frac{d[\text{B}]}{dt} = k[\text{A}][\text{B}] \qquad [7\text{-}3]$$

Here, k is a second-order rate constant and has units of $M^{-1} \cdot s^{-1}$. The velocity of a second-order reaction is therefore proportional to the product of the two reactant concentrations (see Sample Calculation 7-1).  Review Exercise 3 includes a discussion of some simple rate equations.

Sample Calculation 7-1

Problem

Determine the velocity of the reaction $X + Y \rightarrow Z$ when the sample contains 3 μM X and 5 μM Y and k for the reaction is 400 $M^{-1} \cdot s^{-1}$.

Solution

Use Equation 7-3 and make sure that all units are consistent:

$$\begin{aligned} v &= k[\text{X}][\text{Y}] \\ &= (400\ \text{M}^{-1} \cdot \text{s}^{-1})(3\ \mu\text{M})(5\ \mu\text{M}) \\ &= (400\ \text{M}^{-1} \cdot \text{s}^{-1})(3 \times 10^{-6}\ \text{M})(5 \times 10^{-6}\ \text{M}) \\ &= 6 \times 10^{-9}\ \text{M} \cdot \text{s}^{-1} \\ &= 6\ \text{nM} \cdot \text{s}^{-1} \end{aligned}$$

REVIEW LEARNING OBJECTIVE 2

- What are the rate equations for first-order and second-order reactions?

What is the rate equation for an enzyme-catalyzed reaction?

In the simplest case, an enzyme binds its substrate (in an enzyme–substrate complex) before converting it to product, so *the overall reaction actually consists of first-order and second-order processes, each with a characteristic rate constant:*

$$\text{E} + \text{S} \underset{k_{-1}}{\overset{k_1}{\rightleftharpoons}} \text{ES} \overset{k_2}{\rightarrow} \text{E} + \text{P} \qquad [7\text{-}4]$$

The initial collision of E and S is a bimolecular reaction with the second-order rate constant k_1. The ES complex can then undergo one of two possible unimolecular reactions: k_2 is the first-order rate constant for the conversion of ES to E and P, and k_{-1} is the first-order rate constant for the conversion of ES back to E and S. The bimolecular reaction that would represent the formation of ES from E and P is not shown because we assume that the formation of product from ES (the step described by k_2) does not occur in reverse.

The rate equation for product formation is

$$v = \frac{d[\mathrm{P}]}{dt} = k_2[\mathrm{ES}] \quad [7\text{-}5]$$

To calculate the rate constant, k_2, for the reaction, we would need to know the reaction velocity and the concentration of ES. The velocity can be measured relatively easily, for example, by using a synthetic substrate that is converted to a light-absorbing or fluorescent product. (The velocity of the reaction is just the rate of appearance of the product as monitored by a spectrophotometer or fluorimeter.) However, measuring [ES] is more difficult because the concentration of the enzyme–substrate complex depends on its rate of formation from E and S and its rate of decomposition to E + S and E + P:

$$\frac{d[\mathrm{ES}]}{dt} = k_1[\mathrm{E}][\mathrm{S}] - k_{-1}[\mathrm{ES}] - k_2[\mathrm{ES}] \quad [7\text{-}6]$$

To simplify our analysis, we choose experimental conditions such that the substrate concentration is much greater than the enzyme concentration ($[\mathrm{S}] \gg [\mathrm{E}]$). Under these conditions, after E and S have been mixed together, *the concentration of ES remains constant until nearly all the substrate has been converted to product.* This is shown graphically in Figure. 7-4. [ES] is said to maintain a **steady state** (it has a constant value) and

$$\frac{d[\mathrm{ES}]}{dt} = 0 \quad [7\text{-}7]$$

The rate of ES formation must therefore balance the rate of ES consumption:

$$k_1[\mathrm{E}][\mathrm{S}] = k_{-1}[\mathrm{ES}] + k_2[\mathrm{ES}] \quad [7\text{-}8]$$

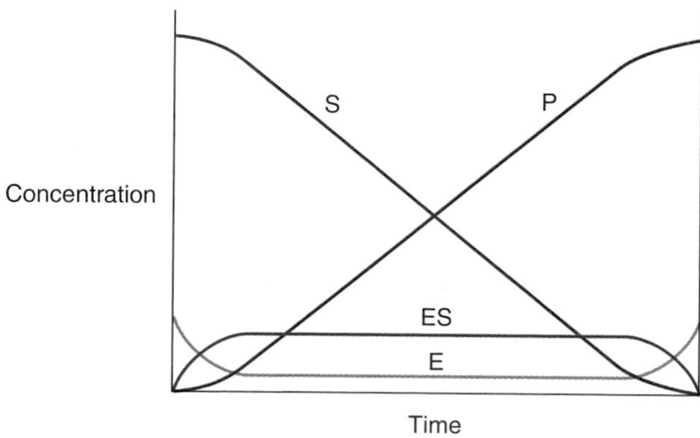

FIGURE 7-4

Changes in concentration for a simple enzyme-catalyzed reaction.
For most of the duration of the reaction, [ES] remains constant while S is converted to P. In this idealized reaction, all the substrate is converted to product.

At any point during the reaction, [E]—like [ES]—is difficult to determine, but the total enzyme concentration, $[E]_T$, is usually known:

$$[E]_T = [E] + [ES] \qquad [7\text{-}9]$$

Thus, $[E] = [E]_T - [ES]$. This expression for [E] can be substituted into Equation 7-8 to give

$$k_1([E]_T - [ES])[S] = k_{-1}[ES] + k_2[ES] \qquad [7\text{-}10]$$

Rearranging (by dividing both sides by [ES] and k_1) gives

$$\frac{([E]_T - [ES])[S]}{[ES]} = \frac{k_{-1} + k_2}{k_1} \qquad [7\text{-}11]$$

At this point, we can define the **Michaelis constant, K_M,** as a collection of rate constants:

$$K_M = \frac{k_{-1} + k_2}{k_1} \qquad [7\text{-}12]$$

Consequently, Equation 7-11 becomes

$$\frac{([E]_T - [ES])[S]}{[ES]} = K_M \qquad [7\text{-}13]$$

or

$$K_M[ES] = ([E]_T - [ES])[S] \qquad [7\text{-}14]$$

Dividing both sides by [ES] gives

$$K_M = \frac{[E]_T[S]}{[ES]} - [S] \qquad [7\text{-}15]$$

or

$$\frac{[E]_T[S]}{[ES]} = K_M + [S] \qquad [7\text{-}16]$$

Solving for [ES] yields

$$[ES] = \frac{[E]_T[S]}{K_M + [S]} \qquad [7\text{-}17]$$

The rate equation for the formation of product (Equation 7-5) is $v = k_2[ES]$, so we can express the reaction velocity as

$$v = k_2[ES] = \frac{k_2[E]_T[S]}{K_M + [S]} \qquad [7\text{-}18]$$

Now we have an equation containing known quantities: $[E]_T$ and [S]. Although some S is consumed in forming the ES complex, we can ignore it because $[S]_T \gg [E]_T$.

Typically, kinetic measurements are made at the start of the reaction, before more than about 10% of the substrate has been converted to product (this is why we can ignore the reverse reaction, E + P $\rightarrow$ ES). Therefore, the velocity at the start of the reaction (at time zero) is expressed as $\boldsymbol{v_0}$ (the **initial velocity**):

$$v_0 = \frac{k_2[\mathrm{E}]_\mathrm{T}[\mathrm{S}]}{K_\mathrm{M} + [\mathrm{S}]} \qquad [7\text{-}19]$$

We can make one additional simplification: When [S] is very high, virtually all the enzyme is in its ES form (it is saturated with substrate) and therefore approaches its point of maximum activity (see Fig. 7-3). The maximum reaction velocity, represented as $\boldsymbol{V_{max}}$, can be expressed

$$V_\mathrm{max} = k_2[\mathrm{E}]_\mathrm{T} \qquad [7\text{-}20]$$

which is similar to Equation 7-5. By substituting Equation 7-20 into Equation 7-19, we obtain

$$\boxed{v_0 = \frac{V_\mathrm{max}[\mathrm{S}]}{K_\mathrm{M} + [\mathrm{S}]}} \qquad [7\text{-}21]$$

This relationship is called the **Michaelis–Menten equation** after Leonor Michaelis and Maude Menten, who derived it in 1913. *It is the rate equation for an enzyme-catalyzed reaction and is the mathematical description of the hyperbolic curve shown in Figure 7-3* (see Sample Calculation 7-2). The graphical presentation of kinetic information is included in Exercise 10.

Sample Calculation 7-2

Problem

An enzyme-catalyzed reaction has a K_M of 1 mM and a V_{max} of 5 nM · s^{-1}. What is the reaction velocity when the substrate concentration is (a) 0.25 mM, (b) 1.5 mM, or (c) 10 mM?

Solution

Use the Michaelis–Menten equation (Equation 7-21):

$$\text{(a)}\ v_0 = \frac{(5\ \mathrm{nM\cdot s^{-1}})(0.25\ \mathrm{mM})}{(1\ \mathrm{mM}) + (0.25\ \mathrm{mM})}$$

$$= \frac{1.25}{1.25}\ \mathrm{nM\cdot s^{-1}}$$

$$= 1\ \mathrm{nM\cdot s^{-1}}$$

$$\text{(b)}\ v_0 = \frac{(5\ \mathrm{nM\cdot s^{-1}})(1.5\ \mathrm{mM})}{(1\ \mathrm{mM}) + (1.5\ \mathrm{mM})}$$

$$= \frac{7.5}{2.5}\ \mathrm{nM\cdot s^{-1}}$$

$$= 3\ \mathrm{nM\cdot s^{-1}}$$

$$\text{(c) } v_0 = \frac{(5 \text{ nM} \cdot \text{s}^{-1})(10 \text{ mM})}{(1 \text{ mM}) + (10 \text{ mM})}$$

$$= \frac{50}{11} \text{ nM} \cdot \text{s}^{-1}$$

$$= 4.5 \text{ nM} \cdot \text{s}^{-1}$$

REVIEW LEARNING OBJECTIVE 3

- Describe the three possible reactions of an ES complex.
- Describe the changes in the concentrations of S, P, E_T, and ES during the course of an enzyme-catalyzed reaction.
- Write the rate equation for an enzyme-catalyzed reaction.

What does K_M represent?

We have seen that the Michaelis constant, K_M, is a combination of three rate constants (Equation 7-12), but it can be fairly easily determined from experimental data. Kinetic measurements are usually made over a range of substrate concentrations. When [S] = K_M, the reaction velocity (v_0) is equal to half its maximum value ($v_0 = V_{max}/2$), as shown in Figure 7-5. Since K_M *is the substrate concentration at which the reaction velocity is half-maximal, it indicates how efficiently an enzyme selects its substrate and converts it to product.* The lower the value of K_M, the more effective the enzyme is at low substrate concentrations; the higher the value of K_M, the less effective the enzyme is. The K_M is unique for each enzyme–substrate pair. Consequently, K_M values are useful for comparing the activities of two enzymes that act on the same substance or for assessing the ability of different substrates to be recognized by a single enzyme.

In practice, K_M *is often used as a measure of an enzyme's affinity for a substrate.* In other words, it approximates the dissociation constant of the ES complex:

$$K_M \approx \frac{[E][S]}{[ES]} \tag{7-22}$$

Note that this relationship is strictly true only when the rate of the ES $\rightarrow$ E + P reaction is much slower than the rate of the ES $\rightarrow$ E + S reaction (that

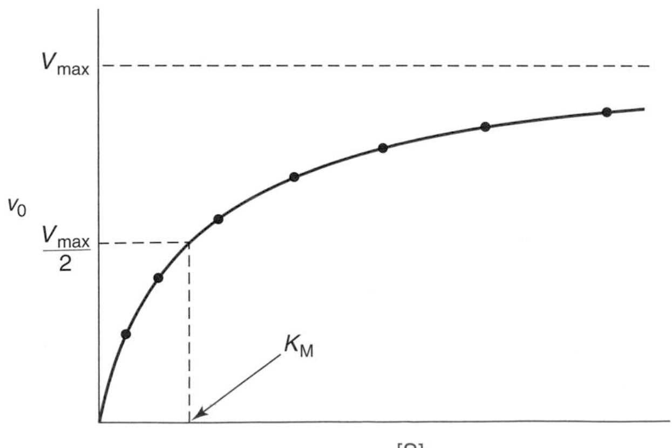

FIGURE 7-5

Graphical determination of K_M. K_M corresponds to the substrate concentration at which the reaction velocity is half-maximal. It can be visually estimated from a plot of v_0 versus [S].

is, when $k_2 \ll k_{-1}$).  Exercise 10 shows how different values of K_M affect an enzymatic reaction.

REVIEW LEARNING OBJECTIVE 3

- What is the formal definition of K_M?
- What other meaning is it often given?

What is the catalytic constant?

It is also useful to know how fast an enzyme operates after it has selected and bound its substrate; that is, how fast does the ES complex proceed to E + P? This quantity is termed the **catalytic constant** and is symbolized $\boldsymbol{k_{cat}}$. For any enzyme-catalyzed reaction,

$$k_{cat} = \frac{V_{max}}{[E]_T} \qquad [7\text{-}23]$$

For a simple reaction, such as the one diagrammed in Equation 7-4,

$$k_{cat} = k_2 \qquad [7\text{-}24]$$

Thus, *k_{cat} is the rate constant of the reaction when the enzyme is saturated with substrate* (when $[ES] \approx [E]_T$ and $v_0 \approx V_{max}$). We have already seen this relationship in Equation 7-20. k_{cat} is also known as the enzyme's **turnover number** because it is the number of catalytic cycles that each active site undergoes per unit time. It is a first-order rate constant and therefore has units of s^{-1}. As shown in Table 7-1, the catalytic constants of enzymes vary over many orders of magnitude.  Exercise 10 shows how different values of V_{max} affect an enzymatic reaction.

REVIEW LEARNING OBJECTIVE 4

- What is the meaning of k_{cat}?
- How can V_{max} be used to calculate k_{cat}?

TABLE 7-1 Catalytic constants of some enzymes

Enzyme	k_{cat} (s^{-1})
Staphylococcal nuclease	95
Cytidine deaminase	299
Triose phosphate isomerase	4300
Cyclophilin	13,000
Ketosteroid isomerase	66,000
Carbonic anhydrase	1,000,000

[Data from Radzicka, A. and Wolfenden, R., *Science* **267**, 90–93 (1995).]

k_{cat}/K_M indicates catalytic efficiency

An enzyme's effectiveness as a catalyst depends on how avidly it binds its substrates and how rapidly it converts them to products. Thus, a measure of catalytic efficiency must reflect both binding and catalytic events. The quantity $\boldsymbol{k_{cat}/K_M}$ satisfies this requirement. At low concentrations of substrate ([S] < K_M), very little ES forms and [E] ≈ $[E]_T$. Equation 7-18 can then be simplified (the [S] term in the denominator becomes insignificant):

$$v_0 = \frac{k_2[E]_T[S]}{K_M + [S]} \quad [7\text{-}25]$$

$$v_0 \approx \frac{k_2}{K_M}[E][S] \quad [7\text{-}26]$$

Equation 7-26 is the rate equation for the second-order reaction of E and S. k_{cat}/K_M, which has units of $M^{-1} \cdot s^{-1}$, is the apparent second-order rate constant. As such, it indicates how the reaction velocity varies according to how often the enzyme and substrate combine with each other. *The value of k_{cat}/K_M, more than either K_M or k_{cat} alone, represents the enzyme's overall ability to convert substrate to product.*

What limits the catalytic power of enzymes? Electronic rearrangements during formation of the transition state occur on the order of 10^{-13} s, the lifetime of a bond vibration. Enzyme turnover numbers are much slower than this (see Table 7-1). An enzyme's overall speed is further limited by how often it collides productively with its substrate. The upper limit for the rate of this second-order reaction (a bimolecular reaction) is about 10^8 to 10^9 $M^{-1} \cdot s^{-1}$, which is the maximum rate at which two freely diffusing molecules can collide with each other in aqueous solution.

This so-called **diffusion-controlled limit** for the second-order reaction between an enzyme and a substrate is matched by several enzymes, including triose phosphate isomerase, whose value of k_{cat}/K_M is 2.4×10^8 $M^{-1} \cdot s^{-1}$. This enzyme is therefore said to have reached **catalytic perfection** because *its overall rate is diffusion-controlled: It catalyzes a reaction as rapidly as it encounters its substrate.*

REVIEW LEARNING OBJECTIVE 4

- Why is k_{cat}/K_M a better indicator of enzyme efficiency than either k_{cat} or K_M alone?
- Why is triose phosphate isomerase considered to be a catalytically perfect enzyme?

Experimental determination of K_M and V_{max}

Velocity versus substrate plots such as Figure 7-5 can be useful for visually estimating the kinetic parameters K_M and V_{max} (from which k_{cat} can be derived by Equation 7-23). However, in practice, hyperbolic curves are prone to misinterpretation because it is difficult to estimate the upper limit of the curve (V_{max}). In order to more accurately determine V_{max} and K_M (the substrate concentration at $V_{max}/2$), it is necessary to perform one of the following steps:

1. Analyze the data by a curve-fitting computer program that mathematically calculates the upper limit for the reaction velocity.

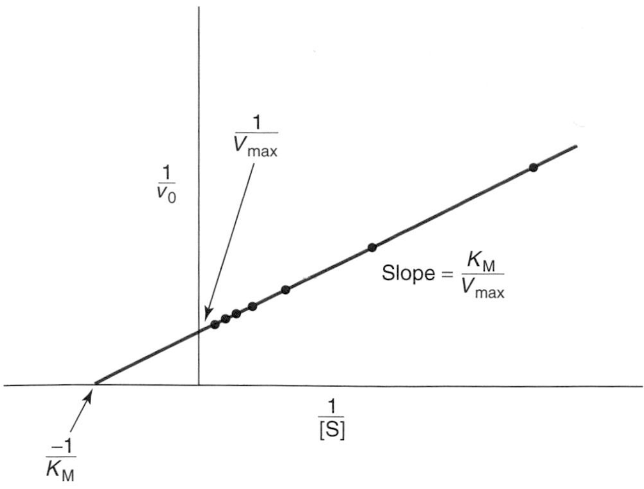

FIGURE 7-6

A Lineweaver–Burk plot.
Plotting the reciprocals of [S] and v_o yields a line whose slope and intercepts yield values of K_M and V_{max}. The plotted points correspond to the points in Figure 7-5.

2. Transform the data to a form that can be plotted as a line. The best-known linear transformation of the velocity versus substrate curve is known as a **Lineweaver–Burk plot,** whose equation is

$$\frac{1}{v_0} = \left(\frac{K_M}{V_{max}}\right)\frac{1}{[S]} + \frac{1}{V_{max}} \qquad [7\text{-}27]$$

Equation 7-27 has the familiar form $y = mx + b$. A plot of $1/v_0$ versus $1/[S]$ gives a straight line whose slope is K_M/V_{max} and whose intercept on the $1/v_0$ axis is $1/V_{max}$. The extrapolated intercept on the $1/[S]$ axis is $-1/K_M$ (Fig. 7-6). A comparison of Figures 7-5 and 7-6, made from the same data, illustrates how evenly spaced points on a velocity versus substrate plot become compressed in a Lineweaver–Burk plot (see Sample Calculation 7-3).

Sample Calculation 7-3

Problem

The velocity of an enzyme-catalyzed reaction was measured at several substrate concentrations. Calculate K_M and V_{max} for the reaction.

[S] (μM)	v_0 (mM · s^{-1})
0.25	0.75
0.5	1.20
1.0	1.71
2.0	2.18
4.0	2.53

Solution

Calculate the reciprocals of the substrate concentration and velocity, then make a plot of $1/v_0$ versus 1/[S] (a Lineweaver–Burk plot).

1/[S] (μM^{-1})	$1/v_0$ ($mM^{-1} \cdot s$)
4.0	1.33
2.0	0.83
1.0	0.58
0.5	0.46
0.25	0.40

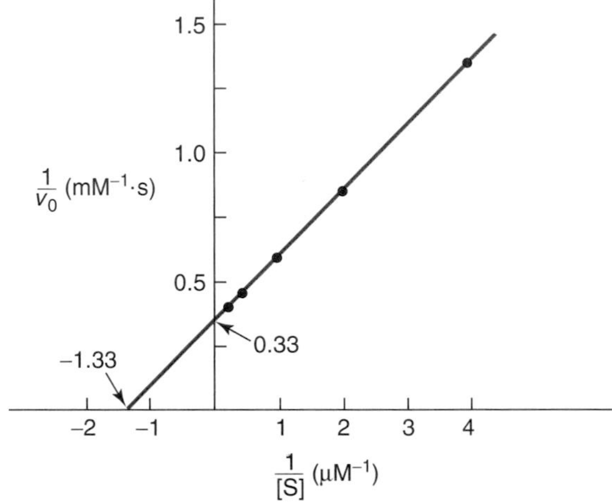

The intercept on the 1/[S] axis (which is equal to $-1/K_M$) is $-1.33\ \mu M^{-1}$. Therefore,

$$K_M = -\left(\frac{1}{-1.33\ \mu M^{-1}}\right) = 0.75\ \mu M$$

The intercept on the $1/v_0$ axis (which is equal to $1/V_{max}$) is $0.33\ mM^{-1} \cdot s$. Therefore,

$$V_{max} = \frac{1}{0.33\ mM^{-1} \cdot s} = 3.0\ mM \cdot s^{-1}$$

Ideally, experimental conditions are chosen so that velocity measurements can be made for substrate concentrations that are both higher and lower than K_M. This yields the most accurate values for K_M and V_{max}. A Lineweaver–Burk plot, whether constructed manually or by computer, offers the advantage that K_M and V_{max} can be quickly estimated by eye. Linear plots are also more convenient than curves for comparing multiple data sets, such as different enzyme preparations, or a single enzyme in the presence of different concentrations of an inhibitor. Examples of Lineweaver–Burk plots are included in Exercise 10.

REVIEW LEARNING OBJECTIVE 5

- Why is a Lineweaver–Burk plot, rather than a velocity versus substrate plot, often used to determine K_M and V_{max}?

Not all enzymes fit the simple Michaelis–Menten model

So far, the discussion has focused on the very simplest of enzyme-catalyzed reactions, namely, a reaction with one substrate and one product. Such reactions represent only a small portion of known enzymatic reactions, many of which involve multiple substrates and products, proceed via multiple steps, or do not meet the assumptions of the Michaelis–Menten kinetic model for other reasons. Nevertheless, the kinetics of these reactions can still be evaluated.

1. Multisubstrate reactions

More than half of all known biochemical reactions involve two substrates. Most of these **bisubstrate reactions** are either oxidation–reduction reactions or transferase reactions. In an oxidation–reduction reaction, electrons are transferred between substrates:

$$X_{\text{oxidized}} + Y_{\text{reduced}} \rightarrow X_{\text{reduced}} + Y_{\text{oxidized}}$$

In a transferase reaction, such as the one catalyzed by transketolase, a group is transferred between two molecules:

$$\begin{array}{c} CH_2OH \\ | \\ C{=}O \\ | \\ HO{-}C{-}H \\ | \\ H{-}C{-}OH \\ | \\ H{-}C{-}OH \\ | \\ CH_2OPO_3^{2-} \\ \text{Fructose-6-phosphate} \end{array} \quad + \quad \begin{array}{c} O \;\; H \\ \backslash\!\backslash \; / \\ C \\ | \\ H{-}C{-}OH \\ | \\ CH_2OPO_3^{2-} \\ \text{Glyceraldehyde-3-phosphate} \end{array}$$

$$\big\downarrow \text{ transketolase}$$

$$\begin{array}{c} O \;\; H \\ \backslash\!\backslash \; / \\ C \\ | \\ H{-}C{-}OH \\ | \\ H{-}C{-}OH \\ | \\ CH_2OPO_3^{2-} \\ \text{Erythrose-4-phosphate} \end{array} \quad + \quad \begin{array}{c} CH_2OH \\ | \\ C{=}O \\ | \\ HO{-}C{-}H \\ | \\ H{-}C{-}OH \\ | \\ CH_2OPO_3^{2-} \\ \text{Xylulose-5-phosphate} \end{array}$$

The transketolase reaction is ubiquitous in nature; it functions in the synthesis and degradation of carbohydrates. As written here, it transforms a 6-carbon sugar and a 3-carbon sugar to a 4-carbon sugar and a 5-carbon sugar. *Each of the substrates interacts with the enzyme with a characteristic* K_M. To experimentally determine each K_M, the reaction velocity is measured at different concentrations of one substrate while the concentration of the other substrate is held constant. V_{max} *is the maximum reaction velocity when both substrates are present at concentrations that saturate their binding sites on the enzyme.*

In some bisubstrate reactions, the substrates can bind in any order, but in the transketolase reaction, fructose-6-phosphate must bind first. It surrenders a two-carbon fragment to the enzyme, and the first product (erythrose-4-phosphate) leaves the active site before the second substrate (glyceraldehyde-3-phosphate) binds and receives the two-carbon fragment to yield the second product (xylulose-5-phosphate).

2. Multistep reactions

As the transketolase reaction illustrates, an enzyme-catalyzed reaction may contain many steps. In this example, the reaction includes an intermediate in which the two-carbon fragment removed from fructose-6-phosphate remains bound to the enzyme while awaiting the arrival of the second substrate. (The chymotrypsin reaction mechanism outlined in Fig. 6-9 similarly requires several steps.) The multistep transketolase reaction can be broken down into a series of simple mechanistic steps and diagrammed as follows:

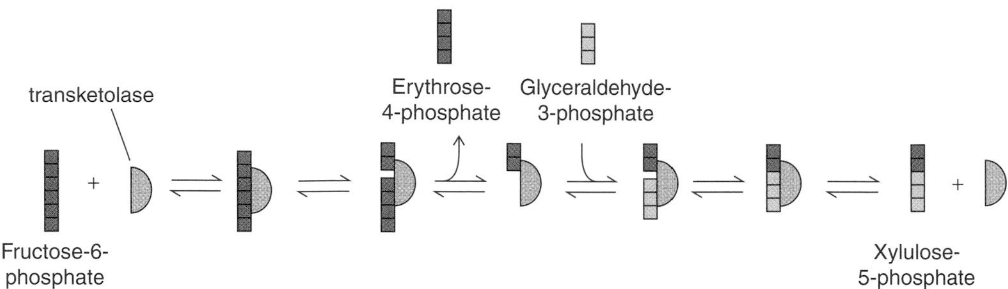

Each step of this process has characteristic forward and reverse rate constants. Consequently, k_{cat} for the overall reaction

$$\text{Fructose-6-phosphate} + \text{glyceraldehyde-3-phosphate} \rightleftharpoons \text{erythrose-4-phosphate} + \text{xylulose-5-phosphate}$$

is a complicated function of many individual rate constants (only for a very simple reaction, for example, Equation 7-4, does $k_{cat} = k_2$). Nevertheless, *the meaning of k_{cat}—the enzyme's turnover number—is the same as for a single-step reaction.*

The rate constants of individual steps in a multistep reaction can sometimes be measured during the initial stages of the reaction, that is, before a steady state is established. This requires instruments that can rapidly mix the reactants and then monitor the mixture on a time scale from 1 s to 10^{-7}s.

3. Nonhyperbolic reactions

Many enzymes, particularly oligomeric enzymes with multiple active sites, do not obey the Michaelis–Menten rate equation and therefore do not yield hyperbolic velocity versus substrate curves. In these enzymes, *the presence of a substrate at one active site can affect the catalytic activity of the other active sites.* This **cooperative** behavior occurs when the enzyme subunits are structurally linked to each other so that a substrate-induced conformational change in one subunit elicits conformational changes in the remaining subunits. (Cooperative behavior also occurs in hemoglobin, when O_2 binding to the heme group in one subunit alters the O_2 affinity of the other subunits; see Section 4-5.) The result is a sigmoidal (S-shaped) velocity versus substrate curve (Fig. 7-7). Although the standard Michaelis–Menten equation does not apply here, K_M and V_{max} can be estimated and used to characterize enzyme activity.

REVIEW LEARNING OBJECTIVE 6

- How many K_M values pertain to a bisubstrate reaction? How many V_{max} values?
- Explain why k_{cat} is not necessarily the rate constant for a single reaction.
- Describe the shape of the velocity versus substrate curve when the enzyme exhibits cooperative behavior.

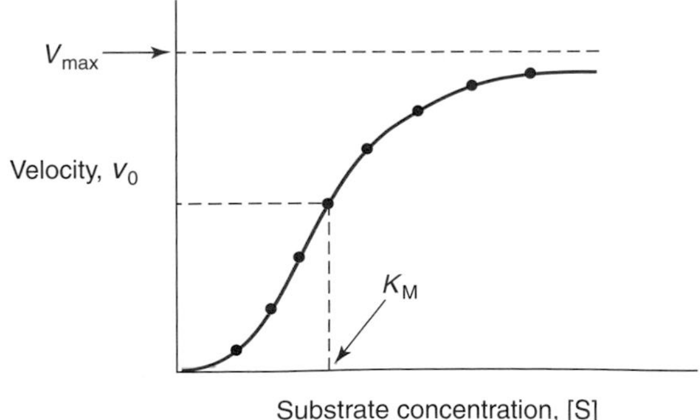

FIGURE 7-7

Effect of cooperative substrate binding.
The velocity versus substrate curve is sigmoidal rather than hyperbolic when substrate binding to one active site in an oligomeric enzyme alters the catalytic activity of the other active sites. The maximum reaction velocity is V_{max}, and K_M is the substrate concentration when the velocity is half-maximal.

3 ENZYME INHIBITION

Inside a cell, an enzyme is subject to a variety of factors that can influence its behavior. Substances that bind to the enzyme can interfere with substrate binding and/or catalysis. Many naturally occurring antibiotics, pesticides, and other poisons are substances that inhibit the activity of essential enzymes. From a strictly scientific point of view, these inhibitors are useful probes of an enzyme's active site structure and catalytic mechanism. Enzyme inhibitors are also used therapeutically as drugs. The ongoing pursuit of more effective drugs requires knowledge of how enzyme inhibitors work and how they can be altered to better inhibit their target enzymes.

Go to **Exercise 11/Enzyme Inhibition** for a closer look at the most common forms of enzyme inhibition, including irreversible inhibition, competitive inhibition, and mixed inhibition. The material in Exercise 11 addresses Learning Objectives 7, 8, and 9.

Irreversible inhibition

Certain compounds interact with enzymes so tightly that their effects are irreversible. For example, diisopropylphosphofluoridate (DIPF), the reagent used to identify the active site Ser residue of chymotrypsin (see Section 6-2), is an **irreversible inhibitor** of the enzyme. When DIPF reacts with chymotrypsin, leaving the DIP group covalently attached to the Ser hydroxyl group (see page 178), the enzyme becomes catalytically inactive.

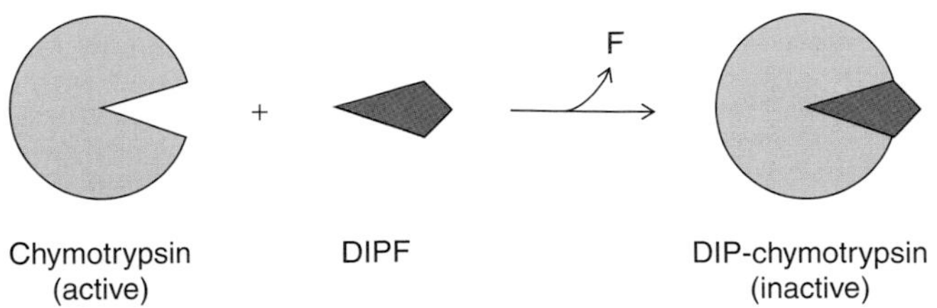

In general, *any reagent that covalently modifies an amino acid side chain in a protein can potentially act as an irreversible enzyme inhibitor.*

Some irreversible enzyme inhibitors are called **suicide substrates** because they enter the enzyme's active site and begin to react, just as a normal substrate would. However, they are unable to undergo the complete reaction and hence become "stuck" in the active site. For example, thymidylate synthase is the enzyme that converts the nucleotide deoxyuridylate (dUMP) to deoxythymidylate (dTMP) by adding a methyl group to C5:

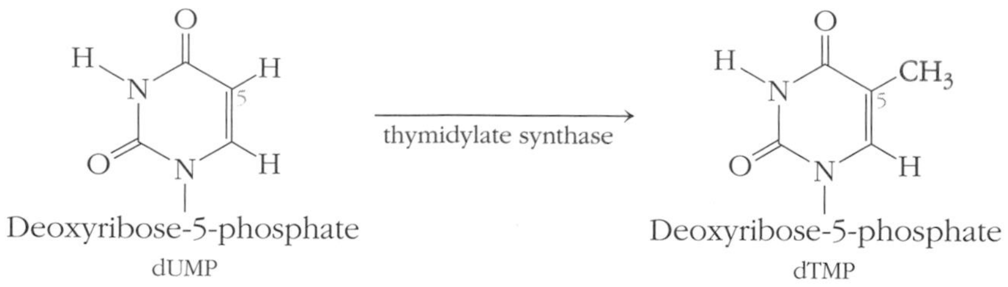

Deoxyribose-5-phosphate
dUMP

thymidylate synthase

Deoxyribose-5-phosphate
dTMP

The synthetic compound 5-fluorouracil

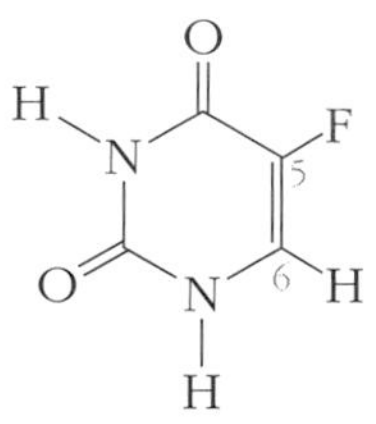

5-Fluorouracil

is readily ribosylated, phosphorylated, and reduced *in vivo* to yield the nucleotide 5-fluorodeoxyuridylate. This compound, like dUMP, enters the active site of thymidylate synthase, where a Cys SH group adds to C6. Normally, this enhances the nucleophilicity (electron richness) of C5 so that it can accept an electron-poor methyl group. However, the presence of the electron-withdrawing F atom prevents methylation. The inhibitor therefore remains in the active site, bound to the Cys side chain, rendering thymidylate synthase inactive. For this reason, 5-fluorouracil is used to disrupt DNA synthesis in rapidly dividing cancer cells.

REVIEW LEARNING OBJECTIVE 7

- Why does an agent that chemically modifies an amino acid residue sometimes act as an irreversible enzyme inhibitor?

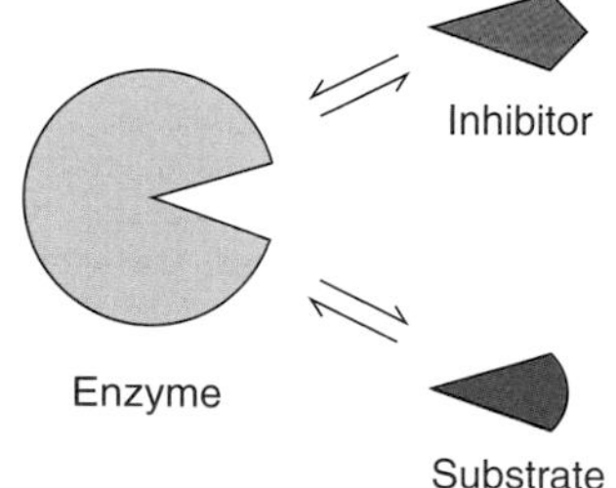

FIGURE 7-8

Competitive enzyme inhibition.
In its simplest form, competitive inhibition of an enzyme occurs when the inhibitor and substrate compete for binding in the enzyme active site. A competitive inhibitor often resembles the substrate in size and shape but cannot undergo a reaction. In all cases, binding of the inhibitor and the substrate is mutually exclusive.

Competitive inhibition is the most common form of reversible enzyme inhibition

As its name implies, reversible enzyme inhibition results when a substance binds reversibly to an enzyme so as to alter its catalytic properties. A reversible inhibitor may affect the enzyme's K_M, k_{cat}, or both. The most common reversible enzyme inhibitors are known as either competitive inhibitors or mixed inhibitors. In the most common form of **competitive inhibition,** *the inhibitor is a substance that directly competes with a substrate for binding to the enzyme's active site* (Fig. 7-8). As expected, the inhibitor usually resembles the substrate in overall size and chemical properties so that it can bind to the enzyme, but it lacks the exact electronic structure that allows it to react.

One well-known competitive inhibitor affects the activity of succinate dehydrogenase, which catalyzes the oxidation (dehydrogenation) of succinate to produce fumarate:

$$\begin{array}{c} COO^- \\ | \\ CH_2 \\ | \\ CH_2 \\ | \\ COO^- \end{array} \xrightarrow{\text{succinate dehydrogenase}} \begin{array}{c} COO^- \\ | \\ CH \\ \| \\ CH \\ | \\ COO^- \end{array}$$

Succinate   Fumarate

The compound malonate

$$\begin{array}{c} COO^- \\ | \\ CH_2 \\ | \\ COO^- \end{array}$$

Malonate

inhibits the reaction because it binds to the dehydrogenase active site but cannot be dehydrogenated. Apparently, the enzyme active site can accommodate either the substrate succinate or the competitive inhibitor malonate.

A plot of an enzyme's reaction velocity in the presence of an inhibitor, as a function of the substrate concentration, is shown in Figure 7-9. Because the inhibitor prevents some of the substrate from reaching the active site, the K_M appears to increase (the enzyme's affinity for the substrate appears to decrease). But because the inhibitor binds reversibly, it constantly dissociates from and reassociates with the enzyme, which allows a substrate molecule to occasionally enter the active site. High concentrations of substrate can overcome the effect of the inhibitor because when [S] ≫ [I], the enzyme is more likely to bind S than I. The presence of a reversible inhibitor does not affect the enzyme's k_{cat}, so as [S] approaches infinity, v_0 approaches V_{max}. To summarize, *a competitive inhibitor increases the apparent K_M of the enzyme but does not affect k_{cat} or V_{max}.*

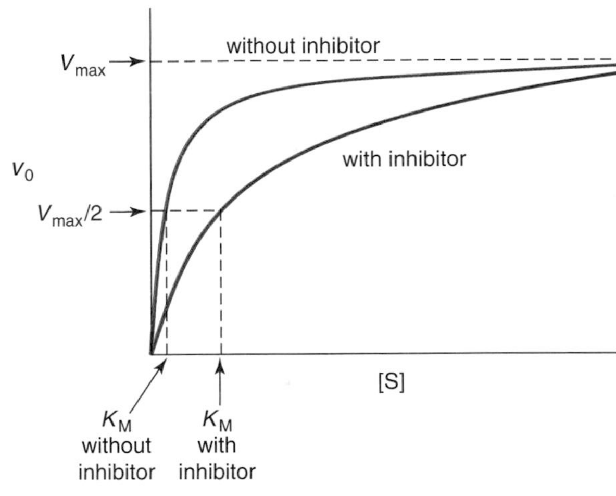

FIGURE 7-9

Effect of a competitive inhibitor on reaction velocity.
In a plot of velocity versus substrate concentration, the inhibitor increases the apparent K_M because it competes with the substrate for binding to the enzyme. The inhibitor does not affect k_{cat}, so at high [S], v_o approaches V_{max}.

The Michaelis–Menten equation for a competitively inhibited reaction has the form

$$v_0 = \frac{V_{max}[S]}{\alpha K_M + [S]} \quad [7\text{-}28]$$

where α is a factor that modifies K_M. The value of α—the degree of inhibition—depends on the inhibitor's concentration and its affinity for the enzyme:

$$\alpha = 1 + \frac{[I]}{K_I} \quad [7\text{-}29]$$

$\boldsymbol{K_I}$ is the **inhibition constant;** it is the dissociation constant for the **enzyme–inhibitor (EI) complex:**

$$K_I = \frac{[E][I]}{[EI]} \quad [7\text{-}30]$$

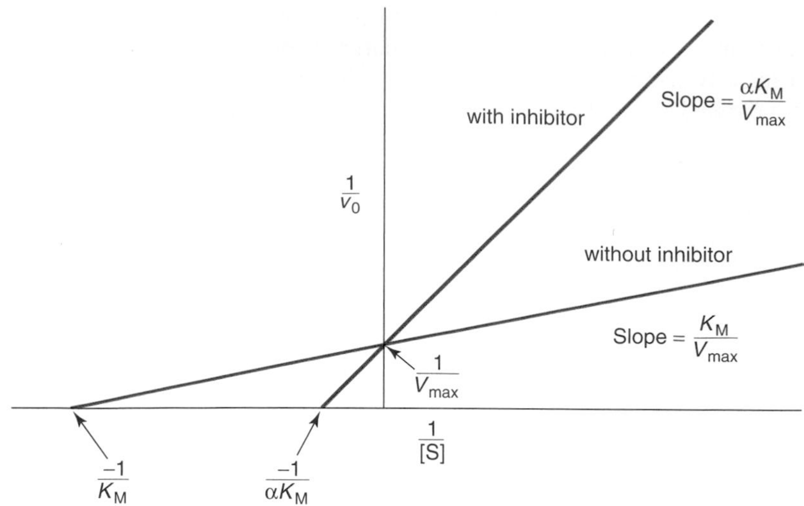

FIGURE 7-10

A Lineweaver–Burk plot for competitive inhibition.
The presence of a competitive inhibitor alters the value of $-1/K_M$, the intercept on the $1/[S]$ axis, by the factor α. Note that the inhibitor does not affect the value of $1/V_{max}$, the intercept on the $1/v_0$ axis.

The lower the value of K_I, the tighter the inhibitor binds to the enzyme. It is possible to derive α (and therefore K_I) by plotting the reaction velocity as a function of substrate concentration in the presence of a known concentration of inhibitor. When the data are replotted in Lineweaver–Burk form, the intercept on the $1/[S]$ axis is $-1/\alpha K_M$ (Fig. 7-10; also see Sample Calculation 7-4). Exercise 11 shows how a competitive inhibitor affects enzyme activity.

Sample Calculation 7-4

Problem

An enzyme has a K_M of 8 μM in the absence of a competitive inhibitor and a K_M of 12 μM in the presence of 3 μM of the inhibitor. Calculate K_I.

Solution

The inhibitor increases K_M by a factor α (Equation 7-28). Since the value of K_M with the inhibitor is 1.5 times greater than the value of K_M without the inhibitor (12 μM ÷ 8 μM), $\alpha = 1.5$. Equation 7-29, which gives the relationship between α, [I], and K_I, can be rearrange to solve for K_I:

$$K_I = \frac{[I]}{\alpha - 1}$$

$$= \frac{3\ \mu M}{1.5 - 1}$$

$$= \frac{3\ \mu M}{0.5}$$

$$= 6\ \mu M$$

K_I values are useful for assessing the inhibitory power of different substances, for example, a series of compounds being tested for usefulness as drugs. Keep in mind that an effective drug is not necessarily the compound with the lowest K_I (the tightest binding), since other factors, such as the drug's solubility or its stability, must also be considered.

Product inhibition occurs when the product of a reaction occupies the enzyme's active site, thereby preventing the binding of additional substrate molecules. This is one reason why measurements of enzyme activity (v_0) are made early in the reaction, before product has significantly accumulated.

Transition state analogs inhibit enzymes

Studies of enzyme inhibitors can reveal information about the chemistry of the reaction and the enzyme's active site. For example, the inhibition of succinate dehydrogenase by malonate, shown above, suggests that the dehydrogenase active site recognizes and binds substances with two carboxylate groups. Similarly, the ability of an inhibitor to bind to an enzyme's active site may confirm a proposed reaction mechanism. The triose phosphate isomerase reaction (introduced on page 199) is believed to proceed through an enediolate transition state (the transition state corresponds to a high-energy structure in which bonds are in the process of breaking and forming):

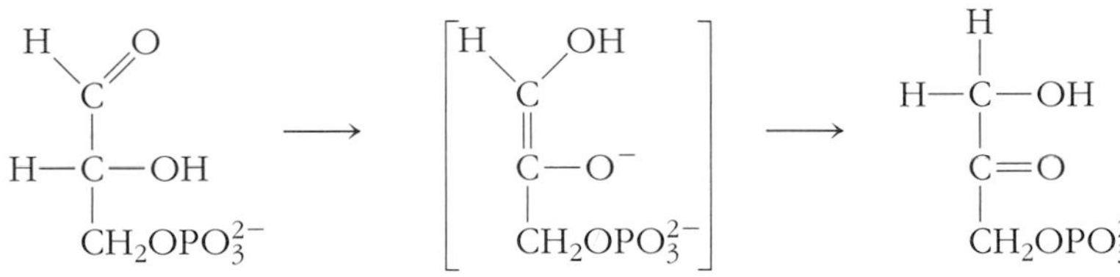

Glyceraldehyde-3-phosphate → [Enediolate] → Dihydroxyacetone phosphate

The compound phosphoglycohydroxamate (at right) resembles the proposed transition state, and in fact binds to triose phosphate isomerase about 300 times more tightly than glyceraldehyde-3-phosphate or dihydroxyacetone phosphate bind.

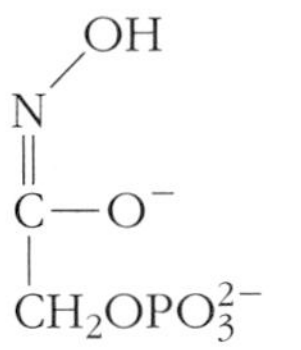

Phosphoglycohydroxamate

Numerous studies demonstrate that *whereas substrate analogs make good competitive inhibitors,* ***transition state analogs*** *make even better inhibitors.* This is because in order to catalyze a reaction, the enzyme must bind to (stabilize) the reaction's transition state (see Section 6-2). A compound that mimics the transition state can take advantage of these interactions in a way that a substrate analog cannot. For example, the nucleoside adenosine is converted to inosine as follows:

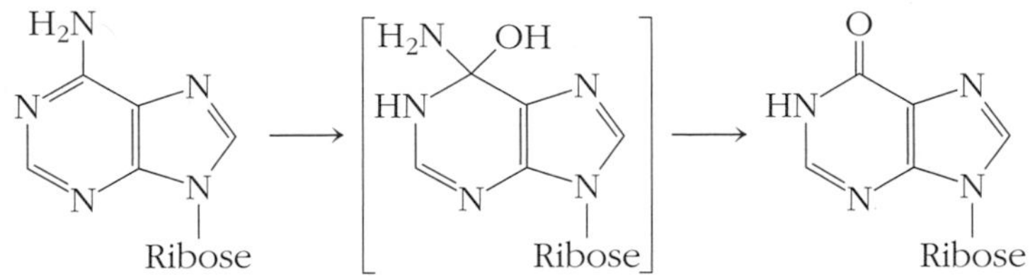

Adenosine → [ ] → Inosine

The K_M of the enzyme for the substrate adenosine is 3×10^{-5} M. The product inosine acts as an inhibitor of the reaction, with a K_I of 3×10^{-4} M. A transition state analog (at right) inhibits the reaction with a K_I of 1.5×10^{-13} M.

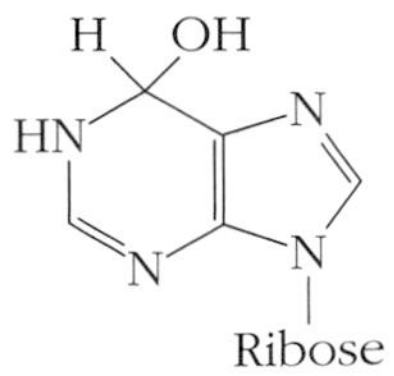

1,6-Dihydroinosine

Not only do such inhibitors shed light on the probable structure of the reaction's transition state, they may provide a starting point for the design of even better inhibitors. Some of the drugs used to treat infection by HIV (the human immunodeficiency virus) were first designed by considering how transition state analogs inhibit viral enzymes (Box 7-A).

REVIEW LEARNING OBJECTIVE 8

- What is the meaning of K_I and how is it determined?
- How does a competitive inhibitor affect K_M and V_{max}?
- Why do some reaction products and transition state analogs act as competitive enzyme inhibitors?

Mixed inhibitors affect both K_M and V_{max}

Some reversible enzyme inhibitors diminish an enzyme's activity not just by interfering with substrate binding (as approximated by K_M), but by directly affecting k_{cat}. This situation most often occurs when the inhibitor binds to a site on the enzyme other than the active site and elicits a conformational change that affects the structure or chemical properties of the active site (Fig. 7-11). As a result, *the apparent* V_{max} *decreases and the apparent* K_M *may increase or decrease.* This phenomenon is called **mixed inhibition** (Fig. 7-12). When the K_M is not affected by the inhibitor—when the inhibitor affects only k_{cat}— inhibition is said to be **noncompetitive.** Because a mixed or noncompetitive inhibitor affects k_{cat} but does not prevent substrate binding to the active site, adding more substrate does not reverse the inhibition, as it does in competitive inhibition. The effects of mixed and noncompetitive inhibitors are illustrated in Exercise 11.

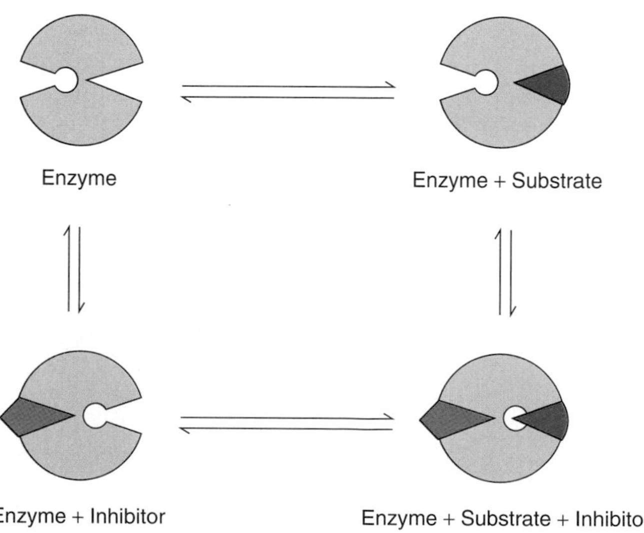

FIGURE 7-11

Mixed enzyme inhibition.
The inhibitor and substrate bind to separate sites. Inhibitor binding alters the catalytic activity of the active site (thereby decreasing V_{max}) and also affects substrate binding so that the K_M may increase or decrease. In noncompetitive inhibition, the inhibitor decreases V_{max} but does not affect K_M.

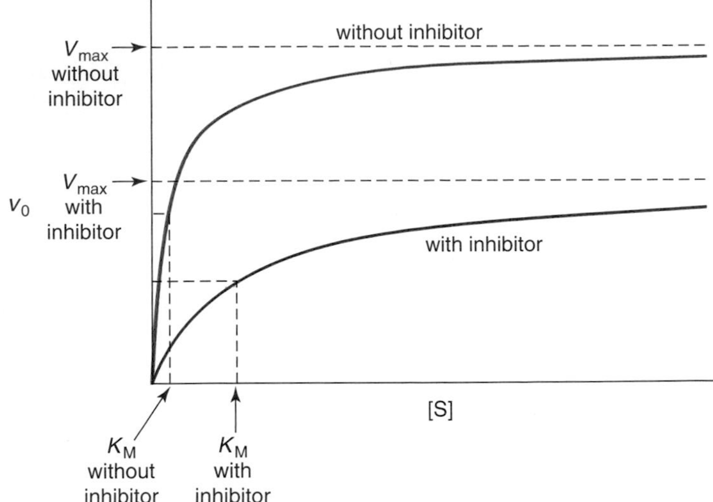

FIGURE 7-12

Effect of a mixed inhibitor on reaction velocity.
As shown here, the inhibitor affects both substrate binding (represented as K_M) and k_{cat} so that the apparent K_M increases and the apparent V_{max} decreases. (In some cases, K_M may not change or may decrease.)

Metal ions frequently act as noncompetitive enzyme inhibitors. For example, trivalent ions such as aluminum (Al^{3+}) inhibit the activity of acetylcholinesterase, which catalyzes the hydrolysis of the neurotransmitter acetylcholine:

$$\underset{\text{Acetylcholine}}{H_3C-\overset{\overset{\large O}{\|}}{C}-O-CH_2-CH_2-\overset{+}{N}(CH_3)_3} + H_2O \xrightarrow[\text{acetylcholinesterase}]{}$$

$$\underset{\text{Acetate}}{H_3C-\overset{\overset{\large O}{\|}}{C}-O^-} + \underset{\text{Choline}}{HO-CH_2-CH_2-\overset{+}{N}(CH_3)_3} + H^+$$

This reaction limits the duration of certain nerve impulses (see Section 8-4). Al^{3+} inhibits acetylcholinesterase noncompetitively by binding to the enzyme at a site distinct from the active site. Consequently, Al^{3+} can bind to the free enzyme or to the enzyme–substrate complex.

Not all reversible enzyme inhibitors are small molecules. The activity of an enzyme may be modulated by proteins that bind to it and thereby alter its conformation in a way that interferes with the chemistry of its active site.

REVIEW LEARNING OBJECTIVE 9

- How does a mixed inhibitor affect K_M and V_{max}?
- Under what conditions is inhibition noncompetitive?

Allosteric inhibition

Oligomeric enzymes—those with multiple active sites in one multisubunit protein—are commonly subject to **allosteric inhibition.** Just as substrate binding to one active site of an oligomeric enzyme may increase the activity of the other active sites (as occurs in hemoglobin; see Section 4-5), inhibitor binding to one subunit of an enzyme may decrease the catalytic activity of all the subunits.

Go to **Exercise 12/Phosphofructokinase Regulation** for a detailed look at the structure of phosphofructokinase and how its conformation is affected by an allosteric inhibitor. The structures shown in Exercise 12 and the questions based on them illuminate Learning Objective 10.

A CLOSER LOOK

BOX 7-A Inhibitors of HIV enzymes

Human immunodeficiency virus (HIV), the causative agent of acquired immune deficiency syndrome (AIDS), consists of a 9 kb RNA genome embedded in a protein core that is in turn surrounded by an outer protein–lipid envelope.

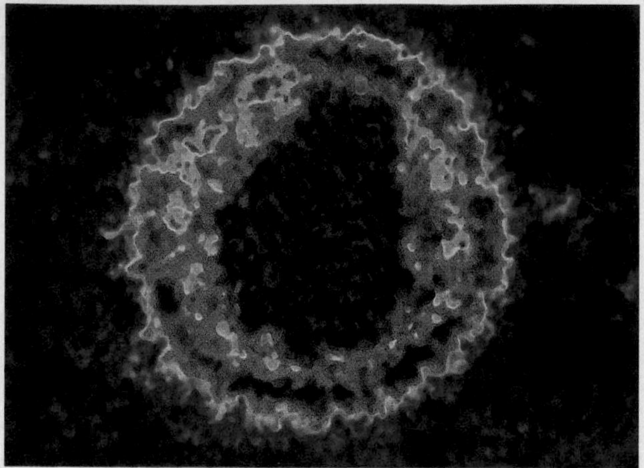

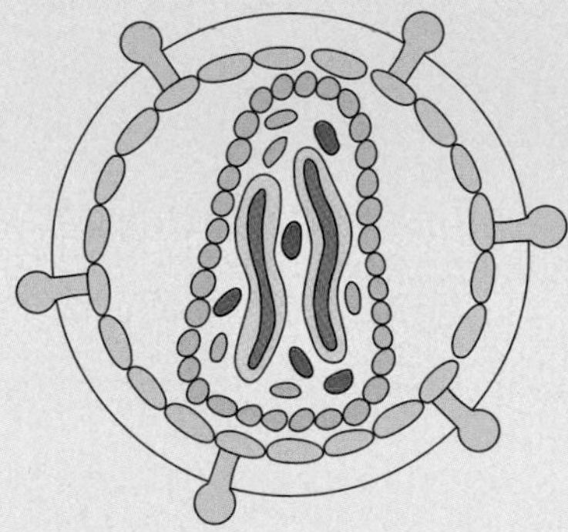

Electron micrograph and diagram of an HIV particle.

The core of the virus includes a few viral enzymes and other proteins that are required for infectivity. After the viral envelope proteins contact certain host cell–surface proteins, the virus enters the cell and its RNA is uncoated. The viral RNA is then transcribed into DNA by the action of the viral enzyme reverse transcriptase. Another viral enzyme, an integrase, incorporates the resulting DNA into the host genome. Expression of the viral genes produces 15 different proteins, some of which must be processed by a viral protease to achieve their mature forms. Eventually, new viral particles are assembled and bud off from the host cell membrane.

The complex life cycle of HIV offers a number of potential points for therapeutic intervention. *Two of the three HIV enzymes, the reverse transcriptase and the protease, were the earliest targets for drug development, in part because enzymes with similar activities had already been extensively studied before HIV was identified in the early 1980s.* These prior studies made it easier to understand the structures and functions of the viral enzymes. Some other aspects of HIV physiology, such as the steps involved in regulating the expression and processing of HIV gene products, are not as well understood as the activities of the reverse transcriptase and protease.

The first anti-HIV drugs were directed against reverse transcriptase, an enzyme with two active sites. One active site catalyzes DNA polymerization to synthesize first a DNA strand complementary to the viral RNA and then a second DNA strand complementary to the first. The other active site is an RNase (an RNA-degrading enzyme) that eliminates the original viral RNA template after the first round of DNA synthesis. These reactions are summarized in the figure. Without reverse transcriptase, the viral genome could not be incorporated into the host DNA and hence the viral genes could not be expressed by the host's transcriptional and translational machinery.

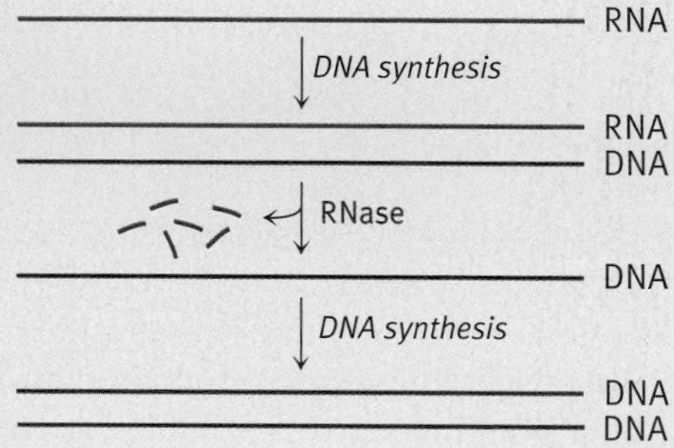

Reverse transcriptase activity can be blocked by two different types of drugs. Nucleoside analogs such as AZT and ddC

HOCH$_2$ O Thymine
H H H H
$\bar{N}=\overset{+}{N}=N$ H

AZT
(3′-Azido-2′,3′-dideoxythymidine, Zidovudine)

HOCH$_2$ O Cytosine
H H H H
H H

ddC
(2′,3′-Dideoxycytidine, Zalcitabine)

(continued)

BOX 7-A *(continued)*

readily enter cells and are phosphorylated. The resulting nucleotides bind in the reverse transcriptase active site and are linked, via their 5′ phosphate group, to the growing DNA chain. However, because they lack a 3′ OH group, further addition of nucleotides is impossible. These inhibitors are therefore called **chain terminators.** (Chain termination is the basis for the DNA sequencing reaction described in Section 3-3.) AZT was originally identified as an anticancer agent and later found to be effective against HIV, setting the stage for the rapid development of additional anti-HIV drugs.

Reverse transcriptase can also be inhibited by non-nucleoside analogs such as nevirapine:

Nevirapine

This noncompetitive inhibitor binds to a hydrophobic patch on the surface of reverse transcriptase near the polymerase active site and blocks its activity.

The 9 kb HIV RNA encodes a total of 15 proteins, including 6 structural proteins (green), 3 enzymes (red), and 6 accessory proteins (blue) that are required for viral gene expression and assembly of new viral particles. Several genes overlap, and two genes are composed of noncontiguous segments of RNA.

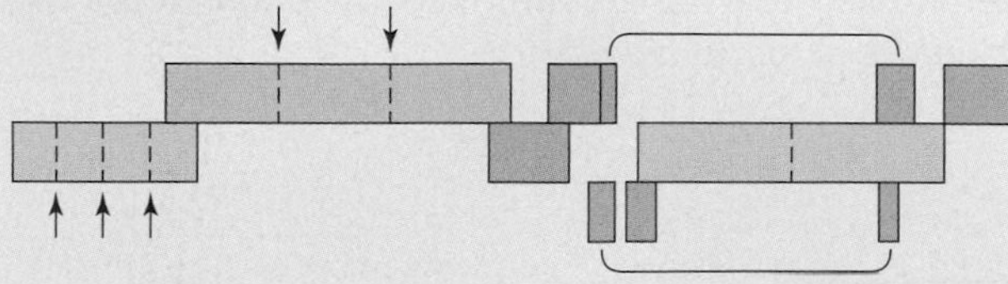

HIV's structural proteins and its three enzymes are initially synthesized as **polyproteins** whose individual members are later separated by proteolysis (at the sites indicated by arrows). HIV protease, itself a product of a polyprotein, carries out the proteolysis reaction by hydrolyzing Tyr—Pro or Phe—Pro peptide bonds in the polyproteins. Catalytic activity is centered on two Asp residues (shown in green in the model above right), each contributed by one subunit of the homodimeric enzyme. The gold structure represents a peptide substrate analog. The side chains of peptide substrates bind in hydrophobic pockets near the active site.

HIV protease inhibitors are the result of **rational drug design**—the development of inhibitors based on detailed knowledge of the enzyme's structure. For example, studies of inhibitor–protease complexes reveal that strong inhibitors must be at least the size of a tetrapeptide but need not be symmetrical (although the enzyme itself is symmetrical). Saquinavir (Invirase) was the first widely used HIV protease inhibitor. It is a transition state analog with bulky side chains that mimics the protease's natural Phe—Pro substrates (the scissile bonds are shown in red).

—Phe—Pro—

Saquinavir

Ritonavir

(continued)

BOX 7-A *(continued)*

Saquinavir acts as a competitive inhibitor with a K_I of 0.15 nM (for comparison, synthetic peptide substrates have K_M values of about 35 μM). Efforts to develop other drugs have focused on improving the solubility (and therefore the bioavailability) of protease inhibitors by adding polar groups without diminishing binding to the protease. Such efforts have yielded, for example, Ritonavir (Norvir) with a K_I of 0.17 nM.

One of the challenges of developing antiviral drugs is to *select a target that is unique to the virus so that the drugs will not disrupt the host's normal metabolic reactions.* For example, HIV protease inhibitors are effective antiviral agents because mammalian proteases do not recognize compounds containing amide bonds to Pro or Pro analogs. Nucleoside analog inhibitors of reverse transcriptase do bind to the host cell's DNA polymerases, but with lower affinity than they bind to HIV reverse transcriptase. Still, antiviral drugs elicit side effects ranging from nausea and rashes to kidney stones and neurological impairment. These problems can be severe because effective treatment of HIV infection requires a multiple-drug approach so that viral particles can be eliminated before they have a chance to develop drug resistance through mutation.

[HIV photo from Cavallini James/Photo Researchers; diagram of HIV genes after Frankel, A.D. and Young, J.A.T., *Annu. Rev. Biochem.* **67,** 3 (1998); structure of HIV protease (pdb 1HXW) determined by C.H. Park, V. Nienaber, and X.P. Kong.]

Allosteric effects are part of the physiological regulation of the enzyme phosphofructokinase, which catalyzes the reaction

$$\text{Fructose-6-phosphate} + \text{ATP} \xrightarrow{\text{phosphofructokinase}} \text{Fructose-1,6-bisphosphate} + \text{ADP}$$

This phosphorylation reaction is Step 3 of glycolysis, the glucose degradation pathway that is an important source of ATP in virtually all cells (see Section 10-2). The phosphofructokinase reaction is inhibited by phosphoenolpyruvate, the product of Reaction 9 of glycolysis:

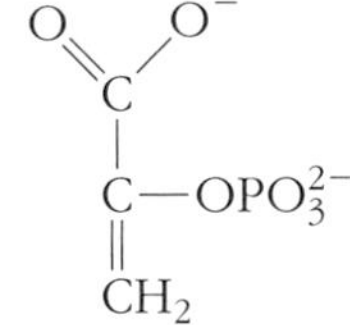

Phosphoenolpyruvate

Phosphoenolpyruvate is an example of a **feedback inhibitor:** When its concentration in the cell is sufficiently high, *it shuts down its own synthesis by blocking an earlier step in its biosynthetic pathway:*

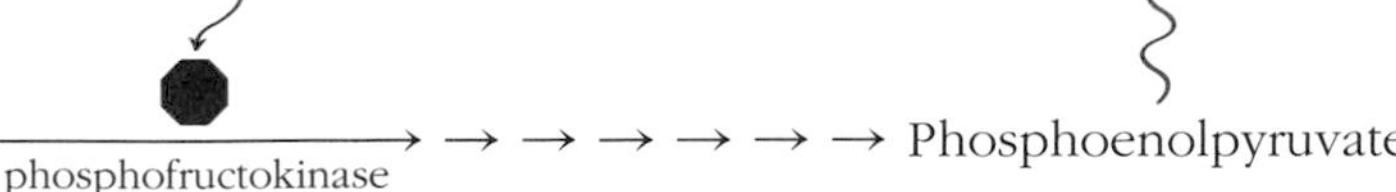

FIGURE 7-13

Structure of phosphofructokinase from *B. stearothermophilus.*
The four identical subunits are arranged as a dimer of dimers (one dimer is shown with blue subunits; the other with purple subunits). [*Structure (pdb 6PFK) determined by P.R. Evans, T. Schirmer, and M. Auer.*]

Phosphofructokinase from the bacterium *Bacillus stearothermophilus* is a tetramer with four active sites. The subunits are arranged as a dimer of dimers (Fig. 7-13). Each of the four fructose-6-phosphate binding sites is made up of residues from both dimers. *B. stearothermophilus* phosphofructokinase

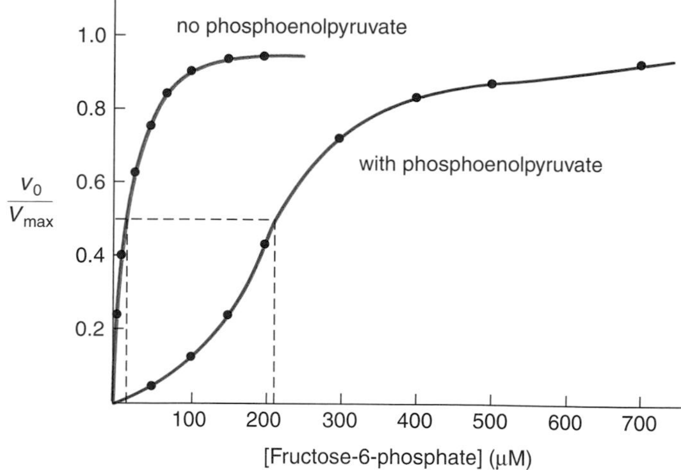

FIGURE 7-14

Effect of phosphoenolpyruvate on phosphofructokinase activity.
In the absence of the inhibitor (green line), *B. stearothermophilus* phosphofructokinase binds the substrate fructose-6-phosphate with a K_M of 23 μM. In the presence of 300 μM phosphoenolpyruvate (red line), the K_M increase to about 200 μM. [*Data from Zhu, X., Byrnes, N., Nelson, J.W., and Chang, S.H.,* Biochemistry ***34,*** *2560–2565 (1995).*]

binds fructose-6-phosphate with hyperbolic kinetics and a K_M of 23 μM. In the presence of 300 μM of the inhibitor phosphoenolpyruvate, fructose-6-phosphate binding becomes sigmoidal and the K_M increases to about 200 μM (Fig. 7-14). The inhibitor does not affect V_{max}, but phosphofructokinase becomes less active because its apparent affinity for fructose-6-phosphate decreases. See Exercise 12 for additional views of phosphofructokinase structure and the locations of its binding sites.

How does phosphoenolpyruvate exert its inhibitory effects? The sigmoidal velocity versus substrate curve (Fig. 7-14) indicates that the phosphofructokinase active sites behave cooperatively in the presence of phosphoenolpyruvate. In each subunit, the inhibitor binds in a pocket that is separated from the fructose-6-phosphate binding site of the neighboring dimer by a loop of protein. When phosphoenolpyruvate occupies its binding site, the protein closes in around it. This causes a conformation change in which two residues in the loop switch positions: Arg 162 moves away from the fructose-6-phosphate binding site of the neighboring subunit and is replaced by Glu 161 (Fig. 7-15). This conformational switch diminishes fructose-6-phosphate binding because the positively charged side chain of Arg 162, which helps stabilize the negatively charged phosphate group of fructose-6-phosphate, is replaced by the negatively charged side chain of Glu 161, which repels the phosphate group. *The effect of phosphoenolpyruvate is communicated to the entire protein (thereby explaining the cooperative effect) because phosphoenolpyruvate binding to one subunit of phosphofructokinase affects fructose-6-phosphate binding to the neighboring subunit in the other dimer.*

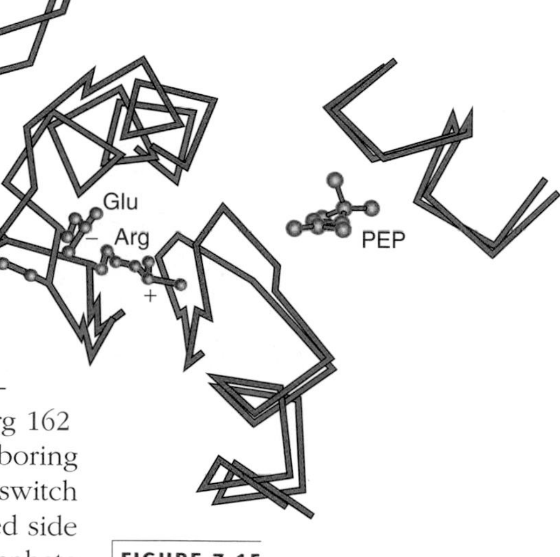

FIGURE 7-15

Conformational change upon phosphoenolpyruvate binding to phosphofructokinase.
The green structure represents the conformation of the enzyme that readily binds the substrate fructose-6-phosphate (labeled F6P). The red structure represents the enzyme with a bound allosteric inhibitor (a phosphoenolpyruvate analog labeled PEP). Phosphoenolpyruvate binding to the enzyme causes a conformational change in which Arg 162 (which forms part of the fructose-6-phosphate binding site of the neighboring subunit) changes places with Glu 161. Because the subunits act cooperatively, the inhibitor diminishes substrate binding to the entire enzyme, causing the K_M to increase. [*After Schirmer, T. and Evans, P.R.,* Nature ***343,*** *142 (1990).*]

Like hemoglobin, phosphofructokinase has two possible quaternary structures. The T (or "tense") state has low activity, and the R (or "relaxed") state has high activity. Because of the interactions between individual subunits, phosphoenolpyruvate binding causes the entire tetramer to switch to the T (low activity) conformation, as measured by fructose-6-phosphate binding affinity. Phosphoenolpyruvate is therefore known as a **negative effector** of the enzyme. Exercise 12 includes views of the conformational changes that occur in phosphofructokinase.

Phosphofructokinase is allosterically inhibited by phosphoenolpyruvate, but it can also be **allosterically activated** by ADP, a **positive effector** of the enzyme. Although ADP is a product of the phosphofructokinase reaction, it is also a general signal for the cell's need for more ATP since the metabolic consumption of ATP yields ADP:

$$ATP + H_2O \rightarrow ADP + P_i$$

Because phosphofructokinase catalyzes Step 3 of the 10-step glycolytic pathway (one of whose ultimate products is ATP), increasing phosphofructokinase activity can increase the rate of ATP produced by the pathway as a whole.

Interestingly, the activator ADP binds not to the active site (where the substrate ATP and the reaction product ADP bind) but to the same site where the inhibitor phosphoenolpyruvate binds. But because ADP is much larger than phosphoenolpyruvate, the enzyme cannot close around it, and the conformational change to the low-activity T state cannot occur. Instead, ADP binding forces Arg 162 to remain where it can stabilize fructose-6-phosphate binding (in other words, it helps keep the enzyme in the high-activity R state). The overall result is that ADP counteracts the inhibitory effect of phosphoenolpyruvate and boosts phosphofructokinase activity.

REVIEW LEARNING OBJECTIVE 10

- How is phosphofructokinase allosterically regulated?

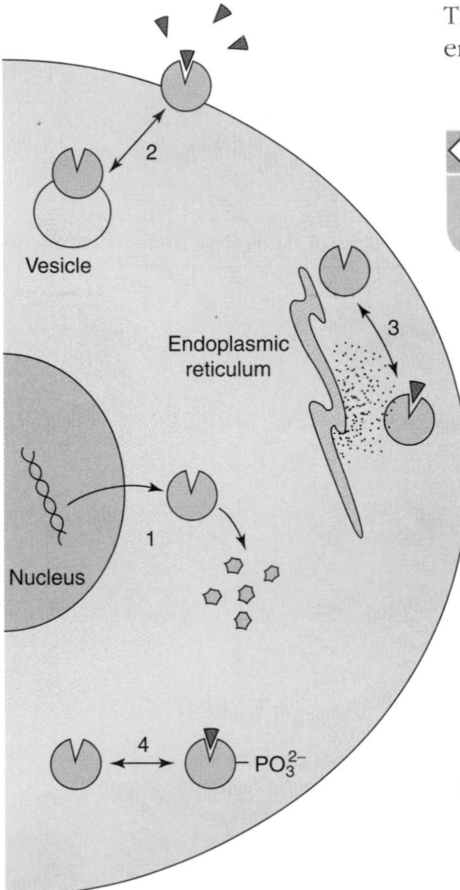

FIGURE 7-16

Some mechanisms for regulating enzyme activity.
The amount of enzyme may depend on its rate of synthesis and degradation (1). The reaction velocity may depend on the enzyme's location (2). A signal such as a burst of Ca^{2+} ions released from the endoplasmic reticulum may affect the enzyme's activity (3). Covalent modification may activate an enzyme; the enzyme may be inactive when dephosphorylated (4).

Other ways to regulate enzyme activity

So far, we have examined how small molecules that bind to an enzyme inhibit (or sometimes activate) that enzyme. These relatively simple phenomena are not the only means for regulating enzyme activity *in vivo*. Listed below and in Figure 7-16 are some additional mechanisms. Keep in mind that several of these mechanisms, along with enzyme inhibition or activation, may operate in concert to precisely adjust the activity of a given enzyme.

1. A change in the rate of an enzyme's synthesis or degradation can alter the amount of enzyme available to catalyze a reaction.
2. A change in subcellular location, for example, from an intracellular membrane to the cell surface, can bring an enzyme into proximity to its substrate and thereby increase reaction velocity. The opposite effect—sequestering an enzyme away from its substrate—dampens the reaction velocity.
3. An ionic "signal" such as a change in pH or the release of stored Ca^{2+} ions can activate or deactivate an enzyme by altering its conformation.
4. Covalent modification of an enzyme can affect the enzyme's K_M or k_{cat}, just as with an allosteric activator or inhibitor. Most commonly, a phosphoryl ($—PO_3^{2-}$) group or a fatty acyl (lipid) group is added to an enzyme so as to alter its catalytic activity. The effects of covalent modification are actually considered to be reversible, since cells contain enzymes that catalyze the removal of the modifying group as well as its addition.

SUMMARY

1. The activity of an enzyme can be expressed as the velocity of the reaction it catalyzes. Rate equations describe the velocity of simple unimolecular (first-order) or bimolecular (second-order) reactions in terms of a rate constant.

2. An enzyme-catalyzed reaction, which is a combination of elementary reactions, can be described by the Michaelis–Menten equation. In the Michaelis–Menten model, the overall rate of the reaction is a function of the rates of formation and breakdown of an enzyme–substrate (ES) complex. This model assumes that the concentration of substrate is much greater than the concentration of enzyme and that the concentration of ES remains constant over the course of the reaction.

3. The Michaelis constant, K_M, is a combination of the three rate constants relevant to the ES complex. It is also equivalent to the substrate concentration at which the enzyme is operating at half-maximal velocity. The maximum velocity is achieved when the enzyme is fully saturated with substrate.

4. The catalytic constant, k_{cat}, for a reaction is the first-order rate constant for the conversion of the enzyme–substrate complex to product. The quotient k_{cat}/K_M, a second-order rate constant for the overall conversion of substrate to product, indicates an enzyme's catalytic efficiency because it accounts for both the binding and catalytic activities of the enzyme.

5. Values for K_M and V_{max} (from which k_{cat} can be calculated) are often derived from Lineweaver–Burk, or double-reciprocal, plots. Not all enzymatic reactions obey the simple Michaelis–Menten model, but their kinetic parameters can still be estimated.

6. Although some substances react irreversibly with enzymes to permanently block catalytic activity, most enzyme inhibitors are reversible. The most common reversible enzyme inhibitors compete with substrate for binding to the active site, thereby increasing the apparent K_M without altering V_{max}. Transition state analogs are often effective competitive inhibitors.

7. Reversible enzyme inhibitors that affect an enzyme's K_M and V_{max} are known as mixed inhibitors, or noncompetitive inhibitors if only V_{max} is decreased.

8. Oligomeric enzymes such as bacterial phosphofructokinase are regulated by inhibitors and activators that bind to the enzyme molecule and affect all the active sites through allosteric mechanisms. The activities of other enzymes may also be regulated by changes in enzyme concentration, location, ion concentrations, and covalent modification.

CHECKLIST

1. Review the Learning Objectives listed on page 198.
2. Complete Review Exercise 3/Elementary Kinetics for background information on chemical reactions.
3. Complete Exercise 10/Enzyme Kinetics to see how K_M and V_{max} are determined and to test how their values influence an enzymatic reaction.
4. Complete Exercise 11/Enzyme Inhibition, which includes reviews of the most common forms of inhibition and their effects on enzyme kinetics.
5. Complete Exercise 12/Phosphofructokinase Regulation, which shows how the binding of an allosteric inhibitor affects the catalytic activity of phosphofructokinase.
6. Apply your knowledge by solving the problems at the end of this chapter. Check your results in the Solutions appendix.
7. Be able to define the boldfaced terms (consult the glossary at the end of the book). Test your understanding by taking the Chapter 7 quiz (www.wiley.com/college/pratt).
8. Explore the websites listed at www.wiley.com/college/pratt.
9. Consult the list of Selected Readings for background information or additional details on enzyme kinetics and inhibition.

GLOSSARY TERMS

ES complex
saturation
unimolecular reaction
bimolecular reaction
rate equation
rate constant (k)
first-order reaction
second-order reaction
steady state
Michaelis constant (K_M)
v_0
V_{max}
Michaelis–Menten equation
catalytic constant (k_{cat})
turnover number
k_{cat}/K_M
diffusion-controlled limit
catalytic perfection
Lineweaver–Burk plot
bisubstrate reaction
cooperativity
irreversible inhibitor
suicide substrate
competitive inhibition
inhibition constant (K_I)
EI complex
product inhibition
transition state analog
chain terminator
polyprotein
rational drug design
mixed inhibition
noncompetitive inhibition
allosteric inhibition
feedback inhibitor
negative effector
allosteric activation
positive effector

PROBLEMS

1. Complete the table for the following one-step reactions:

Reaction	Molecularity	Rate equation	Units of k
A $\rightarrow$ B + C			
A + B $\rightarrow$ C			
2 A $\rightarrow$ B			
2 A $\rightarrow$ B + C			

Reaction	Reaction velocity proportional to...	Order
A $\rightarrow$ B + C		
A + B $\rightarrow$ C		
2 A $\rightarrow$ B		

2. At about the time scientists began analyzing the hyperbolic velocity versus substrate curve (Fig. 7-3), Emil Fischer was formulating his lock-and-key hypothesis of enzyme action (Section 6-3). Explain why kinetic data are consistent with Fischer's model of enzyme action.

3. What portions of the velocity versus substrate curve (Fig. 7-3) correspond to zero-order and first-order processes?

4. Explain why it is usually easier to calculate an enzyme's reaction velocity from the rate of appearance of product rather than the rate of disappearance of a substrate.

5. Draw curves that show the appropriate relationship between the variables of each plot.

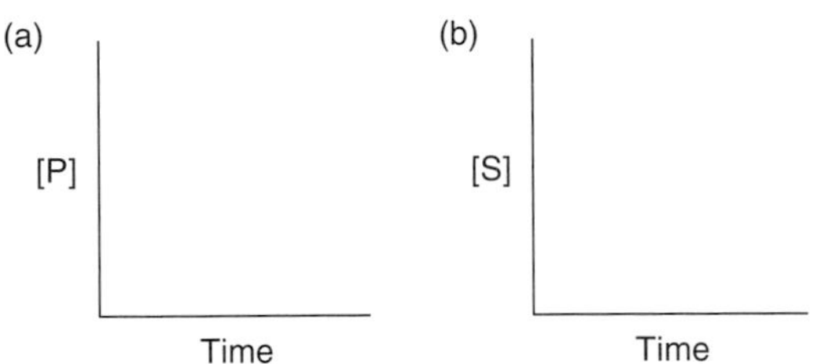

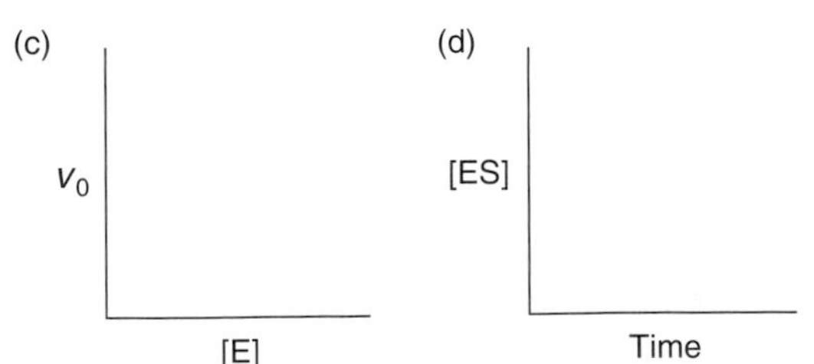

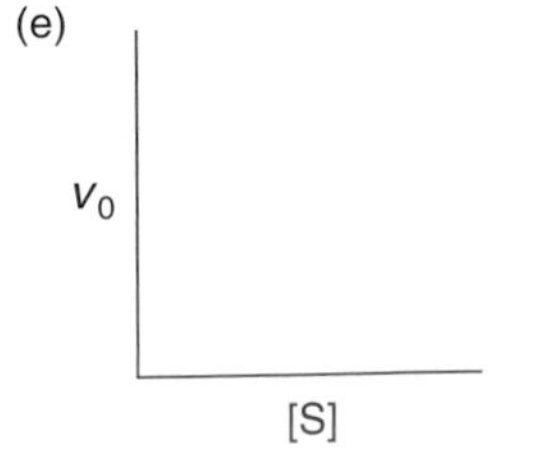

6. You are attempting to determine K_M by measuring the reaction velocity at different concentrations, but you do not realize that the substrate tends to precipitate under the experimental conditions you have chosen. How would this affect your measurement of K_M?

7. You are constructing a velocity versus substrate curve for an enzyme whose K_M is believed to be about 2 μM. The enzyme concentration is 200 nM and the substrate concentrations range from 0.1 μM to 10 μM. What is wrong with this experimental setup and how could you fix it?

8. Is it necessary for measurements of reaction velocity to be expressed in units of concentration per unit time (M/s, for example), in order to calculate an enzyme's K_M?

9. Use the plot provided to estimate values of K_M and V_{max} for an enzyme-catalyzed reaction.

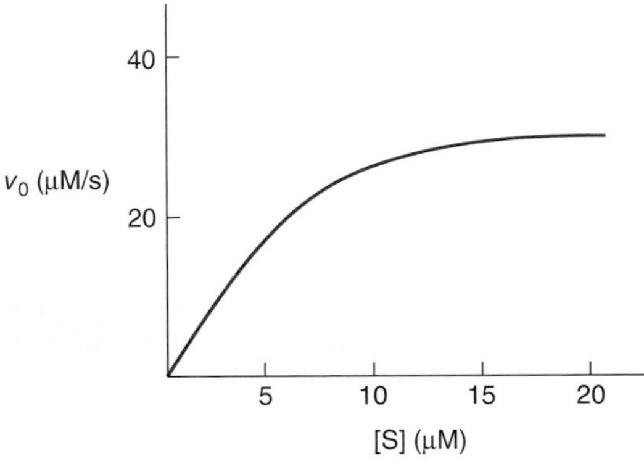

10. Estimate the K_M for each enzyme from the plot shown. Which enzyme would generate product more rapidly when (a) [S] = 1 μM and (b) [S] = 10 μM?

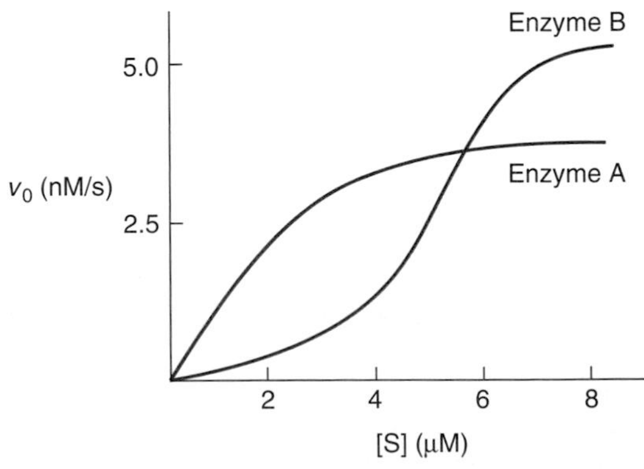

11. Is it necessary to know $[E]_T$ in order to determine (a) K_M, (b) V_{max}, and (c) k_{cat}?

12. The V_{max} of an enzyme-catalyzed reaction was 4.77 mM/s in the presence of 9.0 μM enzyme. What is the value of k_{cat}? What is the meaning of k_{cat}?

13. A Lineweaver–Burk plot is shown for three different enzyme-catalyzed reactions.
(a) Which reaction has the highest K_M value?
(b) Which reaction has the highest V_{max} value?

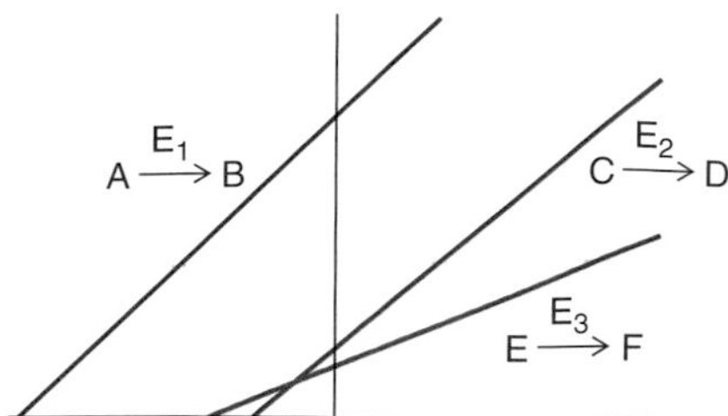

14. The K_M values for the reaction of chymotrypsin with two different substrates are given.

Substrate	K_M (M)
N-acetylvaline ethyl ester	8.8×10^{-2}
N-acetyltyrosine ethyl ester	6.6×10^{-4}

(a) Which substrate has the higher apparent affinity for the enzyme? Use what you have learned about chymotrypsin in Chapter 6 to explain why.
(b) Which substrate is likely to give a higher value for V_{max}?

15. The enzyme hexokinase acts on both glucose and fructose. The K_M and V_{max} values are given in the table. Using these data, compare and contrast the interaction of hexokinase with both of these substrates.

Substrate	K_M (M)	V_{max} (relative)
Glucose	1.0×10^{-4}	1.0
Fructose	7.0×10^{-4}	1.8

16. Consider the following reaction information about substrate S:

Reaction	Enzyme	K_M	V_{max}
$S \rightarrow P$	A	50 μM	100 nM/s
$S \rightarrow Q$	B	5 mM	120 nM/s

A reaction is carried out in which 100 μM S is added to a mixture containing equivalent amounts of enzymes A and B. After 1 minute, which reaction product will be more abundant: P or Q?

17. Use the data provided to determine whether the reactions catalyzed by enzymes A, B, and C are diffusion-controlled.

Enzyme	K_M	k_{cat}
A	0.3 mM	$5000\ s^{-1}$
B	1 nM	$2\ s^{-1}$
C	2 μM	$850\ s^{-1}$

18. The formation of an enzyme–substrate complex for a one-substrate reaction can be expressed as $E + S \rightleftharpoons ES$. For an enzymatic reaction with two substrates, A and B, the expression would be $E + A + B \rightleftharpoons EAB$. Explain why this process is approached as two bimolecular reactions rather than as a single termolecular (three-reactant) reaction.

19. What relationship exists between K_M and [S] when an enzyme-catalyzed reaction proceeds at (a) 75% V_{max} and (b) 90% V_{max}?

20. When [S] = 5 K_M, how close is v_0 to V_{max}? When [S] = 20 K_M, how close is v_0 to V_{max}? What do these results tell you about the accuracy of estimating V_{max} from a plot of v_0 versus [S]?

21. The compound indole is an inhibitor of the enzyme chymotrypsin. Using what you know about the substrate specificity of chymotrypsin and enzyme inhibition, answer the following questions:

(a) What kind of inhibitor is indole?
(b) Compare the values of K_M and V_{max} in the presence and absence of the inhibitor and explain your answer.

Indole

22. The enzyme γ-glutamylcysteine synthetase (γ-GCS) from the protozoan *Trypanosoma brucei*, the parasite that causes African sleeping sickness, catalyzes the first step of the biosynthesis of trypanothione, a compound the parasite requires to maintain proper redox balance.

$$\text{Aminobutyric acid} + \text{Glutamate} + \text{ATP} \xrightarrow{\gamma\text{-GCS}} \longrightarrow \longrightarrow \text{Trypanothione}$$

The K_M values for each of the substrates have been measured.

Substrate	K_M (mM)
Glutamate	5.9
Aminobutyric acid	6.1
ATP	1.4

(a) Does this reaction obey Michaelis–Menten kinetics?
(b) Describe how the K_M values for each of the three substrates were determined.
(c) How would V_{max} be achieved for the γ-GCS reaction?

[From Brekken, D.L. and Phillips, M.A., *J. Biol. Chem.* **273,** 26317–26322 (1998).]

23. How would diisopropylphosphofluoridate (DIPF) affect the apparent K_M and V_{max} of a sample of chymotrypsin?

24. Decamethonium (a muscle relaxant) is an inhibitor of acetylcholinesterase. The normal substrate for acetylcholinesterase is acetylcholine. Structures of both molecules are shown.

Acetylcholine

Decamethonium

(a) What kind of inhibitor is decamethonium? Explain.
(b) Can inhibition by decamethonium be overcome *in vitro* by adding large amounts of substrate? Explain.
(c) Does this inhibitor bind reversibly or irreversibly to the enzyme? Explain.

25. Explain why there are so few examples of inhibitors that decrease V_{max} but do not affect K_M.

26. Based on some preliminary measurements, you suspect that a sample of enzyme contains an irreversible inhibitor. You decide to dilute the sample 100-fold and remeasure the enzyme's activity. What would your results show if the inhibitor in the sample is (a) irreversible or (b) reversible?

27. Glucose-6-phosphate dehydrogenase catalyzes the following reaction:

Glucose-6-phosphate + $NADP^+$ $\longrightarrow$

6-Phosphogluconolactone + NADPH

The K_M for glucose-6-phosphate in yeast is 2.0×10^{-5} M. The K_M for $NADP^+$ is 2.0×10^{-6} M. The glucose-6-phosphate dehydrogenase enzyme can be inhibited by a number of cellular agents whose K_I values are listed in the table below.

Inhibitor	**K_I (M)**
Inorganic phosphate	1.0×10^{-1}
Glucosamine-6-phosphate	7.2×10^{-4}
NADPH	2.7×10^{-5}

(a) Which is the most effective inhibitor? Explain.
(b) Which inhibitor(s) is(are) likely to be completely ineffective under normal cellular conditions? Explain.

28. A reaction is carried out in the presence and absence of an inhibitor, and initial velocities are plotted versus substrate concentration. The results are shown in the graph. What type of inhibitor is this? Explain.

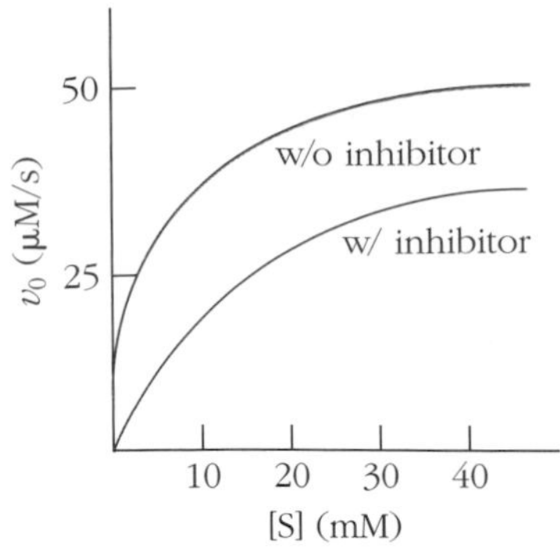

29. The redox state of the cell (the likelihood that certain groups will be oxidized or reduced) is believed to regulate the activities of some enzymes. Explain how the reversible formation of an intramolecular disulfide bond (—S—S—) from two Cys SH groups could affect the activity of an enzyme.

30. The bacterial enzyme proline racemase catalyzes the interconversion of two isomers of proline:

The compound shown below is an inhibitor of proline racemase. Explain why.

31. Cytidine deaminase catalyzes the following reaction:

Cytidine → Uridine

Both of the compounds shown below inhibit the reaction.

A B

The compounds have K_I values of 3×10^{-5} M and 1.2×10^{-12} M. Assign the appropriate K_I value to each inhibitor. Which compound is the more effective inhibitor? Give a structural basis for your answer.

32. Aspartate transcarbamoylase (ATCase) catalyzes the formation of *N*-carbamoyl aspartate from carbamoyl phosphate and aspartate, a step in the multienzyme process that synthesizes cytidine triphosphate (CTP).

Carbamoyl phosphate + Aspartate $\xrightarrow{\text{ATCase}}$
N-Carbamoylaspartate $\longrightarrow \longrightarrow \longrightarrow$
UMP $\longrightarrow \longrightarrow$ UTP $\longrightarrow$ CTP

Kinetic studies of ATCase activity as a function of aspartate concentration yield the results shown in the graph.

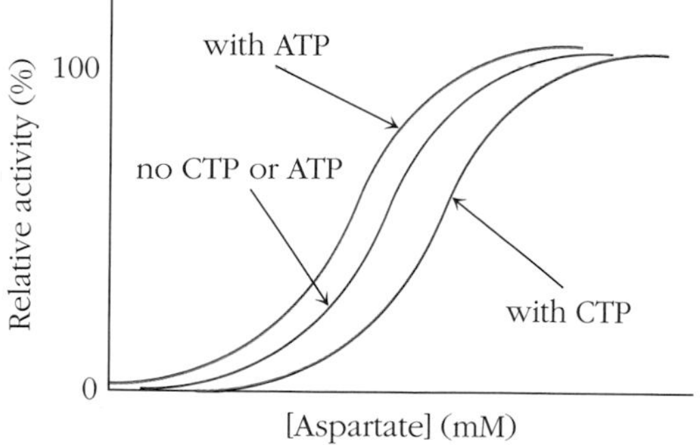

(a) Is ATCase an allosteric enzyme? How do you know?
(b) What kind of an effector is CTP? Explain. What is the biological significance of CTP's effect on ATCase?
(c) What kind of an effector is ATP? Explain. What is the biological significance of ATP's effect on ATCase?

33. Subtilisin is an alkaline serine protease produced by bacteria of the *Bacillus* species and has important industrial applications in the areas of detergents and food processing. Protein engineers have carried out site-directed mutagenesis studies on subtilisin E in an attempt to improve its catalytic activity. In one mutant, the isoleucine at position 31 was replaced with leucine. (Ile 31 was chosen because X-ray diffraction studies had shown that Ile 31 was adjacent to the Asp residue in the Asp–His–Ser catalytic triad.)

The catalytic activity of the Leu 31 mutant was compared to the Ile 31 wild-type enzyme. Both enzymes were assayed for activity using an artificial substrate called N-succinyl-Ala-Ala-Pro-Phe-*p*-nitroanilide (AAPF). The results are shown in the table.

Enzyme	K_M (mM)	k_{cat} (s^{-1})	k_{cat}/K_M ($mM^{-1} \cdot s^{-1}$)
Ile 31 wild-type subtilisin E	1.9 ± 0.2	21 ± 4	11
Leu 31 mutant subtilisin E	2.0 ± 0.3	120 ± 15	60

(a) Explain how the experimenters measured the velocity of the reaction using the AAPF substrate.
(b) What effect did the replacement of Leu for Ile at position 31 have on the catalytic activity of subtilisin E for the hydrolysis of AAPF? Comment on the K_M and k_{cat} values.
(c) Both mutants were also tested for their ability to cleave peptide bonds in the "natural" substrate casein, a protein isolated from milk. The results are shown in the table. Compare the activities of the two enzymes for the casein substrate.

Enzyme	Specific activity (units/mg)
Ile 31 wild-type subtilisin E	109 ± 9
Leu 31 mutant subtilisin E	297 ± 30

(d) Why do you think the substitution of Leu for Ile caused the catalytic efficiency of the enzyme to change?
(e) Why do you think subtilisin would be an effective additive to a detergent to remove protein-based stains from clothing such as milk or blood?

[From Takagi, H., Morinaga, Y., Ikemura, H., Inouye, M., *J. Biol. Chem.* **263,** 19592–19596 (1988).]

34. Aspartate aminotransferase (AspAT) catalyzes the following reaction:

NH_3^+—CH—COO^- (with CH_2—COO^- side chain) Aspartate $\xrightarrow{\text{Asp-AT}}$ COO^-—C=O—CH_2—COO^- Oxaloacetate
(α-Keto acid → α-Amino acid)

The AspAT enzyme has two active-site arginines, Arg 386 and Arg 292, which interact with the α-carboxylate and β-carboxylate groups on the aspartate substrate, respectively. Investigators studied the mechanism of AspAT in more detail by constructing mutant AspAT enzymes in which either or both of the essential arginines was replaced with a lysine residue. The kinetic parameters for the wild-type enzyme and mutant enzymes are shown in the table.

Enzyme	K_M Aspartate (mM)	k_{cat} (s^{-1})
Wild-type Asp AT(Arg 292 Arg 386)	4	530
Mutant Asp AT(Lys 292 Arg 386)	326	4.5
Mutant Asp AT(Arg 292 Lys 386)	72	9.6
Mutant Asp AT(Lys 292 Lys 386)	300	0.055

(a) Compare the ability of the aspartate substrate to bind to the wild-type and mutant enzymes.
(b) Why does replacement of an arginine for a lysine have an effect on substrate binding to AspAT?
(c) Evaluate the catalytic efficiency of the wild-type and mutant enzymes.
(d) Why does the replacement of arginine for lysine affect the catalytic activity of the enzyme?

[From Vacca, R.A., Giannattasio, S., Graber, R., Sandmeier, E., Marra, E., and Christen, P., *J. Biol. Chem.* **272,** 21932–21937 (1997).]

35. Protein phosphatase 1 (PP1) catalyzes a reaction which yields products that are important in regulating cell division. Consequently, PP1 is a possible drug target to treat certain types of cancers. The PP1 enzyme acts to hydrolyze a phosphate group from a specific substrate. One of PP1's substrates is myelin basic protein (MBP). The reaction is shown below:

$$\text{MBP-phosphate} \xrightarrow{\text{PP1}} \text{MBP} + \text{P}_i$$

The activity of PP1 was measured in the presence and absence of the inhibitor phosphatidic acid (PA). The concentration of PA was 300 nM.

[MBP] (mg/mL)	v_0 without PA (nmol P_i released · mL^{-1} · min^{-1})	v_0 with PA (nmol P_i released · mL^{-1} · min^{-1})
0.010	0.0209	0.00381
0.015	0.0335	0.00620
0.025	0.0419	0.00931
0.050	0.0838	0.0140

(a) Use the data provided to construct a Lineweaver–Burk plot for the PP1 enzyme in the presence and absence of PA. What kind of inhibitor is PA?
(b) Report the K_M and V_{max} values for PP1 in the presence and absence of the inhibitor.
(c) Calculate the value of K_I.

[From Kishikawa, K., Chalfant, C.E., Perry, D.K., Bielawska, A., and Hannun, Y.A., *J. Biol. Chem.* **274,** 21335–21341 (1999).]

36. A detailed understanding of the mechanism of viral replication of the human immunodeficiency virus-1 (HIV-1) will allow investigators to design more effective therapeutic treatment regimens for persons with HIV-1 and AIDS. One study has shown that a protein produced by the virus, called p6*, is an inhibitor of the HIV-1 protease, an enzyme important in the viral life cycle. The investigators measured the activity of the HIV-1 protease in the presence and absence of p6* by using an assay involving an artificial substrate, as shown here.

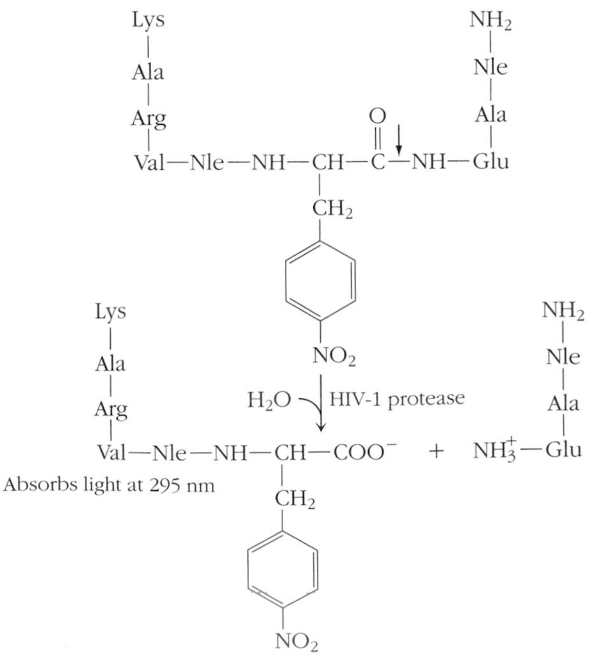

Lys–Ala–Arg–Val—Nle—NH—CH(CH₂–C₆H₄–NO₂)—C(=O)↓NH—Glu–Ala–Nle–NH₂

H_2O ↓ HIV-1 protease

Lys–Ala–Arg–Val—Nle—NH—CH(CH₂–C₆H₄–NO₂)—COO⁻ + NH_3^+—Glu–Ala–Nle–NH₂

Absorbs light at 295 nm

Kinetic assays were carried out in the presence (10 μM) and in the absence of p6*. The data are shown in the table.

[S] (μM)	v_0 without p6* (nmol · min^{-1})	v_0 with p6* (nmol · min^{-1})
10	4.63	2.70
15	5.88	3.46
20	6.94	4.74
25	9.26	6.06
30	10.78	6.49
40	12.14	8.06
50	14.93	9.71

(a) Construct a Lineweaver–Burk plot and determine the K_M and V_{max} in the presence and in the absence of inhibitor.
(b) What type of inhibitor is p6*? Explain.
(c) Calculate the K_I for the inhibitor.

[From Paulus, C., Hellebrand, S., Tessmer, U., Wolf, H., Kräusslich, H.-G., and Wagner, R., *J. Biol. Chem.* **274,** 21539–21543 (1999).]

37. Protein-tyrosine phosphatase enzymes are important regulatory enzymes, for example, in the mechanism of action of insulin. Clinical trials show that inhibitors of these enzymes are effective at treating diabetes. Phosphatase enzymes catalyze the removal of phosphate groups from specific proteins. In this study, the researchers studied the ability of vanadate to inhibit protein tyrosine phosphatase 1B (PTP1B). They measured the activity of PTP1B in the presence and absence of vanadate. They used an artificial substrate, fluorescein diphosphate (FDP), because the product of the reaction, fluorescein monophosphate (FMP), absorbs light at 450 nm. The reaction is shown:

$$\text{Fluorescein diphosphate} \xrightarrow{\text{PTP1B}} \text{Fluorescein monophosphate} + HPO_4^{2-}$$

(absorbs light at 450 nm)

The investigators measured the activity of the PTP1B enzyme in the presence and in the absence of vanadate. The data are shown in the table.

[FDP] (μM)	v_0 without vanadate (nM · s^{-1})	v_0 with vanadate (nM · s^{-1})
6.67	5.7	0.71
10	8.3	1.06
20	12.5	2.04
40	16.7	3.70
100	22.2	8.00
200	25.4	12.5

(a) Construct a Lineweaver–Burk plot using the data provided. Calculate K_M and V_{max} for PTP1B in the absence and in the presence of vanadate.

(b) What kind of inhibitor is vanadate? Explain.

(c) An alternative way to calculate K_I for an inhibitor is to measure the velocity of the enzyme-catalyzed reaction in the presence of increasing amounts of inhibitor and a constant amount of substrate. These data are shown in the table for a substrate concentration of 6.67 μM. To calculate K_I, rearrange Equation 7-28 and solve for α. Then construct a graph plotting α versus [I]. Since $\alpha = 1 + [I]/K_I$, the slope of the line is equal to $1/K_I$. Determine K_I for vanadate using this method.

[vanadate] (μM)	v_0 (nM · s^{-1})
0	5.70
0.2	3.83
0.4	3.07
0.7	2.35
1	2.04
2	1.18
4	0.71

[From Huyer, G., Liu, S., Kelly, J., Moffat, J., Payette, P., Kennedy, B., Tsaprailis, G., Gresser, M., and Ramachandran, C., *J. Biol. Chem.* **272,** 843–851 (1997).]

SELECTED READINGS

Cornish-Bowden, A. and Wharton, C.W., *Enzyme Kinetics,* IRL Press (1988). [Briefly reviews the principles of enzyme kinetics.]

Fersht, A., *Structure and Mechanism in Protein Science: A Guide to Enzyme Catalysis and Protein Folding,* W.H. Freeman (1999). [Explains how enzyme structure relates to catalysis.]

Schramm, V.L., Enzymatic transition states and transition state analog design, *Annu. Rev. Biochem.* **67,** 693–720 (1998). [Describes how transition-state analog inhibitors shed light on enzyme mechanisms.]

Segel, I.H., *Enzyme Kinetics,* Wiley (1975). [Focuses on the equations that describe enzyme activity and various forms of inhibition.]

Wlodawer, A. and Vondrasek, J., Inhibitors of HIV-1 protease: A major success of structure-assisted drug design, *Annu. Rev. Biophys. Biomol. Struct.* **27,** 249–284 (1998). [Reviews the development of some HIV-1 protease inhibitors.]

CHAPTER 8

Biological Membranes

Hand-shaking a mixture of water and amphipathic lipids produces a cloudy suspension of lipid vesicles of varying size that tend to coalesce into larger structures surrounded by multiple layers of lipid molecules. But if the mixing is done at pressures of about 1000 atm, the resulting vesicles, or **liposomes,** are much smaller (<200 nm diameter), uniformly sized, and stable for months. Most importantly, the liposomes can encapsulate water-soluble substances, which makes them attractive as delivery systems for drugs. When applied to the skin, the liposomes penetrate the dead outer layers and then fuse with the underlying cells, releasing their contents directly into the cells. The high local concentration of a drug delivered by liposome offers a clear advantage for treating skin disorders or inducing local anesthesia, particularly when the drug would have adverse effects if administered systemically. Liposomes could potentially deliver packets of drugs to other tissues, for example, a dose of toxins to a tumor. However, a great challenge lies in formulating the surface of the liposome so that it is addressed only to the desired target cells.

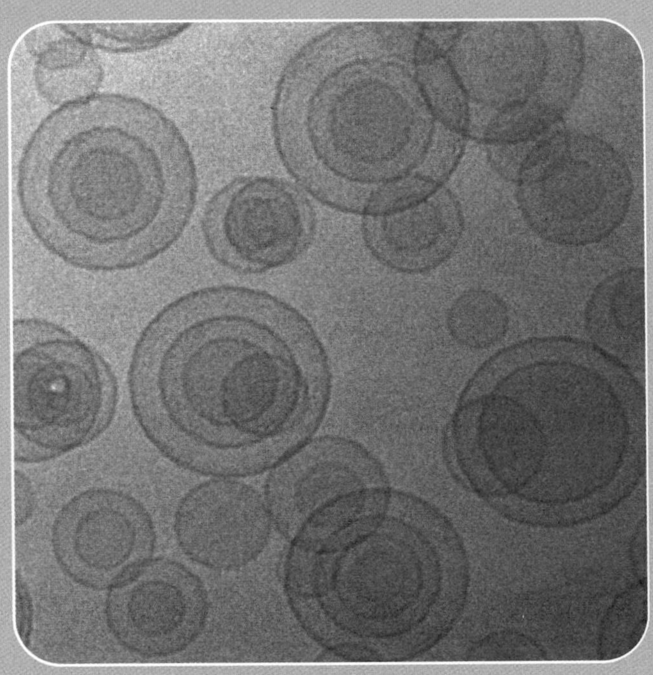

[Image acquired using an FEI Tecnai F20 Cryo TEM. Courtesy of NOF Corporation.]

LEARNING OBJECTIVES

1. Understand the structures of the major types of membrane lipids.
2. Understand bilayer fluidity and the factors that influence the melting points of fatty acyl chains.
3. Understand why transverse diffusion is much slower than lateral diffusion.
4. Understand the different ways that a polypeptide segment can traverse a membrane.
5. Understand that there are different types of membrane proteins.
6. Understand the fluid mosaic model and its refinements.
7. Understand that the distribution of ions on either side of a membrane generates a membrane potential.
8. Understand how ion channels participate in membrane depolarization.
9. Understand the overall structures of different types of transport proteins.
10. Understand how a neurotransmitter is released at a synapse.
11. Understand the role of proteins in membrane fusion.
12. Understand the bilayer changes that occur during exocytosis.

THIS CHAPTER IN CONTEXT

Chapter 8 completes part one of this book, which describes the macromolecular components of cells. We have introduced nucleic acid structure and function, and we have examined proteins, including their functions as oxygen carriers, as cytoskeletal and motor proteins, and as enzymes. In this chapter, we examine another major cell component, the membrane, which consists of both lipids and proteins. We will use nerve signaling to illustrate some of the major functions of biological membranes, since virtually every step involved in generating and transmitting a nerve impulse is tied to a physical property of the lipid bilayer or to the action of membrane-associated proteins. A firm understanding of membrane structure and function also serves as an introduction to other processes that occur largely on membranes, including mitochondrial oxidation reactions and the capture of solar energy in photosynthesis (covered in Chapters 12 and 13).

OPERATION OF THE NERVOUS SYSTEM RELIES ON THE ABILITY OF neurons to transmit signals from cell to cell. This phenomenon is fundamentally electrical in nature and results from the regulated flow of charged particles across the cells' plasma membranes. As noted in Section 2-2, all animal cells—including neurons—maintain intracellular ion concentrations that differ from those outside the cell (see Fig. 2-13). For example, intracellular sodium ion concentrations are much lower than extracellular sodium ion concentrations. The opposite is true for potassium ions. The positive charges of these cations are balanced mostly by extracellular chloride ions and by intracellular proteins (many of which have a net negative charge).

The distributions of Na^+ and K^+ ions are not at equilibrium. To reach equilibrium, Na^+ would have to enter the cell, by spontaneously moving down its concentration gradient. Likewise, K^+ would have to exit the cell, also moving down its concentration gradient (Fig. 8-1). However, the distributions of the ions do *not* change, because the plasma membrane presents a barrier to their diffusion. This impermeability is due to the hydrophobic nature of the lipids that make up the membrane. The ion gradients themselves are established and maintained by proteins that are associated with the membrane. In the following sections, we will explore the structures of lipids, membranes, and membrane proteins. We will also examine how membrane proteins in neurons can mediate changes in ion distribution and membrane shape that are essential for neuronal signaling.

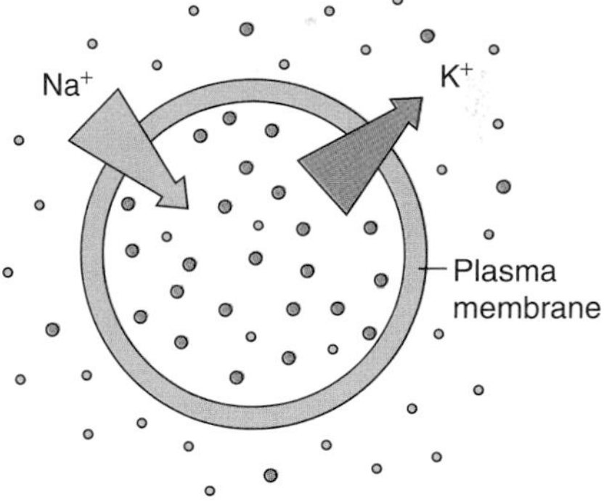

FIGURE 8-1

Distribution of Na^+ and K^+ ions in an animal cell.
The extracellular Na^+ concentration (about 150 mM) is much greater than the intracellular concentration (about 12 mM), whereas the extracellular K^+ concentration (about 4 mM) is much less than the intracellular concentration (about 140 mM). If the plasma membrane were completely permeable to ions, Na^+ would flow into the cell down its concentration gradient (purple arrow), and K^+ would flow out of the cell down its concentration gradient (orange arrow).

1 THE LIPID BILAYER

The fundamental component of a biological membrane is the **lipid bilayer,** a two-dimensional array of lipid molecules. Loosely defined, a **lipid** is a molecule that is soluble in nonpolar media but poorly soluble in water (that is, a hydrophobic molecule). However, the lipids that occur in membranes are not completely hydrophobic: They are amphipathic with hydrophobic tails attached to polar or charged head groups. *This allows them to form a two-*

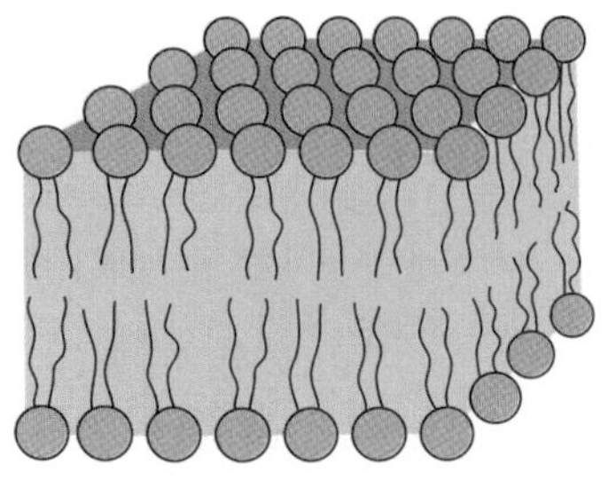

FIGURE 8-2
A lipid bilayer.

layer structure in which the tails associate with each other, out of contact with water, and the head groups interact with the aqueous solution (Fig. 8-2).

The beauty of the bilayer as a barrier for biological systems is that it forms spontaneously (without the input of free energy) due to the influence of the hydrophobic effect, which favors the aggregation of nonpolar groups that are energetically costly to individually hydrate. And because a bilayer self-seals to enclose an aqueous compartment (see page 33), it is a feature of all cells and eukaryotic organelles.

Bilayers contain different kinds of lipids

Most membrane lipids contain derivatives of **fatty acids,** which are long-chain carboxylic acids (at physiological pH, they are ionized to the carboxylate form). These molecules may contain up to 24 carbon atoms, but the most common fatty acids in plants and animals are the even-numbered C_{16} and C_{18} species such as palmitate and stearate:

Palmitate   Stearate

Such molecules are called **saturated fatty acids** since all their tail carbons are "saturated" with hydrogen. **Unsaturated fatty acids** (which contain one or more double bonds) such as oleate and linoleate are also common in biological membranes:

Oleate   Linoleate

TABLE 8-1 Some common fatty acids

Number of carbon atoms	Common name	Systematic name[a]	Structure
Saturated fatty acids			
12	Lauric acid	Dodecanoic acid	$CH_3(CH_2)_{10}COOH$
14	Myristic acid	Tetradecanoic acid	$CH_3(CH_2)_{12}COOH$
16	Palmitic acid	Hexadecanoic acid	$CH_3(CH_2)_{14}COOH$
18	Stearic acid	Octadecanoic acid	$CH_3(CH_2)_{16}COOH$
20	Arachidic acid	Eicosanoic acid	$CH_3(CH_2)_{18}COOH$
22	Behenic acid	Docosanoic acid	$CH_3(CH_2)_{20}COOH$
24	Lignoceric acid	Tetracosanoic acid	$CH_3(CH_2)_{22}COOH$
Unsaturated fatty acids			
16	Palmitoleic acid	9-Hexadecenoic acid	$CH_3(CH_2)_5CH{=}CH(CH_2)_7COOH$
18	Oleic acid	9-Octadecenoic acid	$CH_3(CH_2)_7CH{=}CH(CH_2)_7COOH$
18	Linoleic acid	9,12-Octadecadienoic acid	$CH_3(CH_2)_4(CH{=}CHCH_2)_2(CH_2)_6COOH$
18	α-Linolenic acid	9,12,15-Octadecatrienoic acid	$CH_3CH_2(CH{=}CHCH_2)_3(CH_2)_6COOH$
18	γ-Linolenic acid	6,9,12-Octadecatrienoic acid	$CH_3(CH_2)_4(CH{=}CHCH_2)_3(CH_2)_3COOH$
20	Arachidonic acid	5,8,11,14-Eicosatetraenoic acid	$CH_3(CH_2)_4(CH{=}CHCH_2)_4(CH_2)_2COOH$
20	EPA	5,8,11,14,17-Eicosapentaenoic acid	$CH_3CH_2(CH{=}CHCH_2)_5(CH_2)_2COOH$
24	Nervonic acid	15-Tetracosenoic acid	$CH_3(CH_2)_7CH{=}CH(CH_2)_{13}COOH$

[a]Numbers indicate the starting position of the double bond; the carboxylate carbon is in position 1.

In these molecules, the double bond usually has the *cis* configuration (in which the two hydrogens are on the same side). Some common saturated and unsaturated fatty acids are listed in Table 8-1.

Free fatty acids are relatively scarce in biological systems. Instead, they are usually esterified, for example, to glycerol:

```
CH2—CH—CH2
 |    |    |
OH   OH   OH
```

Glycerol

The fats and oils found in animals and plants are **triacylglycerols** (sometimes called **triglycerides**) in which the **acyl groups** (the R—CO— groups) of three fatty acids are esterified to the three hydroxyl groups of glycerol:

```
CH2—CH—CH2   ]
 |    |    |    ] Glycerol
 O    O    O    ]
 |    |    |    ]
C=O  C=O  C=O   ]
 |    |    |    ]
CH2  CH2  CH2   ] 3 fatty
 :    :    :    ] acyl groups
CH3  CH3  CH3   ]
```

The three fatty acids of a given triacylglycerol may be the same or different. For reasons that are outlined below, triacylglycerols do not form bilayers and so are not important components of biological membranes. However, they do aggregate in large globules, serving as a storage depot for fatty acids that can be broken down to release metabolic energy (these reactions are described in Section 14-1).

Among the major lipids of biological membranes are the **glycerophospholipids,** which contain a glycerol backbone with fatty acyl groups esterified at positions 1 and 2 and a phosphate derivative, called a head group, esterified at position 3. As in triacylglycerols, the fatty acyl components of glycerophospholipids may vary from molecule to molecule. These lipids are usually named according to their head group, for example,

$$\underset{\underset{R}{|}\,C=O\ \ \underset{R}{|}\,C=O}{CH_2(O{-})-CH(O{-})-CH_2}{-}O{-}P(=O)(O^-){-}O{-}CH_2{-}CH_2{-}\overset{+}{N}(CH_3)_3$$

Phosphatidylcholine

$$\underset{\underset{R}{|}\,C=O\ \ \underset{R}{|}\,C=O}{CH_2(O{-})-CH(O{-})-CH_2}{-}O{-}P(=O)(O^-){-}O{-}CH_2{-}CH_2{-}\overset{+}{N}H_3$$

Phosphatidylethanolamine

$$\underset{\underset{R}{|}\,C=O\ \ \underset{R}{|}\,C=O}{CH_2(O{-})-CH(O{-})-CH_2}{-}O{-}P(=O)(O^-){-}O{-}CH_2{-}CH(OH){-}CH_2OH$$

Phosphatidylglycerol

$$\underset{\underset{R}{|}\,C=O\ \ \underset{R}{|}\,C=O}{CH_2(O{-})-CH(O{-})-CH_2}{-}O{-}P(=O)(O^-){-}O{-}CH_2{-}CH(NH_3^+){-}COO^-$$

Phosphatidylserine

The relatively large and polar head groups give most glycerophospholipids the appropriate geometry and amphipathic nature to easily form bilayers, as shown schematically in Figure 8-3.

Other lipids with similar properties also occur in bilayers. For example, **sphingomyelins,** with phosphocholine or phosphoethanolamine head

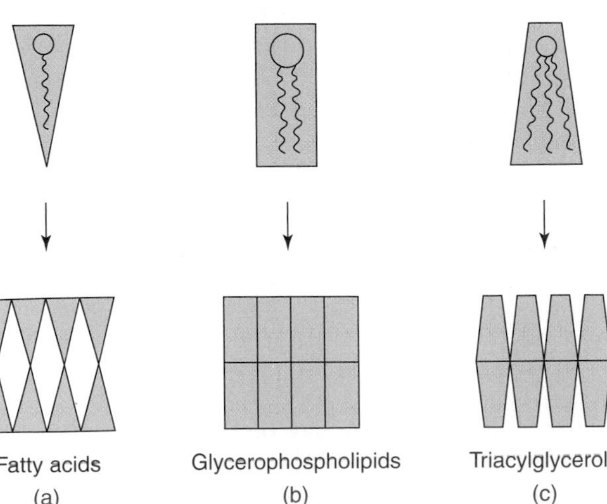

FIGURE 8-3

Bilayer-forming abilities of some lipids. (a) Free fatty acids have a relatively large head group attached to their single hydrophobic tail and therefore cannot align side-by-side to form a bilayer. (b) In most glycerophospholipids, the width of the head group is comparable to the width of the two acyl chains. The lipids can therefore form a bilayer with no voids between lipid molecules. (c) Triacylglycerols, with a small and weakly polar head group, do not form a bilayer.

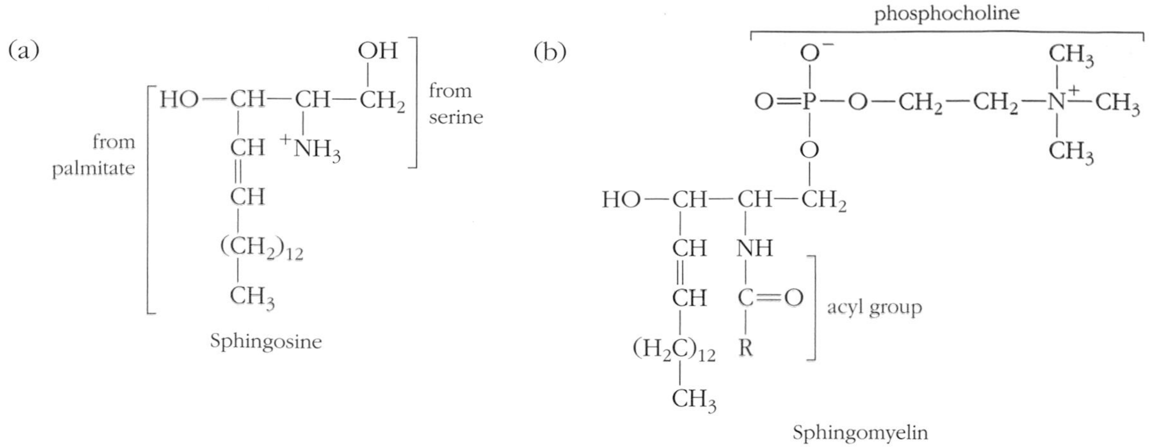

(a)

from palmitate

$HO-CH-CH(^{+}NH_3)-CH_2OH$; $CH=CH-(CH_2)_{12}-CH_3$

from serine

Sphingosine

(b)

phosphocholine

$HO-CH(-CH=CH-(H_2C)_{12}-CH_3)-CH(-NH-C(=O)-R)-CH_2-O-P(=O)(O^-)-O-CH_2-CH_2-N^{+}(CH_3)_3$

acyl group

Sphingomyelin

FIGURE 8-4

Sphingomyelin structure.
(a) The sphingosine backbone is derived from serine and palmitate. (b) The attachment of a second acyl group and a phosphocholine (or phosphoethanolamine) head group yields a sphingomyelin. Compare this structure to those of the glycerophospholipids shown on the facing page.

groups, are sterically similar to their glycerophospholipid counterparts. The major difference is that sphingomyelins are not built on a glycerol backbone. Instead, their basic component is sphingosine, a derivative of serine and the fatty acid palmitate. In a sphingolipid, a second fatty acyl group is attached via an amide bond to the serine nitrogen (Fig. 8-4).

In addition to glycerophospholipids and sphingomyelins, many other types of lipids occur in a mammalian membrane. One of the most abundant of these is cholesterol, a 27-carbon, four-ring molecule:

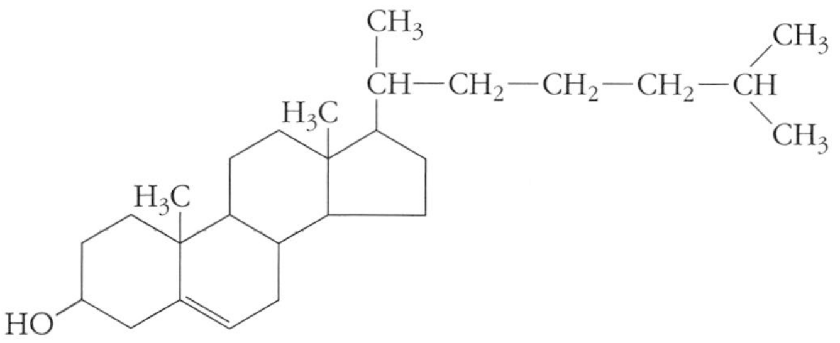

CH_3 CH_3 $CH-CH_2-CH_2-CH_2-CH$ CH_3 H_3C H_3C HO

Cholesterol

Pure cholesterol cannot form a bilayer on its own, because it is entirely hydrophobic except for a single polar hydroxyl group. It is found mostly buried in the hydrophobic region of a membrane, where its planar ring structure inserts among the acyl chains of other lipids. Cholesterol is also a metabolic precursor of steroid hormones such as estrogen and testosterone.

Other lipids are not really structural components of membranes at all but are soluble in the lipid bilayer because of their hydrophobicity. Many of these molecules are **isoprenoids,** which are built from 5-carbon units with the same carbon skeleton as isoprene:

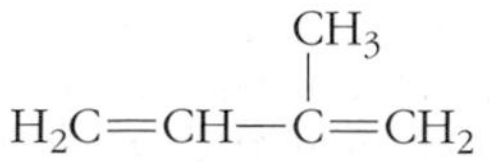

$$H_2C=CH-C(CH_3)=CH_2$$

Isoprene

For example, the isoprenoid ubiquinone is a compound that is reversibly reduced and oxidized in the mitochondrial membrane (it is described further in Section 9-1):

H_3CO, CH_3, CH_3, H_3CO, $(CH_2{-}CH{=}C{-}CH_2)_{10}H$

Isoprenoid units

Ubiquinone

The molecules known as vitamins A, D, and K are all isoprenoids that perform a variety of physiological roles not related to membrane structure (Box 8-A). We will see in Section 16-3 that some lipids serve dual roles as structural components of membranes and as signaling molecules.

REVIEW LEARNING OBJECTIVE 1

- What are the structural features that distinguish triacylglycerols, glycerophospholipids, sphingomyelins, and cholesterol?

The bilayer is a fluid structure

Naturally occurring bilayers are mixtures of many different lipids. This is one reason why *a lipid bilayer has no clearly defined geometry.* Most lipid bilayers have a total thickness of between 30 and 40 Å, with a hydrophobic core about 25 to 30 Å thick. The exact thickness varies according to the lengths of the acyl chains and how they bend and interdigitate. In addition, the head groups of membrane lipids are not of uniform size, and their distance from the membrane center depends on how they nestle in with neighboring head groups (Fig. 8-5).

Finally, the lipid bilayer is impossible to describe in precise terms because it is not a static structure. Rather, it is a dynamic assembly: The head groups bob up and down and the hydrocarbon tails of the lipids are in constant rapid motion. At its very center, the bilayer is as fluid as a sample of pure hydrocarbon.

It is useful to describe the fluidity of a given membrane lipid in terms of its **melting point,** the temperature of transition from an ordered crystalline

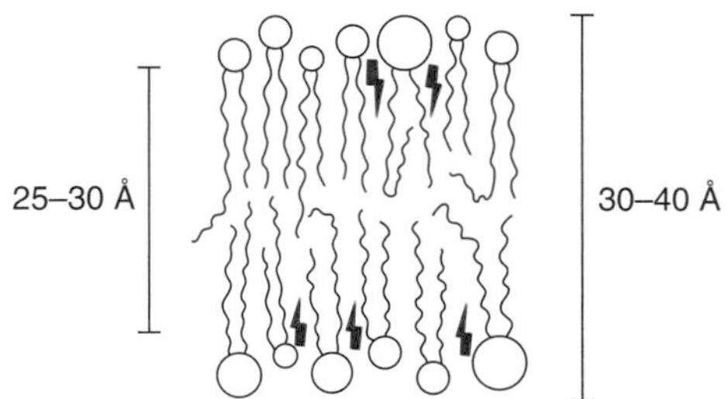

FIGURE 8-5

Dimensions of a lipid bilayer.
The thickness of the hydrophobic core depends in part on the length and bending of the acyl chains of the lipids. The geometry and charge distribution of the head groups influence their packing to form the polar outer surfaces. Note how cholesterol molecules (red) are almost entirely buried within the bilayer core.

A CLOSER LOOK

BOX 8-A The lipid vitamins A, D, and K

The plant kingdom is rich in isoprenoid compounds, which serve as pigments, molecular signals (hormones and pheromones), and defensive agents. During the course of evolution, vertebrate metabolism has co-opted several of these compounds for other purposes. The compounds have become **vitamins,** which are substances that an animal cannot synthesize but must obtain from its food. Vitamins A, D, and K are lipids, but many other vitamins are water-soluble. Other than being lipids, vitamins A, D, and K have little in common.

The first vitamin to be discovered was vitamin A, or retinol.

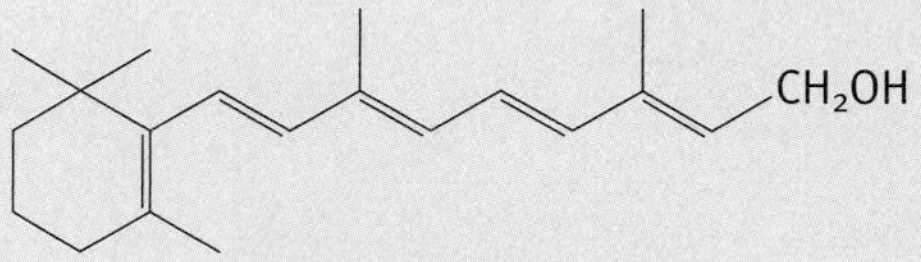

CH_2OH

Retinol
(vitamin A)

It is derived mainly from plant pigments such as β carotene (a red pigment that is present in green vegetables as well as carrots and tomatoes). Retinol is oxidized to retinal, an aldehyde, which functions as a light receptor in the eye. Light causes the retinal to isomerize, triggering an impulse through the optic nerve. A severe deficiency of vitamin A can lead to blindness. Retinal also behaves like a hormone by stimulating tissue repair. It is sometimes used to treat severe acne and skin ulcers.

The steroid derivative vitamin D is actually two similar compounds—one (vitamin D_2) derived from a plant compound, and the other (vitamin D_3) from endogenously produced cholesterol.

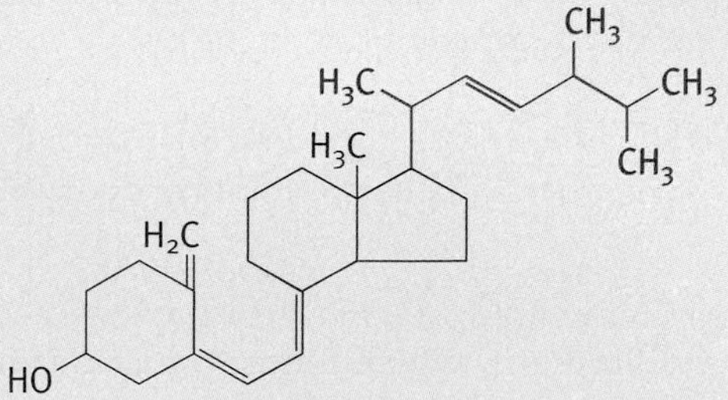

CH_3 H_3C CH_3 H_3C CH_3 H_2C HO

Vitamin D_2

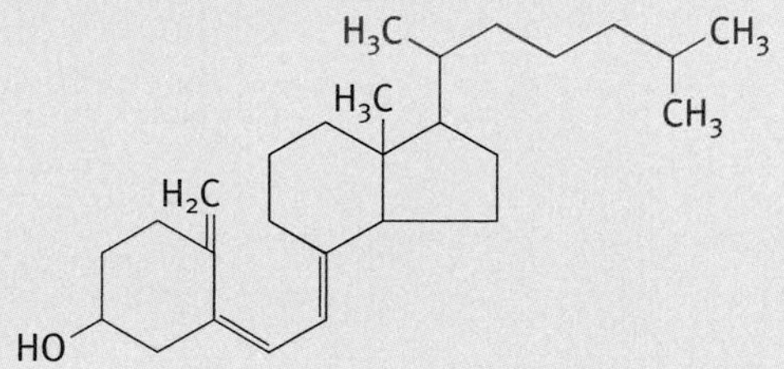

H_3C CH_3 H_3C CH_3 H_2C HO

Vitamin D_3

Ultraviolet light is required for the formation of vitamins D_2 and D_3, giving rise to the saying that sunlight makes vitamin D. Calcium absorption in the intestine is stimulated by vitamin D. The resulting high concentration of Ca^{2+} in the bloodstream promotes Ca^{2+} deposition in the bones and teeth. Rickets, a vitamin D deficiency disease characterized by stunted growth and deformed bones, is easily prevented by good nutrition and exposure to sunlight.

Vitamin K is named for the Danish word *koagulation*. It participates in the enzymatic carboxylation of Glu residues in some of the proteins involved in blood clotting (see Box 6-A). A vitamin K deficiency prevents Glu carboxylation, which inhibits the normal function of the clotting proteins, leading to excessive bleeding. Vitamin K can be obtained from green plants, as phylloquinone,

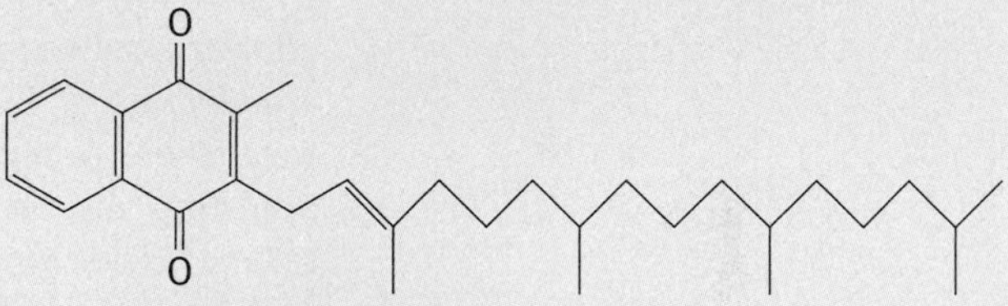

O O

Phylloquinone
(vitamin K)

But about half the daily uptake of the vitamin is supplied by intestinal bacteria. Long-term use of antibiotics, which suppresses the growth of these bacteria, may lead to a vitamin K deficiency.

Because vitamins A, D, and K are water-insoluble, they can accumulate in fatty tissues over time. Excessive vitamin D accumulation can lead to kidney stones and abnormal calcification of soft tissues. High levels of vitamin K have few adverse effects, but extremely high levels of vitamin A can produce a host of nonspecific symptoms as well as birth defects. In general, vitamin toxicities are rare and usually result from overconsumption of commercial vitamin supplements rather than from natural causes.

state to a more fluid state. In the crystalline phase, all the acyl chains in the sample pack together in van der Waals contact. In the fluid phase, the methylene (—CH_2—) groups of the acyl chains in the sample can rotate freely.

The melting point of an individual acyl chain depends on its length and degree of saturation. For a saturated acyl chain, the melting point increases with increasing chain length. This is because more free energy (a higher temperature) is required to disrupt the more extensive van der Waals interactions in a longer chain. A shorter acyl chain melts at a lower temperature because it has less surface area to make van der Waals contacts. A double bond introduces a kink into the acyl chain, so an unsaturated acyl chain is less able to pack efficiently against its neighbors.

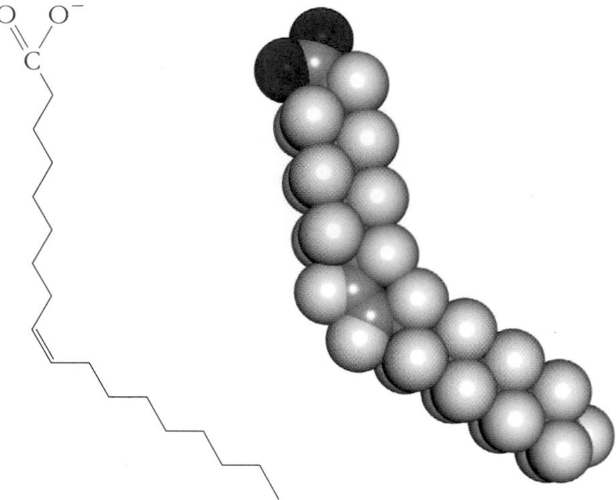

Consequently, the melting point of an acyl chain decreases as the degree of unsaturation increases.

What does all this mean for a mixed bilayer at a constant temperature? In general, *longer acyl chains tend to be less mobile (more crystalline) than shorter acyl chains, and saturated acyl chains are less mobile than unsaturated chains.* Because a fluid membrane is essential for many metabolic processes, organisms endeavor to maintain constant membrane fluidity by adjusting the lipid composition of the bilayer. For example, during adaptation to lower temperatures, an organism may increase its production of lipids with shorter and less saturated acyl chains.

The membranes of most organisms remain fluid over a range of temperatures. This is partly because biological membranes include a variety of different lipids (with different melting points) and do not undergo a sharp transition between liquid and crystalline phases, as would a sample of pure lipid. In addition, cholesterol helps maintain constant membrane fluidity over a range of temperatures through two opposing mechanisms:

1. In a bilayer of mixed lipids, cholesterol's rigid and planar ring system restricts the movements of nearby acyl chains, thereby decreasing membrane fluidity.

2. By inserting between membrane lipids, cholesterol prevents their close packing (that is, their crystallization), which tends to increase membrane fluidity.

Interestingly, different areas of a naturally occurring membrane may be characterized by different degrees of fluidity. Regions known as membrane **rafts** appear to have a near-crystalline consistency. The rafts contain tightly packed cholesterol and **sphingolipids.** Sphingomyelins, with phosphate-derivative head groups (see Fig. 8-4) are a type of sphingolipid. This group also includes cerebrosides and gangliosides, which resemble sphingomyelins but have carbohydrate head groups (and are therefore types of **glycolipids**):

A cerebroside

A ganglioside

REVIEW LEARNING OBJECTIVE 2

- Describe the overall structure of a lipid bilayer.
- How do chain length and saturation influence the melting point of a fatty acid?
- What is the role of cholesterol in maintaining bilayer fluidity?

Natural bilayers are asymmetric

The two leaflets of the bilayer in a biological membrane seldom have identical compositions. For example, sphingolipids—with their carbohydrate head groups—occur almost exclusively on the outer leaflet of the plasma membrane, facing the extracellular space. The polar head groups of phosphatidylcholine also usually face the cell exterior, whereas phosphatidylserine is usually found in the inner leaflet.

The distinct compositions of the inner and outer leaflets are preserved by the extremely slow rate at which most membrane lipids undergo **transverse diffusion** or **flip-flop** (the movement from one leaflet to the other):

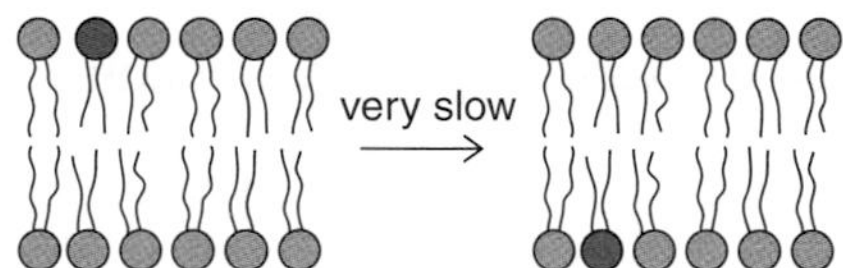

This movement is thermodynamically unfavorable since it would require the passage of a solvated polar head group through the hydrophobic interior of the bilayer. However, cells can and do move certain lipids between leaflets with the assistance of enzymes called **translocases** or **flippases.** Lipid molecules undergo rapid **lateral diffusion,** that is, movement within one leaflet:

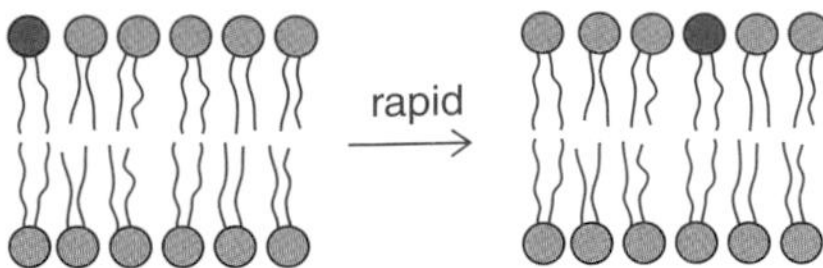

In a membrane bilayer, a lipid changes places with its neighbors as often as 10^7 times per second.

REVIEW LEARNING OBJECTIVE 3

- What are the differences between transverse and lateral diffusion?

2 MEMBRANE PROTEINS

Go to **Exercise 13/Membrane Proteins** to see how the structures of two types of membrane proteins are specialized for interaction with the lipid bilayer. These structures and the questions based on them address Learning Objective 4.

Biological membranes consist of proteins as well as lipids. On average, a membrane is about 50% protein by weight, but this value varies widely, depending on the source of the membrane. Some bacterial plasma membranes and organelle membranes are as much as three-quarters protein. By itself, a lipid bilayer serves as a barrier to the diffusion of polar substances, but virtually all the additional functions of a biological membrane depend on membrane proteins. For example, some membrane proteins sense exterior conditions and communicate them to the cell interior. Other membrane proteins carry out specific metabolic reactions. Still other proteins subvert the lipid barrier by providing passageways for the movement of ions and molecules from one side of the membrane to the other. Finally, some membrane proteins mediate changes in membrane structure associated with vesicle budding and fusion.

Membrane proteins fall into different groups, depending on how they are specialized for interaction with the hydrophobic interior of the lipid bilayer (Fig. 8-6). Their three-dimensional structures have been slow to come to light because they are often difficult to crystallize, even in the presence of lipid molecules.

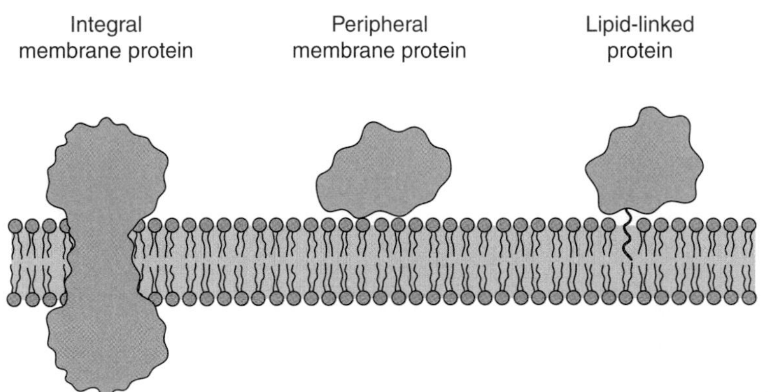

FIGURE 8-6

Types of membrane proteins. This schematic cross-section of a membrane shows an integral protein spanning the width of the membrane, a peripheral membrane protein associated with the membrane surface, and a lipid-linked protein whose attached hydrophobic tail is incorporated into the lipid bilayer.

Integral membrane proteins span the bilayer

In the membrane proteins known as **integral** or **intrinsic membrane proteins,** a portion of structure is fully buried in the lipid bilayer. These proteins are customarily contrasted with **peripheral** or **extrinsic membrane proteins,** which are more loosely associated with the membrane via interactions with lipid head groups or integral membrane proteins (see Fig. 8-6). Except for their weak affiliation with the membrane, peripheral proteins are not notably different from ordinary soluble proteins.

(a) Pro–Glu–Trp–Ile–Trp–Leu–Ala–Leu–Gly–Thr–Ala–Leu–Met–Gly–
Leu–Gly–Thr–Leu–Tyr–Phe–Leu–Val–Lys–Gly–Met

(b)

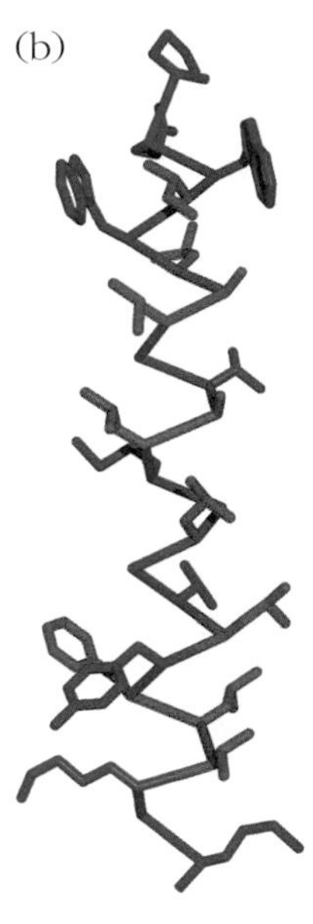

FIGURE 8-7

A membrane-spanning α helix.
(a) A portion of the amino acid sequence from the protein bacteriorhodopsin. (b) Three-dimensional structure of the same sequence. Hydrophobic side chains are colored green.

All but a few integral membrane proteins completely span the lipid bilayer, so they are exposed to the hydrophobic interior as well as the aqueous environment on each side of the membrane. The solvent-exposed portions of an integral membrane protein are typical of other proteins: a polar surface surrounding a hydrophobic core. However, the portion of the protein that penetrates the lipid bilayer must have a hydrophobic surface, since the energetic cost of burying a polar protein group (and its solvating water molecules) is too great.

An α helix can cross the bilayer

One way for a polypeptide chain to cross a lipid bilayer is by forming an α helix whose side chains are all hydrophobic. The hydrogen-bonding tendencies of the amino and carboxyl groups of the backbone are satisfied through hydrogen bonding in the α helix (see Fig. 4-7). The hydrophobic side chains project outward from the helix to mingle with the acyl chains of the lipids.

To span a 30-Å hydrophobic bilayer core, an α helix must contain at least 20 amino acids. A transmembrane helix is often easy to spot by its sequence: It is rich in highly hydrophobic amino acids such as Ile, Leu, Val, and Phe. Polar aromatic groups (Trp and Tyr) and Asn and Gln occur where the helix approaches the more polar lipid head groups. Highly polar residues such as Asp, Glu, Lys, and Arg often mark the point where the polypeptide leaves the membrane and is exposed to the solvent (Fig. 8-7).

Many integral membrane proteins contain several membrane-spanning α helices bundled together (Fig. 8-8). These α helices interact much like the left-

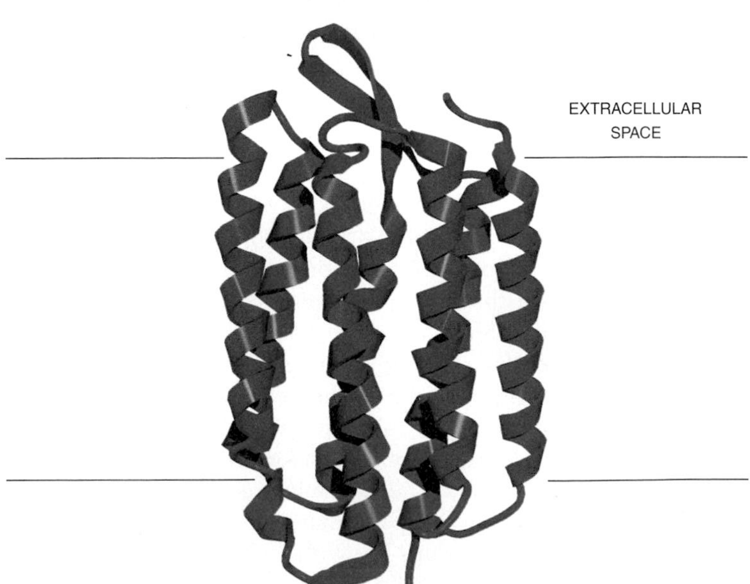

FIGURE 8-8

Bacteriorhodopsin.
This integral membrane protein consists of a bundle of seven membrane-spanning α helices connected by loops that project into solution on each side of the membrane. The horizontal lines approximate the outer surfaces of the membrane. [*Structure (pdb 1QHJ) determined by H. Belrhali, P. Nollert, A. Royant, C. Menzel, J.P. Rosenbusch, E.M. Landau, and E. Pebay-Peyroula.*]

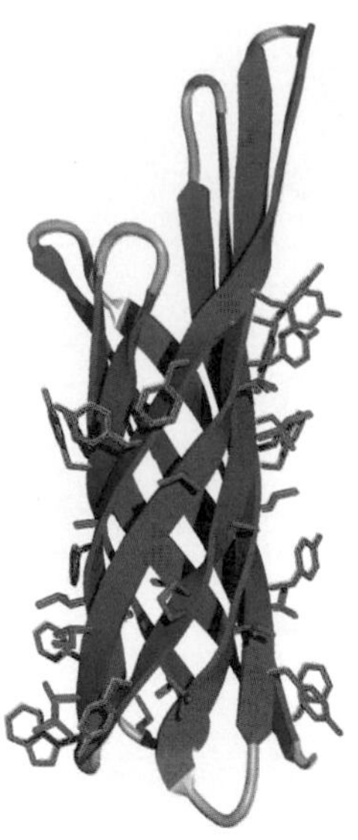

FIGURE 8-9

A membrane-spanning β barrel. The eight strands of this protein, known as *E. coli* OmpX, are fully hydrogen bonded where they span the width of the bilayer. Hydrophobic side chains (green) on the barrel exterior face the bilayer core. Aromatic residues (gold) are located mostly near the lipid head groups. [*Structure (pdb 1QJ9) determined by J. Vogt and G.E. Schulz.*]

handed coiled coils in keratin (see Section 5-5). Some of the helix–helix interactions involve the electrostatic pairing of polar residues, but the surface of the helix bundle—where it contacts the lipid tails—is exclusively hydrophobic. The structure of the α-helical membrane protein bacteriorhodopsin is explored in greater detail in Exercise 13.

The transmembrane β barrel

A polypeptide that crossed the membrane as a β strand would leave its hydrogen-bonding backbone groups unsatisfied. However, if several β strands together form a fully hydrogen-bonded β sheet, they can cross the membrane in an energetically favorable way. In order to maximize hydrogen bonding, the β sheet must close up on itself to form a **β barrel.**

The smallest possible β barrel contains eight strands, although larger barrels (up to 22 strands) also occur. The exterior surface of the barrel includes a band, about 22 Å wide, of hydrophobic side chains. This band is flanked on each side by aromatic side chains, which are more polar and form an interface with the lipid head groups (Fig. 8-9). Exercise 13 shows a β barrel membrane protein.

Because the side chains in a β sheet point alternately to each face, some side chains in a β barrel point into the barrel interior, and others face the lipid bilayer. The absence of a discrete stretch of hydrophobic residues, as in a membrane-spanning α helix, makes it difficult to detect membrane-spanning β strands by examining a protein's sequence.

REVIEW LEARNING OBJECTIVE 4

- What are the sequence requirements for a membrane-spanning α helix and β barrel?

Lipid-linked proteins are anchored in the membrane

A second group of membrane proteins consists of **lipid-linked proteins.** Many of these are otherwise soluble proteins that are anchored in the lipid bilayer by a covalently attached lipid group. A few lipid-linked proteins also contain membrane-spanning polypeptide segments. Some lipid-linked proteins contain a fatty acyl group. For example, a myristoyl residue (from the 14-carbon saturated fatty acid myristate) may be attached via an amide bond to the N-terminal Gly residue of a protein (Fig. 8-10a). Other proteins contain a palmitoyl group (from the 16-carbon palmitate) attached to the sulfur of a Cys side chain via a thioester bond (Fig. 8-10b). Palmitoylation, in contrast to myristoylation, is reversible *in vivo*. Consequently, proteins with a myristoyl group are permanently anchored to the membrane, but proteins with a palmitoyl group may become soluble if the acyl group is removed.

Other lipid-linked proteins in eukaryotes are prenylated; that is, they contain a 15- or 20-carbon isoprenoid group linked to a C-terminal Cys residue via a thioether bond (Fig. 8-10c). The C-terminal is also usually carboxymethylated.

Finally, many eukaryotes, particularly protozoans, contain proteins linked to a lipid–carbohydrate group, known as glycosylphosphatidylinositol, at the C-terminus (Fig. 8-10d). These lipid-linked proteins almost always face the external surface of the cell and are often found in sphingolipid–cholesterol rafts.

(a) HN—CH_2—C(=O)—··· ; C=O

(b) ···—NH—CH—C(=O)—··· ; CH_2 ; S ; C=O

(c) ···—NH—CH—C(=O)—O—CH_3 ; CH_2 ; S

(d) ···—NH—CH—C(=O)—NH—CH_2—CH_2 ; R ; O—P(=O)(O$^-$)—O

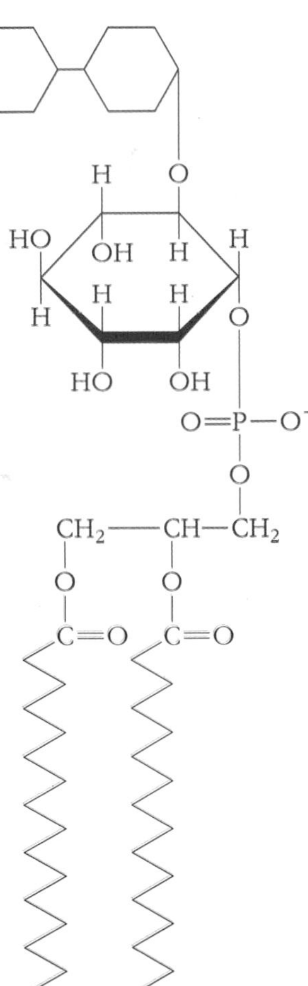

FIGURE 8-10

Attachment of lipids in some lipid-linked proteins.
Proteins are shown in blue, and the lipid anchors are in green. Other groups are purple. (a) Myristoylation. (b) Palmitoylation. (c) Prenylation. The lipid anchor is the 15-carbon farnesyl group. (d) Linkage to a glycosylphosphatidylinositol group. The hexagons represent different monosaccharide residues.

REVIEW LEARNING OBJECTIVE 5

- What are the similarities and differences between integral membrane proteins, peripheral membrane proteins, and lipid-linked proteins?

Many membrane proteins have limited mobility

A given membrane protein has a characteristic orientation; that is, it faces one side of the membrane or the other. After it has assumed its mature conformation and orientation, it does not undergo flip-flop because this would require the passage of large polar polypeptide regions through the hydrophobic bilayer core. However, lateral movement is still possible. An integral membrane protein or a lipid-linked protein can diffuse within the plane of the bilayer, albeit more slowly than a membrane lipid. This sort of movement is a key feature of the **fluid mosaic model** of membrane structure described in 1972 by S. Jonathan Singer and Garth Nicolson. According to their model, membrane proteins are like icebergs floating in a lipid sea (Fig. 8-11).

Over the years, the fluid mosaic model has remained generally valid, although it has been refined. For example, *many membrane proteins do not dif-*

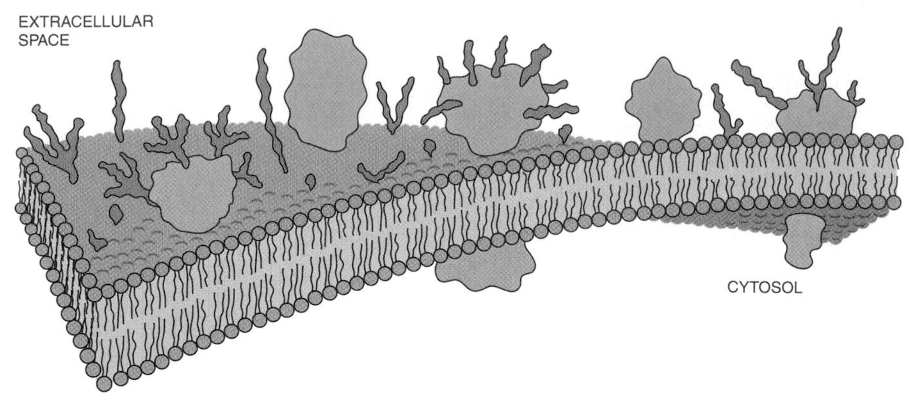

FIGURE 8-11

The fluid mosaic model of membrane structure.
According to this model, integral membrane proteins (blue) float in a sea of lipids and can move laterally but cannot undergo transverse movement (flip-flop). The orange structures are the carbohydrate chains of glycolipids and glycoproteins.

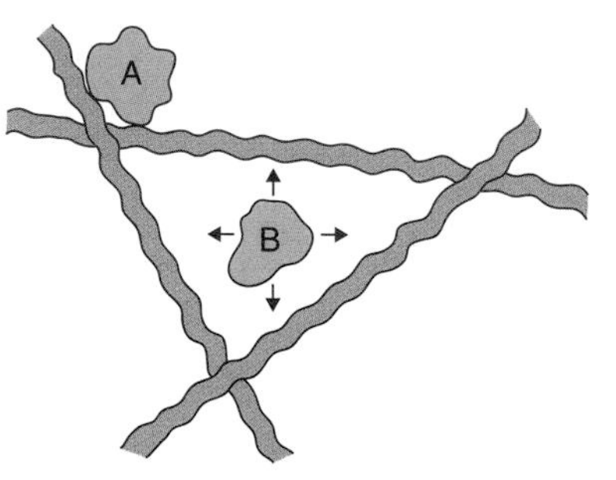

FIGURE 8-12

Limitations on the mobility of membrane proteins.
A protein (labeled A) that interacts tightly with the underlying cytoskeleton appears to be immobile. Another protein (B) can diffuse within a space defined by cytoskeletal proteins. Some proteins appear to diffuse throughout the membrane with no constraints.

fuse as freely as first imagined. Their movements are hindered to some degree according to whether they interact with other membrane proteins or with cytoskeletal elements that lie just beneath the membrane. Thus, a given membrane protein may be virtually immobile (if it is firmly attached to the cytoskeleton), mobile within a small area (if it is confined within a space defined by other membrane and cytoskeletal proteins), or fully free to diffuse (Fig. 8-12).

The presence of lipid rafts, another feature not described in the original fluid mosaic model, may further define the boundaries for a membrane protein. In fact, *certain lipids appear to associate specifically with certain proteins,* possibly to stabilize the protein's structure or modulate its function.

Membrane glycoproteins face the cell exterior

Like membrane lipids, membrane proteins are distributed asymmetrically between the two leaflets. For example, most lipid-linked proteins face the cell interior (glycosylphosphatidylinositol-linked proteins are an exception). The exterior face of the membrane in vertebrate cells is rich in carbohydrate-bearing glycolipids (including cerebrosides and gangliosides, shown on page 241) and **glycoproteins.** The oligosaccharide chains (polymers of monosaccharide residues) that are covalently attached to membrane lipids and proteins shroud the cell in a fuzzy coat (Fig. 8-13). When fully solvated, the highly hydrophilic carbohydrates tend to occupy a large volume.

A single protein may bear several carbohydrate chains that are attached as either ***N*-linked** or ***O*-linked oligosaccharides** (Fig. 8-14). Glycoproteins are not just associated with membranes. In many soluble proteins, particularly extracellular ones, oligosaccharides may help stabilize the protein under harsh extracellular conditions.

Monosaccharide residues can be linked to each other in different ways and in potentially unlimited sequences. This diversity, present in both glycolipids and glycoproteins, is a form of biological information. For example, the well-known ABO blood group system is based on differences in the composition of the carbohydrate components of glycolipids and glycoproteins on red blood cells (Box 8-B). Many other cells appear to recognize each other based on the carbohydrates they display on their surfaces.

REVIEW LEARNING OBJECTIVE 6

- Describe the fluid mosaic model of membrane structure.
- What factors limit membrane protein mobility?
- Why do glycoproteins and glycolipids face the cell exterior?

A CLOSER LOOK

BOX 8-B The ABO blood group system

The carbohydrates on the surface of red blood cells form 15 different blood group systems. The best known and one of the clinically important carbohydrate-classification schemes is the ABO blood group system, which has been known for about a century. Biochemically, the ABO system involves the oligosaccharides attached to sphingolipids and proteins on red blood cells and other cells.

In individuals with type A blood, the oligosaccharide has a terminal *N*-acetylated galactose group. In type B individuals, the terminal sugar is galactose. Neither of these groups appears in the oligosaccharides of type O individuals.

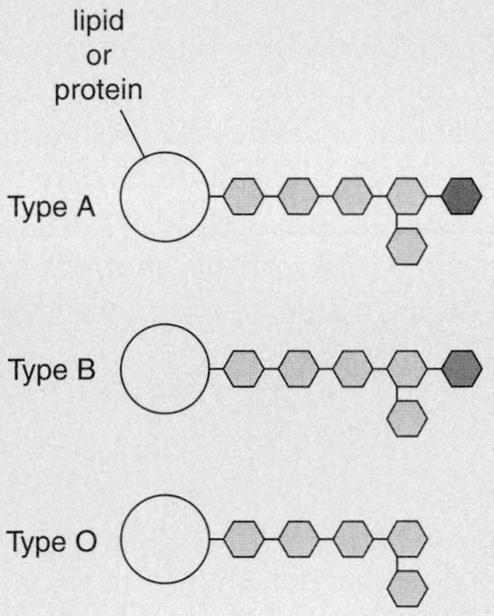

lipid or protein

Type A

Type B

Type O

Blood groups are genetically determined: Type A and B individuals have slightly different versions of an enzyme that adds the final monosaccharide residue to the oligosaccharide. Type O individuals lack the enzyme entirely and therefore lack the final residue.

Type A individuals develop antibodies that recognize and cross-link red blood cells bearing the type B oligosaccharide. Type B individuals develop antibodies to the type A oligosaccharide. Therefore, a transfusion of type B blood cannot be given to a type A individual, and *vice versa*. Individuals with type AB blood bear both types of oligosaccharides and therefore do not develop antibodies to either type. They can receive transfusions of either type A or type B blood. Type O individuals develop both anti-A and anti-B antibodies. If they receive a transfusion of type A, type B, or type AB blood, their antibodies will react with the transfused cells, which causes them to precipitate and block blood vessels. On the other hand, type O individuals are universal donors: Type A, B, or AB individuals can safely receive type O blood (these individuals do not develop antibodies to the O-type oligosaccharide because it occurs naturally in these individuals as a precursor of the A-type and B-type oligosaccharides).

FIGURE 8-13

Electron micrograph of the surface of a red blood cell. Carbohydrate chains attached to membrane lipids and proteins give the cell a fuzzy outer surface. [*Courtesy Harrison Latta, UCLA.*]

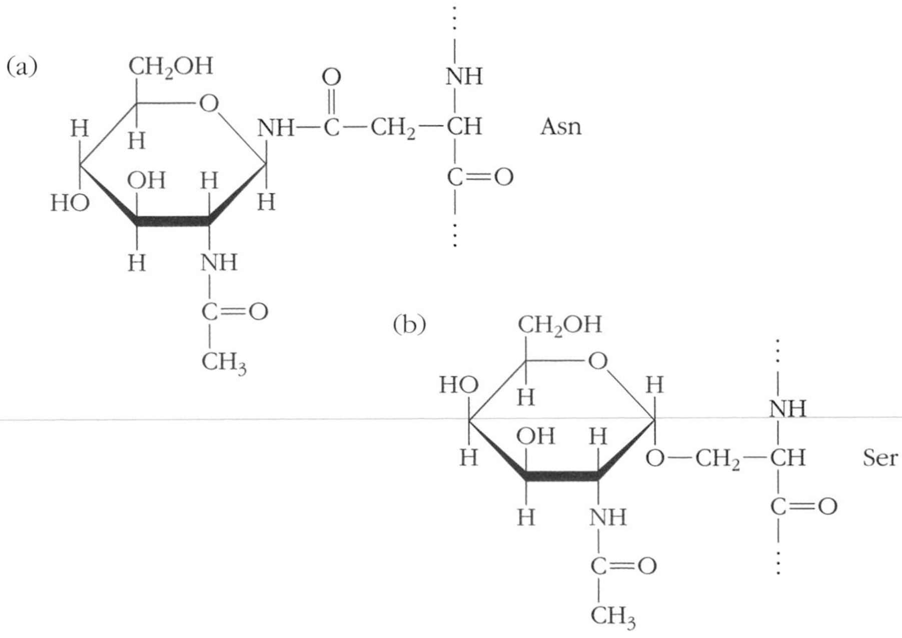

(a) Asn

(b) Ser

FIGURE 8-14

Oligosaccharide linkages in glycoproteins. (a) In *N*-linked oligosaccharides, an *N*-acetylglucosamine residue (a derivative of glucose; purple) is covalently attached to an Asn side chain. The oligosaccharide typically contains several additional monosaccharide residues linked in sequence to one of the glucosamine OH groups. (b) In *O*-linked oligosaccharides, the galactose derivative *N*-acetylgalactosamine (purple) is typically attached to the side chain of a Ser residue (as shown here) or a Thr residue. *O*-linked oligosaccharides are often much larger than *N*-linked oligosaccharides and contain fewer different types of sugar residues.

3 MEMBRANE TRANSPORT

Some of the best-understood membrane-related events occur during production of a nerve impulse, when certain integral membrane proteins act as channels for the passage of ions into and out of neurons. We begin this section by summarizing these ion movements and then examine the types of proteins that mediate transmembrane movements of ions and other substances.

Ion movements alter membrane potential

Although mammalian membranes are largely impermeable to ions, a small percentage of K^+ ions do leak out of the cell. This movement places relatively more positive charges outside the cell and leaves relatively more negative charges inside the cell. The resulting charge imbalance, though small, generates a voltage across the membrane, which is called the **membrane potential** and is symbolized $\Delta\psi$. In the simplest case, $\Delta\psi$ is a function of the ion concentration on each side of the membrane,

$$\Delta\psi = \frac{RT}{Z\mathcal{F}} \ln \frac{[\text{ion}_{in}]}{[\text{ion}_{out}]} \qquad [8\text{-}1]$$

where $\boldsymbol{R}$ is the **gas constant** ($8.3145\ \text{J} \cdot \text{K}^{-1} \cdot \text{mol}^{-1}$), T is temperature in Kelvin (20°C = 293K), $\boldsymbol{Z}$ is the net charge per ion, and $\mathcal{F}$ is the **Faraday constant,** the charge of one mole of electrons ($96{,}485\ \text{coulombs} \cdot \text{mol}^{-1}$ or $96{,}485\ \text{J} \cdot \text{V}^{-1} \cdot \text{mol}^{-1}$). $\Delta\psi$ is expressed in units of volts (V) or millivolts (mV). For a monovalent ion ($Z = 1$) at 20°C, the equation reduces to

$$\Delta\psi = 0.058 \log_{10} \frac{[\text{ion}_{in}]}{[\text{ion}_{out}]} \qquad [8\text{-}2]$$

(see Sample Calculation 8-1). In a neuron, the membrane potential is actually a more complicated function of the concentrations and membrane permeabilities of several different ions (although K^+ is the most important).

Sample Calculation 8-1

Problem

Calculate the intracellular concentration of Na^+ when the extracellular concentration is 160 mM and the membrane protential is −50 mV at 20°C.

Solution

Use Equation 8-2 and solve for $[Na^+_{in}]$:

$$\Delta\psi = 0.058 \log \frac{[\text{Na}^+_{in}]}{[\text{Na}^+_{out}]}$$

$$\frac{\Delta\psi}{0.058} = \log [\text{Na}^+_{in}] - \log [\text{Na}^+_{out}]$$

$$\log [\text{Na}^+_{in}] = \frac{\Delta\psi}{0.058} + \log [\text{Na}^+_{out}]$$

$$\log [\text{Na}^+_{in}] = \frac{-0.050}{0.058} + \log (160)$$

$$\log [Na^+_{in}] = -0.862 + 2.204$$

$$\log [Na^+_{in}] = 1.342$$

$$Na^+_{in} = 22 \text{ mM}$$

Most animal cells maintain a membrane potential of about −70 mV. The negative sign indicates that the inside (the cytosol) is more negative than the outside (the extracellular fluid). A sudden flux of ions across the cell membrane can dramatically alter the membrane potential, and this is exactly what happens when a neuron fires.

When a nerve is stimulated, either mechanically or by a signal ultimately derived from one of the sensory organs, Na^+ channels in the plasma membrane open. Sodium ions immediately move into the cell since their concentration inside is much less than outside. The inward movement of Na^+ makes the membrane potential more positive, increasing it from its resting value of −70 mV to as much as +50 mV. This reversal of membrane potential, or depolarization, is called the **action potential.**

The Na^+ channels remain open for less than a millisecond. However, the action potential has already been generated, and it has two effects. First, *it triggers the opening of nearby* ***voltage-gated*** *K^+ channels* (these channels open only in response to the change in membrane potential). The open K^+ channels allow K^+ to diffuse out of the cell, following its concentration gradient. This action restores the membrane potential to about −70 mV.

The action potential also stimulates the opening of additional Na^+ channels farther along the axon. This induces another round of depolarization and repolarization, and then another. In this way, the action potential travels down the axon. The signal cannot travel backward because once the voltage-sensitive Na^+ channels have shut, they remain closed for a few milliseconds. These events are summarized in Fig. 8-15.

In mammals, action potentials propagate extremely rapidly because the axons are insulated by a so-called **myelin sheath.** This structure consists of several layers of membrane, derived from another cell, coiled around the axon (Fig. 8-16). The myelin sheath is rich in sphingomyelins and contains relatively little protein, about 18%, compared to other membranes, which contain up to 75% protein. Because the myelin sheath prevents ion movements except at the points, or nodes, in between myelinated segments of the axon, the action potential appears to jump from node to node, propagating about 20 times faster than it would in an unwrapped axon. Deterioration of the myelin sheath in diseases such as multiple sclerosis results in the progressive loss of motor control.

REVIEW LEARNING OBJECTIVES 7 AND 8

- What is the membrane potential?
- What is the role of the cell membrane in maintaining membrane potential?
- Describe how an action potential is generated and propagated.

Go to **Exercise 14/Membrane Transport** for a look at the structures, solute selectivity, and mechanisms of several different types of membrane transport proteins. The material presented in Exercise 14 addresses Learning Objective 9. Exercise 14 also includes a general review of membrane structure and permeability.

Overview of transporters

The Na^+ and K^+ channels that participate in the propagation of an action potential are just two members of a large group of transport proteins that occur in the plasma membranes of all cells and in the internal membranes of

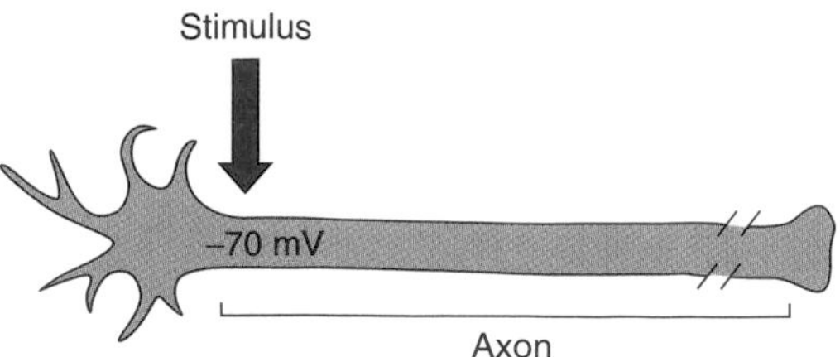

For simplicity, the stimulus is assumed to originate where the axon leaves the neuronal cell body. The stimulus opens Na^+ channels in the membrane.

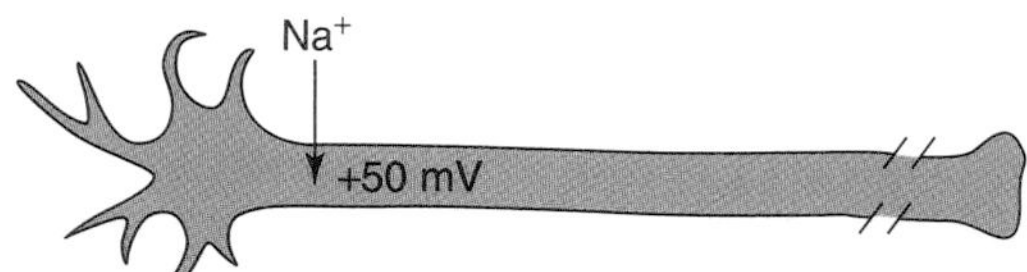

The influx of Na^+ causes depolarization, which triggers the opening of voltage-sensitive K^+ channels. The efflux of K^+ restores the resting potential.

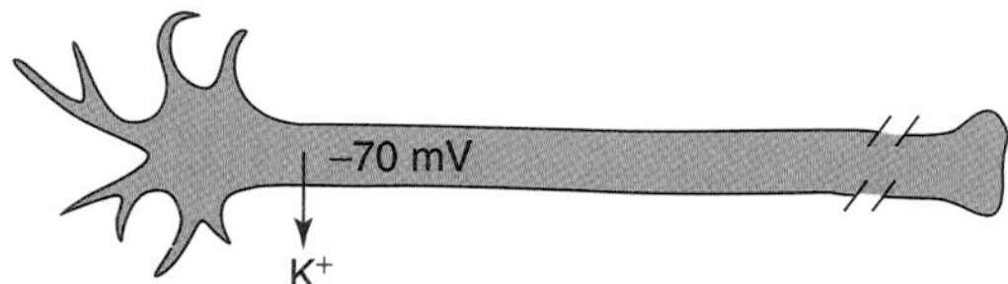

The initial depolarization also triggers the opening of additional Na^+ channels, and then K^+ channels, farther along the axon.

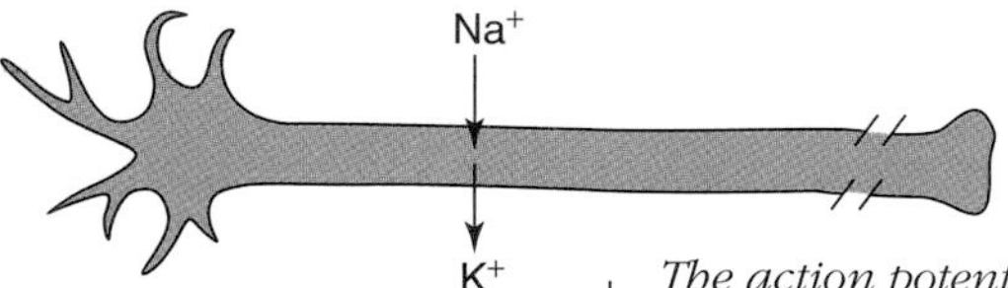

The action potential moves down the length of the axon as a wave of depolarization and repolarization. The action potential travels in only one direction because previously open Na^+ channels remain closed.

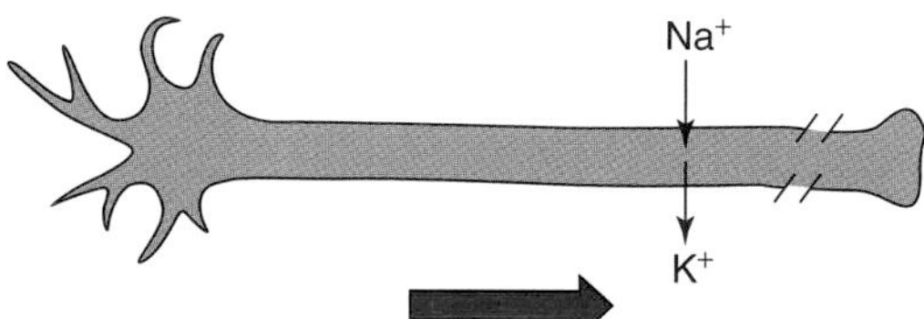

FIGURE 8-15

Propagation of a nerve impulse.

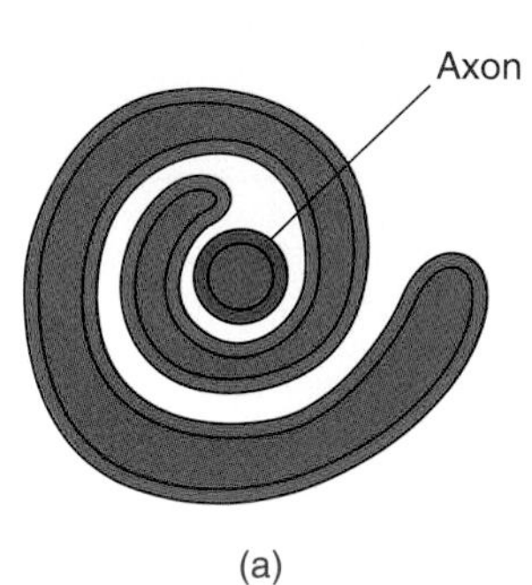

(a)

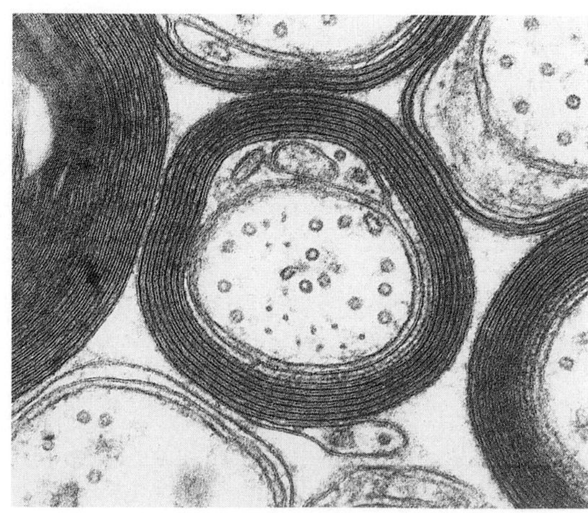

(b)

FIGURE 8-16

Myelination of an axon. (a) Cross-sectional diagram showing how an accessory cell coils around the axon so that multiple layers of its plasma membrane coat the axon. (b) Electron micrograph of myelinated axons. The myelin sheath may be 10 to 15 layers thick. [*Courtesy Cedric S. Raines, Albert Einstein College of Medicine.*]

eukaryotes. Transport proteins go by many different names, depending somewhat arbitrarily on their mode of action: transporters, translocases, permeases, pores, channels, and pumps, to list a few. These proteins are sometimes described by the type of substance they transport across the membrane and by whether they are always open or gated (open only when stimulated). The most important distinction among transport proteins, however, is whether they require a source of free energy to operate. The neuronal Na^+ and K^+ channels are considered **passive transporters** because they provide a means for ions to move down a concentration gradient, a thermodynamically favorable event.

For any transport protein operating independently of the effects of membrane potential, the free energy change for the transmembrane movement of a substance X is

$$\Delta G = RT \ln \frac{[X_{in}]}{[X_{out}]} \qquad [8\text{-}3]$$

Consequently, *the free energy change is negative (the process is spontaneous) only when X moves from an area of high concentration to an area of low concentration* (see Sample Calculation 8-2). The presence of a membrane potential adds a term to the equation:

$$\Delta G = RT \ln \frac{[X_{in}]}{[X_{out}]} + Z\mathcal{F}\Delta\psi \qquad [8\text{-}4]$$

If the transported substance is an ion with charge Z, its transport may not be thermodynamically favored, depending on the membrane potential $\Delta\psi$, even if the concentration gradient alone favors transport.

Sample Calculation 8-2

Problem

Show that $\Delta G < 0$ when Ca^{2+} ions move from the endoplasmic reticulum (where $[Ca^{2+}]$ = 1 mM) to the cytosol (where $[Ca^{2+}]$ = 0.1 μM).

Solution

The cytosol is *in* and the endoplasmic reticulum is *out*.

$$\Delta G = RT \ln \frac{[Ca^{2+}_{in}]}{[Ca^{2+}_{out}]}$$

$$\Delta G = RT \ln \frac{(10^{-7})}{(10^{-3})}$$

$$\Delta G = RT(-9.2)$$

Because the logarithm of $(10^{-7}/10^{-3})$ is a negative quantity, ΔG is also negative.

In contrast to the passive ion channels in neurons, the protein that initially establishes and maintains the cell's Na^+ and K^+ gradients is an **active transporter** that needs the free energy of ATP to move ions against their concentration gradients. In the remainder of this section we will examine various types of transport proteins, beginning with the simplest. Keep in mind that small nonpolar substances can cross a membrane without the aid of any transport protein; they simply diffuse through the lipid bilayer.

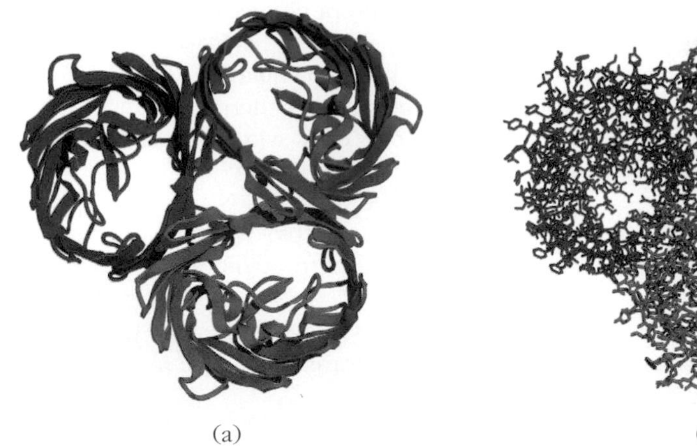

FIGURE 8-17

The *E. coli* OmpF porin.
Each subunit of the trimeric protein forms a transmembrane β barrel that permits the passage of ions or small molecules. (a) Ribbon model, viewed from the extracellular side of the membrane. (b) Stick model. *[Structure (pdb 1OPF) determined by S.W. Cowan, T. Schirmer, R.A Pauptit, and J.N. Jansonius.]*

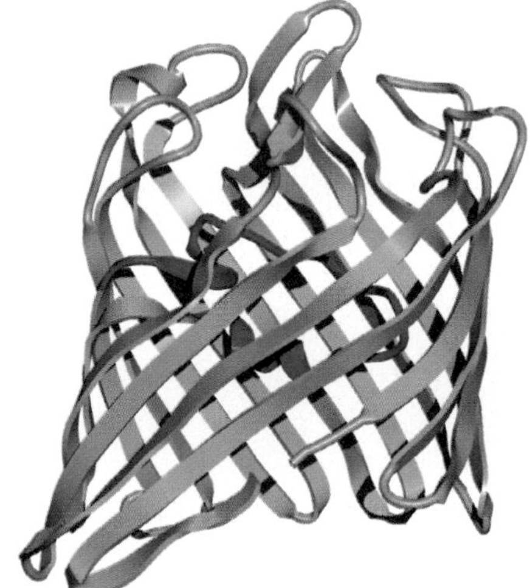

FIGURE 8-18

Side view of an OmpF subunit.
The 16 β strands are connected by loops, one of which (blue) constricts the barrel core and makes the porin specific for small cationic solutes.

Porins

The simplest membrane transporters are the porins, which are located in the outer membranes of bacteria, mitochondria, and chloroplasts (some bacteria and the organelles descended from them have a second outer membrane in addition to the membrane that encloses the cytosol). *All known porins are trimers in which each subunit forms a 16- or 18-stranded membrane-spanning β barrel* (Fig. 8-17). A β barrel of this size has a water-filled core lined with hydrophilic side chains, which forms a passageway for the transmembrane movement of ions or molecules with a molecular mass up to about 1000 D. In the eight-stranded β barrel shown in Figure 8-9, the protein core is too tightly packed with amino acid side chains for it to function as a pore.

In the 16-stranded OmpF protein, long loops connect the β strands (Fig. 8-18). One of these loops in each monomer folds down into the β barrel and constricts its diameter to about 7 Å at one point, thereby preventing the passage of substances larger than 600 D. The loop bears several carboxylate side chains, which make the porin weakly selective for cationic substances. Other porins exhibit a greater degree of solute selectivity, depending on the geometry of the barrel interior and the nature of the side chains that project into it. A porin is considered to be always open, and a solute can travel in either direction through it. The structure and function of a porin are summarized in Exercise 14.

Ion channels

The ion channels in the neuron and in other eukaryotic and prokaryotic cells are more complicated proteins than the porins. Many are multimers of identical or similar subunits with α-helical membrane-spanning segments. The ion passageway itself lies along the central axis of the protein, where the subunits meet. One of the best-known of these proteins is the K^+ channel from the bacterium *Streptomyces lividans.* Each subunit of this tetrameric protein includes two long α helices. One helix forms part of the wall of the transmembrane pore, and the other helix faces the hydrophobic membrane interior (Fig. 8-19). A third, smaller helix is located on the extracellular side of the protein.

The K^+ channel is about 10,000 times more permeant to K^+ than to Na^+, even though Na^+ is smaller and should easily pass through the 6-Å-diameter pore. The high selectivity for K^+ reflects the geometry of the selectivity filter, an arrangement of protein groups that define the extracellular mouth of the pore. At one point, the four polypeptide backbones fold so that their side

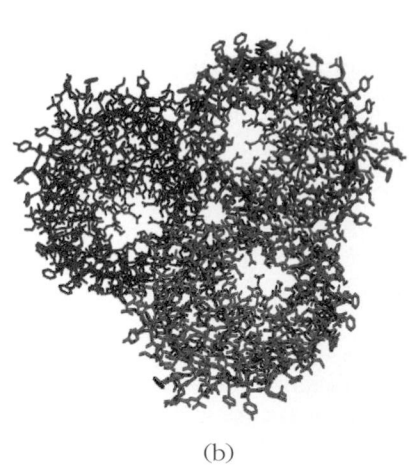

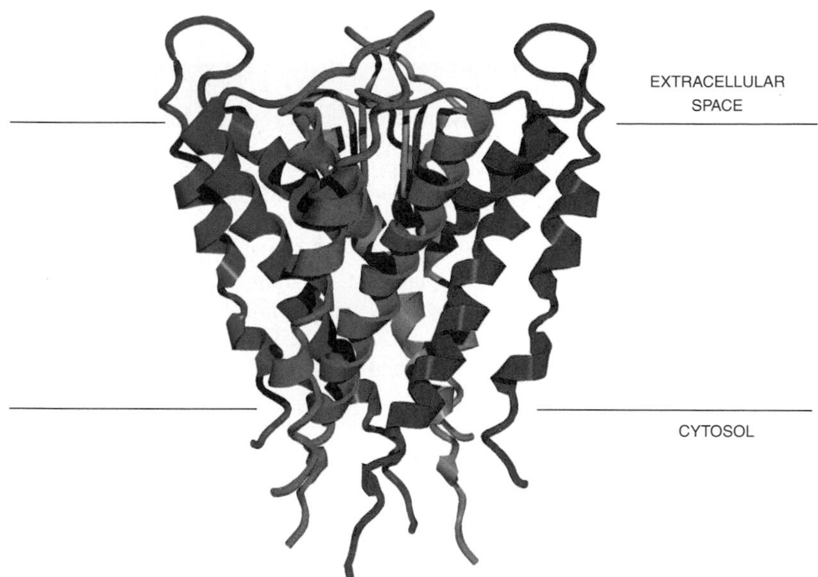

FIGURE 8-19

Structure of the K^+ channel from *S. lividans.*
The four subunits are shown in different colors. Each subunit consists mostly of an inner helix that forms part of the central pore and an outer helix that contacts the membrane interior. [*Structure (pdb 1BL8) determined by D.A. Doyle, J.M. Cabral, R.A. Pfuetzner, A. Kuo, J.M. Gulbis, S.L. Cohen, B.T. Chait, and R. MacKinnon.*]

chains point into the protein interior and their carbonyl groups project into the pore. The carbonyl oxygen atoms are arranged with a geometry suitable for coordinating desolvated K^+ ions (diameter 2.67 Å) as they move through the pore. A desolvated Na^+ ion (diameter 1.90 Å) is too small to coordinate with the carbonyl groups and is therefore excluded from the pore (Fig. 8-20).

Most of the pore is narrow and hydrophobic, except for a hydrophilic cavity located about midway, corresponding to the highest point of the energy barrier in the ion's transmembrane journey. This cavity is about 10 Å in diameter, large enough for a K^+ ion to be surrounded by a layer of water molecules (Fig. 8-21). This aqueous oasis helps decrease the energetic barrier to transporting a K^+ ion through the pore in either direction and therefore promotes the rapid movement of K^+ across the membrane.

The voltage-gated K^+ channel in neurons probably operates similarly to the *S. lividans* K^+ channel. For example, all mammalian K^+ channels are tetramers containing amino acid sequences that could form a selectivity filter. Other ion channels, such as those for Na^+ and Ca^{2+}, necessarily have different filtering mechanisms. Membrane channels that are specific for water molecules form an entirely different family of proteins (Box 8-C).

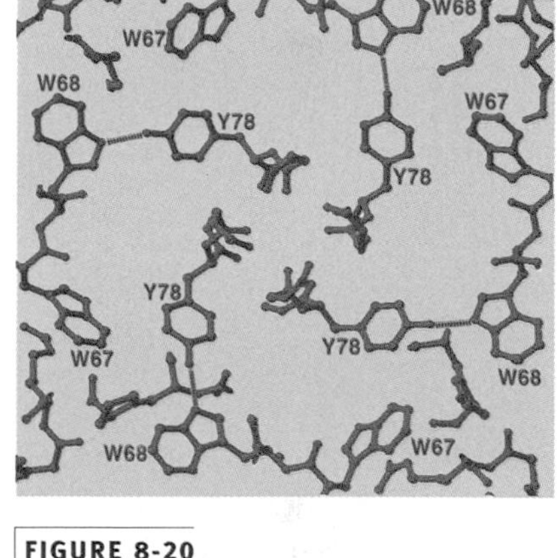

FIGURE 8-20

The K^+ channel selectivity filter.
This view is a cross-section of the pore at the level of the selectivity filter. The pore (center) is lined by backbone carbonyl groups with a geometry suitable for coordinating K^+ ions. A rigid network of Tyr (Y) and Trp (W) residues prevents the pore from contracting to accommodate the smaller Na^+ ions. [*Courtesy Roderick MacKinnon, Rockefeller University.*]

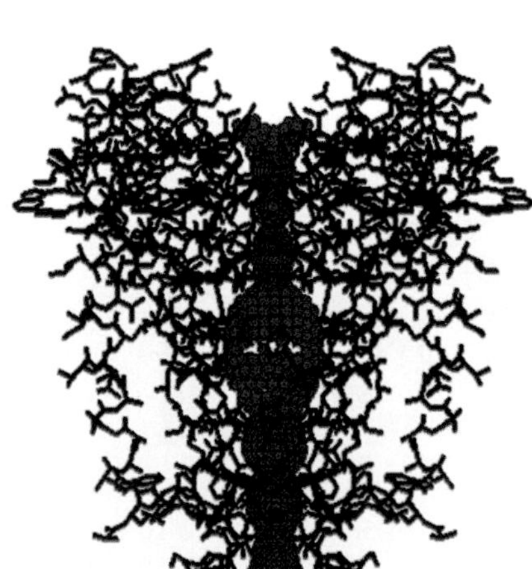

FIGURE 8-21

Surface contour of the K^+ channel pore.
The molecular surface of the pore is visible within a stick model of the protein. The selectivity filter, the narrowest part of the pore, is at the top. A wider cavity, lined with hydrophilic groups, helps decrease the free energy barrier for transporting an ion through the long hydrophobic pore. [*Courtesy Roderick MacKinnon, Rockefeller University.*]

A CLOSER LOOK

BOX 8-C Aquaporins are water-specific pores

For many years, water molecules were assumed to cross membranes by simple diffusion (technically **osmosis,** the movement of water from regions of low solute concentration to regions of high solute concentration). Because water is present in large amounts in biological systems, this premise seemed reasonable. However, certain cells, such as in the kidney, can sustain unexpectedly rapid rates of water transport, which suggested the existence of a previously unrecognized pore for water. The elusive protein was discovered in 1992 by Peter Agre, who coined the term aquaporin.

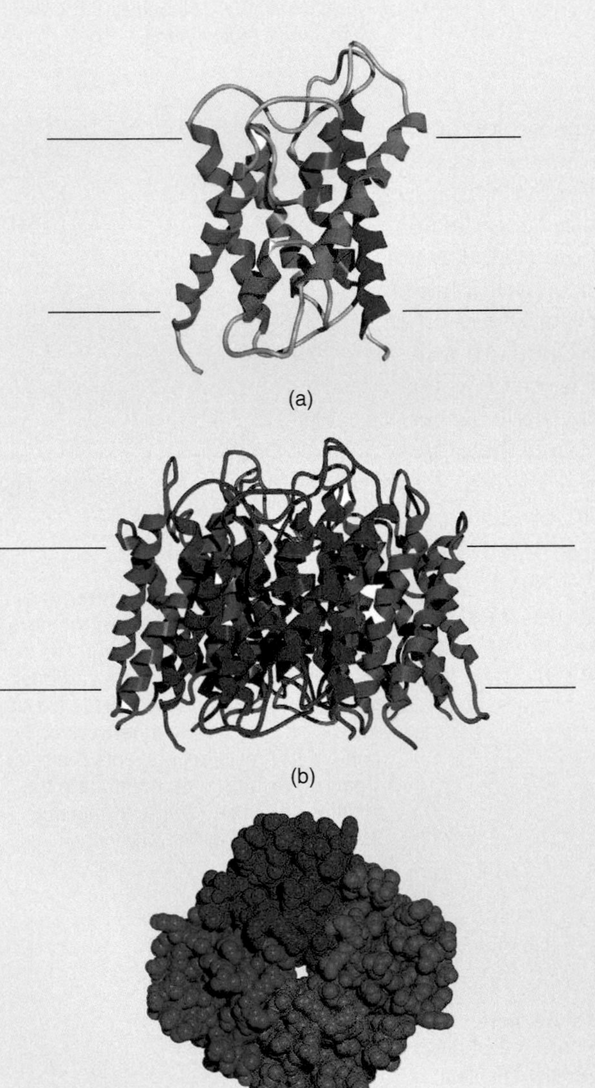

(a)

(b)

(c)

Aquaporins are widely distributed in nature; plants may have as many as 50 different aquaporins. The 10 mammalian aquaporins are expressed at high levels in tissues where fluid transport is important, including the kidney, salivary glands, and lacrimal glands (which produce tears). Most aquaporins are extremely specific for water molecules and do not permit the transmembrane passage of other polar molecules such as glycerol or urea.

$$H_2N-\overset{\displaystyle O}{\overset{\|}{C}}-NH_2$$

Urea

The best-defined member of the aquaporin family (aquaporin 1 or AQP1) is a homotetramer with carbohydrate chains on its extracellular surface. Each subunit consists mostly of six membrane-spanning α helices plus two shorter helices that lie within the dimensions of the bilayer.

The upper figure shows one of the four subunits, viewed from within the membrane. The middle figure is a side view of the aquaporin tetramer, with each subunit a different color. The lower figure is a top view of the tetramer. The subunits associate via hydrogen bonding between helices and through interactions among the loops outside the membrane.

Unlike the K^+ channel, the center of the aquaporin tetramer is narrow and hydrophobic and therefore cannot serve as a passageway for water. Instead, each subunit contains a pore. At its narrowest, the pore is about 3 Å in diameter (the diameter of a water molecule is 2.8 Å). The dimensions of the pore clearly restrict the passage of larger molecules. The pore is lined with hydrophobic residues except for two Asn side chains which have an important function.

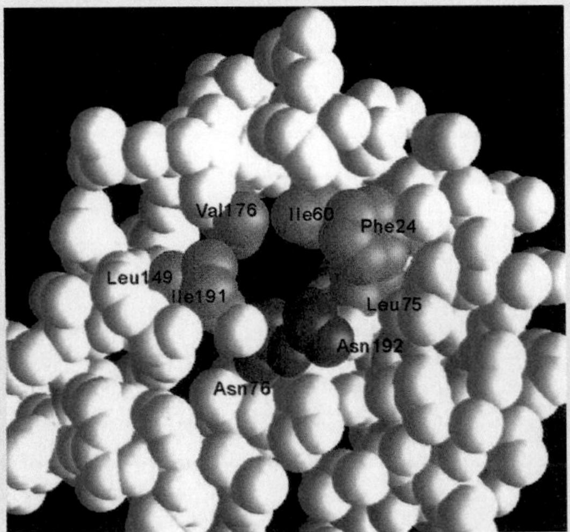

If water were to pass through aquaporin as a chain of hydrogen-bonded molecules, then protons could also easily

(continued)

BOX 8-C *(continued)*

pass through (recall from Section 2-3 that a proton is equivalent to H_3O^+ and that a proton can appear to jump rapidly through a network of hydrogen-bonded water molecules). However, aquaporin does not transport protons (other proteins that do transport protons play important roles in energy metabolism). To prevent proton transport, aquaporin must interrupt the hydrogen-bonded chain of water molecules in its pores, which occurs when the Asn side chains transiently form hydrogen bonds to a water molecule passing by.

[Structure (pdb 1FQY) determined by K. Murata, K. Mitsuoka, T. Hirai, T. Walz, P. Agre, J.B. Heymann, A. Engel, and Y. Fujiyoshi; Photo courtesy Yoshinori Fujiyoshi, Kyoto University. From Murata, et al., *Nature* **407**, 603 (2000).]

Gated channels undergo conformational changes

The mechanisms for gating channels that transport ions are incompletely understood. The opening of some ion channels, such as in the neuron, depends on changes in membrane potential. Other ion channels open and close in response to changes in pH, to the binding of a specific ligand, or to mechanical pressure. The cystic fibrosis Cl^- channel (the CFTR protein, described in Section 3-5), opens when it is phosphorylated. The available evidence suggests that gated channels undergo some small conformational change, such as the movement of a short peptide segment that reversibly blocks the mouth of the transmembrane passageway (Fig. 8-22a). For example, depolarization rapidly opens neuronal Na^+ channels but also causes a peptide segment to move slowly into the channel to block it. This helps explain why the channel remains functional for less than a millisecond and does not immediately reopen. Alternatively, a gated ion channel may undergo rotation that opens the channel without significantly disrupting protein secondary structure (Fig. 8-22b). Exercise 14 includes animated views of gating in an ion channel.

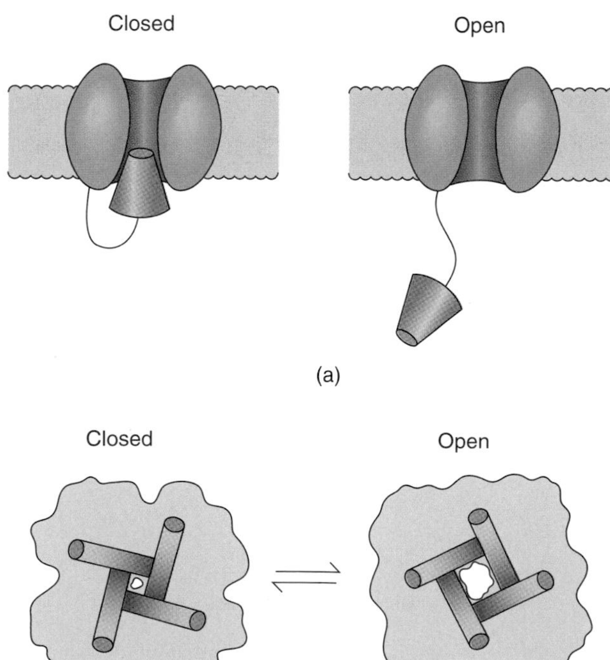

FIGURE 8-22

Models for ion channel gating.
(a) A polypeptide segment that blocks the passageway may move when it binds a ligand or is covalently modified. (b) The subunits may rotate in order to open a central passageway for ions, as depicted by the positions of transmembrane α helices (rods) in this axial view. [*After Perozo, E., Cortes, D.M., and Cuello, L.G., Structural rearrangements underlying K^+-channel activation gating,* Science **285**, *73–78 (1999).*]

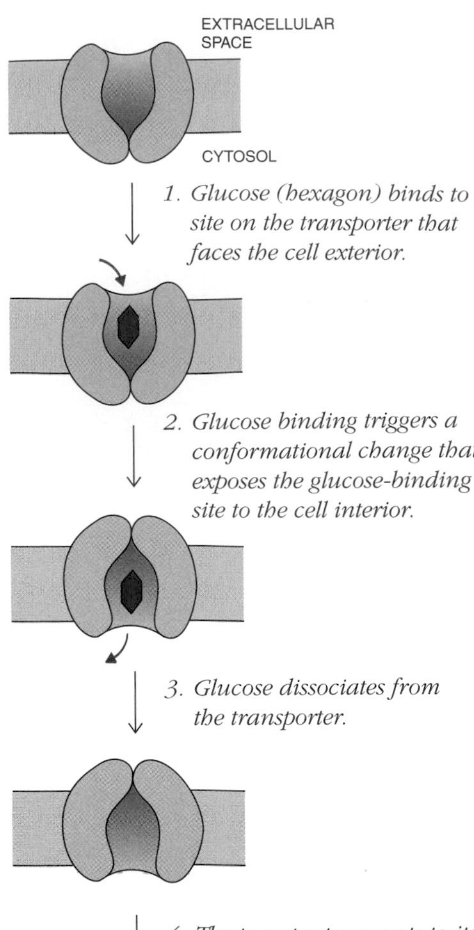

EXTRACELLULAR SPACE
CYTOSOL
1. Glucose (hexagon) binds to a site on the transporter that faces the cell exterior.
2. Glucose binding triggers a conformational change that exposes the glucose-binding site to the cell interior.
3. Glucose dissociates from the transporter.

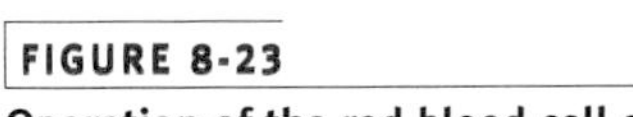

FIGURE 8-23
Operation of the red blood cell glucose transporter.

Some transport proteins alternate between conformations

Not all proteins that mediate transmembrane traffic have an obvious membrane-spanning pore, as in the porins and the *S. lividans* K^+ channel. A protein such as the glucose transporter from red blood cells undergoes a conformational change in order to move a solute from one side of the membrane to the other. Sequence analysis suggests that the glucose transporter has 12 membrane-spanning α helices. Experimental evidence indicates that the protein has a glucose-binding site that alternately faces the cell interior and exterior. When glucose binds to the protein on one face of the membrane, it triggers a conformational change that exposes the bound glucose to the other face (Fig. 8-23). Because the two conformational states of this passive transporter are in equilibrium, it can move glucose in either direction across the cell membrane, depending on the relative concentrations of glucose inside and outside the cell.

Other transport proteins resemble the glucose transporter in that *they are all transmembrane proteins that alternate between conformations in order to bind and release a ligand on opposite sides of the membrane.* For obvious reasons, transport proteins tend to be more solute-selective than porins or ion channels. Their great variety reflects the need to transport many different kinds of metabolic fuels and building blocks into and out of cells and organelles. An estimated 10% of the genes in microorganisms encode transport proteins. The mechanism of these transport proteins is shown schematically in Exercise 14.

Some transport proteins can bind more than one type of ligand, so it is useful to classify them according to how they operate (Fig. 8-24):

1. A **uniport** such as the glucose transporter moves a single substance at a time.
2. A **symport** transports two different substances across the membrane.
3. An **antiport** moves two different substances in opposite directions across the membrane.

Active transport

The differing Na^+ and K^+ concentrations inside and outside of eukaryotic cells are maintained largely by an antiport protein known as the Na,K-ATPase. This active transporter pumps Na^+ out of and K^+ into the cell, working against the ion concentration gradients. As its name implies, ATP is its source of free energy.

The Na,K-ATPase consists of a catalytic α subunit with 10 transmembrane α helices and a β subunit with a single membrane-spanning segment. *In vivo,*

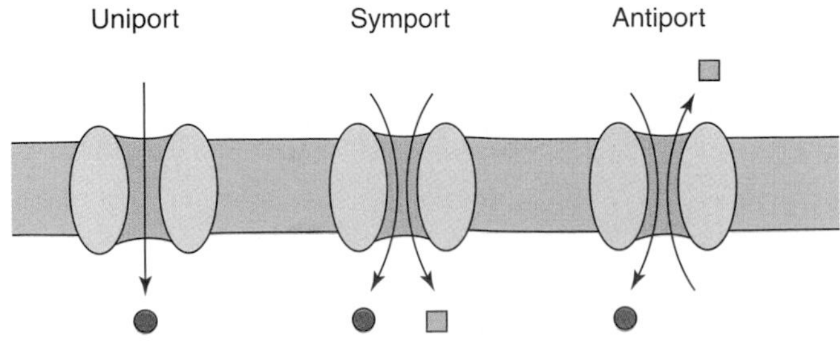

FIGURE 8-24
Some types of membrane transport systems.

several α subunits may associate, possibly along with a peripheral membrane protein, the γ subunit. With each reaction cycle, the ATPase hydrolyzes 1 ATP, pumps 3 Na^+ ions out, and pumps two K^+ ions in:

$$3\ Na^+_{in} + 2\ K^+_{out} + ATP + H_2O \rightarrow 3\ Na^+_{out} + 2\ K^+_{in} + ADP + P_i$$

Like other membrane transport proteins, the Na,K-ATPase has two conformations that alternately expose the Na^+- and K^+-binding sites to each side of the membrane. As diagrammed in Figure 8-25, the protein pumps out 3 Na^+ ions at a time, then transports in 2 K^+ ions at a time, using the free energy released by the hydrolysis of ATP. In effect, *the energetically favorable*

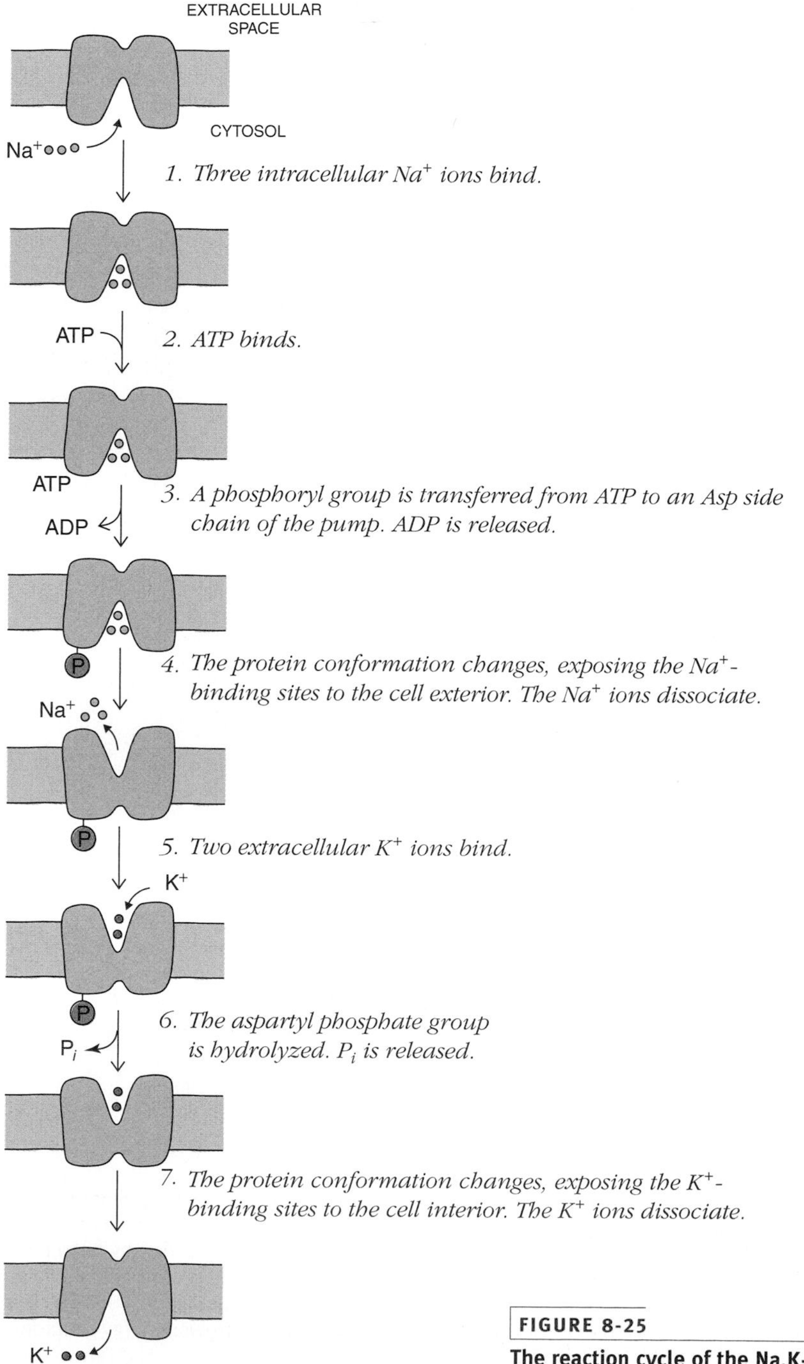

EXTRACELLULAR SPACE
CYTOSOL
Na+
1. Three intracellular Na+ ions bind.
ATP
2. ATP binds.
ATP
ADP
3. A phosphoryl group is transferred from ATP to an Asp side chain of the pump. ADP is released.
P
4. The protein conformation changes, exposing the Na+-binding sites to the cell exterior. The Na+ ions dissociate.
Na+
P
5. Two extracellular K+ ions bind.
K+
P
Pi
6. The aspartyl phosphate group is hydrolyzed. Pi is released.
7. The protein conformation changes, exposing the K+-binding sites to the cell interior. The K+ ions dissociate.
K+

FIGURE 8-25
The reaction cycle of the Na,K-ATPase.

reaction of converting ATP to ADP + P_i drives the energetically unfavorable transport of Na^+ and K^+. The ATP hydrolysis reaction is coupled to ion transport so that phosphoryl group transfer from ATP to the protein triggers one conformational change (Steps 3 and 4) and the subsequent release of the phosphoryl group as P_i triggers another conformational change (Steps 5 and 6). This multistep process (which involves a phosphorylated protein intermediate) ensures that the transporter operates in only one direction and prevents Na^+ and K^+ from diffusing back down their concentration gradients. A similar mechanism operates in myosin (Section 5-2), where ADP and P_i are released separately and so are unlikely to recombine to re-form ATP and drive the reaction cycle in reverse. Other ATP-requiring transport proteins pump substances such as protons and Ca^{2+} ions against their concentration gradients. Exercise 14 describes the operation of these pumps.

In some cases, the "uphill" transmembrane movement of a substance is not directly coupled to the conversion of ATP to ADP + P_i. Instead, *the transporter takes advantage of a gradient already established by an ATPase pump*. For example, the high Na^+ concentration outside of intestinal cells (a gradient established by the Na,K-ATPase) helps drive glucose into the cells via a symport protein (Fig. 8-26). In other words, the free energy released by the movement of Na^+ into the cells (down its concentration gradient) drives the inward movement of glucose (against its gradient). This mechanism allows the intestine to collect glucose from digested food and then release it into the bloodstream. Exercise 14 includes an overview of the Na-glucose transporter.

The free energy of ATP (or its equivalent, GTP) is also required indirectly for the processes that move large substances, such as intact proteins, from one side of the membrane to the other. These phenomena do not involve discrete transport proteins like the ones described above, although they do require a variety of other proteins.

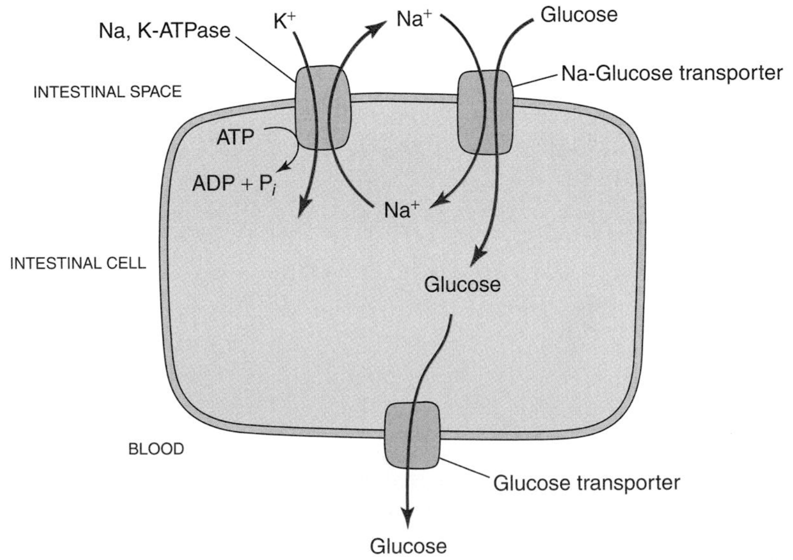

FIGURE 8-26

Glucose transport into intestinal cells.
The Na,K-ATPase establishes a concentration gradient in which $[Na^+_{out}] > [Na^+_{in}]$. Sodium ions move into the cell, down their concentration gradient, along with glucose molecules via a symport protein that transports Na^+ and glucose simultaneously. Glucose thereby becomes more concentrated inside the cell, which it then exits, down its concentration gradient, via a passive uniport glucose transporter similar to the red blood cell transporter described in Figure 8-23. Energetically favorable movements are indicated by green arrows; energy-requiring movements are indicated by red arrows.

REVIEW LEARNING OBJECTIVE 9

- Compare the overall structure, solute selectivity, general mechanism, and energy requirement of porins, the K^+ channel, the red blood cell glucose transporter, and the Na,K-ATPase.

4 MEMBRANE FUSION

The final steps in the transmission of a signal from one neuron to the next, or to a gland or muscle cell, culminate in the release of substances known as **neurotransmitters** from the end of the axon. Common neurotransmitters include amino acids and compounds derived from them. In the case of a synapse linking a motor neuron and its target muscle cell, the neurotransmitter is acetylcholine:

$$H_3C-\overset{\overset{\displaystyle O}{\|}}{C}-O-CH_2-CH_2-\overset{\overset{\displaystyle CH_3}{|}}{\underset{\underset{\displaystyle CH_3}{|}}{N^+}}-CH_3$$

Acetylcholine

Acetylcholine is stored in membrane-bounded compartments, called **synaptic vesicles,** about 400 Å in diameter. When the action potential reaches the axon terminus, voltage-gated Ca^{2+} channels open and allow the influx of extracellular Ca^{2+} ions. The increase in the local intracellular Ca^{2+} concentration, from less than 1 μM to as much as 100 μM, triggers **exocytosis** of the vesicles (exocytosis is the fusion of the vesicle with the plasma membrane such that the vesicle contents are released to the extracellular space). The acetylcholine diffuses across the synaptic cleft, the narrow space between the axon terminus and the muscle cell, and binds to integral membrane protein receptors on the muscle cell surface. This binding initiates a sequence of events that result in muscle contraction (Fig. 8-27). In general, the response of the cell that receives a neurotransmitter depends on the nature of the neurotransmitter and the cellular proteins that are activated when the neurotransmitter binds to its receptor. We will explore other receptor systems in Chapter 16.

The events at the nerve–muscle synapse occur rapidly, within about 1 millisecond, but the effects of a single action potential are limited. First, the Ca^{2+} that triggers neurotransmitter release is quickly pumped back out of the cell by a Ca^{2+}-ATPase. Second, acetylcholine in the synaptic cleft is degraded within a few milliseconds by a lipid-linked or soluble acetylcholinesterase, which catalyzes the reaction

$$\underset{\text{Acetylcholine}}{H_3C-\overset{\overset{\displaystyle O}{\|}}{C}-O-CH_2-CH_2-\overset{+}{N}(CH_3)_3} + H_2O \xrightarrow[\text{acetylcholinesterase}]{} \underset{\text{Acetate}}{H_3C-\overset{\overset{\displaystyle O}{\|}}{C}-O^-} + \underset{\text{Choline}}{HO-CH_2-CH_2-\overset{+}{N}(CH_3)_3} + H^+$$

However, a neuron may contain hundreds of synaptic vesicles, only a small percentage of which undergo exocytosis at a time, so the cell can release acetylcholine repeatedly (as often as 50 times per second).

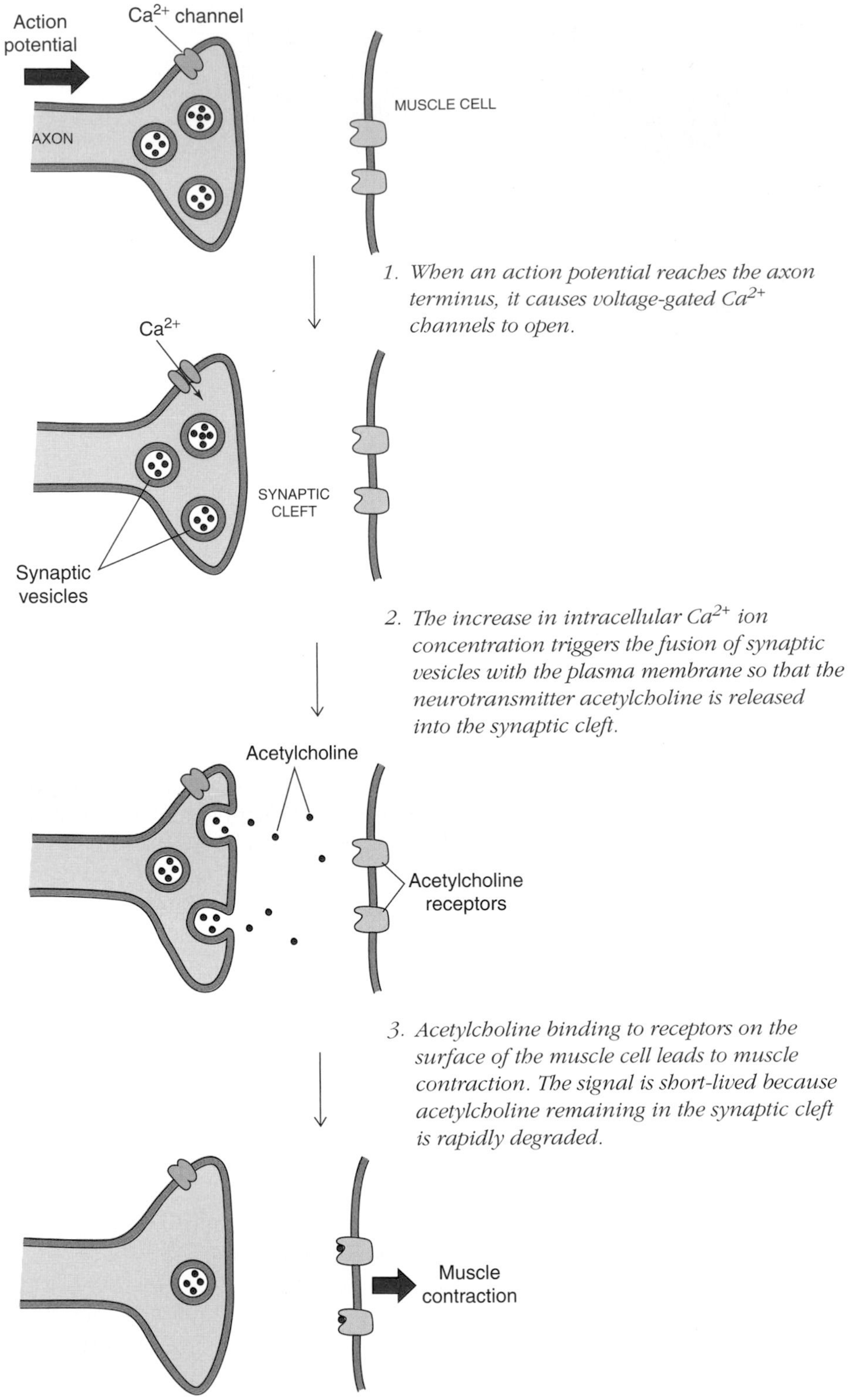

1. When an action potential reaches the axon terminus, it causes voltage-gated Ca^{2+} channels to open.

2. The increase in intracellular Ca^{2+} ion concentration triggers the fusion of synaptic vesicles with the plasma membrane so that the neurotransmitter acetylcholine is released into the synaptic cleft.

3. Acetylcholine binding to receptors on the surface of the muscle cell leads to muscle contraction. The signal is short-lived because acetylcholine remaining in the synaptic cleft is rapidly degraded.

FIGURE 8-27

Events at the nerve–muscle synapse.

REVIEW LEARNING OBJECTIVE 10

- How does acetylcholine inside a synaptic vesicle reach a target muscle cell?

SNAREs link vesicle and plasma membranes

Fusion is a multistep process that begins with the targeting of one membrane (for example the vesicle) to another (for example, the plasma membrane). A number of proteins appear to participate in tethering the two membranes and readying them for fusion. However, many of these proteins may be only accessory factors for the SNAREs, the proteins that physically pair the two membranes and induce them to fuse.

SNAREs are integral membrane proteins (their name is coined from the term "soluble *N*-ethylmaleimide-sensitive-factor attachment protein receptor"). Two SNAREs from the plasma membrane and one from the synaptic vesicle form a complex that includes a 120-Å-long coiled-coil structure containing four helices (two of the SNAREs contribute one helix each, and one SNARE contributes two helices). The four helices, each with about 70 residues, line up in parallel fashion (Fig. 8-28). Unlike other coiled-coil proteins such as keratin (Section 5-5), the four-helix bundle is not a perfectly geometric structure but varies irregularly in diameter. Most of the residues facing the inside of the bundle are nonpolar except at one point, where four polar residues occur. Three glutamine carbonyl groups stabilize an arginine guanidinium group in an otherwise hydrophobic environment. This probably helps the helices assemble in register.

Not only do SNAREs help link a vesicle to its target membrane, they necessarily bring the two membranes close together by forming the four-helix complex (Fig. 8-29). In experimental systems using synthetic vesicles and purified SNAREs, the close approach of the membranes is sufficient for fusion to occur, provided that Ca^{2+} is present. The nature of the Ca^{2+} sensitivity of the fusion machinery is not clear but may involve a Ca^{2+}-induced conformational change in the SNAREs. The rapid rate of acetylcholine release *in vivo* indicates that at least some synaptic vesicles are already docked at the plasma membrane, awaiting the Ca^{2+} signal for fusion. SNAREs are destroyed by the botulinum

FIGURE 8-28

X-Ray structure of the four-helix bundle of the SNARE complex.
The four helices (two of which belong to the same protein) are in different colors. Portions of the SNAREs that do not form the helix bundle were cleaved off before crystallization. [*Structure (pdb 1SFC) determined by R.B. Sutton and A.T. Brunger.*]

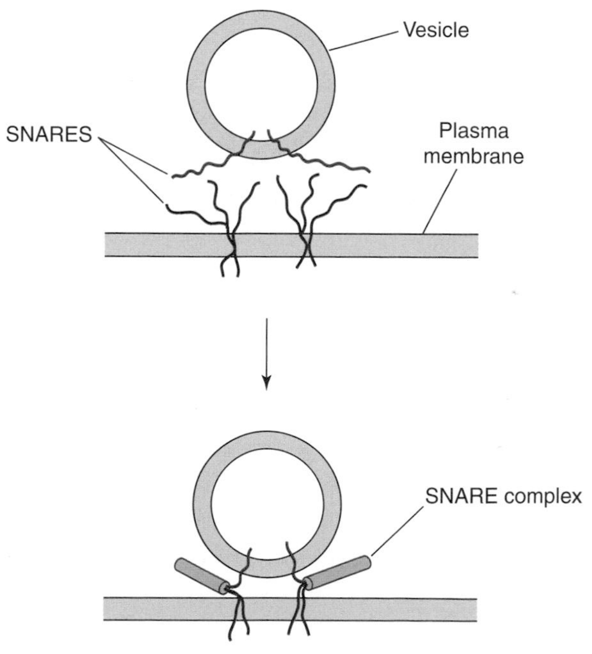

FIGURE 8-29

Model for SNARE-mediated membrane fusion.
Formation of the complex of SNAREs from the vesicle and plasma membranes brings the membranes close together so that they can spontaneously fuse.

and tetanus toxins, which are proteases. Cleavage of the SNAREs prevents vesicle fusion and therefore interrupts the activity of motor neurons, causing paralysis.

REVIEW LEARNING OBJECTIVE 11

- How is the structure of the SNARE complex related to its function?

Membrane fusion requires changes in bilayer curvature

The lipid bilayers of the two membranes must undergo rearrangement as the membranes fuse. The lipids first mix and then form a pore that allows the contents of the synaptic vesicle to enter the extracellular space (Fig. 8-30). The formation of the exceptionally stable SNARE complex may help drive the changes in membrane curvature required for fusion. This could occur if formation of the four-helix bundle caused the transmembrane segments of the SNAREs to bend the bilayers. No external source of free energy is required for this step (although ATP is required later to drive the disassembly of the SNARE complex).

Bilayer curvature could also be affected by structural changes among the lipids themselves, possibly mediated by enzymes that are active only in the presence of the Ca^{2+} trigger for membrane fusion. For example, a calcium-sensitive phospholipid translocase could transfer lipids from one leaflet to the other or even between membranes. Bilayer curvature could also be modified by enzymes that chemically alter specific membrane lipids. For example, the enzymatic removal of an acyl chain would convert a cylindrical lipid to a cone-shaped lipid:

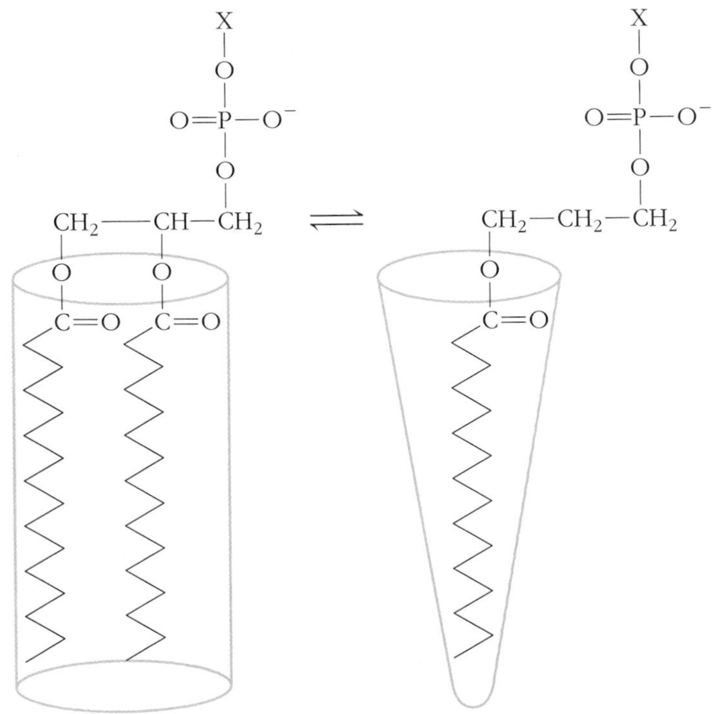

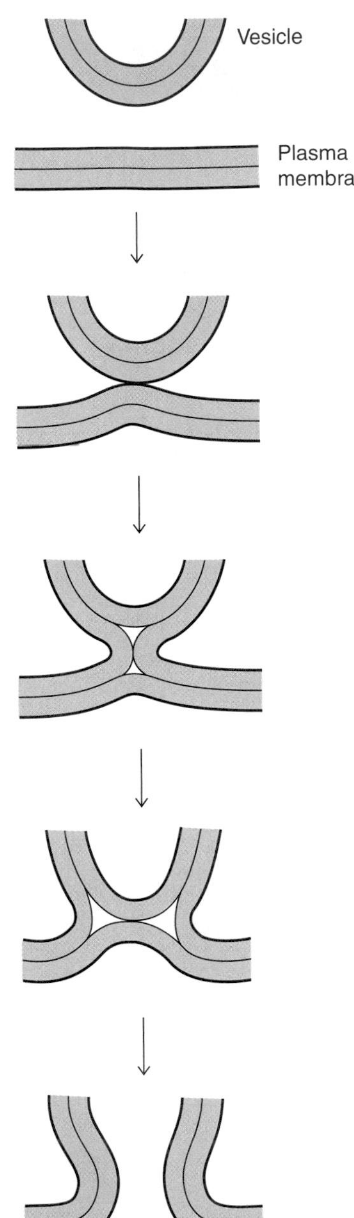

FIGURE 8-30

Schematic view of membrane fusion. For simplicity, the vesicle and plasma membranes are depicted as bilayers.

The neuron eventually recycles some of its plasma membrane material (which has been augmented by material from the fused synaptic vesicle membranes) in the formation of new synaptic vesicles. A new vesicle forms by budding off of an existing membrane, a process known as **endocytosis.** It is the opposite of exocytosis, and follows, in reverse, the scheme diagrammed in Figure 8-30.

REVIEW LEARNING OBJECTIVE 12

- Why are changes in bilayer curvature required for exocytosis?

SUMMARY

1. Biological membranes are based on a lipid bilayer whose components include fatty acid chains. Glycerophospholipids contain two fatty acyl groups attached to a glycerol backbone that bears a phosphate derivative head group. Sphingomyelins are functionally similar but lack a glycerol backbone. Membranes also contain cholesterol and other lipids.
2. Bilayers are dynamic structures. Their fluidity depends on the length and degree of saturation of their fatty acyl groups: Shorter and less saturated chains are more fluid. Cholesterol helps maintain membrane fluidity over a range of temperatures.
3. Membrane lipids can freely diffuse laterally but undergo transverse diffusion very slowly. Membranes may contain crystalline rafts composed of cholesterol and sphingolipids.
4. An integral membrane protein spans the lipid bilayer as one or a bundle of α helices or as a β barrel. Some membrane proteins are anchored in the bilayer by a covalently linked lipid group.
5. According to the fluid mosaic mode, membrane proteins diffuse within the plane of the bilayer. The mobility of proteins may be limited by their interaction with cytoskeletal proteins. Membrane glycolipids and glycoproteins face the cell exterior.
6. The transmembrane movements of ions generate changes in membrane potential during neuronal signaling.
7. Passive transport proteins such as porins allow the transmembrane movement of substances according to their concentration gradient. Ion channels such as the K^+ channel have a selectivity filter that allows passage of one type of ion. Gated channels open and close in response to some other event.
8. Membrane proteins such as the passive glucose transporter undergo conformational changes that alternately expose ligand-binding sites on each side of the membrane.
9. Active transporters such as the Na,K-ATPase require the free energy of ATP to drive the transmembrane movement of substances against their concentration gradient.
10. During the release of neurotransmitters, intracellular vesicles first dock with and then fuse with the cell membrane. SNARE proteins in the vesicle and target membranes form a four-helix structure that brings the fusing membranes close together. Changes in bilayer curvature are necessary for fusion to occur.

CHECKLIST

1. Review the Learning Objectives listed on page 232.
2. Complete Exercise 13/Membrane Proteins to see how the structures of integral membrane proteins are specialized for interaction with the lipid bilayer through formation of membrane-spanning α helices or β barrels.
3. Complete Exercise 14/Membrane Transport, which illustrates the structures and mechanisms of several different types of membrane transport proteins.
4. Apply your knowledge by solving the problems at the end of this chapter. Check your results in the Solutions appendix.
5. Be able to define the boldfaced terms (consult the glossary at the end of the book). Test your understanding by taking the Chapter 8 quiz (www.wiley.com/college/pratt).
6. Explore the websites listed at www.wiley.com/college/pratt.
7. Consult the list of Selected Readings for background information or additional details on lipids, membrane structure, membrane proteins, and transport processes.

GLOSSARY TERMS

liposome
lipid bilayer
lipid
fatty acid
saturated fatty acid
unsaturated fatty acid
triacylglycerol (triglyceride)
acyl group
glycerophospholipid
sphingomyelin
isoprenoid
vitamin
melting point
raft
sphingolipid
glycolipid
transverse diffusion (flip-flop)
translocase (flippase)
lateral diffusion
integral (intrinsic) protein
peripheral (extrinsic) protein
β barrel
lipid-linked protein
fluid mosaic model
glycoprotein
N-linked oligosaccharide
O-linked oligosaccharide
membrane potential ($\Delta\psi$)
gas constant (R)
Z
Faraday constant ($\mathcal{F}$)
action potential
voltage-gated channel
myelin sheath
passive transport
active transport
osmosis
uniport
symport
antiport
neurotransmitter
synaptic vesicle
exocytosis
endocytosis

PROBLEMS

1. Fatty acids are often referred to in a shorthand form which consists of two numbers separated by a colon. The first number is the number of carbons; the second number is the number of double bonds. Therefore, palmitate would be represented by the shorthand 16:0. An unsaturated fatty acid is indicated by a Δ sign, with the positions of the double bonds represented by superscripts (the carboxylate carbon is #1). For example, oleate would be represented by the shorthand 18:1 Δ^9 and linoleate as 18:1 $\Delta^{9,12}$. Draw the structures of the following fatty acids:

(a) Myristate (14:0)
(b) Palmitoleate (16:1 Δ^9)
(c) Linolenate (18:1 $\Delta^{9,12,15}$)

2. Draw the structure of tripalmitin, a triacylglycerol containing three palmitate groups.

3. Phosphatidylinositols (PI) are glycerophospholipids important in cell signaling. Draw the structure of a PI, given the structure of *myo*-inositol. The hydroxyl group involved in the formation of the bond between the inositol and the phosphate group is circled.

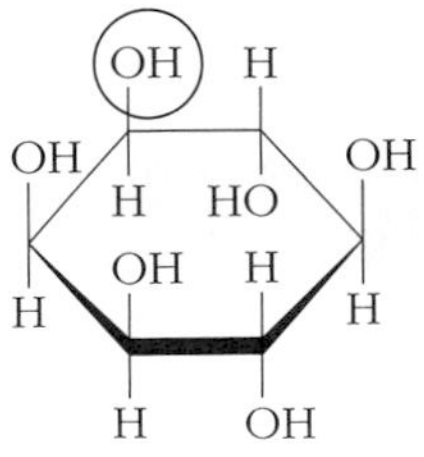

OH H
OH H HO OH
OH H
H H
H OH

myo-Inositol

4. Dipalmitoylphosphatidylcholine (DPPC) is the major lipid of lung surfactant, a protein–lipid mixture essential for pulmonary function. Surfactant production in the developing fetus is low until just before birth, so infants may develop respiratory difficulties if born prematurely. Draw the structure of DPPC.

5. The lipid shown below is known as a plasmalogen.

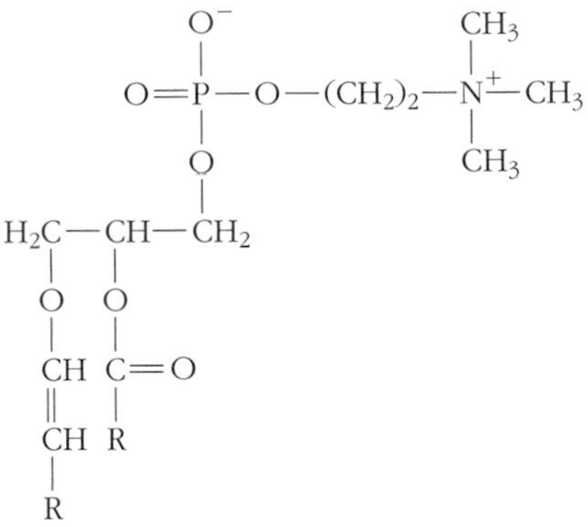

O^- CH_3
$O{=}P{-}O{-}(CH_2)_2{-}\overset{+}{N}{-}CH_3$
O CH_3
$H_2C{-}CH{-}CH_2$
O O
CH C=O
CH R
R

(a) How does it differ from a glycerophospholipid?
(b) Would the presence of this lipid have a dramatic effect on a bilayer that contained only phosphatidylcholine?

6. An unusual sphingosine variant has recently been isolated from the nerve of the squid *Loligo pealei.* Its chemical name is 2-amino-9-methyl-4,8,10-octadecatriene-1,3-diol. Draw the structure of this sphingosine variant.

7. Which of the glycerophospholipids shown on page 236 have hydrogen-bonding head groups?

8. In some autoimmune diseases, an individual develops antibodies that recognize cell constituents such as DNA and phospholipids. Some of the antibodies actually react with both DNA and phospholipids. What is the biochemical basis for this cross-reactivity?

9. Use a simple diagram such as the one in Figure 8-3 to show why bilayer curvature would be affected by replacing a glycerophospholipid bearing two saturated acyl chains with one bearing two highly unsaturated acyl chains.

10. Why can't triacylglycerols form a lipid bilayer?

11. Olestra® is a synthetic lipid which passes through the intestinal tract undigested and is used to make "low-calorie" chips and snacks. Prior to its approval, the FDA required Proctor & Gamble, which markets Olestra®, to add vitamins A, D, and K to products containing the synthetic lipid. Explain.

12. Enzymes known as phospholipases hydrolyze phospholipids. The points of attack of several of these enzymes are indicated in the diagram.

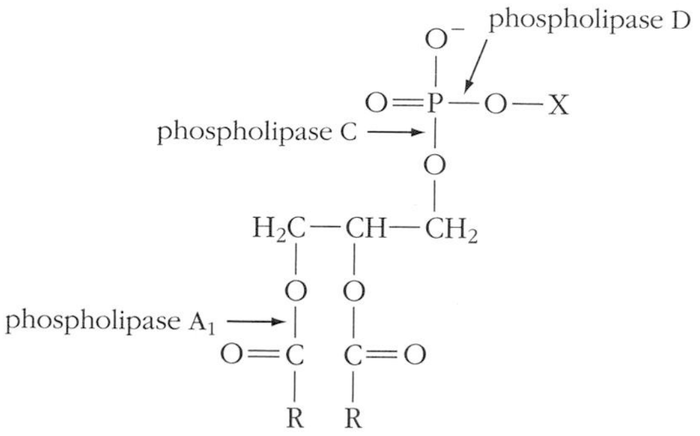

phospholipase D → between P and O—X; phospholipase C → between O and P; phospholipase A_1 → between O and C=O

O^- | O=P—O—X | O | H_2C—CH—CH_2 | O O | O=C C=O | R R

Draw the products of the following reactions:

(a) Phosphatidylserine + phospholipase A_1
(b) Phosphatidylcholine + phospholipase C
(c) Phosphatidylglycerol + phospholipase D

13. The melting points of some common saturated and unsaturated fatty acids are shown in the table. What important factors influence a fatty acid's melting point?

Fatty acid	Melting point (°C)
Laurate (12:0)	44.2
Linoleate (18:2)	−9
Linolenate (18:3)	−17
Myristate (14:0)	52
Oleate (18:1)	13.2
Palmitate (16:0)	63.1
Stearate (18:0)	69.1

14. Rank the melting points of the following fatty acids:

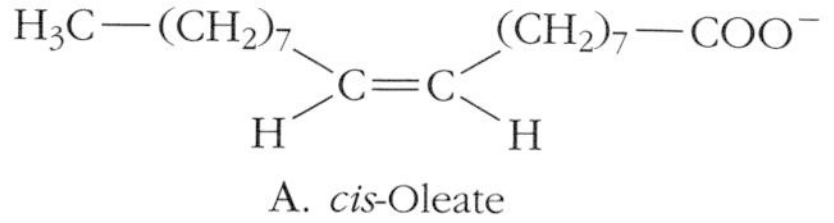

$H_3C—(CH_2)_7$—CH=CH—$(CH_2)_7—COO^-$ (both H on same side)

A. *cis*-Oleate

$H_3C—(CH_2)_7$—CH=CH—$(CH_2)_7—COO^-$ (H on opposite sides)

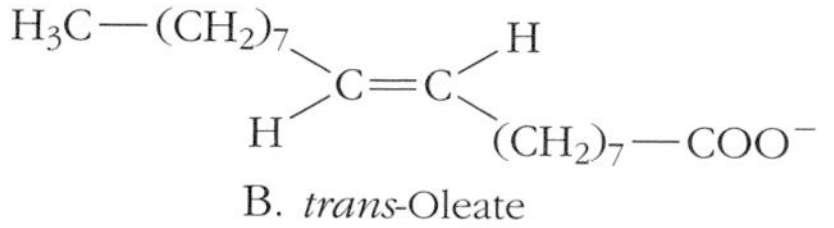

B. *trans*-Oleate

C. Linoleate (18:2)

15. The triacylglycerols of animals tend to be solids (fats), whereas the triacylglycerols of plants tend to be liquids (oils) at room temperature. What can you conclude about the nature of the fatty acyl chains in animal and plant triacylglycerols?

16. Peanut oil contains a high percentage of monounsaturated triacylglycerols (having acyl chains with only one double bond), whereas vegetable oil contains a higher percentage of polyunsaturated triacylglycerols (having acyl chains with more than one double bond). A bottle of peanut oil and a bottle of vegetable oil are stored in a pantry with an outside wall. During a cold spell, the peanut oil freezes whereas the vegetable oil remains liquid. Explain why.

17. Membrane lipids in tissue samples obtained from different parts of the leg of a reindeer show different fatty acid compositions. The proportion of unsaturated fatty acyl chains increases from the top of the leg to the hoof. Provide an explanation for this observation.

18. Phytol is an alcohol found in plants that also becomes part of the diet of mammals consuming the plant. Phytol is converted to phytanic acid, then oxidized to obtain metabolic energy. Inborn errors of metabolism exist in which one of the enzymes in this oxidation pathway is defective. In such individuals, the phytanic acid accumulates in the membranes of nerve cells and causes neurological disorders. What effect would the presence of phytanic acid have on nerve cell membrane fluidity?

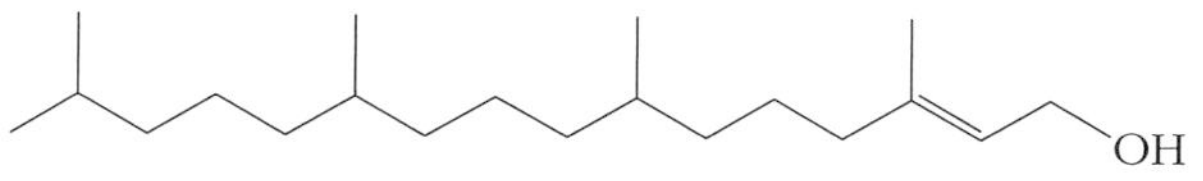

OH

Phytol (3,7,11,15-Tetramethyl-2-hexadecen-1-ol)

↓

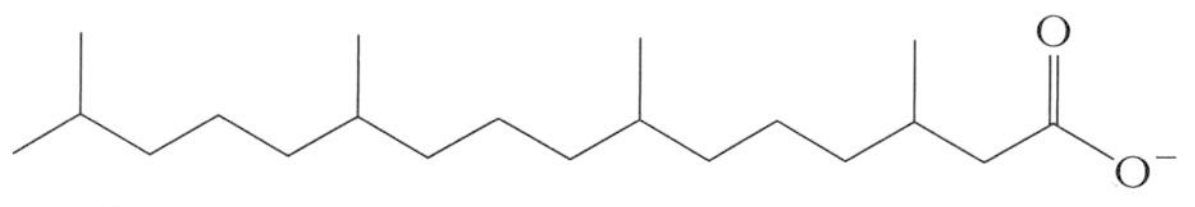

O O^-

Phytanic acid (3,7,11,15-Tetramethylhexadecanoic acid)

19. Bacteria of the genus *Lactobacillus* produce lactobacillic acid, a 19-carbon fatty acid containing a cyclopropane ring. Is the melting point of this fatty acid closer to the melting point of stearate (18:0) or oleate (18:1)? Explain.

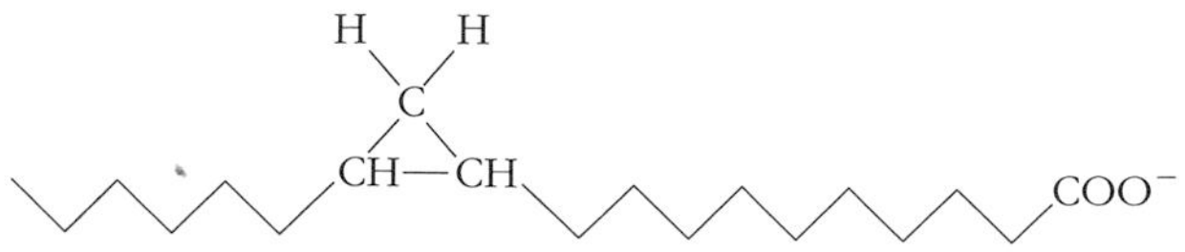

H H | C | CH—CH | COO^-

20. Bacteria are typically grown at a temperature of 37°C. What happens to the membrane lipid composition if the temperature is increased to 42°C?

21. A membrane consisting only of phospholipids undergoes a sharp transition from the crystalline form to the fluid form as it is heated. However, a membrane containing 80% phospholipid and 20% cholesterol undergoes a more gradual change from crystalline to fluid form when heated over the same temperature range. Explain why.

22. Plants can synthesize trienoic acids (fatty acids with three double bonds) by introducing another double bond into a dienoic acid. Would you expect plants growing at higher temperatures to convert more of their dienoic acids into trienoic acids?

23. Why is fluidity greatest at the center of a lipid bilayer?

24. Use the simplified diagram of the plasma membrane to answer the following questions by choosing component A, B, C, D, or E.

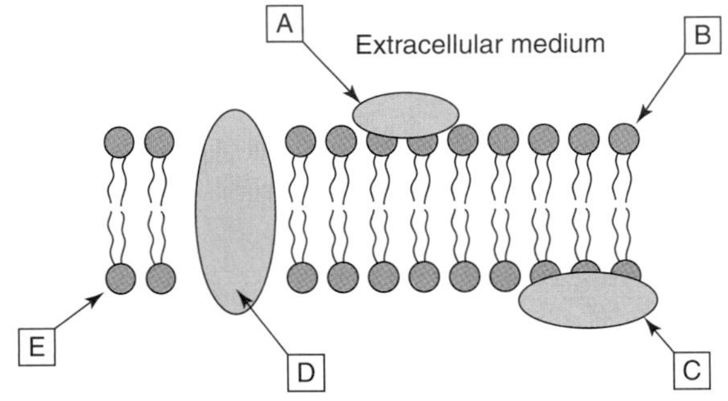

(a) This component could be a glycoprotein.
(b) This component can probably be separated from the others by simply washing the membrane with neutral salt solutions (mild conditions).
(c) In order to separate this component from the others, harsh conditions, such as strong detergents, are needed.
(d) This component is the only component that might bind and transport sodium ions across the membrane.
(e) This component could be a ceramide or a ganglioside.
(f) This component might be able to flip-flop transversely with the assistance of a flippase.

25. Cytochrome *c*, a protein of the electron transport chain in the inner mitochondrial membrane, can be removed by relatively mild means, such as extraction with salt solution. In contrast, cytochrome oxidase from the same source can be removed only by extraction into detergent solutions or organic solvents. What kind of membrane proteins are cytochrome *c* and cytochrome oxidase? Explain. Draw a schematic diagram of what each protein looks like in the membrane.

26. Glycophorin A is a 131-residue integral membrane protein that includes one bilayer-spanning segment. Identify that segment in the glycophorin A amino acid sequence (which uses one-letter abbreviations):

LSTTEVAMHTTTSSSVSKSYISSQTNDTHKRDTYAATPRAHE

VSEISVRTVYPPEEETGERVQLAHHFSEPEITLIIFGVMAGVIG

TILLISYGIRRLIKKSPSDVKPLPSPDTDVPLSSVEIENPETSDQ

27. Proteins that form a transmembrane β barrel always have an even number of β strands. (a) Explain why. (b) Why are the strands antiparallel? (c) Could some of them possibly be parallel?

28. Peptide hormones must bind to receptors on the extracellular surface of their target cells before their effects are communicated to the cell interior. In contrast, receptors for steroid hormones such as estrogen are intracellular proteins. Why is this possible?

29. In a famous experiment, Michael Edidin labeled the proteins on the surface of mouse and human cells with green and red fluorescent markers, respectively. The two types of cells were induced to fuse, forming hybrid cells. Immediately after fusion, green markers could be seen on the surface of one half of a hybrid cell, and red markers on the other half. After a 40-minute incubation at 37°C, the green and red markers became intermingled over the entire surface of the hybrid cell. If the hybrid cells were instead incubated at 15°C, this mixing did not occur. Explain these observations and why they supported the fluid mosaic model of membrane structure.

30. Identify each type of lipid-linked protein in the drawing.

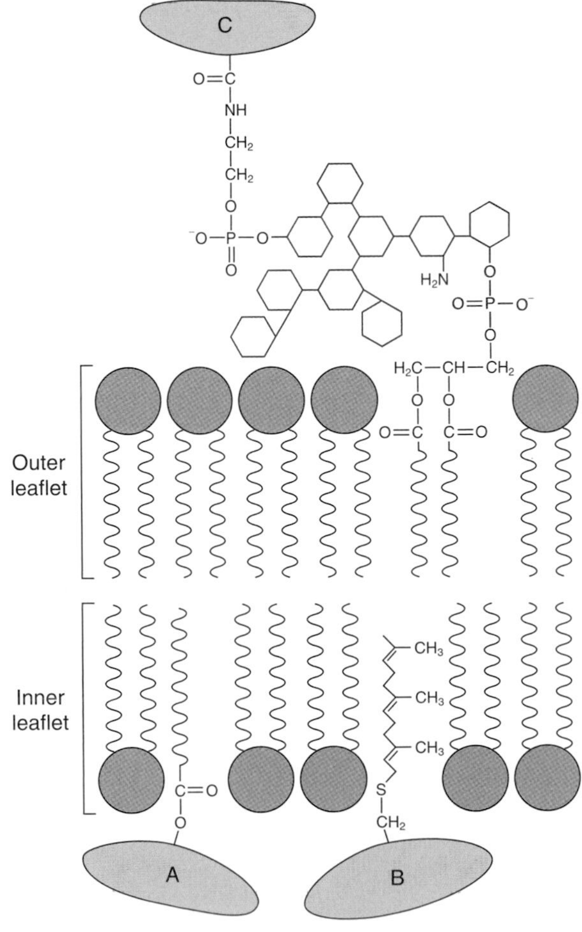

31. In fluorescence photobleaching recovery studies, fluorescent groups are attached to membrane components in a cell. An intense laser beam pulse focused on a very small area destroys (bleaches) the fluorophores in that area. What happens to the fluorescence in that area over time?

32. Explain why carbon dioxide can cross a cell membrane without the aid of a transport protein.

33. Rank the rate of transmembrane diffusion of the following compounds:

$H_3C-C(=O)-NH_2$ A. Acetamide

$H_3C-CH_2-CH_2-C(=O)-NH_2$ B. Butyramide

$HO-CH_2-CH_2-OH$ C. Ethylene glycol

$H_2N-C(=O)-NH_2$ D. Urea

34. A plot of the glucose transport rate versus glucose concentration for the passive glucose transporter of red blood cells is shown below. (a) Explain why the plot yields a hyperbolic curve. (b) What quantitative information does the curve reveal about the transporter?

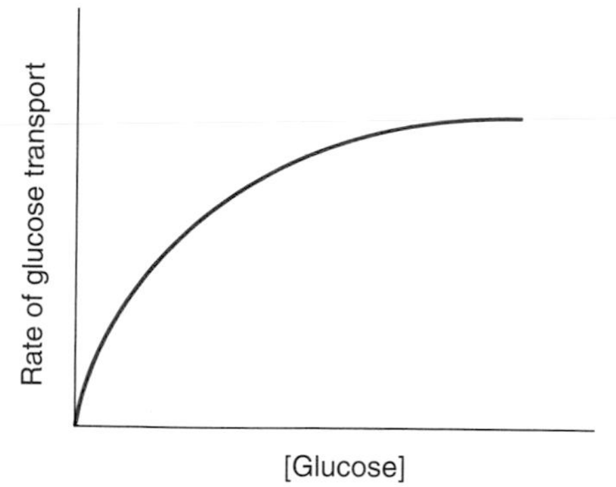

35. Using the data in Sample Calculation 8-2, calculate ΔG at 37°C when the membrane potential is (a) −50 mV (cytosol negative) and (b) +50 mV. In which case is Ca^{2+} movement thermodynamically favorable?

36. In eukaryotes, ribosomes (approximate mass 2.5×10^6 D) are synthesized inside the nucleus, which is enclosed by a double membrane. Protein synthesis occurs in the cytosol. (a) Would you expect that a protein similar to a porin or the glucose transporter would be responsible for transporting ribosomes into the cytoplasm? Explain. (b) Do you think that free energy would be required to move ribosomes from the nucleus to the cytoplasm? Why or why not?

37. A high fever can interfere with normal neuronal activity. Since temperature is one of the terms in Equation 8-1, which defines membrane potential, a fever could potentially alter a neuron's resting membrane potential. (a) Calculate the effect of a change in temperature from 98°F to 104°F (37°C to 40°C) on a neuron's membrane potential. Assume that the normal resting potential is −70 mV and that the distribution of ions does not change.

(b) How else might an elevated temperature affect neuronal activity?

38. In addition to neurons, muscle cells undergo depolarization, although smaller and slower than in the neuron, as a result of the activity of the acetylcholine receptor.

(a) The acetylcholine receptor is also a gated ion channel. What triggers the gate to open?
(b) The acetylcholine receptor/ion channel is specific for Na^+ ions. Do Na^+ ions flow in or out?
(c) How does the Na^+ flow through the ion channel change the membrane potential?

39. The drug known as Botox is a preparation of botulinum toxin. Describe the biochemical basis for its use by plastic surgeons, who inject small amounts of it to alleviate wrinkling in areas such as around the eyes.

40. Phosphatidylinositol is a membrane glycerophospholipid whose head group includes a monosaccharide (inositol) group. A certain kinase adds another phosphate group to a phosphorylated phosphatidylinositol:

$\xrightarrow{\text{kinase}}$

Why might this activity be required during the production of a new vesicle, which forms by budding from an existing membrane?

41. It has been well established that the cellular phospholipid membrane bilayer is asymmetric. Choline-containing lipids such as phosphatidylcholine (PC) and sphingomyelin (SM) are more likely to be found in the outer leaflet of the bilayer, whereas aminophospholipids such as phosphatidylserine (PS) and phosphatidylethanolamine (PE) are more likely to be found in the inner leaflet of the bilayer. Several investigators have used a variety of methods to examine the cellular mechanism that maintains this asymmetry. Several studies have indicated that a transmembrane protein nicknamed a "flippase" is partially responsible for maintaining phospholipid asymmetry in the bilayer. It has been hypothesized that the flip-

pase enzyme acts by "flipping" a phospholipid from one leaflet of the membrane to the other. Specific transport of some types of phospholipids but not others would create the asymmetric distribution that has been observed.

The investigators sonicated (disrupted with sound waves) solutions of a single kind of phospholipid in order to form phospholipid vesicles. The vesicles were then added to red blood cells. The phospholipids migrated from the vesicles to the red blood cells. The investigators then used a microscope to observe the shape of the red blood cells, which normally have a biconcave disk shape. If excess phospholipids are added to the outer membrane leaflet, the red blood cells become echinocytic or "spiky." But if the added phospholipids are acted upon by the flippase enzyme, they are flipped from the outer leaflet to the inner leaflet of the membrane. This results in an excess of phospholipids in the inner leaflet of the membrane. As a result, the red blood cell surface is stomatocytic (covered with "craters"). By observing which phospholipids could be flipped and which could not, the investigators were able to ascertain the properties of the flippase enzyme. In this case study, we examine some of these experiments and use our observations to describe the flippase enzyme in more detail. The experimental design is shown in the figure.

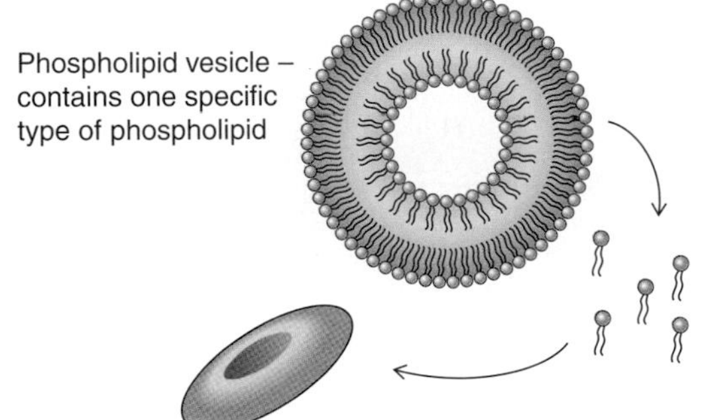

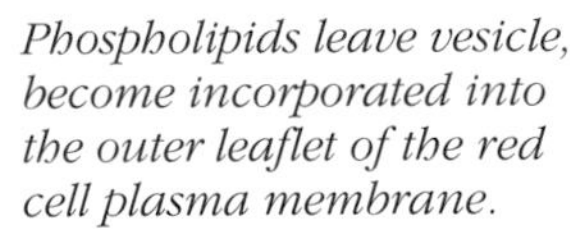

Phospholipids leave vesicle, become incorporated into the outer leaflet of the red cell plasma membrane.
Red blood cell becomes crenated (spiky) when exogenous phospholipids are incorporated into outer leaflet.
If the exogenous phospholipids are acted upon by the flippase, the phospholipids are translocated from the outer leaflet of the membrane to inner leaflet.

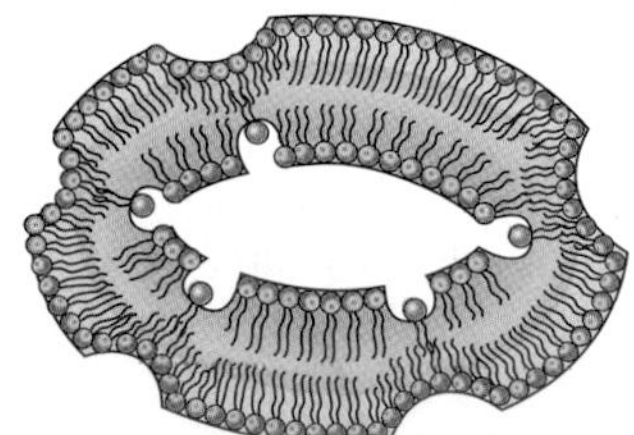

An excess of phospholipid in the inner leaflet of the membrane causes cells to become stomatocytic– they develop "craters."

In order to quantitate their results, the investigators used a numbering system.

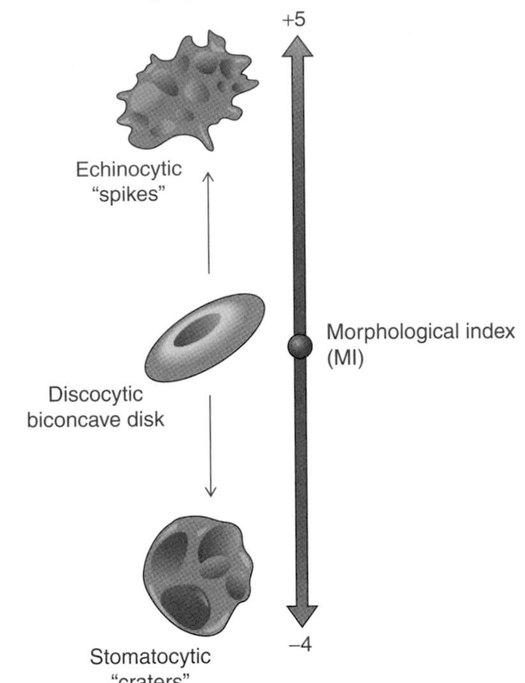

The morphological index (MI) scale for assessing red blood cell shape change

Biconcave red blood cells were given a score of zero. If the red blood cell became echinocytic, it was given a score from +1 to +5, indicating an increasing number of "spikes." Similarly, stomatocytic cells were given scores from −1 to −4, indicating an increasing number of "craters." In each experiment, a field of 100 cells was observed, and their scores were averaged to yield a value called the morphological index (MI). In this manner, the ability of the flippase enzyme to translocate phospholipids from the outer leaflet to the inner leaflet could be ascertained. These studies are important because it is known that the asymmetric distribution of lipids is characteristic of healthy cells. Under some circumstances, however, the membrane phospholipids may become "scrambled" and the asymmetry is lost. This occurs in cells about to undergo a process called *apoptosis*, or programmed cell death. An understanding of the apoptotic process has implications for the development of therapeutic drugs that might encourage cells such as cancer cells to undergo apoptotic death.

Experiments with phospholipid vesicles were carried out as shown above. First, phospholipid vesicles containing one specific type of phospholipid were incubated with red blood cells at 4°C (at this temperature, the flippase is inactive). Then the temperature was increased to 37°C to activate the flippase. The ability of the flippase to translocate three different types of phospholipids was measured. The phospholipids tested were dilauroylphosphatidylserine (DLPS), dilauroylphosphatidylethanolamine (DLPE), and dilauroylphosphatidylcholine (DLPC). The structures of these lipids are shown here.

$CH_2-O-C(=O)-(CH_2)_{10}-CH_3$
$CH-O-C(=O)-(CH_2)_{10}-CH_3$
$CH_2-O-P(=O)(O^-)-O-CH_2-CH(NH_3^+)-COO^-$

Dilauroylphosphatidylserine (DLPS)

$CH_2-O-C(=O)-(CH_2)_{10}-CH_3$
$CH-O-C(=O)-(CH_2)_{10}-CH_3$
$CH_2-O-P(=O)(O^-)-O-CH_2-CH_2-N^+(CH_3)_3$

Dilauroylphosphatidylcholine (DLPC)

$CH_2-O-C(=O)-(CH_2)_{10}-CH_3$
$CH-O-C(=O)-(CH_2)_{10}-CH_3$
$CH_2-O-P(=O)(O^-)-O-CH_2-CH_2-NH_3^+$

Dilauroylphosphatidylethanolamine (DLPE)

$CH_2-O-C(=O)-(CH_2)_{12}-CH_3$
$CH-O-C(=O)-(CH_2)_{12}-CH_3$
$CH_2-O-P(=O)(O^-)-O-CH_2-CH(NH_3^+)-COO^-$

Dimyristoylphosphatidylserine (DMPS)

The results are shown below.

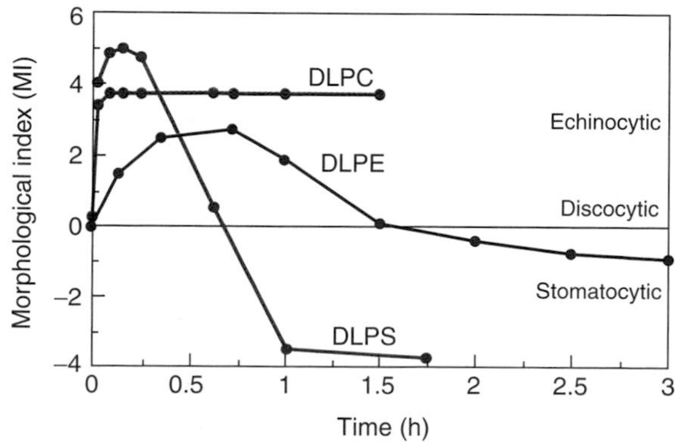

Shape changes induced by incubating red blood cells with phospholipid vesicles

(a) What type(s) of phospholipids is/are preferentially translocated by the flippase? Given the distribution of lipids in the membrane, what can you say about the distribution of charge on each side of the membrane?

(b) The ability of the flippase to translocate DLPS and lyso-PS was compared, and the results are shown here.

$CH_2-O-C(=O)-(CH_2)_{10}-CH_3$
$HO-CH$
$CH_2-O-P(=O)(O^-)-O-CH_2-CH(NH_3^+)-COO^-$

Lyso-phosphatidylserine (Lyso-PS)

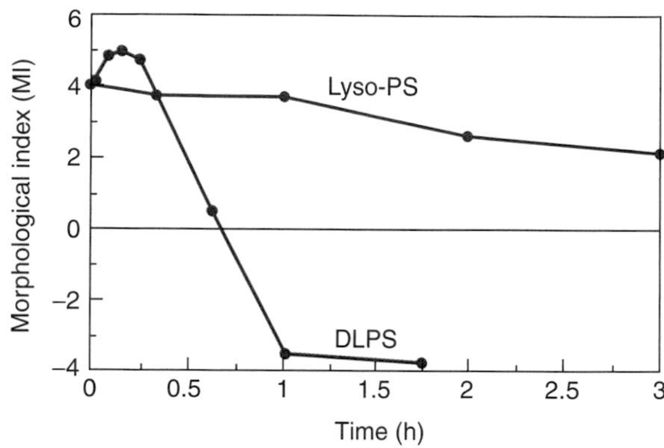

Shape changes in red blood cells induced by lyso-PS

What do these results tell you about the structural requirement for translocation activity?

(c) Additional experiments were carried out to ascertain the involvement of ATP. Previous experiments have shown that incubating red blood cells with iodoacetamide and inosine causes ATP levels to decrease by 60%. Experiments similar to those diagrammed above were carried out, but one batch of red blood cells was pretreated with iodoacetamide and inosine for 30 minutes, then phospholipid vesicles containing purified dimyristoylphosphatidylserine (DMPS) were added. The control batch of red blood cells was not treated with iodoacetamide and inosine. What is your interpretation of the results shown here?

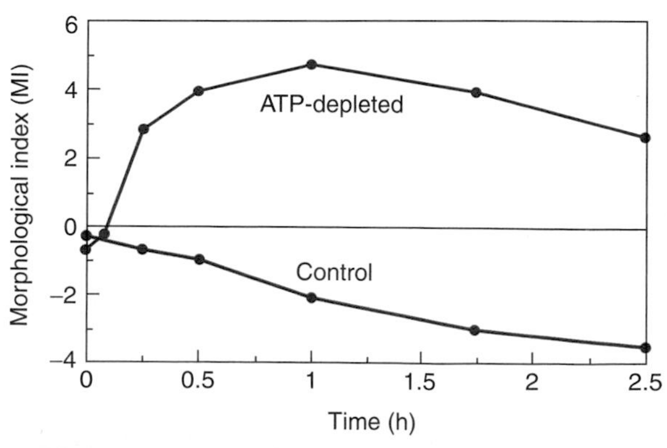

Effect of ATP depletion on red blood cell shape change

(d) Experiments were carried out to determine which amino acid side chains in the flippase enzyme were essential to its translocation activity. Red blood cells were pretreated with diamide, a reagent that modifies sulfhydryl groups.

$$\sim NH-CH(CH_2-SH)-C(=O)\sim \; + \; (CH_3)_2N-C(=O)-N{=}N-C(=O)-N(CH_3)_2 \text{ (Diamide)}$$

$$\downarrow$$

$$\sim NH-CH-C(=O)\sim \text{ with } CH_2-S-N(-C(=O)-N(CH_3)_2)-N{=}C(OH)-N(CH_3)_2$$

After the pretreatment, the shape change experiments were carried out in the usual manner. The results are shown below. What is your interpretation of these results?

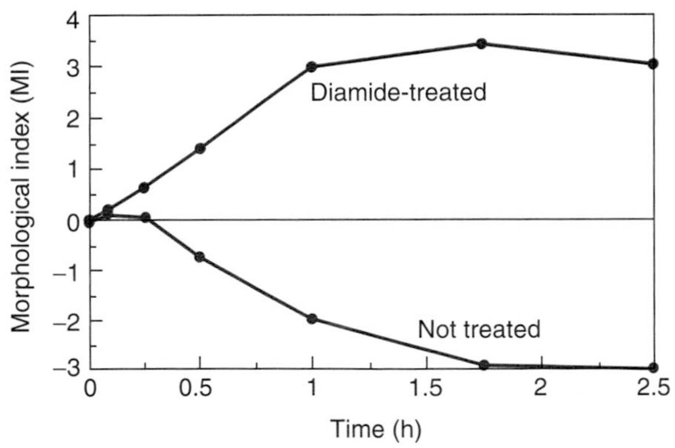

Effect of diamide treatment on DMPS-induced red blood cell shape change

(e) In the next series of experiments, red blood cells were treated with a mixture of an ionophore and a chelating agent (EDTA). The ionophore "pokes holes" in the membrane, so that ions leak out. The chelating agent binds the ions so they can't go back inside the cell. Following this treatment, the shape change experiments were carried out as described above. After 1.5 hours, Mg^{2+} ions were added back to the red blood cells. The results are shown below. What is your interpretation of these results?

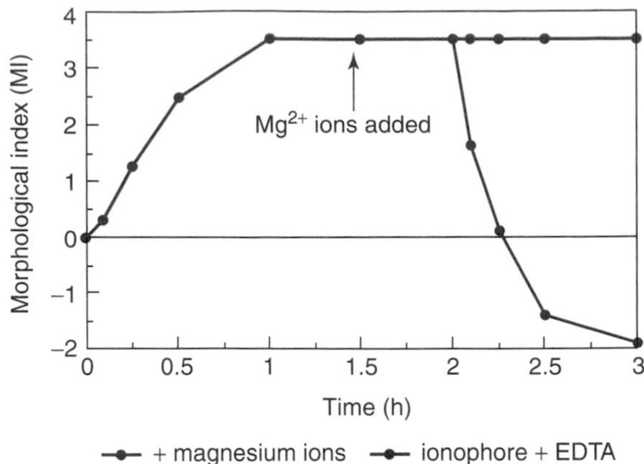

Effect of ion depletion on red blood cell shape change

(f) The experiments such as the ones described here give a good picture of the characteristics of an enzyme that has yet to be completely purified. Write a paragraph that summarizes the characteristics of the flippase enzyme. Include in your description an explanation of how the flippase functions to maintain phospholipid asymmetry.

[From Daleke, D.L. and Heustis, W.H., *Biochemistry* **24,** 5406–5416 (1985).]

42. The retina of the eye contains equal amounts of endothelial cells and cells called pericytes. Pericyte damage and basement membrane thickening occur during the early stages of diabetic retinopathy, which eventually leads to blindness. The investigators were interested in the factors responsible for causing pericyte damage. They noted that cells in the kidney that were similar in function to the pericytes possessed a sodium-coupled glucose transporter (SGLT), and they hypothesized that the pericytes might also have such a transporter that would facilitate and perhaps regulate the entry of glucose into the cell. Next, the investigators determined whether entry of glucose into the cell was correlated with synthesis of collagen, a component of the basement membrane. The cellular concentration of collagen increases in the basement membrane thickening process and contributes to pericyte cell damage. They also investigated several inhibitors of the glucose transport process.

(a) The investigators cultured pericytes and endothelial cells separately and measured their ability to take up glucose. Glucose uptake was measured both in the presence of sodium ions (provided by 145 mM sodium chloride) and in the absence of sodium (145 mM choline chloride was added to maintain constant osmolarity). What is your interpretation of the results shown here?

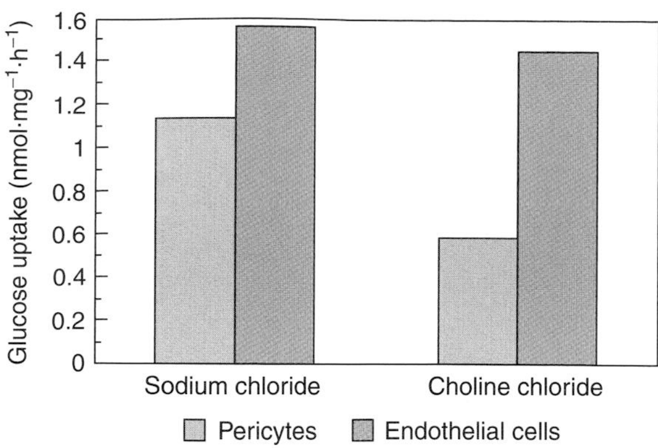

Effect of Na^+ on glucose uptake by two different types of retinal cells

(b) Next, the investigators measured the effect of increasing sodium concentration on glucose uptake in both pericytes and endothelial cells. What is your interpretation of the results shown? What information is conveyed by the shape of the curves?

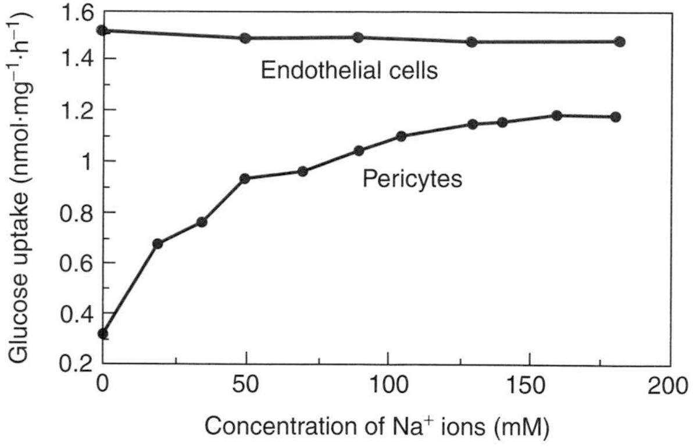

A dose-dependent response for the sodium-dependent uptake of glucose by pericytes and endothelial cells

(c) The kinetics of transport through protein transporters can be described using the language of Michaelis and Menten. The transported substance binding to its protein transporter is analogous to the substrate binding to an enzyme. Thus, K_M and V_{max} values can be determined for protein transporters. Use the information provided in the figure to estimate the K_M and V_{max} for glucose uptake by pericytes in the presence and absence of sodium. What information is conveyed by these values? Have the investigators demonstrated convincingly that an SGLT exists in pericytes? What about endothelial cells? Explain, using the above data to support your answers.

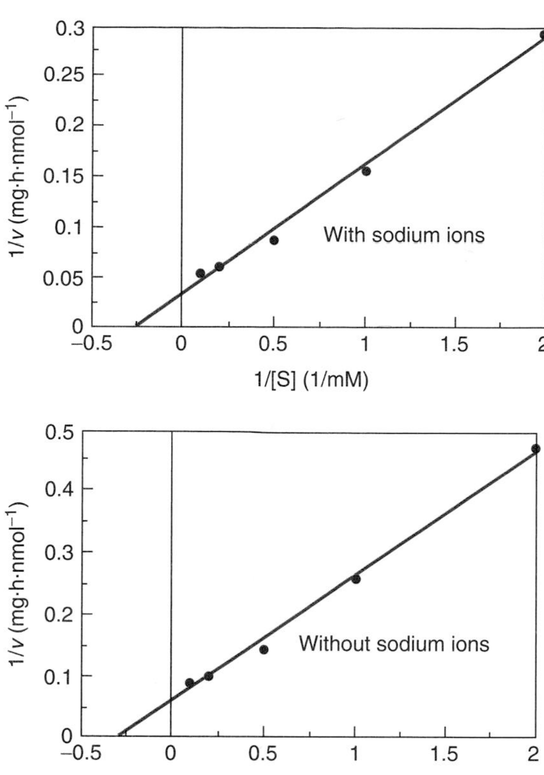

Lineweaver-Burk plots for the uptake of glucose by pericytes in the presence (145 mM NaCl) or absence (145 mM choline chloride) of sodium ions

(d) The activity of the SGLT in pericytes was investigated in the presence of inhibitors. Radioactively labeled glucose was added to cultured pericytes, in the absence (control) or presence of several potential inhibitors. The transport experiments were carried out both in the presence and absence of sodium ions. The results are shown below. What is the effect of

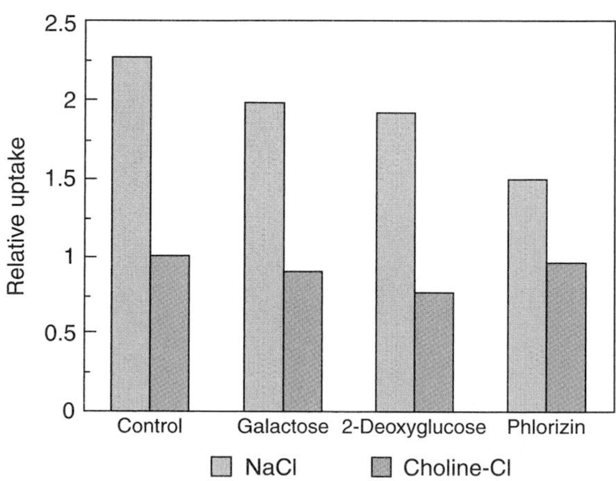

Effect of inhibitors on glucose uptake by pericytes in the presence and absence of sodium ions. The sugars were added at a concentration of 5 mM, while the phlorizin was added at a concentration of 0.2 mM.

sugars such as galactose and 2-deoxyglucose on glucose transport? What is the effect of 0.2 mM phlorizin on glucose transport?

(e) Next, the investigators wanted to explore the relationship between glucose uptake and synthesis of Type IV and Type VI collagen. They added increasing amounts of glucose to cultured pericytes and then measured the relative density of the cells as a result. (They confirmed in a separate immunoblotting experiment that increased collagen synthesis was responsible for the increased cellular density.) The results are shown below. How do you interpret these data?

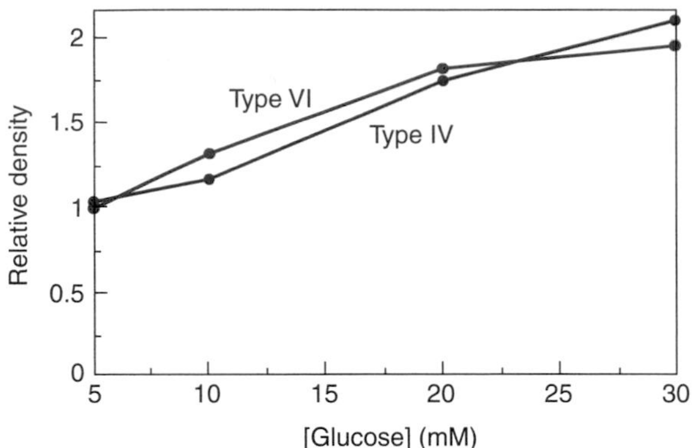

Effect of increasing glucose concentrations on the synthesis of Type IV and Type VI collagen

(f) The effect of the drug phlorizin on glucose transport and relative cellular density was examined. The results are shown below. What effect does phlorizin have on glucose transport at the two different concentrations of glucose? What effect does phlorizin have on the relative density of the cells at the two glucose concentrations?

(g) Diabetes is characterized by high levels of circulating blood glucose. Why do you think that retinopathy is more likely to occur in the diabetic patient? Describe the sequence of biochemical events leading to pericyte cell damage. What steps might be taken to decrease the incidence of diabetic retinopathy? Explain your answer, using the experimental results presented here.

[From Wakisaka, M., Yoshinari, M., Yamamoto, M., Nakamura, S., Asano, T., Himeno, T., Ichikawa, K., Doi, Y., and Fujishima, M., *Biochim. Biophys. Acta* **1362,** 87–96 (1997).]

43. You are a physician and your patients are three sisters, ages 21, 24, and 29, who are short of stature and obese. (There is a fourth sister who appears to be normal, as she is taller than the others. The parents also appear to be normal.) As children, the symptoms of the three sisters were similar: delayed mental and physical development, muscle weakness, and renal tubular acidosis. They frequently suffered bone fractures as children. X-Rays showed cerebral calcification and other skeletal abnormalities. After reviewing the sisters' medical histories, you draw samples of blood and send them to the laboratory for analysis. The laboratory reports that your patients all have a carbonic anhydrase II deficiency.

Humans have three isozymes of carbonic anhydrase, designated I, II, and III, which are all monomeric zinc metalloenzymes. The zinc ion is located at the bottom of the active-site cleft and is coordinated to the imidazole rings of three histidine residues.

The carbonic anhydrase II isozyme is found in bone, kidney, and brain, which is why these tissues are affected when the enzyme is deficient or nonfunctional. The carbonic anhydrase II enzyme is highly active, with one of the highest turnover rates of any known enzyme, and is critical in maintaining proper acid–base balance.

(a) Carbonic anhydrase catalyzes the reaction between water and carbon dioxide to yield carbonic acid. The carbonic acid then undergoes dissociation. Write the two equations that describe these processes.

(b) A genetic analysis of one of the sisters' genes indicates that a His→Tyr mutation at amino acid 107 is responsible for the carbonic anhydrase defect. Using what you know about amino acid structure, explain

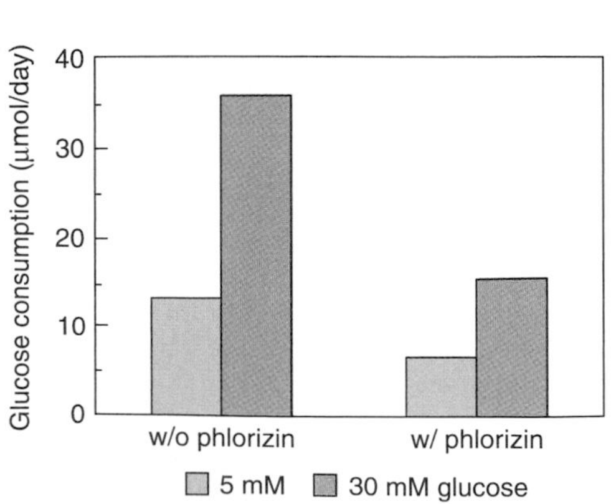

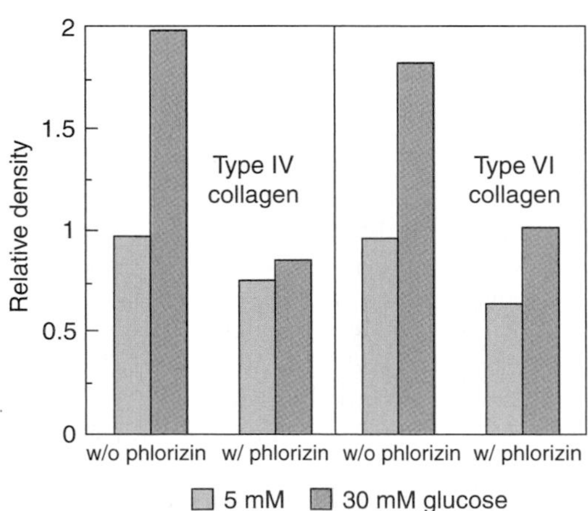

Effect of phlorizin on glucose transport (left) and collagen synthesis (right)

why such a mutation might result in an inactive enzyme, even though His 107 is not one of the residues involved in binding to the zinc ion.

(c) Osteoclasts in bone tissue are particularly rich in carbonic anhydrase II, and a proper functioning enzyme is critical to the development of healthy tissue. For proper bone development, the osteoclast must acidify its extracellular environment, referred to as the bone-resorbing compartment. Several transporters are involved in the acidification: a Na^+/H^+ exchanger, a Cl^-/HCO_3^- exchanger, and the Na^+,K^+-ATPase, which exchanges Na^+ and K^+ ions. A partial diagram of the osteoclast is shown below. Fill in the blanks in the diagram indicating the roles of carbonic anhydrase II and the exchangers in the acidification of the bone-resorbing compartment. Include the reactants and products of the appropriate intracellular reaction(s) and note in which direction each ion is transported in the osteoclast.

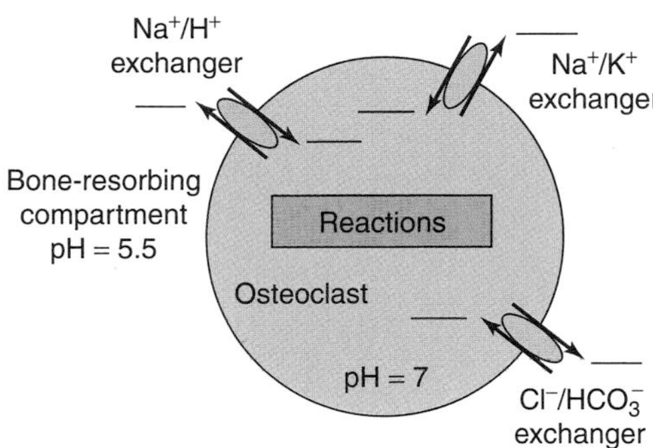

[From Sly, W.S. and Hu, P.Y., *Annu. Rev. Biochem.* **64,** 375–401 (1995), and Whyte, M.P., *Clin. Orthop. Relat. Res.* **294,** 52–63 (1993).]

44. The gene for a high-affinity liver choline transport protein in humans has recently been isolated. The choline transporter is an integral membrane protein that transports choline from the portal circulation to the cytosol of the liver cell. Choline is required for the synthesis of the phospholipid phosphatidylcholine.

$$HO—(CH_2)_2—\overset{CH_3}{\underset{CH_3}{N^+}}—CH_3$$

Choline

Experimenters transfected mouse cells with the gene for the choline transport protein (termed *mOct/Slc22a1*) and used this system to measure the kinetics of hepatic choline transport.

(a) The uptake of radioactively labeled choline by the transfected cells was measured at increasing choline concentrations. Using the language of Michaelis and Menten, K_M and V_{max} values can be obtained from the curve. The K_M value indicates the affinity of the choline for its transport protein whereas V_{max} indicates the maximal velocity of transport. Estimate K_M and V_{max} from the curve shown.

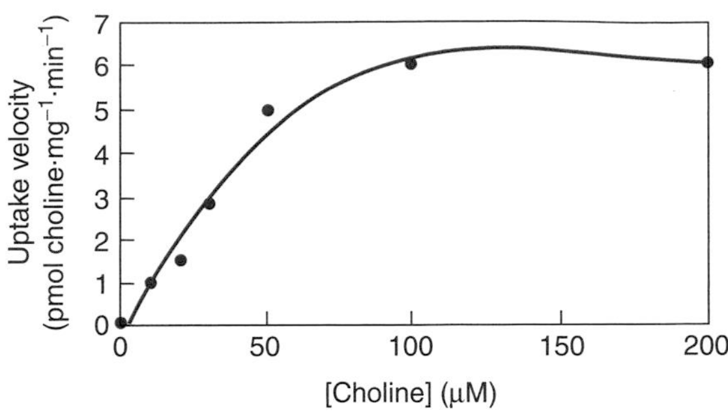

[^{3}H] Choline uptake by *mOct1/Slc221a1*-transfected cells

(b) Plasma concentrations of choline range from 10 to 80 μM, although concentrations may be higher in portal circulation after ingestion of choline. Does the mOct/Slc22a1 transporter operate effectively at these concentrations?

(c) The tetramethylammonium (TEA) ion inhibits the transport of choline. Explain how this is possible.

$$H_3C—CH_2—\overset{CH_3—CH_2}{\underset{CH_2—CH_3}{N^+}}—CH_2—CH_3$$

Tetraethylammonium (TEA)

(d) The mOct/Slc22a1 transporter is inhibited by low external pH and stimulated by high external pH. What role might protons play in the transport of choline?

[From Sinclair, C.J., Chi, K.D., Subramanian, V., Ward, K.L., and Green, R.M., *J. Lipid Res.* **41,** 1841–1847 (2000).]

45. The enzyme lecithin:cholesterol acyltransferase (LCAT) is an important enzyme in lipid metabolism. The enzyme is associated with high-density lipoprotein particles (HDL), which are complexes of phospholipid, cholesterol, cholesteryl esters, and protein. The HDL particle is organized so that the nonpolar cholesterol and cholesteryl esters are located inside the core of the particle, while the polar domains of the phospholipids and proteins are located on the particle surface. The LCAT enzyme has an important role in a process called "reverse cholesterol transport" in which excess cholesterol is moved from the peripheral tissues to the liver for excretion. For this reason, high plasma levels of HDL have been correlated with a low risk of heart disease.

The LCAT enzyme catalyzes the synthesis of cholesteryl ester (CE) from cholesterol and phosphatidylcholine ("lecithin" is another name for this lipid).

Phosphatidylcholine + Cholesterol

↓ LCAT

Lyso-phosphatidylcholine + Cholesteryl ester

Investigators conducting a study of the LCAT enzyme hypothesized that the fluidity of the HDL might influence LCAT activity. Since fluidity of the HDL particle is determined by its constituent phospholipids, the investigators synthesized HDL particles with different phosphatidylcholine compositions. For all phospholipids, the acyl chain in the #1 position was palmitate. The acyl chain in the #2 position was varied. LCAT activity was then measured for each type of synthetic HDL particle.

Examine the data in the table provided. Do these data support a link between LCAT activity and HDL fluidity?

LCAT kinetic constants for synthetic HDL containing phosphatidylcholine species with 20-carbon fatty acyl chains at the #2 position

PC type	Apparent K_M (mM cholesterol)	Apparent V_{max} (nmol CE · h^{-1} · mL^{-1})
20:4	2.0	1038
20:5	3.9	1469
20:3$\Delta^{5,8,11}$	1.1	623
20:3$\Delta^{8,11,14}$	1.6	584
20:3$\Delta^{11,1,17}$*	3.9	270

*Separate experiments indicate that this phospholipid is less fluid than either 20:3$\Delta^{5,8,11}$ or 20:3$\Delta^{8,11,14}$.

SELECTED READINGS

Chen, Y.A. and Scheller, R.H., SNARE-mediated membrane fusion, *Nature Rev. Mol. Cell Biol.* **2,** 98–106 (2001). [Reviews SNARE protein structure and the role of SNAREs in lipid bilayer fusion.]

Doyle, D.A., Cabral, J.M., Pfuetzner, R.A., Kuo, A., Gulbis, J.M., Cohen, S.L., Chait, B.T., and MacKinnon, R., The structure of the potassium channel: Molecular basis of K^+ conduction and selectivity, *Science* **280,** 69–77 (1998). [Describes the structure and mechanism of ion transport through a transmembrane channel protein.]

Drickamer, K. and Taylor, M., Evolving views of protein glycosylation, *Trends Biochem. Sci.* **23,** 321–324 (1998). [Summarizes the synthesis of protein-linked oligosaccharides and their evolution as recognition markers.]

Edidin, M., Shrinking patches and slippery rafts: Scale of domains in the plasma membrane, *Trends Cell Biol.* **11,** 492–496 (2001). [Summarizes current views of the structures, dynamics, and functions of membrane domains.]

Nagle, J.F. and Tristram-Nagle, S., Lipid bilayer structure, *Curr. Opin. Struct. Biol.* **10,** 474–480 (2000). [Explains why it is difficult to quantitatively describe the structure of the lipid bilayer.]

Popot, J.-L. and Engelman, D.M., Helical membrane protein folding, stability, and evolution, *Annu. Rev. Biochem.* **69,** 881–922 (2000). [Shows a number of protein structures and discusses many features of transmembrane proteins.]

Schulz, G.E., β-Barrel membrane proteins, *Curr. Opin. Struct. Biol.* **10,** 443–447 (2000). [Reviews the basic principles of construction for transmembrane β barrels, including the smallest, an eight-stranded barrel.]

Sui, H., Han, B.-G., Lee, J.K., Walian, P., and Jap, B.K., Structural basis of water-specific transport through the AQP1 water channel, *Nature* **414,** 872–878 (2001). [Describes the X-ray structure of the water pore and explains why it does not transport protons.]

PART II

METABOLIC REACTIONS

CHAPTER 9

Overview of Mammalian Metabolism and Free Energy

Malnutrition is not just a matter of inadequate caloric intake but is related to the chemical content of the available food. For example, many of the world's poorest individuals subsist largely on maize (corn)—which is deficient in the essential amino acids lysine and tryptophan—and do not consume adequate amounts of other foods (such as legumes or milk) that could provide the missing nutrients. A maize variety dubbed Quality Protein Maize can alleviate malnutrition by supplying up to twice the amounts of Lys and Trp present in standard maize varieties. This is the result of a mutation in a gene that controls the synthesis of several seed storage proteins. Normally, about half of this protein is zein, which is almost devoid of Lys. The mutation reduces zein synthesis, thereby increasing the proportion of other seed storage proteins that have higher levels of Lys and Trp. After 30 years of selective breeding and field testing, the improved maize also meets other criteria essential for its long-term success: It is pest-resistant, it yields up to 10% more grain than other maize varieties, and—perhaps most importantly—its taste and texture are acceptable to consumers.

[Alan Pitcairn/Grant Heilman Photography.]

LEARNING OBJECTIVES

1. Understand how food molecules are digested, absorbed, stored, and mobilized.
2. Understand that catabolic pathways produce a few kinds of intermediates that are also used in anabolic pathways.
3. Understand that electrons released in oxidative processes can be transferred to cofactors.
4. Understand the meaning of free energy changes and their relationship to equilibrium constants.
5. Understand why ATP often participates in coupled reactions.
6. Understand how reduced cofactors, phosphorylated compounds, and thioesters serve as energy currency in the cell.

THIS CHAPTER IN CONTEXT

This chapter begins a set of eight that explore some of the major themes of metabolism, the processes by which organisms consume and produce matter and energy in the synthesis and degradation of biomolecules. A catalog of all the metabolic reactions undertaken by plants, animals, and bacteria is far beyond the scope of this book. Instead, we will examine a few common metabolic processes, focusing primarily on mammalian systems. In this chapter, we will present an overview of how the molecules we have already introduced—amino acids, nucleotides, saccharides, and lipids—are broken down, rebuilt, and transformed into other substances. We will also look at the meaning and role of free energy in metabolic reactions.

THE MICROBE KNOWN AS *METHANOCOCCUS JANNASCHII* IS A strange creature. It colonizes sulfur-rich hydrothermal vents, growing at temperatures between 48 and 94°C and at pressures of over 200 atm. It is known as a **chemoautotroph** (from the Greek *trophe,* nourishment) because it obtains virtually all its metabolic building material and free energy from the simple inorganic compounds CO_2, N_2, H_2, and S_2. **Photoautotrophs,** such as the familiar green plants, need little more than CO_2, H_2O, a source of nitrogen, and sunlight. In contrast, **heterotrophs,** a group that includes animals, directly or indirectly obtain all their building materials and free energy from organic compounds produced by chemo- or photoautotrophs. Despite their different trophic strategies, all organisms have remarkably similar cellular structures, make the same types of biomolecules, and use similar enzymes to build and break down those molecules.

In all cases, cells break down or **catabolize** large molecules to release free energy and small molecules. The cells then use the free energy and small molecules to rebuild larger molecules, a process called **anabolism** (Fig. 9-1). The set of all catabolic and anabolic activities constitutes an organism's **metabolism.** In the next few chapters, we will examine some catabolic processes that release free energy and some anabolic processes that consume free energy. But first we will introduce a few of the major molecular players in metabolism, including their precursors and degradation products, and further explore the meaning of free energy in biological systems.

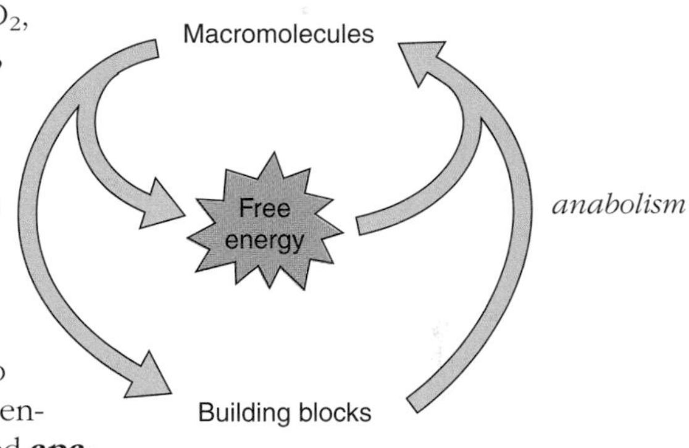

FIGURE 9-1

Catabolism and anabolism. Catabolic (degradative) reactions yield free energy and small molecules that can be used for anabolic (synthetic) reactions. Metabolism is the sum of all catabolic and anabolic processes.

1 FOOD, FUEL, AND ELECTRONS

As heterotrophs, mammals rely on food produced by other organisms. After food is digested and absorbed, it becomes a source of metabolic energy and materials to support the animal's growth and other activities. The human diet includes the four types of biological molecules introduced in Section 1-2 and described in more detail in subsequent chapters. A typical meal consists largely of polymers, including proteins, nucleic acids, polysaccharides, and triacylglycerols (technically, a triacylglycerol, or fat molecule, is not a polymer since it consists of only three monomeric units—fatty acids—linked not to each other as in the other polymers, but to a different molecule, glycerol). *Digestion reduces the polymers to their monomeric components: amino acids, nucleotides, monosaccharides, and fatty acids.* The breakdown of nucleotides does not yield significant amounts of metabolic free energy, so we will devote more attention to the catabolism of other types of biomolecules.

The first stage of digestion takes place extracellularly in the mouth, stomach, and intestine and is catalyzed by hydrolytic enzymes (Fig. 9-2). For example, salivary amylase begins to break down starch, which is a combination of linear polymers of glucose residues (called amylose) and branched polymers (called amylopectin). Gastric and pancreatic proteases (including trypsin, chymotrypsin, and elastase; see Chapter 6) degrade proteins to small peptides and amino acids. Lipases synthesized by the pancreas and secreted into the small intestine catalyze the release of fatty acids from triacylglycerols.

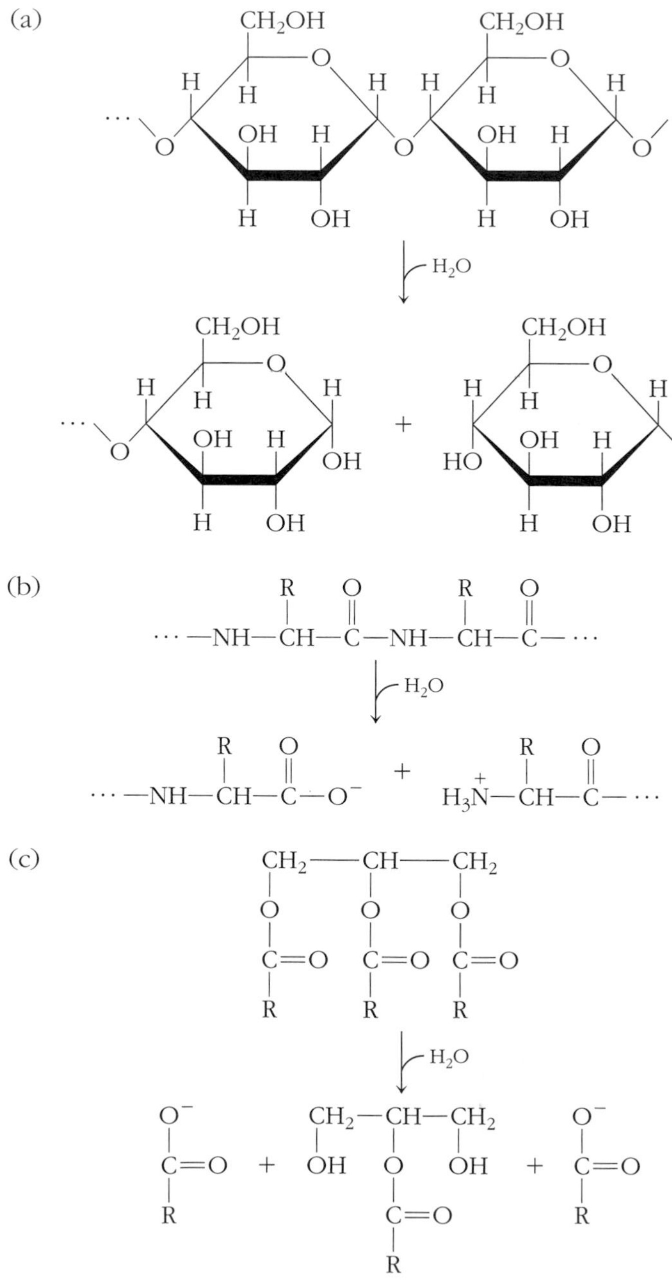

FIGURE 9-2

Digestion of biopolymers. These hydrolytic reactions are just a few of those that occur during food digestion. In each example, the bond to be cleaved is colored red. (a) The linear chains of glucose residues in starch are hydrolyzed by amylases. (b) Proteases catalyze the hydrolysis of peptide bonds in proteins. (c) Lipases hydrolyze the ester bonds linking fatty acids to the glycerol backbone of triacylglycerols.

Water-insoluble lipids do not freely mix with the other digested molecules but instead form micelles (see Fig. 2-10).

Cells take up the products of digestion

The products of digestion are absorbed by the cells lining the intestine. Monosaccharides enter the cells via active transporters such as the Na^+-glucose system diagrammed in Figure 8-26. Similar symport systems bring amino acids and di- and tripeptides into the cells. Some highly hydrophobic lipids diffuse through the cell membrane; others require transporters. Inside the cell, the triacylglycerol digestion products re-form triacylglycerols, and some fatty acids are linked to cholesterol to form cholesteryl esters, for example,

Cholesteryl stearate

Triacylglycerols and cholesteryl esters are packaged, together with specific proteins, to form **lipoproteins.** These particles, known specifically as chylomicrons, are released into the lymphatic circulation before entering the bloodstream for delivery to tissues.

Soluble substances such as amino acids and monosaccharides leave the intestinal cells and enter the portal vein, which drains the intestine and other visceral organs, and are delivered directly to the liver. *The liver therefore receives the bulk of a meal's nutrients and catabolizes them, stores them, or releases them back into the bloodstream.* The liver also takes up chylomicrons and repackages the lipids with different proteins to form other lipoproteins, which circulate throughout the body, carrying cholesterol, triacylglycerols, and other lipids (lipoproteins are discussed in greater detail in Chapter 14).

REVIEW LEARNING OBJECTIVE 1

- Review the steps by which nutrients from food molecules reach the body's tissues.

Monomers are stored as polymers

Immediately following a meal, the concentrations of monomeric compounds are high. Fatty acids are used to build triacylglycerols, many of which travel in the form of lipoproteins to adipose tissue and are stored in fat globules in the adipocytes (Fig. 9-3). Some monosaccharides are immediately catabolized to produce free energy. Some are used to synthesize glycogen, the storage polymer of glucose, primarily in liver and muscle. Glycogen resembles the amylopectin component of starch (a plant product) but is more highly branched (Fig. 9-4). This branched structure means that a single glycogen molecule can be expanded quickly, by adding glucose residues to its many ends, and de-

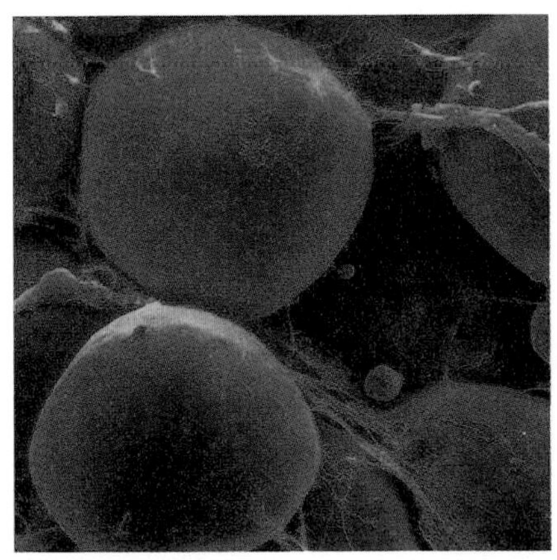

FIGURE 9-3

Adipocytes. These cells, which make up adipose tissue, contain a small amount of cytoplasm surrounding a large globule of triacylglycerols (fat). [*CNRI/Phototake.*]

(a)

Branch point

(b)

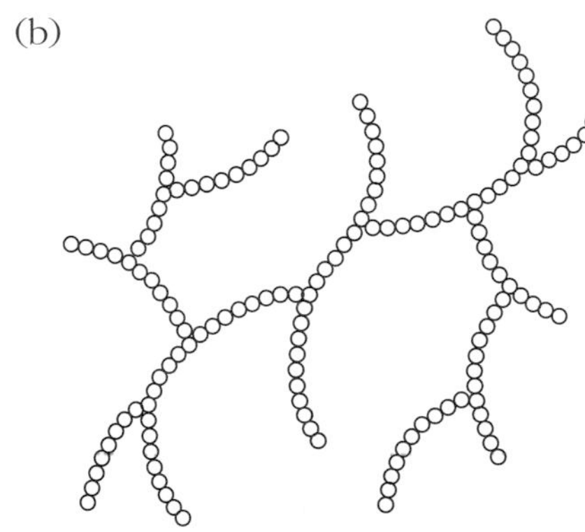

FIGURE 9-4

Structure of glycogen.
(a) Glycogen is a polymer of glucose residues linked at their C1 and C4 positions, with occasional branch points linking C1 and C6 positions. (b) With branches occurring every 8 to 14 residues, a glycogen molecule can store a large amount of glucose in a compact form.

graded quickly, by simultaneously removing glucose from the ends of many branches. Glucose that does not become part of glycogen can be partially catabolized and converted into fatty acids for storage as triacylglycerols.

Amino acids can be used to build polypeptides. A protein is not a dedicated storage molecule for amino acids, as glycogen is for glucose, so excess amino acids cannot be saved for later. However, in certain cases, such as during starvation, proteins are catabolized to supply the body's energy needs. If the intake of amino acids exceeds the body's immediate protein-building needs, the excess amino acids can be broken down, converted to carbohydrate, and stored as glycogen.

Amino acids and glucose are both required to synthesize nucleotides. Asp, Gln, and Gly supply some of the carbon and nitrogen atoms used to build the purine and pyrimidine bases. The glucose derivative glucose-6-phosphate is converted to ribose-5-phosphate, the precursor of the monosaccharide component of nucleotides:

Glucose-6-phosphate

Ribose-5-phosphate

Fuels are mobilized as needed

Amino acids, monosaccharides, and fatty acids are known as ***metabolic fuels*** *because they can be broken down by processes that make free energy available for the cell's activities.* After a meal, free glucose and amino acids are catabolized to release their free energy. When these fuel supplies are exhausted, the body **mobilizes** its stored resources; that is, it converts its polysaccharide and triacylglycerol storage molecules (and sometimes proteins) to their respective monomeric units. Most of the body's tissues prefer to use glucose as their primary metabolic fuel, and the central nervous system can run on almost nothing else. In response to this demand, the liver mobilizes glucose by breaking down glycogen.

In general, depolymerization reactions are hydrolytic, but in the case of glycogen, the molecule that breaks the bonds between glucose residues is not water but phosphate. Thus, the degradation of glycogen is called **phosphorolysis.** The reaction is catalyzed by glycogen phosphorylase, which releases residues from the ends of branches in the glycogen polymer.

$$^{-}O-P(=O)(O^{-})-OH \;+\; \text{Glycogen} \;\longrightarrow\; \text{Glucose-1-phosphate} \;+\; \cdots$$

Glycogen

Glucose-1-phosphate

The phosphate group of glucose-1-phosphate is removed before glucose is released from the liver into the circulation.

Only when the supply of glucose runs low does adipose tissue mobilize its fat stores. A lipase hydrolyzes triacylglycerols so that fatty acids can be released into the bloodstream. These free fatty acids are not water-soluble and therefore bind to circulating proteins. Except for the heart, which uses fatty acids as its primary fuel, the body does not have a budget for burning fatty acids. In general, as long as dietary carbohydrates and amino acids can meet the body's energy needs, stored fat will not be mobilized, even if the diet includes almost no fat. This feature of mammalian fuel metabolism is a source of misery for many dieters!

Amino acids are not mobilized to generate energy except during a fast, when glycogen stores are depleted (in this situation, the liver can also convert amino acids into glucose). However, cellular proteins are continuously degraded and rebuilt with the changing demand for particular enzymes, transporters, cytoskeletal elements, and so on. There are two major mechanisms for degrading unneeded proteins:

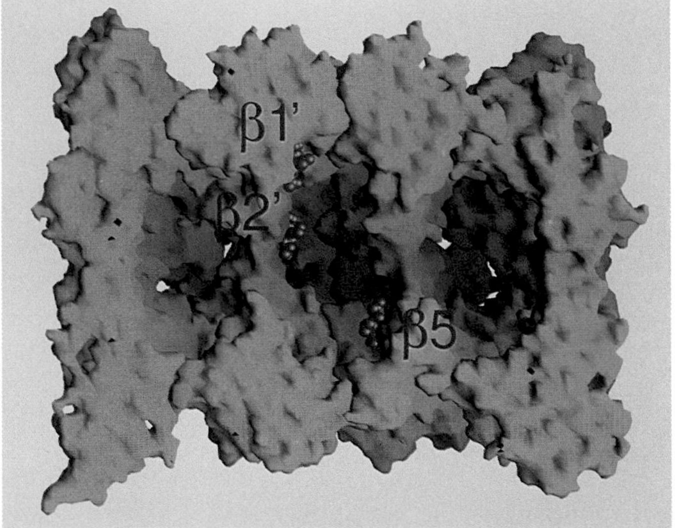

FIGURE 9-5

Structure of the yeast proteasome core. This cutaway view shows the inner chamber, where proteolysis occurs. Additional protein complexes (not shown) assist the entry of proteins into the proteasome. The red structures mark the locations of three protease active sites. [*Courtesy Robert Huber, Max-Planck-Institut fur Biochemie, Germany.*]

1. The **lysosome,** an organelle containing proteases and other hydrolytic enzymes, breaks down proteins that are enclosed in a membranous vesicle, including extracellular proteins taken up by endocytosis.

2. Some proteins are broken down inside a barrel-shaped structure known as a **proteasome.** The 700-kD core of this multiprotein complex encloses an inner chamber with multiple active sites that carry out peptide bond hydrolysis (Fig. 9-5). A protein can enter the proteasome only after it has been covalently tagged with a small protein called

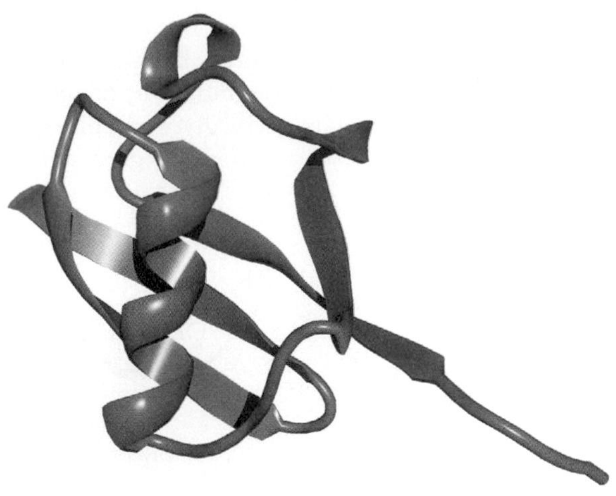

FIGURE 9-6

Ubiquitin.

Several copies of this 76-residue protein are linked to Lys residues in proteins that are to be degraded by a proteasome. [*Structure (pdb 1UBQ) determined by S. Vijay-Kumar, C.E. Bugg, and W.J. Cook.*]

ubiquitin (Fig. 9-6). The structural features that allow a protein to be ubiquitinated, and therefore targeted for destruction, are not completely understood, but the system is sophisticated enough to allow unneeded or defective proteins to be destroyed while sparing essential proteins.

Glucose

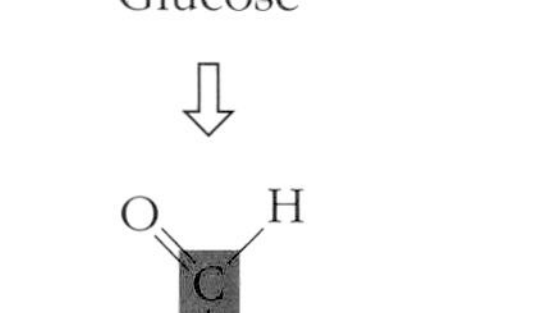

Glyceraldehyde-3-phosphate

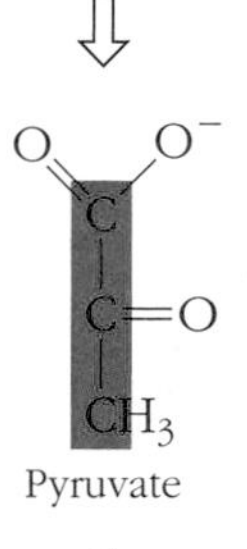

Pyruvate

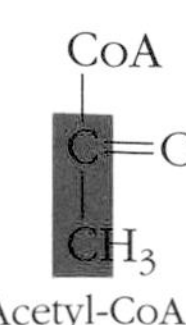

Acetyl-CoA

FIGURE 9-7

Some intermediates resulting from glucose catabolism.

REVIEW LEARNING OBJECTIVE 1

- What are metabolic fuels and how are they stored?
- How and when are they mobilized?

The major catabolic pathways yield a few common intermediates

The interconversion of a biopolymer and its monomeric units is usually accomplished in just one or a few enzyme-catalyzed steps. In contrast, many steps are required to break down monomeric compounds or build them up from smaller precursors. These series of reactions are known as **metabolic pathways.** *A handful of small molecules appear as precursors or products in the pathways that lead to or from virtually all other types of biomolecules.* These small molecules, or **intermediates,** are worth examining at this point, since they will reappear several times in the coming chapters.

In **glycolysis,** the pathway that degrades the monosaccharide glucose, the six-carbon sugar is phosphorylated and split in half, yielding two molecules of glyceraldehyde-3-phosphate (Fig. 9-7). This compound is then converted in several more steps to another three-carbon molecule, pyruvate. The decarboxylation of pyruvate (removal of a carbon atom as CO_2) yields acetyl-CoA, in which a two-carbon acetyl group is linked to a carrier molecule (called coenzyme A or CoA).

Glyceraldehyde-3-phosphate, pyruvate, and acetyl-CoA are key players in other metabolic pathways. For example, glyceraldehyde-3-phosphate is the metabolic precursor of the three-carbon glycerol backbone of triacylglycerols. In plants, it is also the entry point for the carbon "fixed" by photosynthesis;

in this case, two molecules of glyceraldehyde-3-phosphate combine to form a six-carbon monosaccharide. Pyruvate can undergo a reversible amino-group transfer reaction to yield alanine:

$$\underset{\text{Pyruvate}}{\begin{array}{c} COO^- \\ | \\ C{=}O \\ | \\ CH_3 \end{array}} \qquad \underset{\text{Alanine}}{\begin{array}{c} COO^- \\ | \\ H_3\overset{+}{N}{-}C{-}H \\ | \\ CH_3 \end{array}}$$

This makes pyruvate both a precursor of an amino acid and the degradation product of one. Pyruvate can also be carboxylated to yield oxaloacetate, a four-carbon precursor of several other amino acids:

$$\underset{\text{Pyruvate}}{\begin{array}{c} COO^- \\ | \\ C{=}O \\ | \\ CH_3 \end{array}} + CO_2 + ATP \xrightarrow[\text{pyruvate carboxylase}]{} \underset{\text{Oxaloacetate}}{\begin{array}{c} COO^- \\ | \\ C{=}O \\ | \\ CH_2 \\ | \\ COO^- \end{array}} + ADP + P_i$$

Fatty acids are built by the sequential addition of two-carbon units derived from acetyl-CoA; fatty acid breakdown yields acetyl-CoA. These relationships are summarized in Figure 9-8. If not used to synthesize other compounds, the two- and three-carbon intermediates can be broken down to CO_2 by the **citric acid cycle,** a metabolic pathway that carries out the final stages of catabolism of metabolic fuels.

REVIEW LEARNING OBJECTIVE 2

- Why are compounds such as glyceraldehyde-3-phosphate, pyruvate, and acetyl-CoA so important in metabolism?

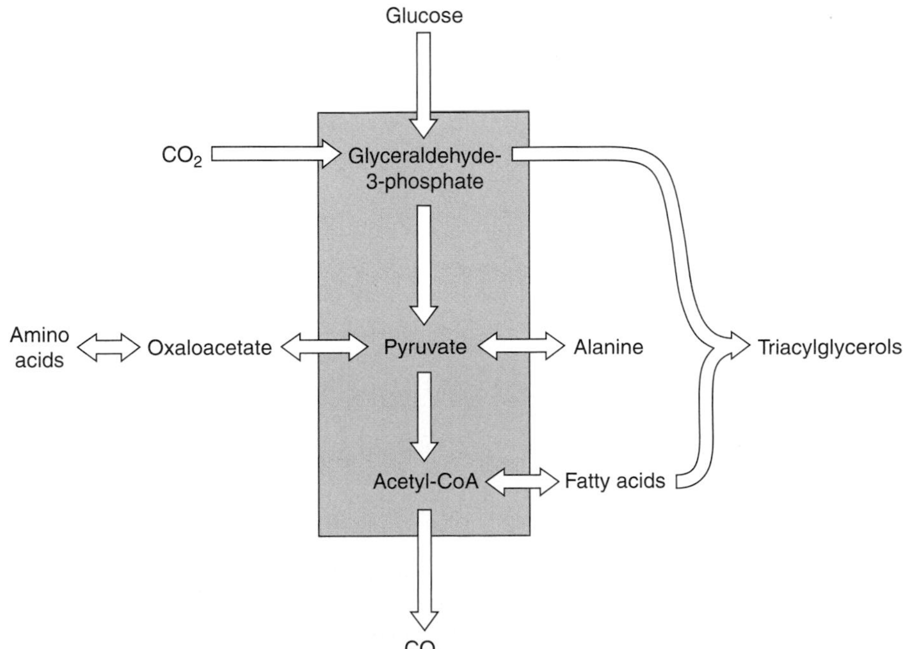

FIGURE 9-8

Some of the metabolic roles of the common intermediates.

Reduced cofactors

In general, *the catabolism of amino acids, monosaccharides, and fatty acids is a process of oxidation* (**oxidation** is the loss of electrons). For example, the saturated (reduced) methylene (—CH_2—) groups of a fatty acid are eventually released as CO_2, in which the carbon is fully oxidized (for a review of oxidation states, see Table 1-3). In contrast, *the synthesis of amino acids, monosaccharides, and fatty acids from smaller precursors is generally a process of reduction* (**reduction** is the gain of electrons). Oxidation–reduction reactions involve the transfer of electrons (symbolized e^-) so that as one compound is oxidized and gives up its electrons, another compound is reduced as it accepts the electrons. When a metabolic fuel is oxidized, its electrons are collected by a coenzyme:

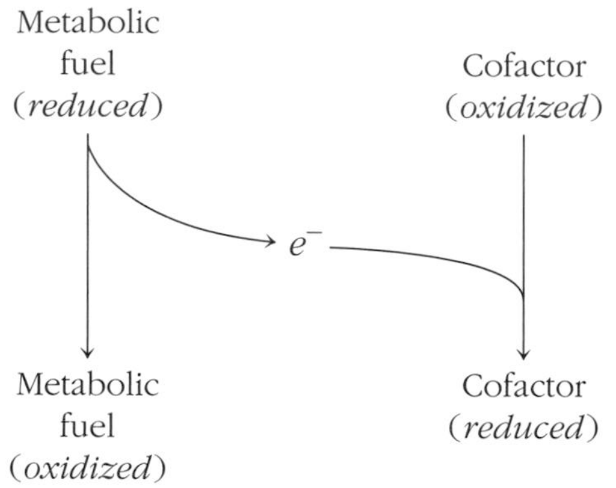

Metabolic fuel (*reduced*) → Metabolic fuel (*oxidized*)

Cofactor (*oxidized*) → Cofactor (*reduced*)

e^-

A **coenzyme** is an organic group that participates in a reaction catalyzed by an enzyme. Some coenzymes are covalently bound to the enzyme as prosthetic groups (for example, heme is a prosthetic group in some enzymes and in myoglobin and hemoglobin; see Section 4-4). Other coenzymes behave more like substrates in an enzyme-catalyzed reaction and diffuse away after the reaction. The term **cofactor** includes coenzymes as well as metal ions that are essential for an enzyme's activity, but the terms *cofactor* and *coenzyme* are often used interchangeably.

The cofactors most relevant to fuel metabolism in mammalian cells are the nicotinamide coenzymes and coenzyme Q. Nicotinamide adenine dinucleotide (NAD^+) and the related compound nicotinamide adenine dinucleotide phosphate ($NADP^+$) consist of two linked nucleotides, one containing an adenine base and the other a base derived from nicotinic acid (also known as the vitamin niacin; Fig. 9-9). In general, NAD^+ participates in catabolic reactions and $NADP^+$ in anabolic reactions, but the two compounds are almost identical from a thermodynamic point of view.

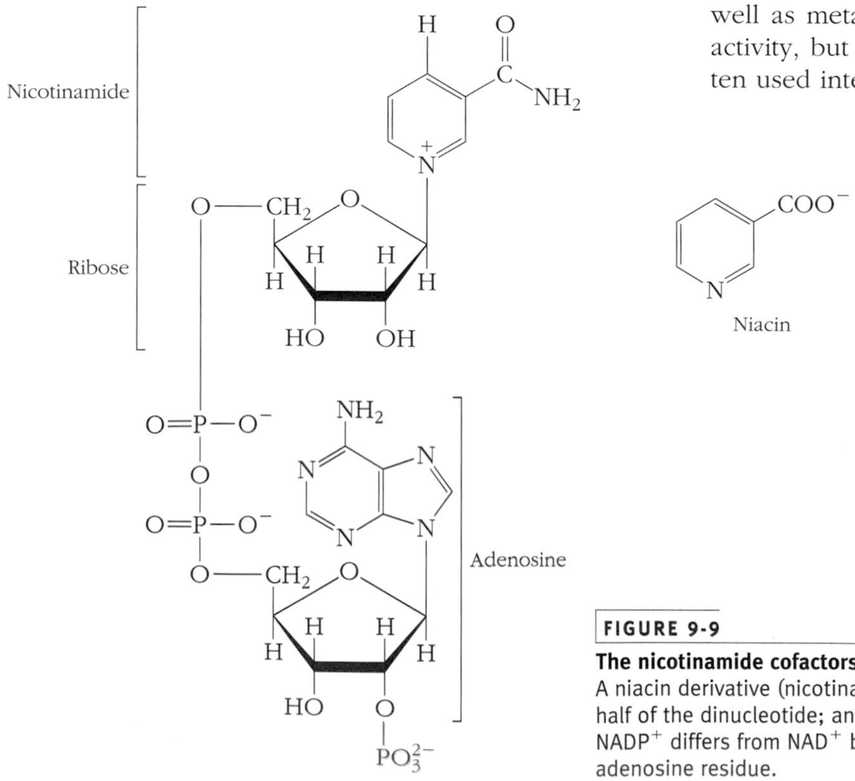

Nicotinamide

Ribose

Adenosine

Niacin

FIGURE 9-9

The nicotinamide cofactors.

A niacin derivative (nicotinamide) and a ribose group together make up one-half of the dinucleotide; an adenine nucleotide makes up the other half. $NADP^+$ differs from NAD^+ by the addition of a phosphate group at C2′ of its adenosine residue.

The active portion of a nicotinamide cofactor is the nicotinamide group. This group is reduced by the enzyme-catalyzed transfer of a hydride ion (a proton with two electrons, H:$^-$) to produce either NADH or NADPH:

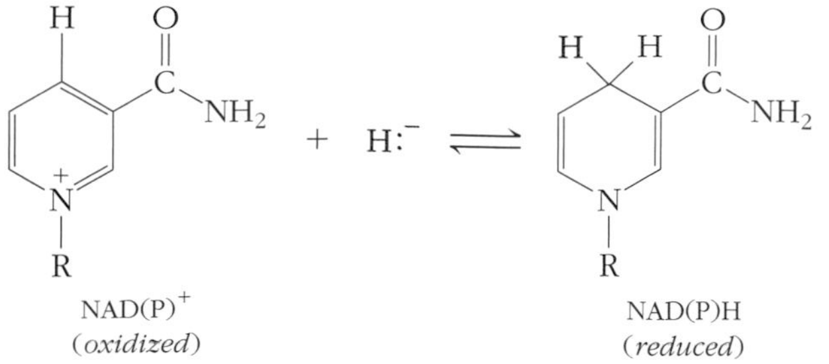

$$\text{NAD(P)}^+ + \text{H:}^- \rightleftharpoons \text{NAD(P)H}$$

NAD(P)$^+$
(oxidized)

NAD(P)H
(reduced)

The reaction is reversible, so *the reduced cofactor can become oxidized by giving up the hydride ion with its two electrons.* Because NAD$^+$ and NADP$^+$ are soluble in aqueous solution, they can travel throughout the cell, shuttling electrons from reduced compounds to oxidized compounds.

Many cellular oxidation–reduction reactions take place at membrane surfaces, for example, in the inner membranes of mitochondria and chloroplasts in eukaryotes and in the plasma membrane of prokaryotes. In these cases, a membrane-associated enzyme may transfer electrons from a substrate to a lipid-soluble electron carrier rather than to a soluble electron carrier such as NAD$^+$. Ubiquinone, or coenzyme Q (abbreviated Q) is such an electron carrier. Its hydrophobic tail, consisting of 10 isoprenoid units in mammals, allows it to diffuse within the membrane. *Ubiquinone can take up one or two electrons* (in contrast to NAD$^+$, which is strictly a one-electron carrier). A one-electron reduction of ubiquinone produces a semiquinone, a stable free radical (shown as QH·). A two-electron reduction yields ubiquinol (QH_2):

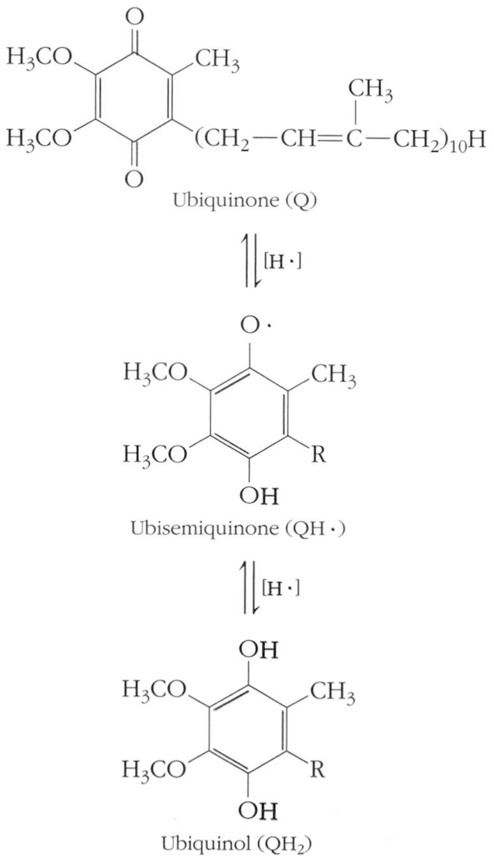

$(CH_2{-}CH{=}C(CH_3){-}CH_2)_{10}H$

Ubiquinone (Q)

[H·]

Ubisemiquinone (QH·)

[H·]

Ubiquinol (QH_2)

However, each electron travels as a hydrogen atom (a proton and an electron). The reduced ubiquinol can then diffuse through the membrane to donate its electrons in another oxidation–reduction reaction.

Among catabolic pathways, the citric acid cycle generates considerable amounts of reduced cofactors. Some of them are reoxidized in anabolic reactions. The rest are reoxidized by a process that is accompanied by the synthesis of ATP from ADP and P_i. In mammals, the reoxidation of NADH and QH_2 and the concomitant production of ATP requires the reduction of O_2 to H_2O and is known as **oxidative phosphorylation.**

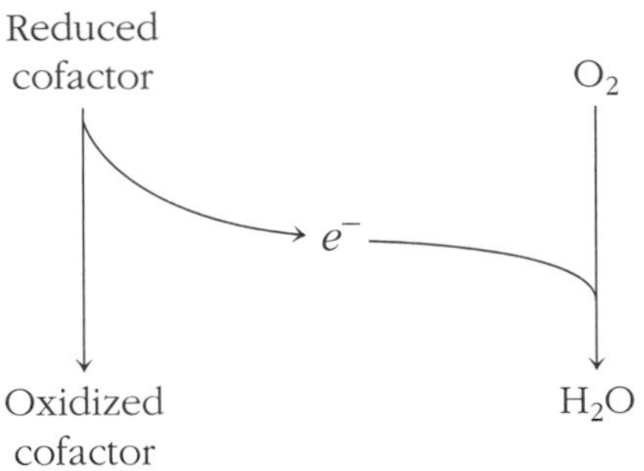

Exclusively anaerobic organisms synthesize ATP without using O_2 as their ultimate oxidizing agent.

REVIEW LEARNING OBJECTIVE 3

- What role do cofactors such as NAD^+ and ubiquinone play in metabolic reactions?
- What is the importance of reoxidizing NADH and QH_2 by molecular oxygen?

Looking at the big picture

So far we have sketched the outlines of mammalian fuel metabolism, in which macromolecules are stored and mobilized so that their monomeric units can be broken down into smaller intermediates. These intermediates can be further degraded (oxidized) and their electrons collected by cofactors. We have also briefly mentioned anabolic (synthetic) reactions in which the common two- and three-carbon intermediates give rise to larger compounds. At this point, we can present this information in schematic form in order to highlight some important features of metabolism (Fig. 9-10).

Figure 9-10 is a composite of a number of metabolic processes, *not all of which occur in every organism or even within every cell of an organism.* For example, mammals do not undertake photosynthesis, and only the liver and kidney can synthesize monosaccharides from noncarbohydrate precursors. You should also keep in mind that Figure 9-10 does not represent the true complexity of metabolic processes: It does not show the individual enzyme-catalyzed steps by which one molecule is transformed into another, and it does not include any of the reactions involved in storing, transmitting, and decoding genetic information (these topics are covered in the final section of this book). However, a diagram such as Figure 9-10 is a useful tool for mapping the relationships among metabolic processes, and we will refer back to it in the coming chapters.

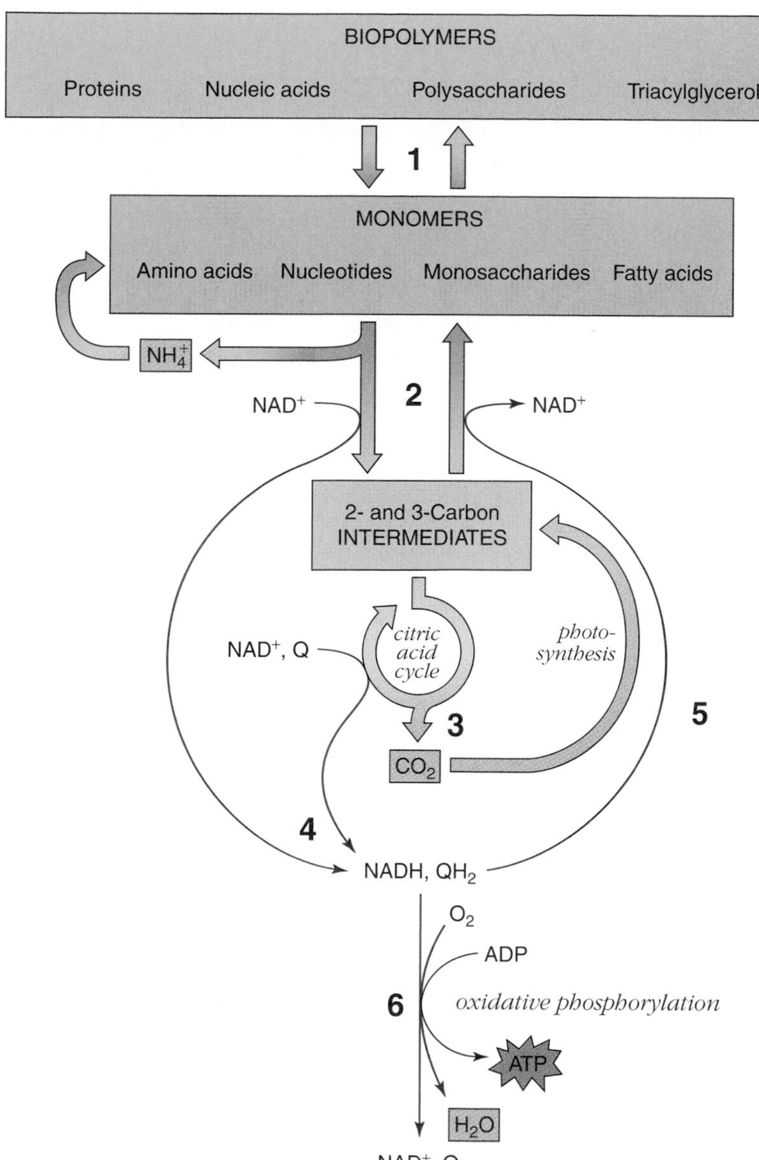

FIGURE 9-10

Outline of metabolism.
In this composite diagram, downward arrows represent catabolic processes, and upward arrows represent anabolic processes. Red arrows indicate some major oxidation–reduction reactions. The major metabolic processes are highlighted: (1) Biological polymers (proteins, nucleic acids, polysaccharides, and triacylglycerols) are built from and are degraded to monomers (amino acids, nucleotides, monosaccharides, and fatty acids). (2) The monomers are broken down into two- and three-carbon intermediates such as glyceraldehyde-3-phosphate, pyruvate, and acetyl-CoA, which are also the precursors of virtually all other biological compounds. (3) The complete degradation of biological molecules yields inorganic compounds such as NH_3, CO_2, and H_2O. These substances are returned to the pool of intermediates by processes such as photosynthesis. (4) Electron carriers (NAD^+ and ubiquinone) accept the electrons released by metabolic fuels (amino acids, monosaccharides, and fatty acids) as they are degraded and then completely oxidized by the citric acid cycle. (5) The reduced cofactors (NADH and QH_2) are required for many biosynthetic reactions. (6) The reoxidation of reduced cofactors drives the production of ATP from ADP + P_i (oxidative phosphorylation).

Each organism has a unique metabolism

In addition to the pathways outlined in Figure 9-10, which are centered on fuel metabolism, cells carry out a plethora of biosynthetic reactions. Many of these reactions are the same in eukaryotes, bacteria, and archaea, in accord with their common evolutionary origins (see Section 1-4). Other pathways are limited to certain species; for example, only methanogens such as *Methanococcus jannaschii* produce CH_4 as a waste product. Each species has a characteristic metabolic repertoire, which reflects its unique genetic makeup. In some organisms, the full complement of metabolic capabilities has come to light only after the organism's genome has been sequenced and the genes for various enzymes have been identified.

Humans lack many of the biosynthetic pathways found in plants and microorganisms. Certain amino acids and unsaturated fatty acids are considered **essential** because the human body cannot synthesize them and must obtain them from food (Table 9-1). **Vitamins** also are compounds that humans need but cannot synthesize (Box 9-A). Presumably, the pathways for syn-

A CLOSER LOOK

BOX 9-A Vitamins

The word *vitamin* comes from *vital amine,* a term coined by Casimir Funk in 1912 to describe organic compounds that are required in small amounts for normal health. It turns out that most vitamins are not amines, but the name has stuck. We have already introduced lipid vitamins (A, D, and K; see Box 8-A). The water-soluble vitamins are a diverse group of compounds, whose discoveries and functional characterization have provided some of the more colorful stories in the history of biochemistry.

Many vitamins have been discovered through studies of nutritional deficiencies. One of the earliest links between nutrition and disease was observed centuries ago in sailors suffering from scurvy, an illness characterized by loose teeth, skin lesions, and poor wound healing. British navy physicians found that citrus juice cured scurvy in sailors whose diet lacked fresh fruit and vegetables. Eventually, the active ingredient was identified as ascorbic acid (vitamin C):

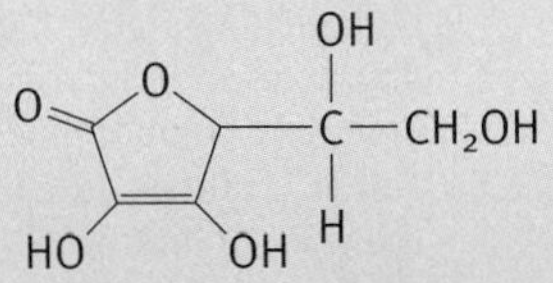

Ascorbic acid (vitamin C)

Ascorbic acid is a cofactor for the enzyme that hydroxylates Pro residues in collagen (see Section 5-6). A deficiency of vitamin C prevents normal formation of collagen fibrils, producing the symptoms of scurvy.

A study of the disease beriberi led to the discovery of the first B vitamin. Beriberi, characterized by leg weakness and swelling, is caused by a deficiency of thiamine (vitamin B_1).

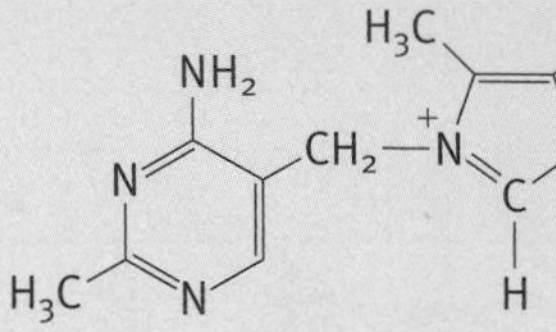

Thiamine (vitamin B_1)

Thiamine acts as a prosthetic group in some essential enzymes, including the one that converts pyruvate to acetyl-CoA. Rice husks are rich in thiamine, and individuals whose diet consists largely of polished (huskless) rice can develop beriberi. The disease was originally thought to be infectious, until the same symptoms were observed in chickens fed a diet of polished rice. Thiamine deficiency can occur in chronic alcoholics and others with a limited diet and impaired nutrient absorption.

Niacin, a precursor of NAD^+ and $NADP^+$ (see Fig. 9-9) was first identified as the factor missing in the vitamin-deficiency disease pellagra. The symptoms of pellagra, including diarrhea and dermatitis, can be alleviated by boosting the intake of the essential amino acid tryptophan, which humans can convert to niacin.

The vitamins mentioned here, as well as others we will encounter in the next few chapters, serve as (or are precursors of) cofactors for enzymatic reactions. However, not all coenzymes have vitamin origins. For example, ubiquinone (coenzyme Q, shown on page 285) is not a vitamin. Most vitamins are readily obtained from a balanced diet, although poor nutrition, particularly in impoverished parts of the world, still causes vitamin-deficiency diseases.

TABLE 9-1 Some essential amino acids and fatty acids

Amino acids	Fatty acids	
Arginine	Linoleate	$CH_3(CH_2)_4(CH{=}CHCH_2)_2(CH_2)_6COO^-$
Histidine		
Isoleucine	Linolenate	$CH_3CH_2(CH{=}CHCH_2)_3(CH_2)_6COO^-$
Leucine		
Lysine		
Methionine		
Phenylalanine		
Threonine		
Tryptophan		
Valine		

thesizing these substances, which require many specialized enzymes, are not necessary for heterotrophic organisms and have been lost through evolution.

2 FREE ENERGY CHANGES IN METABOLIC REACTIONS

We have introduced the idea that catabolic reactions tend to release free energy and anabolic reactions tend to consume it (see Fig. 9-1), but in fact, *all reactions in vivo occur with a net decrease in free energy; that is, ΔG is always less than zero* (free energy is introduced in Section 1-3). In a cell, these reactions are not isolated but are linked, so that the free energy of a thermodynamically favorable reaction can be transferred to a second, unfavorable reaction to allow it to proceed. What is the nature of this free energy and how is it transferred? Free energy is not a substance or the property of a single molecule, so it is misleading to refer to a molecule or a bond within that molecule as having a large amount of free energy. Rather, *free energy is a property of a system, and it changes when the system undergoes a chemical reaction.*

Review Exercise 1/Thermodynamics defines free energy and explains its importance in biochemical reactions. This review material and the associated questions (also relevant to Chapter 1) address Learning Objective 4 of this chapter.

Free energy and concentration

The change in free energy of a system is related to the concentrations of the reacting substances. When a reaction, such as A + B → C + D, is at equilibrium, the concentrations of the four reactants define the **equilibrium constant, K_{eq},** for the reaction:

$$K_{eq} = \frac{[C]_{eq}[D]_{eq}}{[A]_{eq}[B]_{eq}} \qquad [9\text{-}1]$$

(the brackets indicate the molar concentrations of each substance). *When the system is not at equilibrium, the reactants experience a driving force to reach their equilibrium values. This force is the* ***standard free energy change for the reaction, $\Delta G°'$,*** *which is defined as*

$$\boxed{\Delta G°' = -RT \ln K_{eq}} \qquad [9\text{-}2]$$

R is the gas constant (8.3145 J · K^{-1} · mol^{-1}) and T is the temperature in Kelvin. By convention, measurements of standard free energy are valid under **standard conditions,** where the temperature is 25°C (298K) and the pressure is 1 atm (these conditions are indicated by the degree symbol). For a chemist, standard conditions specify concentrations of 1 M for each reactant, but this is impractical for biochemists since most biochemical reactions occur near neutral pH (where $[H^+] = 10^{-7}$ M rather than 1 M) and in aqueous solution (where $[H_2O]$ = 55.5 M). Hence, biochemists add a prime symbol to indicate the standard free energy change for a reaction under *biochemical* standard conditions.

Like K_{eq}, $\Delta G°'$ is a constant for a particular reaction. It may be a positive or negative value, and it indicates whether the reaction proceeds spontaneously ($\Delta G°' < 0$) or not ($\Delta G°' > 0$) *when the reactants are all present at their standard concentrations.* These conditions almost never pertain in a living cell, so *it is important to distinguish the standard free energy change of* *a reaction from its actual free energy change, ΔG.* ΔG is a function of the actual concentrations of the reactants and the temperature (37°C or 310K in humans). It is related to the standard free energy change for the reaction:

$$\Delta G = \Delta G^{\circ\prime} + RT \ln \frac{[\mathrm{C}][\mathrm{D}]}{[\mathrm{A}][\mathrm{B}]} \qquad [9\text{-}3]$$

Here, the bracketed quantities represent the actual, nonequilibrium concentrations of the reactants. The concentration term in Equation 9-3 is sometimes called the **mass action ratio.**

When the reaction is at equilibrium, $\Delta G = 0$ and

$$\Delta G^{\circ\prime} = -RT \ln \frac{[\mathrm{C}]_{\mathrm{eq}}[\mathrm{D}]_{\mathrm{eq}}}{[\mathrm{A}]_{\mathrm{eq}}[\mathrm{B}]_{\mathrm{eq}}} \qquad [9\text{-}4]$$

which is equivalent to Equation 9-2. Note that Equation 9-3 shows that *the criterion for spontaneity for a reaction is ΔG, a property of the actual concentrations of the reactants, and not the constant $\Delta G^{\circ\prime}$*. Thus, a reaction with a positive standard free energy change (a reaction that cannot occur when the reactants are present at standard concentrations) may proceed *in vivo,* depending on the concentrations of reactants in the cell (see Sample Calculation 9-1). Keep in mind that thermodynamic spontaneity does not imply a rapid reaction. Even a substance with a strong tendency to undergo reaction ($\Delta G \ll 0$) will usually not react until acted upon by an enzyme that catalyzes the reaction.

Sample Calculation 9-1

Problem

The standard free energy change for the reaction catalyzed by phosphoglucomutase

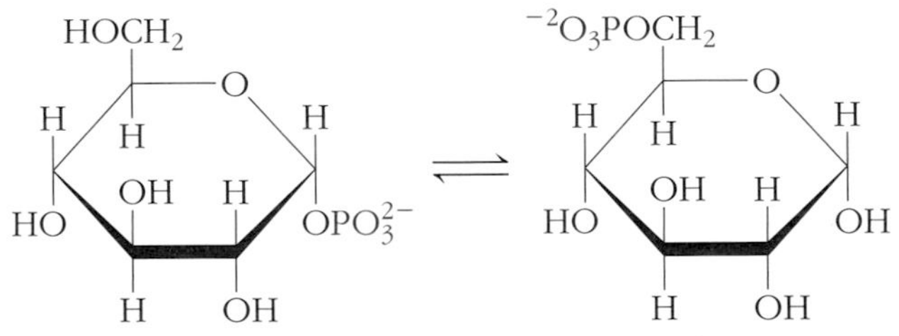

Glucose-1-phosphate ⇌ Glucose-6-phosphate

is $-7.1 \text{ kJ} \cdot \text{mol}^{-1}$. Calculate the equilibrium constant for the reaction. Calculate ΔG at 37°C when the concentration of glucose-1-phosphate is 1 mM and the concentration of glucose-6-phosphate is 25 mM. Is the reaction spontaneous under these conditions?

Solution

The equilibrium constant K_{eq} can be derived by rearranging Equation 9-2:

$$\begin{aligned} K_{\mathrm{eq}} &= e^{-\Delta G^{\circ\prime}/RT} \\ &= e^{-(-7100 \text{ J} \cdot \text{mol}^{-1})/(8.3145 \text{ J} \cdot \text{K}^{-1} \cdot \text{mol}^{-1})(298\text{K})} \\ &= e^{2.87} = 17.6 \end{aligned}$$

At 37°C, $T = 310\text{K}$.

$$\Delta G = \Delta G^{\circ\prime} + RT \ln \frac{[\text{glucose-6-phosphate}]}{[\text{glucose-1-phosphate}]}$$

$$= -7.1 \text{ kJ} \cdot \text{mol}^{-1} + (8.3145 \text{ J} \cdot \text{K}^{-1} \cdot \text{mol}^{-1})(310\text{K}) \ln (0.025/0.001)$$

$$= -7.1 \text{ kJ} \cdot \text{mol}^{-1} + 8.3 \text{ kJ} \cdot \text{mol}^{-1}$$

$$= +1.2 \text{ kJ} \cdot \text{mol}^{-1}$$

The reaction is not spontaneous, because ΔG is greater than zero.

REVIEW LEARNING OBJECTIVE 4

- Why must free energy changes be negative for reactions *in vivo*?
- What is the standard free energy change for a reaction and how is it related to the reaction's equilibrium constant?
- Distinguish ΔG and $\Delta G^{\circ\prime}$. How are they related?

ATP and coupled reactions

A biochemical reaction may at first seem to be forbidden because its free energy change is greater than zero. Yet the reaction can proceed *in vivo* when it is coupled to a second reaction whose value of ΔG is very large and negative, so that the net change in free energy for the combined reactions is still less than zero. ATP is often involved in such coupled reactions, because its reactions can release a large amount of free energy.

Adenosine triphosphate (ATP) contains two phosphoanhydride bonds (Fig. 9-11). Cleavage of either of these bonds—that is, transfer of one or two of its phosphoryl groups to another molecule—is a reaction with a large neg-

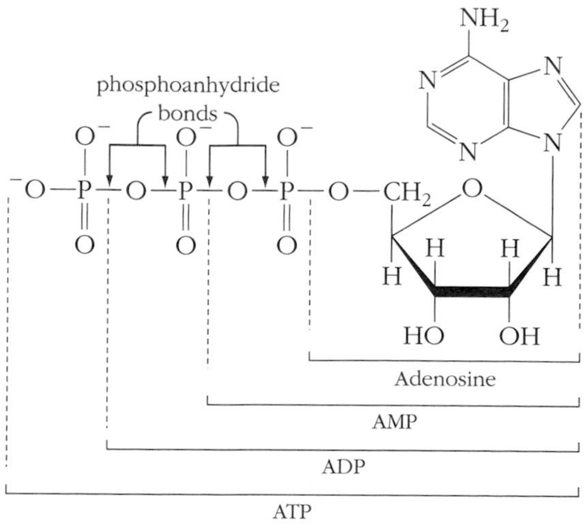

phosphoanhydride bonds

Adenosine

AMP

ADP

ATP

FIGURE 9-11

Adenosine triphosphate.
The second and third phosphoryl groups of ATP are linked via phosphoanhydride bonds. A reaction in which one or two phosphoryl groups are transferred to another compound (a reaction in which a phosphoanhydride bond is cleaved) has a large negative value of $\Delta G^{\circ\prime}$.

TABLE 9-2 Standard free energy changes for cleavage of some phosphoanhydride bonds

Reaction	$\Delta G^{\circ\prime}$ ($kJ \cdot mol^{-1}$)
$ATP + H_2O \rightarrow ADP + P_i$	−30.5
$ATP + H_2O \rightarrow AMP + PP_i$	−32.2
$PP_i + H_2O \rightarrow 2\ P_i$	−33.5

ative standard free energy change (under physiological conditions, ΔG is even more negative). As a reference point, biochemists use the reaction in which a phosphoryl group is transferred to water—in other words, hydrolysis of the phosphoanhydride bond. The free energy changes for hydrolyzing ATP's phosphoanhydride bonds are given in Table 9-2.

The following example illustrates the role of ATP in a coupled reaction. The phosphorylation of glucose by inorganic phosphate (HPO_4^{2-} or P_i) is thermodynamically unfavorable ($\Delta G^{\circ\prime} = +13.8\ kJ \cdot mol^{-1}$):

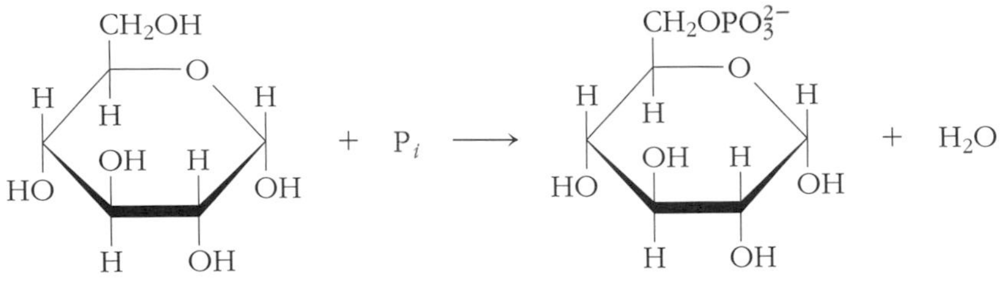

$$\text{Glucose} + P_i \longrightarrow \text{Glucose-6-phosphate} + H_2O$$

Glucose — Glucose-6-phosphate

but the hydrolysis of ATP is a spontaneous reaction ($\Delta G^{\circ\prime} = -30.5\ kJ \cdot mol^{-1}$):

$$ATP + H_2O \rightarrow ADP + P_i$$

When the two reactions are combined, the values of $\Delta G^{\circ\prime}$ are added:

	$\Delta G^{\circ\prime}$
Glucose + P_i → glucose-6-phosphate + H_2O	$+13.8\ kJ \cdot mol^{-1}$
$ATP + H_2O \rightarrow ADP + P_i$	$-30.5\ kJ \cdot mol^{-1}$
Glucose + ATP → glucose-6-phosphate + ADP	$-16.7\ kJ \cdot mol^{-1}$

Thus, the overall reaction for the phosphorylation of glucose is thermodynamically favorable. *In vivo,* this reaction is catalyzed by hexokinase (introduced in Section 6-3), and a phosphoryl group is transferred from ATP directly to glucose. The ATP is not actually hydrolyzed. However, writing out the two coupled reactions, as shown above, makes it easier to see what's going on thermodynamically.

Some biochemical processes appear to occur with the concomitant hydrolysis of ATP to ADP + P_i, for example, the operation of myosin and kinesin (Sections 5-2 and 5-4) or the Na,K-ATPase ion pump (Section 8-3). But a closer look reveals that in all these processes, ATP actually transfers a phosphoryl group to a protein. Later, the phosphoryl group is transferred to water, so that the net reaction takes the form of ATP hydrolysis. The same ATP "hydrolysis" effect applies to some reactions in which the AMP moiety of ATP (rather than a phosphoryl group) is transferred to a substance, leaving inorganic pyrophosphate (PP_i). Cleavage of the phosphoanhydride bond of PP_i also has a large negative value of $\Delta G^{\circ\prime}$ (see Table 9-2).

What's so special about ATP?

Because ATP appears to drive many thermodynamically unfavorable reactions, it is tempting to think of ATP as an agent that transfers a packet of free energy from a reaction that produces ATP to a reaction that consumes ATP. This is one reason why ATP is commonly called the energy currency of the cell. The general role of ATP in linking exergonic catabolic processes to endergonic anabolic processes can be diagrammed as

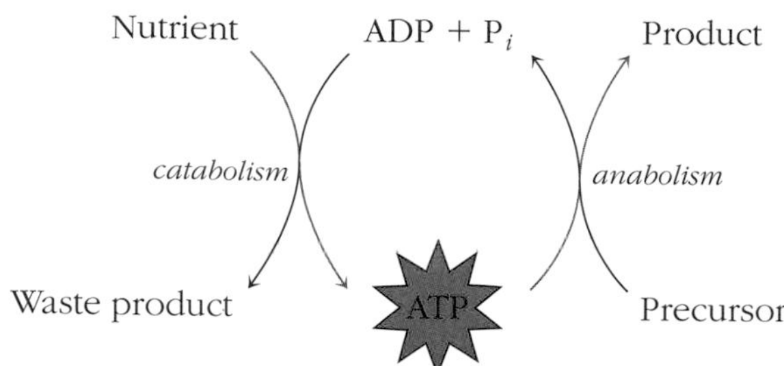

However, there is nothing inherently special about ATP, and its two phosphoanhydride bonds (sometimes called "high-energy" bonds) are no different from other covalent bonds. All that matters is that *cleavage of these bonds has a large negative free energy change.* Using the simple example of ATP hydrolysis, we can state that a large amount of free energy is released when ATP is hydrolyzed because the products of the reaction have less free energy than the reactants. It is worth examining two reasons why this is so:

1. *The ATP hydrolysis products are more stable than the reactants.* At physiological pH, ATP has three to four negative charges (the pK is close to neutral), and the anionic groups exert mutual electrostatic repulsion. In the products ADP and P$_i$, separation of the charges relieves some of this unfavorable repulsion.

2. *A compound with a phosphoanhydride bond experiences less resonance stabilization than its hydrolysis products.* For example, there are fewer equivalent ways of arranging the bonds of the terminal phosphoryl group of ATP than there are in free P$_i$.

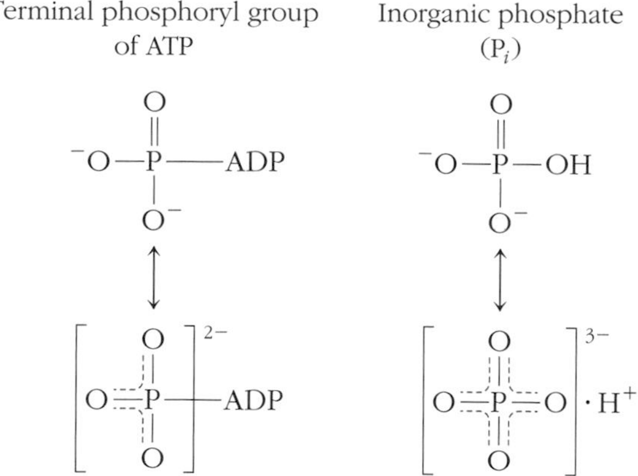

 **REVIEW LEARNING OBJECTIVE 5**

- Why is it misleading to refer to ATP as a high-energy molecule?
- What is meant when ATP is said to transfer free energy from one reaction to another?
- Explain why cleavage of one of ATP's phosphoanhydride bonds releases large amounts of free energy.

Other forms of energy currency

In addition to ATP, other molecules can serve as energy currency in the cell. These substances appear to transfer free energy in coupled reactions, just as ATP does.

Reduced cofactors

We have already seen that the electron carriers NAD^+ and ubiquinone collect electrons from oxidized substrates. The reduced cofactors can then transfer their electrons to another, oxidized substrate. In this way, some of the free energy released in the oxidation of the first substrate can be used to drive the reduction of a second substrate.

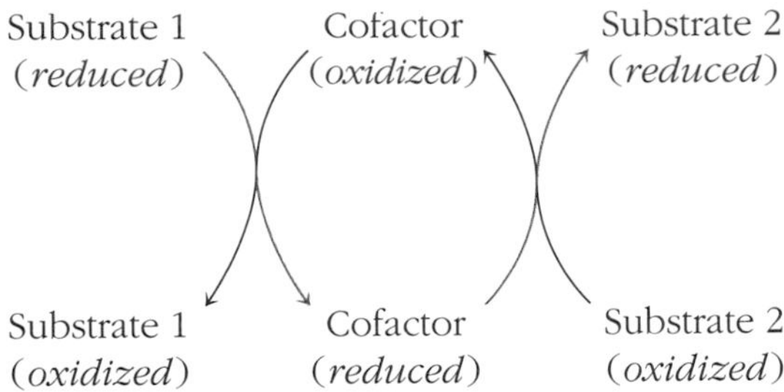

Alternatively, when the reduced cofactor is subsequently oxidized by transferring its electrons to O_2, the free energy is recovered and used to drive the formation of ATP by oxidative phosphorylation.

Phosphorylated compounds

Some phosphorylated compounds other than ATP release large amounts of free energy when they are cleaved (that is, the cleavage reaction has a large negative value of ΔG). For example, phosphocreatine has a standard free energy of hydrolysis of $-43.1\ \text{kJ} \cdot \text{mol}^{-1}$.

$$\mathrm{H_2\overset{+}{N}{=}C(NH{-}PO_3^{2-})(N(CH_3){-}CH_2{-}COO^-)} \xrightarrow{\;H_2O \;\rightarrow\; P_i\;} \mathrm{H_2\overset{+}{N}{=}C(NH_2)(N(CH_3){-}CH_2{-}COO^-)}$$

Phosphocreatine → Creatine

Creatine has lower free energy than phosphocreatine since it has two, rather than one, resonance forms; this resonance stabilization contributes to the large negative free energy change when phosphocreatine transfers its phosphoryl group to another compound. In muscles, phosphocreatine transfers a phosphoryl group to ADP to produce ATP (Box 9-B).

Like ATP, other nucleoside triphosphates have large negative standard free energies of hydrolysis. For this reason, GTP rather than ATP can serve as the energy currency for many of the reactions required for protein synthesis. In the cell, nucleoside triphosphates are freely interconverted by reactions such as the one catalyzed by nucleoside diphosphate kinase, which transfers a phosphoryl group from ATP to a nucleoside diphosphate (NDP):

$$\text{ATP} + \text{NDP} \rightleftharpoons \text{ADP} + \text{NTP}$$

Because the reactants and products are energetically equivalent, $\Delta G^{\circ\prime}$ values for these reactions are near zero.

A CLOSER LOOK

BOX 9-B Powering human muscles

In resting muscles, when the demand for ATP is low, creatine kinase catalyzes the transfer of a phosphoryl group from ATP to creatine to produce phosphocreatine:

$$\text{ATP} + \text{creatine} \rightleftharpoons \text{phosphocreatine} + \text{ADP}$$

This reaction runs in reverse when ADP concentrations rise, as they do when muscle contraction converts ATP to ADP + P_i. Phosphocreatine therefore acts as a sort of phosphoryl-group reservoir to maintain the supply of ATP. Cells cannot stockpile ATP; its concentration remains remarkably stable (between 2 and 5 mM in most cells) under widely varying levels of demand. Without phosphocreatine, a muscle would exhaust its ATP supply before it could be replenished by other, slower processes.

The potential power-boosting function of phosphocreatine has attracted some commercial interest. Administering creatine intravenously appears to improve heart muscle function in individuals with congestive heart failure, but there is no strong evidence that oral creatine supplements enhance muscle performance in athletes.

Different types of physical activity make different demands on a muscle's ATP-generating mechanisms. A short burst of activity is powered by the available ATP. Activities lasting up to a few seconds require phosphocreatine to maintain the ATP supply. Phosphocreatine itself is limited, so continued muscle contraction must rely on ATP produced by catabolizing glucose (obtained from the muscle's store of glycogen) via glycolysis. The end product of this pathway is lactate, the conjugate base of a weak acid, and muscle pain sets in as the acid accumulates and the pH begins to drop. Up to this point, the muscle functions anaerobically. To continue its activity, it must switch to aerobic metabolism and further oxidize glucose via the citric acid cycle. The muscle also catabolizes fatty acids, whose products also enter the citric acid cycle. Recall that the citric acid cycle generates reduced cofactors that must be reoxidized by molecular oxygen. Aerobic metabolism of glucose and fatty acids is slower than anaerobic glycolysis, but it also generates considerably more ATP. Some forms of physical activity and the systems that power them are diagrammed below.

A casual athlete can detect the shift from anaerobic to aerobic metabolism after about a minute and a half. In world-class athletes, the breakpoint occurs at about 150 to 170 seconds, which corresponds roughly to the finish line in a 1000-meter race.

The muscles of sprinters have a high capacity for anaerobic ATP generation, whereas the muscles of distance runners are better adapted to produce ATP aerobically. Such differences in energy metabolism are visibly manifest in the flight muscles of birds. Migratory birds such as geese, which power their long flights primarily with fatty acids, have large numbers of mitochondria to carry out oxidative phosphorylation. The reddish-brown color of the mitochondria gives the flight muscles a dark color. Birds that rarely fly, such as chickens, have fewer mitochondria and lighter-colored muscles. When these birds do fly, it is usually only a short burst of activity that is powered by anaerobic mechanisms.

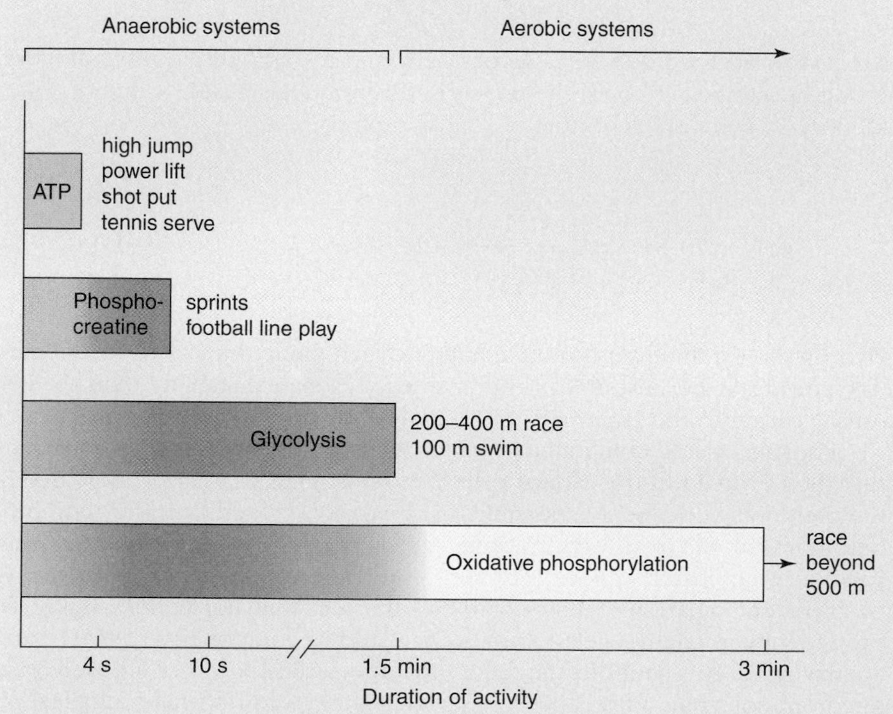

[Figure adapted from McArdle, W.D., Katch, F.I., and Katch, V.L., *Exercise Physiology* (2nd ed.), Lea & Febiger (1986), p. 348.]

Thioesters

A third class of compounds that can release a large amount of free energy upon hydrolysis are **thioesters,** such as acetyl-CoA. Coenzyme A is a nucleotide derivative containing pantothenic acid (also known as vitamin B_3) with a side chain ending in a sulfhydryl (SH) group:

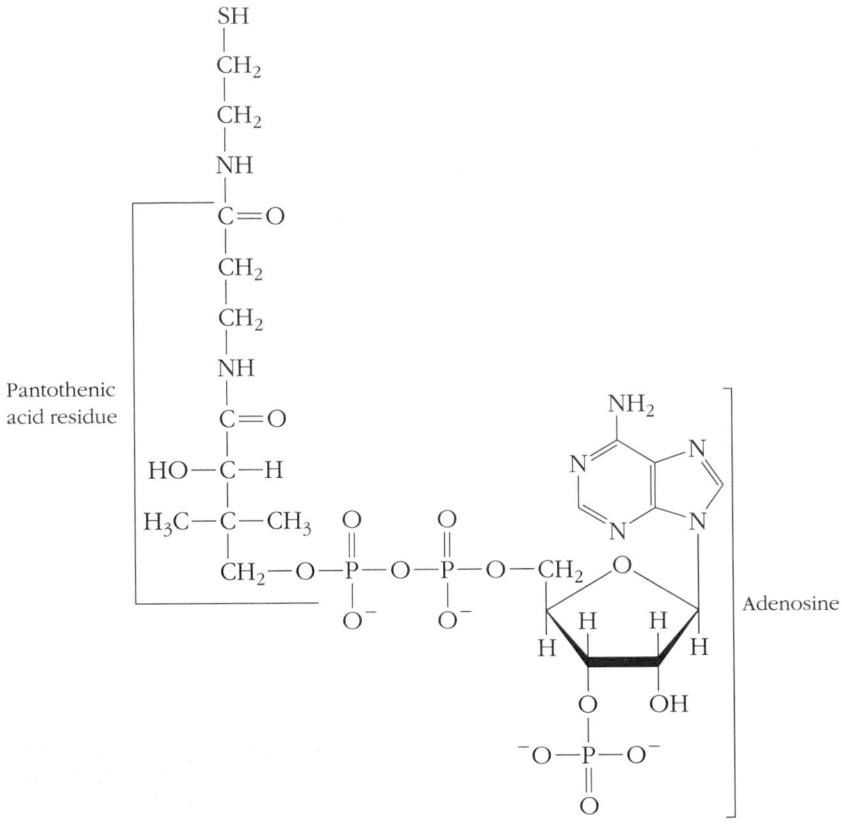

Pantothenic acid residue

Adenosine

Coenzyme A (CoA)

An acyl or acetyl group (the "A" of coenzyme A) is linked to the sulfhydryl group by a thioester bond. Hydrolysis of this bond has a $\Delta G^{\circ\prime}$ value of -31.5 $kJ \cdot mol^{-1}$, comparable to that of ATP hydrolysis:

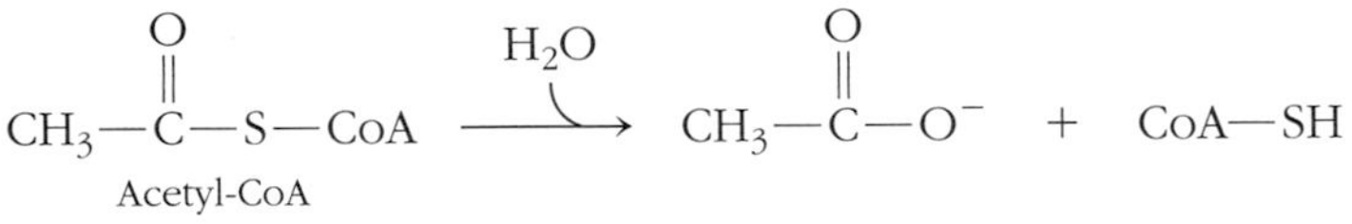

$$\underset{\text{Acetyl-CoA}}{CH_3-\overset{\overset{\displaystyle O}{\|}}{C}-S-CoA} \xrightarrow{H_2O} CH_3-\overset{\overset{\displaystyle O}{\|}}{C}-O^- + CoA-SH$$

Hydrolysis of a thioester is more exergonic than the hydrolysis of an ordinary (oxygen) ester because thioesters have less resonance stability than oxygen esters, owing to the larger size of an S atom relative to an O atom.

Phosphorylated compounds such as ATP are ubiquitous, which suggests that they evolved in the earliest cells. Yet phosphate was probably scarce in the prebiotic world, so it is possible that thioesters served as energy currency before metabolic pathways became specialized for phosphorylated compounds. Significantly, thioesters function in the central metabolic pathways of all extant organisms. Also notable is the fact that nucleotides figure so prominently in energy metabolism, as ATP and as components of NAD^+ and coenzyme A. This supports the theory that the earliest forms of life were centered on molecules such as RNA, which can store and transmit biological information, catalyze chemical reactions, and transfer metabolic free energy.

REVIEW LEARNING OBJECTIVE 6

- How do reduced cofactors, phosphorylated compounds, and thioesters appear to transfer free energy?

SUMMARY

1. Polymeric food molecules such as starch, proteins, and triacylglycerols are broken down to their monomeric components (glucose, amino acids, and fatty acids), which are catabolized. These metabolic fuels are stored as polymers and are mobilized as needed.
2. Series of reactions known as metabolic pathways break down and synthesize biological molecules. Several pathways make use of the same small molecule intermediates.
3. During the oxidation of amino acids, monosaccharides, and fatty acids, electrons are transferred to carriers such as NAD^+ and ubiquinone. Reoxidation of the reduced cofactors drives the synthesis of ATP by oxidative phosphorylation.
4. Not all cells or organisms carry out all possible metabolic processes.
5. The free energy change for a reaction depends on the equilibrium constant for the reaction and on the actual concentrations of the reacting species.
6. A thermodynamically unfavorable reaction may proceed when it is coupled to a favorable process involving ATP, whose phosphoanhydride bonds release a large amount of free energy when cleaved.
7. Other forms of cellular energy currency include reduced cofactors, certain phosphorylated compounds, and thioesters.

CHECKLIST

1. Review the Learning Objectives listed on page 276.
2. Complete Review Exercise 1/Thermodynamics for a review of free energy and its relevance for metabolic reactions.
3. Apply your knowledge by solving the problems at the end of this chapter. Check your results in the Solutions appendix.
4. Be able to define the boldfaced terms (consult the glossary at the end of the book). Test your understanding by taking the Chapter 9 quiz (www.wiley.com/college/pratt).
5. Explore the websites listed at www.wiley.com/college/pratt.
6. Consult the list of Selected Readings for background information or additional details on fuel acquisition and metabolism, metabolic pathways, and the thermodynamics of metabolism.

GLOSSARY TERMS

chemoautotroph
photoautotroph
heterotroph
catabolism
anabolism
metabolism
lipoprotein
metabolic fuel
mobilization
phosphorolysis
lysosome
proteasome
metabolic pathway
intermediate
glycolysis
citric acid cycle
oxidation
reduction
coenzyme
cofactor
oxidative phosphorylation
essential compound
vitamin
equilibrium constant (K_{eq})
standard free energy change ($\Delta G^{\circ\prime}$)
standard conditions
mass action ratio
thioester

PROBLEMS

1. (a) The $\Delta G°'$ value for a hypothetical reaction is 10 kJ · mol^{-1}. Compare the K_{eq} for this reaction with the K_{eq} for a reaction whose $\Delta G°'$ value is twice as large.
   (b) Carry out the same exercise for a hypothetical reaction whose $\Delta G°'$ value is -10 kJ · mol^{-1}.

2. The ΔG for the hydrolysis of ATP under standard conditions at pH 7 and in the presence of magnesium ions is -30.5 kJ · mol^{-1}.
   (a) How would this value change if ATP hydrolysis were carried out at a pH of less than 7? Explain.
   (b) How would this value change if magnesium ions were not present?

3. The $\Delta G°'$ for the formation of UDP–glucose from glucose-1-phosphate and UTP is about zero. Yet the production of UDP–glucose is highly favorable. What is the driving force for this reaction?

   $$\text{Glucose-1-phosphate} + \text{UTP} \rightleftharpoons \text{UDP–glucose} + \text{PP}_i$$

4. (a) The complete oxidation of glucose releases a considerable amount of energy. The $\Delta G°'$ for the reaction shown below is -2850 kJ · mol^{-1}.

   $$C_6H_{12}O_6 + 6\ O_2 \rightarrow 6\ CO_2 + 6\ H_2O$$

   How many moles of ATP could be produced under standard conditions from the oxidation of one mole of glucose, assuming about 40% efficiency?
   (b) The oxidation of palmitate, a 16-carbon saturated fatty acid, releases 9781 kJ · mol^{-1}.

   $$C_{16}H_{32}O_2 + 23\ O_2 \rightarrow 16\ CO_2 + 16\ H_2O$$

   How many moles of ATP could be produced under standard conditions from the oxidation of one mole of palmitate, assuming 40% efficiency?
   (c) Calculate the number of ATP molecules produced per carbon for glucose and palmitate. Explain the reason for the difference.

5. (a) Consider the physical properties of a polar glycogen molecule and an aggregation of hydrophobic triacylglycerols. On a per-weight basis, why is fat a more efficient form of energy storage than glycogen?
   (b) Explain why there is an upper limit to the size of a glycogen molecule, but there is no upper limit to the amount of triacylglycerols that an adipocyte can store.

6. An adult female weighing 125 lbs must consume 2000 Calories of food daily. (Note: A nutritional "large calorie" or "Calorie" is equal to a kilocalorie; 1 calorie = 4.184 J.)
   (a) If this energy is used to synthesize ATP, calculate the number of moles of ATP that would be synthesized each day under standard conditions (assuming 41% efficiency).
   (b) Calculate the number of grams of ATP that would be synthesized each day. The molar mass of ATP is 505 g · mol^{-1}. What is the mass of ATP in pounds? (2.2 kg = 1 lb.)
   (c) There is approximately 40 g of ATP in the adult 125-lb female. Considering this fact and your answer to part b, suggest an explanation that is consistent with these findings.

7. Citrate is isomerized to isocitrate in the citric acid cycle (Chapter 11). The reaction is catalyzed by the enzyme aconitase. The $\Delta G°'$ of the reaction is 5 kJ · mol^{-1}. The kinetics of the reaction are studied *in vitro* where 1 M citrate and 1 M isocitrate are added to an aqueous solution of the enzyme at 25°C.
   (a) What is the K_{eq} for the reaction?
   (b) What are the equilibrium concentrations of the reactant and product?
   (c) What is the preferred direction of the reaction under standard conditions?
   (d) The aconitase reaction is the second step of an eight-step pathway and occurs in the direction shown in the figure. How can you reconcile these facts with your answer to part c?

   Citrate: $CH_2{-}COO^-$ / $HO{-}C{-}COO^-$ / $CH_2{-}COO^-$ $\xrightleftharpoons{\text{aconitase}}$ Isocitate: $CH_2{-}COO^-$ / $HC{-}COO^-$ / $HO{-}CH{-}COO^-$

   Citrate — Isocitate

8. The equilibrium constant and the $\Delta G°'$ values were recently determined for the arginine kinase reaction.

   $$\text{Phosphoarginine} + \text{ADP} \xrightleftharpoons{\text{arginine kinase}} \text{Arginine} + \text{ATP}$$

   (a) What is the value of the K_{eq} for the reaction at 37°C when equilibrium concentrations of the reactants and products are as follows: [phosphoarginine] = 0.737 mM, [ADP] = 0.750 mM, [arginine] = 4.78 mM and [ATP] = 3.87 mM?
   (b) What is the value of the $\Delta G°'$ for this reaction? Is it spontaneous under standard conditions?

9. Calculate ΔG for the hydrolysis of ATP under cellular conditions, where [ATP] = 3 mM, [ADP] = 1 mM, and [P_i] = 5 mM.

10. The standard free energy change for the reaction catalyzed by triose phosphate isomerase is 7.9 kJ · mol^{-1}.

   Glyceraldehyde-3-phosphate: H / $C{=}O$ / $H{-}C{-}OH$ / $CH_2OPO_3^{2-}$ $\rightleftharpoons$ Dihydroxyacetone phosphate: H / $H{-}C{-}OH$ / $C{=}O$ / $CH_2OPO_3^{2-}$

   Glyceraldehyde-3-phosphate — Dihydroxyacetone phosphate

   (a) Calculate the equilibrium constant for the reaction.
   (b) Calculate ΔG at 37°C when the concentration of glyceraldehyde-3-phosphate is 0.1 mM and the concentration of dihydroxyacetone phosphate is 0.5 mM.
   (c) Is the reaction spontaneous under these conditions? Would the reverse reaction be spontaneous?

11. The equilibrium constant for the conversion of glucose-6-phosphate to fructose-6-phosphate is 0.41. The reaction

is reversible and is catalyzed by the enzyme phosphoglucose isomerase.

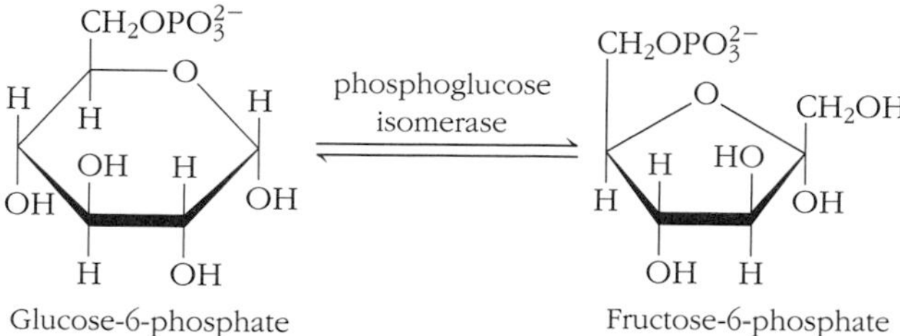

Glucose-6-phosphate   Fructose-6-phosphate

(a) What is the $\Delta G°'$ for this reaction? Would this reaction proceed in the direction written under standard conditions?

(b) What is the ΔG for this reaction at 37°C when the concentration of glucose-6-phosphate is 2.0 mM and the concentration of the fructose-6-phosphate is 0.5 mM? Would the reaction proceed in the direction written under these cellular conditions?

12. The conversion of glutamate to glutamine is unfavorable. In order for this transformation to occur in the cell, it must be coupled to the hydrolysis of ATP. Consider two possible mechanisms:

Mechanism 1: Glutamate + NH_3 $\rightleftharpoons$ glutamine
ATP + H_2O $\rightleftharpoons$ ADP + P_i

Mechanism 2:
Glutamate + ATP $\rightleftharpoons$ γ-glutamylphosphate + ADP
γ-Glutamylphosphate + H_2O + NH_3 $\rightleftharpoons$ glutamine + P_i

Write the overall equation for the reaction for each mechanism. Is one mechanism more likely than the other? Or are both mechanisms equally feasible for the conversion of glutamate to glutamine? Explain.

13. The phosphorylation of glucose to glucose-6-phosphate is the first step of glycolysis (Chapter 10). The phosphorylation of glucose by phosphate is described by the following equation.

Glucose + P_i $\rightleftharpoons$ glucose-6-phosphate + H_2O
$\Delta G°' = +13.8\ \text{kJ} \cdot \text{mol}^{-1}$

(a) Calculate the equilibrium constant for the above reaction.

(b) What would the equilibrium concentration of glucose-6-phosphate be under cellular conditions of [glucose] = [P_i] = 5 mM if glucose were phosphorylated according to the reaction above? Does this reaction provide a feasible route for the production of glucose-6-phosphate for the glycolytic pathway?

(c) One way to increase the amount of a product of a particular reaction is to increase the concentrations of the reactants. This would shift the equilibrium to the right, according to LeChatelier's principle. If the cellular concentration of phosphate were 5 mM, what concentration of glucose would be required in order to achieve the normal physiological concentration of glucose-6-phosphate of 250 μM? Would this route be a physiologically feasible approach to increasing the amount of glucose-6-phosphate product, given that the maximum solubility of glucose in aqueous medium is less than 1 M?

(d) Another way to promote the formation of glucose-6-phosphate is to couple the phosphorylation of glucose to the hydrolysis of ATP:

Glucose + P_i $\rightleftharpoons$ glucose-6-phosphate + H_2O
$\Delta G°' = 13.8\ \text{kJ} \cdot \text{mol}^{-1}$

ATP + H_2O $\rightleftharpoons$ ADP + P_i $\Delta G°' = -30.5\ \text{kJ} \cdot \text{mol}^{-1}$

Glucose + ATP $\rightleftharpoons$ glucose-6-phosphate + ADP

Calculate K_{eq} for the overall reaction.

(e) When the ATP-dependent phosphorylation of glucose is carried out, what concentration of glucose is needed to achieve a 250 μM intracellular concentration of glucose-6-phosphate when the concentrations of ATP and ADP are 5.0 mM and 1.25 mM, respectively? (The actual cellular concentrations may differ from these values.)

(f) Which route is more feasible to accomplish the phosphorylation of glucose to glucose-6-phosphate: the direct phosphorylation by P_i, or the coupling of this phosphorylation to ATP hydrolysis? Explain.

14. Fructose-6-phosphate is phosphorylated to fructose-1,6-bisphosphate as part of the glycolytic pathway. The phosphorylation of fructose-6-phosphate by phosphate is described by the following equation:

Fructose-6-phosphate + P_i $\rightleftharpoons$ Fructose-1,6-bisphosphate
$\Delta G°' = 47.7\ \text{kJ} \cdot \text{mol}^{-1}$

(a) What is the ratio of fructose-1,6-bisphosphate to fructose-6-phosphate at equilibrium if the concentration of phosphate in the cell is 5 mM?

(b) Suppose that the phosphorylation of fructose-6-phosphate is coupled to the hydrolysis of ATP.

ATP + H_2O $\rightleftharpoons$ ADP + P_i $\Delta G°' = -30.5\ \text{kJ} \cdot \text{mol}^{-1}$

Write the new equation that describes the phosphorylation of fructose-6-phosphate coupled with ATP hydrolysis. Calculate the $\Delta G°'$ for the reaction.

(c) What is the ratio of fructose-1,6-bisphosphate to fructose-6-phosphate at equilibrium for the reaction you wrote in part b if the equilibrium concentration of [ATP] = 3 mM and [ADP] = 1 mM?

(d) Write a concise paragraph that summarizes your findings above.

(e) One can envision two mechanisms for coupling ATP hydrolysis to the phosphorylation of fructose-6-phosphate, yielding the same overall reaction:

Mechanism 1: ATP is hydrolyzed as fructose-6-phosphate is transformed to fructose-1,6-bisphosphate:

Fructose-6-phosphate + P_i $\rightleftharpoons$ Fructose-1,6-bisphosphate
ATP + H_2O $\rightleftharpoons$ ADP + P_i

Mechanism 2: ATP transfers its γ phosphate directly to fructose-6-phosphate in one step, producing fructose-1,6-bisphosphate.

Fructose-6-phosphate + ATP + H_2O $\rightleftharpoons$
Fructose-1,6-bisphosphate + ADP

Choose one of the above mechanisms as the more biochemically feasible and provide a rationale for your choice.

15. Glyceraldehyde-3-phosphate (GAP) is eventually converted to 3-phosphoglycerate (3PG) in the glycolytic pathway.

$$\begin{array}{c} H \\ | \\ C{=}O \\ | \\ H{-}C{-}OH \\ | \\ CH_2OPO_3^{2-} \end{array} \rightleftharpoons \begin{array}{c} O \\ \| \\ C{-}O^- \\ | \\ H{-}C{-}OH \\ | \\ CH_2OPO_3^{2-} \end{array}$$

Glyceraldehyde-3-phosphate 3-Phosphoglycerate

$$\begin{array}{c} O \\ \| \\ C{-}OPO_3^{2-} \\ | \\ H{-}C{-}OH \\ | \\ CH_2OPO_3^{2-} \end{array}$$

1,3-Bisphosphoglycerate (1,3-BPG)

Consider these two scenarios:

I. GAP is oxidized to 1,3-BPG ($\Delta G°' = 6.7$ kJ · mol^{-1}), which is subsequently hydrolyzed to yield 3PG ($\Delta G°' = -49.3$ kJ · mol^{-1})

II. GAP is oxidized to 1,3-BPG, which then transfers its phosphate to ADP, yielding ATP ($\Delta G°' = -18.8$ kJ · mol^{-1}).

Write the overall equations for the two scenarios. Which is more likely to occur in the cell, and why?

16. Palmitate is activated in the cell by forming a thioester bond to coenzyme A. The $\Delta G°'$ for the synthesis of palmitoyl-CoA from palmitate and coenzyme A is 31.5 kJ · mol^{-1}.

$$H_3C{-}(CH_2)_{14}{-}\overset{\overset{O}{\|}}{C}{-}O^- + CoA \rightleftharpoons H_3C{-}(CH_2)_{14}{-}\overset{\overset{O}{\|}}{C}{-}S{-}CoA + H_2O$$

(a) What is the ratio of products to reactants at equilibrium for the reaction? Is the reaction favorable? Explain.

(b) Suppose the synthesis of palmitoyl-CoA were coupled with ATP hydrolysis to ADP. The standard free energy for the hydrolysis of the γ phosphate linkage of ATP is -30.5 kJ · mol^{-1}.

$$ATP + H_2O \rightarrow ADP + P_i \quad \Delta G°' = -30.5 \text{ kJ} \cdot \text{mol}^{-1}$$

Write the new equation for the activation of palmitate when coupled with ATP hydrolysis to ADP. Calculate $\Delta G°'$ for the reaction. What is the ratio of products to reactants at equilibrium for the reaction? Is the reaction favorable? Compare your answer to the answer you obtained in part a.

(c) Suppose the reaction described in part a were coupled with ATP hydrolysis to AMP. The standard free energy for the hydrolysis of the β phosphate linkage of ATP is -32.2 kJ · mol^{-1}.

$$ATP + H_2O \rightarrow AMP + PP_i \quad \Delta G°' = -32.2 \text{ kJ} \cdot \text{mol}^{-1}$$

Write the new equation for the activation of palmitate when coupled with ATP hydrolysis to AMP. Calculate $\Delta G°'$ for the reaction. What is the ratio of products to reactants at equilibrium for the reaction? Is the reaction favorable? Compare your answer to the answer you obtained in part b.

(d) The enzyme pyrophosphatase hydrolyzes pyrophosphate, PP_i:

$$PP_i + H_2O \rightarrow 2\ P_i \quad \Delta G°' = -33.5 \text{ kJ} \cdot \text{mol}^{-1}$$

The activation of palmitate, as described in part c, is coupled to the hydrolysis of pyrophosphate. Write the equation for this coupled reaction. Calculate $\Delta G°'$ for the reaction. What is the ratio of products to reactants at equilibrium for the reaction? Is the reaction favorable? Compare your answer to the answers you obtained in parts b and c.

17. DNA containing broken phosphodiester bonds ("nicks") can be repaired by the action of a ligase enzyme. Formation of a new phosphodiester bond in DNA requires the free energy of ATP phosphoanhydride bond cleavage. In the ligase-catalyzed reaction, ATP is hydrolyzed to AMP:

$$\text{ATP} + \text{Nicked bond} \xrightleftharpoons{\text{ligase}} \text{AMP} + PP_i + \text{Phosphodiester bond}$$

The equilibrium constant expression for this reaction can be rearranged to define a constant, C, as follows:

$$K_{eq} = \frac{[\text{Phosphodiester bond}][\text{AMP}][PP_i]}{[\text{Nick}][\text{ATP}]}$$

$$\frac{[\text{Nick}]}{[\text{Phosphodiester bond}]} = \frac{[\text{AMP}][PP_i]}{K_{eq}[\text{ATP}]}$$

$$C = \frac{[PP_i]}{K_{eq}[\text{ATP}]}$$

$$\frac{[\text{Nick}]}{[\text{Phosphodiester bond}]} = C\,[\text{AMP}]$$

The investigators determined the ratio of nicked bonds to phosphodiester bonds at various concentrations of AMP.

(a) Using the data provided, construct a plot of [nick]/[phosphodiester bond] versus [AMP] and determine the value of C from the plot.

[AMP] (mM)	**[Nick]/[phosphodiester bond]**
10	4.0×10^{-5}
15	4.3×10^{-5}
20	5.47×10^{-5}
25	6.67×10^{-5}
30	8.67×10^{-5}
35	9.47×10^{-5}
40	9.30×10^{-5}
45	1.0×10^{-4}
50	1.13×10^{-4}

(b) Determine the value of K_{eq} for the reaction, given the information that the concentrations of PP_i and ATP were held constant at 1.0 mM and 14 μM, respectively.

(c) What is the value of $\Delta G°'$ for the reaction?

(d) What is the value of $\Delta G°'$ for the following reaction?

$$\text{Nicked bond} \rightleftharpoons \text{Phosphodiester bond}$$

Note that the $\Delta G°'$ for the hydrolysis of ATP to AMP and PP_i is $-48.5\ kJ \cdot mol^{-1}$ in the presence of 10 mM Mg^{2+}, the conditions used in these experiments.

(e) The $\Delta G°'$ for the hydrolysis of a typical phosphomonoester to yield P_i and an alcohol is $-13.8\ kJ \cdot mol^{-1}$. Compare the stability of the phosphodiester bond in DNA to the stability of a typical phosphomonoester bond.

[From Dickson, K., Burns, C.M., and Richardson, J., *J. Biol. Chem.* **275**, 15828–15831 (2000).]

SELECTED READINGS

Hanson, R.W., The role of ATP in metabolism, *Biochem. Ed.* **17,** 86–92 (1989). [Provides an excellent explanation of why ATP is an energy transducer rather than an energy store.]

Scriver, C.R., Beaudet, A.L., Sly, W.S., Valle, D., Childs, B., Kinzler, K., and Vogelstein, B. (Eds.), *The Metabolic and Molecular Bases of Inherited Disease* (8th ed.), McGraw-Hill (2001). [Many key metabolic pathways are reviewed in this encyclopedic work.]

CHAPTER 10

Glucose Metabolism

The protozoan *Trypanosoma brucei* causes trypanosomiasis, or sleeping sickness, a debilitating and almost invariably fatal disease in humans. Combating this parasitic disease is difficult for several reasons. First, the protozoan spends part of its life cycle in different tissues of the tsetse fly before finding its way (via the fly's saliva) into the bloodstream of a human host. From there, it invades all organs, including the central nervous system. The host cannot mount an effective immune response because each time the host builds up antibodies to the trypanosome's surface glycoproteins, a few trypanosomes replace their coat proteins with different ones and thereby escape detection. This moving-target mechanism also makes it impractical to develop a vaccine. Finally, unlike pathogenic bacteria, a eukaryotic parasite such as a trypanosome has a metabolism similar to that of its host, so it is difficult to identify drugs that interfere with the parasite's metabolism without harming the host. However, investigations of trypanosome metabolism have provided some promising leads. Catabolism of glucose is the sole energy-producing pathway of trypanosomes in the bloodstream, so blocking this process would effectively kill the parasites. Close examinations of the structures of some of the trypanosome's enzymes have revealed subtle differences that can be exploited in the design of new parasite-specific enzyme inhibitors that could suppress parasite infection.

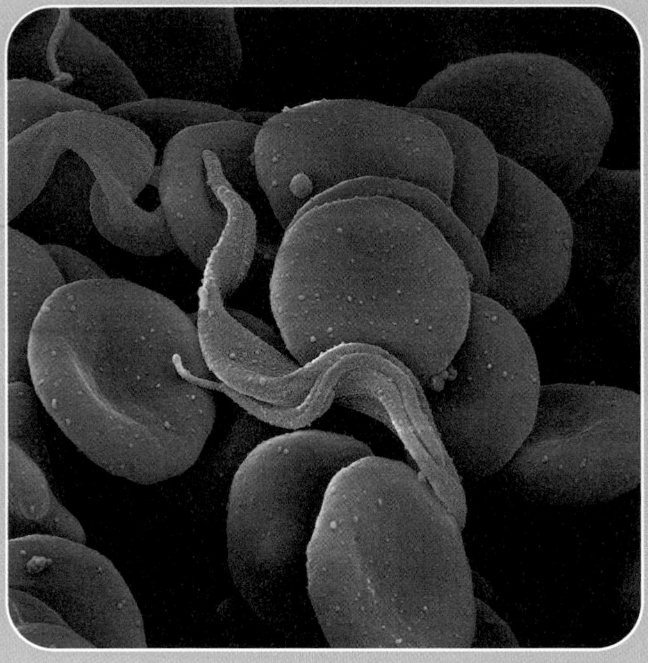

[Science Photo Library/Photo Researchers.]

LEARNING OBJECTIVES

1. Understand the structure of glucose in its monomeric and polymeric forms.
2. Understand the sequence of the 10 enzyme-catalyzed reactions of glycolysis, including their substrates, products, and cofactors.
3. Understand that reactions with large negative free energy changes are irreversible and may function as regulated, rate-determining steps.
4. Understand that the first phase of glycolysis requires the investment of free energy, whereas the second phase produces free energy.

5. Understand that pyruvate may be further oxidized, or used as a precursor of other molecules.
6. Understand why the reactions of gluconeogenesis are not the exact reverse of the reactions of glycolysis.
7. Understand how glycogen synthesis differs from glycogenolysis.
8. Understand that the pentose phosphate pathway is an oxidative catabolic pathway for glucose.

THIS CHAPTER IN CONTEXT

Beginning a study of metabolic pathways with the reactions of glucose is more than just a tradition—it reflects the central position of glucose metabolism in virtually all cells. In this chapter we will use the major glucose pathways (glycolysis, glycogen synthesis and degradation, gluconeogenesis, and the pentose phosphate pathway) to illustrate the first of several general themes of metabolism. Specifically, we will show that pathways are series of chemical reactions in which specific enzymes catalyze the transformation of specific metabolic intermediates. Some of these molecular transformations are metabolically irreversible and serve as regulatory points for the entire pathway.

GLUCOSE OCCUPIES A CENTRAL POSITION IN THE METABOLISM OF most cells. It is a source of metabolic energy (in some cells, such as the parasitic trypanosome, it is the only source) and it provides the precursors for the synthesis of other biomolecules. Yeast and other microorganisms can convert glucose to ethanol and CO_2, a metabolic feat that has been exploited for centuries for the manufacture of alcoholic beverages and the baking of bread (Fig. 10-1).

The catabolism of glucose was also one of the first metabolic pathways to be studied in detail, beginning with Louis Pasteur's observations in the 1850s. This was an era when biological processes were thought to be the result of a "vital force"; enzymes were not yet known. Over the next 100 years or so, with the development of techniques for preparing cell extracts and then for isolating individual enzymes, it became clear that the conversion of glucose to other substances was the result of a series of enzyme-catalyzed chemical reactions.

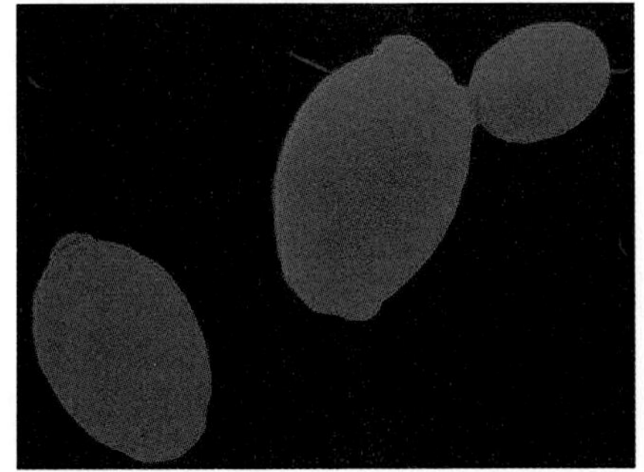

FIGURE 10-1
Yeast cells.
The oxidation of glucose by these cells produces CO_2 and ethanol. Some early biochemists used yeast to test their ideas about how chemical transformations occur in living cells. [*Dr. Dennis Kunkel/ Phototake.*]

The conversion of the six-carbon glucose to the three-carbon pyruvate, a pathway we now call **glycolysis,** occurs in 10 steps. As a result of many years of research, we know a great deal about the pathway's nine intermediates and the enzymes that effect their chemical transformations. One important discovery, in 1905, was that glucose degradation requires inorganic phosphate. Later, ATP was identified as another essential factor. The identification of the pathway intermediates and the enzymes that use them as substrates was further aided by the use of enzyme inhibitors. For example, blocking an enzyme's activity in a cell preparation causes the substrate of that enzyme to accumulate. A simple experiment with a two-step metabolic pathway illustrates how an enzyme inhibitor can reveal the order of enzymes and intermediates in the pathway:

$$A \xrightarrow{X} B \xrightarrow{Y} C$$

Adding an inhibitor of enzyme X causes substrate A to accumulate because it cannot be converted to intermediate B. If B is added to the experimental system, enzyme Y can convert it to product C.

The steps of a metabolic pathway can ultimately be confirmed by purifying the individual enzymes and reassembling them, along with the necessary cofactors, into a functional pathway. For a pathway as long as glycolysis, this was no easy task. The first attempts to reconstitute the pathway from extracts of yeast cells used two crude preparations, one containing a mixture of enzymes and the other containing cofactors such as NAD^+ and metal ions. Eventually, experimental evidence revealed that glycolysis and all other metabolic pathways exhibit the following properties:

1. Each step of the pathway is catalyzed by a distinct enzyme.
2. The free energy consumed or released in certain reactions is transferred by molecules such as ATP and NADH.
3. The rate of the pathway can be controlled by altering the activity of individual enzymes.

If metabolic processes did not occur via multiple enzyme-catalyzed steps, cells would have little control over the amount and type of reaction products and no way to manage free energy. For example, the combustion of glucose and O_2 to CO_2 and H_2O—if allowed to occur in one grand explosion—would release about 2850 $kJ \cdot mol^{-1}$ of free energy all at once. In the cell, glucose combustion requires many steps so that the cell can recover its free energy in smaller, more useful quantities.

In this chapter, we will examine the major metabolic pathways involving glucose. Figure 10-2 shows how these pathways relate to the general metabolic scheme outlined in Figure 9-10. The highlighted pathways include the interconversion of the monosaccharide glucose with its polymeric form glycogen, the degradation of glucose to the three-carbon intermediate pyruvate (the glycolytic pathway), the synthesis of glucose from smaller compounds (**gluconeogenesis**), and the conversion of glucose to the five-carbon monosaccharide ribose. For all the pathways, we will present the intermediates and some of the relevant enzymes. We will also examine the thermodynamics of reactions that release or consume large amounts of free energy and discuss how some of these reactions are regulated.

1 GLUCOSE: WHENCE AND WHEREFORE

Glucose is the most plentiful organic molecule on earth, produced at a rate of about 50 billion tons per year, mainly by photosynthetic organisms. Glucose is a six-carbon sugar (or **hexose**) containing an aldehyde group and five hydroxyl groups. It is convenient to draw it as a linear **Fischer projection,** as shown on the left side of Figure 10-3. However, the aldehyde group readily reacts with one of the hydroxyl groups, so that *glucose actually exists as a circular molecule,* which can be represented by a **Haworth pro-**

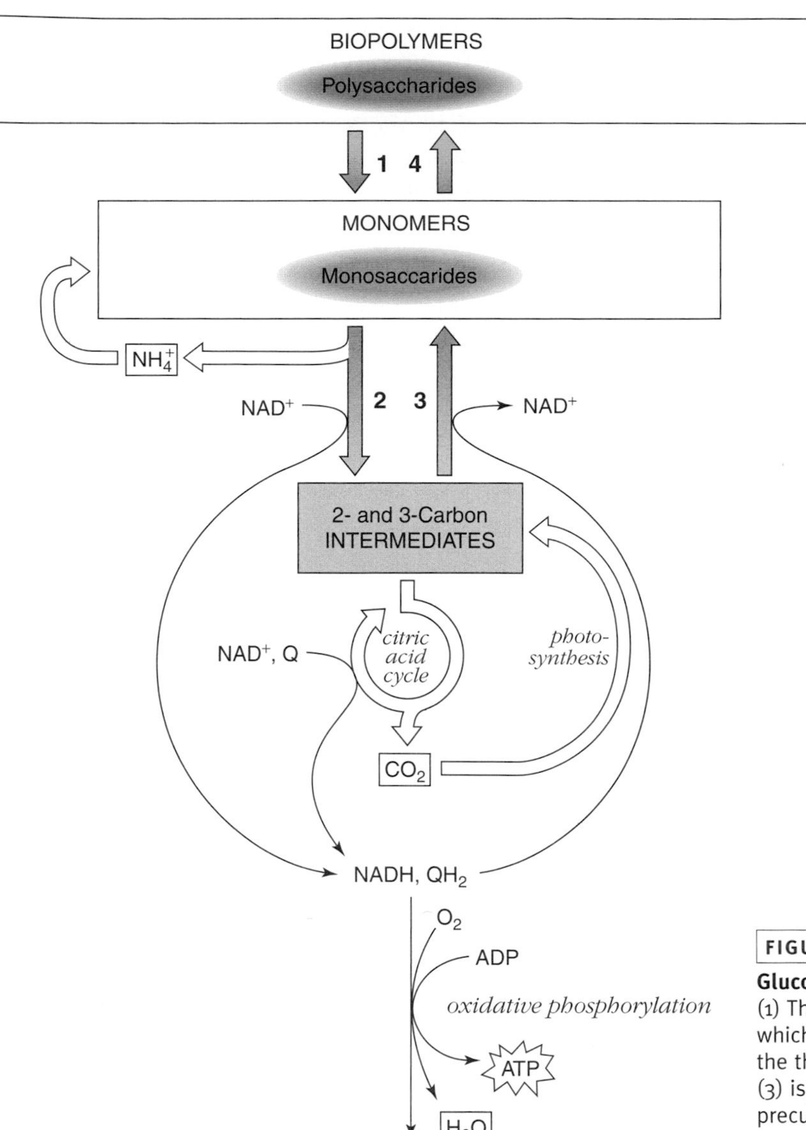

FIGURE 10-2

Glucose metabolism in context.
(1) The polysaccharide glycogen is degraded to glucose, which is then catabolized by the glycolytic pathway (2) to the three-carbon intermediate pyruvate. Gluconeogenesis (3) is the pathway for the synthesis of glucose from smaller precursors. Glucose can then be reincorporated into glycogen (4). The conversion of glucose to ribose, a component of nucleotides, is not shown in this diagram.

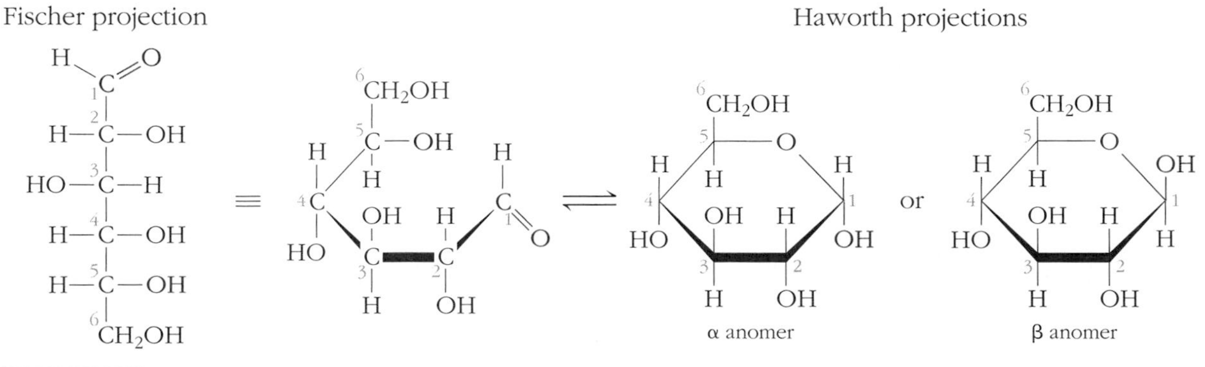

Fischer projection

Haworth projections

α anomer

β anomer

FIGURE 10-3

Representation of glucose structure.
In a linear Fischer projection, the horizontal bonds point out of the page and the vertical bonds point below the page. Glucose cyclizes to form a six-membered ring represented by a Haworth projection, in which the heaviest bonds point out from the plane of the page. Both the α and β anomers are shown.

jection (Fig. 10-3, right side). When the glucose molecule cyclizes, the hydroxyl group at C1 may point either down (in this arrangement, the molecule is known as the **α anomer**) or up (the **β anomer**). Unless the hydroxyl group at C1 is derivatized in some way, *the α and β forms of glucose freely interconvert, although the β anomer is more stable and therefore predominates.*

Of course, the hexose does not form the planar structure suggested by a Haworth projection but puckers so that it resembles the chair conformation of cyclohexane:

Cyclohexane

This allows each carbon atom to retain its tetrahedral geometry. The substituents of these carbons may point either above the ring (axial positions) or outward (equatorial positions). Glucose can adopt a chair conformation in which all its bulky ring substituents (the OH and CH_2OH groups) occupy equatorial positions:

Glucose

In all other hexoses, some of these groups must occupy the more crowded—and therefore less stable—axial positions. The greater stability of glucose may be one reason for its abundance among monosaccharides.

Glucose residues are linked by glycosidic bonds

As introduced in Section 9-1, glucose is stored in polymeric form as starch in plants and as glycogen in animals. The breakdown of these polymers provides glucose monomers that can be catabolized to release energy. Plants also contain significant amounts of glucose polymerized in the form of cellulose, which is a major structural component of plant cell walls (Fig. 10-4). Most animals cannot digest cellulose and therefore cannot use its glucose as an energy source. This is because the amylases that digest dietary starch and glycogen cannot cleave the **glycosidic bonds** between glucose residues in cellulose. In starch and glycogen, most of the glucose residues are linked by $\alpha(1 \rightarrow 4)$ glycosidic bonds; that is, the bond results from the condensation of the OH group at C4 with the α-OH group at C1. In cellulose, glucose residues

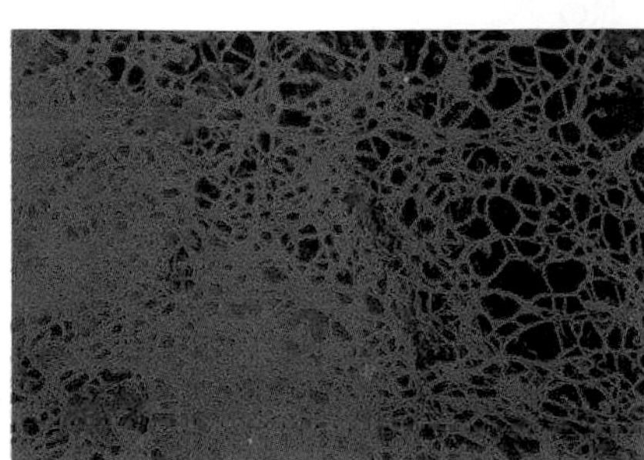

FIGURE 10-4

Cellulose fibers in a plant cell wall. The removal of other polymeric substances leaves cellulose fibers and cross-linking molecules. [Dr. Jeremy Burgess/SPL/Photo Researchers.]

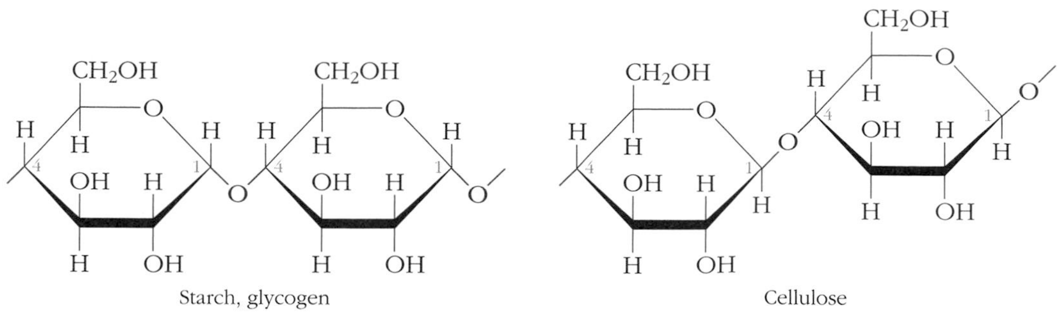

FIGURE 10-5

Comparison of glycosidic bonds in starch, glycogen, and cellulose.
In starch and glycogen, glucose residues are linked in linear chains by $\alpha(1 \rightarrow 4)$ glycosidic bonds. In cellulose, the bonds between glucose residues are $\beta(1 \rightarrow 4)$ glycosidic bonds.

are linked by $\beta(1 \rightarrow 4)$ glycosidic bonds, in which the C1 oxygen has the β configuration (Fig. 10-5).

Herbivores and termites, which appear to digest cellulose, do so only with the aid of cellulases produced by microorganisms in the digestive tract. Humans lack these microorganisms, so although up to 80% of the dry weight of plants consists of glucose, much of this is not available for metabolism in humans (although its bulk is required for normal function of the digestive system).

Glycogenolysis

In humans, the sources of glucose include free glucose liberated from dietary starch and glycogen and glucose released by **glycogenolysis** in the liver. *In glycogenolysis, glycogen is phosphorolyzed to yield glucose-1-phosphate* (see Section 9-1). Because mobilization of glucose must be tailored to meet the body's energy demands when other fuels are in short supply, the activity of glycogen phosphorylase is carefully regulated by a variety of mechanisms linked to hormonal signals (these regulatorymechanisms are described in greater detail in Chapter 16). The enzyme phosphoglucomutase converts glucose-1-phosphate to glucose-6-phosphate, which is then hydrolyzed by glucose-6-phosphatase to release free glucose:

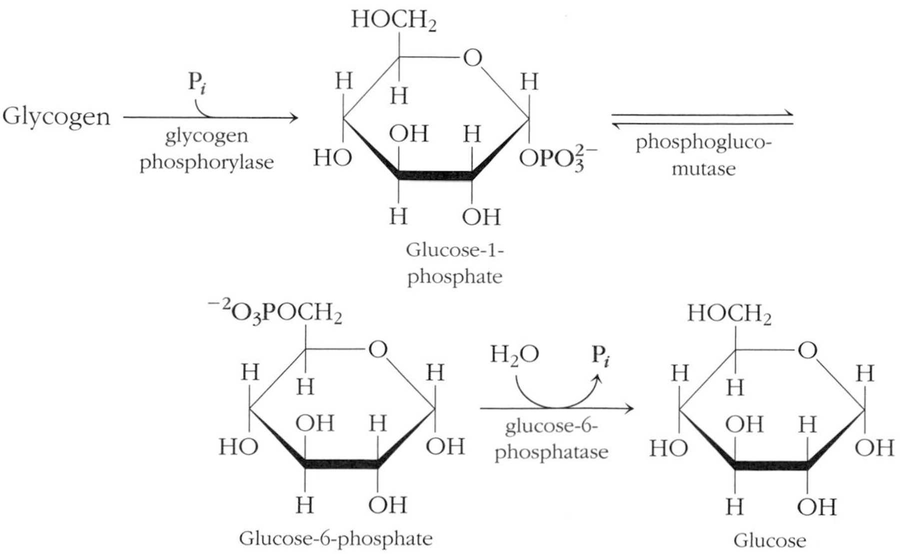

Only the liver can make glucose available to the body at large. Other organs that synthesize glycogen, such as muscle, lack glucose-6-phosphatase and break down glycogen only for their own needs.

REVIEW LEARNING OBJECTIVE 1

- Why is the monosaccharide glucose so stable?
- How does glycogen differ from cellulose?
- What is the product of glycogenolysis?

2 GLYCOLYSIS

Go to **Exercise 15/Glycolytic Enzymes** for an overview of glycolysis as a series of enzyme-catalyzed reactions and for a closer look at these enzymes and their mechanisms. The material in this exercise and the questions related to it address Learning Objective 2.

Glycolysis appears to be an ancient metabolic pathway. The fact that it does not require molecular oxygen suggests that it evolved before photosynthesis increased the level of atmospheric O_2. Overall, glycolysis is a series of 10 enzyme-catalyzed steps in which a six-carbon glucose molecule is broken down into two three-carbon pyruvate molecules. This catabolic pathway is accompanied by the phosphorylation of two molecules of ADP (to produce 2 ATP) and the reduction of two molecules of NAD^+. The net equation for the pathway (ignoring water and protons) is

$$\text{Glucose} + 2\ NAD^+ + 2\ ADP + 2\ P_i \rightarrow 2\ \text{pyruvate} + 2\ NADH + 2\ ATP$$

It is convenient to divide the 10 reactions of glycolysis into two phases. In the first (Reactions 1–5), the hexose is phosphorylated and cleaved in half. In the second (Reactions 6–10), the three-carbon molecules are converted to pyruvate (Fig. 10-6). As you examine each of the reactions of glycolysis described in the following pages, note how the reaction substrates are converted to products by the action of an enzyme (and note how the enzyme's name often reveals its purpose). Pay attention also to the free energy change of each reaction. Details about the active sites and catalytic mechanisms of the glycolytic enzymes are included in Exercise 15.

REVIEW LEARNING OBJECTIVE 2

- Draw the structures of the substrates and products of the 10 glycolytic reactions, name the enzyme that catalyzes each step, and indicate whether ATP or NADH is involved.

Reactions 1–5: energy investment

The first phase of glycolysis can be considered a preparatory phase for the second, energy-producing phase. In fact, the first phase requires the *investment* of free energy in the form of two ATP molecules.

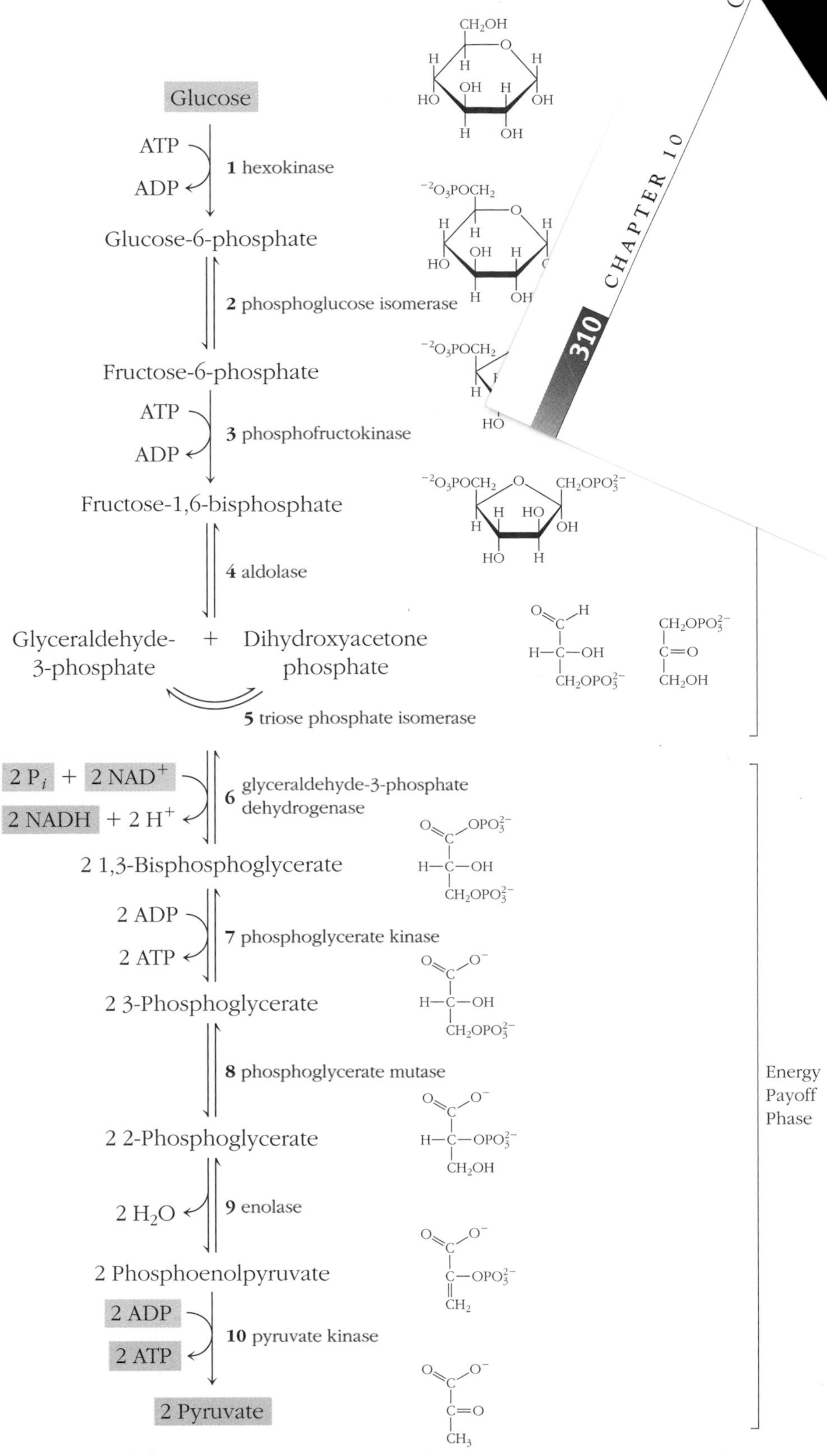

FIGURE 10-6

The reactions of glycolysis.
The substrates, products, and enzymes corresponding to the 10 steps of the pathway are shown. Shading indicates the substrates (purple) and products (green) of the pathway as a whole.

1. Hexokinase

In the first step of glycolysis, the enzyme hexokinase transfers a phosphoryl group from ATP to the C6 OH group of glucose to form glucose-6-phosphate:

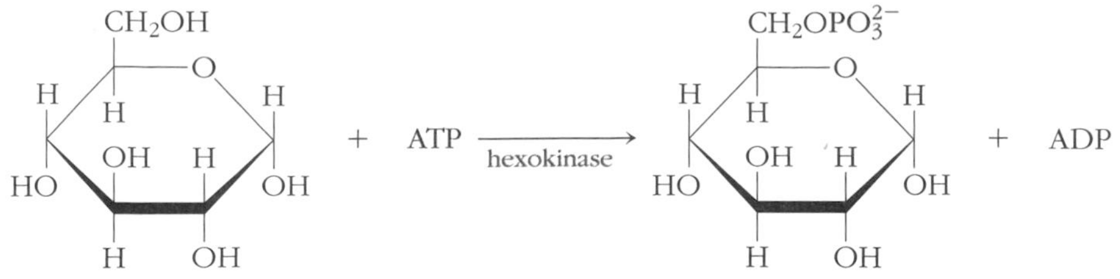

Glucose + ATP → (hexokinase) Glucose-6-phosphate + ADP

A **kinase** is an enzyme that transfers a phosphoryl group from ATP (or another nucleoside triphosphate) to another substance. Hexokinase was the active component of a yeast extract used in some of the first studies of glucose metabolism, which led to the discovery that phosphorylated sugars are intermediates in glycolysis.

Recall from Section 6-3 that the hexokinase active site closes around its substrates so that a phosphoryl group is efficiently transferred from ATP to glucose. The standard free energy change for this reaction, which cleaves one of ATP's phosphoanhydride bonds, is $-16.7\ \text{kJ} \cdot \text{mol}^{-1}$ (ΔG, the actual free energy change for the reaction, has a similar value). The magnitude of this free energy change means that the reaction proceeds in only one direction; the reverse reaction is extremely unlikely since its standard free energy change would be $+16.7\ \text{kJ} \cdot \text{mol}^{-1}$. Consequently, hexokinase is said to catalyze a **metabolically irreversible reaction** that prevents glucose from backing out of glycolysis. *Many metabolic pathways have a similar irreversible step near the start that commits a metabolite to proceed through the pathway.*

Phosphorylation of glucose in the first step of glycolysis also adds a negatively charged group to the electrically neutral glucose molecule. This prevents the sugar from exiting the cell (the glucose transporters that mediate glucose movement across cell membranes do not recognize phosphorylated glucose derivatives).

When glycogen is broken down intracellularly, it yields phosphorylated glucose (Section 9-1). This process bypasses the hexokinase step, thereby sparing the consumption of ATP. As a result, glycolysis from glycogen-derived glucose has a higher net ATP yield than glycolysis from glucose supplied by the bloodstream.

2. Phosphoglucose isomerase

The second reaction of glycolysis is an isomerization reaction in which glucose-6-phosphate is converted to fructose-6-phosphate:

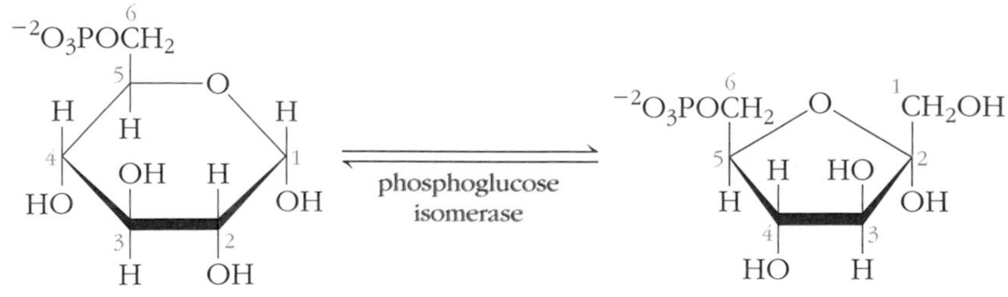

Glucose-6-phosphate ⇌ (phosphoglucose isomerase) Fructose-6-phosphate

Fructose, like glucose, is a six-carbon sugar (a hexose), but it differs from glucose in the position of its carbonyl group so that it is a ketone (called a

ketose) rather than an aldehyde (or **aldose**). This is obvious when the sugars are shown as Fischer projections:

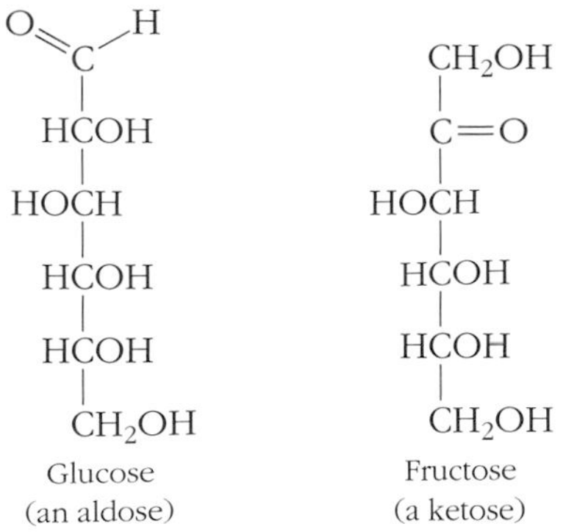

Glucose (an aldose)	Fructose (a ketose)
$O{=}C{-}H$	CH_2OH
HCOH	C=O
HOCH	HOCH
HCOH	HCOH
HCOH	HCOH
CH_2OH	CH_2OH

Glucose (an aldose)   Fructose (a ketose)

The ketone group also accounts for the fact that fructose forms a five-membered ring rather than a six-membered ring.

The standard free energy change for the phosphoglucose isomerase reaction is $+2.2\ \text{kJ} \cdot \text{mol}^{-1}$, but the reactant concentrations *in vivo* yield a ΔG value of about $-1.4\ \text{kJ} \cdot \text{mol}^{-1}$. A value of ΔG near zero indicates a reaction that operates close to equilibrium (at equilibrium, $\Delta G = 0$). *Such* ***near-equilibrium reactions*** *are considered to be freely reversible, since a slight excess of products can easily drive the reaction in reverse by mass-action effects.* In a metabolically irreversible reaction, such as the hexokinase reaction, the concentration of product could never increase enough to compensate for the reaction's large value of ΔG.

3. Phosphofructokinase

The third reaction of glycolysis consumes a second ATP molecule in the phosphorylation of fructose-6-phosphate to yield fructose-1,6-bisphosphate.

$$\text{Fructose-6-phosphate} + \text{ATP} \xrightarrow{\text{phosphofructokinase}} \text{Fructose-1,6-bisphosphate} + \text{ADP}$$

Fructose-6-phosphate   Fructose-1,6-bisphosphate

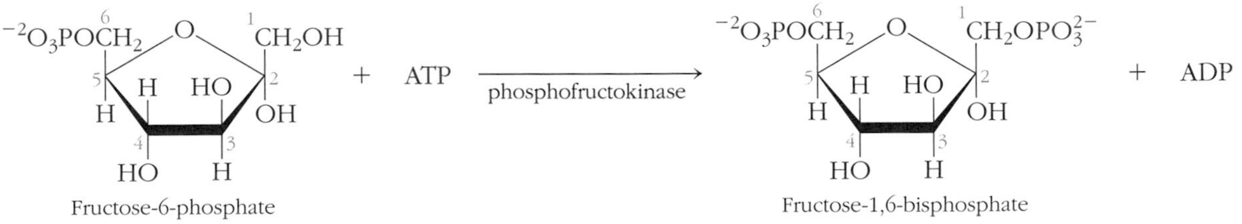

This "hexose diphosphate" was the first glycolytic intermediate to be isolated, in 1905. Phosphofructokinase operates in much the same way as hexokinase, and the reaction it catalyzes is irreversible, with a $\Delta G°'$ value of $-17.2\ \text{kJ} \cdot \text{mol}^{-1}$.

In cells, the activity of phosphofructokinase is regulated. We have already seen how the activity of a bacterial phosphofructokinase responds to allosteric effectors (Section 7-3). ADP binds to the enzyme and causes a conformational change that promotes fructose-6-phosphate binding, which in turn promotes catalysis. This mechanism is useful because the concentration of ADP in the cell is a good indicator of the need for ATP, which is a product of glycolysis. Phosphoenolpyruvate, the product of Step 9 of glycolysis, binds to bacterial phosphofructokinase and causes it to assume a conformation that destabilizes fructose-6-phosphate binding, thereby diminishing catalytic activity. Thus, when the glycolytic pathway is producing plenty of phosphoenolpyruvate and ATP, the phosphoenolpyruvate can act as a feedback inhibitor to slow the pathway by decreasing the rate of the reaction catalyzed by phosphofructokinase (Fig. 10-7).

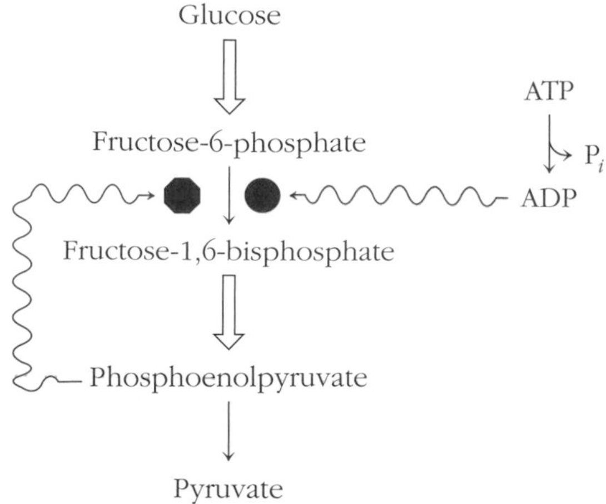

FIGURE 10-7

Regulation of bacterial phosphofructokinase.
ADP, produced when ATP is consumed elsewhere in the cell, stimulates the activity of phosphofructokinase (green symbol). Phosphoenolpyruvate, a late intermediate of glycolysis, inhibits phosphofructokinase (red symbol), thereby decreasing the rate of the entire pathway.

The most potent activator of phosphofructokinase in mammals, however, is the compound fructose-2,6-bisphosphate, which is synthesized from fructose-6-phosphate by an enzyme known as phosphofructokinase-2:

$^{-2}O_3POH_2C$ O CH_2OH H HO H OH HO H

Fructose-6-phosphate

ATP ADP

phosphofructokinase-2

$^{-2}O_3POH_2C$ O CH_2OH H HO H $O{-}PO_3^{2-}$ HO H

Fructose-2,6-bisphosphate

The activity of phosphofructokinase-2 is hormonally stimulated when the concentration of glucose in the blood is high. The resulting increase in fructose-2,6-bisphosphate concentration activates phosphofructokinase to increase the flux of glucose through the glycolytic pathway.

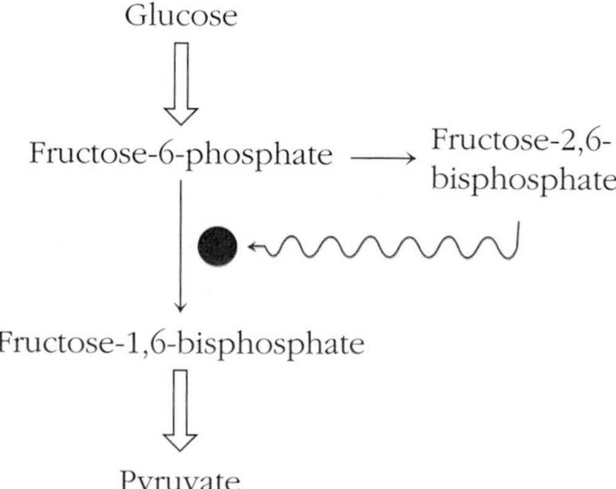

The phosphofructokinase reaction is the primary control point for glycolysis. It is the slowest reaction of glycolysis, so the rate of this reaction largely determines the ***flux*** *(rate of flow) of glucose through the entire pathway.* In general, a **rate-determining reaction**—such as the phosphofructokinase reaction—operates far from equilibrium; that is, it has a large negative free energy change and is irreversible under metabolic conditions. The rate of the reaction can be altered by allosteric effectors but not by fluctuations in the concentrations of its substrates or products. Thus, it acts as a one-way valve. In contrast, a near-equilibrium reaction—such as the phosphoglucose isomerase reaction—cannot serve as a rate-determining step for a pathway, because it can respond to small changes in reactant concentrations by operating in reverse.

Although it too is irreversible and a potential control point, the hexokinase step cannot serve as the rate-determining step of glycolysis because glucose can also enter the pathway as glucose-6-phosphate without the activity of hexokinase.

4. Aldolase

Reaction 4 converts the hexose fructose-1,6-bisphosphate to two three-carbon molecules, each of which bears a phosphate group.

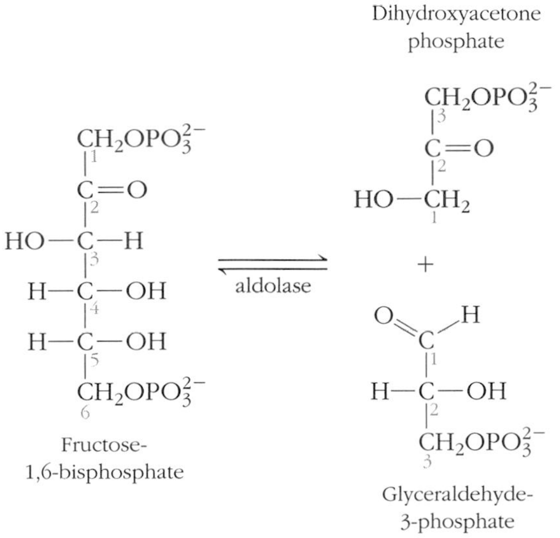

Fructose-1,6-bisphosphate: $^{1}CH_2OPO_3^{2-}$—$^{2}C{=}O$—$HO{-}^{3}C{-}H$—$H{-}^{4}C{-}OH$—$H{-}^{5}C{-}OH$—$^{6}CH_2OPO_3^{2-}$

$\rightleftharpoons$ (aldolase)

Dihydroxyacetone phosphate: $^{3}CH_2OPO_3^{2-}$—$^{2}C{=}O$—$HO{-}^{1}CH_2$

+

Glyceraldehyde-3-phosphate: $O{=}^{1}C{-}H$—$H{-}^{2}C{-}OH$—$^{3}CH_2OPO_3^{2-}$

This reaction is the reverse of an aldol (aldehyde–alcohol) condensation, so the enzyme that catalyzes the reaction is called aldolase. It is worth examining its mechanism.

The active site of mammalian aldolase contains two catalytically important residues: a Lys residue that forms a Schiff base (imine) with the substrate, and an ionized Tyr residue that acts as a base catalyst (Fig. 10-8).

Early studies of aldolase implicated a Cys residue in catalysis, because iodoacetate, a reagent that reacts with the Cys side chain, also inactivates the enzyme:

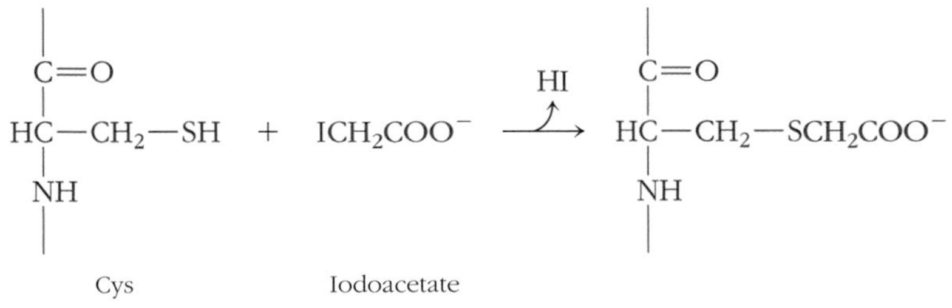

$$\underset{\text{Cys}}{-C{=}O\,/\,HC{-}CH_2{-}SH\,/\,NH-} + \underset{\text{Iodoacetate}}{ICH_2COO^-} \xrightarrow{\;HI\;} -C{=}O\,/\,HC{-}CH_2{-}SCH_2COO^-\,/\,NH-$$

Iodoacetate was one of the inhibitors used by researchers to identify the intermediates of glycolysis: In the presence of iodoacetate, fructose-1,6-bisphosphate

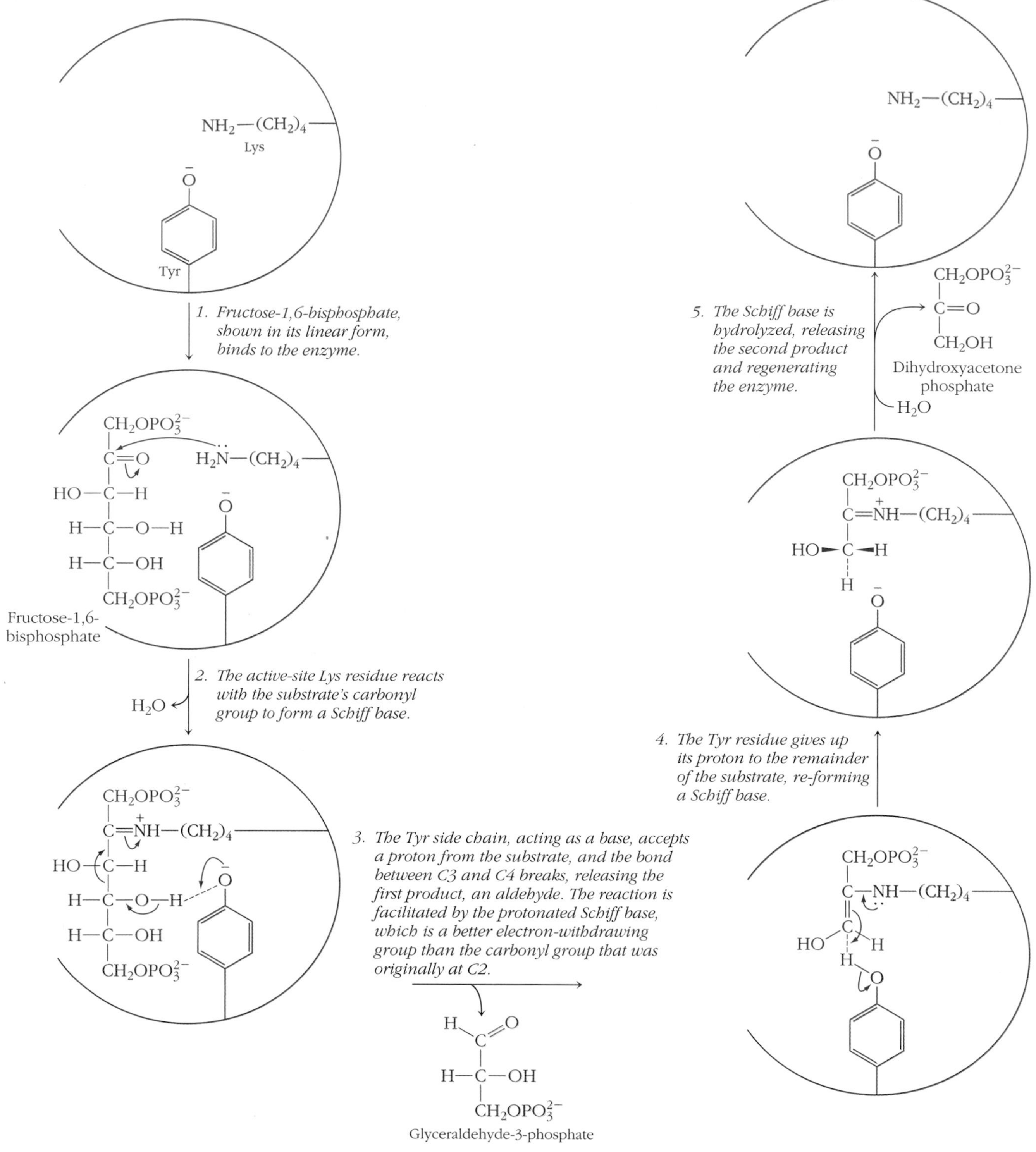

FIGURE 10-8

The aldolase reaction.

accumulates because the next step is blocked. Acetylation of the Cys residue, which is not part of the active site, probably interferes with a conformational change that is necessary for aldolase activity.

The $\Delta G^{\circ\prime}$ value for the aldolase reaction is $+22.8\ \text{kJ} \cdot \text{mol}^{-1}$, indicating that the reaction is unfavorable under standard conditions. However, the reaction proceeds *in vivo* (ΔG is actually less than zero) because the products of the reaction are quickly whisked away by subsequent reactions. In essence, the rapid consumption of glyceraldehyde-3-phosphate and dihydroxyacetone phosphate "pulls" the aldolase reaction forward.

5. Triose phosphate isomerase

The products of the aldolase reaction are both phosphorylated three-carbon compounds, but only one of them—glyceraldehyde-3-phosphate—proceeds through the remainder of the pathway. Dihydroxyacetone phosphate is converted to glyceraldehyde-3-phosphate by triose phosphate isomerase:

$$\underset{\text{Dihydroxyacetone phosphate}}{\mathrm{H{-}C(H){-}OH,\ C{=}O,\ CH_2OPO_3^{2-}}} \xrightleftharpoons{\text{triose phosphate isomerase}} \underset{\text{Glyceraldehyde-3-phosphate}}{\mathrm{O{=}C{-}H,\ H{-}C{-}OH,\ CH_2OPO_3^{2-}}}$$

Triose phosphate isomerase was introduced in Section 7-2 as an example of a catalytically perfect enzyme, one whose rate is limited only by the rate at which its substrates can diffuse to its active site. The catalytic mechanism of triose phosphate isomerase may involve low-barrier hydrogen bonds (which also help stabilize the transition state in serine proteases; see Section 6-3). In addition, the catalytic power of triose phosphate isomerase depends on a protein loop that closes over the active site (Fig. 10-9).

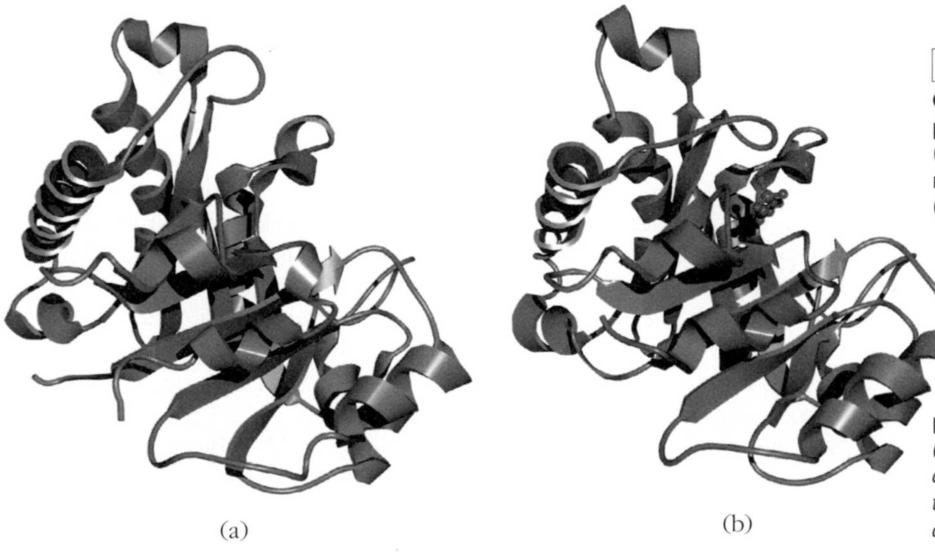

FIGURE 10-9

Conformational changes in yeast triose phosphate isomerase. (a) One loop of the protein, comprising residues 166–176, is highlighted in green. (b) When a substrate binds to the enzyme, the loop closes over the active site to stabilize the reaction's transition state. In this model, the transition-state analog 2-phosphoglycolate occupies the active site. Triose phosphate isomerase is actually a homodimer; only one subunit is pictured. [*Structure of the enzyme alone (pdb 1YPI) determined by T. Alber, E. Lolis, and G.A. Petsko; structure of the enzyme with the analog (pdb 2YPI) determined by E. Lolis and G.A. Petsko.*]

The standard free energy change for the triose phosphate isomerase reaction is slightly positive, even under physiological conditions ($\Delta G^{\circ\prime} = +7.9$ kJ · mol^{-1} and $\Delta G = +4.4$ kJ · mol^{-1}), but the reaction proceeds because glyceraldehyde-3-phosphate is quickly consumed in the next reaction, so that more dihydroxyacetone phosphate is constantly being converted to glyceraldehyde-3-phosphate.

REVIEW LEARNING OBJECTIVE 3

- Distinguish a near-equilibrium reaction from a metabolically irreversible reaction.
- Which reaction is the major rate-determining step of glycolysis and how is its rate regulated in mammals?

Reactions 6–10: energy payoff

So far, the reactions of glycolysis have consumed 2 ATP, but this investment pays off in the second phase of glycolysis when 4 ATP are produced, for a net gain of 2 ATP. All of the reactions of the second phase involve three-carbon intermediates, but keep in mind that *each glucose molecule that enters the pathway yields two of these three-carbon units.*

Some species convert glucose to glyceraldehyde-3-phosphate by different pathways than the one presented above. However, the second phase of glycolysis, which covers the conversion of glyceraldehyde-3-phosphate to pyruvate, is the same in all organisms. This suggests that glycolysis may have evolved from the "bottom up"; that is, it first evolved as a pathway for extracting free energy from abiotically produced small molecules, before cells developed the ability to synthesize larger molecules such as hexoses.

6. Glyceraldehyde-3-phosphate dehydrogenase

In the sixth reaction of glycolysis, glyceraldehyde-3-phosphate is both oxidized and phosphorylated:

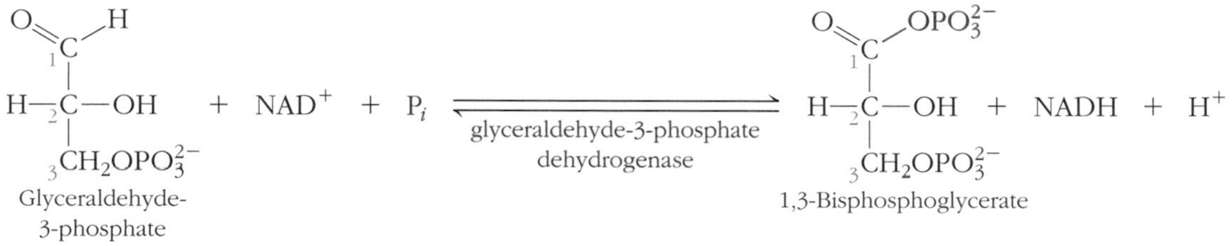

$$\underset{\text{Glyceraldehyde-3-phosphate}}{\overset{{}^1\text{C(=O)H}}{\text{H}-{}^2\text{C}-\text{OH}}\ \underset{{}^3\text{CH}_2\text{OPO}_3^{2-}}{}} + \text{NAD}^+ + \text{P}_i \underset{\text{dehydrogenase}}{\overset{}{\underset{\text{glyceraldehyde-3-phosphate}}{\rightleftharpoons}}} \underset{\text{1,3-Bisphosphoglycerate}}{\overset{{}^1\text{C(=O)OPO}_3^{2-}}{\text{H}-{}^2\text{C}-\text{OH}}\ \underset{{}^3\text{CH}_2\text{OPO}_3^{2-}}{}} + \text{NADH} + \text{H}^+$$

Unlike the kinases that catalyze Reactions 1 and 3, glyceraldehyde-3-phosphate dehydrogenase does not use ATP as a phosphoryl-group donor; it adds inorganic phosphate to the substrate. This reaction is also an oxidation–reduction reaction in which the aldehyde group of glyceraldehyde-3-phosphate is oxidized and the cofactor NAD^+ (see Section 9-1) is reduced to NADH. In effect, glyceraldehyde-3-phosphate dehydrogenase catalyzes the removal of an H atom, hence the name "dehydrogenase." Note that the reaction product NADH must eventually be reoxidized to NAD^+, or else glycolysis will halt. In fact, the reoxidation of NADH, which is a form of "energy currency" (see Section 9-2), can generate ATP (Chapter 12).

An active-site Cys residue participates in the glyceraldehyde-3-phosphate dehydrogenase reaction (Fig. 10-10). The enzyme is inhibited by arsenate (AsO_4^{3-}), which competes with $P_i(PO_4^{3-})$ for binding in the enzyme active site. Arsenate was one of the inhibitors used to elucidate the steps of glycolysis.

7. Phosphoglycerate kinase

The product of Reaction 6, 1,3-bisphosphoglycerate, is an acyl phosphate:

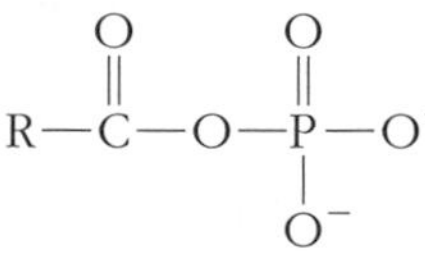

$$\text{R}-\overset{\overset{\text{O}}{\|}}{\text{C}}-\text{O}-\underset{\underset{\text{O}^-}{|}}{\overset{\overset{\text{O}}{\|}}{\text{P}}}-\text{O}^-$$

An acyl phosphate

The subsequent removal of its phosphoryl group releases a large amount of free energy in part because the reaction products are more stable (the same principle contributes to the large negative value of ΔG for reactions involving

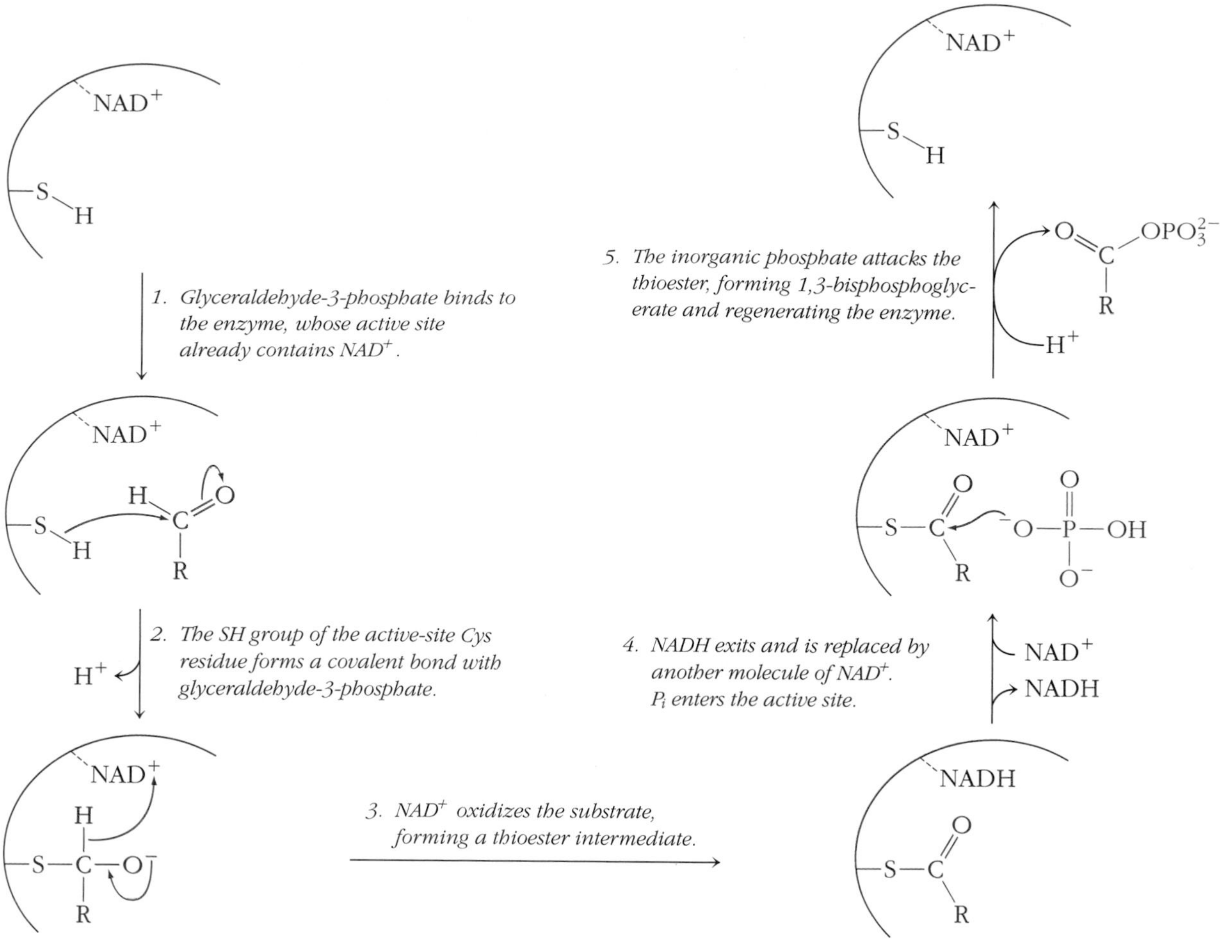

NAD$^+$

1. *Glyceraldehyde-3-phosphate binds to the enzyme, whose active site already contains NAD$^+$.*

2. *The SH group of the active-site Cys residue forms a covalent bond with glyceraldehyde-3-phosphate.*

3. *NAD$^+$ oxidizes the substrate, forming a thioester intermediate.*

4. *NADH exits and is replaced by another molecule of NAD$^+$. P_i enters the active site.*

5. *The inorganic phosphate attacks the thioester, forming 1,3-bisphosphoglycerate and regenerating the enzyme.*

FIGURE 10-10

The glyceraldehyde-3-phosphate dehydrogenase reaction.

cleavage of ATP's phosphoanhydride bonds; see Section 9-2). *The free energy released in this reaction is used to drive the formation of ATP, as 1,3-bisphosphoglycerate donates its phosphoryl group to ADP:*

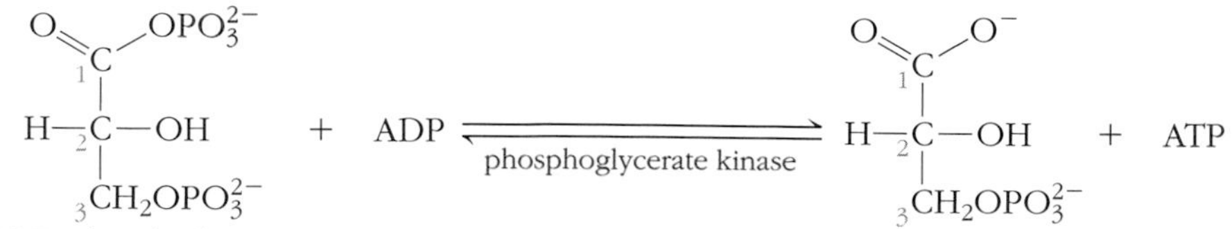

$$\underset{\text{1,3-Bisphosphoglycerate}}{\text{O=}{}^1\text{C(OPO}_3^{2-})\text{–H–}{}^2\text{C–OH–}{}^3\text{CH}_2\text{OPO}_3^{2-}} + \text{ADP} \underset{\text{phosphoglycerate kinase}}{\rightleftharpoons} \underset{\text{3-Phosphoglycerate}}{\text{O=}{}^1\text{C(O}^-)\text{–H–}{}^2\text{C–OH–}{}^3\text{CH}_2\text{OPO}_3^{2-}} + \text{ATP}$$

Note that the enzyme that catalyzes this reaction is called a kinase since it transfers a phosphoryl group between ATP and another molecule.

The standard free energy change for the phosphoglycerate kinase reaction is $-18.8\ \text{kJ} \cdot \text{mol}^{-1}$. This strongly exergonic reaction helps pull the glyceraldehyde-3-phosphate dehydrogenase reaction forward, since its standard free energy change is greater than zero ($\Delta G°' = +6.7\ \text{kJ} \cdot \text{mol}^{-1}$). This pair of reactions provides a good example of the coupling of a thermodynamically favorable and unfavorable reaction so that both proceed with a net decrease in free energy: $-18.8\ \text{kJ} \cdot \text{mol}^{-1} + 6.7\ \text{kJ} \cdot \text{mol}^{-1} = -12.1\ \text{kJ} \cdot \text{mol}^{-1}$. Under physiological conditions, ΔG for the paired reactions is close to zero.

8. Phosphoglycerate mutase

In the next reaction, 3-phosphoglycerate is converted to 2-phosphoglycerate:

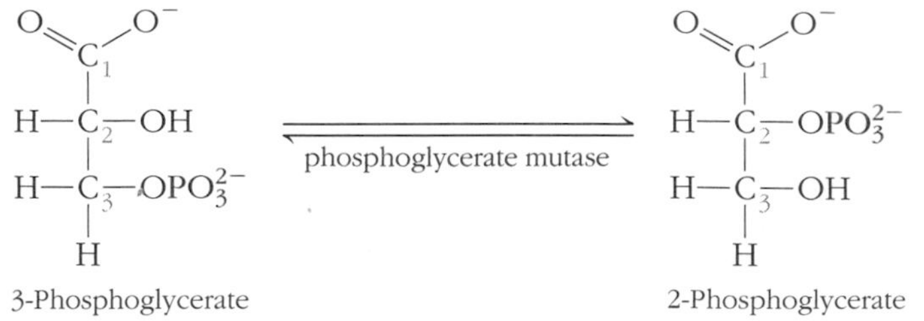

$$\text{3-Phosphoglycerate } (\text{O}^-\text{OC}_1\text{–H–C}_2\text{–OH–H–C}_3\text{–OPO}_3^{2-}\text{–H}) \underset{\text{phosphoglycerate mutase}}{\rightleftharpoons} \text{2-Phosphoglycerate } (\text{O}^-\text{OC}_1\text{–H–C}_2\text{–OPO}_3^{2-}\text{–H–C}_3\text{–OH–H})$$

3-Phosphoglycerate

2-Phosphoglycerate

Although the reaction appears to involve the simple intramolecular transfer of a phosphoryl group, the reaction mechanism is a bit more complicated and requires an enzyme active site that contains a phosphorylated His residue. The phospho-His transfers its phosphoryl group to 3-phosphoglycerate to generate 2,3-bisphosphoglycerate, which then gives a phosphoryl group back to the enzyme, leaving 2-phosphoglycerate and phospho-His:

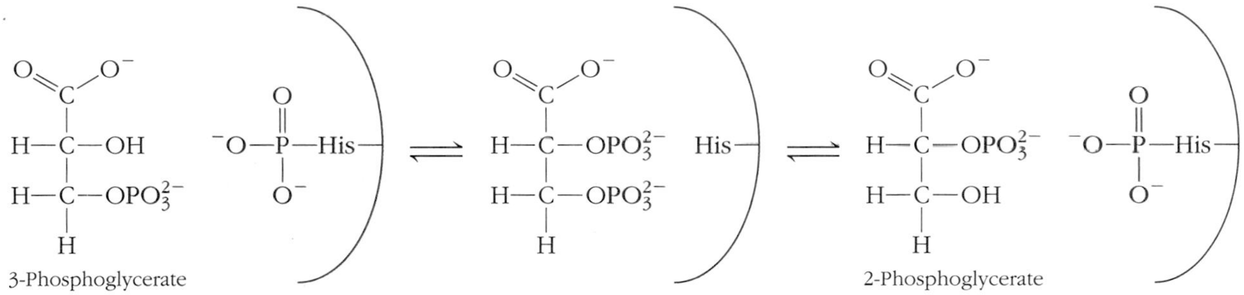

$$\text{3-Phosphoglycerate} + {}^-\text{O–P(=O)(O}^-\text{)–His–E} \rightleftharpoons \text{2,3-bisphosphoglycerate } (\text{H–C–OPO}_3^{2-},\ \text{H–C–OPO}_3^{2-}) + \text{His–E} \rightleftharpoons \text{2-Phosphoglycerate} + {}^-\text{O–P(=O)(O}^-\text{)–His–E}$$

3-Phosphoglycerate

2-Phosphoglycerate

As can be guessed from its mechanism, the phosphoglycerate mutase reaction is freely reversible *in vivo*.

9. Enolase

Enolase catalyzes a dehydration reaction, in which water is eliminated:

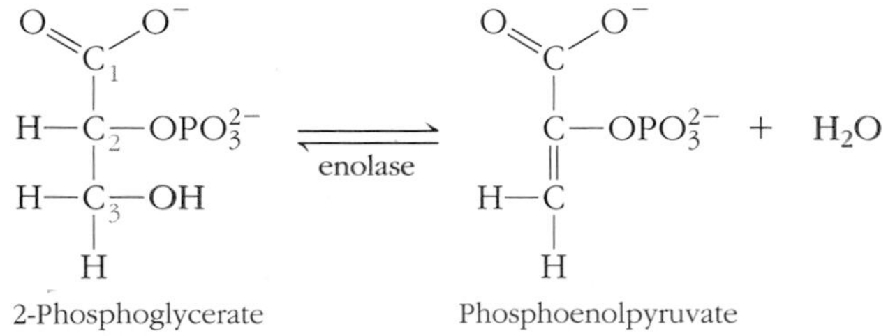

$$\text{2-Phosphoglycerate } (\text{O}^-\text{OC}_1\text{–H–C}_2\text{–OPO}_3^{2-}\text{–H–C}_3\text{–OH–H}) \underset{\text{enolase}}{\rightleftharpoons} \text{Phosphoenolpyruvate } (\text{O}^-\text{OC–C(–OPO}_3^{2-})\text{=CH}_2) + H_2O$$

2-Phosphoglycerate

Phosphoenolpyruvate

The enzyme active site includes a Mg^{2+} ion that apparently coordinates with the OH group at C3 and makes it a better leaving group. A complex of fluoride ion and P_i can form a complex with the Mg^{2+} and thereby inhibit the enzyme. In early studies demonstrating the inhibition of glycolysis by F^-, 2-phosphoglycerate, the substrate of enolase, accumulated. The concentration of 3-phosphoglycerate also increased in the presence of F^- since phosphoglycerate mutase readily converted the excess 2-phosphoglycerate back to 3-phosphoglycerate.

10. Pyruvate kinase

The tenth reaction of glycolysis is catalyzed by pyruvate kinase, which converts phosphoenolpyruvate to pyruvate and transfers a phosphoryl group to ADP to produce ATP:

$$\underset{\text{Phosphoenolpyruvate}}{\mathrm{CH_2{=}C(OPO_3^{2-})\text{—}COO^-}} + \mathrm{ADP} \xrightarrow[\text{pyruvate kinase}]{} \underset{\text{Pyruvate}}{\mathrm{CH_3\text{—}C({=}O)\text{—}COO^-}} + \mathrm{ATP}$$

The reaction actually occurs in two parts. First, ADP attacks the phosphoryl group of phosphoenolpyruvate to form ATP and enolpyruvate:

$$\underset{\text{Phosphoenolpyruvate}}{\mathrm{CH_2{=}C(COO^-)\text{—}O\text{—}P(O)(O^-)\text{—}O^-}} + \underset{\text{ADP}}{\mathrm{{}^-O\text{—}P(O)(O^-)\text{—}O\text{—}AMP}} \xrightarrow[\mathrm{H^+}]{\mathrm{ATP}} \underset{\text{Enolpyruvate}}{\mathrm{CH_2{=}C(OH)\text{—}COO^-}}$$

Removal of phosphoenolpyruvate's phosphoryl group is not a particularly exergonic reaction: When written as a hydrolytic reaction (transfer of the phosphoryl group to water), the $\Delta G^{\circ\prime}$ value is $-16\ \mathrm{kJ\cdot mol^{-1}}$. This is not enough free energy to drive the synthesis of ATP from ADP + P_i (this reaction requires $+30.5\ \mathrm{kJ\cdot mol^{-1}}$). However, the second half of the pyruvate kinase reaction is highly exergonic. This is the **tautomerization** (isomerization through the shift of an H atom) of enolpyruvate to pyruvate:

$$\underset{\text{Enolpyruvate}}{\mathrm{CH_2{=}C(OH)\text{—}COO^-}} \longrightarrow \underset{\text{Pyruvate}}{\mathrm{CH_3\text{—}C({=}O)\text{—}COO^-}}$$

$\Delta G^{\circ\prime}$ for this step is $-46\ \mathrm{kJ\cdot mol^{-1}}$, so that $\Delta G^{\circ\prime}$ for the net reaction (hydrolysis of phosphoenolpyruvate followed by tautomerization of enolpyruvate to pyruvate) is $-61.9\ \mathrm{kJ\cdot mol^{-1}}$, more than enough free energy to drive the synthesis of ATP. The free energy changes for the 10 reactions of glycolysis are shown graphically in Figure 10-11.

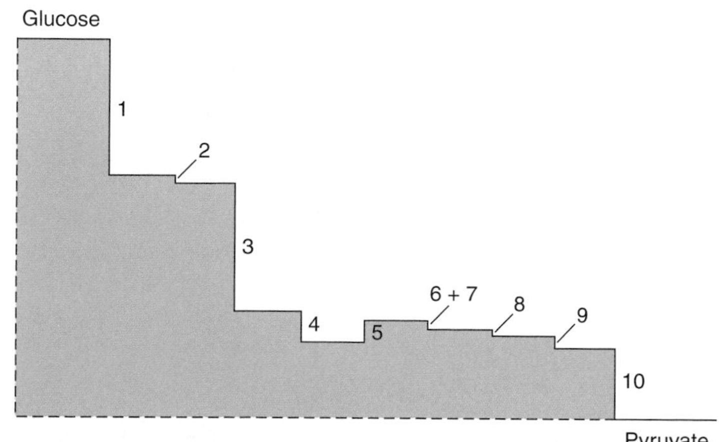

FIGURE 10-11

Graphical representation of the free energy changes of glycolysis. Three steps (catalyzed by hexokinase, phosphofructokinase, and pyruvate kinase) have large negative values of ΔG; the remaining steps are near equilibrium ($\Delta G \approx 0$). The height of each step corresponds to its ΔG value in heart muscle and the numbers correspond to glycolytic enzymes. Keep in mind that ΔG values vary slightly among tissues. [*Data from Newsholme, E.A. and Start, C.,* Regulation in Metabolism, *Wiley (1973), p. 97.*]

Although pyruvate kinase catalyzes an irreversible reaction and is therefore a potential rate-determining reaction for the pathway, it would not be efficient for it to be the primary regulatory step because it occurs at the very end of the 10-step pathway. In yeast, fructose-1,6-bisphosphate activates pyruvate kinase at an allosteric site. This is an example of **feed-forward activation:** Once a monosaccharide has entered glycolysis, fructose-1,6-bisphosphate helps ensure rapid flux through the pathway.

To sum up the second phase of glycolysis: Glyceraldehyde-3-phosphate is converted to pyruvate with the synthesis of 2 ATP (in Reactions 7 and 10). Since each molecule of glucose yields two molecules of glyceraldehyde-3-phosphate, the reactions of the second phase of glycolysis must be doubled, so the yield is 4 ATP. Two molecules of ATP are invested in Phase 1, bringing the net yield to 2 ATP produced per glucose molecule. Other monosaccharides are metabolized in a similar fashion to yield ATP (Box 10-A).

REVIEW LEARNING OBJECTIVE 4

- Which glycolytic reactions consume ATP? Which generate ATP?
- What is the net yield of ATP per glucose molecule?
- Which reactions serve as control points?

The fate of pyruvate

What happens to the pyruvate generated by the catabolism of glucose? It can be further broken down or used to synthesize other compounds. The fate of pyruvate depends on the cell type and the need for metabolic free energy and molecular building blocks.

Lactate synthesis

In a highly active muscle cell, glycolysis rapidly provides ATP to power muscle contraction, but the pathway also consumes NAD^+ at the glyceraldehyde-3-phosphate dehydrogenase step. The two NADH molecules generated for each glucose molecule catabolized can be reoxidized in the presence of oxygen. However, this process is too slow to replenish the NAD^+ needed for the rapid production of ATP by glycolysis. *To regenerate NAD^+, the enzyme lactate dehydrogenase reduces pyruvate to lactate:*

$$\text{Pyruvate} + \text{NADH} + H^+ \underset{\text{lactate dehydrogenase}}{\rightleftharpoons} \text{Lactate} + NAD^+$$

Pyruvate NADH Lactate NAD^+

This reaction, sometimes called the eleventh step of glycolysis, allows the muscle to function anaerobically for a minute or so (see Box 9-B). The net reaction for anaerobic glucose catabolism is

$$\text{Glucose} + 2\ \text{ADP} + 2\ P_i \rightarrow 2\ \text{lactate} + 2\ \text{ATP}$$

Lactate represents a sort of metabolic dead end: Its only options are to be eventually converted back to pyruvate (the lactate dehydrogenase reaction is

A CLOSER LOOK

BOX 10-A Metabolism of other sugars

A typical human diet contains many carbohydrates other than glucose and its polymers; the disaccharides lactose and sucrose are major sources of carbohydrates. Lactose (also called milk sugar) is synthesized in mammary glands and is a major source of energy for newborn mammals:

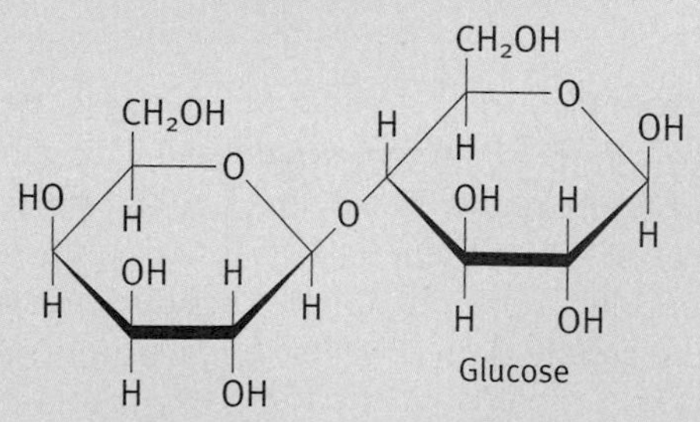

Galactose Glucose

Lactose

This disaccharide, which contains glucose and its isomer galactose, is cleaved in the intestine by the enzyme lactase, and the two monosaccharides are absorbed, transported to the liver, and metabolized. Galactose undergoes phosphorylation and isomerization and enters the glycolytic pathway as glucose-6-phosphate, so its energy yield is the same as that of glucose.

Adult humans who ingest milk sometimes experience diarrhea and flatulence, because they do not express lactase and are unable to digest lactose. Although lactose intolerance or lactase "deficiency" is often considered to be a disorder, it is the natural condition for the majority of the world's adults. Only a few groups of people retain the ability to digest lactose throughout life.

Undigested lactose remains in the small intestine, where it exerts an osmotic effect leading to frequent watery stools. In addition, intestinal bacteria catabolize some of the lactose, ultimately producing gases such as H_2, CO_2, and CH_4, which cause bloating and flatulence. Similar symptoms result from the ingestion of other indigestible oligosaccharides, such as those in some fruits and legumes.

Sucrose, the other major dietary disaccharide, is composed of glucose and fructose.

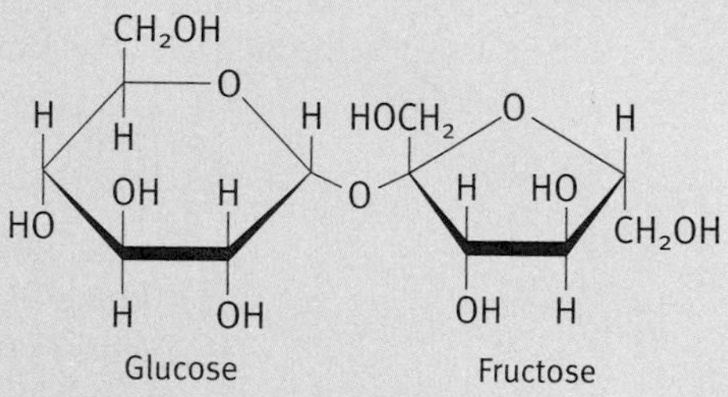

Glucose Fructose

Sucrose

Sucrose (table sugar) is found in a variety of foods of plant origin. Like lactose, it is hydrolyzed in the small intestine, and its component glucose and fructose are absorbed. The monosaccharide fructose is also present in many foods, particularly fruits and honey. It tastes sweeter than sucrose, is more soluble, and is inexpensive to produce in the form of high-fructose corn syrup—all of which make fructose attractive to the manufacturers of soft drinks. This is the primary reason why the consumption of fructose in the United States has increased at least tenfold in the last quarter century.

One effect of the overconsumption of fructose is that the sugar may not be completely absorbed, leading to symptoms similar to those of lactose intolerance. Additional problems may result from the catabolism of fructose, which differs somewhat from the catabolism of glucose. Fructose is metabolized primarily by the liver, but the form of hexokinase present in the liver (called glucokinase) has very low affinity for fructose. Fructose therefore enters glycolysis by a different route.

First, fructose is phosphorylated to yield fructose-1-phosphate. The enzyme fructose-1-phosphate aldolase then splits the six-carbon molecule into two three-carbon molecules: glyceraldehyde and dihydroxyacetone phosphate, as shown here.

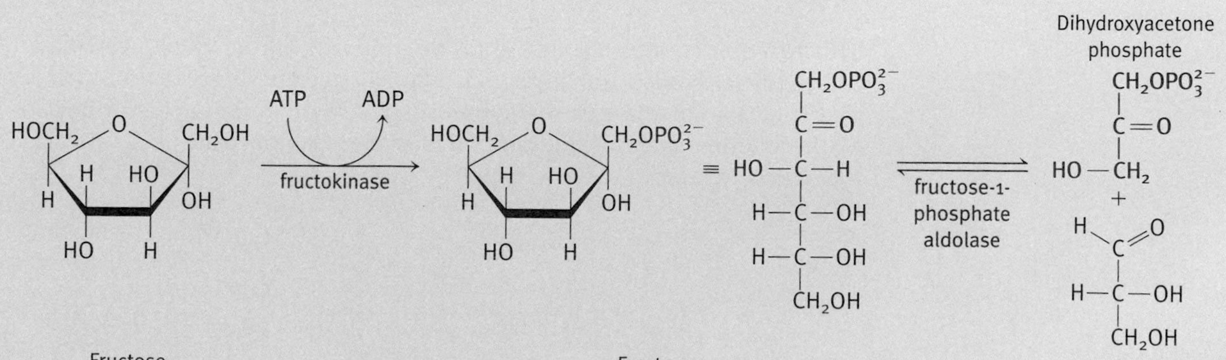

Fructose —(ATP → ADP; fructokinase)→ Fructose-1-phosphate ⇌ (fructose-1-phosphate aldolase) Dihydroxyacetone phosphate + Glyceraldehyde

(continued)

BOX 10-A *(continued)*

Dihydroxyacetone phosphate is converted to glyceraldehyde-3-phosphate by triose phosphate isomerase and can proceed through the second phase of glycolysis. The glyceraldehyde can be phosphorylated to glyceraldehyde-3-phosphate, but it can also be converted to glycerol-3-phosphate, a precursor of the backbone of triacylglycerols. This may contribute to an increase in fat. A second hazard of the fructose catabolic pathway is that fructose-1-phosphate may accumulate faster than it can be cleaved by the aldolase. This may tie up the liver's supply of phosphate, with disastrous consequences for ATP production from ADP. A third notable problem is that fructose catabolism bypasses the phosphofructokinase-catalyzed step of glycolysis and thus avoids a major regulatory point. This somehow disrupts fuel metabolism so that fructose catabolism leads to greater production of lipid than does glucose catabolism. The metabolic consequences of consuming high-fructose foods therefore go beyond their caloric (energetic) content.

reversible) or to be exported from the cell. The liver takes up lactate, oxidizes it back to pyruvate, and then uses it for gluconeogenesis. The glucose produced in this manner may eventually make its way back to the muscle to help fuel continued muscle contraction. When the muscle is functioning aerobically, NADH produced by the glyceraldehyde-3-phosphate dehydrogenase reaction is reoxidized by oxygen and the lactate dehydrogenase reaction is not needed.

Pyruvate oxidation

Although glycolysis is an oxidative pathway, its end product pyruvate is still a relatively reduced molecule. The complete oxidation of its carbons to 3 CO_2 can potentially release a large amount of free energy (Table 10-1). Much of this energy is recovered in the synthesis of ATP by the reactions of the citric acid cycle (Chapter 11) and oxidative phosphorylation (Chapter 12).

The further oxidation of pyruvate by these processes requires molecular oxygen. Organisms such as yeast growing under anaerobic conditions can dispose of pyruvate by converting it to alcohol (Box 10-B). This process was dubbed **fermentation** by Pasteur to describe the phenomenon of life without oxygen.

Pyruvate as a metabolic precursor

Pyruvate's carbon atoms provide the raw material for synthesizing a variety of molecules, including, in the liver, more glucose (discussed below). Fatty acids, the precursors of triacylglycerols and many membrane lipids, are synthesized from the two-carbon unit of acetyl-CoA, which is derived from pyruvate:

$$\underset{\text{Pyruvate}}{CH_3{-}\overset{\overset{\displaystyle O}{\|}}{C}{-}COO^-} \xrightarrow{\text{CoASH} \quad CO_2} \underset{\text{Acetyl-CoA}}{CH_3{-}\overset{\overset{\displaystyle O}{\|}}{C}{-}SCoA}$$

This reaction is one step in the production of fat from excess carbohydrate.

Pyruvate is also the precursor of oxaloacetate, a four-carbon molecule that is an intermediate in the synthesis of several amino acids. It is also one of the intermediates of the citric acid cycle. Oxaloacetate is synthesized by the action of pyruvate carboxylase:

$$\underset{\text{Pyruvate}}{\begin{array}{c} COO^- \\ | \\ C{=}O \\ | \\ CH_3 \end{array}} + CO_2 + ATP \xrightarrow[\text{pyruvate carboxylase}]{} \underset{\text{Oxaloacetate}}{\begin{array}{c} COO^- \\ | \\ C{=}O \\ | \\ CH_2 \\ | \\ COO^- \end{array}} + ADP + P_i$$

TABLE 10-1 Standard free energy changes for glucose catabolism

Catabolic process	$\Delta G°'$ ($kJ \cdot mol^{-1}$)
$C_6H_{12}O_6 \rightarrow 2\ C_3H_5O_3^- + 2\ H^+$ (Glucose) (lactate)	−196
$C_6H_{12}O_6 + 6\ O_2 \rightarrow 6\ CO_2 + 6\ H_2O$ (Glucose)	−2850

A CLOSER LOOK

BOX 10-B Alcohol metabolism

Alcohol is synthesized by yeast under anaerobic conditions. This two-step process eliminates the pyruvate end product of glycolysis and regenerates the NAD^+ required for further glucose catabolism. First, pyruvate decarboxylase (an enzyme not expressed in animals) catabolizes the removal of pyruvate's carboxylate group to produce acetaldehyde. Next, alcohol dehydrogenase reduces acetaldehyde to ethanol.

Pyruvate $\xrightarrow[\text{pyruvate decarboxylase}]{CO_2}$ (Acetaldehyde)

Acetaldehyde $\xrightarrow[\text{alcohol dehydrogenase}]{NADH \rightarrow NAD^+}$ Ethanol

Ethanol is considered to be a waste product of sugar metabolism; its accumulation is toxic to the organisms that produce it. This is one reason why the alcohol content of yeast-fermented beverages such as wine is limited to about 13%. "Hard" liquor must be distilled to increase its ethanol content.

Mammals synthesize alcohol dehydrogenase not to produce ethanol but to metabolize the ethanol that is produced by intestinal bacteria. This liver enzyme also metabolizes the exogenous ethanol from alcoholic drinks. Alcohol dehydrogenase oxidizes ethanol back to acetaldehyde, using NAD^+ as a cofactor. This reaction is followed by one catalyzed by acetaldehyde dehydrogenase, which produces acetate and also requires NAD^+:

Ethanol $\xrightleftharpoons[\text{alcohol dehydrogenase}]{NAD^+ \rightarrow NADH, H^+}$ (Acetaldehyde)

Acetaldehyde $\xrightleftharpoons[\text{acetaldehyde dehydrogenase}]{OH^-\ NAD^+ \rightarrow NADH, H^+}$ Acetate

Both acetaldehyde and acetate are mildly toxic, contributing to the hangover that follows alcohol consumption. In addition, the production of NADH disrupts the cell's balance of NAD^+ and NADH. A shortage of NAD^+ slows glycolysis at the NAD^+-requiring glyceraldehyde-3-phosphate dehydrogenase step, thereby reducing the cell's ability to produce ATP. The excess NADH drives the lactate dehydrogenase reaction toward lactate production, which diverts pyruvate from other pathways such as gluconeogenesis, thereby impairing the ability of the liver to supply glucose to tissues such as the brain.

The pyruvate carboxylase reaction is interesting because of its unusual chemistry. The enzyme has a biotin prosthetic group that acts as a carrier of CO_2. Biotin is considered a vitamin, but a deficiency is rare because it is present in many foods and is synthesized by intestinal bacteria. The biotin group is covalently linked to an enzyme Lys residue:

```
            O
            ‖
            C
          /   \
        HN     NH                                              |
         |      |                                            O=C
     H—C—————C—H                        O                        |
         |      \                        ‖                        |
       H2C       C—H                                             
          \     /  ---CH2—CH2—CH2—CH2—C—NH—(CH2)4—CH
            S                                                     |
                                                                 NH
                                                                  |
 |_____________________________________________| |____________________|
                     Biotin                          Lys residue
```

The Lys side chain and its attached biotin group form a 14-Å-long flexible arm that swings through the enzyme's active site. During the multistep reaction, a CO_2 molecule is first "activated" by its reaction with ATP, then transferred to biotin, and finally added to pyruvate to produce oxaloacetate (Fig. 10-12).

REVIEW LEARNING OBJECTIVE 5

- What are the possible fates of pyruvate?
- What is the metabolic function of lactate dehydrogenase?

3 GLUCONEOGENESIS

We have already alluded to the ability of the liver to synthesize glucose from noncarbohydrate precursors via the pathway of gluconeogenesis. This pathway, which also occurs to a limited extent in the kidneys and pancreas, operates when the liver's supply of glycogen is exhausted. Certain tissues, such as the central nervous system and red blood cells, which burn glucose as their primary metabolic fuel, cannot produce their own glucose and therefore must rely on the liver to supply them with newly synthesized glucose.

Gluconeogenesis is considered to be the reversal of glycolysis, that is, the conversion of two molecules of pyruvate to one molecule of glucose. Although some of the steps of gluconeogenesis are catalyzed by glycolytic enzymes operating in reverse, the gluconeogenic pathway contains several unique enzymes that bypass the three irreversible steps of glycolysis—the steps catalyzed by pyruvate kinase, phosphofructokinase, and hexokinase (Fig. 10-13). This principle applies to all pairs of opposing metabolic pathways: *The pathways may share some near-equilibrium reactions but cannot use the same enzymes to catalyze thermodynamically favorable irreversible reactions.* The three irreversible reactions of glycolysis are clearly visible in the "waterfall" diagram (Fig. 10-11).

Pyruvate to phosphoenolpyruvate

First, pyruvate cannot be converted directly back to phosphoenolpyruvate, because pyruvate kinase catalyzes an irreversible reaction (Reaction 10 of glycolysis). To get around this thermodynamic barrier, pyruvate is carboxylated

ATP + HCO_3^-

ADP

1. CO_2 (as bicarbonate, HCO_3^-) reacts with ATP such that some of the free energy released in the removal of ATP's phosphoryl group is conserved in the formation of the "activated" compound carboxyphosphate.

Carboxyphosphate

P_i

Biotinyl-Lys

2. Like ATP, carboxyphosphate releases a large amount of free energy when its phosphoryl group is liberated. This free energy drives the carboxylation of biotin.

Pyruvate

3. The enzyme abstracts a proton from pyruvate, forming a carbanion.

4. The carbanion attacks the carboxyl group attached to biotin, generating oxaloacetate.

Biotinyl-Lys

Oxaloacetate

FIGURE 10-12

The pyruvate carboxylase reaction.

by pyruvate carboxylase to yield the four-carbon compound oxaloacetate (the same reaction shown in Fig. 10-12). Next, phosphoenolpyruvate carboxykinase catalyzes the decarboxylation of oxaloacetate to form phosphoenolpyruvate:

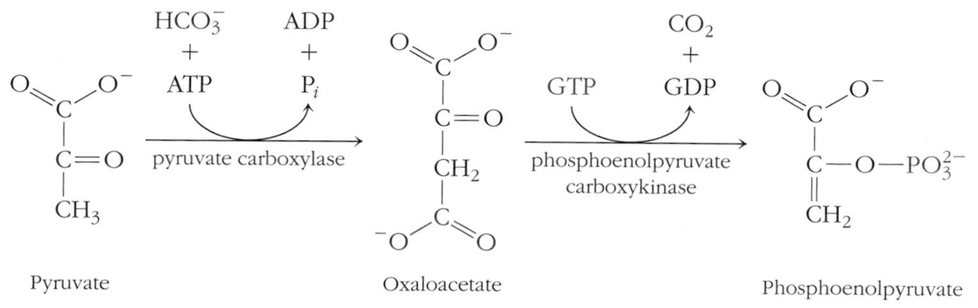

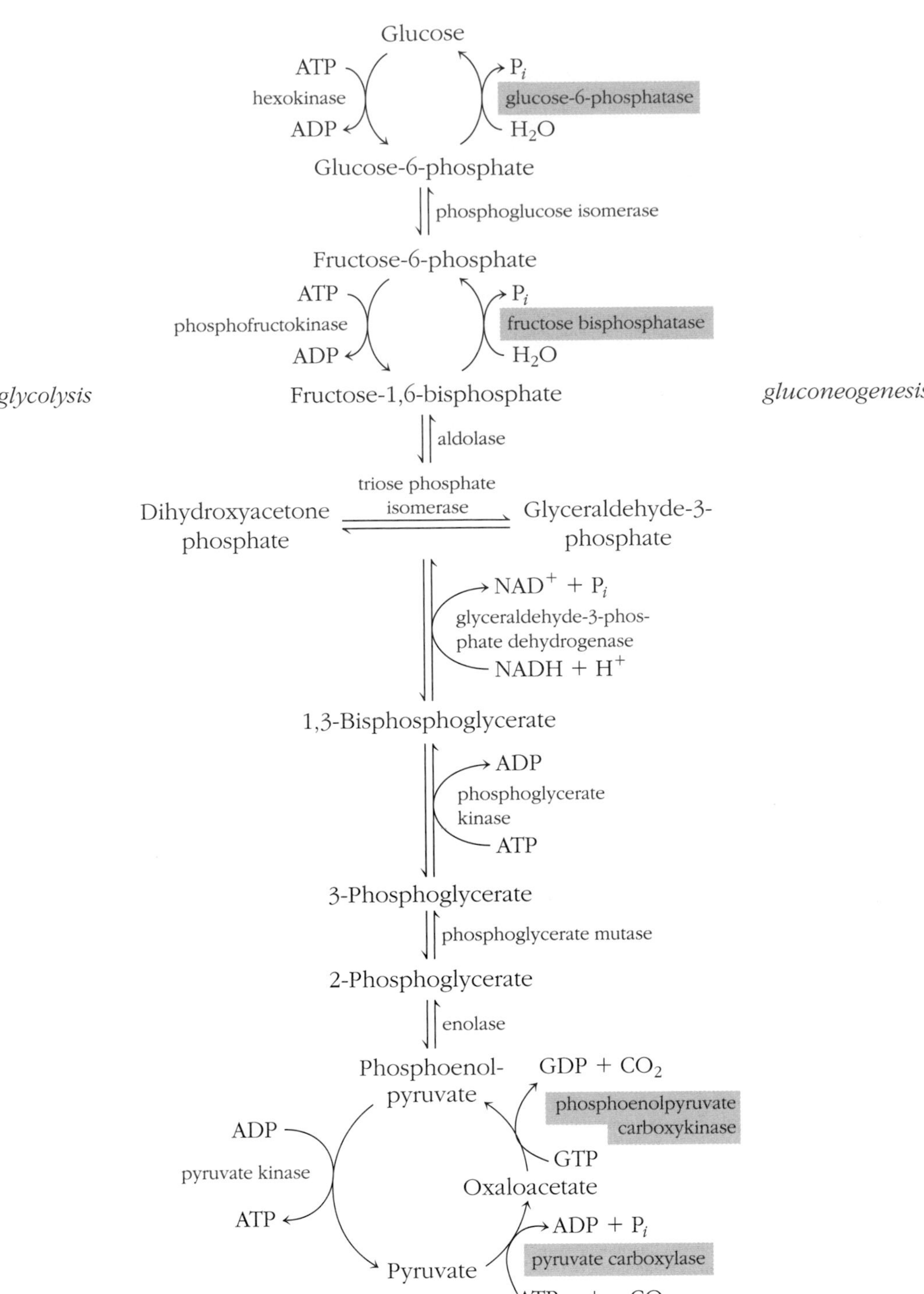

FIGURE 10-13

The reactions of gluconeogenesis.
The pathway uses the seven glycolytic enzymes that catalyze reversible reactions. The three irreversible reactions of glycolysis are bypassed in gluconeogenesis by the four enzymes that are highlighted in blue.

Note that the carboxylate group added in the first reaction is released in the second. The two reactions are energetically costly: Pyruvate carboxylase consumes ATP, and phosphoenolpyruvate carboxykinase consumes GTP (which is interchangeable with and energetically equivalent to ATP). Cleavage of two phosphoanhydride bonds is required to supply enough free energy to "undo" the highly exergonic pyruvate kinase reaction.

Amino acids (except for Leu and Lys) are the main sources of gluconeogenic precursors, because they can all be converted to oxaloacetate and then to phosphoenolpyruvate. Thus, during starvation, proteins can be broken down and used to produce glucose to fuel the central nervous system. Fatty acids cannot serve as gluconeogenic precursors because they cannot be converted to oxaloacetate. (However, the three-carbon glycerol backbone of triacylglycerols is a gluconeogenic precursor.)

Phosphoenolpyruvate to fructose-1,6-bisphosphate

Two molecules of phosphoenolpyruvate are converted to fructose-1,6-bisphosphate in a series of six reactions that are all catalyzed by glycolytic enzymes (Steps 4–9 in reverse order). These reactions are reversible because they are near equilibrium ($\Delta G \approx 0$), and the direction of flux is determined by the concentrations of substrates and products. Note that the phosphoglycerate kinase reaction consumes ATP when it operates in the direction of gluconeogenesis. NADH is also required to reverse the glyceraldehyde-3-phosphate dehydrogenase reaction.

Fructose-1,6-bisphosphate to glucose

The final three reactions of gluconeogenesis require two enzymes unique to this pathway. The first step undoes the phosphofructokinase reaction, the irreversible reaction that is the major control point of glycolysis. In gluconeogenesis, the enzyme fructose bisphosphatase hydrolyzes the C1 phosphate of fructose-1,6-bisphosphate to yield fructose-6-phosphate. This reaction is thermodynamically favorable, with a ΔG value of $-8.6 \text{ kJ} \cdot \text{mol}^{-1}$.

Next, the glycolytic enzyme phosphoglucose isomerase catalyzes the reverse of Step 2 of glycolysis to produce glucose-6-phosphate.

Finally, the gluconeogenic enzyme glucose-6-phosphatase catalyzes a hydrolytic reaction that yields glucose and P_i. Note that the hydrolytic reactions catalyzed by fructose bisphosphatase and glucose-6-phosphatase undo the work of two kinases in glycolysis (phosphofructokinase and hexokinase).

Regulation of gluconeogenesis

Gluconeogenesis is energetically expensive. Producing 1 glucose from 2 pyruvate consumes 6 ATP, 2 each at the steps catalyzed by pyruvate carboxylase, phosphoenolpyruvate carboxykinase, and phosphoglycerate kinase. If glycolysis occurred simultaneously with gluconeogenesis, there would be a net consumption of ATP:

Glycolysis	Glucose + 2 ADP + 2 $P_i \rightarrow$ 2 pyruvate + 2 ATP
Gluconeogenesis	2 Pyruvate + 6 ATP $\rightarrow$ glucose + 6 ADP + 6 P_i
Net	4 ATP $\rightarrow$ 4 ADP + 4 P_i

To avoid this waste of metabolic free energy, gluconeogenic cells (mainly liver cells) carefully regulate the opposing pathways of glycolysis and gluconeogenesis according to the cell's energy needs. *The major regulatory point is centered on the interconversion of fructose-6-phosphate and fructose-1,6-bisphosphate.* We have already mentioned that fructose-2,6-bisphosphate is a potent allosteric activator of phosphofructokinase (PFK), which catalyzes the glycolytic reaction (Section 10-2). Not surprisingly, fructose-2,6-bisphosphate is a potent *inhibitor* of fructose bisphosphatase (FBPase), which catalyzes the opposing gluconeogenic reaction.

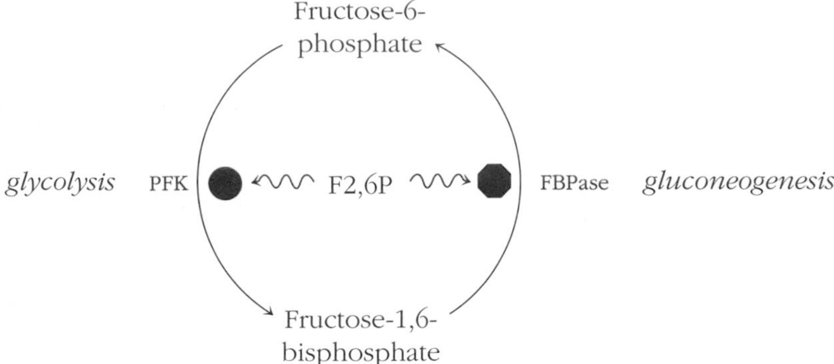

This mode of allosteric regulation is efficient because *a single compound can control flux through two opposing pathways in a reciprocal fashion.* Thus, when the concentration of fructose-2,6-bisphosphate is high, glycolysis is stimulated and gluconeogenesis is inhibited, and vice versa.

Muscle, a tissue that does not undertake gluconeogenesis, does not express the gluconeogenic enzyme glucose-6-phosphatase. Consequently, the phosphorylated glucose produced by glycogenolysis in muscle cells cannot be converted to glucose and released into the circulation. However, many nongluconeogenic cells do contain the gluconeogenic enzyme fructose bisphosphatase. What is the reason for this?

When both fructose bisphosphatase and phosphofructokinase are active, the net result is the hydrolysis of ATP:

PFK	Fructose-6-phosphate + ATP $\rightarrow$ fructose-1,6-bisphosphate + ADP
FBPase	Fructose-1,6-bisphosphate + H_2O $\rightarrow$ fructose-6-phosphate + P_i
Net	ATP + H_2O $\rightarrow$ ADP + P_i

This combination of metabolic reactions is called a **futile cycle** since it seems to have no useful result. However, Eric Newsholme realized that such futile cycles could actually provide a means for fine-tuning the output of a metabolic pathway. For example, flux through the phosphofructokinase step of glycolysis is diminished by the activity of fructose bisphosphatase. An allosteric compound such as fructose-2,6-bisphosphate modulates the activity of both enzymes so that as the activity of one enzyme increases, the activity of the other one decreases. This dual regulatory effect results in a greater possible range of net flux than if the regulator merely activated or inhibited a single enzyme.

REVIEW LEARNING OBJECTIVE 6

- Which reactions of gluconeogenesis are catalyzed by glycolytic enzymes?
- Why are some enzymes unique to gluconeogenesis?
- What is a futile cycle and what is its purpose?

Glycogen synthesis

Some of the glucose produced by gluconeogenesis is stored in the liver and other tissues as glycogen. Recall that glucose units are removed from the glycogen polymer by phosphorolysis (see Section 9-1). This reaction is catalyzed by an enzyme that operates far from equilibrium (that is, it catalyzes a metabolically irreversible reaction). The synthesis of glycogen from glucose must therefore use a different sequence of reactions that are thermodynamically favorable. In mammalian cells, glycogen synthesis requires the free energy released by the nucleotide uridine triphosphate (like GTP, UTP is energetically equivalent to ATP).

The monosaccharide unit that is incorporated into glycogen is glucose-1-phosphate, which is produced from glucose-6-phosphate (the penultimate product of gluconeogenesis) by the action of phosphoglucomutase (introduced on page 307). Glucose-1-phosphate is then "activated" by reacting with UTP to form UDP–glucose:

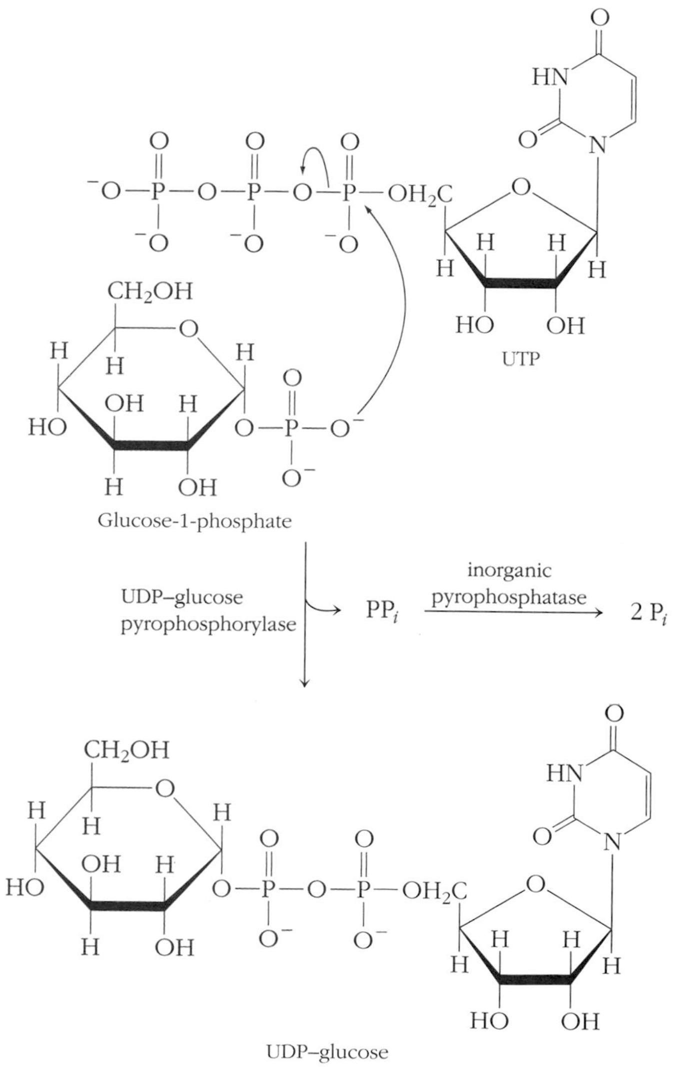

UTP

Glucose-1-phosphate

UDP–glucose pyrophosphorylase → PP_i $\xrightarrow{\text{inorganic pyrophosphatase}}$ 2 P_i

UDP–glucose

This process is a reversible phosphoanhydride exchange reaction ($\Delta G \approx 0$). Note that the two phosphoanhydride bonds of UTP are conserved, one in the product PP_i and one in UDP–glucose. However, the PP_i is rapidly hydrolyzed by inorganic pyrophosphatase to 2 P_i in a highly exergonic reaction ($\Delta G^{\circ\prime} = -33.5\ \text{kJ} \cdot \text{mol}^{-1}$). Thus, cleavage of a phosphoanhydride bond makes the formation of UDP–glucose an exergonic, irreversible process—that is, *PP_i hydrolysis "drives" a reaction that would otherwise be near equilibrium.* The hydrolysis of PP_i by inorganic pyrophosphatase is a common strategy of biosynthetic reactions; we will see this reaction again in the synthesis of other polymers, namely DNA, RNA, and polypeptides.

Finally, glycogen synthase transfers the glucose unit to the C4 OH group at the end of one of glycogen's branches:

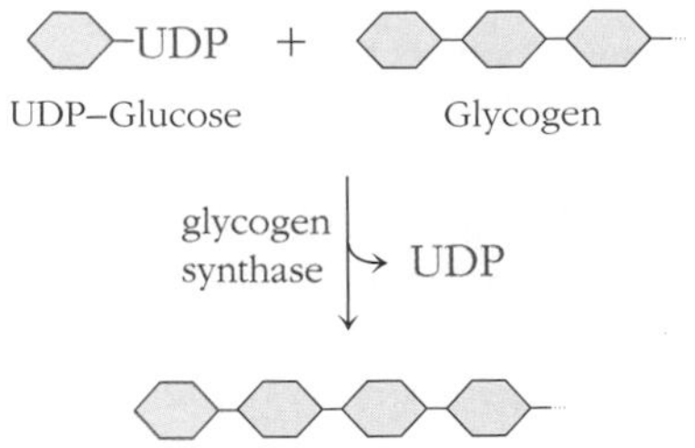

UDP + (glycogen)

UDP–Glucose   Glycogen

glycogen synthase ↓ → UDP

(glycogen, one residue longer)

The steps of glycogen synthesis can be summarized as follows:

UDP–glucose pyrophosphorylase	Glucose-1-phosphate + UTP $\rightleftharpoons$ UDP–glucose + PP_i
Pyrophosphatase	$PP_i + H_2O \rightarrow 2\ P_i$
Glycogen synthase	UDP–glucose + glycogen (*n* residues) $\rightarrow$ glycogen (*n* + 1 residues) + UDP

Glucose-1-phosphate + glycogen + UTP + $H_2O \rightarrow$ glycogen + UDP + 2 P_i

The energetic price for adding one glucose unit to glycogen is the cleavage of one phosphoanhydride bond of UTP. Nucleotides are also required for the synthesis of other oligosaccharides and polysaccharides (Box 10-C). In Chapter 16 we will examine some of the hormonal mechanisms for regulating the opposing pathways of glycogen synthesis and degradation.

REVIEW LEARNING OBJECTIVE 7

- What is the role of UTP in glycogen synthesis?

4 THE PENTOSE PHOSPHATE PATHWAY

We have already seen that glucose catabolism can lead to pyruvate, which can be further oxidized to generate more ATP or used to synthesize amino acids and fatty acids. Glucose is also a precursor for nucleotide synthesis. We will end this chapter on glucose metabolism by examining the pathway that produces the ribose component of nucleotides.

The **pentose phosphate pathway,** like glycolysis, is an oxidative pathway, but it generates NADPH rather than NADH. The two cofactors are not interchangeable and are easily distinguished by degradative enzymes (which use NAD^+) and biosynthetic enzymes (which use $NADP^+$). The pentose phosphate pathway is by no means a minor feature of glucose metabolism. As much as 30% of glucose in the liver may be catabolized by the pentose phosphate pathway. This pathway can be divided into two phases: a series of oxidative reactions followed by a series of reversible interconversion reactions.

A CLOSER LOOK

BOX 10-C The synthesis of other saccharides

The standard free energy change for the formation of a glycosidic bond is about 16 kJ · mol^{-1}, so the reaction must be coupled to a thermodynamically favorable process such as the cleavage of a phosphoanhydride bond ($\Delta G^{\circ\prime} \approx 30$ kJ · mol^{-1}). In glycogen synthesis, the phosphoanhydride bond originates in the nucleotide UTP, and UDP–glucose is an intermediate. Other nucleotide sugars participate in the synthesis of other saccharides.

Lactose, a glucose–galactose disaccharide, is synthesized by the transfer of galactose from UDP–galactose to an OH group of glucose. UDP–galactose itself is the product of an epimerization reaction (**epimers** are sugars, such as glucose and galactose, that differ in configuration at one of their carbons). The enzyme UDP–galactose 4-epimerase catalyzes the isomerization of UDP–glucose to UDP–galactose in a reaction that requires NAD^+:

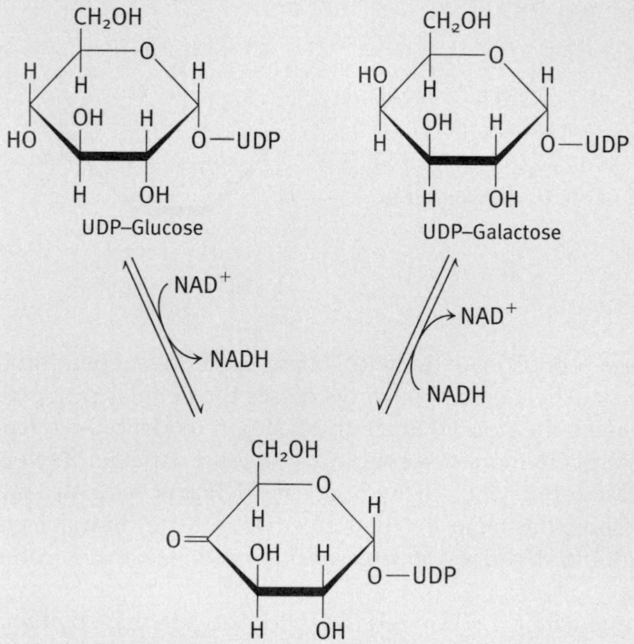

The *O*-linked oligosaccharides of glycoproteins are also assembled from nucleotide sugars. The monosaccharide units are added one at a time to build a chain anchored to the Ser side chain of a polypeptide. These additions are catalyzed by glycosyltransferases that recognize specific nucleotide sugars. In one type of *O*-linked oligosaccharide, monosaccharides are donated by UDP–*N*-acetylgalactosamine, UDP–galactose, and CMP–sialic acid.

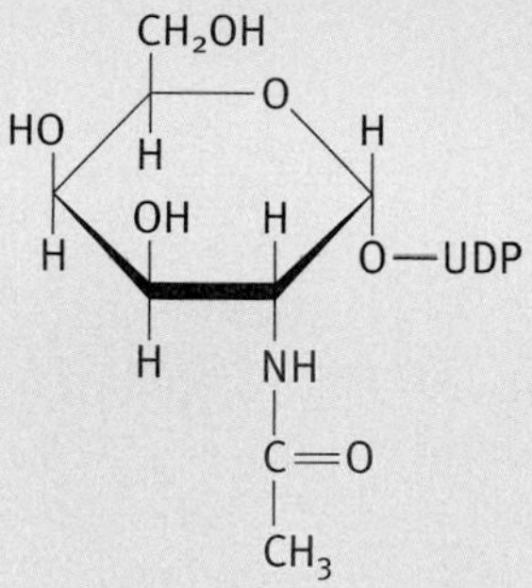

UDP–*N*-acetylgalactosamine

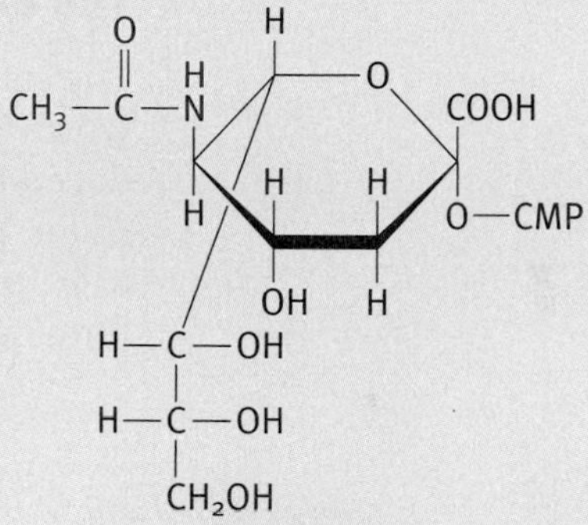

CMP–sialic acid

N-linked oligosaccharides are also built from monosaccharides ultimately derived from nucleotide sugars such as UDP–galactose and GDP–mannose.

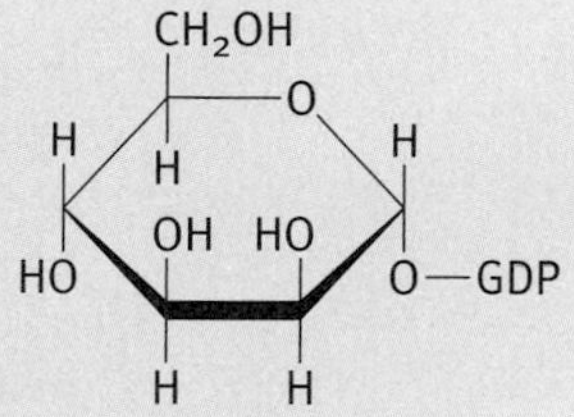

GDP–mannose

However, in this case, a 15-residue oligosaccharide is first assembled on a lipid carrier molecule before being transferred to an Asn side chain of a polypeptide. The oligosaccharide is subsequently trimmed by glycosidases, but additional monosaccharides may be added by specific glycosyltransferases, as in *O*-linked oligosaccharides.

In the enormous molecules known as **proteoglycans,** *O*-linked oligosaccharides are extended by the addition of hundreds of monosaccharide residues. These polysaccharide chains, called **glycosaminoglycans,** consist of an amino sugar such as *N*-acetylgalactosamine alternating with an acidic sugar such as glucuronic acid.

(continued)

BOX 10-C *(continued)*

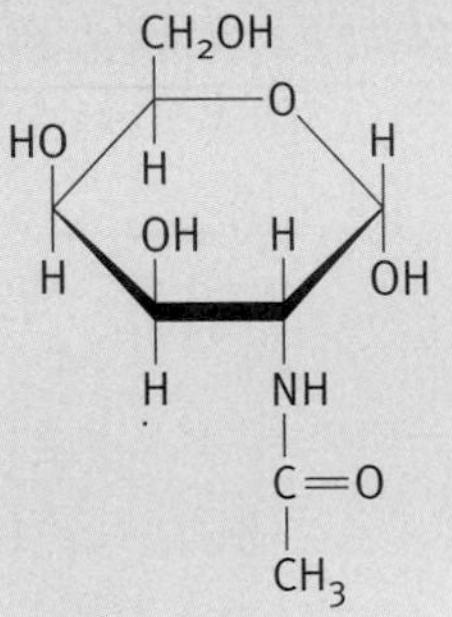

N-Acetylgalactosamine

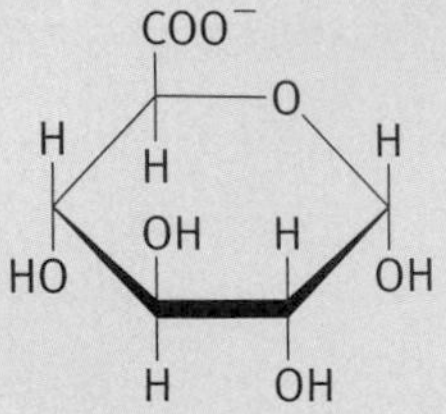

Glucuronic acid

The glycosaminoglycan chain may be sulfated, so in addition to its numerous OH and COO^- groups, it contains negatively charged SO_3^- groups. A proteoglycan bearing dozens or hundreds of glycosaminoglycan chains is therefore heavily hydrated and exists as a gel-like substance. The elasticity and incompressibility of connective tissue is due largely to the glycosaminoglycan chains of proteoglycans.

The oxidative reactions of the pentose phosphate pathway

The starting point of the pentose phosphate pathway is glucose-6-phosphate, which can be derived from free glucose, from the glucose-1-phosphate produced by glycogen phosphorolysis, or from gluconeogenesis. In the first step of the pathway, glucose-6-phosphate dehydrogenase catalyzes the metabolically irreversible transfer of a hydride ion from glucose-6-phosphate to $NADP^+$, forming a lactone and NADPH:

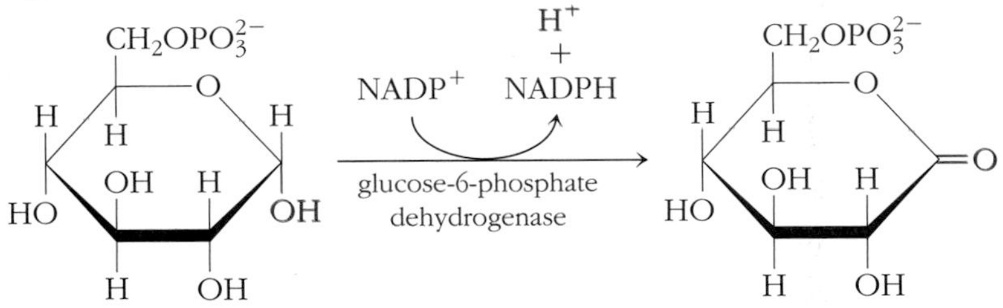

Glucose-6-phosphate $\xrightarrow[\text{glucose-6-phosphate dehydrogenase}]{NADP^+ \;\rightarrow\; NADPH + H^+}$ 6-Phosphoglucono-δ-lactone

A deficiency of glucose-6-phosphate dehydrogenase is the most common human enzyme deficiency. This defect, which decreases the cellular production of NADPH, interferes with the normal function of certain oxidation–reduction processes and makes the cells more susceptible to oxidative damage. However, individuals with glucose-6-phosphate dehydrogenase deficiency are also more resistant to malaria. Thus, the gene for the defective enzyme (like the gene for sickle-cell hemoglobin described in Box 4-D) persists because it confers a selective advantage.

The lactone intermediate is hydrolyzed to 6-phosphogluconate by the action of 6-phosphogluconolactonase:

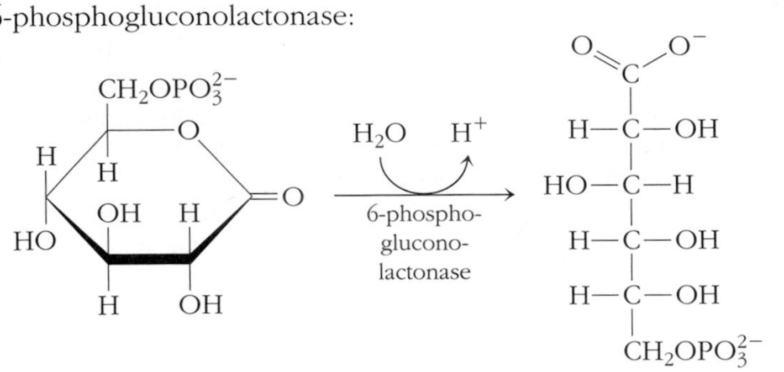

6-Phosphoglucono-δ-lactone $\xrightarrow[\text{6-phosphogluconolactonase}]{H_2O \;\rightarrow\; H^+}$ 6-Phosphogluconate

This reaction can also occur in the absence of the enzyme.

In the third step of the pentose phosphate pathway, 6-phosphogluconate is oxidatively decarboxylated in a reaction that converts the six-carbon sugar (a hexose) to a five-carbon sugar (a **pentose**) and reduces a second $NADP^+$ to NADPH:

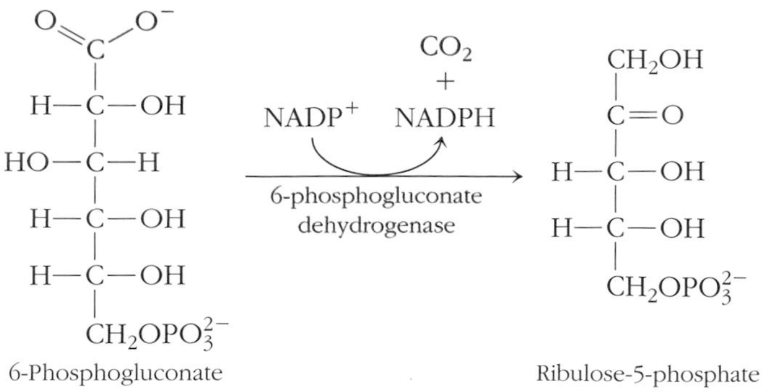

$$\underset{\text{6-Phosphogluconate}}{\begin{array}{c} COO^- \\ | \\ H-C-OH \\ | \\ HO-C-H \\ | \\ H-C-OH \\ | \\ H-C-OH \\ | \\ CH_2OPO_3^{2-} \end{array}} \xrightarrow[\text{6-phosphogluconate dehydrogenase}]{NADP^+ \quad CO_2 + NADPH} \underset{\text{Ribulose-5-phosphate}}{\begin{array}{c} CH_2OH \\ | \\ C=O \\ | \\ H-C-OH \\ | \\ H-C-OH \\ | \\ CH_2OPO_3^{2-} \end{array}}$$

The two molecules of NADPH produced for each glucose molecule that enters the pathway are used primarily for biosynthetic reactions, such as fatty acid synthesis and the synthesis of deoxynucleotides.

Isomerization and interconversion reactions of the pentose phosphate pathway

The ribulose-5-phosphate product of the oxidative phase of the pentose phosphate pathway can isomerize to ribose-5-phosphate:

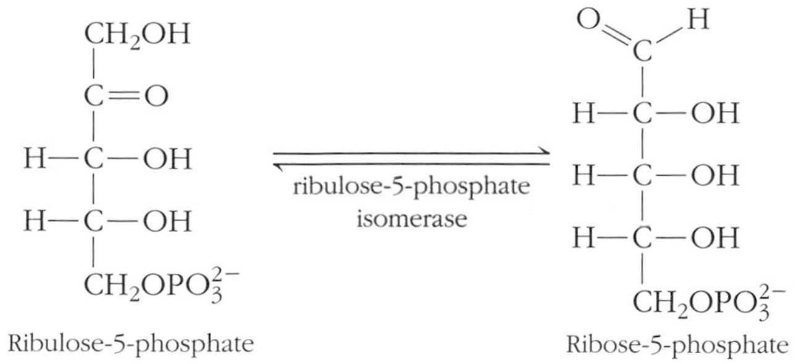

$$\underset{\text{Ribulose-5-phosphate}}{\begin{array}{c} CH_2OH \\ | \\ C=O \\ | \\ H-C-OH \\ | \\ H-C-OH \\ | \\ CH_2OPO_3^{2-} \end{array}} \underset{\text{ribulose-5-phosphate isomerase}}{\rightleftharpoons} \underset{\text{Ribose-5-phosphate}}{\begin{array}{c} CHO \\ | \\ H-C-OH \\ | \\ H-C-OH \\ | \\ H-C-OH \\ | \\ CH_2OPO_3^{2-} \end{array}}$$

Ribose-5-phosphate is the precursor of the ribose unit of nucleotides. In many cells, this marks the end of the pentose phosphate pathway, which has the net equation

$$\text{Glucose-6-phosphate} + 2\ NADP^+ + H_2O \rightarrow \text{ribose-5-phosphate} + 2\ NADPH + CO_2 + 2\ H^+$$

Not surprisingly, the activity of the pentose phosphate pathway is high in rapidly dividing cells that must synthesize large amounts of DNA. In fact, the pentose phosphate pathway not only produces ribose, it also provides a reducing agent (NADPH) required for the reduction of ribose to deoxyribose.

Ribonucleotide reductase carries out the reduction of nucleotide diphosphates (NDPs):

NDP

↓

dNDP

The enzyme, which is oxidized in the process, is restored to its original state by a series of reactions in which NADPH is reduced.

In some cells, however, the need for NADPH for other biosynthetic reactions is greater than the need for ribose-5-phosphate. In this case, the excess carbons of the pentose are recycled into intermediates of the glycolytic pathway so that they can be degraded to pyruvate or used in gluconeogenesis, depending on the cell type and its metabolic needs.

A set of reversible reactions transforms five-carbon ribulose units into six-carbon units (fructose-6-phosphate) and three-carbon units (glyceraldehyde-3-phosphate). This transformation is accomplished mainly by the enzymes transketolase and transaldolase, which transfer two- and three-carbon units among various monosaccharide intermediates (the reaction catalyzed by transketolase was introduced in Section 7-2). Figure 10-14 is a schematic view of this process. Because all of these interconversions are reversible, glycolytic intermediates can also be siphoned from glycolysis to synthesize ribose-5-phosphate.

REVIEW LEARNING OBJECTIVE 8

- What are the main products of the pentose phosphate pathway and how does the cell use them?

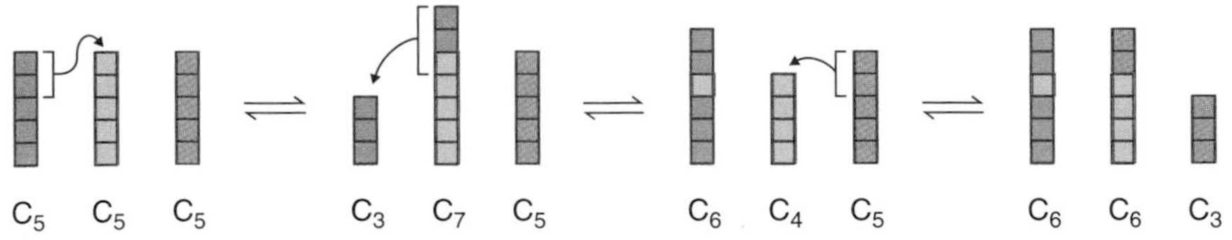

FIGURE 10-14

Rearrangements of the products of the pentose phosphate pathway. Three of the five-carbon products of the oxidative phase of the pentose phosphate pathway are converted to two fructose-6-phosphate and one glyceraldehyde-3-phosphate by reversible reactions involving the transfer of two- and three-carbon units.

5 SUMMARY OF GLUCOSE METABOLISM

The central position of glucose metabolism in all cells warrants its close study. Indeed, the enzymes of glycogen metabolism, glycolysis, gluconeogenesis, and the pentose phosphate pathway are among the best-studied proteins. In nearly all cases, detailed knowledge of their molecular structures has provided insight into their catalytic mechanisms and mode of regulation.

Although our coverage of glucose metabolism is far from exhaustive, this chapter describes quite a few enzymes and reactions, which are compiled in Figure 10-15. As you examine this diagram, keep in mind the following points, which also apply to the metabolic pathways we will encounter in subsequent chapters:

1. A metabolic pathway is a series of enzyme-catalyzed reactions, so that the pathway's substrate is converted to its product in discrete steps.

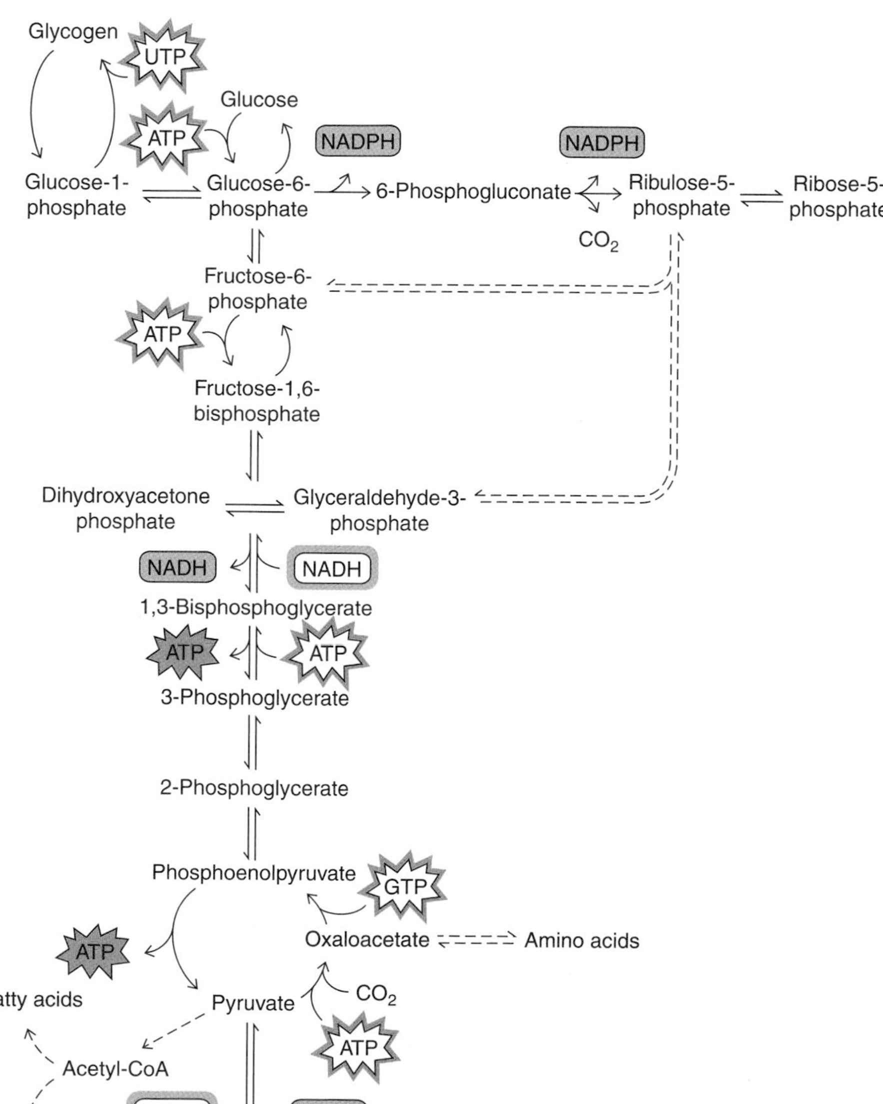

FIGURE 10-15

Summary of glucose metabolism. This diagram includes the pathways of glycogen synthesis and degradation, glycolysis, gluconeogenesis, and the pentose phosphate pathway. Dotted lines are used where the individual reactions are not shown. Filled gold symbols indicate ATP production; shadowed gold symbols indicate ATP consumption. Filled and shadowed red symbols represent the production and consumption of the reduced cofactors NADH and NADPH.

2. A monomeric compound such as glucose is interconverted with its polymeric form (glycogen), with other monosaccharides (fructose-6-phosphate and ribose-5-phosphate, for example), and with smaller metabolites such as the three-carbon pyruvate.

3. Although anabolic and catabolic pathways may share some steps, their irreversible steps are catalyzed by enzymes unique to each pathway.

4. Certain reactions consume or produce free energy in the form of ATP. In most cases, these are phosphoryl-group transfer reactions.

5. Some steps are oxidation–reduction reactions that require or generate a reduced cofactor such as NADH or NADPH.

SUMMARY

1. The pathway of glucose catabolism, or glycolysis, is a series of enzyme-catalyzed steps in which free energy is conserved as ATP or NADH.

2. Glucose, the substrate for glycolysis, can be obtained by breaking the glycosidic bonds in glycogen (glycogenolysis).

3. The 10 reactions of glycolysis convert the six-carbon glucose to two molecules of pyruvate and produce two molecules of NADH and two molecules of ATP. The first phase (reactions catalyzed by hexokinase, phosphoglucose isomerase, phosphofructokinase, aldolase, and triose phosphate isomerase) requires the investment of 2 ATP. The irreversible reaction catalyzed by phosphofructokinase is the rate-determining step and the major control point for glycolysis. The second phase of the pathway (reactions catalyzed by glyceraldehyde-3-phosphate dehydrogenase, phosphoglycerate kinase, phosphoglycerate mutase, enolase, and pyruvate kinase) generates 4 ATP per glucose.

4. Pyruvate may be reversibly reduced to lactate, further oxidized by the citric acid cycle, or converted to other compounds.

5. The pathway of gluconeogenesis converts two molecules of pyruvate to one molecule of glucose at a cost of 6 ATP. The pathway uses seven glycolytic enzymes, and the activities of pyruvate carboxylase, phosphoenolpyruvate carboxykinase, fructose bisphosphatase, and glucose-6-phosphatase bypass the three irreversible steps of glycolysis. A futile cycle involving phosphofructokinase and fructose bisphosphatase helps regulate the flux through glycolysis and gluconeogenesis.

6. Glucose residues are incorporated into glycogen after first being activated by attachment to UDP.

7. The pentose phosphate catabolic pathway for glucose yields NADPH and ribose groups. The five-carbon sugar intermediates can be converted to glycolytic intermediates.

CHECKLIST

1. Review the Learning Objectives listed on page 302.

2. Complete Exercise 15/Glycolytic Enzymes to review the pathway of glycolysis and to explore the structures and mechanisms of the 10 enzymes.

3. Apply your knowledge by solving the problems at the end of this chapter. Check your results in the Solutions appendix.

4. Be able to define the boldfaced terms (consult the glossary at the end of the book). Test your understanding by taking the Chapter 10 quiz (www.wiley.com/college/pratt).

5. Explore the websites listed at www.wiley.com/college/pratt.

6. Consult the list of Selected Readings for background information or additional details on the pathways of glucose metabolism.

GLOSSARY TERMS

glycolysis
gluconeogenesis
hexose
Fischer projection
Haworth projection
α anomer
β anomer
glycosidic bond
glycogenolysis
kinase
metabolically irreversible reaction
ketose
aldose
near-equilibrium reaction
flux
rate-determining reaction
tautomerization
feed-forward activation
fermentation
futile cycle
epimers
proteoglycan
glycosaminoglycan
pentose phosphate pathway
pentose

PROBLEMS

1. The structures of the α and β anomers of mannose are shown. Which of the two is more stable? Draw the structure of its chair conformation.

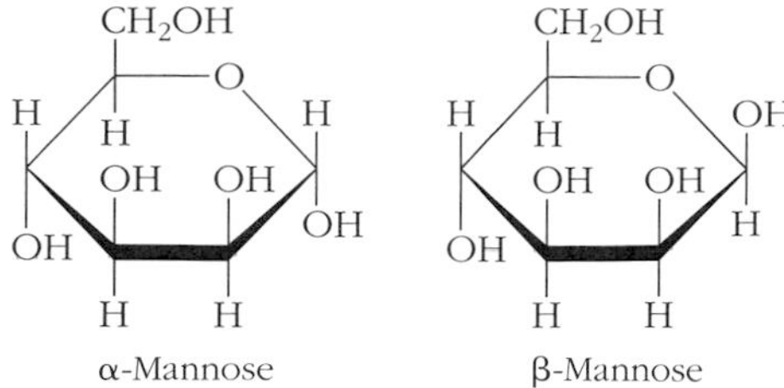

2. The exoskeleton of invertebrates such as crustaceans, insects, and spiders is composed principally of chitin, a polymer of a single type of modified monosaccharide. What is the monosaccharide and how has it been modified? Specify the location of the glycosidic bond.

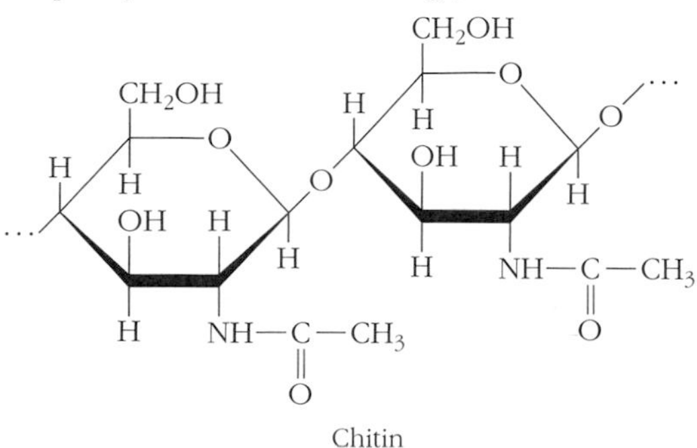

Chitin

3. Except during starvation, the brain burns glucose as its sole metabolic fuel and consumes up to 40% of the body's circulating glucose.
   (a) Why would hexokinase be the primary rate-determining step of glycolysis in the brain? (In tissues such as muscle, phosphofructokinase rather than hexokinase catalyzes the rate-determining step.)
   (b) Brain hexokinase has a K_M for glucose that is 100 times lower than the concentration of circulating glucose (5 mM). What is the advantage of this low K_M?

4. Individuals who seek to reduce their daily caloric intake rely on artificial sweeteners, compounds that have a sweet taste but are indigestible and thus contribute no calories to the diet. Many artificial sweeteners have been discovered by accident, since the structural requirements for sweet taste have remained elusive.
   (a) Over 30 years ago, Shallenberger and Acree suggested that the sensation of a sweet taste occurs when the functional groups —B and —AH on the sweet compound form hydrogen bonds with protein receptor sites on the tongue. Describe how ordinary table sugar, sucrose (a disaccharide of fructose and glucose; see Box 10-A), would interact with protein receptors in this way.

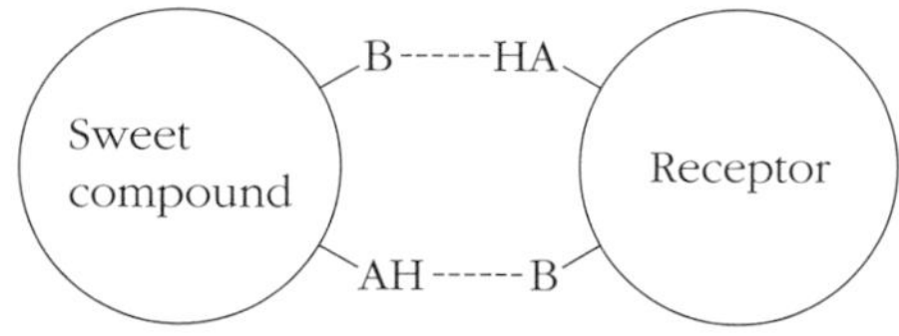

   (b) Because many molecules have —B and —AH groups yet are not sweet, Kier hypothesized that in addition to the —B and —AH groups, a third hydrophobic group, —X, is required to interact with the taste buds. The structure of the sucrose derivative sucralose (Splenda®) is shown. Which functional groups can fulfill the roles of —B, —AH, and —X?

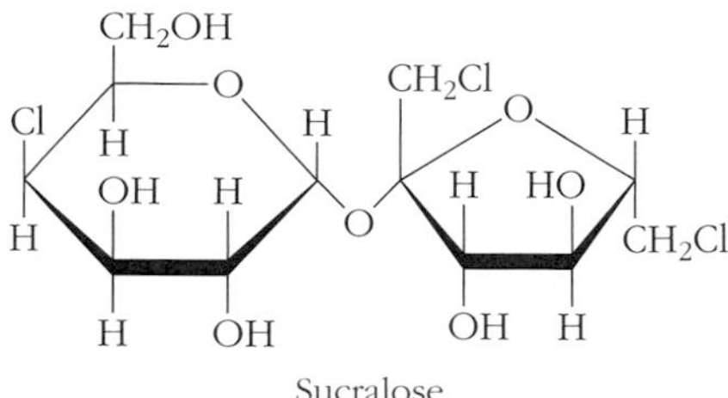

Sucralose

5. Explain why iodoacetate was useful for determining the order of intermediates in glycolysis but provided misleading information about an enzyme's active site.

6. Which of the 10 reactions of glycolysis are (a) phosphorylations, (b) isomerizations, (c) oxidation–reductions, (d) dehydrations, and (e) carbon–carbon bond cleavages?

7. Assuming a standard free energy change of 30.5 $kJ \cdot mol^{-1}$ for the synthesis of ATP from ADP and P_i, how many molecules of ATP can be theoretically produced by the catabolism of glucose to (a) lactate or to (b) CO_2 (see Table 10-1)?

8. The $\Delta G°'$ value for the hexokinase reaction is -16.7 $kJ \cdot mol^{-1}$, and the ΔG value under cellular conditions is similar.
   (a) What is the ratio of glucose-6-phosphate to glucose under standard conditions if the ratio of [ATP] to [ADP] is 10:1?
   (b) How high would the ratio of glucose-6-phosphate to glucose have to be in order to reverse the hexokinase reaction by mass action?

9. What is the ratio of fructose-6-phosphate to glucose-6-phosphate under (a) standard conditions and (b) cellular conditions? In which direction does the reaction proceed under cellular conditions?

10. What is the ratio of glyceraldehyde-3-phosphate (GAP) to dihydroxyacetone phosphate (DHAP) in cells at 37°C under nonequilibrium conditions? Considering your answer to this question, how do you account for the fact that the conversion of DHAP to GAP occurs readily in cells?

11. Glucose is frequently administered intravenously (injected directly into the bloodstream) to patients as a food source. A new resident at a hospital where you are doing one of your rotations suggests administering glucose-6-phosphate instead. You recall from biochemistry class that the transformation of glucose to glucose-6-phosphate requires ATP and you consider the possibility that

administering glucose-6-phosphate might save the patient energy. Should you use the resident's suggestion?

12. Trypanosomes living in the human bloodstream obtain all their free energy from glycolysis. They take up glucose from the host's blood and excrete pyruvate as a waste product. In this part of their life cycle, trypanosomes do not carry out any oxidative phosphorylation but they do use another oxygen-dependent pathway, which is absent in mammals, to oxidize NADH.

(a) Why is this other pathway necessary?

(b) Would the pathway be necessary if the trypanosome excreted lactate rather than pyruvate?

(c) Why would this pathway be a good target for antiparasitic drugs?

13. Red blood cells synthesize and degrade 2,3-BPG as a detour from the glycolytic pathway, as shown in the figure.

Glyceraldehyde-3-P_i

⇅ GAPDH

1,3-Bisphosphoglycerate → 2,3-Bisphosphoglycerate

⇅ PGK

3-Phosphoglycerate ← 2,3-Bisphosphoglycerate

2,3-BPG decreases the oxygen affinity of hemoglobin by binding in the central cavity of the deoxygenated form of hemoglobin. This encourages delivery of oxygen to tissues. A defect in one of the glycolytic enzymes may affect levels of 2,3-BPG. The plot below shows oxygen-binding curves for normal erythrocytes and for hexokinase- and pyruvate kinase-deficient erythrocytes. Identify which curve corresponds to which enzyme deficiency.

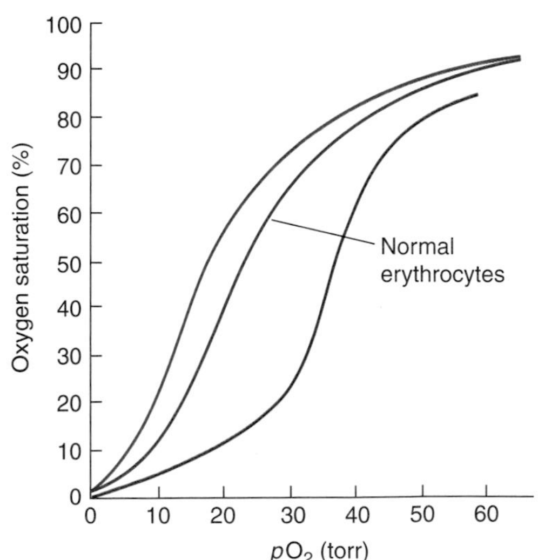

14. Arsenate, AsO_4^{3-}, acts as a phosphate analog and can replace phosphate in the GAPDH reaction. The product of this reaction is 1-arseno-3-phosphoglycerate. It is unstable and spontaneously hydrolyzes to form 3-phosphoglycerate, as shown above right. What is the effect of arsenate on cells undergoing glycolysis?

$$\mathrm{HC(=O)\text{—}HCOH\text{—}H_2COPO_3^{2-}} + AsO_4^{3-} \xrightarrow[\text{GAPDH}]{NAD^+ \rightarrow NADH + H^+} \mathrm{(O=)C\text{—}OAsO_3^{2-}\text{—}HCOH\text{—}H_2COPO_3^{2-}}$$

$$\mathrm{(O=)C\text{—}OAsO_3^{2-}\text{—}HCOH\text{—}H_2COPO_3^{2-}} \xrightarrow{H_2O \rightarrow AsO_4^{3-}} \mathrm{(O=)C\text{—}O^-\text{—}HCOH\text{—}H_2COPO_3^{2-}}$$

15. Vanadate, VO_4^{3-}, inhibits GAPDH, not by acting as a phosphate analog, but by interacting with essential —SH groups on the enzyme. What happens to cellular levels of phosphate, ATP, and 2,3-bisphosphoglycerate (see Problem 13) when red blood cells are incubated with vanadate?

16. Cancer cells have elevated levels of GAPDH, which may account for the high rate of glycolysis seen in cancer cells. The compound methylglyoxal has been shown to inhibit GAPDH in cancer cells but not in normal cells. This observation may lead to the development of rapid screening assays for cancer cells and to the development of drugs for treating cancerous tumors.

(a) What mechanisms might be responsible for the elevated levels of GAPDH in cancer cells?

(b) Why might methylglyoxal inhibit GAPDH in cancer cells but not in normal cells?

[From Ray, M., Basu, N., and Ray, S., *Mol. Cell. Biochem.* **177**, 21–26 (1997).]

17. The mechanism of plant phosphoglycerate mutase is different from the mechanism of mammalian phosphoglycerate mutase presented in the text. 3-PG binds to the plant enzyme, transfers its phosphate to the enzyme, and then the enzyme transfers the phosphate group back to the substrate to form 2-PG. When [^{32}P]-labeled 3-PG is added to cultured (a) hepatocytes or (b) plant cells, what is the fate of the [^{32}P] label?

18. Which intermediates of glycolysis accumulate if fluoride ions are present?

19. Organisms such as yeast growing under anaerobic conditions can convert pyruvate to alcohol in a process called fermentation, as described in the text. Instead of being converted to lactate, pyruvate produced in glycolysis can be converted to ethanol in a two-step reaction. Step 1 is catalyzed by pyruvate decarboxylase; Step 2 is catalyzed by alcohol dehydrogenase. The product ethanol diffuses out of the cell. What is the importance of Step 2 to the yeast cell?

$$\underset{\text{Pyruvate}}{\mathrm{H_3C\text{—}C(=O)\text{—}COO^-}} \xrightarrow[1]{CO_2} \underset{\text{Acetaldehyde}}{\mathrm{H_3C\text{—}CH(=O)}}$$

$$\xrightarrow[2]{NADH \rightarrow NAD^+} \underset{\text{Ethanol}}{\mathrm{H_3C\text{—}CH_2\text{—}OH}}$$

20. Teusink and coworkers have introduced the term "turbo design" to describe pathways such as glycolysis which have one or more ATP-consuming steps followed by one or more ATP-producing steps and a net yield of ATP production for the pathway overall. These investigators have developed mathematical models, which are supported by experimental evidence, showing that such "turbo" pathways have the risk of substrate-accelerated death unless there is a "guard at the gate," that is, a mechanism for inhibiting an early step of the pathway. In yeast, hexokinase is inhibited by a complex mechanism mediated by trehalose-6-phosphate synthase (TPSI). Mutant yeast in which TPSI is defective (there is no "guard at the gate") die if grown under conditions of high glucose concentration. Explain why.

[From Teusink, B., Walsh, M.C., van Dam, K., and Westerhoff, H.V., *Trends Biochem. Sci.* **23,** 162–169 (1998).]

21. Recent studies have shown that the halophilic organism *Halococcus saccharolyticus* degrades glucose via the Entner-Doudoroff pathway rather than by the glycolytic pathway presented in this chapter. A modified scheme of the Entner-Doudoroff pathway is shown here.

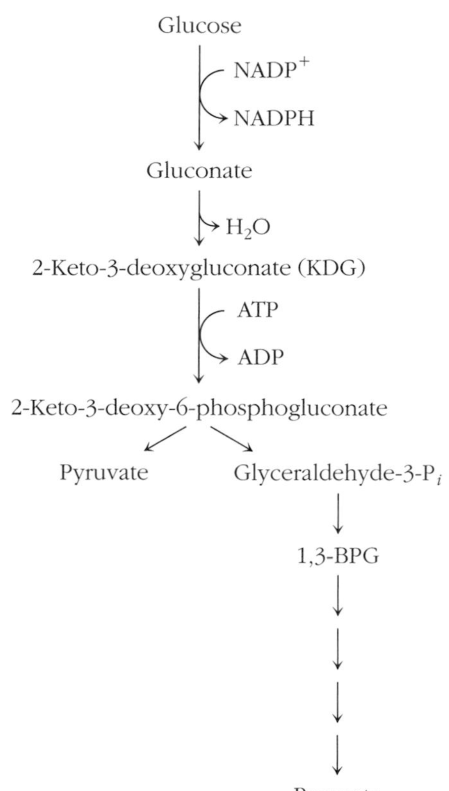

(a) What is the ATP yield per mole of glucose for this pathway?
(b) Describe (in general) what kinds of reactions would need to follow the Entner-Doudoroff pathway in this organism.

[From Johnsen, U., Selig, M., Xavier, K.B., Santos, H., and Schönheit, P., *Arch. Microbiol.* **175,** 52–61 (2001).]

22. During even mild exertion, individuals with McArdle's disease experience painful muscle cramps due to a genetic defect in glycogen phosphorylase, the enzyme that breaks down glycogen. Yet the muscles in these individuals contain normal amounts of glycogen. What did this observation tell researchers about the pathways for glycogen degradation and glycogen synthesis?

23. A given metabolite may follow more than one metabolic pathway. List all the possible fates of glucose-6-phosphate in (a) a liver cell and (b) a muscle cell.

24. Most metabolic pathways include an enzyme-catalyzed reaction that commits a metabolite to continue through the pathway.

(a) Identify the first committed step of the pentose phosphate pathway. Explain your reasoning.
(b) Hexokinase catalyzes an irreversible reaction at the start of glycolysis. Does this step commit glucose to continue through glycolysis?

25. Milk is the sole source of nourishment for nursing kittens. But adult cats suffer from diarrhea if they drink milk. Why?

26. Trehalose is one of the major sugars in the insect hemolymph (the fluid that circulates through the insect's body). It is a disaccharide consisting of two linked glucose residues. In the hemolymph, trehalose serves as a storage form of glucose and also helps protect the insect from desiccation and freezing. Its concentration in the hemolymph must be closely regulated. Trehalose is synthesized in the insect fat body, which plays a role in metabolism analogous to the vertebrate liver. Recent studies on the insect *Manduca sexta* show that during starvation, hemolymph glucose concentration decreases, which results in an increase in fat body glycogen phosphorylase activity and a decrease in the concentration of fructose-2,6-bisphosphate. What effect do these changes have on hemolymph trehalose concentration in the fasted insect?

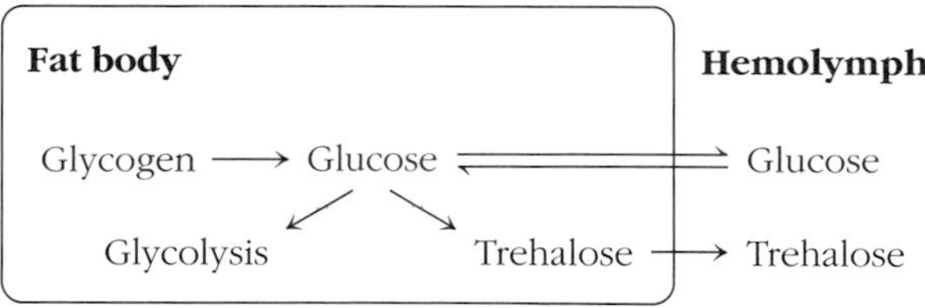

[From Meyer-Fernandes, J.R., Clark, C.P., Gondim, K.C., and Wells, M.A., *Insect Biochem. Mol. Biol.* **31,** 165–170 (2001).]

27. Glycogen is degraded via a phosphorolysis process, which produces glucose-1-phosphate. What advantage does this process have over a simple hydrolysis, which would produce glucose instead of phosphorylated glucose?

$$(\text{Glycogen})_n + P_i \xrightarrow{\textit{phosphorolysis}} (\text{Glycogen})_{n-1} + \text{Glucose-1-}P_i$$

$$(\text{Glycogen})_n + H_2O \xrightarrow{\textit{hydrolysis}} (\text{Glycogen})_{n-1} + \text{Glucose}$$

28. Reduced glutathione, a tripeptide containing a Cys residue, is found in red blood cells, where it reduces organic peroxides formed in cellular structures exposed to high concentrations of reactive oxygen.

$$2\ \gamma\text{-Glu—}\underset{\substack{|\\ \text{SH}}}{\text{Cys}}\text{—Gly} + \text{R—O—OH}$$

Reduced glutathione   Organic peroxide

↓

$$\gamma\text{-Glu—Cys—Gly} \;(\text{S—S bridged to})\; \gamma\text{-Glu—Cys—Gly} + \text{R—OH} + H_2O$$

Oxidized glutathione

Reduced glutathione also plays a role in maintaining normal red blood cell structure and keeping the iron ion of hemoglobin in the +2 oxidation state.

Glutathione is regenerated as shown in the following reaction:

$$\gamma\text{-Glu—Cys—Gly} \;(\text{S—S bridged to})\; \gamma\text{-Glu—Cys—Gly} + \text{NADPH} + H^+$$

↓

$$2\ \gamma\text{-Glu—}\underset{\substack{|\\ \text{SH}}}{\text{Cys}}\text{—Gly} + \text{NADP}^+$$

Use this information to predict the physiological effects of a glucose-6-phosphate dehydrogenase deficiency.

29. A liver biopsy of a four-year-old boy indicated that the fructose-1,6-bisphosphatase enzyme activity was 20% of normal. The patient's blood glucose levels were normal at the beginning of a fast but then decreased suddenly. Pyruvate and alanine concentrations were also elevated, as was the glyceraldehyde-3-phosphate/dihydroxyacetone phosphate ratio. Explain the reason for these symptoms.

30. Patients with von Gierke's disease have a deficiency of glucose-6-phosphatase. One of the most prominent symptoms of the disease is a protruding abdomen due to an enlarged liver. Explain why the liver is enlarged in patients with von Gierke's disease.

31. Write a mechanism for the nonenzymatic hydrolysis of 6-phosphogluconolactone to 6-phosphogluconate.

32. An isozyme of PFK called phosphofructokinase-1 C (PFK-1C) has been purified from brain tissue. There are three isozymes of PFK-1, designated A, B, and C. The A isozyme (M_r = 84,000 D) is found in the muscle and the brain; the B isozyme (M_r = 80,000 D) is found in the liver and the brain; and the C isozyme (M_r = 86,000 D) is found only in the brain. Because the brain contains all three isozymes and there isn't a location where the C isozyme is found exclusively, the enzyme had been difficult to purify. In this case, the investigators purified the desired enzyme to homogeneity and conducted experiments to demonstrate that the C isozyme is distinct from the A and B isozymes. The availability of a pure C preparation means that antibodies can be generated and then used to detect the isozyme. Since the levels of PFK-1 isozymes change during malignant transformation of cells, the availability of a C antibody might be a valuable diagnostic tool.

The investigators carried out kinetic studies using their newly purified PFK-1C isozyme. They studied the catalytic behavior of the enzyme in the presence of the metabolites AMP, inorganic phosphate (P_i), and fructose-2,6-bisphosphate (F2,6BP). The results are shown below.

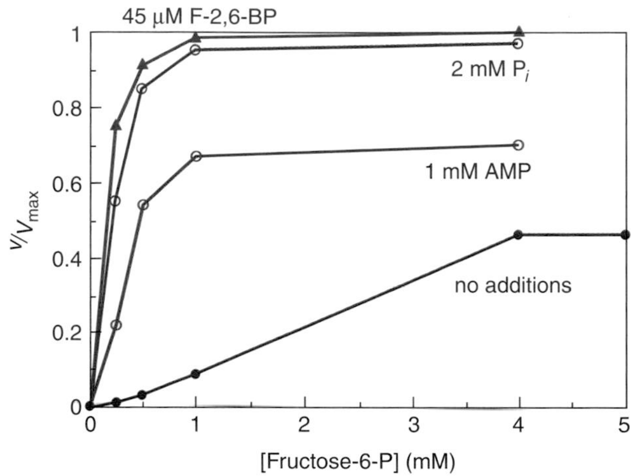

In addition, the relative potency of the allosteric effector citrate on PFK isozymes was measured as the concentration of citrate required to inhibit 50% of the enzyme activity:

Isozyme	Citrate (μM)
A	100
B	>2000
C	750

Three other allosteric effectors were tested. The numbers indicate the concentrations of each effector required to achieve 50% of the maximal velocity.

Isozyme	Phosphate (μM)	AMP (μM)	Fructose-2,6-BP (μM)
A	80	10	0.05
B	200	10	0.05
C	350	75	4.5

(a) Compare the ability of PFK-1C to catalyze the phosphorylation of fructose-6-phosphate in the absence and presence of AMP, F-2,6-BP, and P_i. How do these allosteric effectors influence the velocity of the reaction?

(b) Evaluate whether the investigators have shown that PFK-1C is different from the PFK-1A and PFK-1B isozymes. Speculate why there might be differences among the isozymes.

[From Foe, L.G. and Kemp, R.G., *J. Biol. Chem.* **260,** 726–730 (1985).]

33. Carbohydrate metabolism in the thermophilic organism *Thermoproteus tenax* is rather peculiar compared to the carbohydrate metabolism in the types of organisms usually studied in introductory biochemistry. For example, the phosphofructokinase reaction in *T. tenax* is reversible and uses pyrophosphate rather than ATP. In addition, *T. tenax* has two different glyceraldehyde-3-phosphate dehydrogenase isozymes. One is well known and requires $NADP^+$ as a cofactor. The second isozyme requires NAD^+ as a cofactor and is less well-studied. In this case, we will consider the properties of the NAD^+-dependent glyceraldehyde-3-phosphate dehydrogenase.

T. tenax glyceraldehyde-3-phosphate dehydrogenase (GAPDH) has interesting regulatory properties, some of which are presented in the table.

Without AMP	
V_{max} (units/mg)	36.5
K_M (mM)	3.3
With AMP	
V_{max} (units/mg)	37.0
K_M (mM)	1.4
Molecular mass	
subunit	55,000 D
native	220,000 D

(a) Name the three enzymes that catalyze irreversible, regulated reactions in glycolysis.
(b) Write the balanced equation for the reaction catalyzed by NAD^+-dependent glyceraldehyde-3-phosphate dehydrogenase. Include structures and cofactors.
(c) What is the importance of the GAPDH reaction to glycolysis?
(d) The GAPDH enzyme from *T. tenax* did not show any activity in the presence of 1,3-bisphosphoglycerate. What is the significance of this observation?
(e) The activity of the GAPDH enzyme was assayed in the presence of a constant amount of glyceraldehyde-3-phosphate and an increasing amount of NAD^+. The activity of the control was compared to the activity in the presence of various metabolites. The results are shown in the plot below. Additional data are given in the table. Classify the metabolites listed in the table as inhibitors or activators. Explain your reasoning. What is the physiological significance of your answer?

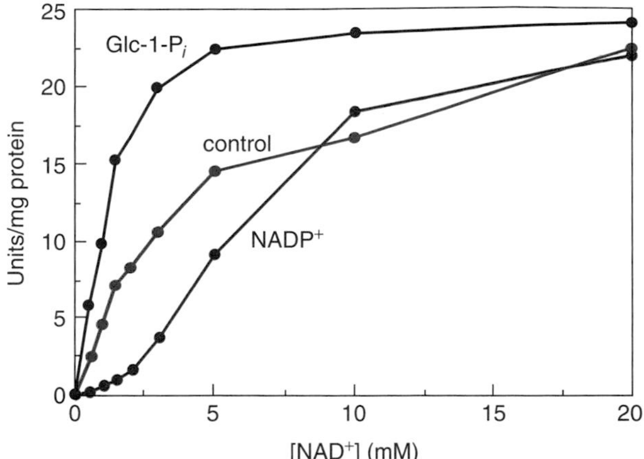

Metabolite	Apparent K_M (mM)
None	3.1
$NADP^+$	4.5
Glucose-1-phosphate	0.4
NADH	8.0
AMP	1.3
ADP	1.7
ATP	30

(f) What does the information in this problem (and other information available to you) tell you about the mechanism of inhibition and activation of GAPDH by these selected metabolites?

[From Brunner, N.A., Brinkmann, H., Siebers, B., and Hensel, R., *J. Biol. Chem.* **273,** 6149–6156 (1998).]

SELECTED READINGS

Brosnan, J.T., Comments on metabolic needs for glucose and the role of gluconeogenesis, *Eur. J. Clin. Nutr.* **53,** S107–S111 (1999). [A very readable review that discusses possible reasons why carbohydrates are used universally as metabolic fuels, why glucose is stored as glycogen, and why the pentose phosphate pathway is important.]

Depre, C., Rider, M.H., and Hue, L., Mechanisms of control of heart glycolysis, *Eur. J. Biochem.* **258,** 277–290 (1998). [Discusses how the control of glycolysis in heart muscle is distributed among several enzymes, transporters, and other pathways.]

CHAPTER 11

The Citric Acid Cycle

In this age of technical marvels, it might come as a surprise to learn that many industrial solvents and other chemicals are not by-products of oil refining but are instead synthesized by microorganisms—albeit on a grand scale. For example, ethanol, acetone, butanol, acetic acid, and citric acid are all the products of bacterial or fungal metabolism. Typically, the organisms are grown in huge fermentation tanks under carefully controlled conditions. The common mold *Aspergillus niger* is the source of citric acid. Under growth conditions first described in 1917, these cells synthesize and excrete enormous amounts of citric acid, directing about 70% of their carbon uptake to this compound. The annual production of about a billion pounds of citric acid is used primarily by the food and pharmaceutical industries. Citric acid—which is stable, water-soluble, palatable, and nontoxic—enhances the tartness of soft drinks, lends acidity to processed foods to minimize spoilage, and buffers pharmaceutical compounds against changes in pH.

[Maximilian Stock, Ltd./Science Photo Library/Photo Researchers]

LEARNING OBJECTIVES

1. Understand how carbon atoms from pyruvate enter the citric acid cycle.
2. Understand the roles of cofactors such as coenzyme A, thiamine pyrophosphate, lipoamide, FAD, and NAD^+.
3. Understand that the citric acid cycle acts as a multistep catalyst to oxidize acetyl groups derived from carbohydrate, amino acid, and lipid catabolism.
4. Understand how flux through the citric acid cycle is regulated.
5. Understand how the citric acid cycle might have evolved.
6. Understand that citric acid cycle intermediates serve as both precursors and products for other metabolic pathways.

THIS CHAPTER IN CONTEXT

Chapter 11, covering the citric acid cycle, logically follows the chapter on glucose metabolism, since the citric acid cycle represents the final stage of carbohydrate metabolism. However, the student should keep in mind that the citric acid cycle is also a pathway for the final catabolism of fatty acids (covered in Chapter 14) and many amino acids (covered in Chapter 15). Rather than focus exclusively on the citric acid cycle's enzymes and reactions, this chapter also addresses how a circular metabolic pathway might have evolved and how the citric acid cycle is part of a network of synthetic and degradative pathways.

THE CITRIC ACID CYCLE IS A METABOLIC PATHWAY THAT CONVERTS carbon atoms to CO_2 and, in doing so, conserves metabolic free energy that ultimately drives the synthesis of ATP. The citric acid cycle represents the final stage in the oxidation of metabolic fuels, including carbohydrates, fatty acids, and amino acids (Fig. 11-1). The eight reactions of the citric acid cycle take place in the cytosol of prokaryotes and in the mitochondria of eukaryotes.

Unlike a linear pathway such as glycolysis (Fig. 10-6) or gluconeogenesis (Fig. 10-13), the citric acid cycle always returns to its starting position, essentially behaving as a multistep catalyst. It is easy to envision how a linear metabolic pathway could evolve through the addition of one step after another, but understanding the evolution of a circular pathway such as the citric acid cycle presents a greater challenge. However, by comparing the metabolic capabilities of organisms that lack some or all of the eight steps of the citric acid cycle, it is possible to postulate how the pathway could have originated.

An examination of the citric acid cycle also illustrates an important feature of metabolic pathways in general, namely, that a pathway is less like an element of plumbing and more like a web or network. In other words, the pathway does not function like a pipeline where substances entering one end emerge transformed at the other end. Instead, the pathway's intermediates can participate in many reactions, depending on the needs of the cell. The citric acid cycle has no equal in the number of metabolic con-

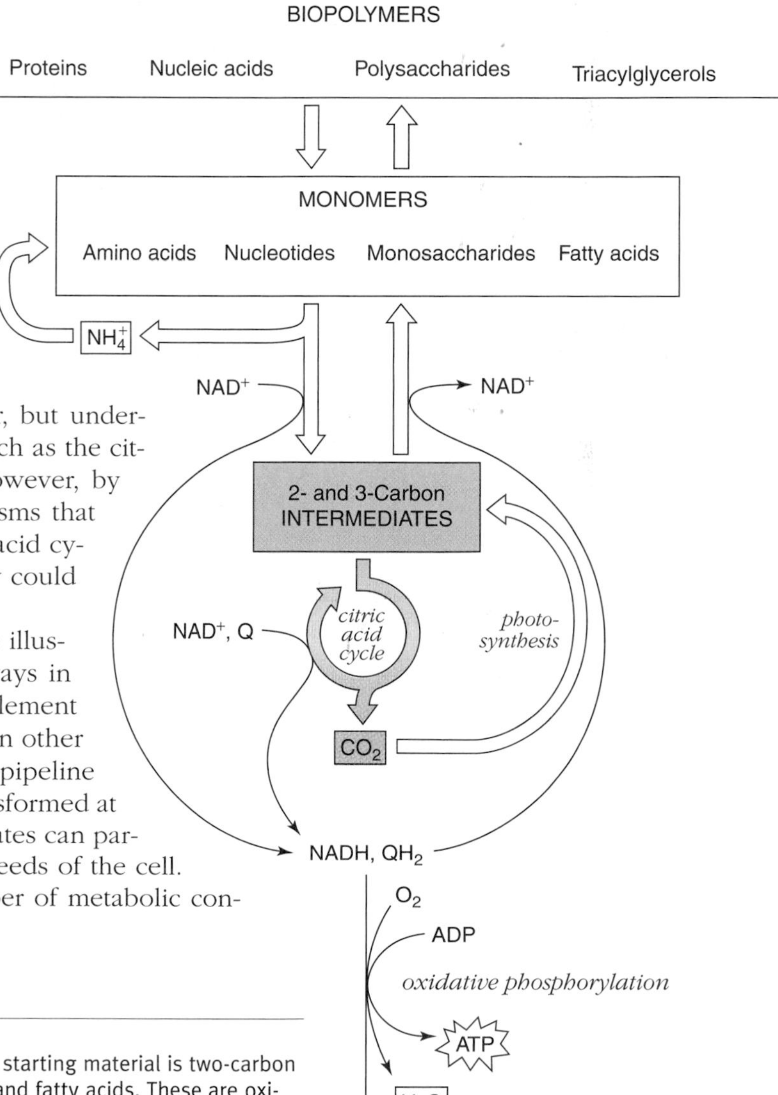

FIGURE 11-1

The citric acid cycle in context.
The citric acid cycle is a central metabolic pathway whose starting material is two-carbon acetyl units derived from amino acids, monosaccharides, and fatty acids. These are oxidized to the waste product CO_2, with the reduction of the cofactors NAD^+ and ubiquinone (Q).

nections it makes: It is a central pathway whose intermediates are both precursors and products of a large variety of biological molecules.

Carbon atoms enter the citric acid cycle in the form of acetyl groups derived from amino acids, fatty acids, and carbohydrates. Since we have already looked at the catabolism of carbohydrates, we will use pyruvate, the end product of glycolysis, as the starting point for our study of the citric acid cycle. We will examine the eight reactions of the citric acid cycle and discuss how this sequence of reactions might have evolved. Finally, we will consider the citric acid cycle as a multifunctional pathway with links to other metabolic processes.

1 THE PYRUVATE DEHYDROGENASE COMPLEX COVERTS PYRUVATE TO ACETYL-CoA

The end product of glycolysis is the three-carbon compound pyruvate. In aerobic organisms, these carbons are ultimately oxidized to 3 CO_2 (although the oxygen atoms come not from molecular oxygen but from water and phosphate). The first molecule of CO_2 is released when pyruvate is decarboxylated to an acetyl unit. The second and third CO_2 molecules are products of the citric acid cycle. The decarboxylation of pyruvate is catalyzed by the pyruvate dehydrogenase complex. This enzyme complex, and the enzymes of the citric acid cycle itself, are located inside the mitochondria (an organelle surrounded by a double membrane and whose interior is called the **mitochondrial matrix**). Accordingly, pyruvate produced by glycolysis in the cytosol must first be transported into the mitochondria.

The pyruvate dehydrogenase complex contains multiple copies of three different enzymes

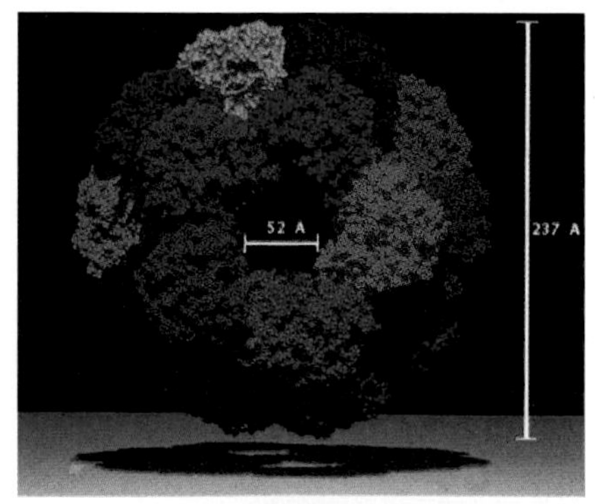

FIGURE 11-2

Structure of the E2 core of the pyruvate dehydrogenase complex from *B. stearothermophilus*.
In this bacterial enzyme complex, 60 E2 subunits form a hollow dodecahedron. The structure has an outer diameter of about 237 Å and an inner diameter of about 118 Å. The opening visible in this structure is about 52 Å wide, large enough to permit the entry of coenzyme A and the exit of acetyl-CoA. [*Courtesy Tina Izard and Wim G. J. Hol. From* Proc. Natl. Acad. Sci. ***96,*** *1241 (1999).*]

For convenience, the three kinds of enzymes that make up the pyruvate dehydrogenase complex are called E1, E2, and E3. Together *they catalyze the oxidative decarboxylation of pyruvate and the transfer of the acetyl unit to coenzyme A:*

$$\text{Pyruvate} + \text{CoA} + \text{NAD}^+ \rightarrow \text{acetyl-CoA} + CO_2 + \text{NADH}$$

The structure of coenzyme A, a nucleotide derivative containing pantothenate, is shown on page 296.

In *E. coli,* the pyruvate dehydrogenase complex contains 60 protein subunits and has a mass of about 4600 kD. In mammals and some other bacteria, the enzyme complex is even larger, with about 140 subunits, including proteins that regulate the complex's enzymatic activity. The pyruvate dehydrogenase complex from *Bacillus stearothermophilus* consists of a core of 60 E2 subunits that form a dodecahedron (a 12-sided polyhedron; Fig. 11-2). The other subunits are arranged over the surface of this structure.

The operation of the pyruvate dehydrogenase complex requires several coenzymes, whose functional roles in the five-step reaction are described below.

1. In the first step, which is catalyzed by E1 (also called pyruvate dehydrogenase), pyruvate is decarboxylated. This reaction requires the cofactor thiamine pyrophosphate (TPP; Fig. 11-3). TPP attacks the carbonyl carbon of pyruvate, and the departure of CO_2 leaves a

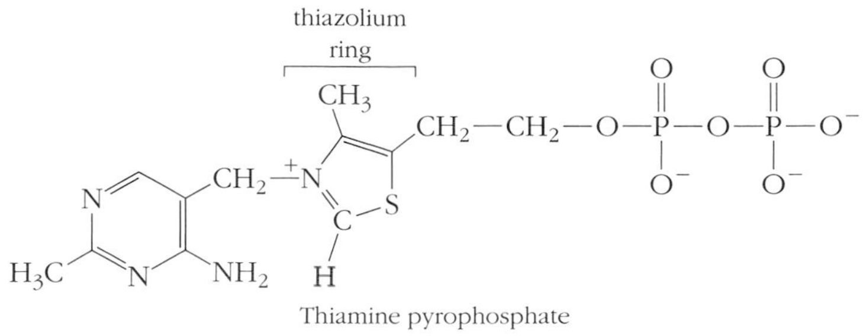

FIGURE 11-3

Thiamine pyrophosphate (TPP). This cofactor is the phosphorylated form of thiamine, also known as vitamin B_1 (see Box 9-A). The central thiazolium ring is the active portion. An acidic proton (red) dissociates, and the resulting carbanion is stabilized by the nearby positively charged nitrogen. TPP is a cofactor for several different decarboxylases.

hydroxyethyl group attached to TPP. This carbanion is stabilized by the positively charged thiazolium ring group of TPP:

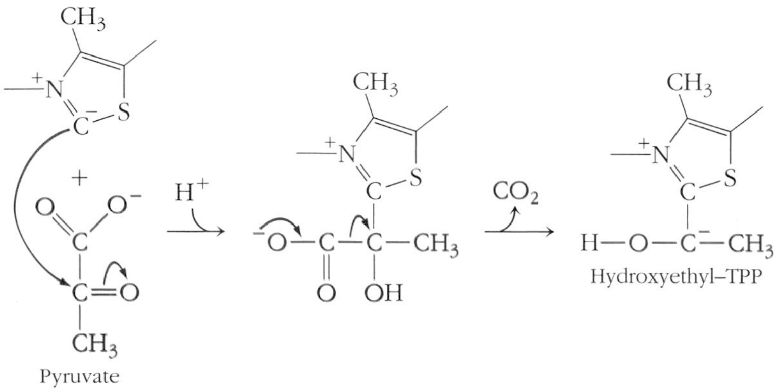

2. The hydroxyethyl group is then transferred to E2 of the pyruvate dehydrogenase complex. The hydroxyethyl acceptor is a lipoamide prosthetic group (Fig. 11-4). The transfer reaction regenerates the TPP cofactor of E1 and oxidizes the hydroxyethyl group to an acetyl group:

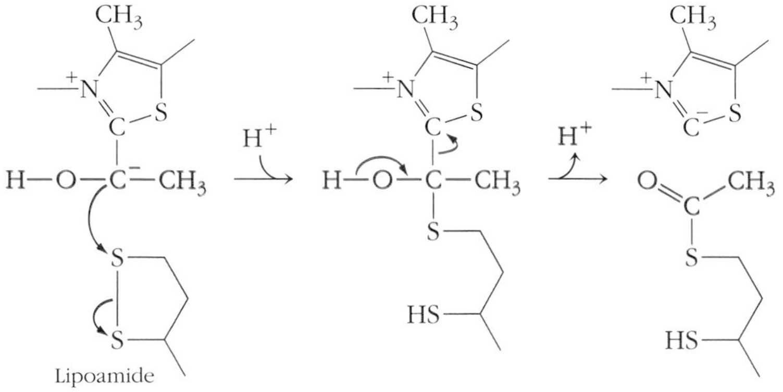

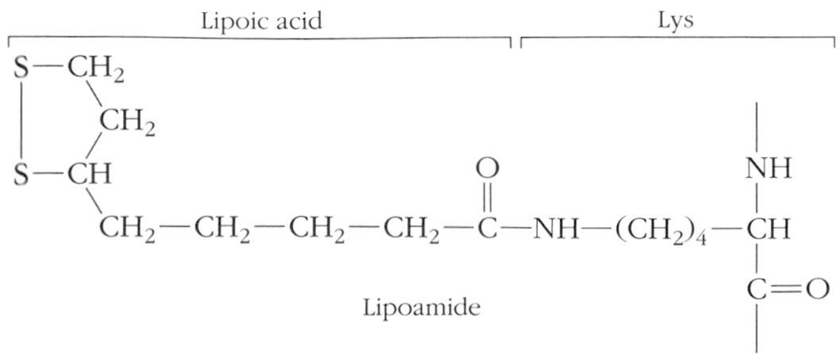

FIGURE 11-4

Lipoamide. This prosthetic group consists of a lipoic acid moiety linked via an amide bond to the ε-amino group of a protein Lys residue. The active portion of the 14-Å-long lipoamide is the disulfide bond (red), which can be reversibly reduced.

2 THE EIGHT REACTIONS OF THE CITRIC ACID CYCLE

Go to **Exercise 16/The Citric Acid Cycle** to explore how the citric acid cycle oxidizes acetyl units and produces CO_2, GTP, and reduced cofactors. The material included in this exercise and the questions based on it address Learning Objectives 3 and 6.

An acetyl-CoA molecule derived from pyruvate is a product of carbohydrate catabolism as well as a product of amino acid catabolism, since the carbon skeletons of many amino acids are broken down to pyruvate. Acetyl-CoA is also a direct product of the degradation of certain amino acids and of fatty acids. In some tissues, the bulk of acetyl-CoA is derived from the catabolism of fatty acids rather than carbohydrates or amino acids.

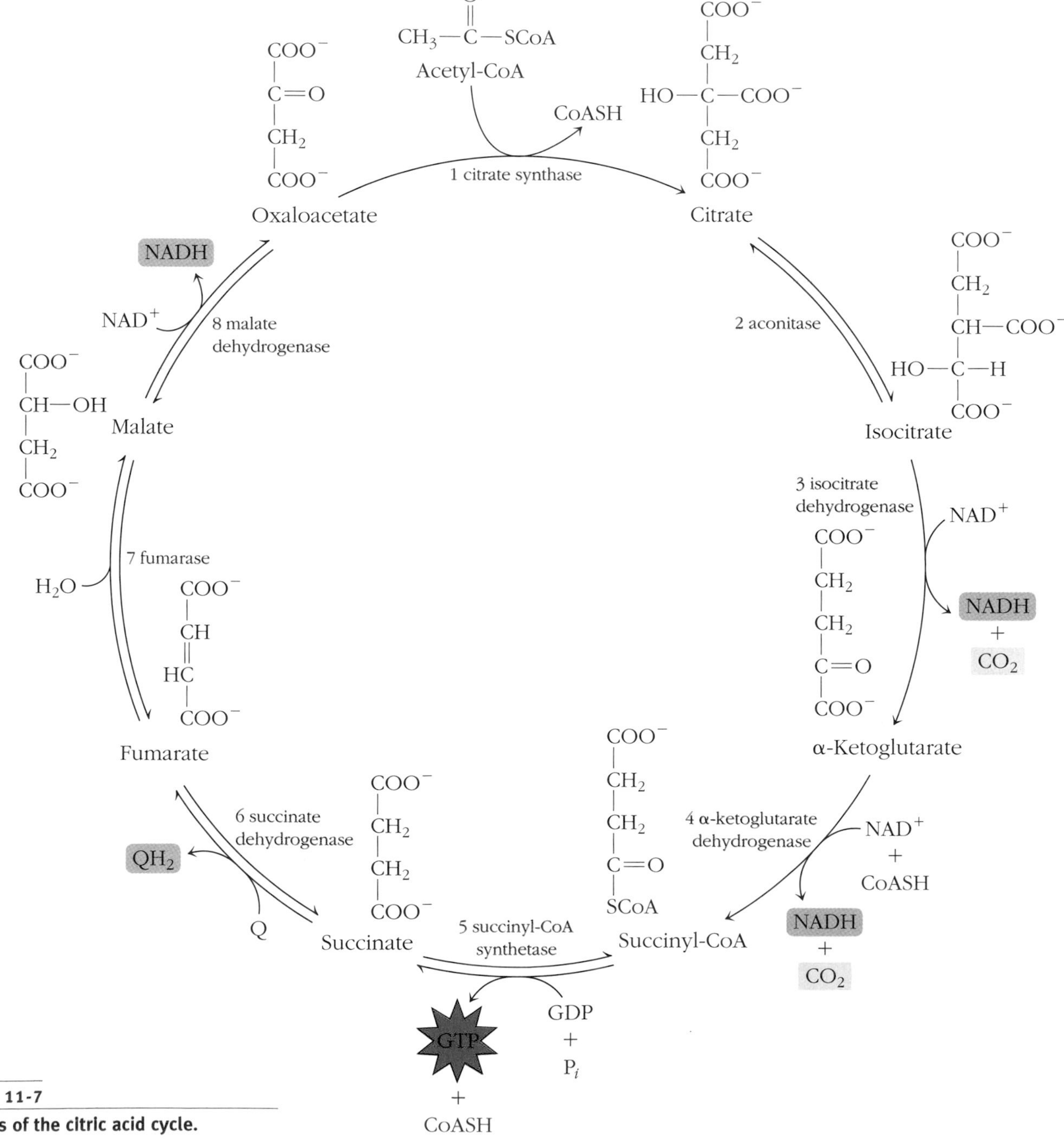

FIGURE 11-7
Reactions of the citric acid cycle.

Whatever its source, acetyl-CoA enters the citric acid cycle for further oxidation. This process is highly exergonic, and free energy is conserved at several steps in the form of a nucleotide triphosphate (GTP) and reduced cofactors. *For each acetyl group that enters the citric acid cycle, two molecules of fully oxidized CO_2 are produced, representing a loss of four pairs of electrons.* These electrons are transferred to 3 NAD^+ and one ubiquinone (Q) to produce 3 NADH and 1 QH_2. The net equation for the citric acid cycle is therefore

$$\text{Acetyl-CoA} + \text{GDP} + P_i + 3\ \text{NAD}^+ + \text{Q} \rightarrow$$
$$2\ CO_2 + \text{CoA} + \text{GTP} + 3\ \text{NADH} + QH_2$$

In this section we examine the sequence of eight enzyme-catalyzed reactions of the citric acid cycle, focusing on a few interesting reactions. The entire pathway is summarized in Figure 11-7. Reviews of each step, including the enzyme, substrate, product, and required cofactor, are shown in Exercise 16.

1. Citrate synthase

In the first reaction of the citric acid cycle, the acetyl group of acetyl-CoA condenses with the four-carbon compound oxaloacetate to produce the six-carbon compound citrate.

$$\underset{\text{Oxaloacetate}}{^{-}OOC{-}C({=}O){-}CH_2{-}COO^{-}} + \underset{\text{Acetyl-CoA}}{CH_3{-}\overset{\overset{\displaystyle O}{\|}}{C}{-}SCoA} \xrightarrow[\text{citrate synthase}]{H_2O \quad HSCoA} \underset{\text{Citrate}}{HO{-}C(COO^{-})(CH_2{-}COO^{-})_2}$$

Citrate synthase is a dimer that undergoes a large conformational change upon substrate binding (Fig. 11-8).

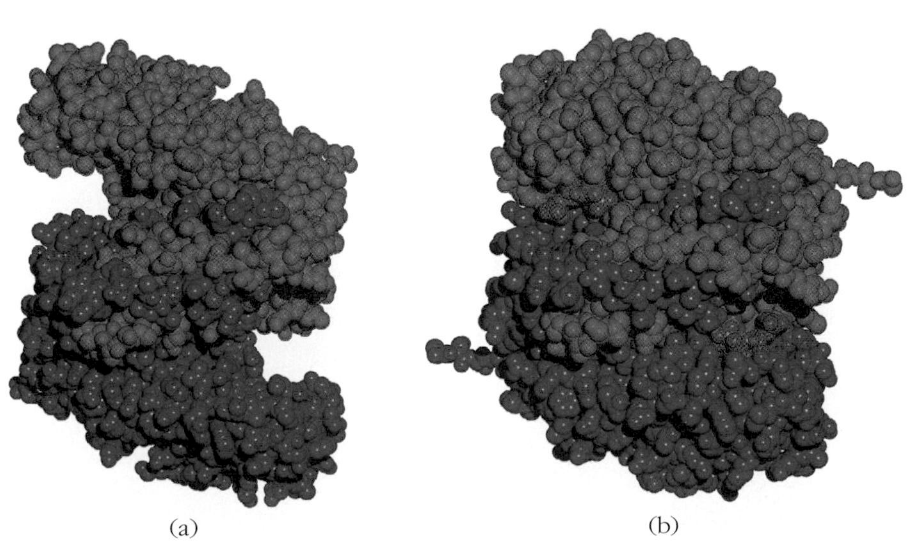

FIGURE 11-8

Conformational changes in citrate synthase from chicken. (a) The enzyme in the absence of substrates. The two subunits of the dimeric enzyme are colored blue and green. (b) When oxaloacetate (red) binds, each subunit undergoes a conformational change that creates a binding site for acetyl-CoA (an acetyl-CoA analog is shown here in orange). This conformational change explains why oxaloacetate must bind to the enzyme before acetyl-CoA can bind. [*Structure of citrate synthase alone (pdb 5CSC) determined by D.-I. Liao, M. Karpusas, and S.J. Remington; structure of citrate synthase with oxaloacetate and carboxymethyl-CoA (pdb 5CTS) determined by M. Karpusas, B. Branchaud, and S.J. Remington.*]

Citrate synthase is one of the few enzymes that can synthesize a carbon–carbon bond without using a metal ion cofactor. Its mechanism is shown in Figure 11-9. The first reaction intermediate is probably stabilized by the formation of low-barrier hydrogen bonds, which are stronger than ordinary hydrogen bonds (see Section 6-3). The coenzyme A released during the final step can be reused by the pyruvate dehydrogenase complex or used later in the citric acid cycle to synthesize the intermediate succinyl-CoA.

The reaction catalyzed by citrate synthase is highly exergonic ($\Delta G^{\circ\prime} = -31.5\ \text{kJ} \cdot \text{mol}^{-1}$, equivalent to the free energy of breaking the thioester bond of acetyl-CoA). We will see later why the efficient operation of the citric acid cycle requires that this step have a large free energy change.

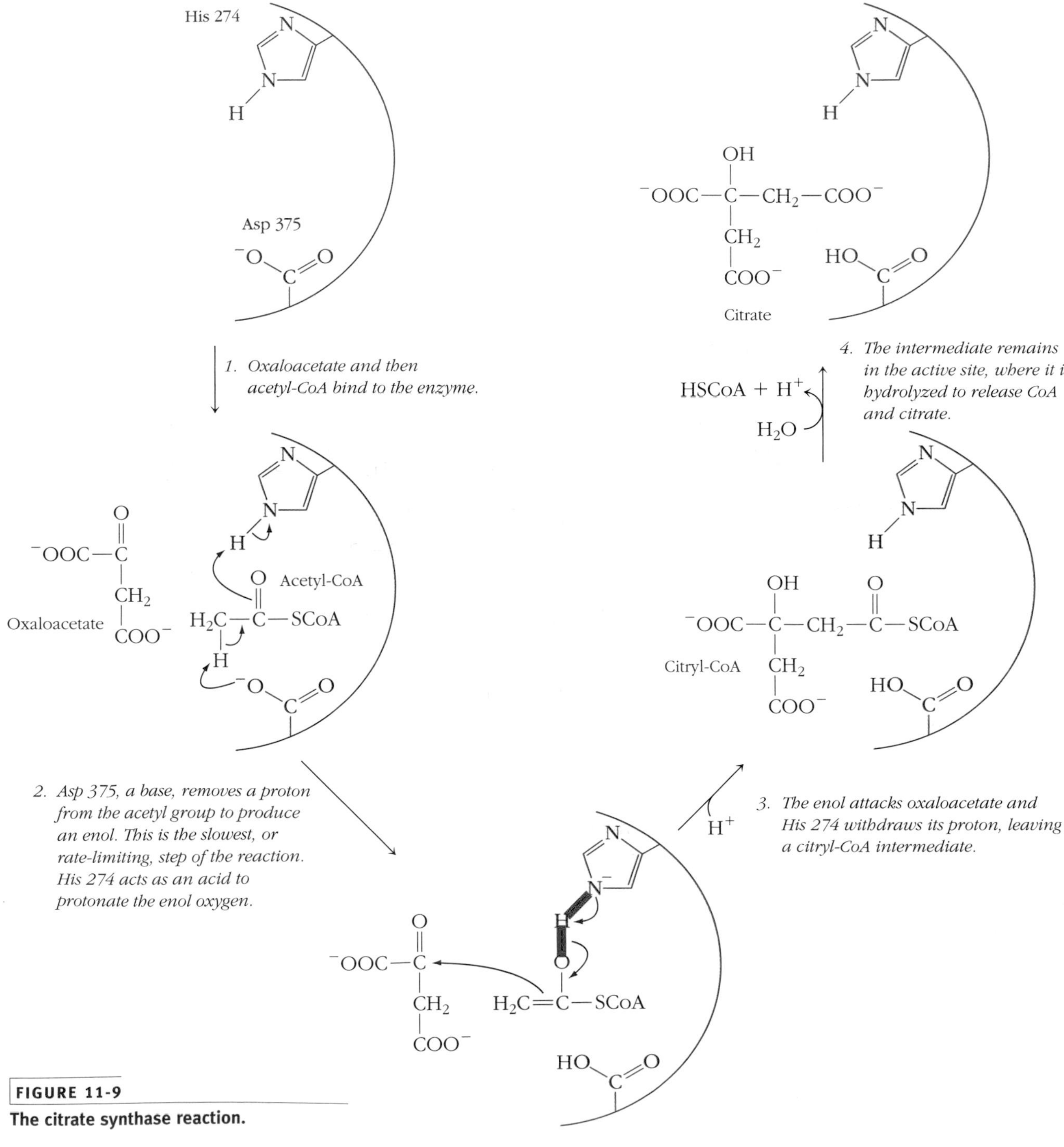

FIGURE 11-9
The citrate synthase reaction.

2. Aconitase

The second enzyme of the citric acid cycle catalyzes the reversible isomerization of citrate to isocitrate:

$$\underset{\text{Citrate}}{\begin{array}{c} COO^- \\ | \\ CH_2 \\ | \\ HO{-}C{-}COO^- \\ | \\ CH_2 \\ | \\ COO^- \end{array}} \underset{}{\overset{H_2O}{\rightleftharpoons}} \underset{\text{Aconitate}}{\left[\begin{array}{c} COO^- \\ | \\ CH_2 \\ | \\ C{-}COO^- \\ \| \\ CH \\ | \\ COO^- \end{array}\right]} \overset{H_2O}{\rightleftharpoons} \underset{\text{Isocitrate}}{\begin{array}{c} COO^- \\ | \\ CH_2 \\ | \\ H{-}C{-}COO^- \\ | \\ HO{-}C{-}H \\ | \\ COO^- \end{array}}$$

The enzyme is named after the reaction intermediate.

Citrate is a symmetrical molecule, yet only one of its two carboxymethyl arms ($-CH_2-COO^-$) undergoes dehydration and rehydration during the aconitase reaction. This stereochemical specificity long puzzled biochemists, but it is a feature of all enzyme-catalyzed reactions (Box 11-A).

3. Isocitrate dehydrogenase

The third reaction of the citric acid cycle is the oxidative decarboxylation of isocitrate to α-ketoglutarate. The substrate is first oxidized in a reaction accompanied by the reduction of NAD^+ to NADH. Then the carboxylate group β to the ketone function (that is, two carbon atoms away from the ketone) is eliminated as CO_2.

$$\underset{\text{Isocitrate}}{\begin{array}{c} COO^- \\ | \\ CH_2 \\ | \\ H{-}C{-}COO^- \\ | \\ HO{-}C{-}H \\ | \\ COO^- \end{array}} \xrightarrow{NAD^+ \;\; H^+ + NADH} \left[\begin{array}{c} COO^- \\ | \\ CH_2 \\ | \\ H{-}C{-}COO^- \\ | \\ C{=}O \\ | \\ COO^- \end{array}\right] \xrightarrow{CO_2} \left[\begin{array}{c} COO^- \\ | \\ CH_2 \\ | \\ H{-}C \\ \| \\ C{-}O^- \\ | \\ COO^- \end{array}\right] \xrightarrow{H^+} \underset{\alpha\text{-Ketoglutarate}}{\begin{array}{c} COO^- \\ | \\ CH_2 \\ | \\ H{-}C{-}H \\ | \\ C{=}O \\ | \\ COO^- \end{array}}$$

A Mn^{2+} ion in the active site helps stabilize the negative charges of the reaction intermediates.

4. α-Ketoglutarate dehydrogenase

α-Ketoglutarate dehydrogenase, like isocitrate dehydrogenase, catalyzes an oxidative decarboxylation reaction. It also transfers the remaining four-carbon fragment to CoA.

$$\underset{\alpha\text{-Ketoglutarate}}{\begin{array}{c} COO^- \\ | \\ CH_2 \\ | \\ CH_2 \\ | \\ C{=}O \\ | \\ COO^- \end{array}} \xrightarrow[NAD^+ \;\; NADH]{CoASH \;\; CO_2} \underset{\text{Succinyl-CoA}}{\begin{array}{c} COO^- \\ | \\ CH_2 \\ | \\ CH_2 \\ | \\ C{=}O \\ | \\ S{-}CoA \end{array}}$$

A CLOSER LOOK

BOX 11-A Asymmetry in the citric acid cycle

In 1937, Hans Krebs described a series of reactions that became known as the citric acid cycle (or the Krebs cycle). In Krebs' original scheme, pyruvate reacts with oxaloacetate to produce CO_2 and citrate (CoA and acetyl-CoA had not yet been discovered).

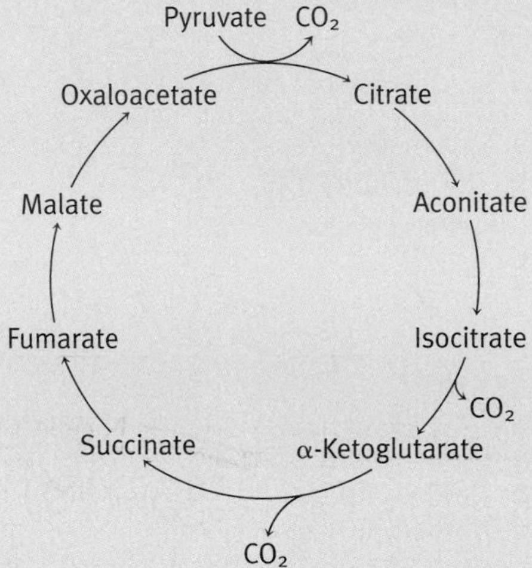

The remainder of the pathway, in which citrate undergoes additional reactions to evolve two more molecules of CO_2 and regenerate oxaloacetate, is similar to the pathway as we know it today.

However, experiments with isotopically labeled starting materials, which became available in the 1940s, cast in doubt Krebs' identification of citrate as the first intermediate of the reaction cycle. One experiment used as the starting material oxaloacetate labeled at one position with the heavy isotope ^{13}C. It was expected that the labeled atom would become part of citrate and then, because citrate is a symmetrical molecule, the label would become distributed symmetrically in subsequent intermediates of the citric acid cycle, such as α-ketoglutarate. This expectation can be diagrammed (labeled carbons are in red):

COO^- — $C{=}O$ — CH_2 — COO^-
Oxaloacetate

↓

COO^- — CH_2 — $HO{-}C{-}COO^-$ — CH_2 — COO^-
Citrate

↙ ↘

COO^- — CH_2 — $C{-}COO^-$ — $\Vert$ — CH — COO^-	Aconitate	COO^- — CH_2 — $C{-}COO^-$ — $\Vert$ — CH — COO^-
↓		↓
COO^- — CH_2 — $CH{-}COO^-$ — $HO{-}C{-}H$ — COO^-	Isocitrate	COO^- — CH_2 — $CH{-}COO^-$ — $HO{-}C{-}H$ — COO^-
↓		↓
COO^- — CH_2 — CH_2 — $C{=}O$ — COO^-	α-Ketoglutarate	COO^- — CH_2 — CH_2 — $C{=}O$ — COO^-

(continued)

BOX 11-A *(continued)*

However, when the α-ketoglutarate was recovered and analyzed, only one of its carbons was labeled (the carboxylate carbon next to the carbonyl group). The conclusion was that citrate could not be an intermediate in the citric acid cycle, and Krebs obligingly renamed the pathway the tricarboxylic cycle and modified it so that citrate is merely the product of a side reaction:

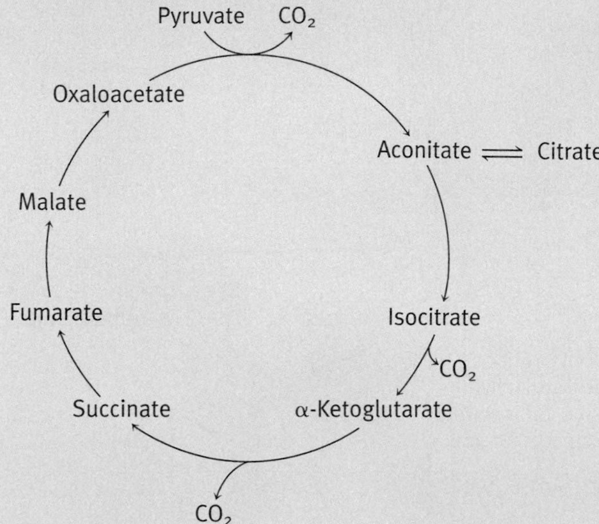

Although it seemed to fit the data, the modified pathway was inadequate, because the interconversion of aconitate with citrate would scramble the labeled carbons just as if citrate were part of the pathway itself, as originally proposed. The truth was exposed in 1948 by Alexander Ogston, who pointed out that although citrate is symmetrical, its two carboxymethyl groups would no longer be identical when it was bound to an asymmetrical enzyme. Ogston showed that a three-point attachment of citrate to an enzyme could result in one carboxymethyl group (green) reacting differently from the other.

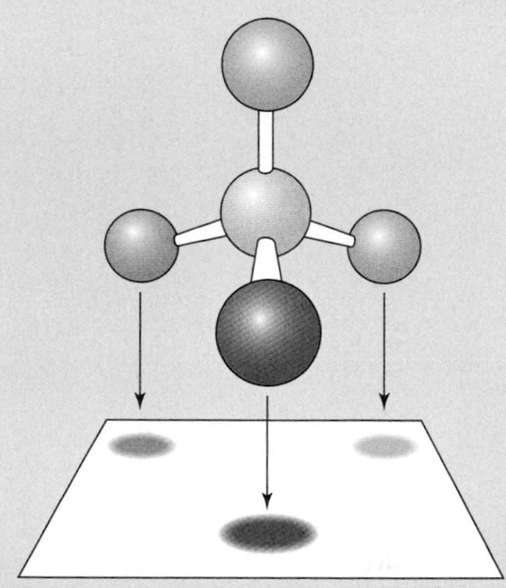

As a result, the labeled citrate would give rise to only one labeled α-ketoglutarate molecule. Further contemplation revealed that a three-point attachment is not even necessary for an enzyme to distinguish two groups in a molecule such as citrate, which are related by mirror symmetry. You can prove this yourself with a simple organic chemistry model kit. By now you should appreciate that biological systems are inherently chiral (also see Box 4-A).

Ogston's finding, which seems obvious in hindsight, is consistent with what we now know about enzyme structure and reaction specificity. In fact, within a few years of Ogston's observation, enzymologists had identified the enzymes of the citric acid cycle and verified that citrate did indeed lie on the circular pathway and was not a side product.

The free energy of oxidizing α-ketoglutarate is conserved in the formation of the thioester succinyl-CoA. α-Ketoglutarate dehydrogenase is a multienzyme complex that resembles the pyruvate dehydrogenase complex in both structure and enzymatic mechanism. In fact, the same E3 enzyme is a member of both complexes.

The isocitrate dehydrogenase and α-ketoglutarate dehydrogenase reactions both release CO_2. These two carbons are *not* the ones that entered the citric acid cycle as acetyl-CoA; those acetyl carbons are released in subsequent rounds of the cycle (Fig. 11-10). However, *the net result of each round of the citric acid cycle is the loss of two carbons as CO_2 for each acetyl-CoA that enters the cycle.* Exercise 16 traces the fates of carbon atoms from acetyl-CoA through two rounds of the citric acid cycle.

5. Succinyl-CoA synthetase

The thioester succinyl-CoA releases a large amount of free energy when it is cleaved ($\Delta G^{\circ\prime} = -32.6 \text{ kJ} \cdot \text{mol}^{-1}$). This is enough free energy to drive the synthesis of a nucleoside triphosphate from a nucleoside diphosphate and

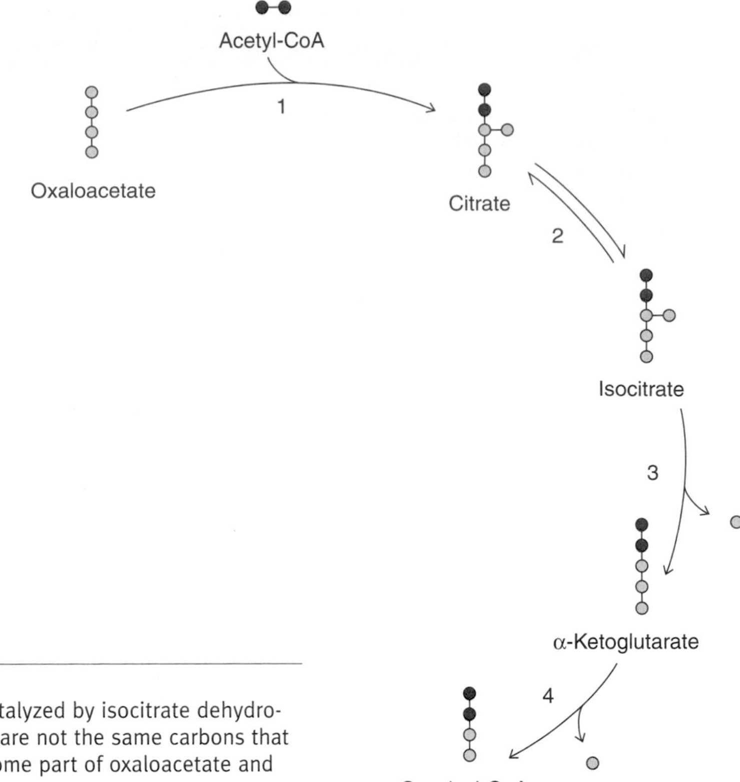

FIGURE 11-10

Fates of carbon atoms in the citric acid cycle. The two carbon atoms that are lost as CO_2 in the reactions catalyzed by isocitrate dehydrogenase (Step 3) and α-ketoglutarate dehydrogenase (Step 4) are not the same carbons that entered the cycle as acetyl-CoA (red). The acetyl carbons become part of oxaloacetate and are lost in subsequent rounds of the cycle.

P_i ($\Delta G^{\circ\prime} = 30.5 \text{ kJ} \cdot \text{mol}^{-1}$). The change in free energy for the net reaction is near zero, so the reaction is reversible. In fact, the enzyme is named for the reverse reaction. Succinyl-CoA synthetase in the mammalian citric acid cycle generates GTP, whereas the plant and bacterial enzymes generate ATP (recall that GTP is energetically equivalent to ATP).

How does succinyl-CoA synthetase couple thioester cleavage to the synthesis of a nucleoside triphosphate? The reaction is a series of phosphoryl group transfers that involves an active-site His residue (Fig. 11-11). The phospho-His reaction intermediate must move a large distance to shuttle the phosphoryl group between the succinyl group and the nucleoside diphosphate (Fig. 11-12).

6. Succinate dehydrogenase

The final three reactions of the citric acid cycle convert succinate back to the cycle's starting substrate, oxaloacetate. Succinate dehydrogenase catalyzes the reversible dehydrogenation of succinate to fumarate. This oxidation–reduction reaction requires an FAD prosthetic group, which is reduced to $FADH_2$ during the reaction:

$$
\underset{\text{Succinate}}{^{-}OOC{-}CH_2{-}CH_2{-}COO^{-}} \xrightleftharpoons[\text{succinate dehydrogenase}]{\text{Enz-FAD} \;\; \text{Enz-FADH}_2} \underset{\text{Fumarate}}{^{-}OOC{-}CH{=}CH{-}COO^{-}}
$$

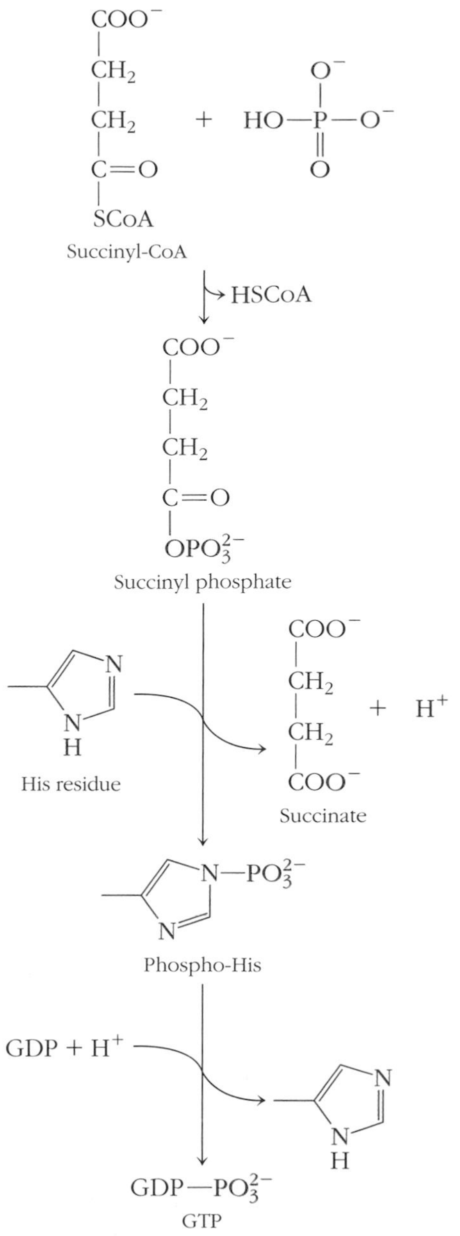

COO^- — CH_2 — CH_2 — C=O — SCoA (Succinyl-CoA) + $HO{-}PO_2{-}O^-$

→ HSCoA

COO^- — CH_2 — CH_2 — C=O — OPO_3^{2-} (Succinyl phosphate)

His residue → COO^- — CH_2 — CH_2 — COO^- (Succinate) + H^+

Phospho-His ($N{-}PO_3^{2-}$)

$GDP + H^+$ → His residue

$GDP{-}PO_3^{2-}$ (GTP)

1. A phosphate group displaces CoA in succinyl-CoA. The product, succinyl phosphate, is an acyl phosphate, which releases a large amount of free energy when cleaved.

2. Succinyl phosphate donates its phosphoryl group to a His residue on the enzyme, producing a phospho-His intermediate and releasing succinate.

3. The phosphoryl group is then transferred to GDP to form GTP.

FIGURE 11-11

The succinyl-CoA synthetase reaction.

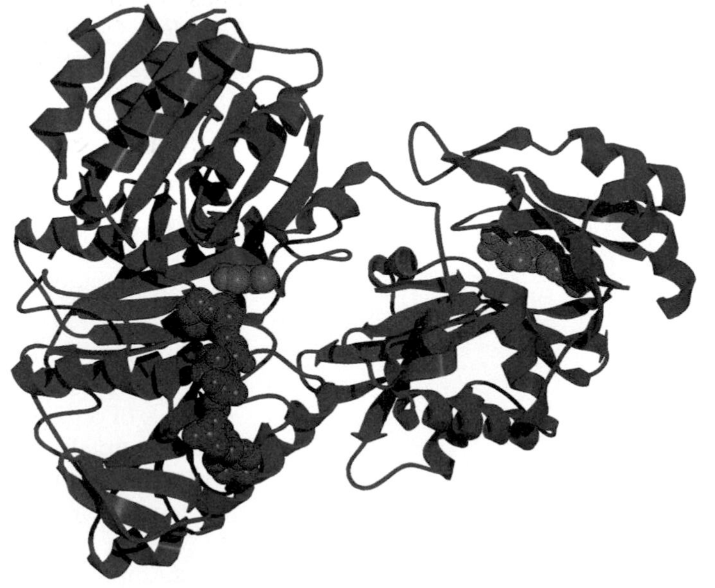

FIGURE 11-12

Substrate binding in succinyl-CoA synthetase.
Succinyl-CoA (represented by coenzyme A, red) binds to the enzyme and its succinyl group is phosphorylated. The succinyl phosphate then transfers its phosphoryl group to the His 246 side chain (green). A protein loop containing the phospho-His side chain must undergo a large movement, because the nucleoside diphosphate awaiting phosphorylation (ADP, orange, in this *E. coli* enzyme) binds to a site about 35 Å away. [*Structure of succinyl-CoA synthetase (pdb 1CQI) determined by M.A. Joyce, M.E. Fraser, M.N.G. James, W.A. Bridger, and W.T. Wolodko.*]

To regenerate the enzyme, the $FADH_2$ group must be reoxidized. Since succinate dehydrogenase is embedded in the inner mitochondrial membrane (it is the only one of the eight citric acid cycle enzymes that is not soluble in the mitochondrial matrix), it can be reoxidized by the lipid-soluble electron carrier ubiquinone (see page 285), rather than by the soluble cofactor NAD^+.

$$\text{Enzyme-FADH}_2 \underset{}{\overset{\text{Q} \quad \text{QH}_2}{\rightleftharpoons}} \text{Enzyme-FAD}$$

7. Fumarase

In the seventh reaction, fumarase catalyzes the reversible hydration of a double bond to convert fumarate to malate:

$$\underset{\text{Fumarate}}{\begin{matrix} COO^- \\ | \\ CH \\ \| \\ HC \\ | \\ COO^- \end{matrix}} \quad \underset{\text{fumarase}}{\overset{H_2O}{\rightleftharpoons}} \quad \underset{\text{Malate}}{\begin{matrix} COO^- \\ | \\ H-C-OH \\ | \\ H-C-H \\ | \\ COO^- \end{matrix}}$$

8. Malate dehydrogenase

The citric acid cycle concludes with the regeneration of oxaloacetate from malate in an NAD^+-dependent oxidation reaction:

$$\underset{\text{Malate}}{\begin{matrix} COO^- \\ | \\ H-C-OH \\ | \\ CH_2 \\ | \\ COO^- \end{matrix}} \quad \underset{\text{malate dehydrogenase}}{\overset{NAD^+ \quad NADH + H^+}{\rightleftharpoons}} \quad \underset{\text{Oxaloacetate}}{\begin{matrix} COO^- \\ | \\ C=O \\ | \\ CH_2 \\ | \\ COO^- \end{matrix}}$$

The standard free energy change for this reaction is $+29.7 \text{ kJ} \cdot \text{mol}^{-1}$, indicating that the reaction has a low probability of occurring as written. However, the product oxaloacetate is a substrate for the next reaction (Reaction 1 of the citric acid cycle). The highly exergonic—and therefore highly favorable—citrate synthase reaction helps pull the malate dehydrogenase reaction forward. This is the reason for the apparent waste of free energy released by cleaving the thioester bond of acetyl-CoA in the first reaction of the citric acid cycle.

The citric acid cycle is an energy-generating catalytic cycle

Because the eighth reaction of the citric acid cycle returns the system to its original state, *the entire pathway acts in a catalytic fashion to dispose of carbon atoms derived from amino acids, carbohydrates, and fatty acids*. Albert Szent-Györgyi discovered the catalytic nature of the pathway by observing that small additions of organic compounds such as succinate, fumarate, and malate stimulated O_2 uptake in a tissue preparation. Because O_2 consumption was much greater than would be required for the direct oxidation of these substances, he inferred that the compounds acted catalytically.

We now know that oxygen is consumed during oxidative phosphorylation, the process that reoxidizes the reduced cofactors (NADH and QH_2) that are produced by the citric acid cycle. Although the citric acid cycle generates one molecule of GTP (or ATP), considerably more ATP is generated when the reduced cofactors are reoxidized by O_2. Each NADH yields approximately 3 ATP, and each QH_2 yields approximately 2 ATP (we will see in Section 12-4 why these values are only approximate). Every acetyl unit that enters the citric acid cycle can therefore generate a total of 12 ATP equivalents. The energy yield of a molecule of glucose, which generates two acetyl units, can be calculated:

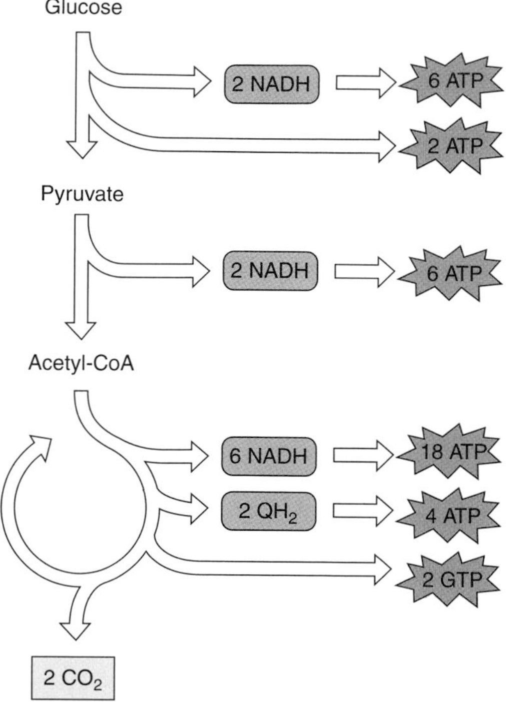

A muscle operating anaerobically produces only 2 ATP per glucose, but under aerobic conditions when the citric acid cycle is fully functional, each glucose molecule generates 38 ATP equivalents. This general phenomenon is called the **Pasteur effect,** after Louis Pasteur, who first observed that the rate of glucose consumption by yeast cells decreased dramatically when the cells were shifted from anaerobic to aerobic growth conditions.  The energetic output of the citric acid cycle is summarized in Exercise 16.

REVIEW LEARNING OBJECTIVE 3

- Describe the sources of the acetyl groups that enter the citric acid cycle.
- List the substrates and products for each of the cycle's eight reactions.
- Which products of the citric acid cycle represent forms of energy currency for the cell?

The citric acid cycle is regulated at three steps

Flux through the citric acid cycle is regulated primarily at the cycle's three metabolically irreversible steps: those catalyzed by citrate synthase (Reaction 1), isocitrate dehydrogenase (Reaction 3), and α-ketoglutarate dehydrogenase (Reaction 4). The major regulators are shown in Figure 11-13.

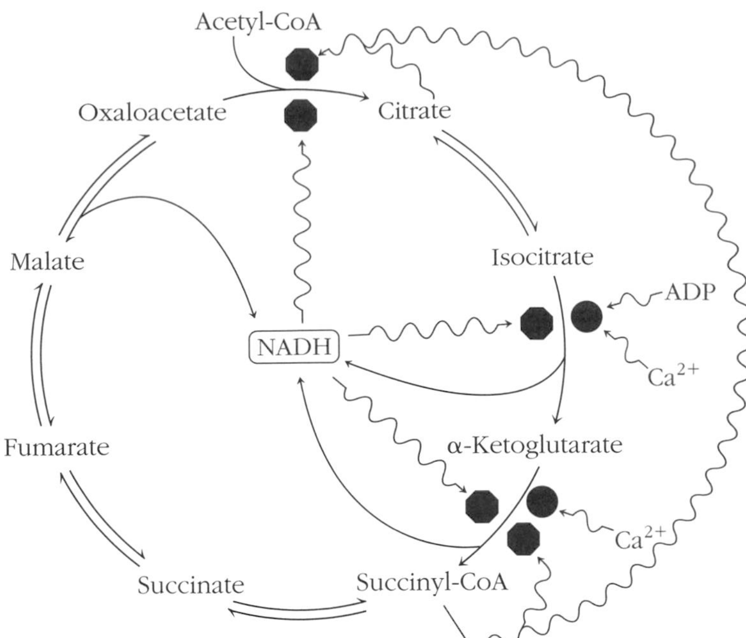

FIGURE 11-13

Regulation of the citric acid cycle. Inhibition is represented by red symbols; activation by green symbols.

Neither acetyl-CoA nor oxaloacetate is present at concentrations high enough to saturate citrate synthase, so flux through the first step of the citric acid cycle depends largely on the substrate concentrations. The product of the reaction, citrate, inhibits citrate synthase (citrate also inhibits phosphofructokinase, thereby decreasing the supply of acetyl-CoA produced by glycolysis). Succinyl-CoA, the product of Reaction 4, inhibits the enzyme that produces it. It also acts as a feedback inhibitor by competing with acetyl-CoA in Reaction 1.

The activity of isocitrate dehydrogenase is inhibited by its reaction product, NADH. NADH also inhibits α-ketoglutarate dehydrogenase and citrate synthase. Both dehydrogenases are activated by Ca^{2+} ions, which generally signify the need to generate cellular free energy. ADP, also representing the need for more ATP, activates isocitrate dehydrogenase.

REVIEW LEARNING OBJECTIVE 4

- Which substrates and products of the citric acid cycle regulate flux through the pathway?

Evolution of the citric acid cycle

A cyclic pathway is efficient because it minimizes waste. Not surprisingly, cyclic processes are common in biology. For example, the circulatory system operates as a cycle: Oxygen is delivered to cells and carbon dioxide and other waste products are removed, all without permanently altering the composition of the blood. The citric acid cycle is just one of several cyclic metabolic pathways.

Like all metabolic pathways, the citric acid cycle evolved from preexisting biochemical reactions. Clues to its origins can be found by examining the metabolism of organisms that resemble earlier life forms. Such organisms emerged before atmospheric oxygen was available and may have used sulfur as their ultimate oxidizing agent, reducing it to H_2S. Their modern-day counterparts are anaerobic autotrophs that harvest free energy by pathways that are independent of the pathways of carbon metabolism. These organisms therefore do not use the citric acid cycle to generate reduced cofactors that are subsequently oxidized by molecular oxygen. However, all organisms must synthesize the small molecules from which they can build proteins, nucleic acids, carbohydrates, lipids, and so on.

The central metabolic pathways of prokaryotes can be reconstructed by assigning functions to their various genes, as identified by genome-sequencing studies (this approach is fruitful because many "housekeeping" genes, which encode enzymes that make free energy and molecular building blocks available to the cell, are highly conserved among different species and are relatively easy to recognize).

Even organisms that do not use the citric acid cycle contain genes for some citric acid cycle enzymes. For example, the cells may condense acetyl-CoA with oxaloacetate, leading to α-ketoglutarate, which is a precursor of several amino acids. They may also convert oxaloacetate to succinate, proceeding through the citric acid cycle intermediates malate and fumarate. These two pathways resemble the citric acid cycle, with the right arm following the usual oxidative sequence of the cycle and the left arm following a reversed, reductive sequence (Fig. 11-14). The reductive sequence of reactions might have evolved as a way to regenerate the cofactors reduced during other catabolic reactions (for example, the NADH produced by the glyceraldehyde-3-phosphate dehydrogenase reaction of glycolysis; see Section 10-2).

It is easy to theorize that the evolution of an enzyme to interconvert α-ketoglutarate and succinate could have created a cyclic pathway similar to the modern citric acid cycle. Interestingly, *E. coli,* which uses the citric acid cycle under aerobic growth conditions, uses an interrupted citric acid cycle like the one diagrammed in Figure 11-14 when it is growing anaerobically.

Since the final four reactions of the modern citric acid cycle are metabolically reversible, the primitive citric acid cycle might easily have accommodated one-way flux in the clockwise direction, forming an oxidative cycle. If the complete cycle proceeded in the counterclockwise direction, the result would have been a reductive biosynthetic pathway (Fig. 11-15). This pathway, which would incorporate, or "fix," atmospheric CO_2 into biological molecules, may have preceded the modern CO_2-fixing pathway found in green plants and some photosynthetic bacteria (described in Section 13-3).

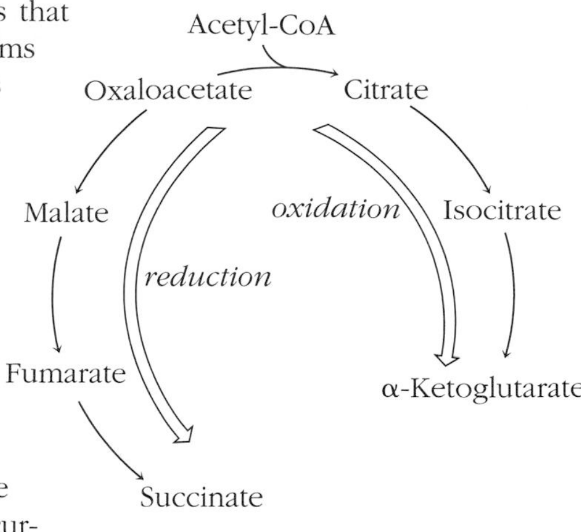

FIGURE 11-14

Pathways that might have given rise to the citric acid cycle. The pathway starting from oxaloacetate and proceeding to the right is an oxidative biosynthetic pathway, whereas the pathway that proceeds to the left is a reductive pathway. The modern citric acid cycle may have evolved by connecting these pathways.

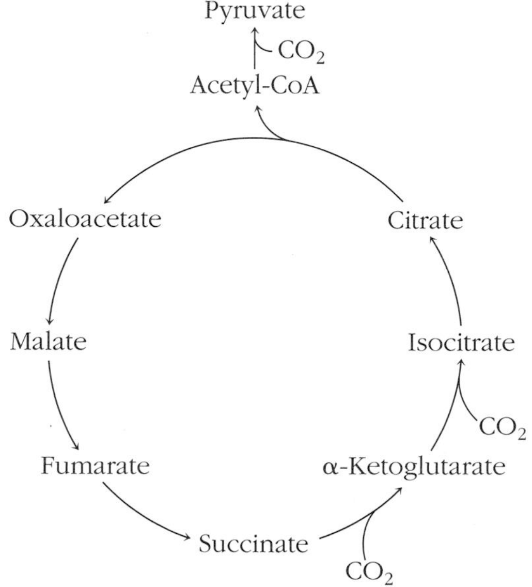

FIGURE 11-15

A proposed reductive biosynthetic pathway based on the citric acid cycle. This pathway might have operated to incorporate CO_2 into biological molecules. [*After Romano, A.H. and Conway, T.,* Res. Microbiol ***147,*** *450 (1996).*]

REVIEW LEARNING OBJECTIVE 5

- Describe how primitive oxidative and reductive biosynthetic pathways might have been combined to generate a circular metabolic pathway.

3 THE CITRIC ACID CYCLE IS BOTH CATABOLIC AND ANABOLIC

In mammals, six of the eight citric acid cycle intermediates (all except isocitrate and succinate) are precursors or products of other substances. For this reason, it is impossible to designate the citric acid cycle as a purely catabolic or anabolic pathway.

Citric acid cycle intermediates are precursors of other molecules

Intermediates of the citric acid cycle can be siphoned off to form other compounds (Fig. 11-16). For example, succinyl-CoA is used for the synthesis of heme. The five-carbon α-ketoglutarate can undergo reductive amination by glutamate dehydrogenase to produce the amino acid glutamate:

$$\underset{\alpha\text{-Ketoglutarate}}{^{-}OOC{-}CH_2{-}CH_2{-}C({=}O){-}COO^{-}} + NH_4^+ \xrightleftharpoons[\text{glutamate dehydrogenase}]{NADH + H^+ \quad NAD^+} \underset{\text{Glutamate}}{^{-}OOC{-}CH_2{-}CH_2{-}CH(NH_3^+){-}COO^{-}} + H_2O$$

Glutamate is a precursor of the amino acids glutamine, arginine, and proline. Glutamine in turn is a precursor for the synthesis of purine and pyrimidine nucleotides. We have already seen that oxaloacetate is a precursor of mono-

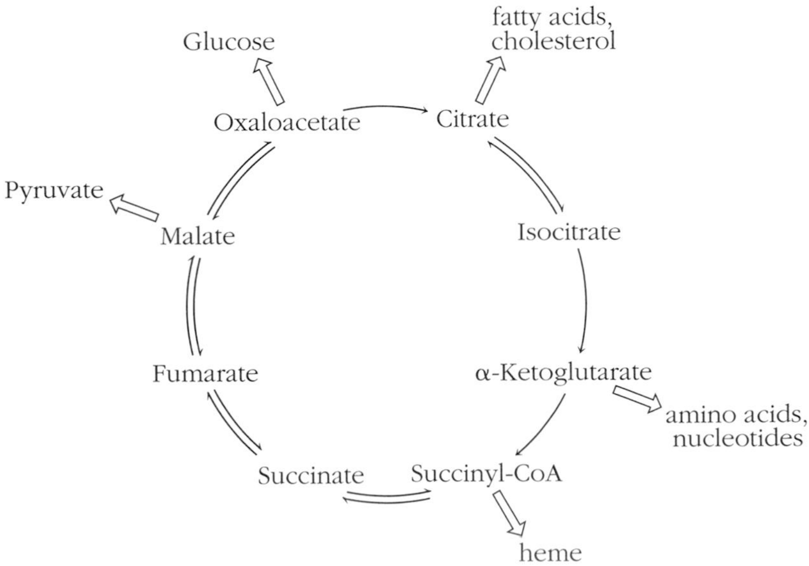

FIGURE 11-16
Citric acid cycle intermediates as biosynthetic precursors.

saccharides (Section 10-3). Consequently, *any of the citric acid cycle intermediates, which can be converted by the cycle to oxaloacetate, can ultimately serve as gluconeogenic precursors.*

Citrate produced by the condensation of acetyl-CoA with oxaloacetate can be transported out of the mitochondria to the cytosol. ATP-citrate lyase then catalyzes the reaction

$$\text{ATP} + \text{citrate} + \text{CoA} \rightarrow \text{ADP} + \text{P}_i + \text{oxaloacetate} + \text{acetyl-CoA}$$

The resulting acetyl-CoA is used for fatty acid and cholesterol synthesis, which take place in the cytosol. The ATP-citrate lyase reaction undoes the work of the exergonic citrate synthase reaction. This seems wasteful, but cytosolic ATP-citrate lyase is essential because acetyl-CoA, which is produced in the mitochondria, cannot cross the mitochondrial membrane to reach the cytosol, whereas citrate can. The oxaloacetate product of the ATP-citrate lyase reaction can be converted to malate by a cytosolic malate dehydrogenase operating in reverse. Malate is then decarboxylated by the action of malic enzyme to produce pyruvate:

$$\underset{\text{Malate}}{\begin{array}{c}\text{COO}^-\\ |\\ \text{CH—OH}\\ |\\ \text{CH}_2\\ |\\ \text{COO}^-\end{array}} \xrightarrow[\text{malic enzyme}]{\text{NADP}^+ \quad \text{NADPH}} \underset{\text{Pyruvate}}{\begin{array}{c}\text{COO}^-\\ |\\ \text{C=O}\\ |\\ \text{CH}_3\end{array}} + \text{CO}_2$$

Pyruvate can reenter the mitochondria and be converted back to oxaloacetate to complete the cycle shown in Figure 11-17. Exercise 16 explores the function of the citric acid cycle as a supplier of metabolic precursors.

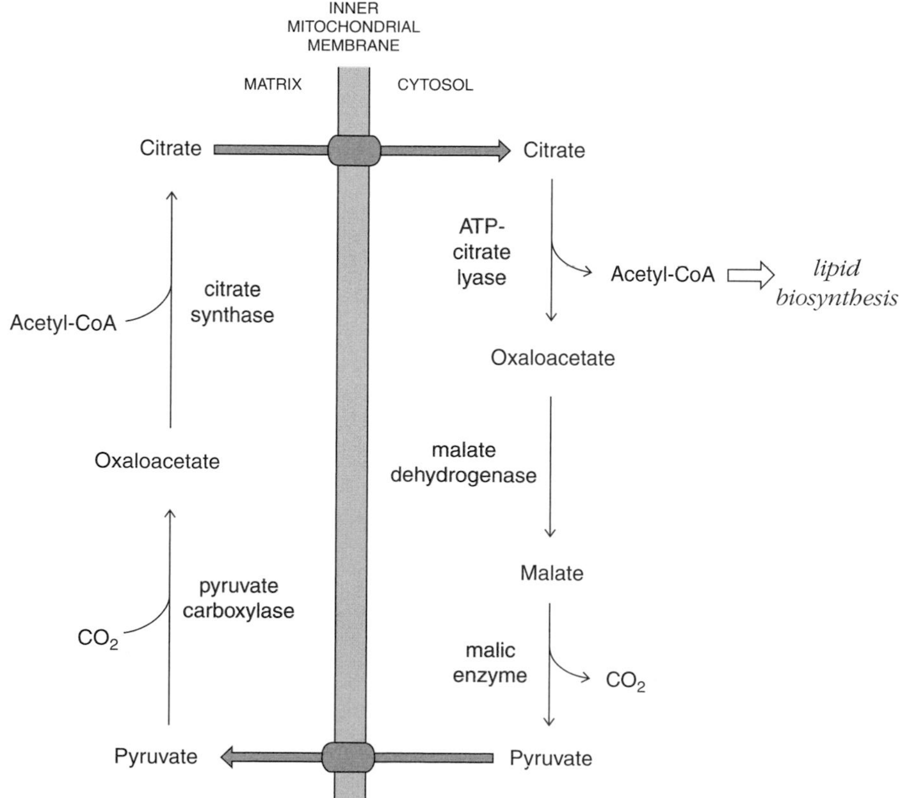

FIGURE 11-17

The citrate transport system. Both citrate and pyruvate cross the inner mitochondrial membrane via specific transport proteins. This system allows carbon atoms from citrate to be released in the cytosol as acetyl-CoA for the synthesis of fatty acids and cholesterol.

In plants, isocitrate is diverted from the citric acid cycle in a biosynthetic pathway known as the **glyoxylate cycle** (Box 11-B).

REVIEW LEARNING OBJECTIVE 6

- Describe how the citric acid cycle supplies the precursors for the synthesis of amino acids, glucose, and fatty acids.
- What does the ATP-citrate lyase reaction accomplish?

Anaplerotic reactions replenish citric acid cycle intermediates

Intermediates that are diverted from the citric acid cycle for other purposes can be replenished through **anaplerotic reactions** (from the Greek *ana,* up and *plerotikos,* to fill; Fig. 11-18). One of the most important of these reactions is catalyzed by pyruvate carboxylase (this is also the first step of gluconeogenesis; Section 10-3):

$$\text{Pyruvate} + CO_2 + \text{ATP} + H_2O \rightarrow \text{oxaloacetate} + \text{ADP} + P_i$$

Acetyl-CoA activates pyruvate carboxylase, so when the activity of the citric acid cycle is low and acetyl-CoA accumulates, more oxaloacetate is produced. The concentration of oxaloacetate is normally low since the malate dehydrogenase reaction is thermodynamically unfavorable and the citrate synthase reaction is highly favorable. The replenished oxaloacetate is converted to citrate, isocitrate, α-ketoglutarate, and so on, so that the concentrations of all the citric acid cycle intermediates increase and the cycle can proceed more quickly. *Since the citric acid cycle acts as a catalyst, increasing the concentrations of its components increases flux through the pathway.*

The degradation of fatty acids with an odd number of carbon atoms yields the citric acid cycle intermediate succinyl-CoA. Other anaplerotic reactions in-

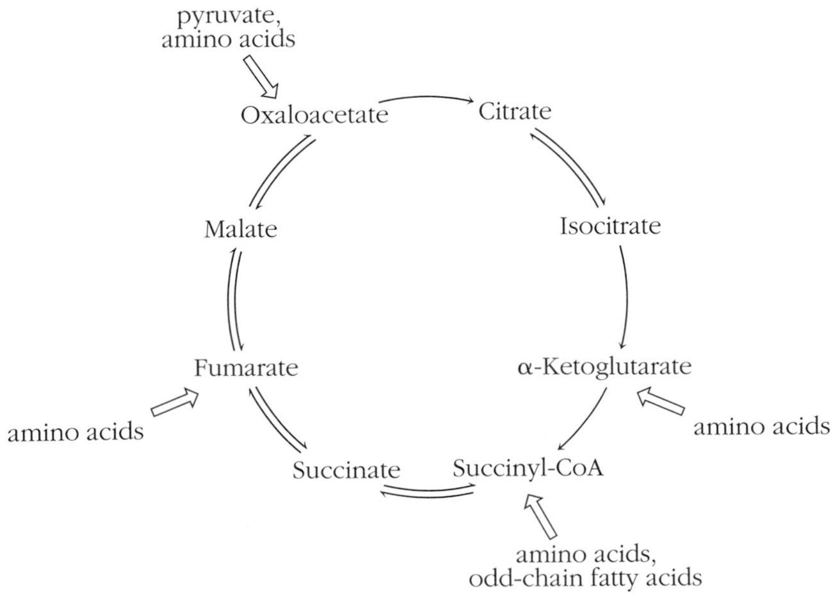

FIGURE 11-18

Anaplerotic reactions of the citric acid cycle.

A CLOSER LOOK

BOX 11-B The glyoxylate cycle

Plants and some bacterial cells contain certain enzymes that act together with some citric acid cycle enzymes to convert acetyl-CoA to oxaloacetate, a gluconeogenic precursor. Animals lack the enzymes to do this and therefore cannot undertake the net synthesis of carbohydrates from two-carbon precursors. In plants, the glyoxylate cycle includes reactions that take place in the mitochondria and the **glyoxisome,** an organelle that, like the peroxisome, contains enzymes that carry out some essential metabolic processes.

In the glyoxisome, acetyl-CoA condenses with oxaloacetate to form citrate, which is then isomerized to isocitrate, as in the citric acid cycle. However, the next step is not the isocitrate dehydrogenase reaction but a reaction catalyzed by the glyoxisome enzyme isocitrate lyase, which converts isocitrate to succinate and the two-carbon compound glyoxylate.

Succinate continues as usual through the mitochondrial citric acid cycle to regenerate oxaloacetate. In the glyoxisome, the glyoxylate condenses with a second molecule of acetyl-CoA in a reaction catalyzed by the glyoxisome enzyme malate synthase to form the four-carbon compound malate. Malate can then be converted to oxaloacetate for gluconeogenesis. The two reactions that are unique to the glyoxylate cycle are shown in green in the figure; reactions that are identical to those of the citric acid cycle are shown in blue.

In essence, the glyoxylate cycle bypasses the two CO_2-generating steps of the citric acid cycle (catalyzed by isocitrate dehydrogenase and α-ketoglutarate dehydrogenase) and incorporates a second acetyl unit (at the malate synthase step). The net result of the glyoxylate cycle is the production of a four-carbon compound that can be used to synthesize glucose. This pathway is highly active in germinating seeds, where stored oils (triacylglycerols) are broken down to acetyl-CoA. The glyoxylate cycle thus provides a route for synthesizing glucose from fatty acids. Because animals lack isocitrate lyase and malate synthase, they cannot undertake the net synthesis of carbohydrates from fat.

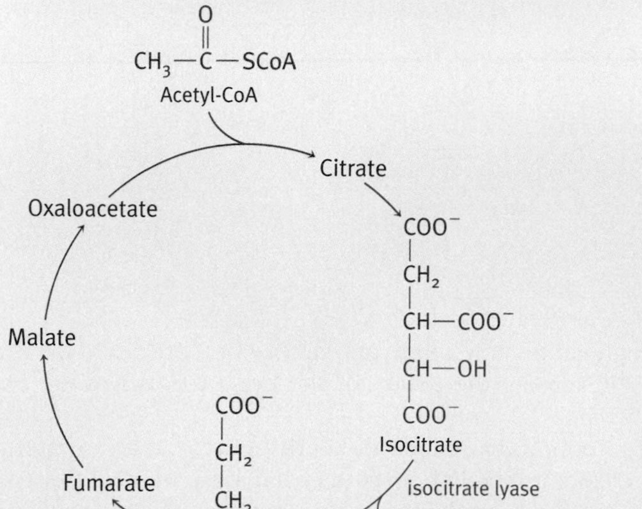

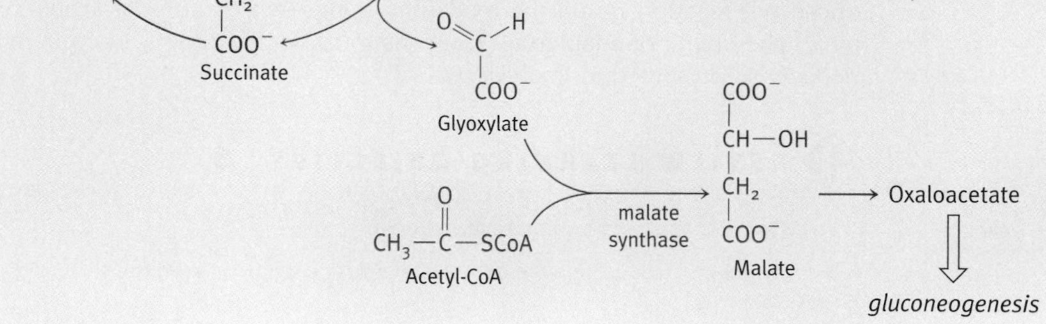

$CH_3-C(=O)-SCoA$
Acetyl-CoA

Citrate

Oxaloacetate

Malate

Fumarate

$^-OOC-CH_2-CH(COO^-)-CH(OH)-COO^-$
Isocitrate

$^-OOC-CH_2-CH_2-COO^-$
Succinate

isocitrate lyase

$O=CH-COO^-$
Glyoxylate

$CH_3-C(=O)-SCoA$
Acetyl-CoA

malate synthase

$^-OOC-CH(OH)-CH_2-COO^-$
Malate

→ Oxaloacetate

⇩

gluconeogenesis

clude the pathways for the degradation of some amino acids, which produce α-ketoglutarate, succinyl-CoA, fumarate, and oxaloacetate. Some of these reactions are transaminations, such as

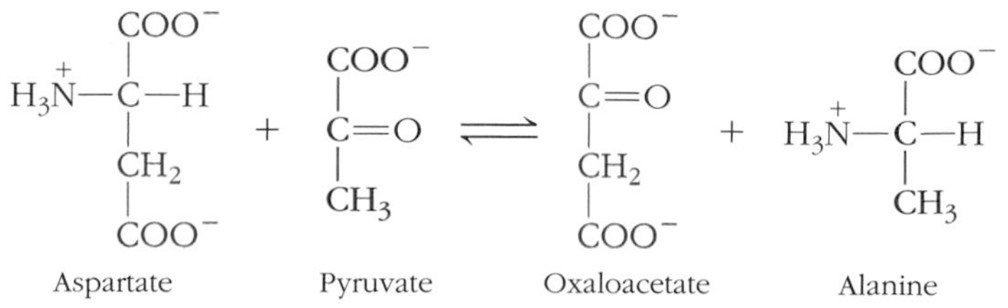

$$\underset{\text{Aspartate}}{\begin{matrix} COO^- \\ | \\ H_3\overset{+}{N}-C-H \\ | \\ CH_2 \\ | \\ COO^- \end{matrix}} + \underset{\text{Pyruvate}}{\begin{matrix} COO^- \\ | \\ C=O \\ | \\ CH_3 \end{matrix}} \rightleftharpoons \underset{\text{Oxaloacetate}}{\begin{matrix} COO^- \\ | \\ C=O \\ | \\ CH_2 \\ | \\ COO^- \end{matrix}} + \underset{\text{Alanine}}{\begin{matrix} COO^- \\ | \\ H_3\overset{+}{N}-C-H \\ | \\ CH_3 \end{matrix}}$$

Because transamination reactions have ΔG values near zero, the direction of flux into or out of the pool of citric acid cycle intermediates depends on the relative concentrations of the reactants.

In vigorously exercising muscle, the concentrations of citric acid cycle intermediates increase about three- to fourfold within a few minutes. This may help boost the energy-generating activity of the citric acid cycle, but it cannot be the sole mechanism, since flux through the citric acid cycle actually increases as much as 100-fold due to the increased activity of the three enzymes at the control points: citrate synthase, isocitrate dehydrogenase, and α-ketoglutarate dehydrogenase. The increase in citric acid cycle intermediates may actually be a mechanism for accommodating the large increase in pyruvate that results from rapid glycolysis at the start of exercise. Rather than converting all the pyruvate to lactate (Section 10-2), some is shunted into the pool of citric acid cycle intermediates via the pyruvate carboxylase reaction. Some pyruvate also undergoes a reversible reaction catalyzed by alanine aminotransferase.

$$\underset{\text{Pyruvate}}{\begin{array}{c} COO^- \\ | \\ C{=}O \\ | \\ CH_3 \end{array}} + \underset{\text{Glutamate}}{\begin{array}{c} COO^- \\ | \\ CH_2 \\ | \\ CH_2 \\ | \\ H{-}C{-}NH_3^+ \\ | \\ COO^- \end{array}} \rightleftharpoons \underset{\text{Alanine}}{\begin{array}{c} COO^- \\ | \\ H_3\overset{+}{N}{-}C{-}H \\ | \\ CH_3 \end{array}} + \underset{\alpha\text{-Ketoglutarate}}{\begin{array}{c} COO^- \\ | \\ CH_2 \\ | \\ CH_2 \\ | \\ C{=}O \\ | \\ COO^- \end{array}}$$

The resulting α-ketoglutarate then augments the pool of citric acid cycle intermediates, thereby increasing the ability of the cycle to oxidize the extra pyruvate.

Note that any compound that enters the citric acid cycle as an intermediate is not itself oxidized; it merely boosts the catalytic activity of the cycle, whose net reaction is still the oxidation of the two carbons of acetyl-CoA.  The ability of anaplerotic reactions to replenish citric acid cycle intermediates is presented in Exercise 16.

REVIEW LEARNING OBJECTIVE 6

- Why is the concentration of oxaloacetate low?
- Why does synthesizing more oxaloacetate increase flux through the citric acid cycle?

SUMMARY

1. In order for pyruvate, the product of glycolysis, to enter the citric acid cycle, it must undergo oxidative decarboxylation catalyzed by the multienzyme pyruvate dehydrogenase complex.
2. The eight reactions of the citric acid cycle function as a multistep catalyst to convert the two carbons of acetyl-CoA to 2 CO_2. The electrons released in this oxidative process are transferred to 3 NAD^+ and to ubiquinone. The reoxidation of the reduced cofactors generates ATP by oxidative phosphorylation. In addition, succinyl-CoA synthetase yields one molecule of GTP.
3. The regulated reactions of the citric acid cycle are its irreversible steps, catalyzed by citrate synthase, isocitrate dehydrogenase, and α-ketoglutarate dehydrogenase.
4. The citric acid cycle most likely evolved from biosynthetic pathways leading to α-ketoglutarate or succinate.

5. Six of the eight citric acid cycle intermediates serve as precursors of other compounds, including amino acids, monosaccharides, and lipids. Anaplerotic reactions convert other compounds into citric acid cycle intermediates, thereby allowing increased flux of acetyl carbons through the pathway.

CHECKLIST

1. Review the Learning Objectives listed on page 342.
2. Complete Exercise 16/The Citric Acid Cycle to explore the fates of carbon atoms that enter the citric acid cycle and to review the eight reactions and the energy output of the cycle.
3. Apply your knowledge by solving the problems at the end of this chapter. Check your results in the Solutions appendix.
4. Be able to define the boldfaced terms (consult the glossary at the end of the book). Test your understanding by taking the Chapter 11 quiz (www.wiley.com/college/pratt).
5. Explore the websites listed at www.wiley.com/college/pratt.
6. Consult the list of Selected Readings for background information or additional details on the pyruvate dehydrogenase complex, the citric acid cycle, anaplerotic reactions, and the glyoxylate cycle.

GLOSSARY TERMS

citric acid cycle
mitochondrial matrix
multienzyme complex
Pasteur effect
glyoxylate cycle
glyoxisome
anaplerotic reaction

PROBLEMS

1. Determine which one of the five steps of the pyruvate dehydrogenase complex reaction is metabolically irreversible and explain why.

2. The product of the pyruvate dehydrogenase complex, acetyl-CoA, is released in Step 3 of the overall reaction. What is the purpose of Steps 4 and 5?

3. Beriberi is a disease that results from a dietary lack of thiamine, the vitamin that serves as the precursor for thiamine pyrophosphate (TPP). There are two metabolites that accumulate in individuals with beriberi, especially after ingestion of glucose. Which metabolites accumulate and why?

4. Arsenite is toxic in part because it binds to sulfhydryl compounds such as lipoamide, as shown below. What effect would the presence of arsenite have on the citric acid cycle?

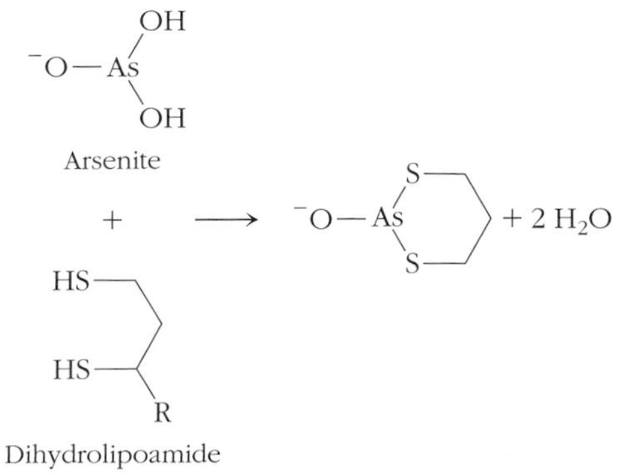

$$^{-}O{-}As(OH)_2 \text{ (Arsenite)} + \text{Dihydrolipoamide (HS, HS, R)} \longrightarrow {}^{-}O{-}As(S{-})_2 \text{ (cyclic)} + 2\,H_2O$$

Arsenite

Dihydrolipoamide

5. Using the pyruvate dehydrogenase complex reaction as a model, reconstruct the TPP-dependent yeast pyruvate decarboxylase reaction in alcoholic fermentation (see Box 10-B).

6. Why is it advantageous for citrate, the product of Reaction 1 of the citric acid cycle, to inhibit phosphofructokinase, which catalyzes the third reaction of glycolysis (see Section 10-2)?

7. Animals that have ingested the leaves of the poisonous South African plant *Dichapetalum cymosum* exhibit a 10-fold increase in levels of cellular citrate. The plant contains fluoroacetate, which is converted to fluoroacetyl-CoA. Describe the mechanism that leads to increased levels of citrate in animals who have ingested this poisonous plant. (Note: Fluoroacetyl-CoA is not an inhibitor of citrate synthase.)

8. Site-directed mutagenesis techniques were used to synthesize a mutant citrate synthase enzyme in which the active site histidine was converted to an alanine. Why did the mutant citrate synthase enzyme exhibit decreased catalytic activity? [From Pereira, D.S., Donald, L.J., Hosfield, D.J., and Duckworth, H.W., *J. Biol. Chem.* **269,** 412–417 (1994).]

9. The compound *S*-acetonyl-CoA can be synthesized from 1-bromoacetone and coenzyme A.

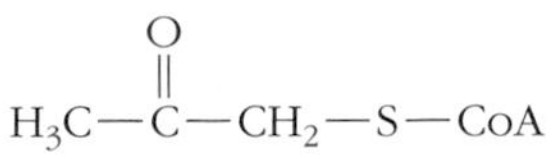

$$H_3C{-}\overset{O}{\overset{\|}{C}}{-}CH_2{-}S{-}CoA$$

S-Acetonyl-CoA

(a) Write the reaction for the formation of *S*-acetonyl-CoA.
(b) The Lineweaver–Burk plot of the inhibition of citrate synthase by *S*-acetonyl-CoA is shown. What type of inhibitor is *S*-acetonyl-CoA? Explain.

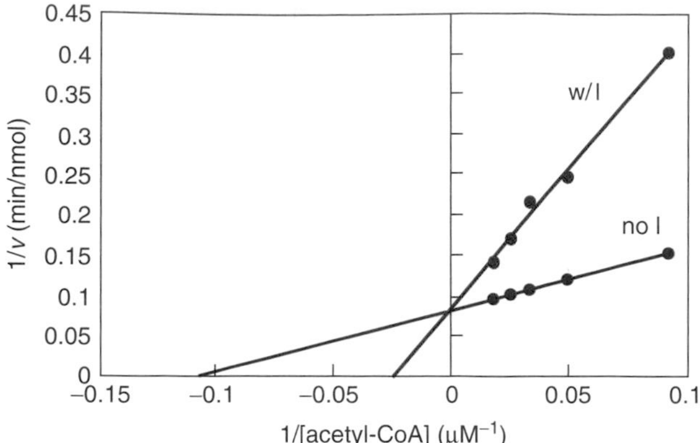

(c) Acetyl-CoA acts as an allosteric activator of pyruvate carboxylase. *S*-acetonyl-CoA does not activate pyruvate carboxylase, and it cannot compete with acetyl-CoA for binding to the enzyme. What does this tell you about the binding requirements for an allosteric activator of pyruvate carboxylase?

[From Rubenstein, P. and Dryer, R., *J. Biol. Chem.* **255,** 7858–7862 (1980).]

10. Administration of high concentrations of oxygen (hyperoxia) is effective in the treatment of lung injuries but at the same time can also be quite damaging.

(a) It has been shown that lung aconitase activity is dramatically decreased during hyperoxia. How would the concentration of citric acid cycle intermediates be affected?
(b) The decreased aconitase activity and decreased mitochondrial respiration in hyperoxia are accompanied by elevated rates of glycolysis and the pentose phosphate pathway. Explain why.

[From Allen, C.B., Guo, X.L., and White, C.W., *Am. J. Physiol.* **274 (3 Pt. 1),** L320–L329 (1998).]

11. Using the pyruvate dehydrogenase complex reaction as a model, draw the intermediates of the α-ketoglutarate dehydrogenase reaction. Describe what happens in each of the five reaction steps.

12. Succinyl-CoA synthetase is also called succinate thiokinase. Why is the enzyme considered to be a kinase?

13. Malonate is a competitive inhibitor of succinate dehydrogenase. What citric acid cycle intermediates accumulate if malonate is present in a preparation of isolated mitochondria?

14. Reaction 8 and Reaction 1 of the citric acid cycle can be considered to be coupled, because the exergonic cleavage of the thioester bond of acetyl-CoA in Reaction 1 drives the regeneration of oxaloacetate in Reaction 8.

(a) Write the equation for the overall coupled reaction and calculate its $\Delta G^{\circ\prime}$.
(b) What is the equilibrium constant for the coupled reaction? Compare this equilibrium constant with the equilibrium constant of Reaction 8 alone.

15. (a) Oxaloacetate labeled at C4 with ^{14}C is added to a suspension of respiring mitochondria. What is the fate of the labeled carbon?
(b) Acetyl-CoA labeled at C1 with ^{14}C is added to a suspension of respiring mitochondria. What is the fate of the labeled carbon?

16. The $\Delta G^{\circ\prime}$ for the fumarase reaction is $-3.4\ kJ \cdot mol^{-1}$, but the ΔG value is close to zero. What is the ratio of fumarate to malate under cellular conditions at 37°C? Is this reaction likely to be a control point for the citric acid cycle?

17. When leavened bread is made, the bread dough is "punched" down and then put in a warm place to "rise" to increase its volume. Give a biochemical explanation for this observation.

18. Flux through the citric acid cycle is regulated by the simple mechanisms of (a) substrate availability, (b) product inhibition, and (c) feedback inhibition. Give examples of each.

19. Certain microorganisms with an incomplete citric acid cycle decarboxylate α-ketoglutarate to produce succinate semialdehyde. A dehydrogenase then converts succinate semialdehyde to succinate. These reactions can be combined with other standard citric acid cycle reactions to create a pathway from citrate to oxaloacetate. How does this alternative pathway compare to the standard citric acid cycle in its ability to make free energy available to the cell?

COO^-–CH_2–CH_2–C=O–COO^-
α-Ketoglutarate

$\xrightarrow{CO_2}$

COO^-–CH_2–CH_2–C=O–H
Succinate semialdehyde

$\xrightarrow{NAD^+ \rightarrow NADH}$

COO^-–CH_2–CH_2–C=O–O^-
Succinate

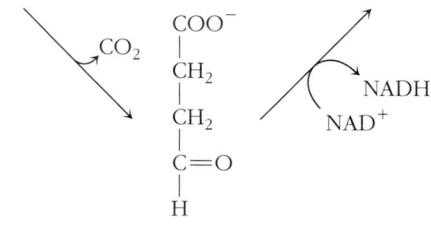

20. Is net synthesis of glucose possible from the following compounds?

(a) The fatty acid palmitate (16:0), which is degraded to 8 acetyl-CoA
(b) Glyceraldehyde-3-phosphate
(c) Leucine, which is degraded to acetoacetate (a compound that is metabolically equivalent to two acetyl-CoA groups) and acetyl-CoA

(d) Tryptophan, which is degraded to alanine and acetoacetate

(e) Phenylalanine, which is degraded to acetoacetate and fumarate

21. Pancreatic islet cells cultured in the presence of 1–20 mM glucose showed increased activities of pyruvate carboxylase and the E1 subunit of the pyruvate dehydrogenase complex proportional to the increase in glucose concentration. Explain why.

22. The udder of a ewe uses the majority of the glucose synthesized by the ewe, both for production of the milk sugar lactose and for the production of triacylglycerols. During the winter, milk production decreases due to a poor food supply. Administration of propionate (which is converted to succinyl-CoA in ruminants) is an appropriate remedy. How does this treatment work?

23. Metabolites in rat muscle were measured before and after exercising. After exercise, the rat muscle showed an increase in oxaloacetate concentration, a decrease in phosphoenolpyruvate concentration, and no change in pyruvate concentration. Explain.

24. The activity of the isocitrate dehydrogenase enzyme in *E. coli* is regulated by the covalent attachment of a phosphate group to the enzyme. Phosphorylated isocitrate dehydrogenase is inactive. When acetate is the food source for a culture of *E. coli,* isocitrate dehydrogenase is phosphorylated.

(a) Draw a diagram showing how acetate is metabolized in *E. coli.*

(b) If glucose is added to the culture, the phosphate group is removed from the isocitrate dehydrogenase enzyme. How does flux through the metabolic pathways change in *E. coli* when glucose is the food source instead of acetate?

25. Animals lack a glyoxylate cycle and cannot convert fats to carbohydrates. If an animal is fed a fatty acid with all its carbons replaced by the isotope ^{14}C, some of the labeled carbons later appear in glucose. How is this possible?

26. Many amino acids are broken down to intermediates of the citric acid cycle.

(a) Why can't these amino acid "remnants" be completely oxidized to CO_2 by the citric acid cycle?

(b) Explain why amino acids that are broken down to pyruvate can be completely oxidized by the citric acid cycle.

27. Describe how the transamination reaction below could function as an anaplerotic reaction for the citric acid cycle.

$$\underset{\text{Oxaloacetate}}{\begin{array}{c}COO^-\\|\\C{=}O\\|\\CH_2\\|\\COO^-\end{array}} + \underset{\text{Alanine}}{\begin{array}{c}COO^-\\|\\{}^+H_3N{-}CH\\|\\CH_3\end{array}} \rightleftharpoons \underset{\text{Aspartate}}{\begin{array}{c}COO^-\\|\\{}^+H_3N{-}CH\\|\\CH_2\\|\\COO^-\end{array}} + \underset{\text{Pyruvate}}{\begin{array}{c}COO^-\\|\\C{=}O\\|\\CH_3\end{array}}$$

28. The plant metabolite hydroxycitric acid (shown in its ionized form) is advertised as an agent that prevents fat buildup.

$$\begin{array}{c}CH_2{-}COO^-\\|\\HO{-}C{-}COO^-\\|\\HO{-}CH{-}COO^-\end{array}$$

Hydroxycitrate

(a) How does this compound differ from citrate?

(b) Hydroxycitrate inhibits the activity of ATP-citrate lyase. What kind of inhibition is likely to occur?

(c) Why might inhibition of ATP-citrate lyase block the conversion of carbohydrates to fats?

(d) The synthesis of what other compounds would be inhibited by hydroxycitrate?

29. *Helicobacter pylori* is a bacterium that colonizes the upper gastrointestinal tract in humans and is the causative agent of chronic gastritis, ulcers, and possibly gastric cancer. Knowledge of the intermediary metabolism of this organism will be helpful in the development of effective drug therapies to treat these diseases.

The citric acid "cycle" in *H. pylori* is a noncyclic, branched pathway that is used to produce biosynthetic intermediates instead of metabolic energy. Succinate is produced in the "reductive branch" whereas α-ketoglutarate is produced in the "oxidative branch." The two branches are linked by the α-ketoglutarate oxidase reaction. The pathway is shown in the diagram below.

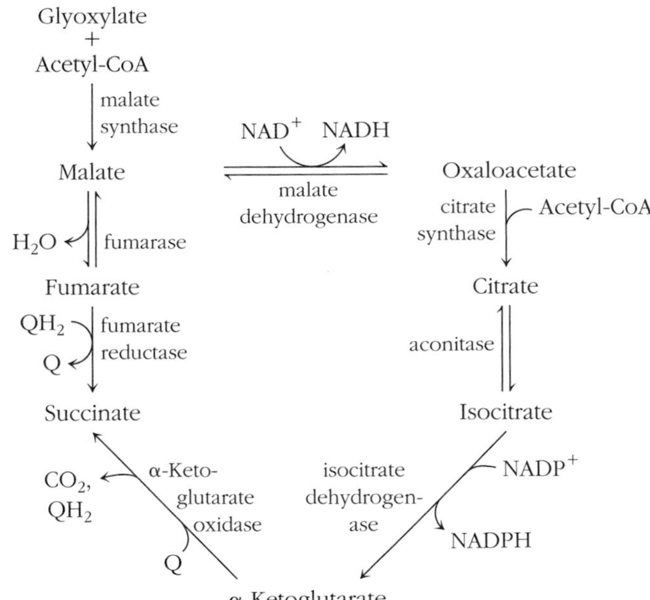

(a) Compare and contrast the citric acid cycle in *H. pylori* with the citric acid cycle in mammals.

(b) The K_M values for the enzymes in the table on the next page are higher than the K_M values for the corresponding enzymes in other species of bacteria. What does this tell you about the conditions under which the citric acid cycle operates in *H. pylori*?

(c) Compare the properties of *H. pylori* citrate synthase and mammalian citrate synthase.

(d) What enzymes might serve to regulate the citric acid cycle in *H. pylori*?

(e) What enzymes might be used as drug targets for persons suffering from gastritis, ulcers, or gastric cancer?

Enzyme	Substrate	Inhibitors	Activators
Citrate synthase	Acetyl-CoA, Oxaloacetate	ATP	
Aconitase	Citrate		
Isocitrate dehydrogenase ($NADP^+$-dependent)	Isocitrate, $NADP^+$	Higher concentrations $NADP^+$, isocitrate	AMP (slight)
α-ketoglutarate oxidase	α-ketoglutarate		CoASH
Malate dehydrogenase	Oxaloacetate, NADH		
Fumarase	Malate		
Fumarate reductase	Fumarate, QH_2		
Malate synthase	Glyoxylate, acetyl-CoA		

[From Pitson, S.M., Mendz, G.L., Srinivasan, S., and Hazell, S.L., *Eur. J. Biochem.* **260,** 258–267 (1999).]

30. As one of the reactants in the first reaction of the citric acid cycle, oxaloacetate is an important cellular metabolite. The concentrations of oxaloacetate are tightly regulated. Different organisms employ different mechanisms to obtain oxaloacetate. In mammals and yeast, oxaloacetate is the product of the pyruvate carboxylase (PC) reaction. In *E. coli,* the enzyme phosphoenolpyruvate carboxylase (PPC) provides oxaloacetate from phosphoenolpyruvate obtained from glucose oxidation in the glycolytic pathway. If glucose is absent and *E. coli* is using acetate as a carbon source, the glyoxylate pathway serves to generate the needed oxaloacetate. Usually an organism will employ PPC or PC, but not both.

Detectable levels of PC in the methanogenic bacterium *Methanobacterium thermoautotrophicum* had previously not been found, and since PPC had been detected, it was believed that *M. thermoautotrophicum* did not possess PC. However, microbiologists found that if they added biotin (a cofactor required by PC) to cultures of the methanogenic bacterium, pyruvate carboxylase activity could be detected. The investigators set out to isolate, purify, and characterize the enzyme.

(a) Write the balanced reaction catalyzed by pyruvate carboxylase.

(b) Write the balanced reaction catalyzed by phosphoenolpyruvate carboxylase.

(c) Why is it so important that oxaloacetate be generated in the cell?

(d) The catalytic properties of the pyruvate carboxylase enzyme were assessed following purification. The activity of the enzyme was assayed in the presence of ATP, pyruvate, bicarbonate, and Mg^{2+} ions as a control. In addition, the dependence of the enzyme on these various metabolites was tested by replacing them with similar compounds. The results are shown in the table below.

Does PC depend on ATP for activity? Can other nucleotides substitute for ATP? What happens when other nucleotides are added to the assay mixture in addition to ATP?

Effector	Activity of *M. thermoautotrophicum* pyruvate carboxylase (% of control)
Control	100
Nucleotide replacing ATP	
AMP	0
ADP	0
CTP	0
GTP	0
ITP	0
UTP	0
Nucleotide in addition to ATP	
AMP	104
ADP	73
CTP	106
GTP	94
ITP	80
UTP	105
Citric acid cycle–related compounds	
Acetyl-CoA	84
Aspartate	91
Glutamate	95
α-Ketoglutarate	73
Divalent cation replacing Mg^{2+}	
Mn^{2+}	17
Co^{2+}	46
Zn^{2+}	0

(e) What is the effect of the other metabolites on PC activity?

(f) What ion or ions are required for PC activity?

[From Mukhopadhyay, B., Stoddard, S.F., and Wolfe, R.S., *J. Biol. Chem.* **273,** 5155–5166 (1998).]

SELECTED READINGS

Barry, M.J., Enzymes and symmetrical molecules, *Trends Biochem. Sci.* **22,** 228–230 (1997). [Recounts how experiments and insight revealed that the symmetrical citrate molecule can react asymmetrically.]

Gibala, M.J., Young, M.E., and Taegtmeyer, H., Anaplerosis of the citric acid cycle: Role in energy metabolism of heart and skeletal muscle, *Acta Physiol. Scand.* **168,** 657–658 (2000). [Reviews anaplerotic mechanisms and discusses their importance in muscle physiology.]

Huynen M.A., Dandekar, T., and Bork, P., Variation and evolution of the citric-acid cycle: A genomic perspective, *Trends Microbiol.* **7,** 281–291 (1999). [Discusses how genome studies can allow reconstruction of metabolic pathways, even when some enzymes appear to be missing.]

CHAPTER 12

Oxidative Phosphorylation

Humans and other obligate aerobes cannot survive in the absence of oxygen. In animals, even a few minutes without breathing causes irreversible brain damage. Yet the same chemical reactivity that makes oxygen essential for life also makes it potentially toxic. One type of hazard is oxygen free radicals, the partially reduced oxygen molecules, such as O_2^-, that are occasionally produced as by-products of normal oxidative metabolism. These highly reactive molecules and their reaction products can damage the DNA, proteins, and lipids of cells. Some of the body's own defensive players, the white blood cells known as neutrophils and macrophages, intentionally produce oxygen radicals to accelerate the inactivation and death of invading microorganisms (such as the bacteria shown here). The oxygen radicals of the initial onslaught quickly convert to other compounds such as H_2O_2 (hydrogen peroxide) and OCl^- (hypochlorite or bleach), which are the body's natural counterparts to the household chemicals commonly used to kill germs.

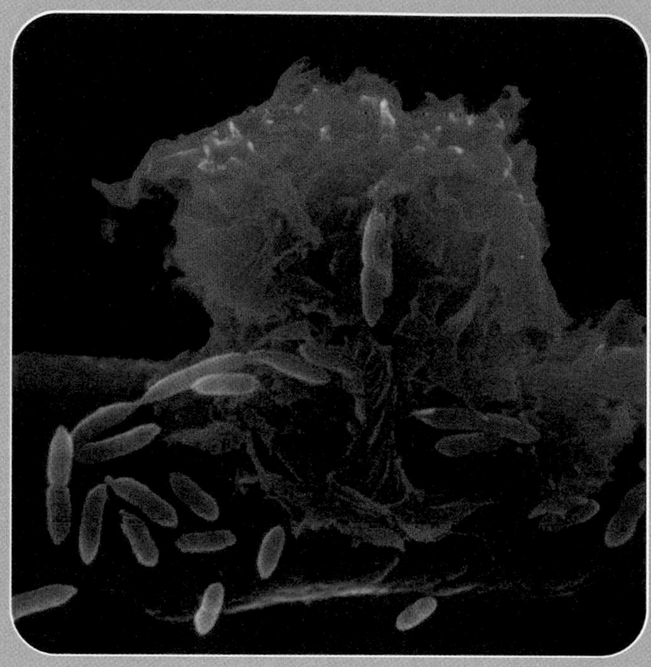

[Dennis Kunkel/Phototake.]

LEARNING OBJECTIVES

1. Understand that an oxidation–reduction reaction involves the transfer of electrons according to the reduction potentials of the reactants.
2. Understand how changes in reduction potential are linked to changes in free energy of a system.
3. Understand that the inner mitochondrial membrane is impermeable to protons and other ionic substances.
4. Understand that Complexes I, III, and IV transfer electrons between redox centers that can undergo reversible one- or two-electron reduction.
5. Understand that the electrons donated by the reduced cofactors NADH and QH_2 are ultimately transferred to O_2.
6. Understand the chemiosmotic theory, which describes how respiration generates a transmembrane proton gradient that supplies free energy to drive ATP synthesis.
7. Understand the structure and function of the two components of the F_1F_o ATP synthase and how they are linked.
8. Understand the binding change mechanism for ATP synthesis.
9. Understand how oxidative phosphorylation is regulated.

THIS CHAPTER IN CONTEXT

This chapter represents a departure from the themes of the preceding two chapters on glucose metabolism (Chapter 10) and the citric acid cycle (Chapter 11), which mostly trace the fates of carbon atoms. This chapter follows the movement of electrons and free energy in oxidative phosphorylation, which is the synthesis of ATP powered by the free energy of reduced compounds produced by other metabolic pathways, including glycolysis and the citric acid cycle. Electron transport and ATP synthesis are carried out by integral membrane proteins (a topic explored in Chapter 8). Understanding the structures and mechanisms of these proteins is the key to understanding the theory of chemiosmosis, a major theme of this chapter. A similar phenomenon underlies photosynthesis, the subject of Chapter 13.

THE OXIDATION OF METABOLIC FUELS SUCH AS GLUCOSE, FATTY acids, and amino acids, and the oxidation of acetyl carbons to CO_2 via the citric acid cycle yields the reduced cofactors NADH and ubiquinol (QH_2). These compounds are forms of energy currency (see Section 9-2), because their reoxidation—ultimately by molecular oxygen in aerobic organisms—is an exergonic reaction. The free energy thereby released is harvested to synthesize ATP, a phenomenon called **oxidative phosphorylation.** In the scheme introduced in Figure 9-10, oxidative phosphorylation represents the final phase of the catabolism of metabolic fuels and the major source of the cell's ATP (Fig. 12-1).

Oxidative phosphorylation differs from the conventional biochemical reactions we have focused on in the previous two chapters. In particular, ATP synthesis is not directly coupled to a single discrete chemical reaction, such as a kinase-catalyzed reaction. Rather, *oxidative phosphorylation is a more indirect process in which free energy is temporarily stored, or conserved, in the form of a gradient of protons before being used to drive ATP synthesis.*

To understand oxidative phosphorylation, we must first consider how the reduced cofactors produced in other metabolic reactions are reoxidized by molecular oxygen. The flow of electrons from reduced compounds such as NADH and QH_2 to an oxidized

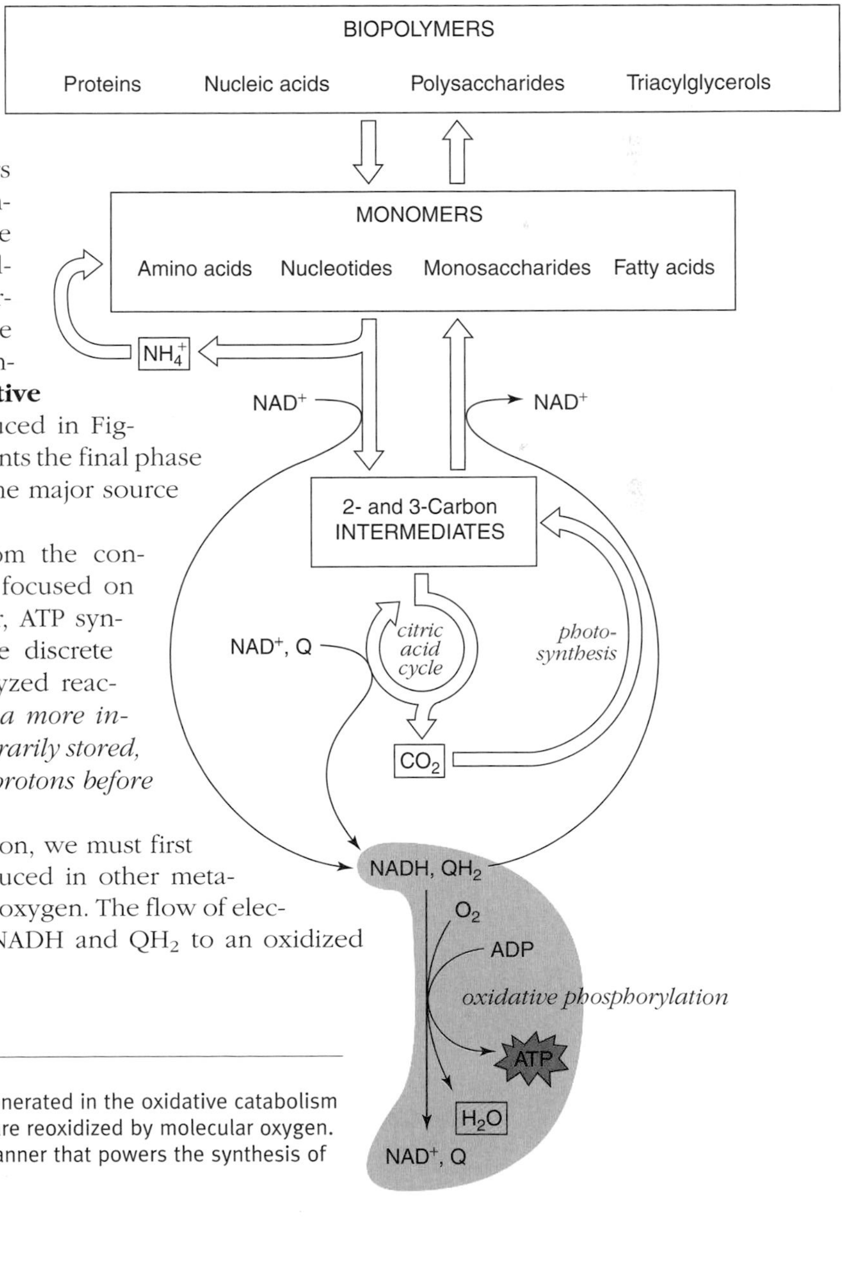

FIGURE 12-1
Oxidative phosphorylation in context.
The reduced cofactors NADH and QH_2, which are generated in the oxidative catabolism of amino acids, monosaccharides, and fatty acids, are reoxidized by molecular oxygen. The free energy of this process is conserved in a manner that powers the synthesis of ATP from ADP + P_i.

compound such as O_2 is a thermodynamically favorable process. We will see that the free energy changes for electron transfer reactions can be quantified by considering the reduction potentials of the chemical species involved. Next, we will track the movements of the electrons through a series of electron carriers that include small molecules as well as prosthetic groups of large integral membrane proteins. In aerobic eukaryotes, these proteins are located in the mitochondrial membrane. As electrons are shuttled from NADH and QH_2 to molecular oxygen, the membrane-bound proteins translocate protons from the inside to the outside of the mitochondria. This process is the first step of chemiosmosis, the formation and dissipation of a transmembrane chemical gradient. Later in this chapter we will consider the thermodynamic implications of proton transport. Finally, we will examine the structure of ATP synthase, the enzyme complex that taps the free energy of the proton gradient to synthesize ATP from ADP + P_i.

1 THE THERMODYNAMICS OF OXIDATION–REDUCTION REACTIONS

Go to **Review Exercise 4/Redox Reactions** for a refresher course in the chemistry of oxidation and reduction and the meaning and use of reduction potentials. The material presented here addresses Learning Objectives 1 and 2.

Oxidation–reduction reactions (or **redox reactions**) are similar to other chemical reactions in which a portion of a molecule—electrons in this case—is transferred. In any oxidation–reduction reaction, one reactant (called the **oxidizing agent** or **oxidant**) is reduced as it gains electrons. The other reactant (called the **reducing agent** or **reductant**) is oxidized as it gives up electrons:

$$A_{oxidized} + B_{reduced} \rightleftharpoons A_{reduced} + B_{oxidized}$$

For example, in the succinate dehydrogenase reaction (Step 6 of the citric acid cycle; see Section 11-2), the two electrons of the reduced $FADH_2$ prosthetic group of the enzyme are transferred to ubiquinone (Q) so that $FADH_2$ is oxidized and ubiquinone is reduced:

$$\underset{(reduced)}{FADH_2} + \underset{(oxidized)}{Q} \rightleftharpoons \underset{(oxidized)}{FAD} + \underset{(reduced)}{QH_2}$$

In this reaction, the two electrons are transferred as H atoms (an H atom consists of a proton and an electron, or H^+ and e^-). In oxidation–reduction reactions involving the cofactor NAD^+, the electron pair takes the form of a hydride ion (H^-, a proton with two electrons). In biological systems, electrons usually travel in pairs, although, as we shall see, they may also be transferred one at a time.

Reduction potential indicates a substance's tendency to accept electrons

The tendency of a substance to accept electrons (to become reduced) or to donate electrons (to become oxidized) can be quantified. Although an oxidation–reduction reaction necessarily requires both an oxidant and a reductant, it is helpful to consider just one substance at a time, that is, a **half-reaction.** Using the example above, the half-reaction for ubiquinone (written as a reduction reaction) is

$$Q + 2\,H^+ + 2\,e^- \rightleftharpoons QH_2$$

(the reverse reaction would describe an oxidation half-reaction).

The affinity of a substance such as ubiquinone for electrons is its **standard reduction potential ($\mathcal{E}^{\circ\prime}$),** which has units of volts (note that the degree and prime symbols indicate a value under standard biochemical conditions where the pressure is 1 atm, the temperature is 25°C, the pH is 7.0, and all species are present at concentrations of 1 M). *The greater the value of $\mathcal{E}^{\circ\prime}$, the greater the tendency of the oxidized form of the substance to accept electrons and become reduced.* The standard reduction potentials of some biological substances are given in Table 12-1.

Like a ΔG value, *the actual reduction potential depends on the concentrations of the oxidized and reduced species.* The actual **reduction potential ($\mathcal{E}$)** is related to the standard reduction potential ($\mathcal{E}^{\circ\prime}$) by the **Nernst equation:**

$$\mathcal{E} = \mathcal{E}^{\circ\prime} + \frac{RT}{n\mathcal{F}} \ln \frac{[\mathrm{A}_{reduced}]}{[\mathrm{A}_{oxidized}]} \qquad [12\text{-}1]$$

R (the gas constant) has a value of 8.3145 $\mathrm{J \cdot K^{-1} \cdot mol^{-1}}$, T is the temperature in Kelvin, n is the number of electrons transferred (1 or 2 in most of

TABLE 12-1 Standard reduction potentials of some biological substances

Half-reaction	$\mathcal{E}^{\circ\prime}$ (V)
$\frac{1}{2}O_2 + 2\,H^+ + 2\,e^- \rightleftharpoons H_2O$	0.815
$SO_4^{2-} + 2\,H^+ + 2\,e^- \rightleftharpoons SO_3^{2-} + H_2O$	0.48
$NO_3^- + 2\,H^+ + 2\,e^- \rightleftharpoons NO_2^- + H_2O$	0.42
Cytochrome a_3 (Fe^{3+}) + e^- $\rightleftharpoons$ cytochrome a_3 (Fe^{2+})	0.385
Cytochrome a (Fe^{3+}) + e^- $\rightleftharpoons$ cytochrome a (Fe^{2+})	0.29
Cytochrome c (Fe^{3+}) + e^- $\rightleftharpoons$ cytochrome c (Fe^{2+})	0.235
Cytochrome c_1 (Fe^{3+}) + e^- $\rightleftharpoons$ cytochrome c_1 (Fe^{2+})	0.22
Cytochrome b (Fe^{3+}) + e^- $\rightleftharpoons$ cytochrome b (Fe^{2+}) (*mitochondrial*)	0.077
Ubiquinone + 2 H^+ + 2 e^- $\rightleftharpoons$ ubiquinol	0.045
Fumarate$^-$ + 2 H^+ + 2 e^- $\rightleftharpoons$ succinate$^-$	0.031
FAD + 2 H^+ + 2 e^- $\rightleftharpoons$ $FADH_2$ (*in flavoproteins*)	~0.
Oxaloacetate$^-$ + 2 H^+ + 2 e^- $\rightleftharpoons$ malate$^-$	−0.166
Pyruvate$^-$ + 2 H^+ + 2 e^- $\rightleftharpoons$ lactate$^-$	−0.185
Acetaldehyde + 2 H^+ + 2 e^- $\rightleftharpoons$ ethanol	−0.197
$S + 2\,H^+ + 2\,e^- \rightleftharpoons H_2S$	−0.23
Lipoic acid + 2 H^+ + 2 e^- $\rightleftharpoons$ dihydrolipoic acid	−0.29
NAD^+ + H^+ + 2 e^- $\rightleftharpoons$ NADH	−0.315
$NADP^+$ + H^+ + 2 e^- $\rightleftharpoons$ NADPH	−0.320
Cystine + 2 H^+ + 2 e^- $\rightleftharpoons$ 2 cysteine	−0.340
Acetoacetate$^-$ + 2 H^+ + 2 e^- $\rightleftharpoons$ 3-hydroxybutyrate$^-$	−0.346
Acetate$^-$ + 3 H^+ + 2 e^- $\rightleftharpoons$ acetaldehyde + H_2O	−0.581

Source: Mostly from Loach, P.A., In Fasman, G.D. (ed.), *Handbook of Biochemistry and Molecular Biology* (3rd ed.), Physical and Chemical Data, Vol. I, pp. 123–130, CRC Press (1976).

the reactions we will encounter), and $\mathcal{F}$ is the **Faraday constant** (96,485 J · V^{-1} · mol^{-1}; it is equivalent to the electrical charge of one mole of electrons). At 25°C (298K), the Nernst equation reduces to

$$\mathcal{E} = \mathcal{E}^{\circ\prime} + \frac{0.026 \text{ V}}{n} \ln \frac{[A_{reduced}]}{[A_{oxidized}]} \quad [12\text{-}2]$$

In fact, for many substances in biological systems, the concentrations of the oxidized and reduced species are similar, so the logarithmic term is small (recall that ln 1 = 0) and $\mathcal{E}$ is close to $\mathcal{E}^{\circ\prime}$.

REVIEW LEARNING OBJECTIVE 1

- Explain why an oxidation–reduction reaction must include both an oxidant and a reductant.
- How does the Nernst equation link $\mathcal{E}$ to $\mathcal{E}^{\circ\prime}$?

Changes in reduction potential and free energy changes

Knowing the reduction potentials of different substances is useful for predicting the movement of electrons between the two substances. When the substances are together in solution or connected by wire in an electrical circuit, *electrons flow spontaneously from the substance with the lower reduction potential to the substance with the higher reduction potential.* For example, in a system containing Q/QH_2 and NAD^+/NADH, we can predict whether electrons will flow from QH_2 to NAD^+ or from NADH to Q. Using the standard reduction potentials given in Table 12-1, we note that $\mathcal{E}^{\circ\prime}$ for NAD^+ (−0.315 V) is lower than $\mathcal{E}^{\circ\prime}$ for ubiquinone (0.045 V). Therefore, NADH will tend to transfer its electrons to ubiquinone; that is, NADH will be oxidized and Q will be reduced.

A complete oxidation–reduction reaction is just a combination of two half-reactions. For the NADH–ubiquinone reaction, the net reaction is the ubiquinone reduction half-reaction (the half-reaction as listed in Table 12-1) combined with the NADH oxidation reaction (the reverse of the half-reaction listed in Table 12-1). Note that because the NAD^+ half-reaction has been reversed to indicate oxidation, the sign of its $\mathcal{E}^{\circ\prime}$ value has been reversed:

$$\text{NADH} \rightleftharpoons \text{NAD}^+ + \text{H}^+ + 2\,e^- \qquad \mathcal{E}^{\circ\prime} = +0.315 \text{ V}$$

$$\text{Q} + 2\,\text{H}^+ + 2\,e^- \rightleftharpoons \text{QH}_2 \qquad \mathcal{E}^{\circ\prime} = 0.045 \text{ V}$$

$$\textit{Net:}\ \ \text{NADH} + \text{Q} + \text{H}^+ \rightleftharpoons \text{NAD}^+ + \text{QH}_2 \qquad \Delta\mathcal{E}^{\circ\prime} = +0.360 \text{ V}$$

When the two half-reactions are added, their reduction potentials are also added, yielding a $\Delta\mathcal{E}^{\circ\prime}$ value. Additional examples are provided in Review Exercise 4.

Not surprisingly, *the larger the difference in $\mathcal{E}$ values (the greater the $\Delta\mathcal{E}$ value), the greater the tendency of electrons to flow from one substance to the other, and the greater the change in free energy of the system.* ΔG is related to $\Delta\mathcal{E}$ as follows:

$$\boxed{\Delta G^{\circ\prime} = -n\mathcal{F}\Delta\mathcal{E}^{\circ\prime}} \quad [12\text{-}3]$$

or

$$\Delta G = -n\mathcal{F}\Delta\mathcal{E} \quad [12\text{-}4]$$

Accordingly, an oxidation–reduction reaction with a large positive $\Delta\mathcal{E}$ value has a large negative value of ΔG (see Sample Calculation 12-1). Depending on the relevant reduction potentials, an oxidation–reduction reaction can release considerable amounts of free energy. This is what happens in the mitochondria, where the reduced cofactors generated by the oxidation of metabolic fuels are reoxidized. The free energy released in this process powers ATP synthesis by oxidative phosphorylation. The relationship between reduction potential and free energy is explained in Review Exercise 4.

Sample Calculation 12-1

Problem

Calculate the standard free energy change for the oxidation of malate by NAD^+. Is this reaction spontaneous under standard conditions?

Solution

First, write the relevant half-reactions, reversing the malate half-reaction (so that it becomes an oxidation reaction) and reversing the sign of its $\mathcal{E}^{\circ\prime}$:

$$\text{Malate} \rightarrow \text{oxaloacetate} + 2\,H^+ + 2\,e^- \qquad \mathcal{E}^{\circ\prime} = +0.166\text{ V}$$

$$NAD^+ + H^+ + 2\,e^- \rightarrow \text{NADH} \qquad \mathcal{E}^{\circ\prime} = -0.315\text{ V}$$

$$\text{Malate} + NAD^+ \rightarrow \text{oxaloacetate} + \text{NADH} + H^+ \qquad \Delta\mathcal{E}^{\circ\prime} = -0.149\text{ V}$$

The $\Delta\mathcal{E}^{\circ\prime}$ for the net reaction is -0.149 V. Use Equation 12-3 to calculate $\Delta G^{\circ\prime}$:

$$\begin{aligned}\Delta G^{\circ\prime} &= -n\mathcal{F}\Delta\mathcal{E}^{\circ\prime}\\ &= -(2)(96{,}485\text{ J}\cdot\text{V}^{-1}\cdot\text{mol}^{-1})(-0.149\text{ V})\\ &= +28{,}750\text{ J}\cdot\text{mol}^{-1}\\ &= +28.8\text{ kJ}\cdot\text{mol}^{-1}\end{aligned}$$

The reaction has a positive value of $\Delta G^{\circ\prime}$ and so is not spontaneous. (*In vivo*, this reaction occurs as Step 8 of the citric acid cycle.)

REVIEW LEARNING OBJECTIVE 2

- Explain how adding the $\mathcal{E}^{\circ\prime}$ values for two half-reactions yields a value of $\Delta\mathcal{E}^{\circ\prime}$ and $\Delta G^{\circ\prime}$ for an oxidation–reduction reaction.

2 MITOCHONDRIAL ELECTRON TRANSPORT

In aerobic organisms, the NADH and ubiquinol produced by glycolysis, the citric acid cycle, and other metabolic pathways are ultimately reoxidized by molecular oxygen, a process called **respiration.** The standard reduction potential of $+0.815$ V for the reduction of O_2 to H_2O indicates that O_2 is a more effective oxidizing agent than any other biological compound (see Table 12-1). The oxidation of NADH by O_2, that is, the transfer of electrons from NADH directly to O_2, would release a large amount of free energy, but this

Go to **Exercise 17/Oxidative Phosphorylation** for a summary of the functions of the electron-transporting protein complexes that make up the mitochondrial respiratory chain. The animations and questions based on them address Learning Objectives 4, 5, and 6.

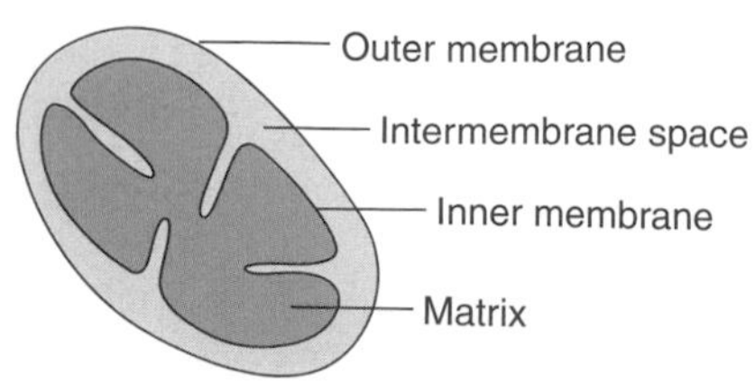

FIGURE 12-2

Model of mitochondrial structure. The relatively impermeable inner mitochondrial membrane encloses the protein-rich matrix. The intermembrane space has an ionic composition similar to that of the cytosol because the outer mitochondrial membrane is permeable to substances with masses less than about 10 kD.

reaction does not occur in a single step. Instead, *electrons are shuttled from NADH to O_2 in a multistep process that offers several opportunities to conserve the free energy of oxidation.* In eukaryotes, all the steps of oxidative phosphorylation are carried out by a series of membrane-bound protein complexes in the mitochondria (in prokaryotes, plasma membrane proteins perform similar functions). The following sections describe how electrons flow through this "respiratory chain" from reduced cofactors to oxygen.

Mitochondrial anatomy

In accordance with its origin as a bacterial symbiont, the **mitochondrion** (plural, mitochondria) has two membranes. The outer membrane, analogous to the outer membrane of some bacteria, is relatively porous due to the presence of porin-like proteins that permit the transmembrane diffusion of substances with masses up to about 10 kD (see Section 8-3 for an example of porin structure and function). The inner membrane has a convoluted architecture that encloses a space called the **mitochondrial matrix.** Because the inner mitochondrial membrane prevents the transmembrane movements of ions and small molecules (except via specific transport proteins), the composition of the matrix differs from that of the space between the inner and outer membranes. In fact, the ionic composition of the **intermembrane space** is considered to be equivalent to that of the cytosol due to the presence of the porins in the outer mitochondrial membrane (Fig. 12-2). Exercise 17 includes a review of mitochondrial structure.

Mitochondria are customarily shown as kidney-shaped organelles with the inner mitochondrial membrane forming a system of baffles called **cristae** (Fig. 12-3a). However, **electron tomography,** a technique for visualizing cellular structures in three dimensions by analyzing micrographs of sequential cell slices, reveals that mitochondria are highly variable structures. For example, the cristae may be irregular and bulbous rather than planar, and may make several tubular connections with the rest of the inner mitochondrial membrane (Fig. 12-3b). Moreover, a cell may contain hundreds to thousands

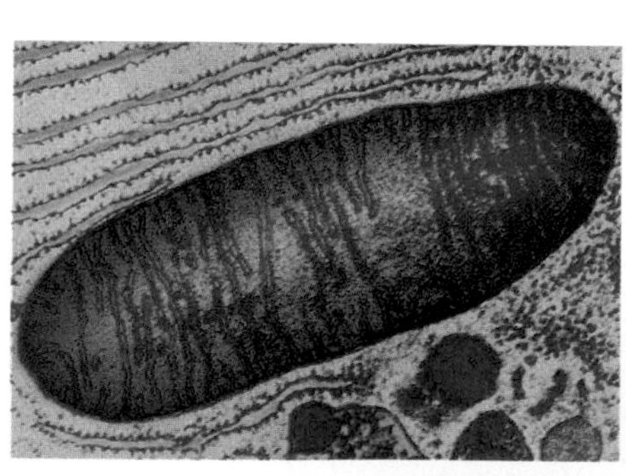

(a)

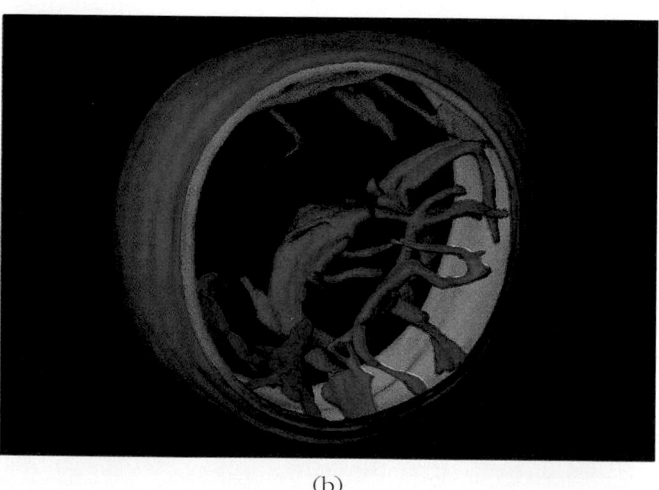

(b)

FIGURE 12-3

Images of mitochondria.
(a) Conventional electron micrograph showing cristae as a system of planar baffles. [*Courtesy K. Porter/Photo Researchers.*] (b) Three-dimensional reconstruction of a mitochondrion by electron tomography, showing irregular tubular cristae. [*Courtesy Carmen Mannella, Resource for the Visualization of Biological Complexity, Wadsworth Center, and NCRR/NIH.*]

of discrete bacteria-shaped mitochondria, or a single tubular organelle may take the form of an extended network with many branches and interconnections (Fig. 12-4).

Reflecting its ancient origin as a free-living organism, the mitochondrion has its own genome and protein-synthesizing machinery consisting of mitochondrially encoded rRNA and tRNA. The mitochondrial genome encodes 13 proteins, all of which are components of the respiratory chain complexes. This is only a small subset of the proteins required for mitochondrial function; the other respiratory chain proteins, matrix enzymes, transporters, and so on are encoded by the cell's nuclear genome, synthesized in the cytosol, and imported into the mitochondria (across one or both membranes) by special mechanisms.

Much of the cell's NADH and QH_2 is generated by the citric acid cycle in the mitochondrial matrix. Fatty acid oxidation also takes place largely in the matrix and yields NADH and QH_2. These reduced cofactors transfer their electrons to the protein complexes of the respiratory electron transport chain, which are tightly associated with the inner mitochondrial membrane. However, NADH produced by glycolysis and other oxidative processes in the cytosol cannot directly reach the respiratory chain. There is no transport protein that can ferry NADH across the inner mitochondrial membrane. Instead, "reducing equivalents" are imported into the matrix by the chemical reactions of systems such as the malate–aspartate shuttle system (Fig. 12-5).

Mitochondria also need a mechanism to export ATP and to import ADP and P_i, since most of the cell's ATP is generated in the matrix by oxidative phosphorylation and is consumed in the cytosol. A transport protein called the adenine nucleotide translocase exports ATP and imports ADP, binding one or the other and changing its conformation to release the bound nu-

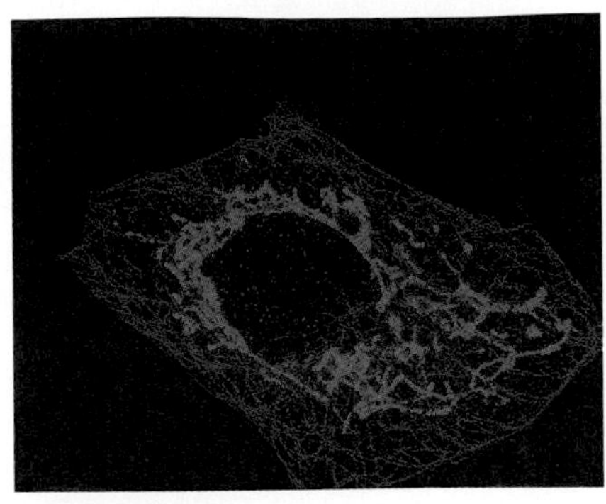

FIGURE 12-4

Micrograph of a mitochondrial network. The tubular mitochondria form a network (labeled with a green fluorescent dye) in a mammalian fibroblast. The remainder of the cytosol is delineated by microtubules (labeled with a red fluorescent dye). [*Courtesy Michael P. Yaffe. From* Science ***283,*** *1493–1497 (1991), Fig. 1.*]

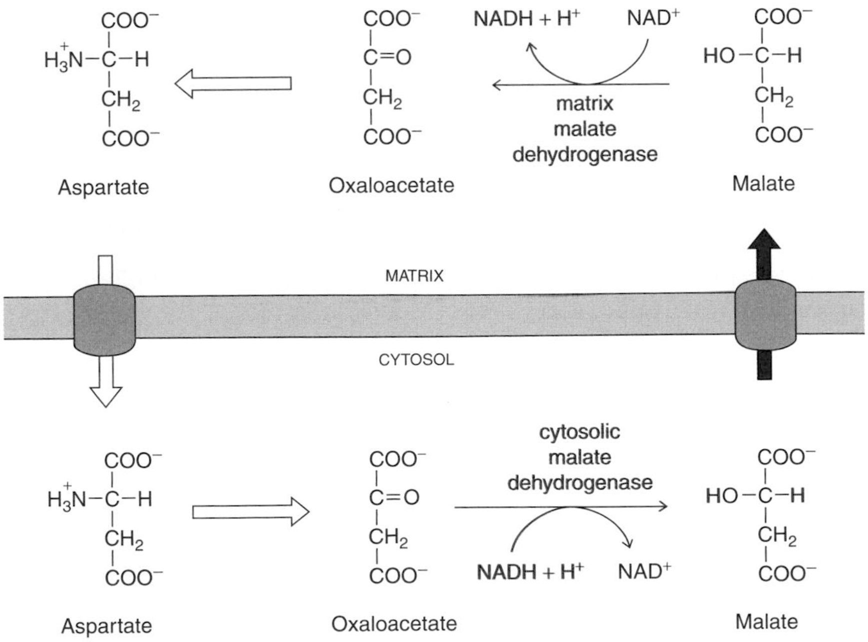

FIGURE 12-5

The malate–aspartate shuttle system.
Cytosolic oxaloacetate is reduced to malate for transport into mitochondria. Malate is then reoxidized in the matrix. The net result is the transfer of "reducing equivalents" from the cytosol to the matrix. Mitochondrial oxaloacetate can be exported back to the cytosol after being converted to aspartate by an aminotransferase.

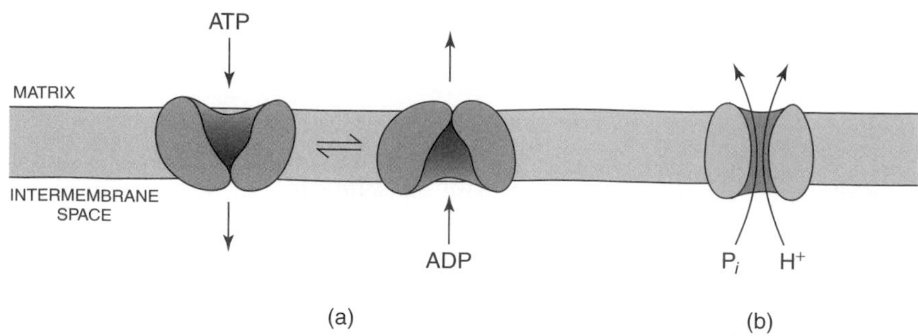

FIGURE 12-6

Mitochondrial transport systems.
(a) The adenine nucleotide translocase binds either ATP or ADP and changes its conformation to release the nucleotide on the opposite side of the inner mitochondrial membrane. This transporter can therefore export ATP and import ADP. (b) A P_i-H^+ symport protein permits the simultaneous movement of inorganic phosphate and a proton into the mitochondrial matrix.

cleotide on the other side of the membrane (Fig. 12-6a). Inorganic phosphate, a substrate for oxidative phosphorylation, is imported from the cytosol in symport with H^+ (Fig. 12-6b).  The functions of these shuttle systems are summarized in Exercise 17.

The protein complexes that carry out electron transport and ATP synthesis are oriented in the inner mitochondrial membrane so that they can bind the NADH, ADP, and P_i present in the matrix. Their orientation in the inner mitochondrial membrane is also essential to their ability to either generate or dissipate a transmembrane gradient of protons, since the membrane itself is impermeable to protons.

REVIEW LEARNING OBJECTIVE 3

- Why are transport proteins present in the inner mitochondrial membrane?

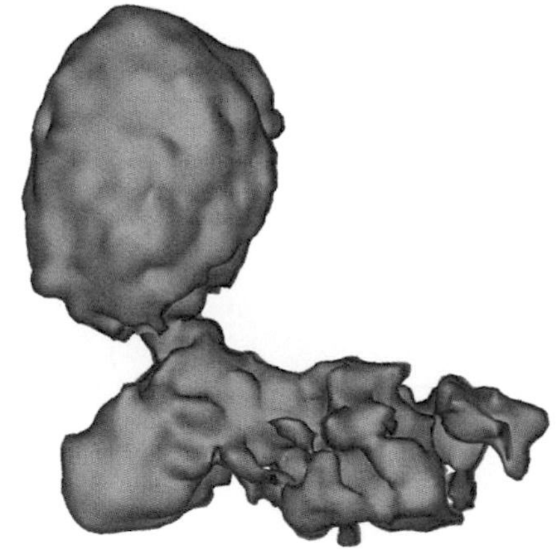

FIGURE 12-7

Structure of Complex I determined by cryoelectron microscopy.
A narrow stalk links the soluble and membrane-embedded portions of the mammalian protein complex. The resolution of this model is about 22 Å. [*Courtesy Nikolaus Grigorieff. From* J. Cell. Biol. ***152****, F1–F10. (2001).*]

Complex I transfers electrons from NADH to ubiquinone

The path electrons travel through the respiratory chain begins with Complex I, also called NADH:ubiquinone oxidoreductase or NADH dehydrogenase. This enzyme catalyzes the transfer of a pair of electrons from NADH to ubiquinone:

$$\text{NADH} + \text{H}^+ + \text{Q} \rightleftharpoons \text{NAD}^+ + \text{QH}_2$$

Complex I is the largest of the electron transport proteins in the mitochondrial respiratory chain, with 43 different subunits and a total mass of about 900 kD. Its three-dimensional shape has been visualized by **cryoelectron microscopy.** This technique analyzes micrographs of single particles embedded in crystalline ice and is useful for constructing models of large complexes when X-ray crystallography is not practical. Overall, the complex is L-shaped, with a soluble globular arm connected by a narrow stalk to a membrane-embedded foot (Fig. 12-7). The stalk is only about 30 Å in diameter, but this is large enough to provide a pathway for electrons to travel between an NADH-binding site in the arm and a quinone-binding site in the foot.

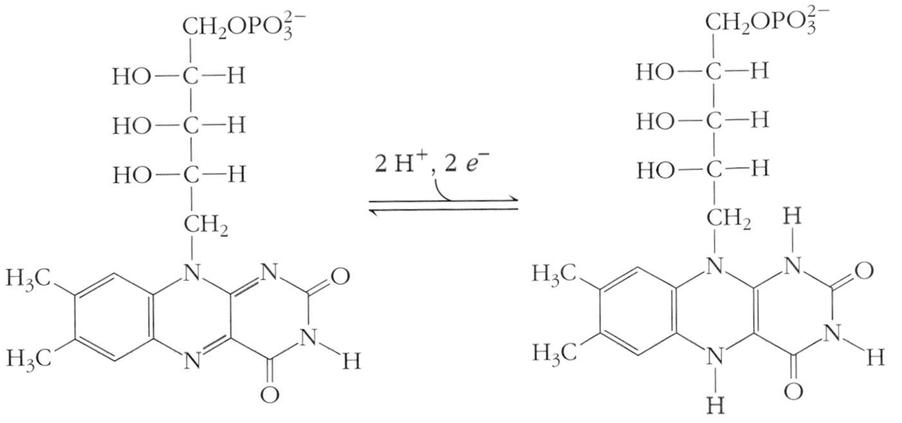

$2\,H^+, 2\,e^-$

Flavin mononucleotide (FMN)

$FMNH_2$

FIGURE 12-8

Flavin mononucleotide (FMN).
This prosthetic group resembles flavin adenine dinucleotide (FAD; see Fig. 11-5) but lacks the ADP group of FAD. The two-electron reduction of FMN yields $FMNH_2$.

Complex I includes several prosthetic groups that participate in electron transfer, undergoing reduction as they receive electrons and becoming oxidized as they give up their electrons to the next group. All these groups, or **redox centers,** appear to have reduction potentials approximately between the reduction potentials of NAD^+ ($\mathcal{E}^{\circ\prime} = -0.315$ V) and ubiquinone ($\mathcal{E}^{\circ\prime} = +0.045$ V). This allows them to form a chain where the electrons travel a path of increasing reduction potential. The redox centers do not need to be in intimate contact with each other, as they would be if the transferred group were a larger chemical entity. An electron can move between redox centers up to 14 Å apart by "tunneling" through the covalent bonds of the protein.

The two electrons donated by NADH are first picked up by flavin mononucleotide (FMN; Fig. 12-8). This noncovalently bound prosthetic group then transfers the electrons, one at a time, to a second type of redox center, an iron-sulfur (Fe-S) cluster. Complex I bears six to eight of these prosthetic groups, which contain equal numbers of iron and sulfide ions (Fig. 12-9). Unlike the electron carriers we have introduced so far, Fe-S clusters are one-electron carriers. They have an oxidation state of either +3 (oxidized) or +2 (reduced), regardless of the number of Fe atoms in the cluster (each cluster is a conjugated structure that functions as a single unit). Electrons travel between several Fe-S clusters before reaching ubiquinone. Like FMN, ubiquinone is a two-electron carrier, but it accepts one electron at a time from an Fe-S donor. Iron-sulfur clusters may be among the most ancient of electron carriers, dating from a time when the earth's abundant iron and sulfur were major players in prebiotic chemical reactions.

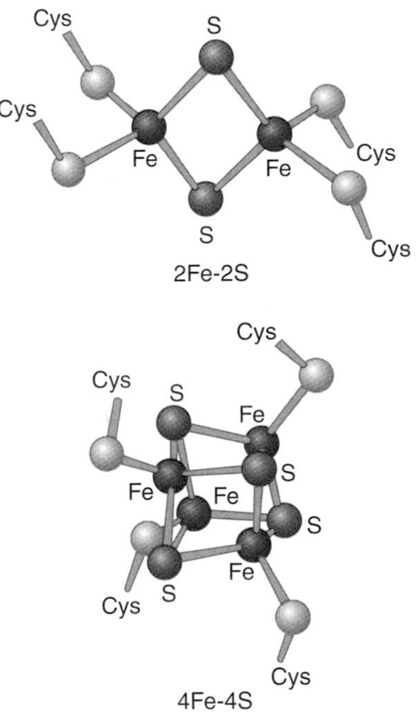

FIGURE 12-9

Iron-sulfur clusters.
Although some Fe-S clusters contain up to eight Fe atoms, the most common are the 2Fe-2S and 4Fe-4S clusters. In all cases, the iron-sulfur clusters are coordinated by the S atoms of Cys side chains.

As electrons are transferred from NADH to ubiquinone, Complex I transfers four protons from the matrix to the intermembrane space. The exact mechanism is not understood, in part because the fine structure of Complex I is not known. Proton movement is probably accomplished by protein conformational changes that occur as redox groups are transiently reduced and reoxidized. The proton transport pathway is not a pore in the conventional sense (see Section 8-3) but probably takes the form of a **proton wire,** a series of hydrogen-bonded protein groups plus water molecules that form a chain through which a proton can be rapidly relayed (the proton jumping among water molecules is shown in Fig. 2-14). Note that in this relay mech-

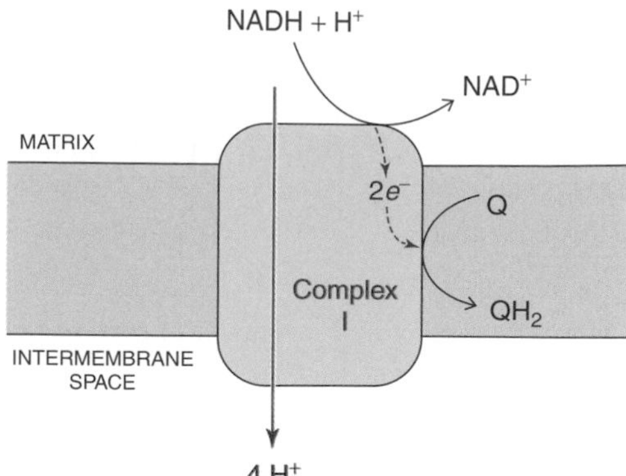

FIGURE 12-10
Complex I function.
As two electrons from the water-soluble NADH are transferred to the lipid-soluble ubiquinone, four protons are translocated from the matrix into the intermembrane space.

anism, the protons taken up from the matrix are not the same ones that are released into the intermembrane space. The reactions of Complex I are summarized in Figure 12-10.

Other oxidation reactions contribute to the ubiquinol pool

The reduced quinone product of the Complex I reaction joins a pool of quinones that are soluble in the inner mitochondrial membrane by virtue of their long hydrophobic isoprenoid tails (see page 285). *The pool of reduced quinones is augmented by the activity of other oxidation–reduction reactions.* One of these is catalyzed by succinate dehydrogenase, which carries out Step 6 of the citric acid cycle (see Section 11-2).

$$\text{Succinate} + \text{Q} \rightleftharpoons \text{fumarate} + \text{QH}_2$$

Succinate dehydrogenase is the only one of the citric acid cycle enzymes that is not soluble in the mitochondrial matrix; it is embedded in the inner membrane. Like the other respiratory complexes, it contains several redox centers, including an FAD group. Succinate dehydrogenase is also called Complex II of the mitochondrial respiratory chain. However, it is more like a tributary because it does not undertake proton translocation and therefore does not directly contribute the free energy of its oxidation–reduction reaction toward ATP synthesis. Nevertheless, it does feed reducing equivalents as ubiquinol into the electron transport chain (Fig. 12-11a).

Electrons from cytosolic NADH can also enter the mitochondrial ubiquinol pool through the actions of a cytosolic and a mitochondrial glycerol-3-phosphate dehydrogenase (Fig. 12-11b). This system, which shuttles electrons from NADH to ubiquinol, bypasses Complex I. In addition, some of the electrons released by the oxidation of fatty acids in the matrix are funneled into the quinone pool in the inner mitochondrial membrane.

Complex III transfers electrons from ubiquinol to cytochrome *c*

Ubiquinol is reoxidized by Complex III, an integral membrane protein with 11 subunits in each of its two monomeric units. Complex III, also called ubiquinol:cytochrome *c* oxidoreductase or cytochrome bc_1, transfers electrons to the peripheral membrane protein cytochrome *c*. **Cytochromes** are proteins with heme prosthetic groups. The name *cytochrome* literally means "cell color"; cytochromes are largely responsible for the purplish-brown color of mitochondria. Cytochromes are commonly named with a letter (*a, b,* or *c*) indicating the exact structure of the porphyrin ring of their heme group (Fig.

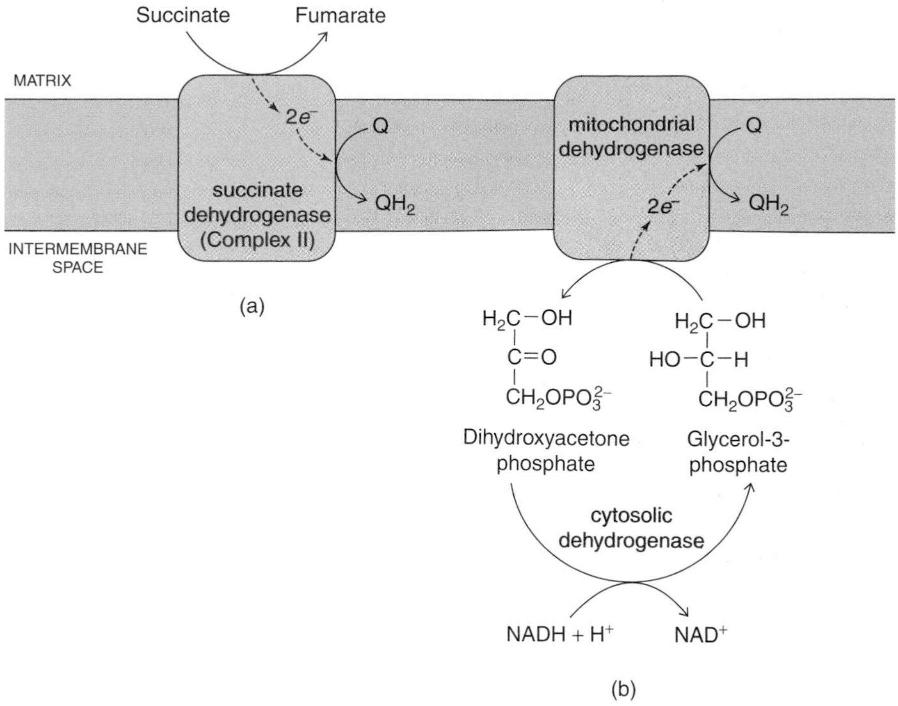

FIGURE 12-11

Reactions that contribute to the ubiquinol pool.
(a) The succinate dehydrogenase (Complex II) reaction transfers electrons to the pool of reduced ubiquinone in the inner mitochondrial membrane. (b) In the glycerol-3-phosphate shuttle system, electrons from cytosolic NADH are used by a cytosolic glycerol-3-phosphate dehydrogenase to reduce dihydroxyacetone phosphate to glycerol-3-phosphate. The mitochondrial enzyme, embedded in the inner membrane, then reoxidizes the glycerol-3-phosphate, ultimately transferring the two electrons to the membrane ubiquinone pool.

12-12). The structure of the heme group and the surrounding protein microenvironment influence the protein's absorption spectrum. They also determine the reduction potentials of cytochromes, which range from about −0.080 V to about +0.385 V.

Unlike the heme prosthetic groups of hemoglobin and myoglobin (see Section 4-4), the heme groups of cytochromes undergo reversible one-electron reduction, with the central Fe atom cycling between the Fe^{3+} (oxidized) and Fe^{2+} (reduced) states. Consequently, the net reaction for Complex III, in which two electrons are transferred, includes two cytochrome *c* proteins:

$$QH_2 + 2 \text{ cytochrome } c^{3+} \rightleftharpoons Q + 2 \text{ cytochrome } c^{2+} + 2\,H^+$$

Complex III itself contains two cytochromes (cytochrome *b* and cytochrome c_1) that are integral membrane proteins. These two proteins, along with an iron-sulfur protein (also called the Rieske protein), form the functional core of Complex III (these same three subunits are the only ones that have homologs in the corresponding bacterial respiratory complex). Altogether, each monomer of Complex III is anchored in the membrane by 13 transmembrane α helices (Fig. 12-13).

The flow of electrons through Complex III is complicated, in part because the two electrons donated by ubiquinol must split up in order to travel through a series of one-electron carriers that includes the 2Fe-2S cluster of the iron-sulfur protein, cytochrome c_1, and cytochrome *b* (which actually contains two heme groups with slightly different reduction potentials). Except for

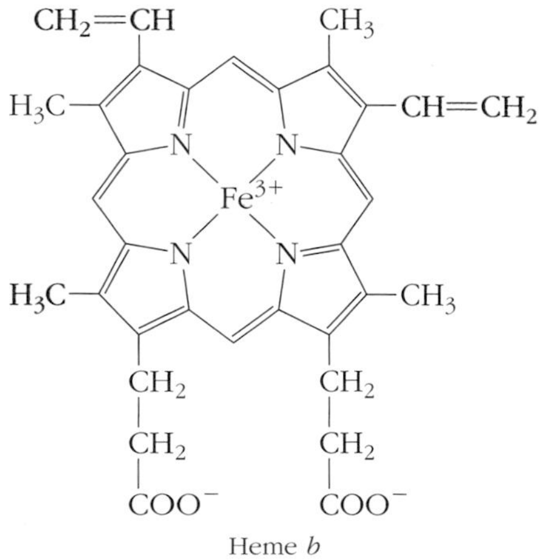

Heme *b*

FIGURE 12-12

The heme group of a *b* cytochrome.
The planar porphyrin ring surrounds a central Fe atom (shown here in its oxidized state). The heme substituent groups that are colored blue differ in the *a* and *c* cytochromes (the heme group of hemoglobin and myoglobin has the *b* structure; see page 116).

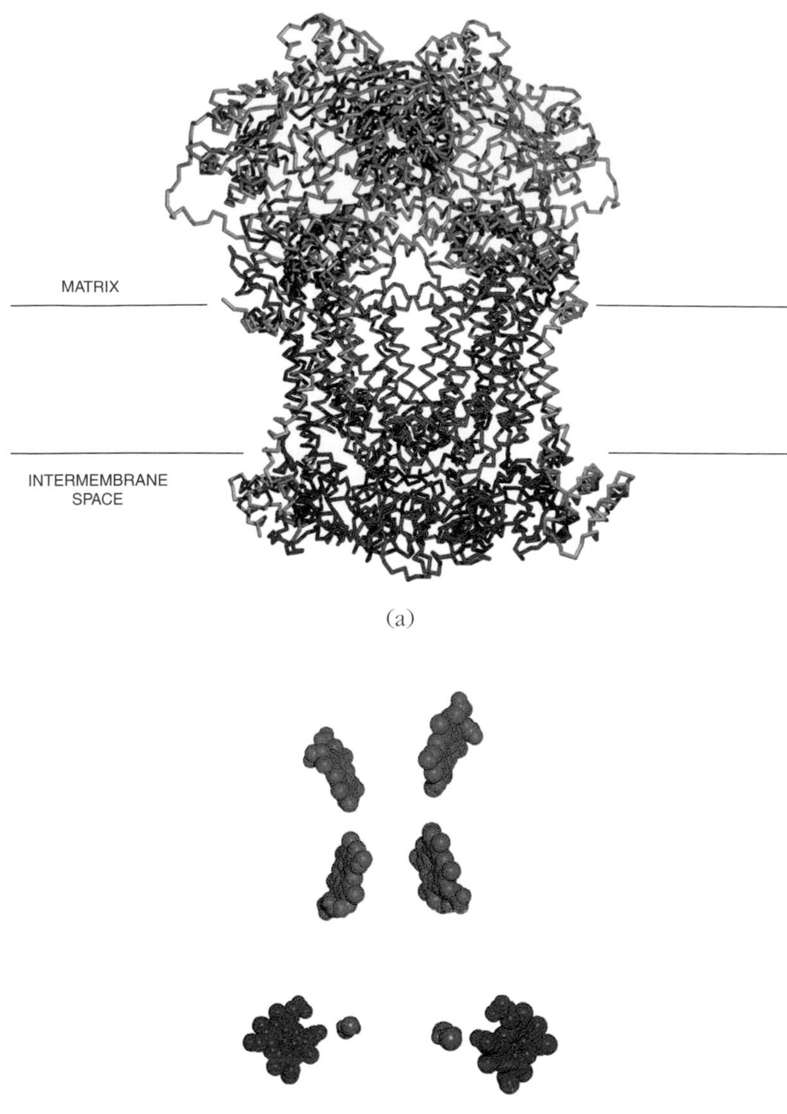

FIGURE 12-13

X-Ray structure of mammalian Complex III.

(a) Backbone model. Eight of the 13 transmembrane helices in each monomer of the dimeric complex are contributed by cytochrome *b* (yellow). The iron-sulfur protein (orange) and cytochrome c_1 (red) project into the intermembrane space. The approximate position of the membrane is indicated. (b) Arrangement of prosthetic groups. The two heme groups of cytochrome *b* (yellow) and the heme group of cytochrome c_1 (red), along with the iron-sulfur cluster (orange), provide a pathway for electrons between ubiquinol (in the membrane) and cytochrome *c* (in the intermembrane space). [*Structure (pdb 1BE3) determined by S. Iwata, J.W. Lee, K. Okada, J.K. Lee, M. Iwata, S. Ramaswamy, and B.K. Jap.*]

the 2Fe-2S cluster, all the redox centers are arranged in such a way that electrons can tunnel from one to another. The iron-sulfur protein must change its conformation by rotating and moving about 22 Å in order to pick up and deliver an electron. Further complicating the picture is the fact that each monomeric unit of Complex III has two active sites where quinone cofactors undergo reduction and oxidation.

The circuitous route of electrons from ubiquinol to cytochrome *c* is described by the two-round **Q cycle,** diagrammed in Figure 12-14. *The net result of the Q cycle is that two electrons from QH_2 reduce two molecules of cytochrome c. In addition, four protons are translocated to the intermembrane space, two from QH_2 in the first round of the Q cycle and two from QH_2 in the second round.* This proton movement contributes to the transmembrane proton gradient. The reactions of Complex III are summarized in Figure 12-15.

Complex IV oxidizes cytochrome *c* and reduces O_2

Just as ubiquinone ferries electrons from Complex I (and other enzymes) to Complex III, cytochrome *c* ferries electrons between Complexes III and IV.

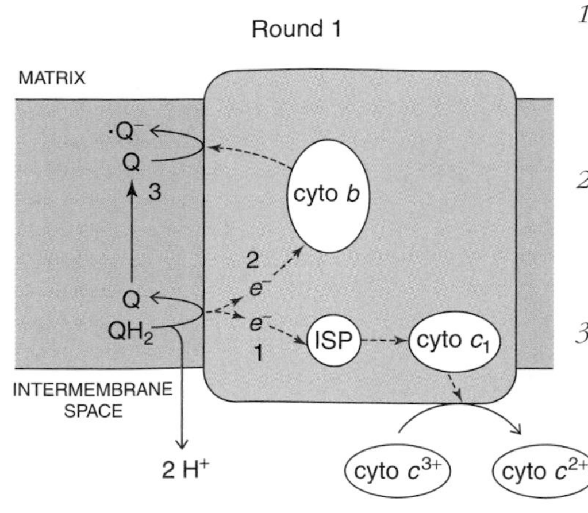

1. *In the first round, QH_2 donates one electron to the iron-sulfur protein (ISP). The electron then travels to cytochrome c_1 and then to cytochrome c.*

2. *QH_2 donates its other electron to cytochrome b. The two protons from QH_2 are released into the intermembrane space.*

3. *The oxidized ubiquinone diffuses to another quinone-binding site, where it accepts the electron from cytochrome b, becoming a half-reduced semiquinone ($\cdot Q^-$).*

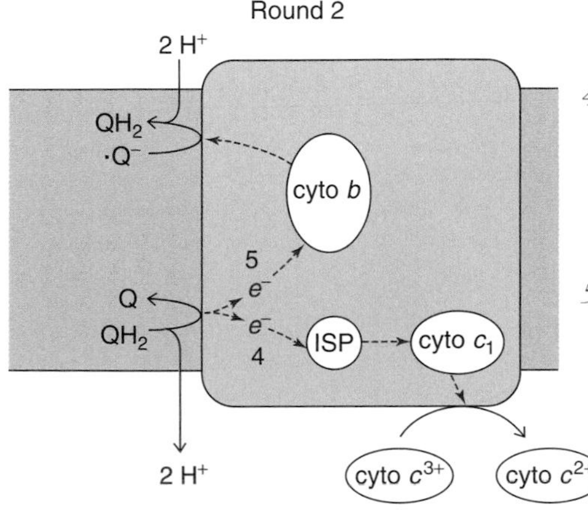

4. *In the second round, a second QH_2 surrenders its two electrons to Complex III and its two protons to the intermembrane space. One electron goes to reduce cytochtome c.*

5. *The other electron goes to cytochrome b and then to the waiting semiquinone produced in the first part of the cycle. This step regenerates QH_2, using protons from the matrix.*

FIGURE 12-14

The Q cycle.

Unlike ubiquinone and the other proteins of the respiratory chain, cytochrome *c* is soluble in the intermembrane space. Several Lys residues on its surface help it bind to Complex III, where it is reduced, and to Complex IV, where it is oxidized (Fig. 12-16).

Complex IV, also called cytochrome *c* oxidase, is the last enzyme to deal with the electrons derived from the oxidation of metabolic fuels. Four electrons

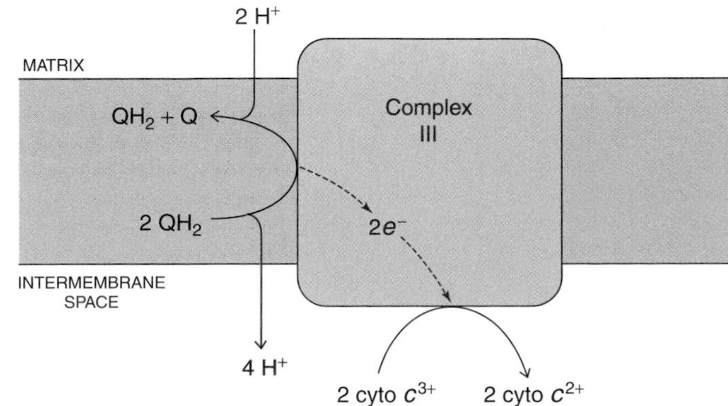

FIGURE 12-15

Complex III function.

For every two electrons that pass from ubiquinol to cytochrome *c*, four protons are translocated to the intermembrane space.

FIGURE 12-16

Cytochrome *c* from tuna.
This peripheral membrane protein transfers one electron at a time from Complex III to Complex IV. Lys side chains (purple) on the protein surface are involved in protein–protein contacts. The heme group is shown in red. [*Structure (pdb 5CYT) determined by T. Takano.*]

delivered by cytochrome *c* are consumed in the reduction of molecular oxygen to water:

$$4\ \text{cytochrome}\ c^{2+} + O_2 + 4\ H^+ \rightarrow 4\ \text{cytochrome}\ c^{3+} + 2\ H_2O$$

The redox centers of mammalian Complex IV include heme groups and copper ions situated among the 13 subunits in each half of the dimeric complex (Fig. 12-17).

Each electron travels from cytochrome *c* to the Cu_A redox center, which has two copper ions, and then to a heme *a* group. From there it travels to a binuclear center consisting of the iron atom of heme a_3 and a copper ion (Cu_B). The four-electron reduction of O_2 occurs at the Fe-Cu binuclear center. One possible sequence of reaction intermediates is shown in Figure 12-18. Note that the chemical reduction of O_2 to H_2O consumes four protons from the mitochondrial matrix.

Cytochrome *c* oxidase also relays four additional protons from the matrix to the intermembrane space (two protons for every pair of electrons).

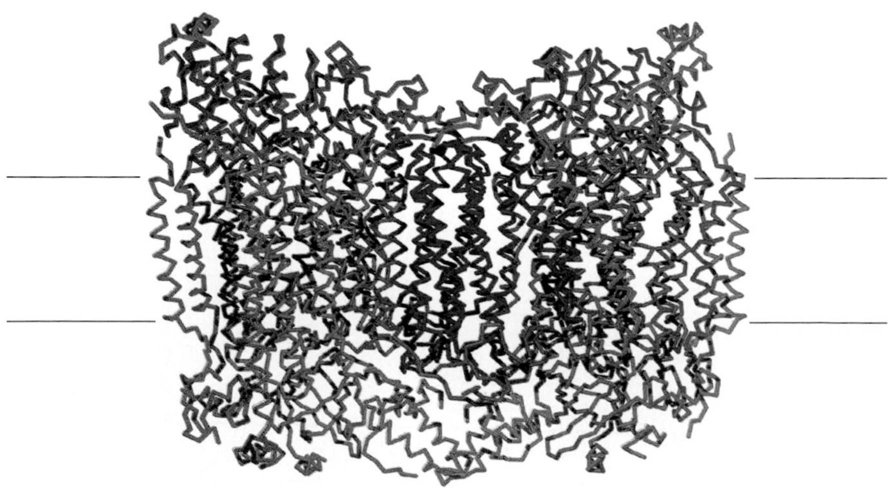

FIGURE 12-17

X-Ray structure of cytochrome *c* oxidase.
The 13 subunits in each monomeric half of the mammalian complex comprise 28 transmembrane α helices. [*Structure (pdb 2OCC) determined by T. Tsukihara and M. Yao.*]

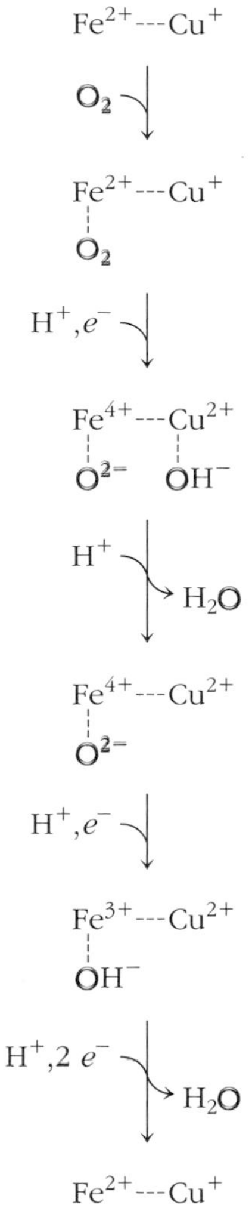

FIGURE 12-18

A proposed model for the cytochrome *c* oxidase reaction.
Although the exact sequence of proton and electron transfers is not known, the reaction intermediates shown here are inferred from spectroscopic and other evidence. An enzyme tyrosine radical probably plays a role in electron transfer.

Fe^{2+}---Cu^{+}

O_2 ↓

Fe^{2+}---Cu^{+}
O_2

H^+, e^- ↓

Fe^{4+}---Cu^{2+}
O^{2-} OH^-

H^+ ↓ H_2O

Fe^{4+}---Cu^{2+}
O^{2-}

H^+, e^- ↓

Fe^{3+}---Cu^{2+}
OH^-

$H^+, 2\,e^-$ ↓ H_2O

Fe^{2+}---Cu^{+}

The protein complex appears to harbor two proton wires spanning the 50 Å distance between the matrix and intermembrane faces of the protein (these proton wires are distinct from the one that delivers H^+ ions to the oxygen-reducing active site). Site-directed mutagenesis has implicated some Arg and Glu side chains as components of the protein wires. Protons are probably relayed when the protein changes its conformation in response to changes in its oxidation state. This notion is not at all far-fetched; recall that the conformational changes in hemoglobin are triggered by changes in the geometry of the heme group upon O_2 binding (see Section 4-5). *The production of water and the proton relays both contribute to the formation of a proton gradient across the inner mitochondrial membrane* (Fig. 12-19). **Exercise 17 traces the movements of electrons and protons through all the mitochondrial electron transport complexes.**

REVIEW LEARNING OBJECTIVES 4 AND 5

- List the different types of redox centers in the respiratory electron transport chain.
- Which are one-electron carriers and which are two-electron carriers?
- How are electrons delivered to Complexes I, III, and IV?
- Why is oxygen the final electron acceptor of the respiratory chain?
- What can you conclude about the reduction potentials of the series of electron carriers that constitute the electron transport chain between NADH and O_2?

3 CHEMIOSMOSIS

The electrons collected from metabolic fuels during their oxidation are now fully disposed of in the reduction of O_2 to H_2O. However, their free energy has been conserved. How much free energy is potentially available? Assuming standard conditions, we can calculate $\Delta\mathcal{E}^{\circ\prime}$ and $\Delta G^{\circ\prime}$ (using Equation 12-3)

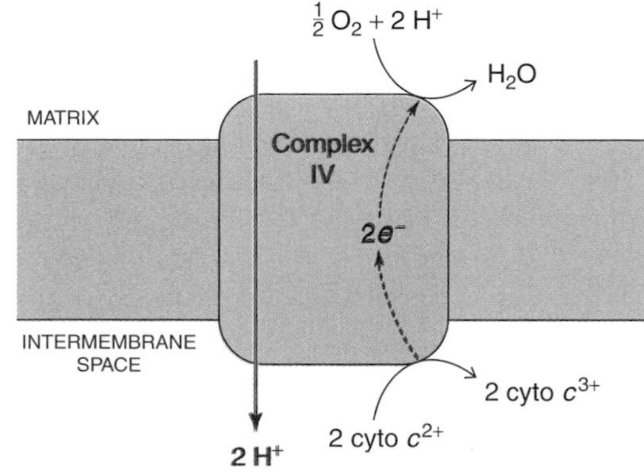

FIGURE 12-19

Complex IV function.
For every two electrons donated by cytochrome *c*, two protons are translocated to the intermembrane space. Two protons from the matrix are also consumed in the reaction $1/2\ O_2 \rightarrow H_2O$ (the full reduction of O_2 requires four electrons).

TABLE 12-2 Free energy available from mitochondrial electron transport reactions

	$\Delta\mathcal{E}^{\circ\prime}$	$\Delta G^{\circ\prime}$
Complex I	0.360 V	$-69.5\ \text{kJ}\cdot\text{mol}^{-1}$
Complex III	0.190 V	$-36.7\ \text{kJ}\cdot\text{mol}^{-1}$
Complex IV	0.580 V	$-112.0\ \text{kJ}\cdot\text{mol}^{-1}$

for the reactions catalyzed by Complexes I, III, and IV. These values are listed in Table 12-2. Each of the three respiratory complexes theoretically releases enough free energy to drive the endergonic phosphorylation of ADP to form ATP ($\Delta G^{\circ\prime} = +30.5\ \text{kJ}\cdot\text{mol}^{-1}$).

Chemiosmosis links electron transport and oxidative phosphorylation

Until the 1960s, the connection between respiratory electron transport (measured as O_2 consumption) and ATP synthesis was a mystery. Credit for discovering the connection belongs primarily to Peter Mitchell, who was inspired by his work on mitochondrial phosphate transport and recognized the importance of compartmentation in biological systems. Mitchell's **chemiosmotic theory** proposed that the proton-translocating activity of the electron transport complexes in the inner mitochondrial membrane generates a proton gradient across the membrane. The protons cannot diffuse back into the matrix, because the membrane is impermeable to ions. *The imbalance of protons represents a source of free energy, also called a* ***protonmotive force,*** *that can drive the activity of an ATP synthase.*

We now know that for each pair of electrons that flow through Complexes I, III, and IV, 10 protons are translocated from the matrix to the intermembrane space (which is ionically equivalent to the cytosol). In bacteria, electron transport complexes in the plasma membrane translocate protons from the cytosol to the cell exterior. Mitchell's theory of chemiosmosis actually explains more than just aerobic respiration. It also applies to systems where the energy from sunlight is used to generate a transmembrane proton gradient (this aspect of photosynthesis is described in Section 13-2).

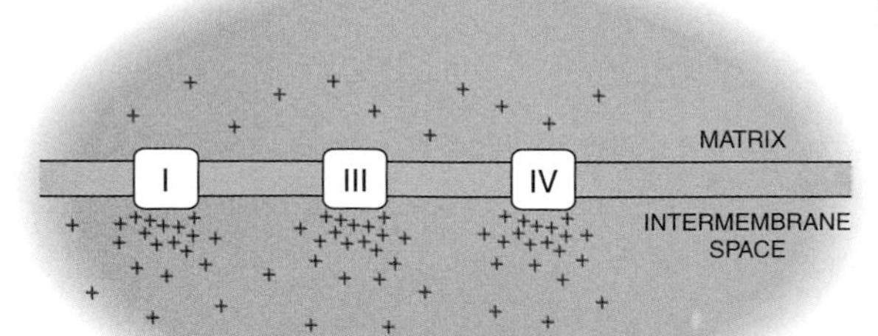

FIGURE 12-20

Generation of a proton gradient. During the oxidation–reduction reactions catalyzed by mitochondrial Complexes I, III, and IV, protons (represented by positive charges) are translocated out of the matrix into the intermembrane space. This creates an imbalance in both proton concentration and electrical charge.

The proton gradient is an electrochemical gradient

When the mitochondrial complexes translocate protons across the inner mitochondrial membrane, the concentration of H^+ outside increases and the concentration of H^+ inside decreases (Fig. 12-20). *This imbalance of protons, a nonequilibrium state, has an associated free energy (the force that would restore the system to equilibrium).* The free energy of the proton gradient has two components, reflecting the difference in the concentration of the chemical species and the difference in electrical charge of the positively charged protons (for this reason, the mitochondrial proton gradient is referred to as an electrochemical gradient rather than a simple concentration gradient). The free energy change for generating the chemical imbalance of protons is

$$\Delta G = RT \ln \frac{[H^+_{out}]}{[H^+_{in}]} \qquad [12\text{-}5]$$

The pH ($-\log [H^+]$) of the intermembrane space (*out*) is typically about 0.75 units less than the pH of the matrix (*in*).

The free energy change for generating the electrical imbalance of protons is

$$\Delta G = Z\mathcal{F}\Delta\psi \qquad [12\text{-}6]$$

where Z is the ion's charge (+1 in this case) and $\Delta\psi$ is the membrane potential caused by the imbalance in positive charges. For mitochondria, $\Delta\psi$ is positive, usually 150 to 200 mV. This value indicates that the intermembrane space or cytosol is more positive than the matrix (recall from Section 8-3 that for a whole cell, the cytosol is more negative than the extracellular space and $\Delta\psi$ is negative).

Combining the chemical and electrical effects gives an overall free energy change of

$$\Delta G = RT \ln \frac{[H^+_{out}]}{[H^+_{in}]} + Z\mathcal{F}\Delta\psi \qquad [12\text{-}7]$$

Typically, the free energy change for translocating one proton out of the matrix is about $+20\ \text{kJ} \cdot \text{mol}^{-1}$ (see Sample Calculation 12-2 for a detailed application of Equation 12-7). This is a thermodynamically costly event. Passage of the proton back *into* the matrix, following its electrochemical gradient, would have a free energy change of about $-20\ \text{kJ} \cdot \text{mol}^{-1}$. This is thermodynamically favorable, but it does not provide enough free energy to drive the synthesis of ATP. However, the 10 protons translocated for each pair of electrons transferred from NADH to O_2 have an associated protonmotive force of about $200\ \text{kJ} \cdot \text{mol}^{-1}$, enough to drive the phosphorylation of several molecules of ADP.

Sample Calculation 12-2

Problem

Calculate the free energy change for translocating a proton out of the mitochondrial matrix, where $\text{pH}_{matrix} = 7.8$, $\text{pH}_{cytosol} = 7.15$, $\Delta\psi = 170$ mV, and $T = 25°\text{C}$.

Solution

Since $\text{pH} = -\log [H^+]$ (Equation 2-4), the logarithmic term of Equation 12-7 can be rewritten. Equation 12-7 then becomes

$$\Delta G = 2.303\ RT(\text{pH}_{in} - \text{pH}_{out}) + Z\mathcal{F}\Delta\psi$$

Substituting known values gives

$$\begin{aligned}\Delta G &= 2.303\ (8.3145\ \text{J} \cdot \text{K}^{-1} \cdot \text{mol}^{-1})(298\text{K})(7.8 - 7.15) \\ &\quad + (1)(96{,}485\ \text{J} \cdot \text{V}^{-1} \cdot \text{mol}^{-1})(0.170\ \text{V}) \\ &= 3700\ \text{J} \cdot \text{mol}^{-1} + 16{,}400\ \text{J} \cdot \text{mol}^{-1} \\ &= +20.1\ \text{kJ} \cdot \text{mol}^{-1}\end{aligned}$$

REVIEW LEARNING OBJECTIVE 6

- What is the source of the protons that form the gradient across the inner mitochondrial membrane?
- Why does the free energy of the proton gradient have a chemical and an electrical component?

Go to **Exercise 18/ATP Synthase** for a closer examination of the structure and function of the enzyme complex that uses the free energy of the proton gradient to phosphorylate ADP. The structures shown in this exercise and the accompanying questions address Learning Objectives 7 and 8.

4 ATP SYNTHASE

The protein that taps the electrochemical proton gradient to phosphorylate ADP is known as the F_1F_0 ATP synthase (or Complex V). One part of the protein, called F_0, functions as a transmembrane channel that permits H^+ to flow back into matrix, following its gradient. The F_1 component catalyzes the reaction ADP + $P_i \rightarrow$ ATP + H_2O (Fig. 12-21). This section describes the structures of the two components of ATP synthase and shows how their activities are linked so that exergonic H^+ transport can be coupled to endergonic ATP synthesis.

The structure of ATP synthase

Not surprisingly, the structure of ATP synthase is highly conserved among different species. We will focus mainly on the *E. coli* protein, which is located in the bacterial plasma membrane and differs from the mammalian (mitochondrial) enzyme only in the number of some of the smaller subunits (Fig. 12-22). The membrane-embedded F_0 component consists of an *a* subunit, two *b* subunits that extend upward to interact with the F_1 component, and 9 or more *c* subunits that form a donut-shaped ring in the membrane. The exact number of *c* subunits varies with the source; the chloroplast ATP synthase, for example, has 14 *c* subunits.

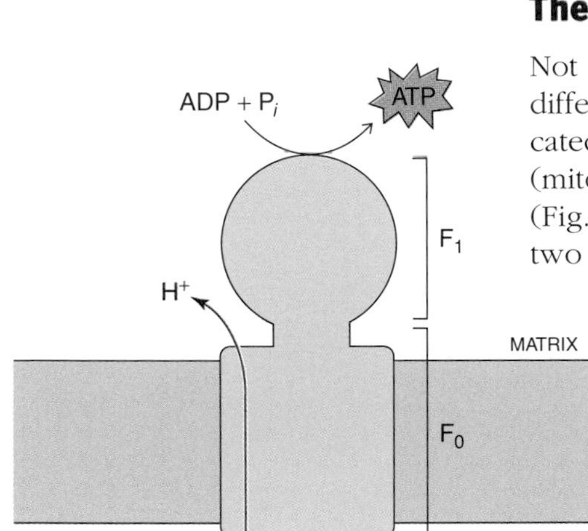

FIGURE 12-21
ATP synthase function.
As protons flow through the F_o transmembrane channel from the intermembrane space to the matrix, the F_1 component catalyzes the synthesis of ATP from ADP + P_i.

Proton transport through F_0 is believed to involve the rotation of the *c* ring past the stationary *a* subunit. According to one model, a highly conserved carboxylate side chain on each *c* subunit serves as a proton binding site (Fig. 12-23). When properly positioned at the *a* subunit, a *c* subunit can take up a proton from the intermembrane space. Rotation of the *c* ring brings another *c* subunit into position so that it can release its bound proton into the matrix (Fig. 12-23). The favorable thermodynamics of proton translocation would tend to keep the *c* ring moving in one direction. Depending on the relative concentrations of protons on the two sides of the membrane, the *c* ring could spin in either direction. Exercise 18 includes animated presentations of *c* ring function.

Attached to the *c* ring and rotating along with it is the γ subunit of the F_1 component. Additional subunits, such as ε in the *E. coli* complex (see Fig. 12-22b), are located near the *c*–γ interface. The bulk of F_1, which is the soluble portion of ATP synthase, consists of three α and three β subunits. These six subunits have similar tertiary structures and are arranged like the sections of an orange. Although all six subunits can bind adenine nucleotides, only the β subunits have catalytic activity (nucleotide binding to the α subunits may play a regulatory role). The γ subunit, which has two long α helices arranged as a bent coiled coil, protrudes into the center of the αβ hexamer (Fig. 12-24). The structure of ATP synthase is explained in detail in Exercise 18.

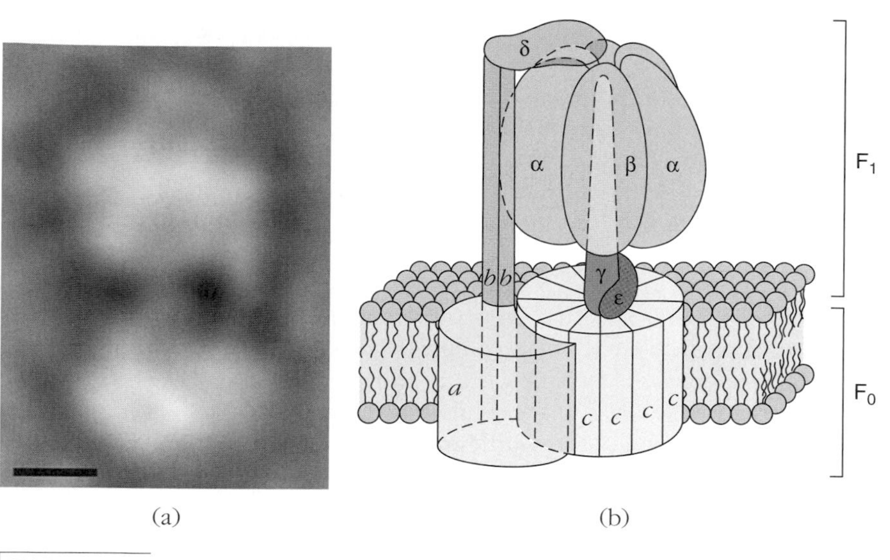

FIGURE 12-22

Structure of ATP synthase.
(a) Cryoelectron microscopy images of *E. coli* ATP synthase, showing the asymmetry of the complex. [*Courtesy Roderick A. Capaldi, University of Oregon.*] (b) Schematic model of the *E. coli* enzyme with individual subunits labeled. The F_o component has the composition ab_2c_{9-12}, and the F_1 component has the composition $\alpha_3\beta_3\gamma\delta\varepsilon$. (c) X-Ray structure of yeast ATP synthase at 3.9 Å resolution. Each residue is represented by a sphere. The proteins corresponding to the *a*, *b*, and δ subunits in part (b) are not included in this model. [*Structure (pdb 1QO1) determined by D. Stock, A.G.W. Leslie, and J.E. Walker.*]

H+
MATRIX
a
H+ H+ H+ H
c
INTERMEMBRANE SPACE
H+

FIGURE 12-23

Model for proton transport by F_o.
When a *c* subunit (purple) binds a proton from one side of the membrane, it moves away from the *a* subunit (blue). Because the *c* subunits form a ring, rotation brings another *c* subunit toward the *a* subunit, where it releases its bound proton to the opposite side of the membrane.

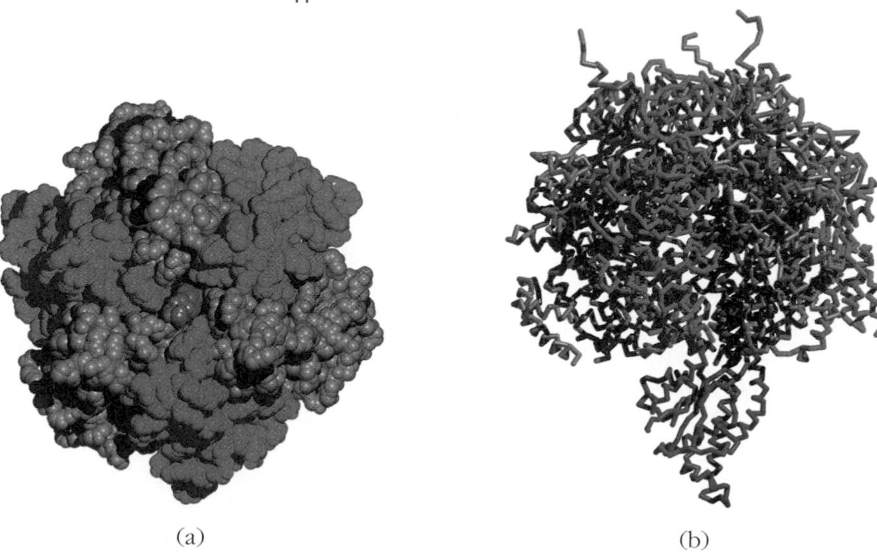

FIGURE 12-24

Structure of the F_1 component of ATP synthase.
The alternating α (blue) and β (green) subunits form a hexameric ring around the end of the γ shaft (purple). (a) Top view (from the matrix). (b) Side view. The other subunits of F_1 are not shown. [*Structure (pdb 1E79) determined by C. Gibbons, M.G. Montgomery, A.G.W. Leslie, and J.E. Walker.*]

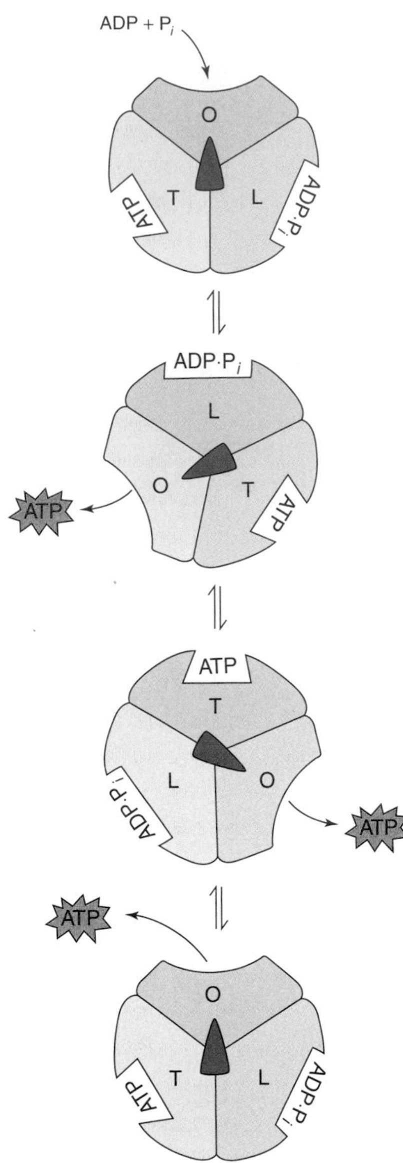

FIGURE 12-25

The binding change mechanism. The diagram shows the catalytic (β) subunits of the F_1 component of ATP synthase. Each of the three β subunits adopts a different conformation: open (O), loose (L), or tight (T). The substrates ADP and P_i bind to a loose site; ATP is synthesized when the site becomes tight; and ATP is released when the subunit becomes open. The conformational shifts are triggered by the 120° rotation of the γ subunit, arbitrarily represented by the purple shape. Because each of the three catalytic sites cycles through the three conformational states, ATP is released from one of the three β subunits with each 120° rotation of the γ subunit.

A close examination of the F_1 assembly reveals that the γ subunit interacts asymmetrically with the three pairs of αβ units. In fact, each αβ unit has a slightly different conformation, and model-building indicates that for steric reasons, the three units cannot simultaneously adopt the same conformation. *The three αβ pairs change their conformations as the γ subunit rotates* (it is like a shaft driven by the *c* ring "rotor"). The αβ hexamer itself does not rotate, since it is held in place by the ab_2 "stator" of F_0 (see Fig. 12-22b).

For an ATP synthase containing 12 *c* subunits, the transmembrane movement of each proton could potentially turn the γ shaft by 30°. However, videomicroscopy indicates that the γ subunit rotates in steps of 120°, apparently interacting with each of the three αβ pairs in one full rotation of 360°. Electrostatic interactions between the γ and β subunits may act as a catch that holds the γ subunit in place while translocation of three protons builds up strain. Translocation of the fourth proton causes the γ subunit to suddenly jump to the next β subunit, a movement of 120°.

The binding change mechanism

How does the rotation of the γ subunit promote ATP synthesis? According to the **binding change mechanism** described by Paul Boyer, *rotation-driven conformational changes alter the affinity of each catalytic β subunit for adenine nucleotides.* At any moment, each catalytic site has a different conformation (and binding affinity), referred to as the open, loose, or tight state. ATP synthesis occurs as follows (Fig. 12-25):

1. The substrates ADP and P_i bind to a β subunit in the loose state.
2. The substrates are converted to ATP as rotation of the γ subunit causes the β subunit to shift to the tight conformation.
3. The product ATP is released after the next rotation, when the β subunit shifts to the open conformation.

Because the three β subunits of ATP synthase interact cooperatively, they all change their conformations simultaneously as the γ subunit turns. A full rotation of 360° is required to restore the enzyme to its initial state, but each rotation of 120° results in the release of ATP from one of the three active sites. Exercise 18 shows how ATP synthesis occurs as proton translocation drives rotation of the γ subunit.

Much of the evidence supporting the binding change mechanism comes from experiments with the isolated F_1 component. In the absence of F_0, F_1 functions as an ATPase, hydrolyzing ATP to ADP + P_i. This is a thermodynamically favorable reaction ($\Delta G^{\circ\prime} = -30.5 \text{ kJ} \cdot \text{mol}^{-1}$). ATP synthesis cannot occur in the absence of F_0 because this endergonic reaction requires the free energy of the proton gradient. In the intact F_1F_0 ATP synthase, dissipation of the gradient via F_0 is tightly coupled to ATP synthesis. Consequently, *in the absence of a proton gradient, no ATP is synthesized because there is no free energy to drive the rotation of the γ subunit.* Agents that dissipate the proton gradient can therefore "uncouple" ATP synthesis from oxidative electron transport, the source of the proton gradient (Box 12-A).

The tight coupling between ATP synthesis and proton translocation also means that under certain experimental conditions, the entire F_1F_0 ATP synthase can operate in reverse. When $[H^+_{out}] < [H^+_{in}]$, protons are translocated in the opposite direction, reversing the rotation of the *c* ring and its attached γ subunit. The β subunits of F_1 then follow the steps of the binding change mechanism (see Fig. 12-25) in reverse order so that ATP is hydrolyzed rather than synthesized. This reversibility indicates that ATP synthase operates at near 100% efficiency (that is, little of the free energy of the proton gradient is wasted).

A CLOSER LOOK

BOX 12-A Uncoupling agents prevent ATP synthesis

When the metabolic need for ATP is low, the oxidation of reduced cofactors proceeds until the transmembrane proton gradient builds up enough to halt further electron transport. When the protons reenter the matrix via the F_o component of ATP synthase, electron transport resumes. However, if the protons leak back into the matrix by a route other than ATP synthase, then electron transport will continue without any ATP being synthesized. ATP synthesis is said to be "uncoupled" from electron transport, and the agent that allows the proton gradient to dissipate in this way is called an **uncoupler.**

One well-known uncoupler is the dinitrophenolate ion, which can take up a proton (its p*K* is near neutral):

$$\text{Dinitrophenolate } (O^-, NO_2, NO_2) \underset{}{\overset{H^+}{\rightleftharpoons}} \text{Dinitrophenol } (OH, NO_2, NO_2)$$

Dinitrophenolate   Dinitrophenol

In its neutral state, dinitrophenol can diffuse through a lipid bilayer and release the proton on the other side. Dinitrophenol can therefore dissipate a proton gradient by providing a way for protons to cross the membrane.

Dinitrophenol, of course, is not usually present in mitochondria; however, physiological uncoupling does occur. Dissipating a proton gradient prevents ATP synthesis, but it allows oxidative metabolism to continue at a high rate. The by-product of this metabolic activity is heat.

Uncoupling for **thermogenesis** (heat production) occurs in specialized adipose tissue known as brown fat (its dark color is due to the relatively high concentration of cytochrome-containing mitochondria; ordinary adipose tissue is white). The inner membrane of the mitochondria in brown fat contains a transmembrane proton channel called a UCP (uncoupling protein). Protons translocated to the intermembrane space during respiration can reenter the mitochondrial matrix via the uncoupling protein, bypassing ATP synthase. The free energy of respiration is therefore lost as heat rather than used to synthesize ATP. Brown fat is present in hibernating mammals and newborn humans, and the activity of the UCP is under the control of hormones that also mobilize the stored fatty acids to be oxidized in the brown fat mitochondria.

Adult humans lack brown fat, but the mitochondria of ordinary adipose tissue and muscle appear to contain one or more uncoupling proteins. These uncouplers may help regulate metabolic rates, and variations in UCP levels might explain why some people seem to have higher or lower metabolic rates. UCPs might be good targets for treating obesity, since increasing the concentration or activity of UCPs could uncouple respiration from ATP synthesis. Metabolic fuel oxidation could proceed at a high rate, literally burning up fat stores. It is also possible that UCPs have nothing to do with metabolic rates or tendencies to lose or gain weight. They may instead function to maintain the body's core temperature and might therefore be poor candidates for manipulation for the purpose of weight loss.

[John Serrao/Photo Researchers.]

Stoichiometric considerations of oxidative phosphorylation

Since the γ shaft of ATP synthase is attached to the *c*-subunit rotor, one ATP molecule is synthesized for every three or four protons translocated through F_0 (the exact number of protons depends on the number of *c* subunits). For yeast ATP synthase, which has 10 *c* subunits, the stoichiometry is clearly nonintegral. Such nonintegral values would be difficult to reconcile with most biochemical reactions, but they are consistent with the chemiosmotic theory: *Chemical energy (from the respiratory oxidation–reduction reactions) is transduced to a protonmotive force, then to the mechanical movement of a rotary engine (the c ring and its attached γ shaft), and finally back to chemical energy in the form of ATP.*

The relationship between respiration (the activity of the electron-transport complexes) and ATP synthesis is traditionally expressed as a **P:O ratio,** that is, the number of phosphorylations of ADP relative to the number of oxygen atoms reduced. For example, the oxidation of NADH by O_2 (carried out by the sequential activities of Complexes I, III, and IV) translocates 10 protons into the intermembrane space. The flow of 10 protons back into the matrix via the F_0 component would drive the synthesis of about 3 ATP (the 10 protons would account for about one complete rotation of the γ subunit through the three ATP-synthesizing sites of F_1). Thus, the P:O ratio would be about 3 (3 ATP per 1/2 O_2 reduced). For an electron pair originating as QH_2, only 6 protons would be translocated (by the activities of Complexes III and IV), and the P:O ratio would be approximately 2.

These sorts of calculations are the basis for our tally of the ATP yield for the complete oxidation of glucose by glycolysis and the citric acid cycle (see page 357). Keep in mind that these values are approximate, due to the nonintegral nature of P:O ratios. *In vivo,* the ratios are actually a bit lower, because some of the protons translocated during electron transport are consumed in other processes, such as the transport of P_i into the mitochondrial matrix (see Fig. 12-6).

REVIEW LEARNING OBJECTIVES 7 AND 8

- How does the F_0 component of ATP synthase dissipate the proton gradient? How is this activity related to the activity of the ATP-synthesizing catalytic sites of F_1?
- Describe how the three conformational states of the β subunits of ATP synthase are involved in ATP synthesis.
- How does the binding change mechanism account for ATP hydrolysis by ATP synthase?
- What is a P:O ratio and why is it often nonintegral?

Regulation of oxidative phosphorylation

In most metabolic pathways, control is exerted at highly exergonic (irreversible) steps. In oxidative phosphorylation, this step would be the reaction catalyzed by cytochrome *c* oxidase (Complex IV; see Table 12-2). However, there are no known effectors of cytochrome *c* oxidase activity. Apparently, *the close coupling between generation of the proton gradient and ATP synthesis allows oxidative phosphorylation to be regulated primarily by the availability of reduced cofactors (NADH and QH_2) produced by other metabolic processes.*

Less important regulatory mechanisms may involve the availability of the substrates ADP and P_i (which depend on the activity of their respective transport proteins). Experiments with ATP synthase show that when ADP and P_i are absent, the β subunits of the F_1 component cannot undergo the conformational changes required by the binding change mechanism. The γ subunit therefore remains immobile, and no protons are translocated though the *c* ring of F_0. This tight coupling between ATP synthesis and proton translocation prevents the waste of the free energy of the proton gradient.

There is also evidence that mitochondria contain a regulatory protein that binds to ATP synthase to inhibit its rate of ATP hydrolysis. The inhibitor is sensitive to pH so that it does not bind to ATP synthase when the matrix pH is high (as it is when electron transport is occurring). However, if the matrix pH drops as a result of a momentary disruption of the proton gradient, the inhibitor binds to ATP synthase. This regulatory mechanism prevents ATP synthase from operating in reverse as an ATPase.

REVIEW LEARNING OBJECTIVE 9

- Explain why the availability of reduced substrates is the primary mechanism for regulating oxidative phosphorylation.

SUMMARY

1. The electron affinity of a substance participating in an oxidation–reduction reaction, which involves the transfer of electrons, is indicated by its reduction potential, $\mathcal{E}°'$.
2. The difference in reduction potential between species undergoing oxidation and reduction is related to the free energy change for the reaction.
3. Oxidation of reduced cofactors generated by metabolic reactions takes place in the mitochondrion. Shuttle systems and transport proteins allow the transmembrane movement of reducing equivalents, ATP, ADP, and P_i.
4. The electron transport chain consists of a series of integral membrane protein complexes that contain multiple redox groups, including iron-sulfur clusters, flavins, cytochromes, and copper ions, and that are linked by mobile electron carriers. Starting from NADH, electrons travel a path of increasing reduction potential through Complex I, ubiquinone, Complex III, cytochrome *c*, and then to Complex IV, where O_2 is reduced to H_2O.
5. As electrons are transferred, protons are translocated to the intermembrane space via proton wires in Complexes I and IV and by the action of the Q cycle associated with Complex III.
6. The chemiosmotic theory describes how proton translocation during mitochondrial electron transport generates an electrochemical gradient whose free energy drives ATP synthesis.
7. The energy of the proton gradient is tapped as protons spontaneously flow through ATP synthase. Proton transport allows rotation of a ring of integral membrane *c* subunits. The linked γ subunit thereby rotates, triggering conformational changes in the F_1 portion of ATP synthase.
8. According to the binding change mechanism, the three functional units of the F_1 portion cycle through three conformational states to sequentially bind ADP and P_i, convert the substrates to ATP, and release ATP.
9. The P:O ratio quantifies the link between electron transport and oxidative phosphorylation in terms of the ATP synthesized and the O_2 reduced. Because these processes are coupled, the rate of oxidative phosphorylation is controlled primarily by the availability of reduced cofactors.

CHECKLIST

1. Review the Learning Objectives listed on page 370.
2. Complete Review Exercise 4/Redox Reactions to review the chemistry of oxidation–reduction reactions and the use of reduction potentials.
3. Complete Exercise 17/Oxidative Phosphorylation to explore the functions of the electron transport complexes of the mitochondrial respiratory chain.
4. Complete Exercise 18/ATP Synthase to examine the structure and function of the enzyme complex that uses the free energy of the proton gradient to synthesize ATP.
5. Apply your knowledge by solving the problems at the end of this chapter. Check your results in the Solutions appendix.
6. Be able to define the boldfaced terms (consult the glossary at the end of the book). Test your understanding by taking the Chapter 12 quiz (**www.wiley.com/college/pratt**).
7. Explore the websites listed at **www.wiley.com/college/pratt**.
8. Consult the list of Selected Readings for background information or additional details on mitochondrial structure, respiratory electron transport, and ATP synthase.

GLOSSARY TERMS

oxidative phosphorylation
redox reaction
oxidizing agent (oxidant)
reducing agent (reductant)
half-reaction
standard reduction potential ($\mathcal{E}°'$)
reduction potential ($\mathcal{E}$)
Nernst equation
Faraday constant ($\mathcal{F}$)
respiration
mitochondrion
mitochondrial matrix
intermembrane space
cristae
electron tomography
cryoelectron microscopy
redox center
proton wire
cytochrome
Q cycle
chemiosmotic theory
protonmotive force
Z
binding change mechanism
uncoupler
thermogenesis
P:O ratio

PROBLEMS

1. Calculate the standard free energy change for the reduction of pyruvate by NADH. Is this reaction spontaneous under standard conditions?

2. Calculate the standard free energy change for the reduction of oxygen by cytochrome a_3. Is this reaction spontaneous under standard conditions?

3. For every two QH_2 that enter the Q cycle, one is regenerated and the other passes its two electrons to two cytochrome c_1 centers. The overall equation is

$$QH_2 + 2\text{ cytochrome } c_1^{3+} + 2\text{ H}^+ \rightarrow Q + 2\text{ cytochrome } c_1^{2+} + 4\text{ H}^+$$

Calculate the free energy change associated with the Q cycle.

4. Acetoacetate may be reduced to 3-hydroxybutyrate. What serves as a better reducing agent, NADH or $FADH_2$? Explain.

5. Why is succinate oxidized by FAD instead of by NAD^+?

6. (a) What is the $\Delta\mathcal{E}$ value for the oxidation of ubiquinol by cytochrome *c* when the ratio of QH_2/Q is 10 and the ratio of cyt c^{3+}/cyt c^{2+} is 5?
(b) Calculate the ΔG for the reaction in part (a).

7. An iron-sulfur protein in Complex III donates an electron to cytochrome c_1. The reduction half reactions and $\mathcal{E}°'$ values are listed below. Write the balanced equation for the reaction and calculate the standard free energy change. How can you account for the fact that this reaction occurs spontaneously in the cell?

$$\text{FeS (ox)} + e^- \rightarrow \text{FeS (red)} \qquad \mathcal{E}°' = 0.280\text{ V}$$

$$\text{Cytochrome } c_1^{3+} + e^- \rightarrow \text{cytochrome } c_1^{2+} \qquad \mathcal{E}°' = 0.215\text{ V}$$

8. Calculate the overall efficiency of oxidative phosphorylation, assuming standard conditions, by comparing the free energy potentially available from the oxidation of NADH by O_2 and the free energy required to synthesize 3 ATP from 3 ADP.

9. Using the percent efficiency calculated in the previous problem, calculate the number of ATP generated by (a) Complex I (where NADH is oxidized by ubiquinone), (b) Complex III (where ubiquinol is oxidized by cytochrome *c*), and (c) Complex IV (where cytochrome *c* is oxidized by molecular oxygen).

10. The succinate dehydrogenase (Complex II) reaction occurs with a negative value of $\Delta G°'$. Show why this does not supply enough free energy to drive ATP synthesis.

11. How much ATP can be obtained by the cell from the complete oxidation of one mole of glucose? Compare this value with the amount of ATP obtained when glucose is anaerobically converted to lactate or ethanol.

12. The complete oxidation of glucose yields $-2850\text{ kJ} \cdot \text{mol}^{-1}$ of free energy. Incomplete oxidation by conversion to lactate yields $-196\text{ kJ} \cdot \text{mol}^{-1}$ and by alcoholic fermentation yields $-235\text{ kJ} \cdot \text{mol}^{-1}$.
(a) Calculate the overall efficiencies of glucose oxidation by these three processes.
(b) Do organisms that can completely oxidize glucose have an advantage over organisms that cannot?

13. A culture of yeast grown under anaerobic conditions is exposed to oxygen, resulting in a dramatic decrease in glucose consumption by the cells. This phenomenon is referred to as the Pasteur effect.
(a) Explain the Pasteur effect.
(b) The [NADH]/[NAD^+] and [ATP]/[ADP] ratios also change when an anaerobic culture is exposed to oxygen. Explain how these ratios change, and what effect this has on glycolysis and the citric acid cycle in the yeast.

14. The sequence of events in electron transport was elucidated in part by the use of inhibitors that block electron transfer at specific points along the chain. For example, adding rotenone (a plant toxin) or amytal (a barbiturate) blocks electron transport in Complex I; antimycin A (an antibiotic) blocks electron transport in Complex III; and cyanide (CN^-) blocks electron transport in Complex IV by binding to the Fe^{2+} in the Fe-Cu binuclear center.
(a) What happens to oxygen consumption if these inhibitors are added to a suspension of respiring mitochondria?
(b) What is the redox state of the electron carriers in the electron transport chain if each of the inhibitors is added separately to the mitochondrial suspension?
(c) What is the effect of added succinate to rotenone-blocked, amytal-blocked, or cyanide-blocked mitochondria? In other words, can succinate help "bypass" the block? Explain.

15. The compound tetramethyl-*p*-phenylenediamine (TMPD) donates a pair of electrons directly to Complex IV. What is the P:O ratio of this compound?

16. Ascorbate (vitamin C) can donate a pair of electrons to cytochrome *c*. What is the P:O ratio for ascorbate?

17. Can tetramethyl-*p*-phenylenediamine (see Problem 15) act as a bypass for the rotenone-blocked, amytal-blocked, or cyanide-blocked mitochondria described in Problem 14? Can ascorbate act as a bypass? Explain.

18. When the antifungal agent myxothiazol is added to a suspension of respiring mitochondria, the QH_2/Q ratio increases. Where in the electron transport chain does myxothiazol inhibit electron transfer?

19. If cyanide poisoning (see Problem 14) is diagnosed immediately, it can be treated by the administration of nitrites that have the ability to oxidize the Fe^{2+} in hemoglobin to Fe^{3+}. Why is this treatment effective?

20. The effect of the drug fluoxetine (Prozac) on isolated rat brain mitochondria was examined by measuring the rate of electron transport (units not given) in the presence of various combinations of substrates and inhibitors (see Problems 14–16).
(a) How do pyruvate and malate serve as substrates for electron transport?
(b) What is the effect of fluoxetine on electron transport? Explain.

(c) Fluoxetine can also inhibit ATP synthase. Why might long-term use of fluoxetine be a concern?

Fluoxetine concentration (mM)	3.5 mM pyruvate + 7.5 mM malate	10 mM succinate +1 μg/mL rotenone	0.1 mM ascorbate +4 mM TMPD
0	163 ± 15.1	145 ± 14.2	184 ± 22.2
0.15	77 ± 7.3	131 ± 13.5	116 ± 13.9

21. Ubiquinone is not anchored in the mitochondrial membrane but is free to diffuse laterally throughout the membrane among the electron transport chain components. What aspects of its structure account for this behavior?

22. Cytochrome *c* is easily dissociated from isolated mitochondrial membrane preparations, but the isolation of cytochrome c_1 requires the use of strong detergents. Explain why.

23. Consider the adenine nucleotide translocase and the P_i-H^+ symport protein that import ADP and P_i, the substrates for oxidative phosphorylation, into the mitochondria (see Fig. 12-6).

(a) How does the activity of the adenine nucleotide translocase affect the electrochemical gradient across the mitochondrial membrane?

(b) How does the activity of the P_i-H^+ symport protein affect the gradient?

(c) What can you conclude about the thermodynamic force that drives the two transport systems?

24. The compounds atractyloside and bongkrekic acid both bind tightly to the ATP/ADP translocase. What is the effect of these compounds on ATP synthesis? On electron transport?

25. Several key experimental observations were important in the development of the chemiosmotic theory. Explain how each observation listed below is consistent with the chemiosmotic theory as described by Peter Mitchell.

(a) The pH of the intermembrane space is lower than the pH of the mitochondrial matrix.

(b) Oxidative phosphorylation does not occur in mitochondrial preparations to which detergents have been added.

(c) Lipid-soluble compounds such as DNP (see Box 12-A) inhibit oxidative phosphorylation while allowing electron transport to continue.

26. Mitchell's original chemiosmotic hypothesis relies on the impermeability of the inner mitochondrial membrane to ions other than H^+, such as Na^+ and Cl^-.

(a) Why was this thought to be important?

(b) Could ATP still be synthesized if the membrane were permeable to other ions?

27. Nigericin is an antibiotic that integrates into membranes and functions as a K^+/H^+ antiporter. Another antibiotic, valinomycin, is similar, but it allows the passage of K^+ ions. When both antibiotics are added simultaneously to suspensions of respiring mitochondria, the electrochemical gradient completely collapses.

(a) Draw a diagram of a mitochondrion in which nigericin and valinomycin have integrated into the inner mitochondrial membrane, in a manner that is consistent with the experimental results.

(b) Explain why the electrochemical gradient dissipates. What happens to ATP synthesis?

28. Calculate the free energy change for translocating a proton out of the mitochondrial matrix, where $pH_{matrix} = 7.6$, $pH_{cytosol} = 7.2$, $\Delta\psi = 200$ mV, and $T = 37°C$.

29. Dicyclohexylcarbodiimide (DCCD) is a reagent that reacts with Asp or Glu residues. Explain why the reaction of DCCD with just one *c* subunit completely blocks both the ATP-synthesizing and ATP-hydrolyzing activity of ATP synthase.

30. Oligomycin is an antibiotic that blocks proton transfer through the F_0 proton channel of ATP synthase. What is the effect on (a) ATP synthesis, (b) electron transport, and (c) oxygen consumption if oligomycin is added to a suspension of respiring mitochondria? (d) What changes occur if dinitrophenol is then added to the suspension?

31. The compound dinitrophenol (DNP) was introduced as a "diet pill" in the 1920s. Its use was discontinued because the side effects were fatal in some cases. What was the rationale for believing that DNP would be an effective diet aid?

32. The compound carbonylcyanide-*p*-trifluoromethoxy phenylhydrazone (FCCP) is an uncoupler similar to DNP. Describe how FCCP acts as an uncoupler.

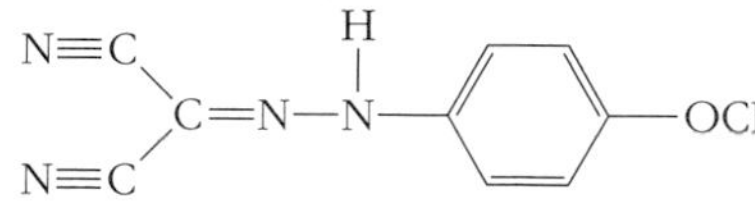

N≡C
C=N—N(H)—[benzene ring]—OCF$_3$
N≡C

33. In the 1950s, experiments with isolated mitochondria showed that organic compounds are oxidized and O_2 is consumed only when ADP is included in the preparation. When the ADP supply runs out, oxygen consumption halts. Explain these results.

34. A patient seeks treatment because her metabolic rate is twice normal and her temperature is elevated. She suspects a thyroid problem but an examination reveals that her thyroid functions normally, so a muscle biopsy is performed for further tests. An examination of the muscle mitochondria indicates that muscle mitochondria are structurally unusual and not subject to normal respiratory controls. Electron transport takes place regardless of the concentration of ADP.

(a) What is the P:O ratio (compared to normal) of NADH that enters the electron transport chain in the mitochondria of this patient?

(b) Why are the patient's metabolic rate and temperature elevated?

(c) What is the ability of this patient to carry out strenuous exercise?

35. In experimental systems, the F_0 component of ATP synthase can be reconstituted into a membrane. F_0 can then act as a proton channel that is blocked when the F_1 component is added to the system. What molecule must be added to the system in order to restore the proton-translocating activity of F_0? Explain.

36. A bacterial ATP synthase has 9 *c* subunits, and a chloroplast ATP synthase has 14 *c* subunits. Would you expect the bacterium or the chloroplast to have a higher P:O ratio?

37. In yeast, pyruvate may be converted to ethanol in a two-step pathway catalyzed by pyruvate decarboxylase and alcohol dehydrogenase (Box 10-B). Pyruvate may also be converted to acetyl-CoA by pyruvate dehydrogenase. Yeast mutants in which the pyruvate decarboxylase gene is missing (*pdc–*) are useful for studying the regulation of pyruvate dehydrogenase enzyme. When wild-type yeast were pulsed with glucose, glycolytic flux increased dramatically and the rate of respiration increased. But when the same experiments were performed with *pdc–* mutants, only a small increase in glycolysis was observed, and pyruvate was the main product excreted by the yeast cells. Explain these results.

38. Hibernating animals and human infants contain brown fat deposits, so-called because of the presence of large numbers of mitochondria, the site of electron transport and oxidative phosphorylation. In brown fat, given the appropriate stimulus, oxidative phosphorylation and electron transport can be uncoupled, causing energy to be dissipated as heat. The protein responsible for the uncoupling is a brown fat inner mitochondrial membrane protein previously named UCP (for uncoupling protein), but now referred to as UCP1, since a second uncoupling protein has since been discovered. Previous experiments have shown that UCP1 protects against cold and is involved in regulation of energy expenditure. The ability of UCP1 to stimulate the consumption of calories solely for the production of heat led some investigators to postulate that UCP1 was involved in regulating body weight. Scientists have always wondered why some people seem to be able to ingest a large number of calories without gaining weight, whereas others eat moderately but are obese. If the UCP1 of brown fat were involved, scientists postulated that obese people would be efficient "burners," whereas humans of moderate weight might burn calories inefficiently, releasing a greater proportion of energy as heat. But the role of UCP1 in humans has always been debated since infants contain a large amount of brown fat but mature adults do not.

In order to examine the biochemical role of UCP1 more fully, the investigators in this case worked with genetically engineered ("knockout") mice that lack the gene for UCP1. By examining the characteristics of knockout mice, the biochemical and physiological roles of a particular protein can be ascertained. The results of an experiment with UCP1-knockout mice, described below, led to the discovery of a second uncoupling protein referred to as UCP2. UCP2 may play a more significant role in obesity, since UCP2 is found in abundance in white fat.

(a) Why is heat produced when UCP1 uncouples oxidative phosphorylation from electron transport?

(b) In their study, the investigators injected a β3 adrenergic agonist (a substance that mimics the effect of a hormone), which stimulated UCP1 in adipose tissue. When they injected the agonist into normal mice, they noted that oxygen consumption increased over twofold. But when the agonist was injected into knockout mice, oxygen consumption increased only slightly. Explain these results.

(c) The investigators who produced the UCP1 knockout mice noted that the knockouts were normal in every way except for increased lipid deposition in their adipose tissue. Explain why.

(d) The investigators carried out an experiment in which normal mice and UCP1 knockout mice were placed in a cold (5°C) room for 24 hours. The normal mice were able to maintain their body temperature at 37°C even after 24 hours in the cold. But when the knockout mice were placed in a cold room, their body temperature decreased 10°C or more. Explain.

(e) Investigators wanted to know whether UCP1 synthesis could be induced by overeating, and if so, if it could be involved in burning excess fat such that the animal's adipose tissue content remained relatively constant. They assumed that UCP1 knockout mice, unable to synthesize UCP1, would become obese if fed a high-fat diet. Interestingly, knockout mice did not become obese. Instead, overfeeding induced the synthesis of a second protein. Separate experiments indicated that this second protein was also able to uncouple oxidative phosphorylation, and this protein was given the name UCP2. How would the induction of UCP2 help animals maintain a constant amount of adipose tissue?

[From Enerbäck, S., Jacobsson, A., Simpson, E.M., Guerra, C., Yamashita, H., Harper, M.-E., and Kozack, L.P., *Nature* **387,** 90–94 (1997); Fleury, C., Neverova, M., Collins, S., Raimbault, S., Champigny, O., Levi-Meyrueis, C., Bouillard, F., Seldin, M.F., Surwit, R.S., Ricquier, D., and Warden, C.H., *Nature Genetics* **15,** 269–272 (1997); Lowell, B.B., S-Susulic, V., Hamann, A., Lawitts, J.A., Himms-Hagen, J., Boyer, B.B., Kozak, L.P., and Flier, J.S., *Nature* **366,** 740–742 (1993); Hirsch, J., *Nature* **387,** 27–28 (1997).]

39. In mammals, body temperature can be regulated by uncoupling proteins, which act by uncoupling electron transport and oxidative phosphorylation. But the mechanism of thermogenesis in plants is not as well understood. Several studies indicate that thermogenesis is an important process in the developing plant. For example, the eastern skunk cabbage has the ability to maintain its temperature 15–35°C higher than ambient temperatures during the months of February and March, when ambient temperatures range from −15 to +15°C. Thermogenesis in the skunk cabbage is critical to the survival of the plant since the spadix tissue (part of the flower) is not frost-resistant. The skunk cabbage may be unique in its ability to carry out thermogenesis for longer than just a few hours. Thermoregulation has also been studied in lotus flowers. Investigators in Australia noted that the temperature of the lotus flower receptacle rises 1 or 2 days before the flower opens. Temperatures fall only after the flower is completely opened. The investigators speculated that heat production in the lotus flower enhances evaporation of the floral scent that attracts pollinating insects. Noting that many pollinating insects require a thoracic temperature of 30°C or higher to initiate flight, the investigators hypothesized that heat production in the lotus flower would assist the departure of the insects from the flower.

Although observations of temperature and oxygen utilization in the skunk cabbage and lotus flower implicated an uncoupler, an uncoupling protein was not isolated until recently. A research team in France identified an uncoupling protein in potatoes (*Solanum tuberosum*) that is involved in thermoregulation in plants in a manner similar to the brown fat uncoupling proteins in mammals. The uncoupling protein is named StUCP and its synthesis is induced by cold temperatures. The authors note that heat production concomitant with a "burst of respiration" occurs during plant flowering and fruit ripening.

(a) In the skunk cabbage, the site of thermogenesis is the spadix. However, the spadix tissue does not store starch. Instead, the spadix relies on the massive root system, which stores appreciable quantities of starch. The large quantity of starch available may explain why the skunk cabbage is able to carry out thermogenesis for weeks rather than hours. Why is starch required for thermogenesis?

(b) Oxygen consumption by the skunk cabbage increases as the temperature decreases. The rate of oxygen consumption nearly doubles with every 10°C drop in ambient temperature. Similarly, in the lotus flower, the receptacle temperature is maintained between 30 and 35°C for a 2-day period during flower opening. The ambient air temperature during this 2-day period may fluctuate between 10 and 30°C. Oxygen consumption was observed to decrease during the day when temperatures were close to 30°C. Explain the biochemical mechanism for these observations.

(c) In the skunk cabbage, lotus flower, and potato, thermogenesis occurs because of the action of an uncoupling protein. The gene that codes for the uncoupling protein in potatoes was recently isolated. The result from a Northern blot analysis is shown below (a Northern blot detects levels of mRNA). What is your interpretation of these results? How is the production of the uncoupling protein regulated?

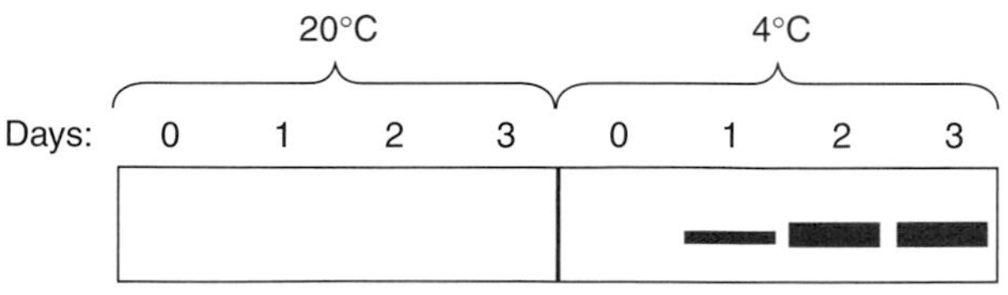

[From Laloi, M., Klein, M., Riesmeier, J.M. Müller-Röber, B., Fleury, C, Bouillaud, F., and Ricquier, D., *Nature* **389,** 135–136 (1997); Knutson, R.M., *Science* **186,** 746–747 (1974); Seymour, R.S., and Schultze-Motel, P., *Nature* **383,** 305 (1996).]

40. Glutamate can be used experimentally as a substrate for mitochondrial respiration, as shown in the diagram below. When the sphingolipid ceramide is added to a mitochondrial suspension respiring in the presence of glutamate, respiration decreases. The authors of the study propose that ceramide could act as a regulator of mitochondrial function *in vivo*.

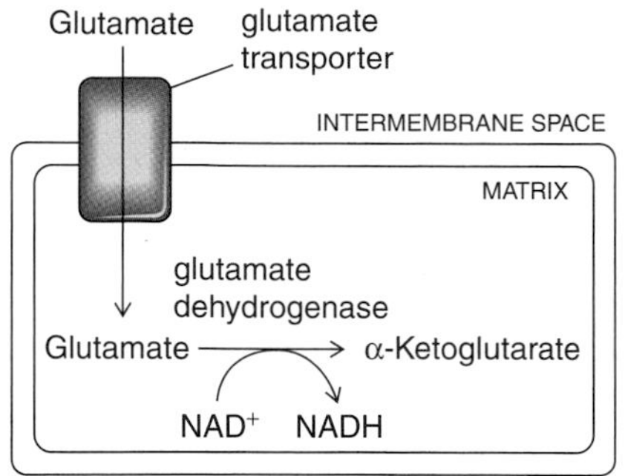

(a) How does glutamate act as a substrate for mitochondrial respiration?

(b) Ceramide-induced inhibition of respiration could be due to several different factors. List all of the possibilities.

(c) Mitochondria treated with ceramide were exposed to an uncoupler. Respiration decreased, and the results were identical to the results in the absence of the uncoupler. What site(s) of inhibition can be ruled out?

(d) In another experiment, mitochondria were subjected to a freeze–thaw cycle that makes the inner mitochondrial membrane permeable to NADH. NADH can then be used as substrate for electron transport in these mitochondria. When NADH is used as a substrate, the ceramide-induced decrease in respiration is identical to that seen when glutamate is used in untreated mitochondria. What site(s) of inhibition can be ruled out?

[From Gudz, T.I., Tserng, K.-Y., and Hoppel, C.L., *J. Biol. Chem.* **272,** 24154–24158 (1997).]

SELECTED READINGS

Boyer, P.D., Catalytic site forms and controls in ATP synthase catalysis, *Biochim. Biophys. Acta* **1458,** 252–262 (2000). [A description of the steps of ATP synthesis and hydrolysis, along with experimental evidence and alternative explanations, by the author of the binding change mechanism.]

Iwata, S., Lee, J.W., Okada, K., Lee, J.K., Iwata, M., Rasmussen, B., Link, T.A., Ramaswamy, S., and Jap, B.K., Complete structure of the 11-subunit bovine mitochondrial cytochrome bc_1 complex, *Science* **281,** 64–71 (1998). [Reports the structure of Complex III, including the locations of all its redox centers.]

Iwata, S., Ostermeier, C., Ludwig, B., and Michel, H., Structure at 2.8 Å resolution of cytochrome *c* oxidase from *Paracoccus denitrificans, Nature* **376,** 660–669 (1995). [Describes the structure of a bacterial Complex IV and discusses mechanisms of proton translocation.]

Oster, G. and Wang, H., ATP synthase: Two motors, two fuels, *Structure* **7,** R67–R72 (1999). [Describes how structural changes (rotation of *c* subunits) in ATP synthase are linked to catalytic conformational changes that result in ATP production.]

Saraste, M., Oxidative phosphorylation at the *fin de siècle, Science* **283,** 1488–1493 (1999). [Provides a succinct overview of the entire process of electron transport and ATP synthesis.]

CHAPTER 13

Photosynthesis

The hot sun of the tropics spells sunburn—and not just for humans with sensitive skin. Corals that inhabit clear shallow waters must also protect themselves from DNA-damaging ultraviolet light, which is strongest near the equator. Coral tissues are mostly transparent, to allow sunlight to reach the photosynthetic algae that live symbiotically with the coral. So what do corals use for sunscreen? They rely on the algae to produce modified amino acids that absorb ultraviolet radiation. These compounds accumulate in the coral host and thereby prevent sun damage. Invertebrates and fish that feed on corals acquire the sunscreen compounds and can take advantage of their photoprotective effects. These natural sunscreens may soon find their way onto human skin, since they are chemically simple and potentially less allergenic than currently available synthetic sunscreens.

[M. McCoy/Photo Researchers.]

LEARNING OBJECTIVES

1. Understand the role of pigments in absorbing and dissipating solar energy.
2. Understand how photooxidation at a reaction center drives electron transfer to groups of increasing reduction potential.
3. Understand the functions of the major proteins and cofactors of the light reactions of photosynthesis.
4. Understand that the light reactions in plants and cyanobacteria produce O_2, NADPH, and ATP.
5. Understand how photophosphorylation resembles mitochondrial oxidative phosphorylation.
6. Understand that rubisco carries out both carboxylation and oxygenation.
7. Understand that the Calvin cycle directs fixed CO_2 into three-carbon intermediates and regenerates the five-carbon CO_2 acceptor.
8. Understand how the light reactions regulate the "dark" reactions of photosynthesis.
9. Understand how phosphoanhydride bonds are consumed in starch and sucrose synthesis.

THIS CHAPTER IN CONTEXT

Although much of this textbook is concerned with mammalian metabolism, Chapter 13 shifts the focus to plants and provides an opportunity to compare mammalian and plant biochemistry. The study of photosynthesis is a logical extension of the presentation of electron transport and oxidative phosphorylation in Chapter 12. It also builds on the reactions presented in Chapter 10, which show the interconversions of small molecules with monosaccharides. Photosynthesis also marks a change in emphasis from catabolic reactions, the main focus of Chapters 10 through 12, to anabolic reactions, which are prominently featured in Chapters 14 and 15.

EVERY YEAR, THE EARTH'S PLANTS CONVERT AN ESTIMATED 6 $\times$ 10^{16} grams of carbon to organic compounds by **photosynthesis.** Most of this activity occurs in forests and savannahs, but it also occurs in the ocean and under ice—wherever water, carbon dioxide, and light are available. The organic materials produced by photosynthetic organisms sustain them as well as the organisms that feed on them.

The ability to use sunlight as an energy source evolved about 3.5 billion years ago. Before that, cellular metabolism probably centered around the inorganic reductive reactions associated with hydrothermal vents. The first photosynthetic organisms produced various pigments (light-absorbing molecules) to capture solar energy and thereby drive the reduction of metabolites. The descendants of some of these organisms are known today as purple bacteria and green sulfur bacteria. By about 2.5 billion years ago, the cyanobacteria had evolved (Fig. 13-1; these organisms are sometimes misleadingly called blue-green algae, although they are prokaryotes and true algae are eukaryotes). Cyanobacteria absorb enough solar energy to undertake the energetically costly oxidation of water to molecular oxygen. In fact, the dramatic increase in the level of atmospheric oxygen (from an estimated 1% to the current level of about 20%) around 2.5 billion years ago is attributed to the rise of cyanobacteria. Modern plants are the result of the symbiosis of early eukaryotic cells with cyanobacteria.

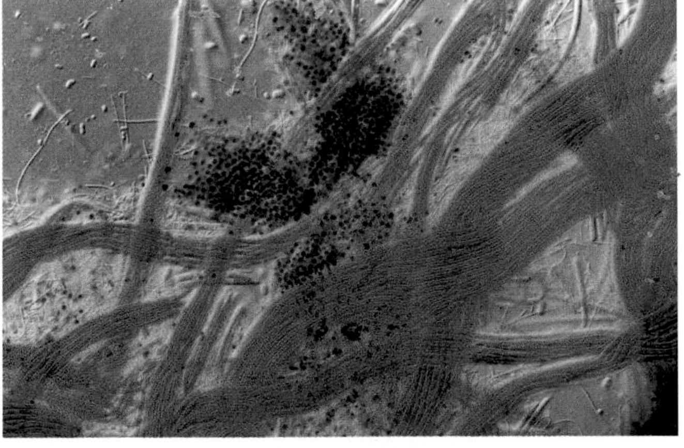

FIGURE 13-1

Cyanobacteria and purple sulfur bacteria. The greenish filaments are cyanobacteria, and the small dark spheres are purple sulfur bacteria (their name reflects their metabolic dependence on sulfur). These organisms use different pigments to absorb solar radiation. [*Courtesy Lee Prufert-Bebout/NASA Ames Research Center, Exobiology Branch.*]

Photosynthesis as a biochemical phenomenon is worth studying not just because of its importance in global carbon and oxygen cycles. It also affords an opportunity to compare the biochemistry of plants and mammals. To begin, we will examine light-absorbing chloroplasts, which are the relics of bacterial symbionts, as are mitochondria. We will then look at the electron-transporting complexes that convert solar energy to biologically useful forms of free energy such as ATP and the reduced cofactor NADPH. Finally, we will see how plants use these energy currencies to carry out biosynthetic reactions—the **fixation** of carbon dioxide into organic compounds.

Although the apparatus and reactions of photosynthesis are not found in all organisms, we can place them in the context of the metabolic scheme outlined in Chapter 9 (Fig. 13-2). As we describe the harvest of solar energy and its use in the incorporation of CO_2 into three-carbon compounds, we

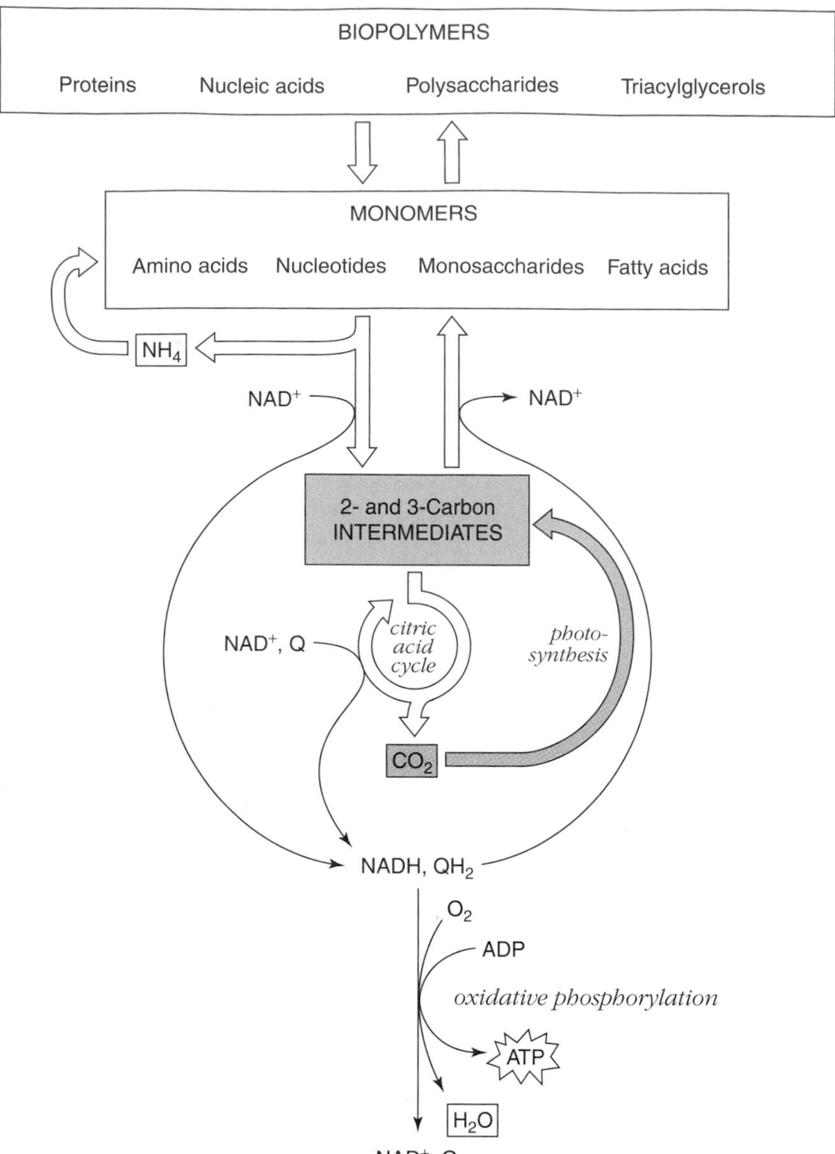

FIGURE 13-2

Photosynthesis in context. Photosynthetic organisms incorporate atmospheric CO_2 into three-carbon compounds that are the precursors of biological molecules such as carbohydrates and amino acids. Photosynthesis requires light energy to drive the production of the ATP and NADPH consumed in biosynthetic reactions.

will emphasize how these processes differ from and resemble the metabolic processes that occur in animal cells.

1 CHLOROPLASTS AND SOLAR ENERGY

Photosynthesis in green plants takes place in **chloroplasts,** discrete organelles that are descended from cyanobacteria. Like mitochondria, chloroplasts contain their own DNA, in this case coding for 100 to 200 chloroplast proteins. DNA in the cell's nucleus contains close to a thousand more genes whose products are essential for photosynthesis.

The chloroplast is enclosed by a porous outer membrane and an ion-impermeable inner membrane (Fig. 13-3). The inner compartment, called the **stroma,** is analogous to the mitochondrial matrix and is rich in enzymes, including those required for carbohydrate synthesis. Within the stroma is a membranous structure called the **thylakoid.** Unlike the planar or tubular mitochondrial cristae (see Fig. 12-3), the thylakoid membrane folds into stacks of flattened vesicles and encloses a compartment called the thylakoid lumen. The energy-transducing reactions of photosynthesis take place in the thylakoid membrane. The analogous reactions in photosynthetic bacteria typically take place in folded regions of the plasma membrane.

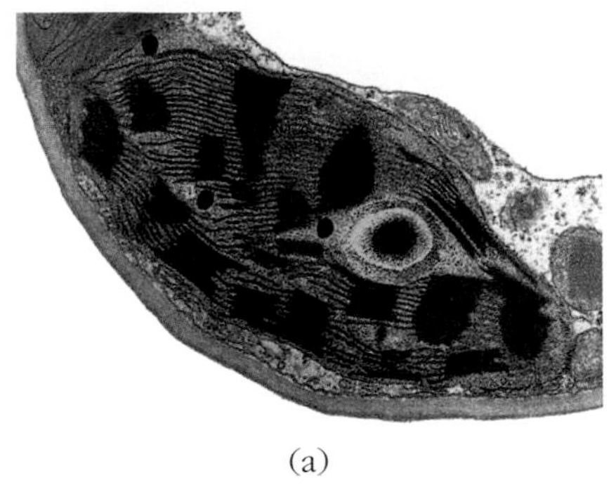

(a)

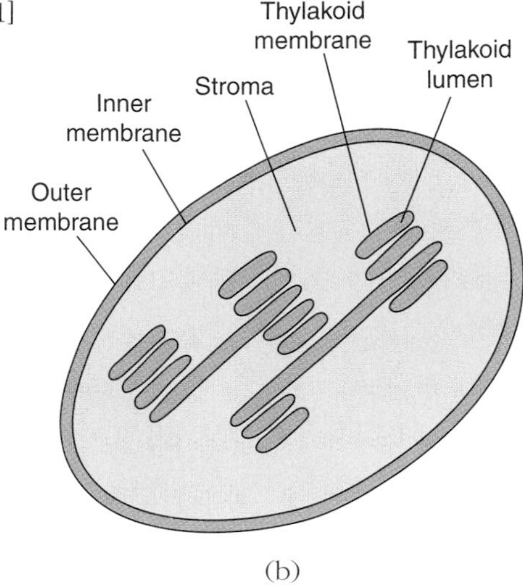

(b)

FIGURE 13-3

The chloroplast.
(a) Electron micrograph of a chloroplast from corn. (b) Model. The stacked thylakoid membranes are known as grana (singular, granum). [*Dr. George Chapman/ Visuals Unlimited.*]

How do plants absorb light?

Light can be considered as both a wave and a particle, the **photon.** The energy (E) of a photon depends on its wavelength, as expressed by **Planck's law,**

$$E = \frac{hc}{\lambda} \qquad [13\text{-}1]$$

where h is Planck's constant (6.626×10^{-34} J · s), c is the speed of light (2.998×10^{8} m · s^{-1}), and λ is the wavelength (about 400 to 700 nm for visible light). *This energy is absorbed by the photosynthetic apparatus of the chloroplast and transduced to chemical energy.*

Chloroplasts contain a variety of light-absorbing groups called pigments or **photoreceptors** (Fig. 13-4). Chlorophyll is the principal pho-

(a)

Chlorophyll *a*

(b)

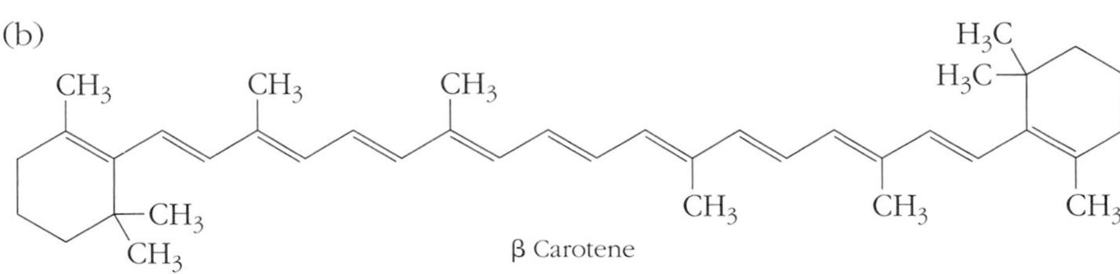

β Carotene

(c)

Phycocyanin

FIGURE 13-4

Some common chloroplast photoreceptors.
(a) Chlorophyll *a*. In chlorophyll *b*, a methyl group (blue) is replaced by an aldehyde group. Chlorophyll resembles the heme groups of hemoglobin and cytochromes (see Fig. 12-12), but it has a central Mg^{2+} rather than an Fe^{2+} ion, it includes a fused cyclopentane ring, and it has a long lipid side chain. (b) The carotenoid β carotene, a precursor of vitamin A (see Box 8-A). (c) Phycocyanin, a linear tetrapyrrole. It resembles an unfolded chlorophyll molecule.

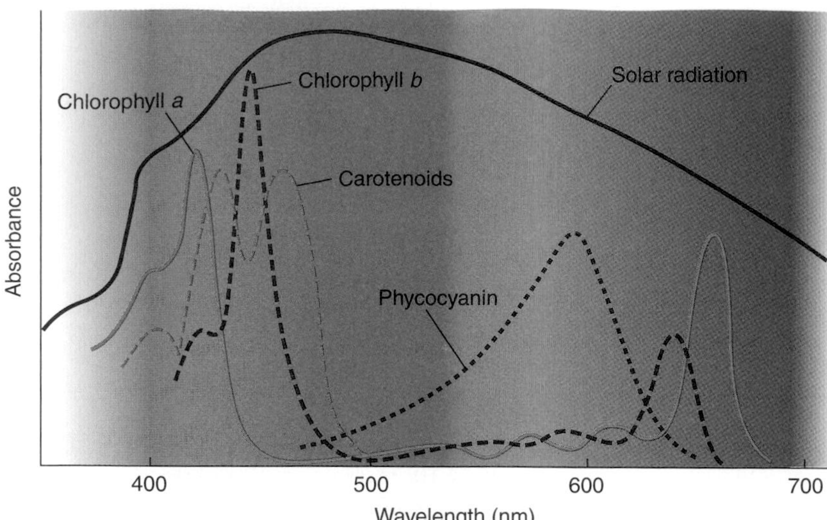

FIGURE 13-5

Visible light absorption by some photosynthetic pigments.
The wavelengths of absorbed light correspond to the peak of the solar energy that reaches the earth.

toreceptor. It appears green because it absorbs both blue and red light. The second most common pigments are the red carotenoids, which absorb blue light. Pigments such as phycocyanin, which absorb longer-wavelength red light, are common in aquatic systems, because water absorbs blue light. Together, these types of pigments absorb all the wavelengths of visible light (Fig. 13-5).

A photosynthetic pigment is a highly conjugated molecule. When it absorbs a photon of the appropriate wavelength, one of its delocalized electrons is promoted to a higher-energy orbital, and the molecule is said to be excited. The excited molecule can return to its low-energy, or ground, state by several mechanisms (Fig. 13-6).

1. The absorbed energy can be lost as heat.
2. The energy can be given off as light, or **fluorescence.** For thermodynamic reasons, the emitted photon has a lower energy (longer wavelength) than the absorbed photon.
3. The energy can be transferred to another molecule. This process is called **exciton transfer** (an exciton is the packet of transferred energy) or resonance energy transfer, since the molecular orbitals of the donor and recipient groups must be oscillating in a coordinated manner in order to transfer energy.
4. An electron from the excited molecule can be transferred to a recipient molecule. In this process, called **photooxidation,** the excited mole-

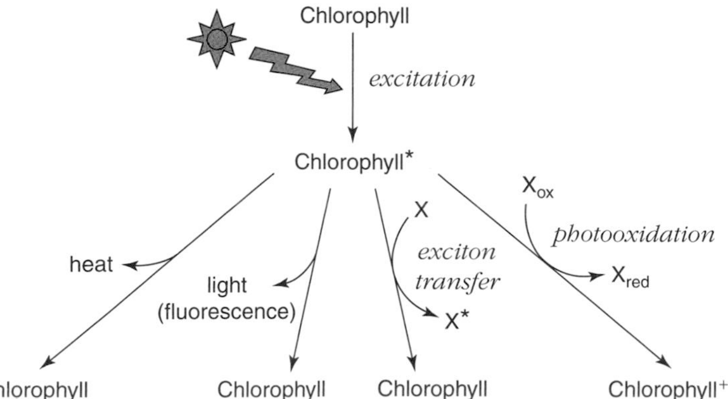

FIGURE 13-6

Dissipation of energy in a photoexcited molecule.
A pigment molecule such as chlorophyll is excited by absorbing a photon. The excited molecule (chlorophyll*) can return to its ground state by one of several mechanisms.

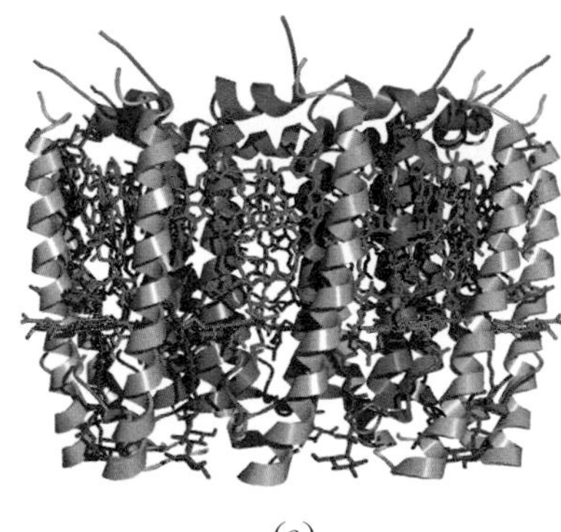
(a)

(b)

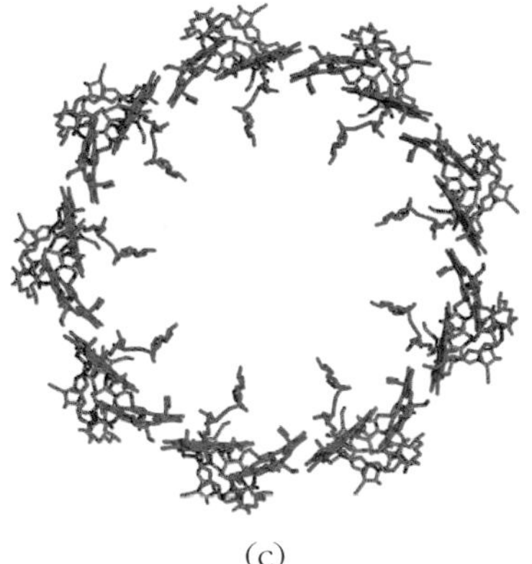
(c)

FIGURE 13-7

A light-harvesting complex from *Rhodopseudomonas acidophila*.
The nine pairs of subunits (light and dark gray) are mostly buried in the membrane and form a scaffold for two rings of chlorophyll molecules (yellow and green) and carotenoids (red). The pigments are all within a few angstroms of each other. (a) Side view. The extracellular side is at the top. (b) Top view. (c) Top view showing only the chlorophyll molecules. The 18 chlorophyll molecules in the inner ring (green) overlap so that excitation energy may be delocalized over the entire ring. [*Structure (pdb 1KZU) determined by R.J. Cogdell, A.A. Freer, N.W. Isaacs, A.M. Hawthornthwaite-Lawless, G. McDermott, M.Z. Papiz, and S.M. Prince.*]

cule becomes oxidized and the acceptor molecule becomes reduced. Another electron-transfer reaction is required to restore the photooxidized molecule to its original reduced state.

All of these energy-transferring processes occur in chloroplasts to some extent, but *exciton transfer and photooxidation are the most important for photosynthesis.*

Light-harvesting complexes

The primary reactions of photosynthesis occur at chlorophyll molecules in complexes called **reaction centers.** However, chloroplasts contain many more chlorophyll molecules and other pigments than reaction centers. *Many of these extra, or* ***antenna,*** *pigments are located in membrane proteins called* ***light-harvesting complexes*****.** Over 30 different kinds of light-harvesting complexes have been characterized, and they are remarkable for their regular geometry. For example, one light-harvesting complex in purple bacteria consists of 18 polypeptide chains holding two concentric rings of chlorophyll molecules, plus carotenoids (Fig. 13-7). This artful arrangement of light-absorbing groups is essential for the function of the light-harvesting complex.

The protein microenvironment of each photoreceptor influences the wavelength (and therefore the energy) of the photon it can absorb (just as cytochrome protein structure influences the reduction potential of its heme group; see Section 12-2). Consequently, *the various light-harvesting complexes with their multiple pigments can absorb light of many different wavelengths.* Within a light-harvesting complex, the precisely aligned pigment molecules can quickly transfer their energy to other pigments. *Exciton transfer eventually brings the energy to the chlorophyll at the reaction center* (Fig. 13-8). Without

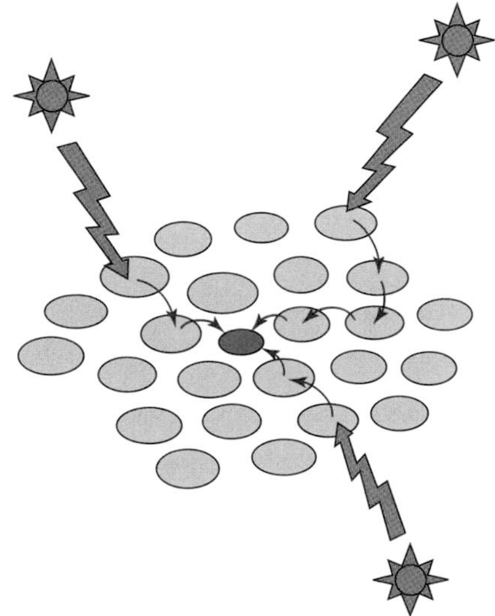

FIGURE 13-8

Function of light-harvesting complexes.
A typical photosynthetic system consists of a reaction center (dark green) surrounded by light-harvesting complexes (light green), whose multiple pigments absorb light of different wavelengths. Exciton transfer funnels this captured solar energy to the chlorophyll at the reaction center. Because the exciton must move from higher-energy to lower-energy states, the antenna pigments farthest from the reaction center have the highest-energy excited states.

light-harvesting complexes to collect and concentrate light, the reaction center chlorophyll could collect only a small fraction of the incoming solar radiation. Even so, a leaf captures only about 1% of the available solar energy.

During periods of high light intensity, some accessory pigments may function to dissipate excess solar energy as heat so that it does not damage the photosynthetic apparatus by inappropriate photooxidation. Various pigment molecules may also act as photosensors to regulate the plant's growth rate and shape and to coordinate the plant's activities—such as germination, flowering, and dormancy—according to daily or seasonal light levels. In fact, photosensors are found in many organisms, not just plants (Box 13-A).

REVIEW LEARNING OBJECTIVE 1

- How is a photon's energy related to its wavelength?
- Why is it advantageous for photosynthetic pigments to absorb different colors of light?
- Describe the four mechanisms by which a photoexcited molecule can return to its ground state.
- What is the function of a light-harvesting complex?

2 THE LIGHT REACTIONS

In plants and cyanobacteria, the energy captured by the antenna pigments of the light-harvesting complexes is funneled to two photosynthetic reaction centers. *Excitation of the reaction centers drives a series of oxidation–reduction reactions whose net results are the oxidation of water, the reduction of $NADP^+$, and the generation of a transmembrane proton gradient that powers ATP synthesis.* These events are known as the **light reactions** of photosynthesis. (Photosynthetic bacteria undertake similar reactions but have a single reaction center and do not produce oxygen.) The two photosynthetic reaction centers that mediate the input of light energy are part of protein complexes called Photosystem I and Photosystem II. These, along with other integral and peripheral proteins of the thylakoid membrane, operate in a series, much like the mitochondrial electron transport chain.

Photosystem II is a light-activated oxidation–reduction enzyme

In plants and cyanobacteria, the light reactions begin with Photosystem II (the number indicates that it was the second to be discovered). This integral membrane protein complex is dimeric, with more bulk on the lumenal side of the thylakoid membrane than on the stromal side. The cyanobacterial Photosystem II contains at least 17 subunits (14 of them integral membrane proteins). Its numerous prosthetic groups include light-absorbing pigments and redox-active cofactors (Fig. 13-9).

At least 50 chlorophyll molecules in Photosystem II function as internal antennas, funneling energy to the two reaction centers, each of which includes a chlorophyll group known as P680 (680 nm is the wavelength of one of its absorption peaks). In some bacteria, the reaction center includes a "special pair" of overlapping chlorophyll molecules that are electronically coupled and function as a single unit. When P680 is excited, as indicated by the

A CLOSER LOOK

BOX 13-A Cryptochromes: animal photosensors

It is not surprising that plants have evolved mechanisms for sensing ambient light, since their ability to sustain anabolic activities depends profoundly on the supply of solar energy. Even nonphotosynthetic organisms whose metabolisms are not so closely tied to the sun modulate their activities according to the daily light–dark cycle. Such **circadian cycles** (from the Latin *circa,* around and *dies,* day) allow organisms to enhance their survival by hunting, hiding, or sleeping at different intervals during a 24-hour day.

In all species, circadian cycles are maintained by an internal "clock" of oscillating biochemical processes, such as synthesis of a protein that inhibits its own production when its concentration rises. The clock is set daily by the sun (or in the laboratory by artificial light). Circadian clocks may also sense seasonal changes in day length to prepare organisms for hibernation, migration, mating, and so on.

In plants, the circadian clock is set by photoreceptors known as phytochromes and cryptochromes, which absorb red and blue light, respectively. Cryptochromes are also present in the eyes of fruit flies and mammals, where they function—in concert with several other proteins—to set the circadian clock. Cryptochromes are flavoproteins containing a flavin (the same group found in FAD and FMN; see Figs. 11-5 and 12-8) and another pigment known as a pterin. The light-absorbing portions of these groups are shown below:

Flavin

Pterin

The flavin absorbs light in the range of 420 to 480 nm. The pterin probably acts as an antenna pigment.

Cryptochromes get their name from the long-hidden nature of their light-absorbing groups. The crystal structure of an algal FAD-containing cryptochrome is shown here.

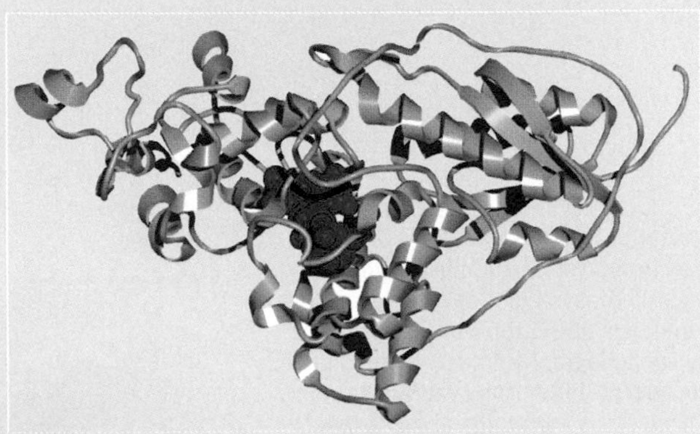

A cryptochrome most likely transduces a light signal to a biochemical signal through electron transfer, although its redox partners have not been identified. There is some evidence that cryptochromes interact with DNA-binding proteins to regulate gene expression. Understanding the role of cryptochromes in setting and maintaining circadian clocks may have practical implications for treating severe jet lag and sleep disorders.

[*Chlamydomonas* cryptochrome F structure (pdb 1CFM) determined by Y.I. Chi, L.S. Huang, Z. Zhang, J.G. Fernandez-Velasco, R. Malkin, and E.A. Berry.]

notation P680*, it quickly gives up an electron, dropping to a lower-energy state, $P680^+$. In other words, light has oxidized P680. The photooxidized chlorophyll molecule must be reduced in order to return to its original state.

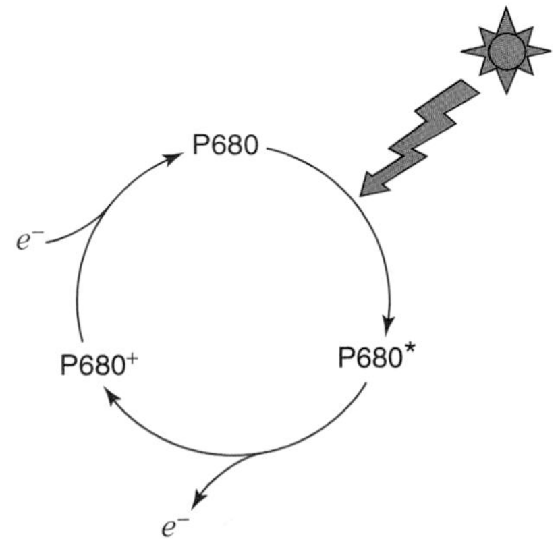

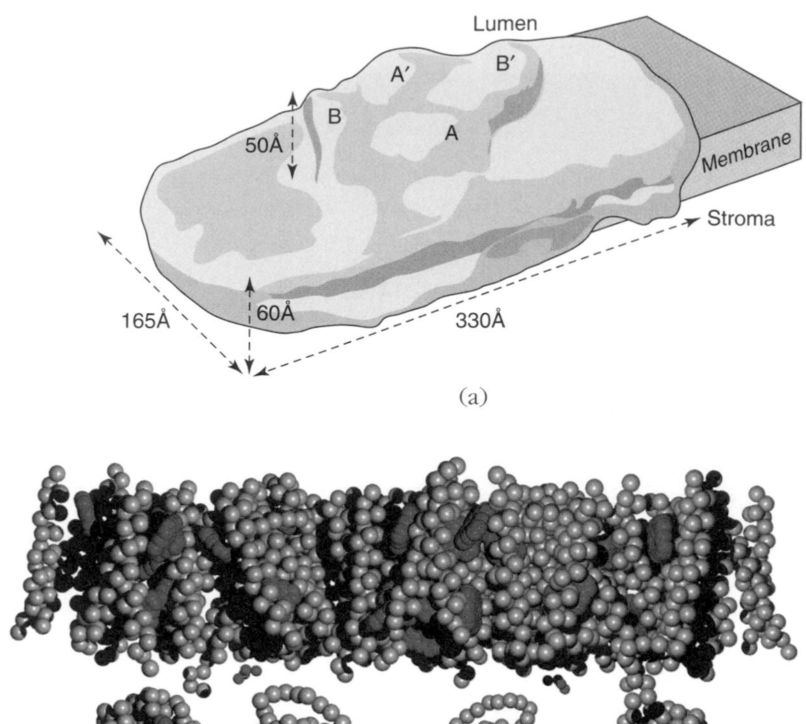

FIGURE 13-9

Structure of Photosystem II. (a) Electron crystallographic image of spinach Photosystem II at 24 Å resolution. The lumen is at the top. (b) Incomplete X-ray crystallographic image of the smaller Photosystem II from the cyanobacterium *Synechococcus elongatus* at 3.8 Å resolution. Polypeptide α carbons are shown in gray, and prosthetic groups are in green. In this image, the lumen is at the bottom. [*Part (a) courtesy James Barber; X-ray structure (pdb 1FE1) determined by A. Zouni, H.-T. Witt, J. Kern, P. Fromme, N. Krauss, W. Saenger, and P. Orth.*]

We will see in the next section that $P680^+$ reduction requires electrons derived from water.

P680 is located near the lumenal side of Photosystem II. The electron given up by photooxidized P680 travels through several redox groups (Fig. 13-10). Although the prosthetic groups in Photosystem II are arranged more

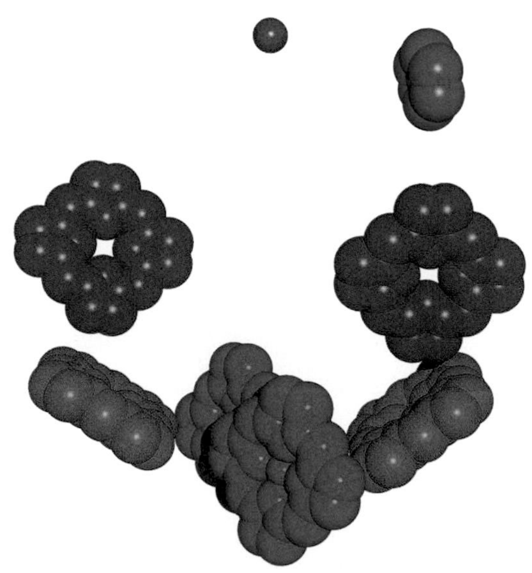

FIGURE 13-10

Arrangement of prosthetic groups in *Synechococcus* Photosystem II. One of the two chlorophyll groups colored green is the photooxidizable P680. The two "accessory" chlorophyll groups, colored yellow, do not undergo oxidation or reduction. An electron from P680 travels to one of the pheophytin groups (orange), which are essentially chlorophyll molecules without the central Mg^{2+} ion. Next, the electron is transferred to a tightly bound plastoquinone molecule (blue) and then to a loosely bound plastoquinone (not shown). An iron atom (red) may assist the final electron transfer. The lipid tails of the prosthetic groups are not shown.

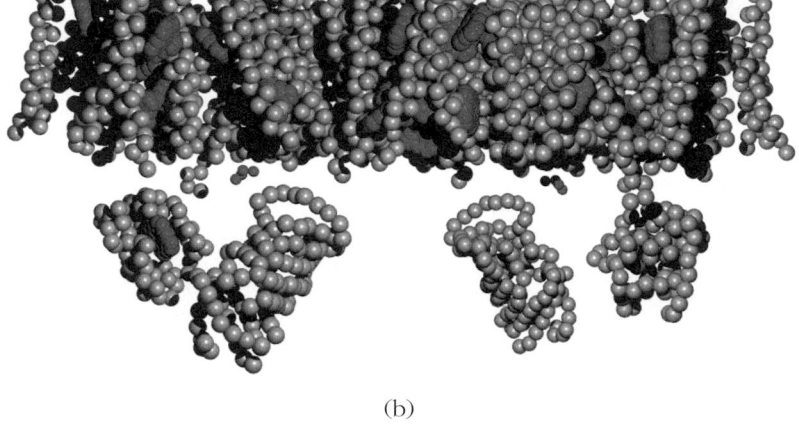

or less symmetrically, they do not all directly participate in electron transfer. The electron eventually reaches a plastoquinone molecule on the stromal side of Photosystem II. Plastoquinone (PQ) is similar to mammalian mitochondrial ubiquinone (see page 285).

$[CH_2{-}CH{=}C(CH_3){-}CH_2]_n{-}H$

Plastoquinone

It functions in the same way as a two-electron carrier. The fully reduced plastoquinol (PQH_2) joins a pool of plastoquinones that are soluble in the thylakoid membrane. Two electrons (two photooxidations of P680) are required to fully reduce plastoquinone to PQH_2. This reaction also consumes two protons, which are taken from the stroma.

The oxygen-evolving complex of Photosystem II oxidizes water

The electrons that reduce photooxidized P680 are derived from the oxidation of water to O_2 by a lumenal portion of Photosystem II called the oxygen-evolving center.

$$2\ H_2O \rightarrow O_2 + 4\ H^+ + 4\ e^-$$

This reaction is rapid, with about 50 O_2 produced per second per Photosystem II, and generates most of the earth's atmospheric O_2 (Fig. 13-11).

The catalyst for the water-splitting reaction is a cofactor with the approximate composition $Mn_4CaCl_{1-2}O_x(HCO_3)_y$. This unusual inorganic cofactor has the same composition in all Photosystem II complexes, which suggests a unique chemistry that has remained unaltered for about 2.5 billion years. No synthetic catalyst can match the manganese cluster in its ability to extract electrons from water to form O_2.

During water oxidation, the manganese cluster undergoes changes in its oxidation state, somewhat reminiscent of the changes in the Fe-Cu binuclear center of cytochrome *c* oxidase (mitochondrial Complex IV; see Fig. 12-18), which carries out the reverse reaction. The four water-derived protons are released into the thylakoid lumen, contributing to a drop in pH relative to the stroma. A tyrosine radical (Y·)

Tyrosine radical

in Photosystem II transfers each of the four water-derived electrons to $P680^+$ (a tyrosine radical also plays a role in electron transfer in cytochrome *c* oxidase; see Section 12-2).

The oxidation of water is a thermodynamically demanding reaction because O_2 has an extremely high reduction potential (+0.815 V) and electrons spontaneously flow from a group with a lower reduction potential to a group with a higher reduction potential (see Section 12-1). In fact, P680 is the most powerful biological oxidant, with a reduction potential of about +1.15 V.

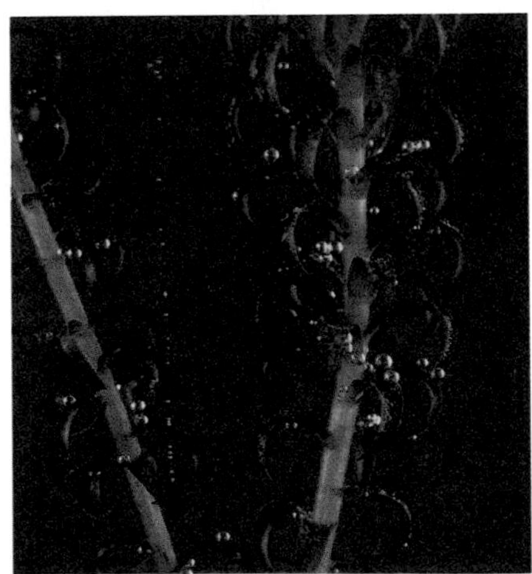

FIGURE 13-11

Photosynthetic oxygen evolution. The O_2 produced by the activity of Photosystem II is visible as "champagne bubbles" emanating from a freshwater plant. [*Nigel Cattlin/Holt Studios International/Photo Researchers.*]

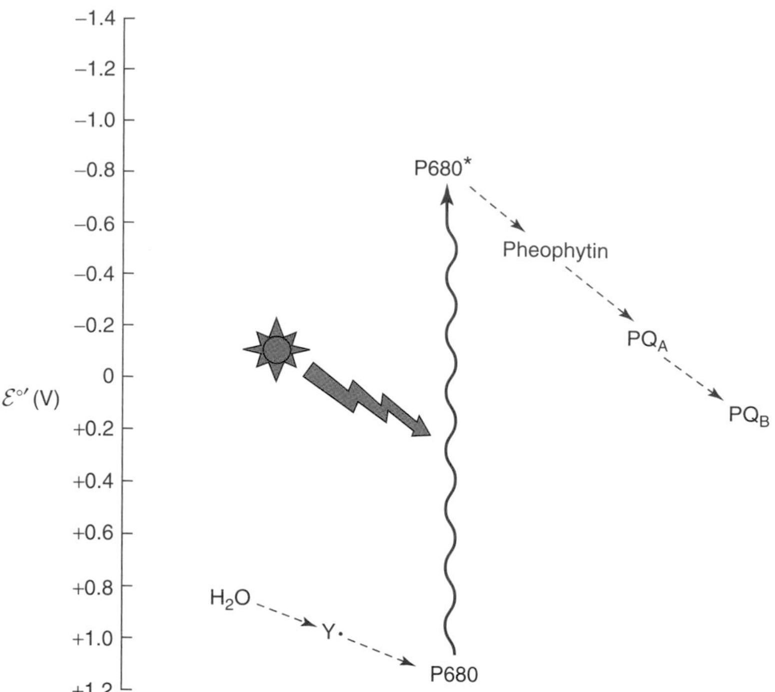

FIGURE 13-12

Reduction potential and electron flow in Photosystem II.
Electrons flow spontaneously from a group with a low reduction potential to a group with a high reduction potential. The transfer of electrons from H_2O to plastoquinone (dashed red lines) is made possible by the excitation of P680 (wavy arrow), which lowers its reduction potential.

Upon photooxidation, the reduction potential of P680 (now P680*) is dramatically diminished, to about −0.8 V. *This low reduction potential allows P680* to surrender an electron to a series of groups with increasingly positive reduction potentials* (Fig.13-12). The overall result is that the input of solar energy allows an electron to travel a thermodynamically favorable path from water to plastoquinone. *Four photooxidation events in Photosystem II are required to oxidize two H_2O molecules and produce one O_2 molecule.* Figure 13-13 summarizes the functions of Photosystem II.

REVIEW LEARNING OBJECTIVE 2

- How does absorption of a photon drive electron transfer from water to plastoquinone?

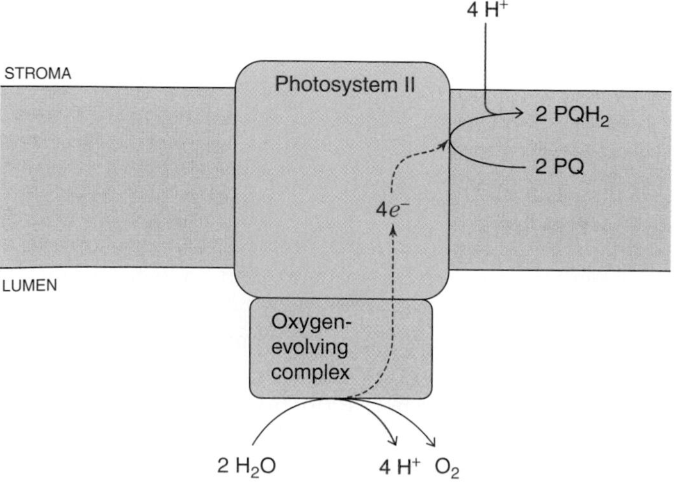

FIGURE 13-13

Photosystem II function.
For every oxygen molecule evolved, two plastoquinone molecules are reduced.

Cytochrome b_6f links Photosystems I and II

After they leave Photosystem II as plastoquinol, electrons reach a second membrane-bound protein complex known as cytochrome b_6f. This complex resembles mitochondrial Complex III (also called cytochrome bc_1)—from the entry of electrons in the form of a reduced quinone, through the circular flow of electrons among its redox groups, to the final transfer of electrons to a mobile electron carrier.

The cytochrome b_6f complex contains eight subunits in each of its monomeric halves. Three subunits bear electron-transporting prosthetic groups. One of these subunits is cytochrome b_6, which is homologous to mitochondrial cytochrome *b*. The second is cytochrome *f*, whose heme group is actually of the *c* type. Although it shares no sequence homology with mitochondrial cytochrome c_1, it functions similarly. The chloroplast complex also contains a Rieske iron-sulfur protein with a 2Fe–2S group. This subunit is structurally and functionally related to its mitochondrial counterpart. In fact, electron crystallography of cytochrome b_6f reveals that the chloroplast and mitochondrial complexes are similar, particularly in the transmembrane region (Fig. 13-14). However, the cytochrome b_6f complex also contains subunits with prosthetic groups that are absent in the mitochondrial complex: a chlorophyll molecule and a β carotene. These light-absorbing molecules do not appear to participate in electron transfer and may instead help regulate the activity of cytochrome b_6f by registering the amount of available light.

Electron flow in the cytochrome b_6f complex follows a cyclic pattern that is probably identical to the Q cycle in mitochondrial Complex III (see Fig. 12-14). However, in chloroplasts, the final electron acceptor is not cytochrome *c* but plastocyanin, a small protein with an active-site copper ion (Fig. 13-15). Plastocyanin functions as a one-electron carrier by cycling between the Cu^+ and Cu^{2+} oxidation states. Like cytochrome *c*, plastocyanin is a peripheral membrane protein; it picks up electrons at the lumenal surface of cytochrome b_6f and delivers them to another integral membrane protein complex, in this case Photosystem I.

The net result of the cytochrome b_6f Q cycle is that for every two electrons emanating from Photosystem II, four protons are released into the thylakoid lumen. Since the oxidation of 2 H_2O is a four-electron reaction, *the production of one molecule of O_2 causes the cytochrome b_6f complex to produce 8 lumenal H^+* (Fig. 13-16). The resulting pH gradient between the stroma and the lumen is a source of free energy that drives ATP synthesis, as described below.

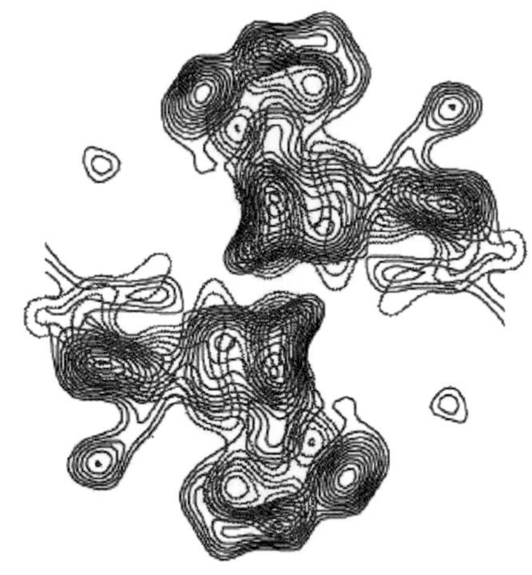

FIGURE 13-14

Comparison of transmembrane domains of cytochrome b_6f and mitochondrial cytochrome bc_1.
The transmembrane region of the chicken mitochondrial cytochrome bc_1 (red) is superimposed on a cryoelectron microscopic model of the transmembrane region of cytochrome b_6f from the alga *Chlamydomonas*. [*Courtesy Cecile Breyton,* Biochim. Biophys. Acta ***1459****, 467–474 (2000).*]

FIGURE 13-15

Plastocyanin.
The redox-active copper ion (green) is coordinated by a Cys, a Met, and two His residues (yellow). [*Structure of plastocyanin from poplar leaves (pdb 1PLC) determined by J.M. Guss and H.C. Freeman.*]

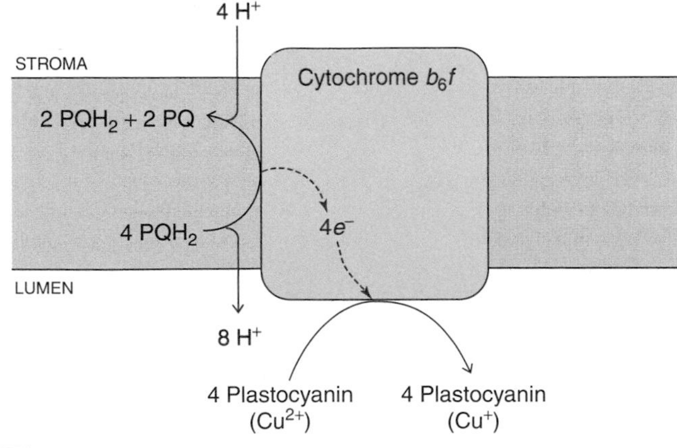

FIGURE 13-16

Cytochrome b_6f function.
The stoichiometry shown for the cytochrome b_6f Q cycle reflects the four electrons released by the oxygen-evolving complex of Photosystem II.

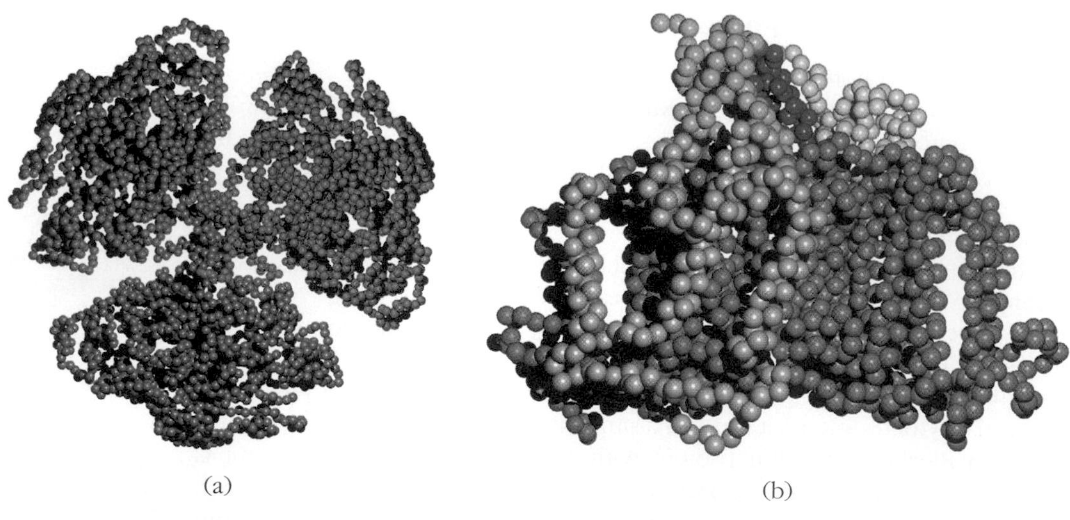

FIGURE 13-17

Structure of Photosystem I from *Synechococcus*.
(a) Top view of the trimeric complex. Only the α carbons are shown. (b) Side view of a monomer, which contains 11 different proteins. The stroma is at the top. [*Structure (pdb 1C51) determined by O. Klukas, W.-D. Schubert, P. Jordan, N. Krauss, P. Fromme, H.T. Witt, and W. Saenger.*]

A second photooxidation occurs at Photosystem I

Photosystem I, like Photosystem II, is a large protein complex containing multiple pigment molecules. The Photosystem I in the cyanobacterium *Synechococcus* is a symmetric trimer with 31 transmembrane helices to anchor each monomer in the membrane (Fig. 13-17). Ninety-six chlorophyll molecules and 22 carotenoids operate as a built-in light-harvesting complex for Photosystem I.

In the core of each monomer, one or both of a pair of chlorophyll molecules constitute the photoactive group known as P700 (it has a slightly longer-wavelength absorbance maximum than P680). Like P680, P700 undergoes exciton transfer from an antenna pigment. P700* gives up an electron to achieve a low-energy oxidized state, $P700^+$. The group is then reduced by accepting an electron donated by plastocyanin.

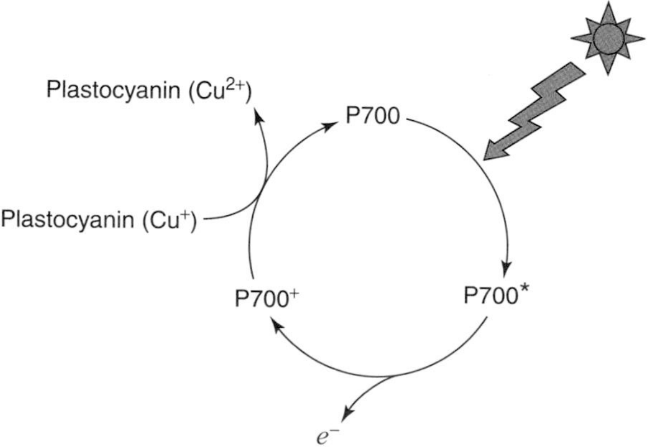

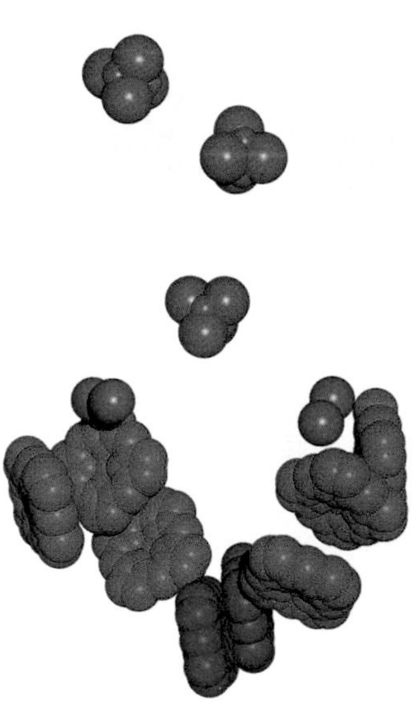

FIGURE 13-18

Prosthetic groups in Photosystem I.
The groups include P700 (one or both of the green chlorophyll molecules), "accessory" chlorophylls (yellow), quinones (marked by blue spheres), and 4Fe–4S clusters (orange).

P700 is not a particularly good reducing agent (its reduction potential is relatively high, about +0.45 V). However, excited P700 (P700*) has an extremely low $\mathcal{E}^{\circ\prime}$ value (about −1.3 V), so electrons can spontaneously flow from P700* to the other redox groups of Photosystem I. These groups include four additional chlorophyll molecules, quinones, and iron-sulfur clusters of the 4Fe–4S type (Fig. 13-18). As in Photosystem II, these prosthetic groups are arranged with approximate symmetry. However, in Photosystem I, all the redox groups appear to undergo oxidation and reduction.

Each electron given up by photooxidized P700 eventually reaches ferredoxin, a small peripheral protein on the stromal side of the thylakoid membrane. Ferredoxin undergoes a one-electron reduction at a 2Fe–2S cluster (Fig. 13-19). Reduced ferredoxin participates in two electron-transport pathways in the chloroplast, which are known as noncyclic and cyclic electron flow.

In **noncyclic electron flow,** ferredoxin serves as a substrate for ferredoxin–$NADP^+$ reductase. This stromal enzyme uses two electrons (from two separate ferredoxin molecules) to reduce $NADP^+$ to NADPH (Fig. 13-20). *The net result of noncyclic electron flow is therefore the transfer of electrons from water, through Photosystem II, cytochrome b_6f, Photosystem I, and then on to $NADP^+$*. Photosystem I does not contribute to the transmembrane proton gradient except by consuming stromal protons in the reduction of $NADP^+$ to NADPH.

When plotted according to reduction potential, the electron-carrying groups of the pathway from water to $NADP^+$ form a diagram called the **Z-scheme** of photosynthesis (Fig. 13-21). The zigzag pattern is due to the two photooxidation events, which markedly decrease the reduction potentials of P680 and P700. Note that the four-electron process of producing one O_2 and two NADPH is accompanied by the absorption of 8 photons (4 each at Photosystem II and Photosystem I).

In **cyclic electron flow,** electrons from Photosystem I do not reduce $NADP^+$ but instead return to the cytochrome b_6f complex. There, the electrons are transferred to plastocyanin and flow back to Photosystem I to reduce photooxidized P700*. Meanwhile, plastoquinol molecules circulate between the two quinone-binding sites of cytochrome b_6f, so that protons are translocated from the stroma to the lumen, as in the Q cycle (Fig. 13-22). Cyclic electron flow requires the input of light energy at Photosystem I but not Photosystem II. During cyclic flow, no free energy is recovered in the form of the reduced cofactor NADPH, but free energy is conserved in the formation of a transmembrane proton gradient by the activity of the cytochrome b_6f complex.

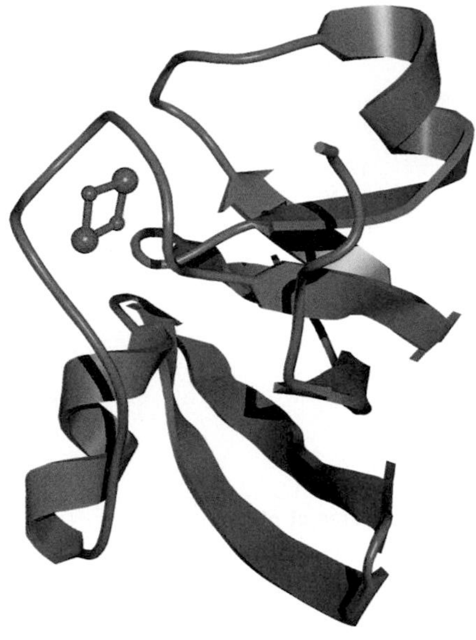

FIGURE 13-19

Ferredoxin.
The 2Fe–2S cluster is shown in orange. [*Structure of ferredoxin from the cyanobacterium* Anabaena *(pdb 1CZP) determined by R. Morales, M.H. Charon, and M. Frey.*]

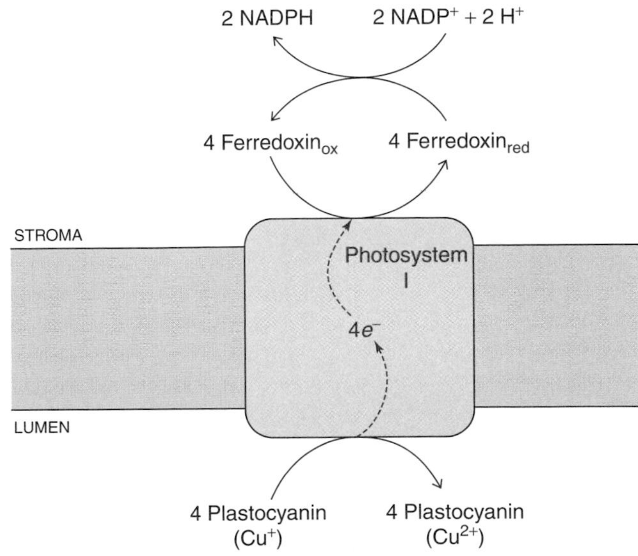

FIGURE 13-20

Noncyclic electron flow through Photosystem I.
Electrons donated by plastocyanin are transferred to ferredoxin and used to reduce $NADP^+$. The stoichiometry reflects the four electrons released by the oxidation of 2 H_2O in Photosystem II. Therefore, 2 NADPH are produced for every molecule of O_2.

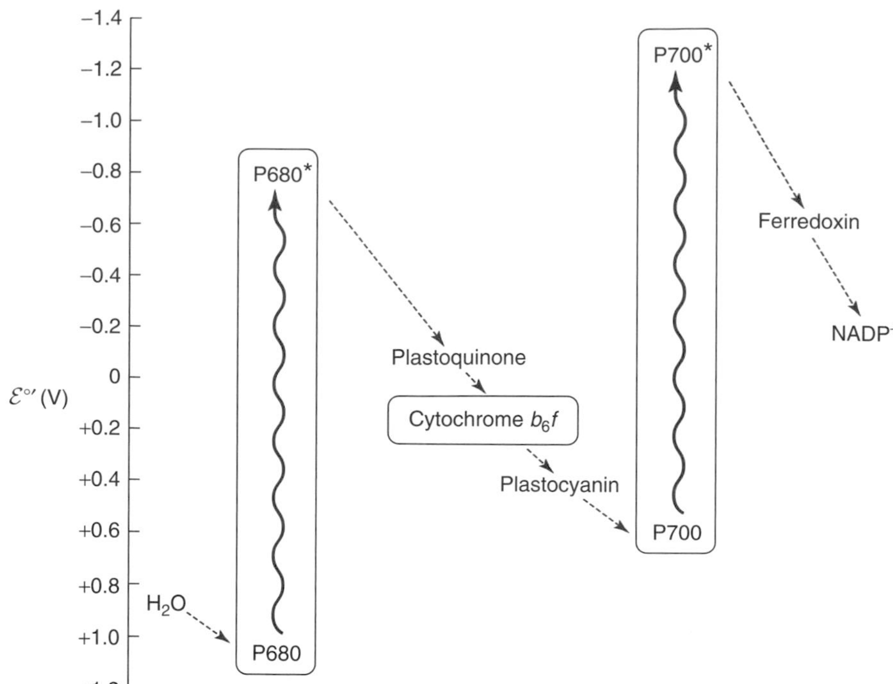

FIGURE 13-21

The *Z*-scheme of photosynthesis. The reduction potentials of the major components are indicated (the individual redox groups within Photosystem II, cytochrome b_6f, and Photosystem I are not shown). Excitation of P680 and P700 ensures that electrons follow a thermodynamically favorable pathway to groups with increasing reduction potential.

Consequently, *cyclic electron flow augments ATP generation by chemiosmosis* (in some bacteria with just a single reaction center, electrons flow through a similar pathway that does not produce O_2 or NADPH). By varying the proportion of electrons that follow the noncyclic and cyclic pathways through Photosystem I, a photosynthetic cell can vary the proportions of ATP and NADPH produced by the light reactions.

REVIEW LEARNING OBJECTIVE 3

- Summarize the functions of Photosystem II, the oxygen-evolving complex, plastoquinone, the cytochrome b_6f complex, plastocyanin, Photosystem I, and ferredoxin.
- Describe the *Z*-scheme of photosynthesis and explain its zigzag shape.

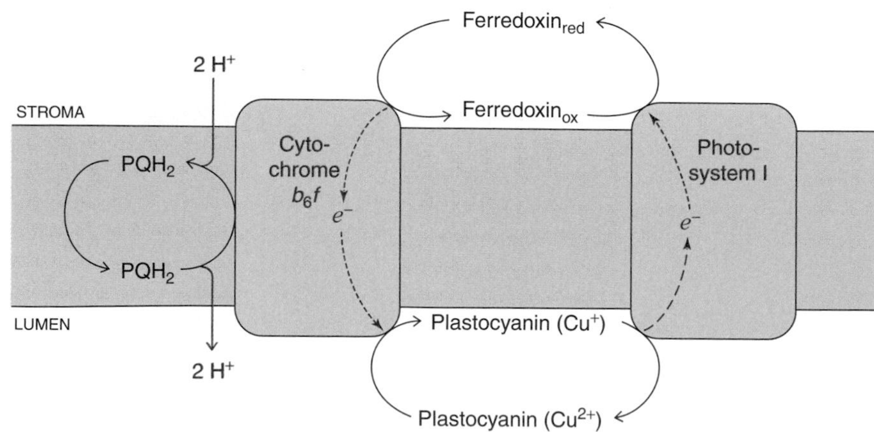

FIGURE 13-22

Cyclic electron flow. Electrons circulate between Photosystem I and the cytochrome b_6f complex. No NADPH or O_2 is produced, but the activity of cytochrome b_6f builds up a proton gradient that drives ATP synthesis.

REVIEW LEARNING OBJECTIVE 4

- Discuss the yields of O_2, NADPH, and ATP in cyclic and noncyclic electron flow.
- Compare and contrast the chloroplast light reactions and mitochondrial electron transport.

Photophosphorylation: ATP synthesis by chemiosmosis

Chloroplasts and mitochondria use the same mechanism to synthesize ATP: *They couple the dissipation of a proton gradient to the phosphorylation of ADP.* In photosynthetic organisms, this process is called **photophosphorylation.** Chloroplast ATP synthase is highly homologous to mitochondrial and bacterial ATP synthases. The CF_1CF_0 complex ("C" indicates chloroplast) consists of a proton-translocating integral membrane component (CF_0) mechanically linked to a soluble CF_1 component where ATP synthesis occurs by a binding change mechanism (as described in Fig. 12-25). The movement of protons from the thylakoid lumen to the stroma provides the free energy to drive ATP synthesis (Fig. 13-23).

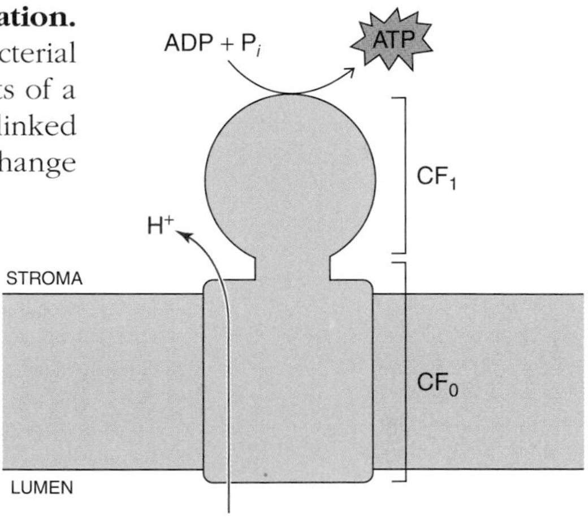

FIGURE 13-23

Photophosphorylation. As protons traverse the CF_0 component of chloroplast ATP synthase (following their concentration gradient from the lumen to the stroma), the CF_1 component carries out ATP synthesis.

As in mitochondria, the proton gradient has both chemical and electrical components. In chloroplasts, the pH gradient (about 3.5 pH units) is much larger than in mitochondria (about 0.75 units). However, in chloroplasts, the electrical component is less than in mitochondria because of the permeability of the thylakoid membrane to ions such as Mg^{2+} and Cl^-. Diffusion of these ions tends to minimize the difference in charge due to protons.

Assuming linear electron flow, 8 photons are absorbed (4 by Photosystem II, and 4 by Photosystem I) to generate 4 lumenal protons from the oxygen-evolving complex and 8 protons from the cytochrome b_6f complex. Theoretically, these 12 protons can drive the synthesis of about 3 ATP, which is consistent with experimental results showing approximately 3 ATP generated for each molecule of O_2.

REVIEW LEARNING OBJECTIVE 5

- How does photophosphorylation resemble oxidative phosphorylation?
- Which reactions contribute to the pH gradient that drives the activity of chloroplast ATP synthase?

3 CARBON FIXATION

The production of ATP and NADPH by the photoactive complexes of the thylakoid membrane (or bacterial plasma membrane) is only part of the story of photosynthesis. The rest of this chapter focuses on the use of the products of the light reactions in the so-called **"dark" reactions.** These reactions, which occur in the chloroplast stroma, fix atmospheric carbon dioxide in biologically useful organic molecules.

Rubisco catalyzes CO_2 fixation

Carbon dioxide is fixed by the action of ribulose bisphosphate carboxylase/oxygenase, or rubisco. *This enzyme adds CO_2 to a five-carbon sugar and*

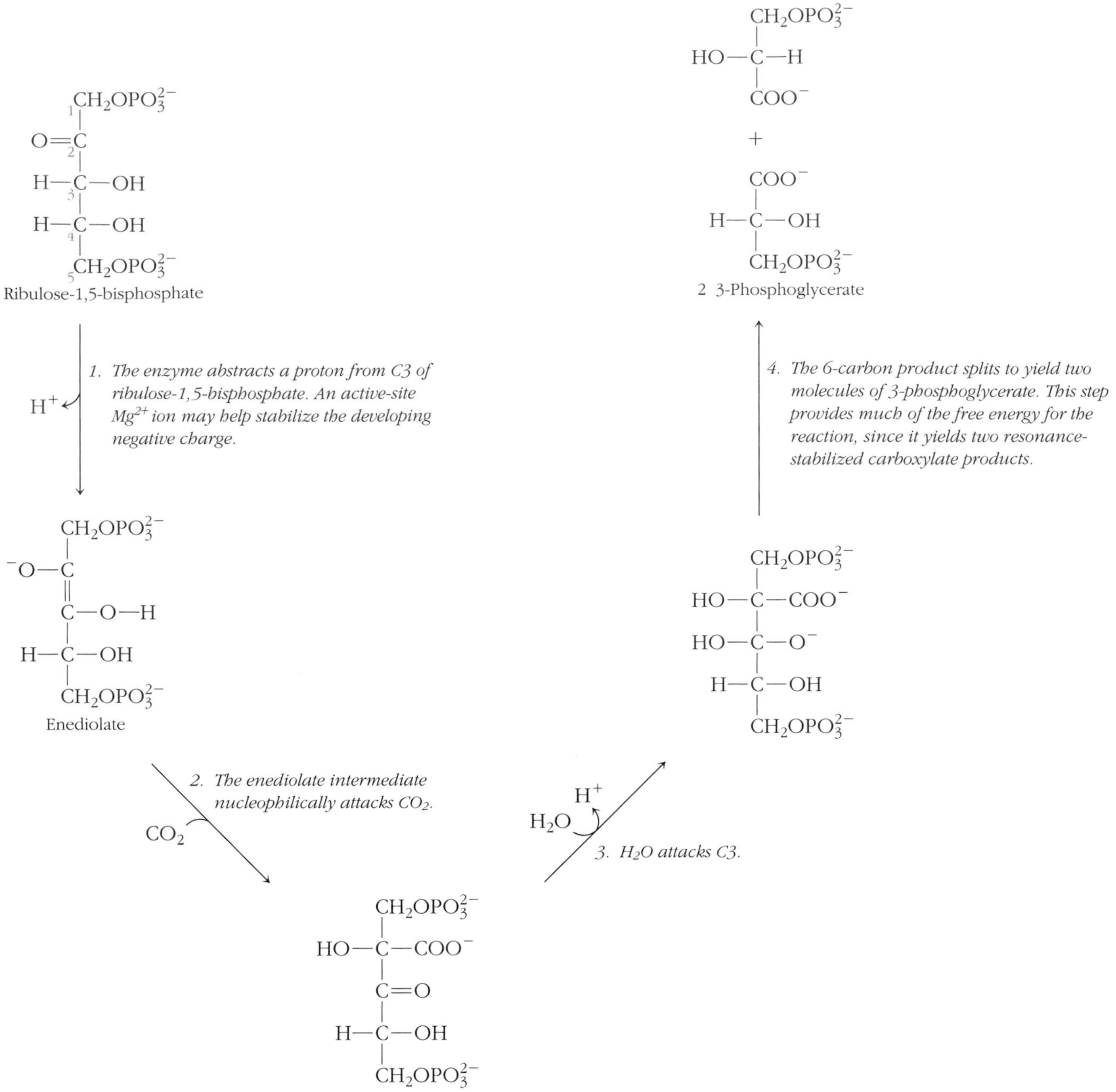

Ribulose-1,5-bisphosphate

1. *The enzyme abstracts a proton from C3 of ribulose-1,5-bisphosphate. An active-site Mg^{2+} ion may help stabilize the developing negative charge.*

Enediolate

2. *The enediolate intermediate nucleophilically attacks CO_2.*

3. *H_2O attacks C3.*

4. *The 6-carbon product splits to yield two molecules of 3-phosphoglycerate. This step provides much of the free energy for the reaction, since it yields two resonance-stabilized carboxylate products.*

2 3-Phosphoglycerate

FIGURE 13-24

The rubisco carboxylation reaction.

then cleaves the product to two three-carbon units (Fig. 13-24). This reaction itself does not require ATP or NADPH, but the reactions that transform the rubisco reaction product, 3-phosphoglycerate, to the three-carbon sugar glyceraldehyde-3-phosphate require both ATP and NADPH, as we shall see below.

Three-carbon compounds are the biosynthetic precursors of monosaccharides, amino acids, and—indirectly—nucleotides. They also give rise to the two-carbon acetyl units used to build fatty acids. The metabolic importance of these small molecular building blocks is one reason why the scheme shown in Figure 13-2 presents photosynthesis as a process in which CO_2 is converted to two- and three-carbon intermediates.

Rubisco is a notable enzyme, in part because its activity directly or indirectly sustains most of the earth's biomass. Plant chloroplasts are packed with the enzyme, which accounts for about half of the chloroplast's protein content. Rubisco is easily the most abundant biological catalyst. One reason for its high concentration is that it is not a particularly efficient enzyme. Its catalytic output is only about three CO_2 fixed per second.

Bacterial rubisco is usually a small dimeric enzyme (Fig. 13-25a), whereas the plant enzyme is a large multimer of eight large and eight small subunits (Fig. 13-25b). In some archaea, rubisco has 10 identical subunits. Enzymes with multiple catalytic sites typically exhibit cooperative behavior and are regulated allosterically, but this does not seem to be true for plant rubisco, whose eight active sites operate independently. Multimerization may simply be an efficient way to pack more active sites into the limited space of the chloroplast.

Despite its metabolic importance, rubisco is not a highly specific enzyme. It also acts as an oxygenase (as reflected in its name) by reacting with O_2, which chemically resembles CO_2. The products of the oxygenase reaction are a three-carbon and a two-carbon compound:

(a)

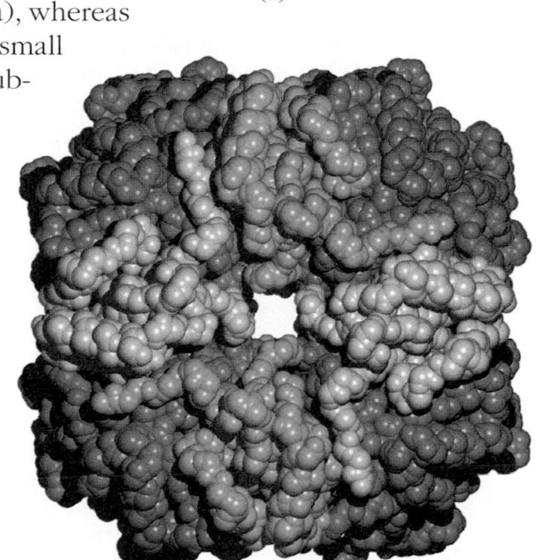

(b)

FIGURE 13-25

Bacterial and plant rubisco.
(a) The enzyme from the bacterium *Rhodospirillum rubrum*. A catalytically essential Mg^{2+} ion (pink) marks the location of each active site. (b) The spinach enzyme. The complex has a mass of approximately 550 kD. The eight catalytic sites are located in the large subunits (dark colors). Only four of eight small subunits (light colors) are visible in this image. [*Structure of bacterial rubisco (pdb 9RUB) determined by T. Lundqvist and G. Schneider; structure of plant rubisco (pdb 1RCX) determined by T.C. Taylor and I. Anderson.*]

$$\underset{\text{Ribulose bisphosphate}}{\begin{array}{c} CH_2OPO_3^{2-} \\ | \\ O{=}C \\ | \\ H{-}C{-}OH \\ | \\ H{-}C{-}OH \\ | \\ CH_2OPO_3^{2-} \end{array}} + O_2 \xrightarrow[\text{rubisco}]{H_2O} \underset{\text{2-Phosphoglycolate}}{\begin{array}{c} CH_2OPO_3^{2-} \\ | \\ {}^{-}O{-}C{=}O \end{array}} + \underset{\text{3-Phosphoglycerate}}{\begin{array}{c} COO^{-} \\ | \\ H{-}C{-}OH \\ | \\ CH_2OPO_3^{2-} \end{array}}$$

The 2-phosphoglycolate product of the rubisco oxygenation reaction is subsequently metabolized by a pathway that consumes ATP and NADPH and produces CO_2. This process, called **photorespiration,** uses the products of the light reactions and therefore wastes some of the free energy of captured photons.

Oxygenase activity is a feature of all known rubisco enzymes and must play an essential role that has been conserved throughout plant evolution. *Photorespiration apparently provides a mechanism for plants to dissipate excess free energy under conditions where the CO_2 supply is insufficient for carbon fixation.* Photorespiration may consume significant amounts of ATP and NADPH at high temperatures, which favor oxygenase activity. Some plants have evolved a mechanism, called the **C_4 pathway,** to minimize photorespiration (Box 13-B).

A CLOSER LOOK

BOX 13-B The C_4 pathway

On hot, bright days, high temperatures favor photorespiration, and CO_2 supplies are low as plants close their stomata (pores in the leaf surface) to avoid evaporative water loss. This combination of events can bring photosynthesis to a halt. Some plants avoid this possibility by stockpiling CO_2 in four-carbon molecules so that photosynthesis can proceed even while stomata are closed (the closed stomata also limit the availability of O_2 for photorespiration).

The mechanism for storing carbon begins with the condensation of bicarbonate (HCO_3^-) with phosphoenolpyruvate to yield oxaloacetate, which is then reduced to malate. These four-carbon acids give the C_4 pathway its name. The subsequent oxidative decarboxylation of malate regenerates CO_2 and NADPH to be used in the Calvin cycle. The three-carbon remnant, pyruvate, is recycled back to phosphoenolpyruvate.

COO^- — $C{-}OPO_3^{2-}$ = CH_2

Phosphoenolpyruvate

CO_2 → (carbonic anhydrase) → HCO_3^-; HCO_3^- in, P_i out (phosphoenolpyruvate carboxylase)

COO^- — $C{=}O$ — CH_2 — COO^-

Oxaloacetate

NADPH in, $NADP^+$ out (malate dehydrogenase)

COO^- — CHOH — CH_2 — COO^-

Malate

$NADP^+$ in, NADPH out (malic enzyme)

Calvin cycle ⟸ CO_2 + NADPH

COO^- — $C{=}O$ — CH_3

Pyruvate

Because the C_4 pathway and the rubisco reaction compete for CO_2, they take place in different types of cells or at different times of day. For example, in some plants, carbon accumulates in mesophyll cells, which are near the leaf surface and lack rubisco. The C_4 compounds then enter bundle sheath cells in the leaf interior, which contain abundant rubisco. In other plants, the C_4 pathway occurs at night, when the stomata are open and water loss is minimal, and carbon is fixed by rubisco during the day.

The C_4 pathway is energetically expensive, so it requires lots of sunlight. Consequently, C_4 plants grow more slowly than conventional, or C_3, plants when light is limited, but have the advantage in hot, dry climates. About 5% of the earth's plants, including the economically important maize (corn), use the C_4 pathway.

[© Terry Brandt/Grant Heilman Photography.]

REVIEW LEARNING OBJECTIVE 6

- What are the products of the two reactions catalyzed by rubisco?
- Compare the physiological implications of carbon fixation and photorespiration.

The Calvin cycle

If rubisco is responsible for fixing CO_2, what is the origin of its other substrate, ribulose bisphosphate? The answer, elucidated over many years by Melvin Calvin, James Bassham, and Andrew Benson, is a metabolic pathway known as the **Calvin cycle.** Early experiments to study the fate of ^{14}C-labeled CO_2 in algae showed that within a few minutes, the cells had synthesized a complex mixture of sugars, all containing the radioactive label. Rearrangements among these sugar molecules generate the five-carbon substrate for rubisco.

The Calvin cycle actually begins with a sugar monophosphate, ribulose-5-phosphate, which is phosphorylated in an ATP-dependent reaction:

$$\begin{array}{c} CH_2OH \\ | \\ C{=}O \\ | \\ H{-}C{-}OH \\ | \\ H{-}C{-}OH \\ | \\ CH_2OPO_3^{2-} \end{array} \xrightarrow[\text{phosphoribulokinase}]{\text{ATP} \quad \text{ADP}} \begin{array}{c} CH_2OPO_3^{2-} \\ | \\ C{=}O \\ | \\ H{-}C{-}OH \\ | \\ H{-}C{-}OH \\ | \\ CH_2OPO_3^{2-} \end{array}$$

Ribulose-5-phosphate → Ribulose-1,5-bisphosphate

The bisphosphate is a substrate for rubisco, as we have already seen:

$$\begin{array}{c} CH_2OPO_3^{2-} \\ | \\ C{=}O \\ | \\ H{-}C{-}OH \\ | \\ H{-}C{-}OH \\ | \\ CH_2OPO_3^{2-} \end{array} \xrightarrow[\text{rubisco}]{CO_2} \begin{array}{c} CH_2OPO_3^{2-} \\ | \\ HO{-}C{-}H \\ | \\ COO^- \\ + \\ COO^- \\ | \\ H{-}C{-}OH \\ | \\ CH_2OPO_3^{2-} \end{array}$$

Ribulose-1,5-bisphosphate → 2 3-Phosphoglycerate

Each 3-phosphoglycerate product of the rubisco reaction is then phosphorylated:

$$\begin{array}{c} O \quad O^- \\ \backslash\!\!\backslash \;\; / \\ C \\ | \\ H{-}C{-}OH \\ | \\ CH_2OPO_3^{2-} \end{array} \xrightarrow[\text{phosphoglycerate kinase}]{\text{ATP} \quad \text{ADP}} \begin{array}{c} O \quad OPO_3^{2-} \\ \backslash\!\!\backslash \;\; / \\ C \\ | \\ H{-}C{-}OH \\ | \\ CH_2OPO_3^{2-} \end{array}$$

3-Phosphoglycerate → 1,3-Bisphosphoglycerate

This reaction is identical to the phosphoglycerate kinase reaction of glycolysis (see Section 10-2). Next, bisphosphoglycerate is reduced by the chloroplast enzyme glyceraldehyde-3-phosphate dehydrogenase, which resembles the glycolytic enzyme but uses NADPH rather than NADH.

$$\begin{array}{c} O \diagdown \; \diagup OPO_3^{2-} \\ C \\ | \\ H{-}C{-}OH \\ | \\ CH_2OPO_3^{2-} \end{array} \xrightarrow[\text{glyceraldehyde-3-phosphate dehydrogenase}]{\text{NADPH} \quad \text{NADP}^+ + P_i} \begin{array}{c} O \diagdown \; \diagup H \\ C \\ | \\ H{-}C{-}OH \\ | \\ CH_2OPO_3^{2-} \end{array}$$

1,3-Bisphosphoglycerate → Glyceraldehyde-3-phosphate

Some glyceraldehyde-3-phosphate is siphoned from the Calvin cycle for metabolic fates such as glucose or amino acid synthesis, and the remainder enters a series of isomerization and group-transfer reactions that regenerate ribulose-5-phosphate. These interconversion reactions are similar to those of the pentose phosphate pathway (Section 10-4). We can represent them simply by showing how carbon atoms are shuffled among three- to seven-carbon sugars:

$$C_3 + C_3 \rightarrow C_6$$
$$C_3 + C_6 \rightarrow C_4 + C_5$$
$$C_3 + C_4 \rightarrow C_7$$
$$C_3 + C_7 \rightarrow C_5 + C_5$$

The net reaction is

$$5\ C_3 \rightarrow 3\ C_5$$

Consequently, if the Calvin cycle starts with three five-carbon ribulose molecules, so that three CO_2 molecules are fixed, the products are six three-carbon glyceraldehyde-3-phosphate molecules, five of which are recycled to form three ribulose molecules, leaving the sixth (representing the three fixed CO_2) as its net product.

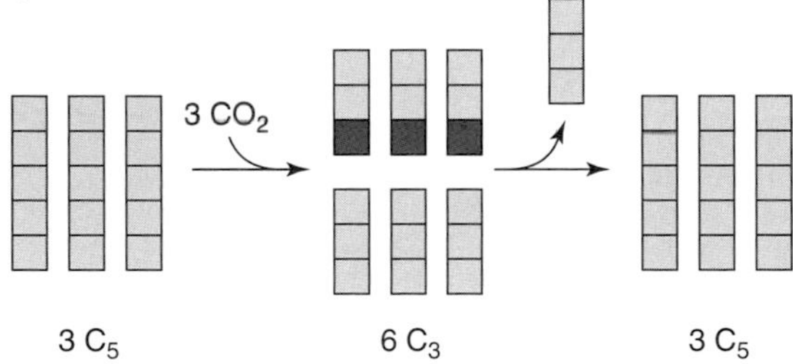

Including the ATP and NADPH cofactors yields a net equation for the Calvin cycle:

$$3\ CO_2 + 9\ ATP + 6\ NADPH \rightarrow \text{Glyceraldehyde-3-phosphate} + 9\ ADP + 8\ P_i + 6\ NADP^+$$

Fixing a single CO_2 therefore requires 3 ATP and 2 NADPH—approximately the same quantity of ATP and NADPH produced by the absorption of 8 photons. The relationship between the number of photons absorbed and the amount of carbon fixed or oxygen released is known as the **quantum yield** of photosynthesis. Keep in mind that the exact number of carbons fixed per photon absorbed depends on factors such as the number of protons translocated per ATP synthesized by the CF_1CF_0 ATP synthase and the ratio of cyclic to noncyclic electron flow in Photosystem I.

REVIEW LEARNING OBJECTIVE 7

- How does the carbon from a molecule of fixed CO_2 become incorporated into other compounds such as monosaccharides?
- What is the source of the ribulose-1,5-bisphosphate used for carbon fixation by rubisco?

Regulation of carbon fixation

Plants must coordinate light availability with carbon fixation. During the day, both processes occur. At night, when the photosystems are inactive, the plant turns off the "dark" reactions to conserve ATP and NADPH while it turns on pathways to regenerate these cofactors by metabolic pathways such as glycolysis and the pentose phosphate pathway. It would be wasteful for these catabolic processes to proceed simultaneously with the Calvin cycle. Thus, the "dark" reactions do not actually occur in the dark!

All the mechanisms for regulating the Calvin cycle are directly or indirectly linked to the availability of light energy. Some of the regulatory mechanisms are highlighted here. For example, a catalytically essential Mg^{2+} ion in the rubisco active site is coordinated in part by a carboxylated Lys side chain that is produced by the reaction of CO_2 with the ε-amino group:

$$\underset{\text{Lys}}{—(CH_2)_4—NH_2} + CO_2 \rightleftharpoons —(CH_2)_4—NH—COO^- + H^+$$

By forming the Mg^{2+}-binding site, this "activating" CO_2 molecule promotes the ability of rubisco to fix additional substrate CO_2 molecules. The carboxylation reaction is favored at high pH, a signal that the light reactions are working (depleting the stroma of protons) and that ATP and NADPH are available for the Calvin cycle.

Magnesium ions also directly activate rubisco and several of the Calvin cycle enzymes. During the light reactions, the rise in stromal pH triggers the flux of Mg^{2+} ions from the lumen to the stroma (this ion movement helps balance the charge of the protons that are translocated in the opposite direction). Some of the Calvin cycle enzymes are also activated when the ratio of reduced ferredoxin to oxidized ferredoxin is high, another signal that the photosystems are active.

REVIEW LEARNING OBJECTIVE 8

- Describe some of the mechanisms for regulating the activity of the "dark" reactions.

Carbohydrate synthesis

Many of the three-carbon sugars produced by the Calvin cycle are converted to sucrose or starch. The polysaccharide starch is synthesized in the chloroplast stroma as a temporary storage depot for glucose. It is also synthesized as a long-term storage molecule elsewhere in the plant, including leaves, seeds, and roots. In the first stage of starch synthesis, two molecules of glyceraldehyde-3-phosphate are converted to glucose-6-phosphate by reactions analogous to those of mammalian gluconeogenesis (see Fig. 10-13). Phosphoglucomutase then carries out an isomerization reaction to produce glucose-1-phosphate. Next, this sugar is "activated" by its reaction with ATP to form ADP–glucose:

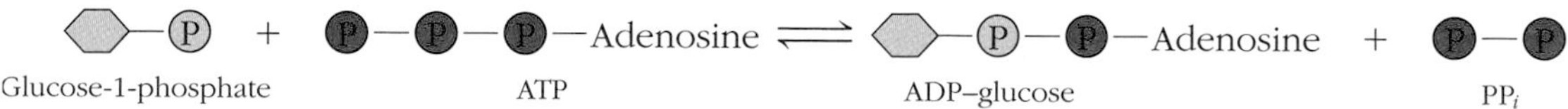

(Recall from Section 10-3 that glycogen synthesis uses the chemically related nucleotide sugar UDP–glucose.) Starch synthase then transfers the glucose residue to the end of a starch polymer, forming a new glycosidic linkage.

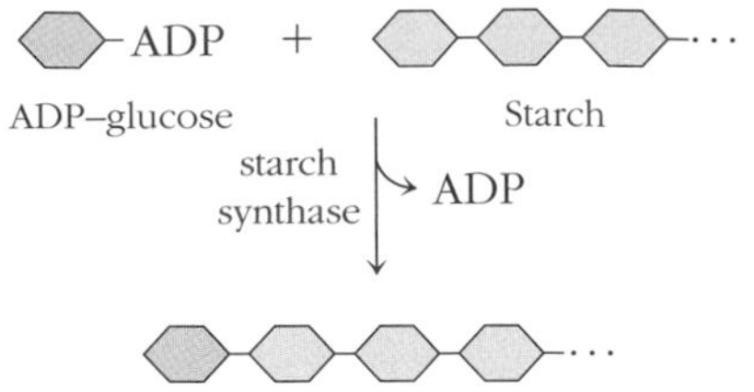

ADP–glucose + Starch → (starch synthase) → ADP

The overall reaction is driven by the exergonic hydrolysis of the PP_i released in the formation of ADP–glucose. Thus, one phosphoanhydride bond is consumed in lengthening a starch molecule by one glucose residue.

Sucrose, a disaccharide of glucose and fructose, is synthesized in the cytosol. Glyceraldehyde-3-phosphate or its isomer dihydroxyacetone phosphate is transported out of the chloroplast by an antiport protein that exchanges phosphate for a phosphorylated three-carbon sugar. Two of these sugars combine to form fructose-6-phosphate and two others combine to form glucose-1-phosphate, which is subsequently activated by UTP. Next, fructose-6-phophate reacts with UDP–glucose to produce sucrose:

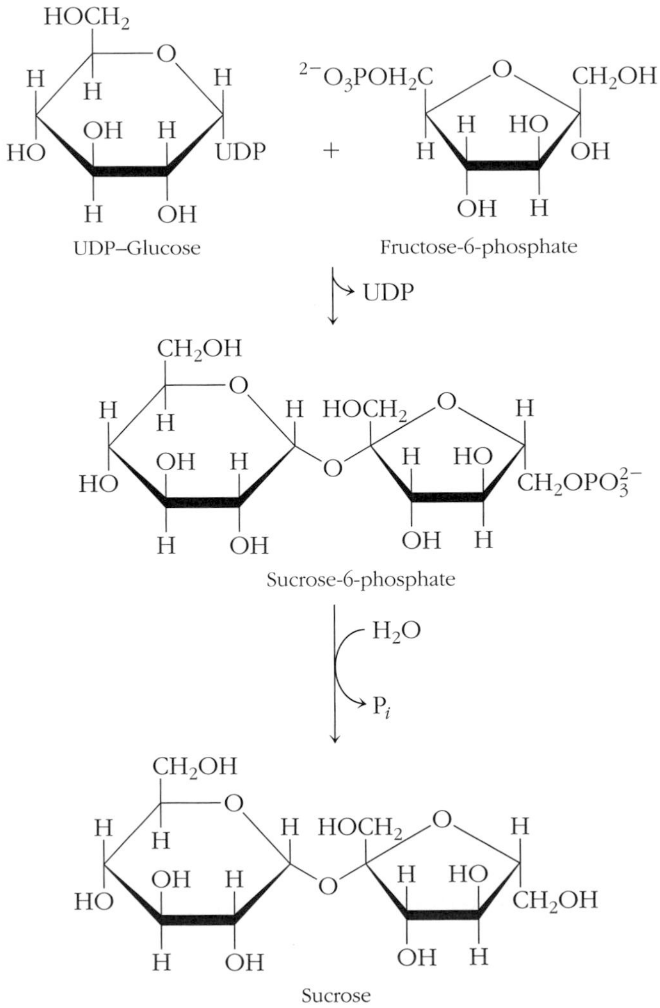

UDP–Glucose + Fructose-6-phosphate → UDP

Sucrose-6-phosphate

H_2O → P_i

Sucrose

Sucrose can then be exported to other plant tissues. This disaccharide probably became the preferred transport form of carbon in plants because its glycosidic linkage is insensitive to amylases (starch-digesting enzymes) and other common hydrolases. Also, its two anomeric carbons are tied up in the glycosidic bond and therefore cannot react nonenzymatically with other substances.

Cellulose, the other major polysaccharide of plants, is also synthesized from UDP–glucose (the structure of cellulose is shown in Fig. 10-5). Plant cell walls consist of almost-crystalline cables containing approximately 36 cellulose polymers, all embedded in an amorphous matrix of other polysaccharides (synthetic materials such as fiberglass are built on the same principle). Unlike starch in plants or glycogen in mammals, cellulose is synthesized by enzyme complexes in the plant plasma membrane and extruded into the extracellular space.

REVIEW LEARNING OBJECTIVE 9

- What is the role of nucleotides in the synthesis of starch and sucrose?

SUMMARY

1. Plant chloroplasts contain pigments that absorb photons and release the energy primarily by transferring it to another molecule (exciton transfer) or giving up an electron (photooxidation). Light-harvesting complexes act to capture and funnel light energy to the photosynthetic reaction centers.
2. In the so-called light reactions of photosynthesis, electrons from the photooxidized P680 reaction center of Photosystem II pass through several prosthetic groups and then to plastoquinone. The P680 electrons are replaced when the oxygen-evolving complex of Photosystem II converts water to O_2, a four-electron oxidation reaction.
3. Electrons flow next to a cytochrome b_6f complex that carries out a proton-translocating Q cycle, and then to the protein plastocyanin.
4. A second photooxidation at P700 of Photosystem I allows electrons to flow to the protein ferredoxin and finally to $NADP^+$ to produce NADPH.
5. The free energy of light-driven electron flow is also conserved in the formation of a transmembrane proton gradient that drives ATP synthesis in the process called photophosphorylation.
6. The enzyme rubisco "fixes" CO_2 by catalyzing the carboxylation of a five-carbon sugar. Rubisco also acts as an oxygenase in the process of photorespiration.
7. The reactions of the Calvin cycle use the products of the light reactions (ATP and NADPH) to convert the product of the rubisco reaction to glyceraldehyde-3-phosphate and to regenerate the five-carbon carboxylate receptor. These "dark" reactions are regulated according to the availability of light energy.
8. Chloroplasts convert the glyceraldehyde-3-phosphate product of photosynthesis into glucose residues for incorporation into starch, sucrose, and cellulose.

CHECKLIST

1. Review the Learning Objectives listed on page 398.
2. Apply your knowledge by solving the problems at the end of this chapter. Check your results in the Solutions appendix.
3. Be able to define the boldfaced terms (consult the glossary at the end of the book). Test your understanding by taking the Chapter 13 quiz (www.wiley.com/college/pratt).
4. Explore the websites listed at www.wiley.com/college/pratt.
5. Consult the list of Selected Readings for background information or additional details on photosynthetic pigments, reaction centers, rubisco, and the Calvin cycle.

GLOSSARY TERMS

photosynthesis
carbon fixation
chloroplast
stroma
thylakoid
photon
Planck's law
photoreceptor
fluorescence
exciton transfer
photooxidation
reaction center
antenna pigment
light-harvesting complex
circadian cycle
light reactions
noncyclic electron flow
Z-scheme
cyclic electron flow
photophosphorylation
"dark" reactions
photorespiration
C_4 pathway
Calvin cycle
quantum yield

PROBLEMS

1. Indicate with a C or an M whether the following occur in chloroplasts or mitochondria:

Proton translocation	________
Photophosphorylation	________
Photooxidation	________
Quinones	________
Oxygen reduction	________
Water oxidation	________
Electron transport	________
Oxidative phosphorylation	________
Carbon fixation	________
NADH oxidation	________
Mn cofactor	________
Heme groups	________
Binding change mechanism	________
Iron-sulfur clusters	________
$NADP^+$ reduction	________

2. Assuming 100% efficiency, calculate the molecules of ATP that could be generated per mole of photons with wavelengths of (a) 400 nm and (b) 700 nm.

3. Red tides are caused by algal blooms that cause seawater to become visibly red. In the photosynthetic process, red algae take advantage of wavelengths not absorbed by other organisms. Describe the photosynthetic pigments of the red algae.

4. Some photosynthetic bacteria live in murky ponds where visible light does not penetrate easily. What light might the photosynthetic pigments in these organisms absorb?

5. You are investigating the functional similarities of chloroplast cytochrome *f* and mitochondrial cytochrome c_1.
   (a) Which would provide more useful information: the amino acid sequences of the proteins or models of their three-dimensional shapes? Explain.
   (b) Would it be better to examine high-resolution models of the two apoproteins (the polypeptides without their heme groups) or low-resolution models of the holoproteins (polypeptides plus heme groups)?

6. Calculate the standard free energy for the oxidation of one molecule of water by $NADP^+$. Compare this value to the energy available in two photons of wavelength 600 nm.

7. Under conditions of very high light intensity, excess absorbed solar energy is dissipated by the action of "photoprotective" proteins in the thylakoid membrane.
   (a) Explain why it is advantageous for these proteins to be activated by a buildup of a proton gradient across the membrane.
   (b) Of the four mechanisms for dissipating light energy shown in Figure 13-6, which would be best for "protecting" the photosystems from excess light energy?

8. The three electron-transporting complexes of the thylakoid membrane can be called plastocyanin–ferredoxin oxidoreductase, plastoquinone–plastocyanin oxidoreductase, and water–plastoquinone oxidoreductase. What are the common names of these enzymes and in what order do they act?

9. Photosystem II is located mostly in the tightly stacked regions of the thylakoid membrane, whereas Photosystem I is located mostly in the unstacked regions (see Fig. 13-3). Why might it be important for the two photosystems to be separated?

10. Calculate the free energy of translocating a proton out of the stroma when the lumenal pH is 3.5 units lower than the stromal pH and $\Delta\psi$ is -50 mV. Compare this value to the free energy of translocating a proton out of a mitochondrion where the pH difference is 0.75 units and $\Delta\psi$ is 200 mV.

11. Predict the effect of an uncoupler such as dinitrophenol (see Box 12-A) on production of (a) ATP and (b) NADPH by a chloroplast.

12. Antimycin A (an antibiotic) blocks electron transport in Complex III of the electron transport chain in mitochondria. How would the addition of antimycin A to chloroplasts affect chloroplast ATP synthesis and NADPH production?

13. The herbicide 3-(3,4-dichlorophenyl)-1,1-dimethylurea (DCMU) blocks electron flow from Photosystem II to Photosystem I. What is the effect on oxygen production and photophosphorylation when DCMU is added to plants?

14. When the antifungal agent myxothiazol is added to a suspension of chloroplasts, the QH_2/Q ratio increases. Where does myxothiazol inhibit electron transfer?

15. Plastoquinone is not firmly anchored to any thylakoid membrane component but is free to diffuse laterally throughout the membrane among the photosynthetic components. What aspects of its structure account for this behavior?

16. The net equation for the light and "dark" reactions of photosynthesis—that is, the incorporation of one molecule of CO_2 into carbohydrate, which has the chemical formula $(CH_2O)_n$—is
$$CO_2 + 2\ H_2O \rightarrow (CH_2O) + O_2 + H_2O$$
How would this equation differ for a bacterial photosynthetic system in which H_2S rather than H_2O serves as a source of electrons?

17. Does the quantum yield of photosynthesis increase or decrease for systems where (a) the CF_0 component of ATP synthase contains more *c* subunits and (b) the proportion of cyclic electron flow through Photosystem I increases?

18. Oligomycin inhibits the proton channel (F_0) of the ATP synthase enzyme in mitochondria but does not inhibit CF_0. When oligomycin is added to plant cells undergoing photosynthesis, the cytosolic ATP/ADP ratio decreases whereas the chloroplastic ATP/ADP ratio is unchanged or even increases. Explain these results. [From Gardeström, P. and Lernmark, U., *J. Bioenergetics Biomembranes* **27,** 415–421 (1995).]

19. Efforts to engineer a more efficient rubisco, one that could fix CO_2 more quickly than 3 per second, could improve farming by allowing plants to grow larger and/or faster. Explain why the engineered rubisco might also decrease the need for nitrogen-containing fertilizers.

20. Some plants synthesize a sugar known as 2-carboxyarabinitol-1-phosphate. This compound inhibits the activity of rubisco.

$$\begin{array}{c} CH_2OPO_3^{2-} \\ | \\ HO-C-CO_2^- \\ | \\ H-C-OH \\ | \\ H-C-OH \\ | \\ CH_2OH \end{array}$$

2-Carboxyarabinitol-1-phosphate

(a) What is the probable mechanism of action of the inhibitor?

(b) Why do plants synthesize the inhibitor at night and break it down during the day?

21. Chloroplasts contain thioredoxin, a small protein with two Cys residues that can form an intramolecular disulfide bond. The sulfhydryl/disulfide interconversion in thioredoxin is catalyzed by an enzyme known as ferredoxin–thioredoxin reductase. This enzyme, along with some of the Calvin cycle enzymes, also includes two Cys residues that undergo sulfhydryl/disulfide transitions. Show how disulfide interchange reactions involving thioredoxin could coordinate the activity of Photosystem I with the activity of the Calvin cycle.

22. The inner chloroplast membrane is impermeable to large polar and ionic compounds such as NADH and ATP. However, the membrane has an antiport protein that facilitates the passage of dihydroxyacetone phosphate or 3-phosphoglycerate in exchange for P_i. This system permits the entry of P_i for photophosphorylation and the exit of the products of carbon fixation. Show how the same antiport could "transport" ATP and reduced cofactors from the chloroplast to the cytosol.

23. An "activating" CO_2 reacts with a Lys side chain on rubisco to carboxylate it. The carboxylation reaction is favored at high pH. Explain why.

24. The $\Delta G°'$ for the ribulose bisphosphate carboxylase reaction is $-35.1\ kJ \cdot mol^{-1}$ and the ΔG is $-41.0\ kJ \cdot mol^{-1}$. What is the ratio of products to reactants under normal cellular conditions?

25. The sedoheptulose bisphosphatase (SBPase) enzyme in the Calvin cycle catalyzes the removal of a phosphate group from C1 of sedoheptulose-1,7-bisphosphate (SBP) to produce sedoheptulose-7-phosphate (S7P). The $\Delta G°'$ for this reaction is $-14.2\ kJ \cdot mol^{-1}$ and the ΔG is $-29.7\ kJ \cdot mol^{-1}$. What is the ratio of products to reactants under normal cellular conditions? Is this enzyme likely to be regulated in the Calvin cycle?

26. Melvin Calvin and his colleagues noted that when $^{14}CO_2$ was added to algal cells, a single compound was radiolabeled within 5 seconds of exposure. What is the compound, and where does the radioactive label appear?

27. The Photosystem II reaction center contains a protein referred to as D1, which contains the PQ_B binding site. The D1 protein in the single-celled alga *Chlamydomonas reinhardtii* is predicted to have five hydrophobic membrane-spanning helical regions. A loop between the fourth and fifth membrane-spanning helical domains is located in the stroma. This loop is believed to lie along the membrane surface, and it contains several highly conserved amino acid residues. Site-directed mutagenesis was used to synthesize variant D1 proteins with mutations at the Ala 251 position. The mutant cells were evaluated for their ability to carry out photosynthesis and their susceptibility to herbicides. The results are shown in the table. What are the essential properties of the amino acid at position 251 in the D1 protein?

Amino acid at position 251	Characteristics
Ala	Wild-type
Cys	Similar to wild-type
Gly, Pro, Ser	Impaired in photosynthesis
Ile, Leu, Val	Impaired in photoautotrophic growth Impaired in photosynthesis Resistant to herbicides
Arg, Gln, Glu, His, Asp	Not photosynthetically competent

[From Lardans, A., Förster, B., Prásil, O., Falkowski, P.G., Sobolev, V., Edelman, M., Osmond, C.B., Gillman, N.W., and Boynton, J.E., *J. Biol. Chem.* **273,** 11082–11091 (1998).]

SELECTED READINGS

Blankenship, R.E. and Hartman, H., The origin and evolution of oxygenic photosynthesis, *Trends Biochem. Sci.* **23,** 94–97 (1998). [Discusses why absorbance of high-energy photons was required for the evolution of an oxygen-evolving photosynthetic system.]

Jordan, P., Fromme, P., Witt, H.T., Klukas, O., Saenger, W., and Krauss, N., Three-dimensional structure of cyanobacterial photosystem I at 2.5 Å resolution, *Nature* **411,** 909–917 (2001). [Provides recent data on the photosynthetic machinery.]

Nugent, J.H.A., Oxygenic photosynthesis: Electron transfer in photosystem I and photosystem II, *Eur. J. Biochem.* **237,** 519–531 (1996). [A brief but thorough review of the basic players in the light reactions.]

Zouni, A., Witt, H.-T., Kern, J., Fromme, P., Krauss, N., Saenger, W., and Orth, P., Crystal structure of photosystem II from *Synechococcus elongatus* at 3.8 Å resolution, *Nature* **409,** 739–743 (2001). [Presents much of the protein structure and positions of the prosthetic groups of a cyanobacterial Photosystem II.]

CHAPTER 14

Lipid Metabolism

About 400 years ago, farmers crossed turnip and cabbage plants to create a hybrid known as rape. This plant with bright yellow flowers (it is a member of the mustard family) can be cultivated almost anywhere and is grown worldwide for the sake of its oil. Rapeseed has the highest yield of oil of any crop; one hectare (10,000 square meters) can produce about 1000 kg of the oil. The original rape oil contains as much as 50% erucic acid, a C_{22} fatty acid with one double bond. Unfortunately, a diet rich in erucic acid contributes to heart disease. Modern varieties of rape contain less than 1% of this undesirable fatty acid and are known as canola. Nearly all of the canola oil harvest is intended for human consumption, but the oil can also be burned in an internal combustion engine. In fact, cars that run on canola oil demonstrate the utility of inexpensive renewable alternatives to fossil hydrocarbons.

[Inga Spence/Visuals Unlimited.]

LEARNING OBJECTIVES

1. Understand how fatty acids are activated and transported into mitochondria for oxidation.
2. Understand that the reactions of β oxidation result in the formation of reduced cofactors and acetyl-CoA molecules that can be further catabolized to release free energy.
3. Understand that the oxidation of unsaturated, odd-chain, very-long-chain, and branched fatty acids requires additional enzymes, some of them in peroxisomes.
4. Understand how fatty acid synthesis resembles and differs from β oxidation.
5. Understand the metabolic importance of the acetyl-CoA carboxylase reaction.
6. Understand the purpose and reactions of ketogenesis.
7. Understand that fatty acids are the precursors of triacylglycerols, glycerophospholipids, and lipids with signaling activity.
8. Understand that cholesterol is synthesized from acetyl-CoA.
9. Understand the role of lipoproteins in cholesterol metabolism.

THIS CHAPTER IN CONTEXT

Lipid metabolism can be conveniently divided into degradative and synthetic reactions, mainly involving fatty acids, the building blocks of lipids such as triacylglycerols and membrane phospholipids. The catabolism of fatty acids, like the catabolism of carbohydrates (covered in Chapter 10), yields acetyl-CoA, which is further metabolized to yield free energy by the citric acid cycle (Chapter 11) and oxidative phosphorylation (Chapter 12). This chapter also presents the biosynthesis of fatty acids, which allows easy comparison to the pathway of fatty acid degradation and illustrates some similarities and differences between opposing metabolic pathways. The related pathways of ketone body and cholesterol synthesis are also included. In addition to the catabolic and anabolic chemical reactions that occur within cells, lipid metabolism involves the transport of lipids via lipoproteins, whose importance in disease is outlined at the start of the chapter.

APPROXIMATELY HALF OF ALL DEATHS IN THE UNITED STATES ARE linked to the vascular disease atherosclerosis (a term derived from the Greek *athero,* paste and *sclerosis,* hardness). Atherosclerosis is a slow progressive disease that begins with the accumulation of lipids in the walls of large blood vessels. The trapped lipids initiate inflammation by triggering the production of chemical signals that attract white blood cells, particularly macrophages. These cells engorge themselves by taking up the accumulated lipids and continue to recruit more macrophages, thereby perpetuating the inflammation.

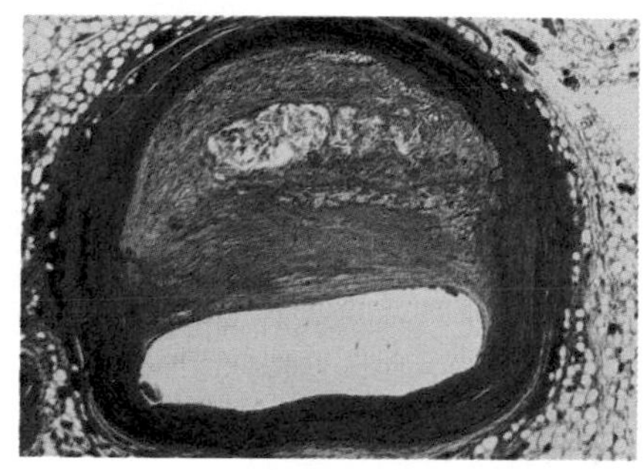

FIGURE 14-1

An atherosclerotic plaque blocking the lumen of an artery. [*Biodisc/Visuals Unlimited.*]

The damaged vessel wall forms a plaque with a core of cholesterol, cholesteryl esters, and remnants of dead macrophages, surrounded by proliferating smooth muscle cells that may undergo calcification, as occurs in bone formation. This accounts for the "hardening" of the arteries. Although a very large plaque can occlude the lumen of the artery (Fig. 14-1), blood flow is usually not completely blocked unless the plaque ruptures, triggering formation of a blood clot that can prevent circulation to the heart (a heart attack) or brain (a stroke).

What is the source of the lipids that accumulate in vessel walls? They are deposited by lipoproteins known as LDL (for low-density lipoproteins). **Lipoproteins** (particles consisting of lipids and specialized proteins) are the primary form of circulating lipid (Fig. 14-2). Recall from Section 9-1 that dietary lipids travel from the intestine to other tissues as chylomicrons. These lipoproteins are relatively large (1000 to 5000 Å in diameter) with a protein

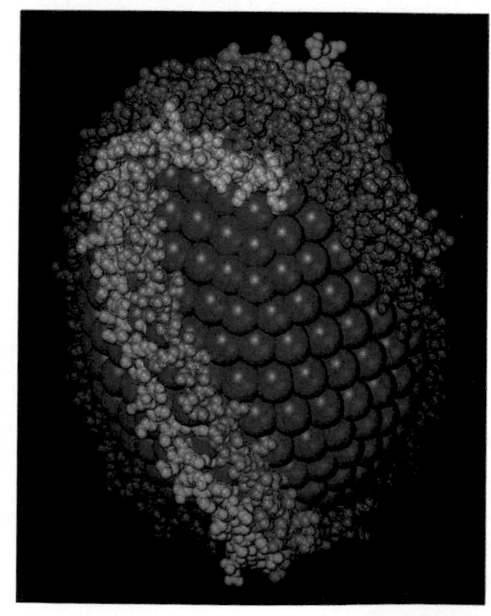

FIGURE 14-2

Theoretical model of lipoprotein structure.
The core of the lipoprotein contains highly hydrophobic lipids such as triacylglycerols and cholesteryl esters. These are surrounded by a monolayer of amphipathic lipids such as cholesterol and phospholipids (whose head groups are represented here as blue spheres). Proteins (colored structures) associated with the lipoprotein surface target the particle to cell surfaces and modulate the activities of enzymes that act on the component lipids. The various types of lipoproteins differ in size, lipid composition, protein composition, and density (a function of the relative proportions of lipid and protein). [*Courtesy David W. Borhani. Reprinted with permission from* Proc. Natl. Acad. Sci. ***94,*** *12291–12296 (1997).*]

TABLE 14-1 Characteristics of lipoproteins

Lipoprotein	Diameter (Å)	Density ($g \cdot cm^{-3}$)	% Protein	% Triacylglycerol	% Cholesterol and cholesteryl ester
Chylomicrons	1000–5000	< 0.95	1–2	85–90	4–8
VLDL	300–800	0.95–1.006	5–10	50–65	15–25
IDL	250–350	1.006–1.019	10–20	20–30	40–45
LDL	180–250	1.019–1.063	20–25	7–15	45–50
HDL	50–120	1.063–1.210	40–55	3–10	15–20

content of only 1 to 2%. Their primary function is to transport dietary triacylglycerols to adipose tissue and cholesterol to the liver. The liver repackages the cholesterol and other lipids—including triacylglycerols, phospholipids, and cholesteryl esters—into other lipoproteins known as VLDL (very low-density lipoproteins). VLDL have a triacylglycerol content of about 50% and a diameter of about 500 Å. As they circulate in the bloodstream, VLDL give up triacylglycerols to the tissues, becoming smaller, denser, and richer in cholesterol and cholesteryl esters. After passing through an intermediate state (IDL, or intermediate-density lipoproteins), they become LDL, about 200 Å in diameter and about 45% cholesteryl ester (Table 14-1).

High concentrations of circulating LDL, measured as serum cholesterol (popularly called "bad cholesterol"), are a prerequisite for atherosclerosis. Some high-fat diets (especially those rich in saturated fats) may contribute to atherosclerosis by boosting LDL levels, but genetic factors, smoking, and infection also increase the risk of atherosclerosis. The disease is less likely to occur in individuals who consume low-cholesterol diets and who have high levels of HDL (high-density lipoproteins; sometimes called "good cholesterol"). HDL particles are even smaller and denser than LDL (see Table 14-1), and their primary function is to transport the body's excess cholesterol back to the liver. HDL therefore counter the atherogenic tendencies of LDL. The roles of the various lipoproteins are summarized in Figure 14-3. The structures and physiological roles of lipoproteins are described in Exercise 20/Lipoproteins.

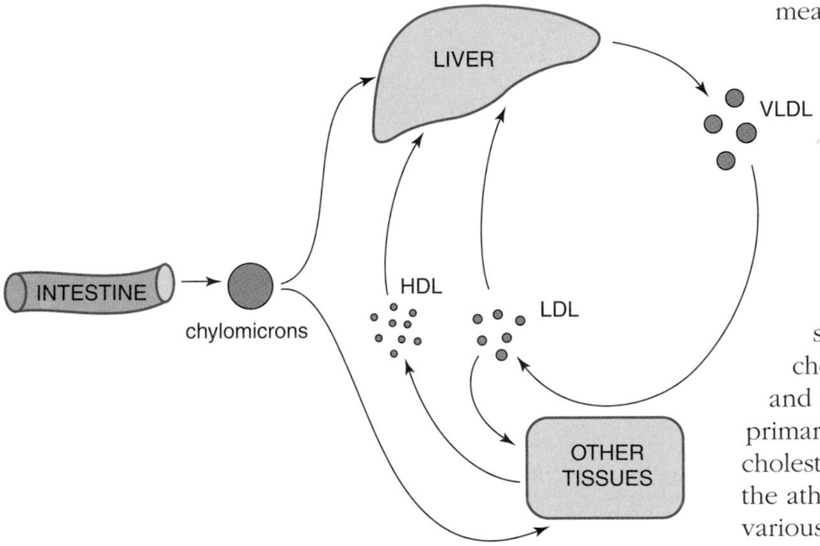

FIGURE 14-3

Lipoprotein function.
Large chylomicrons, which are mostly lipid, transport dietary lipids to the liver and other tissues. The liver produces triacylglycerol-rich very-low-density lipoproteins (VLDL). As they circulate in the tissues, VLDL give up their triacylglycerols, becoming cholesterol-rich low-density lipoproteins (LDL), which are taken up by tissues. High-density lipoproteins (HDL), the smallest and densest of the lipoproteins, transport cholesterol from the tissues back to the liver.

The opposing actions of LDL and HDL are just one part of the body's efforts to regulate lipid metabolism, which consists of multiple pathways. For example, lipids are obtained by digesting food; they are synthesized from smaller precursors; they are used by cells as a source of free energy, as building materials, and as signaling molecules; they are stored in adipose tissue; and they are transported between tissues via lipoproteins. Most of this chapter focuses on the opposing pathways of lipid synthesis and degradation. We will begin by looking at the oxidative degradation of fatty acids to acetyl-CoA. Next, we will examine fatty acid synthesis from acetyl-CoA and, finally, the synthesis of other lipids. For the most part, these reactions fall within the shaded portions of the metabolic map shown in Figure 14-4.

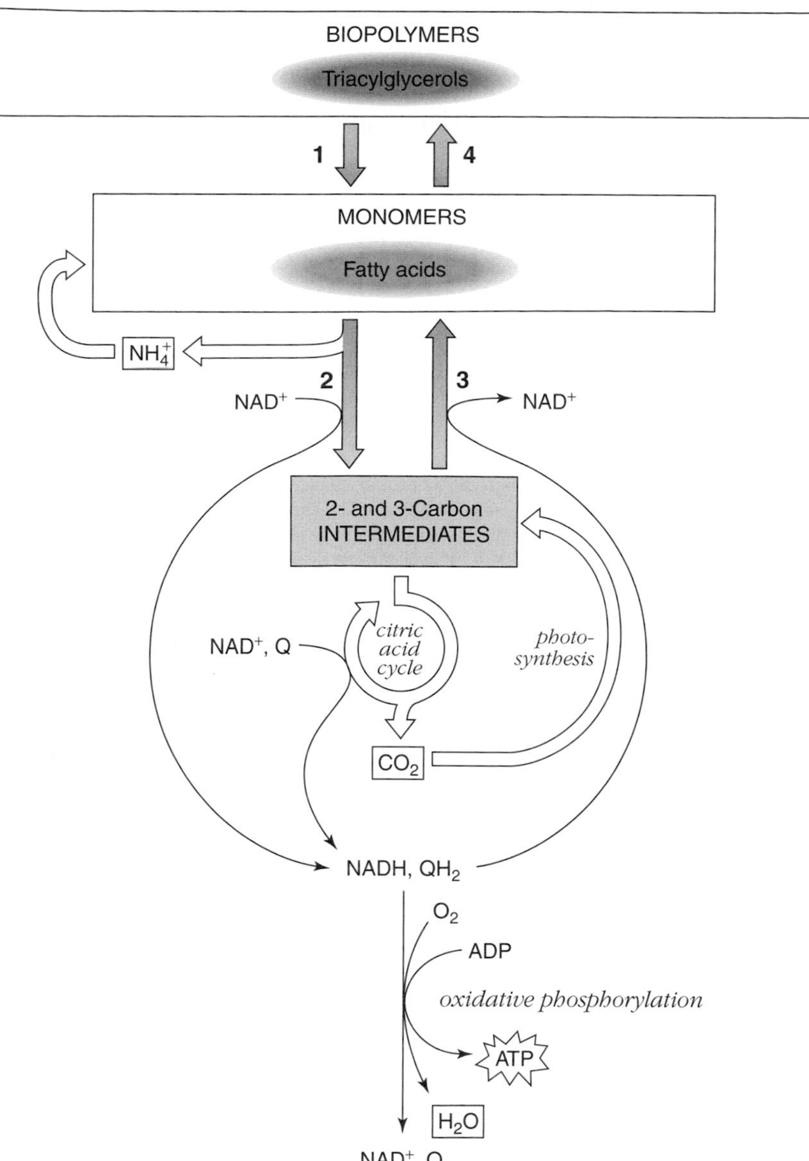

FIGURE 14-4

Lipid metabolism in context. Triacylglycerols, the "polymeric" form of fatty acids, are hydrolyzed to release fatty acids (1) that are oxidatively degraded to the two-carbon intermediate acetyl-CoA (2). Acetyl-CoA is also the starting material for the reductive biosynthesis of fatty acids (3), which can then be stored as triacylglycerols (4) or used in the synthesis of other lipids. The pathways of fatty acid or triacylglycerol degradation and synthesis offer insights into how cells carry out highly similar but opposing pathways. Acetyl-CoA is also the precursor of lipids that are not built from fatty acids (these pathways are not shown here).

1 FATTY ACID OXIDATION

The degradation (oxidation) of fatty acids is a source of metabolic free energy. In this section we describe how cells obtain, activate, and oxidize fatty acids. In humans, *dietary triacylglycerols are the primary source of fatty acids used as metabolic fuel.* The triacylglycerols are carried by lipoproteins to tissues, where hydrolysis releases their fatty acids from the glycerol backbone.

Go to **Exercise 19/Fatty Acid Metabolism** for a detailed look at the reactions and free energy yield of fatty acid oxidation as well as fatty acid synthesis. The material presented here addresses Learning Objectives 1, 2, and 3.

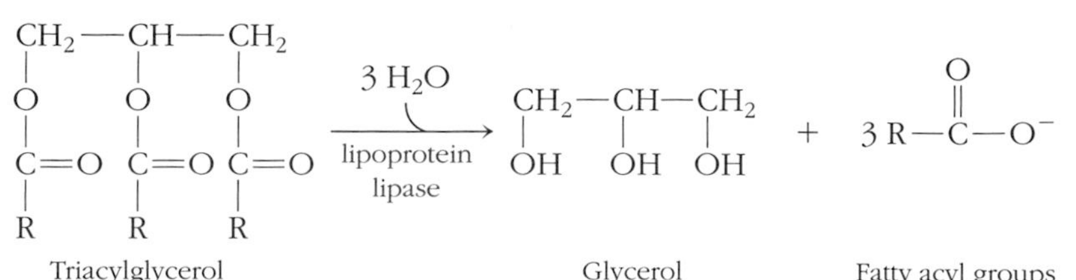

Hydrolysis occurs extracellularly, catalyzed by lipoprotein lipase, an enzyme associated with the outer surface of cells.

Triacylglycerols that are stored in adipose tissue are mobilized (their fatty acids are released to be used as fuel) by a hormone-sensitive lipase. The fatty acids travel through the bloodstream, not as part of lipoproteins, but bound to albumin, a 66-kD protein that accounts for about half of the serum protein (it also binds metal ions and hormones, serving as an all-purpose transport protein).

The concentration of free fatty acids in the body is very low, because these molecules are detergents (which form micelles; see Section 2-2) and can disrupt cell membranes. After they enter cells, probably with the assistance of proteins, the fatty acids are either broken down for energy or re-esterified to form triacylglycerols or other complex lipids (as described in Section 14-3). Many free fatty acids are deployed to the liver and muscle cells, especially heart muscle, which prefers to burn fatty acids even when carbohydrate fuels are available.

Fatty acids are "activated" before they are degraded

To be oxidatively degraded, a fatty acid must first be activated. Activation is a two-step reaction catalyzed by acyl-CoA synthetase. First, the fatty acid displaces the diphosphate group of ATP, then coenzyme A (HSCoA) displaces the AMP group to form an acyl-CoA:

$$\underset{\text{Fatty Acid}}{R{-}C({=}O){-}O^-} + \underset{\text{ATP}}{{}^-O{-}P(O)(O^-){-}O{-}P(O)(O^-){-}O{-}P(O)(O^-){-}O{-}\text{Adenosine}}$$

$$\downarrow \rightarrow PP_i \xrightarrow[\text{inorganic pyrophosphatase}]{H_2O} 2\,P_i$$

$$\underset{\text{Acyladenylate}}{R{-}C(O){-}O{-}P(O)(O^-){-}O{-}\text{Adenosine}}$$

$$\downarrow \leftarrow H{-}SCoA$$

$$\underset{\text{Acyl-CoA}}{R{-}C(O){-}SCoA} + \underset{\text{AMP}}{{}^-O{-}P(O)(O^-){-}O{-}\text{Adenosine}}$$

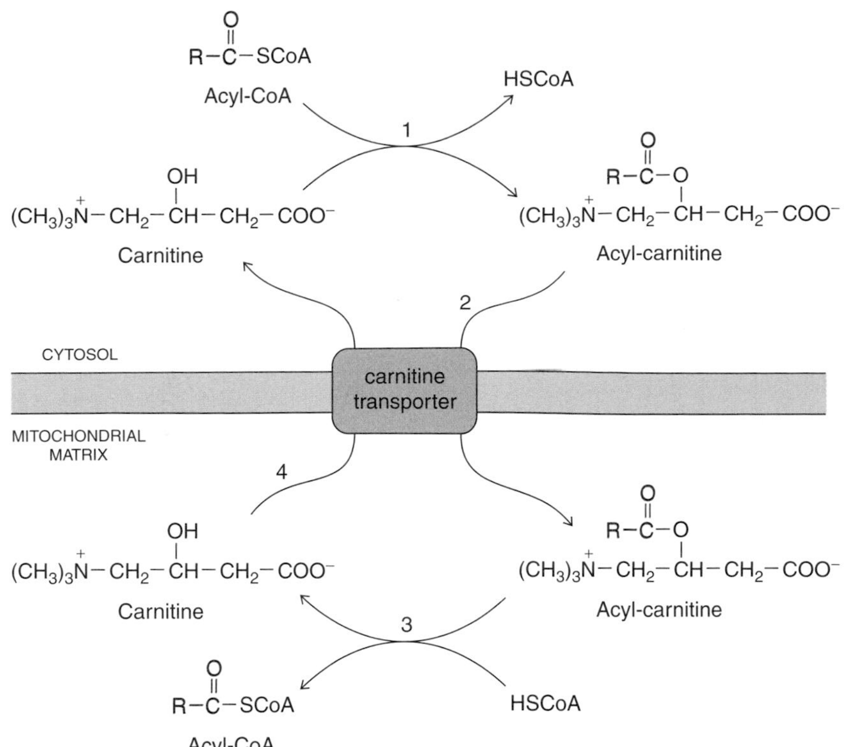

FIGURE 14-5

The carnitine shuttle system. 1. A cytosolic carnitine acyltransferase transfers an acyl group from CoA to carnitine. 2. The carnitine transporter allows the acyl-carnitine to enter the mitochondrial matrix. 3. A mitochondrial carnitine acyltransferase transfers the acyl group to a mitochondrial CoA molecule. 4. Free carnitine returns to the cytosol via the transport protein.

The acyladenylate product of the first step has a large free energy of hydrolysis (in other words, its cleavage would release a large amount of free energy), so it conserves the free energy of the cleaved phosphoanhydride bond in ATP. The second step, transfer of the acyl group to CoA (the same molecule that carries acetyl groups as acetyl-CoA), likewise conserves free energy in the formation of a thioester bond (see Section 9-2). Consequently, the overall reaction

$$\text{Fatty acid} + \text{CoA} + \text{ATP} \rightleftharpoons \text{Acyl-CoA} + \text{AMP} + \text{PP}_i$$

has a free energy change near zero. However, subsequent hydrolysis of the product PP_i (by the ubiquitous enzyme inorganic pyrophosphatase) is highly exergonic, and *this reaction makes the formation of acyl-CoA spontaneous and irreversible.*

Fatty acids are activated in the cytosol, but oxidation occurs in the mitochondria. Because there is no transport protein for CoA adducts, *acyl groups must enter the mitochondria via a shuttle system involving the small molecule carnitine* (Fig. 14-5). The acyl group is now ready to be oxidized.  Fatty acid activation and transport into the mitochondria are summarized in Exercise 19.

REVIEW LEARNING OBJECTIVE 1

- What is the source of fatty acids that enter the β oxidation pathway?
- Why must a fatty acid be activated by formation of a thioester bond?

β Oxidation: a pathway with four reactions

The pathway known as **β oxidation** degrades an acyl-CoA in a way that produces acetyl-CoA molecules for further oxidation and energy production by the citric acid cycle. In fact, in some tissues or under certain conditions, β oxidation supplies far more acetyl groups to the citric acid cycle than does glycolysis. β oxidation also feeds electrons directly into the mitochondrial electron transport chain, which generates ATP by oxidative phosphorylation.

β Oxidation is a spiral pathway. *Each round consists of four enzyme-catalyzed steps that yield one molecule of acetyl-CoA and an acyl-CoA shortened by two carbons, which becomes the starting substrate for the next round.* Seven rounds of β oxidation degrade a C_{16} fatty acid to eight molecules of acetyl-CoA:

Figure 14-6 shows the reactions of β oxidation. The "β" indicates that oxidation occurs at the β position, which is the carbon atom that is two away from the carbonyl carbon (C3 is the β carbon). Note that acetyl units are not lost from the methyl end of the fatty acid but from the activated, CoA end. **Exercise 19 includes animations of the four reactions of β oxidation.**

The oxidation of fatty acids by the successive removal of two-carbon units was discovered around 100 years ago, and the enzymatic steps were elucidated about 40 years ago. But β oxidation still offers surprises in the details. For example, many enzymes are required to fully degrade an acyl-CoA to acetyl-CoA. Each of the four steps shown in Figure 14-6 (plus the acyl-CoA synthetase reaction) appears to be catalyzed by two to five different enzymes with different chain-length specificities. The existence of some of these isozymes was inferred from studies of patients with disorders of fatty acid oxidation. One of these often-fatal diseases is due to a deficiency of medium-chain acyl-CoA dehydrogenase; affected individuals cannot degrade acyl-CoAs having 4 to 12 carbons, and derivatives of these molecules accumulate in the liver and are excreted in the urine.

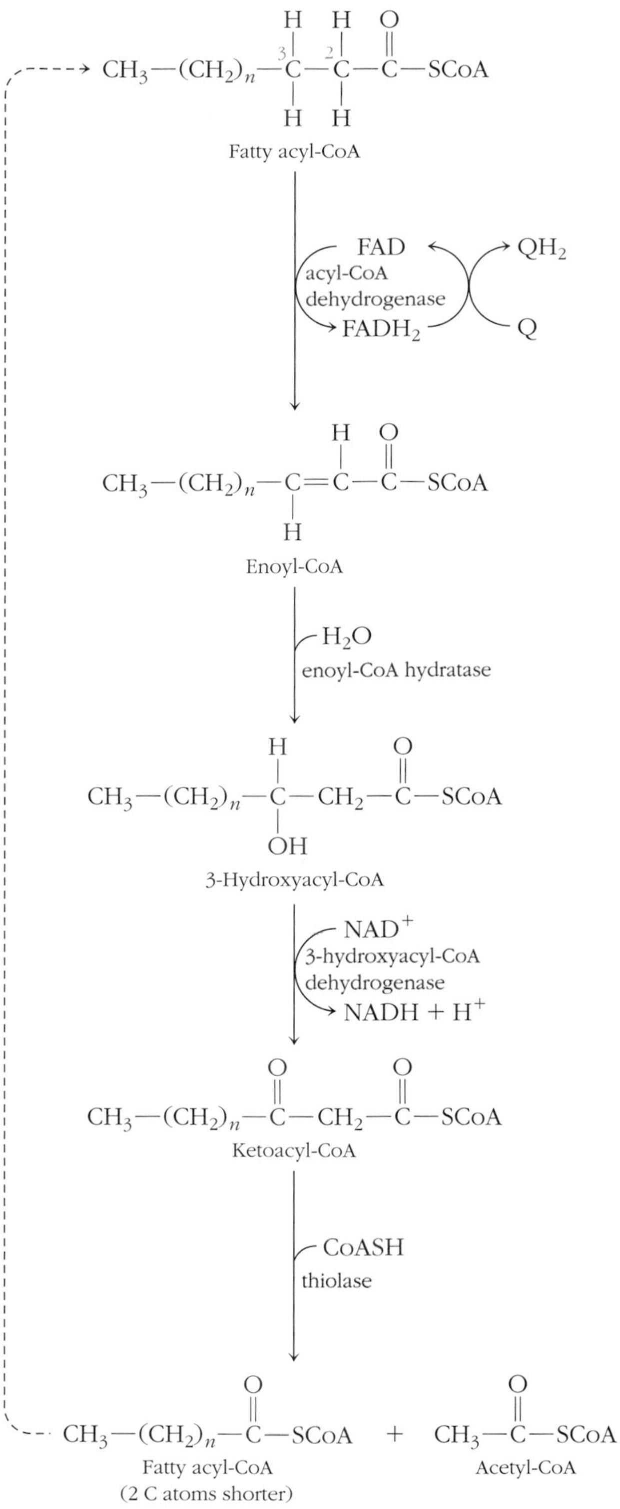

$CH_3-(CH_2)_n-CH_2-CH_2-C(=O)-SCoA$

Fatty acyl-CoA

FAD → $FADH_2$ (acyl-CoA dehydrogenase); Q → QH_2

1. *Oxidation of acyl-CoA at the 2,3 position is catalyzed by an acyl-CoA dehydrogenase to yield a 2,3-enoyl-CoA. The two electrons removed from the acyl group are transferred to an FAD prosthetic group. A series of electron transfer reactions eventually transfers the electrons to ubiquinone (Q).*

$CH_3-(CH_2)_n-CH=CH-C(=O)-SCoA$

Enoyl-CoA

H_2O (enoyl-CoA hydratase)

2. *The second step is catalyzed by a hydratase, which adds the elements of water across the double bond produced in the first step.*

$CH_3-(CH_2)_n-CH(OH)-CH_2-C(=O)-SCoA$

3-Hydroxyacyl-CoA

NAD^+ → $NADH + H^+$ (3-hydroxyacyl-CoA dehydrogenase)

3. *The hydroxyacyl-CoA is oxidized by another dehydrogenase. In this case, NAD^+ is the cofactor.*

$CH_3-(CH_2)_n-C(=O)-CH_2-C(=O)-SCoA$

Ketoacyl-CoA

CoASH (thiolase)

4. *The final step, thiolysis, is catalyzed by a thiolase and releases acetyl-CoA. The remaining acyl-CoA, two carbons shorter than the starting substrate, undergoes another round of the four reactions (dotted line).*

$CH_3-(CH_2)_n-C(=O)-SCoA + CH_3-C(=O)-SCoA$

Fatty acyl-CoA (2 C atoms shorter) + Acetyl-CoA

FIGURE 14-6

The reactions of β oxidation.

Energy yield of β oxidation

β Oxidation is a major source of cellular free energy, especially during a fast, when carbohydrates are not available. Each round of β oxidation produces one QH_2, one NADH, and one acetyl-CoA. The citric acid cycle oxidizes the acetyl-CoA to produce an additional three NADH, one QH_2, and one GTP. Oxidation of all the reduced cofactors yields approximately 16 ATP (4 from the two QH_2 and 12 from the four NADH; see Section 12-4). Therefore, considerable free energy is made available for the cell. The ATP equivalents obtained from fatty acid oxidation are also tallied in Exercise 19.

One round of β oxidation		*Citric acid cycle*		*Oxidative phosphorylation*
1 QH_2	⟶		⟶	2 ATP
1 NADH	⟶		⟶	3 ATP
1 Acetyl-CoA	⟶	3 NADH	⟶	9 ATP
		1 QH_2	⟶	2 ATP
		1 GTP	⟶	1 ATP
			Total	17 ATP

β Oxidation is regulated primarily by the availability of free CoA (to make acyl-CoA) and by the ratios of NAD^+/NADH and Q/QH_2 (these reflect the state of the oxidative phosphorylation system). Some individual enzymes are also regulated by product inhibition.

REVIEW LEARNING OBJECTIVE 2

- Summarize the chemical reactions of β oxidation.
- How do these reactions lead to ATP production in the cell?

Oxidation of unsaturated fatty acids

Common fatty acids such as oleate and linoleate

Oleate

Linoleate

contain *cis* double bonds that present obstacles to the enzymes that catalyze β oxidation. For linoleate, the first three rounds of β oxidation proceed as usual. But the acyl-CoA that begins the fourth round has a 3,4 double bond (originally the 9,10 double bond). Furthermore, this molecule is a *cis* enoyl-CoA, but enoyl-CoA hydratase (the enzyme that catalyzes Step 2 of β oxidation) recognizes only the *trans* configuration. This metabolic obstacle is removed by the enzyme enoyl-CoA isomerase, which converts the *cis* 3,4 double bond to a *trans* 2,3 double bond so that β oxidation can continue.

enoyl-CoA isomerase

A second obstacle arises after the first reaction of the fifth round of β oxidation. Acyl-CoA dehydrogenase introduces a 2,3 double bond as usual, but the original 12,13 double bond of linoleate is now at the 4,5 position. The resulting dienoyl-CoA is not a good substrate for the next enzyme, enoyl-CoA hydratase. The dienoyl-CoA must undergo an NADPH-dependent reduction to convert its two double bonds to a single *trans* 3,4 double bond. This product must then be isomerized to produce the *trans* 2,3 double bond that is recognized by enoyl-CoA hydratase.

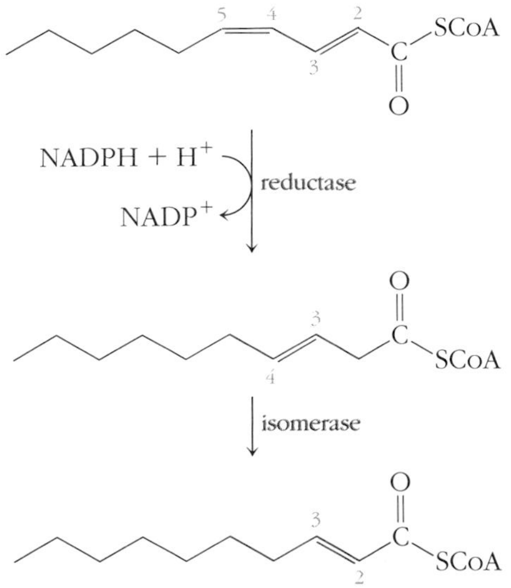

5 4 2 3 SCoA C O

$NADPH + H^+$ → $NADP^+$ reductase

O C SCoA 3 4

isomerase

O C SCoA 3 2

Unsaturated fatty acids yield less free energy than saturated fatty acids, as a result of these bypass reactions. First, the enoyl-CoA isomerase reaction bypasses the QH_2-producing acyl-CoA dehydrogenase step, so two less ATP are produced. Second, the NADPH-dependent reductase consumes three ATP equivalents in the form of NADPH (which is energetically equivalent to NADH). **The energy yield from oxidation of unsaturated fatty acids is explored in Exercise 19.**

Oxidation of odd-chain fatty acids

Most fatty acids have an even number of carbon atoms (this is because they are synthesized by the addition of two-carbon acetyl units, as we shall see later in this chapter). However, some plant and bacterial fatty acids that make their way into the human system have an odd number of carbon atoms. *The final round of β oxidation of these molecules leaves a three-carbon fragment, propionyl-CoA, rather than the usual acetyl-CoA.*

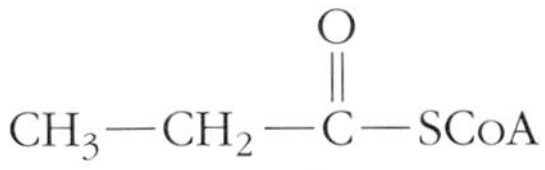

$$CH_3-CH_2-\overset{O}{\overset{\|}{C}}-SCoA$$

Propionyl-CoA

This intermediate can be further metabolized by the sequence of steps outlined in Figure 14-7. At first, this pathway seems longer than necessary. For example, adding a carbon to C3 of the propionyl group would immediately generate succinyl-CoA. However, such a reaction is not chemically favored, because C3 is too far from the electron-delocalizing effects of the CoA thioester. Consequently, propionyl-CoA carboxylase must add a carbon to C2, and then methylmalonyl-CoA mutase must rearrange the carbon skeleton to produce succinyl-CoA. Note that succinyl-CoA is not the end point of the pathway. Because it is a citric acid cycle intermediate it acts catalytically and is not consumed by the cycle (see Section 11-2). *The complete catabolism of the*

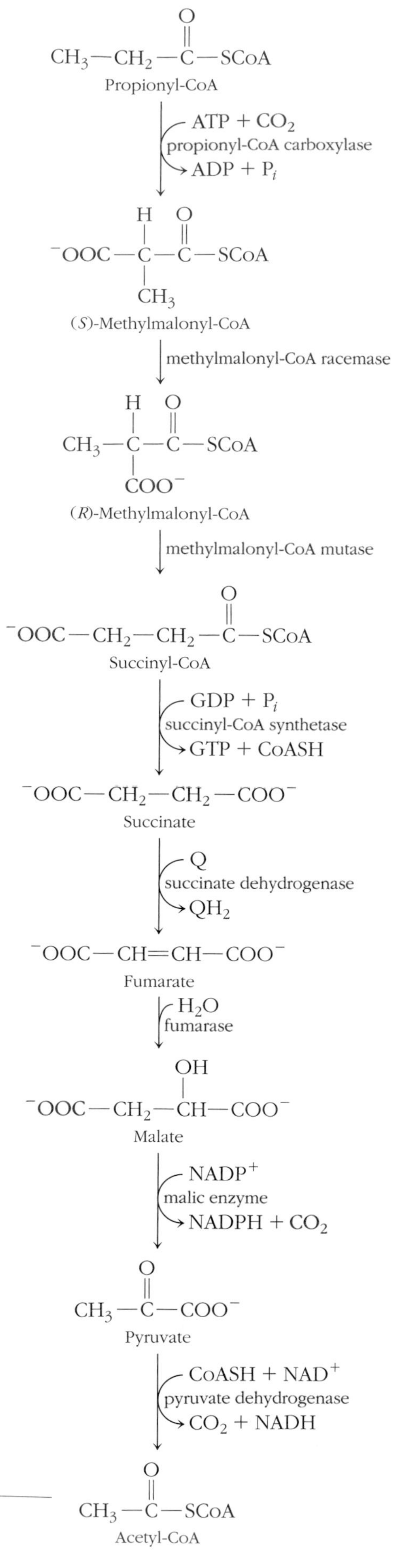

$$CH_3-CH_2-\overset{O}{\overset{\|}{C}}-SCoA$$

Propionyl-CoA

↓ $ATP + CO_2$ → propionyl-CoA carboxylase → $ADP + P_i$

1. *Propionyl-CoA carboxylase adds a carboxyl group at C2 of the propionyl group to form a four-carbon methylmalonyl group.*

$$^{-}OOC-\underset{CH_3}{\underset{|}{\overset{H}{\overset{|}{C}}}}-\overset{O}{\overset{\|}{C}}-SCoA$$

(*S*)-Methylmalonyl-CoA

↓ methylmalonyl-CoA racemase

2. *A racemase interconverts the two different methylmalonyl-CoA stereoisomers (the two configurations are indicated by the R and S symbols).*

$$CH_3-\underset{COO^-}{\underset{|}{\overset{H}{\overset{|}{C}}}}-\overset{O}{\overset{\|}{C}}-SCoA$$

(*R*)-Methylmalonyl-CoA

↓ methylmalonyl-CoA mutase

3. *Methylmalonyl-CoA mutase rearranges the carbon skeleton to generate succinyl-CoA.*

$$^{-}OOC-CH_2-CH_2-\overset{O}{\overset{\|}{C}}-SCoA$$

Succinyl-CoA

↓ $GDP + P_i$ → succinyl-CoA synthetase → $GTP + CoASH$

4–6. *Succinyl-CoA, a citric acid cycle intermediate, is converted to malate by reactions 5–7 of the citric acid cycle (see Fig. 11-7).*

$$^{-}OOC-CH_2-CH_2-COO^-$$

Succinate

↓ Q → succinate dehydrogenase → QH_2

$$^{-}OOC-CH{=}CH-COO^-$$

Fumarate

↓ H_2O → fumarase

$$^{-}OOC-CH_2-\overset{OH}{\overset{|}{C}H}-COO^-$$

Malate

↓ $NADP^+$ → malic enzyme → $NADPH + CO_2$

7. *After being exported from the mitochondria, malate is decarboxylated by malic enzyme to produce pyruvate in the cytosol.*

$$CH_3-\overset{O}{\overset{\|}{C}}-COO^-$$

Pyruvate

↓ $CoASH + NAD^+$ → pyruvate dehydrogenase → $CO_2 + NADH$

8. *Pyruvate, imported back into the mitochondria, can then be converted to acetyl-CoA by the pyruvate dehydrogenase complex.*

$$CH_3-\overset{O}{\overset{\|}{C}}-SCoA$$

Acetyl-CoA

FIGURE 14-7

Catabolism of propionyl-CoA.

carbons derived from propionyl-CoA requires that the succinyl-CoA be converted to pyruvate and then to acetyl-CoA, which enters the citric acid cycle as a substrate. These reactions are summarized in Exercise 19.

Methylmalonyl-CoA mutase (see Step 3 of Fig. 14-7) is an unusual enzyme because it catalyzes a rearrangement of carbon atoms and requires a prosthetic group derived from the vitamin cobalamin (vitamin B_{12}; Fig. 14-8). Only about a dozen enzymes are known to use cobalamin cofactors. The small amounts of cobalamin required for human health are usually easily obtained from the diet. A disorder of vitamin B_{12} absorption causes the disease pernicious anemia.

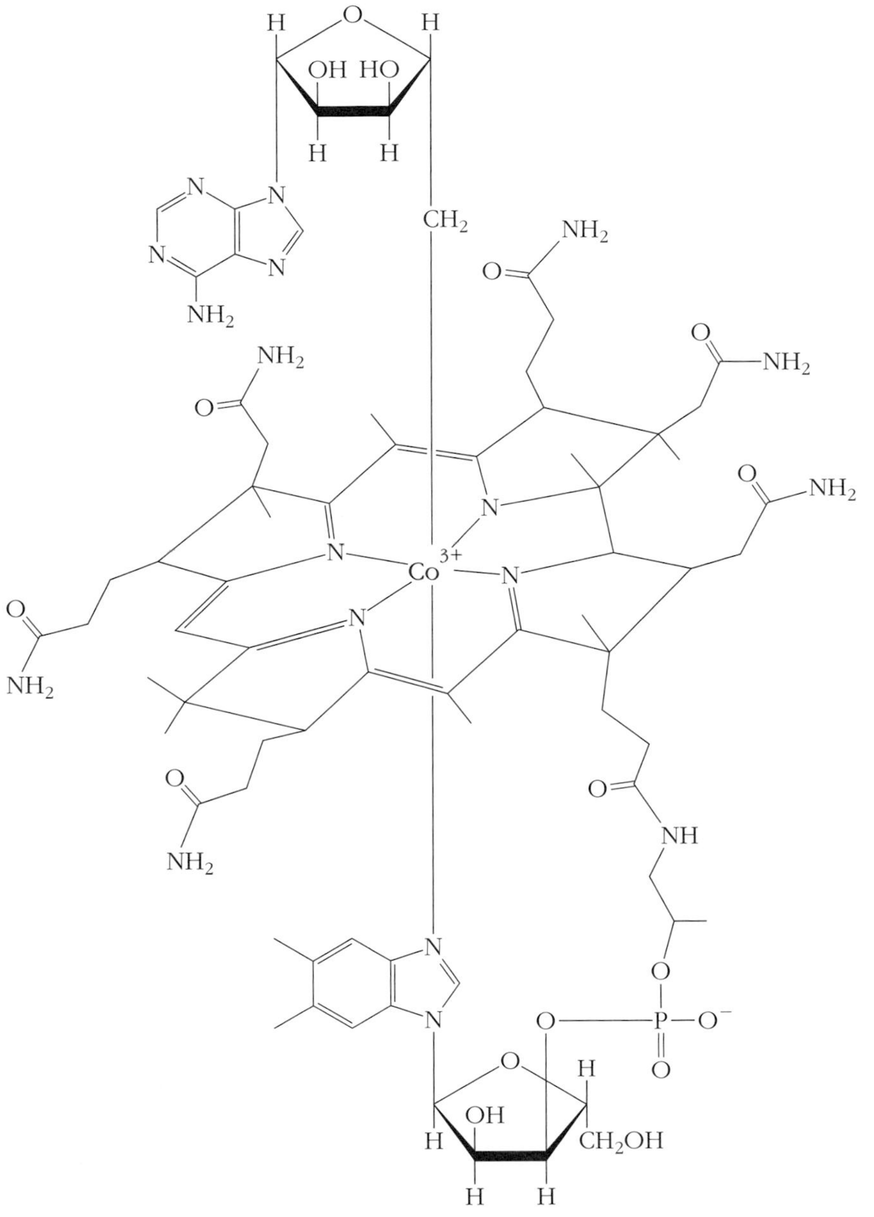

FIGURE 14-8

The cobalamin-derived cofactor.
The prosthetic group of methylmalonyl-CoA mutase is a derivative of the vitamin cobalamin. The structure includes a hemelike ring structure with a central cobalt ion. Note that one of the Co ligands is a carbon atom, an extremely rare instance of a carbon–metal bond in a biological system.

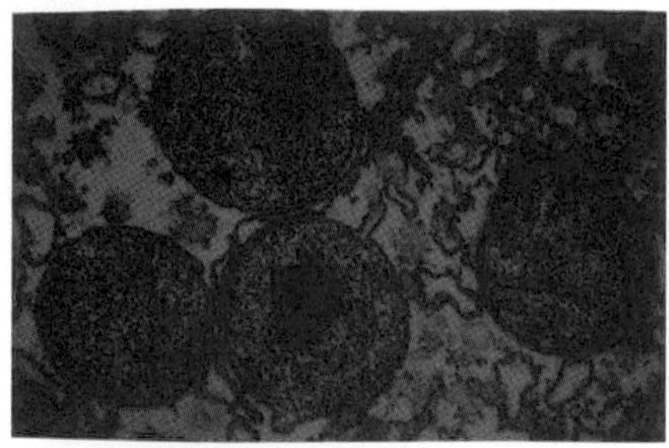

FIGURE 14-9

Peroxisomes.
Nearly all eukaryotic cells contain these single-membrane-bound organelles, which are similar to plant glyoxisomes (see Box 11-B). [*D. Friend and D. Fawcett/Visuals Unlimited.*]

Fatty acid oxidation in peroxisomes

The majority of the cell's fatty acid oxidation occurs in mitochondria, but a small percentage is carried out in organelles known as **peroxisomes** (Fig. 14-9). These single membrane–bound compartments contain a variety of degradative and biosynthetic enzymes. The peroxisomal β oxidation pathway differs from the mitochondrial pathway in the first step. An acyl-CoA oxidase catalyzes the reaction:

$$CH_3-(CH_2)_n-\underset{H}{\overset{H}{C}}-\underset{H}{\overset{H}{C}}-\overset{O}{\overset{\|}{C}}-SCoA$$

Acyl-CoA

$$\xrightarrow[\text{acyl-CoA oxidase}]{FAD \rightarrow FADH_2;\ O_2 \rightarrow H_2O_2}$$

$$CH_3-(CH_2)_n-\underset{H}{C}=\overset{H}{C}-\overset{O}{\overset{\|}{C}}-SCoA$$

Enoyl-CoA

The enoyl-CoA product of the reaction is identical to the product of the mitochondrial acyl-CoA dehydrogenase reaction (see Fig. 14-6), but the electrons removed from the acyl-CoA are transferred not to ubiquinone but directly to molecular oxygen to produce hydrogen peroxide, H_2O_2. This reaction product, which gives the peroxisome its name, is subsequently broken down by the peroxisomal enzyme catalase:

$$2\ H_2O_2 \rightarrow 2\ H_2O + O_2$$

The second, third, and fourth reactions of fatty acid oxidation are the same as in mitochondria.

Because the peroxisomal oxidation enzymes are specific for very-long-chain fatty acids (for example, over 20 carbons) and bind short-chain fatty acids with low affinity, *the peroxisome serves as a chain-shortening system.* The partially degraded fatty acyl-CoAs then make their way to the mitochondria for complete oxidation. Peroxisomal β oxidation is included in Exercise 19.

The peroxisome is also responsible for degrading some branched-chain fatty acids, which are not recognized by the mitochondrial enzymes. One such nonstandard fatty acid is phytanate,

3 COO^-

Phytanate

which is derived from the side chain of chlorophyll molecules (see Fig. 13-4) and is present in all plant-containing diets. Phytanate must be degraded by peroxisomal enzymes because the methyl group at C3 prevents dehydrogenation by 3-hydroxyacyl-CoA dehydrogenase (Step 3 of the standard β oxidation pathway). A deficiency of any of the phytanate-degrading enzymes results in Refsum's disease, a degenerative neuronal disorder characterized by an accumulation of phytanate in the tissues. The importance of peroxisomal enzymes in lipid metabolism (both catabolic and anabolic) is confirmed by the fatal outcome of most diseases stemming from deficient peroxisomal enzymes or improper synthesis of the peroxisomes themselves.

REVIEW LEARNING OBJECTIVE 3

- Summarize the enzyme activities that are required to fully metabolize unsaturated and odd-chain fatty acids.
- What is the role of peroxisomes in fatty acid catabolism?

2 FATTY ACID SYNTHESIS

Go to **Exercise 19/Fatty Acid Metabolism** to explore how fatty acid synthesis differs from and resembles β oxidation. The material presented in the exercise addresses Learning Objective 4.

At first glance, fatty acid synthesis appears to be the exact reverse of fatty acid oxidation. For example, fatty acyl groups are built and degraded two carbons at a time, and several of the reaction intermediates in the two pathways are similar or identical. However, *the pathways for fatty acid synthesis and degradation must differ for thermodynamic reasons,* as we saw for glycolysis and gluconeogenesis. Since fatty acid oxidation is a thermodynamically favorable process, simply reversing the steps of this pathway would be energetically unfavorable.

In cells, the opposing metabolic pathways of fatty acid synthesis and degradation are entirely separate. β Oxidation takes place in the mitochondrial matrix, and synthesis occurs in the cytosol. Furthermore, the two pathways use different cofactors. In β oxidation, the acyl group is attached to coenzyme A, but a growing fatty acyl chain is bound by an acyl carrier protein (ACP; Fig. 14-10). β Oxidation funnels electrons to ubiquinone and NAD^+, but in fatty acid synthesis, NADPH is the source of reducing power. Finally, β oxidation requires two ATP equivalents (two phosphoanhydride bonds) to "activate" the acyl group, but the biosynthetic pathway consumes one ATP

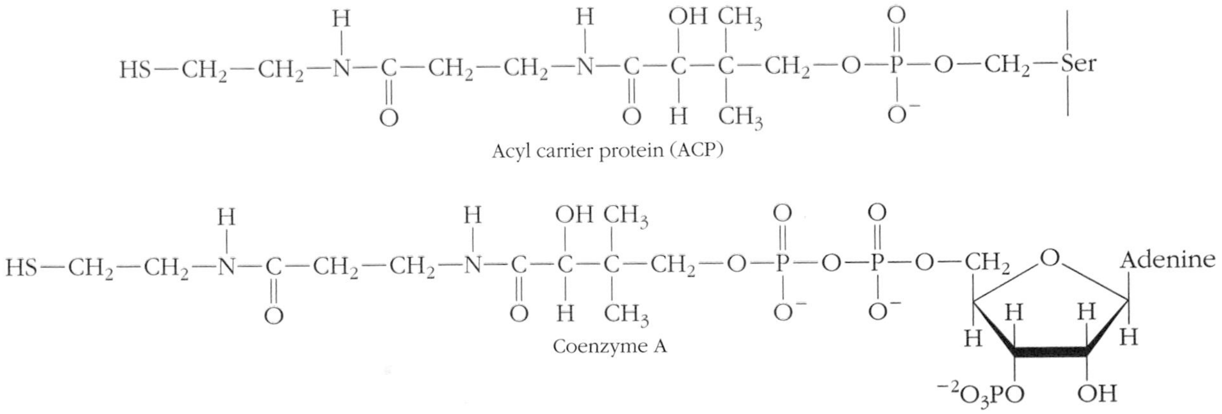

$HS{-}CH_2{-}CH_2{-}N(H){-}C({=}O){-}CH_2{-}CH_2{-}N(H){-}C({=}O){-}C(OH)(H){-}C(CH_3)_2{-}CH_2{-}O{-}P({=}O)(O^-){-}O{-}CH_2{-}Ser$

Acyl carrier protein (ACP)

$HS{-}CH_2{-}CH_2{-}N(H){-}C({=}O){-}CH_2{-}CH_2{-}N(H){-}C({=}O){-}C(OH)(H){-}C(CH_3)_2{-}CH_2{-}O{-}P({=}O)(O^-){-}O{-}P({=}O)(O^-){-}O{-}CH_2$–ribose(Adenine, $^{-2}O_3PO$, OH)

Coenzyme A

FIGURE 14-10

Acyl carrier protein and coenzyme A.
Both acyl carrier protein (ACP) and coenzyme A (CoA) include a pantothenate (vitamin B_3) derivative ending with a sulfhydryl group that forms a thioester with an acyl or acetyl group. In CoA, the pantothenate derivative is esterified to an adenine nucleotide, and in ACP, the group is esterified to a Ser OH group of a polypeptide (in mammals, ACP is part of a larger multifunctional protein, fatty acid synthase).

for every two carbons incorporated into a fatty acid. In this section, we focus on the reactions of fatty acid synthesis, comparing and contrasting them to β oxidation.

The source of cytosolic acetyl-CoA

Just as acyl-CoAs cannot directly enter the mitochondria to be oxidized, acetyl-CoA generated in the mitochondria cannot exit to the cytoplasm for biosynthetic reactions. *The transport of acetyl groups to the cytosol involves citrate, which has a transport protein.* Citrate synthase (the enzyme that catalyzes the first step of the citric acid cycle; see Fig. 11-7) combines acetyl-CoA with oxaloacetate to produce citrate, which then leaves the mitochondria. ATP-citrate lyase "undoes" the citrate synthase reaction to produce acetyl-CoA and oxaloacetate in the cytosol (Fig. 14-11). Note that ATP is consumed in the ATP-citrate lyase reaction to drive the formation of a thioester bond.

Acetyl-CoA carboxylase catalyzes the first step of fatty acid synthesis

The first step of fatty acid synthesis is the carboxylation of acetyl-CoA, an ATP-dependent reaction carried out by acetyl-CoA carboxylase. This enzyme catalyzes the rate-controlling step for the fatty acid synthesis pathway. The acetyl-CoA carboxylase mechanism is similar to that of propionyl-CoA carboxylase (Step 1 in Fig. 14-7) and pyruvate carboxylase (see Fig. 10-12). First, CO_2 (as bicarbonate, HCO_3^-) is "activated" by its attachment to a biotin prosthetic group in a reaction that converts ATP to ADP + P_i:

$$\text{Biotin} + HCO_3^- + \text{ATP} \rightarrow \text{Biotin—COO}^- + \text{ADP} + P_i$$

Next, the carboxybiotin prosthetic group transfers the carboxylate group to acetyl-CoA to form the three-carbon malonyl-CoA and regenerate the enzyme.

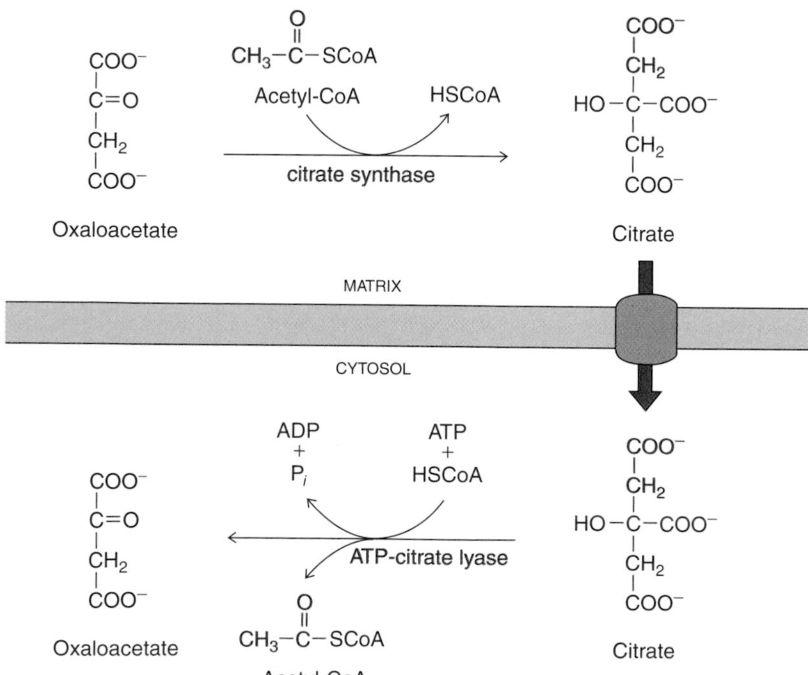

FIGURE 14-11

The citrate transport system. The citrate transport protein, along with mitochondrial citrate synthase and cytoplasmic ATP-citrate lyase, provides a route for transporting acetyl units from the mitochondrial matrix to the cytosol. The complete transport system showing how oxaloacetate returns to the matrix is presented in Figure 11-17.

$$\text{Biotin—COO}^- + \text{CH}_3\text{—}\overset{\overset{\text{O}}{\|}}{\text{C}}\text{—SCoA} \longrightarrow {}^-\text{OOC—CH}_2\text{—}\overset{\overset{\text{O}}{\|}}{\text{C}}\text{—SCoA} + \text{Biotin}$$

Acetyl-CoA → Malonyl-CoA

Malonyl-CoA is the donor of the two-carbon acetyl units that are used to build a fatty acid. The carboxylate group added by the carboxylation reaction is lost in a subsequent decarboxylation reaction. This sequence of carboxylation followed by decarboxylation also occurs in the conversion of pyruvate to phosphoenolpyruvate in gluconeogenesis (see Section 10-3). Note that fatty acid synthesis requires a C_3 intermediate, whereas β oxidation involves only two-carbon acetyl units.

The reactions of fatty acid synthase

The protein that carries out the main reactions of fatty acid synthesis in animals is a 530-kD **multifunctional enzyme** made of two identical polypeptides (Fig. 14-12). Each polypeptide of this fatty acid synthase has active sites to carry out multiple discrete reactions, which are summarized in Fig. 14-13.

Reactions 1 and 2 are transacylation reactions that serve to prime or load the enzyme with the reactants for the condensation reaction (Step 3). In the condensation reaction, decarboxylation of the malonyl group allows C2 to attack the acetyl thioester group to form acetoacetyl-ACP:

$$\underset{\text{Malonyl-ACP}}{\text{ACP—S—C(=O)—CH}_2\text{—COO}^-} + \underset{\text{Acetyl-Cys}}{\text{Cys—S—C(=O)—CH}_3} \xrightarrow{CO_2} \underset{\text{Acetoacetyl-ACP}}{\text{ACP—S—C(=O)—CH}_2\text{—C(=O)—CH}_3} + \text{Cys—SH}$$

This chemistry explains the necessity for carboxylating an acetyl group to a malonyl group: C2 of an acetyl group would not be sufficiently reactive.

The hydroxyacyl product of Reaction 4 is chemically similar to the hydroxyacyl product of Step 2 of β oxidation, but the intermediates of the two pathways have opposite configurations (see Box 4-A):

$$\text{CH}_3\text{—(CH}_2)_n\text{—}\overset{\text{OH}}{\underset{\text{H}}{\text{C}}}\text{—CH}_2\text{—}\overset{\overset{\text{O}}{\|}}{\text{C}}\text{—SACP} \qquad \text{CH}_3\text{—(CH}_2)_n\text{—}\overset{\text{H}}{\underset{\text{OH}}{\text{C}}}\text{—CH}_2\text{—}\overset{\overset{\text{O}}{\|}}{\text{C}}\text{—SCoA}$$

3-Hydroxyacyl-ACP intermediate of fatty acid synthesis (D configuration)

3-Hydroxyacyl-ACP intermediate of β oxidation (L configuration)

Note also that growth of the acyl chain—like chain shortening in β oxidation—occurs at the thioester end of the molecule. Exercise 19 presents the reactions of fatty acid synthesis.

The NADPH required for the two reduction steps of fatty acid synthesis (Steps 4 and 6) is supplied mostly by the pentose phosphate pathway (see Section 10-4). The synthesis of one molecule of palmitate (the usual product of fatty acid synthase) requires the production of 7 malonyl-CoA, at a cost of 7 ATP. The seven rounds of fatty acid synthesis consume 14 NADPH, which

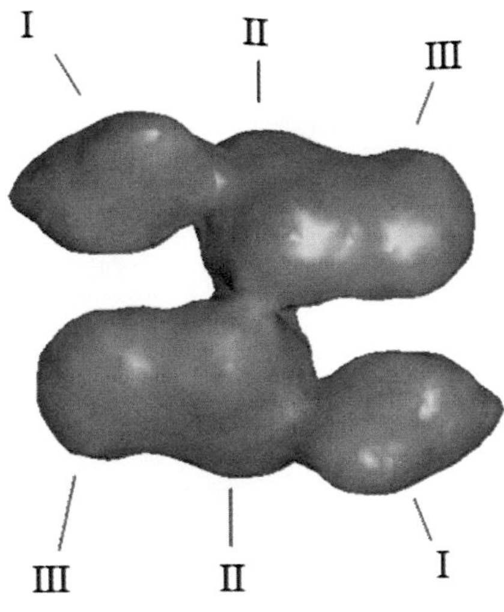

FIGURE 14-12

Electron micrograph of fatty acid synthase.
This electron cryomicroscopic reconstruction of human fatty acid synthase, with a resolution of 19 Å, indicates that the two monomers are arranged in an antiparallel fashion so that the dimer has dimensions of 180 × 130 × 75 Å. The three structural domains of each monomer comprise seven different enzymatic activities. [*Courtesy Dr. J. Brink and Prof. Wah Chiu, Verna & Mars McLean Department of Biochemistry and Molecular Biology, Baylor College of Medicine, Houston, TX. From* Proc. Natl. Acad. Sci. ***99***, *138–43 (2002).*]

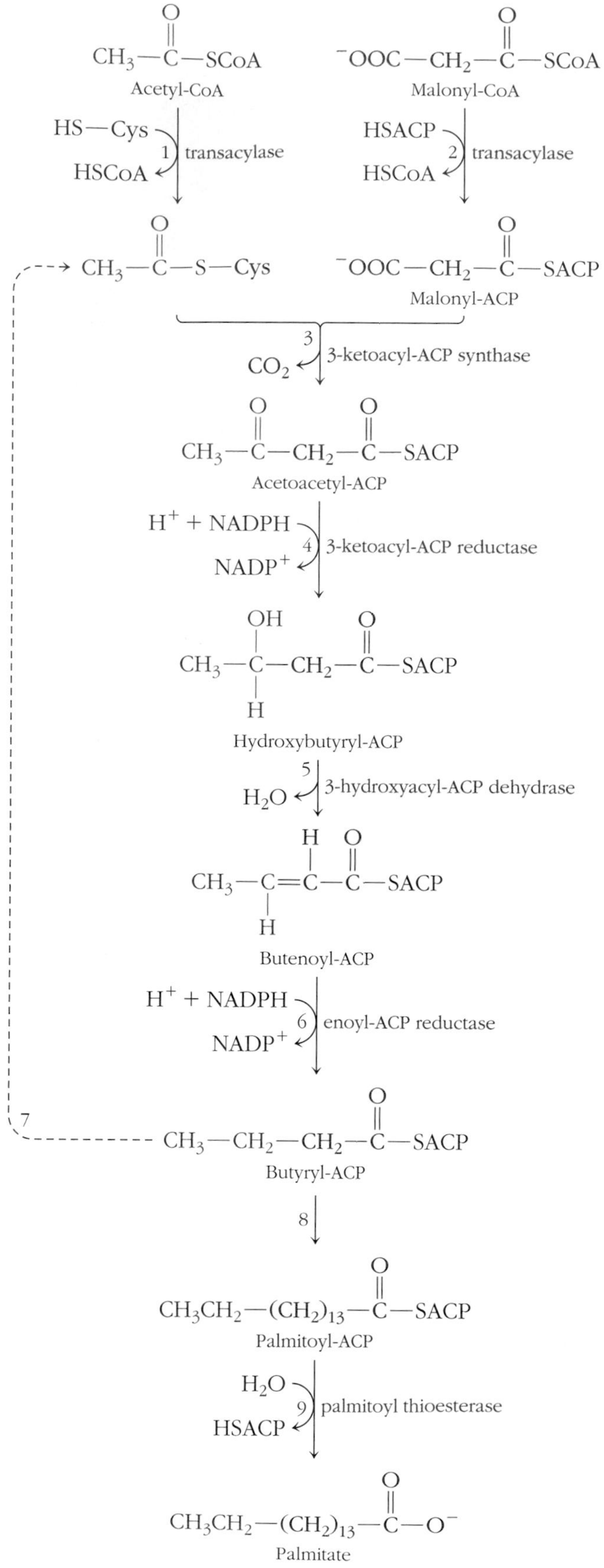

$$CH_3-\overset{O}{\overset{\|}{C}}-SCoA$$
Acetyl-CoA

$$^{-}OOC-CH_2-\overset{O}{\overset{\|}{C}}-SCoA$$
Malonyl-CoA

HS—Cys → (1) transacylase → HSCoA

HSACP → (2) transacylase → HSCoA

$$CH_3-\overset{O}{\overset{\|}{C}}-S-Cys$$

$$^{-}OOC-CH_2-\overset{O}{\overset{\|}{C}}-SACP$$
Malonyl-ACP

(3) 3-ketoacyl-ACP synthase → CO_2

$$CH_3-\overset{O}{\overset{\|}{C}}-CH_2-\overset{O}{\overset{\|}{C}}-SACP$$
Acetoacetyl-ACP

H^+ + NADPH → (4) 3-ketoacyl-ACP reductase → $NADP^+$

$$CH_3-\overset{OH}{\underset{H}{C}}-CH_2-\overset{O}{\overset{\|}{C}}-SACP$$
Hydroxybutyryl-ACP

(5) 3-hydroxyacyl-ACP dehydrase → H_2O

$$CH_3-\underset{H}{C}=\overset{H}{C}-\overset{O}{\overset{\|}{C}}-SACP$$
Butenoyl-ACP

H^+ + NADPH → (6) enoyl-ACP reductase → $NADP^+$

$$CH_3-CH_2-CH_2-\overset{O}{\overset{\|}{C}}-SACP$$
Butyryl-ACP

(7) (returns to the Cys)

(8)

$$CH_3CH_2-(CH_2)_{13}-\overset{O}{\overset{\|}{C}}-SACP$$
Palmitoyl-ACP

H_2O → (9) palmitoyl thioesterase → HSACP

$$CH_3CH_2-(CH_2)_{13}-\overset{O}{\overset{\|}{C}}-O^-$$
Palmitate

1. *The two-carbon acetyl group that will be lengthened is transferred from CoA to a Cys side chain of fatty acid synthase.*
2. *The malonyl group that will donate an acetyl group to the growing fatty acyl chain is transferred from CoA to the ACP domain of the enzyme.*
3. *In this condensation reaction, the malonyl group is decarboxylated and the resulting two-carbon fragment attacks the acetyl group to form a four-carbon product.*
4. *The 3-ketoacyl product of Step 3 is reduced.*
5. *A dehydration introduces a 2,3 double bond.*
6. *A second NADPH-dependent reduction completes the conversion of the condensation product to an acyl group.*
7. *The acyl group is transferred from ACP to the enzyme Cys group, and another malonyl group is loaded onto the free ACP, ready for another condensation reaction.*
8. *Steps 3–6 are repeated six times to build a C_{16} fatty acid.*
9. *A thioesterase hydrolyzes the thioester bond to release palmitate.*

FIGURE 14-13

Fatty acid synthesis.
The steps show how fatty acid synthase carries out the synthesis of the C_{16} fatty acid palmitate, starting from acetyl-CoA.

A CLOSER LOOK

BOX 14-A Fats, diet, and heart disease

Years of study have established a link between elevated LDL levels and atherosclerosis and indicate that certain diets contribute to the formation of fatty deposits that clog arteries and produce cardiovascular disease. Considerable research has been devoted to showing how dietary lipids influence serum lipid levels. For example, early studies showed that diets rich in saturated fats increase blood cholesterol (that is, LDL), whereas diets in which unsaturated vegetable oils replaced the saturated fats had the opposite effect. These and other findings led to recommendations that individuals at risk for atherosclerosis avoid butter, which is rich in saturated fat as well as cholesterol, and instead use margarine, which is prepared from cholesterol-free vegetable oils.

[Yoav Levy/Phototake.]

The production of semisolid margarine from liquid plant oils (triacylglycerols containing unsaturated fatty acids) often includes a hydrogenation step to chemically saturate the carbons of the fatty acyl chains. In this process, some of the original *cis* double bonds are converted to *trans* double bonds. In clinical studies, *trans* fatty acids are comparable to saturated fatty acids in their tendency to increase LDL levels and decrease HDL levels. Dietary guidelines now warn against the excessive intake of *trans* fatty acids in the form of hydrogenated vegetable oils (small amounts of *trans* fatty acids also occur naturally in some animal fats). This would mean avoiding donuts, french fries, and processed foods whose list of ingredients includes the term "partially hydrogenated vegetable oil."

So should you consume butter or margarine or neither, or does it really matter? Linking specific types of dietary fats to human health and disease has always been a risky venture, because quantitative information comes mainly from epidemiological and clinical studies, which are typically time-consuming and often inconclusive or downright contradictory. For example, since 1970, Americans have decreased their fat intake from about 40% of total calories to about 34%, and average serum cholesterol levels have also dropped. The death rate from heart disease has dropped, but—surprise!—the incidence of heart disease has not declined. And scientists still do not fully understand *how* the consumption of specific fatty acids—saturated or unsaturated, *cis* or *trans*—influences lipoprotein metabolism. Likewise, the metabolic effects of other dietary components are poorly understood. One consequence of low-fat diets, for example, is that individuals consume relatively more carbohydrates. And when they reduce their meat intake (an obvious source of fat), people eat more fruits and vegetables, which may have health-enhancing effects of their own in the form of vitamins and antioxidant compounds.

is equivalent to 14×3 or 42 ATP, bringing the total cost to 49 ATP. Still, this energy investment is much less than the free energy yield of oxidizing palmitate. Exercise 19 tallies the free energy required to synthesize a fatty acid.

During fatty acid synthesis, the long flexible arm of the pantothenate derivative in ACP (see Fig. 14-10) shuttles intermediates between the active sites of fatty acid synthase (the lipoamide group in the pyruvate dehydrogenase complex functions similarly; see Section 11-1). In the synthase dimer, two fatty acids can be built simultaneously.

Packaging several enzyme activities into one multifunctional protein like mammalian fatty acid synthase allows the enzymes to be synthesized and controlled in a coordinated fashion. Also, the product of one reaction can quickly diffuse to the next active site. These advantages are lacking in bacteria and chloroplasts, where fatty acid synthesis is carried out by separate proteins. These processes are presented schematically in Exercise 19.

Other enzymes elongate and desaturate newly synthesized fatty acids

Some sphingolipids contain C_{22} and C_{24} fatty acyl groups. *These and other long-chain fatty acids are generated by enzymes known as elongases,* which extend the C_{16} fatty acid produced by fatty acid synthase. Elongation can occur in either the endoplasmic reticulum or mitochondria. The endoplasmic reticulum reactions use malonyl-CoA as the acetyl-group donor and are chemically similar to those of fatty acid synthase. In the mitochondria, fatty acids are elongated by reactions that more closely resemble the reversal of β oxidation but use NADPH.

Desaturases introduce double bonds into saturated fatty acids. These reactions take place in the endoplasmic reticulum, catalyzed by membrane-bound enzymes. The electrons removed in the dehydrogenation of the fatty acid are eventually transferred to molecular oxygen to produce H_2O. The most common unsaturated fatty acids in animals are palmitoleate (a C_{16} molecule) and oleate (a C_{18} fatty acid; see page 432), both with one *cis* double bond at the 9,10 position. *Trans* fatty acids are relatively rare in plants and animals, but they are abundant in some prepared foods, which has produced confusion among individuals concerned with eating the "right" kinds of fats (Box 14-A).

Elongation can follow desaturation (and vice versa), so animals can synthesize a variety of fatty acids with different chain lengths and degrees of unsaturation. However, mammals cannot introduce double bonds at positions beyond C9 and therefore cannot synthesize fatty acids such as linoleate and linolenate. These molecules are precursors of the C_{20} fatty acid arachidonate and other lipids with specialized biological activities (Fig. 14-14). *Mammals must therefore obtain linoleate and linolenate from their diet.* These **essential fatty acids** are abundant in fish and plant oils. A deficiency of essential fatty acids resulting from a very-low-fat diet may elicit symptoms such as slow growth and poor wound healing.

REVIEW LEARNING OBJECTIVE 4

- Why must the pathways of fatty acid synthesis and degradation differ?
- Compare these two pathways with respect to location, cofactors, ATP requirement, and reaction intermediates.
- What is the function of elongases and desaturases?

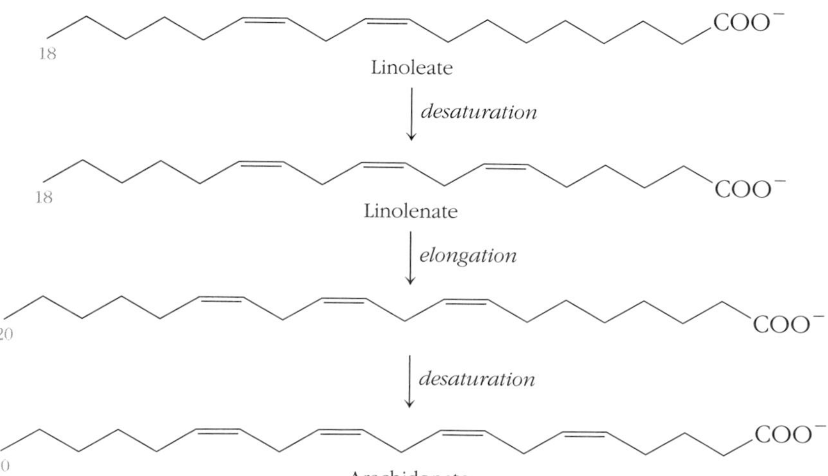

FIGURE 14-14

Synthesis of arachidonate. Linoleate (or linolenate) is elongated and desaturated to produce arachidonate, a C_{20} fatty acid with four double bonds.

Regulation of fatty acid synthesis

Under conditions of abundant metabolic fuel, the products of carbohydrate and amino acid catabolism are directed toward fatty acid synthesis and the resulting fatty acids are stored as triacylglycerols. The rate of fatty acid synthesis is controlled by acetyl-CoA carboxylase, which catalyzes the first step of the pathway. This enzyme is inhibited by a pathway product (palmitoyl-CoA) and is allosterically activated by citrate (which signals abundant acetyl-CoA). The enzyme is also subject to allosteric regulation by hormone-stimulated phosphorylation and dephosphorylation.

The concentration of malonyl-CoA is also critical for preventing the wasteful simultaneous activity of fatty acid synthesis and fatty acid oxidation. Malonyl-CoA is the source of acetyl groups that are incorporated into fatty acids, and it also blocks β oxidation by inhibiting carnitine acyltransferase, the enzyme involved in shuttling acyl groups into the mitochondria (see Fig. 14-5). Consequently, *when fatty acid synthesis is under way, no acyl groups are transported into the mitochondria for oxidation.* Some of the mechanisms that regulate fatty acid metabolism are summarized in Figure 14-15.

There are both natural and synthetic inhibitors of fatty acid synthase, such as the widely used antibacterial agent triclosan (Box 14-B). Fatty acid synthase inhibitors are of great scientific and popular interest, given that excess body weight (due to fat) is a major health problem, affecting over half the population of the United States. And because many tumors sustain high levels of fatty acid synthesis, fatty acid synthase inhibitors may also be useful for treating cancer.

REVIEW LEARNING OBJECTIVE 5

- What is the role of malonyl-CoA in fatty acid synthesis?

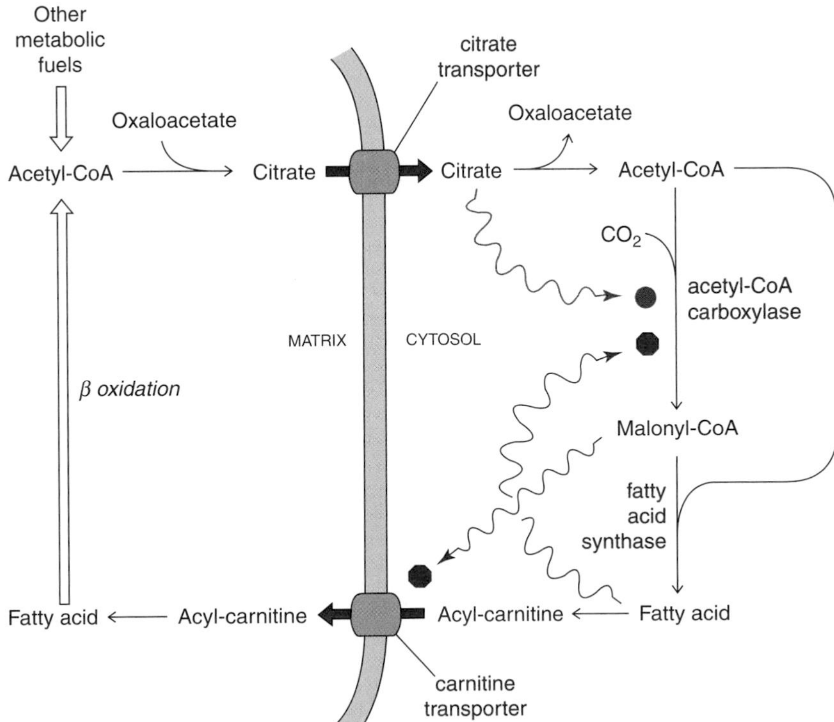

FIGURE 14-15

Some control mechanisms in fatty acid metabolism.
Red symbols indicate inhibition, and green symbols indicate activation.

A CLOSER LOOK

BOX 14-B Triclosan, an inhibitor of fatty acid synthesis

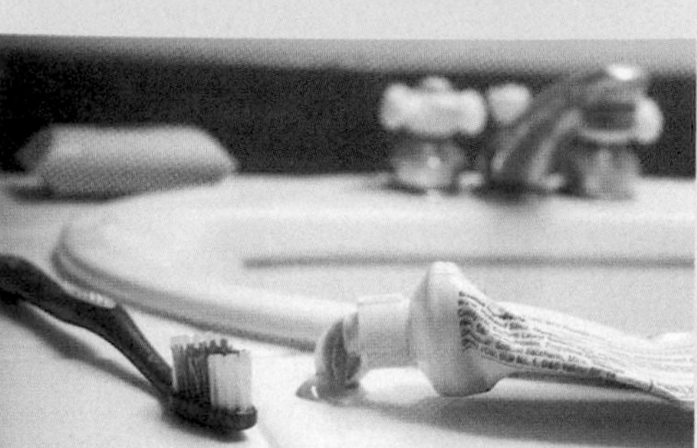

[Bruce Burkhardt/Corbis Images.]

Many cosmetics, toothpastes, antiseptic soaps, and even plastic toys and kitchenware contain the compound 5-chloro-2-(2,4-dichlorophenoxy)phenol, better known as triclosan:

Cl OH O Cl Cl

Triclosan

This compound has been used for over 30 years as an antibacterial agent, although its mechanism of action was not understood until 1998.

Triclosan was long believed to act as a general microbicide, which meant that it killed nonspecifically, much like household bleach or ultraviolet light. Such nonspecific microbicides are effective because it is difficult for bacteria to evolve specific resistance mechanisms. However, triclosan actually operates more like an antibiotic with a specific biochemical target, in this case, one of the enzymes of fatty acid synthesis (in bacteria, the enzymes are separate proteins, not part of a multifunctional protein). In bacteria, triclosan inhibits enoyl-ACP reductase (which catalyzes Step 6 as shown in Fig. 14-13).

The enzyme's natural substrate has a K_M of about 22 μM, but the dissociation constant for the inhibitor is 20 to 40 pM, indicating extremely tight binding. In the active site, one of the phenyl rings of triclosan, whose structure mimics the structure of the reaction intermediate, stacks on top of the nicotinamide ring of the NADH cofactor (similar to the stacking of bases in a polynucleotide; see Section 3-1). Triclosan also binds through van der Waals interactions and hydrogen bonds with amino acid residues in the active site. Some *E. coli* strains that are resistant to triclosan have mutations in one of these contact residues.

The specific action of triclosan as an inhibitor of a fatty acid synthesis enzyme and the existence of resistant bacterial strains indicate that triclosan is subject to the same drawbacks as other antibiotics—the eventual emergence of resistance through gene mutation. The widespread use of triclosan in the home and workplace increases the chances for resistance genes to spread. It is probably only a matter of time before resistance genes are ubiquitous, rendering triclosan useless as an antimicrobial agent.

Ketogenesis

During a prolonged fast, when glucose is unavailable from the diet and liver glycogen has been depleted, many tissues depend on fatty acids released from stored triacylglycerols to meet their energy needs. However, the brain does not burn fatty acids, because they pass poorly through the blood–brain barrier. Gluconeogenesis helps supply the brain's energy needs, but the liver also produces **ketone bodies** to supplement gluconeogenesis. The ketone bodies—acetoacetate and 3-hydroxybutyrate (also called β-hydroxybutyrate)—are synthesized from acetyl-CoA in liver mitochondria by a process called **ketogenesis.** Because ketogenesis uses fatty acid–derived acetyl groups, it helps spare amino acids that would otherwise be diverted to gluconeogenesis.

The assembly of ketone bodies is somewhat reminiscent of the synthesis of fatty acids or the oxidation of fatty acids in two-carbon steps (Fig. 14-16). In fact, the hydroxymethylglutaryl-CoA intermediate is chemically similar to the 3-hydroxyacyl intermediates of β oxidation and fatty acid synthesis.

Because they are small and water-soluble, ketone bodies are transported in the bloodstream without specialized lipoproteins, and they can easily pass into the central nervous system. During periods of high ketogenic activity, such as in diabetes, ketone bodies may be produced faster than they are con-

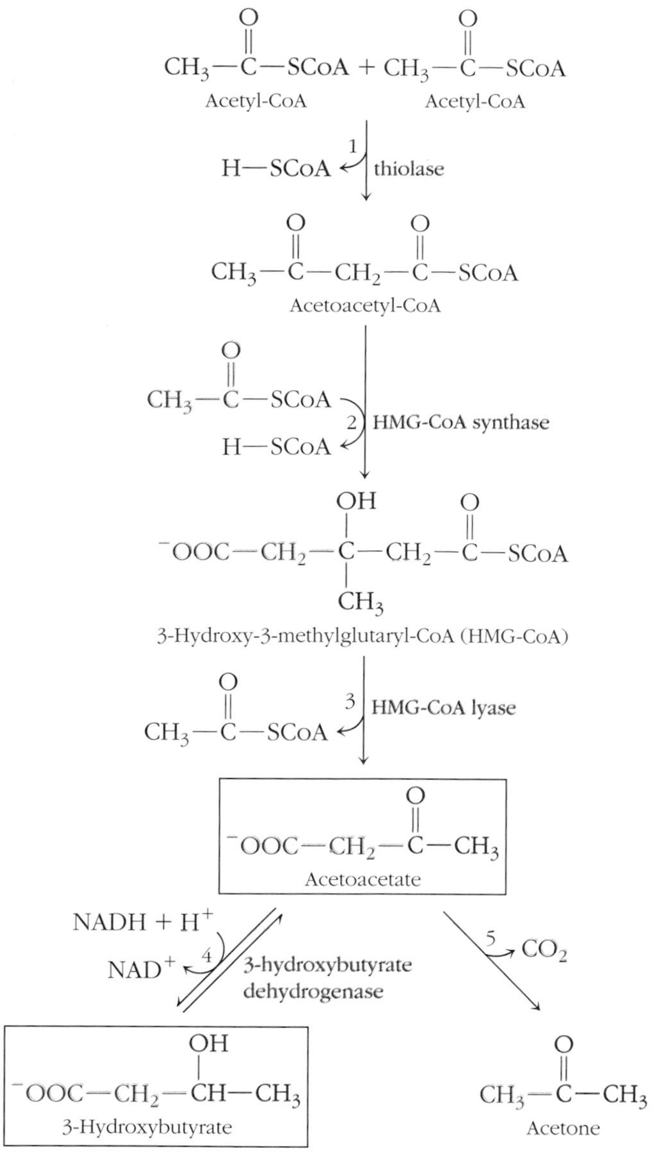

$CH_3-C(=O)-SCoA + CH_3-C(=O)-SCoA$

Acetyl-CoA Acetyl-CoA

1 thiolase; $H-SCoA$

$CH_3-C(=O)-CH_2-C(=O)-SCoA$

Acetoacetyl-CoA

1. *Two molecules of acetyl-CoA condense to form acetoacetyl-CoA. The reaction is catalyzed by a thiolase, which breaks a thioester bond.*

$CH_3-C(=O)-SCoA$; 2 HMG-CoA synthase; $H-SCoA$

$^{-}OOC-CH_2-C(OH)(CH_3)-CH_2-C(=O)-SCoA$

3-Hydroxy-3-methylglutaryl-CoA (HMG-CoA)

2. *The four-carbon acetoacetyl group condenses with a third molecule of acetyl-CoA to form the six-carbon 3-hydroxymethylglutaryl-CoA (HMG-CoA).*

3 HMG-CoA lyase; $CH_3-C(=O)-SCoA$

$^{-}OOC-CH_2-C(=O)-CH_3$

Acetoacetate

3. *HMG-CoA is then degraded to the ketone body acetoacetate and acetyl-CoA.*

$NADH + H^+$; 4 3-hydroxybutyrate dehydrogenase; NAD^+

$^{-}OOC-CH_2-CH(OH)-CH_3$

3-Hydroxybutyrate

5 CO_2

$CH_3-C(=O)-CH_3$

Acetone

4. *Acetoacetate undergoes reduction to produce another ketone body, 3-hydroxybutyrate.*
5. *Some acetoacetate may also undergo nonenzymatic decarboxylation to acetone and CO_2.*

FIGURE 14-16

Ketogenesis.
The ketone bodies are boxed.

sumed. Some of the excess acetoacetate breaks down to acetone, which gives the breath a characteristic sweet smell. Ketone bodies are also acids, with a pK of about 3.5. Their overproduction can lead to a drop in the pH of the blood, a condition called ketoacidosis.

Ketone bodies produced by the liver are used by other tissues as metabolic fuels after being converted back to acetyl-CoA (Fig. 14-17). The liver itself cannot catabolize ketone bodies because it lacks one of the required enzymes, 3-ketoacyl-CoA transferase.

REVIEW LEARNING OBJECTIVE 6

- What are ketone bodies and when are they synthesized?

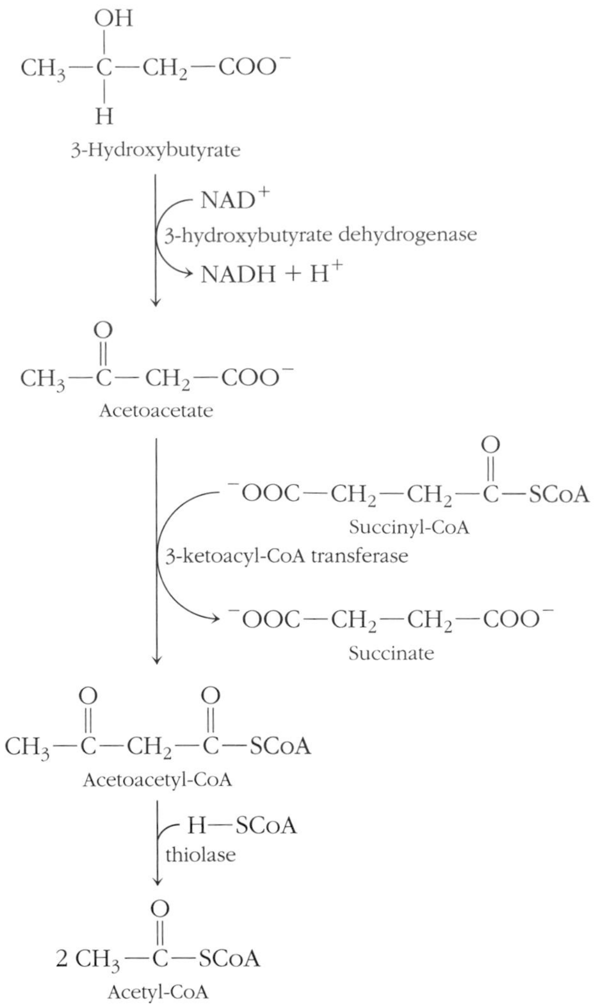

$$CH_3-\underset{\displaystyle H}{\overset{\displaystyle OH}{\underset{|}{\overset{|}{C}}}}-CH_2-COO^-$$

3-Hydroxybutyrate

$\downarrow$ NAD^+ → $NADH + H^+$ (3-hydroxybutyrate dehydrogenase)

1. *3-Hydroxybutyrate is oxidized back to acetoacetate (this is the reverse of Reaction 4 in Fig. 14-16).*

$$CH_3-\overset{\displaystyle O}{\overset{\|}{C}}-CH_2-COO^-$$

Acetoacetate

$\downarrow$ $^-OOC-CH_2-CH_2-\overset{O}{\overset{\|}{C}}-SCoA$ (Succinyl-CoA) → $^-OOC-CH_2-CH_2-COO^-$ (Succinate) (3-ketoacyl-CoA transferase)

2. *Succinyl-CoA donates its CoA group to produce acetoacetyl-CoA.*

$$CH_3-\overset{\displaystyle O}{\overset{\|}{C}}-CH_2-\overset{\displaystyle O}{\overset{\|}{C}}-SCoA$$

Acetoacetyl-CoA

$\downarrow$ $H-SCoA$ (thiolase)

3. *Thiolase then uses a free CoA group to cleave the four-carbon unit to two molecules of acetyl-CoA.*

$$2\ CH_3-\overset{\displaystyle O}{\overset{\|}{C}}-SCoA$$

Acetyl-CoA

FIGURE 14-17

Catabolism of ketone bodies.

3 SYNTHESIS OF OTHER LIPIDS

Lipid metabolism encompasses many chemical reactions, but many of these involve fatty acids, which are structural components of other lipids such as triacylglycerols, glycerophospholipids, and sphingolipids. Certain fatty acids are also the precursors of lipids with specialized biological roles as signaling molecules. This section covers the biosynthesis of some of the major types of lipids, including the synthesis of cholesterol from acetyl-CoA.

Triacylglycerol synthesis

Cells have a virtually unlimited capacity for storing fatty acids in the form of triacylglycerols. These highly hydrophobic molecules aggregate in the cytoplasm to form droplets surrounded by a layer of amphipathic phospholipids. In an adipose tissue cell, triacylglycerols occupy almost the entire cell volume (Fig. 14-18).

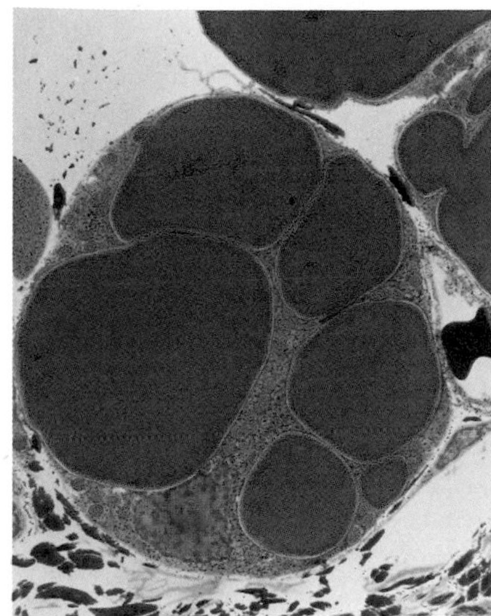

FIGURE 14-18

Electron micrograph of an adipocyte. Droplets of triacylglycerol occupy most of the cell's volume. [*Science Photo Library/Photo Researchers.*]

Triacylglycerols are synthesized by attaching fatty acyl groups to a glycerol backbone derived from phosphorylated glycerol or from glycolytic intermediates, for example, dihydroxyacetone phosphate:

$$\underset{\text{Dihydroxyacetone phosphate}}{\begin{array}{l} CH_2{-}OH \\ | \\ C{=}O \\ | \\ CH_2{-}O{-}PO_3^{2-} \end{array}} \xrightarrow[\text{glycerol-3-phosphate dehydrogenase}]{NADH + H^+ \quad NAD^+} \underset{\text{Glycerol-3-phosphate}}{\begin{array}{l} CH_2{-}OH \\ | \\ CH{-}OH \\ | \\ CH_2{-}O{-}PO_3^{2-} \end{array}}$$

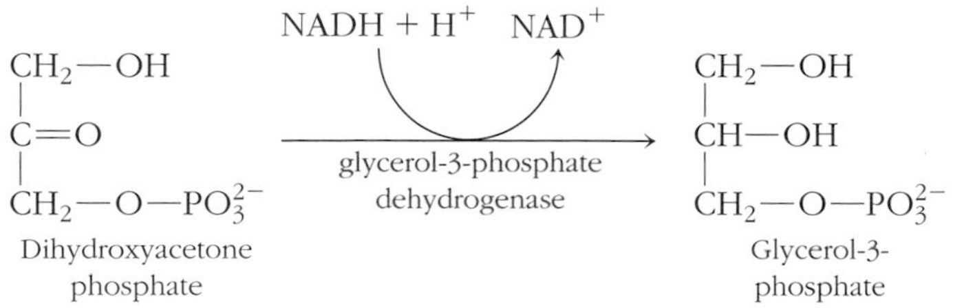

The fatty acyl groups are first activated to CoA thioesters in an ATP-dependent manner:

$$\text{Fatty acid} + \text{CoA} + \text{ATP} \rightleftharpoons \text{Acyl-CoA} + \text{AMP} + \text{PP}_i$$

This reaction is catalyzed by acyl-CoA synthetase, the same enzyme that activates fatty acids for oxidation. Triacylglycerols are assembled as shown in Figure 14-19. The acyltransferases that add fatty acids to the glycerol backbone are not highly specific with respect to chain length or degree of unsaturation of the fatty acyl group, but human triacylglycerols usually contain palmitate at C1 and unsaturated oleate at C2.

Phospholipid synthesis

The triacylglycerol biosynthetic pathway also provides the precursors for glycerophospholipids. *These amphipathic phospholipids are synthesized from phosphatidate or diacylglycerol by pathways that include an activating step in*

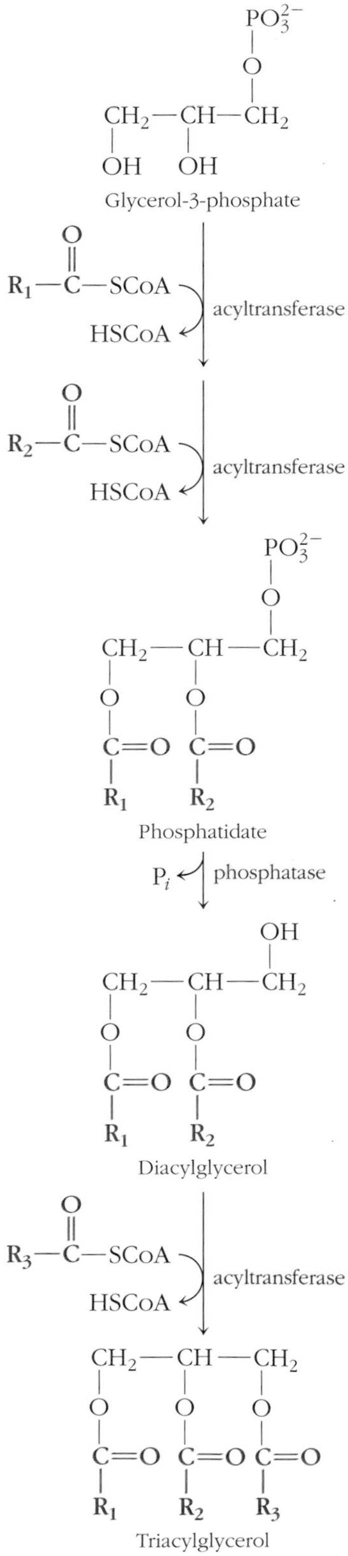

Glycerol-3-phosphate: $CH_2(OH){-}CH(OH){-}CH_2{-}O{-}PO_3^{2-}$

$\downarrow$ acyltransferase ($R_1{-}C(=O){-}SCoA \rightarrow HSCoA$)

$\downarrow$ acyltransferase ($R_2{-}C(=O){-}SCoA \rightarrow HSCoA$)

Phosphatidate: $CH_2(O{-}C(=O){-}R_1){-}CH(O{-}C(=O){-}R_2){-}CH_2{-}O{-}PO_3^{2-}$

$\downarrow$ phosphatase (P_i released)

Diacylglycerol: $CH_2(O{-}C(=O){-}R_1){-}CH(O{-}C(=O){-}R_2){-}CH_2{-}OH$

$\downarrow$ acyltransferase ($R_3{-}C(=O){-}SCoA \rightarrow HSCoA$)

Triacylglycerol: $CH_2(O{-}C(=O){-}R_1){-}CH(O{-}C(=O){-}R_2){-}CH_2(O{-}C(=O){-}R_3)$

FIGURE 14-19

Triacylglycerol synthesis.
An acyltransferase appends a fatty acyl group to C1 of glycerol-3-phosphate. A second acyltransferase reaction adds an acyl group to C2, yielding phosphatidate. A phosphatase removes P_i to produce diacylglycerol. The addition of a third acyl group yields a triacylglycerol.

which the nucleotide cytidine triphosphate (CTP) is cleaved. In some cases, the phospholipid head group is activated, and in other cases, the lipid tail portion is activated.

Figure 14-20 shows how the head groups ethanolamine and choline are activated before being added to diacylglycerol to produce phosphatidylethanolamine and phosphatidylcholine. Similar chemistry involving nucleotide sugars is used in the synthesis of glycogen from UDP–glucose (see Section 10-3) and starch from ADP–glucose (see Section 13-3).

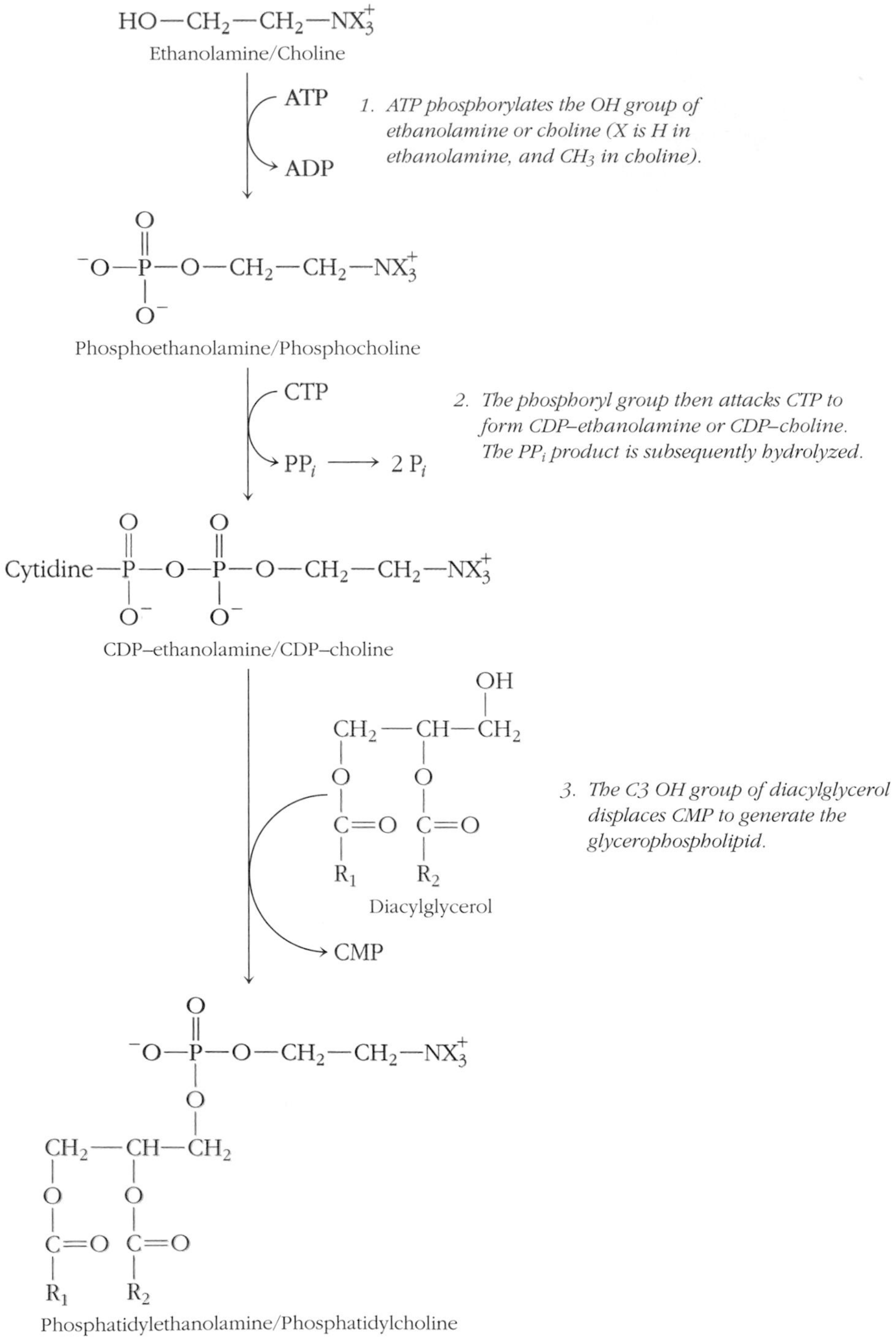

$HO{-}CH_2{-}CH_2{-}NX_3^+$

Ethanolamine/Choline

ATP → ADP

1. *ATP phosphorylates the OH group of ethanolamine or choline (X is H in ethanolamine, and CH_3 in choline).*

$^-O{-}P(=O)(O^-){-}O{-}CH_2{-}CH_2{-}NX_3^+$

Phosphoethanolamine/Phosphocholine

CTP → PP_i ⟶ 2 P_i

2. *The phosphoryl group then attacks CTP to form CDP–ethanolamine or CDP–choline. The PP_i product is subsequently hydrolyzed.*

$Cytidine{-}P(=O)(O^-){-}O{-}P(=O)(O^-){-}O{-}CH_2{-}CH_2{-}NX_3^+$

CDP–ethanolamine/CDP–choline

Diacylglycerol: $R_1{-}C(=O){-}O{-}CH_2{-}CH(O{-}C(=O){-}R_2){-}CH_2{-}OH$

→ CMP

3. *The C3 OH group of diacylglycerol displaces CMP to generate the glycerophospholipid.*

$^-O{-}P(=O){-}O{-}CH_2{-}CH_2{-}NX_3^+$, with P bonded through O to $CH_2{-}CH(O{-}C(=O){-}R_2){-}CH_2{-}O{-}C(=O){-}R_1$

Phosphatidylethanolamine/Phosphatidylcholine

FIGURE 14-20

Synthesis of phosphatidylethanolamine and phosphatidylcholine.

Phosphatidylserine is synthesized from phosphatidylethanolamine by a head-group exchange reaction in which serine displaces the ethanolamine head group.

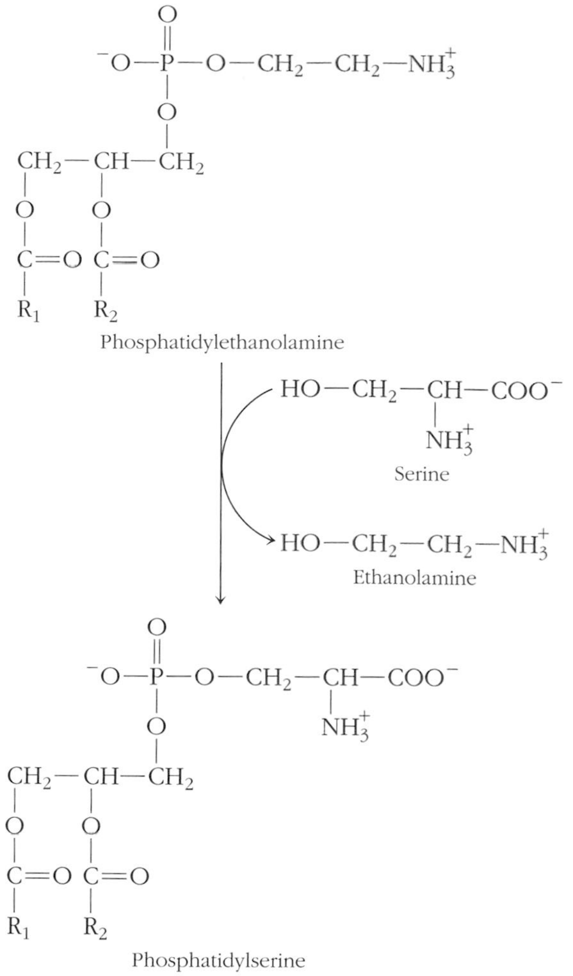

Phosphatidylethanolamine

Serine

Ethanolamine

Phosphatidylserine

In the synthesis of phosphatidylinositol, the diacylglycerol component is activated, rather than the head group, so that the inositol head group adds to CDP–diacylglycerol (Fig. 14-21).

Glycerophospholipids (and sphingolipids) are components of cellular membranes. New membranes are formed by inserting proteins and lipids into preexisting membranes, mainly in the endoplasmic reticulum. The newly synthesized membrane components reach their final cellular destinations primarily via vesicles that bud off the endoplasmic reticulum, and in some cases, by diffusing at points where two membranes make physical contact. Glycerophospholipids may undergo remodeling through the action of phospholipases and acyltransferases that remove and reattach different fatty acyl groups.

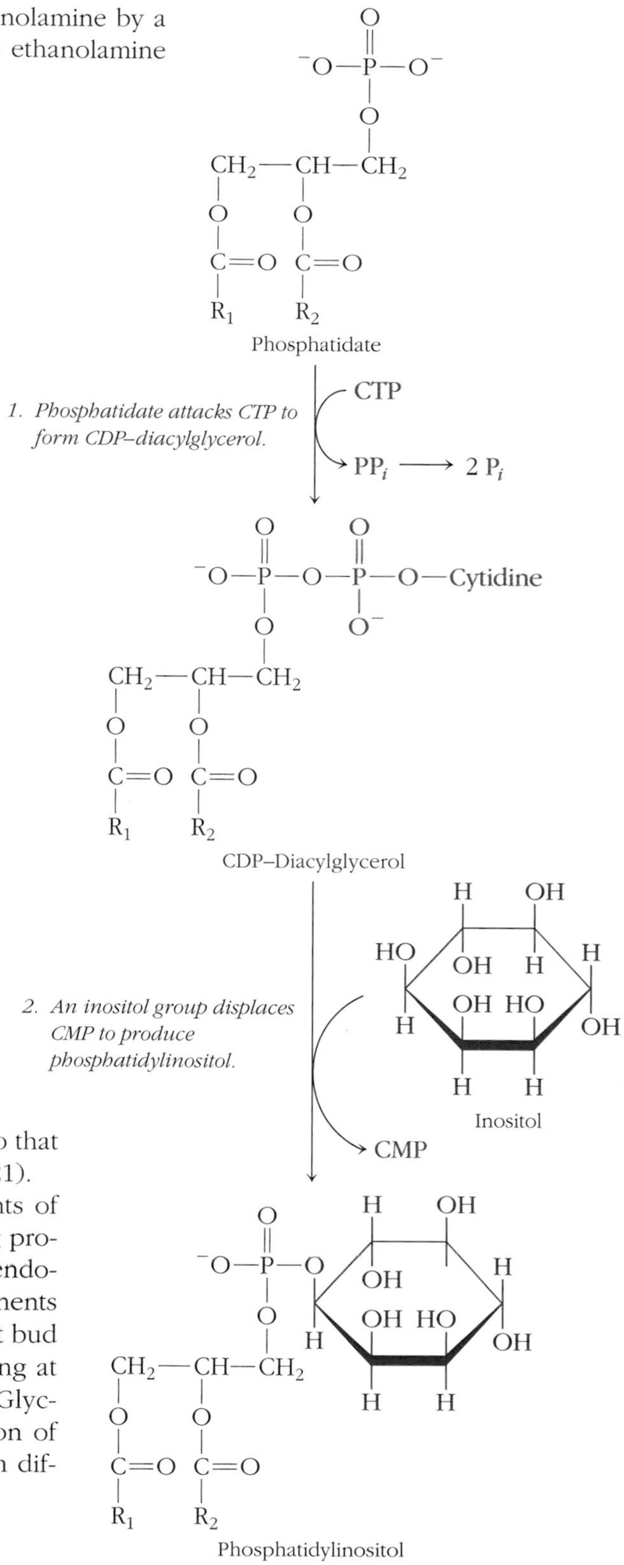

Phosphatidate

1. Phosphatidate attacks CTP to form CDP–diacylglycerol.

CTP

$PP_i \longrightarrow 2\ P_i$

CDP–Diacylglycerol

2. An inositol group displaces CMP to produce phosphatidylinositol.

Inositol

CMP

Phosphatidylinositol

FIGURE 14-21
Phosphatidylinositol synthesis.

Lipids as biological signals

Many membrane lipids are the precursors of molecules with signaling functions. For example, a hormone binding to a receptor protein on the surface of the cell may activate a phospholipase that hydrolyzes a membrane phospholipid to release its head group, which then diffuses through the cytosol and interacts with other proteins (we will investigate some hormone signaling pathways in Chapter 16).

Some membrane lipids may contain arachidonate, a C_{20} fatty acid with four double bonds, at the C2 position (see Fig. 14-14). Arachidonate released by the action of phospholipases is used in the synthesis of **eicosanoids.** These compounds (named for the Greek *eikosi,* twenty) have highly diverse structures and regulate a wide variety of physiological systems, including blood pressure, smooth muscle contraction, and inflammatory reactions. They induce pain, fever, and blood clotting. Most eicosanoids are short-lived, breaking down within seconds to minutes, and they usually act in or near the cells that produce them. One group of eicosanoids are called prostaglandins, reflecting an early theory that these compounds originated in the prostate gland. In humans, almost all tissues produce eicosanoids, with the exact product depending on the tissue and the synthetic enzymes present.

The pathway from arachidonate to prostaglandins begins with an enzyme commonly called cyclooxygenase (abbreviated COX), which actually has two catalytic activities. The cyclooxygenase activity catalyzes the addition of two oxygen molecules, with cyclization of the arachidonate. Next, a peroxidase activity converts the resulting peroxy group (—OOH) to a hydroxyl group (—OH):

COO^-

Arachidonate

$2\ O_2$ cyclooxygenase

O O COO^- OOH

peroxidase

O O COO^- OH

The product of the COX reactions undergoes additional modifications, depending on the tissue.

Blocking the production of prostaglandins by inhibiting COX has obvious implications for alleviating the pain and inflammation that are mediated by prostaglandins. One of the oldest and best-known COX inhibitors is aspirin. A new generation of COX inhibitors may be even more effective (Box 14-C).

REVIEW LEARNING OBJECTIVE 7

- What is the role of coenzyme A in triacylglycerol biosynthesis?
- What are the roles of CTP in glycerophospholipid biosynthesis?
- Describe the function of cyclooxygenase in eicosanoid biosynthesis.

Cholesterol synthesis

Cholesterol molecules, like fatty acids, are built from two-carbon acetyl units. In fact, *the first steps of cholesterol synthesis resemble those of ketogenesis.* How-

A CLOSER LOOK

BOX 14-C Aspirin and other inhibitors of cyclooxygenase

The bark of the willow *Salix alba* has been used since ancient times to relieve pain and fever. The active ingredient, when acetylated, is known as acetylsalicylate, or aspirin.

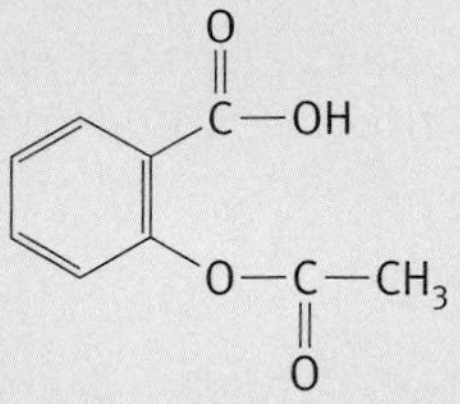

Acetylsalicylic acid (aspirin)

Aspirin was first prepared in 1853, but it was not used clinically for another 50 years or so. Effective promotion by the Bayer chemical company around the turn of the last century marked the beginning of the modern pharmaceutical industry.

Despite its universal popularity, aspirin's mode of action was not discovered until 1971. It inhibits the production of prostaglandins (which induce pain and fever, among other things) by inhibiting the cyclooxygenase activity of the COX enzyme. Inhibition results from acetylation of a Ser residue located near the active site in a cavity that accommodates the arachidonate substrate. Other pain-relieving substances such as ibuprofen and acetaminophen

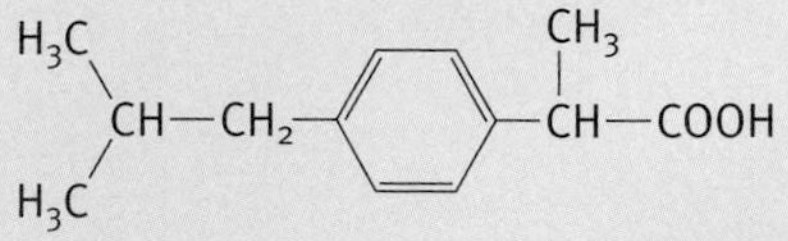

Ibuprofen

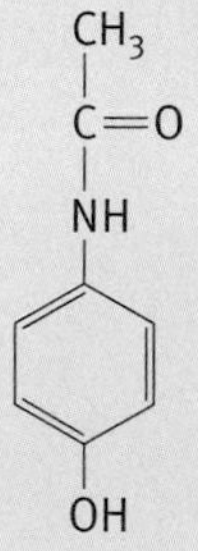

Acetaminophen

also bind to COX to prevent the synthesis of prostaglandins, although neither of these drugs acetylates the enzyme.

Since the 1970s, low doses of aspirin, which is a relatively inexpensive drug, have been used to prevent blood clotting (thrombosis) that could precipitate a heart attack. Aspirin is effective since it blocks the first step in the conversion of arachidonate to the eicosanoids known as thromboxanes, which trigger platelet aggregation (one of the events of blood clotting, or thrombosis). However, aspirin is not without side effects, notably stomach irritation. Gastric ulcers may develop because inhibiting COX prevents the synthesis of a prostaglandin that stimulates the gastric mucosa to secrete its protective coating.

Researchers discovered that the human body produces several isozymes of COX. COX-1 is expressed at all times and is required for the production of prostaglandins that protect the stomach and other organs from damage. COX-2 is induced by inflammatory stimuli and leads to production of prostaglandins that mediate the pain and swelling of inflammation. Aspirin and other older drugs inhibit both COX-1 and COX-2, but newer inhibitors that are selective for COX-2 and ineffective against COX-1 are valuable anti-inflammatory agents that do not produce gastric side effects.

COX-1 and COX-2 have similar sequences and three-dimensional structures, but they differ in the size and shape of the inhibitor binding site. In COX-2, the binding cavity is slightly larger. The newer drugs such as Celebrex® and Vioxx® can therefore bind to and inhibit COX-2 but not COX-1.

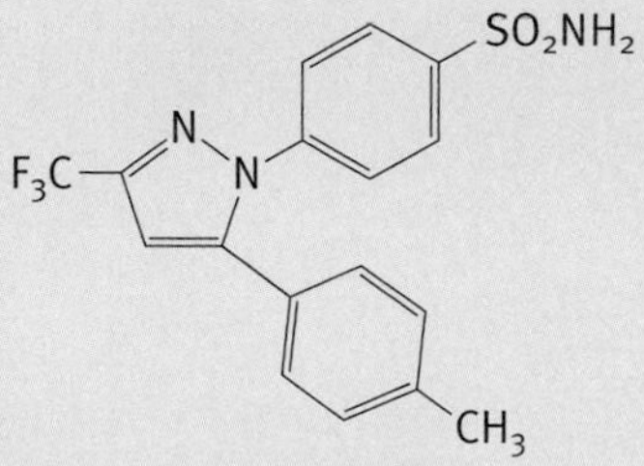

Celebrex®

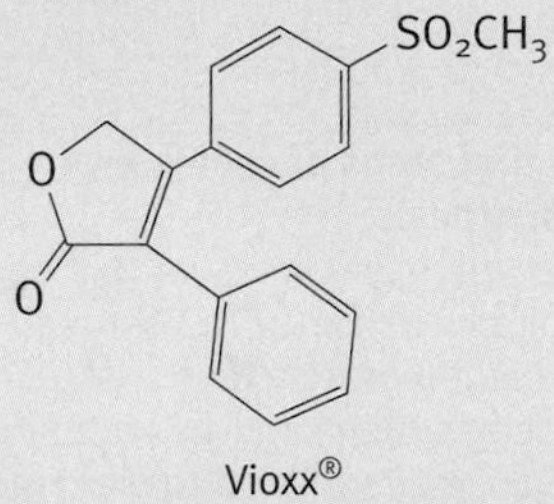

Vioxx®

These drugs are particularly effective for relieving the symptoms of chronic inflammatory diseases such as rheumatoid arthritis, since they can be used for long periods without eliciting harmful side effects.

Acetaminophen targets a third isozyme, COX-3. Blocking this enzyme relieves pain and fever but has little anti-inflammatory effect. The existence of multiple COX isozymes offers the possiblity of developing additional drugs that are tailored to specific ailments.

ever, ketone bodies are synthesized in the mitochondria (and only in the liver), and cholesterol is synthesized in the cytosol. The reactions of cholesterol biosynthesis and ketogenesis diverge after the production of HMG-CoA. In ketogenesis, this compound is cleaved to produce acetoacetate (see Fig. 14-16). In cholesterol synthesis, the thioester group of HMG-CoA is reduced to an alcohol, releasing the six-carbon compound mevalonate (Fig. 14-22).

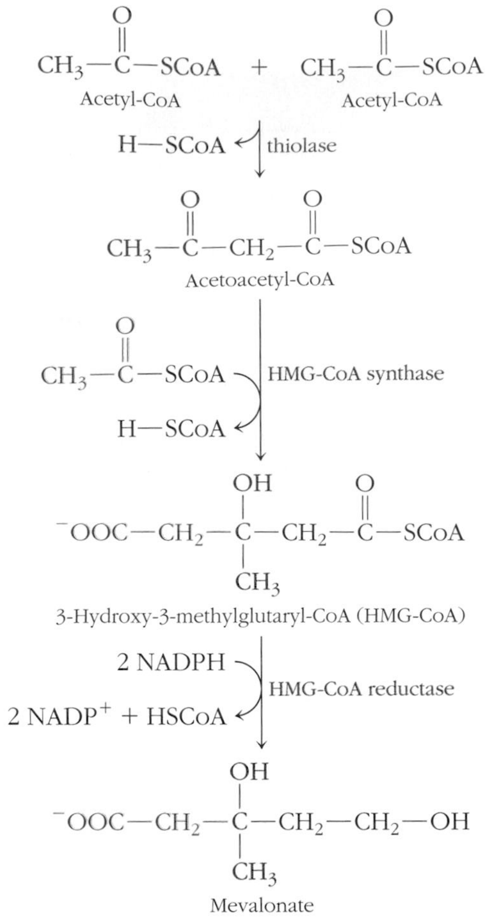

$$CH_3-\overset{O}{\overset{\|}{C}}-SCoA + CH_3-\overset{O}{\overset{\|}{C}}-SCoA$$

Acetyl-CoA   Acetyl-CoA

H—SCoA ← thiolase

$$CH_3-\overset{O}{\overset{\|}{C}}-CH_2-\overset{O}{\overset{\|}{C}}-SCoA$$

Acetoacetyl-CoA

$CH_3-\overset{O}{\overset{\|}{C}}-SCoA$ → HMG-CoA synthase → H—SCoA

$$^{-}OOC-CH_2-\underset{CH_3}{\overset{OH}{C}}-CH_2-\overset{O}{\overset{\|}{C}}-SCoA$$

3-Hydroxy-3-methylglutaryl-CoA (HMG-CoA)

2 NADPH → HMG-CoA reductase → $2\ NADP^+ + HSCoA$

$$^{-}OOC-CH_2-\underset{CH_3}{\overset{OH}{C}}-CH_2-CH_2-OH$$

Mevalonate

FIGURE 14-22

The first steps of cholesterol biosynthesis.
Note the resemblance of this pathway to ketogenesis (Fig. 14-16) through the production of HMG-CoA. HMG-CoA reductase then catalyzes a four-electron reductive deacylation to yield mevalonate.

In the next four steps of cholesterol synthesis, mevalonate acquires two phosphoryl groups and is decarboxylated to produce the five-carbon compound isopentenyl pyrophosphate:

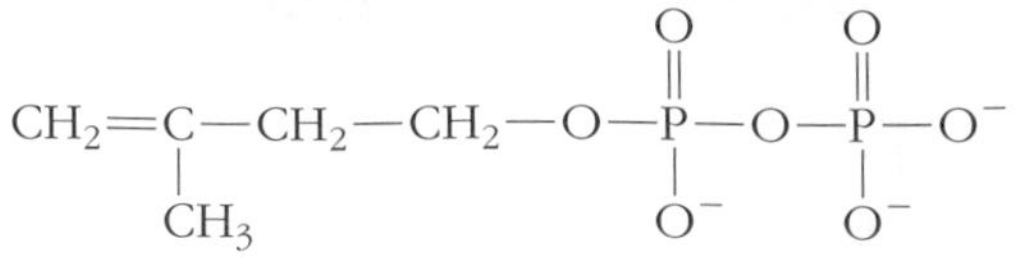

$$CH_2=\underset{CH_3}{C}-CH_2-CH_2-O-\underset{O^-}{\overset{O}{\overset{\|}{P}}}-O-\underset{O^-}{\overset{O}{\overset{\|}{P}}}-O^-$$

Isopentenyl pyrophosphate

This isoprene derivative is the precursor of cholesterol as well as other isoprenoids, such as ubiquinone, the C_{15} farnesyl group that is attached to some lipid-linked membrane proteins, and carotenoid pigments.

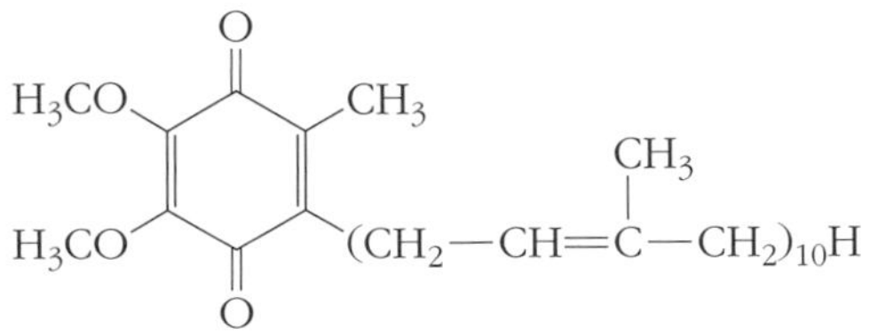

H_3CO, H_3CO, CH_3, $(CH_2-CH=\overset{CH_3}{C}-CH_2)_{10}H$

Ubiquinone

$$---NH-\underset{\underset{S}{CH_2}}{CH}-\overset{O}{\overset{\|}{C}}-O-CH_3$$

farnesyl group

CH_3 CH_3 CH_3 H_3C H_3C CH_3 CH_3 CH_3 CH_3 CH_3

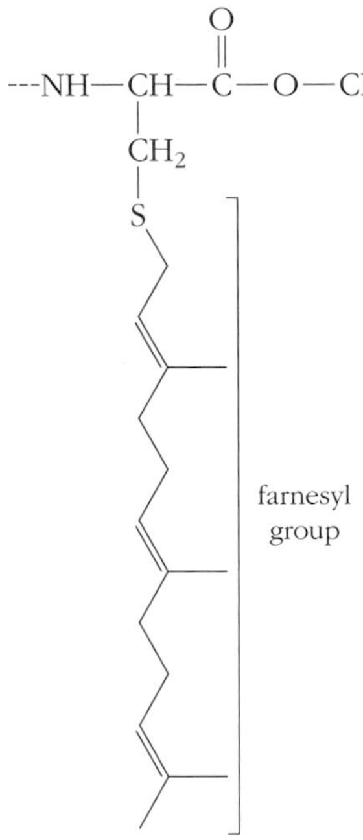

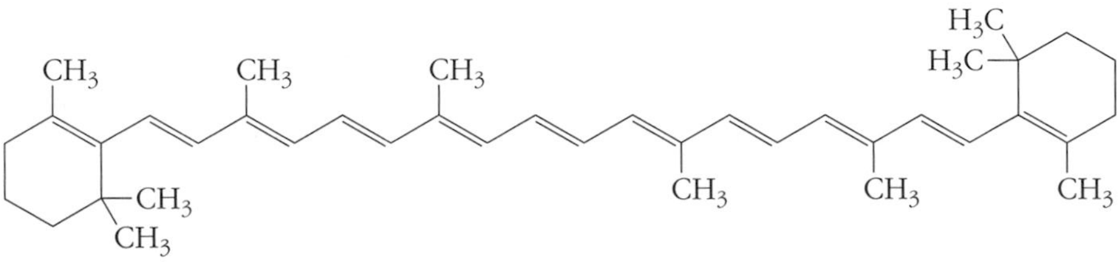

β Carotene

Isoprenoids are an extremely diverse group of compounds, with about 25,000 characterized to date.

In cholesterol synthesis, six isoprene units condense to form the C_{30} compound squalene. Cyclization of this linear molecule leads to a structure with four rings, resembling cholesterol (Fig. 14-23). A total of 21 reactions are re-

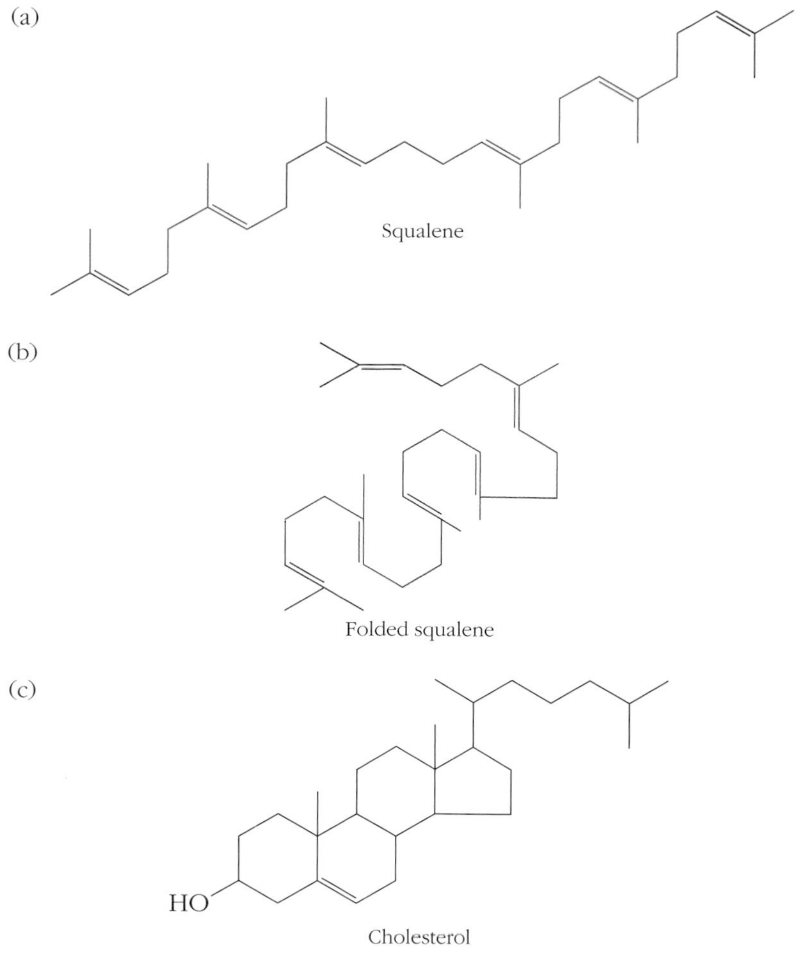

(a)

(b)

(c)

FIGURE 14-23

Structure of squalene, a precursor of cholesterol.
(a) Squalene, with its six isoprene units shown in different colors. (b) A folded squalene molecule before cyclization. (c) Cholesterol, a C_{27} derivative of the C_{30} squalene.

quired to convert squalene to cholesterol. These steps include cyclization, oxidation, and the loss of three methyl groups. NADH or NADPH is required for several steps.

The rate-determining step of cholesterol synthesis (a pathway with over 30 steps) and the major control point is the conversion of HMG-CoA to mevalonate by HMG-CoA reductase. This enzyme is one of the most highly regulated enzymes known. For example, the rates of its synthesis and degradation are tightly controlled, and the enzyme is subject to inhibition by phosphorylation of a Ser residue.

Synthetic inhibitors known as statins bind extremely tightly to HMG-CoA reductase, with K_I values in the nanomolar range. The substrate HMG-CoA has a K_M of about 4 μM. All the statins have an HMG-like group that acts as a competitive inhibitor of HMG-CoA binding to the enzyme (Fig. 14-24). Their rigid hydrophobic groups also prevent the enzyme from forming a structure that would accommodate the pantothenate moiety of CoA. The physiological effect of the statins is to lower serum cholesterol levels by blocking mevalonate synthesis. Cells must then obtain cholesterol from circulating lipoproteins. But since mevalonate is also the precursor of other isoprenoids, the long-term use of statins can have negative side effects.

FIGURE 14-24

Some statins.
These inhibitors of HMG-CoA reductase have a bulky hydrophobic group plus an HMG-like group (colored red).

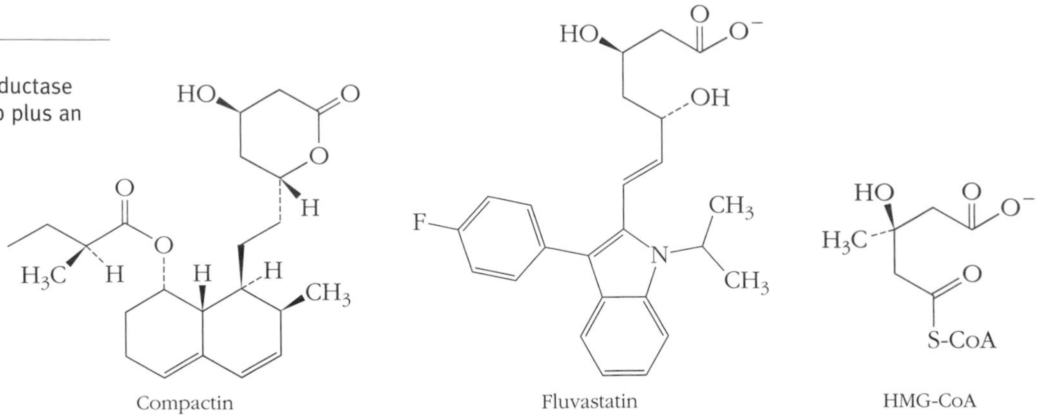

Compactin

Fluvastatin

HMG-CoA

The fate of cholesterol

Newly synthesized cholesterol has several fates:

Go to **Exercise 20/Lipoproteins** for an overview of the structure and physiological functions of the lipoproteins that transport cholesterol to and from cells. The material included here addresses Learning Objective 9.

1. It can be incorporated into a cell membrane.
2. It may be acylated to form a cholesteryl ester for storage or, in liver, for packaging in VLDL.

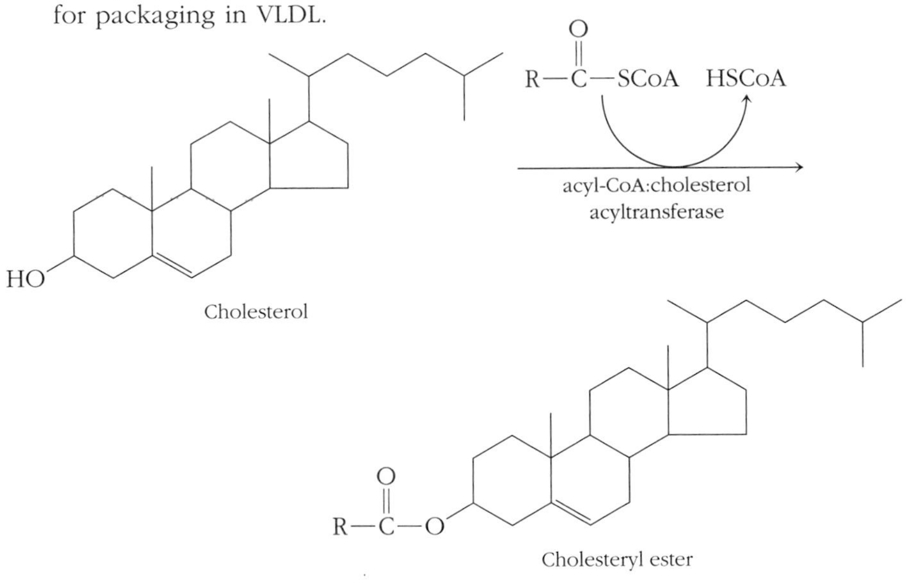

Cholesterol

R—C(=O)—SCoA → HSCoA

acyl-CoA:cholesterol acyltransferase

Cholesteryl ester

3. It is a precursor of steroid hormones such as testosterone and estrogen in the appropriate tissues.
4. It is a precursor of bile acids such as cholate:

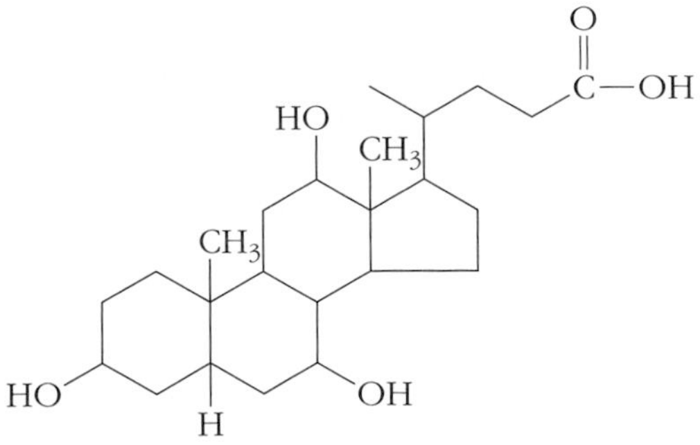

Cholate

Bile acids are synthesized in the liver, stored in the gallbladder, and secreted into the small intestine. There, they aid digestion by acting as detergents to solubilize dietary fats and make them more susceptible to lipases. Although bile acids are mostly reabsorbed and recycled through the liver for reuse, some are excreted from the body. *This is virtually the only route for cholesterol disposal.*

Cells can synthesize cholesterol as well as obtain it from circulating LDL. When LDL proteins dock with the LDL receptor on the cell surface, the lipoprotein–receptor complex undergoes endocytosis. Inside the cell, the lipoprotein is degraded and cholesterol enters the cytosol.

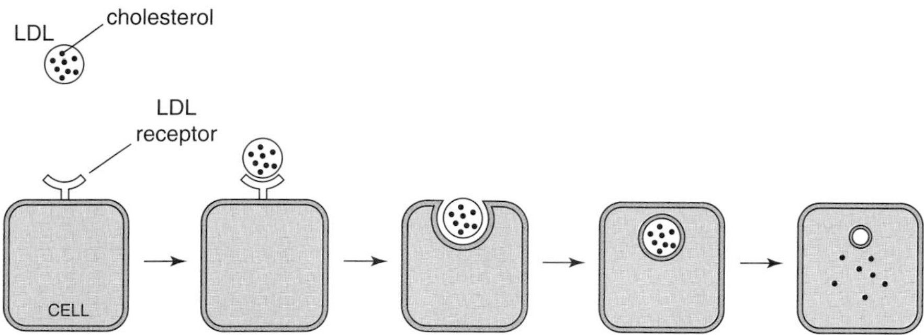

The role of LDL in delivering cholesterol to cells is dramatically illustrated by the disease familial hypercholesterolemia, which is due to a genetic defect in the LDL receptor. The cells of homozygotes are unable to take up LDL, so the concentration of serum cholesterol is about three times higher than normal. This contributes to atherosclerosis, and many individuals die of the disease before age 30. Exercise 20 illustrates cholesterol transport by lipoproteins, including receptor-mediated endocytosis.

High-density lipoproteins (HDL) are essential for removing excess cholesterol from cells. The efflux of cholesterol requires the close juxtaposition of the cell membrane and an HDL particle as well as specific cell surface proteins. One of these appears to be a transporter or flippase that moves cholesterol from the cytosolic leaflet to the extracellular leaflet, from which it can diffuse into the HDL particle.

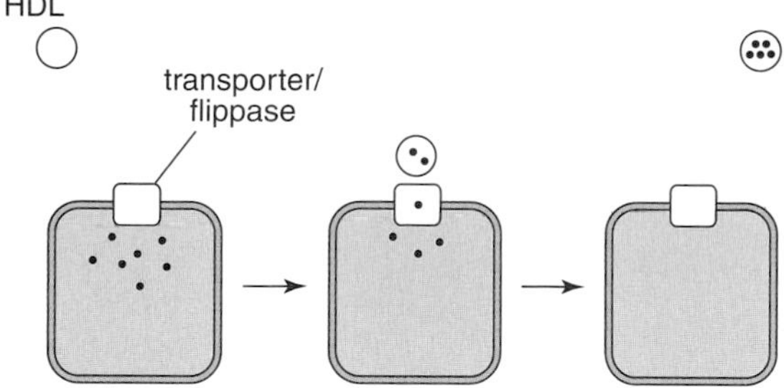

Defects in the gene for the transporter cause Tangier disease, which is characterized by accumulations of cholesterol in tissues and a high risk of heart attack.

Because cells do not break down cholesterol and because the accumulation of cholesterol is potentially toxic (it could disrupt membrane structure), the body must coordinate cholesterol synthesis and transport among tissues. For example, cholesterol shuts down its own synthesis by inhibiting the synthesis of enzymes such as HMG-CoA reductase. Cellular cholesterol also represses transcription of the gene for the LDL receptor. In contrast to fatty acid metabolism, where the two opposing pathways of synthesis and degradation operate in balance to meet the cell's needs, cholesterol metabolism in many cells is characterized by a balance between influx and efflux.

REVIEW LEARNING OBJECTIVES 8 AND 9

- How does cholesterol synthesis resemble ketogenesis?
- List the metabolic fates of newly synthesized cholesterol.
- Summarize the roles of LDL and HDL in cholesterol metabolism.
- What is the link between lipoproteins and atherosclerosis?

4 SUMMARY OF LIPID METABOLISM

The processes of breaking down and synthesizing lipids illustrate some general principles related to how the cell carries out opposing metabolic pathways. The diagram in Figure 14-25 includes the major lipid metabolic pathways covered in this chapter. Several features are worth noting:

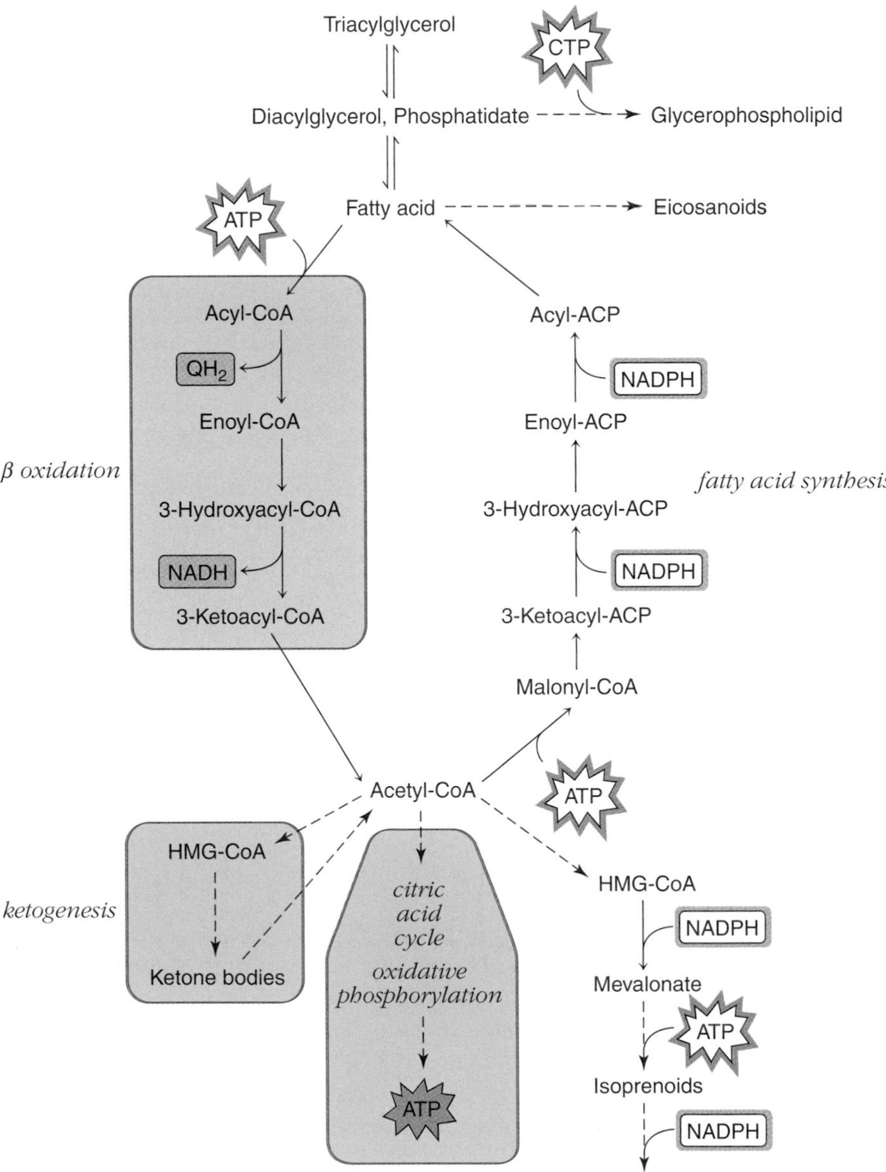

FIGURE 14-25

Summary of lipid metabolism. Only the major pathways covered in this chapter are included. Open gold symbols indicate ATP consumption; filled gold symbols indicate ATP production. Open and filled red symbols represent the consumption and production of reduced cofactors (NADH, NADPH, and QH_2). The shaded portions of the diagram indicate reactions that occur in mitochondria.

1. The pathways for fatty acid catabolism and synthesis, and the synthesis of some other compounds all converge at the common intermediate acetyl-CoA, which is also a product of carbohydrate metabolism (see Section 11-1) and a key player in amino acid metabolism (which will be covered in Chapter 15).

2. The pathways for fatty acid degradation and fatty acid synthesis have a certain degree of symmetry, with similar intermediates and a role for thioesters, but the pathways have very different free energy considerations. β Oxidation produces reduced cofactors and requires only one ATP; fatty acid synthesis consumes NADPH and requires the input of ATP in each round. Other metabolic pathways, including cholesterol synthesis, consume reduced cofactors generated by catabolic reactions.

3. The catabolic pathway of β oxidation, the conversion of acetyl-CoA to ketone bodies, the oxidation of acetyl-CoA via the citric acid cycle, and the reoxidation of reduced cofactors occur in mitochondria (although some lipid metabolic reactions also occur in peroxisomes). In contrast, many lipid biosynthetic reactions take place in the cytosol or associated with the endoplasmic reticulum. Various pathways therefore require transmembrane transport systems and/or separate pools of substrates and cofactors.

4. Although the central pathways of lipid metabolism, as outlined in Figure 14-25, comprise just a few different reactions, complexity is introduced in the form of isozymes with different acyl-chain length specificities, in additional enzymes to deal with odd-chain, branched, and unsaturated fatty acids, and in tissue-specific reactions leading to particular products such as eicosanoids or isoprenoids.

Understanding lipid metabolism is essential for diagnosing and treating certain human diseases, including those that reflect deficiencies of lipid-metabolizing enzymes or abnormalities among the proteins involved in transporting lipids via lipoproteins.

SUMMARY

1. Lipoproteins transport lipids, including cholesterol, in the bloodstream. High levels of LDL are associated with the development of atherosclerosis.

2. Fatty acids released from triacylglycerols by the action of lipases are activated by their attachment to CoA in an ATP-dependent reaction.

3. In the process of β oxidation, a series of four enzymatic reactions degrades a fatty acyl-CoA two carbons at a time, producing one QH_2, one NADH, and one acetyl-CoA, which can be further oxidized by the citric acid cycle. Reoxidation of the reduced cofactors generates considerable ATP.

4. Oxidation of unsaturated and odd-chain fatty acids requires additional enzymes. Very-long-chain and branched fatty acids are oxidized in peroxisomes.

5. Fatty acids are synthesized by a pathway that resembles the reverse of β oxidation. In the first step of fatty acid synthesis, acetyl-CoA carboxylase catalyzes an ATP-dependent reaction that converts acetyl-CoA to malonyl-CoA, which becomes the donor of two-carbon groups.

6. Mammalian fatty acid synthase is a multifunctional enzyme in which the growing fatty acyl chain is attached to acyl carrier protein rather than CoA. Elongases and desaturases may modify newly synthesized fatty acids.

7. The liver can convert acetyl-CoA to ketone bodies to be used as metabolic fuels in other tissues.

8. Triacylglycerols are synthesized by attaching three fatty acyl groups to a glycerol backbone. Intermediates of the triacylglycerol pathway are the starting materials for the synthesis of phospholipids. Phospholipid-derived eicosanoids are signaling molecules with diverse physiological roles.

9. Cholesterol is synthesized from acetyl-CoA. The rate-determining step of this pathway is the target of drugs known as statins. Excess cholesterol circulates via HDL.

CHECKLIST

1. Review the Learning Objectives listed on page 424.
2. Complete Exercise 19/Fatty Acid Metabolism for a closer look at the reactions of β oxidation and fatty acid synthesis.
3. Complete Exercise 20/Lipoproteins for a summary of the structures and functions of lipoproteins and their role in cholesterol metabolism and disease.
4. Apply your knowledge by solving the problems at the end of this chapter. Check your results in the Solutions appendix.
5. Be able to define the boldfaced terms (consult the glossary at the end of the book). Test your understanding by taking the Chapter 14 quiz (www.wiley.com/college/pratt).
6. Explore the websites listed at www.wiley.com/college/pratt.
7. Consult the list of Selected Readings for background information or additional details on lipoproteins and the metabolism of fatty acids, membrane lipids, and cholesterol.

GLOSSARY TERMS

lipoprotein
β oxidation
peroxisome
multifunctional enzyme
essential fatty acid
ketone bodies
ketogenesis
eicosanoids
bile acids

PROBLEMS

1. The overall reaction of the activation of a fatty acid to fatty acyl-CoA, with concomitant hydrolysis of ATP to AMP, has a free energy change of about zero. The reaction is favorable because of subsequent hydrolysis of pyrophosphate to orthophosphate ($\Delta G^{\circ\prime}$ for this reaction is given in Table 9-2). Write the equation for the coupled reaction, and calculate (a) $\Delta G^{\circ\prime}$ and (b) the equilibrium constant for the reaction.
2. Fatty acid activation catalyzed by acyl-CoA synthetase begins with nucleophilic attack of the negatively charged oxygen of the fatty acid carboxylate group on the α-phosphate (the innermost phosphate) of ATP. An acyladenylate mixed anhydride is formed. Write the mechanism of the reaction.
3. During β oxidation, methylene ($—CH_2—$) groups in a fatty acid are oxidized to carbonyl ($C{=}O$) groups, yet no oxygen is consumed by the reactions of β oxidation. How is this possible?
4. The first three reactions of the β oxidation pathway are similar to three reactions of the citric acid cycle. Which reactions are these, and why are they similar?
5. The β oxidation pathway was elucidated in part by Franz Knoop in 1904. He fed dogs fatty acid phenyl derivatives and then analyzed the dogs' urine for the resulting metabolites. What metabolite was produced when the dogs were fed (a) phenylpropionate and (b) phenylbutyrate?

$C_6H_5—CH_2—CH_2—COO^-$

Phenylpropionate

$C_6H_5—CH_2—CH_2—CH_2—COO^-$

Phenylbutyrate

6. (a) A deficiency of carnitine results in muscle cramps, which are exacerbated by fasting or exercise. Give a biochemical explanation for the muscle cramping and explain why cramping increases during fasting and exercise.
   (b) Muscle biopsy and enzyme assays of a carnitine-deficient individual demonstrate that medium-chain (C_8–C_{10}) fatty acids can be metabolized normally, despite the carnitine deficiency. What does this tell you about the role of carnitine in fatty acid transport across the inner mitochondrial membrane?
7. Some individuals suffer from medium-chain acyl-CoA dehydrogenase (MCAD) deficiency. Which intermediates would accumulate in individuals with MCAD deficiency?
8. A deficiency of phytanate-degrading enzymes in peroxisomes results in Refsum's disease, a neuronal disorder caused by phytanate accumulation. Patients with Refsum's disease cannot convert phytanate to pristanate because they lack the enzymes involved in this α oxidation reaction. Pristanate normally enters the peroxisomal β oxidation pathway. The α oxidation of phytanate to pristanate is shown below. Show how pristanate is oxidized via β oxidation, and list the products of pristanate oxidation.

Phytanate (COO^-, β, α)

α oxidation → CO_2

Pristanate (COO^-, α, β)

9. How many molecules of ATP are generated when (a) palmitate and (b) stearate are completely oxidized via the β oxidation pathway in mitochondria?

10. How many molecules of ATP are generated when oxidation of a fully saturated C_{24} fatty acid begins in the peroxisome and is completed by the mitochondrion when 12 carbons remain?

11. How many molecules of ATP are generated by the complete oxidation of (a) oleic acid (9-*cis*-octadecenoic acid) and (b) linoleic acid (9,12-*cis*-octadecadienoic acid)? Compare these values with the amount of ATP generated by the complete oxidation of stearate (18:0).

12. How many ATP are generated when a fully saturated 17-carbon fatty acid is oxidized via the β oxidation pathway?

13. Both fatty acid oxidation and glucose oxidation by glycolysis lead to the formation of large amounts of ATP. Explain why a cell preparation containing all the enzymes required for either pathway cannot generate ATP when a fatty acid or glucose is added, unless a small amount of ATP is also added to the preparation.

14. The complete oxidation to CO_2 of glucose and palmitate releases considerable free energy: $\Delta G^{\circ\prime} = -2850\ \text{kJ}\cdot\text{mol}^{-1}$ for glucose oxidation, and $\Delta G^{\circ\prime} = -9781\ \text{kJ}\cdot\text{mol}^{-1}$ for palmitate. For each fuel molecule, compare the ATP yield per carbon atom (a) in theory and (b) *in vivo*. (c) What do these results tell you about the relative efficiency of oxidizing carbohydrates and fatty acids?

15. Compare fatty acid degradation and fatty acid synthesis by filling in the table below.

	Fatty acid degradation	**Fatty acid synthesis**
Cellular location		
Acyl group carrier		
Electron carrier(s)		
ATP requirement		
Unit product/unit donor		
Configuration of hydroxyacyl intermediate		
Shortening/growth occurs at which end of the fatty acyl chain?		

16. Write the mechanism for the carboxylation of acetyl-CoA to malonyl-CoA catalyzed by acetyl-CoA carboxylase.

17. During fatty acid synthesis, why is the condensation of an acetyl group and a malonyl group energetically favorable, whereas the condensation of two acetyl groups would be unfavorable?

18. Compare the ATP yield of palmitate oxidation and the ATP cost of palmitate synthesis.

19. Compare the carboxylation/decarboxylation sequence of reactions in gluconeogenesis (Fig. 10-13) and fatty acid synthesis. Discuss the source of free energy for these steps in each pathway.

20. On what carbon atoms does the $^{14}CO_2$ used to synthesize malonyl-CoA from acetyl-CoA appear in palmitate?

21. Which of the following fatty acids are essential fatty acids for humans?

$O{=}C(-O^-)-(CH_2)_7-CH{=}CH-(CH_2)_7-CH_3$

Oleate

$O{=}C(-O^-)-(CH_2)_7-CH{=}CH-CH_2-CH{=}CH-(CH_2)_4-CH_3$

Linoleate

$O{=}C(-O^-)-(CH_2)_7-CH{=}CH-CH_2-CH{=}CH-CH_2-CH{=}CH-CH_2-CH_3$

α-Linolenate

$O{=}C(-O^-)-(CH_2)_7-CH{=}CH-(CH_2)_5-CH_3$

Palmitoleate

22. Does activity increase or decrease for the given fatty acid synthase under the following conditions? Explain.

(a) High-carbohydrate diet (liver fatty acid synthase)
(b) High-fat diet (liver fatty acid synthase)
(c) Mid- to late pregnancy (mammary gland fatty acid synthase)

23. Isolated heart cells undergo contraction even in the absence of glucose and fatty acids if they are supplied with acetoacetate.

(a) How does this compound act as a metabolic fuel?
(b) Even with plentiful acetoacetate, the rate of flux through the citric acid cycle gradually drops off unless pyruvate is added to the cells. Explain.

24. One of the symptoms of untreated diabetes is ketosis. Diabetics lack functional pancreatic β cells and are unable to produce the insulin required to stimulate cellular uptake of glucose. Blood glucose levels rise, while the cells are literally "starved" for glucose. Explain why blood concentrations of ketone bodies rise in the untreated diabetic patient.

25. Discuss the energetic costs of converting 2 acetyl-CoA to the ketone body 3-hydroxybutyrate in the liver and then converting the 3-hydroxybutyrate back to 2 acetyl-CoA in the muscle.

26. Premenopausal women typically have higher HDL levels than men.

(a) Why would this tend to decrease the risk of heart disease in these women?
(b) Why is HDL level alone not a good indicator of the risk of developing heart disease?

27. Cholesterol is poorly soluble in aqueous solution, yet cells must be able to sense the cholesterol level in order to regulate uptake and biosynthesis, in part by altering the expression of genes for HMG-CoA reductase and the LDL receptor. The cellular cholesterol sensors appear to be proteins called SREBPs (sterol regulatory element binding proteins). In the absence of cholesterol, an SREBP residing in the endoplasmic reticulum is proteolytically cleaved to release a large soluble N-terminal domain that includes a structural motif found in many DNA-binding proteins.

(a) Why is it important that the SREBP be an integral membrane protein?
(b) Why is proteolysis of the SREBP required?
(c) How might the SREBP regulate the transcription of enzymes related to cholesterol metabolism?

28. As a nutritionist, you have persuaded one of your patients with atherosclerosis to avoid animal fats, which are rich in saturated fatty acids. He loves buttered toast but finally agrees to switch from butter to margarine. Would your recommend that he use a soft (spreadable) margarine or a hard (butter-like) margarine? Explain.

29. Why does triclosan inhibit bacterial fatty acid synthase but not mammalian fatty acid synthase?

30. Cancer cells exhibit greater-than-normal levels of fatty acid synthesis, which has been attributed to enhanced expression of the fatty acid synthase enzyme (FAS). Rather than synthesizing lipids for storage, as normal cells do, the malignant cells use the fatty acids to fuel tumor growth. This observation has led cancer researchers to investigate the use of fatty acid synthesis inhibitors as antitumor agents. It has been hypothesized that the increased levels of malonyl-CoA that result when FAS activity is inhibited cause apoptosis, or programmed cell death. An effective design of a fatty acid synthase inhibitor requires a detailed knowledge of the enzymatic mechanism of the enzyme. A portion of the proposed mechanism is shown below.

Elongating fatty acid — Malonyl-ACP

FAS—CH_2—S—C(=O)—R + $^-$OOC—CH_2—C(=O)—S-ACP

Elongating fatty acid ↓ → CO_2

FAS—CH_2—S—C(=O)—R + H_2C=C(O^-)—S-ACP

Enolate anion

↓

FAS—CH_2—S—C(R)(O^-)—CH_2—C(=O)—S-ACP

↓

FAS—CH_2—S^- + R—C(=O)—CH_2—C(=O)—S-ACP

(a) A series of inhibitors were synthesized with varying alkyl chain (R) lengths. Explain why these compounds were effective inhibitors of FAS.

HOOC, O, =O, O, R

(b) Next, the investigators tested each compound's ability to inhibit FAS activity in normal cells and in SKBr3 cells, a breast-cancer cell line that expresses the highest levels of FAS activity of all known cancer cells. For each compound, an ID_{50} was calculated (this is the concentration of inhibitor required to inhibit the growth of 50% of the cells). The results are shown in the table. Which inhibitors are most effective? What are the characteristics of the effective inhibitors? (An effective inhibitor must also be soluble in aqueous solution, so consider solubility as an additional factor in your answer.)

Compound	Alkyl side chain (R =)	SKBr3 cells ID_{50} (μg/mL)	Normal cells ID_{50} (μg/mL)
C83	—$C_{13}H_{27}$	3.9 ± 0.2	10.6 ± 0.3
C81	—$C_{11}H_{23}$	4.8 ± 0.2	29.0 ± 5.0
C77	—C_9H_{19}	5.2 ± 0.3	12.8 ± 1.2
C75	—C_8H_{17}	5.0 ± 0.1	21.6 ± 1.4
C49	—C_7H_{15}	4.8 ± 0.5	21.7 ± 0.5
C73	—C_6H_{13}	8.4 ± 0.2	12.4 ± 0.8

(c) Slow-binding inhibitors are inhibitors for which equilibration of the enzyme, inhibitor, and enzyme–inhibitor complex takes place on the order of seconds or minutes (rather than fractions of a second). Slow-binding inhibitors are effective pharmacological agents because they minimize rapid substrate buildup that occurs when an enzyme is inhibited. Is C75 a slow-binding inhibitor? Explain.

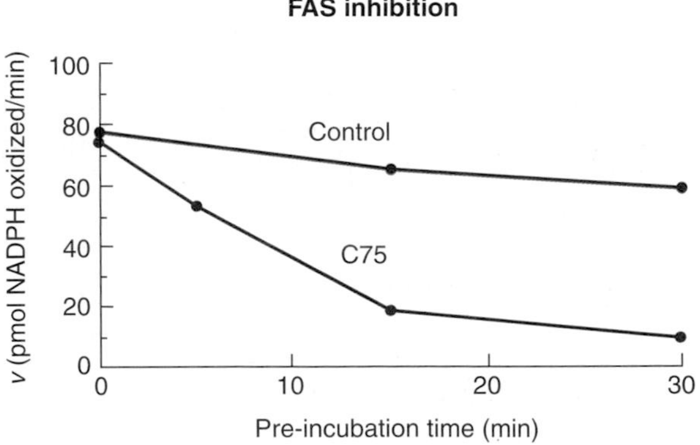

(d) Interpret the data in the graph shown below.

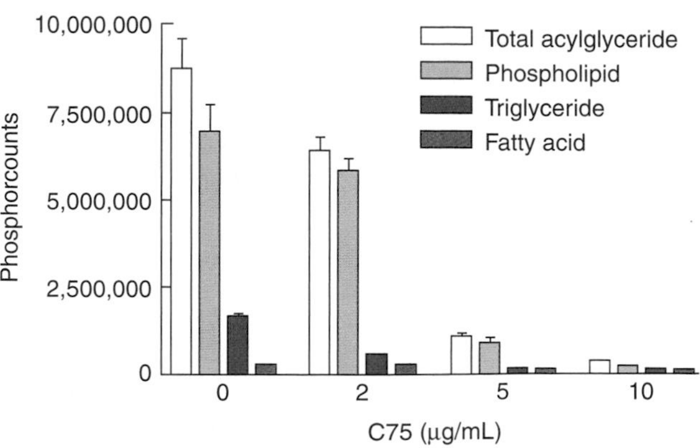

Inhibition of FAS by C75 in HL60 leukemia cells as measured by incorporation of radioactively labeled acetate into acylglycerides.

[From Kuhajda, F.P., Pizer, E.S., Li, J.N., Mani, N.S., Frehywot, G.L., and Townsend, C., *Proc. Natl. Acad. Sci.* **97,** 3450–3454 (2000).]

31. Fumonisins are mycotoxins isolated from the fungus *Fasarum moniliforme,* which is commonly found on corn and other grains. Fumonisins are both toxic and carcinogenic and can cause disease in animals grazing on fungus-contaminated grain. The structure of fumonisin B_1 is shown here. Note the structural similarity to sphingosine.

OH, CH_2OH, NH_3^+

Sphingosine

OR, OH OH, CH_3, CH_3 OR CH_3 OH, NH_3^+

Fumonisin B_1

R= C(=O), COOH, COOH

Fumonisins inhibit one of the enzymes in the ceramide synthetic pathway, which is outlined below. Ceramide is an important cell-signaling molecule, and its regulation is critical to cell survival.

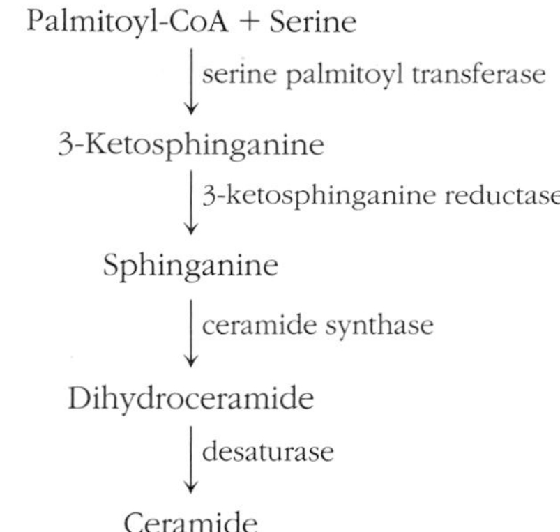

(a) Deduce which enzyme is inhibited in the pathway based on the following clues:
   1. The addition of fumonisin B_1 to rat hepatocytes almost completely inhibited ceramide synthesis. The synthesis of other phospholipids was not affected.
   2. The activity of serine palmitoyl transferase in the presence and absence of fumonsin B_1 was 88 ± 3 and 61 ± 15 pmol of 3-ketosphinganine formed per min per mg protein, respectively.
   3. There was no accumulation of 3-ketosphinganine.
   4. When radioactively labeled serine was added to culture medium containing fumonisin B, there was an increase in the amount of label in sphinganine as compared to controls.

(b) How does fumonisin inhibit the target enzyme identified in part a?

[From Wang, E., Norred, W.P., Bacon, C.W., Riley, R.T., and Merrill, A.H., *J. Biol. Chem.* **266,** 14486–14490 (1991).]

SELECTED READINGS

Eaton, S., Bartlett, K., and Pourfarzam, M., Mammalian mitochondrial β-oxidation, *Biochem. J.* **320,** 345–357 (1996). [Discusses the reactions and enzymes of mitochondrial β oxidation as well as their regulation and diseases associated with their deficiencies.]

Simons, K. and Ikonen, E., How cells handle cholesterol, *Science* **290,** 1721–1726 (2000). [Discusses mechanisms of cholesterol influx and efflux from cells.]

Steinberg, D., Atherogenesis in perspective: hypercholesterolemia and inflammation as partners in crime, *Nature Medicine* **8,** 1211–1216 (2002). [Summarizes current hypotheses linking cholesterol levels to the development of atherosclerosis.]

Vane, J.R. and Botting, R.M., Anti-inflammatory drugs and their mechanism of action, *Inflamm. Res.* **47, Suppl 2,** S78–87 (1998). [Describes the action of prostaglandins, the COX isozymes, and selective inhibitors of COX-2.]

CHAPTER 15

Nitrogen Metabolism

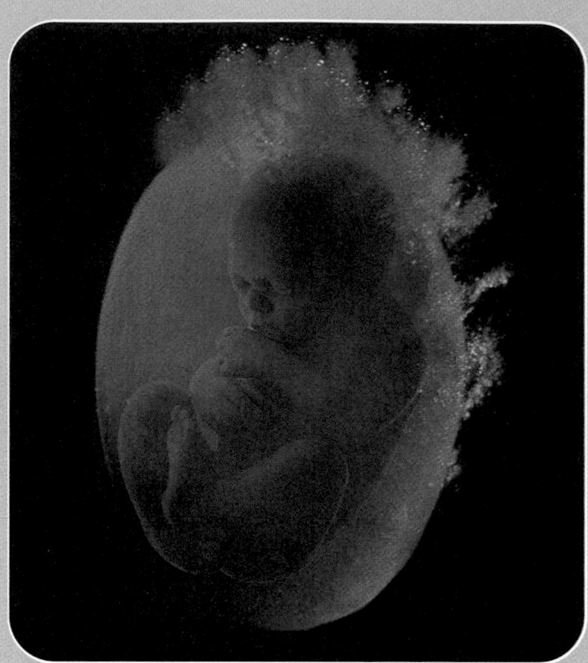

The growth of a fetus presents an immunological challenge to its mother. Half of its genes are derived from the father, so it expresses proteins that the mother's immune system recognizes as foreign. What prevents the mother from mounting an immune attack against the fetus? The answer may lie in the placenta, a fetal tissue, which expresses the enzyme indoleamine 2,3-dioxygenase (IDO). IDO destroys the amino acid tryptophan, which the mother's immune cells require in order to proliferate and attack a foreign tissue. Consequently, the mother does not somehow "tolerate" the fetus; the fetus actively defends itself from attack by producing IDO.

[Yoav Levy/Phototake]

LEARNING OBJECTIVES

1. Understand how nitrogen is assimilated into biological molecules.
2. Understand how the pyridoxal phosphate cofactor participates in the transaminase reaction.
3. Understand that mammals can synthesize nonessential amino acids from common metabolites but must obtain essential amino acids from food.
4. Understand that amino acids are the precursors of other substances, such as neurotransmitters.
5. Understand that nucleotides are synthesized largely from amino acids.
6. Understand the advantages of channeling.
7. Understand how amino acids are catabolized as metabolic fuels.
8. Understand that the urea cycle disposes of nitrogen.

THIS CHAPTER IN CONTEXT

This chapter is the last to cover metabolic pathways. In the preceding chapters we examined the metabolism of carbohydrates and lipids—compounds containing primarily carbon, hydrogen, and oxygen. We will now look at the metabolism of amino acids and nucleotides—compounds that contain nitrogen as well. Rather than simply examining how these compounds are synthesized and broken down, we will focus on the metabolism of nitrogen: how it is fixed, how it makes its way into nitrogen-containing amino acids and nucleotides, and how it is disposed of. Along the way, we will examine some of the interesting chemistry of amino acid and nucleotide metabolism.

1 NITROGEN FIXATION AND ASSIMILATION

Approximately 80% of the air we breathe is nitrogen (N_2), but we cannot use this form of nitrogen for the synthesis of amino acids, nucleotides, and other nitrogen-containing biomolecules. Instead, we—along with most macroscopic and many microscopic life forms—depend on the activity of a few types of microorganisms that can "fix" gaseous N_2 by transforming it into biologically useful forms. The availability of fixed nitrogen, in the form of nitrite, nitrate, and ammonia, is believed to limit the biological productivity in much of the world's oceans. It also limits the growth of terrestrial organisms, which is why farmers use fertilizer (a source of fixed nitrogen, among other things) to promote crop growth.

The known **nitrogen-fixing** organisms include certain marine cyanobacteria and bacteria that colonize the root nodules of leguminous plants (Fig. 15-1). These bacteria make the enzyme nitrogenase, which carries out the energetically expensive reduction of N_2 to NH_3. Nitrogenase is a metalloprotein containing iron–sulfur centers and a cofactor with both iron and molybdenum, which resembles an elaborate Fe–S cluster (Fig. 15-2). The industrial fixation of nitrogen also involves metal catalysts, but this process requires temperatures of 300 to 500°C and pressures of over 300 atm in order to break the triple bond between the two nitrogen atoms.

Biological N_2 reduction consumes large amounts of ATP and requires a strong reducing agent such as ferredoxin (see Section 13-2) to donate electrons. The net reaction is

$$N_2 + 8\ H^+ + 8\ e^- + 16\ ATP + 16\ H_2O \rightarrow 2\ NH_3 + H_2 + 16\ ADP + 16\ P_i$$

Note that eight electrons are required for the nitrogenase reaction, although N_2 reduction formally requires only six electrons; the two extra electrons are used to produce H_2. *In vivo,* the inefficiency of the reaction boosts the ATP toll to about 20 or 30 per N_2 reduced. Oxygen inactivates nitrogenase, so many nitrogen-fixing bacteria are confined to anaerobic habitats or carry out nitrogen fixation when O_2 is scarce.

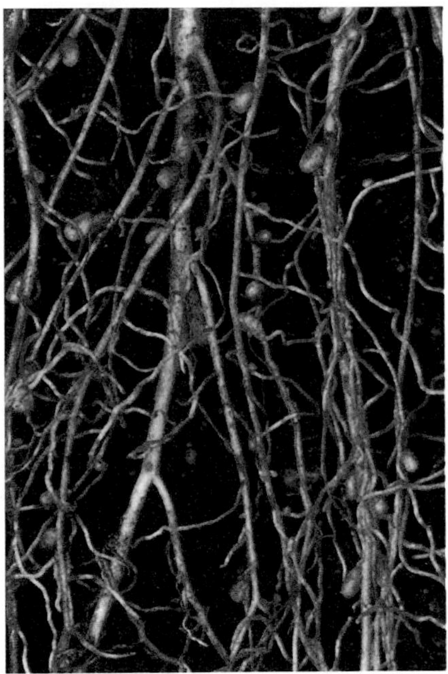

FIGURE 15-1

Root nodules from clover. Legumes (such as beans, clover, and alfalfa) and some other plants harbor nitrogen-fixing bacteria in root nodules. The symbiotic relationship revolves around the ability of the bacteria to fix nitrogen and the ability of the plant to make other nutrients available to the bacteria. [*Dr. Jeremy Burgess/Science Photo Library/Photo Researchers.*]

Other sources of fixed nitrogen

Biologically useful nitrogen also originates from nitrate (NO_3^-), which is naturally present in water and soils. *Nitrate is reduced to NH_3 by plants, fungi,*

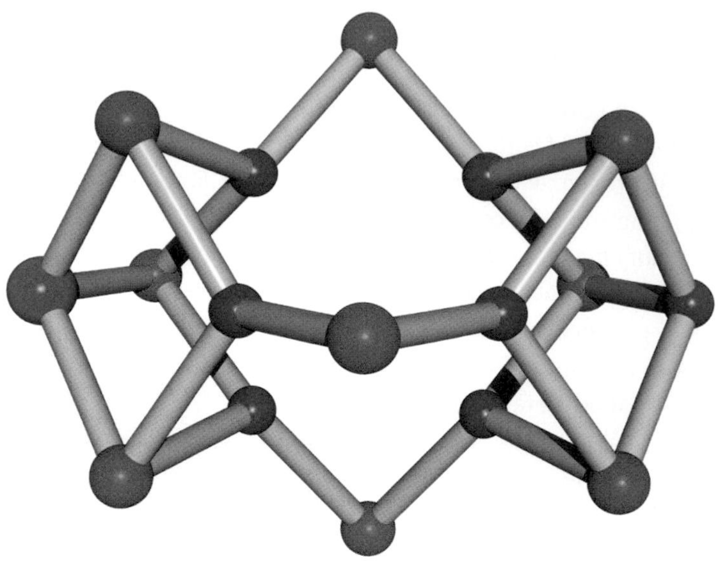

FIGURE 15-2

Model of the FeMo cofactor of nitrogenase.
This prosthetic group in the enzyme nitrogenase consists of iron atoms (red), sulfur atoms (orange), and a molybdenum atom (green). The central cavity may include an N atom coordinated with the six iron atoms. The manner in which N_2 interacts with the FeMo cofactor is not understood. [*Structure of the FeMo cofactor in nitrogenase (pdb 1QGU) determined by S.M. Mayer, D.M. Lawson, C.A. Gormal, S.M. Roe, and B.E. Smith.*]

and many bacteria. First, nitrate reductase catalyzes the two-electron reduction of nitrate to nitrite (NO_2^-):

$$NO_3^- + 2\ H^+ + 2\ e^- \rightarrow NO_2^- + H_2O$$

Next, nitrite reductase converts nitrite to ammonia:

$$NO_2^- + 8\ H^+ + 6\ e^- \rightarrow NH_4^+ + 2\ H_2O$$

Under physiological conditions, ammonia exists primarily in the protonated form, NH_4^+ (the ammonium ion), which has a pK of 9.25.

Nitrate is also produced by certain bacteria that oxidize NH_4^+ to NO_2^- and then NO_3^-, a process called **nitrification.** Still other organisms convert nitrate back to N_2, which is called **denitrification.** All of the reactions we have discussed so far constitute the earth's **nitrogen cycle** (Fig. 15-3).

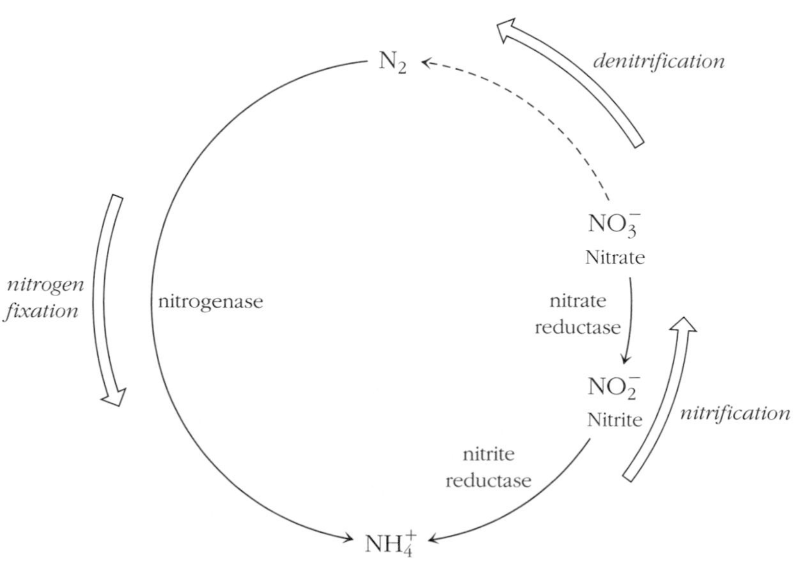

FIGURE 15-3

The nitrogen cycle.
Nitrogen fixation converts N_2 to the biologically useful NH_4^+. Nitrate can also be converted to NH_4^+. Ammonia is transformed back to N_2 by nitrification followed by denitrification.

Assimilation of ammonia

The enzyme glutamine synthetase is found in all organisms. In microorganisms, it is a metabolic entry point for fixed nitrogen. In animals, it helps mop up excess ammonia, which is toxic. In the first step of the reaction, ATP donates a phosphoryl group to glutamate. Then ammonia reacts with the reaction intermediate, displacing P_i to produce glutamine.

$$\underset{\text{Glutamate}}{H-\overset{\overset{COO^-}{|}}{\underset{\underset{^+NH_3}{|}}{C}}-CH_2-CH_2-C(=O)O^-} \xrightarrow{ATP \rightarrow ADP} \left[H-\overset{\overset{COO^-}{|}}{\underset{\underset{^+NH_3}{|}}{C}}-CH_2-CH_2-C(=O)OPO_3^{2-}\right] \xrightarrow{NH_4^+ \rightarrow P_i} \underset{\text{Glutamine}}{H-\overset{\overset{COO^-}{|}}{\underset{\underset{^+NH_3}{|}}{C}}-CH_2-CH_2-C(=O)NH_2}$$

The name *synthetase* indicates that ATP is consumed in the reaction.

Glutamine, along with glutamate, is usually present in organisms at much higher concentrations than the other amino acids, which is consistent with its role as a carrier of amino groups. Not surprisingly, *the activity of glutamine synthetase is tightly regulated to maintain a supply of accessible amino groups.* For example, the dodecameric glutamine synthetase from *E. coli* is regulated allosterically and by covalent modification (Fig. 15-4).

The glutamine synthetase reaction that introduces fixed nitrogen (ammonia) into biological compounds requires a nitrogen-containing compound (glutamate) as a substrate. So what is the source of the nitrogen in glutamate?

In bacteria and plants, the enzyme glutamate synthase catalyzes the reaction

$$\underset{\alpha\text{-Ketoglutarate}}{^-OOC-C(=O)-CH_2-CH_2-COO^-} + \underset{\text{Glutamine}}{H_3\overset{+}{N}-CH(COO^-)-CH_2-CH_2-C(=O)NH_2} \xrightarrow[\text{glutamate synthase}]{NADPH + H^+ \rightarrow NADP^+} \underset{\text{Glutamate}}{2\; H_3\overset{+}{N}-CH(COO^-)-CH_2-CH_2-COO^-}$$

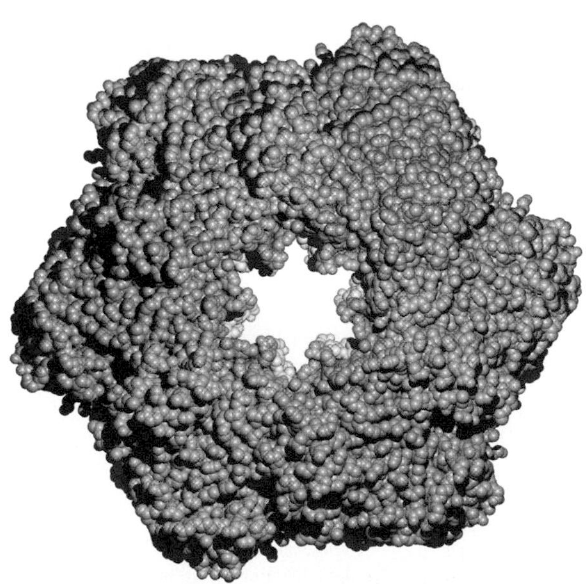

FIGURE 15-4

***E. coli* glutamine synthetase.** The 12 identical subunits of this enzyme are arranged in two stacked rings of six subunits (only one ring is visible here). The symmetrical arrangement of subunits is a general feature of enzymes that are regulated by allosteric effectors: Changes in activity at one of the active sites can be efficiently communicated to the other active sites. [*Structure (pdb 2GLS) determined by D. Eisenberg, R.J. Almassy, and M.M. Yamashita.*]

(a reaction catalyzed by a synthase does not require ATP). The net result of the glutamine synthetase and glutamate synthase reactions is

$$\alpha\text{-Ketoglutarate} + NH_4^+ + NADPH + ATP \rightarrow \text{glutamate} + NADP^+ + ADP + P_i$$

In other words, *the combined action of these two enzymes assimilates fixed nitrogen (NH_4^+) into an organic compound (α-ketoglutarate, a citric acid cycle intermediate) to produce an amino acid (glutamate).* Mammals lack glutamate synthase, but glutamate concentrations are relatively high because glutamate is produced by other reactions.

REVIEW LEARNING OBJECTIVE 1

- What does the nitrogenase reaction accomplish?
- What other compounds give rise to ammonia?
- Describe the reactions catalyzed by glutamine synthetase and glutamate synthase.

2 TRANSAMINATION

Because reduced nitrogen is so precious, amino groups are transferred from molecule to molecule, with glutamate often serving as an amino-group donor. We saw some of these **transamination** reactions in Section 11-3 when we examined how citric acid cycle intermediates participate in other metabolic pathways.

A transaminase (also called an aminotransferase) catalyzes the transfer of an amino group to an α-keto acid. For example,

$$\underset{\substack{\text{Glutamate}\\\text{(amino acid)}}}{^{-}OOC{-}CH_2{-}CH_2{-}CH(\overset{+}{N}H_3){-}COO^{-}} + \underset{\substack{\text{Pyruvate}\\\text{(α-keto acid)}}}{CH_3{-}C({=}O){-}COO^{-}} \underset{\text{transaminase}}{\rightleftharpoons} \underset{\substack{\text{α-Ketoglutarate}\\\text{(α-keto acid)}}}{^{-}OOC{-}CH_2{-}CH_2{-}C({=}O){-}COO^{-}} + \underset{\substack{\text{Alanine}\\\text{(amino acid)}}}{CH_3{-}CH(\overset{+}{N}H_3){-}COO^{-}}$$

During such an amino-group transfer reaction, the amino group is transiently attached to a prosthetic group of the enzyme. This group is pyridoxal-5′-phosphate (PLP), a derivative of pyridoxine (an essential nutrient also known as vitamin B_6):

Pyridoxal-5′-phosphate (PLP) [structure: pyridinium ring with 4′ CHO at C4, $^{-2}O_3P{-}O{-}H_2C$ (5′) at C5, OH at C3, CH_3 at C2, $\overset{+}{N}{-}H$ at position 1]

Pyridoxine (vitamin B_6) [structure: pyridinium ring with $H_2C{-}OH$ at C4, $HO{-}H_2C$ at C5, OH at C3, CH_3 at C2, $\overset{+}{N}{-}H$]

PLP is covalently attached to the enzyme via a Schiff base (imine) linkage to the ε-amino group of a Lys residue:

Enzyme–PLP
Schiff base

The amino acid substrate of the transaminase displaces this Lys amino group, which then acts as an acid–base catalyst. The steps of the reaction are diagrammed in Figure 15-5.

The transamination reaction is freely reversible, so transaminases participate in pathways for amino acid synthesis as well as degradation. Note that if the α-keto acid produced in Step 4 re-enters the active site, then the amino group that was removed from the starting amino acid is restored. However, most transaminases accept only α-ketoglutarate or oxaloacetate as the α-keto acid substrate for the second part of the reaction (Steps 5 to 7). This means that most transaminases generate glutamate or aspartate. Lysine is the only amino acid that cannot be transaminated.

The presence of transaminases in muscle and liver cells makes them useful markers of tissue damage. Assays of the enzymes' activities in the blood are the basis of the well-known clinical measurements known as SGOT (serum glutamate-oxaloacetate transaminase, also known as aspartate aminotransferase, AST) and SGPT (serum glutamate-pyruvate transaminase, or alanine transaminase, ALT). The concentrations of these enzymes in the blood increase after a heart attack, when damaged heart muscle leaks its intracellular contents. Similarly, liver damage can be monitored by SGOT and SGPT readings.

REVIEW LEARNING OBJECTIVE 2

- What is the function of the PLP cofactor?
- Explain why transaminases catalyze reversible reactions.

3 AMINO ACID BIOSYNTHESIS

Amino acids are synthesized from intermediates of glycolysis, the citric acid cycle, and the pentose phosphate pathway. Their amino groups are derived from the nitrogen-carrier molecules glutamate or glutamine. Using the meta-

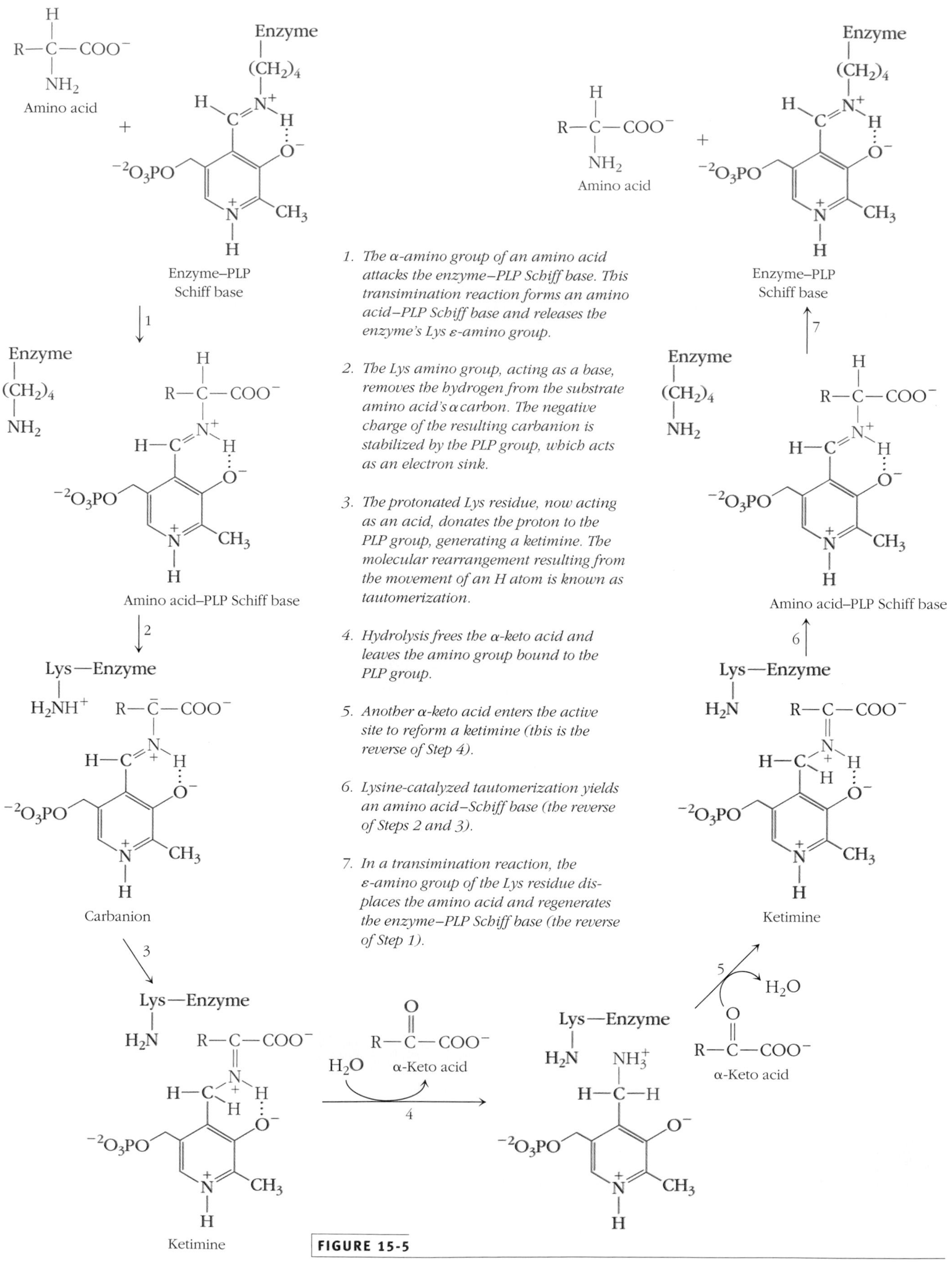

1. *The α-amino group of an amino acid attacks the enzyme–PLP Schiff base. This transimination reaction forms an amino acid–PLP Schiff base and releases the enzyme's Lys ε-amino group.*

2. *The Lys amino group, acting as a base, removes the hydrogen from the substrate amino acid's α carbon. The negative charge of the resulting carbanion is stabilized by the PLP group, which acts as an electron sink.*

3. *The protonated Lys residue, now acting as an acid, donates the proton to the PLP group, generating a ketimine. The molecular rearrangement resulting from the movement of an H atom is known as tautomerization.*

4. *Hydrolysis frees the α-keto acid and leaves the amino group bound to the PLP group.*

5. *Another α-keto acid enters the active site to reform a ketimine (this is the reverse of Step 4).*

6. *Lysine-catalyzed tautomerization yields an amino acid–Schiff base (the reverse of Steps 2 and 3).*

7. *In a transimination reaction, the ε-amino group of the Lys residue displaces the amino acid and regenerates the enzyme–PLP Schiff base (the reverse of Step 1).*

FIGURE 15-5

PLP-catalyzed transamination.

bolic scheme introduced in Chapter 9, we can show how amino acid biosynthesis and other reactions of nitrogen metabolism are related to the other pathways we have examined (Fig. 15-6).

Humans can synthesize only 10 of the 20 amino acids that are commonly found in proteins. These 10 are known as **nonessential amino acids.** The other 10 amino acids are said to be **essential** because humans must obtain them from their food. The ultimate sources of the essential amino acids are plants and microorganisms, which produce all the enzymes necessary to undertake the synthesis of these compounds. The essential and nonessential amino acids for humans are listed in Table 15-1. Their structures are shown in Figure 4-2.

TABLE 15-1 Essential and nonessential amino acids in humans

Essential	Nonessential
Arginine	Alanine
Histidine	Asparagine
Isoleucine	Aspartate
Leucine	Cysteine
Lysine	Glutamate
Methionine	Glutamine
Phenylalanine	Glycine
Threonine	Proline
Tryptophan	Serine
Valine	Tyrosine

Synthesis of nonessential amino acids

We have already seen that *some amino acids can be produced by transamination reactions*. In this way, alanine is produced from pyruvate, aspartate from oxaloacetate, and glutamate from α-ketoglutarate. Glutamine synthetase cat-

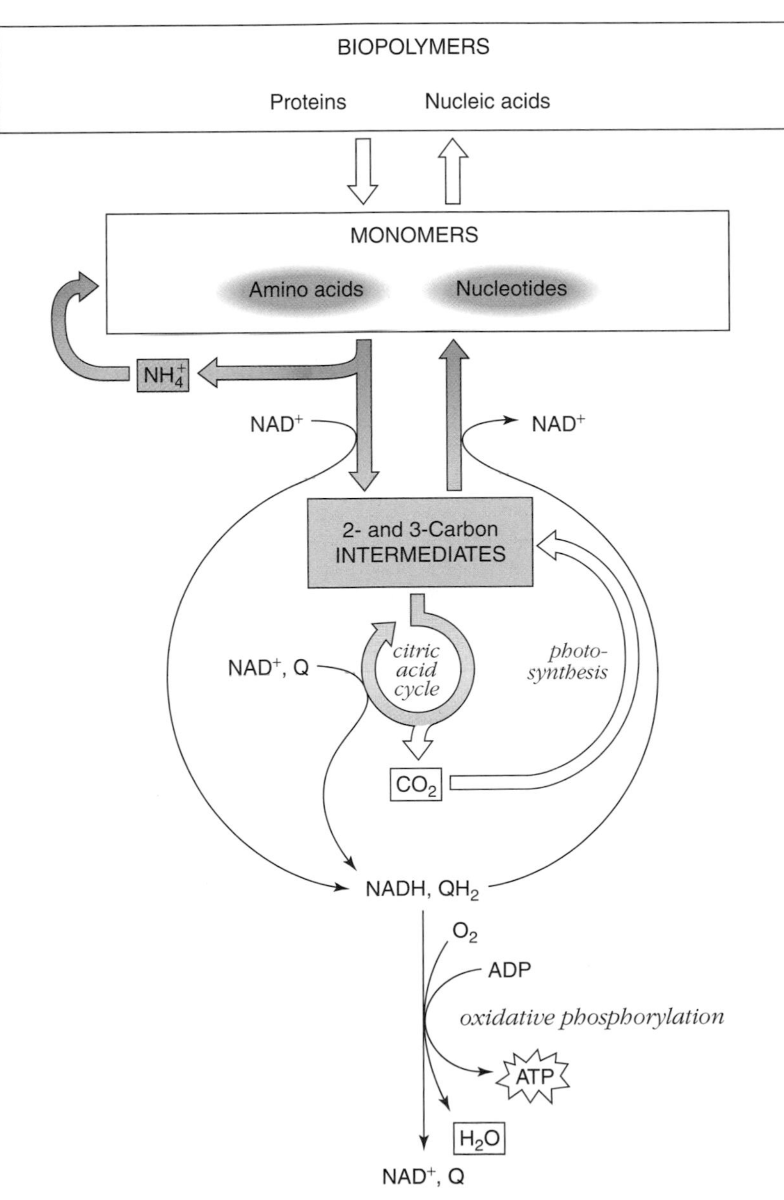

FIGURE 15-6
Nitrogen metabolism in context. Amino acids are synthesized mostly from three-carbon intermediates of glycolysis and from intermediates of the citric acid cycle. Amino acid catabolism yields some of the same intermediates, as well as the two-carbon acetyl-CoA. Amino acids are also the precursors of nucleotides. Both types of molecules contain nitrogen, so a discussion of amino acid metabolism includes pathways for obtaining, using, and disposing of ammonia.

alyzes the amidation of glutamate to produce glutamine, and asparagine synthetase, which uses glutamine as an amino-group donor rather than ammonia, converts aspartate to asparagine:

$$\underset{\text{Aspartate}}{\begin{array}{c}COO^-\\ |\\ H_3\overset{+}{N}-C-H\\ |\\ CH_2\\ |\\ {}^{-}O-C=O\end{array}} \xrightarrow[\text{asparagine synthetase}]{\text{Glutamine + ATP} \;\to\; \text{Glutamate + AMP + PP}_i} \underset{\text{Asparagine}}{\begin{array}{c}COO^-\\ |\\ H_3\overset{+}{N}-C-H\\ |\\ CH_2\\ |\\ H_2N-C=O\end{array}}$$

So far, we have seen that three common metabolic intermediates (pyruvate, oxaloacetate, and α-ketoglutarate) give rise to 5 of the 10 nonessential amino acids by simple transamination and amidation reactions.

Slightly longer pathways convert glutamate to proline and arginine:

$$\underset{\text{Glutamate}}{\begin{array}{c}COO^-\\ |\\ H_3\overset{+}{N}-C-H\\ |\\ CH_2\\ |\\ CH_2\\ |\\ COO^-\end{array}} \dashrightarrow \underset{\text{Arginine}}{\begin{array}{c}COO^-\\ |\\ H_3\overset{+}{N}-C-H\\ |\\ CH_2\\ |\\ CH_2\\ |\\ CH_2\\ |\\ NH\\ |\\ H_2N-C=\overset{+}{N}H_2\end{array}} \qquad \dashrightarrow \underset{\text{Proline}}{\begin{array}{c}COO^-\\ |\\ HN-C-H \;\; CH_2\\ H_2C-CH_2\end{array}}$$

Serine is derived from the glycolytic intermediate 3-phosphoglycerate in three steps:

$$\underset{\text{3-Phosphoglycerate}}{\begin{array}{c}COO^-\\ |\\ H-C-OH\\ |\\ CH_2OPO_3^{2-}\end{array}} \xrightarrow{NAD^+ \;\to\; NADH} \underset{\text{3-Phospho-hydroxypyruvate}}{\begin{array}{c}COO^-\\ |\\ C=O\\ |\\ CH_2OPO_3^{2-}\end{array}} \xrightarrow{\text{Glutamate} \;\to\; \alpha\text{-Ketoglutarate}} \underset{\text{3-Phosphoserine}}{\begin{array}{c}COO^-\\ |\\ H_3\overset{+}{N}-C-H\\ |\\ CH_2OPO_3^{2-}\end{array}} \xrightarrow{P_i} \underset{\text{Serine}}{\begin{array}{c}COO^-\\ |\\ H_3\overset{+}{N}-C-H\\ |\\ CH_2\\ |\\ OH\end{array}}$$

Serine, a three-carbon amino acid, gives rise to the two-carbon glycine in a reaction catalyzed by serine hydroxymethyltransferase (the reverse reaction converts glycine to serine). This enzyme uses a PLP-dependent mechanism to remove the hydroxymethyl ($—CH_2OH$) group attached to the α carbon of serine; this one-carbon fragment is then transferred to the cofactor tetrahydrofolate:

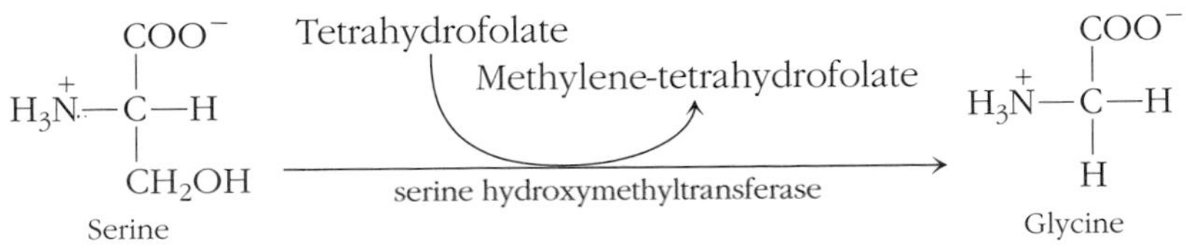

$$\text{Serine} \;(\overset{+}{H_3N}-CH(COO^-)-CH_2OH) \xrightarrow[\text{serine hydroxymethyltransferase}]{\text{Tetrahydrofolate} \to \text{Methylene-tetrahydrofolate}} \text{Glycine} \;(\overset{+}{H_3N}-CH(COO^-)-H)$$

Tetrahydrofolate functions as a carrier of one-carbon units in several reactions of amino acid and nucleotide metabolism (Fig. 15-7). Mammals cannot synthesize folate (the oxidized form of tetrahydrofolate) and must therefore obtain it as a vitamin from their diet. Folate is abundant in foods such as fortified cereal, fruits, and vegetables. The requirement for folate increases during the first few weeks of pregnancy, when the fetal nervous system begins to develop. Supplemental folate appears to prevent certain neural tube defects such as spina bifida, in which the spinal cord remains exposed.

Finally, cysteine is derived from serine by a pathway that requires the nonstandard amino acid homocysteine (a sulfur-containing compound that is a product of methionine catabolism).

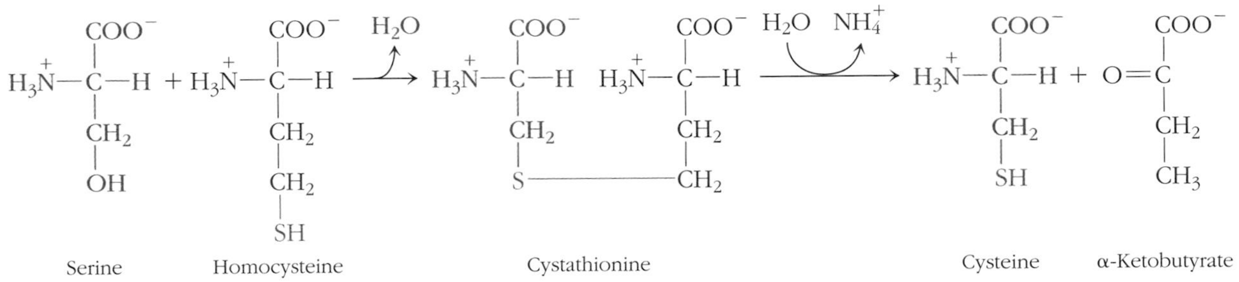

$$\text{Serine} + \text{Homocysteine} \xrightarrow{H_2O} \text{Cystathionine} \xrightarrow{H_2O \;\; NH_4^+} \text{Cysteine} + \alpha\text{-Ketobutyrate}$$

Serine: $\overset{+}{H_3N}-CH(COO^-)-CH_2-OH$; Homocysteine: $\overset{+}{H_3N}-CH(COO^-)-CH_2-CH_2-SH$; Cystathionine: $\overset{+}{H_3N}-CH(COO^-)-CH_2-S-CH_2-CH_2-CH(COO^-)-\overset{+}{N}H_3$; Cysteine: $\overset{+}{H_3N}-CH(COO^-)-CH_2-SH$; α-Ketobutyrate: $O=C(COO^-)-CH_2-CH_3$

Synthesis of essential amino acids

Although the essential amino acids—like the nonessential amino acids—are derived from common metabolites, their biosynthetic pathways tend to require multiple steps. At some point in their evolution, animals lost the ability to synthesize these amino acids, probably because the pathways were energetically expensive and the compounds were already available in food. To sum-

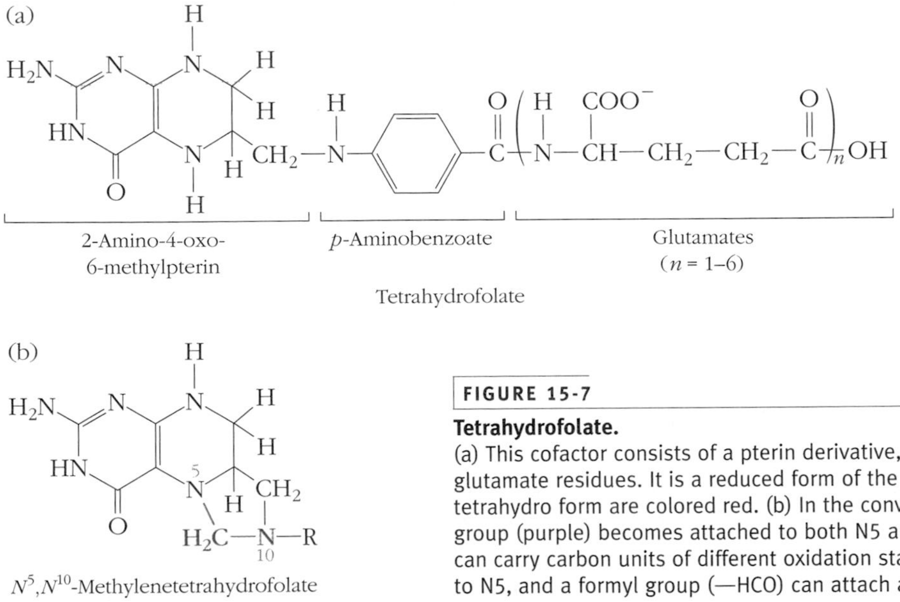

(a) 2-Amino-4-oxo-6-methylpterin — *p*-Aminobenzoate — Glutamates (n = 1–6)

Tetrahydrofolate

(b) N^5,N^{10}-Methylenetetrahydrofolate

FIGURE 15-7

Tetrahydrofolate.

(a) This cofactor consists of a pterin derivative, a *p*-aminobenzoate residue, and up to six glutamate residues. It is a reduced form of the vitamin folate. The four H atoms of the tetrahydro form are colored red. (b) In the conversion of serine to glycine, a methylene group (purple) becomes attached to both N5 and N10 of tetrahydrofolate. Tetrahydrofolate can carry carbon units of different oxidation states. For example, a methyl group can attach to N5, and a formyl group (—HCO) can attach at N5 or N10.

marize, humans cannot synthesize branched-chain amino acids or aromatic amino acids and cannot incorporate sulfur into compounds such as methionine. In this section, we will focus on a few interesting points related to the synthesis of essential amino acids.

As we saw in the preceding section, the mammalian pathway for synthesizing cysteine requires a sulfur atom ultimately derived from methionine. In bacteria, methionine synthesis requires a sulfur atom derived from cysteine. *This sulfur comes from inorganic sulfide in the conversion of serine to cysteine in bacteria:*

$$\underset{\text{Serine}}{H_3\overset{+}{N}{-}CH(COO^-){-}CH_2{-}OH} \xrightarrow{H_3C{-}C(=O){-}SCoA \;\to\; CoASH} \underset{O\text{-Acetylserine}}{H_3\overset{+}{N}{-}CH(COO^-){-}CH_2{-}O{-}C(=O){-}CH_3} \xrightarrow{S^{2-} + H^+ \;\to\; CH_3COO^-} \underset{\text{Cysteine}}{H_3\overset{+}{N}{-}CH(COO^-){-}CH_2{-}SH}$$

Cysteine can then donate its sulfur atom to homocysteine, whose four-carbon skeleton is derived from aspartate. The final step of methionine synthesis is catalyzed by methionine synthase, which adds to homocysteine a methyl group carried by tetrahydrofolate:

$$\underset{\text{Homocysteine}}{H_3\overset{+}{N}{-}CH(COO^-){-}CH_2{-}CH_2{-}SH} \xrightarrow{\text{Methyl-tetrahydrofolate} \;\to\; \text{Tetrahydrofolate}} \underset{\text{Methionine}}{H_3\overset{+}{N}{-}CH(COO^-){-}CH_2{-}CH_2{-}S{-}CH_3}$$

In humans, high levels of homocysteine in the blood are associated with cardiovascular disease. The link was first discovered in individuals with homocystinuria, a disorder in which excess homocysteine is excreted in the urine. These individuals develop atherosclerosis as children, probably because the homocysteine directly damages the walls of blood vessels even in the absence of elevated LDL levels (see Chapter 14). Increasing the intake of folate, the vitamin precursor of tetrahydrofolate, helps decrease the level of homocysteine by promoting its conversion to methionine.

Aspartate, the precursor of methionine, is also the precursor of the essential amino acids threonine and lysine. Since these amino acids are derived from another amino acid, they already have an amino group. The aliphatic amino acids (valine, leucine, and isoleucine) are synthesized by pathways that use pyruvate as the starting substrate. These amino acids require a step catalyzed by a transaminase (with glutamate as a substrate) to introduce an amino group.

In plants and bacteria, the pathway for synthesizing the aromatic amino acids (phenylalanine, tyrosine, and tryptophan) begins with the condensation of the C_3 compound phosphoenolpyruvate (a glycolytic intermediate) and erythrose-4-phosphate (a four-carbon intermediate of the pentose phosphate pathway). The seven-carbon reaction product then cyclizes and undergoes additional modifications, including the addition of three more carbons from phosphoenolpyruvate, before becoming chorismate, the last common intermediate in the synthesis of the three aromatic amino acids.

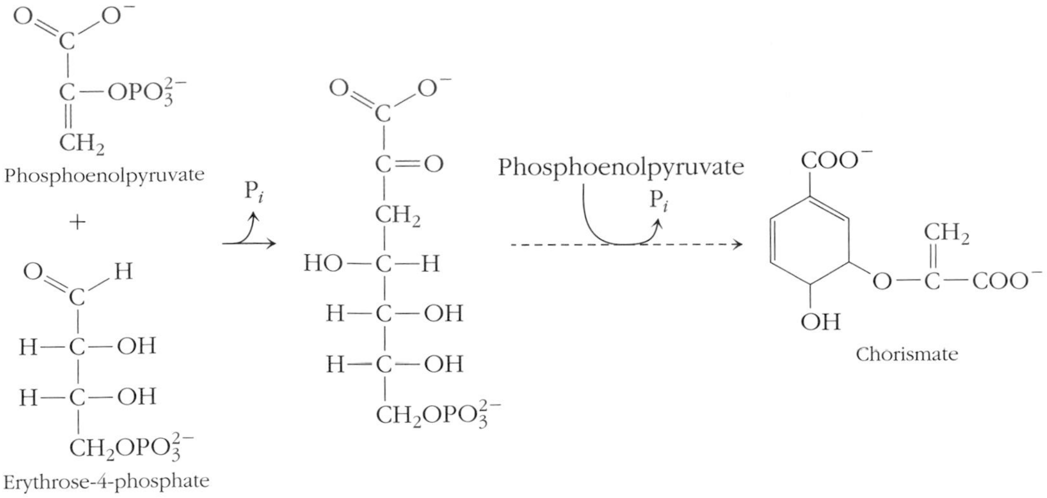

Phosphoenolpyruvate

+

Erythrose-4-phosphate

P_i

Phosphoenolpyruvate

P_i

Chorismate

Because animals do not synthesize chorismate, this pathway is an obvious target for agents that can inhibit plant metabolism without affecting animals. For example, the herbicide glyphosate (Roundup®) competes with the second phosphoenolpyruvate in the pathway leading to chorismate.

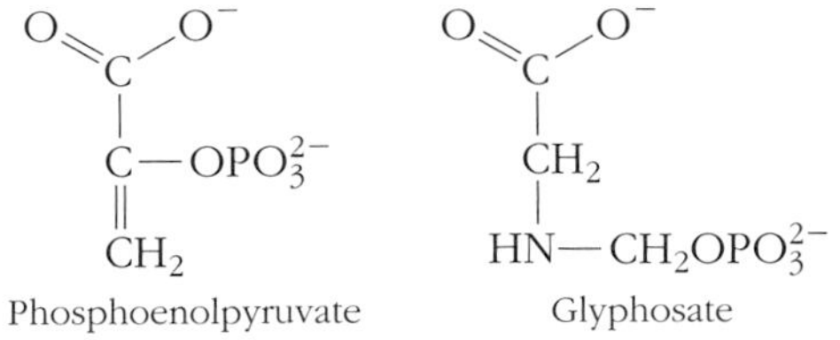

Phosphoenolpyruvate

Glyphosate

Go to **Exercise 21/Enzymes of Nitrogen Metabolism** to explore the role of channeling in two enzymes involved in amino acid and nucleotide metabolism. The structures and processes highlighted here address Learning Objective 6.

The final two reactions of the tryptophan biosynthetic pathway (which has 13 steps altogether) are catalyzed by tryptophan synthase, a bifunctional enzyme with an $\alpha_2\beta_2$ quaternary structure. The α subunit cleaves indole-3-glycerol phosphate to indole and glyceraldehyde-3-phosphate, then the β subunit adds serine to indole to produce tryptophan.

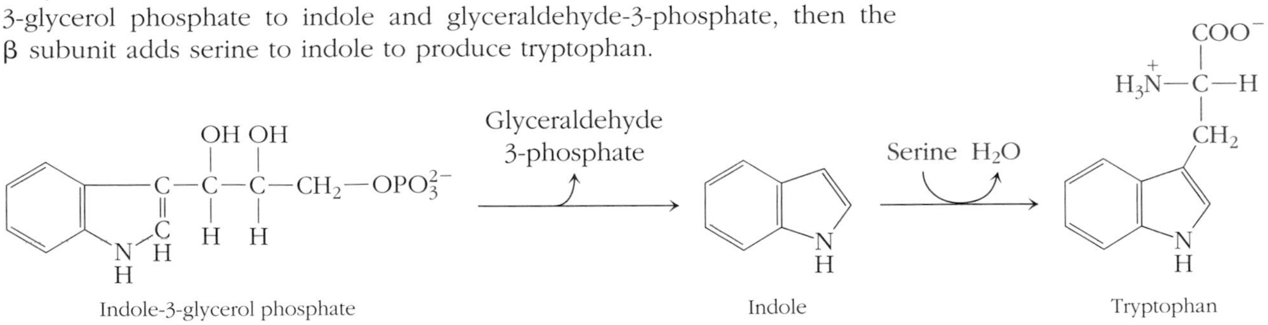

Glyceraldehyde 3-phosphate

Serine H_2O

Indole-3-glycerol phosphate

Indole

Tryptophan

Indole, the product of the α-subunit reaction and the substrate for the β-subunit reaction, never leaves the enzyme. Instead, it diffuses directly from one active site to the other without entering the surrounding solvent. The X-ray structure of the enzyme reveals that the active sites in adjacent α and β subunits are 25 Å apart but are connected by a tunnel through the protein that is large enough to accommodate indole (Fig. 15-8). *The movement of a reactant between two active sites is called* ***channeling****, and it increases the rate of a metabolic process by preventing the loss of intermediates.* Channeling can occur in multisubunit enzymes like tryptophan synthase as well as in polypeptides with multiple active sites, such as fatty acid synthase (see Section 14-2). The structure of the channel in tryptophan synthase is explored in Exercise 21.

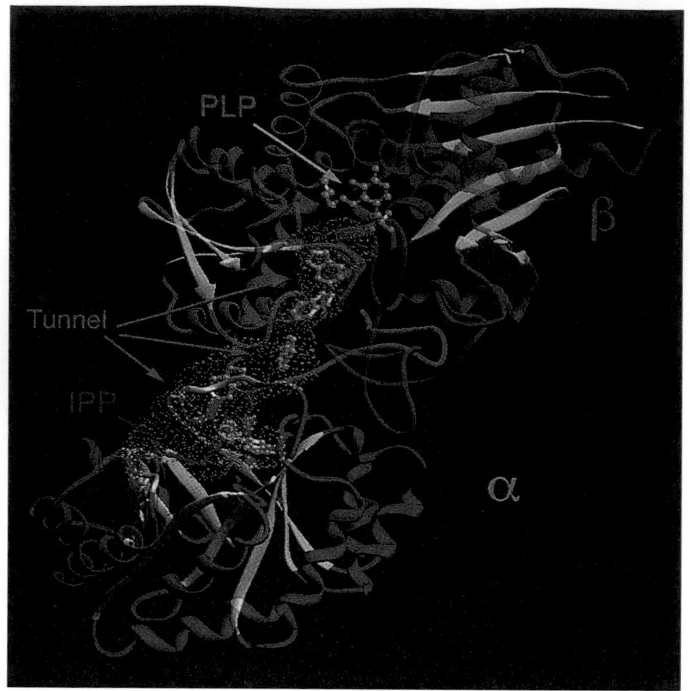

FIGURE 15-8

Tryptophan synthase.
Only one α subunit (blue and tan) and one β subunit (yellow, orange, and tan) are shown. Indolepropanol phosphate (IPP; red) marks the active site of the α subunit. The β active site is marked by its PLP cofactor (yellow). The surface of the tunnel between the two active sites is outlined with yellow dots. Several indole molecules (green) are included in the model to show how this intermediate can pass between the active sites. [*Courtesy Craig Hyde, National Institutes of Health.*]

All but one of the 20 standard amino acids are synthesized entirely from precursors produced by the main carbohydrate-metabolizing pathways. The exception is histidine, to which ATP provides one nitrogen and one carbon atom. Glutamate and glutamine donate the other two nitrogen atoms, and the remaining five carbons are derived from a phosphorylated monosaccharide, 5-phosphoribosyl pyrophosphate:

NH_2
Ribose triphosphate
ATP
+
$^{-2}O_3P—O—CH_2$
HO OH
5-Phosphoribosyl pyrophosphate

Glutamine
Glutamate

COO^-
$H_3\overset{+}{N}—C—H$
CH_2
HN CH
HC=N
Histidine

5-Phosphoribosyl pyrophosphate is also the source of the ribose group of nucleotides. This suggests that histidine might have been one of the first amino acids synthesized by an early life form making the transition from an all-RNA to an RNA-and-protein-based metabolism.

REVIEW LEARNING OBJECTIVE 3

- Which amino acids are synthesized by simple transamination reactions?
- Describe the role of tetrahydrofolate in amino acid biosynthesis.
- Why do herbicides target the pathways for synthesizing aromatic amino acids?
- How does His differ from other amino acids in its synthesis?

4 AMINO ACIDS AS METABOLIC PRECURSORS

Many amino acids that are ingested or built from scratch find their way into a cell's proteins, but some also have essential functions as precursors of other compounds, such as nucleotides, heme groups, and neurotransmitters. In this section we examine the synthesis of some of these molecules.

Many neurotransmitters are amino acid derivatives

Communication in the complex neuronal circuitry of the nervous system relies on small chemical signals that are released by one neuron and taken up by another (see Section 8-4). Common **neurotransmitters** include the amino acids glycine and glutamate. (For this reason, overuse of the flavor enhancer MSG, monosodium glutamate, is suspected of contributing to some psychological disorders, although this has never been proved). Glutamate from which a carboxylate group has been removed is known as γ-aminobutyric acid (GABA).

$$H_3\overset{+}{N}-CH_2-CH_2-CH_2-COO^-$$

γ-Aminobutyrate

It is the major inhibitory neurotransmitter, and its concentration in the brain is up to a thousand times greater than those of the other neurotransmitters.

Several other amino acid derivatives also function as neurotransmitters. For example, tyrosine gives rise to dopamine, norepinephrine, and epinephrine. These compounds are called catecholamines, reflecting their resemblance to catechol:

Tyrosine ⟹ Dopamine, Norepinephrine, Epinephrine; Catechol

A deficiency of dopamine produces the symptoms of Parkinson's disease: tremor, rigidity, and slow movements. Catecholamines are also produced by other tissues and function as hormones.

Tryptophan is the precursor of the neurotransmitter serotonin,

Tryptophan ⟹ Serotonin

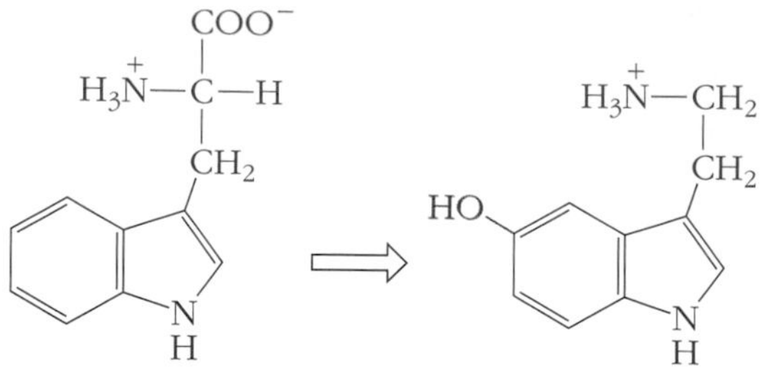

Low levels of serotonin in the brain have been linked to conditions such as depression, aggression, and hyperactivity. The antidepressive effect of drugs such as Prozac® results from their ability to increase serotonin levels by blocking the reabsorption of the released neurotransmitter. Serotonin is the precursor of melatonin:

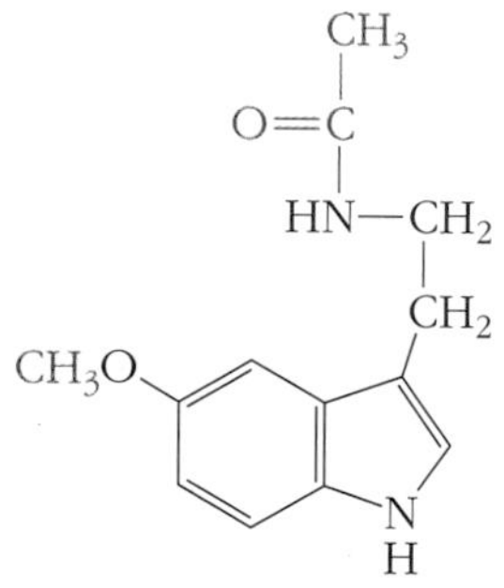

Melatonin

This tryptophan derivative is synthesized in the pineal gland and retina. Its concentration is low during the day, rising during darkness. Because melatonin appears to govern the synthesis of some other neurotransmitters that control circadian (daily) rhythms, it has been touted as a cure for sleep disorders and jet lag.

Arginine is also the precursor of a signaling molecule that was discovered only a few years ago to be the free radical gas nitric oxide (NO; Box 15-A).

REVIEW LEARNING OBJECTIVE 4

- Which amino acids are neurotransmitters?
- List some amino acid derivatives that act as signaling molecules.

Nucleotide biosynthesis

Nucleotides are synthesized from precursors that include several amino acids. The human body can also recycle nucleotides from nucleic acids and nucleotide cofactors that are broken down. Although food supplies nucleotides,

A CLOSER LOOK

Box 15-A Nitric oxide

In the 1980s, vascular biologists were investigating the nature of an endothelial cell-derived "relaxation factor" that caused blood vessels to dilate. This substance diffused quickly, acted locally, and disappeared within seconds. To the surprise of many, the mysterious factor turned out to be the free radical nitric oxide (·NO). Although NO was known to elicit vasodilation, it had not been considered a good candidate for a biological signaling molecule because its unpaired electron makes it extremely reactive and it breaks down to yield the corrosive nitric acid.

NO is a signaling molecule in the nervous, muscular, cardiovascular, and immune systems. At low concentrations it induces blood vessel dilation; at high concentrations it kills pathogens, much as oxygen radicals do (see page 370). NO is synthesized from arginine by nitric oxide synthase, an enzyme whose cofactors include FMN, FAD, tetrahydrobiopterin, and a heme group. A portion of murine nitric oxide synthase is shown here. Each of the two monomeric structures has a heme group (red), tetrahydrobiopterin (orange), and an arginine substrate (purple).

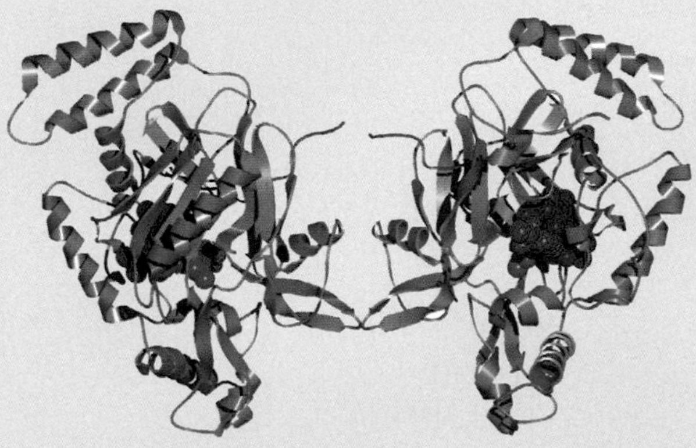

The first step of NO production is a hydroxylation reaction. In the second step, one electron oxidizes *N*-hydroxyarginine.

NO is unusual among signaling molecules for several reasons: It cannot be stockpiled for later release; it diffuses into cells so it does not need a receptor protein; and it needs no degradative enzyme because it breaks down on its own. NO is produced only when and where it is needed. A free radical gas such as NO cannot be directly introduced into the body, but an indirect source of NO has been clinically used for over a century. Individuals who suffer from angina pectoris, a painful condition caused by obstruction of the coronary blood vessels, can relieve their symptoms by taking nitroglycerin.

$$\begin{array}{ccc} CH_2 - & CH - & CH_2 \\ | & | & | \\ O & O & O \\ | & | & | \\ NO_2 & NO_2 & NO_2 \end{array}$$

Nitroglycerin

In vivo, nitroglycerin yields NO, which rapidly stimulates vasodilation, temporarily relieving the symptoms of angina.

Interestingly, naturally produced NO reacts with hemoglobin, not by binding to the heme Fe atom, as might be expected, but by reacting with certain Cys side chains to form a stable *S*-nitrosothiol group. NO reacts preferentially with hemoglobin in the oxy conformation. Consequently, in the lungs, NO is loaded onto hemoglobin along with O_2. When O_2 dissociates in the tissues and hemoglobin reverts to the deoxy state, the NO is also released. This dilates the small blood vessels, thereby facilitating O_2 delivery to the tissues.

[Structure of nitric oxide synthase (pdb 1NOD) determined by B.R. Crane, A.S. Arvai, E.D. Getzoff, D.J. Stuehr, and J.A. Tainer.]

$$\text{Arginine } (COO^-\text{–}C(H)(\overset{+}{H_3N})\text{–}CH_2\text{–}CH_2\text{–}CH_2\text{–}NH\text{–}C(=\overset{+}{NH_2})NH_2) \xrightarrow[O_2 \;\to\; H_2O]{NADPH \;\to\; NADP^+} N\text{-Hydroxyarginine } (\ldots\text{–}NH\text{–}C(=N\text{–}OH)NH_2) \xrightarrow[O_2 \;\to\; H_2O]{\frac{1}{2}NADPH \;\to\; \frac{1}{2}NADP^+} \text{Citrulline } (\ldots\text{–}NH\text{–}C(=O)NH_2) + \cdot NO$$

Arginine — *N*-Hydroxyarginine — Citrulline

the biosynthetic and "salvage" pathways are so efficient that there is no true dietary requirement for purines and pyrimidines. In this section we will take a brief look at the biosynthetic pathways for purine and pyrimidine nucleotides in mammals.

Purine nucleotides (AMP and GMP) are synthesized by building the purine base onto a ribose-5-phosphate molecule. In fact, the first step of the pathway is the production of 5-phosphoribosyl pyrophosphate (which is also a precursor of histidine):

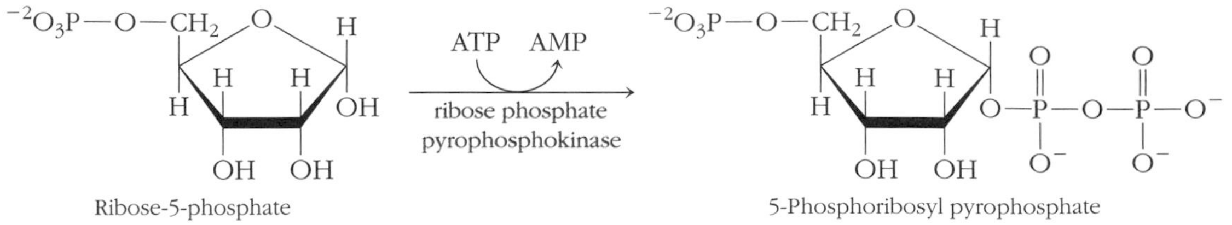

Ribose-5-phosphate → (ATP → AMP; ribose phosphate pyrophosphokinase) → 5-Phosphoribosyl pyrophosphate

The subsequent 10 steps of the pathway require as substrates glutamine, glycine, aspartate, bicarbonate, plus one-carbon formyl (—HC=O) groups donated by tetrahydrofolate. The product is inosine monophosphate (IMP), a nucleotide whose base is the purine hypoxanthine:

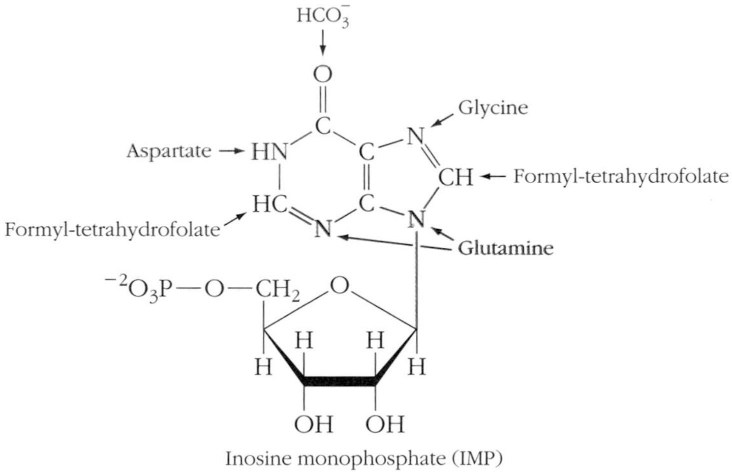

HCO_3^- → C; Glycine → N; Formyl-tetrahydrofolate → CH; Aspartate → HN; Formyl-tetrahydrofolate → HC; Glutamine → N, N

Inosine monophosphate (IMP)

IMP is the substrate for two short pathways that yield AMP and GMP. In AMP synthesis, an amino group from aspartate is transferred to the purine; in GMP synthesis, glutamate is the source of the amino group (Fig. 15-9). Kinases then catalyze phosphoryl-group transfer reactions to convert the nucleoside monophosphates to diphosphates and then triphosphates (ATP and GTP).

Figure 15-9 indicates that GTP participates in AMP synthesis and ATP participates in GMP synthesis. High concentrations of ATP therefore promote GMP production, and high concentrations of GTP promote AMP production. *This reciprocal relationship is one mechanism for balancing the production of adenine and guanine nucleotides.* (Because most nucleotides are destined for DNA or RNA synthesis, they are required in roughly equal amounts.) The pathway leading to AMP and GMP is also regulated by feedback inhibition at several points, including the first step, the production of 5-phosphoribosyl pyrophosphate from ribose-5-phosphate, which is inhibited by both ADP and GDP.

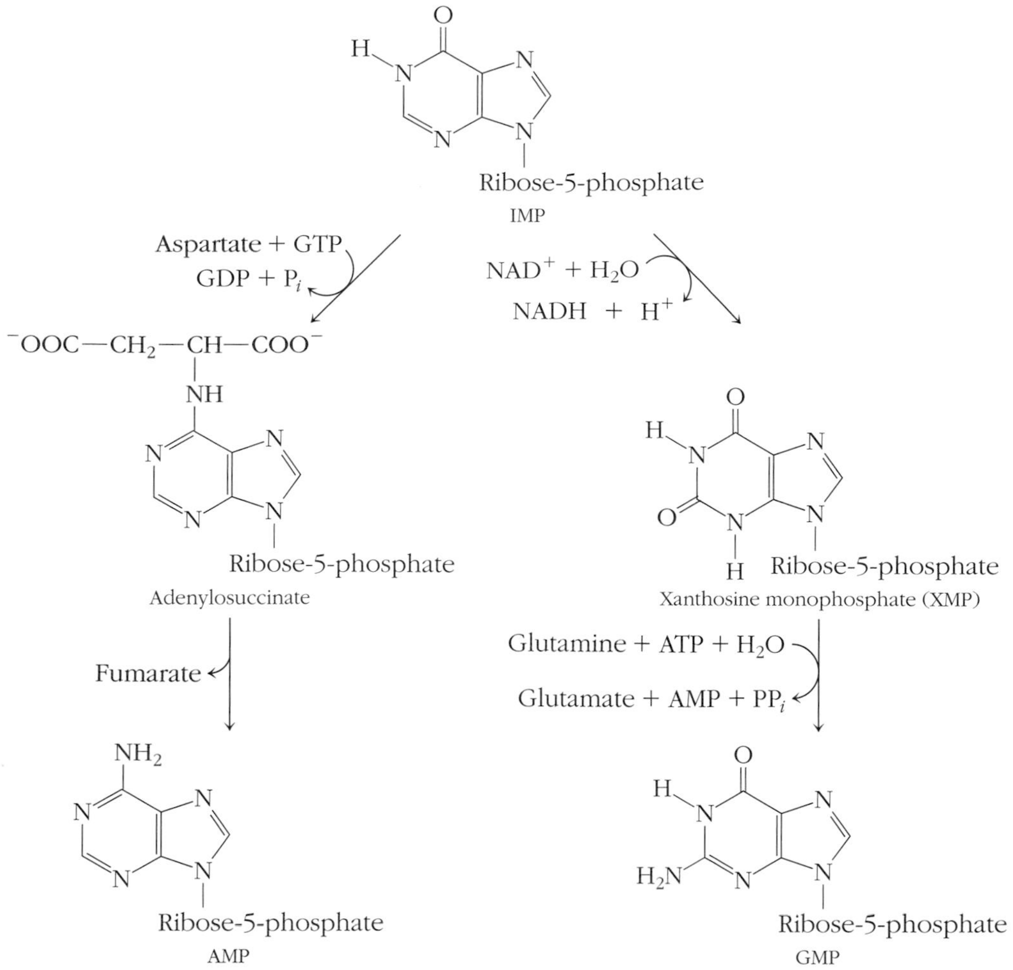

FIGURE 15-9

AMP and GMP synthesis from IMP.

In contrast to purine nucleotides, pyrimidine nucleotides are synthesized as a base that is subsequently attached to 5-phosphoribosyl pyrophosphate to form a nucleotide. The six-step pathway that yields uridine monophosphate (UMP) requires glutamine, aspartate, and bicarbonate.

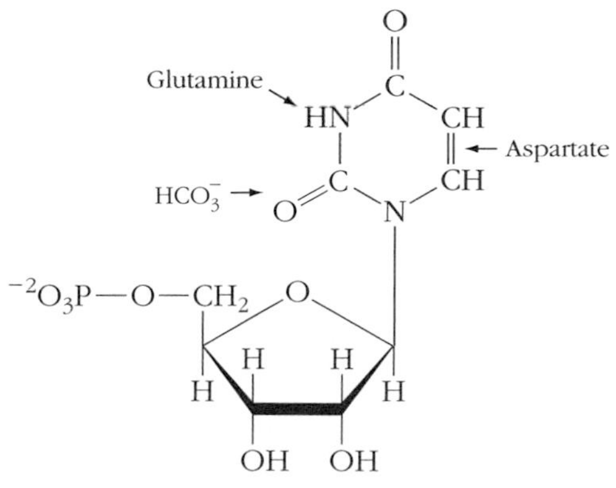

Uridine monophosphate (UMP)

UMP is phosphorylated to yield UDP and then UTP. CTP synthase catalyzes the amination of UTP to CTP, using glutamate as the donor:

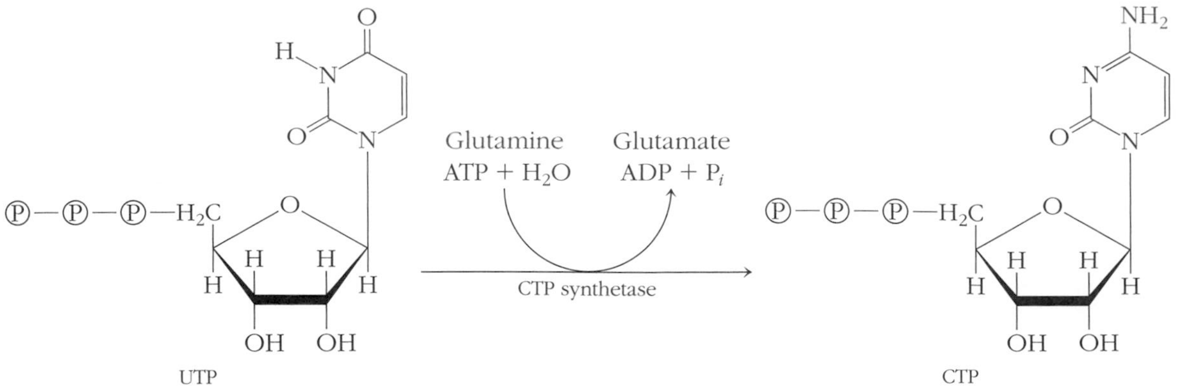

UTP —(Glutamine, ATP + H_2O → Glutamate, ADP + P_i; CTP synthetase)→ CTP

The UMP synthetic pathway in mammals is regulated primarily through feedback inhibition by UMP, UDP, and UTP. ATP activates the enzyme that catalyzes the first step; this helps balance the production of purine and pyrimidine nucleotides. Exercise 21 highlights channeling in *E. coli* carbamoyl phosphate synthetase, the enzyme that catalyzes the first step of pyrimidine synthesis.

So far, we have accounted for the synthesis of ATP, GTP, CTP, and UTP, which are substrates for the synthesis of RNA. DNA, of course, is built from deoxynucleotides, which are formed by the reduction of the ribonucleoside diphosphates ADP, GDP, CDP, and UDP (Box 15-B) followed by phosphorylation of the deoxynucleoside diphosphates to deoxynucleoside triphosphates.

dUTP is not used for DNA synthesis. Instead, *it is rapidly converted to thymine nucleotides* (which prevents the accidental incorporation of uracil into DNA). First, dUTP is hydrolyzed to dUMP. Next, thymidylate synthase adds a methyl group to dUMP to produce dTMP, using methylene-tetrahydrofolate as a one-carbon donor.

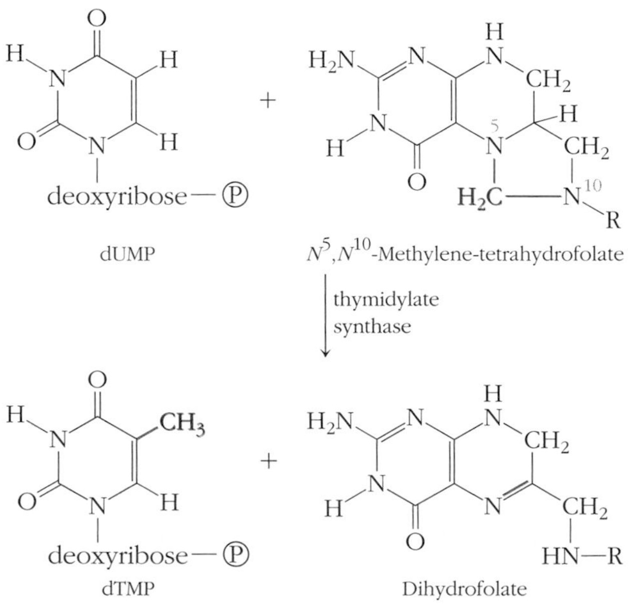

dUMP + N^5,N^{10}-Methylene-tetrahydrofolate —(thymidylate synthase)→ dTMP + Dihydrofolate

The serine hydroxymethyltransferase reaction is the main source of methylene-tetrahydrofolate.

In converting the methylene group (—CH_2—) to a methyl group (—CH_3), thymidylate synthase converts the tetrahydrofolate cofactor to dihydrofolate.

An NADPH-dependent enzyme called dihydrofolate reductase must then regenerate the reduced tetrahydrofolate cofactor. Finally, dTMP is phosphorylated to produce dTTP, the substrate for DNA polymerase.

Because cancer cells undergo rapid cell division, the enzymes of nucleotide synthesis, including thymidylate synthase and dihydrofolate reductase, are highly active. Compounds that inhibit either of these reactions can therefore act as anticancer agents. For example, the dUMP analog 5-fluorodeoxyuridylate, introduced in Section 7-3, inactivates thymidylate synthase. "Antifolates" such as methotrexate (at right) are competitive inhibitors of dihydrofolate reductase because they compete with dihydrofolate for binding to the enzyme. In the presence of methotrexate, a cancer cell cannot regenerate the tetrahydrofolate required for dTMP production, and the cell dies. Most noncancer cells, which grow much more slowly, are not as sensitive to the effect of the drug.

Methotrexate

In plants and some protozoa, thymidylate synthase and dihydrofolate reductase activities are contained in a single bifunctional enzyme. The dihydrofolate intermediate is channeled between the active sites—not through a solvent-inaccessible tunnel as in tryptophan synthase (see Fig. 15-8) but via a 40-Å-long "electrostatic highway" on the enzyme surface (Fig. 15-10). The positively charged groups of Lys and Arg side chains guide the anionic dihydrofolate. Electrostatic guidance may be a common feature of enzyme active sites.

REVIEW LEARNING OBJECTIVE 5

- Why don't humans require purines and pyrimidines in their diet?
- What are the metabolic fates of IMP and UTP?
- Describe the mechanisms that help balance the production of purine and pyrimidine nucleotides.
- Explain the importance of the thymidylate synthase and dihydrofolate reductase reactions.

REVIEW LEARNING OBJECTIVE 6

- Describe the ways reactants can be channeled between active sites.

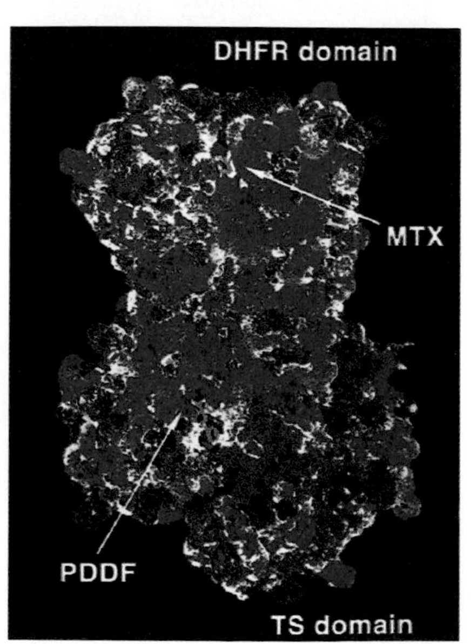

FIGURE 15-10

The bifunctional thymidylate synthase–dihydrofolate reductase of the protozoan *Leishmania*.
The enzyme consists of two identical chains, each of which has a thymidylate synthase active site (in the lower domain) and a dihydrofolate reductase active site (in the upper domain). The active sites are indicated by the yellow bonds to substrate analogs. In each subunit, the dihydrofolate product of the thymidylate synthase active site is channeled to the dihydrofolate reductase active site via a positively charged "electrostatic highway" on the enzyme surface. Red indicates negatively charged regions, and blue positively charged regions. [*Courtesy Edith Miles. From* Nat. Struct. Biol. ***1**, 186–194 (1994).*]

A CLOSER LOOK

Box 15-B Ribonucleotide reductase

Ribonucleotide reductase converts ribonucleotides to deoxyribonucleotides. This essential enzyme carries out the chemically difficult replacement of ribose 2′ OH groups with H, by a mechanism involving free radical chemistry. The well-studied *E. coli* ribonucleotide reductase is an $\alpha_2\beta_2$ tetramer with one active site. Reducing power comes from a cofactor such as thioredoxin, a 108-residue protein with two Cys residues that undergo disulfide exchange. Thioredoxin is reduced in turn by NADPH. The active site of ribonucleotide reductase itself contains iron atoms, Cys residues, and an unusually stable tyrosine radical (most free radicals are highly reactive and short-lived).

Tyrosine radicals are also features of cytochrome *c* oxidase and Photosystem II. The tyrosine radical of ribonucleotide reductase reacts with a Cys side chain to generate a thiyl radical, which then abstracts the 3′ H from the ribose ring (as shown at right). The activity of ribonucleotide reductase is tightly regulated so that the cell can balance the levels of ribo- and deoxyribonucleotides as well as the proportions of each of the four deoxyribonucleotides. Control of the enzyme is exerted at two regulatory sites that are distinct from the substrate-binding site. For example, ATP binding to the so-called activity site activates the enzyme. Binding of the deoxyribonucleotide dATP decreases enzyme activity. Several nucleotides bind to the so-called substrate specificity site. Here, ATP binding induces the enzyme to act on pyrimidine nucleotides, and dTTP binding causes the enzyme to prefer GDP as a substrate. These mechanisms, in concert with other mechanisms for balancing the amounts of the various nucleotides, help make all four deoxynucleotides available for DNA synthesis.

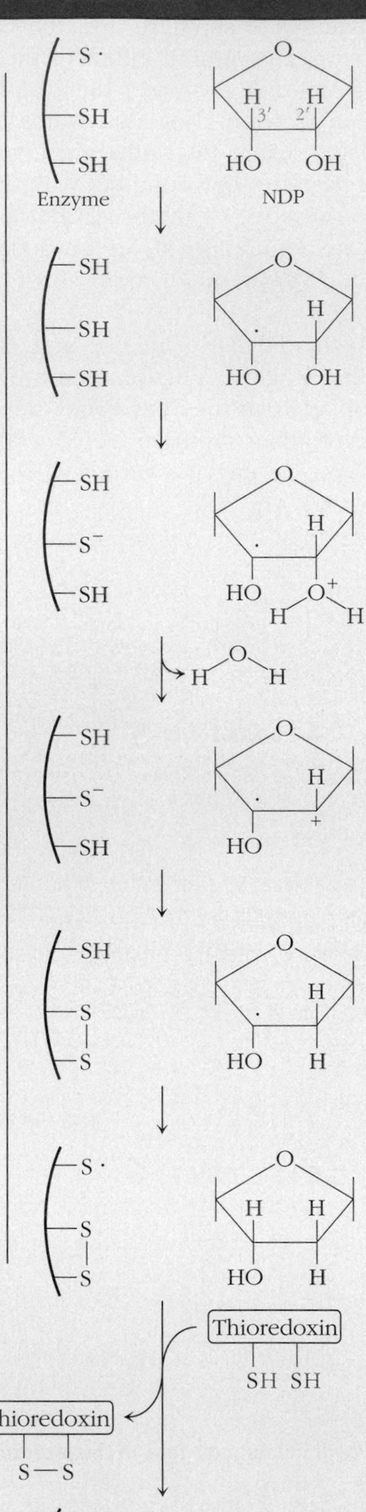

1. *The Cys free radical (—S·) reacts with the ribose of the NDP substrate (only a partial ribose ring is shown) to create a radical at C3′.*
2. *An enzyme Cys SH group donates a proton to the oxygen at C2′.*
3. *The radical helps stabilize the carbocation at C2′ formed by the loss of H_2O.*
4. *Transfer of a proton and electron reduces the cation and produces a disulfide bond between two enzyme Cys residues.*
5. *A second proton and electron transfer (the reverse of Step 1) generates the deoxyribose group and regenerates the thiyl radical.*
6. *The oxidized Cys groups of the enzyme are reduced in a disulfide exchange reaction with thioredoxin.*

5 AMINO ACID CATABOLISM

Like monosaccharides and fatty acids, *amino acids are metabolic fuels that can be broken down to release free energy.* In fact, dietary amino acids, not serum glucose, are the major fuel for the cells lining the small intestine. These cells absorb dietary amino acids and break down almost all of the available glutamate and aspartate and a good portion of the glutamine supply (note that these are all nonessential amino acids).

Other tissues, mainly the liver, also catabolize amino acids originating from the diet, from biosynthetic reactions, and from the normal turnover of intracellular proteins. During periods when dietary amino acids are not available, such as during a prolonged fast, amino acids are mobilized through the breakdown of muscle tissue, which accounts for about 40% of the total protein in the body. The amino acids undergo transamination reactions to remove their α-amino groups, and their carbon skeletons then enter the central pathways of energy metabolism (principally the citric acid cycle). However, *the catabolism of amino acids in the liver is not complete.* There is simply not enough oxygen available for the liver to completely oxidize all the carbon to CO_2. And even if there were, the liver would not need all the ATP that would be produced as a result. Instead, the amino acids are partially oxidized to substrates for gluconeogenesis (or ketogenesis). Glucose can then be exported to other tissues or stored as glycogen.

The reactions of amino acid catabolism, like those of amino acid synthesis, are too numerous to describe in full here, and the catabolic pathways do not necessarily mirror the anabolic pathways, as they do in carbohydrate and fatty acid metabolism. In this section, we will focus on some general principles and a few interesting chemical aspects of amino acid catabolism. In the following section we will see how organisms dispose of the nitrogen component of catabolized amino acids.

Amino acids are glucogenic, ketogenic, or both

It is useful to classify amino acids in humans as **glucogenic** (giving rise to gluconeogenic precursors such as citric acid cycle intermediates) or **ketogenic** (giving rise to acetyl-CoA, which can be used for ketogenesis or fatty acid synthesis, but not gluconeogenesis), or both glucogenic and ketogenic. As shown in Table 15-2, *all but leucine and lysine are at least partly glucogenic; all the nonessential amino acids are glucogenic; and the large skeletons of the aromatic amino acids are both glucogenic and ketogenic.*

Three amino acids are converted to gluconeogenic substrates by simple transamination (the reverse of their biosynthetic reactions): alanine to pyruvate, aspartate to oxaloacetate, and glutamate to α-ketoglutarate. Glutamate can also be deaminated in an oxidation reaction that we will examine in the following section. Asparagine undergoes deamidation to aspartate, which is then transaminated to oxaloacetate:

$$\underset{\text{Asparagine}}{H_3\overset{+}{N}-CH(COO^-)-CH_2-C(=O)NH_2} \xrightarrow[\text{asparaginase}]{H_2O \;\; NH_4^+} \underset{\text{Aspartate}}{H_3\overset{+}{N}-CH(COO^-)-CH_2-COO^-} \xrightarrow[\text{transaminase}]{\alpha\text{-Ketoglutarate} \;\; \text{Glutamate}} \underset{\text{Oxaloacetate}}{O=C(COO^-)-CH_2-COO^-}$$

TABLE 15-2 Catabolic fates of amino acids

Glucogenic	Both glucogenic and ketogenic	Ketogenic
Alanine	Isoleucine	Leucine
Arginine	Phenylalanine	Lysine
Aspartate	Threonine	
Asparagine	Tryptophan	
Cysteine	Tyrosine	
Glutamate		
Glutamine		
Glycine		
Histidine		
Methionine		
Proline		
Serine		
Valine		

Similarly, glutamine is deamidated to glutamate and then deaminated to α-ketoglutarate. Serine is converted to pyruvate:

$$\underset{\text{Serine}}{H_3\overset{+}{N}-\underset{\underset{CH_2-OH}{|}}{\overset{\overset{COO^-}{|}}{C}}-H} \xrightarrow{\;NH_4^+\;} \underset{\text{Pyruvate}}{\underset{\underset{CH_3}{|}}{\overset{\overset{COO^-}{|}}{C}}=O}$$

Note that in this reaction and in the conversion of asparagine and glutamine to their acid counterparts, the amino group is released as NH_4^+ rather than being transferred to another compound.

Proline and arginine (which are synthesized from glutamate) and histidine are catabolized to glutamate, which is then converted to α-ketoglutarate. The glutamate "family" of amino acids, which includes glutamine, proline, arginine, and histidine, constitute about 25% of dietary amino acids, so their potential contribution to energy metabolism is significant.

Cysteine is converted to pyruvate by a process that releases ammonia as well as sulfur:

$$\underset{\text{Cysteine}}{H_3\overset{+}{N}-\underset{\underset{CH_2-SH}{|}}{\overset{\overset{COO^-}{|}}{C}}-H} \xrightarrow{\;H_2O \quad NH_4^+,\ H_2S\;} \underset{\text{Pyruvate}}{\underset{\underset{CH_3}{|}}{\overset{\overset{COO^-}{|}}{C}}=O}$$

The products of the reactions listed so far—pyruvate, oxaloacetate, and α-ketoglutarate—are all gluconeogenic precursors:

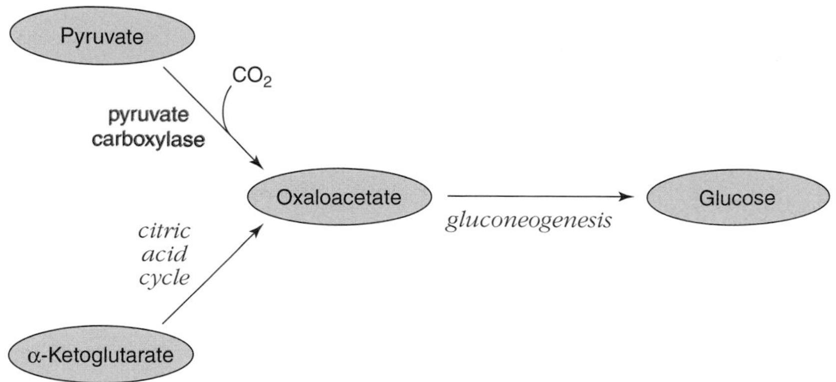

Threonine is both glucogenic and ketogenic because it is broken down to acetyl-CoA and glycine:

$$\underset{\text{Threonine}}{H_3\overset{+}{N}-\underset{\underset{\underset{CH_3}{|}}{CH-OH}}{|}{\overset{\overset{COO^-}{|}}{C}}-H} \xrightarrow{NAD^+ \;\curvearrowright\; NADH + H^+} \underset{\alpha\text{-Amino-}\beta\text{-ketobutyrate}}{H_3\overset{+}{N}-\underset{\underset{\underset{CH_3}{|}}{C=O}}{|}{\overset{\overset{COO^-}{|}}{C}}-H} \xrightarrow{CoASH} \underset{\text{Glycine}}{H_3\overset{+}{N}-\underset{\underset{H}{|}}{\overset{\overset{COO^-}{|}}{C}}-H} + \underset{\text{Acetyl-CoA}}{H_3C-\overset{\overset{O}{\|}}{C}-SCoA}$$

The acetyl-CoA is a precursor of ketone bodies (see Section 14-2), and the glycine is potentially glucogenic—if it is first converted to serine by the action of serine hydroxymethyltransferase. The major route for glycine disposal, however, is catalyzed by a multiprotein complex known as the glycine cleavage system:

$$H_3\overset{+}{N}-\overset{\overset{COO^-}{|}}{CH_2} + \text{Tetrahydrofolate} \xrightarrow[\text{glycine cleavage system}]{NAD^+ \;\curvearrowright\; NADH} \textbf{Methylene}\text{-tetrahydrofolate} + NH_4^+ + CO_2$$

The degradation pathways for the remaining amino acids are more complicated. For example, the branched-chain amino acids—valine, leucine, and isoleucine—undergo transamination to their α-keto acid forms and are then linked to coenzyme A in an oxidative decarboxylation reaction. This step is catalyzed by the branched-chain α-keto-acid dehydrogenase complex, a multienzyme complex that resembles the pyruvate dehydrogenase complex (see Section 11-1) and even shares some of the same subunits. A deficiency of the branched-chain α-keto-acid dehydrogenase complex leads to maple syrup urine disease, in which high concentrations of the branched-chain amino acids are excreted in the urine and give it a characteristic smell. The disease is rapidly fatal unless treated with a diet low in branched-chain amino acids.

The initial reactions of valine catabolism are shown in Figure 15-11. Subsequent steps yield the citric acid cycle intermediate succinyl-CoA. Isoleucine is degraded by a similar pathway that yields succinyl-CoA and acetyl-CoA. Leucine degradation yields acetyl-CoA and the ketone body acetoacetate. Lysine degradation, which follows a different pathway from the branched-chain amino acids, also yields acetyl-CoA and acetoacetate. The degradation of methionine produces succinyl-CoA.

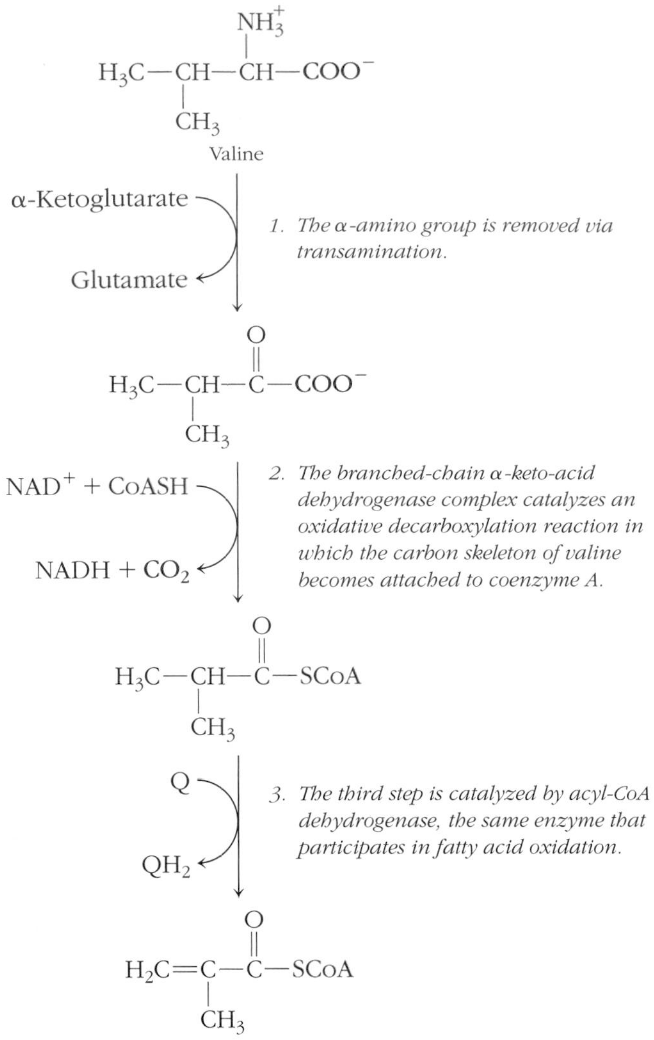

$$H_3C-CH(CH_3)-CH(NH_3^+)-COO^-$$

Valine

α-Ketoglutarate → Glutamate

1. The α-amino group is removed via transamination.

$$H_3C-CH(CH_3)-C(=O)-COO^-$$

$NAD^+ + CoASH$ → $NADH + CO_2$

2. The branched-chain α-keto-acid dehydrogenase complex catalyzes an oxidative decarboxylation reaction in which the carbon skeleton of valine becomes attached to coenzyme A.

$$H_3C-CH(CH_3)-C(=O)-SCoA$$

Q → QH_2

3. The third step is catalyzed by acyl-CoA dehydrogenase, the same enzyme that participates in fatty acid oxidation.

$$H_2C=C(CH_3)-C(=O)-SCoA$$

FIGURE 15-11

The initial steps of valine degradation.

Finally, the cleavage of the aromatic amino acids—phenylalanine, tyrosine, and tryptophan—yields the ketone body acetoacetate as well as a glucogenic compound (alanine or fumarate). The first step of phenylalanine degradation is a hydroxylation reaction that produces tyrosine (this is why tyrosine is considered nonessential). The reaction is worth noting because it uses the cofactor tetrahydrobiopterin (which, like folate, contains a pterin moiety). The tetrahydrobiopterin is oxidized to dihydrobiopterin in the phenylalanine hydroxylase reaction (above right). This cofactor must be subsequently reduced to the tetrahydro form by a separate NADH-dependent enzyme. Another step of the phenylalanine (and tyrosine) degradation pathway is also notable because a deficiency of the enzyme was one of the first-characterized "inborn errors of metabolism" (Box 15-C).

REVIEW LEARning OBJECTIVE 7

- What is the role of the liver in catabolizing amino acids?
- Distinguish the glucogenic and ketogenic amino acids.
- What is the role of coenzyme A in amino acid catabolism?

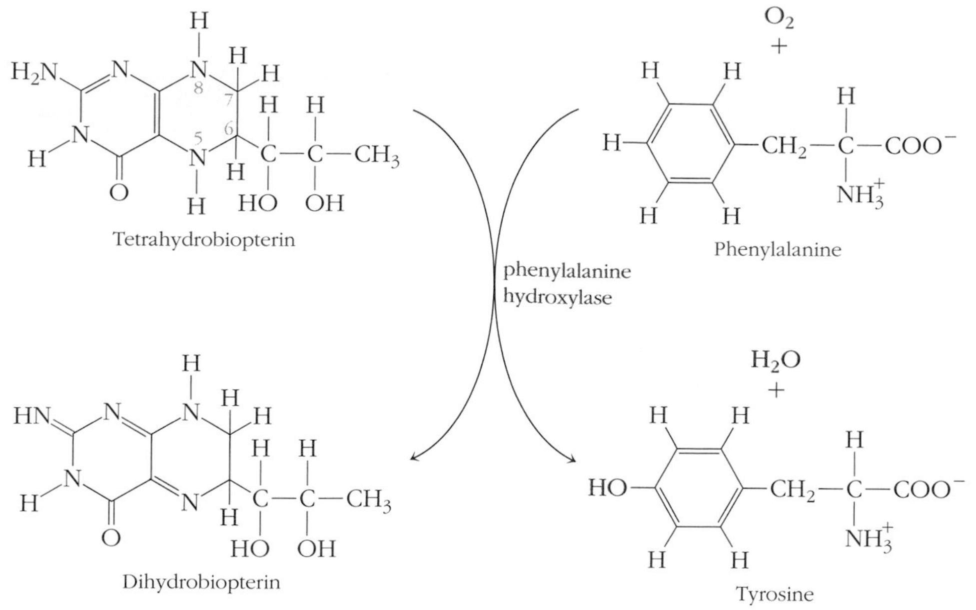

Tetrahydrobiopterin

Phenylalanine

phenylalanine hydroxylase

Dihydrobiopterin

Tyrosine

6 NITROGEN DISPOSAL: THE UREA CYCLE

When the supply of amino acids exceeds the cell's immediate needs for protein synthesis or other amino acid–consuming pathways, the carbon skeletons are broken down and the nitrogen disposed of. All amino acids except lysine can be deaminated by the action of transaminases, but this merely transfers the amino group to another molecule; it does not eliminate it from the cell.

Some catabolic reactions release free ammonia, which can be excreted as a waste product in the urine. In fact, the kidney is a major site of glutamine catabolism, and the resulting NH_4^+ facilitates the excretion of metabolic acids such as H_2SO_4 that arise from the catabolism of methionine and cysteine. However, ammonia production is not a practical means for disposing of excess nitrogen. First, high concentrations of NH_4^+ in the blood cause alkalosis and second, ammonia is highly toxic. It easily enters the brain, where it combines with α-ketoglutarate to form glutamate. This depletion of a citric acid cycle intermediate may be the cause of ammonia's neurotoxicity.

Approximately 80% of the body's excess nitrogen is excreted in the form of urea,

$$H_2N-\overset{\overset{\displaystyle O}{\|}}{C}-NH_2$$

Urea

which is produced in the liver by the reactions of the ***urea cycle.*** (This catabolic cycle was elucidated in 1932 by Hans Krebs and Kurt Henseleit; Krebs went on to outline another circular pathway—the citric acid cycle—in 1937.)

Glutamate supplies nitrogen to the urea cycle

Because many transaminases use α-ketoglutarate as the amino-group acceptor, glutamate is one of the most abundant amino acids inside cells. Glutamate can be deaminated to regenerate α-ketoglutarate and release NH_4^+ in an oxidation–reduction reaction catalyzed by glutamate dehydrogenase:

A CLOSER LOOK

Box 15-C Inborn errors of metabolism

Today we understand that inherited diseases result from defective genes. We are also beginning to understand that malfunctioning genes underlie many noninherited diseases as well. The link between genes and disease was first recognized about 100 years ago by the physician Archibald Garrod, who coined the term "inborn error of metabolism" in 1902. Garrod's insights came from his studies of individuals with alcaptonuria. Their urine turned black upon exposure to air because it contained homogentisate, a product of tyrosine catabolism. Garrod concluded that this inherited condition resulted from the lack of a specific enzyme. We now know that homogentisate is excreted because the enzyme that breaks it down, homogentisate dioxygenase, is missing or defective. The pathway is shown below.

Garrod's findings built on Mendel's discovery of the laws of inheritance but went largely unappreciated for about half a century. In the 1950s, George Beadle and Edward Tatum popularized the "one gene, one enzyme" theory based on their work with *Neurospora* mutants. Around the same time, Vernon Ingram discovered the molecular defect responsible for sickle-cell hemoglobin. At this point, Garrod's work finally seemed relevant.

Garrod also described a number of other inborn errors of metabolism, including albinism, cystinuria (excretion of cysteine in the urine), and several other disorders that were not life-threatening and left easily detected clues in the patient's urine. Of course, many "inborn errors" are catastrophic. For example, phenylketonuria (PKU) results from a deficiency of phenylalanine hydroxylase, the first enzyme in the pathway shown below. Phenylalanine cannot be broken down, although it can undergo transamination. The resulting α-keto acid derivative phenylpyruvate accumulates and is excreted in the urine, giving it a mousy odor. If not treated, PKU causes mental retardation. Fortunately, the disease can be detected in newborns. Afflicted individuals develop normally if they consume a diet that is low in phenylalanine. Sadly, the biochemical defects behind many other diseases are not as well understood, making them difficult to identify and treat.

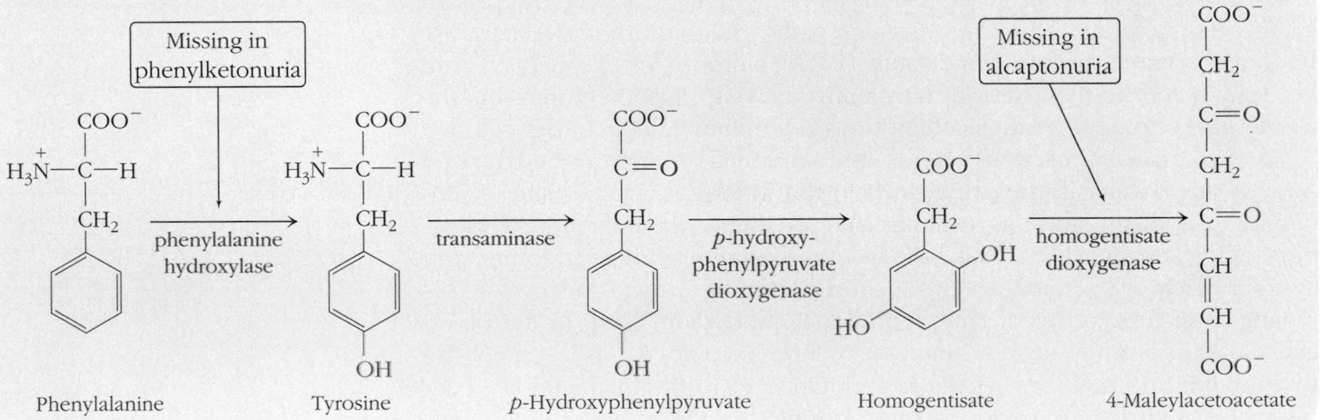

Missing in phenylketonuria

Missing in alcaptonuria

Phenylalanine $\xrightarrow{\text{phenylalanine hydroxylase}}$ Tyrosine $\xrightarrow{\text{transaminase}}$ *p*-Hydroxyphenylpyruvate $\xrightarrow{p\text{-hydroxyphenylpyruvate dioxygenase}}$ Homogentisate $\xrightarrow{\text{homogentisate dioxygenase}}$ 4-Maleylacetoacetate

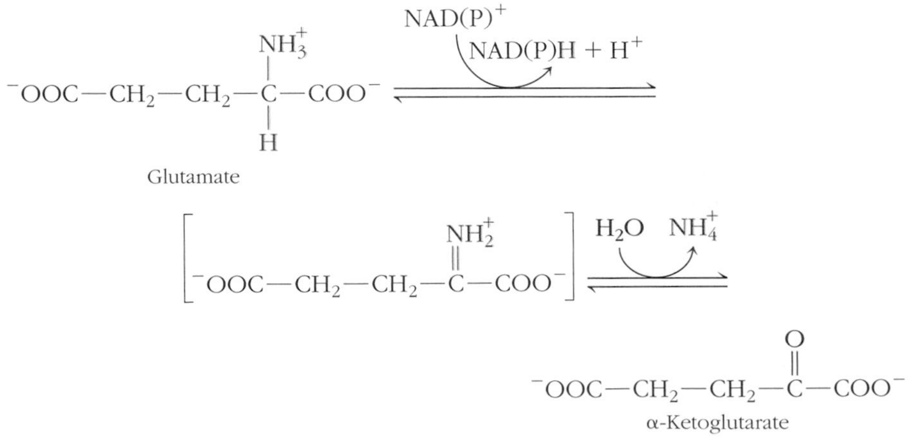

$$^{-}OOC-CH_2-CH_2-\overset{\overset{NH_3^+}{|}}{\underset{\underset{H}{|}}{C}}-COO^- \xrightleftharpoons[]{NAD(P)^+ \;\; NAD(P)H + H^+} \left[^{-}OOC-CH_2-CH_2-\overset{\overset{NH_2^+}{\|}}{C}-COO^- \right] \xrightleftharpoons[]{H_2O \;\; NH_4^+} {}^{-}OOC-CH_2-CH_2-\overset{\overset{O}{\|}}{C}-COO^-$$

Glutamate

α-Ketoglutarate

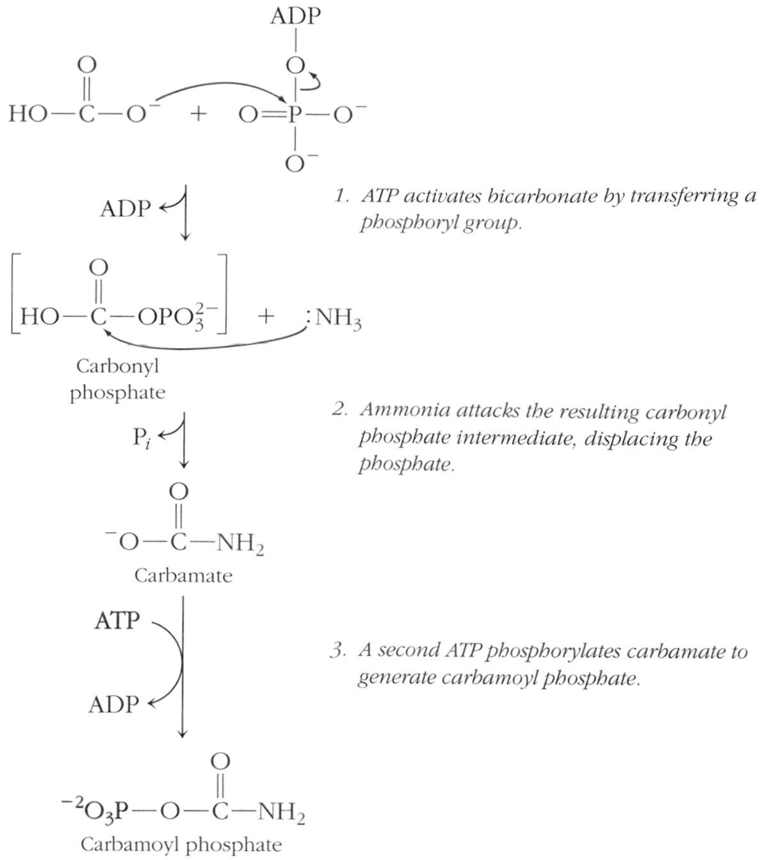

$$\text{HO—}\overset{\overset{\text{O}}{\|}}{\text{C}}\text{—O}^- + \text{ATP} \xrightarrow{\text{ADP}} \left[\text{HO—}\overset{\overset{\text{O}}{\|}}{\text{C}}\text{—OPO}_3^{2-}\right]$$

1. ATP activates bicarbonate by transferring a phosphoryl group.

Carbonyl phosphate

$$\left[\text{HO—}\overset{\overset{\text{O}}{\|}}{\text{C}}\text{—OPO}_3^{2-}\right] + :\text{NH}_3 \xrightarrow{P_i} {}^-\text{O—}\overset{\overset{\text{O}}{\|}}{\text{C}}\text{—NH}_2$$

2. Ammonia attacks the resulting carbonyl phosphate intermediate, displacing the phosphate.

Carbamate

$${}^-\text{O—}\overset{\overset{\text{O}}{\|}}{\text{C}}\text{—NH}_2 \xrightarrow{\text{ATP} \rightarrow \text{ADP}} {}^{-2}\text{O}_3\text{P—O—}\overset{\overset{\text{O}}{\|}}{\text{C}}\text{—NH}_2$$

3. A second ATP phosphorylates carbamate to generate carbamoyl phosphate.

Carbamoyl phosphate

FIGURE 15-12

The carbamoyl phosphate synthetase reaction.

This mitochondrial enzyme is unusual: It is the only known enzyme that can use either NAD^+ or $NADP^+$ as a cofactor. The reaction is reversible *in vivo,* and *the direction of flux depends on the relative concentrations of substrates and products.* Consequently, glutamate dehydrogenase can assimilate free NH_4^+ into an amino acid (the reverse reaction). As written above, the reaction is a major route for feeding amino acid–derived amino groups into the urea cycle.

The starting substrate for the urea cycle is an "activated" molecule produced by the condensation of bicarbonate and ammonia, as catalyzed by carbamoyl phosphate synthetase (Fig. 15-12). The NH_4^+ may be contributed by the glutamate dehydrogenase reaction or another process that releases ammonia. The bicarbonate is the source of the urea carbon. Note that the phosphoanhydride bonds of two ATP molecules are consumed in the energetically costly production of carbamoyl phosphate.

The urea cycle consists of four reactions

The four enzyme-catalyzed reactions of the urea cycle proper are shown in Figure 15-13. The cycle also provides a means for synthesizing arginine: The five-carbon ornithine is derived from glutamate, and the urea cycle converts it to arginine. However, the arginine needs of children exceed the biosynthetic capacity of the urea cycle, so arginine is classified as an essential amino acid.

The fumarate generated in Step 3 of the urea cycle can be converted back to the aspartate substrate for Step 2. Combining these ancillary reactions with those of the urea cycle, the carbamoyl phosphate synthetase reaction, and the glutamate dehydrogenase reaction yields the pathway outlined in Fig-

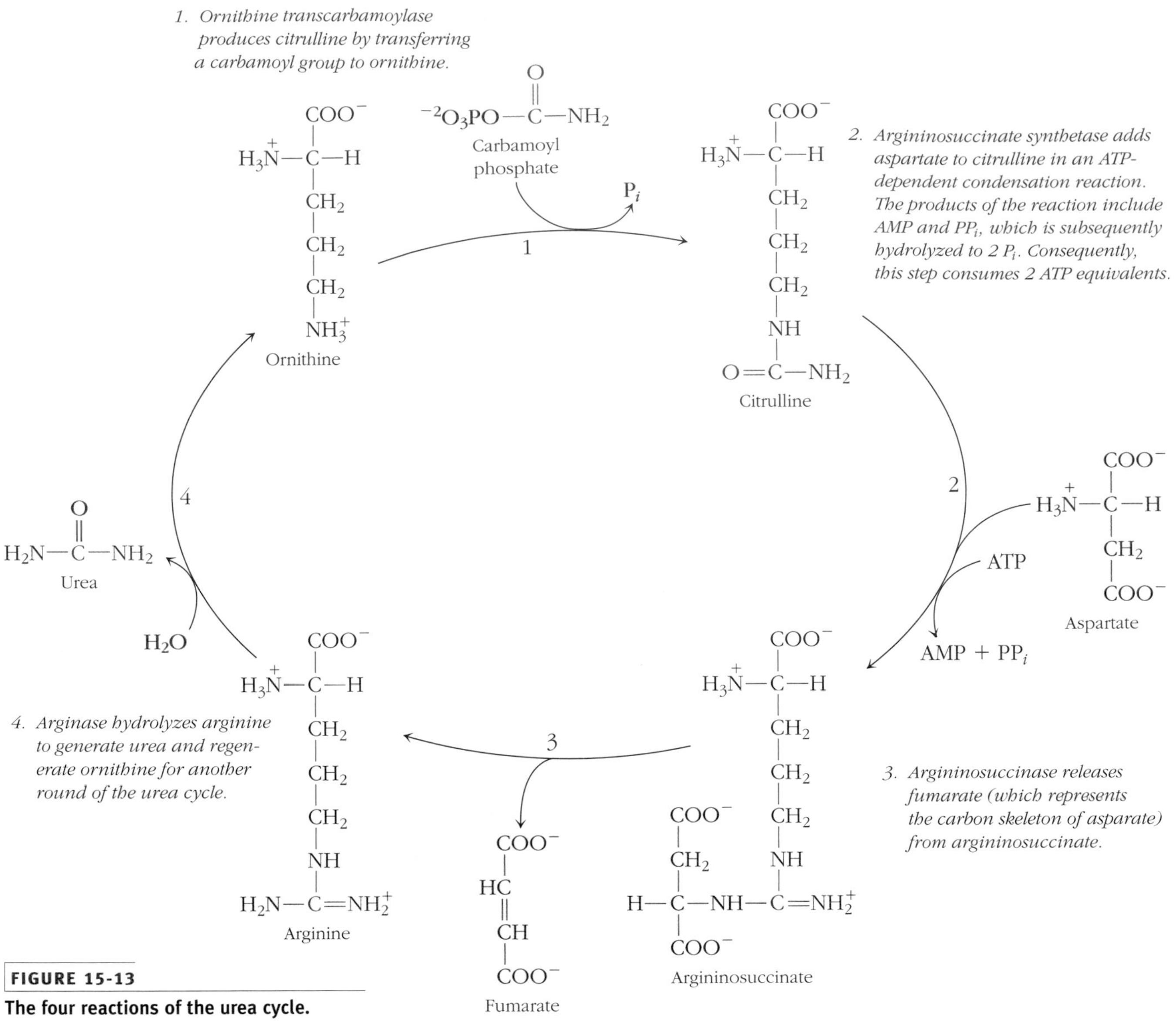

1. Ornithine transcarbamoylase produces citrulline by transferring a carbamoyl group to ornithine.

2. Argininosuccinate synthetase adds aspartate to citrulline in an ATP-dependent condensation reaction. The products of the reaction include AMP and PP_i, which is subsequently hydrolyzed to 2 P_i. Consequently, this step consumes 2 ATP equivalents.

3. Argininosuccinase releases fumarate (which represents the carbon skeleton of asparate) from argininosuccinate.

4. Arginase hydrolyzes arginine to generate urea and regenerate ornithine for another round of the urea cycle.

Ornithine — Carbamoyl phosphate — P_i — 1 — Citrulline — 2 — Aspartate — ATP — AMP + PP_i — Argininosuccinate — 3 — Fumarate — Arginine — 4 — H_2O — Urea

FIGURE 15-13

The four reactions of the urea cycle.

ure 15-14. *The overall effect is that transaminated amino acids donate amino groups, via glutamate and aspartate, to urea synthesis.*

Like many other metabolic loops, the urea cycle involves enzymes located in both the mitochondria and cytosol. Glutamate dehydrogenase, carbamoyl phosphate synthetase, and ornithine transcarbamoylase are mitochondrial, whereas argininosuccinate synthetase, argininosuccinase, and arginase are cytosolic. Consequently, citrulline is produced in the mitochondria but must be transported to the cytosol for the next step, and ornithine produced in the cytosol must be imported into the mitochondria to begin a new round of the cycle.

Although the carbamoyl phosphate synthetase reaction and the argininosuccinate synthetase reactions each consume 2 ATP equivalents, for a total cost of 4 ATP per urea, *the urea cycle and its ancillary reactions actually generate ATP:* The glutamate dehydrogenase reaction produces NADH (or NADPH), whose free energy is conserved in the synthesis of 3 ATP by oxidative phosphorylation. The conversion of malate to oxaloacetate by malate dehydrogenase also generates NADH (equivalent to 3 ATP), for a total yield of 6 ATP and a net yield of 2 ATP per urea.

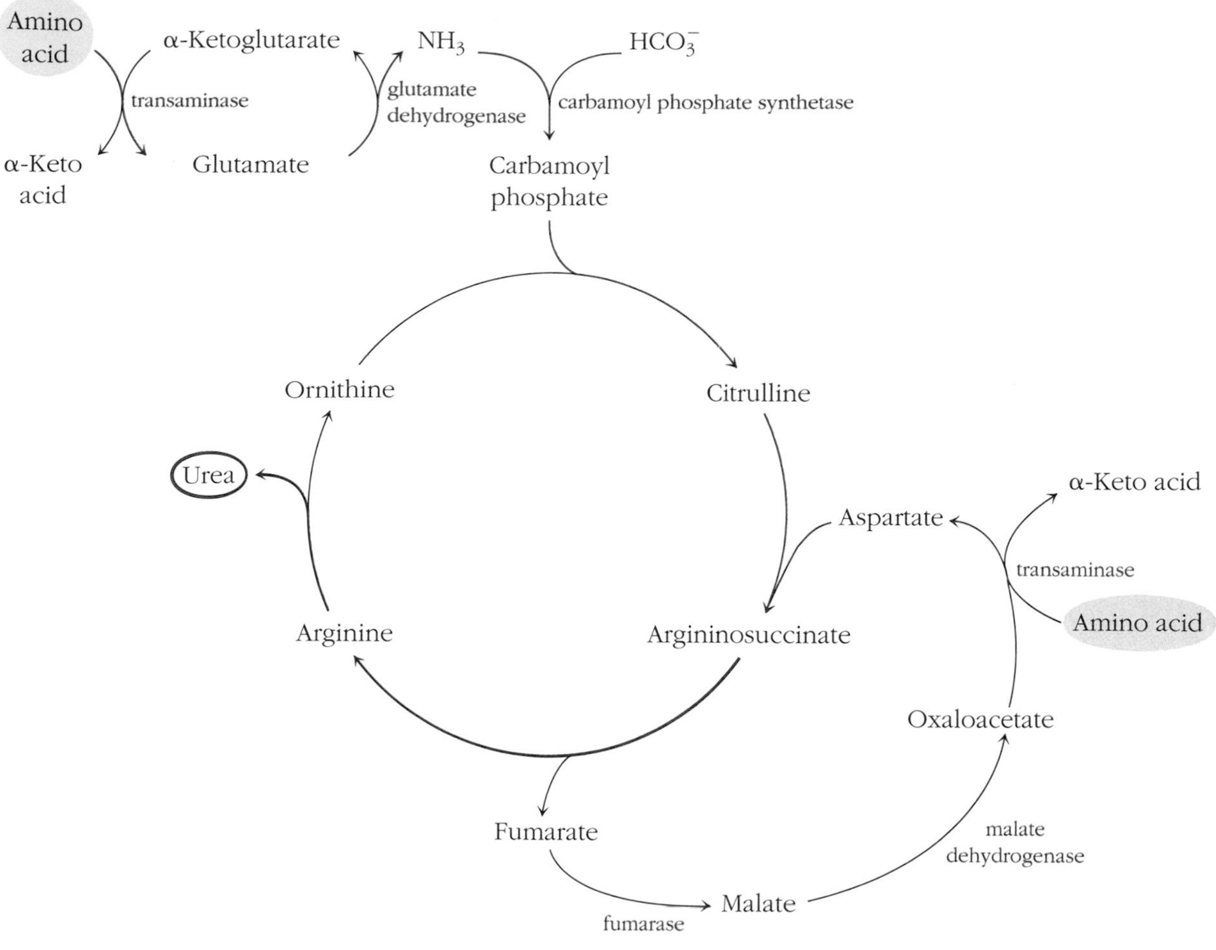

FIGURE 15-14

The urea cycle and related reactions.
Two routes for the disposal of amino groups are highlighted. The blue pathway shows how an amino group from an amino acid enters the urea cycle via glutamate and carbamoyl phosphate. The red pathway shows how an amino group from an amino acid enters via aspartate. The fumarase and malate dehydrogenase at the lower right are cytosolic isozymes of the mitochondrial citric acid cycle enzymes.

The rate of urea production is controlled largely by the activity of carbamoyl phosphate synthetase. This enzyme is allosterically activated by *N*-acetylglutamate, which is synthesized from glutamate and acetyl-CoA:

$$\underset{\text{Glutamate}}{\mathrm{{}^{+}H_3N{-}CH(COO^-){-}CH_2{-}CH_2{-}COO^-}} + \underset{\text{Acetyl-CoA}}{\mathrm{H_3C{-}\overset{O}{\overset{\|}{C}}{-}SCoA}} \xrightarrow{\text{HSCoA}} \underset{N\text{-Acetylglutamate}}{\mathrm{H_3C{-}\overset{O}{\overset{\|}{C}}{-}NH{-}CH(COO^-){-}CH_2{-}CH_2{-}COO^-}}$$

When amino acids are undergoing transamination and being catabolized, the resulting increase in the cellular glutamate concentration boosts production of *N*-acetylglutamate. This stimulates carbamoyl phosphate synthetase activity, and flux through the urea cycle increases. Such a regulatory system allows the cell to efficiently dispose of the nitrogen released from amino acid degradation.

Other mechanisms for nitrogen disposal

Urea is relatively nontoxic and easily transported through the bloodstream to the kidneys. However, the polar urea molecule requires large amounts of water for its efficient excretion. This presents a problem for flying vertebrates such as birds and for reptiles that are adapted to arid habitats. These organisms produce uric acid as a nitrogenous waste product and excrete it as a semisolid slurry.

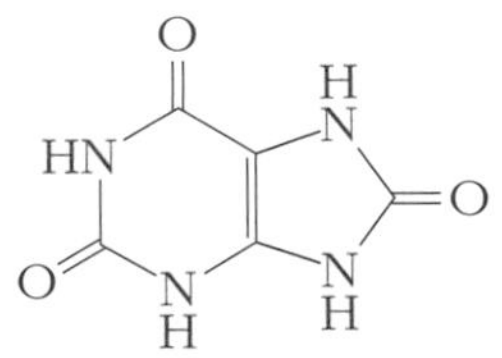

Uric acid

Uric acid is also the end product of purine catabolism in primates and some other animals. Purine nucleotides are broken down by nucleotidases (which remove phosphate groups to yield nucleosides), deaminases (which remove the ring amino substituents), and purine nucleoside phosphorylase (which removes the ribose group). These reactions yield the purine xanthine, which xanthine oxidase converts to uric acid:

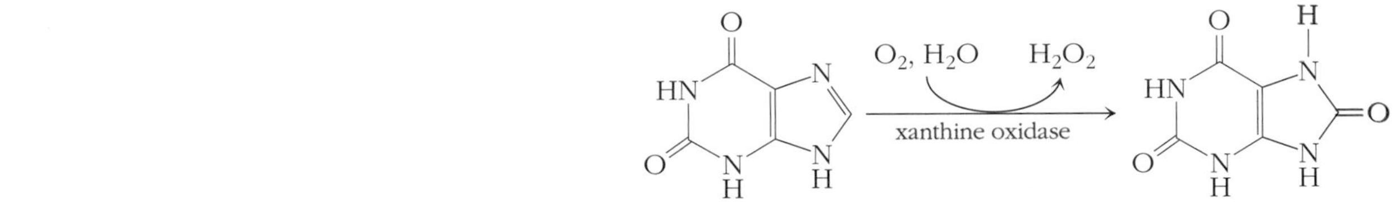

$$\text{Xanthine} \xrightarrow[\text{xanthine oxidase}]{O_2,\ H_2O \quad H_2O_2} \text{Uric acid}$$

Keep in mind that the intermediates of nucleotide catabolism are substrates for **salvage reactions,** which use the compounds in the synthesis of new nucleotides.

An excess of uric acid, which is not highly soluble in water, may lead to the precipitation of uric acid crystals as kidney stones. Uric acid may also precipitate in the joints, primarily the knees or toes, a painful condition known as gout. Uric acid accumulation can be treated with a purine analog that blocks the activity of xanthine oxidase. The preceding intermediates in the purine catabolic pathways, which are more soluble than uric acid, are then excreted.

Pyrimidine nucleotides, like purine nucleotides, undergo deamination and removal of phosphate and ribose groups, and the reaction products are often salvaged for the synthesis of new nucleotides. However, unlike purines, the pyrimidine bases (uracil and thymine) can be further degraded as CoA adducts. Consequently, *cellular pyrimidine catabolism contributes slightly to the pool of metabolic fuels* (in contrast, excess purines are not used as fuels but are excreted).

Urease breaks down urea

To complete the story of nitrogen disposal, we turn to urease, an enzyme that allows bacteria, fungi, and some other organisms to break down urea.

$$H_2N{-}\overset{\overset{\displaystyle O}{\|}}{C}{-}NH_2 + H_2O \xrightarrow[\text{urease}]{} 2\,NH_3 + CO_2$$

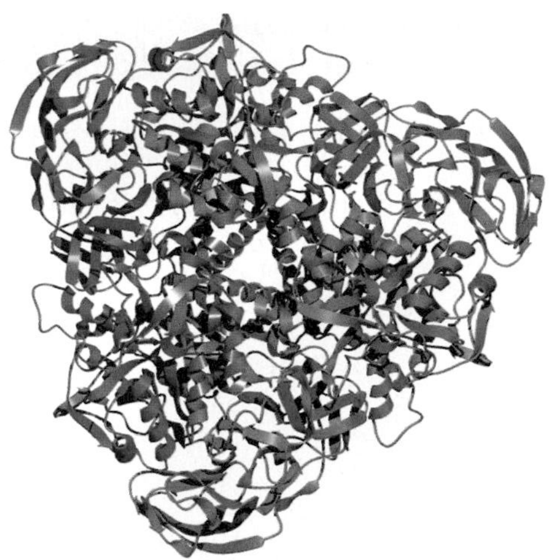

FIGURE 15-15

Urease.
This enzyme, from the bacterium *Klebsiella aerogenes*, contains nine polypeptides arranged as a trimer of trimers. Two nickel atoms (dark purple spheres) mark the active site in each trimeric unit. [*Structure (pdb 2KAU) determined by E. Jabri, M.B. Carr, R.P. Hausinger, and P.A. Karplus.*]

Urease has the distinction of being the first enzyme to be crystallized (in 1926). It helped promote the theory that catalytic activity was a property of proteins. This premise is only partly true, as we have seen, since many enzymes contain metal ions or inorganic cofactors (urease itself contains two catalytic nickel atoms). The detailed X-ray crystallographic structure of urease was determined only in 1995, some 70 years after the first crystal studies (Fig. 15-15).

REVIEW LEARNING OBJECTIVE 8

- Why do mammals eliminate waste nitrogen as urea rather than ammonia? Why do some organisms excrete uric acid?
- What does the carbamoyl phosphate synthetase reaction accomplish?
- Describe the four reactions of the urea cycle.
- How are amino groups from amino acids incorporated into urea?

SUMMARY

1. Nitrogen-fixing organisms convert N_2 to NH_3 in an ATP-consuming reaction. Nitrate and nitrite can also be reduced to NH_3.
2. Ammonia is incorporated into glutamine by the action of glutamine synthetase.
3. Transaminases use a PLP prosthetic group to catalyze the reversible interconversion of α-amino acids and α-keto acids.
4. The 10 nonessential amino acids are synthesized from common metabolic intermediates such as pyruvate, oxaloacetate, and α-ketoglutarate. The 10 essential amino acids, which include the branched-chain and aromatic amino acids, are synthesized by more elaborate pathways in bacteria and plants.
5. Amino acids are the precursors of some neurotransmitters.
6. The synthesis of nucleotides requires glutamate, glycine, and aspartate as well as ribose-5-phosphate. The pathways for purine and pyrimidine biosynthesis are regulated to balance the production of the various nucleotides. A ribonucleotide reductase converts the nucleotides to deoxynucleotides. Thymidine production requires a methyl group donated by the cofactor tetrahydrofolate.
7. Following removal of their amino groups by transamination, amino acids are broken down to intermediates that can be used as metabolic fuels or converted to glucose (glucogenic amino acids) or ketone bodies (ketogenic amino acids).
8. In mammals, excess amino groups are converted to urea for disposal. The urea cycle is regulated at the carbamoyl phosphate synthetase step, an entry point for ammonia. Other organisms convert excess nitrogen to compounds such as uric acid.

CHECKLIST

1. Review the Learning Objectives listed on page 462.
2. Complete Exercise 21/Enzymes of Nitrogen Metabolism to explore channeling in the operation of two of the enzymes involved in amino acid and nucleotide metabolism.
3. Apply your knowledge by solving the problems at the end of this chapter. Check your results in the Solutions appendix.
4. Be able to define the boldfaced terms (consult the glossary at the end of the book). Test your understanding by taking the Chapter 15 quiz (www.wiley.com/college/pratt).
5. Explore the websites listed at www.wiley.com/college/pratt.
6. Consult the list of Selected Readings for background information or additional details on nitrogen fixation, amino acid catabolism, nucleotide metabolism, and the urea cycle.

GLOSSARY TERMS

nitrogen fixation
nitrification
denitrification
nitrogen cycle
transamination
nonessential amino acid
essential amino acid
channeling
neurotransmitter
glucogenic amino acid
ketogenic amino acid
urea cycle
salvage reaction

PROBLEMS

1. The $\mathcal{E}^{\circ\prime}$ for the half-reaction

$$N_2 + 6\ H^+ + 6\ e^- \rightleftharpoons 2\ NH_3$$

is −0.34 V. The reduction potential of the nitrogenase component that donates electrons for nitrogen reduction is about −0.29 V. ATP hydrolysis apparently induces a conformational change in the protein that alters its reduction potential (a change with a magnitude of about 0.11 V). Does this change increase or decrease the $\mathcal{E}^{\circ\prime}$ of the electron donor and why is this change necessary?

2. The highly versatile prokaryotic cell can incorporate ammonia into amino acids using two different mechanisms, depending on the concentration of available ammonia.
   (a) One method involves coupling glutamine synthetase, glutamate synthase, and transamination reactions. Write the overall balanced equation for this process.
   (b) A second method involves coupling the freely reversible glutamate dehydrogenase reaction and a transamination reaction. Write the overall balanced equation for this process.
   (c) Which process is used when the concentration of available ammonia is low? When the concentration of ammonia is high? Why is the prokaryotic cell at a disadvantage when the concentration of ammonia is low? Explain. (Hint: The K_M of glutamine synthetase for ammonia is lower than the K_M of glutamate dehydrogenase for the ammonium ion.)

3. Plants whose root nodules contain nitrogen-fixing bacterial symbionts synthesize a heme-containing protein, called leghemoglobin, that resembles myoglobin in its structure. What is the function of this protein in the root nodules?

4. Which three mammalian enzymes can potentially "mop up" excess NH_4^+?

5. Describe how the reversible glutamate dehydrogenase reaction (a) contributes to amino acid biosynthesis and (b) functions as an anaplerotic reaction for the citric acid cycle.

6. Draw the products of the following transamination reactions:
   (a) Glycine + α-ketoglutarate → glutamate + ________
   (b) Arginine + α-ketoglutarate → glutamate + ________
   (c) Serine + α-ketoglutarate → glutamate + ________
   (d) Phenylalanine + α-ketoglutarate → glutamate + ________

7. Draw the products of the following transamination reactions. What do all of the products have in common?
   (a) Aspartate + α-ketoglutarate → glutamate + ________
   (b) Alanine + α-ketoglutarate → glutamate + ________
   (c) Glutamate + oxaloacetate → aspartate + ________

8. Name the amino acids that serve as reactants in the following transamination reactions.

(a) $COO^- - C{=}O - CH_2 - CH_2 - COO^-$ (α-Ketoglutarate) + [ ] $\xrightarrow{\text{Glu}}$ $COO^- - C{=}O - CH_2 - CH(-CH_3) - CH_3$

(b) $COO^- - C{=}O - CH_2 - CH_2 - COO^-$ (α-Ketoglutarate) + [ ] $\xrightarrow{\text{Glu}}$ $COO^- - C{=}O - CH_2 - CH_2 - S - CH_3$

(c) $COO^- - C{=}O - CH_2 - CH_2 - COO^-$ (α-Ketoglutarate) + [ ] $\xrightarrow{\text{Glu}}$ $COO^- - C{=}O - CH_2 - C_6H_4 - OH$

(d) $COO^- - C{=}O - CH_2 - CH_2 - COO^-$ (α-Ketoglutarate) + [ ] $\xrightarrow{\text{Glu}}$ $COO^- - C{=}O - CH(-CH_3) - CH_3$

9. Serine hydroxymethyltransferase catalyzes the conversion of threonine to glycine in a PLP-dependent reaction. The mechanism is slightly different from that shown for transamination in Figure 15-5. The degradation of threonine to glycine begins with a threonine $C_\alpha{-}C_\beta$ bond cleavage. Draw the structure of the threonine–Schiff base intermediate that forms in this reaction and show how $C_\alpha{-}C_\beta$ bond cleavage occurs.

10. Why is cysteine sometimes listed as an essential amino acid?

11. Sulfonamides (sulfa drugs) act as antibiotics by inhibiting the synthesis of folate in bacteria.

$H_2N - C_6H_4 - S(=O)_2 - NH - R$

A sulfonamide

(a) Which portion of the folate molecule does the sulfonamide resemble?
(b) Why do sulfonamides kill bacteria without harming their mammalian host?

12. Purine nucleotide synthesis is a highly regulated process. The main objective is to provide the cell with approximately equal concentrations of ATP and GTP for DNA synthesis. The purine nucleotide synthesis pathway is outlined in the figure below.

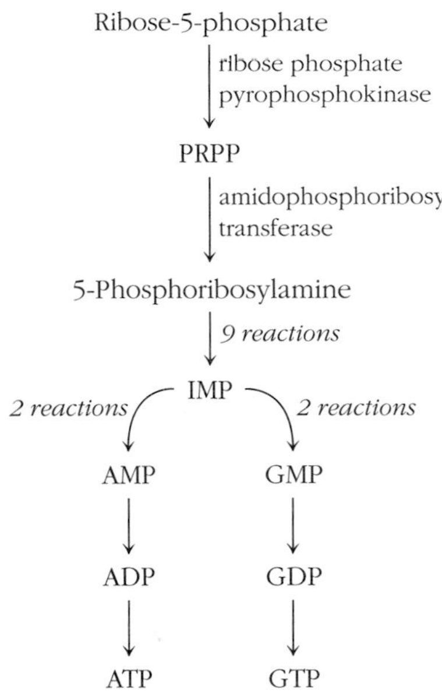

(a) How do ADP and GDP affect ribose phosphate pyrophosphokinase?
(b) Amidophosphoribosyl transferase catalyzes the committed step of the IMP synthetic pathway. How might PRPP, AMP, ADP, ATP, GMP, GDP, and GTP affect the activity of this enzyme? Explain your reasoning.

13. Hypoxanthine phosphoribosyl transferase (HPRT) catalyzes the following reaction:

Hypoxanthine + 5-phosphoribosyl pyrophosphate → inosine monophosphate + PP_i

The concentration of hypoxanthine in human cells is relatively high, about 8 μM. Some intracellular protozoan parasites have high levels of HPRT, and inhibitors of this enzyme are being studied for their effectiveness in blocking parasite growth. What would be the metabolic effect of inhibiting the parasite's HPRT, and what does this tell you about the parasite's metabolic capabilities?

14. Lesch–Nyhan syndrome is a disease caused by a severe deficiency in HPRT activity. In the absence of HPRT, hypoxanthine and guanine are not converted to IMP and GMP, respectively, in the salvage reactions. The disease is characterized by the accumulation of excessive amounts of uric acid, a product of nucleotide degradation, which causes neurological abnormalities and destructive behavior, including self-mutilation. Explain why the absence of HPRT causes uric acid to accumulate.

15. Although amino acids are classified as glucogenic, ketogenic, or both, it is possible for all their carbon skeletons to be broken down to acetyl-CoA. Explain.

16. Vigorous exercise is known to break down muscle proteins. What is the probable metabolic fate of the resulting free amino acids?

17. In mammals, metabolic fuels can be stored: glucose as glycogen and fatty acids as triacylglycerols. What type of molecule could be considered as a sort of storage depot for amino acids? How does it differ from other fuel-storage molecules?

18. Isoleucine is degraded to acetyl-CoA and propionyl-CoA by a pathway in which the first steps are identical to those of valine degradation (Fig. 15-11) and the last steps are identical to those of fatty acid oxidation.

(a) Draw the intermediates of isoleucine degradation and indicate the enzyme that catalyzes each step.
(b) Which reaction in the degradation scheme is analogous to the reaction catalyzed by pyruvate dehydrogenase?
(c) What reaction is analogous to the reaction catalyzed by acyl-CoA dehydrogenase in fatty acid oxidation?

19. What is the fate of the propionyl-CoA that is produced upon degradation of isoleucine?

20. Phenylketonuria (PKU) is an inherited disease that results from the lack of the enzyme phenylalanine hydroxylase (PAH). PAH catalyzes the first step in the degradation of phenylalanine. Individuals with phenylketonuria cannot break down phenylalanine, which accumulates in the blood and is eventually transaminated to phenylpyruvate, a phenylketone compound. The accumulation of phenylpyruvate causes irreversible brain damage if the disease is not treated.

(a) Draw the structure of phenylpyruvate, the product of phenylalanine transamination.
(b) Why do children with a deficiency of tetrahydrobiopterin excrete large quantities of this compound?
(c) Individuals diagnosed with PKU are placed on a low-phenylalanine diet. Why would a phenylalanine-free diet be undesirable?
(d) Why should patients with PKU avoid the artificial sweetener Aspartame?
(e) Patients with PKU tend to have blue eyes, fair hair, and very light skin. Explain why.
(f) Explain why individuals on a low-phenylalanine diet may need to increase their tyrosine intake.

21. A person whose diet is poor in just one of the essential amino acids may enter a state of negative nitrogen balance, in which nitrogen excretion is greater than nitrogen intake. Explain why this occurs, even when the supply of other amino acids is high.

22. Identify the source of the two nitrogen atoms in urea.

23. A complete deficiency of a urea cycle enzyme usually causes death soon after birth, but a partial deficiency may be tolerated.

(a) Explain why hyperammonemia (high levels of ammonia in the blood) accompanies a urea cycle enzyme deficiency.
(b) What dietary adjustments might minimize the possibility of ammonia toxicity?
(c) Compounds that covalently bind glycine and glutamine and are then excreted in the feces are sometimes used to treat hyperammonemia. One of these compounds is the drug Ucephan, which consists of a mixture of sodium salts of phenylacetate and benzoate. What is the biochemical rationale for this treatment?

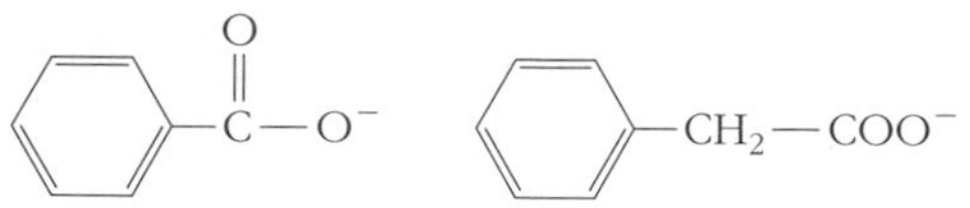

Benzoate Phenylacetate

(d) If there is a deficiency of argininosuccinase, what compound could be added to the diet to boost urea production?

24. Incorporation of ammonium ions into urea for excretion removes the toxic ions from the bloodstream. Elevated blood ammonium ion concentrations result in coma and eventual death. Increased NH_4^+ concentrations may shift the equilibrium of the glutamate dehydrogenase reaction. How does this explain the toxicity of ammonium ions?

25. Production of the enzymes that catalyze the reactions of the urea cycle can increase or decrease according to the metabolic needs of the organism. High levels of these enzymes are associated with high-protein diets as well as starvation. Explain this paradox.

26. The C8 ketone of uric acid undergoes tautomerization to produce an enol form that can undergo ionization with a p*K* of 5.4 (this is why the compound is called uric *acid*). Draw the structure of the conjugate base, urate.

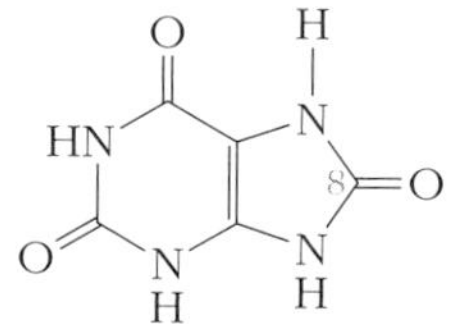

Uric acid (keto tautomer)

27. Gastric ulcers result from infection by *Helicobacter pylori.* To survive in the extreme acidity of the stomach, the bacteria express high levels of the enzyme urease. The concentration of urea in gastric juice is approximately 1–5 mM.

(a) Why is urease activity essential for *H. pylori* survival?
(b) Why is it important for at least some urease to be associated with the bacterial cell surface?

28. Because glutamine is the amino group donor in many biosynthetic reactions, as well as being a storage form of ammonia, control of the glutamine synthetase reaction is vitally important. Bacterial glutamine synthetase has an extremely elaborate control system, which includes both allosteric and covalent control. The bacterial enzyme consists of 12 identical subunits arranged at the corners of a hexagonal prism (see Fig. 15-4). Its activity is regulated by covalent modification by adenylylation (attachment of an AMP group) as shown in the figure. The enzyme has 12 adenylylation sites, one on each subunit. Adenylylated glutamine synthetase is less enzymatically active; the deadenylylated enzyme is more active. The level of adenylylation is controlled by the activity of a uridylyltransferase (UTase), an enzyme that attaches a UMP group on a regulatory protein termed P_{II}. A uridylyl removing enzyme (UR) removes the UMP group from P_{II}. The P_{II} protein is associated with the adenylyltransferase (ATase), which catalyzes both adenylylation and deadenylylation of the glutamine synthetase enzyme. When P_{II} is uridylylated, the ATase enzyme catalyzes deadenylylation of glutamine synthetase; when the UMP group is removed from P_{II}, adenylylation of glutamine synthetase takes place.

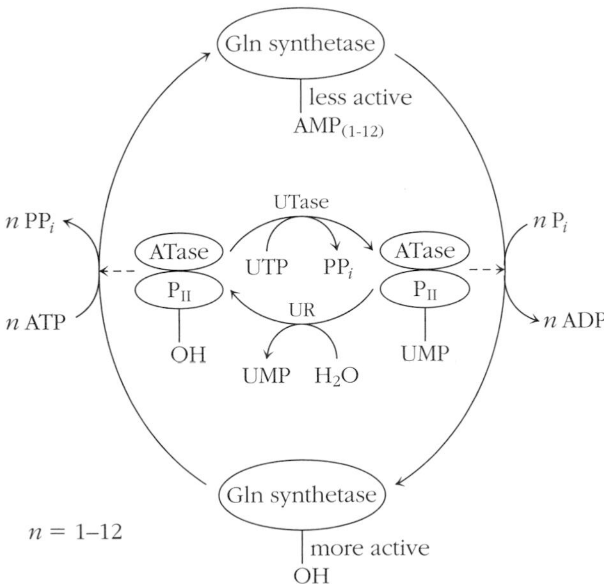

(a) The activity of the UTase enzyme is itself regulated, since its activity determines the amount of uridylylation of P_{II}, and consequently, the activity of the ATase. The activity of UTase is affected by cellular concentrations of α-ketoglutarate, ATP, glutamine, and inorganic phosphate. The activity of UR is affected by cellular concentrations of glutamine. Classify each of these substances as inhibitors or activators of UTase and UR.

(b) Increased levels of adenylylation of glutamine synthetase render the enzyme more susceptible to allosteric inhibition by the following metabolites: histidine, tryptophan, carbamoyl phosphate, glucosamine-6-phosphate, AMP, CTP, NAD^+, alanine, serine, and glycine. Why do these substances inhibit glutamine synthetase?

(c) When Earl Stadtman and his colleagues first purified glutamine synthetase from *E. coli,* they isolated the enzyme from cells grown in media containing glycerol as the only carbon source and glutamate as the only nitrogen source. The enzyme prepared from these cells was sensitive to the allosteric inhibitors listed above. A second batch of enzyme was prepared from *E. coli* cells grown in a medium containing glucose and ammonium chloride. The glutamine synthetase enzyme purified from these cells was insensitive to the inhibitors listed above. Explain these observations.

[From Stadtman, E.R., *J. Biol. Chem.* **276,** 44357–44363 (2001).]

29. Threonine deaminase catalyzes the committed step of the biosynthetic pathway leading to the branched-chain amino acid isoleucine. The enzyme catalyzes the dehydration and deamination of threonine to α-ketobutyrate.

$^+H_3N-CH-COO^-$ | CH—OH | CH_3

Threonine

threonine deaminase ↓ H_2O, NH_3

COO^- | C=O | CH_2 | CH_3

α-Ketobutyrate

4 reactions ↓

$^+H_3N-CH-COO^-$ | CH—CH_3 | CH_2 | CH_3

Isoleucine

COO^- | C=O | CH_3

Pyruvate

4 reactions ↓

$^+H_3N-CH-COO^-$ | CH—CH_3 | CH_3

Valine

Threonine deaminase is a tetramer and has a low affinity, or "T," form and a high affinity, or "R," form. The enzyme has been shown to be allosterically regulated. The activity of threonine deaminase was measured at increasing concentrations of its substrate, threonine (Curve A next page). Measurements were also made in the presence of isoleucine (Curve B) and valine (Curve C). Velocity was measured in terms of the μmol of α-ketobutyrate produced per mg of enzyme protein per minute. The kinetic constants obtained from the graphs are shown in the table.

(a) What information can you determine about threonine deaminase from these data?

	No modulators present (Thr only)	**Isoleucine (50 mM) added**	**Valine (0.5 mM) added**
V_{max} ($\mu mol \cdot mg^{-1} \cdot min^{-1}$)	214	180	225
K_M (mM)	8.0	74	5.7

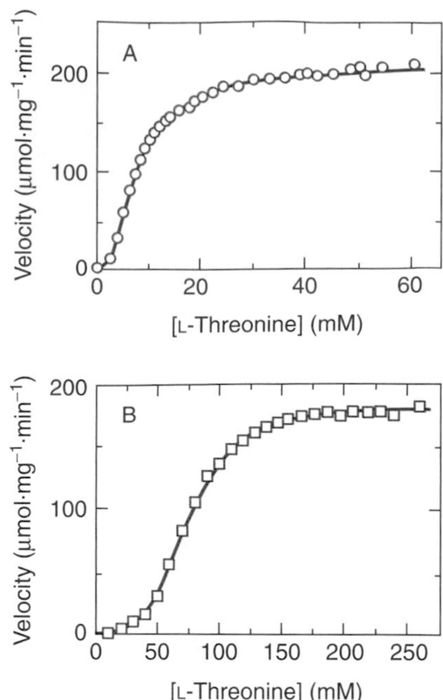

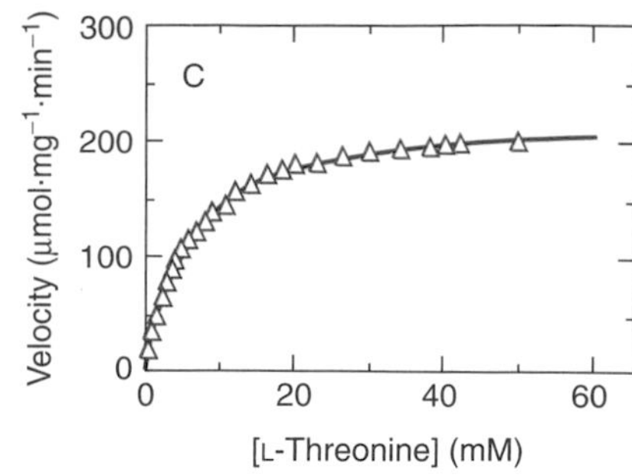

(b) What is the effect of isoleucine on threonine deaminase activity? To which form of the enzyme does isoleucine bind?

(c) What is the effect of valine (a product of a parallel pathway) on threonine deaminase activity? To which form of the enzyme does valine bind?

[From Eisenstein, E., *J. Biol. Chem.* **266,** 5801–5807 (1991), and Eisenstein, E., Yu, H.D., and Schwartz, F.P., *J. Biol. Chem.* **269,** 29423–29429 (1994).]

SELECTED READINGS

Brosnan, J.T., Glutamate, at the interface between amino acid and carbohydrate metabolism, *J. Nutr.,* **130,** 988S–990S (2000). [Summarizes the metabolic roles of glutamate as a fuel and as a participant in deamination and transamination reactions.]

Miles, E.W., Rhee, S., and Davies, D.R., The molecular basis of substrate channeling, *J. Biol. Chem.* **274,** 12193–12196 (1999). [Reviews the mechanisms of channeling inside proteins and on the protein surface in several enzymes involved in amino acid and nucleotide metabolism.]

Reichard, P., The evolution of ribonucleotide reduction, *Trends Biochem. Sci.* **22,** 81–85 (1997). [Speculates on the evolutionary divergence of ribonucleotide reductases, which all use a free-radical mechanism.]

Richardson, D.J. and Watmough, N.J., Inorganic nitrogen metabolism in bacteria, *Curr. Opin. Chem. Biol.* **3,** 207–219 (1999). [Describes the structural and mechanistic features of some of the enzymes involved in bacterial nitrogen fixation and other steps of the nitrogen cycle.]

Withers, P.C., Urea: Diverse functions of a "waste product," *Clin. Exp. Pharm. Physiol.* **25,** 722–727 (1998). [Discusses the various roles of urea, contrasting it with ammonia with respect to toxicity, acid–base balance, and nitrogen transport.]

CHAPTER 16

Regulation of Mammalian Fuel Metabolism

Impure drinking water may contain *Vibrio cholerae* cells (shown at right), which are responsible for the often-fatal disease cholera. The bacteria colonize the small intestine, but they do not directly damage the host's tissues. Instead, the bacteria produce a protein toxin that interferes with the normal functioning of intestinal cells. The cholera toxin enters the host cells, where it is activated by reduction of a disulfide bond. The active toxin is an enzyme that catalyzes the transfer of an ADP-ribose group from NAD^+ to a cell-signaling component known as a G protein. Once it is ADP-ribosylated, the G protein cannot be turned off. As a result, the intestinal cells secrete unusually large amounts of Na^+ ions. As the ions are pumped out of the cells, water follows. The ensuing diarrhea—up to several liters per hour—can cause death by dehydration. Many cases of cholera can be cured by oral rehydration therapy, which gives the immune system time to combat the underlying bacterial infection.

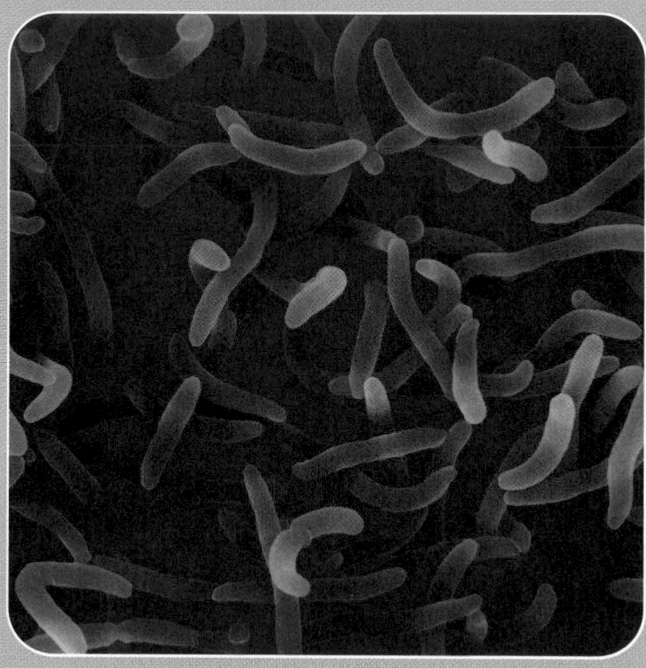

[Dennis Kunkel/Phototake.]

LEARNING OBJECTIVES

1. Understand that the metabolism of different organs is linked.
2. Understand the functions of the main components of signal transduction pathways.
3. Understand how insulin affects carbohydrate and fatty acid metabolism.
4. Understand how glucagon and epinephrine act in opposition to insulin.
5. Understand the metabolic defects in diabetes.

THIS CHAPTER IN CONTEXT

Chapter 16 completes Part 2 of this book by summarizing some of the metabolic relationships among different tissues and describing how hormones regulate the allocation and use of metabolic fuels. The pathways by which insulin and other hormones exert their effects are presented in order to illustrate some of the general features of receptors, G proteins, kinases, second messengers, and other components of signal transduction pathways. The chapter ends with a brief discussion of diabetes, a disease of carbohydrate and lipid fuel metabolism.

1 WHY ARE REGULATORY SYSTEMS NEEDED?

In examining the various pathways for the catabolism and anabolism of the major metabolic fuels and building blocks in mammals—carbohydrates, fatty acids, amino acids, and nucleotides—we have seen that biosynthetic and degradative pathways differ for thermodynamic reasons. These pathways are also regulated so that their simultaneous operation does not waste resources. One form of regulation is the compartmentation of opposing processes. For example, fatty acid oxidation takes place in mitochondria whereas fatty acid synthesis takes place in the cytosol. The locations of the major metabolic pathways are shown in Figure 16-1.

Compartmentation also takes the form of organ specialization: *Different tissues have different roles in energy storage and use.* For example, the liver carries out most metabolic processes as well as liver-specific functions such as gluconeogenesis, ketogenesis, and urea production. Adipose tissue is specialized to store about 95% of the body's triacylglycerols. Some tissues, such as the brain and red blood cells, do not store glycogen or fat but rely primarily on glucose supplied by the liver. The various organs are linked via

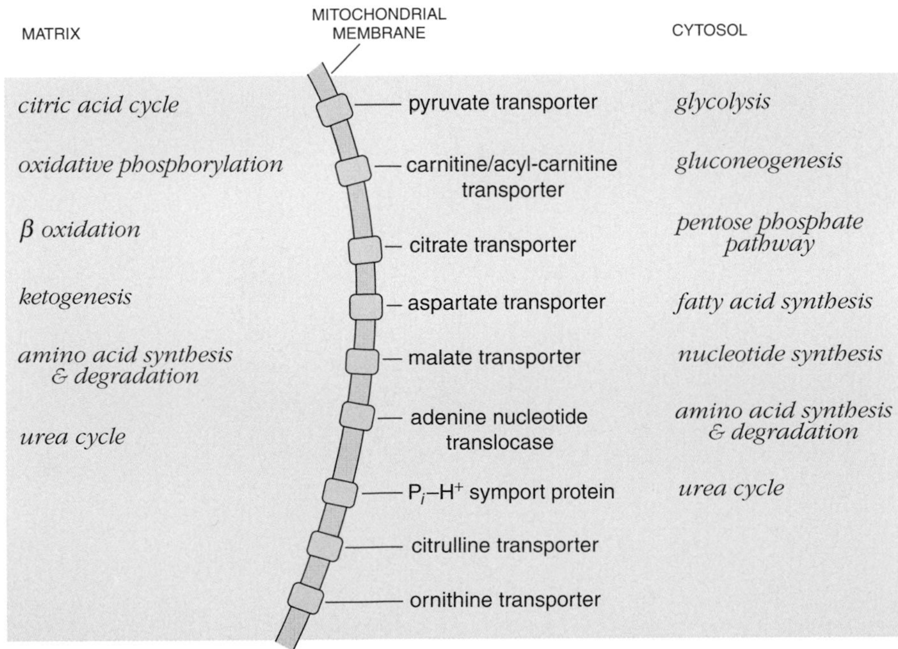

FIGURE 16-1

Cellular locations of major metabolic pathways.
In mammalian cells, most metabolic reactions occur in either the cytosol or the mitochondrial matrix. The urea cycle requires enzymes located in the matrix and cytosol. Amino acid degradation also occurs in both compartments. Other reactions, not pictured here, occur in the peroxisome, endoplasmic reticulum, Golgi apparatus, and lysosome. The diagram includes some transport proteins that transfer substrates and products between the mitochondria and cytosol.

the bloodstream, which transports certain metabolites between tissues. In fact, at least two metabolic pathways require interorgan transport.

Interorgan pathways: the Cori cycle and the glucose–alanine cycle

The **Cori cycle** (named after Carl and Gerty Cori, who first described it) is a metabolic pathway involving the muscles and liver. During periods of high activity, muscle glycogen is broken down to glucose, which is catabolized via glycolysis to produce the ATP required for muscle contraction. The rapid catabolism of muscle glucose, which exceeds the ability of the mitochondria to reoxidize the reduced cofactors, generates lactate as an end product. This three-carbon molecule is excreted from the muscle cells and travels via the bloodstream to the liver, where it serves as a substrate for gluconeogenesis. The resulting glucose can then return to the muscle cells to sustain ATP production even after the muscle glycogen has been depleted (Fig. 16-2). The free energy to drive gluconeogenesis in the liver is derived from ATP produced by the oxidation of fatty acids. In effect, *the Cori cycle transfers free energy from the liver to the muscles.*

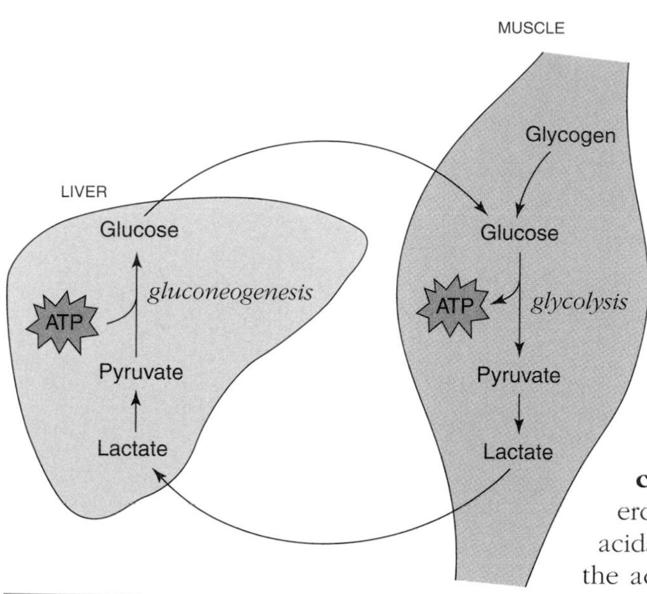

FIGURE 16-2

The Cori cycle. The product of muscle glycolysis is lactate, which travels to the liver. Lactate dehydrogenase converts the lactate back to pyruvate, which can then be used to synthesize glucose by gluconeogenesis. The input of free energy in the liver (in the form of ATP) is recovered when glucose returns to the muscles to be catabolized.

A second interorgan pathway, the **glucose–alanine cycle,** also links the muscles and liver. During vigorous exercise, muscle proteins break down, and the resulting amino acids undergo transamination to generate intermediates to boost the activity of the citric acid cycle (see Section 11-3). Transamination reactions convert pyruvate, a product of glycolysis, to alanine, which travels by the blood to the liver. There, the amino group is used for urea synthesis (see Section 15-6) and the resulting pyruvate is converted back to glucose by the reactions of gluconeogenesis. As in the Cori cycle, the glucose travels back to the muscle cells to complete the metabolic loop (Fig. 16-3). *The net effect of the glucose–alanine cycle is to transport nitrogen from muscles to the liver.* Both the Cori cycle and the glucose–alanine cycle redistribute metabolic fuels, but the allocation of resources at the level of the whole body is more complicated.

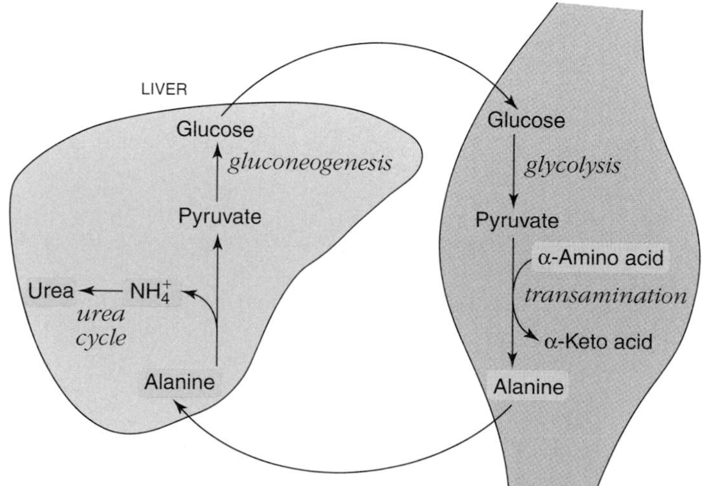

FIGURE 16-3

The glucose–alanine cycle. The pyruvate produced by muscle glycolysis undergoes transamination to alanine, which delivers amino groups to the liver. The carbon skeleton of alanine is converted back to glucose to be used by the muscles, and the nitrogen is disposed of as urea.

REVIEW LEARNING OBJECTIVE 1

- Describe the individual steps and the net effect of the Cori cycle and the glucose–alanine cycle.

Hormones coordinate metabolic functions

Individual cells or organs must regulate the activities of their respective pathways according to their metabolic needs and the availability of fuel and building materials. In mammals, these are supplied intermittently, as meals alternate with periods of fasting. The body buffers itself against fluctuations in the fuel supply by storing carbohydrates as glycogen, fatty acids as triacylglycerols, and amino acids as proteins. The stored metabolic fuels can be mobilized as needed and replenished after the next meal.

Go to **Exercise 22/Signal Transduction** to explore some of the events that occur in hormone signaling pathways. The material in this exercise and the questions based on it address Learning Objective 2.

Coordination among the organs during the course of a day is accomplished by the action of **hormones,** which are substances produced by one tissue that affect the functions of other tissues throughout the body. The most important hormones involved in fuel metabolism are insulin, glucagon, and the catecholamines (epinephrine and norepinephrine). The ability of a cell to respond to these extracellular signals depends on a set of proteins that recognize the hormonal signal, transmit the signal to the cell interior, and mediate changes in enzyme activity and gene expression. This process is known as **signal transduction.**

Signal transduction pathways vary enormously in their molecular details, but all share certain elements (Fig. 16-4).

1. *A cell surface* ***receptor*** *binds an extracellular molecule (called a* ***ligand****) that does not penetrate through the cell membrane.* The ligand may be a growth factor or another peptide hormone, a neurotransmitter, or an odorant molecule. The interaction between a receptor and its ligand is highly specific, with binding involving any combination of hydrophobic or electrostatic interactions.
2. *The receptor is a transmembrane protein communicating with both the cell exterior and interior.* Ligand binding alters the receptor's conformation, thereby transmitting the signal across the membrane.
3. *The conformational change may alter the receptor's interactions with other proteins or it may stimulate the receptor's own catalytic*

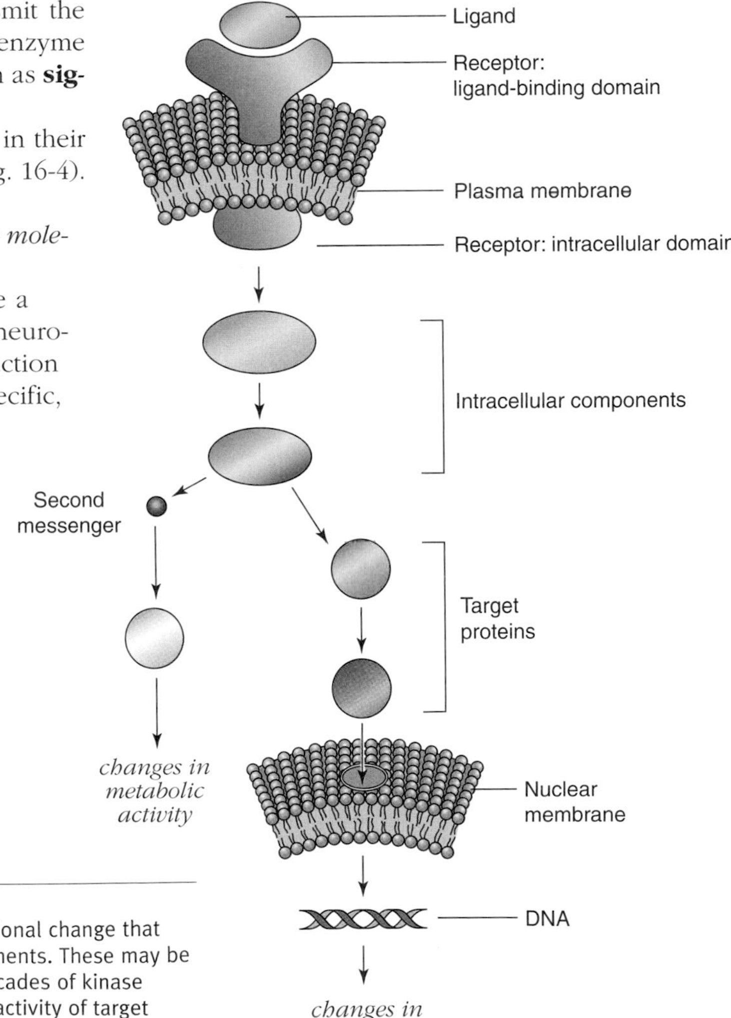

FIGURE 16-4

Elements of a signal transduction pathway.
Ligand binding to a cell surface receptor triggers a conformational change that alters the interaction of the receptor with intracellular components. These may be kinases or enzymes that give rise to second messengers. Cascades of kinase activity as well as the action of second messengers alter the activity of target proteins, which may lead to changes in metabolism or gene expression. Individual signal transduction pathways vary in the number and nature of their components.

activity. For example, some receptors are enzymes, and some are ion channels that open or close in response to ligand binding.

4. *Ligand binding to a receptor initiates a series of events that occur in a cascading fashion as one component of the signaling pathway activates others*. These steps usually serve to amplify the initial signal. Many signaling pathways involve a series of **protein kinases,** enzymes that transfer a phosphoryl group from ATP to a substrate polypeptide. These kinases are of two types: Tyrosine kinases transfer a phosphoryl group to the OH group of a Tyr side chain. Ser/Thr kinases transfer the group to a Ser or Thr side chain.

$$—CH_2—C_6H_4—O—PO_3^{2-} \qquad —CH_2—O—PO_3^{2-} \qquad —CH(CH_3)—O—PO_3^{2-}$$

Phospho-Tyr Phospho-Ser Phospho-Thr

5. *A signaling pathway may also include enzymes that generate an intracellular signal called a* ***second messenger*** (to distinguish it from the first, or extracellular messenger). Second messengers include nucleotide and lipid derivatives and calcium ions, all of which can diffuse some distance within the cytosol. The movement of second messengers or other components of the signaling pathway may be necessary for an extracellular signal to elicit changes in a metabolic pathway whose enzymes are located far from the cell surface or inside an organelle. If the signaling pathway alters gene expression, the signal must travel to the nucleus, where it can be recognized by the nuclear import machinery.

6. *Signaling pathways may have a built-in "off" switch*. The same signal that initiates an intracellular response may also activate a mechanism for shutting down that response. For example, the activation of a protein kinase often triggers the activation of a **phosphatase,** an enzyme that removes the phosphoryl group and thereby restores signaling components to their resting state. This mechanism tends to limit both the extent and duration of a hormone's effects.  The general features of signal transduction pathways are summarized in Exercise 22.

REVIEW LEARNING OBJECTIVE 2

- How can hormones regulate different processes in different tissues?
- How does extracellular ligand binding to a receptor trigger intracellular events?
- Describe the reaction catalyzed by kinases.
- How does a signal transduction pathway permit amplification of a hormonal signal?

2 THE ACTION OF INSULIN

Let us take a closer look at how hormones regulate fuel metabolism in mammals. Most tissues in the body use glucose as their preferred fuel and turn to fatty acids when the glucose supply dwindles. Except in organs such

TABLE 16-1 Source of metabolic fuels under different conditions

	Carbohydrates (%)	Fatty acids (%)	Amino acids (%)
Immediately after a meal	50	33	17
After an overnight fast	12	70	18
After a 40-day fast	0	95*	5

*This value reflects a high concentration of fatty acid–derived ketone bodies.

as the intestine, amino acids are not the primary fuel. Ketone bodies do not play a role until other fuels have been depleted, many hours after the last meal. These patterns of fuel use are summed up in Table 16-1.

During a 40-day fast, the concentrations of circulating fatty acids vary about 15-fold, and the concentration of ketone bodies increases about 100-fold. In contrast, the concentration of glucose in the blood varies by no more than threefold. A constant glucose supply is critical for the brain, which exerts a large and relatively constant demand for glucose, regardless of how the dietary intake of carbohydrates varies or how much carbohydrate is oxidized to support physical activities.

How does the body control glucose levels so closely? The hormone insulin plays a large role. Insulin stimulates cells to take up glucose and suppresses glycogen breakdown when dietary glucose is abundant. Insulin also suppresses fatty acid release from adipose tissue and stimulates triacylglycerol synthesis. In short, *insulin signals fuel abundance: It decreases the metabolism of stored fuel while promoting fuel storage* (Table 16-2). A lack of insulin or an inability to respond to it results in the disease **diabetes mellitus.**

TABLE 16-2 Summary of insulin action

Target tissue	Metabolic effect
Muscle and other tissues	Promotes glucose transport into cells Stimulates glycogen synthesis Suppresses glycogen breakdown
Adipocytes	Activates extracellular lipoprotein lipase Increases level of acetyl-CoA carboxylase Stimulates triacylglycerol synthesis

Pancreatic β cells release insulin

Immediately following a meal, blood glucose concentrations may rise to about 8 mM, from a normal concentration of about 3.6 to 5.8 mM. The increase in circulating glucose triggers the release of the hormone insulin, a 51-amino acid polypeptide (Fig. 16-5). Insulin is synthesized in the β cells of pancreatic islets, which are small clumps of cells that produce hormones rather than digestive enzymes (Fig. 16-6). The hormone is named after the Latin *insula,* island.

The mechanism that triggers the release of insulin from the β cells is not well understood. The pancreatic cells do not express a glucose receptor on their surface, as might be expected. Instead, *the cellular metabolism of glucose itself seems to generate the signal to release insulin.*

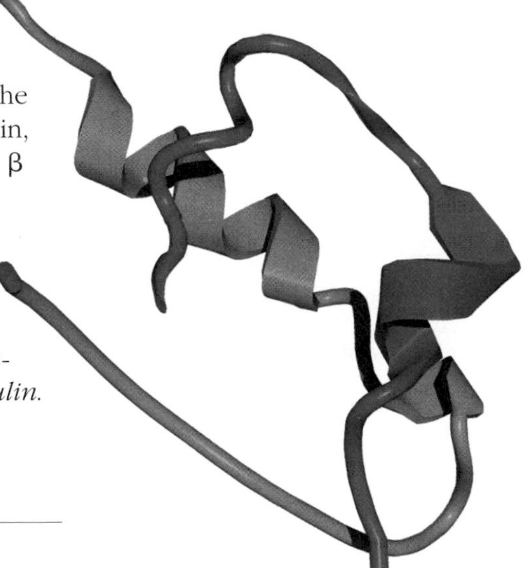

FIGURE 16-5

Structure of human insulin.
[*Structure (pdb 1AIo) determined by X. Chang, A.M.M. Jorgensen, P. Bardrum, and J.J. Led.*]

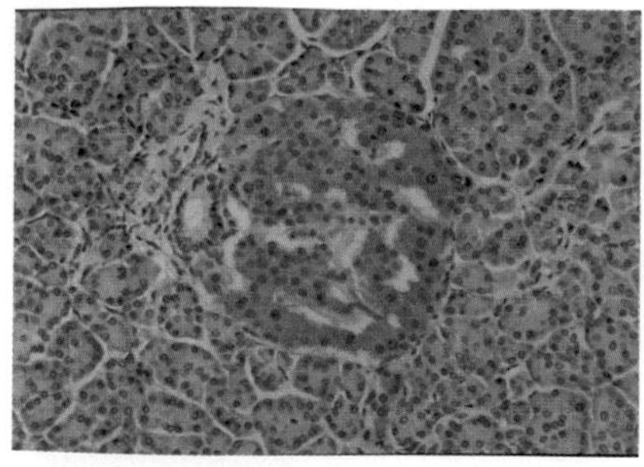

FIGURE 16-6

Pancreatic islet cells.
The pancreatic islets of Langerhans (named for their discoverer) consist of two types of cells. The β cells produce the hormone insulin, and the α cells produce glucagon. Most other pancreatic cells produce digestive enzymes. [*Carolina Biological Supply Co./Phototake.*]

In liver and pancreatic β cells, the glycolytic degradation of glucose begins with a reaction catalyzed by glucokinase (an isozyme of hexokinase; see Section 10-2):

$$\text{Glucose} + \text{ATP} \rightarrow \text{glucose-6-phosphate} + \text{ADP}$$

The hexokinases in other cell types have a relatively low K_M for glucose (less than 0.1 mM), which means that the enzymes are saturated with substrate at physiological glucose concentrations. Glucokinase, in contrast, has a high K_M of 5–10 mM, *so it is never saturated and its activity is maximally sensitive to the concentration of available glucose* (Fig. 16-7).

Interestingly, the velocity versus substrate curve for glucokinase is not hyperbolic, as might be expected for a monomeric enzyme such as glucokinase. Instead, the curve is sigmoidal, which is typical of allosteric enzymes with multiple active sites operating cooperatively (see Section 7-2). The sigmoidal kinetics of glucokinase, which has only one active site, may be due to a substrate-induced conformational change such that at the end of the catalytic cycle, the enzyme briefly maintains a high affinity for the next glucose molecule. At high glucose concentrations, this would mean a high reaction velocity; at low glucose concentrations, the enzyme would operate more slowly because it reverts to a low-affinity conformation before binding another glucose substrate.

The role of glucokinase as a pancreatic glucose sensor is supported by the fact that mutations in the glucokinase gene cause a rare form of diabetes. However, other cellular factors may be involved, particularly in the mitochondria of the β cells. The glucose sensor responsible for triggering insulin release may also depend on the mitochondrial NAD^+/NADH or ADP/ATP ratios. For this reason, age-related declines in mitochondrial function may be a factor in the development of diabetes in the elderly.

Insulin binds to a specific receptor

Once released into the bloodstream, insulin can bind to its receptors on the surface of cells in muscle and other tissues. The insulin receptor is a protein with an $(\alpha\beta)_2$ quaternary structure. The two large α subunits (135 kD each) are extracellular and each β subunit (95 kD) includes a transmembrane domain as well as extra- and intracellular domains. The three-dimensional structure

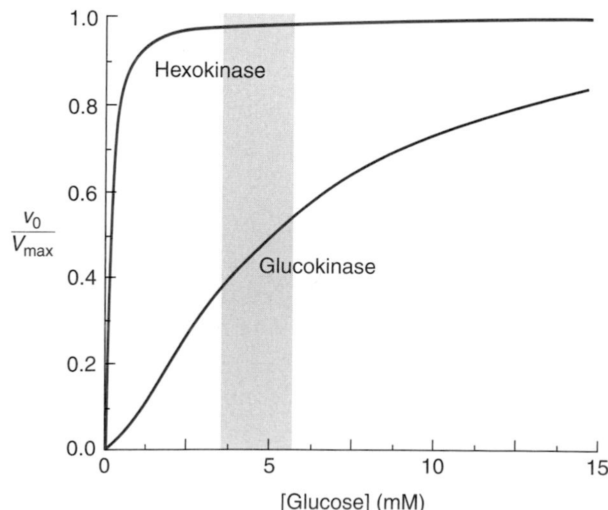

FIGURE 16-7

Activities of glucokinase and hexokinase.
Both enzymes catalyze the ATP-dependent phosphorylation of glucose as the first step of glycolysis. Glucokinase has a high K_M, so its reaction velocity changes in response to changes in glucose concentrations. In contrast, hexokinase is saturated with glucose at physiological concentrations (shaded region).

of the insulin receptor is known from electron microscopy (Fig. 16-8) and from detailed X-ray crystallographic analysis of some of its domains. Although the symmetric insulin receptor has two potential insulin-binding sites, the binding of one insulin molecule prevents binding of the second (an example of negative cooperativity among binding sites). Binding depends mostly on electrostatic interactions between insulin and the α subunits.

In the absence of the hormone, the two intracellular domains of the β subunits are separated. Insulin binding triggers a conformational change in the receptor that brings the domains closer together. Each β subunit intracellular domain is a tyrosine kinase that phosphorylates its partner in the receptor. This mutual phosphorylation is known as **autophosphorylation,** because the receptor appears to phosphorylate itself. The tyrosine kinase also phosphorylates one or more tyrosine residues in other proteins, including one known as IRS-1 (insulin receptor substrate-1). IRS-1 then triggers additional events in the cell, not all of which have been fully characterized.

Many of the receptors for polypeptide growth factors are also tyrosine kinases. Unlike the insulin receptor, most of these proteins are monomeric until their ligand binds. Binding induces dimerization (sometimes oligomerization) of the receptor. When brought together in this way, the intracellular tyrosine kinase domains become active and phosphorylate each other and other substrates. Additional components of the signal transduction pathways contain a conserved region known as an SH2 domain that includes an Arg side chain to interact with the phospho-Tyr groups of the receptor and other proteins. *Such interactions allow a single hormone-binding event to alter the activities of several intracellular proteins.* In at least one case, a portion of the receptor itself is cleaved off by a protease and travels to the nucleus, where it modulates gene transcription.

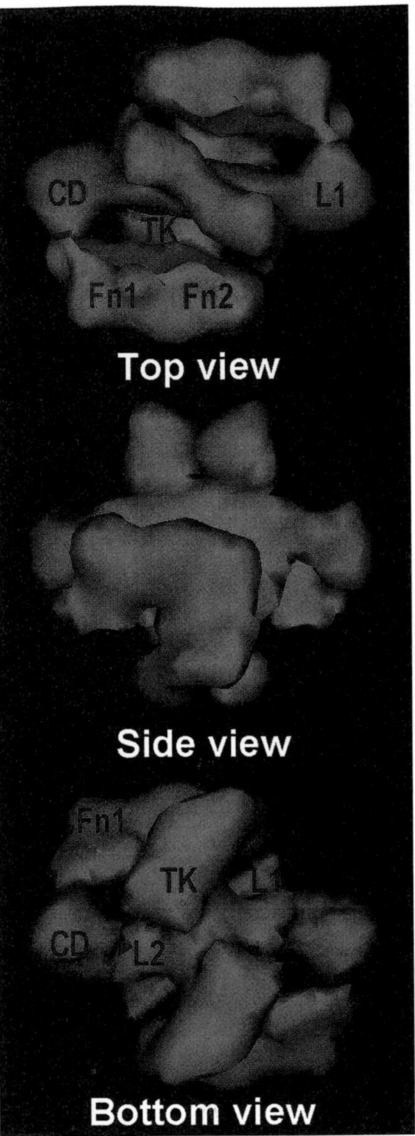

FIGURE 16-8

The insulin receptor.
These images are reconstructed from electron microscopic data. (a) Top (extracellular) view. (b) Side view (from within the membrane). (c) Bottom (intracellular) view. [*Courtesy Cecil Yip and Peter Ottensmeyer, University of Toronto. Adapted from Luo, et al.,* Science ***285****, 1077–80 (1999).*]

Insulin alters metabolic processes

Only cells that bear insulin receptors can respond to the hormone, and the response varies by cell type. In insulin-responsive tissues such as muscles and adipose tissue, insulin stimulates glucose transport into cells by severalfold. The V_{max} for glucose transport increases, not because insulin alters the intrinsic catalytic activity of the transporter, but because insulin increases the number of transporters at the cell surface. These transporters, named GLUT4 to distinguish them from other glucose-transport proteins, are situated in the membranes of intracellular vesicles. When insulin binds to the cell, the vesicles fuse with the plasma membrane. *This translocation of transporters to the cell surface increases the rate at which glucose enters the cell* (Fig. 16-9). GLUT4 is a passive transporter, operating similarly to the erythrocyte glucose trans-

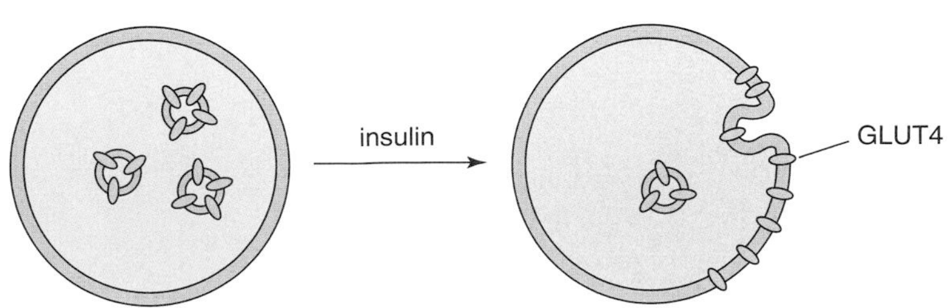

FIGURE 16-9

Effect of insulin on GLUT4.
Insulin triggers vesicle fusion so that the glucose transport protein GLUT4 is translocated from intracellular vesicles to the plasma membrane. This increases the rate at which the cells take up glucose.

porter (see Fig. 8-23). When the insulin stimulus is removed, endocytosis returns the transporters to intracellular vesicles.

Insulin stimulates fatty acid uptake as well as glucose uptake. When the hormone binds to its receptors in adipose tissue, it activates the extracellular protein lipoprotein lipase, which helps remove fatty acids from circulating lipoproteins so that they can be taken up for storage by adipocytes.

Insulin binding to its cell surface receptor also modulates glycogen-metabolizing enzymes. Glycogen metabolism is characterized by a balance between glycogen synthesis and glycogen degradation. Synthesis is carried out by the enzyme glycogen synthase, which adds glucose units donated by UDP–glucose to the ends of the branches of a glycogen polymer (see Section 10-3):

$$\text{UDP–glucose} + \text{glycogen}_{(n\ \text{residues})} \rightarrow \text{UDP} + \text{glycogen}_{(n+1\ \text{residues})}$$

Glycogen phosphorylase mobilizes glucose residues from glycogen by phosphorolysis (cleavage through addition of a phosphoryl group rather than water):

$$\text{Glycogen}_{(n\ \text{residues})} + \text{P}_i \rightarrow \text{glycogen}_{(n-1\ \text{residues})} + \text{glucose-1-phosphate}$$

This reaction, followed by an isomerization reaction, yields glucose-6-phosphate, the first intermediate of glycolysis.

Glycogen synthase is a homodimer, and glycogen phosphorylase is a heterodimer. Both enzymes are regulated by allosteric effectors. For example, glycogen synthase is activated by glucose-6-phosphate. AMP activates glycogen phosphorylase and ATP inhibits it—effects that are consistent with the role of glycogen phosphorylase in making glucose available to boost cellular ATP production. However, *the primary mechanism for regulating glycogen synthase and glycogen phosphorylase is covalent modification (phosphorylation and dephosphorylation) that is under hormonal control.* Both enzymes undergo reversible phosphorylation at specific Ser residues. Phosphorylation deactivates glycogen synthase and activates glycogen phosphorylase. Removal of the phosphoryl groups has the opposite effect: Dephosphorylation activates glycogen synthase and deactivates glycogen phosphorylase (Fig. 16-10).

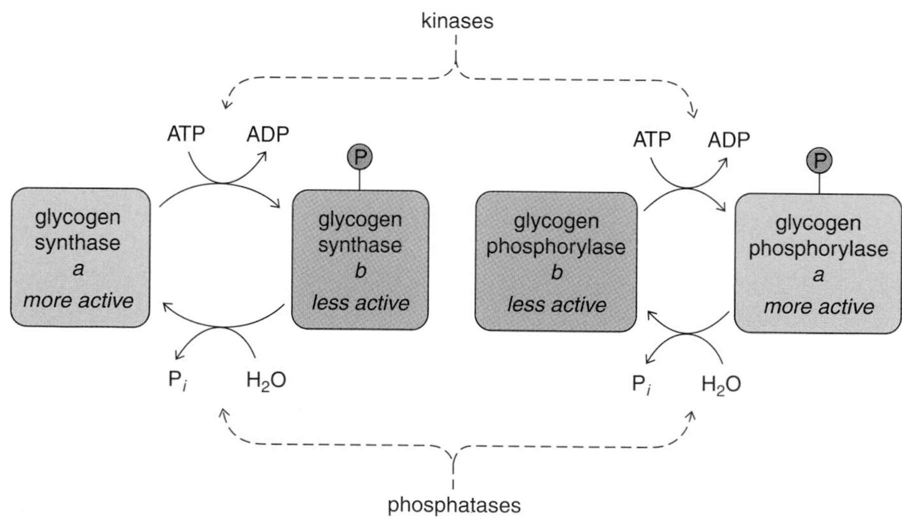

FIGURE 16-10

The reciprocal regulation of glycogen synthase and glycogen phosphorylase. Phosphorylation (transfer of a phosphoryl group from ATP) deactivates glycogen synthase and activates glycogen phosphorylase. Dephosphorylation has the opposite effect. The more active form of each enzyme is known as the *a* form (indicated in green), and the less active form is known as the *b* form (in red).

Phosphorylation/dephosphorylation is a type of allosteric regulation

Glycogen synthase and glycogen phosphorylase are examples of enzymes whose active sites are influenced by an allosteric event, in this case, a covalent modification (phosphorylation or dephosphorylation) elsewhere on the protein. The attachment or removal of the highly anionic phosphoryl group triggers a conformational shift between a more active (*a* or R) state and a less active (*b* or T) state. *The reciprocal regulation of glycogen synthase and glycogen phosphorylase promotes metabolic efficiency, since the two enzymes catalyze key reactions in opposing metabolic pathways.* The advantage of this regulatory system is that a single kinase can tip the balance between glycogen synthesis and degradation. Similarly, a single phosphatase can tip the balance in the opposite direction. Covalent modifications, such as phosphorylation and dephosphorylation, permit a much wider range of enzyme activities than could be accomplished solely through the allosteric effects of metabolites whose cellular concentrations do not vary much.

Insulin participates in the regulation of glycogen synthase and glycogen phosphorylase indirectly. The insulin receptor tyrosine kinase activates other proteins, including phosphatases that dephosphorylate (activate) glycogen synthase and dephosphorylate (deactivate) glycogen phosphorylase. As a result, glycogen synthesis accelerates and the rate of glycogenolysis decreases. In fact, multiple kinases and phosphatases interact with the enzymes of glycogen metabolism—and with each other—to provide an exquisitely fine-tuned regulatory network ultimately controlled by signals introduced by extracellular hormones.

REVIEW LEARNING OBJECTIVE 3

- Summarize the metabolic effects of insulin signaling on muscle cells and adipocytes.
- How does glucokinase differ from hexokinase?
- How does insulin trigger tyrosine phosphorylation?
- How does insulin increase the rate of glucose entry into cells?
- Explain how phosphorylation and dephosphorylation reciprocally regulate glycogen synthase and glycogen phosphorylase.

3 THE ACTION OF GLUCAGON

Within hours of a meal, dietary glucose has been taken up by cells and consumed as fuel, stored as glycogen, or converted to fatty acids for long-term storage. At this point, the liver must begin mobilizing glucose in order to keep the blood glucose concentration constant. This phase of fuel metabolism is governed not by insulin but by other hormones, mainly glucagon and the catecholamines epinephrine and norepinephrine.

Glucagon, a 29-residue peptide hormone, is synthesized and released by the α cells of pancreatic islets when the blood glucose concentration begins to drop below about 5 mM (Fig. 16-11). In contrast to insulin, glucagon stimulates the liver to release glucose produced by glycogenolysis and gluconeogenesis, and it stimulates adipose tissue to undergo **lipolysis** to

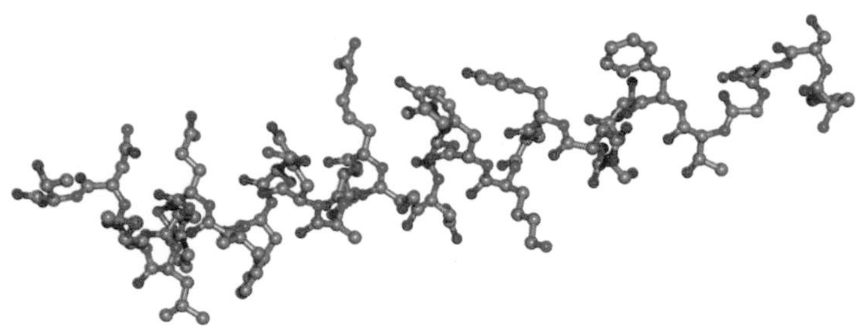

FIGURE 16-11

Structure of glucagon.
The atoms of the 29-residue peptide are colored by type. [*Structure (pdb 1GCN) determined by T.L. Blundell, K. Sasaki, S. Dockerill, and I.J. Tickle.*]

FIGURE 16-12

Structure of rhodopsin.
This protein, with seven transmembrane helices, is a major component of the membrane of rod cells of the eye. Rhodopsin contains a retinal prosthetic group (red) that triggers a series of conformational changes when it absorbs light energy. Hormone receptors have the same overall protein structure and undergo conformational changes in response to hormone binding. [*Structure of bovine rhodopsin (1HZX) determined by D.C. Teller, T. Okada, C.A. Behnke, K. Palczewski, and R.E. Stenkamp.*]

release fatty acids to the circulation. Muscle cells do not express a glucagon receptor and therefore do not respond to the hormone.

Although their signal transduction pathways differ somewhat, catecholamines elicit the same overall effects as glucagon. For example, epinephrine stimulation of muscle cells activates glycogenolysis, which makes more glucose available to power muscle contraction. Recall that the catecholamines are tyrosine derivatives (see page 475) that are synthesized by the central nervous system as neurotransmitters and by the adrenal glands as hormones. Catecholamines bind to cell surface proteins known as adrenergic receptors.

The glucagon receptor and adrenergic receptors such as the epinephrine-binding β-adrenergic receptor are transmembrane proteins containing seven membrane-spanning α helices. These proteins are members of a large family of structurally similar proteins, exemplified by the light receptor rhodopsin (Fig. 16-12). *The glucagon and catecholamine receptors are not tyrosine kinases, but hormone binding elicits a conformational change that affects the intracellular portion of the protein.* Such receptors are thought to act like allosteric proteins that alternate between two conformations, switching in response to ligand binding.

G proteins are intracellular mediators

Inside the cell, a trimeric protein known as a **G protein** interacts with the intracellular portion of the receptor. G proteins are named for their ability to bind guanine nucleotides (either GDP or GTP). The three subunits of the G protein are designated α, β, and γ (Fig. 16-13). In the resting state, the G

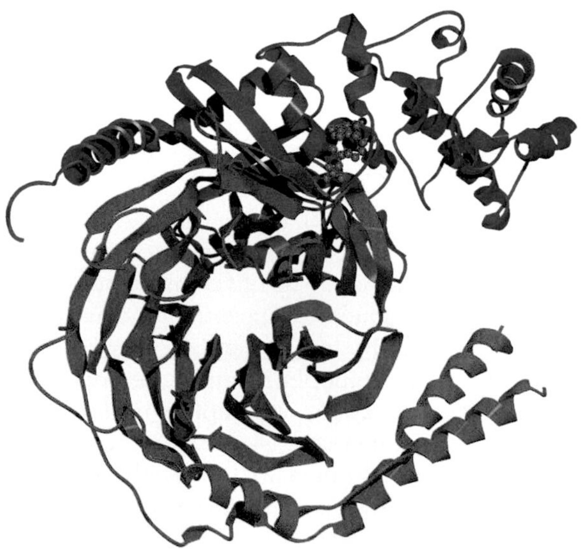

FIGURE 16-13

A G protein.
The β subunit (green) has a propeller-like structure. The small γ subunit (yellow) associates tightly with the β subunit and has an attached prenyl group, which may help anchor the G protein to the inside surface of the plasma membrane near a receptor. The α subunit (blue), where the guanine nucleotide (orange) binds, may also bear a fatty acyl membrane anchor. [*Structure (pdb 1GP2) determined by M.A. Wall and S.R. Sprang.*]

protein exists as a trimer with GDP bound to the α subunit. Ligand binding to the associated receptor triggers a conformational change in the G protein, which is probably mediated by a C-terminal extension of the α subunit. As a result of the conformational change, the α subunit releases its bound GDP and binds GTP in its place. The third phosphate group of GTP is not easily accommodated in the αβγ trimer, so the α subunit dissociates from the β and γ subunits, which remain tightly associated as a dimer. *Once dissociated, the α_{GTP} subunit and the βγ dimer are both active; that is, they interact with additional cellular components in the signal transduction pathway.*

The signaling activity of the G protein is limited by the intrinsic GTPase activity of the α subunit, which converts the bound trinucleotide to GDP:

$$\text{GTP} + \text{H}_2\text{O} \rightarrow \text{GDP} + \text{P}_i$$

Hydrolysis of the GTP allows the three G protein subunits to reassociate as an inactive trimer ($\alpha_{GDP}\beta\gamma$; Fig. 16-14).

The GDP nucleotide bound to the trimeric G protein dissociates with a basal rate of about $10^{-5}\ s^{-1}$, which means that physiologically relevant activation of the G protein (the release of GDP and binding of GTP) would take hours. *In vivo,* the exchange of GDP for GTP occurs in less than one second due to the assistance of proteins known as GEFs (guanine nucleotide exchange factors). Similarly, the GTPase catalytic rate is often enhanced by accessory proteins known as GAPs (GTPase activating proteins), which boost the intrinsic GTPase activity of about 2–4 min^{-1} to about 10^2–$10^3\ min^{-1}$.

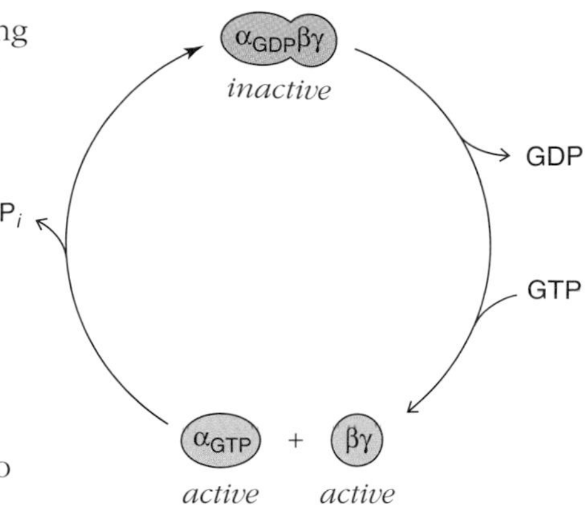

FIGURE 16-14

The G protein cycle. The αβγ trimer, with GDP bound to the α subunit, is inactive. Ligand binding to a receptor associated with the G protein triggers a conformational change that causes GTP to replace GDP and the α subunit to dissociate from the βγ dimer. Both portions of the G protein are active in the signaling pathway. The GTPase activity of the α subunit returns the G protein to its inactive trimeric state.

The adenylyl cyclase pathway

The separated α_{GTP} and βγ units exert different effects on cellular proteins. For example, the α_{GTP} unit binds to and allosterically activates an enzyme called adenylyl cyclase. This heterodimeric enzyme is an integral membrane protein with 12 membrane-spanning α helices. The intracellular catalytic domains (Fig. 16-15) convert ATP to a molecule known as cyclic AMP (cAMP):

$$\text{ATP} \xrightarrow[\text{adenylyl cyclase}]{\text{PP}_i} \text{Cyclic AMP (cAMP)}$$

The adenylyl cyclase catalytic site includes two Asp side chains as well as metal ions (probably Mg^{2+}) that coordinate with the two Asp residues and the phosphate groups of ATP. Activation and inhibition of adenylyl cyclase are considered in Exercise 22.

The product cAMP is a second messenger, a small and highly soluble substance that can freely diffuse in the cell. cAMP activates a Ser/Thr kinase called protein kinase A. In the absence of cAMP, this kinase is an inactive tetramer

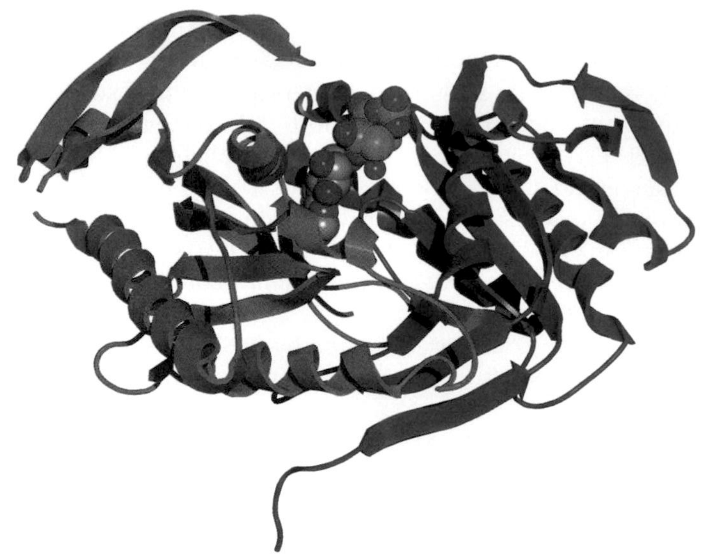

FIGURE 16-15

Structure of the adenylyl cyclase catalytic domains.
The two intracellular domains (blue and green) of the integral membrane protein form a cleft where the substrate ATP (space-filling model) binds. Mg^{2+} (purple) and Mn^{2+} (green) are also present in the active site in this model. [*Structure (pdb 1CJK) determined by J.J.G. Tesmer and S.R. Sprang.*]

of two regulatory (R) and two catalytic (C) subunits. cAMP binds to the regulatory subunits, causing the tetramer to release its two active catalytic subunits.

$$\underset{\textit{inactive}}{\text{R}_2\text{C}_2} \xrightarrow{\text{cAMP}} \text{R}_2 + \underset{\textit{active}}{\text{C}} + \underset{\textit{active}}{\text{C}}$$

Consequently, *the level of the second messenger cAMP determines the level of activity of protein kinase A.* The cAMP signal (and hence protein kinase A activity) is shut off by the action of cAMP phosphodiesterase, which converts cAMP to AMP:

$$\text{cAMP} \xrightarrow[\text{phosphodiesterase}]{H_2O} \text{AMP}$$

cAMP AMP

One of the intracellular targets of protein kinase A is phosphorylase kinase, the enzyme that phosphorylates (deactivates) glycogen synthase and phosphorylates (activates) glycogen phosphorylase. This series of signaling events explains how hormones such as glucagon and epinephrine, which lead to cAMP production, promote glycogenolysis and inhibit glycogen synthesis (Fig. 16-16). Although phosphorylase kinase is activated by protein kinase A, it is maximally active only when Ca^{2+} ions are present as the result of other signaling pathways (described below).

In adipocytes, protein kinase A phosphorylates an enzyme known as hormone-sensitive lipase, thereby activating it. This lipase catalyzes the rate-limiting step of lipolysis, the conversion of stored triacylglycerols to diacylglycerols and then to monoacylglycerols (see Section 9-1). *Hormone stimulation*

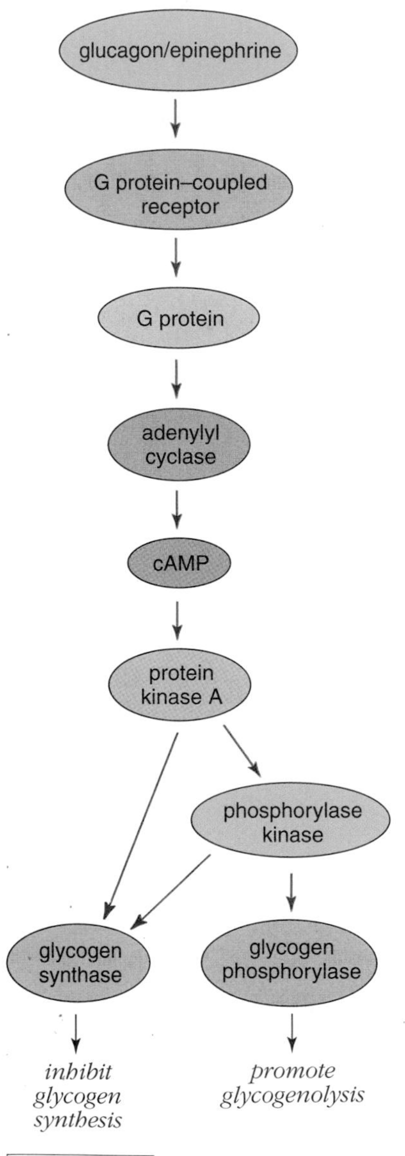

FIGURE 16-16

Effect of glucagon and epinephrine on glycogen metabolism.
All but one of the arrows linking the components of the signal transduction pathway represent activation events. Both glucagon and epinephrine inhibit glycogen synthesis and promote the mobilization of glucose from glycogen.

not only increases the lipase catalytic activity, it also relocates the lipase from the cytosol to the fat droplet of the adipocyte. Co-localization with its substrate, possibly through binding to a lipid-binding protein, boosts the rate at which fatty acids are mobilized (Fig. 16-17). Thus, glucagon and epinephrine promote the breakdown of both glycogen and fat.

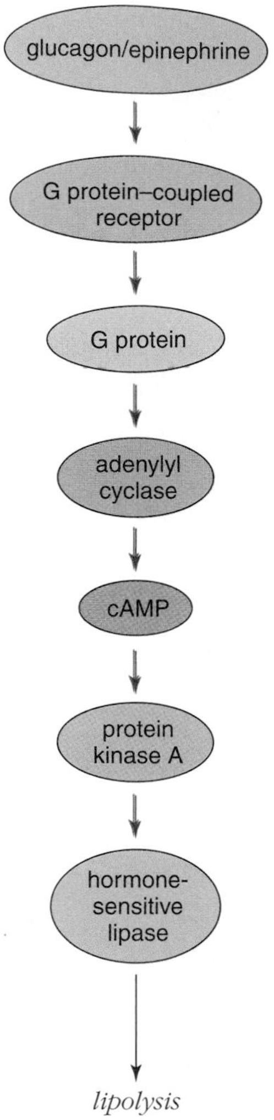

FIGURE 16-17

Effect of glucagon and epinephrine on fatty acid mobilization.
Hormone binding to receptors on adipocytes leads to increased lipolysis.

G proteins participate in other signaling pathways

G protein–coupled receptors, such as the glucagon and catecholamine receptors, represent the largest family of cell surface receptors involved in signal transduction. Some G protein–coupled receptors are activated by light (for example, the eye pigment rhodopsin; shown in Fig. 16-12) or by odorant molecule (Box 16-A). Over 1% of the human genome appears to code for G protein–coupled receptors, and these proteins are the targets of about half of all drugs currently in clinical use.

G proteins themselves also exhibit variety. For example, in addition to the G proteins that activate adenylyl cyclase, there are inhibitory G proteins that lead to a decrease in cellular cAMP levels by binding to different sites on the adenylyl cyclase protein. Other G proteins activate cAMP phosphodiesterase, with similar effects on cAMP-dependent processes. Consequently, *a cell's response to a hormone signal depends not only on whether the appropriate receptor is present, but also on whether the associated G protein is activating or inhibitory.* Because a single type of hormone may activate both types of G protein, the signaling system may be active for only a brief time before it is turned off.

The diversity of receptors, coupled with the diversity of G proteins, creates almost unlimited possibilities for producing second messengers and altering the activities of cellular enzymes. The targets of various α_{GTP} and $\beta\gamma$ components of G proteins include adenylyl cyclase, guanylyl cyclase (which produces the second messenger cyclic GMP), and an assortment of kinases and phospholipases. Some of the second messengers generated by these enzymes are lipids derived from membrane glycerophospholipids and sphingolipids.

For example, epinephrine binds to a receptor known as the α-adrenergic receptor, which is part of the **phosphoinositide signaling system.** The G protein associated with the α-adrenergic receptor activates the cellular enzyme phospholipase C, which converts the membrane lipid phosphatidylinositol bisphosphate to inositol trisphosphate and diacylglycerol.

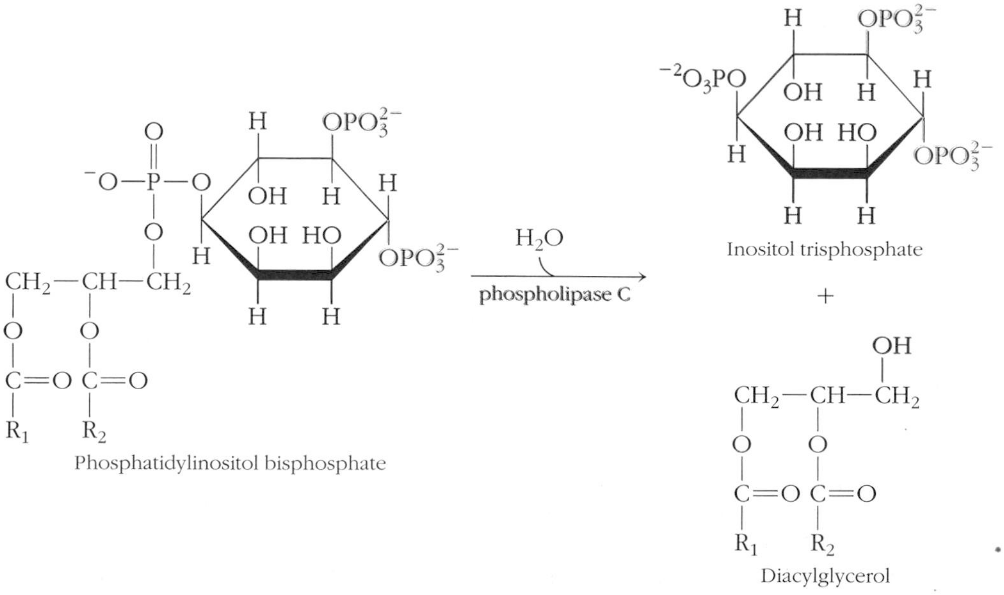

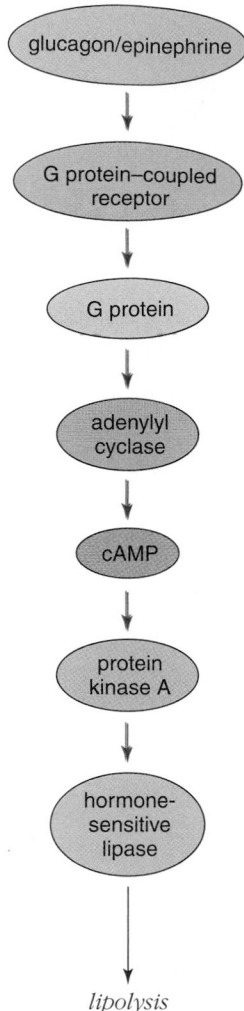

A CLOSER LOOK

BOX 16-A Olfaction

Humans are not celebrated for their superior sense of smell, but the human olfactory system can detect and discriminate among thousands of low-molecular-weight organic compounds. Complex organisms from flies to mammals actually have two olfactory systems. One, called the vomeronasal system, is geared toward recognizing species-specific olfactory signals (pheromones) primarily related to reproductive behavior (the importance of this system in human physiology is not well understood). The other olfactory system, which we will discuss here, is primarily responsible for the sense of smell. Like the immune system, it has evolved to contend with a large number of potential ligands (odorants) that the organism might encounter.

Olfactory sensory neurons line the nasal passages and communicate directly with the brain. The sensory region of each neuron terminates in small "cilia" that are covered with mucus. Odor molecules bind to cell surface receptors, which resemble other receptor proteins in that they contain seven membrane-spanning α helices and are coupled to G proteins. The olfactory receptors expressed by different cells are similar except for a region of hypervariability localized to a bundle of three transmembrane helices. This is the putative odorant-binding site (colored red in the model at right; the most conserved regions of the olfactory receptor are shown in blue).

Odorant binding to a receptor activates the associated G protein, which stimulates adenylyl cyclase to produce cAMP. The cAMP second messenger opens a cAMP-gated ion channel. The resulting influx of Na^+ and Ca^{2+} ions causes depolarization (the inside of the cell becomes less negative). If the stimulus persists, an action potential propagates along the length of the neuron and enters the brain (see Section 8-3 for a summary of neural cell function). Thus, binding of the odorant molecule is transduced to an electrical signal.

Odorant signaling also activates a negative feedback mechanism. The increase in cellular Ca^{2+} decreases the sensitivity of the ion channel to cAMP, so that a stronger odor stimulus is required to keep the channel open and trigger an action potential. This mechanism allows the olfactory neuron to adjust its sensitivity so that it can respond to a wide range of odorant concentrations.

The nose can also detect a wide variety of odor molecules, because most odorants are recognized by more than one type of receptor, and most receptors can accommodate more than one type of odorant. This cross-reactivity means that even a modest complement of different receptors (perhaps just a few hundred in humans) could work in concert to detect an almost infinite array of odors.

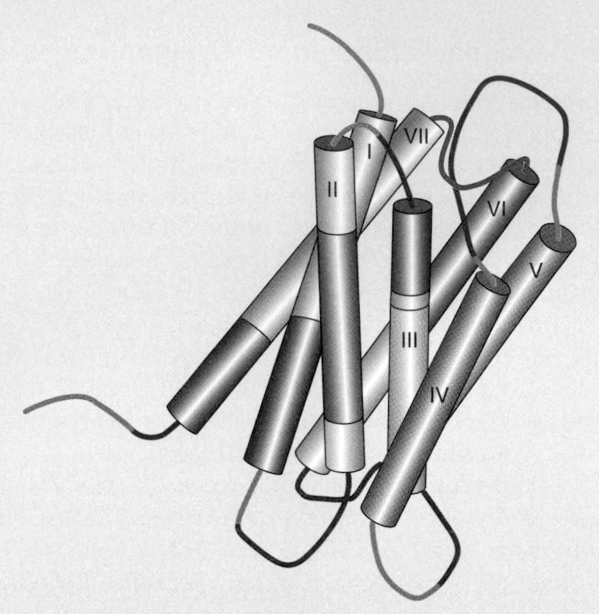

[Drawing of olfactory receptor based on an image by Stuart Firestein, Columbia University.]

The highly polar inositol trisphosphate second messenger activates calcium channel proteins, causing them to open and allow Ca^{2+} ions into the cytosol. This increase in Ca^{2+} leads to full activation of phosphorylase kinase. Calcium also activates a Ser/Thr kinase known as protein kinase B (Fig. 16-18).

The hydrophobic diacylglycerol product of the phospholipase C reaction is also a second messenger. Although it remains in the cell membrane, it can diffuse laterally to activate protein kinase C, which phosphorylates its target proteins at Ser or Thr residues.

The phosphoinositide signaling system depends closely on the activity of kinases and phosphatases, which can modify the phosphorylation state of the parent phosphatidylinositol or its inositol second messenger. In fact, *kinases and phosphatases form a close partnership in all known signal transduction pathways,* including the receptor tyrosine kinase system. Thus, when kinase-catalyzed phosphorylation alters the activity of a protein, phosphatase-catalyzed

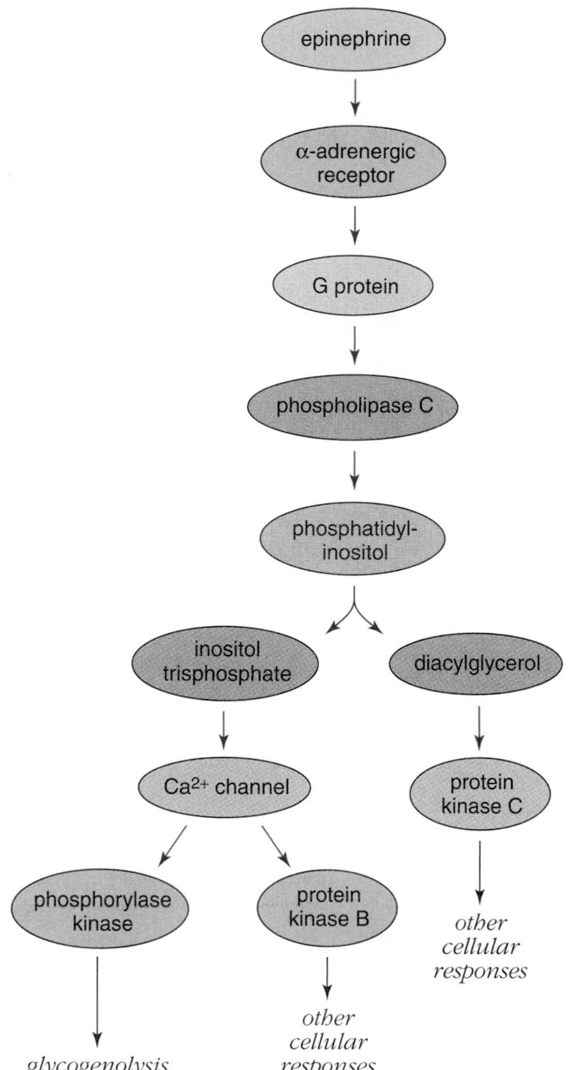

FIGURE 16-18
The phosphoinositide signaling pathway.

dephosphorylation restores it to its resting state. Both types of enzymes exhibit similar diversity and specificity with respect to their substrate proteins.

The various components of a signal transduction pathway—receptors, G proteins, kinases, phosphatases, and second-messenger-generating or -destroying enzymes—may be preassembled in a complex. As a result, the cellular response to an extracellular signal is not the result of random diffusion and activation of components scattered throughout the cell. Rather, *the response is rapid and localized.* Of course, other signaling events elsewhere in the cell may augment or cancel the effect of one signal transduction complex. In addition, signal transduction pathways often interconnect by sharing the same second messengers or other components. This communication between signaling pathways is known as **cross-talk.**

The ultimate targets of some signal transduction pathways are nuclear proteins known as transcription factors. These proteins help control gene expression by regulating the ability of the enzyme RNA polymerase to produce messenger RNA corresponding to particular genes. For example, many growth factors stimulate pathways that govern the activation of transcription factors. Disruptions in such pathways may lead to uncontrolled growth, as in cancer (discussed further in Chapter 18). In addition, many pathogenic bacteria, including the one that causes anthrax, interfere with their host's signaling pathways (Box 16-B).

A CLOSER LOOK

BOX 16-B Anthrax

Anthrax is a bacterial disease that is widespread among animals. It is rare in humans but potentially deadly, leading to its exploitation as an agent of biological warfare. The effectiveness of anthrax as a weapon was first confirmed through the accidental release of anthrax spores from a laboratory in the Soviet Union in 1979, when 68 people died.

Bacillus anthracis is a nonmotile aerobic bacterium that quickly dies outside of host tissues. However, it forms spores, particles about 1 μm in diameter, that can survive for decades.

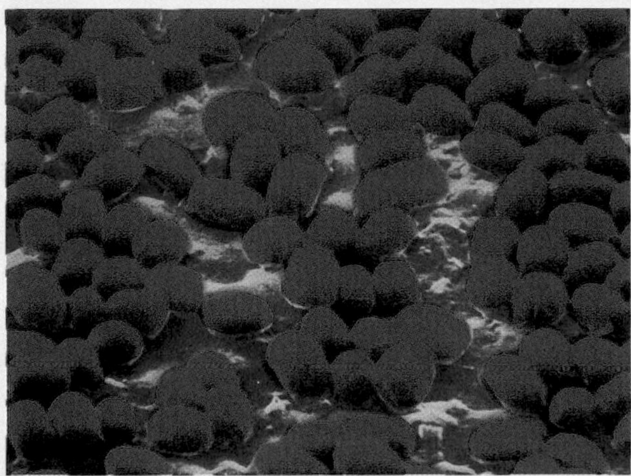

Such spores are naturally present in soils worldwide and are ingested by herbivores. The mechanisms whereby the spores germinate to form full-sized bacterial cells is not well understood. If released into the air, the odorless and invisible spores can travel long distances and easily find their way indoors. These spores can cause inhalation anthrax in humans, which is an extremely rare disease. A cluster of inhalation anthrax cases almost certainly indicates that the spores have been specifically targeted to humans, for example, in letters or packages. As recent events in the United States have shown, spores from one envelope can easily contaminate others.

Historically, anthrax infections in humans have been of the cutaneous variety and are easily recognized by the black lesions that result (cutaneous anthrax was understood to be an occupational hazard for woolsorters and others who worked with animal hides). Unfortunately, the early symptoms of inhalation anthrax are nonspecific and resemble the flu. The later stages of the disease are rapidly fatal, with an average interval of only 3 days between the onset of symptoms and death.

Like many deadly microbes, *B. anthracis* synthesizes a protein toxin. In this case, the toxin consists of three parts, known as protective antigen (PA), lethal factor (LF), and edema factor (EF). Although the toxin enters cells of many tissues, it is particularly lethal for macrophages, the immune system cells that engulf and destroy invading bacteria. Thus, the toxin prevents the immune system from destroying the bacteria.

The three components of the anthrax toxin operate in concert. Protective antigen, which is used in the preparation of vaccines (hence its name), binds to a host cell surface protein whose normal function is not known. After protective antigen has bound, a cell surface protease cleaves it, and its N-terminal fragment diffuses away. The remaining membrane-bound portion combines with six others to form a heptameric complex that binds the other two toxin proteins, lethal factor and edema factor. The resulting complex then undergoes endocytosis. Once it is enclosed in an intracellular vesicle, protective antigen forms a 14-stranded β-barrel pore that allows lethal factor and edema factor to enter the cytosol.

Lethal factor is a four-domain protease whose active site contains a Zn^{2+} ion. The substrates for lethal factor are protein kinases known as MAPKKs (mitogen-activated protein kinase kinases) that phosphorylate and thereby activate another kinase involved in a signaling pathway that regulates cell growth. Lethal factor is extremely specific for MAPKKs because it interacts with a 40-Å-long 16-residue segment of MAPKK. Cleavage of the kinase by lethal factor blocks its signaling activity.

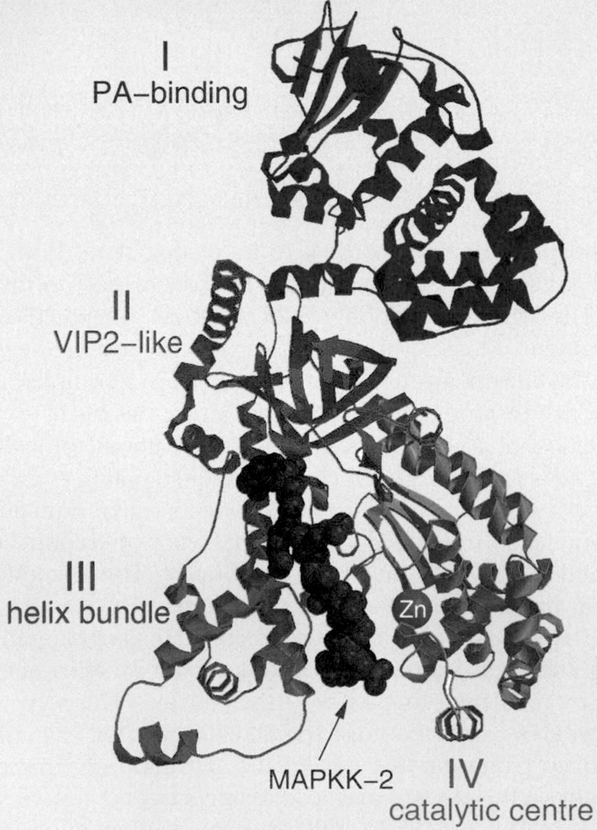

(continued)

BOX 16-B *(continued)*

In the ribbon model of lethal factor shown at left, the four domains of the protein are in different colors. A purple Zn^{2+} ion marks the protease active site. A 16-residue peptide corresponding to a MAPKK substrate is shown in orange.

Early in infection, lethal factor prevents macrophages from effectively combating the bacteria; later in infection, lethal factor kills the macrophages, which then release large quantities of inflammatory mediators (including nitric oxide; see Box 15-A). This presumably triggers the massive septic shock that causes death.

The third anthrax toxin component, edema factor, is an adenylyl cyclase that interferes with signal transduction, particularly in macrophages. The toxic effects of edema factor result from the increased concentrations of cAMP, which lead to edema (abnormal buildup of extracellular fluid), a hallmark of cutaneous anthrax infections. Edema factor is active only when it forms a complex with a host Ca^{2+}-binding protein known as calmodulin. This function of edema factor action may protect the bacteria themselves from the toxin, since bacterial cells lack calmodulin. By binding calmodulin, edema factor also enhances cAMP production indirectly, since calmodulin normally activates cAMP phosphodiesterase, the enzyme that breaks down cAMP.

Once the bacterial toxin has entered cells, it cannot be stopped, so early identification of anthrax infection—or even just exposure—is essential to prevent death. Most naturally occurring strains of *B. anthracis* are sensitive to penicillin, but it is feared that "weaponized" anthrax may have been engineered for resistance to common antibiotics. The drug of choice is therefore the newer, broad-spectrum antibiotic ciprofloxacin. Treatment for about 60 days is required to prevent infection by spores that have delayed germination. The good news is that anthrax—unlike smallpox and bubonic plague, for example—is not highly contagious.

[Photo of spores from NIBSC/Science Photo Library/Photo Researchers. Photo of lethal factor courtesy Robert C. Liddington, The Burnham Institute; from *Nature* **441,** 229–233 (2001).]

REVIEW LEARNING OBJECTIVE 4

- Summarize the metabolic effects of glucagon and epinephrine on liver cells and adipocytes.
- Describe the structure and function of G proteins.
- How is the second messenger cAMP generated and destroyed?
- Describe the phosphoinositide signaling system.

4 DIABETES, A DISORDER OF FUEL METABOLISM

The multifaceted regulation of mammalian fuel metabolism offers many opportunities for things to go wrong. One of the best known and most common disorders is diabetes mellitus, which affects as much as 5% of the population of the United States. The words *diabetes* (meaning "to run through") and *mellitus* ("honey") describe an obvious symptom of the disease. Diabetics excrete large amounts of urine containing high concentrations of glucose (the kidneys work to eliminate excess circulating glucose by excreting it in urine, a process that requires large amounts of water).

There are two main types of diabetes

Type I diabetes (juvenile onset or insulin-dependent diabetes) is an autoimmune disease in which the immune system destroys pancreatic β cells. Symptoms first appear in childhood as insulin production begins to drop off. At one time, the disease was invariably fatal. This changed dramatically in 1922, when Frederick Banting and Charles Best administered a pancreatic

FIGURE 16-19

Banting and Best.
Frederick Banting (right) and Charles Best (left) surgically removed the pancreas of dogs to induce diabetes. When preparations of the pancreatic tissue were administered to the animals, their symptoms improved. This work laid the foundation for treating human diabetes with pancreatic extracts, which contain insulin. [*Hulton Archive/Getty Images.*]

extract to save the life of a severely ill diabetic boy (Fig. 16-19). Since then, the treatment of type I diabetes with purified insulin has been refined, but one remaining challenge lies in tailoring the delivery of insulin to the body's needs over the course of a typical 24-hour cycle of eating and fasting. Gene therapy to treat diabetes is an elusive goal, because the insulin gene must be introduced into the body in such a way that the gene's expression is glucose-sensitive.

By far the most common form of diabetes, accounting for approximately 90% of all cases, is type II diabetes (also known as adult-onset or non-insulin-dependent diabetes). These cases are characterized by **insulin resistance,** which is the failure of the body to respond to normal or even elevated concentrations of the hormone. Only a small fraction of patients with type II diabetes bear genetic defects in the insulin receptor, as might be expected; in the majority of cases, the underlying cause is not known.

The metabolic effects of diabetes

The primary feature of untreated diabetes is chronic **hyperglycemia** (high levels of glucose in the blood). The loss of responsiveness of tissues to insulin means that cells fail to take up glucose. The body's metabolism responds as if no glucose were available, so liver gluconeogenesis increases, thereby further promoting hyperglycemia. Glucose circulating at high concentrations can participate in nonenzymatic glycosylation of proteins. This process is slow, but the modified proteins may gradually accumulate and damage tissues with low turnover rates, such as neurons and the lens of the eye.

Tissue damage also results from the metabolic effects of hyperglycemia. Since muscle and adipose tissue are unable to increase their uptake of glucose in response to insulin, glucose tends to enter other tissues. Inside these cells, aldose reductase catalyzes the conversion of glucose to sorbitol:

$$\underset{\text{Glucose}}{\begin{array}{c} \text{H}-\text{C}{=}\text{O} \\ | \\ \text{H}-\text{C}-\text{OH} \\ | \\ \text{HO}-\text{C}-\text{H} \\ | \\ \text{H}-\text{C}-\text{OH} \\ | \\ \text{H}-\text{C}-\text{OH} \\ | \\ \text{CH}_2\text{OH} \end{array}} \xrightarrow[\text{aldose reductase}]{\text{NADPH} + \text{H}^+ \;\to\; \text{NADP}^+} \underset{\text{Sorbitol}}{\begin{array}{c} \text{CH}_2\text{OH} \\ | \\ \text{H}-\text{C}-\text{OH} \\ | \\ \text{HO}-\text{C}-\text{H} \\ | \\ \text{H}-\text{C}-\text{OH} \\ | \\ \text{H}-\text{C}-\text{OH} \\ | \\ \text{CH}_2\text{OH} \end{array}}$$

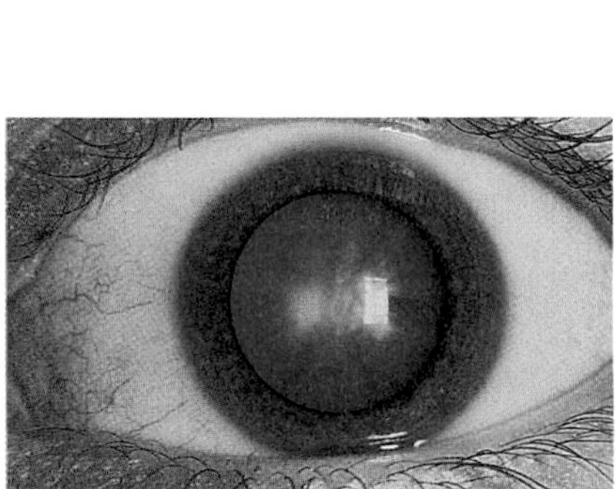

FIGURE 16-20

Photo of a diabetic cataract.
The accumulation of sorbitol in the lens leads to swelling and precipitation of lens proteins. The resulting opacification can cause blurred vision or complete loss of sight. [*Courtesy Dr. Manuel Datiles III, Cataract and Cornea Section, OGCSB, National Eye Institute, National Institutes of Health.*]

Because aldose reductase has a relatively high K_M for glucose (about 100 mM), flux through this reaction is normally very low. But under hyperglycemic conditions, sorbitol accumulates and may alter the cell's osmotic balance. This may alter kidney function and may trigger protein precipitation in other tissues. Aggregation of lens proteins leads to cataracts (Fig. 16-20). Neurons and cells lining blood vessels may be similarly damaged, increasing the likelihood of neuropathies and circulatory problems that in severe cases result in heart attack, stroke, or the amputation of extremities.

Although traditionally considered a disorder of glucose metabolism, *diabetes is also a disorder of fat metabolism,* since insulin normally stimulates triacylglycerol synthesis and suppresses lipolysis in adipocytes. Uncontrolled diabetics tend to metabolize fatty acids rather than carbohydrates, and the resulting production of ketone bodies may give the breath a sweet odor. Overproduction of ketone bodies leads to diabetic ketoacidosis.

In diabetes, the body behaves as if it were starving. Paradoxically, about 80% of patients with type II diabetes are obese, and obesity—particularly when abdominal fat deposits are large—is strongly correlated with the development of the disease. It was once thought that large stockpiles of triacylglycerols would increase the levels of circulating fatty acids, which would somehow interfere with the metabolism of the body's other major fuel, carbohydrates. According to this view, cells would be less likely to take up glucose to use as a fuel. In response, the pancreatic β cells would increase insulin production (accounting for the high levels of insulin in some type II diabetics) but would eventually fail.

However, it now appears that adipose tissue itself plays an active role in the development of type II diabetes. Adipocytes are not inert fat stores but actively synthesize peptide and steroid hormones that act on other organs, including the brain. For example, adipocytes secrete leptin, a hormone that helps regulate appetite and may contribute to obesity (Box 16-C).

Adipocytes also produce a 108-residue hormone called resistin, which may be the missing link between obesity and type II diabetes. Resistin appears to block the action of insulin, at least in other adipocytes, and levels of resistin increase in obesity. Certain drugs used to treat type II diabetes appear to decrease the amount of resistin secreted by adipocytes (in fact, studies of the drugs' action led to the discovery of the hormone). However, the role of resistin under normal (nondiabetic) conditions remains obscure.

REVIEW LEARNING OBJECTIVE 5

- Distinguish type I and type II diabetes.
- Explain why diabetes is a disorder of fat metabolism as well as carbohydrate metabolism.
- Describe some of the complications of diabetes.

SUMMARY

1. The specialized metabolic functions of mammalian organs require coordination. Pathways such as the Cori cycle and the glucose–alanine cycle link different organs, and hormones can regulate metabolic activities throughout the body.
2. A cell's response to a hormone involves a signal transduction pathway that may include a cell surface receptor that can transmit a signal to the cell interior, intracellular kinases and phosphatases, and second messengers.
3. Insulin, which is synthesized by the pancreas in response to glucose, binds to a receptor tyrosine kinase. Cellular responses to insulin include increased uptake of glucose and fatty acids.
4. The balance between glycogen synthesis and degradation depends on the relative activities of glycogen synthase and glycogen phosphorylase, which are controlled by hormone-triggered phosphorylation and dephosphorylation.
5. Glucagon and catecholamines bind to seven-helix transmembrane receptors that are associated with GTP-binding G proteins. Activation of a G protein triggers additional intracellular events such as the production of the second messenger cAMP by adenylyl cyclase.
6. Other G proteins have inhibitory effects or activate signal transduction pathways that lead to the formation of second messengers derived from phosphatidylinositol.
7. The most common form of diabetes is characterized by insulin resistance, the inability to respond to insulin. The resulting hyperglycemia can lead to tissue damage. Paradoxically, many individuals with insulin resistance are obese, suggesting that abnormal fat metabolism causes this form of diabetes.

A CLOSER LOOK

BOX 16-C Obesity

Obesity has become an enormous public health problem. In addition to its impact on the quality of life, it is physiologically costly: Masses of fat prevent the lungs from fully expanding, and the heart must work harder to circulate blood through a larger body. Obesity is frequently accompanied by diabetes and heart disease. And it is not an exaggeration to describe obesity as an epidemic, since it affects an estimated one-third of the adult population of the United States.

Like many conditions, obesity has no single cause. It is a complex disorder involving appetite and metabolism and reflecting environmental as well as genetic factors. There is an obvious link between overeating and the deposition of fat in adipose tissue, but several organs are involved, including those that store fuel (principally the liver and adipose tissue), those that primarily use fuel (skeletal muscle, for example), and the brain, which controls appetite. Regulation of this entire system is so complex that—as dieters can attest—a simple adjustment such as eating less food may not be sufficient to correct a tendency to gain weight.

In human beings, energy intake is highly irregular, with the quantity and type of food varying widely from one meal to the next and from one day to the next. Nevertheless, the body balances energy intake and energy expenditure over a period of decades. Thus, there appears to be a set-point for body weight that remains constant and relatively independent of energy intake and expenditure.

The polypeptide hormone leptin may be partly responsible. Leptin, which is synthesized by adipocytes, was first identified as the product of the gene lacking in genetically obese mice.

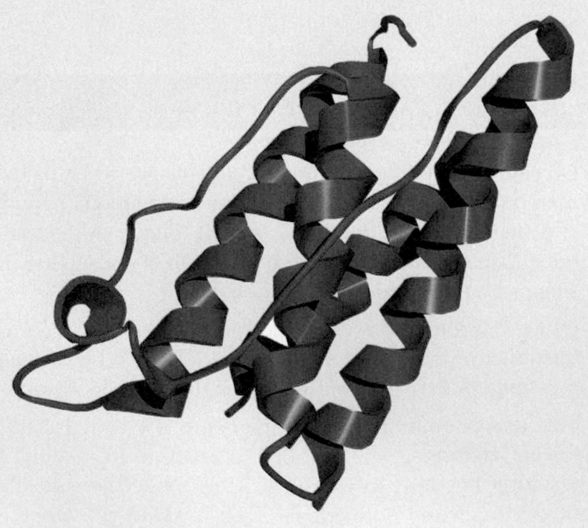

Leptin acts on the central nervous system as a "satiety" signal and suppresses the appetite. The absence of leptin in rodents causes severe obesity. However, in obese humans, the problem seems to be leptin resistance, which reflects a defect in the leptin-sensing mechanism. Consequently, leptin is less effective at suppressing the appetite and the individual gains weight. Eventually, the increase in leptin concentration resulting from the increase in adipose tissue mass succeeds in signaling satiety, but the result is a high set-point (a higher body weight that must be maintained). This may be one reason why overweight people who manage to shed a few pounds often regain the lost weight, returning to the original set-point. Decreased sensitivity to leptin may also account for the failure of supplemental leptin to cure obesity in humans.

Whatever the set-point, the body exhibits remarkable resilience in maintaining it. For example, during starvation, metabolic rates may decrease by as much as 40% (hence dieting may simply train the body to reduce its fuel expenditure, preventing any substantial weight loss). Feeding increases metabolic rates by as much as 25%. Still, the allocation of calories varies among individuals: Some efficiently store the excess as fat, and some tend to burn it off.

One mechanism for eliminating excess calories is **thermogenesis,** or heat production. For example, by uncoupling oxidative phosphorylation from electron transport, the free energy of fuel oxidation can be given off as heat without being linked to the cell's need for ATP (see Box 12-A). Another mechanism for wasting free energy is fidgeting, the voluntary and involuntary muscle activity that maintains posture and governs small body movements over the course of the day.

Although considerable research into the mechanisms that regulate metabolic fuel use and storage can provide some guidelines to those who are anxious to maintain a healthy body weight, modern humans must struggle against millions of years of evolution. Clearly, animals experienced a selective advantage in developing efficient mechanisms for fuel storage in the face of unpredictable food supplies. A certain amount of adipose tissue probably allowed early humans to survive seasonal food shortages and periodic famines. However, such thrift, operating in a modern era of food abundance (or overabundance), predisposes to obesity.

[Structure of human leptin (pdb 1AX8) determined by F. Zhang, J.M. Beals, S.L. Briggs, D.K. Clawson, J.-P. Wery, and R.W. Schevitz.]

CHECKLIST

1. Review the Learning Objectives listed on page 500.
2. Complete Exercise 22/Signal Transduction for an overview of signaling pathways and the roles of specific signaling proteins.
3. Apply your knowledge by solving the problems at the end of this chapter. Check your results in the Solutions appendix.
4. Be able to define the boldfaced terms (consult the glossary at the end of the book). Test your understanding by taking the Chapter 16 quiz (www.wiley.com/college/pratt).
5. Explore the websites listed at www.wiley.com/college/pratt.
6. Consult the list of Selected Readings for background information or additional details on hormone signaling pathways.

GLOSSARY TERMS

Cori cycle
glucose–alanine cycle
hormone
signal transduction
receptor
ligand
protein kinase
second messenger
phosphatase
diabetes mellitus
autophosphorylation
lipolysis
G protein
phosphoinositide signaling system
cross-talk
insulin resistance
hyperglycemia
thermogenesis

PROBLEMS

1. What is the "energy cost" in ATP of running the Cori cycle?
2. In the Cori cycle, the muscle converts pyruvate into lactate, which then diffuses out of the muscle and travels via the bloodstream to the liver, where the reaction is reversed and lactate is converted back to pyruvate. Why is this extra step necessary? Why wouldn't it be possible for the muscle to simply release pyruvate for uptake by the liver?
3. Explain how the reactions of the glucose–alanine cycle would operate during starvation.
4. Explain why fasting results in an increase in the liver concentrations of phosphoenolpyruvate carboxykinase and glucose-6-phosphatase.
5. Why does the sigmoidal behavior of glucokinase, the liver isozyme of hexokinase, help the liver to adjust its metabolic activities to the amount of available glucose?
6. Some bacterial signaling systems involve kinases that transfer a phosphoryl group to a His side chain. Draw the structure of phospho-His.
7. Inexperienced athletes might consume a meal high in glucose just before a race, but veteran marathon runners know that doing so would impair their performance. Explain.
8. Explain why a tyrosine phosphatase might be involved in limiting the signaling effect of insulin.
9. Why is insulin required for triacylglycerol synthesis in adipocytes?
10. Explain why some drugs used to treat type I diabetes are compounds that diffuse into the cell and activate tyrosine kinases.
11. Insulin activates cAMP phosphodiesterase. Explain why this augments insulin's metabolic effect.
12. Among the cell's many kinases involved in signal transduction pathways is one that is activated by AMP. Would this AMP-dependent kinase lead to activation or inhibition of anabolic processes such as glycogen and triacylglycerol synthesis?
13. Glycogen phosphorylase cleaves glucose residues from glycogen via a phosphorolytic rather than a hydrolytic cleavage:

    Phosphorolytic cleavage

    $$\text{Glycogen}_{(n \text{ residues})} + P_i \rightarrow \text{glycogen}_{(n-1 \text{ residues})} + \text{glucose-1-phosphate}$$

    Hydrolytic cleavage

    $$\text{Glycogen}_{(n \text{ residues})} + H_2O \rightarrow \text{glycogen}_{(n-1 \text{ residues})} + \text{glucose}$$

    What is the advantage of phosphorolytic cleavage?
14. Addition of the nonhydrolyzable analog GTPγS to cultured cells is a common practice in signal transduction experiments. What effect does GTPγS have on cellular cAMP levels?

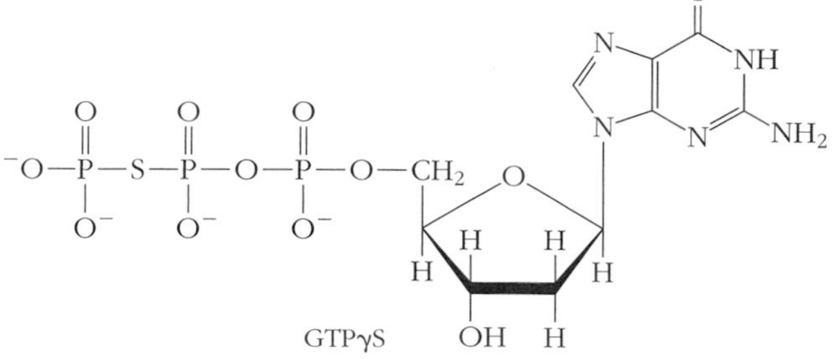

    GTPγS

15. After several days of starvation, the ability of the liver to metabolize acetyl-CoA via the citric acid cycle is severely compromised. Explain why.

16. Glycogen storage diseases are so named because a hallmark of the diseases is abnormal glycogen storage due to a deficiency of one of the enzymes involved in either glycogen synthesis or glycogen degradation. At least 10 kinds of glycogen storage diseases have been identified so far. Each type is characterized by the lack of a specific enzyme. An understanding of glycogen metabolism is essential for the proper treatment of these diseases, and identification of the deficient enzyme is required before a treatment protocol can be designed.

You are a physician and your patient is a 15-year-old Caucasian male named Alex K. Alex's mother has brought him to see you because she is concerned about his inability to perform any kind of strenuous exercise. During his physical education classes, Alex cannot keep up with his classmates and often suffers painful muscle cramps if he does attempt to exercise. He appears to be normal while at rest or performing light to moderate exercise. A physical examination reveals that his liver appears to be normal in size, but his muscles are flabby and poorly developed. A fasting glucose test shows that Alex is not hypo- or hyperglycemic. A number of biochemical tests are carried out to identify the type of glycogen storage disease in this patient.

(a) You decide to test Alex's response to glucagon by injecting a high dose of glucagon intravenously, drawing samples of blood periodically, and measuring their glucose content. After the glucagon injection, Alex's blood sugar rises dramatically. Is this the response you would expect in a normal person? Explain.

(b) Liver and muscle biopsies are taken from Alex and analyzed. The biopsies reveal that glycogen content in the liver is normal, but muscle glycogen content is elevated. The biochemical structure of glycogen in both tissues appears to be normal. Suggest some possible explanations for these observations.

(c) Next, you have Alex perform ischemic exercise for as long as he is able to do so. Blood is withdrawn every few minutes or so during the exercise period. Alex's blood samples are tested for lactate and compared with a control sample from a patient who does not suffer from a glycogen storage disease. The results are shown below. Why does lactate concentration increase in the normal patient? Why is there no corresponding increase in Alex's lactate concentration?

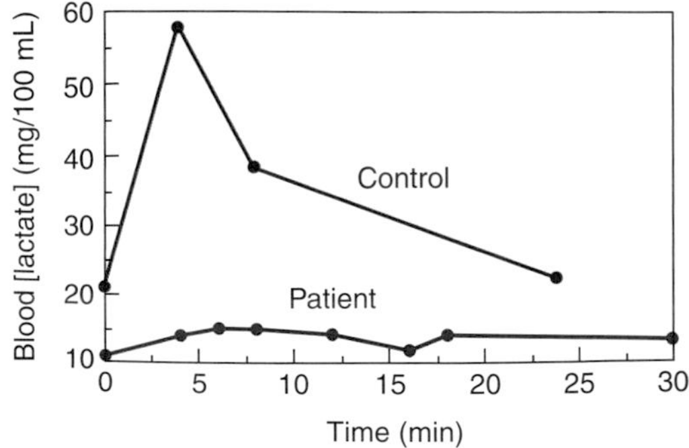

Blood lactate concentration following ischemic exercise in a patient with a glycogen storage disease and in a control

(d) Tests performed after Alex has completed his exercise reveal the presence of myoglobin in his urine. Myoglobin isn't normally found in the urine but in muscle cells. Why does Alex suffer from myoglobinuria following ischemic exercise?

(e) Alex's blood samples are also tested for alanine content. In a normal person, blood alanine increases during ischemic exercise. But in Alex's blood samples, you see a decrease in alanine concentrations, leading you to believe that Alex's muscle cells are taking up alanine rather than releasing it. Why do blood alanine concentrations increase in a normal person? Why do blood alanine concentrations decrease in your patient?

(f) Alex's enzyme deficiency does not cause him to suffer from either hypo- or hyperglycemia. Explain why.

(g) As a treatment, you tell Alex that the best thing he can do is to avoid strenuous exercise. If he does wish to exercise, you advise him to consume sports drinks containing glucose or fructose frequently while exercising. Why would this help alleviate Alex's suffering during exercise?

[From Stanbury, J.B., Wyngaarden, J.B., and Fredrickson, D.S., *The Metabolic Basis of Inherited Disease,* McGraw-Hill, pp. 151–153 (1978).]

17. Because of its importance in the disease phenylketonuria (PKU; see Problem 15–20), the enzyme phenylalanine hydroxylase (PAH) has been fairly well characterized. PAH has been isolated and purified from both rats and humans.

(a) Once the enzyme was purified, the investigators set out to determine its properties. They wanted to see if phenylalanine, in addition to serving as a substrate, had an additional role in the regulation of the enzyme. Polyacrylamide gel electrophoresis (under denaturing and nondenaturing conditions) was carried out with rat liver PAH. The results are shown below. How do you interpret these data?

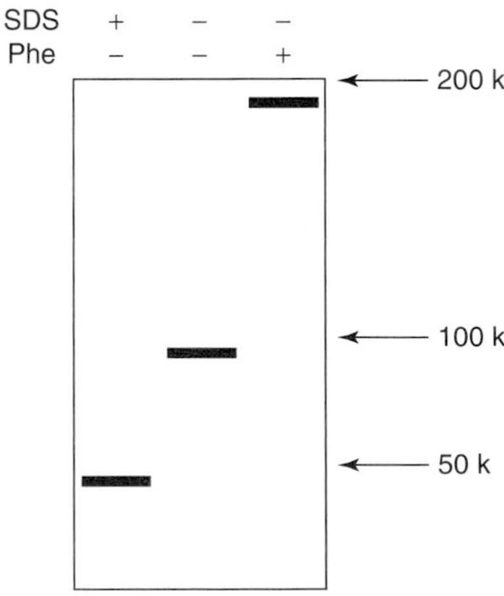

(b) Next, kinetic studies were carried out with the enzymes. A plot of reaction velocity versus phenylalanine concentration yields a sigmoidal curve. What does this tell you about the enzyme?

(c) PAH activity with and without preincubation with phenylalanine is shown below. Provide a structural explanation of these data.

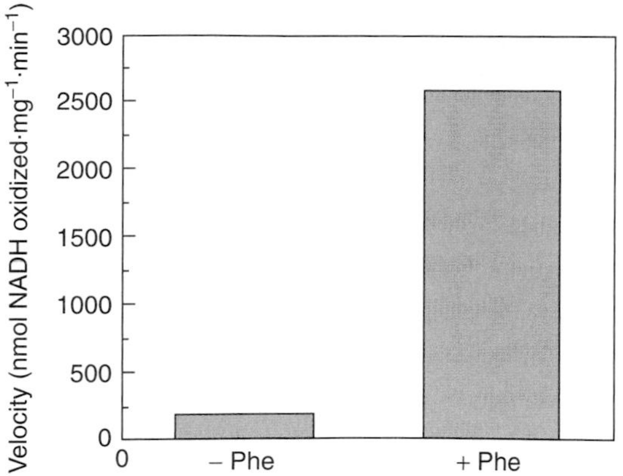

(d) The effects of the hormones glucagon and insulin on PAH activity were investigated. The results are shown here. In addition, the amount of radioactive phosphate incorporated into PAH upon glucagon treatment was found to be nearly sevenfold greater than in controls. How would you interpret these data? Draw a diagram demonstrating the mechanism for hormonal activation of PAH. Which hormone activates PAH and why?

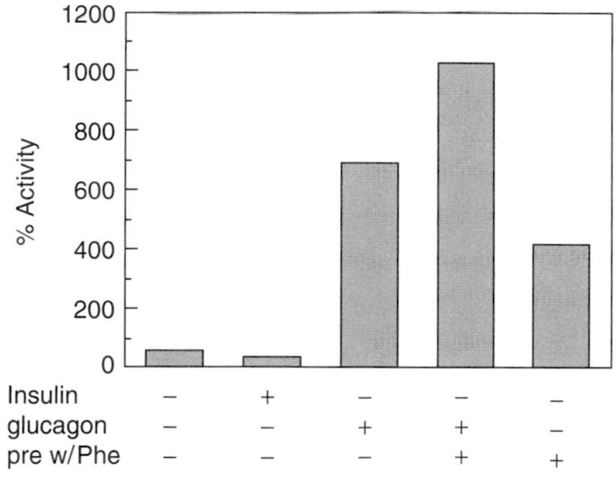

[From Agranoff, B.W., and Aprison, M.H., *Advances in Neurochemistry,* Plenum Press, pp. 1–132 (1977); Kaufman, S., *Tetrahydrobiopterin: Basic Biochemistry and Role in Human Disease,* Johns Hopkins University Press, pp. 31–139 and 262–322 (1997); DiLella, A.G., Marvit, J., and Woo, S.L.C., The Molecular Genetics of Phenylketonuria *in Amino Acids in Health and Disease: New Perspectives,* Alan R. Liss, Inc., pp. 553–564 (1987).]

18. Glucagon is a 29-amino acid peptide hormone secreted by the pancreatic α cells in response to low glucose concentrations. Its primary amino acid sequence is shown on the next page. Glucagon acts primarily on the liver, where binding to specific extracellular receptors stimulates glycogenolysis and gluconeogenesis with subsequent release of glucose from the liver for the benefit of other body tissues. Glucagon is counterregulatory to insulin, which is secreted by pancreatic β cells and stimulates cellular uptake of exogenous glucose from the blood. During feeding, insulin levels are high and glucagon levels are low. The opposite is true during fasting: Glucagon levels rise and insulin concentrations decrease.

The glucagon hormone has been the subject of much research interest in decades past, not just because of its importance in carbohydrate metabolism, but also because its mechanism of action, via the activation of a G protein–linked enzyme, is a model for signal transduction. But more recently, attention has focused on the role of glucagon in the disease diabetes mellitus. Several studies have shown that in type I diabetics, the lack of insulin is accompanied by hypersecretion of glucagon. The excess glucagon secretion leads to release of glucose from the liver, which exacerbates the high blood glucose concentrations in the untreated diabetic.

Diabetics are currently treated with exogenous insulin. But some investigators have suggested that the treatment regimen of the diabetic should address the glucagon hypersecretion as well as the lack of insulin. One way to do this would be to administer a glucagon antagonist along with insulin. A glucagon antagonist is a molecule that could bind to extracellular receptors on liver cells but could not carry out the signal transduction process. The glucagon antagonist would compete for binding with endogenous glucagon. If the antagonist bound instead of the endogenous glucagon, glycogenolysis would not occur.

In order to construct a glucagon antagonist, it is necessary to determine exactly which amino acids contribute to receptor binding and which amino acids are involved in signal transduction. These experiments were first carried out in the mid-1970s, but recent advances in biotechnology have facilitated the process. For example, the glucagon receptor gene has been cloned and sequenced, and studies have shown that an aspartate residue near the C-terminus of the receptor is essential for glucagon binding. Retaining the amino acid residues important for binding while modifying those amino acids involved in signal transduction would result in a glucagon antagonist. Many such compounds have been synthesized, but the search for the ideal antagonist has been complicated by the fact that several amino acid residues in glucagon are important for both receptor binding and signal transduction.

In the current study, the investigators used solid-phase peptide synthesis to produce modified glucagon molecules. They carried out two separate studies. The first study examined the role of amino acid residues at positions 9, 15, and 21. The second study examined the role of amino acid residues at positions 1, 12, 17, and 18. In each study, certain amino acid residues were replaced by amino acids with different properties, and the resulting analogs were tested for their ability to bind to liver membrane receptors and initiate signal transduction. A true antagonist would bind to receptors but elicit no response whatsoever. Analogs capable of binding, with diminished (but not abolished) activity are referred to as partial agonists.

Sequence of human glucagon

1	2	3	4	5	6	7	8	9	10
His	Ser	Gln	Gly	Thr	Phe	Thr	Ser	Asp	Tyr
11	12	13	14	15	16	17	18	19	20
Ser	Lys	Tyr	Leu	Asp	Ser	Arg	Arg	Ala	Gln
21	22	23	24	25	26	27	28	29	
Asp	Phe	Val	Gln	Trp	Leu	Met	Asn	Thr	

(a) Glucagon carries out its biological function by binding to extracellular hepatic receptors and then putting into motion a series of events that lead to glycogenolysis. Draw a diagram showing the steps of this process.

(b) Why did the investigators choose to modify the amino acids at positions 1, 12, 17, and 18?

(c) The investigators synthesized a number of glucagon analogs, which are listed in the table. The ability of the glucagon analogs to bind to receptors and elicit a biological response was measured and compared to native glucagon. Binding affinity refers to the ability of the glucagon analog to bind to hepatic membrane receptors. Activity was measured by testing each analog's ability to stimulate cAMP production as compared to native glucagon. The des prefix indicates that the specified amino acid has been deleted. Use the information provided in the table to answer the following: What is the effect of substituting or eliminating the amino acid at position 9? Of replacement or modification at position 12? Of replacement at position 17 and 18? What is the role of the histidine at position 1?

Glucagon analog	Binding affinity (%)	% of maximum activity
Glucagon	100	100
Des-Asp9	45	8.3
Lys9	54	0
Nε-acetyl-Lys12	47	90
Ala12	17	60
Gly12	11	86
Glu12	1.0	80
Ala17	38	29
Leu17	30	88
Glu17	21	95
Ala18	13	94
Leu18	56	95
Glu18	6.2	100
Des-His1	63	44
Des-His1-Des-Asp9	7	0
Des-His1-Lys9	70	0
Des-His1-Glu12	0.11	28
Des-His1-Glu17	1.7	22
Des-His1-Glu18	0.44	18

(d) Write a paragraph summarizing the main findings of this study.

(e) Of the glucagon analogs presented here, which is the best glucagon antagonist? Could you design a better glucagon antagonist than the analogs presented here? Explain the rationale for your design.

[From Unson, C.G., Macdonald, D., Ray, K., Durrah, T.L., and Merrifield, R.B., *J. Biol. Chem.* **266,** 2763–2766 (1991); Unson, C.G., Wu, C.-R., Cheung, C.P., and Merrifield, R.B., *J. Biol. Chem.* **273,** 10308–10312 (1998).]

19. The simple one-celled eukaryotic organism *Saccharomyces cerevisiae* (yeast) has been well studied because extensive research has shown that the metabolic pathways in the yeast are subject to the same types of controls as cells of higher organisms. Yeast is also easy to culture in a laboratory setting. Yeast has been employed for literally thousands of years in human history in the production of leavened bread and alcoholic beverages. To the biochemist, yeast is important because of Buchner's experiments, which disproved the theory of vitalism by demonstrating that metabolic pathways could still occur in cell-free yeast extracts. This pioneering work allowed scientists to study metabolic pathways in yeast in great detail. Once the pathways had been elucidated, scientists turned their attention to the various mechanisms that regulate the pathways. These studies have been carried out with yeast mutants in which a single enzyme associated with a specific pathway is deficient or nonfunctional. Observing the phenotype of the mutants allows scientists to pinpoint the mutated enzyme and increases our understanding of yeast metabolic pathways. In this case, we will focus on carbohydrate metabolism in yeast.

(a) Yeast is used in the production of wine (see Box 10-B). Describe how yeast converts the sugar in the grape (mainly fructose) to ethanol. Explain why concentrations of acetate often rise during alcoholic fermentation.

(b) Cells lacking alcohol dehydrogenase accumulate large amounts of glycerol during anaerobic fermentation. Explain why.

(c) Yeast is unusual in that it can use ethanol as a gluconeogenic substrate. Ethanol is converted to glucose using the assistance of the glyoxylate pathway (see Box 11-B). Describe how the ethanol → glucose conversion takes place.

(d) A yeast mutant was isolated that was deficient in the enzyme phosphofructokinase. The mutant yeast was able to use glycerol as an energy source but not glucose. Explain why.

(e) Yeast is used as a leavening agent in making bread. Explain, in biochemical terms, why the bread dough rises when placed in a warm place.

(f) Yeast cells that are growing on nonfermentable substrates and are then abruptly switched to glucose exhibit substrate-induced inactivation of several enzymes. Which enzymes would glucose cause to be inactivated, and why?

(g) During anaerobic fermentation, the majority of the available glucose is oxidized via the glycolytic pathway and the rest enters the pentose phosphate pathway to generate NADPH and ribose. This occurs during aerobic respiration as well, except that the percentage of glucose entering the pentose phosphate pathway is much greater in aerobic respiration than during anaerobic fermentation. Explain why.

[From Wills, C. *Critical Reviews in Biochemistry and Molecular Biology* **25,** 245–280 (1990).]

20. *Helicobacter pylori* is a bacterium that colonizes the upper gastrointestinal tract in humans and causes chronic gastritis, ulcers, and possibly gastric cancer. Knowledge of the intermediary metabolism of this organism will be helpful in the development of effective drug therapies to treat these diseases.

The citric acid "cycle" in *H. pylori* is a noncyclic branched pathway that produces biosynthetic intermediates instead of metabolic energy (see Problem 11-29). Succinate is produced in the "reductive branch," whereas α-ketoglutarate is produced in the "oxidative branch." The two branches are linked by the α-ketoglutarate oxidase reaction, as shown below. *H. pylori* uses amino acids and fatty acids present in the gastrointestinal tract as a source of biosynthetic intermediates, rather than synthesizing these compounds *de novo.* Describe the fate of (a) acetyl-CoA (from fatty acid breakdown) and (b) aspartate.

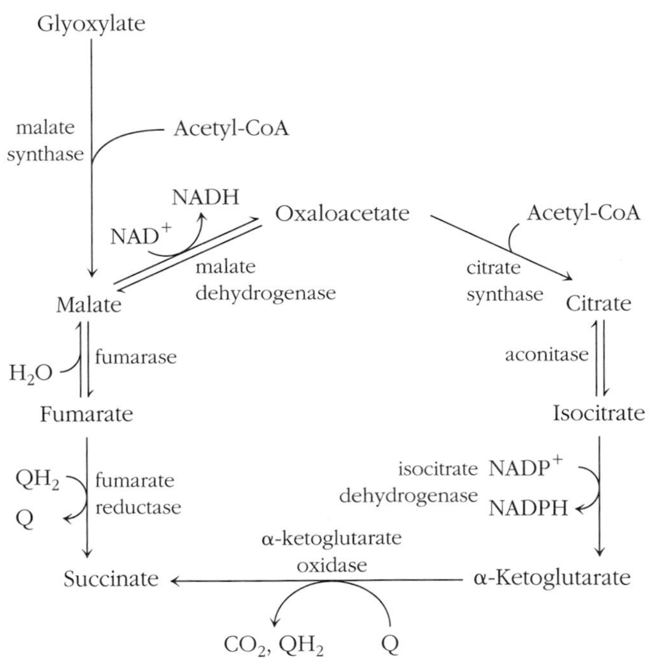

[From Pitson, S.M., Mendz, G.L., Srinivasan, S., and Hazell, S.L., *Eur. J. Biochem.* **260,** 258–267 (1999).]

21. A three-month-old patient has just been diagnosed with a pyruvate carboxylase deficiency. Fibroblasts obtained from the infant were cultured and studied in order to make the diagnosis. The patient's symptoms include psychomotor retardation, poor head control, hypotonia, lactic acidosis, and ketosis. She seemed normal at birth but had recently begun having seizures.

(a) Which metabolites would be elevated in this patient? Which metabolites would be deficient? Why does the patient suffer from lactic acidosis and ketosis?

(b) Psychomotor retardation arises from the lack of the neurotransmitter amino acids glutamate, aspartate, and γ-aminobutyric acid (GABA). Why would a pyruvate carboxylase deficiency result in decreased synthesis of these neurotransmitters?

(c) The scientists who cultured the patient's fibroblasts added acetyl-CoA to the culture medium to see if they could detect pyruvate carboxylase activity. What was the rationale behind this experiment?

[From Stanbury, J.B., Wyngaarden, J.B., Fredrickson, D.S., Goldstein, J.L., and Brown, M.S., *The Metabolic Basis of Inherited Disease,* McGraw-Hill, pp. 196–198 (1983).]

22. The enzymes of fatty acid synthetic pathways have been examined in the hopes that a more thorough understanding of these pathways might lead to better treatments for obesity. A recent study focused on acetyl-CoA carboxylase (ACC), the enzyme that catalyzes the highly regulated first step of fatty acid synthesis, the production of malonyl-CoA from acetyl-CoA (see Section 14-2). Humans have two forms of ACC, termed ACC1 and ACC2, whose properties are summarized in the table.

	ACC1	**ACC2**
Molecular mass (M_r)	265,000	280,000
Tissue expression	liver and adipose tissue	heart and muscle
Cellular location	cytosol	mitochondrial matrix
Sensitive to regulation by malonyl-CoA	yes	yes

(a) Review Section 14-2 and explain how malonyl-CoA regulates fatty acid synthesis.

(b) Lipid is not consumed in a fat-free diet, or during starvation or diabetes. However, ACC is stimulated by a fat-free diet and inhibited by starvation and diabetes. Explain these observations.

(c) The investigators generated ACC2 "knockout" mice, which were missing both maternal and paternal copies of the ACC2 gene. ACC1 functioned normally in these mice. The investigators noted several differences between the wild-type and knockout mice, including 20% less glycogen in the liver of the knockout mice. Explain this observation.

(d) In the knockout mice, the concentration of fatty acids in the blood was lower but the concentration of triacylglycerols was higher than in the wild-type mice. Explain.

(e) Fatty acid oxidation was measured in muscle tissue samples collected from both types of mice. The addition of insulin caused a 45% decrease in palmitate oxidation in muscle tissue from normal mice, but there was no change in the rate of palmitate oxidation in the knockout mice. Explain these results.

(f) Groups of both knockout and wild-type mice were allowed access to as much food as they cared to eat. At the end of a 27-week period, the knockout mice had consumed 20–30% more food than the wild-type mice. Interestingly, despite the increased food intake, the mice weighed about 10% less and accumulated less fat in adipose tissue. Explain these results.

(g) How might you design the next new "diet pill," based on these results?

[From Abu-Elheiga, L., Matzuk, M.M., Abo-Hashema, K.A.H., and Wakhil, S.J., *Science* **291,** 2613–2616 (2001).]

23. Fatty acid synthase inhibitors have been investigated as possible candidates for weight-loss drugs. A fatty acid synthase (FAS) inhibitor termed C75, a cerulenin derivative, has been synthesized by Kuhajda and colleagues (see Problem 14–30). Cerulenin, a fungal epoxide, inhibits FAS but is unstable, so derivatives with greater stability were synthesized as more suitable drug candidates.

(a) Mice were injected intraperitoneally (IP) with C75 and radioactively labeled acetate. What is the fate of the label?

(b) Mice receiving IP injections of C75 reduced their food intake by more than 90% and lost nearly one-third of their body weight, although they gained back the weight when the drug was withdrawn. The investigators measured brain concentrations of neuropeptide Y (NPY), a compound known to act on the hypothalamus to increase appetite during starvation. Based on the results presented here, predict the effect of C75 on brain levels of NPY.

(c) Because hepatic malonyl-CoA levels were high in the C75-treated mice but not in control mice, the investigators hypothesized that malonyl-CoA inhibits feeding. If their hypothesis is correct, predict what would happen if mice were pretreated with the acetyl-CoA carboxylase inhibitor TOFA prior to injection with C75.

(d) What other cellular metabolites accumulate when concentrations of malonyl-CoA rise? (These molecules are candidates for signaling molecules, which could stimulate a biochemical pathway that decreases appetite.)

[From Loftus, T.M., Jaworksy, D.E., Grehywot, G.L., Townsend, C.A., Ronnett, G.V., Lane, M.D., and Kuhajda, F.P., *Science* **288,** 2379–2381 (2000).]

24. Adipocytes secrete leptin, a hormone that suppresses appetite (see Box 16-C). Leptin exerts its effects through the central nervous system and also directly on target tissues via leptin receptors. Leptin can inhibit insulin secretion but can also act as an insulin mimic by binding to its own receptors and activating the same signaling pathway as insulin. For example, leptin can induce tyrosine phosphorylation of the insulin receptor substrate-1 (IRS-1). Using this information, predict leptin's effect on the following:

(a) Glucose uptake by skeletal muscle.

(b) Hepatic glycogenolysis and liver glycogen phosphorylase activity.

(c) cAMP phosphodiesterase, the enzyme that catalyzes a ring-opening reaction and converts cAMP to AMP.

25. Glucagon regulates blood glucose concentrations during fasting by increasing glycogenolysis and gluconeogenesis in the liver. Glucagon also acts on other enzymes involved in carbohydrate metabolism. For example, the signal transduction pathway put into motion by glucagon also leads to the phosphorylation and subsequent activation of the glycolytic enzyme pyruvate kinase.

(a) The investigators of one study found that glucagon is responsible for an acute stimulation of glucose-6-phosphate hydrolysis by a temperature-sensitive mechanism. How could this explain the following results: Phosphoenolpyruvate concentration increased twofold, glucose-6-phosphate concentration decreased by 60%, and hepatic glucose concentration was increased twofold in the presence of glucagon and an exogenously administered substrate (dihydroxyacetone phosphate).

(b) The investigators state that activation of glucose-6-phosphate hydrolysis results in both activation of gluconeogenesis and inhibition of glycolysis. Explain.

[From Ichai, C., Guignot, L., El-Mir, M.Y., Nogueira, V., Guigas, B., Chauvin, C., Fontaine, E., Mithieux, G., and Leverve, X.M., *J. Biol. Chem.* **276,** 28126–28133 (2001).]

26. The ATP-generating glycolytic pathway is controlled in part by the activities of the allosteric enzymes phosphofructokinase and pyruvate kinase. The activities of these two enzymes are influenced by cellular concentrations of ATP and AMP, as discussed in Chapter 10. But the relative concentrations of ATP and AMP alone are not sufficient to account for the dramatic changes in enzyme activities measured under a variety of cellular situations. Ammonium ions have been shown to influence the activities of the two enzymes, thus "fine-tuning" the glycolytic pathway's responses to changes in the energy state of the cell. The enzyme activities are under hormonal control as well.

AMP deaminase, an enzyme involved in the purine nucleotide cycle, generates ammonium ions as a reaction product. Describe how changes in AMP deaminase activity can control flux through glycolysis and gluconeogenesis.

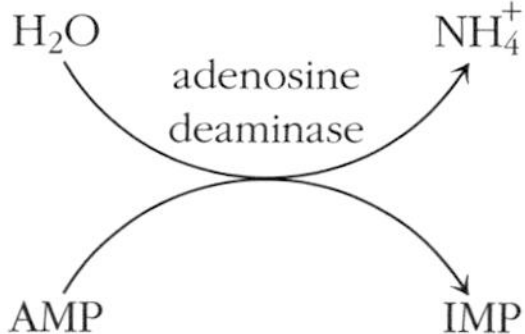

[From Yoshino, M. and Murakami, K., *J. Biol. Chem.* **260,** 4729–4732 (1985).]

SELECTED READINGS

Bollen, M., Keppens, S., and Stalmans, W., Specific features of glycogen metabolism in the liver, *Biochem. J.* **336,** 19–31 (1998). [Summarizes the roles of some of the many proteins that regulate glucose storage and mobilization in liver cells.]

Cantley, L.C., The phosphoinositide 3-kinase pathway, *Science* **296,** 1655–1657 (2002). [Briefly summarizes the lipid signaling pathway.]

Firestein, S., How the olfactory system makes sense of scents, *Nature* **413,** 211–218 (2001). [Briefly reviews the neuronal and signal-transducing components of the olfactory system.]

Holman, G.D. and Sandoval, I.V., Moving the insulin-regulated glucose transporter GLUT4 into and out of storage, *Trends Cell. Biol.* **11,** 173–178 (2001). [Summarizes current knowledge and controversies about how GLUT4 uses cellular machinery to move between the cell interior and the cell surface in response to insulin.]

Iiri, T., Farfel, Z., and Bourne, H.R., G-protein diseases furnish a model for the turn-on switch, *Nature* **394,** 35–38 (1998). [Proposes a mechanism whereby binding of GTP to Gα promotes dissociation of the G protein from the receptor and dissociation of Gα from G$\beta\gamma$.]

Marinissen, M.J. and Gutkind, J.S., G-protein-coupled receptors and signaling networks: Emerging paradigms, *Trends Pharm. Sci.* **22,** 368–375 (2001). [Discusses some of the pathways in which G-proteins mediate signal transduction.]

Saltiel, A.R. and Kahn, C.R., Insulin signaling and the regulation of glucose and lipid metabolism, *Nature* **414,** 799–806 (2001). [Summarizes the mechanisms of insulin action and insulin resistance.]

Steppan, C., Bailey, S.T., Bhat, S., Brown, E.J., Banerjee, R.R., Wright, C.M., Patel, H.R., Ahima, R.S., and Lazar, M.A., The hormone resistin links obesity to diabetes, *Nature* **409,** 307–312 (2001). [Describes the discovery and possible role of resistin in linking obesity and type II diabetes.]

PART III

MANAGEMENT OF GENETIC INFORMATION

CHAPTER 17

DNA Replication

Viruses, such as adenovirus (shown here), are small particles containing a nucleic acid surrounded by protein. The adenovirus genome is a linear double-stranded DNA molecule of about 35 kb. This long, thin molecule is packed inside an icosahedral (20-sided) protein shell, called a capsid. The assembled virus has a diameter of 65 to 80 nm, with fibrous projections that help it attach to a host cell but are easily broken off during preparation for electron microscopy. Adenovirus multiplies inside animal cells, causing relatively mild illnesses. One of the mysteries of adenovirus and other viruses containing DNA as well as RNA is how a new viral particle is assembled, that is, how a newly synthesized nucleic acid is inserted into the capsid without becoming tangled or broken. Answers to this riddle might lead to new ways to treat viral infections.

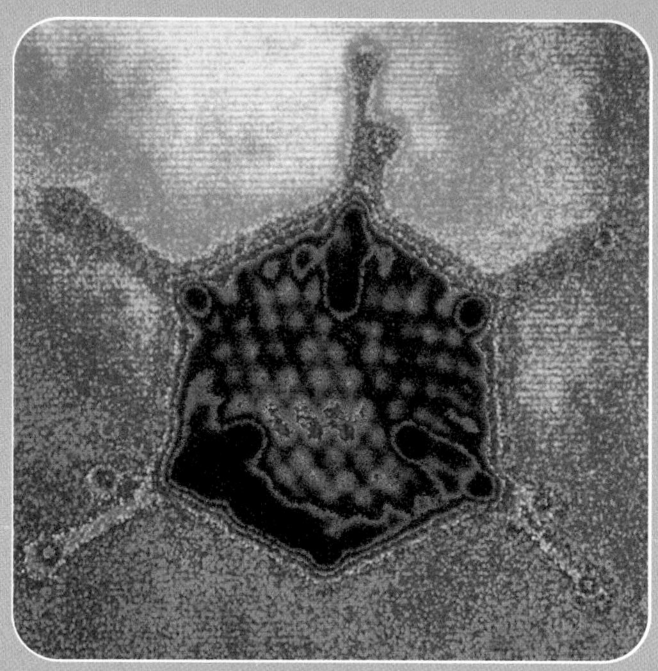

[Alfred Pasieka/Science Photo Library/Photo Researchers.]

LEARNING OBJECTIVES

1. Understand how DNA helix twisting relates to supercoiling.
2. Understand how topoisomerases alter supercoiling.
3. Understand that stationary protein "factories" carry out DNA replication.
4. Understand that the leading strand is synthesized continuously, and the lagging strand discontinuously.
5. Understand the structure and functions of DNA polymerase.
6. Understand the functions of helicase, SSB, primase, the sliding clamp, RNase H, and DNA ligase.
7. Understand the structure, function, and origin of telomeres.
8. Understand the basic structure of a nucleosome.

THIS CHAPTER IN CONTEXT

This chapter is the first in a series of four that look at processes involving nucleic acids, including DNA replication, RNA transcription, and protein synthesis (translation). This part of the book extends the material on nucleic acids that was introduced in Chapter 3. It is also a logical sequel to the metabolism section of the book, since the management of genetic information, or nucleic acid metabolism, includes synthetic as well as degradative reactions. Chapter 17 discusses the proteins and reactions of DNA replication. It also describes DNA topology and packaging in nucleosomes, features that affect the cell's ability to store, duplicate, and express genetic information.

WHEN WATSON AND CRICK DESCRIBED THE COMPLEMENTARY, double-stranded nature of DNA in 1953, they recognized that DNA could be duplicated by a process involving separation of the strands followed by the assembly of two new complementary strands. This mechanism of copying, or **replication,** was elegantly demonstrated by Matthew Meselson and Franklin Stahl in 1958. They grew bacteria in a medium containing the heavy isotope ^{15}N in order to label the cells' DNA. The bacteria were then transferred to fresh medium containing only ^{14}N, and the newly synthesized DNA was isolated and sedimented according to its density in an ultracentrifuge. Meselson and Stahl found that the first generation of replicated DNA had a lower density than the parental DNA but a higher density than DNA containing only ^{14}N. From this, they concluded that newly synthesized DNA is a hybrid containing one parental (heavy) strand and one new (light) strand. In other words, DNA is replicated **semiconservatively.** Because Meselson and Stahl did not observe any all-heavy DNA in the first generation, they were able to discount the possibility that DNA was copied in a way that left intact—or conserved—the original double-stranded molecule (Fig. 17-1).

Although simple in principle, DNA replication is a process of many steps, involving many enzymes and accessory factors. This complexity reflects the relatively large size of DNA molecules and the need to copy them quickly and accurately. In addition, the very structure of DNA—its helical twisting and its extraordinary length—present challenges to the cellular machinery for replication and other processes. In this chapter we examine the topology of DNA, the enzymes that replicate DNA, and the packaging of DNA molecules in eukaryotic cells.

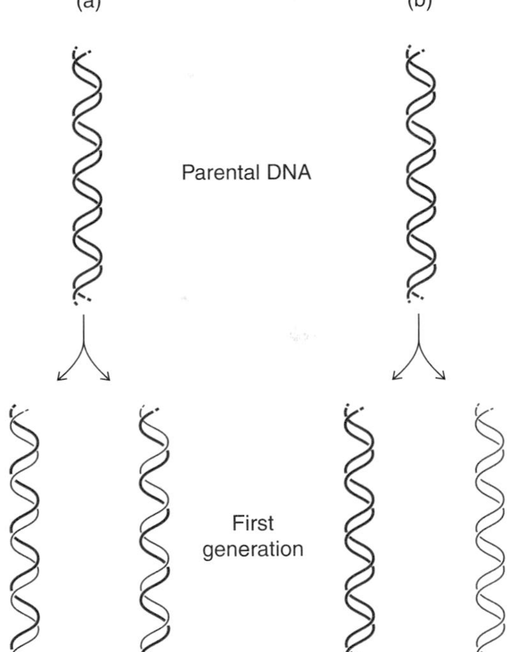

FIGURE 17-1

Semiconservative versus conservative DNA replication.
(a) Experiments performed by Meselson and Stahl demonstrated that new DNA molecules contain one parental (heavy) and one new (light) polynucleotide strand. Thus, DNA replication is semiconservative. (b) If DNA replication were conservative, the parental DNA (both strands heavy) would persist while new DNA would consist of two light strands.

1 DNA TOPOLOGY

Before DNA can be replicated, its two strands, which are coiled around each other in the form of a double helix, must be separated. Inside cells, the DNA helix is already slightly unwound. However, in order to maintain a conformation close to the stable B form, the DNA molecule twists up on itself, much like the cord of an old-fashioned telephone. This phenomenon, termed **supercoiling,** is readily apparent in small circular DNA molecules (Fig. 17-2).

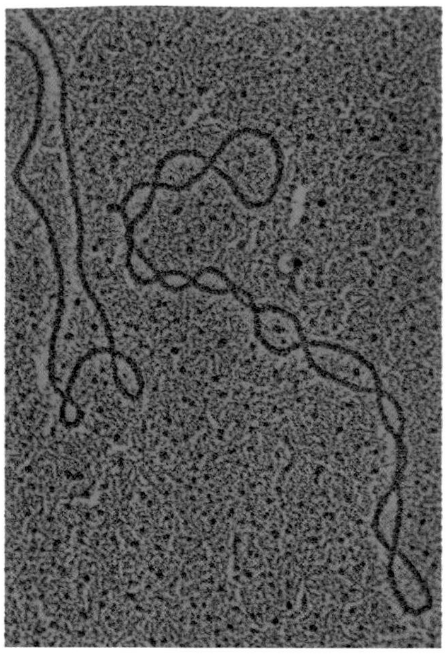

FIGURE 17-2

Supercoiled DNA molecules.
A slightly unwound circular DNA molecule coils up on itself, forming supercoils. [*Gopal Murti/Visuals Unlimited.*]

The geometry of a DNA molecule can be described by the branch of mathematics known as topology. For example, consider a strip of paper. When the strip is looped once, it is said to have one writhe. If the ends of the paper are gently pulled in opposite directions, the strip is deformed into a shape called a twist.

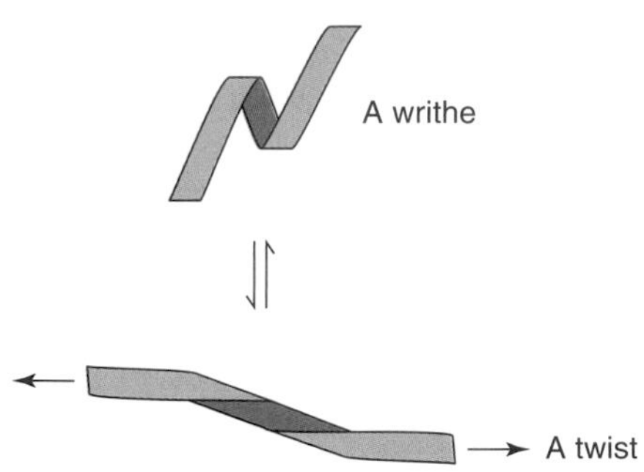

Twisting the strip further introduces more writhes; twisting it in the opposite direction removes writhes, or introduces negative writhes. The same topological terms apply to DNA: *Each writhe, or supercoil, in DNA is the result of overtwisting or undertwisting the DNA helix.* The twisted molecule, like the piece of paper, prefers to writhe, since this is more energetically favorable.

To demonstrate supercoiling for yourself, cut a flat rubber band to obtain a piece a few inches long. Hold the ends apart, twist them, and then bring the ends closer together. You will see the twists collapse into writhes (supercoils). The same thing happens if you let the rubber band relax and then twist it in the opposite direction. Note that the twisted rubber band (representing a double-stranded DNA molecule) adopts a more compact shape. In fact, supercoiling is essential for packaging DNA efficiently into the small space of a bacterial cell or a eukaryotic nucleus.

Naturally occurring DNA molecules are negatively supercoiled. This means that if the DNA were stretched out (that is, its writhes converted to twists), the two strands would unwind, in effect loosening the double helix. Consequently, *the negative supercoiling of DNA helps the replication machinery gain access to the two template strands during synthesis of two new complementary DNA strands.* In the absence of negative supercoils, strand separation would force additional twisting ahead of the separation point:

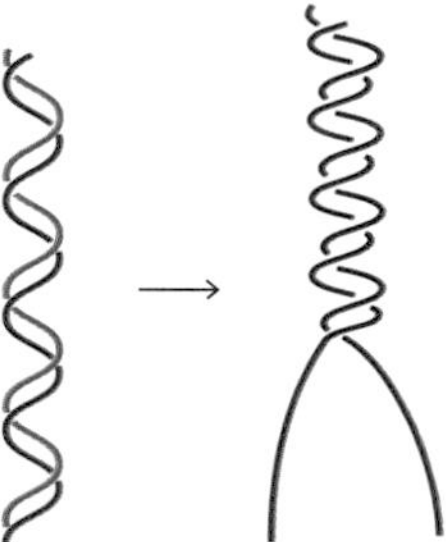

REVIEW LEARNING OBJECTIVE 1

- What is the relationship between a writhe (supercoil) and a twist? How does twisting alter supercoiling?
- Explain why negative supercoiling of DNA assists processes such as replication.

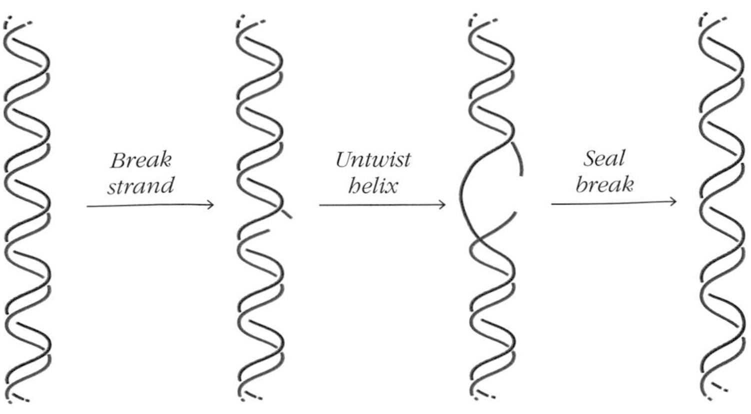

FIGURE 17-3

Action of type I topoisomerase.
In this scheme, the supercoiling of a DNA segment is decreased when a DNA strand is broken, untwisted, and then sealed. Some topoisomerases can increase the degree of supercoiling by twisting rather than untwisting the DNA.

Topoisomerases alter DNA supercoiling

Normal replication and transcription require that cells actively maintain the supercoiling state of DNA by adding or removing supercoils. It is difficult to imagine how the molecule could be pulled taut and its ends twisted or untwisted. Instead, topological changes in DNA molecules occur when an enzyme called a topoisomerase cuts one or both strands of the supercoiled DNA, alters the structure, and then seals the broken strands. Type I topoisomerases cut one DNA strand; the type II enzymes cut both strands of DNA and require ATP.

A type I topoisomerase alters supercoiling by altering the DNA's helical twisting (Fig. 17-3). This process can be carried out by a relatively small protein. Human topoisomerase I, for example, is a monomeric protein with four domains that encircle double-stranded DNA (Fig. 17-4). The surface of the protein that contacts the DNA is rich in positively charged groups that can

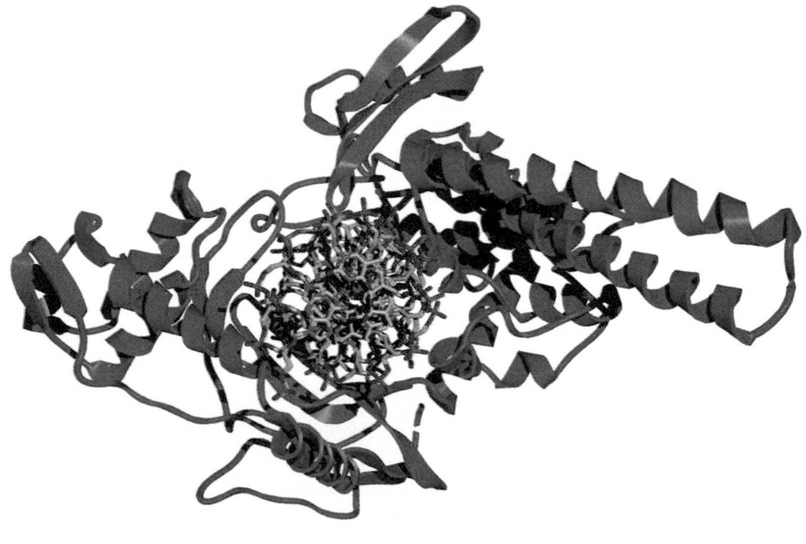

FIGURE 17-4

Structure of human topoisomerase I.
The protein (blue) surrounds a DNA molecule (shown end-on in this model). [*Structure (pdb 1A36) determined by L. Stewart, M.R. Redinbo, X. Qiu, J.J. Champoux, and W.G.J. Hol.*]

interact with the backbone phosphate groups of about 10 base pairs. When one strand of the DNA is cleaved, an active-site Tyr residue forms a covalent bond with the backbone phosphate at one side of the break:

Formation of this diester linkage conserves the free energy of the broken phosphodiester bond of the DNA strand, so strand cleavage and subsequent sealing do not require any other source of free energy. While the DNA is thus **nicked** (one strand broken), the molecule may undergo more than one rotation (see Fig. 17-3) to add or remove one or more supercoils.

Anticancer drugs such as camptothecin

Camptothecin

bind to topoisomerase I and prolong the lifetime of the nicked DNA intermediate. DNA containing strand breaks is a poor substrate for the enzyme complexes that carry out replication and transcription, so camptothecin kills cells that are rapidly dividing, such as cancer cells.

Type II topoisomerases directly alter the number of writhes in a supercoiled DNA molecule by passing one DNA segment through another, a process that requires a double-strand break (Fig. 17-5). The type II enzymes are dimeric, with two active-site Tyr residues that form covalent bonds to the 5′ phosphate groups of the broken DNA strands. ATP hydrolysis powers the mechanical rearrangement of DNA, most likely by driving the protein conformational changes required for cleaving a DNA molecule and holding the ends apart while another portion of the DNA segment passes through the break.

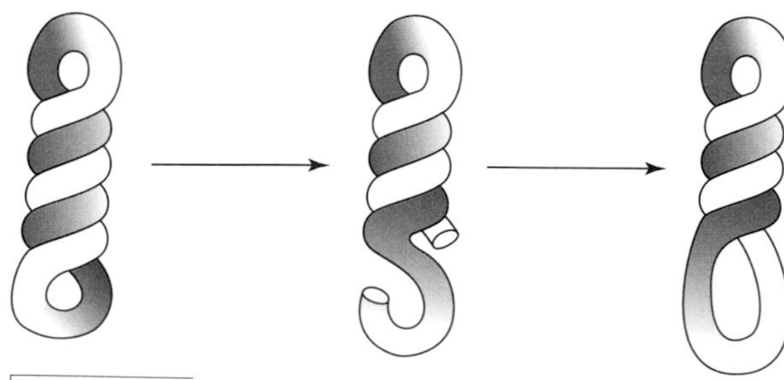

FIGURE 17-5

Action of type II topoisomerase. The twisted-worm shape represents a supercoiled double-stranded DNA molecule. Type II topoisomerase breaks both strands of the DNA, passes another segment of DNA through the break, then seals the break. In this diagram, the degree of supercoiling decreases.

Bacterial type II topoisomerase is called DNA gyrase, and it introduces additional negative supercoils into DNA (the net effect is to further underwind the DNA helix). A number of antibiotics inhibit DNA gyrase without affecting eukaryotic type II topoisomerases. Drugs such as ciprofloxacin

Ciprofloxacin

act on DNA gyrase to enhance the rate of DNA cleavage or reduce the rate of sealing broken DNA. The result is a large number of DNA breaks that prevent normal cell growth and division. Ciprofloxacin has been widely prescribed for its ability to treat infections by *Bacillus anthracis* (anthrax), the spores of which have been used as biological weapons (see Box 16-B).

REVIEW LEARNING OBJECTIVE 2

- Describe how type I and type II topoisomerases operate. When is a source of free energy required?
- What is the use of drugs that interfere with topoisomerase action?

2 THE PROTEINS OF DNA REPLICATION

DNA polymerase, the enzyme that catalyzes the polymerization of deoxynucleotides, is just one of the proteins involved in replicating double-stranded DNA. The entire process—separating the two template strands, initiating new polynucleotide chains, and extending them—is carried out by a complex of enzymes and other proteins. In this section we examine the structures and functions of the major players in DNA replication.

Go to **Exercise 23/DNA Replication** to explore the roles of the various replication proteins and to see how the replication machinery can simultaneously synthesize the leading and lagging strands. The material in this exercise addresses Learning Objectives 4 and 6.

Replication occurs in factories

In the circular chromosomes of bacteria, DNA replication begins at a particular site, known as the origin. Here, proteins bind to the DNA and melt it open in an ATP-dependent manner. Polymerization then proceeds in both directions from this point until the entire chromosome (4.6×10^6 bp in *E. coli*) has been replicated. The point where the parental strands separate and the new strands are synthesized is known as the **replication fork.**

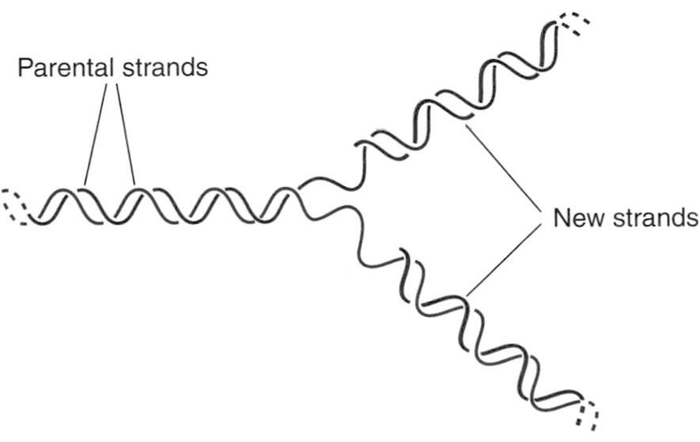

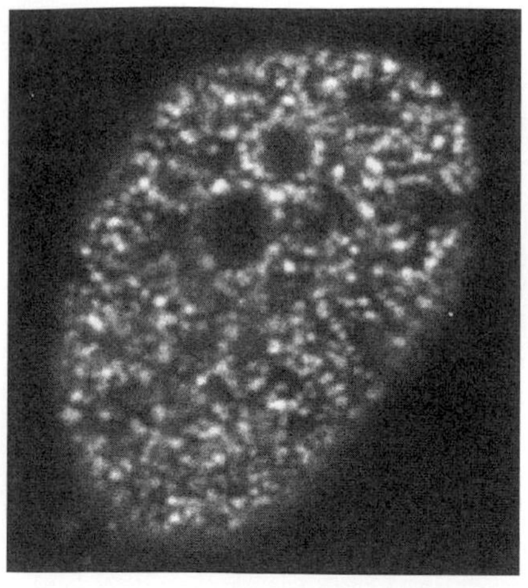

FIGURE 17-6

Replication foci.
The fluorescent patches (foci) in the nucleus of a eukaryotic cell mark the presence of newly synthesized DNA. These sites of DNA replication remain stationary, probably due to attachment of the replication machinery to the nucleoskeleton. [*Courtesy A. Pombo. From* Science ***284***, *1790–1795 (1999).*]

Replication begins at numerous sites in the 46 chromosomes (2.9×10^9 bp of DNA) in the nucleus of each human cell.

At one time, complexes of DNA polymerase and other replication proteins were thought to move along DNA like a train on tracks. This "locomotive" model of DNA replication requires that the large replication proteins move along the relatively thin template strand, rotating around it while generating a double-stranded helical product. In fact, cytological studies indicate that DNA replication (and transcription likewise) occur in "factories" at discrete sites. For example, in bacteria, DNA polymerase and associated factors appear to be immobilized in one or two complexes near the plasma membrane. In the eukaryotic nucleus, newly synthesized DNA appears at 100 to 150 spots, each one representing several hundred replication forks (Fig. 17-6). *According to the* ***factory model*** *of replication, the protein machinery is stationary and the DNA is reeled through it.* In eukaryotes, this organization presumably facilitates the synchronous elongation of many DNA segments, which allows for efficient replication of enormous eukaryotic genomes.

REVIEW LEARNING OBJECTIVE 3

- Explain why the factory model for DNA replication is superior to the locomotive model.

Helicases convert double-stranded DNA to single-stranded DNA

As replication proceeds, the parental DNA strands are continuously separated by the action of ATP-dependent helicases (this unwinding is also aided by the negative supercoiling of DNA). One family of helicases are hexameric proteins that bind to a single strand of DNA (Fig. 17-7). As the helicase pulls the strand through its ring, it unwinds the double-stranded DNA ahead of it. For each ATP hydrolyzed, an estimated one to five base pairs of the helix are separated. This helicase may operate in a rotary fashion, with conformational changes driven by ATP hydrolysis. This mechanism is reminiscent of the binding change mechanism of the F_1 component of the ATP synthase, another hexameric ATP-binding protein (see Section 12-4).

In *E. coli,* DNA appears to move through the helicase at a rate of about 35 nucleotides per second, but the rate increases to about 750 nucleotides per second when DNA polymerase is also present. A second helicase, which may be a dimer rather than a hexamer, probably binds to the other strand of DNA to help open up the replication fork.

As single-stranded DNA is exposed at the replication fork, it associates with a protein known as single-strand binding protein (SSB). SSB forms a tetramer that coats the DNA strands so that they cannot reanneal or form secondary structures that might impede replication (Fig. 17-8). As DNA polymerase converts the single-stranded template to a double-stranded DNA molecule, the SSB is displaced. The roles of helicase and SSB are summarized in Exercise 23.

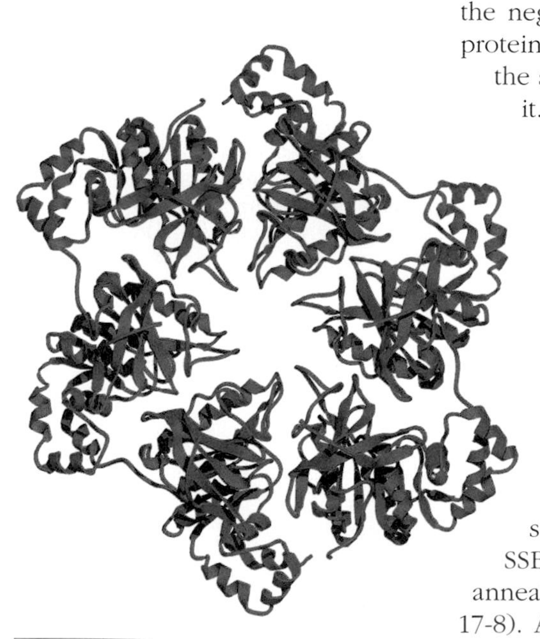

FIGURE 17-7

Structure of a hexameric helicase.
This helicase, from the bacteriophage T7, forms a hexameric ring around single strand of DNA and pushes double-stranded DNA apart. [*Structure (pdb 1EOJ) determined by M.R. Singleton, M.R. Sawaya, T. Ellenberger, and D.B. Wigley.*]

DNA polymerase faces two problems

The mechanism of DNA polymerase and the double-stranded structure of DNA present two serious obstacles to the efficient replication of DNA. First,

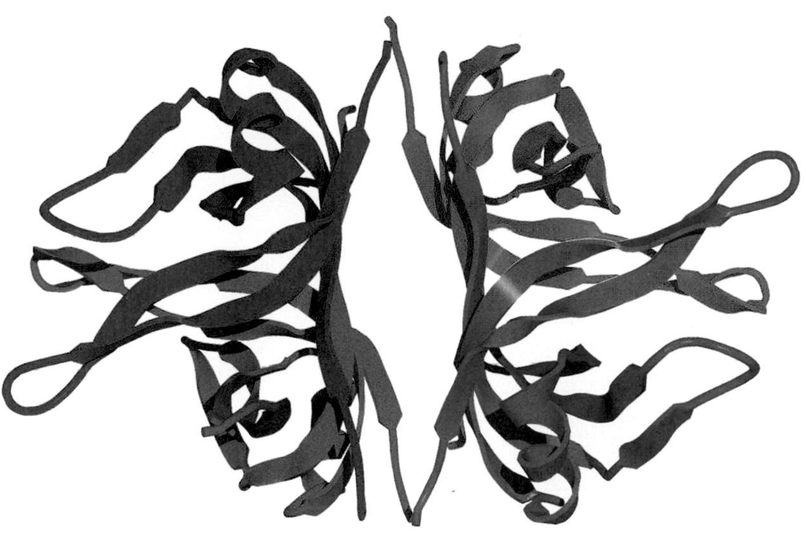

FIGURE 17-8

Structure of SSB.
Each SSB tetramer binds to a stretch of single-stranded DNA. Binding is mediated by Lys side chains that interact with the phosphates of the DNA backbone and by Phe and Trp side chains that stack with the purine and pyrimidine bases. In this model, the DNA would lie diagonally across the face of the protein. [*Structure (pdb 3ULL) determined by C. Yang, U. Curth, C. Urbanke, and C. Kang.*]

DNA polymerase can only extend a preexisting chain; it cannot initiate polynucleotide synthesis. However, RNA polymerase can do this, so DNA chains *in vivo* begin with a short stretch of RNA that is later removed and replaced with DNA.

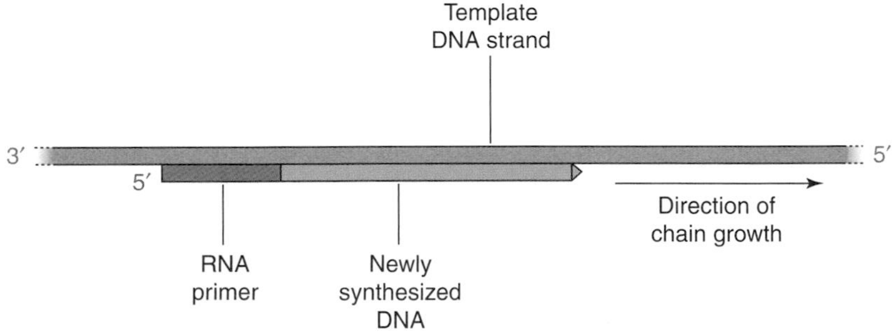

This stretch of RNA, up to 60 nucleotides long, is known as a **primer,** and the enzyme that produces it during DNA replication is known as a primase. As we shall see below, primase is required throughout DNA replication, not just at the start. The active site of the primase is narrow (about 9 Å in diameter) at one end, where the single-stranded DNA template is threaded through. The other end of the active site is wide enough to accommodate a DNA–RNA hybrid helix (which has an A-DNA–like conformation; see Fig. 3-6).

The second problem facing DNA polymerase is that the antiparallel template DNA strands are replicated simultaneously by a pair of polymerase enzymes. But each DNA polymerase catalyzes a reaction in which the 3′ OH group at the end of the growing DNA chain attacks the phosphate group of a free nucleotide that base-pairs with the template DNA strand (Fig. 17-9). For this reason, *a polynucleotide chain is said to be synthesized in the 5′→3′ direction.* Because the two template DNA strands are antiparallel, the synthesis of two new DNA strands would require that the template strands be pulled in opposite directions through the replication machinery so that the DNA

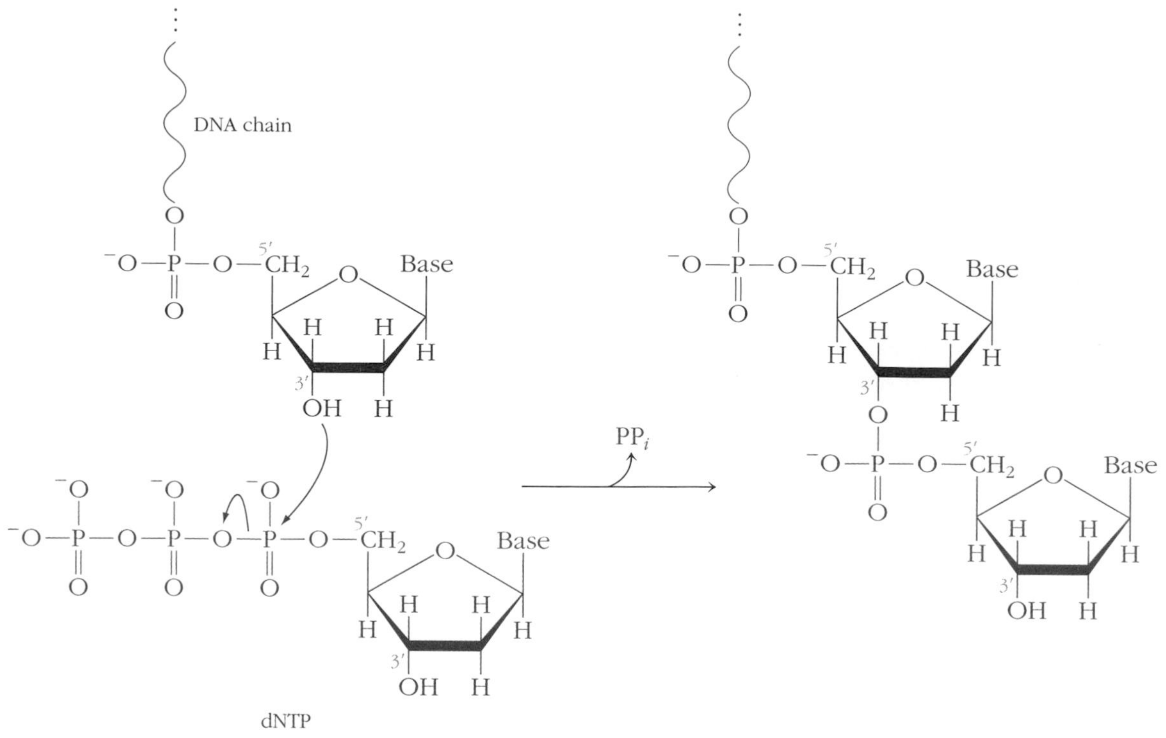

FIGURE 17-9

Mechanism of DNA polymerase.
The 3′ OH group (at the 3′ end of a growing polynucleotide chain) is a nucleophile that attacks the phosphate group of an incoming deoxynucleoside triphosphate (dNTP) that base-pairs with the template DNA strand. Formation of a new phosphodiester bond eliminates PP_i. The reactants and products of this reaction have similar free energies, so the polymerization reaction is reversible. However, the subsequent hydrolysis of PP_i makes the reaction irreversible *in vivo*. RNA polymerase follows the same mechanism.

polymerases could continually add nucleotides to the 3′ end of each new strand.

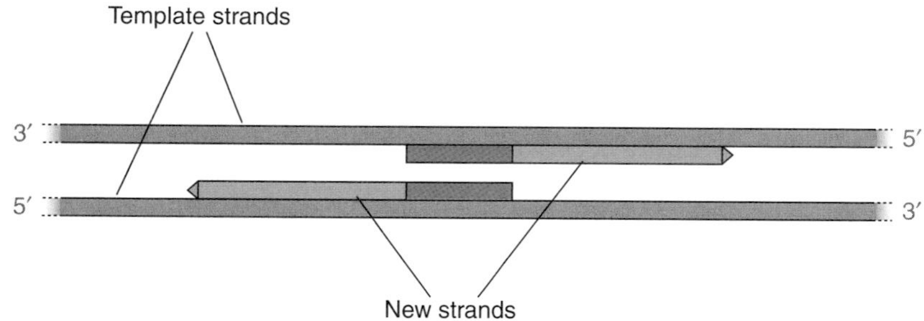

This awkward situation does not occur in cells. Instead, *the two polymerases work side-by-side, and one template DNA strand periodically loops out.* In this scenario, one strand of DNA, called the **leading strand,** can be synthesized in one continuous piece. It is initiated by the action of a primase, then extended in the 5′→3′ direction by the action of one DNA polymerase enzyme. The other strand, called the **lagging strand,** is synthesized in pieces, or **discontinuously.** Its template is repeatedly looped out so that its polymerase can also operate in the 5′→3′ direction. Thus, *the lagging strand consists of a series of polynucleotide segments, which are called* ***Okazaki fragments*** *after their discoverer* (Fig. 17-10).

Bacterial Okazaki fragments are about 500 to 2000 nucleotides long; in eukaryotes, they are about 100 to 200 nucleotides long. Each Okazaki fragment

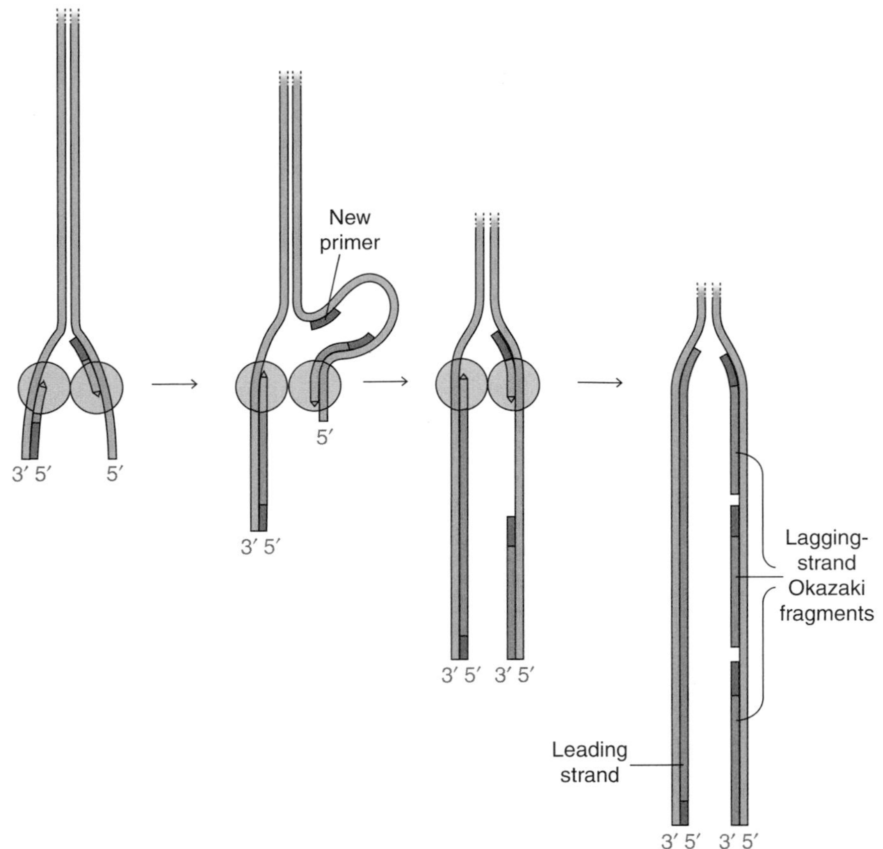

FIGURE 17-10

A model for DNA replication.
Two DNA polymerase enzymes (green) are positioned at the replication fork to make two complementary strands of DNA. The leading and lagging strands both start with RNA primers (red) and are extended by DNA polymerase in the 5′→3′ direction. The replication machinery is stationary, and the template DNA is reeled through it. Because the two template strands are antiparallel, the lagging-strand template loops out (the single-stranded DNA becomes coated with SSB). The leading strand is therefore continuously synthesized while the lagging strand is synthesized as a series of Okazaki fragments.

has a short stretch of RNA at its 5′ end, since each segment is initiated by a separate priming event. This explains why primase is required throughout replication: Although the leading strand theoretically requires only one priming event, the discontinuously synthesized lagging strand requires multiple primers.

The mechanism for continuous synthesis of the leading strand and discontinuous synthesis of the lagging strand requires that the lagging-strand template be periodically repositioned. Each time an Okazaki fragment is completed, the polymerase must begin extending the RNA primer of the next Okazaki fragment. Presumably, other protein components of the replication complex assist in this repositioning in order to coordinate the activities of the two DNA polymerases at the replication fork.  Exercise 23 shows how the replication machinery synthesizes the leading and lagging strands of DNA.

REVIEW LEARNING OBJECTIVE 4

- Explain why DNA polymerization must be primed by RNA.
- Describe how two DNA polymerases, operating in the 5′→3′ direction, replicate the two antiparallel template strands.

DNA polymerases share a common structure and mechanism

All known DNA polymerases are shaped somewhat like a hand, with domains corresponding to palm, fingers, and thumb. These structure are likely the result of convergent evolution, as only the palm domains exhibit strong

Go to **Exercise 24/DNA Polymerase** for a closer look at the structure of the enzyme and its polymerizing and proofreading activities. This material addresses Learning Objective 5.

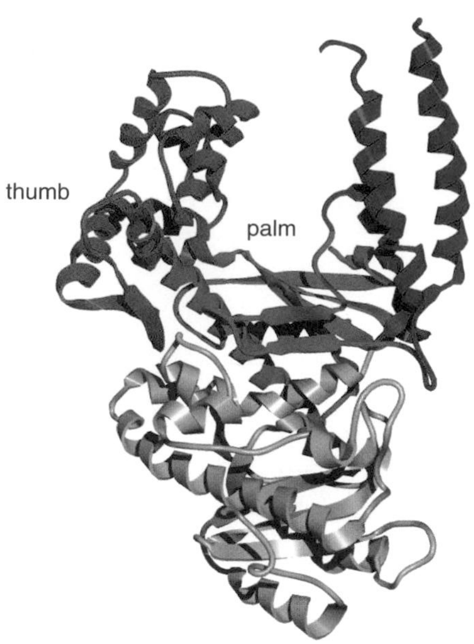

FIGURE 17-11

Structure of *E. coli* DNA polymerase I. This model shows the so-called Klenow fragment of DNA polymerase I (residues 324 to 928). The palm, fingers, and thumb domains are labeled. A loop at the end of the fingers is missing in this model. [*Structure (pdb 1KFD) determined by L.S. Beese, J.M. Friedman, and T.A. Steitz.*]

homology. One of the best known polymerases is *E. coli* DNA polymerase I, the first such enzyme to be characterized (Fig. 17-11). The template strand and the newly synthesized DNA strand, which form a double helix, lie across the palm, in a cleft lined with basic residues.

The polymerase active site, at the bottom of the cleft, contains two metal ions (probably Mg^{2+}) about 3.6 Å apart. These metal ions are coordinated by basic side chains of the protein, by water molecules, and by an oxygen atom of the incoming nucleoside triphosphate. These interactions help position the nucleotide for nucleophilic attack. The deoxy carbon at the 2′ position of the ribose ring lies in a hydrophobic pocket. This binding site allows the polymerase to discriminate against ribonucleotides, which bear a 2′ OH group. The mechanism of DNA polymerization is included in Exercise 24.

After each polymerization event, the enzyme must advance the template strand by one nucleotide. This translocation occurs as the polymerase momentarily relaxes its grip, possibly by opening and closing the fingers domain against the palm.

Most DNA polymerases are **processive** enzymes, which means that they undergo several catalytic cycles (about 10 to 15 for *E. coli* DNA polymerase *in vitro*) before dissociating from their substrates. *E. coli* DNA polymerase is even more processive *in vivo,* polymerizing as many as 5000 nucleotides before releasing the DNA. *This enhanced processivity is due to an accessory protein that forms a sliding clamp around the DNA and helps hold DNA polymerase in place.* In *E. coli,* the clamp is a dimeric protein and in yeast, the clamp is a trimer. Both proteins have a hexagonal ring structure and similar dimensions, consistent with a common function (Fig. 17-12). Additional proteins help assemble the clamp around the DNA. For the lagging-strand polymerase, the clamp must be reloaded at the start of each Okazaki fragment. This occurs about once every second in *E. coli.* The role of the sliding clamp is illustrated in Exercise 23.

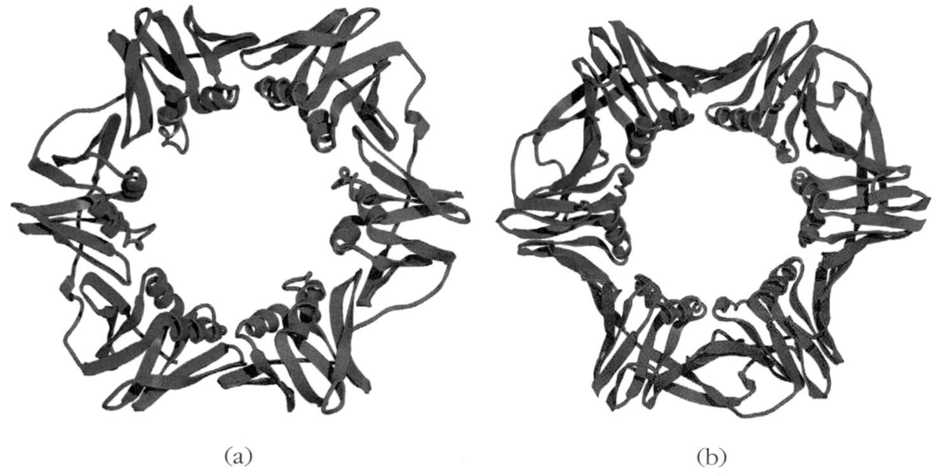

FIGURE 17-12

DNA polymerase–associated clamps.
(a) In *E. coli,* the β subunit of DNA polymerase III forms a dimeric clamp. (b) In yeast, the clamp is a trimer, called the proliferating cell nuclear antigen (PCNA). The inner surface of each clamp is positively charged and encloses a space with a diameter of about 35 Å, more than large enough to accommodate double-stranded DNA or a DNA–RNA hybrid helix with a diameter of 26 Å. Both structures enhance the processivity of their respective DNA polymerases, thereby increasing the efficiency of DNA replication. [*Structure of the β clamp (pdb 2POL) determined by X.-P. Kong and J. Kuriyan; structure of yeast PCNA (pdb 1AXC) determined by J.M. Gulbis and J. Kuriyan.*]

E. coli has at least five different DNA polymerase enzymes, and there are at least nine different eukaryotic DNA polymerases, not including those found in mitochondria and chloroplasts. Why so many? In eukaryotes and at least some prokaryotes, two different types of polymerases work in concert to synthesize the leading and lagging strands. For example, the leading-strand polymerase must be extremely processive, whereas the lagging-strand polymerase need not be. Many polymerases appear to function primarily in DNA repair pathways, some of which involve the excision and replacement of damaged DNA. In all cases, DNA polymerases function as part of multiprotein complexes that often include such proteins as a helicase, primase, and clamp.

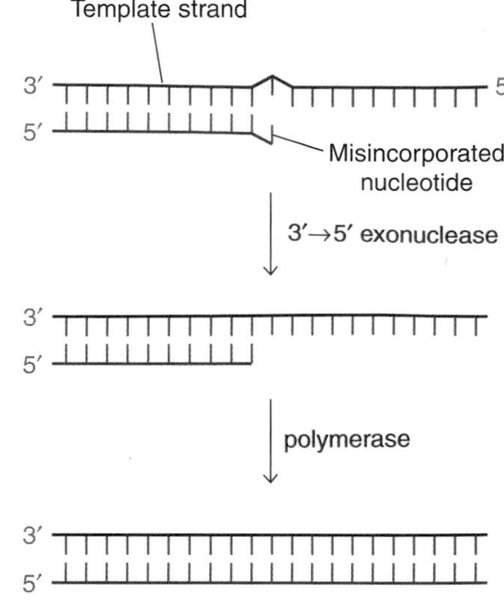

FIGURE 17-13

Proofreading during polymerization. DNA polymerase detects a distortion in double-stranded DNA that results from the incorporation of a mismatched nucleotide. The 3′→5′ exonuclease activity hydrolyzes the nucleotide at the 3′ end of the new strand (an exonuclease removes residues from the end of a polymer; an endonuclease cleaves within the polymer). The polymerase then resumes its activity, generating an accurately base-paired DNA product.

DNA polymerase proofreads newly synthesized DNA

During polymerization, the incoming nucleotide base-pairs with the template DNA so that the new strand will be complementary to the original. The polymerase accommodates base pairs snugly—recall that all possible pairings (A:T, T:A, C:G, and G:C) have the same overall geometry (see page 57). This minimizes the chance of mispairings. If the wrong nucleotide does become covalently linked to the growing chain, the polymerase can detect the distortion it creates in the newly generated double helix.

Many DNA polymerases contain a second active site that catalyzes hydrolysis of the nucleotide at the 3′ end of the growing DNA strand. *This 3′→5′ exonuclease excises misincorporated nucleotides, thereby acting as a* ***proofreader*** *for DNA polymerase* (Fig. 17-13). In *E. coli* DNA polymerase I, the 3′→5′ exonuclease active site is located about 25 Å from the polymerase active site, indicating that the enzyme–DNA complex must undergo a large conformational change in order to shift from polymerization to nucleotide hydrolysis. Exercise 24 explains the relationship between the polymerization and proofreading sites of DNA polymerase.

Proofreading during polymerization limits the error rate of DNA polymerase to about one in 10^5 to 10^6 bases. Misincorporated bases can also be removed following replication, through various DNA repair mechanisms, which further reduces the error rate of replication. This high degree of fidelity is absolutely essential for the accurate transmission of biological information from one generation to the next.

An RNase and a ligase are required to complete the lagging strand

Replication is not complete until the lagging strand—which is synthesized one Okazaki fragment at a time—is made whole. *After the completion of an Okazaki fragment, its RNA primer, along with some of the adjoining DNA, is hydrolytically removed and replaced with DNA; then the nick is sealed to generate a continuous DNA strand.* This process also increases the accuracy of DNA replication: Primase has low fidelity, so the RNA primers tend to contain errors, as do the first few deoxynucleotides added to the primer by the action of DNA polymerase.

In many cells, an exonuclease known as RNase H (H stands for hybrid) operates in the 5′→3′ direction to excise nucleotides at the primer end of an Okazaki fragment. Nucleotide hydrolysis may continue until DNA polymerase, now in the process of completing the next Okazaki fragment, "catches up" with RNase H (the polymerase is faster than the exonuclease). In extending the next Okazaki fragment, the polymerase replaces the excised ribonucleotides with deoxyribonucleotides, leaving a single-strand gap (called a nick) between the two lagging-strand segments (Fig. 17-14).

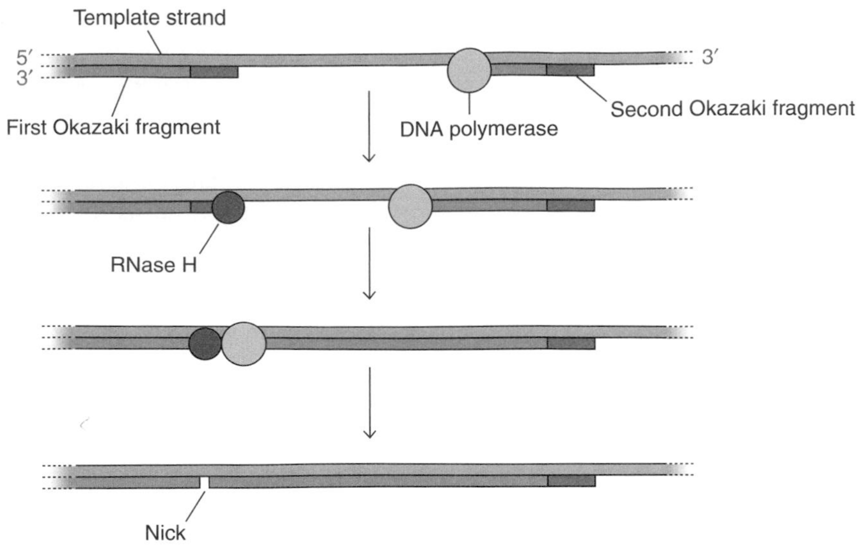

FIGURE 17-14

Primer excision.
RNase H removes the RNA primer and some of the adjoining DNA, allowing DNA polymerase to accurately replace these nucleotides. The nick can then be sealed.

In the event that DNA polymerase reaches the first Okazaki fragment before its primer has been completely removed, the polymerase displaces the primer, creating a single-stranded "flap." RNase H or another protein then acts as an endonuclease to cut away the primer (Fig. 17-15). This flap endonuclease appears to recognize the junction between the RNA and DNA portions of a single polynucleotide strand.

The *E. coli* DNA polymerase I polypeptide actually includes a 5′→3′ exonuclease activity (this is in addition to the 3′→5′ proofreading endonuclease), so that, at least *in vitro,* a single protein can remove ribonucleotides from the previous Okazaki fragment as it extends the next Okazaki fragment. The combined activities of removing and replacing nucleotides have the net effect of moving the nick in the 5′→3′ direction. This phenomenon is known as **nick translation.** DNA polymerase cannot seal the nick; this is the function of yet another enzyme.

The discontinuous segments of the lagging strand are joined by the action of DNA ligase. The reaction, which results in the formation of a phosphodiester bond, consumes the free energy of a similar bond in a nucleotide

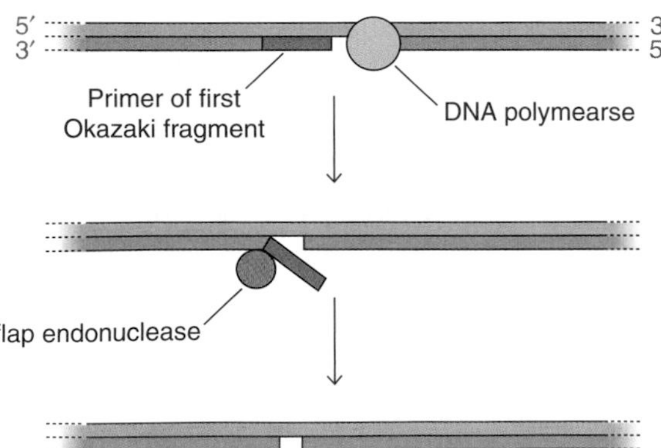

FIGURE 17-15

Function of the flap endonuclease.
If DNA polymerase displaces the RNA primer of the preceding Okazaki fragment, an endonuclease can remove the single-stranded flap.

cofactor. Depending on the species, either ATP or NAD^+ fulfils this role, yielding as products AMP and PP_i or AMP and nicotinamide mononucleotide.

$$\underset{\text{ATP}}{\text{Adenine—Ribose}}\text{—O—}\overset{\overset{\displaystyle O}{\|}}{\underset{\underset{\displaystyle O^-}{|}}{P}}\text{—O—}\overset{\overset{\displaystyle O}{\|}}{\underset{\underset{\displaystyle O^-}{|}}{P}}\text{—O—}\overset{\overset{\displaystyle O}{\|}}{\underset{\underset{\displaystyle O^-}{|}}{P}}\text{—O}^-$$

$$\downarrow$$

$$\underset{\text{AMP}}{\text{Adenine—Ribose}}\text{—O—}\overset{\overset{\displaystyle O}{\|}}{\underset{\underset{\displaystyle O^-}{|}}{P}}\text{—O}^- \quad + \quad PP_i$$

$$\text{Adenine—Ribose—O—}\overset{\overset{\displaystyle O}{\|}}{\underset{\underset{\displaystyle O^-}{|}}{P}}\text{—O—}\overset{\overset{\displaystyle O}{\|}}{\underset{\underset{\displaystyle O^-}{|}}{P}}\text{—O—Ribose—Nicotinamide}$$

Nicotinamide adenine dinucleotide (NAD)

$$\downarrow$$

$$\underset{\text{AMP}}{\text{Adenine—Ribose}}\text{—O—}\overset{\overset{\displaystyle O}{\|}}{\underset{\underset{\displaystyle O^-}{|}}{P}}\text{—O}^- \quad + \quad {}^-\text{O—}\overset{\overset{\displaystyle O}{\|}}{\underset{\underset{\displaystyle O^-}{|}}{P}}\text{—O—Ribose—Nicotinamide}$$

Nicotinamide mononucleotide

Ligation of Okazaki fragments yields a continuous lagging strand, completing the process of DNA replication. The roles of the RNase and DNA ligase are summarized in Exercise 23.

REVIEW LEARNING OBJECTIVE 5

- What protein enhances the processivity of DNA polymerase?
- How does the polymerase proofread its mistakes?
- Explain why an RNase, a polymerase, and a ligase are required to complete lagging-strand synthesis.

3 TELOMERES

In a circular bacterial DNA molecule, replication terminates where the two replication forks meet, at a point more or less opposite the replication origin. In a linear eukaryotic chromosome, replication must proceed to the very end of the chromosome. Copying the 5′ ends of the two parental DNA strands is straightforward, since DNA polymerase extends a new complementary strand in the 5′→3′ direction.

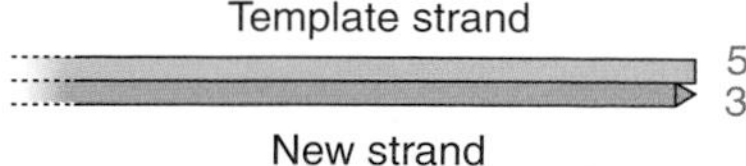

However, replication of the extreme 3′ ends of the parental DNA strands presents a problem, for the same reason. Even if an RNA primer were paired with the 3′ end of a template strand, DNA polymerase would not be able to replace the ribonucleotides with deoxynucleotides.

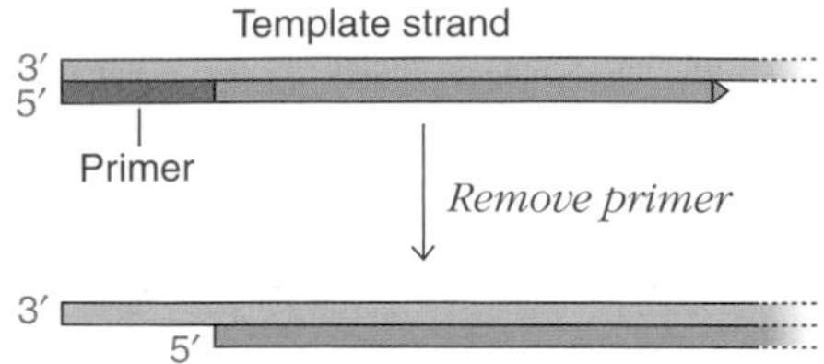

The 3′ end of each parental (template) DNA strand would then extend past the end of each new strand and would be susceptible to nucleases. Consequently, *each round of DNA replication would lead to chromosome shortening.*

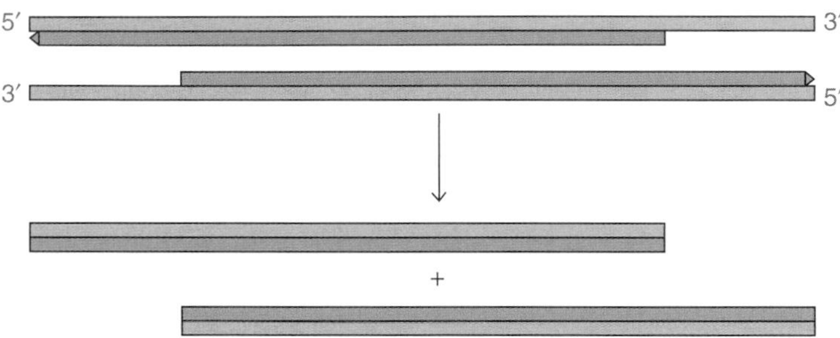

Eukaryotic cells counteract this potential chromosomal shortening by actively extending the ends of chromosomes. The resulting structure, which consists of short tandem (side-by-side) repeats, is called a **telomere.** A telomere—and its associated proteins—protects the end of the DNA molecule from nucleolytic degradation and prevents the joining of two chromosomes through end-to-end ligation (a reaction that normally occurs to repair broken DNA molecules).

Telomerase extends chromosomes

The telomere proteins include an enzymatic activity known as telomerase, which was first described by Elizabeth Blackburn. Telomerase adds the repeating sequence of nucleotides to the 3′ end of a DNA strand, using an enzyme-associated RNA molecule as a template. The catalytic subunit of telomerase is a **reverse transcriptase,** a homolog of the viral enzyme that copies the viral RNA genome into DNA (Box 17-A).

The RNA subunit of telomerase ranges from several hundred to over a thousand nucleotides, depending on the species. In humans, the RNA is a sequence of 450 bases, including the six that serve as a template for the addition of DNA repeats with the sequence TTAGGG. Similar G-rich sequences extend the 3′ ends of chromosomes in all eukaryotes. In order to synthesize telomeric DNA, the template RNA must be repeatedly realigned with the end of the preceding hexanucleotide extension. The precise alignment probably depends on base-pairing between telomeric DNA and a nontemplate portion of the telomerase RNA. When the 3′ end has been extended, it can then serve as a template for the conventional synthesis of a C-rich extension of the complementary DNA strand (Fig. 17-16). In humans, telomeric DNA is 5 to 15 kb long.

FIGURE 17-16

Synthesis of telomeric DNA. Telomerase extends the 3′ end of one strand by adding TTAGGG repeats (yellow segments). DNA polymerase can then extend the complementary strand by the normal mechanism for lagging-strand synthesis (orange segments). Note that this process still leaves a 3′ overhang (about 300 nucleotides in humans), but the chromosome has been lengthened.

A CLOSER LOOK

BOX 17-A HIV and reverse transcriptase

When HIV, the virus that causes AIDS, was discovered in 1983, much was already known about its biology. HIV is one of the retroviruses, originally called RNA tumor viruses because of their ability to cause tumors in certain animals. HIV kills cells of the immune system rather than causing cancer, but its RNA genome resembles those of the tumor viruses.

Studies of RNA viruses in the 1960s redefined the meaning of genetic information and rewrote the Central Dogma, which until that point held that biological information flows unidirectionally from DNA to RNA to protein. RNA viruses copy their genes to DNA by a "reverse transcription" process (hence the name "retrovirus") using reverse transcriptase, an enzyme that synthesizes DNA from an RNA template.

Viral reverse transcriptase proved to be more than a biological curiosity. It became a valuable laboratory tool, allowing researchers to purify messenger RNA transcripts from cells, transform them to DNA (called cDNA), and then clone them for sequencing or genetic engineering experiments.

The structure of reverse transcriptase includes domains corresponding to the fingers, thumb, and palm regions of other polymerases. This part of the protein (colored red below) contains a polymerase active site that can actually use either DNA or RNA as a template (no other enzyme has this dual specificity). A separate domain (green) contains an RNase active site that degrades the RNA template.

The entire process of reverse transcription occurs as follows: The enzyme binds to the RNA template and generates a complementary DNA strand. Next, the RNase active site degrades the RNA strand of the RNA–DNA hybrid molecule, leaving a single DNA strand that then serves as a template for the reverse transcriptase polymerase active site to use in synthesizing a second strand of DNA. The result is a double-stranded DNA molecule.

Chemical inhibitors of reverse transcriptase include the active-site-directed dideoxynucleotides derived from nucleosides such as AZT (3′-azido-2′,3′-dideoxythymidine; see Box 7-A). These compounds were originally developed as anticancer agents (although they were not very effective) and were already on the shelf when HIV came along. Non-nucleoside inhibitors of HIV reverse transcriptase bind to the enzyme near the base of the thumb domain. This does not interfere with RNA or nucleotide binding, but it does inhibit polymerase activity, probably by restricting thumb movement. HIV reverse transcriptase is also presented in Exercise 24.

[Structure of HIV reverse transcriptase (pdb 1BQN) determined by Y. Hsiou, K. Das, and E. Arnold.]

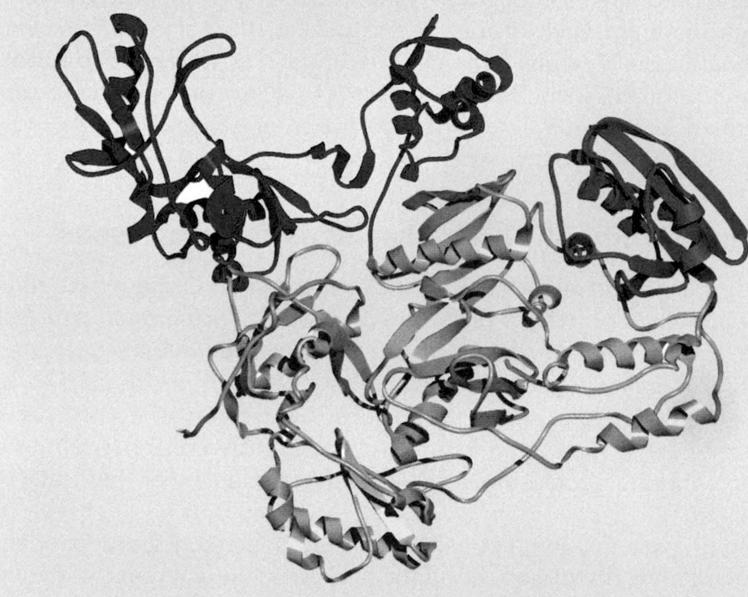

Is telomerase activity linked to cell immortality?

Cells that normally undergo a limited number of cell divisions appear to contain no active telomerase. Consequently, the size of the telomeres decreases with each replication cycle, until the cells reach a senescent stage and no longer divide. In contrast, cells that are "immortal" appear to have active telomerase. These findings are consistent with the role of telomerase in maintaining the ends of chromosomes over many cycles of replication. In fact, telomerase appears to be activated in cancer cells derived from tissues that are normally senescent. This suggests that shutting down telomerase activity might be an effective treatment for cancer. However, for many cells, telomerase activity by itself is not an indicator of the cells' potential for immortality. Rather, the ability of the cells to undergo repeated divisions without chromosome shortening appears to be a more complicated function of telomerase activity, telomere length, and the integrity of the telomere (which may depend on telomere proteins other than telomerase).

REVIEW LEARNING OBJECTIVE 6

- Why does DNA replication lead to chromosome shortening?
- Describe the structure of telomeric DNA.
- What is the function of the RNA component of telomerase?

4 DNA PACKAGING

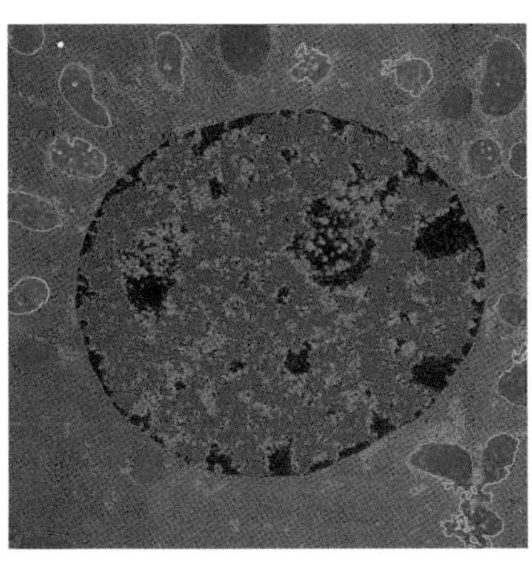

FIGURE 17-17

A eukaryotic nucleus.
Heterochromatin is the darkly staining material; euchromatin stains more lightly. [*Dr. Gopal Murti/Science Photo Library/ Photo Researchers*]

At some point after it is replicated, eukaryotic DNA assumes a highly condensed form. This is advantageous for a cell about to divide, since elongated DNA molecules would become hopelessly tangled rather than segregating neatly to form two equivalent sets of chromosomes (fully extended human chromosomes would have a collective length of about 2 meters). Even when the cell is not dividing, much of the DNA is packaged in a form that compresses its length considerably. This DNA is known as **heterochromatin,** which is transcriptionally silent. **Euchromatin** is less condensed and appears to be transcribed at a higher rate. The two forms of chromatin are distinguishable by electron microscopy (Fig. 17-17).

The fundamental unit of DNA packaging is the nucleosome

Both heterochromatin and euchromatin contain structural units known as **nucleosomes,** which are complexes of DNA and protein. *The core of a nucleosome consists of eight histone proteins: two each of the histones known as H2A, H2B, H3, and H4. Approximately 146 base pairs of DNA wind around the histone octamer* (Fig. 17-18). A complete nucleosome contains the core structure plus histone H1, which appears to bind outside the core, between the DNA coils. Neighboring nucleosomes are separated by short stretches of DNA of variable length.

The histone proteins interact with DNA in a sequence-independent manner, primarily via hydrogen bonding and ionic interactions with the sugar–phosphate backbone. Although prokaryotes lack histones, other DNA-binding proteins may help package DNA in bacterial cells.

The winding of DNA in the nucleosome (it makes approximately 1.65 turns around the histone octamer) generates positive supercoils (in other words, the DNA in nucleosomes becomes slightly overwound). This conformation is stabilized by interactions with the histone octamer. During DNA replication, nucleosomes are disassembled as the DNA spools through the replication machinery. A protein complex called the replication-coupling as-

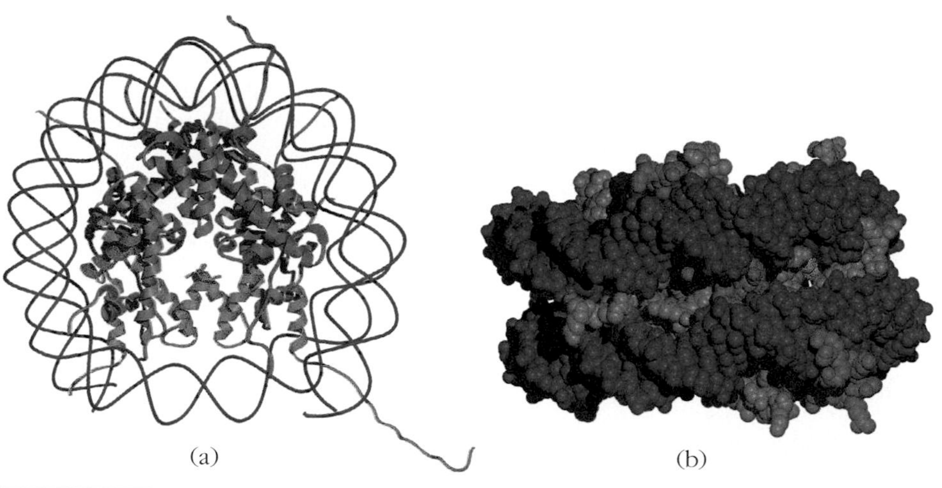

FIGURE 17-18

Structure of the nucleosome core.
(a) Top view. (b) Side view (space-filling model). The DNA (dark blue) winds around the outside of the histone octamer. [*Structure (pdb 1AOI) determined by K. Luger, A.W. Maeder, R.K. Richmond, D.F. Sargent, and T.J. Richmond.*]

sembly factor helps reassemble nucleosomes on newly replicated DNA, using the displaced histones as well as newly synthesized histones imported from the cytoplasm to the nucleus.

Nucleosomes compact the DNA only by a factor of about 30 to 40, but the chain of nucleosomes itself coils into a solenoid with a diameter of about 300 Å (Fig. 17-19). A portion of histone H4 in one nucleosome appears to

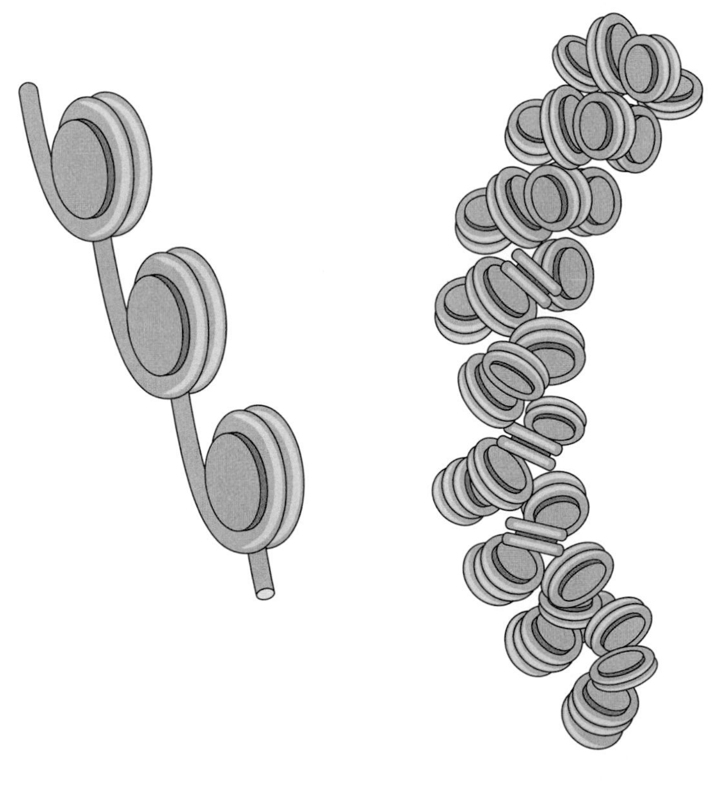

FIGURE 17-19

Two levels of chromatin structure.
(a) Nucleosomes give DNA the appearance of beads on a string. (b) The packing of nucleosomes into a solenoid further condenses the DNA molecule.

interact with the H2A–H2B pair of histones in a neighboring nucleosome. *The solenoid is irregular in structure, owing to the slight differences in the length of the linker DNA between nucleosomes.* This form of chromatin is presumably well-protected from nuclease attack and is inaccessible to the proteins that carry out replication and transcription.

Just prior to cell division, each chromosome condenses even further, to form a rod with an average length of about 10 μm and a diameter of 1 μm (for comparison, the DNA helix has a diameter of about 20 Å, or 2 nm). The mechanical feat of compressing DNA into this structure requires additional proteins and the free energy of ATP hydrolysis.

Histones are covalently modified

The histones are among the most highly conserved proteins known, in keeping with their function in packaging the genetic material in all eukaryotic cells. Each histone pairs with another in a sort of handshake (Fig. 17-20), and the set of eight forms a compact structure (see Fig. 17-18). However, the tails of the histones, which are flexible and charged, extend outward from the core of the nucleosome. These histone tails are subject to covalent modification, including acetylation of Lys residues, phosphorylation of Ser and His side chains, and methylation of Lys and Arg side chains.

The addition or removal of the various histone-modifying groups can potentially introduce considerable variation in the fine structure of chromatin, offering a mechanism for promoting or preventing gene expression. For example, a Lys residue of a histone is positively charged and could interact strongly with the negatively charged DNA backbone. Acetylation of the Lys side chain would neutralize it and weaken its interaction with the DNA, possibly destabilizing the nucleosome or allowing other proteins access to the DNA. In fact, acetylation of histones is associated with transcriptionally active chromatin, and deacetylation appears to repress transcription. Histone phosphorylation appears to be a prelude to the chromosome condensation that occurs during cell division. Recall that phosphorylation catalyzed by kinases is a general feature of growth factor signaling pathways (see Section 16-3).

Some of the enzymes that modify histones act on many residues in different histone proteins; others are highly specific for one residue in one histone. The modifications themselves are interdependent. For example, in histone H3, methylation of Lys 9 inhibits phosphorylation of Ser 10, but phosphorylation of Ser 10 promotes acetylation of Lys 14. Such factors may explain why there is sometimes no clear correlation between the histone

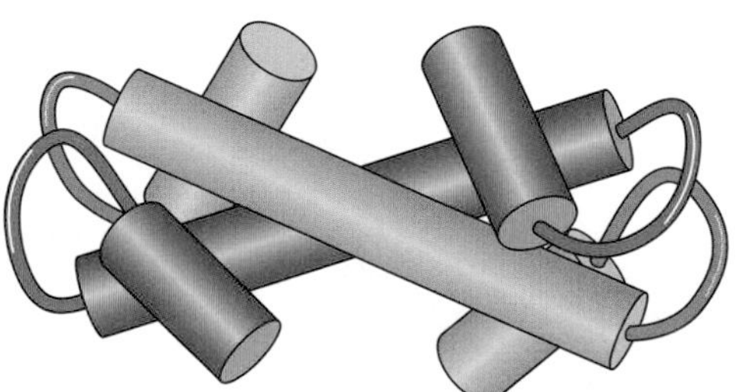

FIGURE 17-20

Interaction of histones.
Each histone contains a structural motif consisting of a long central helix flanked by two shorter helices. The long helices of two histones (shown here in green and yellow) overlap.

acetylation state and readiness for transcription. The different combinations of histone modification may be a sort of "histone code" that can be "read" by various proteins. *Alterations in histone–histone and histone–DNA contacts could alter the structure of the nucleosomes as well as provide binding sites for proteins such as transcription factors* (these are discussed in greater detail in Section 19-1).

DNA also undergoes covalent modification

In many organisms, including plants and animals, DNA methyltransferases add methyl groups to cytosine residues:

5-Methylcytosine residue

The methyl group projects out into the major groove of DNA and could potentially alter interactions with DNA-binding proteins.

In mammals, the methyltransferases target C residues next to G residues, so that about 80% of CG sequences (formally represented as CpG) are methylated. CpG sequences occur much less frequently than statistically predicted, but clusters of CpG (called **CpG islands**) are often located near the starting points of genes. Interestingly, these CpG islands are usually unmethylated. This suggests that *methylation may be part of the mechanism for marking or "silencing" DNA that contains no genes.*

After DNA is replicated, methyltransferases can modify the new strand, using the methylation pattern of the parent DNA strand as a guide. Cells also contain enzymes that methylate DNA without a template, and enzymes that demethylate DNA.

Variations in DNA methylation may be responsible for **imprinting,** in which the level of expression of a gene depends on its parental origin. Recall that an individual receives one copy of a gene from each parent. Normally, both copies are expressed, but a gene that has been methylated may not be expressed. The gene behaves as if it has been "imprinted" and knows its parentage. Imprinting explains why some traits are transmitted in a maternal or paternal fashion, in opposition to the Mendelian laws of inheritance. Imprinted genes may also account for the abnormalities in cloned mammals, whose genes all originate from a single maternal cell. In female mammals, some form of imprinting may help compensate for the double dose of X-chromosome genes (Box 17-B).

REVIEW LEARNING OBJECTIVE 7

- How does heterochromatin differ from euchromatin in structure and function?
- What is the role of histones in packaging DNA?
- How could covalent modification of histones alter gene expression?
- What are the possible functions of DNA methylation?

A CLOSER LOOK

BOX 17-B X chromosome inactivation

The cells of female mammals contain two X chromosomes; the males have one X and one Y. The Y chromosome is small and contains mostly male-determining genes, but the X chromosome is much larger and contains more genes. To restore balance, females must somehow shut off half of their X-chromosome genes.

Early in female development, one of the X chromosomes is inactivated and remains inactive throughout all subsequent cell divisions. This early inactivation affects the maternal or paternal X chromosome at random, so females are said to be mosaics of tissues expressing either the maternal or paternal X-chromosome genes. Genetic studies have shown that X inactivation affects the entire chromosome (with one exception, explained below); it is not the result of randomly shutting down one or the other of each pair of genes. The inactive X chromosome is often visible as a Barr body, a highly condensed, darkly staining mass in the cell nucleus.

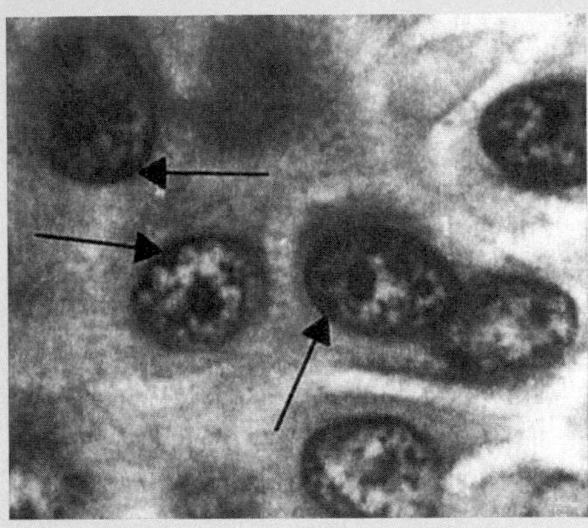

[Courtesy Murray L. Barr and Brian Blumerfelt, University of Western Ontario.]

X inactivation is a multifactorial event. The inactive chromosome is enriched in the histone H2A variant known as macro H2A1, which is otherwise randomly distributed throughout the cell's chromatin. In addition, the X histones are hypoacetylated and its CpG islands are methylated. The only gene that escapes inactivation is called *Xist* in humans (for X-inactive specific transcript). *Xist* encodes a relatively large RNA molecule of about 15 to 17 kb. This RNA is extremely stable and coats the inactive chromosome by an unknown mechanism (the *Xist* gene is inactive on the active X chromosome). The *Xist* RNA is essential for maintaining the inactive state of the chromosome. This appears to explain why the inactivated X chromosome is the last to be replicated and why it remains in its inactive condensed state following cell division.

SUMMARY

1. In order to be replicated, the two strands of DNA must be separated, or unwound. This unwinding is facilitated by the negative supercoiling (underwinding) of DNA molecules. Enzymes called topoisomerases can add or remove supercoils by temporarily introducing breaks in one or both DNA strands.

2. DNA replication requires a host of enzymes and other proteins located in a stationary factory. Helicases separate the two DNA strands at a replication fork, and SSB binds to the exposed single strands.

3. DNA polymerase can only extend a preexisting chain and therefore requires an RNA primer synthesized by a primase. Two polymerases operate side-by-side to replicate DNA, so the leading strand of DNA is synthesized continuously while the lagging strand is synthesized discontinuously as a series of Okazaki fragments. The RNA primers of the Okazaki fragments are removed, the gaps are filled in by DNA polymerase, and the nicks are sealed by DNA ligase.

4. DNA polymerase and other polymerases catalyze a reaction in which the 3′ OH group of the growing chain nucleophilically attacks the phosphate group of an incoming nucleotide that base-pairs with the template strand. A sliding clamp promotes the processive activity of DNA polymerase.

5. Many DNA polymerases contain a second active site that hydrolytically excises a mismatched nucleotide.

6. In eukaryotes, the extreme 3′ end of a DNA strand cannot be replicated, so the enzyme telomerase adds repeating sequences to the 3′ end to create a structure known as a telomere.

7. After it is replicated, eukaryotic DNA is wound around a core of eight histone proteins to form a nucleosome, which represents the first level of DNA compaction in the nucleus.

8. The histone components of nucleosomes may be modified by acetylation, phosphorylation, and methylation as a potential mechanism for regulating gene expression. The DNA may also be covalently modified, often by methylation of C residues in "silent" regions of the genome.

CHECKLIST

1. Review the Learning Objectives listed on page 530.
2. Complete Exercise 23/DNA Replication for an overview of the process of DNA synthesis and the functions of some replication proteins.
3. Complete Exercise 24/DNA Polymerase for a closer look at the polymerization and proofreading activities of DNA polymerase.
4. Apply your knowledge by solving the problems at the end of this chapter. Check your results in the Solutions appendix.
5. Be able to define the boldfaced terms (consult the glossary at the end of the book). Test your understanding by taking the Chapter 17 quiz (www.wiley.com/college/pratt).
6. Explore the websites listed at www.wiley.com/college/pratt.
7. Consult the list of Selected Readings for background information or additional details on DNA topology, DNA replication, and nucleosomes.

GLOSSARY TERMS

replication
semiconservative replication
supercoiling
nick
replication fork
factory model of replication
primer
leading strand
lagging strand
discontinuous synthesis
Okazaki fragment
processivity
proofreading
nick translation
telomere
reverse transcriptase
heterochromatin
euchromatin
nucleosome
CpG island
imprinting

PROBLEMS

1. Semiconservative replication is shown in Figure 17-1. Draw a diagram that illustrates the composition of DNA daughter molecules for the second, third, and fourth generations.

2. In bacteria, replication initiates at a single site called the origin. Would the origin likely be richer in G:C or A:T base pairs?

3. Explain how DNase (an endonuclease that cleaves the backbone of one strand of a DNA molecule), *E. coli* DNA polymerase I (which includes 5′→3′ exonuclease activity), and DNA ligase could be used in the laboratory to incorporate radioactive nucleotides into a DNA molecule.

4. A variety of compounds inhibit topoisomerases. Novobiocin (like ciprofloxacin) inhibits DNA gyrase, while doxorubicin and etoposide are anticancer drugs. What properties distinguish the antibiotics from the anticancer drugs?

5. The percentages of arginine and lysine residues in the histones of calf thymus DNA are shown in the table below. Why do histones have a large number of Lys and Arg residues?

	% Arg	% Lys
H1	1	29
H2A	9	11
H2B	6	16
H3	13	10
H4	14	11

6. Draw the structures of the side chains that correspond to the following histone modifications: (a) acetylation of Lys, (b) phosphorylation of Ser, (c) phosphorylation of His, (d) methylation of Lys, and (e) methylation of Arg.

7. In the chain termination method of DNA sequencing (described in Section 3-3), the Klenow fragment of DNA polymerase I (see Fig. 17-11) is used to synthesize a complementary strand using single-stranded DNA as a template. Along with a primer and the four dNTP substrates, a small amount of 2′,3′-dideoxynucleoside triphosphate

(ddNTP) is added to the reaction mixture. When the ddNTP is incorporated into the growing nucleotide chain, polymerization stops. Explain.

Ⓟ—Ⓟ—Ⓟ—OCH_2 O Base H H H H H H

2′,3′-Dideoxynucleoside triphosphate (ddNTP)

8. The mechanism of *E. coli* DNA ligase involves the transfer of the adenylyl group of NAD^+ to the ε-amino group of the side chain of an essential Lysine residue on the enzyme. The adenylyl group is subsequently transferred to the 5′ phosphate group of the nick. The first step is shown in the figure below. Draw the mechanism of *E. coli* DNA ligase.

Enzyme—$(CH_2)_4$—NH_2

Nicotinamide—Ribose—O—P(=O)(O^-)—O—P(=O)(O^-)—O—Ribose—Adenine

9. You have discovered a drug that inhibits the activity of the enzyme inorganic pyrophosphatase. What effect would this drug have on DNA synthesis?

10. Based on your knowledge of replication proteins, compare DNA polymerase and single-strand binding protein (SSB) with respect to (a) affinity for DNA and (b) cellular concentration.

11. Describe the ways in which a cell minimizes the incorporation of mispaired nucleotides during DNA replication.

12. The "flap" endonuclease recognizes the junction between RNA and DNA near the 5′ end of an Okazaki fragment. What feature(s) of this structure might the enzyme recognize?

13. Why would it not make sense for the cell to wait to combine Okazaki fragments into one continuous lagging strand until the entire DNA molecule had been replicated?

14. The enzyme deoxyribonuclease I (DNase I) is an endonuclease that hydrolyzes the phosphodiester bonds of the double-stranded DNA backbone to yield small oligonucleotide fragments. DNase I is used therapeutically to treat patients with cystic fibrosis (CF). The DNase I enzyme is inhaled into the lungs, where it acts on the DNA contained in the viscous sputum secreted by the lungs in these patients. Hydrolysis of the DNA decreases sputum viscosity and improves lung function. Animal studies also have shown that DNase I is effective in treating the autoimmune disease systemic lupus erythematosus (SLE). In this disease, DNA present in blood serum provokes an immune response. DNase I prevents the immune response by degrading the DNA to smaller fragments that are not recognized by the immune system.

Genentech, Inc., a company that produces recombinant DNase I, was interested in improving the efficiency of DNase I so that less drug would be needed to achieve the same results. Scientists in the protein-engineering lab constructed hyperactive DNase variants that actually worked better than the wild-type enzyme. DNase I acts by processively nicking the phosphodiester backbone, so the scientists reasoned that a variant that could create more nicks in a shorter period of time would act more efficiently than the wild-type enzyme. In this case, we will examine the engineered hyperactive variants and use the results to make some conclusions about the mechanism of DNase I.

(a) The DNase I variants engineered by the Genentech scientists are listed in the table. (A note on nomenclature: Q9R means that the glutamine at position 9 in the wild-type DNase I has been changed to an arginine.) What structural feature do all of the DNase I variants have in common? Explain the meaning of the abbreviation in the table. Why do you suppose that the protein engineers thought that these changes would improve the catalytic efficiency of DNase I?

DNase variant	**Abbreviation**
Q9R	+1
E13R	+1
T14K	+1
H44K	+1
N74K	+1
T205K	+1
E13R/N74K	+2
Q9R/E13R/N74K	+3
E13R/N74K/T205	+3
Q9R/E13R/N74K/T205K	+4
E13R/H44K/N74K/T205K	+4
T14K/H44K/N74K/T205K	+4
E13R/T14K/N74K/T205K	+4
Q9R/E13R/H44K/N74K/T205K	+5
Q9R/E13R/T14K/H44R/N74K/T205K	+6

(b) The enzymatic activity of the DNase I variants was tested using a DNA hyperchromicity assay. The absorbance of a solution of intact DNA was measured at 260 nm. Then the enzyme was added, and the solution was monitored for an increase in absorbance. (The increase in absorbance at 260 nm is referred to as the hyperchromic effect.) Why was a hyperchromicity assay effective in assessing the activity of the DNase I variants?

(c) The DNA hyperchromicity assay was used to measure the K_M and V_{max} values for each variant. The results are shown in the following table. Explain the significance of the K_M and V_{max} values. How do the amino acid changes affect the activity of the enzyme variants as compared to the wild type? Which variant(s) is(are) most efficient?

DNase variant	K_M (μg/mL DNA)	V_{max} (A_{260} units·min^{-1}·mg^{-1})
wild type	1.0	1.0
Q9R	1.1	2.8
E13R	0.23	1.5
T14K	0.43	1.1
H44K	0.43	1.1
N74K	0.77	3.6
T205K	0.42	2.1
E13R/N74K	0.20	5.3
Q9R/E13R/N74K	0.20	7.0
E13R/N74K/T205K	0.18	7.7
Q9R/E13R/N74K/T205K	0.09	2.8
E13R/H44K/N74K/T205K	0.17	6.4
T14K/H44K/N74K/T205K	0.18	7.7
E13R/T14K/N74K/T205K	0.37	3.5
Q9R/E13R/H44K/N74K/T205K	0.11	2.4
Q9R/E13R/T14K/H44R/N74K/T205K	0.20	2.5

(d) Next, the protein engineers characterized the DNase I variants in terms of their ability to cut or nick DNA. A *cut* refers to the hydrolysis of phosphodiester bonds on both strands, whereas a *nick* is the hydrolysis of just one strand. This was assessed by using the circular plasmid pBR322. The plasmid is most stable in the supercoiled form. If the phosphodiester backbone of one strand is nicked, the plasmid forms a relaxed circle, but if the backbones of both strands are cut, the circle linearizes, as shown at right.

Supercoiled, relaxed circular, and linear DNA can be detected by their different rates of migration during electrophoresis through agarose gels. In a series of experiments, the pBR322 substrate was incubated with DNase I for 45 minutes, then the products were analyzed by agarose gel electrophoresis. The results are shown here. Describe the results for each lane. Compare the selected variants to the wild-type DNase I with regard to their ability to cut or nick the DNA. (C, control; WT, wild-type; +3, 13R/74K/205K variant; +4, 13R/14K/74K/205K variant; +6, 9R/13R/14K/44R/74K/205K variant.)

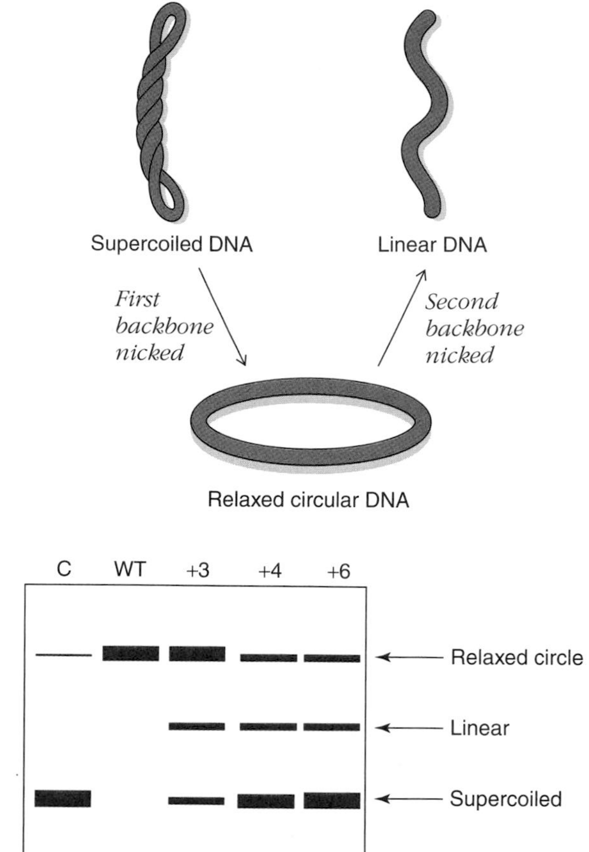

(e) The investigators next tested the DNase I variants' enzymatic activity with high- and low-molecular-weight DNA at high and low concentrations. High-molecular-weight (MW) DNA is present in the lung secretions of CF patients in fairly high concentrations, but the DNA present in the serum of mice with SLE is present in one-tenth the concentration. The data are shown in the table. What is your interpretation of these data?

	Nicking activity (compared to wild type)			
	Low DNA concentration		High DNA concentration	
	Low MW	High MW	Low MW	High MW
Wild type	1	1	1	1
N74K	26		10.4	
E13R/N74K	211	31	24.3	13
E13R/N74K/T205K	7		1.3	
E13R/T14K/N74K/T205K	7		0.7	

(f) Use the experimental evidence presented here to compare the action of the wild-type DNase I with the variant DNase I enzymes.

[From Pan, C.Q. and Lazarus, R.A., *J. Biol. Chem.* **273,** 11701–11708 (1998).]

15. DNA helicases can be considered to be molecular motors that convert the chemical energy of NTP hydrolysis into mechanical energy for separating DNA strands. The bacteriophage T7 genome encodes a protein that assembles into a hexameric ring with helicase activity.

(a) In order for the helicase to unwind DNA, it requires two single-stranded DNA tails at one end of a double-stranded DNA segment. However, the helicase appears to bind to only one of the single strands, apparently by encircling it. Is this consistent with its ability to unwind a double-stranded DNA helix?

(b) Kinetic measurements indicate that the T7 helicase moves along the DNA at a rate of 132 bases per second. The protein hydrolyzes 49 dTTP per second. What is the relationship between dTTP hydrolysis and DNA unwinding?

(c) What does the structure of T7 helicase suggest about its processivity?

[From Kim, D.-E., Narayan, M., and Patel, S.S., *J. Mol. Biol.* **321,** 807–819 (2002).]

16. DNA polymerases include two Mg^{2+} ions in the active site.

(a) Describe how Mg^{2+} could enhance the nucleophilic attack of the DNA chain on an incoming nucleotide.

(b) The polymerization mechanism includes the formation of a pentacovalent phosphorus. Sketch this structure. How could a Mg^{2+} ion contribute to transition-state stabilization during DNA polymerization?

17. A reaction mixture contains purified DNA polymerase, the four dNTPs, and one of the DNA molecules whose structures are represented below. Which reaction mixture produces PP_i?

(a)

(b)

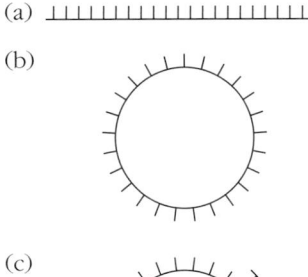

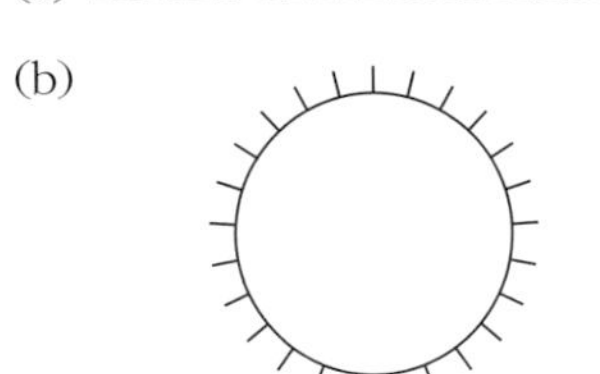

(c)

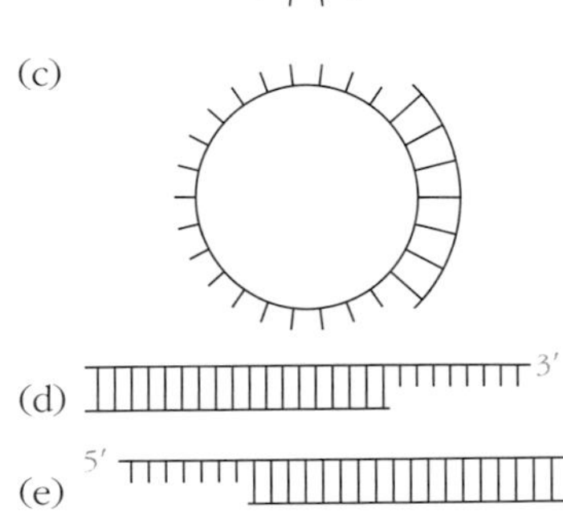

(d)

(e) 5′

(f)

18. Eukaryotic cells contain a number of DNA polymerases. Several of these enzymes were tested for their ability to cleave nucleotides from the 3′ end of a DNA chain (3′→5′ exonuclease activity). The enzymes were also tested for the accuracy of DNA polymerization, expressed as the rate of base substitution. The results are summarized in the table.

Polymerase	3′→5′ exonuclease activity	Base substitution rate ($\times 10^{-5}$)
α	no	16
β	no	67
δ	yes	1
ε	yes	1
η	no	3500

(a) Is there a correlation between the presence of 3′→5′ exonuclease activity and the error rate (base substitutions) during DNA polymerization?

(b) Express the error rate of each polymerase in terms of how often a wrong base is incorporated.

(c) Polymerization errors can result from the ability to insert an incorrect (mispaired) base or from the inability to efficiently insert the correct (template-matched) base. To tell which mechanism accounts for the high error rate of polymerase η, the catalytic efficiency (k_{cat}/K_M) of the polymerization reaction was measured for matched and unmatched bases. The results were compared to the catalytic efficiency of another polymerase, HIV reverse transcriptase (HIV RT). The data are summarized in the table.

Polymerase	Template base	Uncoming base	k_{cat}/K_M ($\mu M \cdot min^{-1} \times 10^3$)
Polymerase η	T	A	420
Polymerase η	T	G	22
Polymerase η	T	C	1.6
Polymerase η	G	C	760
Polymerase η	G	G	8.7
HIV RT	T	A	800
HIV RT	T	G	0.07

Compare the efficiency of DNA polymerase η and HIV RT in incorporating correct bases and incorrect bases. What do these results reveal about the cause of errors during polymerization by polymerase η?

(d) The overexpression of genes that code for DNA polymerases similar to polymerase η has been observed in bacteria. What effect would this have on the mutation rate in these bacteria?

[From Matsuda, T., Bebenek, K., Masutani, C., Hanaoka, F., and Kunkel, T.A., *Nature* **404,** 1011–1013 (2000).]

19. The G-rich telomeric sequences of eukaryotic chromosomes terminate in a single-stranded overhang that can fold up on itself to form a four-stranded structure. In this structure, four guanine residues assume a hydrogen-bonded planar arrangement with an overall geometry that can be represented as

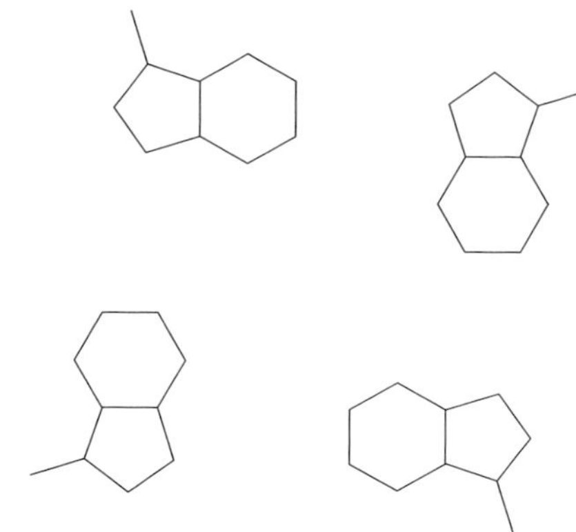

(a) Draw the complete structure of this "G quartet," including the hydrogen bonds between the purine bases.

(b) Show schematically how a single strand of four repeating TTAGGG sequences can fold to generate a structure with three stacked G quartets linked by TTA loops.

20. DNA methylation requires the methyl group donor *S*-adenosylmethionine, which is produced by the condensation of methionine with ATP. The sulfonium ion's methyl group is used in methyl group transfer reactions.

$$CH_3-\overset{+}{S}(CH_2\text{-ribose-Adenine})-CH_2-CH_2-\underset{NH_3^+}{\overset{H}{C}}-COO^-$$

H H H H, HO OH, O, Adenine

S-Adenosylmethionine

(a) The demethylated *S*-adenosylmethionine is then hydrolyzed to produce adenosine and a nonstandard amino acid. Draw the structure of this amino acid. How does the cell convert this compound back to methionine to regenerate *S*-adenosylmethionine?

(b) The proper regulation of gene expression requires methylation as well as demethylation of cytosine residues in DNA. If a demethylase carries out a hydrolytic reaction to restore cytosine residues, what is the other reaction product?

SELECTED READINGS

Cook, P.R., The organization of replication and transcription, *Science* **284,** 1790–1795 (1999). [Argues for the factory model of replication.]

Felsenfeld, G. and Groudine, M., Controlling the double helix, *Nature* **421,** 448–453 (2003). [Describes the structure and dynamics of chromatin.]

Hübscher, U., Nasheuer, H.-P., and Syväoja, J.E., Eukaryotic DNA polymerases, a growing family, *Trends Biochem. Sci.* **25,** 143–147 (2000). [Lists the known eukaryotic polymerases and their probable functions in replication or DNA repair.]

Inglesby, T.V., et al., Anthrax as a biological weapon: Medical and public health management, *JAMA* **281,** 1735–1745 (1999). [Describes the history, symptoms, and treatment of anthrax infections.]

O'Reilly, M.O., Teichmann, S.A., and Rhodes, D., Telomerases, *Curr. Opin. Struct. Biol.* **9,** 56–65 (1999). [Reviews the structures and functions of the RNA and reverse transcriptase components of telomerase.]

Strahl, B.D. and Allis, C.D., The language of covalent histone modifications, *Nature* **403,** 41–45 (2000). [Describes how combinations of histone modifications could function as a code to regulate replication and gene expression.]

CHAPTER 18

Cancer and DNA Repair

Despite their tiny size and unassuming appearance, these cells of *Deinococcus radiodurans* may well be the world's toughest organisms. These bacteria have been isolated from X-irradiated meat and medical instruments. They can also survive more mundane insults—such as ultraviolet light and hydrogen peroxide—that routinely kill other bacteria. *Deinococcus* owes its success to its tightly compacted genome and highly efficient DNA-repair systems. Not only does its genome contain a full suite of genes for DNA repair enzymes, but these genes are present as multiple isozymes in multiple copies. A rich complement of genes for other metabolic processes indicates a high capacity for recovering from extreme conditions, including desiccation and starvation. As if this weren't enough, genetic engineers have introduced additional genes that allow the bacteria to detoxify mercury compounds. The resulting "superbugs" can potentially be used at nuclear waste dumps—where no other bacteria can survive—to clean up the heavy metals that commonly pollute the ground and water at these sites.

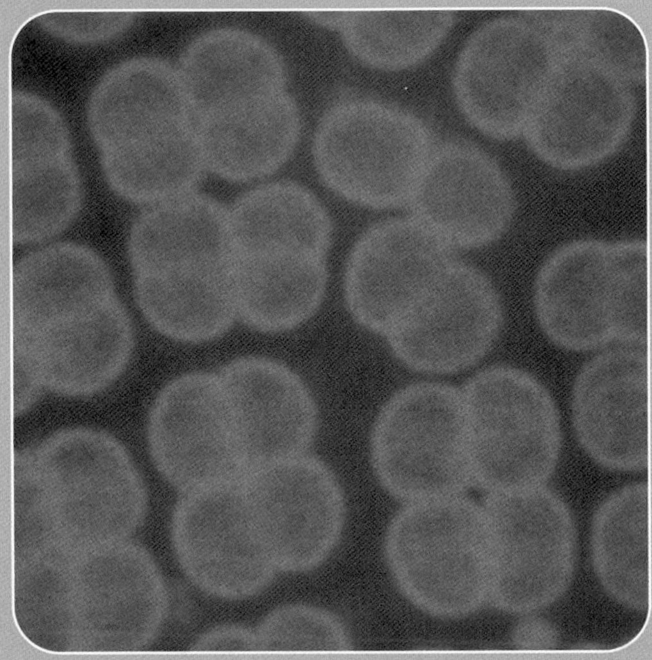

[John D. Cunningham/Visuals Unlimited.]

LEARNING OBJECTIVES

1. Understand how cancer can result from genetic changes.
2. Understand that cancer affects the regulation of the cell cycle, apoptosis, and other processes.
3. Understand the causes and forms of DNA damage.
4. Understand the main pathways for repairing damaged DNA.
5. Understand that the cell cycle and its checkpoints are signal transduction pathways.
6. Understand why p53 mutations are strongly associated with cancer.

THIS CHAPTER IN CONTEXT

Cancer describes a variety of diseases characterized by uncontrolled cell proliferation, but cancer is basically a disease of damaged DNA. This chapter builds on the previous chapter covering DNA replication by describing how replication errors and other forms of DNA damage are repaired. If the damage is not repaired and if it affects a gene controlling cell division or cell death, cancer may result. Similarly, certain genetic defects predispose to cancer. This chapter presents a few of the proteins involved in recognizing DNA damage, repairing it, and signaling the cell to stop growing. These processes illustrate that DNA is not inert but is a biochemically active molecule.

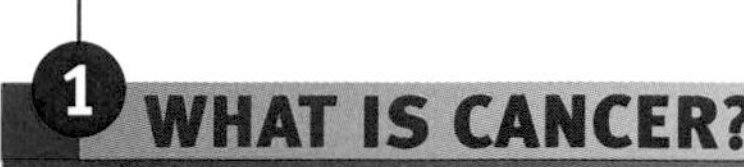

1 WHAT IS CANCER?

Cancer is one of the most common diseases, affecting one in three people and killing one in four. The clinical picture of cancer varies tremendously, depending on the tissue affected, but all cancer cells share an ability to proliferate uncontrollably and ignore the usual signals to differentiate or undergo **apoptosis** (programmed cell death; from the Greek *apo*, away from, and *ptosis*, a falling). Eventually, cancer may kill by invading and eroding the surrounding normal tissues (Fig. 18-1).

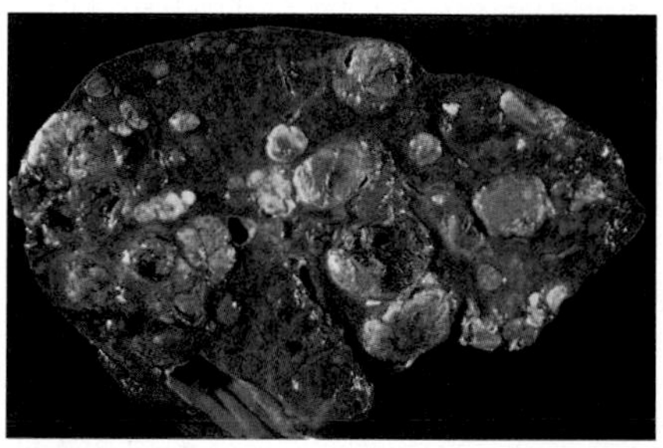

FIGURE 18-1

Tumors in human liver.
The white bulges are masses of cancerous cells growing in the liver. [*CNRI/Science Photo Library/Photo Researchers.*]

Many cancers appear to result from genetic predisposition, environmental factors, or infections. Only about 1% of cancers are classified as hereditary, but there are over 20 different kinds of these diseases, and they shed considerable light on specific molecular mechanisms of **carcinogenesis** (cancer development). The linkage between environmental factors and cancer comes mostly from epidemiological studies, which have shown, for example, that sunlight increases the risk of cutaneous melanoma (skin cancer) and smoking and asbestos exposure promote the development of lung cancer. Viral infections contribute to certain types of cancer, for example, liver cancer from hepatitis B virus and cervical cancer from human papillomaviruses. Chronic bacterial infections may also lead to cancer.

Cancer is a genetic disease

All the known causes of cancer are linked in some way to damage of the cell's DNA. Thus, *cancer is fundamentally a disease of the genes.* Damage to a cell's genetic material can potentially activate or inactivate genes. Evidence for activating events first came from studies of viruses that cause cancer in experimental animals. Certain viral genes, called **oncogenes,** were responsible for **transforming** cells, that is, converting them to a cancerous form. These genes turned out to be activated versions of normal cellular genes, accordingly called **proto-oncogenes.** Proto-oncogenes appear to function in controlling cell growth and differentiation, leading to a theory that *the inappropriate activation of proto-oncogenes could cause cancer.*

Inactivating genetic events also contribute to cancer. For example, the childhood cancer known as retinoblastoma (which is characterized by retinal tumors) occurs in an inherited and a sporadic (noninherited) form. The age of tumor onset varies widely, which is consistent with a scenario in which both copies (alleles) of a gene must be inactivated in order to develop the

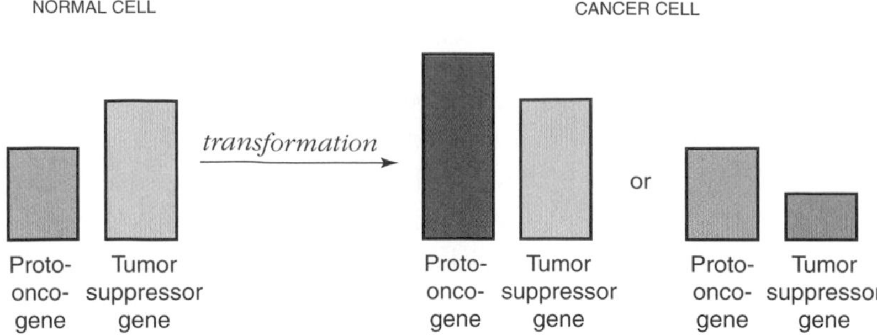

FIGURE 18-2

Action of proto-oncogenes and tumor suppressor genes.
In a normal cell, a proto-oncogene has relatively low activity and the tumor suppressor gene has relatively high activity. Cancer may result when either the proto-oncogene is activated or the tumor suppressor gene is inactivated.

disease. Children with the inherited form of retinoblastoma succumb early because they already bear one defective allele (inherited from one of their parents). Inactivation of the second allele leads to transformation. The retinoblastoma gene is known as a **tumor suppressor gene,** since *its loss leads to the development of cancer* (Fig. 18-2).

Tumor suppressor genes appear to outnumber proto-oncogenes, principally because DNA damage more frequently leads to loss of function (inactivation of a tumor suppressor gene) than to gain of function (activation of a proto-oncogene). DNA damage can take the form of substitutions of one nucleotide for another, large or small deletions of chromatin, or inappropriate methylation that leads to gene silencing (see Section 17-4). If the affected gene is involved in regulating the cell cycle or apoptosis, the result could be the uncontrolled growth of a **tumor** (a mass of cancer cells).

Nearly all cancers in adults arise from epithelial cells, which normally divide repeatedly (for example, skin cells and the lining of the intestine). This observation initially suggested that cancer stems from overly rapid cell division. However, loss of control over the cell cycle is just one hallmark of cancerous cells. In general, several regulatory pathways are overridden by separate genetic events. This is known as the multiple-hit theory of carcinogenesis.

REVIEW LEARNING OBJECTIVE 1

- What are three major causes of cancer?
- Explain how alterations in proto-oncogenes and tumor suppressor genes can lead to cancer.

Multiple defects contribute to cancer

There is no defined sequence of events for transforming a normal cell to a cancerous tumor. Rather, *several different genetic defects occurring at different times can all produce the same final outcome.* Some physiological processes that play a part in the development and growth of tumors are outlined below:

1. *Cell division and differentiation.* Most normal cells undergo differentiation unless they receive a distinct signal (such as a mitogen) to continue dividing. A cancer cell may produce its own mitogen, or its growth-signaling pathway may be permanently set in the "on" position, so that it can proliferate in the absence of the normal signal.

2. *Apoptosis.* Cells can be induced to die when certain survival factors or nutrients are exhausted or when the cell is damaged beyond repair. Cancer cells, in contrast, may resist apoptotic signals.

3. *Telomere erosion.* Repeated cycles of DNA replication can lead to the loss of telomeric DNA, which normally triggers senescence or apoptosis. In cancer cells, telomerase may be activated to counter this effect (see Section 17-3).

4. *Contact inhibition.* Most cells stop dividing when they form tight contacts with their neighbors, but many cancer cells continue to grow (Fig. 18-3). Moreover, some normal cells die when detached from a surface, but cancer cells tend to detach and resettle elsewhere, a property that allows them to invade other tissues and grow into new tumors (a process called **metastasis**).

5. *Angiogenesis.* Cells must be located within 100 to 200 μm of a capillary in order to obtain nutrients and oxygen. Cancer cells produce signals that stimulate **angiogenesis,** the growth of new blood vessels, to support the growth of a tumor.

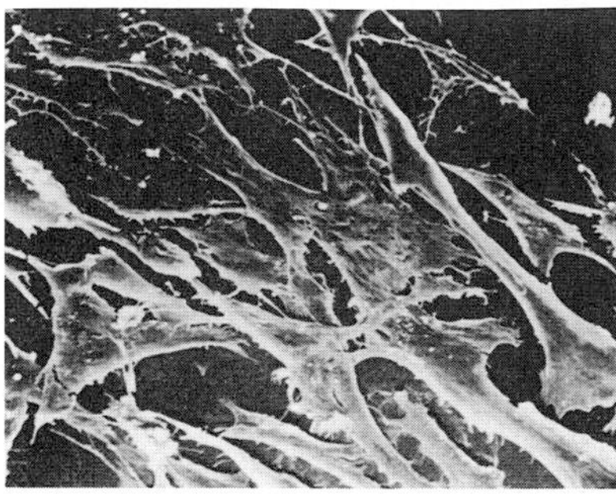

(a)

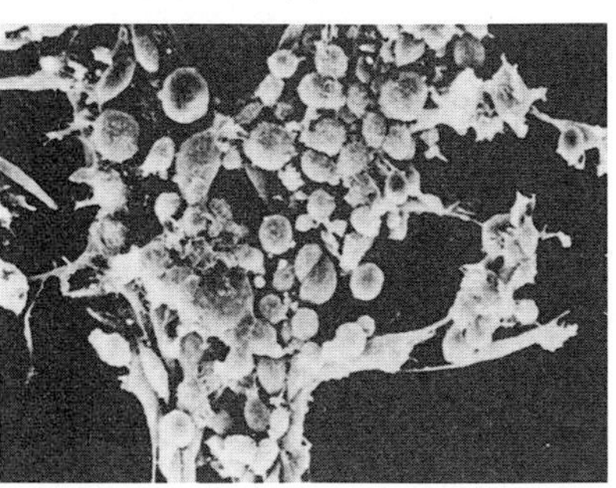

(b)

FIGURE 18-3

Contact inhibition.
(a) Normal cells stop growing when they reach confluence (form a single layer of cells in contact with each other).
(b) Transformed cells continue to grow, piling up on top of each other. [*Photos courtesy G. Steven Martin.*]

The pathways that regulate the processes mentioned above intersect to some degree. Unfortunately, we know relatively little about how these pathways function in normal cells or how well they inhibit or prevent transformation. By necessity, we see only the failures of these control systems, that is, the cases where a cell is able to proliferate to the point where it causes disease. Not surprisingly, tumors exhibit different combinations of genetic defects. Techniques for profiling a tumor at the molecular level can potentially identify its propensities and provide information on how best to treat it (Box 18-A).

In the absence of specific information about a tumor, cancer therapy can only target the features common to all forms of the disease. At present, this mainly means suppressing cell proliferation. Most drugs that interfere with mitosis and cell division are useful but are not specific for cancer cells; they also kill tissues that normally undergo regular renewal. Radiation is effective in killing cancer cells, although it may act indirectly by destroying the radiation-sensitive endothelial cells of the blood vessels that provide nutrients to the rapidly growing cancer cells. Drugs that inhibit angiogenesis or promote apoptosis hold promise but are not yet practical.

Two drugs that have proved their usefulness in the clinic target specific alterations in cancer cell biology. For example, about one-quarter of breast cancers overexpress a growth factor receptor called HER2. A monoclonal antibody called Herceptin binds specifically to HER2, blocking its growth-signaling activity. As a result, the tumor stops growing or even shrinks. Herceptin may be even more effective if the antibody is conjugated to another drug, such as one that interferes with microtubule assembly necessary for cell division (see Section 5-3).

A second clinical success story is Gleevec,

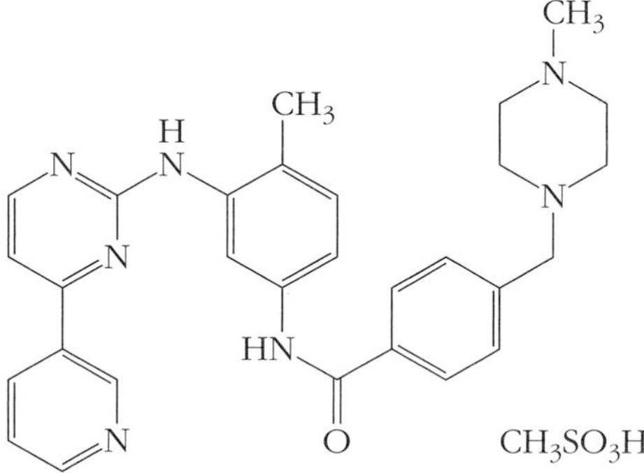

Gleevec

A CLOSER LOOK

Box 18-A Gene expression profiling of cancers

Each tumor consists primarily of cloned cells, that is, cells descended from a single cancer cell. Yet each tumor differs with respect to the set of genes it expresses, so its responsiveness to certain drugs or other treatments may differ from other tumors, even ones that arise from the same type of tissue. Differences in gene expression can be assessed using DNA microarrays, which are glass slides or chips on which are dotted thousands of DNA sequences corresponding to different genes (see Box 3-E). To profile a tumor, mRNA from cancer cells is isolated and then converted to cDNA using reverse transcriptase (see Box 17-A). The single-stranded cDNA incorporates a fluorescent tag so that it can be detected. The cDNA strands are allowed to hybridize with complementary sequences in the DNA microarray, and unbound cDNAs are washed away. The pattern of fluorescent spots (each one representing an mRNA present in the tumor) paints a picture of gene expression that can be quantified and analyzed.

Part of a gene expression profile of breast tumors is shown here. Each horizontal band corresponds to a different gene, and the columns represent different tumors. The green spots indicate levels of gene expression below the mean; red spots indicate levels above the mean; and intermediate shades indicate levels closer to the mean (gray spots indicate no data).

Gene expression profiling reveals differences not visible in standard histological examinations of cancer cells and offers a holistic snapshot of cancer cell biology that would be impossible to obtain through a gene-by-gene search. Microarray analysis has been successfully applied to melanomas and breast cancers. Ideally, the technique should be able to identify patterns of gene expression that suggest how best to treat a cancer, perhaps by correlating observed gene expression patterns with known susceptibilities of similar cancers. This would save time and minimize suffering, since current cancer treatments often progress through a series of regimens of increasing potency.

Eventually, gene expression profiling could be used to pinpoint specific genes that contribute to certain cancer phenotypes, for example, high metastatic potential (ability to spread to other tissues). Profiling precancerous cells—which have undergone some genetic changes but have not yet formed tumors—might suggest treatments to prevent full-blown cancer.

[*Courtesy David Botstein and Therese Sorlie, Stanford University. From* Nature ***406****, 747–52 (2000).*]

an inhibitor of a tyrosine kinase known as Bcr-Abl. This enzyme, an overactive form of the protein Abl, is produced by a chromosomal translocation that leads to chronic myelogenous leukemia. Gleevec inhibits cell proliferation and induces apoptosis. Because it also inhibits other receptor tyrosine kinases, Gleevec or its relatives may be useful drugs for other forms of cancer.

Because cancer is so difficult to cure, preventative efforts should also be considered. Epidemiological studies suggest that deaths from cancer could be significantly reduced (by as much as one-half) by eliminating known risk factors, including smoking and exposure to certain chemicals and microorganisms. The remaining cases of cancer might someday be treated with agents designed to rectify specific defects that are currently being elucidated at the molecular level. In the following sections we will examine the biochemical details of some of the pathways involved in cancer development, including those for repairing damaged DNA and linking DNA repair to the cell cycle.

REVIEW LEARNING OBJECTIVE 2

- Describe the differences between normal cells and cancer cells with respect to cell division, apoptosis, telomeres, contact inhibition, and angiogenesis.
- Which of these activities is the target of most anticancer therapies?

DNA DAMAGE AND REPAIR

Since cancer arises as a result of genetic alterations, it is worthwhile to see how such changes arise and how cells attempt to prevent or minimize them. We will present some examples of DNA damage and describe the cellular processes for repairing DNA. In general, different types of damage demand different repair enzymes. If not repaired, the damage may become manifest as a **mutation,** a permanent and heritable alteration in the cell's genetic material.

DNA can be damaged in different ways

Environmental agents such as ultraviolet light, ionizing radiation, and certain chemicals can physically damage DNA. For example, ultraviolet (UV) light induces covalent linkages between adjacent thymine bases:

UV light

DNA backbone

Thymine residues

Thymine dimer

This brings the bases closer together, which distorts the helical structure of DNA. Thymine dimers can thereby interfere with normal replication and transcription.

Ionizing radiation also damages DNA either through its direct action on the DNA molecule or indirectly by inducing the formation of free radicals, particularly the hydroxyl radical (·OH), in the surrounding medium. This can lead to strand breakage.

Many thousands of chemicals can potentially react with the DNA molecule. Compounds that cause mutations are known as **mutagens,** or as **carcinogens** if the mutations lead to cancer. Cigarette smoke contains mutagens, particularly benzo(*a*)pyrene:

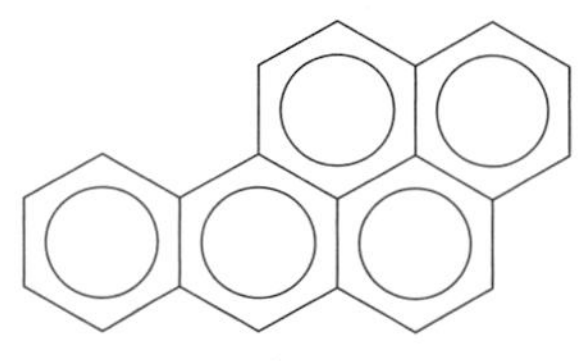

Benzo(*a*)pyrene

Another common mutagen is aflatoxin,

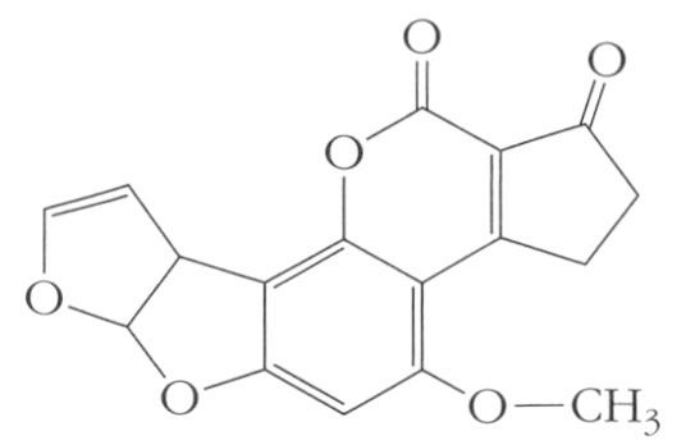

Aflatoxin

a carcinogen that leads to liver cancer. This compound is produced by the mold *Aspergillus* and may contaminate agricultural products (peanuts, corn, and rice, among others) either before or after harvest.

Cellular metabolism itself exposes DNA to the damaging effects of reactive oxygen species (for example, the superoxide anion $\cdot O_2^-$, the hydroxyl radical, or H_2O_2) that are normal by-products of oxidative metabolism. Over 100 different oxidative modifications of DNA have been catalogued. For example, guanine can be oxidized to 8-oxoguanine (oxoG):

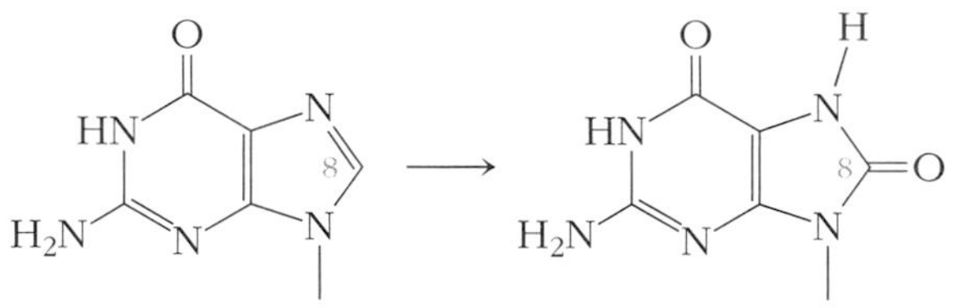

Guanine → 8-Oxoguanine (oxoG)

When the modified DNA strand is replicated, the oxoG can base-pair with either an incoming C or A. Ultimately, the original G:C base pair can become a T:A base pair. A nucleotide substitution such as this is called a **point mutation.**

In addition to the chemical insults described above, *nonenzymatic reactions disintegrate DNA under physiological conditions.* For example, hydrolysis of the *N*-glycosidic bond linking a base to a deoxyribose group yields an **abasic site** (also called an apurinic or apyrimidinic or AP site).

Deamination reactions can alter the identities of bases. This is particularly dangerous in the case of cytosine, since deamination (actually an oxidative deamination) yields uracil:

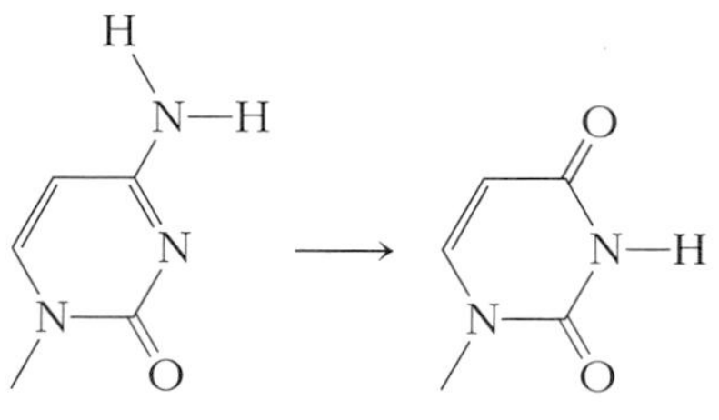

Cytosine → Uracil

Recall that uracil has the same base-pairing propensities as thymine, so the original C:G base pair could give rise to a T:A base pair after DNA replication. Since DNA has evolved to contain thymine rather than uracil, a uracil base resulting from cytosine deamination can be recognized and corrected before the change becomes permanent.

Finally, the normal process of DNA replication introduces errors such as mismatched bases and the occasional incorporation of uracil rather than thymine. DNA polymerase may occasionally add or delete nucleotides, producing bulges or other irregularities in the DNA. Such errors can give rise to mutations as insertions or deletions.

The unavoidable nature of many DNA lesions has driven the evolution of mechanisms to detect and remedy errors. A cell containing a point mutation or a small insertion or deletion may suffer no ill effects, particularly if the mutation is in a part of the genome that does not contain an essential gene. However, more serious lesions, such as single- or double-strand breaks, usually bring replication or translation to a halt. The cell must deal with these show-stoppers immediately. If the damage cannot be repaired, the cell may undergo apoptosis. *Cells that fail to properly repair damaged DNA or that resist apoptosis are good candidates for transformation.*

REVIEW LEARNING OBJECTIVE 3

- Provide examples of how environmental and intracellular agents can damage DNA.
- Can all forms of DNA damage be attributed to one of these agents?
- How does a point mutation arise?

Repair enzymes restore damaged DNA

In a few cases, repair of damaged DNA is a simple process involving one enzyme. For example, in bacteria and some other organisms (but not mammals), UV-induced thymine dimers can be restored to their monomeric form by the action of a light-activated enzyme called DNA photolyase.

Mammals can reverse other simple forms of DNA damage, such as the rare methylation of a guanine residue, which yields O^6-methylguanine (this modified base can pair with either cytosine or thymine):

CH_3 O N 6 N H_2N N N

O^6-Methylguanine

A methyltransferase removes the offending methyl group, transferring it to one of its Cys residues. This permanently inactivates the protein. Apparently, the expense of sacrificing the methyltransferase is justified by the highly mutagenic nature of O^6-methylguanine.

In bacteria as well as eukaryotes, *nucleotide mispairings are corrected shortly after DNA replication by a* ***mismatch repair*** *system.* A protein, called MutS in bacteria, monitors newly synthesized DNA and binds to the mispair. Although it binds only 20 times more tightly to the mispair than to a normal base pair, MutS undergoes a conformational change and causes the DNA to bend (Fig. 18-4). These changes apparently induce an endonuclease to cleave the strand with the incorrect base at a site as far as 1000 bases away. A third protein then unwinds the helix so that the defective segment of DNA can be destroyed and replaced with accurately paired nucleotides by DNA polymerase. How does the endonuclease know which strand contains the incorrect

(a)

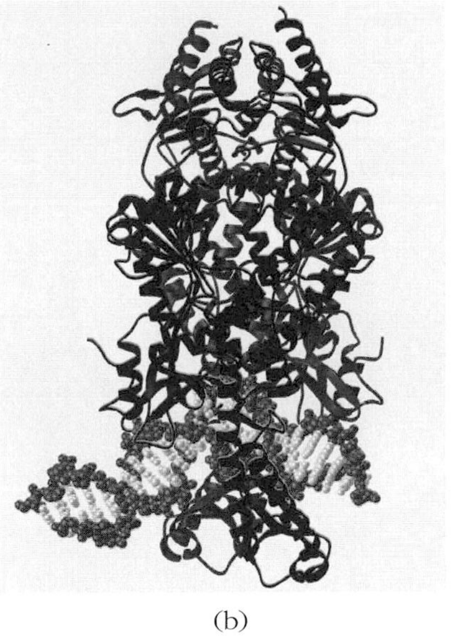

(b)

FIGURE 18-4

The mismatch repair protein MutS bound to DNA.

Two subunits of the comma-shaped MutS protein encircle the DNA at the site of a mispaired nucleotide. Binding causes the DNA to bend sharply, compressing the major groove and widening the minor groove. (a) Front view. (b) Side view. [*Photos courtesy Wei Yang/NIH. From* Nature ***407****, 703–10 (2000).*]

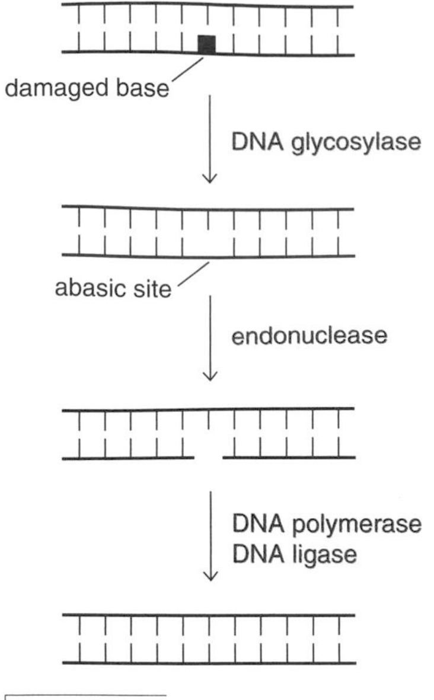

FIGURE 18-5
Base-excision repair.

base? Recall that cellular DNA is normally methylated (see Section 17-4). The endonuclease can select the newly synthesized strand because it has not yet been methylated. A defect in the human homolog of MutS increases the mutation rate and predisposes to a form of hereditary colon cancer.

Base excision repair corrects the most frequent DNA lesions

Modified bases that cannot be directly repaired are removed and replaced in a process known as ***base excision repair.*** This pathway begins with a glycosylase, which removes the damaged base. An endonuclease then cleaves the backbone, and the gap is filled in by DNA polymerase (Fig. 18-5).

The structures and mechanisms of several DNA glycosylases have been described in detail. When a glycosylase binds DNA, the damaged base flips outward from the helix, so that it can bind in a cavity on the protein surface. For example, the glycosylase that recognizes oxoG binds the modified base by stacking it between a Phe and a Cys side chain (Fig. 18-6a). Arg and Asn side chains, taking the place of the extruded guanine, form hydrogen bonds with the unpaired cytosine (Fig. 18-6b).

One of the best known repair glycosylases is uracil-DNA glycosylase. This enzyme removes the uracil bases that are mistakenly incorporated into DNA during replication or that result from cytosine deamination. The reaction mechanism is somewhat unusual. Recall that most enzymes lower the activation energy for a reaction by stabilizing the transition state (see Section 6-3). In the uracil-DNA glycosylase reaction, the substrate itself (DNA) contributes about 80% of the transition state stabilization, a phenomenon termed **substrate catalysis.** The rate-enhancing effect is due to the positioning of DNA

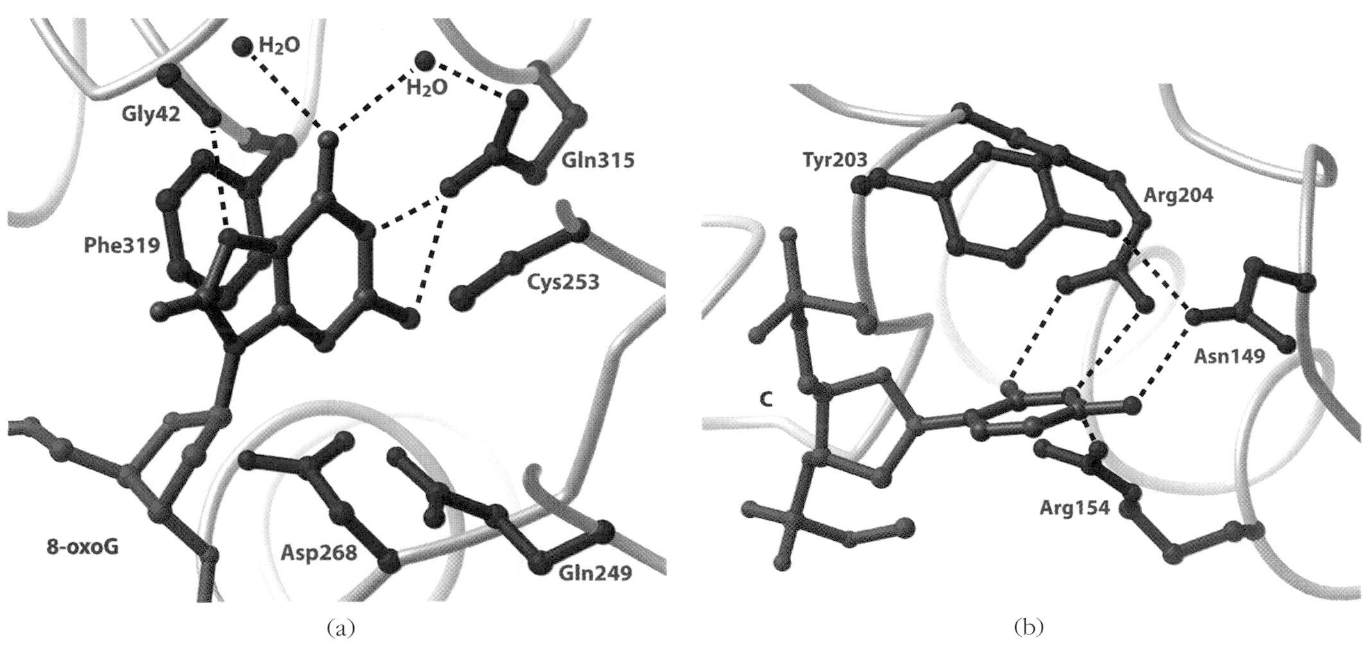

FIGURE 18-6
Glycosylase binding to oxoG.
The 8-oxoguanine-DNA glycosylase binds to DNA containing an oxoG residue. (a) The damaged base (red) twists out of the DNA helix and binds in a pocket on the enzyme surface, between a Phe and a Cys side chain (blue). (b) The cytosine (pink) that had been paired with the oxoG forms hydrogen bonds with several protein side chains. [*Photos courtesy Greg L. Verdine, Erving Professor of Chemistry, Harvard College. From* Nature ***403****, 859–66 (2000).*]

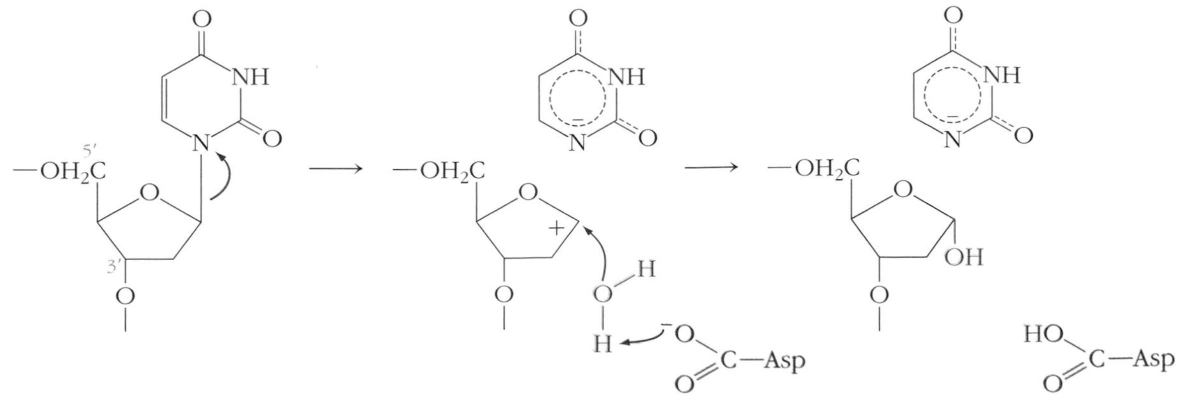

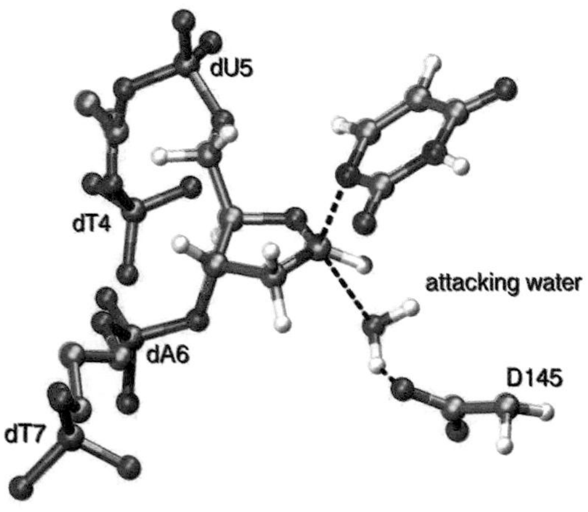

(a)

(b)

FIGURE 18-7

The uracil-DNA glycosylase reaction. (a) The two-step reaction mechanism. Cleavage of the glycosidic bond yields a cationic ribose group and an anionic uracil derivative. In the second step, water attacks the ribose group with the aid of an Asp side chain. (b) Stabilization of the transition state for Step 1. The negatively charged phosphate groups (gold and red) of DNA residues 4 to 7 stabilize the positively charged ribose group of residue 5 (center). The anionic uracil leaving group is at the upper right. [*Photo courtesy Aaron Dinner. Reproduced with permission. From* Nature ***413***, *752–55 (2001).*]

phosphate groups that stabilize the developing positive charge on the deoxyribose group and repel the anionic leaving group (Fig. 18-7).

DNA glycosylases appear to remain bound to the DNA at the abasic site, possibly to help recruit the next enzyme of the repair pathway. This enzyme, usually called the AP endonuclease (APE1 in humans), nicks the DNA backbone at the 5′ side of the abasic ribose. The nuclease inserts two protein loops into the major and minor grooves of the DNA and bends the DNA by about 35° to expose the abasic site. A backbone structure with a base attached cannot enter the active-site pocket. During the hydrolysis reaction, a Mg^{2+} ion in the active site stabilizes the anionic leaving group (Fig. 18-8). Like the glycosylases, the AP endonuclease remains with its product. For most enzymes, rapid product dissociation is the rule, but in DNA repair, the continued association of the endonuclease with the broken DNA strand may be favored because it prevents unwanted side reactions.

In the final step of base excision repair, a DNA polymerase fills in the one-nucleotide gap, and a DNA ligase seals the nick. In some cases, DNA polymerase may replace as many as 10 nucleotides. The displaced single strand can then be cleaved off by a flap endonuclease (see Section 17-2).

Nucleotide excision repair targets the second most common form of DNA damage

Nucleotide excision repair, as its name suggests, is similar to base excision repair but mainly targets DNA damage resulting from insults such as ultraviolet light or oxidation. In nucleotide excision repair, a segment containing the

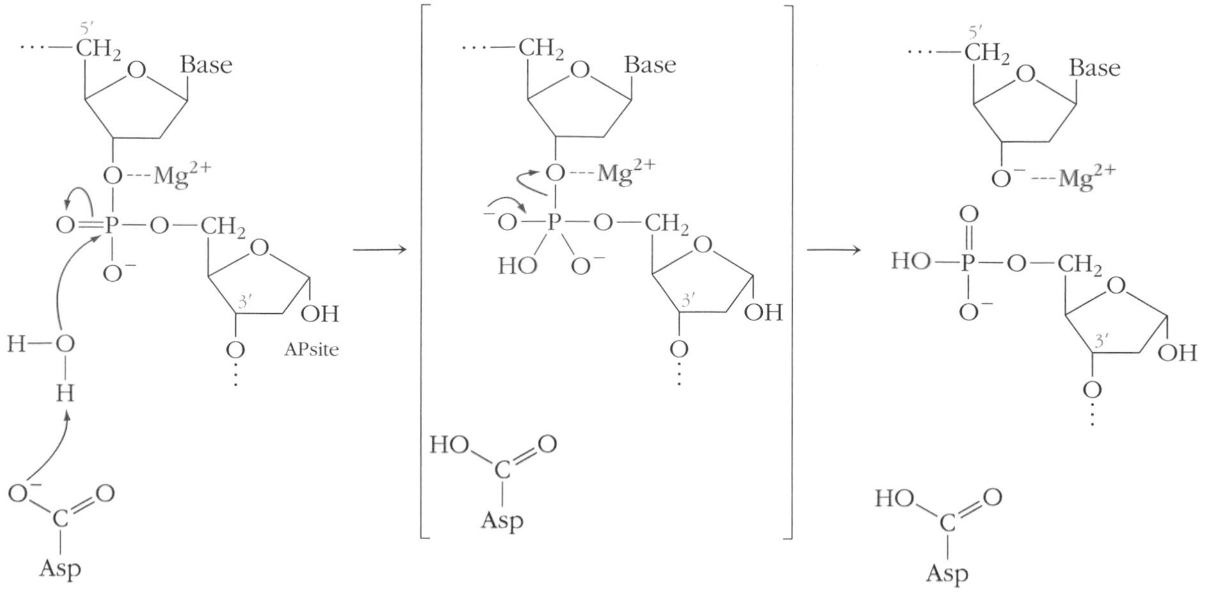

FIGURE 18-8

The AP endonuclease reaction.

damaged nucleotide and about 30 of its neighbors is removed and the resulting gap is filled in by a DNA polymerase, using the intact complementary strand as a template (Fig. 18-9). Many of the enzymes involved in this pathway have been identified through mutations that are manifest as two genetic diseases.

The rare hereditary disease Cockayne syndrome is characterized by neural underdevelopment, failure to grow, and sensitivity to sunlight. It results from a mutation in any one of five genes. These genes encode proteins participating in a pathway that recognizes an RNA polymerase that has stalled in the act of transcribing a gene into messenger RNA. This happens when the DNA template is damaged and distorted so that it blocks the progress of the RNA polymerase. The polymerase must be removed so that the damage can be addressed by the nucleotide excision repair system. Defects in the Cockayne syndrome genes prevent the cell from recognizing and removing the stalled RNA polymerase. Consequently, *the DNA never has a chance to*

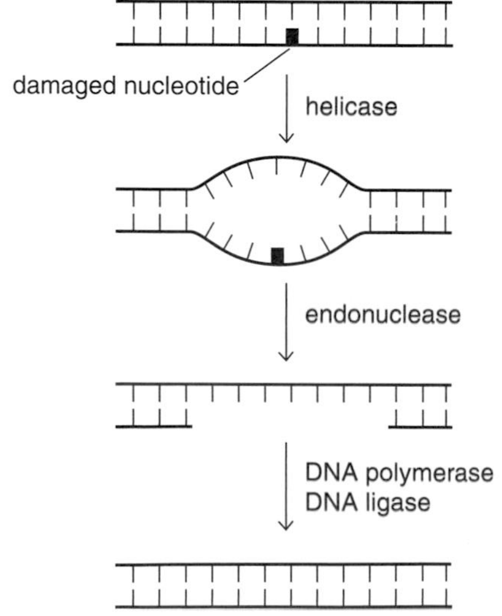

FIGURE 18-9

Nucleotide excision repair.

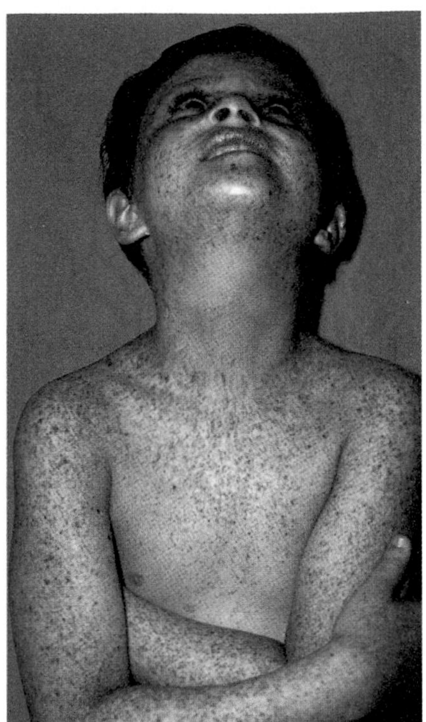

FIGURE 18-10

Xeroderma pigmentosum.
A defect in one of the genes responsible for nucleotide excision repair, a pathway for restoring UV-damaged DNA, allows the development of numerous skin cancers. [*Ken Gree/Visuals Unlimited.*]

be repaired, and the cell undergoes apoptosis. The death of transcriptionally active cells may account for the developmental symptoms of Cockayne syndrome.

Like Cockayne syndrome, the disease xeroderma pigmentosum is characterized by high sensitivity to sunlight, but individuals with xeroderma pigmentosum are about 1000 times more likely to develop skin cancer and do not suffer from developmental problems (Fig. 18-10). Xeroderma pigmentosum is caused by a mutation in any of seven genes that participate directly in nucleotide excision repair. Whereas Cockayne syndrome gene products appear to detect DNA damage that prevents transcription, the xeroderma pigmentosum proteins are responsible for repairing the damage. Thus, *in xeroderma pigmentosum, apoptosis is not triggered, but damaged DNA cannot be repaired.* The failure to repair UV-induced lesions explains the high incidence of skin cancer.

A variant form of xeroderma pigmentosum exhibits increased rates of skin cancer even though the nucleotide excision repair proteins are normal. The defect in this disorder has been pinpointed to an enzyme known as DNA polymerase η. When functioning normally, this polymerase can bypass DNA lesions such as UV-induced thymine dimers, incorporating two adenine bases in the new strand. Although it is useful as a translesion polymerase, DNA polymerase η is relatively inaccurate and has no proofreading exonuclease activity (see Section 17-2). It inserts an incorrect base on average every 30 nucleotides. This may not be problematic, as the errors can be detected and corrected by the mismatch repair system described above.

The existence of error-prone polymerases provides a fail-safe mechanism for replicating stretches of DNA that cannot be navigated by the standard replication machinery. In fact, such alternative DNA polymerases typically outnumber the polymerases devoted to normal replication. For example, synthesis of three of *E. coli*'s five DNA polymerases increases when the cell experiences DNA damage.

There is some evidence that bacterial cells stressed by starvation can preferentially use error-prone polymerases and thereby increase the rate of mutation, leading to altered gene products and possibly increasing the cells' chances for survival. This hypothesis is highly controversial. Although starvation-triggered mutations do not seem to occur specifically in the genes for food-utilizing enzymes (as was once proposed), they do not appear to occur entirely at random. Error-prone DNA synthesis may offer a selective advantage in times of stress, but it is difficult to see how this would be effective in the long term, since most mutations are detrimental.

Some damage can be repaired through recombination

A single-strand gap presents an obvious obstacle to standard DNA polymerases, but replication can continue through **recombination** and the lesion can be repaired later. This process resembles homologous recombination, which occurs in the absence of DNA damage as a mechanism for shuffling genes (the term *homologous* emphasizes that recombination occurs between DNA segments with similar sequences). Recombination as a mechanism for replicating DNA with a single-strand break is diagrammed in Figure 18-11.

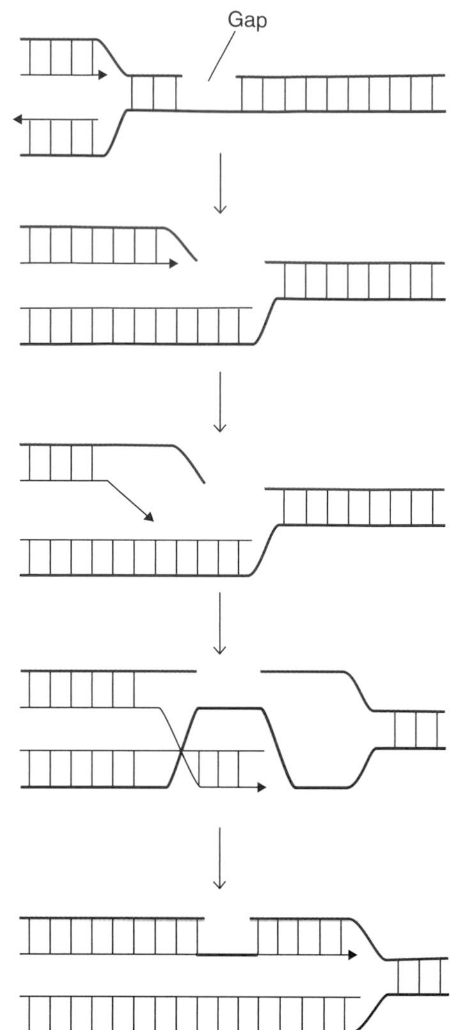

1. *Replication halts when a single-strand gap is encountered.*

2. *A helicase unwinds the newly synthesized helix at the gap.*

3. *The 3′ end of the freed single strand invades the double-stranded molecule on other arm of the replication fork, pairing with complementary bases. The 3′ end serves as a primer for DNA polymerase.*

4. *A protein complex that includes an endonuclease resolves the crossed-over strands, and replication can continue. Note that the lesion has been bypassed but not yet repaired.*

FIGURE 18-11
Recombination to bypass a single-strand gap.

Double-strand breaks can be similarly repaired through recombination, which requires pairing with another intact homologous double-stranded molecule that can act as a template (Fig. 18-12). However, double-strand breaks may be more commonly repaired by a straightforward **end-joining** process (Fig. 18-13). End joining is the only way to repair a double-strand break when a homologous DNA molecule is not available (for example, before the DNA has been replicated). Not surprisingly, end-joining may link two wrong pieces, since it does not depend on homology between double-stranded segments. Consequently, *end-joining is potentially mutagenic.*

In *E. coli,* recombination requires a protein known as RecA, which binds to double-stranded DNA and extends and untwists the helix in an ATP-dependent manner (Fig. 18-14). This presumably makes it easier for the invading single strand to form base pairs with the complementary segment. The importance of recombination as a repair mechanism is illustrated in humans by the disease ataxia telangiectasia, which results from defective recombination and is characterized by neuronal abnormalities, predisposition to cancer, and extreme sensitivity to X-rays (radiation commonly generates double-strand breaks).

The mechanism of recombination is highly conserved among prokaryotes and eukaryotes, which argues for its essential role in maintaining the integrity of DNA as a vehicle of genetic information. The proteins that carry out re-

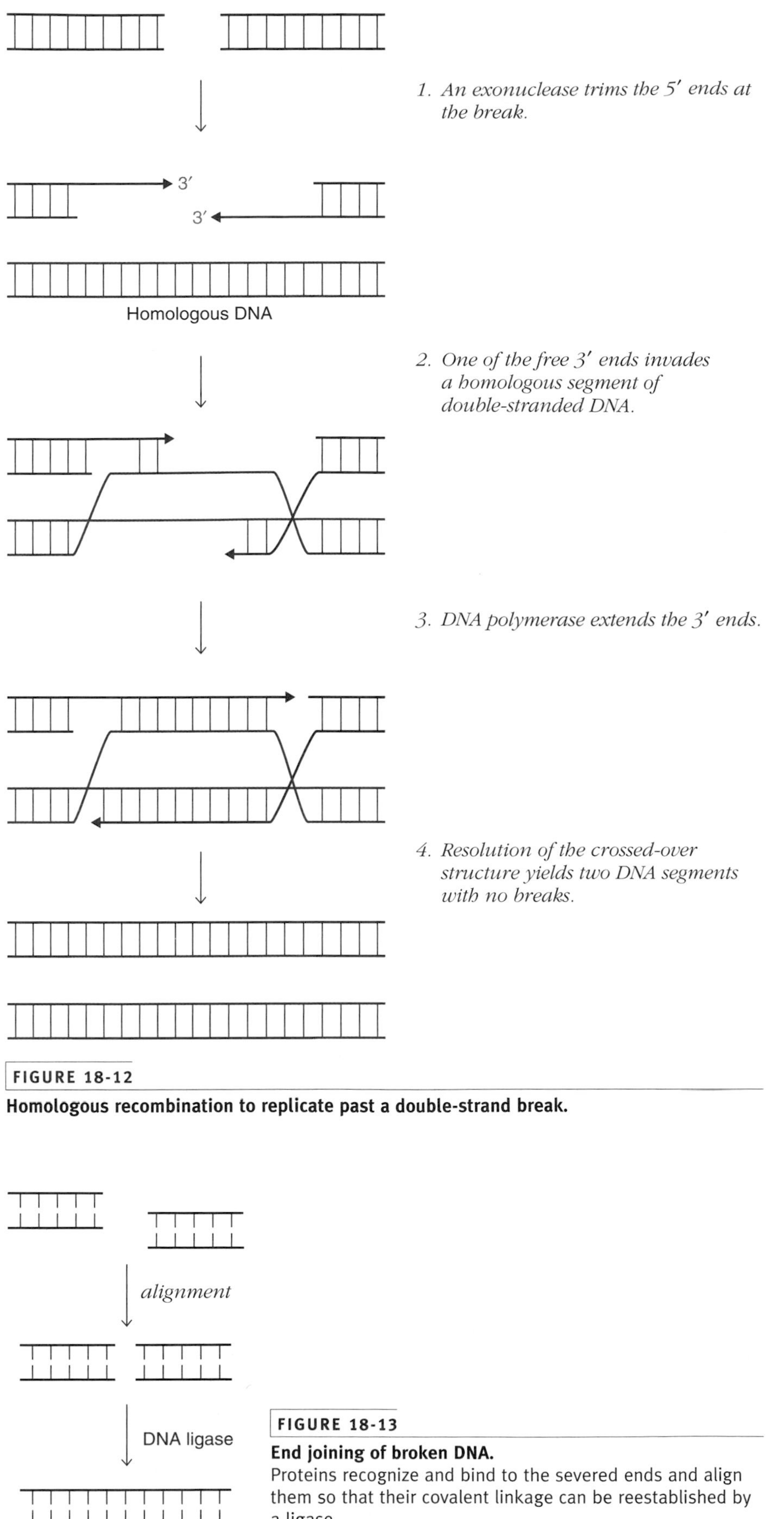

1. An exonuclease trims the 5′ ends at the break.
3′
3′
Homologous DNA
2. One of the free 3′ ends invades a homologous segment of double-stranded DNA.
3. DNA polymerase extends the 3′ ends.
4. Resolution of the crossed-over structure yields two DNA segments with no breaks.
alignment
DNA ligase

FIGURE 18-12

Homologous recombination to replicate past a double-strand break.

FIGURE 18-13

End joining of broken DNA.

Proteins recognize and bind to the severed ends and align them so that their covalent linkage can be reestablished by a ligase.

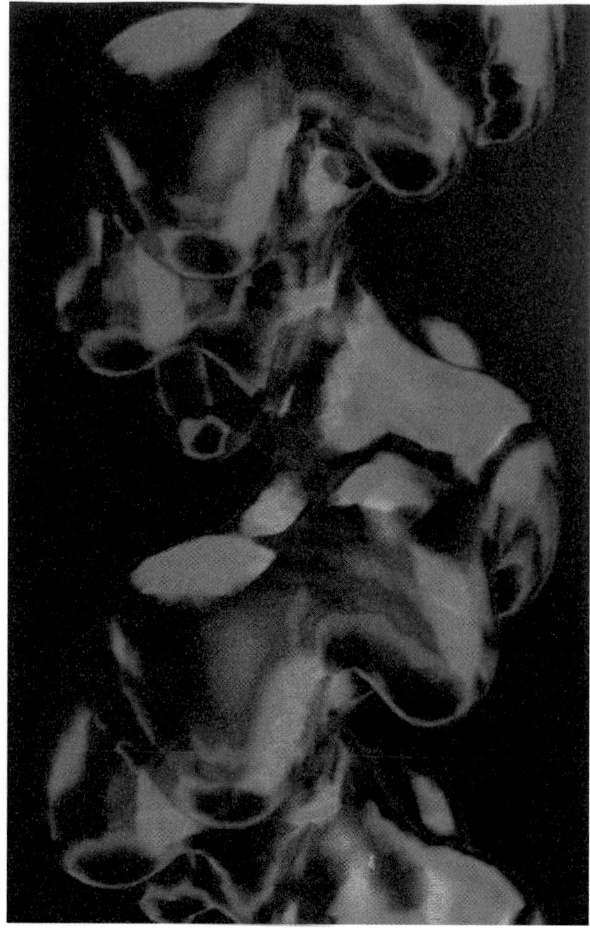

FIGURE 18-14
Model of a RecA–DNA complex.
In this model, the DNA is red and the protein is transparent. RecA coats the DNA, extending and untwisting the helix. [*Courtesy Edward H. Egelman, University of Virginia.*]

combination appear to function **constitutively** (that is, the genes are always expressed), which is consistent with proposals that DNA strand breaks are unavoidable. In contrast, *many other DNA repair mechanisms are induced only when the relevant form of DNA damage is detected.* This makes sense, since the repair enzymes might otherwise interfere with normal replication. In fact, activation of the repair pathways usually halts DNA synthesis, an advantage when error-prone DNA polymerases—which would be a liability in normal replication—are induced.

 **REVIEW LEARNING OBJECTIVE 4**

- Describe the general pathways and enzymes required for mismatch repair, base excision repair, nucleotide excision repair, recombination, and end joining.
- Describe the advantages and disadvantages of error-prone DNA polymerases.
- Why is it usually necessary for DNA replication and DNA repair to take place in a mutually exclusive fashion?

3 CELL CYCLE CONTROL

In the preceding section we surveyed different types of DNA damage and some of the mechanisms that correct the damage. Defects in the repair pathways allow DNA lesions to become permanently established in the cell and its progeny. We know that mutations in certain enzyme-coding genes lead to metabolic diseases. Can we identify the genes whose mutations cause cells to grow uncontrollably?

Identifying specific genetic lesions in cancer cells is tricky, in part because cancers are genetically unstable. In addition to small mutations (substitutions, insertions, and deletions of a few nucleotides), entire chromosomes may be lost or gained, and parts of chromosomes may be translocated to other chromosomes. Any of these changes could disrupt the expression of genes that help control the cell cycle.

The life cycle of a cell can be diagrammed as shown in Figure 18-15. Progression from one phase to the next is controlled by proteins called cyclins, which exert their effects through cyclin-dependent kinases. The levels of cyclins fluctuate as they are tagged with the protein ubiquitin and then degraded by proteasomes (see Section 9-1). The cyclin-dependent kinases form signaling pathways that commit the cell to **S phase** (when DNA is synthesized) and to **M phase** (when mitosis occurs).

Additional signaling pathways that intersect with the cyclin-dependent pathways serve as checkpoints during the G1 and G2 phases. These checkpoints monitor such things as the presence of growth-promoting signals, DNA damage, telomere length, and proper spindle formation (for the segregation of chromosomes during mitosis). If all is not well, *the checkpoints can arrest the cell cycle or initiate apoptosis* (Box 18-B). A checkpoint that detects irregularities in DNA could prevent the perpetuation of damaged genes by halting the cell cycle, promoting DNA repair, or killing the entire cell (Fig. 18-16).

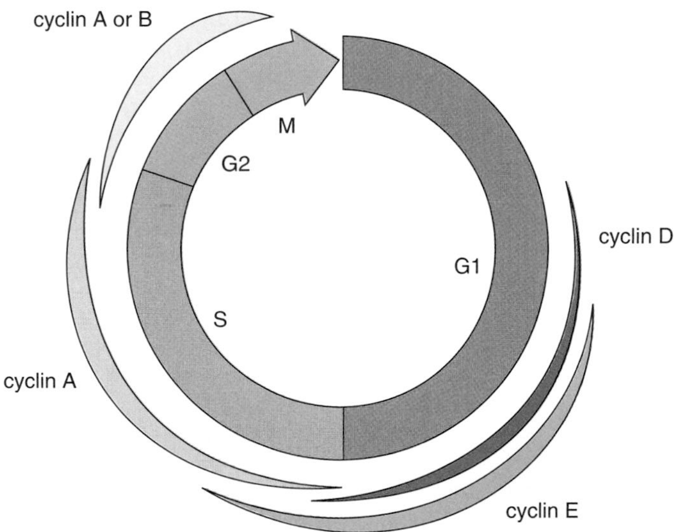

FIGURE 18-15

The cell cycle and cyclin activity.
DNA synthesis occurs during S phase and mitosis during M phase. The duration of each phase may vary. Transition from one phase to the next is regulated by the appearance and disappearance of cyclins. The width of each colored band represents the activity of a cyclin and its associated kinase.

A CLOSER LOOK

BOX 18-B Apoptosis

Apoptosis, or programmed cell death, comprises a number of events that culminate in the death of the cell. Apoptosis occurs within minutes (for comparison, mitosis takes about 20 times longer), so it is difficult to catch a cell in the act of dying. This may be one reason why programmed cell death has only recently become an active area of research. Cells that die by other means—from mechanical injury, bacterial infection, or lack of oxygen—typically leak their contents and trigger inflammation. In contrast, apoptotic cells follow an orderly process in which internal structures collapse and membrane blebbing occurs.

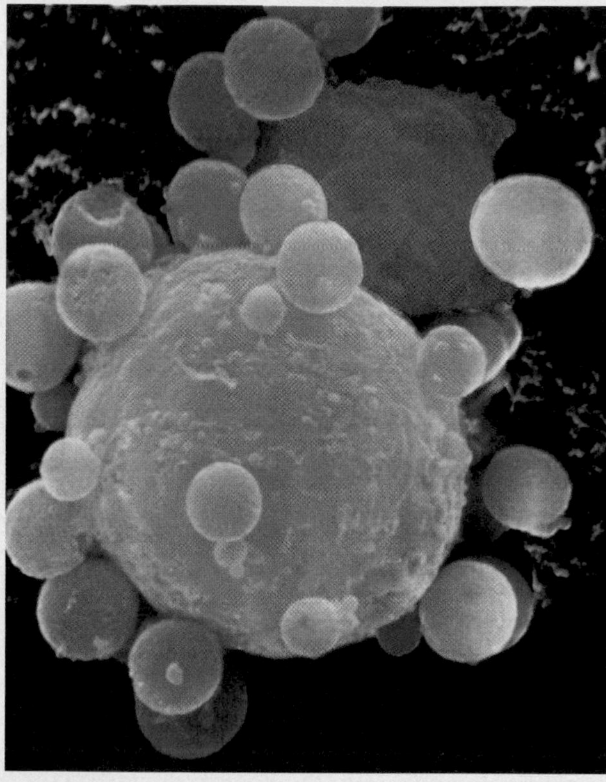

Eventually, the entire cell contents become packaged in membranous compartments and are engulfed by macrophages.

Apoptosis is an essential event in embryogenesis, serving to eliminate excess cells and sculpt organs and structures such as fingers and toes. It also occurs throughout life, to balance mitotic increases in cell number and as a mechanism for disposing of damaged cells. Cell death may be initiated from inside or outside the cell. Extracellular signals include tumor necrosis factor, a polypeptide produced by macrophages and T lymphocytes in response to infection. Depending on the target cell, tumor necrosis factor elicits additional inflammatory responses or apoptosis (hence its name). Intracellular signals for apoptosis are linked to the checkpoints for DNA replication and mitosis. Both types of signaling mechanisms converge on a set of proteolytic enzymes called caspases, which have an active-site Cys residue and cleave their substrates after Asp residues. Human cells contain at least 12 caspases, about two-thirds of them involved in apoptosis. Like most proteases, they are synthesized as zymogens and are activated by proteolysis. Active caspases are derived from two polypeptide chains and have two active sites.

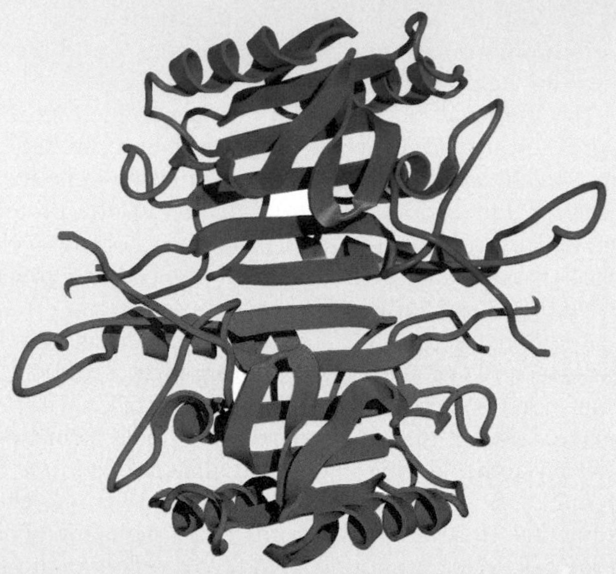

Interestingly, caspase activation sequences include Asp–X peptide bonds, indicating that they are sequentially activated by others of their kind, just like digestive enzymes (see Section 6-4) and the coagulation proteases (see Box 6-A). Caspase activation presumably begins early in the apoptotic pathway, when zymogens are induced to aggregate. The low intrinsic activity of one zymogen may activate another, triggering an amplifying cascade of protease activity.

The advantage of a cascade of caspases is that it helps ensure an all-or-none response to an apoptotic signal. Without this amplification mechanism, the cell might be only crippled rather than killed.

The caspases are selective proteases, cleaving a few bonds in as many as 100 different target proteins. Caspases activate a DNase that cuts between nucleosomes, generating DNA fragments with an average length of about 180 bp. Other caspase substrates are cytoskeletal and nucleoskeletal proteins. Cleavage of these proteins contributes to the structural collapse of the cell. Additional signals probably govern membrane shape changes.

A host of other proteins appear to participate in apoptotic signaling as cofactors or inhibitors. Malfunctions in any of these could theoretically interfere with normal apoptosis and promote cancerous growth. Intriguingly, one of the proteins involved in apoptotic signaling is cytochrome *c*, a pe-

(*continued*)

BOX 18-B (continued)

ripheral membrane protein located between the inner and outer mitochondrial membranes (see Fig. 12-16). The involvement of mitochondria, and specifically of cytochrome *c*, is well-documented, although it is not clear whether the mitochondria play a governing role or merely amplify an apoptotic stimulus though cytochrome *c* release. The exit of cytochrome *c* from the mitochondria, probably via an outer membrane pore protein, represents a point of no return. Once the oxidative machinery of the cell is compromised, cell death is inevitable.

[Photo © Dr. Andrejs Liepins/Science Photo Library/Photo Researchers. Structure of caspase-3 (pdb 1GFW) determined by N.O. Concha and C.A. Janson.]

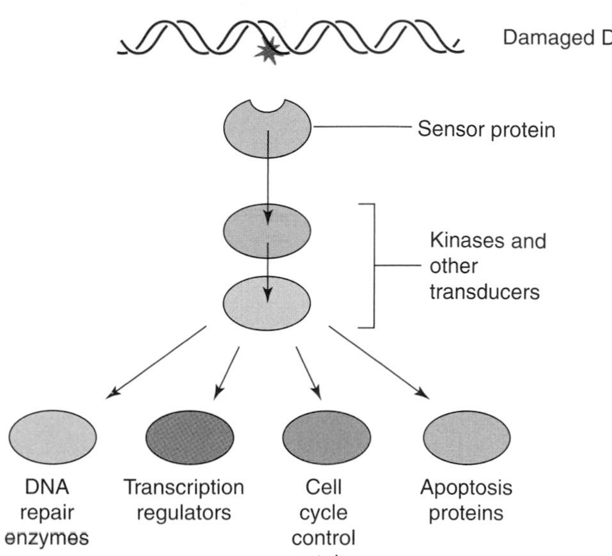

FIGURE 18-16

Model of a DNA damage response pathway. As in other signal-transduction pathways, an initial signal (a damaged DNA molecule) is transduced (via kinases and other factors) to one or more cellular effectors that carry out different tasks.

Cellular signaling networks, including the cell-cycle checkpoints, include kinases, phosphatases, and other components (see Section 16-1). In fact, an estimated 20% of the protein-coding genes in the human genome are related to signal transduction. Genetic studies have implicated a number of kinases and phosphatases in cancer and cell-cycle control. However, the cell-cycle and checkpoint signaling pathways operate as networks, with some degree of redundancy, so biochemical studies have been slow to assign specific functions to particular proteins. Nevertheless, it has been possible to link certain protein defects in these pathways to the development of cancer.

The retinoblastoma protein, for example, participates in cell-cycle control. It maintains cells in G1 phase by binding to and inhibiting the activity of a protein that promotes the synthesis of proteins, including some cyclins, that allow the cell to proceed to S phase. When the level of the retinoblastoma protein drops, the cell begins DNA replication (Fig. 18-17). The loss of the retinoblastoma gene could therefore contribute to the rapid cell division characteristic of some tumors.

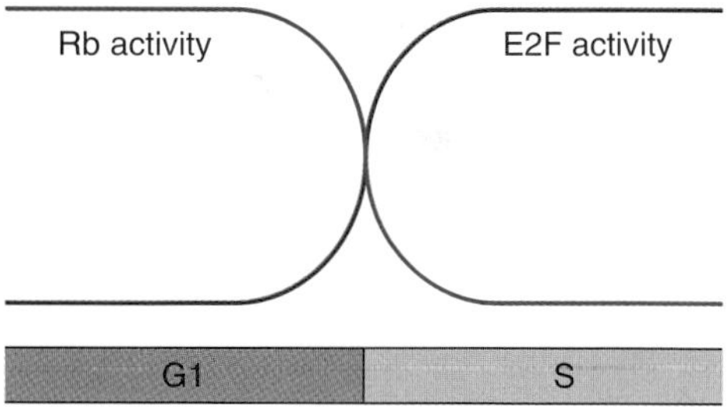

FIGURE 18-17

Action of the retinoblastoma protein. High levels of the retinoblastoma protein (Rb) inhibit the activity of the transcription factor E2F. When Rb levels drop, E2F can stimulate the expression of cyclins and other proteins required for S phase.

Some members of the DNA-damage checkpoint pathway have been identified

One of the checkpoint proteins that senses damaged DNA may resemble PCNA, the eukaryotic sliding-clamp protein that enhances the processivity of DNA polymerase (see Fig. 17-12). Such a protein could assemble at a site of

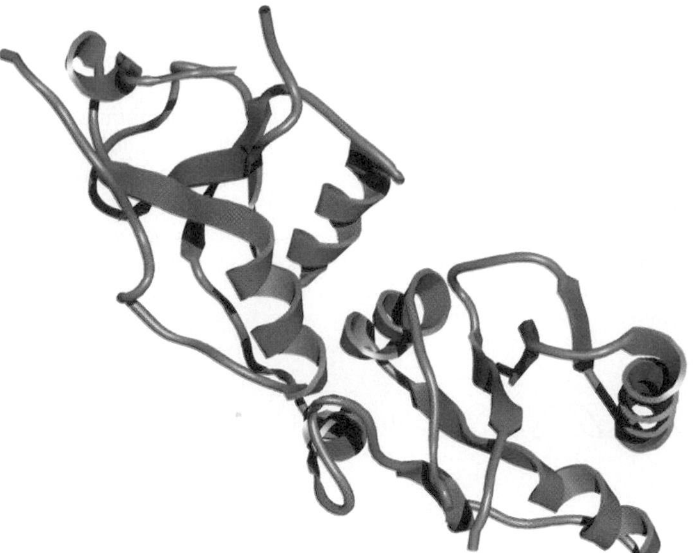

FIGURE 18-18

Partial structure of the *BRCA1* protein. This C-terminal structure, comprising residues 1646 to 1859, represents one of several protein domains in *BRCA1*. The C-terminal domain consists of two repeated motifs (called Brct repeats). Mutations in this region of the protein likely destabilize the structure shown here, which is important for protein–protein contacts. [*Structure (pdb 1JNX) determined by R.S. Williams, R. Green, and J.N.M. Glover.*]

DNA damage (for example, when a DNA or RNA polymerase is stalled) and activate a signaling cascade to arrest the cell cycle until the damage is repaired.

A likely instigator of this signaling cascade is the protein kinase known as ATM, which is defective in ataxia telangiectasia. ATM is a large protein (350,000 D) that includes a DNA-binding domain. It appears to monitor DNA that is damaged or incompletely replicated. It may also bind to double-strand breaks and activate repair by recombination. Among the substrates for ATM's kinase activity are the proteins *BRCA1* (whose gene is commonly mutated in breast cancer) and the tumor suppressor known as p53.

Breast cancer strikes about one in nine women in developed countries. About 10% of breast cancers have a familial form, and about half of these exhibit mutations in the *BRCA1* or *BRCA2* genes. A woman bearing one of these mutated genes has a 70% chance of developing breast cancer. The *BRCA1* protein is an 1863-residue multidomain protein whose structure is partially known (Fig. 18-18). When phosphorylated by ATM or another kinase, it may function as a sort of scaffolding protein, bringing together proteins involved in DNA repair and recombination. There is some evidence that the *BRCA1* protein binds to double-strand breaks, perhaps to discourage error-prone end joining in favor of the more accurate repair by recombination. The *BRCA1* protein may also interact with cell-cycle-related kinases. The loss of *BRCA1* function could sever a link between DNA repair and cell-cycle control. This would allow cells to divide when they would normally be arrested in G1 or G2 phase.

REVIEW LEARNING OBJECTIVE 5

- Why is it necessary for the cell cycle to have checkpoints before proceeding to S phase or M phase?
- How can mutations in a cell cycle or checkpoint gene lead to cancer?

p53 plays a central role in cancer

The tumor suppressor gene p53 is found to be mutated in at least half of all human tumors (melanoma cells are an interesting exception; they usually have a functional p53). The level of p53 in the cell is controlled by its rate of degradation (like the cyclins, it is ubiquitinated and targeted to a proteasome

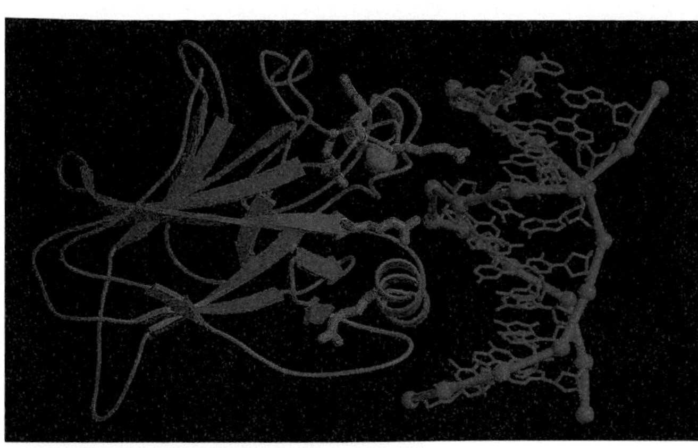

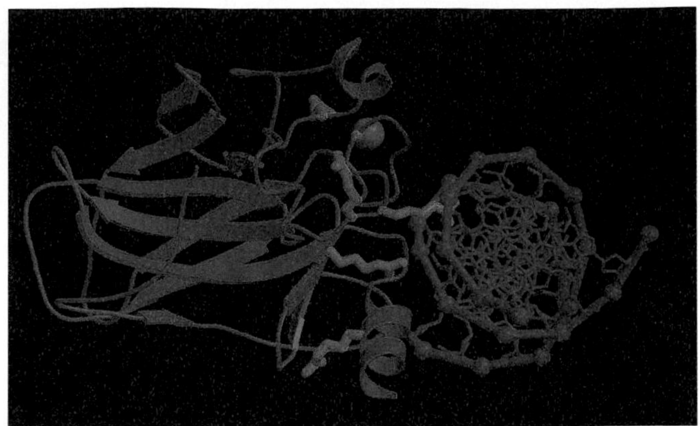

FIGURE 18-19

Structure of the core domain of p53.
p53 forms a tetramer and stimulates the transcription of genes related to apoptosis and cell-cycle control. Residues 325 to 356, which are required for tetramerization, are not included in this model. Mutations that affect p53 binding to DNA block its transcription-stimulating activity. The six most frequently mutated residues in p53 are highlighted in yellow. These groups (five Arg and one Gly side chain) are at or near the surface of the protein that contacts DNA (blue). The two views are related by a 90° rotation. [*Photos courtesy Nikola Pavletich. From* Science ***274****, 1001–1005 (1996).*]

for destruction). The concentration of p53 increases when its degradation is slowed. This can occur indirectly via the loss of an enzyme that attaches ubiquitin to the p53 protein. It also occurs when p53 is phosphorylated by the action of a kinase such as ATM. Thus, *DNA damage, which activates ATM, leads to an increase in the cellular concentration of p53.*

Phosphorylation—and possibly other modifications, such as acetylation and glycosylation—also increase the activity of p53. Covalent modification triggers a conformational change that allows the protein to bind to specific DNA sequences in order to promote the transcription of several dozen different genes. The six residues that are most commonly mutated in p53 from cancer cells map to the region of the protein that interacts with DNA (Fig. 18-19).

Activated p53 stimulates the synthesis of a protein that inhibits cyclin-dependent kinases, thereby blocking progression of the cell cycle. This regulatory mechanism would conceivably buy time for the cell to repair DNA using enzymes whose synthesis is also stimulated by p53. Moreover, activated p53 apparently turns on the gene for a subunit of ribonucleotide reductase (see Section 15-4). Thus, activation of p53 could promote synthesis of the deoxynucleotides required for DNA repair.

Some of p53's other target genes encode proteins that carry out apoptosis. It is possible that p53's ultimate effects are dose-dependent, with lower doses (reflecting mild DNA damage) halting the cell cycle, and higher doses (reflecting severe DNA damage) leading to cell death. *The position of p53 at the interface of pathways related to DNA repair, cell cycle control, and apoptosis indicate why the loss of the p53 gene is so strongly associated with the development of cancer.* Still, p53 is not the only player in cellular transformation, and unraveling the steps of carcinogenesis continues to be a formidable task.

REVIEW LEARNING OBJECTIVE 6

- Explain how p53 links the cell cycle and DNA repair.

SUMMARY

1. Cancer can result from genetic events that activate proto-oncogenes or inactivate tumor suppressor genes.
2. Cancer cells are characterized by abnormalities in cell division, differentiation, apoptosis, telomere maintenance, contact inhibition, and angiogenesis. Most cancer therapies target cell-proliferation pathways.
3. Radiation, chemical mutagens, spontaneous deamination, and normal replication errors can cause mutations in DNA.
4. Mechanisms for repairing damaged DNA include mismatch repair, base excision repair, nucleotide excision repair, and recombination. Defects in these repair pathways increase the risk of cancer.
5. DNA repair pathways are linked to pathways that control the cell cycle so that damaged DNA can be detected and repaired before it is replicated.

CHECKLIST

1. Review the Learning Objectives listed on page 556.
2. Apply your knowledge by solving the problems at the end of this chapter. Check your results in the Solutions appendix.
3. Be able to define the boldfaced terms (consult the glossary at the end of the book). Test your understanding by taking the Chapter 18 quiz (www.wiley.com/college/pratt).
4. Explore the websites listed at www.wiley.com/college/pratt.
5. Consult the list of Selected Readings for background information or additional details on DNA mutation, DNA repair, and cancer.

GLOSSARY TERMS

apoptosis
carcinogenesis
oncogene
transformation
proto-oncogene
tumor suppressor gene
tumor
metastasis
angiogenesis
mutation
mutagen
carcinogen
point mutation
abasic site
mismatch repair
base excision repair
substrate catalysis
nucleotide excision repair
recombination
end joining
constitutive
S phase
M phase

PROBLEMS

1. Many proteins that participate in intracellular signaling pathways have been linked to cancer. Explain why a kinase encoded by a proto-oncogene could be complemented by a phosphatase encoded by a tumor suppressor gene.
2. From what you know about the causes of cancer, explain why the risk of developing cancer increases with age.
3. In eukaryotic cells, a specific triphosphatase cleaves deoxy-8-oxoguanosine triphosphate (oxo-dGTP) to oxo-dGMP + PP_i. What is the advantage of the reaction?
4. Why do unicellular organisms have no need for apoptosis?
5. Draw the structure of an oxoguanine:adenosine base pair. (Hint: The oxoguanine base pivots around the glycosidic bond in order to form two hydrogen bonds with adenine.)
6. What change does the methylation of a guanine residue make in the succeeding generations of DNA?
7. Compounds such as ethidium bromide, proflavin, and acridine orange can be used to stain DNA bands in a gel. These compounds are known as intercalating agents, and they interact with the DNA by slipping in between stacked base pairs. This interaction with the DNA increases the fluorescence of the intercalating agent and allows for visualization of the DNA bands in a gel under ultraviolet light. Care must be taken when working with these com-

pounds, however, because they are powerful mutagens. Explain why.

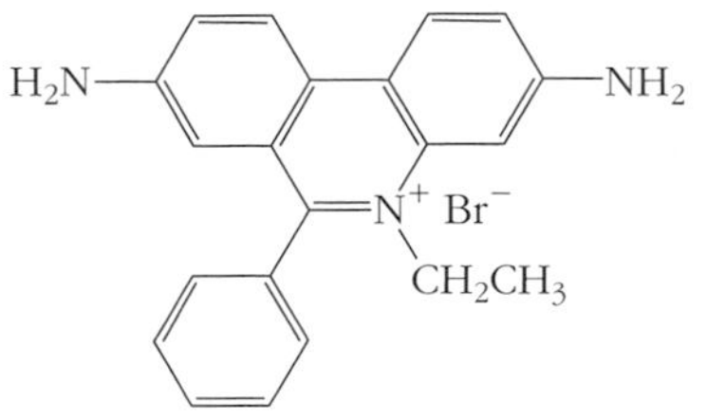

Ethidium bromide

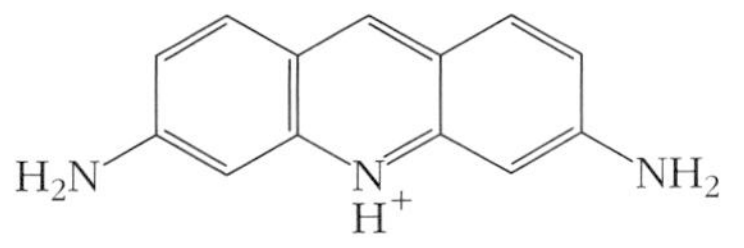

Proflavin

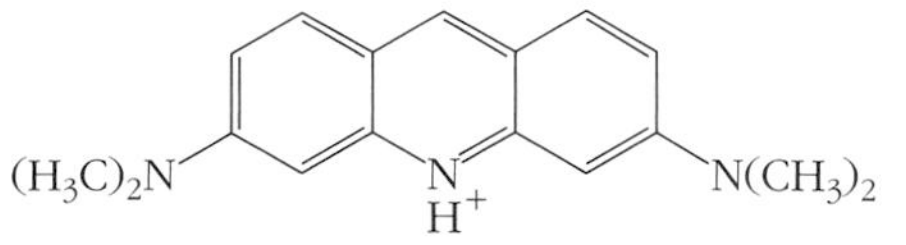

Acridine orange

8. Chemical mutagens, such as the compounds discussed in Problem 7, are substances that induce mutations, or changes in the DNA sequence. They fall into two classes:

(1) Point mutations, in which one base replaces another. The point mutation is referred to as a transition if one purine (or pyrimidine) replaces another; a transversion if a purine replaces a pyrimidine, or the reverse.
(2) Insertion/deletion mutations result from inserting or deleting bases in the DNA such that the reading frame is shifted.

The compound 5-bromouracil is a thymine analog and can be incorporated into DNA in the place of thymine. 5-Bromouracil readily converts to an enol tautomer, which can base pair with guanine. (The tautomers freely interconvert through the movement of a hydrogen between an adjacent nitrogen and oxygen.) Draw the structure of the base pair formed by the enol form of 5-bromouracil and guanine. What kind of mutation can 5-bromouracil induce?

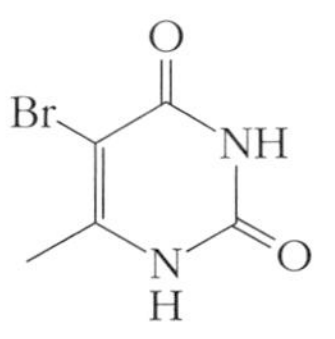

5-Bromouracil

9. Researchers often perform experiments on cells in a tissue culture hood, which filters the air to minimize the chance of contaminating the experimental cells with airborne bacteria. When the experiments have been completed, the cells are removed from the hood, and an ultraviolet light in switched on in the hood and remains on until the hood is used again. What is the rationale for this procedure?

10. As discussed in the text, deamination of cytosine produces uracil. Deamination of other DNA bases can also occur. For example, deamination of adenine produces hypoxanthine.

(a) Draw the structure of hypoxanthine.
(b) Hypoxanthine can base-pair with cytosine. Draw the structure of this base pair.
(c) What is the consequence to the DNA if this deamination is not repaired?

11. Deamination of guanine produces xanthine.

(a) Draw the structure of xanthine.
(b) Xanthine base-pairs with cytosine. Would this cause a mutation? Explain.

12. The fact that DNA has evolved to contain the bases A, C, G, and T makes the DNA molecule easy to repair. For example, deamination of adenine produces hypoxanthine, deamination of guanine produces xanthine, and deamination of cytosine produces uracil. Why are these deaminations repaired quickly?

13. Studies of bacteria and other organisms indicate that mutations occur more frequently at certain positions. These "hotspots" are due to the presence of 5-methycytosine (a naturally occurring methylated base), which may undergo oxidative deamination.

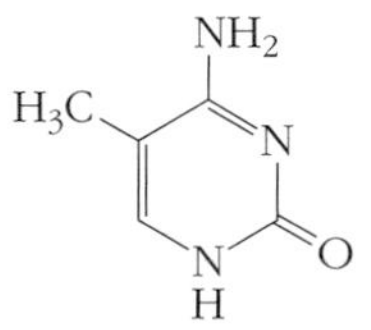

5-Methylcytosine

(a) Draw the structure of deaminated 5-methylcytosine.
(b) By what other name is the base known?
(c) What kind of mutation results from 5-methylcytosine deamination?
(d) Why is the cell unable to repair the altered base?

14. The adenine analog 2-aminopurine is a potent mutagen in bacteria. The 2-aminopurine substitutes for adenine during DNA replication and gives rise to mutations because it pairs with cytosine instead of thymine. Ultraviolet, fluorescence, and nuclear magnetic resonance studies have been used to determine the structure of the base pair formed by 2-aminopurine and cytosine. The results indicate that there is an equilibrium between a "neutral wobble" base pair (so-called because it is the dominant structure at neutral pH) and a "protonated Watson–Crick" structure, which forms at lower pH.

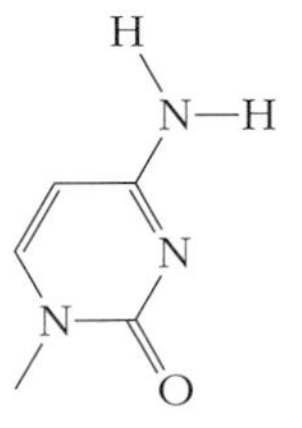

Cytosine 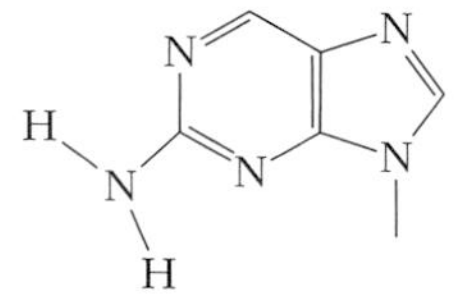 2-Aminopurine

(a) Two hydrogen bonds form between cytosine and 2-aminopurine in the "neutral wobble" pair. Draw the structure of this base pair.

(b) At lower pH, either the cytosine or the 2-aminopurine can become protonated. The two protonated forms are in equilibrium in the "protonated Watson–Crick" base-pair structure; the proton essentially "shuttles" from one base to the other in the pair, and the hydrogen bond is maintained. Draw the two possible structures for the base pair, one with a protonated 2-aminopurine, and one with a protonated cytosine.

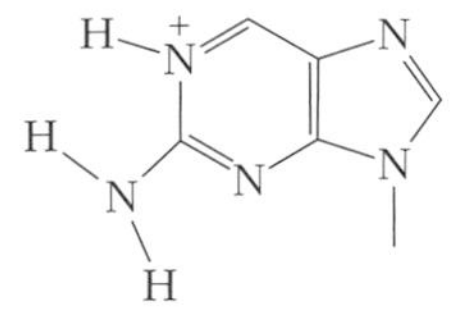

Protonated 2-aminopurine

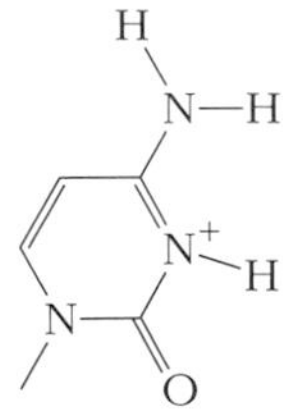

Protonated cytosine

[From Sowers, L.C., Boulard, Y., and Fazakerley, G.V., *Biochemistry* **29,** 7613–7620 (2000).]

15. Chemical methylating agents can react with guanine residues in DNA to produce O^6-methylguanine, which causes mutations because it can pair with thymine. This type of chemical damage is difficult to repair, because the O^6-methylguanine:T base pair is structurally similar to a normal G:C base pair and the mismatch repair system has difficulty recognizing it. Mutations resulting from this type of damage can convert proto-oncogenes into oncogenes. Ultraviolet melting studies have shown that the O^6-methylguanine can form a "wobble base pair" with thymine (similar to the 2-aminopurine:C base pair in Problem 14). The methylated guanine can also form a base pair with protonated cytosine at lower pH values.

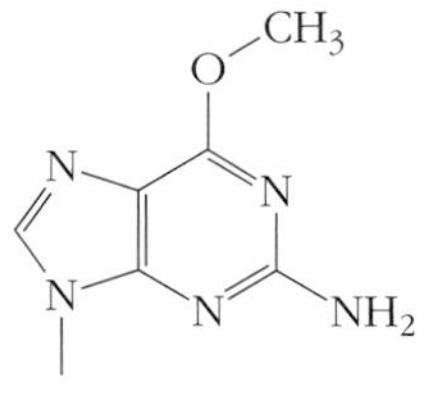

O^6-Methylguanine

(a) At physiological pH, the O^6-methylguanine:T wobble base pair predominates. Two hydrogen bonds form between the two bases. Draw the structure of this base pair.

(b) What kind of mutation results?

(c) At lower pH values, cytosine becomes protonated (see Problem 14b). The ability of O^6-methylguanine to base-pair with protonated cytosine explains why methylation of guanine residues does not always lead to a mutation (a mutation occurs only when O^6-methylguanine pairs with T). Draw the structure of the O^6-methylguanine:protonated cytosine base pair (three hydrogen bonds form between the two bases).

[From Leonard, G.A., Thomson, J., Watson, W.P., and Brown, T., *Proc. Natl. Acad. Sci.* **87,** 9573–9576 (1990).]

16. The compound 5-fluorouracil is used topically to treat skin cancer. Why is this treatment effective?

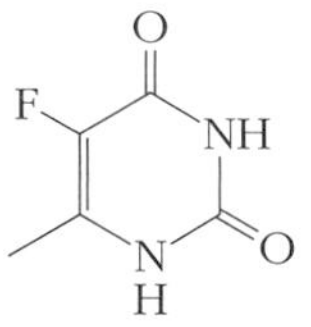

5-Fluorouracil

17. A strain of mutant bacterial cells lacks the enzyme uracil-DNA glycosylase. What is the consequence for the organism?

18. What is the most common DNA lesion in individuals with the disease xeroderma pigmentosum?

19. In many cases, a point mutation in DNA has no effect on the encoded amino acid sequence. Explain.

20. During replication in *E. coli,* certain types of DNA damage, such as thymine dimers, can be bypassed by DNA polymerase V. This polymerase tends to incorporate guanine residues opposite the damaged thymine residues and has a higher overall error rate than other polymerases. Thymine dimers can be bypassed by other polymerases, such as DNA polymerase III (which carries out most DNA replication in *E. coli*). DNA polymerase III incorporates adenine residues opposite the damaged thymine residues, but much more slowly than DNA polymerase V. Polymerase III is a highly processive enzyme, whereas polymerase V adds only 6–8 nucleotides before dissociating from the DNA. Explain how DNA polymerases III and V together carry out the efficient replication of UV-damaged DNA with minimal errors.

SELECTED READINGS

Friedberg, E.C., Wagner, R., and Radman, M., Specialized DNA polymerasese, cellular survival, and the genesis of mutations, *Science* **296,** 1627–1630 (2002). [Reviews the roles of error-prone DNA polymerases in DNA repair and antibody generation.]

Hengartner, M.O., The biochemistry of apoptosis, *Nature* **407,** 770–776 (2000). [Discusses the structures and possible roles of cellular proteins involved in apoptosis, particularly the caspases.]

Kowalczykowski, S.C., Initiation of genetic recombination and recombination-dependent replication, *Trends Biochem. Sci.* **25,** 156–165 (2000). [Presents models for and describes the proteins involved in recombination/replication events in bacteria.]

Rouse, J. and Jackson, S.P., Interfaces between the detection, signaling, and repair of DNA damage, *Science* **297,** 547–551 (2002). [Summarizes current understanding of how signaling pathways triggered by DNA damage lead to DNA repair.]

Vogelstein, B., Lane, D., and Levine, A., Surfing the p53 network, *Nature* **408,** 307–310 (2000). [Briefly reviews the function of p53 as a central node in the pathways that regulate cell growth and apoptosis.]

CHAPTER 19

Transcription and RNA

Amanita phalloides has earned its nickname, death cap. This mushroom produces a cyclic octapeptide known as α-amanitin that binds with high affinity to RNA polymerase and inhibits its ability to transcribe DNA into messenger RNA. As a result, protein synthesis is severely curtailed. Amanitin poisoning has a fatality rate of about 50%. Part of its deadliness lies in the fact that symptoms of poisoning—vomiting and diarrhea—do not appear until about a day after the mushroom has been ingested. By that time, all the toxin has been absorbed by the body, and there is no antidote. Victims often appear to recover but then relapse and die of kidney and liver failure within a few days.

[Vaughan Fleming/Science Photo Library/Photo Researchers.]

LEARNING OBJECTIVES

1. Understand why *gene* is a loose term for a sequence of transcribed DNA.
2. Understand the functions of promoters and general transcription factors.
3. Understand how DNA-binding regulatory proteins can affect transcription.
4. Understand how RNA polymerase transcribes a DNA sequence.
5. Understand how phosphorylation promotes transcript elongation by RNA polymerase.
6. Understand that mRNA transcripts are processed by capping, polyadenylation, and splicing.
7. Understand that rRNA and tRNA transcripts are processed by nucleases and covalently modified.
8. Understand that some RNA molecules have catalytic activity.

THIS CHAPTER IN CONTEXT

This chapter continues the story begun in Chapter 17 by showing how genetic information in DNA is transcribed into RNA so that it can be translated into protein. Transcription is mechanistically similar to replication but is more complex in its regulation because only specific portions of the genome—the genes—are transcribed. Most of this chapter focuses on eukaryotic messenger RNA transcription and its processing and ends with a brief discussion of RNA catalytic function.

TRANSCRIPTION IS THE FUNDAMENTAL MECHANISM BY WHICH A gene is expressed. It is the conversion of stored genetic information (DNA) to a more active form (RNA). The information contained in the sequence of deoxynucleotides in DNA is preserved in the sequence of ribonucleotides in the RNA transcript. Like DNA replication, transcription is characterized by template-directed nucleotide polymerization that requires a certain degree of fidelity. However, unlike DNA synthesis, RNA synthesis takes place selectively, on a gene-by-gene basis. This requires mechanisms for selecting which genes to transcribe and for identifying the sites where RNA polymerase must initiate polymerization. These regulatory aspects of RNA transcription are the responsibility of numerous proteins that interact with each other, with RNA polymerase, and with the DNA template.

In describing RNA transcription, we invoke the idea of a **gene,** a segment of DNA that is transcribed for the purpose of expressing the encoded genetic information, or transforming it to a form that is more useful to the cell. This definition of a gene requires some qualification:

1. *For protein-coding genes, the RNA transcript (called* ***messenger RNA*** *or* ***mRNA****) includes all the information specifying the sequence of amino acids in a polypeptide.* But keep in mind that not all RNA molecules are translated into protein. **Ribosomal RNA (rRNA), transfer RNA (tRNA),** and other types of small RNA molecules carry out their functions without undergoing translation.

2. *Most RNA transcripts correspond to a single functional unit,* for example, one polypeptide. But some RNAs, particularly in prokaryotes, code for multiple proteins and result from the transcription of an **operon,** a set of contiguous genes whose products have related metabolic functions. In a few rare cases, a single mRNA carries information for two proteins in overlapping sequences of nucleotides.

3. *RNA transcripts typically undergo* ***processing****—the addition, removal, and modification of nucleotides—before becoming fully functional.* For most eukaryotic protein-coding genes, processing includes splicing out **introns.** These long segments of noncoding nucleotides are not expressed but are nevertheless part of the gene. Because of variations in mRNA splicing and posttranslational modification, *several different forms of a protein may be derived from a single gene.*

4. Finally, *the proper transcription of a gene may depend on DNA sequences that are not transcribed but help position the RNA polymerase at the transcription start site or are involved in the regulation of gene expression.*

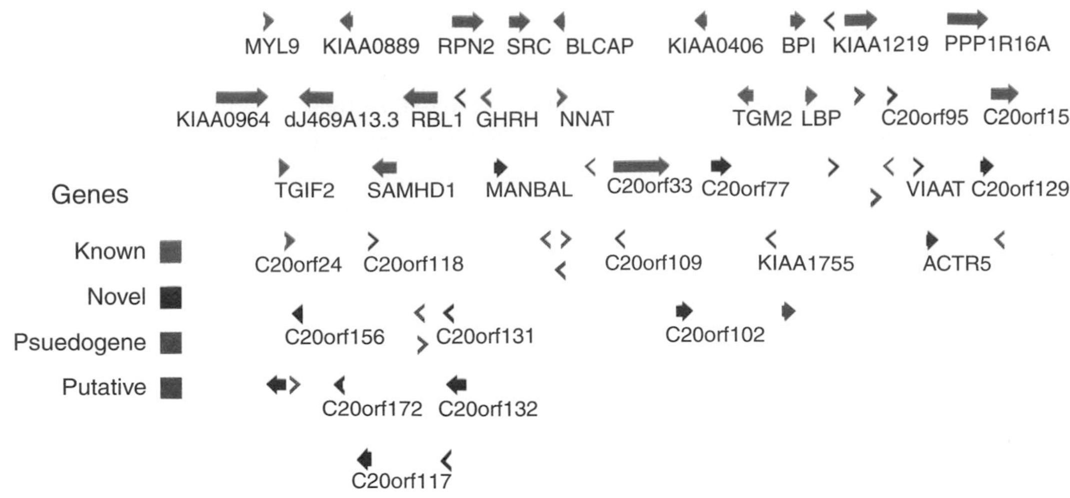

FIGURE 19-1

Genes in a segment of human chromosome 20.
This portion of the genome, representing a "finished" (>99.99% accurate) euchromatic sequence, exhibits typical gene density. Genes are shown as short arrows and are classified as *known* (identical to known protein sequences), *novel* (with ORFs or having sequences that are homologous to known genes), or *putative* (containing an expressed sequence whose gene or ORF is not known). Pseudogenes are sequences that are homologous to known genes or proteins but have a disrupted ORF and therefore cannot be expressed. [*After P. Deloukas et al., Nature* **414**, *866 (2001).*]

Most of this chapter focuses on the transcription of eukaryotic protein-coding genes. The human genome includes an estimated 30,000 to 35,000 such genes, which have an average length of about 27,000 bp. The actual fraction of protein-coding sequences is quite small, however (about 1% of the genome), because a typical gene consists of eight **exons** (protein-coding segments) with an average length of 145 bp that are separated by introns with an average length of 3365 bp. Many genes are known only as **open reading frames (ORFs;** see Section 3-5), that is, the DNA sequences can potentially be transcribed but their protein products have not yet been identified. Other gene sequences that have been cataloged as **expressed sequence tags (ESTs;** see Section 3-5) may represent novel genes (Fig. 19-1).

The sequence of the human genome reveals far fewer genes than were once expected. Similar trends have been observed in the genomic sequences for *Drosophila* (with about 13,000 genes) and the nematode *Caenorhabditis elegans* (with about 19,000 genes). Apparently, organismal complexity is not a straightforward function of gene number (or of DNA content, for that matter, as many plants have considerably more DNA than animals), but probably depends instead on the regulation of gene expression. This would relate to how many different combinations of genes can be expressed and at what levels. Humans have approximately the same number of genes as worms or flies for basic metabolic processes but have relatively more genes for higher-level processes such as hormone-signaling pathways. Not surprisingly, the accurate transcription of such genes and the processing of their mRNAs to a translation-ready form occurs in many steps and requires many enzymes and protein cofactors.

REVIEW LEARNING OBJECTIVE 1

- Describe some of the limitations of the term *gene*.
- What is the significance of ORFs and ESTs in an organism's genome?

1 TRANSCRIPTION INITIATION

Like DNA replication, *RNA transcription is most likely carried out by immobile protein complexes that reel in the DNA.* These transcription factories in eukaryotic nuclei can be visualized by immunofluorescence microscopy and are distinct from the replication factories where DNA is synthesized (Fig. 19-2). If the RNA polymerase were free to track along the length of a DNA molecule, rotating around the helical template, the newly synthesized RNA strand would become tangled with the DNA. In fact, except for a short 8- to 9-bp hybrid DNA–RNA helix at the polymerase active site, the newly synthesized RNA is released as a single-stranded molecule.

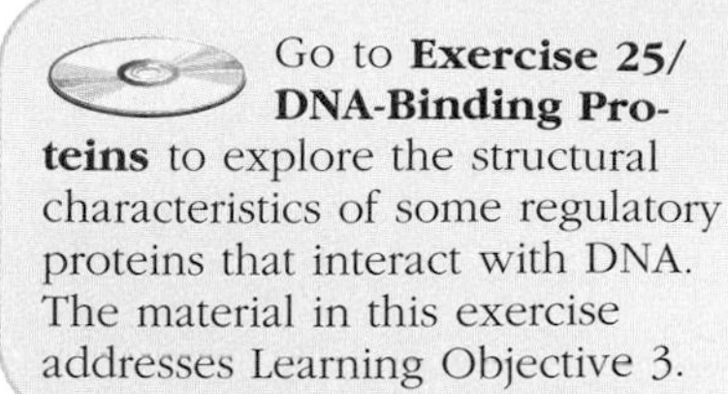

Go to **Exercise 25/ DNA-Binding Proteins** to explore the structural characteristics of some regulatory proteins that interact with DNA. The material in this exercise addresses Learning Objective 3.

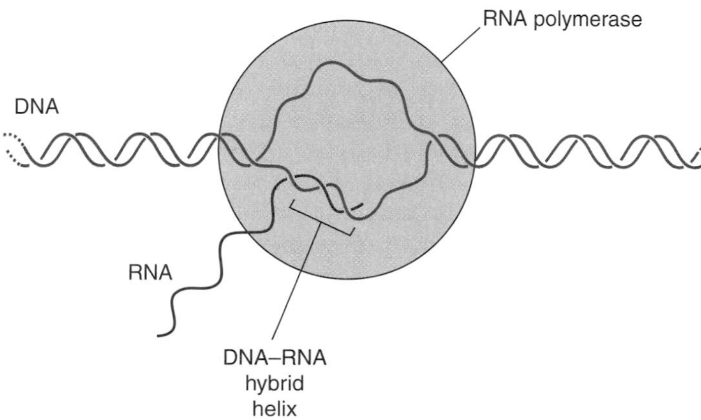

Synthesis of a new RNA molecule is a multistep process, which includes locating the transcription start site, melting apart the double-stranded DNA, initiating RNA synthesis, and extending the RNA molecule. We will examine these steps below and then will examine how the RNA transcript is modified (a process that begins even before its synthesis is complete).

Transcription begins at promoters

In both prokaryotes (which typically have compact genomes without much nontranscribed DNA) and eukaryotes (where genes may be separated by large tracts of "junk" DNA), identifying the transcription start site for a particular gene is essential for the efficient expression of that portion of genetic information. In all cells, *transcription begins at a site known as a* ***promoter.*** The DNA sequence at this site is recognized by specific proteins, which either are part of the RNA polymerase protein or subsequently recruit the appropriate RNA polymerase to the promoter to begin RNA synthesis.

In bacteria such as *E. coli,* the promoter comprises a sequence of about 40 bases on the 5′ side of the transcription start site. By convention, such sequences are written for the coding, or nontemplate, DNA strand (see page

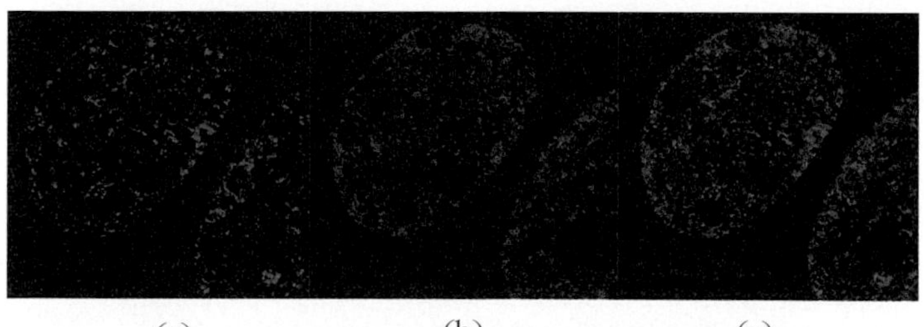

FIGURE 19-2

Spatial separation of transcription and replication. In these fluorescence microscopy images of mouse cells in early S phase, sites of DNA replication are green (a) and sites of RNA transcription are red (b). The merged images are shown in (c). A single nucleus may contain 2000 to 3000 transcription sites or "factories." [*Courtesy Ronald Berezney. From Wei et al.,* Science ***281,*** *1502–1505 (1998).*]

Transcription start site
−35 region −10 region +1
5′··· NNNTTGACANNNNNNNNNNNNNNNNNNTATAATNNNNNNNNNNNN ···3′

FIGURE 19-3

The *E. coli* promoter.
The first nucleotide to be transcribed is position +1. The two consensus promoter sequences, centered around positions −10 and −35, are shaded. N represents any nucleotide.

64) so that the DNA sequence has the same sense and 5′→3′ directionality as the transcribed RNA. The *E. coli* promoter includes two conserved segments, whose **consensus sequences** (sequences indicating the nucleotides found most frequently at each position) are centered at positions −35 and −10 relative to the transcription start site (position +1; Fig. 19-3).

E. coli RNA polymerase is a five-subunit enzyme with a mass of about 450 kD and a subunit composition of $\alpha_2\beta\beta'\sigma$. The σ subunit, or σ factor, recognizes the promoter, thereby precisely positioning the RNA polymerase enzyme to begin transcribing. Although bacterial cells contain only one core RNA polymerase ($\alpha_2\beta\beta'$), they contain multiple σ factors, each specific for a different promoter sequence. Since different genes may have similar promoters, *bacterial cells can regulate patterns of gene expression through the use of different σ factors.* Once transcription is under way, the σ factor is jettisoned, and the remaining subunits of RNA polymerase extend the transcript.

Eukaryotic promoters for protein-coding genes often include a conserved AT-rich sequence, called a **TATA box,** just upstream of (on the 5′ side of) the transcription start site (Fig. 19-4). This sequence resembles the −10 region of a prokaryotic promoter but is located farther upstream. Eukaryotic promoters typically include additional upstream sequences, some of which may be hundreds of nucleotides away. A promoter may also include sequences within the coding region of the gene. Eukaryotic RNA polymerases do not include a subunit corresponding to the prokaryotic σ factor. Instead, RNA polymerase is recruited to the promoter through a series of more complex protein–protein and protein–DNA interactions.

Transcription factors recognize eukaryotic promoters

Eukaryotic promoters must be distinguished from vast tracts of meaningless DNA and may be partially hidden by the structure of chromatin, in which DNA is wound around histone proteins to form disk-shaped nucleosomes that pack together to form larger structures (see Section 17-4). The accessibility of chromatin to transcription depends on its condensation state, which is regulated in part through covalent modification of the histone proteins. Transcriptionally "silent" chromatin is highly condensed. In "active" chromatin,

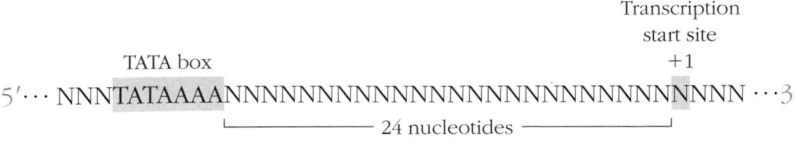

Transcription start site
TATA box +1
5′··· NNNTATAAAANNNNNNNNNNNNNNNNNNNNNNNNNNNNNNN ···3′
24 nucleotides

FIGURE 19-4

A eukaryotic promoter.
The consensus sequence of the conserved TATA box is shaded. The nucleotide at the transcription start site (+1) is often T, so that the mRNA often begins with A.

which is less condensed than silent chromatin, histones have been acetylated. The enzymes known as histone acetyltransferases add acetyl groups from acetyl-CoA to the side chains of Lys residues.

$$-CH_2-CH_2-CH_2-CH_2-NH-C(=O)-CH_3$$

Acetyl-Lys

This modification can be reversed later by the action of a histone deacetylase. *Acetylation and other covalent modifications are believed to constitute a sort of code that allows other proteins to recognize certain segments of DNA in order to transcribe them.* The roles of a number of these proteins are known.

In eukaryotes, the initiation of transcription requires a set of highly conserved proteins known as ***general transcription factors.*** They are abbreviated as TFIIA, TFIIB, TFIID, TFIIE, TFIIF, and TFIIH (the II indicates that these transcription factors are specific for RNA polymerase II, the enzyme that transcribes protein-coding genes). TFIID is one of the most important. It consists of a TATA-binding protein (TBP) and as many as 12 additional proteins called TBP-associated factors (TAFs).

TBP is a saddle-shaped protein about 32 × 45 × 60 Å that binds to the TATA box of a eukaryotic promoter. TBP's two structurally similar domains sit astride a double-stranded-DNA molecule at an angle (Fig. 19-5). This protein–DNA interaction introduces two sharp kinks into the DNA. The kinks are caused by the insertion of two Phe side chains like a wedge between a T and an A residue at each end of the TATA box. There are other sequence-specific interactions based on hydrogen bonding and van der Waals interactions. DNA–TBP binding is explored in Exercise 25.

The TAFs, the other subunits of TFIID, range in size from 15 to 250 kD. Different TFIID complexes appear to share a set of seven or eight TAFs, with the additional subunits conferring promoter specificity by interacting with DNA in a highly sequence-specific manner. Transcription from some promoters appears to occur without the participation of TAFs, although TBP is

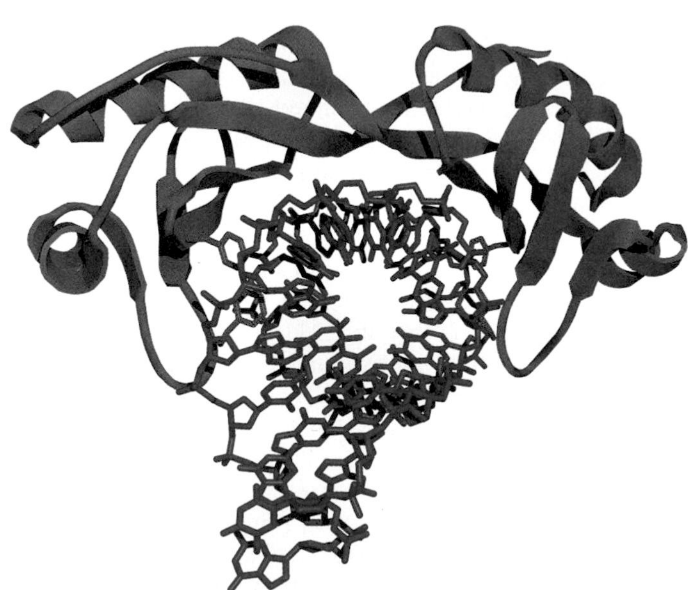

FIGURE 19-5

Structure of TBP bound to DNA. The TBP polypeptide (green) forms a pseudosymmetrical structure that straddles a segment of DNA containing a TATA box (the DNA is shown in blue and viewed end-on). The insertion of TBP Phe residues (gold) bends the DNA in two places. As a result, the DNA unwinds slightly, widening the major groove so that it can make contacts with eight strands of TBP's 10-stranded β sheet. [*Structure (pdb 1YTB) determined by Y. Kim, J.H. Geiger, S. Hahn, and P.B. Sigler.*]

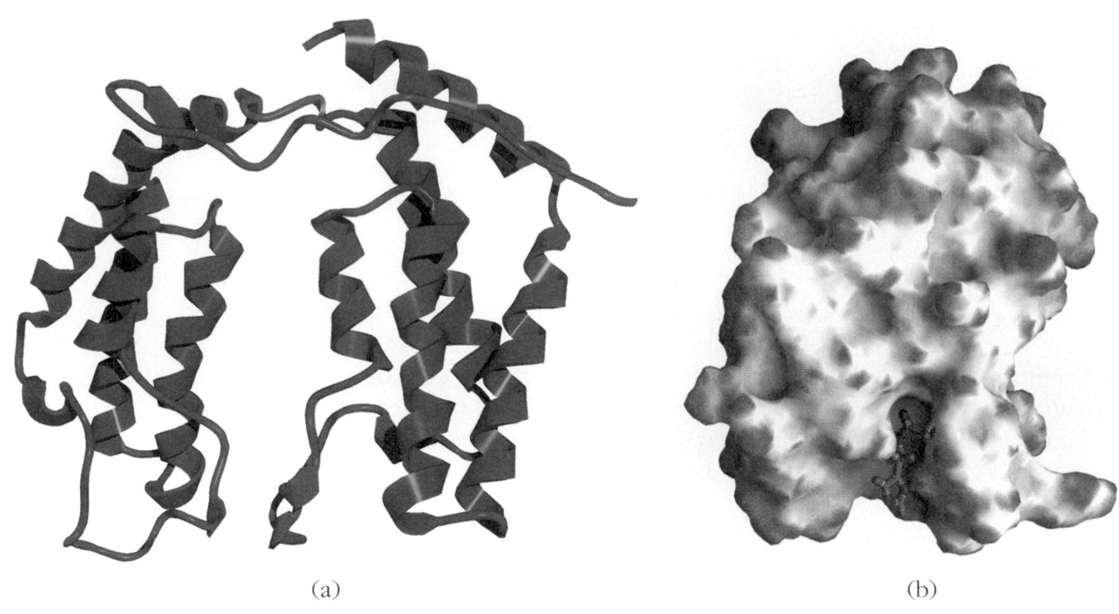

FIGURE 19-6

The double bromodomain of TAFII250.

(a) This crystal structure represents residues 1359 to 1638 of TAFII250. A bromodomain consists of about 120 amino acids arranged in a bundle of four α helices. [*Structure (pdb 1EQF) determined by R.H. Jacobson, A.G. Ladurner, D.S. King, and R. Tjian.*] (b) Model of a single bromodomain showing the cavity for an acetyl-Lys group (ball-and-stick model). [*Courtesy Ming-Ming Zhou, Mt. Sinai School of Medicine. From* Science ***285,*** *1201 (1999).*]

required. In general, TAFs appear to act by receiving regulatory signals that govern when and how often a particular gene is transcribed (see below).

The TFIID component known as TAFII250 (it is the largest of the TAFs) has several enzymatic activities. It is a kinase and a histone acetyltransferase, which may allow it to modify histones and thereby alter the structure of nucleosomes or their ability to interact with regulatory proteins. In addition to loosening up chromatin by neutralizing Lys groups, TAFII250 may diminish histone H1 cross-linking of neighboring nucleosomes by helping to link the small protein ubiquitin to H1, thereby marking it for proteolytic destruction by a proteasome (see Section 9-1).

TAFII250 and some other transcription factors include structural motifs known as bromodomains. These consist of a bundle of four antiparallel α helices, which are arranged with an unusual left-handed twist. In the center of the bundle is a hydrophobic pocket that binds an acetylated Lys side chain. In TAFII250, two bromodomains are arranged side by side, which might allow the protein to bind cooperatively to a multiply acetylated histone protein (Fig. 19-6). The histone acetyltransferase activity of the TAF itself may lead to localized hyperacetylation, a positive feedback mechanism for promoting the transcription of particular genes.

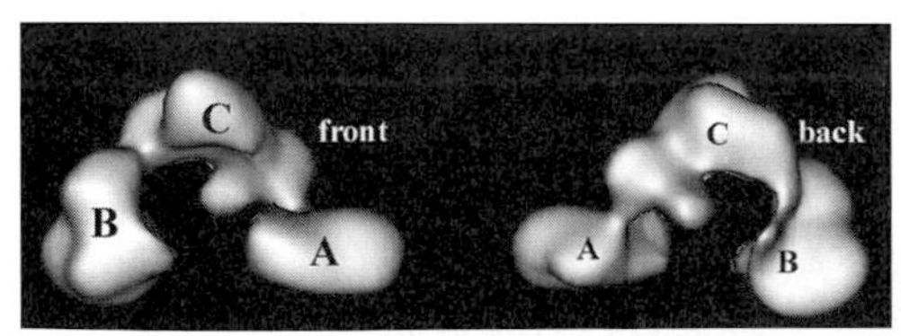

FIGURE 19-7

Three-dimensional reconstruction of human TFIID.

This 35-Å-resoution structure determined by electron microscopy shows that TFIID forms a horseshoe shape with three lobes (labeled A, B, and C). [*Courtesy Eva Nogales. Used with permission of American Association for the Advancement for Science. From F. Andel et al.,* Science ***286,*** *2153–156 (1999).*]

The structure of the entire TFIID complex, consisting of TBP and TAFs, has been visualized by electron microscopy (Fig. 19-7). The horseshoe-shaped particle is about 200 Å in its largest dimension, with a central cavity large enough for a double-stranded DNA molecule to fit inside. The TBP component of TFIID probably binds to the DNA at the top of the horseshoe (near lobe C in Fig. 19-7).

Once TBP is in place at the promoter, the conformationally altered DNA in the TBP–DNA complex may serve as a stage for the assembly of additional proteins, including the general transcription factors TFIIA and TFIIB. A model

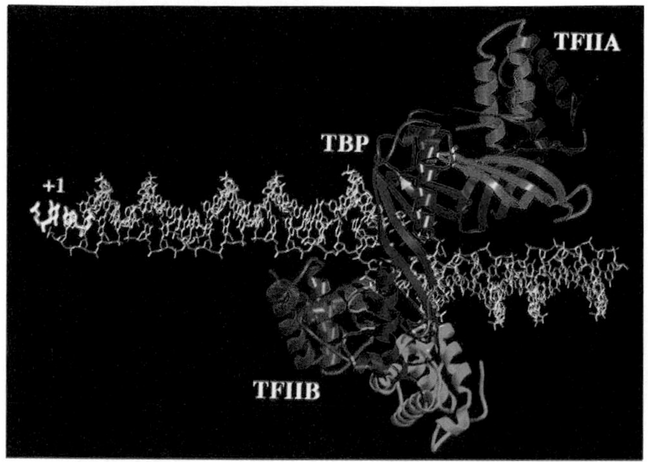

FIGURE 19-8

Model of eukaryotic transcription factors bound to DNA.
This model shows the TBP component of TFIID (blue and cyan; the TAF subunits of TFIID are not included), TFIIA (green and yellow), and TFIIB (pink and red) bound to DNA (white). Note how TBP bends the DNA helix. The transcription start site is at the left (marked as +1). [*Courtesy Stephen Burley, Rockefeller University.*]

of the TBP–TFIIA–TFIIB–DNA complex is shown in Figure 19-8. Additional transcription factors can then bind to the complex, including TFIIH, which is a helicase that unwinds the DNA in an ATP-dependent manner. The result is an open structure called a transcription bubble.

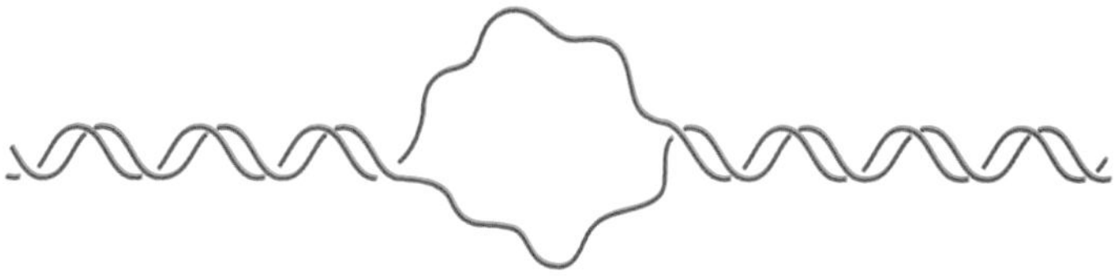

Most importantly, *the assembled general transcription factors provide a docking site for RNA polymerase.* Altogether, as many as 50 polypeptides may be positioned on the DNA at the start of transcription, including the 12 subunits of RNA polymerase.

The spacing between the TATA box (where TBP binds) and the transcription start site (where the catalytic site of RNA polymerase is positioned) tends to be about 25 bp, corresponding to about 2.5 turns of the DNA helix. Interestingly, some promoters that lack a TATA box still require TBP and other general transcription factors to position RNA polymerase properly for transcription.

 **REVIEW LEARNING OBJECTIVE 2**

- How do prokaryotic and eukaryotic promoters differ?
- How are eukaryotic transcription factors analogous to the bacterial σ factor?
- How is histone acetylation related to transcription?
- Describe the roles of TBP, TAFs, and TFIIH in initiating transcription.
- How many different enzymes activities occur among the general transcription factors?

Enhancers and silencers act at a distance from the promoter

Additional sets of protein–protein and protein–DNA interactions may participate in the highly regulated expression of many eukaryotic genes. Whereas the rate of prokaryotic gene transcription varies about 1000-fold between the

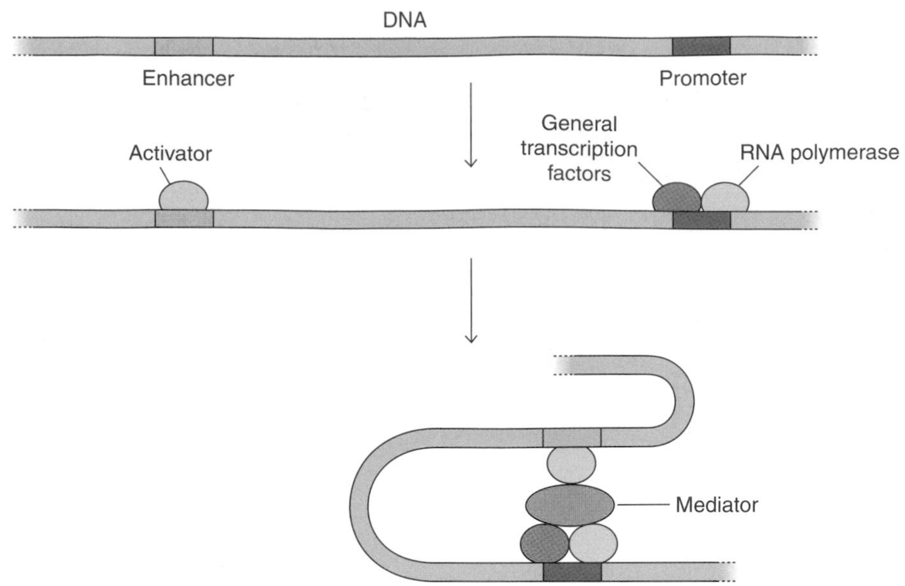

FIGURE 19-9

Overview of enhancer function. An activator protein binds to a gene's enhancer sequence. Yeast Mediator, binding to the activator as well as to the general transcription factors and RNA polymerase at the gene's promoter, transmits a transcription-activating signal to RNA polymerase, thereby promoting gene expression. The negative regulation of gene expression may occur through the binding of a repressor protein to a gene's silencer sequence. In both cases, protein–protein interactions cause the DNA to loop out between the regulatory sequence and the promoter. This simple diagram does not convey the complexity of many activator and repressor pathways, which may involve competition for binding sites among the many protein factors. Multicellular organisms contain numerous complexes with Mediator-like activity.

most and least frequently expressed genes, the gene transcription rate in eukaryotes may vary by as much as 10^9. Some of this fine control is due to **enhancers,** DNA sequences that range from 50 to 1500 bp and are located up to 85 kb upstream of a promoter or 69 kb downstream or even within the coding region of the gene. A single gene may have more than one enhancer functionally associated with it.

A protein ***activator*** *binds to the enhancer, and a protein complex (known as Mediator in yeast) may link the enhancer-bound activator to the transcription machinery poised at the promoter.* Note that this interaction requires that the DNA form a loop in order for the enhancer and promoter to be linked (Fig. 19-9). The packaging of DNA in nucleosomes may facilitate this long-range interaction by minimizing the length of the intervening DNA loop. In addition to enhancers, *a gene may have associated* ***silencer*** *sequences that bind proteins known as* ***repressors.*** Mediator or a similar complex may also relay silencer–repressor signals to the transcription machinery in order to repress gene transcription. Exercise 25 examines the structure and DNA binding of the *E. coli lac* repressor protein.

Yeast Mediator, which contains about 20 polypeptides, is a discrete particle visible by electron microscopy (Fig. 19-10). Mediator appears to interact with the C-terminal domain of the largest subunit of RNA polymerase as well as with individual TAFs or the general transcription factors. Mediator complexes with different polypeptide compositions could potentially recognize different activators and repressors. In more complicated organisms, the variability among Mediator-like complexes, coupled with multiple enhancers and silencers, could constitute a sophisticated system for fine-tuning gene expression, leading to the different patterns of gene expression that distinguish the 200 or so different cell types found in the human body.

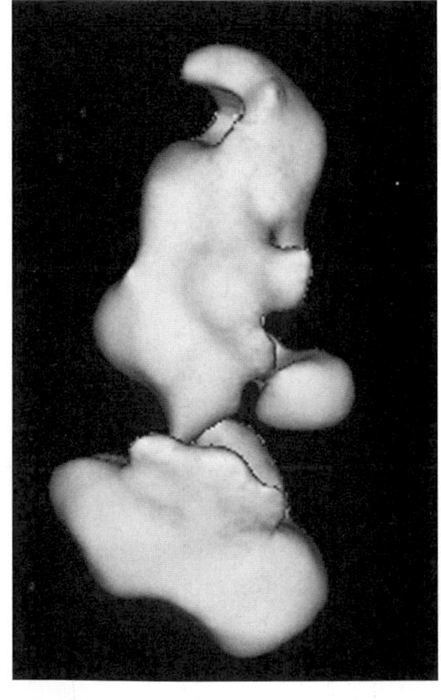

FIGURE 19-10

The yeast Mediator complex. Electron microscopy reveals the structure of this protein complex to about 40-Å resolution. The irregular Mediator particle is about 400 Å long. [*Courtesy Francisco J. Asturias. From M.R. Dotson et al.,* Proc. Natl. Acad. Sci. ***97,*** *14307 (2000).*]

Changes in gene expression may be accomplished through transcription-regulating proteins that are themselves regulated by a signal-transduction pathway or another mechanism. For example, steroid hormones diffuse through cell membranes, enter the nucleus, and bind to a nuclear hormone receptor. The hormone–receptor complex, acting as an activator, binds to a specific DNA sequence called a hormone-response element. Other hormones that bind to cell surface receptors trigger intracellular signaling cascades involving kinase-catalyzed phosphorylations. The ultimate targets of some of

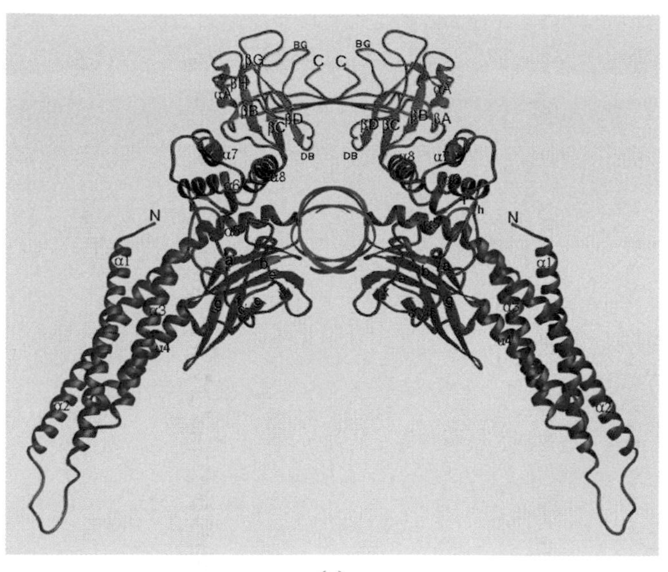

(a)

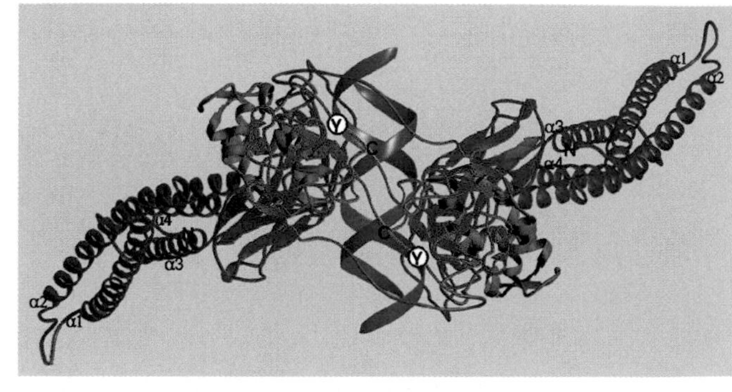

(b)

FIGURE 19-11

Structure of a STAT dimer bound to DNA.
The STAT known as Stat3β dimerizes when phosphorylated (the phosphorylated Tyr residues are located in the yellow dimerization domains). DNA binding is mediated by backbone and side chain groups of the red and green domains. The DNA is shown as a purple ribbon. (a) View along the axis of the DNA, which is bent about 40°. (b) View from the side, with the phospho-Tyr residues marked with a Y. [*Photos courtesy Chistoph W. Mueller, EMBL, Grenoble Outstation, France. From* Nature ***394****, 145–51 (1998).*]

these pathways are known as STAT proteins (for *s*ignal *t*ransducer and *a*ctivator of *t*ranscription). When a tyrosine residue of a STAT is phosphorylated, the STAT dimerizes, translocates to the nucleus, and binds to a specific DNA sequence. The STAT dimer grips the DNA like a pair of pliers (Fig. 19-11). Binding kinks the DNA in two places, generating a 40° bend overall. This conformational change may alter the activities of other proteins required for transcription initiation. Transcription-regulating proteins such as TBP and the STATs belong to a large group of DNA-binding proteins that interact with the nucleic acid in diverse ways (Box 19-A).

Some genes are coordinately expressed

In prokaryotes, the organization of functionally related genes in operons ensures their coordinated expression in response to some metabolic signal. For example, in *E. coli,* six genes involved in synthesizing the amino acid tryptophan form the *trp* operon. The genes are expressed when tryptophan is scarce and are silent when tryptophan is abundant (Fig. 19-12). About 13% of prokaryotic genes are found in operons, including the three genes of the well-studied *lac* operon (Box 19-B).

P	*trpL*	*trpE*	*trpD*	*trpC*	*trpB*	*trpA*

FIGURE 19-12

The *E. coli trp* operon.
The products of five genes (*trpA* through *trpE*) are polypeptides that catalyze several reactions in the tryptophan biosynthetic pathway. The genes are transcribed as a single unit from the promoter, *P*.

A CLOSER LOOK

BOX 19-A DNA binding proteins

The proteins that directly participate in or regulate processes such as DNA replication, repair, and transcription must interact intimately with the DNA double helix. In fact, many of the proteins that promote or suppress transcription recognize and bind to particular sequences in the DNA. However, there do not seem to be any strict rules by which certain amino acid side chains pair with certain nucleotide bases. In general, interactions are based on van der Waals interactions and hydrogen bonds, often with intervening water molecules. Exercise 25 explores the structures of four representative DNA-binding proteins.

Some of the best-understood DNA-binding proteins are found in prokaryotic systems. For example, the expression of a prokaryotic gene is often prevented by a protein called a repressor, which binds near the gene's promoter. A decrease in the concentration of the repressor, or some alteration of its activity, allows transcription to proceed. Although transcription regulation in prokaryotes is less complex than in eukaryotes, both types of organisms rely on a variety of DNA-binding proteins. As much as 3% of a typical prokaryote's genes encode DNA-binding proteins. The proportion may be as high as 7% in eukaryotes.

An examination of the structures of a large variety of protein–DNA complexes in prokaryotic and eukaryotic cells reveals that the DNA-binding proteins fall into a limited number of classes, depending on the structural motif that contacts the DNA. Many of these motifs are likely the result of convergent evolution and may therefore represent the most stable and evolutionarily adaptable ways for proteins to interact with DNA.

By far the most prevalent mode of protein–DNA interaction involves a protein α helix that binds in the major groove of DNA. This DNA-binding motif may take the form of a helix–turn–helix (HTH) structure in which two perpendicular α helices are connected by a small loop of at least four residues. The HTH motifs are colored red in the structure shown here, which is from the bacteriophage λ repressor. The DNA is blue.

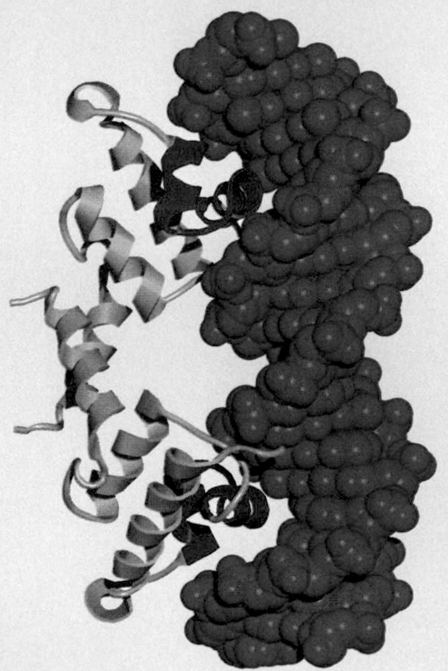

In most cases, the side chains of the DNA-binding helix insert into the major groove and directly contact the exposed edges of bases in the DNA. Residues in the other helix and the turn may interact with the DNA backbone.

In prokaryotic and eukaryotic proteins, the HTH helices are usually part of a larger bundle of several α helices, which form a stable domain with a hydrophobic core. Prokaryotic transcription factors tend to be homodimeric proteins (as in the example above) that recognize palindromic DNA sequences. In contrast, eukaryotic transcription factors more commonly are heterodimeric or contain multiple domains that recognize a nonsymmetrical series of binding sites. For this reason, eukaryotic DNA-binding proteins are able to interact with a wider variety of target DNA sequences.

Many eukaryotic transcription factors include a DNA-binding motif in which one zinc ion (sometimes two) is tetrahedrally coordinated by Cys or His side chains. The metal ion stabilizes a small protein domain (which is sometimes involved in protein–protein rather than protein–DNA interactions). In most cases, the DNA-binding motif, known as a zinc finger (see Fig. 4-17), consists of two antiparallel β strands followed by an α helix. The Zn^{2+} coordinating groups are part of the helix and the second β strand. The structure shown here is from a mouse transcription factor. One of its three zinc fingers is shown in red. The Zn^{2+} ions are represented by purple spheres.

As in the HTH proteins, the helix of the zinc finger motif inserts into the major groove of DNA, where it interacts with a three-base-pair sequence.

Some homodimeric DNA binding proteins in eukaryotes include a leucine zipper motif that mediates protein dimerization. Each subunit has an α helix about 60 residues long, which forms a coiled coil with its counterpart in the other subunit (see Section 5-5). Leucine residues, appearing at every eighth position, or about every two turns of the α helix, mediate hydrophobic contacts between the two helices (they do not actually interdigitate, as the term *zipper* might

(continued)

BOX 19-A *(continued)*

suggest). The DNA-binding portions of a leucine zipper protein are extensions of the dimerization helices that bind in the major groove. The leucine zipper shown here is part of a yeast transcription factor.

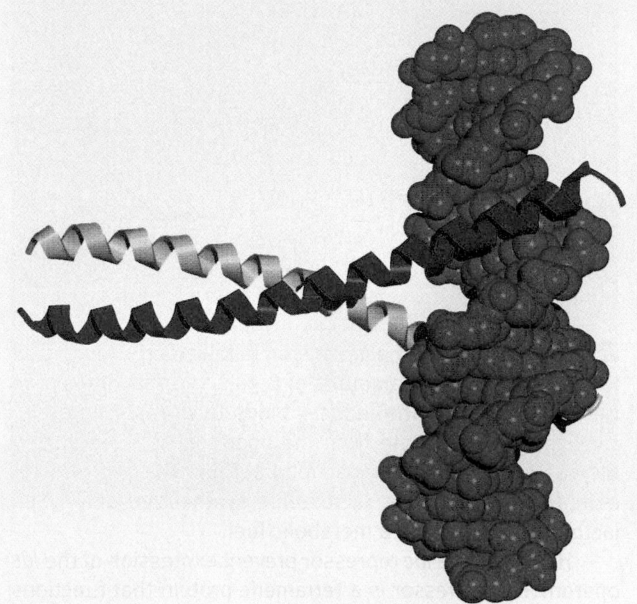

In a few proteins, a β sheet interacts with the DNA (TBP is one such protein; see Fig. 19-5). In a few cases, two antiparallel β strands constitute the DNA-binding segment, fitting into the major groove so that protein side chains can form hydrogen bonds with the functional groups on the edges of the DNA bases. The STAT proteins are among the few DNA-binding proteins that interact with both the major and minor groove of DNA.

The DNA-binding proteins described here interact with a limited portion of the DNA (typically just a few base pairs), marking the positions of regulatory DNA sequences and making additional protein–protein contacts that control processes such as gene transcription. Proteins that carry out catalytic functions (polymerases, for example) interact with the DNA much more extensively—but in a sequence-independent manner—and they tend to envelope the entire DNA helix.

[Structure of the λ repressor/operator complex (pdb 1LMB) determined by L.J. Beamer and C.O. Pabo; structure of the Zif268 zinc finger–DNA complex (pdb 1AAY) determined by M. Elrod-Erickson, M.A. Rould, and C.O. Pabo; structure of the yeast GCN4 leucine zipper–DNA complex (pdb 1DGC) determined by W. Keller, P. Konig, and T.J. Richmond.]

Although some eukaryotic protein-coding genes are found in clusters, they are not transcribed as a single unit, as are the genes in prokaryotic operons. Nevertheless, the expression of eukaryotic genes can be coordinated if they share similar control sequences (enhancers, silencers, or promoters) and regulatory proteins (activators, repressors, or transcription factors). Groups of coregulated genes in eukaryotes are called **synexpression groups;** they account for an estimated 5 to 10% of eukaryotic genes. *Both operons and synexpression groups promote efficient transcription* and, in some cases, help ensure the stoichiometric synthesis of related gene products.

REVIEW LEARNING OBJECTIVE 3

- Summarize the roles of enhancers, silencers, activators, repressors, and Mediator in regulating gene transcription.
- How is a bacterial operon similar to a eukaryotic synexpression group?

2 RNA POLYMERASE

Go to **Exercise 24/ DNA Polymerase** to review the chemistry of the polymerization reaction, which is similar for DNA and RNA polymerases.

Bacterial cells contain just one type of RNA polymerase, but eukaryotic cells contain three (plus additional polymerases for chloroplasts and mitochondria). Eukaryotic RNA polymerase I is responsible for transcribing rRNA genes, which are present in multiple copies. RNA polymerase III mainly synthesizes tRNA

8 or 9 base pairs. The conformation of the hybrid is intermediate to the A form (as in double-stranded RNA) and the B form (as in double-stranded DNA).

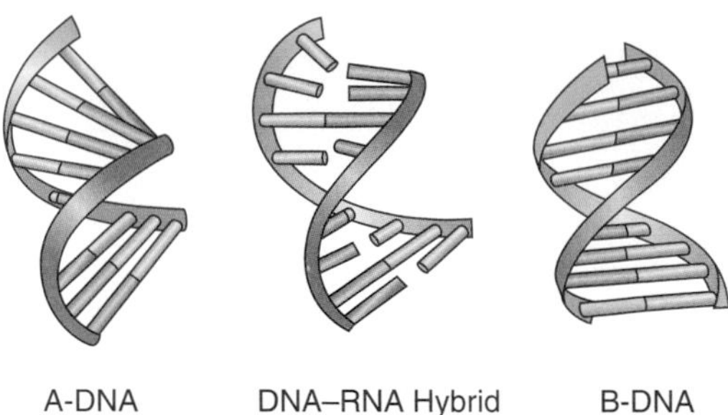

The DNA to be transcribed enters the active site cleft of RNA polymerase between two mobile protein "jaws." The two DNA strands separate to form a transcription bubble a few bases in advance of the point where the DNA–RNA hybrid starts. The transcription bubble extends to 12 or 14 nucleotides—just past the end of the hybrid helix (Fig. 19-14).

At the polymerization site, the DNA template strand makes an abrupt right angle where it encounters a wall of protein. Here, the template base points away from the standard B-DNA conformation and toward the floor of the active site cleft. This geometry allows the deoxynucleotide residue to base-pair with an incoming ribonucleoside triphosphate, which enters the active site through a channel in the floor (Fig. 19-14). Note that the incoming nucleotide has only a 25% chance of pairing correctly (since there are four possible ribonucleotides: ATP, CTP, GTP, and UTP). α-Amanitin probably binds in the nucleotide entry tunnel, since mutations that confer resistance to the toxin map to this region of the polymerase.

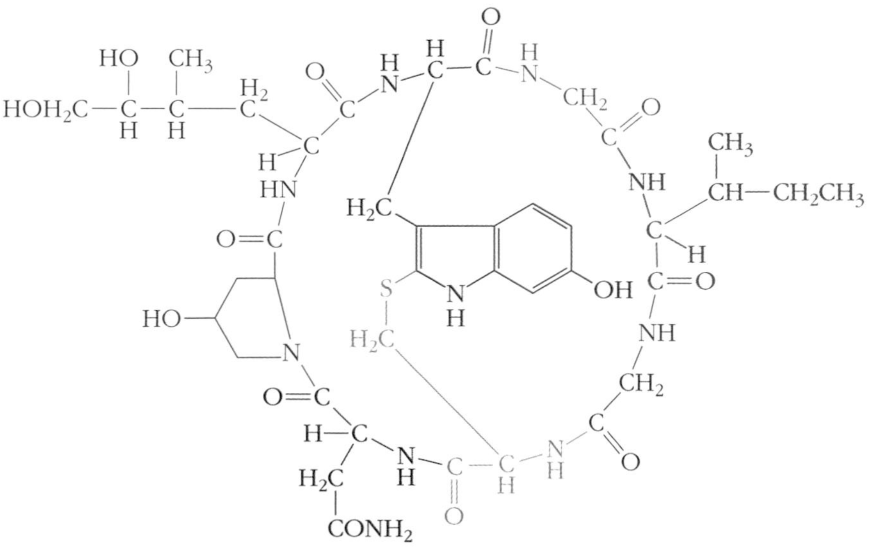

RNA polymerase is a processive enzyme

During transcription, a portion of RNA polymerase known as the clamp (the orange structure in Figs. 19-13 and 19-14) rotates by about 30° to close down snugly over the DNA template and RNA transcript. Clamp closure probably depends on specific interactions with the RNA in the hybrid helix and therefore does not occur in the presence of double-stranded DNA. *The clamp appears to ensure the high processivity of RNA polymerase.* In experiments where RNA

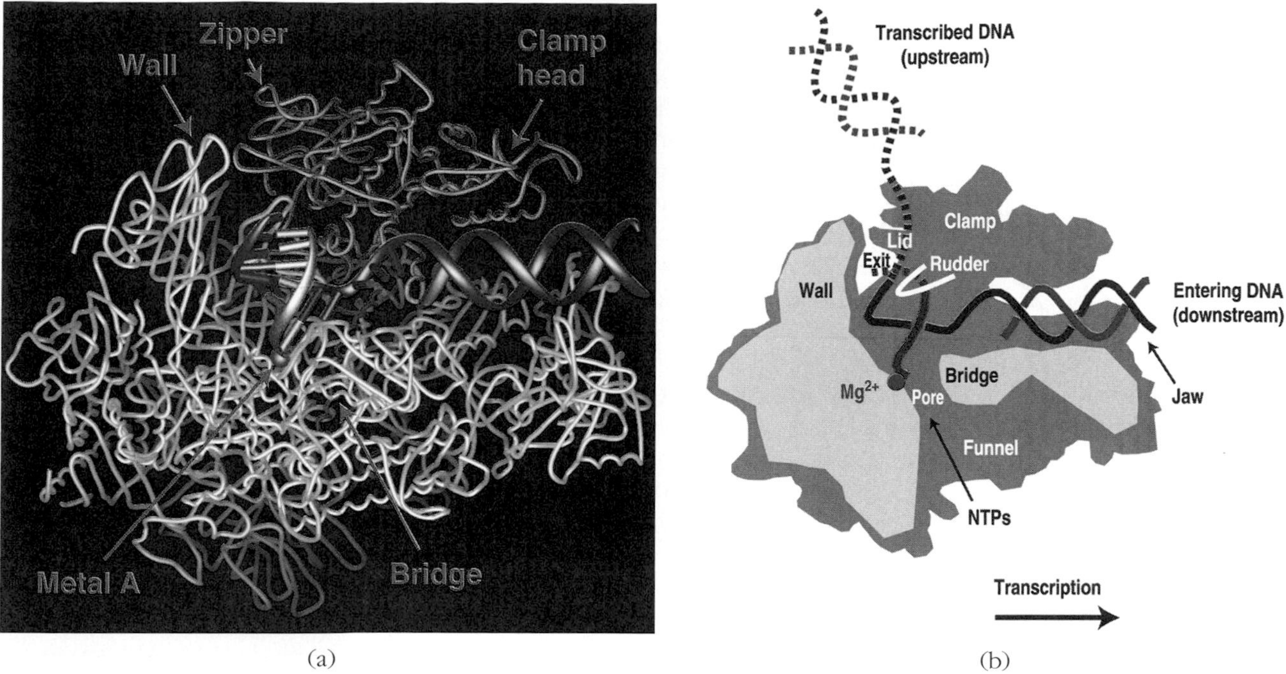

(a) (b)

FIGURE 19-14

RNA polymerase with bound DNA and RNA.
The protein is shown rotated relative to the structure in Fig. 19-13. In both structures, the region called the clamp is orange. (a) X-Ray structure of RNA polymerase (gray and orange) with DNA (coding strand green and template strand blue) and RNA (red). (b) Cutaway diagram. DNA enters the enzyme between jaws at the right. A magenta sphere marks the position of one of the catalytic Mg^{2+} ions. Newly synthesized RNA forms a short hybrid helix with the template strand before exiting the enzyme. The two DNA strands separate in advance of the polymerization site and reanneal just beyond the RNA exit site. The exact position of the nontemplate DNA strand has not been determined. [*Courtesy Roger Kornberg. (a) from* Science ***292,*** *1876 (2001). (b) from* Science ***292,*** *1844 (2001).*]

polymerase was immobilized and a magnetic bead was attached to the DNA, up to 180 rotations (representing thousands of base pairs at 10.4 bp per turn) were observed before the RNA polymerase slipped. This processivity is essential, since genes are usually thousands—sometimes millions—of nucleotides in length.

During transcription, a helix located near the active site (the long green helix visible in Fig. 19-14a) appears to oscillate between a straight and bent conformation. This alternating movement may act as a ratchet to aid translocation of the template so that the next nucleotide can be added to the growing RNA chain. But how can translocation—which requires some degree of protein mobility—be reconciled with the tight fit of the clamp over the DNA–RNA hybrid? One possible explanation is that several protein side chains may interact simultaneously with the sugar–phosphate backbones so that binding is tight but not sequence-specific, allowing the slippage necessary for translocation. Alternatively, a set of about 20 positively charged side chains may form a sort of shell about 4 to 8 Å from the hybrid helix. This shell would attract and hold the negatively charged hybrid helix without restraining its movement.

Like DNA polymerase, *RNA polymerase carries out proofreading.* If a deoxynucleotide or a mispaired ribonucleotide is mistakenly incorporated into RNA, the DNA–RNA hybrid helix becomes distorted. This causes polymer-

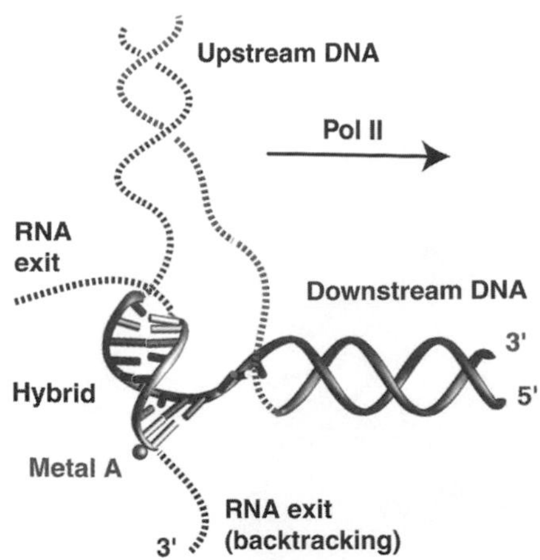

FIGURE 19-15

Schematic view of backtracking RNA in RNA polymerase.

If polymerization stops due to a polymerization error, the 3′ end of the RNA transcript may back up into the channel for incoming nucleotides. An exonuclease can then cleave the 3′ end of the RNA as it emerges from the channel. [*Courtesy Roger Kornberg. From* Science ***292,*** *1879 (2001).*]

ization to cease, and the newly synthesized RNA "backs out" of the active site through the channel by which ribonucleotides enter (Fig. 19-15). The exposed RNA containing the error is then trimmed by the nuclease action of a transcription factor known as TFIIS. Transcription may resume if the 3′ end of the truncated transcript is then repositioned at the active site.

Throughout transcription, the sizes of the transcription bubble and the DNA–RNA hybrid helix remain constant. The mechanism that maintains the transcription bubble is not understood. A protein loop known as the rudder (see Fig. 19-14b) may help separate the RNA and DNA strands so that a single RNA strand is extruded from the enzyme.

REVIEW LEARNING OBJECTIVE 4

- Describe the overall structure of RNA polymerase and the arrangement of DNA and RNA within the protein.
- In what ways does RNA polymerase differ from DNA polymerase?
- Why is it important that the DNA–RNA hybrid helix be short?
- Describe the role of the clamp in RNA synthesis.

What allows RNA polymerase to elongate a transcript?

One of the puzzles of RNA polymerase action is that the enzyme appears to initiate RNA synthesis repeatedly, producing many short transcripts before committing to elongating a transcript. This suggests that *the transcription machinery must undergo a transition from initiation mode to elongation mode.* In eukaryotes, the switch appears to involve the C-terminal domain of the largest subunit of RNA polymerase (a structure that is disordered and therefore not visible in the models shown in Figs. 19-13 and 19-14). The C-terminal domain of mammalian RNA polymerase contains 52 seven-amino acid pseudorepeats with the consensus sequence

Tyr–Ser–Pro–Thr–Ser–Pro–Ser
1 2 3 4 5 6 7

Serine residues 2 and 5 of each heptad can potentially be phosphorylated. During the initiation phase of transcription, the C-terminal domain of RNA polymerase is not phosphorylated, but elongating RNA polymerase bears numerous phosphate groups. The phosphorylation that accompanies the switch from an initiating to an elongating RNA polymerase may be the result of the kinase activity of TFIIH or a cyclin-dependent kinase (see Section 18-3).

When RNA polymerase becomes phosphorylated at its C-terminal domain, it can no longer bind a Mediator complex. This may allow the polymerase to abandon other transcription-initiating factors and advance the template DNA beyond the promoter region. In fact, when RNA polymerase "clears" the promoter, it leaves behind some general transcription factors, including TFIID (Fig. 19-16). These proteins, along with Mediator, can reinitiate transcription by recruiting another RNA polymerase to the promoter. Consequently, *the first RNA polymerase to transcribe a gene acts as a "pioneer" polymerase that helps pave the way for additional rounds of transcription.* Histone acetyltransferases associated with the pioneer RNA polymerase may alter the nucleosomes of a gene undergoing transcription. However, the histone octamer never entirely dissociates from the transcribed DNA (Fig. 19-17).

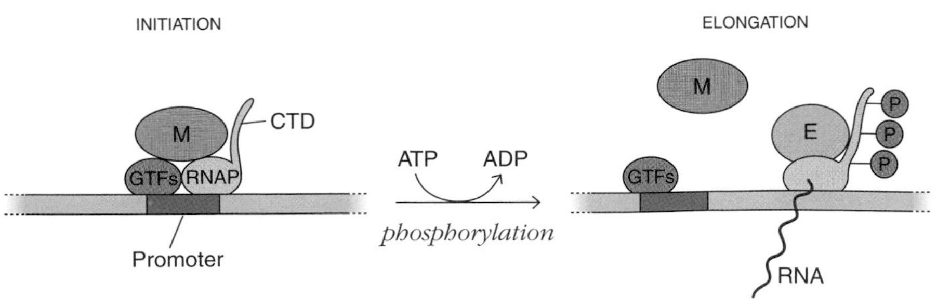

FIGURE 19-16

The transition from transcription initiation to elongation.
During initiation, the nonphosphorylated C-terminal domain (CTD) of RNA polymerase (RNAP) serves as a binding site for a Mediator-like complex (M). When the C-terminal domain undergoes phosphorylation, RNA polymerase switches to an elongation mode, dissociating from the Mediator complex and the general transcription factors (GTFs) that remain at the promoter. The Elongator complex (E) binds to the phosphorylated CTD during elongation.

During transcription elongation, various proteins, including a six-protein complex (called Elongator in yeast), bind to the phosphorylated C-terminal domain of RNA polymerase II, taking the place of the jettisoned initiation factors. Although RNA polymerase by itself can transcribe a DNA sequence *in vitro,* the presence of these additional factors accelerates transcription. Interestingly, the general transcription factors TFIIF and TFIIH, which participate in transcription initiation, remain associated with RNA polymerase during elongation (TFIIH also participates in nucleotide excision repair; its deficiency causes xeroderma pigmentosum; see Section 18-2). The phosphorylated domain of an elongating RNA polymerase also serves as a docking site for proteins that begin processing the nascent RNA transcript (see below). Transcription may not terminate until after these enzymes have completed their tasks.

During elongation, RNA polymerase may pause or even cease transcribing in response to certain DNA template signals. *In vitro,* RNA polymerase may remain paused for a long period without terminating transcription (for example, if one of its four nucleotide substrates is missing; it resumes transcription when the missing nucleotide is supplied). The task of terminating transcription is not simple, since RNA polymerase is highly processive.

In *E. coli,* about half of all transcription termination sites are defined by DNA sequences that code for an RNA sequence that forms a base-paired hairpin followed by a series of U residues, a motif that somehow destabilizes RNA polymerase. Eukaryotic genes do not include such termination signals,

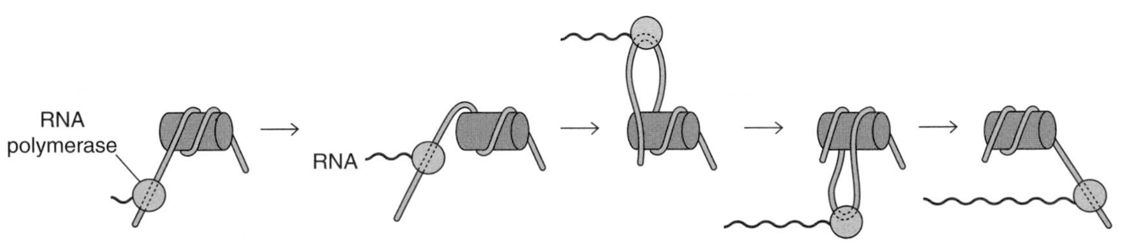

FIGURE 19-17

Transcription through a nucleosome.
In this model, which shows RNA polymerase advancing along a DNA molecule (purple) and synthesizing RNA (red), the DNA loops out so that it never entirely leaves the histone octamer (yellow). [*After Orphanides, G. and Reinberg, D., Nature* ***407****, 472 (2000).*]

and it remains unclear how the RNA polymerase clamp opens up, the RNA–DNA hybrid helix unwinds, and the transcription bubble seals.

REVIEW LEARNING OBJECTIVE 5

- Describe the events that accompany the switch from transcription initiation to elongation.
- What does the "pioneer" polymerase accomplish?

RNA PROCESSING

RNA transcripts undergo a number of modifications that affect their ability to direct translation (or participate in other processes, if they do not encode proteins). Messenger RNA processing includes the covalent modification of both ends of the transcript, splicing to remove introns, export from the nucleus, delivery to ribosomes for translation, and eventually, degradation of the message. All of these events play a role in the degree to which a gene is expressed in the context of a living cell.

mRNA processing begins well before transcription is complete, as soon as the transcript begins to emerge from RNA polymerase. Many of the various enzymes required for capping the 5′ end of the mRNA, for extending the 3′ end, and for splicing are recruited to the phosphorylated domain of RNA polymerase, so that *processing is closely linked to transcription.* In fact, the presence of processing enzymes may actually promote transcriptional elongation.

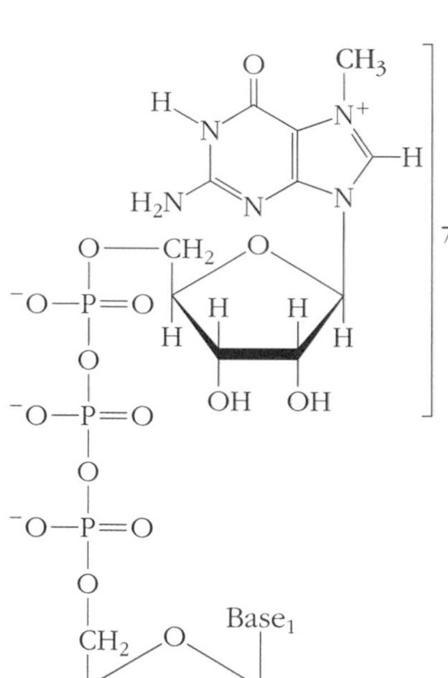

FIGURE 19-18
An mRNA 5′ cap.

5′ capping

At least three enzyme activities modify the 5′ end of the emerging mRNA to produce a structure called a ***cap*** *that protects the polynucleotide from 5′ exonucleases.* First, a triphosphatase removes the terminal phosphate from the 5′ triphosphate end of the mRNA. Next, a guanylyltransferase transfers a GMP unit from GTP to the remaining 5′ diphosphate. These two reactions, which are carried out by a bifunctional enzyme in mammals, create a 5′–5′ triphosphate linkage between two nucleotides. Finally, methyltransferases add a methyl group to the guanine and to the 2′ OH group of ribose residues (Fig. 19-18).

3′ polyadenylation

The 3′ end of an mRNA is also modified. Processing begins following the synthesis of the RNA sequence AAUAAA, which is a signal for a protein complex to cleave the transcript and extend it by adding adenosine residues. In fact, the RNA cleavage reaction may occur while the RNA polymerase is still operating, causing it to terminate transcription.

The enzyme poly(A) polymerase generates a 3′ **poly(A) tail** (also called a polyadenylate tail) of 20 to 50 residues. The enzyme structurally resembles other polymerases (Fig. 19-19), and its active site includes conserved Asp residues that coordinate two metal ions required for catalysis. No template is needed to direct the addition of nucleotides. Multiple copies of a poly(A) binding protein associate with the tail and help protect the 3′ end of the mRNA transcript from 3′ exonucleases. In addition, *the poly(A) tail may serve as a handle for the proteins that deliver mRNA to ribosomes.*

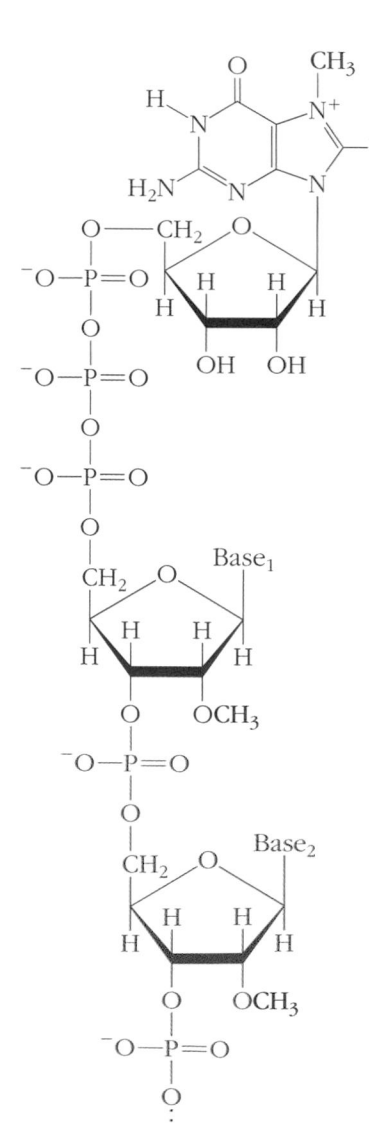

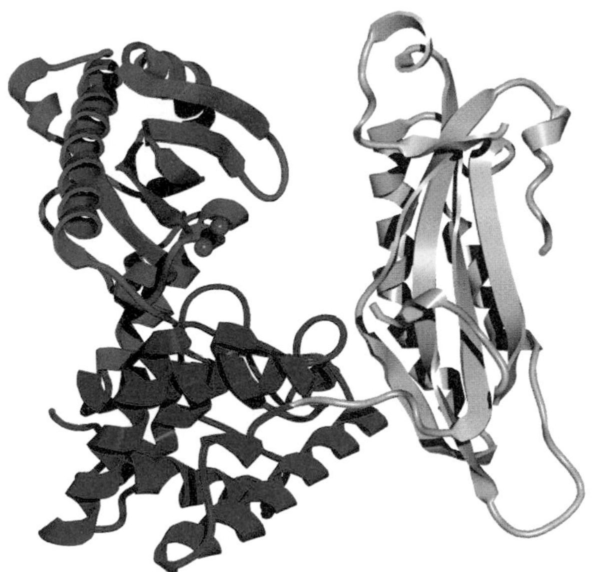

FIGURE 19-19

Yeast poly(A) polymerase.
This small enzyme structurally resembles other polymerases, with domains corresponding to the palm (green) and fingers (red) domains (see Fig. 17-11). The blue spheres represent Mn^{2+} ions in the active site. [*Structure (pdb 1FAo) determined by J. Bard, A.M. Zhelkovsky, S. Helmling, C.L. Moore, and A. Bohm.*]

Splicing

Splicing is the removal of an intervening sequence (intron) and the joining of the ends of the two adjacent expressed sequences (exons) in an RNA transcript. Like capping, splicing commences before RNA polymerase has finished transcribing a gene, and some components of the splicing machinery assemble on the phosphorylated C-terminal domain of RNA polymerase. Most mRNA splicing is carried out by a **spliceosome,** a complex of five small RNA molecules (called **snRNAs,** for **small nuclear RNAs**) and their associated proteins (numbering over 50). This spliceosome recognizes conserved sequences at the 5′ intron/exon junction and at a conserved A residue within the intron, called the branch point (Fig. 19-20). Recognition depends on base-pairing between the conserved mRNA sequences and snRNA sequences.

Splicing is a two-step transesterification process (Fig. 19-21). Each step requires an attacking nucleophile (a ribose OH group) and a leaving group (a phosphoryl group). A catalytically essential metal ion, most likely Mg^{2+}, may enhance the nucleophilicity of the attacking hydroxyl group and stabilize the phosphate leaving group.

Some types of introns, particularly in protozoan rRNA genes, undergo self-splicing; that is, they catalyze their own transesterification reactions without the aid of proteins. These rRNA molecules were the first RNA enzymes

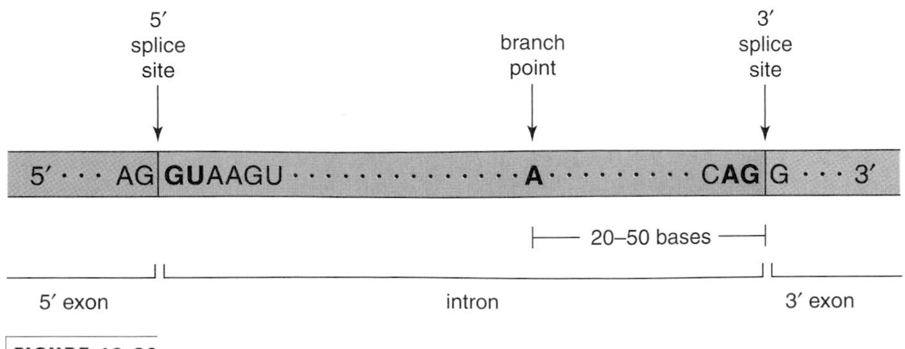

FIGURE 19-20

Consensus sequence at eukaryotic mRNA splice sites.
Nucleotides shown in bold are invariant.

A CLOSER LOOK

BOX 19-C RNA interference

Mechanisms for regulating gene expression *in vivo* include the unusual and the surprising. One of these is a posttranscriptional RNA-dependent gene-silencing mechanism known as **RNA interference (RNAi).** RNA interference was first described in studies of plant gene expression. For example, the introduction of a gene encoding an enzyme for flower pigment synthesis occasionally suppressed rather than enhanced flower pigmentation.

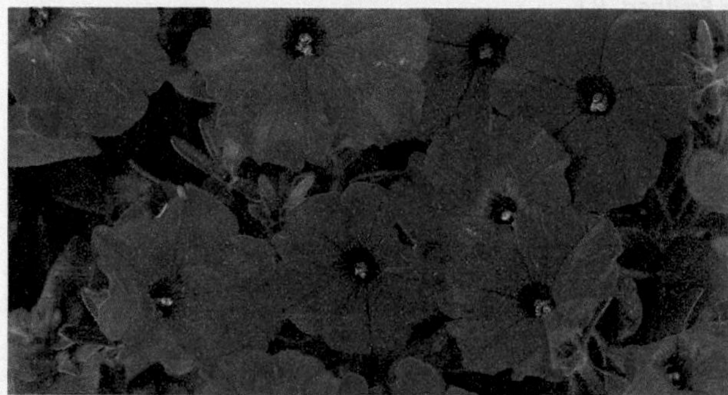

[© Barry L. Runk/Grant Heilman Photography.]

Experiments with other types of genes suggested that an additional gene could silence itself as well as its existing cellular counterpart. *In vitro* experiments indicated that injecting double-stranded RNA had the same suppressor effect (hence the name *RNA interference*).

This sort of gene silencing is distinct from that achieved by antisense RNA, where RNA molecules that are complementary to an mRNA molecule prevent translation of that mRNA through the formation of a double-stranded RNA. A well-known example of the usefulness of antisense RNA is the transgenic tomato known as Flavr Savr. The plants were engineered to express an artificial gene whose transcripts were complementary to the mRNA of a polygalactouronase (this enzyme helps break down cell walls, leading to softening of the fruit). The transgenic tomatoes—despite their long shelf-life and a flavor superior to chemically ripened fruits—never caught on with consumers.

Other experiments attempting to regulate gene expression with antisense RNAs sometimes produced dramatic gene silencing. The effects were further enhanced when the RNA was administered in the form of double-stranded RNA—results consistent with RNA interference rather than antisense RNA. This suggested that the double-stranded RNA interfered with gene expression, but how?

Double-stranded RNA can arise when an RNA molecule folds up on itself to form an extensively base-paired segment. In cells, this RNA is degraded by a nuclease (called Dicer in *Drosophila*) that recognizes the end of a double-stranded RNA molecule and cleaves it into 20–25 bp fragments. These double-stranded RNA fragments, called small interfering RNAs (siRNAs), then direct the degradation of single-stranded RNA that is homologous to one of its strands.

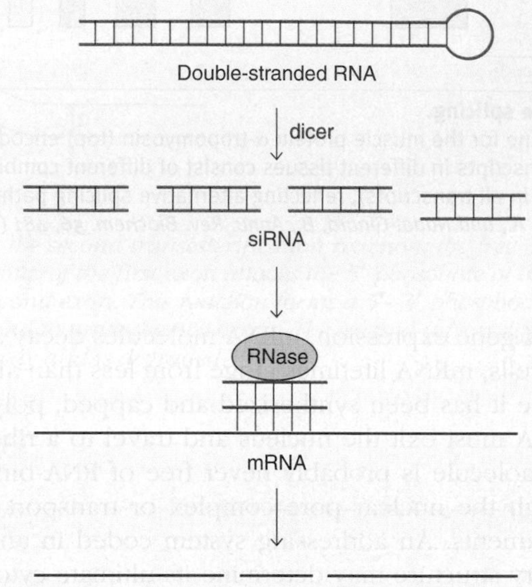

In other words, the siRNA confers sequence-specificity to the RNase. Thus, the presence of an siRNA can lead to the posttranscriptional destruction of an mRNA molecule. Experiments in *Caenorhabditis elegans* suggest that RNAi requires an RNA helicase, which might act by promoting RNA polymerase–catalyzed extension of a single-stranded RNA, using the RNA itself as a template, to produce a double-stranded RNA molecule.

RNA interference may have evolved as a defense mechanism. For example, posttranscriptional gene silencing in plants may help quell the activity of RNA viruses. In animals, the target might be retrotransposons. These viral-derived mobile genetic elements may act by a mechanism in which an RNA-dependent RNA polymerase synthesizes a complementary strand to make a double-stranded RNA molecule.

RNA interference may also help control normal gene expression during development. For example, in the nematode *Caenorhabditis elegans,* 22-nucleotide species known as small temporal RNAs (stRNAs) contain sequences complementary to specific mRNAs that encode developmental timing proteins. Although the stRNAs and siRNAs appear to be derived by similar mechanisms, stRNAs do not promote mRNA degradation, as occurs in RNAi.

RNA interference has been observed in mammalian cells in the laboratory, but its importance *in vivo* is not known. Further exploration will be required to ascertain whether RNAi could be exploited to treat diseases associated with excessive expression of particular genes.

thereby avoiding the waste of synthesizing a nonfunctional truncated polypeptide. Sequence-specific degradation of certain RNAs, a phenomenon called RNA interference, provides another mechanism for regulating gene expression (Box 19-C).

REVIEW LEARNING OBJECTIVE 6

- Describe the reactions that modify the 5′ and 3′ ends of an mRNA molecule.
- What is the purpose of these modifications?
- How does splicing occur and what is its purpose?

rRNA and tRNA processing

rRNA transcripts, which are generated mainly by RNA polymerase I in eukaryotes, must be processed to produce mature rRNA molecules. rRNA processing and ribosome assembly take place in the **nucleolus,** a discrete region in the nucleus. The initial eukaryotic rRNA transcript is cleaved and trimmed by endo- and exonucleases to yield three rRNA molecules (called 18S, 5.8S, and 28S; Fig. 19-24).

rRNA transcripts may be covalently modified (in both prokaryotes and eukaryotes) by the conversion of some uridine residues to pseudouridine

Ribose — Uridine

Ribose — Pseudouridine (ψ)

and by the methylation of certain bases and ribose 2′ OH groups. This last type of modification is guided by a multitude of **small nucleolar RNA molecules** (called **snoRNAs**) that recognize and pair with specific 15-base segments in the rRNA sequences, thereby directing an associated protein methylase to each site. *Without the snoRNAs to mediate sequence-specific ribose methylation, the cell would require many different methylases in order to recognize all the different nucleotide sequences to be modified.*

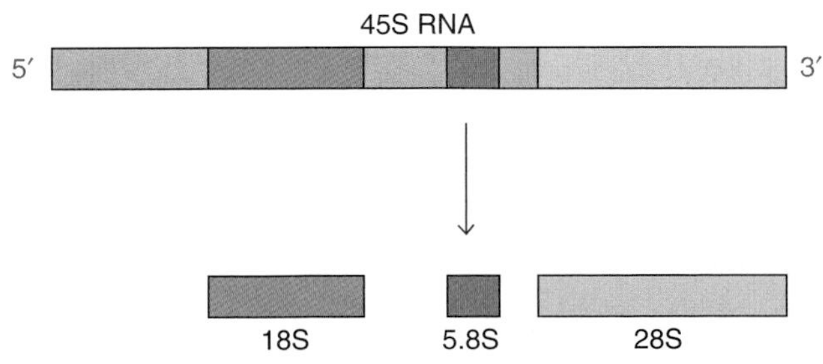

FIGURE 19-24

Eukaryotic rRNA processing.
The initial transcript of about 13.7 kb has a sedimentation coefficient of 45S. Three smaller rRNA molecules (18S, 5.8S, and 28S) are derived from it by the action of nucleases.

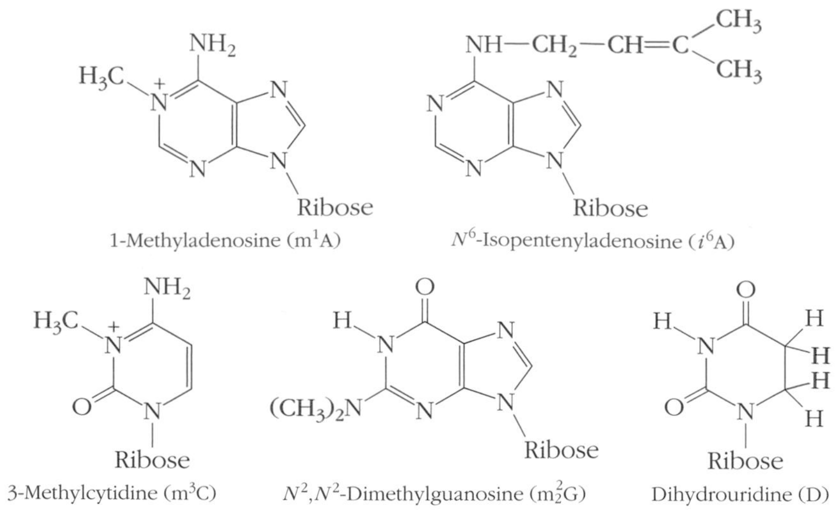

Ribose
1-Methyladenosine (m^1A)

Ribose
N^6-Isopentenyladenosine (i^6A)

Ribose
3-Methylcytidine (m^3C)

Ribose
N^2,N^2-Dimethylguanosine (m_2^2G)

Ribose
Dihydrouridine (D)

FIGURE 19-25

Some modified nucleotides in tRNA molecules.
The parent nucleotide is in black; the modification is shown in red.

A rapidly growing mammalian cell may synthesize as many as 7500 rRNA transcripts each minute, each of which associates with about 150 different snoRNAs. The processed rRNAs then combine with some 80 different ribosomal proteins to generate fully functional ribosomes, a task that requires careful coordination between RNA synthesis and ribosomal protein synthesis.

tRNA molecules, produced by the action of RNA polymerase III in eukaryotes, undergo nucleolytic processing and covalent modification. In bacteria, the initial tRNA transcripts are trimmed by ribonuclease P (RNase P), a ribonucleoprotein enzyme whose catalytic activity resides in its 377-nucleotide RNA component, not in its 14-kD polypeptide component.

Approximately 10 to 25% of the nucleotides in tRNA molecules are covalently modified, for example, by the methylation of ribose 2′ OH groups or the addition of methyl or isopentyl groups to bases (Fig. 19-25). tRNA molecules often contain pseudouridine or dihydrouridine and sometimes ribothymidine. Some tRNA transcripts also undergo splicing to remove introns. In eukaryotes, a CCA sequence is added to the 3′ end of the immature tRNA; this is the end to which an amino acid becomes attached for protein synthesis (see Section 20-1).

REVIEW LEARNING OBJECTIVE 7

- Summarize the reactions involved in rRNA and tRNA processing.
- Compare the extent to which mRNA, rRNA, and tRNA transcripts undergo nuclease digestion, covalent modification of nucleotides, splicing, and addition of nucleotides.

4 RNA: A VERSATILE MACROMOLECULE

We have already alluded to the catalytic activity of RNA, particularly as it relates to other aspects of RNA metabolism (for example, splicing and nucleolytic processing). There are at least eight naturally occurring types of catalytic RNAs and many more synthetic ribozymes. Although the discovery of catalytic RNA

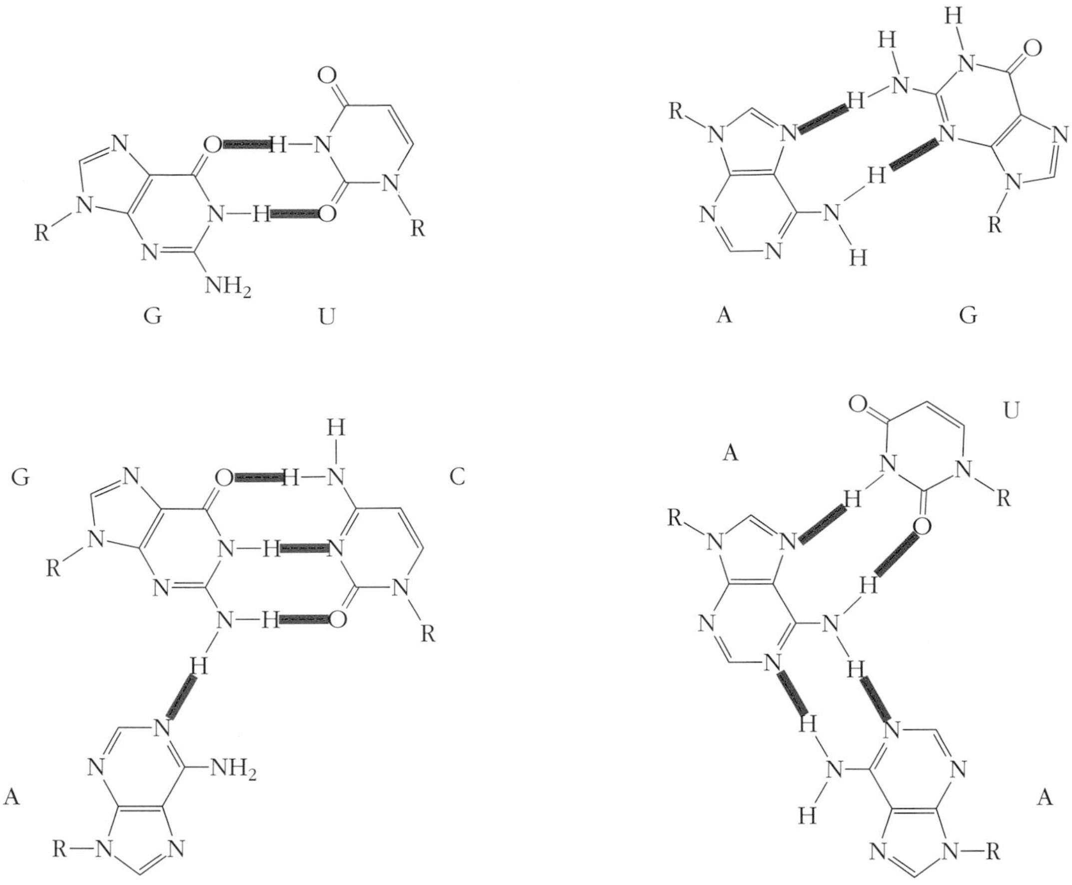

FIGURE 19-26

Some nonstandard base pairs in RNA.
These interactions can bring segments of RNA together to form complex tertiary structures. R represents the ribose–phosphate backbone.

was initially greeted with skepticism, subsequent studies have amply demonstrated that *RNA has the same properties that allow proteins to function as catalysts, namely, complex tertiary structure and reactive functional groups.*

Unlike DNA, whose conformational flexibility is considerably constrained by its double-stranded nature, a single-stranded RNA molecule can adopt highly convoluted shapes through base-pairing between different segments. In addition to the standard (Watson–Crick) types of base pairs, RNA accommodates nonstandard base pairs as well as hydrogen-bonding interactions among three bases (Fig. 19-26). Base stacking stabilizes the tertiary structure, achieving the same sort of balance between rigidity and flexibility exhibited by protein enzymes. A folded RNA molecule can then bind substrates, orient them, and stabilize the transition state of a chemical reaction.

The first ribozyme structure to be determined was that of the hammerhead ribozyme, an RNA from plant viruses that catalyzes its own cleavage. The secondary structure of the ribozyme, with three base-paired stems, resembles a hammer, but the three-dimensional structure of the molecule is more heart-shaped (Fig. 19-27). The tertiary structure depends on Watson–Crick as well as nonstandard base pairs, and it positions a catalytic Mg^{2+} ion close to the scissile bond.

It was once thought that all ribozymes must contain metal ions (as the hammerhead ribozyme does), since purine and pyrimidine bases appear to lack reactive functional groups. However, some ribozymes include groups

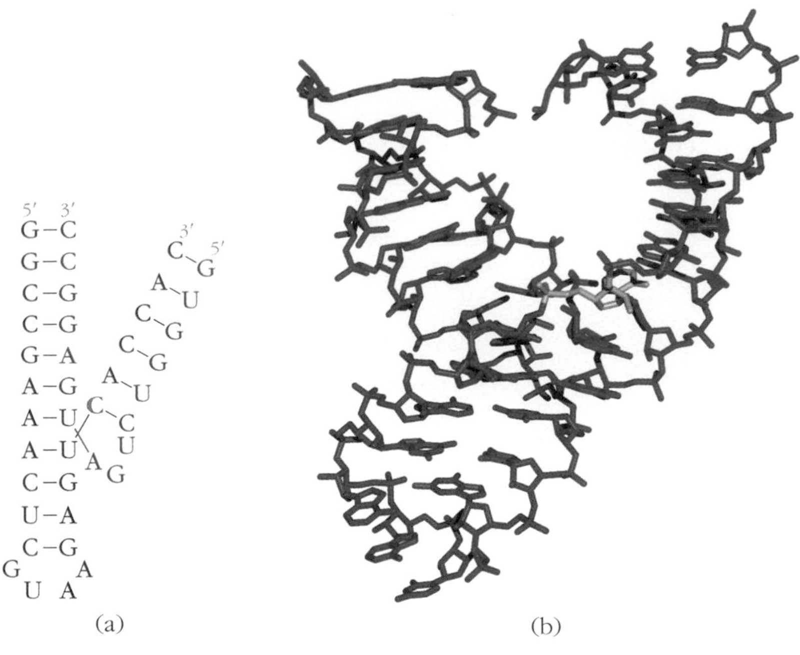

FIGURE 19-27

The hammerhead ribozyme.
The two polynucleotide strands of the ribozyme are blue and purple. The C residue at the cleavage site is shown in gray. (a) Secondary structure. (b) Three-dimensional structure. The catalytic Mg^{2+} ion is not shown. [*Structure (pdb 1MME) determined by W.G. Scott, J. Finch, and A. Klug.*]

that can function catalytically as acid or base catalysts and therefore do not require any metal ion cofactor. For example, the amino groups of adenine and cytosine can act as bases, particularly in a nonaqueous ribozyme active site.

The existence of RNA catalytic activity lends support to the theory of an early "RNA world" when RNA functioned as a repository of biological information (like modern DNA) as well as a catalyst (like modern proteins). Experiments with synthetic RNAs *in vitro* have demonstrated that RNA can catalyze a wide variety of chemical reactions, including the biologically relevant synthesis of glycosidic bonds (the type of bond that links the base and ribose in a nucleoside) and RNA-template-directed RNA synthesis. In nature, protein cofactors frequently enhance the catalytic activity of ribozymes. Presumably, most ribozymes that originated in an early RNA world were later supplanted by protein catalysts, leaving only a few examples of RNA's catalytic abilities.

REVIEW LEARNING OBJECTIVE 8

- In what ways are ribozymes similar to protein enzymes?

SUMMARY

1. Transcription is the process of converting a segment of DNA into RNA. An RNA transcript may represent a protein-coding gene or it may participate in protein synthesis or other activities, including RNA processing.

2. Transcription begins at a DNA sequence known as a promoter. A gene to be transcribed must be recognized by a regulatory factor such as the σ factor in prokaryotes.

3. In eukaryotes, general transcription factors such as the TATA-binding protein and other factors interact with DNA to form a complex that includes enzymes for modifying histones and unwinding DNA.
4. DNA-binding proteins known as activators and repressors may help regulate gene transcription.
5. Eukaryotic RNA polymerase II transcribes protein-coding genes. It requires no primer and polymerizes ribonucleotides to generate an RNA chain that forms a short double helix with the template DNA.
6. The polymerase slides along the DNA template but reverses to allow the excision of a mispaired nucleotide.
7. The elongation phase of transcription is triggered by phosphorylation of the C-terminal domain of RNA polymerase II. Elongation requires accessory factors.
8. mRNA transcripts undergo processing that includes the addition of a 5′ cap structure and a 3′ poly(A) tail. mRNA splicing, carried out by RNA–protein complexes called spliceosomes, joins exons and eliminates introns.
9. rRNA and tRNA transcripts are processed by nucleases and enzymes that modify particular bases.
10. The conformational flexibility and binding activity of some RNA molecules allows them to act as catalysts.

CHECKLIST

1. Review the Learning Objectives listed on page 580.
2. Complete Exercise 25/DNA-Binding Proteins for a look at the structures of some regulatory proteins that interact with DNA.
3. Apply your knowledge by solving the problems at the end of this chapter. Check your results in the Solutions appendix.
4. Be able to define the boldfaced terms (consult the glossary at the end of the book). Test your understanding by taking the Chapter 19 quiz (www.wiley.com/college/pratt).
5. Explore the websites listed at www.wiley.com/college/pratt.
6. Consult the list of Selected Readings for background information or additional details on transcription and RNA structure.

GLOSSARY TERMS

transcription
gene
mRNA
rRNA
tRNA
operon
RNA processing
intron
exon
ORF
EST
promoter
consensus sequence
TATA box
general transcription factor
enhancer
activator
silencer
repressor
synexpression group
cap
poly(A) tail
spliceosome
small nuclear RNA (snRNA)
ribozyme
RNAi
nucleolus
small nucleolar RNA (snoRNA)

PROBLEMS

1. Why is it effective for a bacterial cell to organize genes for related functions as an operon? How do eukaryotes achieve the same benefits?

2. RNA synthesis is much less accurate than DNA synthesis. Why does this not harm the cell?

3. The promoters for genes transcribed by eukaryotic RNA polymerase I exhibit little sequence variation, yet the promoters for genes transcribed by eukaryotic RNA polymerase II are highly variable. Explain.

4. Proteins can interact with DNA through relatively weak forces, such as hydrogen bonds and van der Waals interactions, as well as through stronger electrostatic interactions such as ion pairs. Which types of interactions predominate for sequence-specific DNA-binding proteins and for sequence-independent binding proteins?

5. The RNA polymerase from bacteriophage T7 recognizes specific promoter sequences and melts open the DNA to form a transcription bubble. No other transcription factors are necessary for T7 RNA polymerase to initiate transcription.

   (a) Dissociation constants (K_d) were measured for the interaction between the polymerase and DNA segments containing the promoter sequences. In some cases, the DNA contained a bulge, caused by a mismatch of one, four, or eight bases, to mimic the intermediates in the formation of a transcription bubble. The results are shown in the table. To which DNA segment does the polymerase bind most tightly? Explain in terms of the DNA structure.

DNA promoter segment	K_d (nM)
Fully base-paired	315
One-base bulge	0.52
Four-base bulge	0.0025
Eight-base bulge	0.0013

(b) How can the data in part a be used to calculate the free energy change for the binding of T7 RNA polymerase to DNA? (Note that a dissociation constant is the inverse of an association constant.)

(c) Compare the ΔG values for T7 RNA polymerase binding to double-stranded DNA and to DNA with an eight-base bulge.

(d) What do these results reveal about the thermodynamics of melting open a DNA helix for transcription? What is the approximate free energy cost of forming a transcription bubble equivalent to eight base pairs?

(e) Which of the following DNA segments most likely represents the sequence where the transcription bubble begins to form during transcription initiation? Explain.

CTATA GGGAG
GATAT CCCTC

[From Bandwar, R.P. and Patel, S.S., *J. Mol. Biol.* **324,** 63–72 (2002).]

6. Explain why the adenosine derivative cordycepin inhibits RNA synthesis.

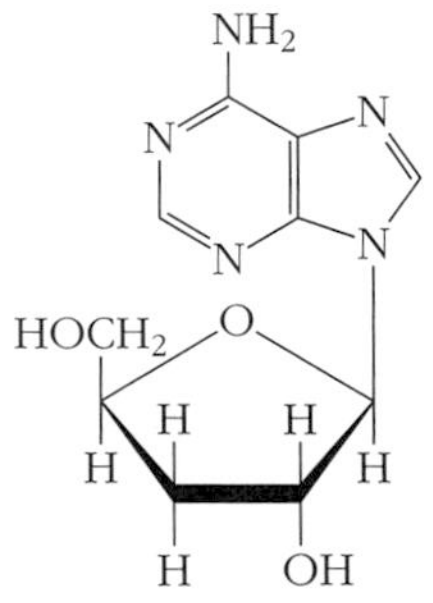

Cordycepin

7. The bacterial enzyme polynucleotide phosphorylase (PNPase) is a 3′→5′ exoribonuclease that degrades mRNA.

(a) The enzyme catalyzes a phosphorolysis reaction, as does glycogen phosphorylase (see Section 9-1), rather than hydrolysis. Write an equation for the mRNA phosphorolysis reaction.

(b) *In vitro,* PNPase also catalyzes the reverse of the phosphorolysis reaction. What does this reaction accomplish and how does it differ from the reaction carried out by RNA polymerase?

(c) PNPase includes a binding site for long ribonucleotides, which may promote the enzyme's processivity. Why would this be an advantage for the primary activity of PNPase *in vivo*?

8. In bacteria, the core RNA polymerase binds to DNA with a dissociation constant of 5×10^{-12} M. The polymerase in complex with its σ factor has a dissociation constant of 10^{-7} M. Explain.

9. The coding strand of a gene has the sequence shown below. Write the sequence of the mRNA that corresponds to this DNA sequence.

GTCCGATCGAATGCATG

10. Certain proteins that stimulate expression of a gene bind to DNA in a sequence-specific manner and also induce conformational changes in the DNA. Describe the purpose of these two modes of interaction with the DNA.

11. The enzyme known as ribonuclease H (RNase H) cleaves the RNA portion of RNA–DNA hybrid helices (see Section 17-2).

(a) Structural studies of RNase H indicate that the active site is nestled in a cluster of Lys residues. What is the probable function of these Lys residues?

(b) RNase H does not bind double-stranded DNA, single-stranded DNA, or single-stranded RNA. It binds to double-stranded RNA but does not cleave it. What does this reveal about the substrate conformation required for binding and hydrolysis?

12. A number of human neurological diseases result from the presence of trinucleotide repeats in certain protein-coding genes. The severity of each disease is correlated with the number of repeats, which may increase due to the slippage of DNA polymerase during replication.

(a) The most common repeated triplet is CAG, which is almost always located within an open reading frame. What amino acid is encoded by this triplet (see Table 3-3) and what effect would the repeats have on the protein?

(b) To test the effect of CAG repeats on transcription, researchers used a yeast expression system with genes engineered to contain CAG repeats. In addition to the expected transcripts corresponding to the known lengths of the genes, RNA molecules up to three times longer were obtained. Based on your knowledge of RNA synthesis and processing, what factors could account for longer-than-expected transcripts of a given gene?

(c) Unexpectedly long transcripts could result from slippage of RNA polymerase II during transcription of the CAG repeats. In this scenario, the polymerase temporarily ceases polymerization, slides backward along the DNA template, then resumes transcription, in effect retranscribing the same sequence. Slippage may be triggered by the formation of secondary structure in the DNA template strand. Draw a diagram showing how a DNA strand containing CAG repeats could form a secondary structure that might prevent the advance of RNA polymerase.

[From Fabre, E., Dujon, B., and Richard, G.-F., *Nucleic Acids Res.* **30,** 3540–3547 (2002).]

13. Explain why capping the 5′ end of an mRNA molecule makes it resistant to 5′→3′ exonucleases. Why is it necessary for capping to occur before the mRNA has been completely synthesized?

14. The poly(A) polymerase that modifies the 3′ end of mRNA molecules differs from other polymerases.

(a) The active sites of DNA and RNA polymerases are large enough to accommodate a double-stranded polynucleotide, but the active site of poly(A) polymerase is much narrower. Explain.

(b) Explain how the substrate specificity of poly(A) polymerase differs from that of a conventional RNA polymerase.

15. In bacteria, the organization of functionally related genes in an operon allows the simultaneous regulation of expression of those genes. If the operon consists of genes encoding the enzymes for a biosynthetic pathway, then the pathway activity as a whole can be feedback inhibited when the concentration of the pathway's final product accumulates.

(a) In one mode of feedback regulation, a repressor protein binds to a site in the operon (called the operator) to decrease the rate of transcription only when the repressor has bound a molecule representing the operon's ultimate metabolic product. Draw a diagram showing how such a regulatory system would work.

(b) Feedback regulation of gene expression can also occur after RNA synthesis has begun. In this case, the presence of the operon's ultimate product causes transcription to terminate prematurely or leads to an mRNA that cannot be translated. Draw a diagram illustrating this control mechanism. Assume that the feedback mechanism includes a protein to which the product binds.

(c) How would the feedback-inhibition system in part b differ if no protein were involved?

(d) In some bacteria, several genes required for the biosynthesis of the redox cofactor flavin adenine dinucleotide (FAD; see Fig. 11-5) are arranged in an operon. Comparisons of the sequences of this operon in different species reveal a conserved sequence in the untranslated region at the 5′ end of the operon's mRNA. The tertiary structure of an RNA molecule typically includes regions of base-pairing and unpaired loops (stem-loop structures; see Fig. 19-27). By examining an RNA sequence and noting which positions are most conserved, it is possible to predict the stem-loop structure of the RNA. A portion of a conserved mRNA sequence called RFN, which regulates the expression of the FAD-synthesizing operons, is shown below. Draw the stem-loop structure for this RNA segment.

···GAUUCAGUUUAAGCUGAAGC···

(e) In order to function as an FAD sensor, the RFN element (which consists of about 165 nucleotides) must alter its conformation when FAD binds. How could researchers assess RNA conformational changes?

(f) FAD can be considered as a derivative of flavin mononucleotide (FMN; see Fig. 12-8), which in turn is derived from riboflavin.

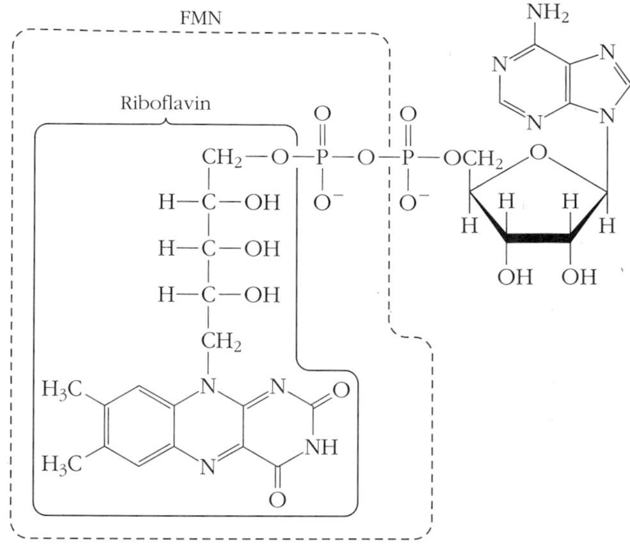

Flavin adenine dinucleotide (FAD)

The ability of FAD, FMN, and riboflavin to bind to the RFN element was measured as a dissociation constant, K_d.

Compound	K_d (nM)
FAD	300
FMN	5
Riboflavin	3000

Which compound would be the most effective regulator of FAD biosynthesis in the cell? What portion of the FAD molecule is likely to be important for interacting with the mRNA?

[From Winkler, W.C., Cohen-Chalamish, S., and Breaker, R.B., *Proc. Natl. Acad. Sci.* **99,** 15908–15913 (2002).]

16. What can you conclude about the free energy change for the splicing reaction diagrammed in Figure 19-21?

17. ATP can be labeled with ^{32}P at any one of its three phosphate groups, designated α, β and γ:

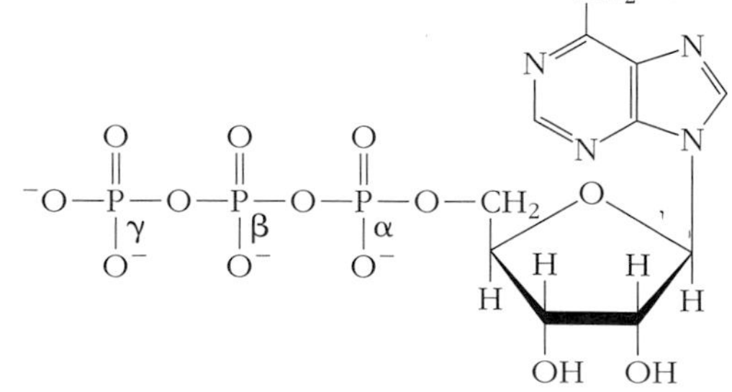

A eukaryotic cell carrying out transcription and RNA processing is incubated with labeled ATP. Where will the radioactive isotope appear in RNA if the ATP is labeled with ^{32}P at the (a) α position, (b) β position, and (c) γ position?

18. Introns in eukaryotic protein-coding genes may be quite large, but almost none are smaller than about 65 bp. What are some reasons for this minimum intron size?

19. Some introns can catalyze their own splicing reactions. Explain why these self-splicing introns are not true catalysts.

20. The endonuclease known as RNase P processes certain immature tRNA and rRNA molecules. In bacteria, RNase P consists of a relatively large RNA of about 400 nucleotides and a small protein of about 120 amino acids.

(a) How could you determine whether the RNA component of purified RNase P contained the endonuclease catalytic activity?

(b) Comparisons of RNase P RNAs from different species reveal conserved features that appear to be involved in the endonuclease reaction. For example, an unpaired uridine at position 69 is universally conserved. This residue does not pair with another nucleotide but forms a bulge in the RNA secondary structure. To test whether the identity or the geometry of U69 is critical for RNase P activity, several mutants were constructed and their endonuclease activity studied. The results for each mutant are given as a rate constant (k) for catalysis and a dissociation constant (K_d) for substrate binding.

RNase P RNA	k (min^{-1})	K_d (nM)
Wild-type U69	0.26	1.7
U69 → A69	0.062	4
U69 → G69	0.0034	73
U69 → C69	0.0056	3
U69 deletion	0.0056	7
U69 + U70	0.0054	181

According to these results, is the U69 bulge more important for substrate binding or catalysis?

(c) What is the effect of increasing the size of the bulge by adding a second U residue (the U69 + U70 mutant)?

[From Kaye, N.M., Zahler, N.H., Christian, E.L., and Harris, M.E., *J. Mol. Biol.* **324,** 429–442 (2002).]

21. Explain why the vast majority of nucleic acids with catalytic activity are RNA rather than DNA.

SELECTED READINGS

Caprara, M.G. and Nilsen, T.W., RNA: versatility in form and function, *Nature Struct. Biol.* **7,** 831–833 (2000). [Summarizes the catalytic functions of ribozymes and the sequence-specific targeting of snoRNAs.]

Cramer, P., Bushnell, D.A., and Kornberg, R.D., Structural basis of transcription: RNA polymerase II at 2.8 Ångstrom resolution, *Science* **292,** 1863–1876 (2001) and Gnatt, A.L., Cramer, P., Fu, J., Bushnell, D.A., and Kornberg, R.D., Structural basis of transcription: An RNA polymerase II elongation complex at 3.3 Å resolution, *Science* **292,** 1876–1882 (2001). [These two papers describe the structure of RNA polymerase II, the binding of RNA and DNA to the enzyme, and the various features of the enzyme that are involved in catalysis, formation of a transcription bubble, and translocation.]

Butcher, S.E., Structure and function of the small ribozymes, *Curr. Opin. Struct. Biol.* **11,** 315–320 (2001). [Summarizes current knowledge of ribozymes, including the hammerhead ribozyme.]

Maniatis, T. and Reed, R., An extensive network of coupling among gene expression machines, *Nature* **416,** 499–506 (2002). [Describes how the events of transcription, RNA processing, and export my be physically linked.]

Proudfoot, N., Furger, A., and Dye, M.J., Integrating mRNA processing with transcription, *Cell* **108,** 501–512 (2002). [Summarizes the links between transcription and capping, splicing, and polyadenylation.]

Storz, G., An expanding universe of noncoding RNAs, *Science* **296,** 1260–1263 (2002). [Briefly describes processes that involve noncoding RNAs, including RNA interference.]

Wilusz, C.J., Wormington, M., and Peltz, S.W., The cap-to-tail guide to mRNA turnover, *Nature Rev. Mol. Cell Biol.* **2,** 237–244 (2001). [Discusses possible mechanisms for regulating the rate of mRNA degradation.]

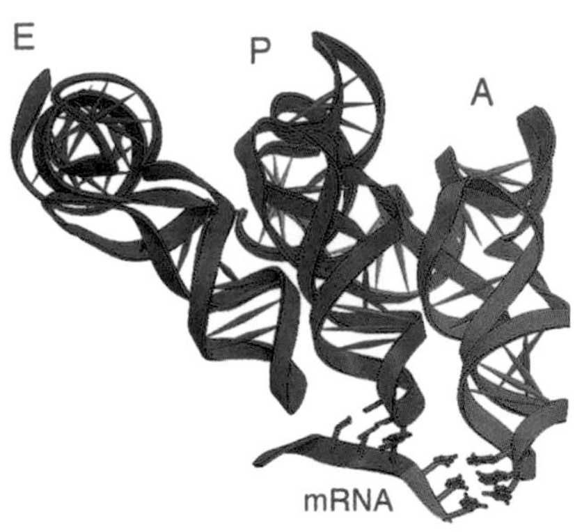

FIGURE 20-13

tRNA–mRNA pairing.
The mRNA bends between the A codon (gold) and the P codon (orange) so that the corresponding anticodons of the aminoacyl–tRNA and peptidyl–tRNA can pair with the mRNA. [*Courtesy Harry Noller. From* Science ***292,*** *883–96 (2001).*]

phate group that forces the mRNA exposed in the ribosomal subunit interface to bend by about 45° between the A and P codons (Fig. 20-13). *This kink helps the tRNAs avoid steric clashes as they read adjacent codons.*

Before the 50S subunit catalyzes peptide bond formation, the ribosome must verify that the correct aminoacyl–tRNA is in place. tRNA binding to the A site induces conformational changes in the 30S subunit that cause two highly conserved residues (A1492 and A1493) of the 16S rRNA to "flip out" of an rRNA loop in order to form hydrogen bonds with two ribose OH groups of the mRNA codon as it pairs with the tRNA anticodon. *These interactions physically link the two rRNA bases with the first two base pairs of the codon–anticodon helix so that they can sense a correctly matched codon and anticodon* (Fig. 20-14). Incorrect base-pairing at the first or second codon position would generate a detectable distortion in the mRNA–tRNA helix. As expected from the wobble hypothesis (see Section 20-1), the A1492/A1493 sensor does not monitor nonstandard base-pairing at the third codon position.

Favorable contacts between the rRNA sensor and the mRNA may alter the conformation of the 30S subunit so that it binds the tRNA more tightly and/or transmits a signal to the active site in the 50S subunit. At the same time, EF-Tu, which is a G-protein, hydrolyzes its bound GTP to GDP + P_i. This chemical reaction alters the conformation of EF-Tu, which allows the protein to leave the ribosome. The departure of EF-Tu–GDP leaves behind the aminoacyl–tRNA, whose acceptor end is now free to occupy the A site of the 50S subunit.

However, if the tRNA anticodon is not properly paired with the A-site codon, the aminoacyl–tRNA dissociates from the ribosome without triggering the 30S conformational changes or GTP hydrolysis by EF-Tu. Because a peptide bond cannot form before GTP is hydrolyzed, *EF-Tu ensures that polymerization does not occur unless the correct aminoacyl–tRNA is positioned in the A site.* The delay between aminoacyl–tRNA binding and GTP hydrolysis by EF-Tu is known as **kinetic proofreading.** This mechanism prevents polymerization of the wrong amino acid and helps limit the error rate of translation to about 10^{-4}. The energetic cost of proofreading at the decoding stage of translation is the free energy of GTP hydrolysis (catalyzed by EF-Tu). The function of EF-Tu is summarized in Figure 20-15.

Peptidyl transferase catalyzes peptide bond formation

The proper docking of an aminoacyl–tRNA in the decoding site of the 30S subunit is communicated to the 50S subunit, most likely by rRNA-mediated conformational changes (recall that the tRNA anticodon and aminoacyl group are separated by about 75 Å). When the A site contains an aminoacyl–tRNA and the P site contains a peptidyl–tRNA (or, prior to formation of the first peptide bond, fMet–$tRNA_f^{Met}$), the peptidyl transferase activity of the 50S subunit catalyzes a **transpeptidation** reaction in which the free amino group of the aminoacyl group of the A-site tRNA attacks the ester bond that links the peptidyl group to the P-site tRNA (Fig. 20-16). This reaction lengthens the peptidyl group by one amino acid at its C-terminal end. Thus, *a polypeptide*

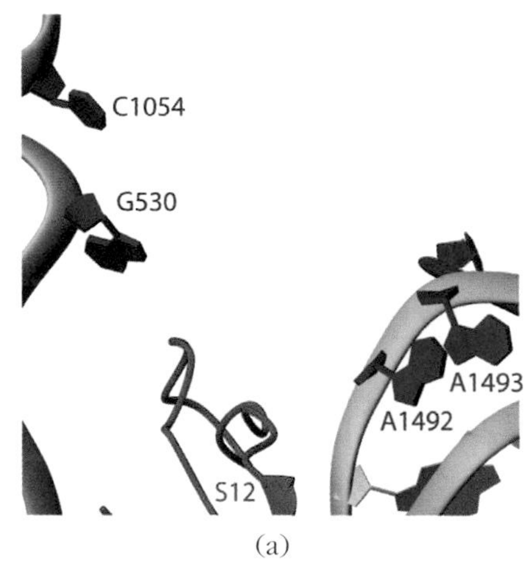

(a)

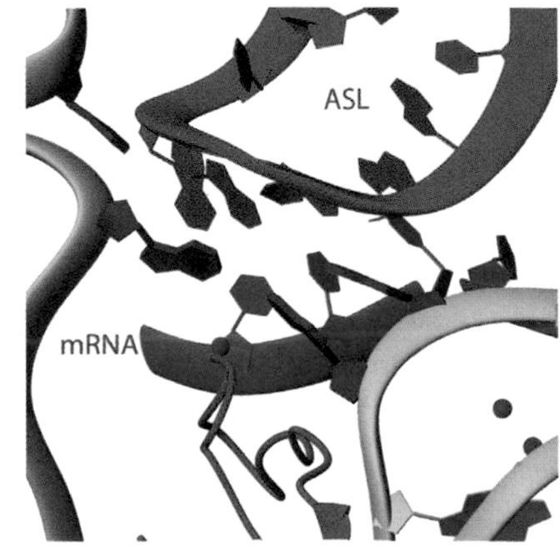

(b)

FIGURE 20-14

The ribosomal sensor for proper codon–anticodon pairing.
These images show the A site of the 30S subunit in the (a) absence and (b) presence of mRNA and tRNA analogs. The rRNA is gray, with the "sensor" bases in red. The mRNA analog, representing the A-site codon, is purple, and the tRNA analog (labeled ASL) is gold. A ribosomal protein (S12) and two Mg^{2+} ions (magenta) are also visible. Note how rRNA bases A1492 and A1493 flip out to sense the codon–anticodon interaction. [*Courtesy Venki Ramakrishnan. From* Science ***292,*** *897–902 (2001).*]

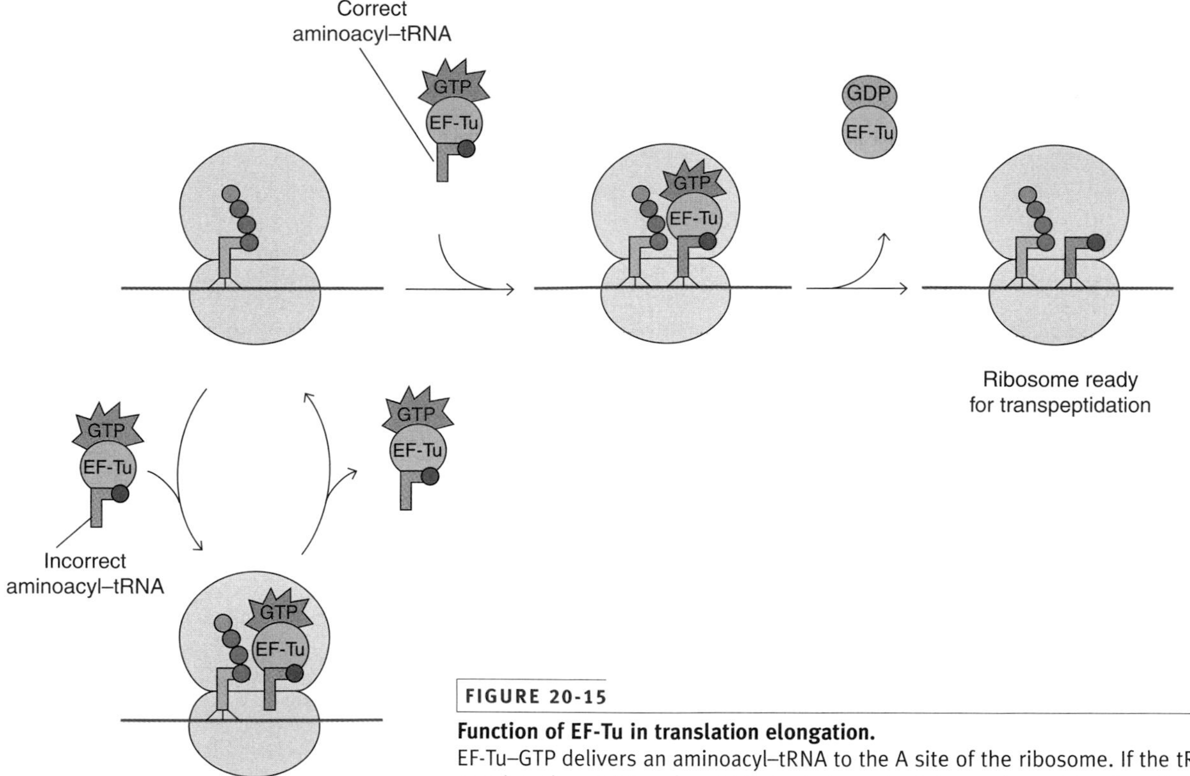

FIGURE 20-15

Function of EF-Tu in translation elongation. EF-Tu–GTP delivers an aminoacyl–tRNA to the A site of the ribosome. If the tRNA anticodon matches the mRNA codon, EF-Tu hydrolyzes its GTP and dissociates from the ribosome, leaving the aminoacyl–tRNA in the A site. If the tRNA anticodon and mRNA codon are mismatched, the aminoacyl–tRNA dissociates before EF-Tu hydrolyzes GTP.

grows in the N → C direction. No external source of free energy is required for transpeptidation because the free energy of the broken ester bond of the peptidyl–tRNA is comparable to the free energy of the newly formed peptide bond.

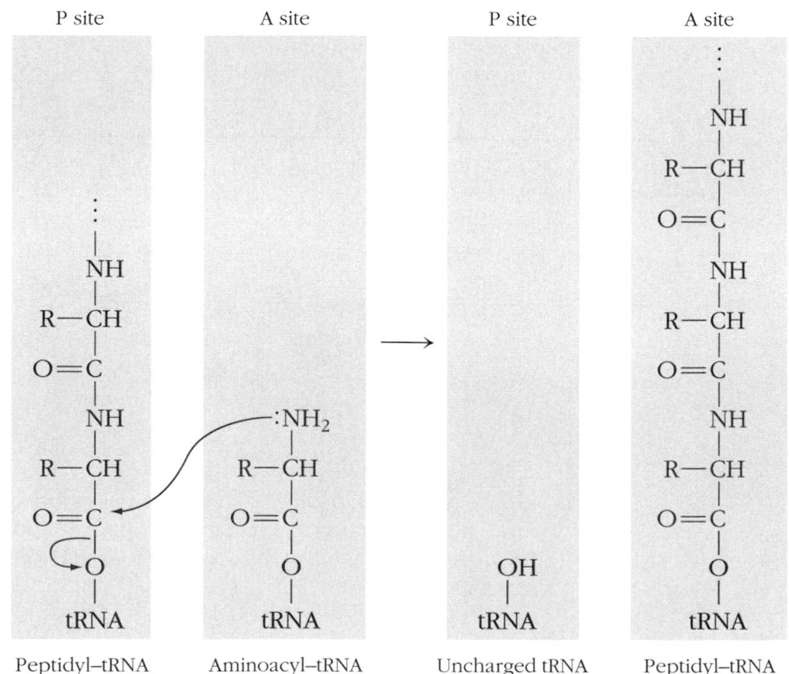

FIGURE 20-16

The peptidyl transferase reaction. Note that the nucleophilic attack of the aminoacyl group on the peptidyl group produces a free tRNA in the P site and a peptidyl–tRNA in the A site.

A CLOSER LOOK

BOX 20-A Antibiotic inhibitors of protein synthesis

Antibiotics interfere with a variety of cellular processes, including cell-wall synthesis, DNA replication, and RNA transcription. Some of the most effective antibiotics, including many in clinical use, target protein synthesis. Because prokaryotic and eukaryotic ribosomes and translation factors differ, these antibiotics can kill bacteria without harming their mammalian hosts.

Translation-blocking antibiotics have also proved useful in the laboratory as probes of ribosomal structure and function. For example, puromycin resembles the 3′ end of Tyr–tRNA and competes with aminoacyl–tRNAs for binding to the ribosomal A site.

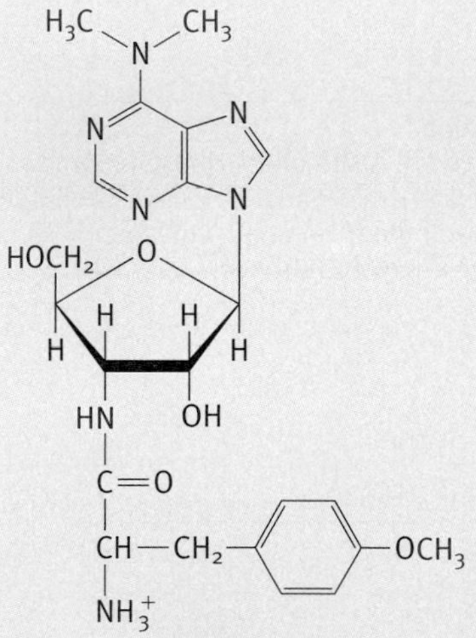

Puromycin

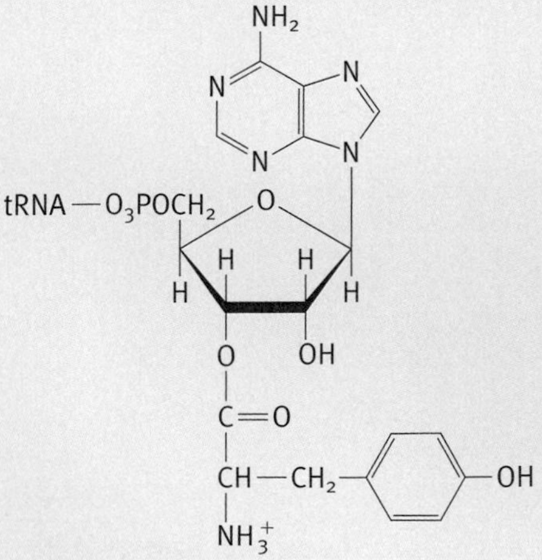

Tyrosyl–tRNA

Transpeptidation generates a puromycin–peptidyl group that cannot be further elongated because the puromycin "amino acid" group is linked by an amide bond rather than an ester bound to its "tRNA" group. As a result, peptide synthesis comes to a halt.

The antibiotic chloramphenicol blocks transpeptidation by binding at the peptidyl transferase active site. X-Ray studies indicate that this relatively small compound

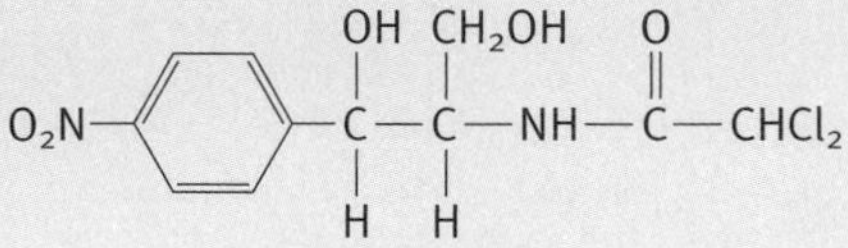

Chloramphenicol

forms complexes with Mg^{2+} ions that interact with the rRNA but not ribosomal proteins. Chloramphenicol may exert its antibacterial effect either by interfering with aminoacyl–tRNA binding in the A site or by preventing formation of the transition state during the transpeptidation reaction.

Erythromycin and related compounds

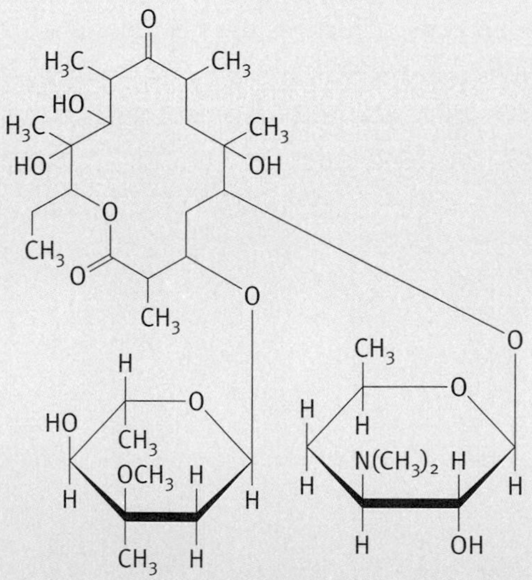

Erythromycin

do not inhibit peptidyl transferase activity but instead physically block the tunnel that conveys the nascent polypeptide away from the active site. Six to eight peptide bonds form before the constriction of the exit tunnel blocks further chain elongation.

(continued)

BOX 20-A *(continued)*

Streptomycin kills cells by an entirely different mechanism.

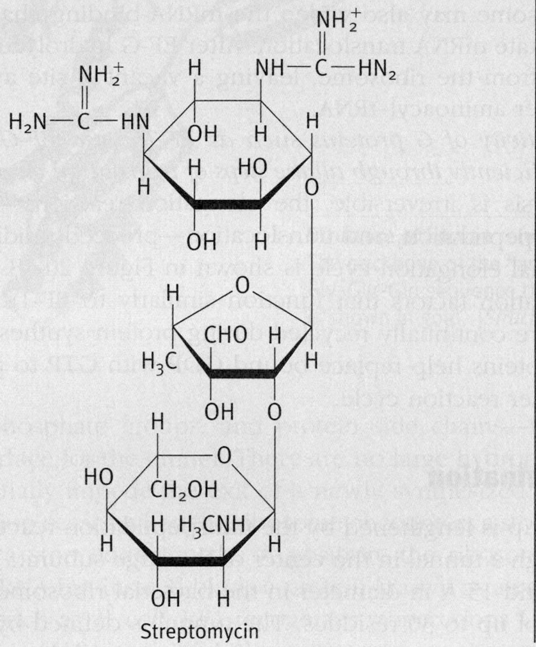

Streptomycin

It binds tightly to the backbone of the 16S rRNA and stabilizes an error-prone conformation of the ribosome. In the presence of the antibiotic, the ribosome's affinity for aminoacyl–tRNAs increases, which increases the likelihood of codon–anticodon mispairing and therefore increases the error rate of translation. Presumably, the resulting burden of inaccurately synthesized proteins kills the cell.

The drugs described here, like all antibiotics, lose their effectiveness when their target organisms become resistant to them. Resistance can be acquired through a variety of mechanisms. For example, mutations in ribosomal components can prevent antibiotic binding. In fact, mapping the locations of such mutations in ribosomal RNA has been instrumental in elucidating ribosomal function. Alternatively, an antibiotic-susceptible organism may acquire a gene, often present on an extrachromosomal plasmid, whose product inactivates the antibiotic or hastens its export from the cell. Acquisition of a gene encoding a particular acetyltransferase, for example, leads to the addition of an acetyl group to chloramphenicol, which prevents its binding to the ribosome.

The peptidyl transferase active site lies in a highly conserved region of the 50S subunit, and the newly formed peptide bond is about 18 Å away from the nearest protein. These observations suggest that the ribosome is a ribozyme (an RNA catalyst). If so, how does rRNA catalyze peptide bond formation? One possibility is that a highly conserved G residue (G2447 in *E. coli*) and an A residue (A2451) that is invariant in all species could participate in an acid–base catalytic mechanism. However, mutagenesis experiments have demonstrated that neither of these two rRNA residues is absolutely required as an acid or base catalyst for protein synthesis by isolated ribosomes *in vitro*. Most likely, the conserved bases promote the transpeptidation reaction by positioning the reactants and stabilizing the reaction's transition state—catalytic activities that still qualify the ribosome as a ribozyme. Some antibiotics exert their effects by binding to the peptidyl transferase active site to directly block protein synthesis (Box 20-A). The peptidyl transferase site is shown in Exercise 27.

During transpeptidation, the peptidyl group is transferred to the tRNA in the A site, and the P-site tRNA becomes deacylated. The new peptidyl–tRNA then moves into the P site, and the deacylated tRNA moves into the E site. The mRNA, which is still base-paired with the peptidyl–tRNA anticodon, advances through the ribosome by one codon. The ribosome can now receive another aminoacyl–tRNA for another round of transpeptidation. *The movement of tRNA and mRNA, which allows the next codon to be translated, is*

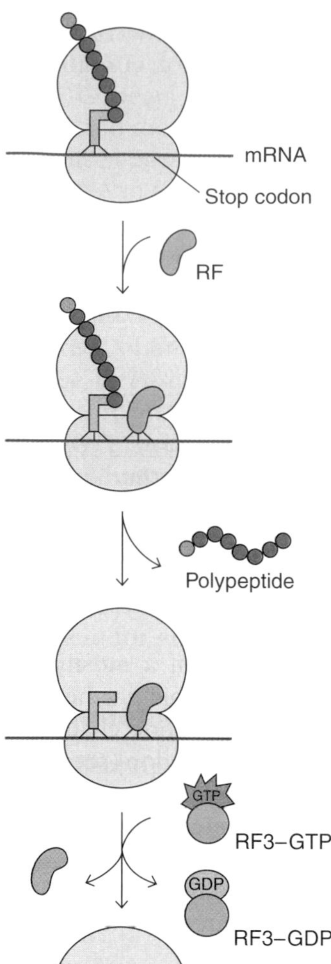

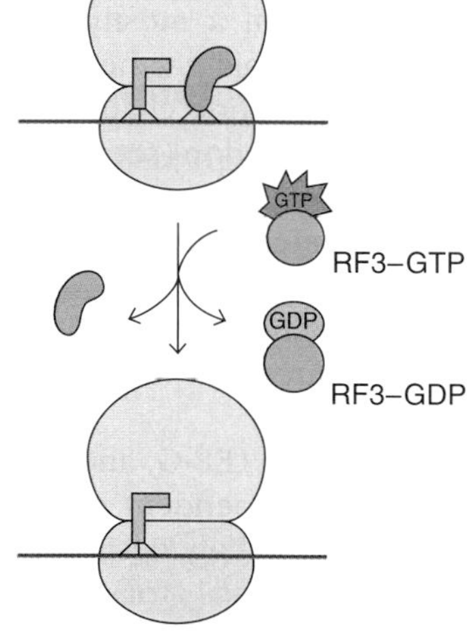

1. A release factor (RF1 or RF2) recognizes a stop codon in the A site.

2. RF1/RF2 causes the ribosome to transfer the peptidyl group to water to release the polypeptide chain.

3. GTP hydrolysis by RF3 allows the release factors to dissociate. Additional steps are required to prepare the ribosome for another round of translation.

FIGURE 20-20

Translation termination in *E. coli*.

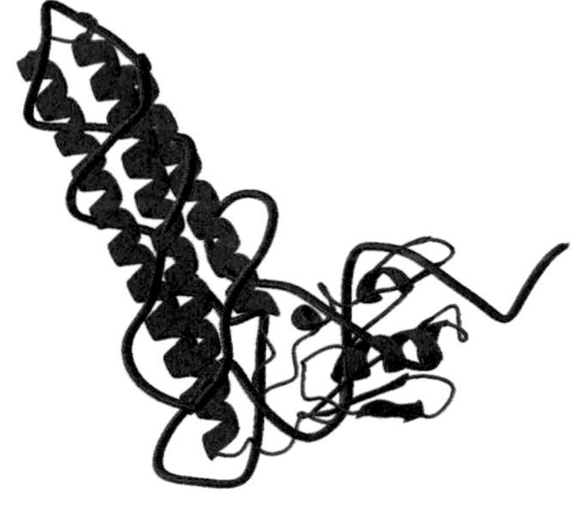

FIGURE 20-21

Structure of RRF from *Thermotoga maritima*.
This two-domain protein (blue) resembles a tRNA molecule. The backbone structure of yeast $tRNA^{Phe}$ (red) is superimposed. The acceptor end of the tRNA is at the lower right. [*Courtesy Anders Liljas. From* Science ***286,*** *2349–2352 (1999)*.]

second ribosome can assemble and begin translating the mRNA. Eukaryotic initiation factors appear to interact with both the 5′ cap and the 3′ poly(A) tail of the mRNA, forming a circular structure that probably enhances the efficacy with which ribosomal subunits are recycled between termination and reinitiation of translation (Fig. 20-23).

REVIEW LEARNING OBJECTIVE 5

- What determines the translation initiation and termination sites in prokaryotes and eukaryotes?
- Summarize the roles of initiation, elongation, and release factors in translation.
- Which of these proteins are G proteins? Which structurally resemble tRNA molecules?

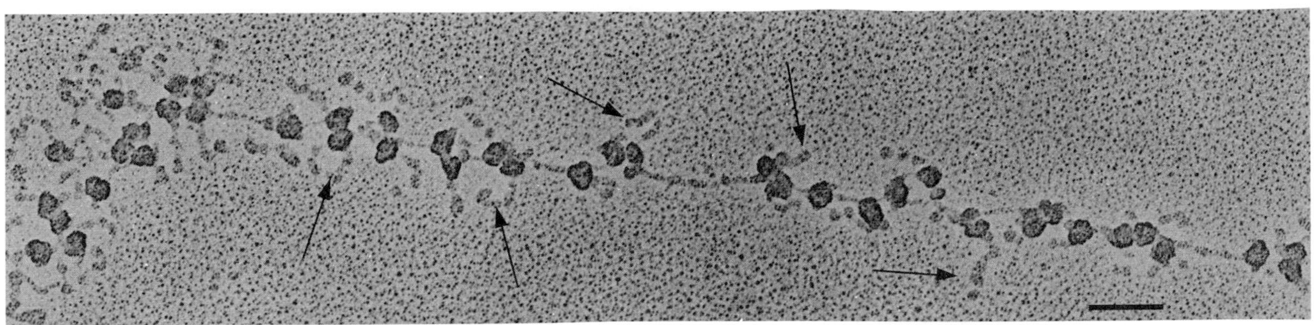

FIGURE 20-22

Electron micrograph of a polysome.
A single mRNA strand encoding silk fibroin (from the silkworm *Bombyx mori*) is studded with ribosomes. Arrows indicate growing fibroin polypeptide chains. The bar represents 0.1 μm. [*Courtesy Oscar L. Miller, Jr., and Steven L. McKnight, University of Virginia.*]

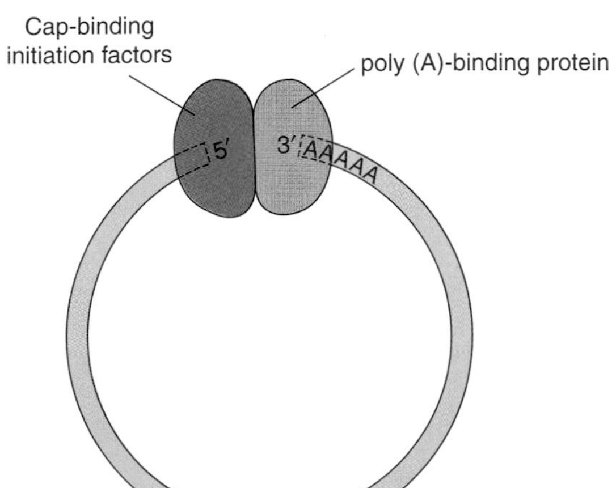

FIGURE 20-23

Circularization of eukaryotic mRNA.
A complex of poly(A)-binding protein and cap-binding initiation factors brings the 3′ and 5′ ends of a eukaryotic mRNA molecule together. The existence of such a structure has been confirmed by electron microscopy. Circularization of the mRNA enhances the efficiency of ribosome recycling.

REVIEW LEARNING OBJECTIVE 6

- How does the ribosome prevent polymerization of the wrong amino acid?
- Explain how polysomes increase the rate of protein synthesis.

4 POSTTRANSLATIONAL EVENTS

The polypeptide released from a ribosome is not yet fully functional. For example, it must fold to its native conformation; it may need to be transported to another location inside or outside the cell; and it may undergo posttranslational modification, or **processing.**

Chaperones assist protein folding

Studies of protein folding *in vitro* have revealed numerous insights into the pathways by which proteins (usually relatively small ones that have been chemically denatured) assume a compact globular shape with a hydrophobic core and a hydrophilic surface (see Section 4-3). Protein folding *in vivo* is only partly understood. For one thing, a protein begins to fold as soon as its N-terminus emerges from the ribosome, even before it has been fully synthesized. In addition, a polypeptide must fold in an environment crowded with other proteins with which it might unfavorably interact. Finally, for proteins with quaternary structure, individual polypeptide chains must assemble with the proper stoichiometry and orientation. All of these processes may be facilitated in a cell by proteins known as **molecular chaperones.**

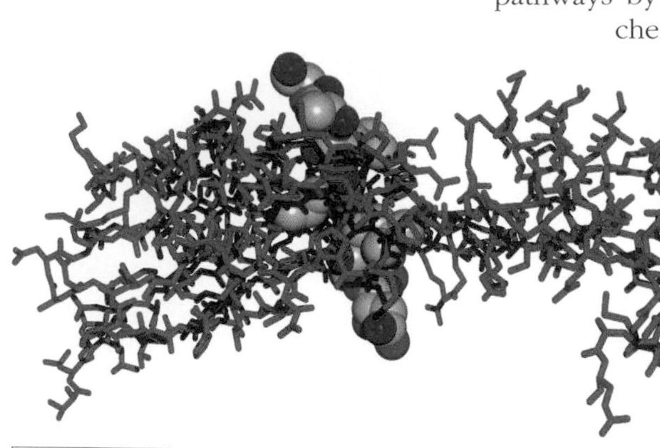

FIGURE 20-24
Structure of a heat-shock protein. *E. coli* DnaK is an example of a monomeric molecular chaperone of the hsp70 class (that is, a heat-shock protein with a mass of about 70 kD). Only the substrate-binding domain of the protein is shown. A polypeptide segment, shown as a space-filling model, binds in a channel through the chaperone. [*Structure (pdb 1DKZ) determined by X. Zhu, X. Zhao, W.F. Burkholder, A. Gragerov, C.M. Ogata, M.E. Gottesman, and W.A. Hendrickson.*]

To prevent improper associations within or between polypeptide chains, chaperones bind to exposed hydrophobic patches on the protein surface (recall that hydrophobic groups tend to aggregate, which could lead to nonnative protein structure or protein aggregation and precipitation). *Many chaperones are ATPases that use the free energy of ATP hydrolysis to drive conformational changes that allow them to bind and release a polypeptide substrate while it assumes its native shape.*

The first chaperones to be identified were known as heat-shock proteins because their synthesis is induced by high temperatures—conditions under which proteins tend to denature (unfold) and aggregate. These monomeric proteins (such as the one shown in Fig. 20-24) bind to newly synthesized polypeptides and to existing cellular proteins and therefore can prevent as well as reverse improper folding. The heat-shock protein, in a complex with ATP, binds to a short extended polypeptide segment with exposed hydrophobic groups (it does not recognize folded proteins, whose hydrophobic groups are sequestered in the interior). ATP hydrolysis causes the chaperone to release the polypeptide. As the polypeptide folds, the heat-shock protein may repeatedly bind and release it. *In vivo,* heat-shock proteins may facilitate the initial folding of a polypeptide, which is completed inside a chaperonin complex.

In addition to monomeric chaperones, all cells contain chaperonins, which form multisubunit cagelike structures that physically sequester a folding polypeptide. The best known chaperonin complex is the GroEL/GroES complex of *E. coli.* Fourteen GroEL subunits form two rings of seven subunits, with each ring enclosing a 45-Å-diameter chamber that is large enough to accommodate a folding polypeptide. Seven GroES subunits form a domelike cap for one GroEL chamber (Fig. 20-25). The GroEL ring nearest the cap is called the cis ring, and the other is the trans ring.

Each GroEL subunit has an ATPase active site. All seven subunits of the ring act in concert, hydrolyzing their bound ATP and undergoing conformational changes. *The two GroEL rings of the chaperonin complex act in a reciprocating fashion to promote the folding of two polypeptide chains*

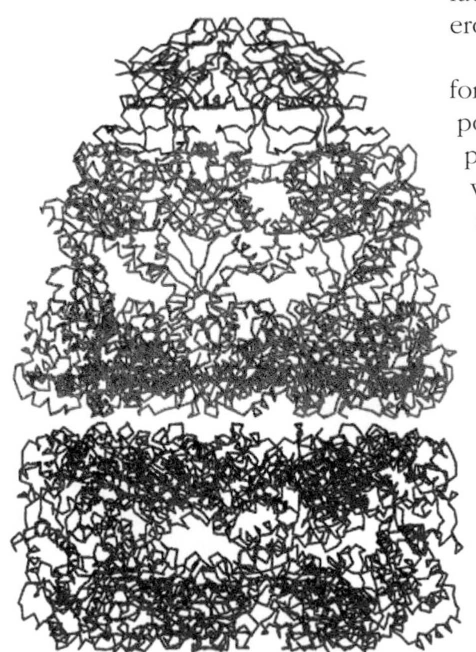

FIGURE 20-25
The GroEL/GroES chaperonin complex. The two seven-subunit GroEL rings, viewed from the side, are colored red and yellow. A seven-subunit GroES complex (blue) caps the so-called cis GroEL ring. [*Structure (pdb 1AON) determined by Z. Xu, A.L. Horwich, and P.B. Sigler.*]

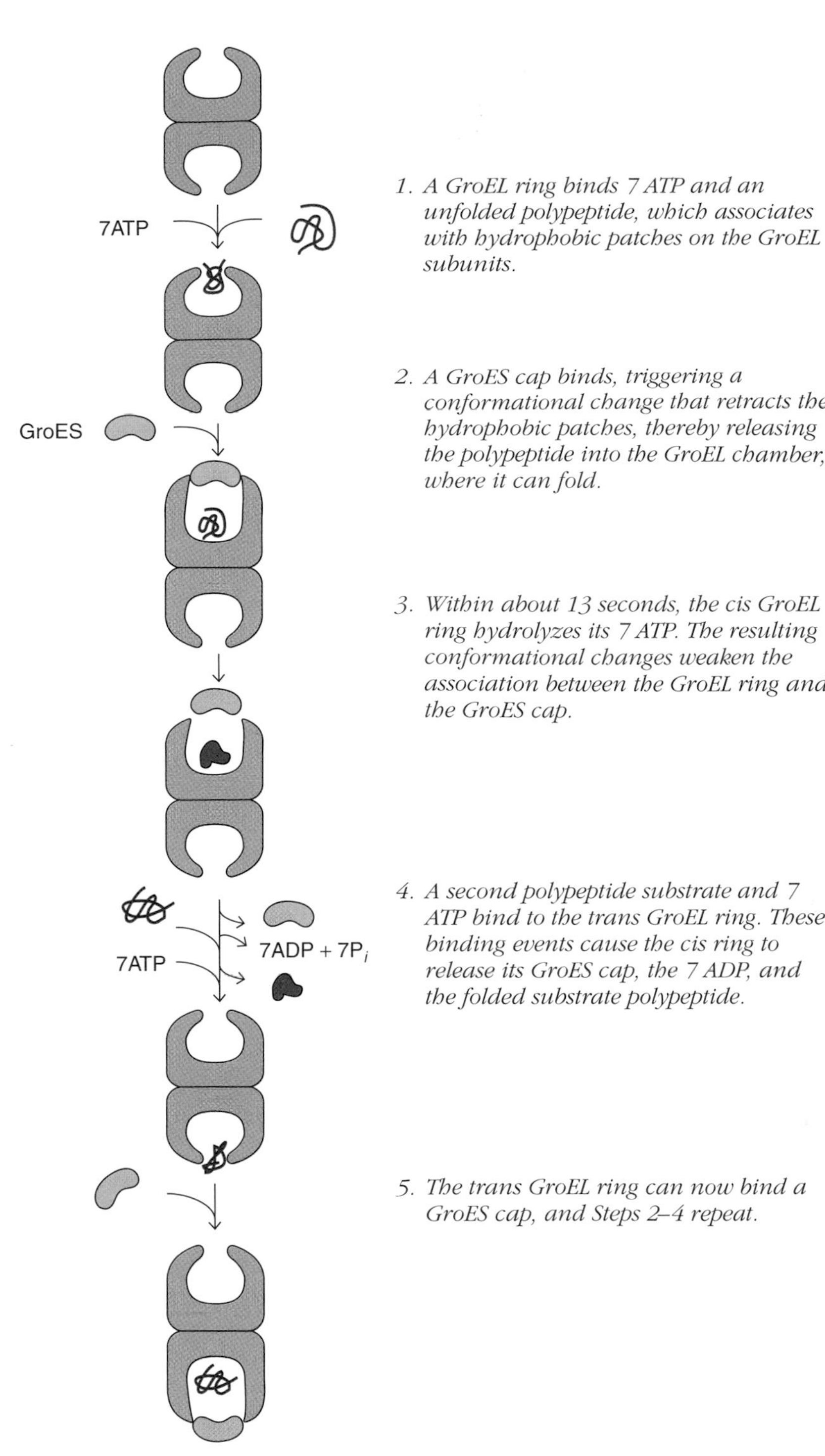

7ATP
GroES
7ATP
7ADP + 7P$_i$
1. A GroEL ring binds 7 ATP and an unfolded polypeptide, which associates with hydrophobic patches on the GroEL subunits.
2. A GroES cap binds, triggering a conformational change that retracts the hydrophobic patches, thereby releasing the polypeptide into the GroEL chamber, where it can fold.
3. Within about 13 seconds, the cis GroEL ring hydrolyzes its 7 ATP. The resulting conformational changes weaken the association between the GroEL ring and the GroES cap.
4. A second polypeptide substrate and 7 ATP bind to the trans GroEL ring. These binding events cause the cis ring to release its GroES cap, the 7 ADP, and the folded substrate polypeptide.
5. The trans GroEL ring can now bind a GroES cap, and Steps 2–4 repeat.

FIGURE 20-26
Chaperonin function.

in a safe environment (Fig. 20-26). Note that 7 ATP are consumed for each 13-second protein-folding opportunity. If the released substrate has not yet achieved its native conformation, it may rebind to the chaperonin complex. Only about 15 to 20% of bacterial proteins seem to require the GroEL/GroES chaperonin complex, and most of these range in size from 10 to 55 kD (a protein larger than about 70 kD probably could not fit inside the protein-folding chamber). Immunocytological studies indicate that some proteins never stray far from a chaperonin complex, perhaps because they tend to unfold and must periodically restore their native structures.

REVIEW LEARNING OBJECTIVE 7

- Compare protein folding as catalyzed by heat-shock proteins and by the chaperonin complex.

The signal recognition particle targets some proteins for membrane translocation

For a cytosolic protein, the journey from a ribosome to the protein's final cellular destination is straightforward (in fact, the journey may be short, since some mRNAs are directed to specific cytosolic locations for translation). In contrast, an integral membrane protein or a protein that is to be secreted from the cell follows a different route since it must pass partly or completely through a membrane.

In prokaryotic cells, membrane and secretory proteins are synthesized by ribosomes in the cytosol and are subsequently ushered to or through the plasma membrane. In eukaryotes, the insertion of most proteins into a membrane occurs cotranslationally, that is, as the polypeptides are being elongated by a ribosome. However, membrane translocation is fundamentally similar in all cells and usually requires a ribonucleoprotein known as the **signal recognition particle (SRP).** *The SRP directs certain proteins to the plasma membrane (in bacteria) or the endoplasmic reticulum (in eukaryotes).*

As in other ribonucleoproteins, including the ribosome and the spliceosome (see Section 19-3), the RNA component of the SRP is highly conserved and is essential for SRP function. The *E. coli* SRP consists of a single multidomain protein and a 4.5S RNA. The mammalian SRP contains a larger RNA and six different proteins, but its core is virtually identical to the bacterial SRP. The RNA component of the SRP includes several nonstandard base pairs, including G:A, G:G, and A:C, and interacts with several protein backbone carbonyl groups (most RNA–protein interactions involve protein side chains rather than the backbone).

How does the SRP recognize membrane and secretory proteins? Such proteins typically have an N-terminal **signal peptide** consisting of an α-helical stretch of 6 to 15 hydrophobic amino acids preceded by at least one positively charged residue. For example, human proinsulin (the polypeptide precursor of the hormone insulin) has the following signal sequence:

MALWMRLLPLLALLALWGPDPAAAFVN····

The hydrophobic segment and a flanking Arg residue are shaded.

The signal peptide binds to the SRP in a pocket formed mainly by a methionine-rich protein domain. The flexible side chains of the hydrophobic Met residues allow the pocket to accommodate helical signal peptides of variable sizes and shapes. In addition to the Met residues, the SRP binding pocket contains a segment of RNA, whose negatively charged backbone interacts electrostatically with the positively charged N-terminus and basic residue of the signal peptide (Fig. 20-27).

Binding of the signal peptide to the SRP probably induces a conformational change in the SRP that allows it to target the polypeptide to a membrane. In eukaryotes, the SRP halts polypeptide elongation until the ribosome is docked at a receptor on the endoplasmic reticulum membrane. When translation resumes, the growing polypeptide is translocated through the membrane. The prokaryotic SRP docks a full-length polypeptide (already free of the ribosome) to the membrane-translocating machinery (called a **translocon**). In all cases, GTP hydrolysis by the SRP is an essential step of this process.

The translocon proteins form a transmembrane channel with a pore of about 20 to 25 Å. Passage of a polypeptide through the translocon requires ATP hydrolysis, which may power a ratchet-like mechanism. When the signal peptide emerges on the far side of the membrane, it may be cleaved off by an integral membrane protein known as a signal peptidase (Fig. 20-28). This enzyme recognizes extended polypeptide segments such as those flanking the hydrophobic segment of a signal peptide, but it does not recognize α-helical structures, which are common in mature membrane proteins. The signal peptides of some integral membrane proteins are not cleaved off, so the proteins remain anchored in the membrane. Proteins with multiple membrane-spanning segments may contain multiple internal signal sequences to direct their insertion into a membrane.

After translocation, *chaperones and other proteins may help the polypeptide fold into its native conformation, form disulfide bonds, and assemble with*

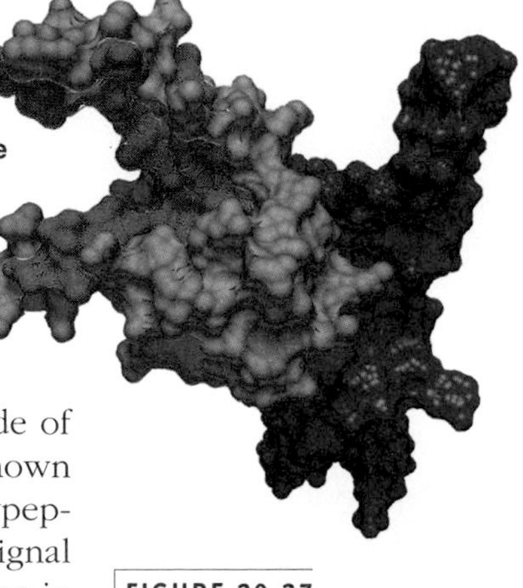
signal sequence
binding groove

FIGURE 20-27

The SRP signal peptide binding domain. This model shows the molecular surface of a portion of the *E. coli* signal recognition particle. The protein is magenta with hydrophobic residues in yellow. Adjacent RNA phosphate groups are red, and the rest of the RNA is dark blue. A signal peptide binds in the SRP groove, making hydrophobic as well as electrostatic contacts with the protein and RNA. [*Courtesy Robert Batey. From* Science ***287***, *1232–39 (2000).*]

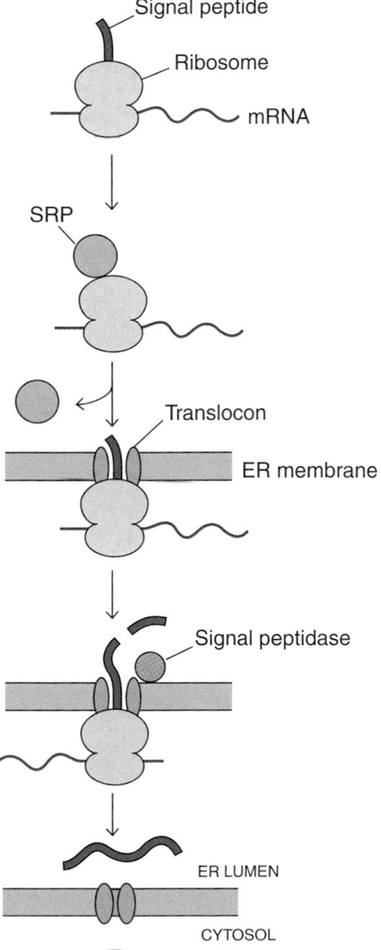
Signal peptide
Ribosome
mRNA
SRP
Translocon
ER membrane
Signal peptidase
ER LUMEN
CYTOSOL
1. The SRP binds to the signal peptide when it emerges from the ribosome. Polypeptide elongation halts.
2. The SRP delivers the ribosome to the endoplasmic reticulum membrane. Translocation resumes, and the polypeptide crosses the membrane via a translocon.
3. Signal peptidase removes the signal peptide from the growing polypeptide.
4. The rest of the polypeptide is translocated into the ER lumen. From here it is transported by vesicles to the extracellular space.

FIGURE 20-28

Membrane translocation of a eukaryotic secretory protein.

other protein subunits. Eukaryotic extracellular proteins are then transported from the endoplasmic reticulum through the Golgi apparatus and to the plasma membrane via vesicles. The proteins may undergo processing (described below) en route to their final destination.

Not all membrane and secretory proteins follow the SRP-mediated pathway described above. For example, some translocated proteins lack signal sequences. In addition, some eukaryotic proteins, particularly mitochondrial proteins, are translocated across membranes posttranslationally. Of the approximately 10,000 different proteins in a eukaryotic cell, about 10% are mitochondrial. The mitochondrion itself synthesizes about a dozen of these; the rest must be imported by translocation complexes in the outer and inner mitochondrial membranes. Similar machinery mediates the import of proteins into chloroplasts (which have three membranes).

REVIEW LEARNING OBJECTIVE 8

- How does the SRP recognize a membrane or secretory protein?
- What are the functions of the translocon and the signal peptidase?

Posttranslational modifications

Proteolysis is part of the maturation pathway of many proteins. For example, after it has entered the endoplasmic reticulum lumen and had its signal peptide removed and its Cys side chains cross-linked as disulfides, the insulin precursor undergoes proteolytic processing. The prohormone is cleaved at two sites to generate the mature hormone (Fig. 20-29).

Many extracellular eukaryotic proteins are **glycosylated** at Asn, Ser, or Thr side chains to generate glycoproteins (see Section 8-2). The short sugar chains (oligosaccharides) attached to secreted glycoproteins may protect the proteins from degradation or denaturation in the relatively harsh extracellular environment and may mediate molecular recognition events, much like cell-surface carbohydrates do (see Box 8-B). A typical *N*-linked oligosaccharide (one that is covalently bonded to an Asn side chain) is shown in Figure 20-30.

In addition to oligosaccharides, other groups such as methyl, phosphoryl, and fatty acyl groups may be covalently attached to a polypeptide at some

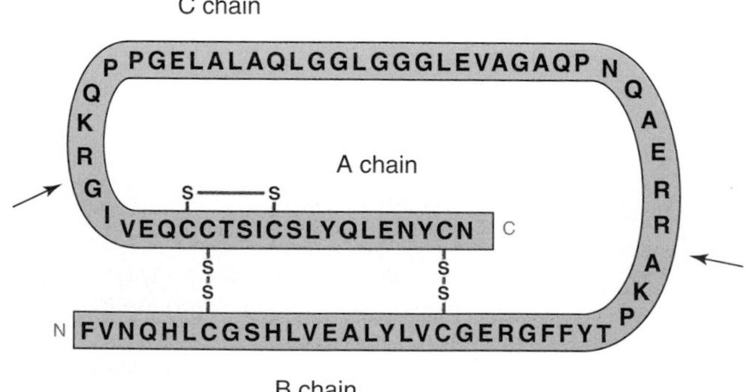

FIGURE 20-29

Conversion of proinsulin to insulin.
The prohormone, with three disulfide bonds, is proteolyzed at two bonds (indicated by arrows) to eliminate the C chain. The mature insulin hormone consists of the disulfide-linked A and B chains.

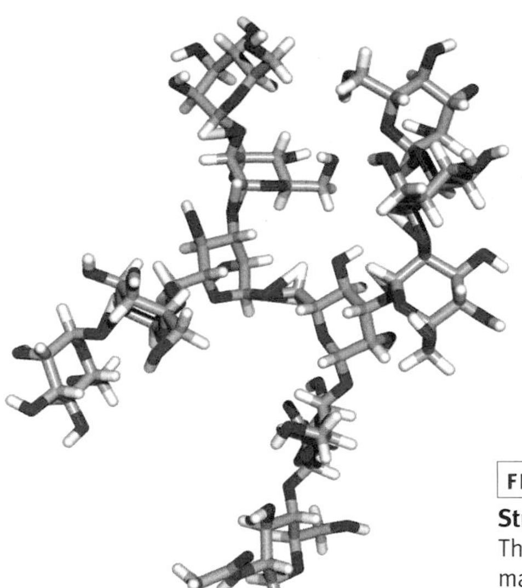

FIGURE 20-30

Structure of an *N*-linked oligosaccharide.
The 11 monosaccharide residues, linked by glycosidic bonds, experience considerable conformational flexibility, so the structure shown here is just one of many possible. [*Structure of the carbohydrate chain of soybean agglutinin determined by A. Darvill and H. Halbeek.*]

point after it has been synthesized. N-terminal groups are frequently modified by acetylation, and C-terminal groups by amidation. More rarely, an amino acid may undergo isomerization to the D configuration. All these posttranslational modifications are catalyzed by specific enzymes, which dramatically increase the potential variation in the ultimate protein products of a given gene.

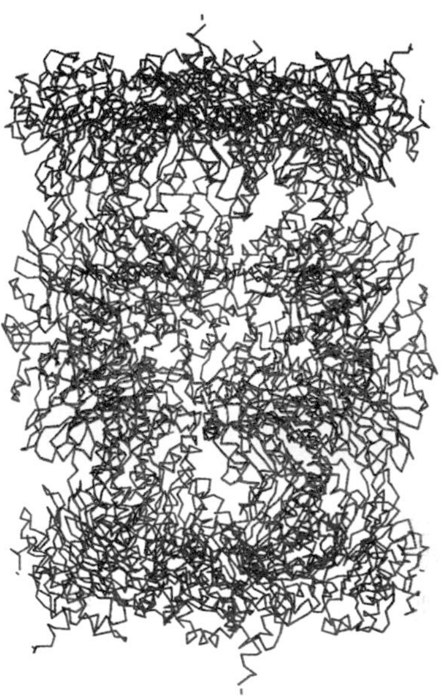

FIGURE 20-31

Structure of the proteasome core.
This barrel-shaped structure is built from four rings with seven subunits in each ring. Additional protein complexes form caps at each end. [*Structure (pdb 1RYP) determined by M. Groll, L. Ditzel, J. Loewe, D. Stock, M. Bochtler, H.D. Bartunik, and R. Huber.*]

Protein turnover

Multistep biological processes must balance speed against accuracy, with high fidelity usually achieved at the expense of speed. Protein synthesis proceeds relatively quickly but with a failure rate of as much as 30% due to inaccurate translation or posttranslational misfolding. *In order to minimize waste, incorrectly synthesized proteins are degraded and their amino acids recycled.* In eukaryotes, the proteins are tagged with the protein ubiquitin and then enter a **proteasome** for degradation (see Section 9-1). Prokaryotes lack ubiquitin but do have proteasomes. The proteasome, a multimeric barrel-shaped structure, superficially resembles the GroEL/GroES chaperonin complex (Fig. 20-31). Intriguingly, both complexes require ATP for their operation: ATP hydrolysis drives chaperonin-mediated protein folding and drives proteasome-mediated protein unfolding that exposes peptide bonds to the proteasome hydrolytic active sites.

The proteasome accounts for about 1% of cellular proteins, which suggests that protein turnover is a normal and essential part of protein metabolism. Even accurately synthesized and correctly folded proteins undergo degradation in the cell. Most enzymes have half-lives in the range of a few hours to a few days. Many regulatory and signaling proteins have even shorter life spans. The continual turnover of cellular proteins is one likely reason that some sloppiness is tolerated during protein synthesis. Unlike a DNA molecule, which must be replicated with extreme accuracy, a protein need not be perfect because it can be readily replaced.

REVIEW LEARNING OBJECTIVE 9

- Describe the types of reactions that a polypeptide may undergo following its synthesis.

GLOSSARY

3′ end. The terminus of a polynucleotide whose C3′ is not esterified to another nucleotide residue.

5′ end. The terminus of a polynucleotide whose C5′ is not esterified to another nucleotide residue.

α anomer. The configuration of a sugar such as glucose, in which the oxygen attached to C1 points down when the molecule is shown as a standard Haworth projection.

α helix. A regular secondary structure of polypeptides, with 3.6 residues per right-handed turn and hydrogen bonds between each backbone C═O group and the backbone N—H group that is four residues further.

α-amino acid. See amino acid.

A site. The ribosomal binding site that accommodates an aminoacyl–tRNA.

Abasic site. The deoxyribose residue remaining after the removal of a base from a DNA strand.

Acid. A substance that can donate a proton.

Acid catalysis. A catalytic mechanism in which partial proton transfer from an acid lowers the free energy of a reaction's transition state.

Acid–base catalysis. A catalytic mechanism in which partial proton transfer from an acid or partial proton abstraction by a base lowers the free energy of a reaction's transition state.

Acid dissociation constant (K_a). The dissociation constant for an acid in water.

Acidic solution. A solution whose pH is less than 7.0 ($[H^+] > 10^{-7}$ M).

Acidosis. A pathological condition in which the pH of the blood drops below its normal value of 7.4.

Action potential. The momentary reversal of membrane potential that occurs during transmission of a nerve impulse.

Activation energy (free energy of activation; $\Delta G^{\ddagger}$). The free energy of the transition state minus the free energies of the reactants in a chemical reaction.

Activator. A protein that binds at or near a gene so as to promote its transcription.

Active site. The region of an enzyme in which catalysis takes place.

Active transport. The transmembrane movement of a substance from low to high concentrations by a protein that couples this endergonic transport to an exergonic process such as ATP hydrolysis.

Acyl group. A portion of a molecule with the formula —COR, where R is an alkyl group.

A-DNA. A conformation of DNA in which the double helix is wider than the standard B-DNA helix and in which base pairs are inclined to the helix axis.

Aerobic. Occurring in or requiring oxygen.

Affinity label. A labeled compound that resembles an enzyme's substrate but reacts irreversibly with and thereby labels a group in the enzyme's active site.

Aldose. A sugar whose carbonyl group is an aldehyde.

Alkalosis. A pathological condition in which the pH of the blood rises above its normal value of 7.4.

Allosteric activation. Binding of an activator to one subunit of a multisubunit enzyme, which increases the catalytic activity of all the subunits.

Allosteric inhibition. Binding of an inhibitor to one subunit of a multisubunit enzyme, which reduces the catalytic activity of all the subunits.

Allosteric protein. A protein in which the binding of ligand at one site affects the binding of other ligands at other sites. See also cooperative binding.

Amido group. A portion of a molecule with the formula —CONH—.

Amino acid (α-amino acid). A compound consisting of a carbon atom to which are attached a primary amino group, a carboxylate group, a side chain (R group), and an H atom.

Amino group. A portion of a molecule with the formula $—NH_2$, —NHR, or $—NR_2$, where R is an alkyl group.

Amino terminus. See N-terminus.

Amphiphilic (amphipathic). Having both polar and nonpolar regions and therefore being both hydrophilic and hydrophobic.

Amyloid deposit. An accumulation of certain types of insoluble protein aggregates in tissues (e.g., in the brain in Alzheimer's disease).

Anabolism. The reactions by which biomolecules are synthesized from simpler components.

Anaerobic. Occurring independently of oxygen.

Anaplerotic reaction. A reaction that replenishes the intermediates of a metabolic pathway.

Angiogenesis. The formation of new blood vessels.

Anneal. To allow base pairing between complementary single polynucleotide strands so that double-stranded segments form.

Anomers. Sugars that differ only in the configuration around the carbonyl carbon that becomes a chiral center when the sugar cyclizes.

Antenna pigment. A molecule that transfers its absorbed energy to other pigment molecules and eventually to a photosynthetic reaction center.

Anticodon. The sequence of three nucleotides in a tRNA that recognizes an mRNA codon through complementary base pairing.

Antiparallel. Running in opposite directions.

Antiparallel β sheet. See β sheet.

Antiport. A transporter that allows the simultaneous transmembrane movement of two molecules in opposite directions.

Antisense strand. See noncoding strand.

Apoptosis. Programmed cell death that results from extracellular or intracellular signals and involves the activation of caspases that selectively degrade cellular structures.

Archaea. One of the two major groups of prokaryotes.

Atherosclerosis. A disease characterized by the formation of cholesterol-containing fibrous plaques in the walls of blood vessels.

ATPase. An enzyme that catalyzes the hydrolysis of ATP to ADP + P_i.

Autoactivation. A process by which the product of an activation reaction also acts as a catalyst for the same reaction, so that it appears that the compound catalyzes its own activation.

Autophosphorylation. The kinase-catalyzed phosphorylation of itself or an identical molecule.

β anomer. The configuration of a sugar such as glucose, in which the oxygen attached to C1 points up when the molecule is shown as a standard Haworth projection.

β barrel. A protein structure consisting of a β sheet rolled into a cylinder.

β oxidation. A series of enzyme-catalyzed reactions in which fatty acids are progressively degraded by the removal of two-carbon units as acetyl-CoA.

β sheet. A regular secondary structure in which extended polypeptide chains form interstrand hydrogen bonds. In parallel β sheets, the polypeptide chains all run in the same direction; in antiparallel β sheets, neighboring chains run in opposite directions.

Backbone. The atoms that form the repeating linkages between successive residues of a polymeric molecule, exclusive of the side chains.

Bacteria. One of the two major groups of prokaryotes.

Bacteriophage. A virus specific for bacteria. Also known as a phage.

Base. (1) A substance that can accept a proton. (2) A purine or pyrimidine component of a nucleoside, nucleotide, or nucleic acid.

Base catalysis. A catalytic mechanism in which partial proton abstraction by a base lowers the free energy of a reaction's transition state.

Base excision repair. A DNA repair pathway in which a damaged base is removed by a glycosylase so that the resulting abasic site can be repaired.

Base pair. The specific hydrogen-bonded association between nucleic acid bases. The standard base pairs are A:T and G:C. See also bp.

Basic solution. A solution whose pH is greater than 7.0 ($[H^+] < 10^{-7}$ M).

B-DNA. The standard conformation of double-helical DNA.

Bilayer. An ordered, two-layered arrangement of amphiphilic molecules in which polar segments are oriented toward the two solvent-exposed surfaces and the nonpolar segments associate in the center.

Bile acid. A cholesterol derivative that acts as a detergent to solubilize lipids for digestion and absorption.

Bimolecular reaction. A reaction involving two molecules, which may be identical or different.

Binding change mechanism. The mechanism whereby the subunits of ATP synthase adopt three successive conformations to convert ADP + P_i to ATP as driven by the dissipation of the transmembrane proton gradient.

Bisubstrate reaction. An enzyme-catalyzed reaction involving two substrates.

Blue-white screening. A technique for distinguishing cells containing a β-galactosidase gene that has been interrupted by a foreign DNA segment such that it cannot give rise to an active enzyme that converts a substrate to a blue-colored product.

Blunt ends. The fully base-paired ends of a DNA fragment that are generated by a restriction endonuclease that cuts both strands at the same point.

Bohr effect. The decrease in O_2 binding affinity of hemoglobin in response to a decrease in pH.

bp. Base pair, the unit of length used for DNA molecules.

Buffer. A solution of a weak acid and its conjugate base, which resists changes in pH upon the addition of acid or base.

C_4 pathway. A photosynthetic process used in some plants to concentrate CO_2 by incorporating it into oxaloacetate (a C_4 compound).

Cα. The alpha carbon, the carbon of an amino acid whose substituents are an amino group, a carboxylate group, an H atom, and a variable R group.

Calvin cycle. The sequence of photosynthetic reactions in which ribulose-5-phosphate is carboxylated, converted to three-carbon carbohydrate precursors, and regenerated.

Cap. A 7-methylguanosine residue that is posttranslationally added to the 5′ end of a eukaryotic mRNA.

Capping protein. A protein that binds to the exposed end of a polymeric molecule, thereby preventing further polymerization or depolymerization.

Carbanion. A compound that bears a negative charge on a carbon atom.

Carbohydrate. A compound with the formula $(CH_2O)_n$ where $n \geq 3$. Also called a saccharide.

Carbon fixation. The incorporation of CO_2 into biologically useful organic molecules.

Carbonyl group. A portion of a molecule with the formula >C=O.

Carboxyl group. A portion of a molecule with the formula —COOH.

Carboxyl terminus. See C-terminus.

Carcinogen. An agent that causes a mutation in DNA that leads to cancer.

Carcinogenesis. The process of developing cancer.

Catabolism. The degradative metabolic reactions in which nutrients and cell constituents are broken down for energy and raw materials.

Catalyst. A substance that promotes a chemical reaction without undergoing permanent change. A catalyst increases the rate at which a reaction approaches equilibrium but does not affect the free energy change of the reaction.

Catalytic constant (k_{cat}). The ratio of the maximal velocity (V_{max}) of an enzyme-catalyzed reaction to the enzyme concentration. Also called a turnover number.

Catalytic perfection. A state achieved by an enzyme that operates at the diffusion-controlled limit.

Catalytic triad. The hydrogen bonded Ser, His, and Asp residues that participate in catalysis in serine proteases.

cDNA. See complementary DNA.

Cell cycle. The sequence of events between eukaryotic cell divisions, which includes mitosis and cell division (M phase), a gap (G_1) phase, DNA synthesis (S phase), and a second gap (G_2) phase.

CF. See cystic fibrosis.

Chain terminator. A nucleotide lacking a 3′ OH group that is incorporated into a polynucleotide but cannot support further polymerization.

Channeling. The transfer of an intermediate product from one enzyme active site to another in such a way that the intermediate remains in contact with the protein.

Chaperone. See molecular chaperone.

Chemical labeling. A technique for identifying functional groups in a macromolecule by treating the molecule with a reagent that reacts with those groups.

Chemiosmotic theory. The postulate that the free energy of electron transport is conserved in the formation of a transmembrane proton gradient that can be subsequently used to drive ATP synthesis.

Chemoautotroph. An organism that obtains its building supplies and free energy from inorganic compounds.

Chirality. The asymmetry or "handedness" of a molecule such that it cannot be superimposed on its mirror image.

Chloroplast. The plant organelle in which photosynthesis takes place.

Chromatin. The complex of DNA and protein that comprises the eukaryotic chromosomes.

Chromosome. The complex of protein and a single DNA molecule that comprises some or all of an organism's genome.

Chylomicrons. Lipoprotein particles that transport dietary lipids from the intestines to the tissues.

Circadian cycle. A daily oscillation in a biochemical process that is coordinated with the daily light–dark cycle.

Citric acid cycle. A set of eight enzymatic reactions, arranged in a cycle, in which energy in the form of ATP, NADH, and QH_2 is recovered from the oxidation of the acetyl group of acetyl-CoA to CO_2.

Clone. An organism or collection of identical cells derived from a single parental cell.

Cloning vector. A DNA molecule, such as a plasmid, that can accommodate a segment of foreign DNA for cloning.

Coagulation. The process of forming a blood clot.

Coding strand. The DNA strand that has the same sequence (except for the replacement of U with T) as the transcribed RNA; it is the nontemplate strand. Also called the sense strand.

Codon. The sequence of three nucleotides in DNA or RNA that specifies a single amino acid.

Coenzyme. A small organic molecule that is required for the catalytic activity of an enzyme. A coenzyme may be tightly associated with the enzyme as a prosthetic group.

Cofactor. A small organic molecule (coenzyme) or metal ion that is required for the catalytic activity of an enzyme.

Coiled coil. An arrangement of polypeptide chains in which two α helices wind around each other.

Collagen. One of a family of extracellular, fibrous proteins found primarily in bone and connective tissue.

Competitive inhibition. A form of enzyme inhibition in which a substance competes with the substrate for binding to the enzyme active site and thereby appears to increase K_M.

Complement. A molecule that pairs in a reciprocal fashion with another.

Complementary DNA (cDNA). A segment of DNA synthesized from an RNA template.

Condensation reaction. The formation of a covalent bond between two molecules, during which the elements of water are lost.

Conformation. The three-dimensional shape of a molecule that it attained through rotation of its bonds.

Conjugate acid. The compound that forms when a base accepts a proton.

Conjugate base. The compound that forms when an acid donates a proton.

Consensus sequence. A DNA or RNA sequence showing the nucleotides most commonly found at each position.

Conservative substitution. A change of an amino acid residue in a protein to one with similar properties (e.g., Leu to Ile or Asp to Glu).

Constitutive. Being expressed at a continuous, steady rate rather than induced.

Contig. A linear stretch of genomic DNA that has been fully sequenced.

Convergent evolution. The independent development of similar characteristics in unrelated species.

Cooperative binding. A situation in which the binding of a ligand at one site on a macromolecule affects the affinity of other sites for the same ligand. See also allosteric protein.

Cori cycle. A metabolic pathway in which lactate produced by glycolysis in the muscles is transported via the bloodstream to the liver, where it is used for gluconeogenesis. The resulting glucose returns to the muscles.

Covalent catalysis. A catalytic mechanism in which the transient formation of a covalent bond between the catalyst and a reactant lowers the free energy of a reaction's transition state.

CpG island. A cluster of CG sequences that often marks the beginning of a gene in a mammalian genome.

Cristae. The invaginations of the inner mitochondrial membrane.

Cross-talk. The interactions of different signal-transduction pathways through activation of the same signaling components.

Cryoelectron microscopy. A technique in visualizing by electron microscopy the shapes of large particles that are embedded in ice so that they retain their native shape to a greater extent than in convention electron microscopy.

C-terminus. The end of a polypeptide that has a free carboxylate group.

Cyclic electron flow. The light-driven circulation of electrons between Photosystem I and cytochrome b_6f, which leads to the production of ATP but not NADPH.

Cystic fibrosis (CF). A genetic disease that is caused by a mutation in a gene for a membrane transport protein and is characterized by thick mucus and bacterial lung infections.

Cytochrome. A protein that carries electrons via a prosthetic Fe-containing heme group.

Cytokinesis. The splitting of the cell into two following mitosis.

Cytoplasm. The entire contents of a cell excluding the nucleus.

Cytoskeleton. The network of intracellular fibers that gives a cell its shape and structural rigidity.

Cytosol. The contents of a cell (cytoplasm) minus its nucleus and other membrane-bounded organelles.

Dark reactions. The photosynthetic reactions in which NADPH and ATP produced by the light reactions are used to incorporate CO_2 into carbohydrates.

ddNTP. A dideoxynucleoside triphosphate.

Deamination. The hydrolytic removal of an amino group.

Degenerate code. A code in which more than one "word" encodes the same entity.

Denaturation. The loss of ordered structure in a polymer, such as the disruption of native conformation in an unfolded polypeptide or the unstacking of bases and separation of strands in a nucleic acid.

Denitrification. The conversion of nitrate (NO_3^-) to nitrogen (N_2).

Deoxyhemoglobin. Hemoglobin that does not contain bound oxygen or is not in the oxygen-binding conformation.

Deoxynucleotide. A nucleotide in which the pentose is 2′-deoxyribose.

Deoxyribonucleic acid. See DNA.

Diabetes mellitus. A disease caused by a deficiency of insulin or the inability to respond to insulin, and characterized by elevated levels of glucose in the blood.

Dideoxy DNA sequencing. A technique for determining the nucleotide sequence of a DNA using dideoxy nucleotides so as to yield a collection of strands of all possible lengths.

Diffraction pattern. The record of the radiation scattered from an object, for example, in X-ray crystallography.

Diffusion-controlled limit. The theoretical maximum rate of an enzymatic reaction in solution, about 10^8 to 10^9 $M^{-1} \cdot s^{-1}$.

Dimer. An assembly consisting of two monomeric units.

Dipole–dipole interaction. A type of van der Waals interaction between two strongly polar groups.

Discontinuous synthesis. A mechanism whereby the lagging strand of DNA is synthesized as a series of fragments that are later joined.

Dissociation constant (*K*). The ratio of the products of the concentrations of the dissociated species to those of their parent compounds at equilibrium.

Disulfide bond. A covalent —S—S— linkage, often between two Cys residues in a protein.

DNA. A polymer of deoxynucleotides whose sequence of bases encodes genetic information in all living cells.

DNA chip. See microarray.

DNA ligase. An enzyme that catalyzes the formation of a phosphodiester bond to join two DNA strands.

DNA marker. An element of DNA structure, such as a gene or other known sequence, whose position on a chromosome is known.

DNA polymerase. An enzyme that catalyzes synthesis of a new DNA strand using an existing strand as a template for the assembly of nucleotides.

dNTP. A deoxyribonucleoside triphosphate.

Domain. A stretch of polypeptide residues that fold into a globular unit with a hydrophobic core.

$\mathcal{E}$. See reduction potential.

$\mathcal{E}^{\circ\prime}$. See standard reduction potential.

E site. The ribosomal binding site that accommodates a deacylated–tRNA before it dissociates from the ribosome.

Edman degradation. A procedure for the stepwise removal and identification of the N-terminal residues of a polypeptide.

EF. See elongation factor.

Ehlers–Danlos syndrome. A genetic disease characterized by elastic skin and joint hyperextensibility, caused by mutations in genes for collagen or collagen-processing proteins.

EI complex. The noncovalent complex that forms between an enzyme and a reversible inhibitor.

Eicosanoids. Compounds derived from the C_{20} fatty acid arachidonic acid, which act in or near the cells that produce them and mediate pain, fever, and other physiological responses.

Electron crystallography. A technique for determining molecular structure, in which the electron beam of an electron microscope is used to elicit diffraction from a two-dimensional crystal of the molecules of interest.

Electron tomography. A technique for reconstructing three-dimensional structures by analyzing electron micrographs of consecutive tissue slices.

Electron transport chain. A series of membrane-associated electron carriers that pass electrons from reduced coenzymes to molecular oxygen so as to recover free energy for the synthesis of ATP.

Electronegativity. A measure of an atom's affinity for electrons.

Electrophile. A compound containing an electron-poor center. An electrophile (electron-lover) reacts readily with a nucleophile (nucleus-lover).

Electrophoresis. A procedure in which macromolecules are separated on the basis of charge or size by their differential migration through a gel-like matrix under the influence of an applied electric field. In polyacrylamide gel electrophoresis (PAGE), the matrix is cross-linked polyacrylamide.

Electrostatic catalysis. A catalytic mechanism in which sequestering the reacting groups away from the aqueous solvent lowers the free energy of a reaction's transition state.

Elongation factor (EF). A protein that interacts with tRNA and/or the ribosome during polypeptide synthesis.

Emphysema. A chronic disease characterized by difficulty breathing due to alveolar degeneration and loss of lung elasticity.

End joining. The ligation process that repairs a double-stranded break in DNA.

Endergonic reaction. A reaction that has an overall positive free energy change (a nonspontaneous process).

Endocytosis. The inward folding and budding of the plasma membrane to form a new intracellular vesicle.

Endonuclease. An enzyme that catalyzes the hydrolysis of the phosphodiester bonds between two nucleotide residues within a polynucleotide strand.

Endopeptidase. An enzyme that catalyzes the hydrolysis of a peptide bond within a polypeptide chain.

Enhancer. A eukaryotic DNA sequence located some distance from the transcription start site, where an activator of transcription may bind.

Enthalpy (*H*). A thermodynamic quantity that is taken to be equivalent to the heat content of a biochemical system.

Entropy (*S*). A measure of the degree of randomness or disorder of a system.

Enzyme. A biological catalyst. Most enzymes are proteins; a few are RNA.

Epimers. Sugars that differ only by the configuration at one C atom (excluding the anomeric carbon).

Equilibrium. The point in a process at which the forward and reverse rates are exactly balanced so that it undergoes no net change.

Equilibrium constant (K_{eq}). The ratio, at equilibrium, of the product of the concentrations of reaction products to that of the reactants.

ES complex. The noncovalent complex that forms between an enzyme and its substrate in the first step of an enzyme-catalyzed reaction.

Essential compound. An amino acid, fatty acid, or other compound that an animal cannot synthesize and must therefore obtain in its diet.

Essential amino acid. An amino acid that an animal cannot synthesize and must therefore obtain in its diet.

Essential fatty acid. A fatty acid that an animal cannot synthesize and must therefore obtain in its diet.

EST. See expressed sequence tag.

Ester group. A portion of a molecule with the formula —COOR, where R is an alkyl group.

Ether. A molecule with the formula ROR′, where R and R′ are alkyl groups.

Euchromatin. The transcriptionally active, relatively uncondensed chromatin in a eukaryotic cell.

Eukarya. See eukaryote.

Eukaryote. An organism consisting of a cell (or cells) whose genetic material is contained in a membrane-bounded nucleus.

Exciton transfer. A mode of decay of an energetically excited molecule, in which electronic energy is transferred to a nearby unexcited molecule.

Exergonic reaction. A reaction that has an overall negative free energy change (a spontaneous process).

Exocytosis. The fusion of an intracellular vesicle with the plasma membrane in order to release the contents of the vesicle outside the cell.

Exon. A portion of a gene that appears in both the primary and mature mRNA transcripts.

Exonuclease. An enzyme that catalyzes the hydrolytic excision of a nucleotide residue from the end of a polynucleotide strand.

Exopeptidase. An enzyme that catalyzes the hydrolytic excision of an amino acid residue from one end of a polypeptide chain.

Expressed sequence tag (EST). A segment of DNA that is synthesized from an mRNA template and which therefore represents a portion of the genome that is expressed.

Extrinsic protein. See peripheral membrane protein.

$\mathcal{F}$. See Faraday.

F-actin. The polymerized form of the protein actin. See also G actin.

Factory model of replication. A model for DNA replication in which DNA polymerase and associated proteins remain stationary while the DNA template is spooled through them.

Faraday ($\mathcal{F}$). The charge of one mole of electrons, equal to 96,485 coulombs $\cdot$ mol^{-1} or 96,485 J $\cdot$ V^{-1} $\cdot$ mol^{-1}.

Fatty acid. A carboxylic acid with a long-chain hydrocarbon side group.

Feedback inhibitor. A substance that inhibits the activity of an enzyme that catalyzes an early step of the substance's synthesis.

Feed-forward activation. The activation of a later step in a reaction sequence by the product of an earlier step.

Fermentation. An anaerobic catabolic process.

Fe–S cluster. See iron–sulfur cluster.

Fibrous protein. A protein characterized by a stiff, elongated conformation, that tends to form fibers.

First-order reaction. A reaction whose rate is proportional to the concentration of a single reactant.

Fischer projection. A graphical convention for specifying molecular configuration in which horizontal lines represent bonds that extend above the plane of the paper and vertical bonds extend below the plane of the paper.

Flip-flop. See transverse diffusion.

Flippase. See translocase.

Fluid mosaic model. A model of biological membranes in which integral membrane proteins float and diffuse laterally in a fluid lipid layer.

Fluorescence. A mode of decay of an excited molecule, in which electronic energy is emitted in the form of a photon.

Fluorophore. A fluorescent group.

Flux. The rate of flow of metabolites through a metabolic pathway.

Fractional saturation (*Y*). The fraction of a protein's ligand-binding sites that are occupied by ligand.

Free energy (*G*). A thermodynamic quantity whose change indicates the spontaneity of a process. For spontaneous processes, $\Delta G < 0$, whereas for a process at equilibrium, $\Delta G = 0$.

Free energy of activation. See activation energy.

Free radical. A molecule with an unpaired electron.

Futile cycle. Two opposing metabolic reactions that function together to provide a control point for regulating metabolic flux.

***G*.** See free energy.

$\Delta G^{\ddagger}$. See activation energy.

$\Delta G^{\circ\prime}$. See standard free energy change.

$\Delta G_{\text{reaction}}$. The difference in free energy between the reactants and products of a chemical reaction; $\Delta G_{\text{reaction}} = \Delta G_{\text{products}} - \Delta G_{\text{reactants}}$.

G-actin. The monomeric form of the protein actin. See also F-actin.

Gas constant (*R*). A thermodynamic constant equivalent to 8.3145 J $\cdot$ K^{-1} $\cdot$ mol^{-1}.

Gene. A unique sequence of nucleotides that encodes a polypeptide or RNA; it may include nontranscribed and nontranslated sequences that have regulatory functions.

Gene expression. The transformation by transcription and translation of the information contained in a gene to a functional RNA or protein product.

Gene therapy. The transfer of genetic material to the cells of an individual in order to produce a therapeutic effect.

General transcription factor. One of a set of eukaryotic proteins that are required for the synthesis of all mRNAs.

Genetic code. The correspondence between the sequence of nucleotides in a nucleic acid and the sequence of amino acids in a polypeptide; a series of three nucleotides (a codon) specifies an amino acid.

Genome. The complete set of genetic instructions in an organism.

Genome map. A reconstruction of an organism's genome, based on DNA sequences and physical DNA markers.

Genomics. The study of the size, organization, and gene content of organisms' genomes.

Genotype. An organism's genetic characteristics.

Globin. The polypeptide component of myoglobin and hemoglobin.

Globular protein. A water-soluble protein characterized by a compact, highly folded structure.

Glucogenic amino acid. An amino acid whose degradation yields a gluconeogenic precursor. See also ketogenic amino acid.

Gluconeogenesis. The synthesis of glucose from noncarbohydrate precursors.

Glucose–alanine cycle. A metabolic pathway in which pyruvate produced by glycolysis in the muscles is converted to alanine and transported to the liver, where it is converted back to pyruvate for gluconeogenesis. The resulting glucose returns to the muscles.

Glycerophospholipid. An amphipathic lipid in which two fatty acyl groups and a polar phosphate derivative are attached to a glycerol backbone.

Glycogenolysis. The enzymatic degradation of glycogen to glucose-1-phosphate.

Glycolipid. A lipid to which carbohydrate is covalently attached.

Glycolysis. The 10-reaction pathway by which glucose is broken down to 2 pyruvate with the concomitant production of 2 ATP and the reduction of 2 NAD^+ to 2 NADH.

Glycoprotein. A protein to which carbohydrate is covalently attached.

Glycosaminoglycan. An unbranched polysaccharide consisting of alternating residues of an amino sugar and a sugar acid.

Glycosidic bond. The covalent linkage between two monosaccharide units in a polysaccharide, or the linkage between the anomeric carbon of a saccharide and an alcohol or amine.

Glycosylation. The attachment of carbohydrate chains to a protein through *N*- or *O*-glycosidic linkages.

Glyoxisome. A membrane-bounded plant organelle in which the reactions of the glyoxylate cycle take place.

Glyoxylate cycle. A variation of the citric acid cycle in plants that allows acetyl-CoA to be converted quantitatively to gluconeogenic precursors.

Gout. An inflammatory disease, usually caused by impaired uric acid excretion and characterized by painful deposition of uric acid in the joints.

G protein. A guanine nucleotide–binding and –hydrolyzing protein, involved in a process such as signal transduction or protein synthesis, that is inactive when it binds GDP and active when it binds GTP.

***H*.** See enthalpy.

Half-reaction. The single oxidation or reduction process, involving the reduced and oxidized forms of a substance, that must be combined with another half-reaction to form a complete oxidation–reduction reaction.

Haworth projection. A drawing of a sugar ring in which ring bonds that project in front of the plane of the paper are represented by heavy lines and ring bonds that project behind the plane of the paper are represented by light lines.

Heat shock protein (Hsp). See molecular chaperone.

Helicase. An enzyme that unwinds DNA.

Heme. A protein prosthetic group that binds O_2 (in myoglobin and hemoglobin) or undergoes redox reactions (in cytochromes).

Henderson–Hasselbalch equation. The mathematical expression of the relationship between the pH of a solution of a weak acid and its p*K*: $pH = pK + \log ([A^-]/[HA])$.

Hetero-. Different. In a heteropolymer, the subunits are not all identical.

Heterochromatin. Highly condensed, nonexpressed eukaryotic DNA.

Heterotroph. An organism that obtains its building materials and free energy from organic compounds produced by other organisms.

Heterozygous. Having one each of two gene variants.

Hexose. A six-carbon sugar.

Histones. Highly conserved basic proteins that form a core to which DNA is bound in a nucleosome.

HIV. The human immunodeficiency virus, the causative agent of acquired immunodeficiency syndrome (AIDS).

Homologous genes. Genes that are related by evolution from a common ancestor.

Homologous proteins. Proteins that are related by evolution from a common ancestor.

Homo-. The same. In a homopolymer, all the subunits are identical.

Homozygous. Having two identical copies of a particular gene.

Hormone. A substance that is secreted by one tissue into the bloodstream and that induces a physiological response in other tissues.

Hybridization. The formation of double-stranded segments consisting of complementary DNA and RNA sequences.

Hydration. The molecular state of being surrounded by and interacting with solvent water molecules; that is, solvated by water.

Hydrogen bond. A partly electrostatic, partly covalent interaction between a donor group such as O—H or N—H and an electronegative acceptor atom such as O or N.

Hydrolase. An enzyme that catalyzes a hydrolytic reaction.

Hydrolysis. The cleavage of a covalent bond accomplished by adding the elements of water; the reverse of a condensation.

Hydronium ion. A proton associated with a water molecule, H_3O^+.

Hydropathy. A measure of the hydrophobicity of an amino acid residue that reflects the likelihood of finding that residue in the protein interior.

Hydrophilic. Having high enough polarity to readily interact with water molecules. Hydrophilic substances tend to dissolve in water.

Hydrophobic. Having insufficient polarity to readily interact with water molecules. Hydrophobic substances tend to be insoluble in water.

Hydrophobic effect. The tendency of water to minimize its contacts with nonpolar substances, thereby inducing the substances to aggregate.

Hydroxyl group. A portion of a molecule with the formula —OH.

Hyperglycemia. Elevated levels of glucose in the blood.

IF. See initiation factor.

Inhibition constant (K_I). The dissociation constant for the complex between an enzyme and a reversible inhibitor.

Imine. A molecule with the formula >C=NH.

Imino group. A portion of a molecule with the formula >C=NH.

Imprinting. A heritable variation in the level of expression of a gene according to its parental origin.

Induced fit. An interaction between a protein and its ligand that induces a conformational change in the protein that enhances the protein's interaction with the ligand.

Inducible enzyme. An enzyme that is synthesized only when required by the cell.

Inhibitor. A substance that reduces an enzyme's activity by affecting its substrate binding or turnover number.

Initiation factor (IF). A protein that interacts with mRNA and/or the ribosome and which is required to initiate translation.

Insulin resistance. The inability of cells to respond to insulin.

Integral membrane protein. A membrane protein that is embedded in the lipid bilayer. Also called an intrinsic protein.

Intermediate. A precursor in the synthesis of another molecule or a product of the degradation of a molecule.

Intermediate filament. A 100-Å-diameter cytoskeletal element consisting of coiled-coil polypeptide chains.

Intermembrane space. The compartment between the inner and outer mitochondrial membranes, which is equivalent to the cytosol in ionic composition.

Intrinsic protein. See integral membrane protein.

Intron. A portion of a gene that is transcribed but excised by splicing prior to translation.

Invariant residue. A residue in a protein that is the same in all evolutionarily related proteins.

***In vitro*.** In the laboratory (literally, in glass).

***In vitro* mutagenesis.** See site-directed mutagenesis.

***In vivo*.** In a living organism.

Ion exchange. A technique in which molecules are selectively bound to a surface bearing oppositely charged groups.

Ion pair. An electrostatic interaction between two ionic groups of opposite charge.

Ionic interaction. An electrostatic interaction between two groups that is stronger than a hydrogen bond but weaker than a covalent bond.

Ionization constant of water (K_w). A quantity that relates the concentrations of H^+ and OH^- in pure water: $K_w = [H^+][OH^-] = 10^{-14}$.

Iron–sulfur cluster. A prosthetic group consisting of iron and sulfur ions ([2Fe–2S] or [4Fe–4S]) and that usually participates in oxidation–reduction reactions.

Irregular secondary structure. A segment of a polymer in which each residue has a different backbone conformation; the opposite of regular secondary structure.

Irreversible inhibitor. An molecule that binds to and permanently inactivates an enzyme.

Isoacceptor tRNA. A tRNA that carries the same amino acid as another tRNA but has a different codon.

Isomerase. An enzyme that catalyzes an isomerization reaction.

Isoprenoid. A lipid constructed from 5-carbon units with an isoprene skeleton.

Isozymes. Different proteins that catalyze the same reaction.

***k*.** See rate constant.

K_a. See acid dissociation constant.

kb. Kilobase pairs; 1000 base pairs.

k_{cat}. See catalytic constant.

k_{cat}/K_M. The apparent second-order rate constant for an enzyme-catalyzed reaction; it indicates the enzyme's overall catalytic efficiency.

K_{eq}. See equilibrium constant.

Ketogenesis. The synthesis of ketone bodies from acetyl-CoA.

Ketogenic amino acid. An amino acid whose degradation yields compounds that can be converted to fatty acids or ketone bodies but not to glucose. See also glucogenic amino acid.

Ketone bodies. Compounds (acetoacetate and 3-hydroxybutyrate) that are produced from acetyl-CoA by the liver and used as metabolic fuels in other tissues when glucose is unavailable.

Ketose. A sugar whose carbonyl group is a ketone.

Ketosis. A pathological condition in which ketone bodies are produced in excess of their utilization.

K_I. See inhibition constant.

Kinase. An enzyme that transfers a phosphoryl group between ATP and another molecule.

Kinetic proofreading. A mechanism for promoting translational accuracy in which a noncognate tRNA dissociates from the ribosome before EF-Tu hydrolyzes its GTP.

K_M. See Michaelis constant.

Knockout. A genetic engineering process that deletes or inactivates a specific gene in an animal.

K_w. See ionization constant of water.

Lagging strand. The DNA strand that is synthesized as a series of discontinuous fragments that are later joined.

Lateral diffusion. The movement of a lipid within one leaflet of a bilayer.

Leading strand. The DNA strand that is synthesized continuously during DNA replication.

Ligand. (1) A small molecule that binds to a larger molecule. (2) A molecule or ion bound to a metal ion.

Ligase. An enzyme that catalyzes bond formation coupled with the hydrolysis of ATP.

Light reactions. The photosynthetic reactions in which light energy is absorbed and used to generate NADPH and ATP.

Light-harvesting complex. A pigment-containing protein that collects light energy in order to transfer it to a photosynthetic reaction center.

Lineweaver–Burk plot. A rearrangement of the Michaelis–Menten equation that permits the determination of K_M and V_{max} from a linear plot.

Lipid. Any member of a broad class of macromolecules that are largely or wholly hydrophobic and therefore tend to be insoluble in water but soluble in organic solvents.

Lipid bilayer. See bilayer.

Lipid-linked protein. A protein that is anchored to a biological membrane via a covalently attached lipid.

Lipolysis. The degradation of a triacylglycerol so as to release fatty acids.

Lipoprotein. A globular particle, containing lipids and proteins, that transports lipids between tissues via the bloodstream.

Liposome. A vesicle bounded by a single lipid bilayer.

Lock-and-key model. An early model of enzyme action, in which the substrate fit the enzyme like a key in a lock.

London dispersion forces. The weak van der Waals interactions between nonpolar groups as a result of fluctuations in their electron distributions that create a temporary separation of charge (polarity).

Loop. A segment of a polypeptide that joins two elements of secondary structure; usually found on the protein surface.

Low-barrier hydrogen bond. A short, strong hydrogen bond in which the proton is equally shared by the donor and acceptor atoms.

Lyase. An enzyme that catalyzes the elimination of a group to form a double bond.

Lysosome. A membrane-bounded organelle in a eukaryotic cell that contains a battery of hydrolytic enzymes and that functions to digest ingested material and to recycle cell components.

Major groove. The wider of the two grooves on a DNA double helix.

Mass action ratio. The ratio of the product of the concentrations of reaction products to that of the reactants.

Melting temperature (T_m). The midpoint temperature of the melting curve for the thermal denaturation of a macromolecule. For a lipid, the temperature of transition from an ordered crystalline state to a more fluid state.

Membrane potential ($\Delta\psi$). The difference in electrical charge across a membrane.

Messenger RNA (mRNA). A ribonucleic acid whose sequence is complementary to that of a protein-coding gene in DNA.

Metabolic acidosis. A low blood pH caused by the overproduction or retention of hydrogen ions.

Metabolic fuel. A molecule that can be oxidized to provide free energy for an organism.

Metabolic pathway. A series of enzyme-catalyzed reactions by which one substance is transformed into another.

Metabolically irreversible reaction. A reaction whose value of ΔG is large and negative so that the reaction cannot proceed in reverse.

Metabolism. The total of all degradative and biosynthetic cellular reactions.

Metabolite. A reactant, intermediate, or product of a metabolic reaction.

Metal ion catalysis. A catalytic mechanism that requires the presence of a metal ion to lower the free energy of a reaction's transition state.

Metalloenzyme. An enzyme that contains a tightly bound metal ion cofactor, typically a transition metal ion such as Fe^{2+}, Zn^{2+}, or Mn^{2+}.

Metastasis. The spread of tumor cells to other sites in the body.

Micelle. A globular aggregate of amphiphilic molecules in aqueous solution that are oriented such that polar segments form the surface of the aggregate and the nonpolar segments form a core that is out of contact with the solvent.

Michaelis constant (K_M). For an enzyme that follows the Michaelis–Menten model, $K_M = (k_{-1} + k_2)/k_1$; K_M is equal to the substrate concentration at which the reaction velocity is half-maximal.

Michaelis–Menten equation. A mathematical expression that describes the activity of an enzyme in terms of the substrate concentration ([S]), the enzyme's maximal velocity (V_{max}), and its Michaelis constant (K_M): $v_0 = V_{max}[S]/(K_M + [S])$.

Microarray. A collection of DNA sequences that hybridize with RNA molecules and that can therefore be used to identify active genes. Also called a DNA chip.

Microenvironment. A group's immediate neighbors, whose chemical and physical properties may affect the group.

Microfilament. A 70-Å-diameter cytoskeletal element composed of polymerized actin subunits.

Microtubule. A 240-Å-diameter cytoskeletal element consisting of a hollow tube of polymerized tubulin subunits.

Minor groove. The narrower of the two grooves on a DNA double helix.

(−) end. The end of a polymeric filament where growth is slower. See also (+) end.

Mismatch repair. A DNA repair pathway that removes and replaces mispaired nucleotides on a newly synthesized DNA strand.

Mitochondrial matrix. The gel-like solution of enzymes, substrates, cofactors, and ions in the interior of the mitochondrion.

Mitochondrion (*pl.* mitochondria). The double-membrane-enveloped eukaryotic organelle in which aerobic metabolic reactions occur, including those of the citric acid cycle, fatty acid oxidation, and oxidative phosphorylation.

Mixed inhibition. A form of enzyme inhibition in which an inhibitor binds to the enzyme such that it affects both K_M and V_{max}. See also noncompetitive inhibition.

Mobilization. The process in which polysaccharides, triacylglycerols, and proteins are degraded to make metabolic fuels available.

Molecular chaperone. A protein that binds to unfolded or misfolded proteins in order to promote their normal folding.

Molecular weight. See M_r.

Monomer. A structural unit from which a polymer is built up.

Monosaccharide. A carbohydrate consisting of a single sugar molecule.

Motor protein. An intracellular protein that couples the free energy of ATP hydrolysis to molecular movement relative to another protein that often acts as a track for the linear movement of the motor protein.

M phase. The phase of the cell cycle when mitosis occurs.

M_r. Relative molecular mass. A dimensionless quantity that is defined as the ratio of the mass of a particle to 1/12th the mass of a ^{12}C atom. Also known as molecular weight.

mRNA. See messenger RNA.

Multienzyme complex. A group of noncovalently associated enzymes that catalyze two or more sequential steps in a metabolic pathway.

Multifunctional enzyme. A protein that carries out more than one chemical reaction.

Mutagen. An agent that induces a mutation in an organism.

Mutase. An enzyme that catalyzes the transfer of a functional group from one position to another on a molecule.

Mutation. A heritable alteration in an organism's genetic material.

Myelin sheath. The multilayer coating of sphingomyelin-rich membranes that insulates a mammalian neuron.

Native structure. The fully folded conformation of a macromolecule.

Natural selection. The evolutionary process by which the continued existence of a replicating entity depends on its ability to survive and reproduce under the existing conditions.

Near-equilibrium reaction. A reaction whose ΔG value is close to zero, so that it can operate in either direction depending on the substrate and product concentrations.

Negative effector. A substance that diminishes an enzyme's activity through allosteric inhibition.

Nernst equation. An expression of the relationship between the actual ($\mathcal{E}$) and standard reduction potential ($\mathcal{E}^{\circ\prime}$) of a substance A: $\mathcal{E} = \mathcal{E}^{\circ\prime} + RT/n\mathcal{F} \ln ([A_{oxidized}]/[A_{reduced}])$.

Neurotransmitter. A substance released by a nerve cell to alter the activity of a target cell.

Neutral solution. A solution whose pH is equal to 7.0 ($[H^+] = 10^{-7}$ M).

Nick. A single-strand break in a double-stranded nucleic acid.

Nick translation. The progressive movement of a single-strand break (nick) in DNA through the actions of an exonuclease that removes residues followed by a polymerase that replaces them.

Nitrification. The conversion of ammonia (NH_3) to nitrate (NO_3^-).

Nitrogen cycle. A set of reactions, including nitrogen fixation, nitrification, and denitrification, for the interconversion of different forms of nitrogen.

Nitrogen fixation. The process by which atmospheric N_2 is converted to a biologically useful form such as NH_3.

***N*-linked oligosaccharide.** An oligosaccharide linked to the amide group of a protein Asn residue.

Noncoding strand. The DNA strand that has a sequence complementary (except for the replacement of U with T) to the transcribed RNA; it is the template strand. Also called the antisense strand.

Noncompetitive inhibition. A special case of mixed inhibition in which the inhibitor reduces k_{cat} but does not alter K_M.

Noncyclic electron flow. The light-driven linear path of electrons from water through Photosystems I and II, which leads to the production of O_2, NADPH, and ATP.

Nonessential amino acid. An amino acid that an organism can synthesize from common intermediates.

Nonspontaneous process. A thermodynamic process that has a net increase in free energy ($\Delta G > 0$) and can occur only with the input of free energy from outside the system. See also endergonic reaction.

N-terminus. The end of a polypeptide that has a free amino group.

Nuclease. An enzyme that degrades nucleic acids.

Nucleic acid. A polymer of nucleotide residues. The major nucleic acids are deoxyribonucleic acid (DNA) and ribonucleic acid (RNA). Also known as a polynucleotide.

Nucleolus. The region of the eukaryotic nucleus where rRNA is processed and ribosomes are assembled.

Nucleophile. A compound containing an electron-rich group. A nucleophile (nucleus-lover) reacts with an electrophile (electron-lover).

Nucleoside. A compound consisting of a nitrogenous base linked to a five-carbon sugar (ribose or deoxyribose).

Nucleosome. The disk-shaped complex of a histone octamer and DNA that represents the fundamental unit of DNA organization in eukaryotes.

Nucleotide. A compound consisting of a nucleoside esterified to one or more phosphate groups. Nucleotides are the monomeric units of nucleic acids.

Nucleotide excision repair. A DNA repair pathway in which a damaged single-stranded segment of DNA is removed and replaced with normal DNA.

Okazaki fragments. The short segments of DNA formed in the discontinuous lagging-strand synthesis of DNA.

Oligonucleotide. A polynucleotide consisting of a few nucleotide residues.

Oligopeptide. A polypeptide consisting of a few amino acid residues.

Oligosaccharide. A polymeric carbohydrate containing a few monosaccharide residues.

***O*-linked oligosaccharide.** An oligosaccharide linked to the hydroxyl group of a protein Ser or Thr side chain.

Oncogene. A viral gene that interferes with the normal regulation of cell growth and contributes to cancer.

Open reading frame (ORF). A portion of the genome that potentially codes for a protein.

Operon. A prokaryotic genetic unit that consists of several genes with related functions that are transcribed as a single mRNA molecule.

Orientation effects. See proximity and orientation effects.

ORF. See open reading frame.

Orphan gene. A gene with no assigned function.

Osmosis. The movement of solvent from a region of low solute concentration to a region of high solute concentration.

Osteogenesis imperfecta. A disease caused by a mutations in collagen genes and characterized by bone fragility and deformation.

Oxidant. See oxidizing agent.

Oxidation. A reaction in which a substance loses electrons.

Oxidative phosphorylation. The process by which the free energy obtained from the oxidation of metabolic fuels is used to generate ATP from ADP + P_i.

Oxidizing agent. A substance that can accept electrons, thereby becoming reduced. Also called an oxidant.

Oxidoreductase. An enzyme that catalyzes an oxidation–reduction reaction.

Oxyanion hole. A cavity in the active site of a serine protease that accommodates the reactants during the transition state and thereby lowers its energy.

Oxyhemoglobin. Hemoglobin that contains bound oxygen or is in the oxygen-binding conformation.

P site. The ribosomal binding site that accommodates a peptidyl–tRNA.

P:O ratio. The ratio of the number of molecules of ATP synthesized from ADP + P_i to the number of atoms of oxygen reduced.

pO_2. The partial pressure of oxygen; the oxygen concentration.

p_{50}. The ligand concentration (or pressure for a gaseous ligand) at which a binding protein such as hemoglobin is half-saturated with ligand.

PAGE. Polyacrylamide gel electrophoresis. See electrophoresis.

Palindrome. A segment of DNA that has the same sequence on each strand when read in the $5' \rightarrow 3'$ direction.

Parallel β sheet. See β sheet.

Partial oxygen pressure (pO_2). The concentration of gaseous O_2 in units of torr.

Passive transport. The thermodynamically spontaneous protein-mediated transmembrane movement of a substance from high to low concentration.

Pasteur effect. The greatly increased sugar consumption of yeast grown under anaerobic conditions compared to that of yeast grown under aerobic conditions.

PCR. See polymerase chain reaction.

Pentose. A five-carbon sugar.

Pentose phosphate pathway. A pathway for glucose degradation that yields ribose-5-phosphate and NADPH.

Peptide. A short polypeptide.

Peptide bond. An amide linkage between the α-amino group of one amino acid and the α-carboxylate group of another. Peptide bonds link the amino acid residues in a polypeptide.

Peptide group. The planar —CO—NH— group that encompasses the peptide bond between amino acid residues in a polypeptide.

Peripheral membrane protein. A protein that is weakly associated with the surface of a biological membrane. Also called an extrinsic protein.

Periplasmic compartment. The space between the cell wall and the outer membrane of gram-negative bacteria.

Peroxisome. A eukaryotic organelle with specialized oxidative functions, including fatty acid degradation.

pH. A quantity used to express the acidity of a solution, equivalent to $-\log[H^+]$.

Phage. See bacteriophage.

Phenotype. An organism's physical characteristics.

Phosphatase. An enzyme that hydrolyzes phosphoryl ester groups.

Phosphodiester bond. The linkage in which a phosphate group is esterified to two alcohol groups (e.g., two ribose units that join the adjacent nucleotide residues in a polynucleotide).

Phosphoinositide signaling system. A signal transduction pathway in which hormone binding to a cell surface receptor induces phospholipase C to catalyze the hydrolysis of phosphatidylinositol bisphosphate to yield the second messengers inositol trisphosphate and diacylglycerol.

Phospholipase. An enzyme that hydrolyzes one or more bonds in a glycerophospholipid.

Phosphorolysis. The cleavage of a chemical bond by the substitution of a phosphate group rather than water.

Phosphoryl group. A portion of a molecule with the formula $—PO_3H_2$.

Photoautotroph. An organism that obtains its building supplies from inorganic compounds and its free energy from sunlight.

Photon. A packet of light energy.

Photooxidation. A mode of decay of an excited molecule, in which oxidation occurs through the transfer of an electron to an acceptor molecule.

Photophosphorylation. The synthesis of ATP from ADP + P_i coupled to the dissipation of a proton gradient that has been generated through light-driven electron transport.

Photoreceptor. A light-absorbing molecule, or pigment.

Photorespiration. The consumption of O_2 and evolution of CO_2 by plants (a dissipation of the products of photosynthesis), a consequence of the competition between O_2 and CO_2 for ribulose-5-phosphate carboxylase.

Photosynthesis. The light-driven incorporation of CO_2 into organic compounds.

P_i. Inorganic phosphate or a phosphoryl group: HPO_3^- or PO_3^{2-}.

p*K*. A quantity used to express the tendency for an acid to donate a proton (dissociate); equal to $-\log K$, where K is the dissociation constant.

Planck's law. An expression for the energy (E) of a photon: $E = hc/\lambda$, where c is the speed of light, λ is its wavelength, and h is Planck's constant (6.626×10^{-34} J · s).

Plasmid. A small circular DNA molecule that autonomously replicates and may be used as a cloning vector.

(+) end. The end of a polymeric filament where growth is faster. See also (−) end.

pO_2. See partial oxygen pressure.

Point mutation. The substitution of one base for another in DNA, arising from mispairing during DNA replication or from chemical alterations of existing bases.

Polarity. Having an uneven distribution of charge.

Poly(A) tail. The sequence of adenylate residues that is posttranslationally added to the 3′ end of eukaryotic mRNAs.

Polyacrylamide gel electrophoresis (PAGE). See electrophoresis.

Polymer. A molecule consisting of numerous smaller units that are linked together in an organized manner.

Polymerase. An enzyme that catalyzes the addition of nucleotide residues to a polynucleotide.

Polymerase chain reaction (PCR). A procedure for amplifying a segment of DNA by repeated rounds of replication centered between primers that hybridize with the two ends of the DNA segment of interest.

Polynucleotide. See nucleic acid.

Polypeptide. A polymer consisting of amino acid residues linked in linear fashion by peptide bonds.

Polyprotein. A polypeptide that undergoes proteolysis after its synthesis to yield several separate protein molecules.

Polysaccharide. A polymeric carbohydrate containing multiple monosaccharide residues.

Polysome. An mRNA transcript bearing multiple ribosomes in the process of translating the mRNA.

Positive effector. A substance that boosts an enzyme's activity through allosteric activation.

Posttranslational modification (processing). The removal or derivatization of amino acid residues following their incorporation into a polypeptide.

PP_i. A pyrophosphoryl group: $H_3P_2O_6$, $H_2P_2O_6^-$, $HP_2O_6^{2-}$, or $P_2O_6^{3-}$.

Primary structure. The sequence of residues in a polymer.

Primer. An oligonucleotide that base-pairs with a template polynucleotide strand and is extended through template-directed polymerization.

Probe. A labeled single-stranded DNA or RNA segment that can hybridize with a DNA or RNA of interest in a screening procedure.

Processing. See RNA processing and posttranslational modification.

Processivity. A property of a motor protein or other enzyme that undergoes many reaction cycles before dissociating from its track or substrate.

Product inhibition. A form of enzyme inhibition in which the reaction product acts as a competitive inhibitor.

Prokaryote. A unicellular organism that lacks a membrane-bounded nucleus. All bacteria are prokaryotes.

Promoter. The DNA sequence at which RNA polymerase binds to initiate transcription.

Proofreading. An additional catalytic activity of an enzyme, which acts to correct errors made by the primary enzymatic activity.

Prosthetic group. An organic group (coenzyme) that is permanently associated with an enzyme.

Protease. An enzyme that catalyzes the hydrolysis of peptide bonds.

Protease inhibitor. An agent, often a protein, that reacts incompletely with a protease so as to inhibit further proteolytic activity.

Proteasome. A multiprotein complex with a hollow cylindrical core in which cellular proteins are degraded to peptides in an ATP-dependent process.

Protein. A macromolecule that consists of one or more polypeptide chains.

Protein kinase. An enzyme that catalyzes the transfer of a phosphoryl group from ATP to the OH group of a protein Ser, Thr, or Tyr residue.

Proteoglycan. An extracellular aggregate of protein and glycosaminoglcyans.

Proteome. The complete set of proteins synthesized by a cell.

Proteomics. The study of all the proteins synthesized by a cell.

Protofilament. One of the 13 linear polymers of tubulin subunits that forms a microtubule.

Proton jumping. The rapid movement of a protons among hydrogen-bonded water molecules.

Proton wire. A series of hydrogen-bonded water molecules and protein groups that can relay protons from one site to another.

Protonmotive force. The free energy of the electrochemical proton gradient that forms during electron transport.

Proto-oncogene. The normal cellular analog of an oncogene, whose mutation may contribute to cancer.

Proximity and orientation effects. A catalytic mechanism in which reacting groups are brought close together in an enzyme active site to accelerate the reaction.

Purine. A derivative of the compound purine, such as the nucleotide base adenine or guanine.

Pyrimidine. A derivative of the compound pyrimidine, such as the nucleotide base cytosine, uracil, or thymine.

Q cycle. The cyclic flow of electrons involving a semiquinone intermediate in Complex III of mitochondrial electron transport and in photosynthetic electron transport.

Quantum (*pl.* quanta). A packet of energy. See also photon.

Quantum yield. The ratio of carbon atoms fixed or oxygen molecules produced to the number of photons absorbed by the photosynthetic machinery.

Quaternary structure. The spatial arrangement of a macromolecule's individual subunits.

***R*.** See gas constant.

R group. A symbol for a variable portion of a molecule, such as the side chain of an amino acid.

R state. One of two conformations of an allosteric protein; the other is the T state.

Raft. An area of a lipid bilayer with a distinct lipid composition and near-crystalline consistency.

Rate constant (*k*). The proportionality constant between the velocity of a chemical reaction and the concentration(s) of the reactant(s).

Rate equation. A mathematical expression for the time-dependent progress of a reaction as a function of reactant concentration.

Rate-determining reaction. The slowest step in a multistep sequence, such as a metabolic pathway, whose rate determines the rate of the entire sequence.

Rational drug design. The use of information about an enzyme's structure, mechanism, and inhibitors to design even more effective inhibitors (drugs).

Reaction center. A chlorophyll-containing protein where photooxidation takes place.

Reaction coordinate. A line representing the progress of a reaction, part of a graphical presentation of free energy changes during a reaction.

Reaction specificity. The ability of an enzyme to discriminate between possible substrates and to catalyze a single type of chemical reaction.

Reading frame. The grouping of nucleotides in sets of three whose sequence corresponds to a polypeptide sequence.

Receptor. A binding protein that is specific for its ligand and elicits a discrete biochemical effect when its ligand is bound.

Recombinant DNA. A DNA molecule constructed by combining DNA from different sources.

Recombination. The exchange of polynucleotide strands between separate DNA segments; recombination is one mechanism for repairing damaged DNA by allowing a homologous segment to serve as a template for replacement of the damaged bases.

Redox center. A group that can undergo an oxidation–reduction reaction.

Redox reaction. A chemical reaction in which one substance is reduced and another substance is oxidized.

Reducing agent. A substance that can donate electrons, thereby becoming oxidized. Also called a reductant.

Reductant. See reducing agent.

Reduction. A reaction in which a substance gains electrons.

Reduction potential ($\mathcal{E}$). A measure of the tendency of a substance to gain electrons.

Regular secondary structure. A segment of a polymer in which the backbone adopts a regularly repeating conformation; the opposite of irregular secondary structure.

Release factor (RF). A protein that recognizes a stop codon and causes a ribosome to terminate polypeptide synthesis.

Renaturation. The refolding of a denatured macromolecule so as to regain its native conformation.

Repetitive DNA. Stretches of DNA containing up to several thousand bases that occur in multiple copies and places in an organism's genome, often arranged in tandem (side by side).

Replication. The process of making an identical copy of a DNA molecule. During DNA replication, the parental polynucleotide strands separate so that each can direct the synthesis of a complementary daughter strand, resulting in two complete DNA double helices.

Replication fork. The point in a replicating DNA molecule where the two parental strands separate in order to serve as templates for the synthesis of new strands.

Repressor. A protein that binds at or near a gene so as to prevent its transcription.

Residue. A term for a monomeric unit after it has been incorporated into a polymer.

Respiration. The metabolic phenomenon whereby organic molecules are oxidized, with the electrons eventually transferred to molecular oxygen.

Respiratory acidosis. A low blood pH caused by insufficient elimination of CO_2 (carbonic acid) by the lungs.

Restriction digest. The generation of a set of DNA fragments by the action of a restriction endonuclease.

Restriction endonuclease. A bacterial enzyme that cleaves a specific DNA sequence.

Restriction fragment. A segment of DNA produced by the action of a restriction endonuclease.

Restriction fragment length polymorphism (RFLP). Differences in a DNA sequence (polymorphism) that are detected by variation in the sizes of the fragments produced when the DNA is digested with a particular restriction endonuclease.

Reverse transcriptase. A DNA polymerase that uses RNA as its template.

RF. See release factor.

RFLP. See restriction fragment length polymorphism.

Ribonucleic acid. See RNA.

Ribosomal RNA (rRNA). The RNA molecules that provide structural support for the ribosome and catalyze peptide bond formation.

Ribosome. The RNA-and-protein organelle that synthesizes polypeptides under the direction of mRNA.

Ribosome recycling factor (RRF). A protein that binds to a ribosome after protein synthesis to prepare it for another round of translation.

Ribozyme. An RNA molecule that has catalytic activity.

RNA. A polymer of ribonucleotides, such as messenger RNA (mRNA), transfer RNA (tRNA), and ribosomal RNA (rRNA).

RNA interference (RNAi). A phenomenon in which short double-stranded RNA molecules direct the degradation of complementary mRNA, thereby inhibiting gene expression.

RNA processing. The addition, removal, or modification of nucleotides in an RNA molecule that is necessary to produce a fully functional RNA.

RRF. See ribosome recycling factor.

rRNA. See ribosomal RNA.

***S*.** See entropy.

Saccharide. See carbohydrate.

Salvage reaction. A reaction that reincorporates an intermediate of nucleotide degradation into a new nucleotide, thereby minimizing the need for the nucleotide biosynthetic pathways.

Saturated fatty acid. A fatty acid that does not contain any double bonds in its hydrocarbon chain.

Saturation. The state in which all of a macromolecule's ligand-binding sites are occupied by ligand.

Schiff base. An imine that forms between an amine and an aldehyde or ketone.

Scissile bond. The bond that is to be cleaved during a proteolytic reaction.

Second messenger. An intracellular ion or molecule that acts as a signal for an extracellular event such as ligand binding to a cell surface receptor.

Secondary structure. The local spatial arrangement of a polymer's backbone atoms without regard to the conformations of its substituent side chains.

Second-order reaction. A reaction whose rate is proportional to the square of the concentration of one reactant or to the product of the concentrations of two reactants.

Selection. A technique for distinguishing cells that contain a particular feature, such as resistance to an antibiotic.

Semiconservative replication. The mechanism of DNA duplication in which each new molecule contains one strand from the parent molecule and one newly synthesized strand.

Sense strand. See coding strand.

Serine protease. A peptide-hydrolyzing enzyme that has a reactive Ser residue in its active site.

Signal peptide. A short sequence in a membrane or secretory protein that binds to the signal recognition particle in order to direct the translocation of the protein across a membrane.

Signal recognition particle (SRP). A complex of protein and RNA that recognizes membrane and secretory proteins and mediates their binding to a membrane for translocation.

Signal transduction. The process by which an extracellular signal is transmitted to the cell interior by binding to a cell surface receptor such that binding triggers a series of intracellular events.

Silencer. A DNA sequence some distance from the transcription start site, where a repressor of transcription may bind.

Single-nucleotide polymorphism. A nucleotide sequence variation in the genomes of two individuals from the same species.

Site-directed mutagenesis. A technique in which a cloned gene is mutated in a specific manner. Also called *in vitro* mutagenesis.

Small nuclear RNA (snRNA). Highly conserved RNAs that participate in eukaryotic mRNA splicing.

Small nucleolar RNA (snoRNA). RNA molecules that direct the sequence-specific methylation of eukaryotic rRNA transcripts.

snoRNA. See small nucleolar RNA.

SNP. See single-nucleotide polymorphism.

snRNA. See small nuclear RNA.

Solvation. The state of being surrounded by solvent molecules.

Specificity pocket. A cavity on the surface of a serine protease, whose chemical characteristics determine the identity of the substrate residue on the N-terminal side of the bond to be cleaved.

S phase. The phase of the cell cycle in which DNA replication occurs.

Sphingolipid. An amphipathic lipid containing an acyl group, a palmitate derivative, and a polar head group attached to a serine backbone. In sphingomyelins, the head group is a phosphate derivative.

Sphingomyelin. See sphingolipid.

Spliceosome. A complex of protein and snRNA that carries out the splicing of immature mRNA molecules.

Splicing. The process by which introns are removed and exons are joined to produce a mature RNA transcript.

Spontaneous process. A thermodynamic process that has a net decrease in free energy ($\Delta G < 0$) and occurs without the input of free energy from outside the system. See also exergonic reaction.

SRP. See signal recognition particle.

Stacking interactions. The stabilizing van der Waals interactions between successive (stacked) bases in a polynucleotide.

Standard free energy change ($\Delta G^{\circ\prime}$). The force that drives reactants to reach their equilibrium values when the system is in its biochemical standard state.

Standard conditions. A set of conditions including a temperature of 25°C, a pressure of 1 atm, and reactant concentrations of 1 M. Biochemical standard conditions include a pH of 7.0 and a water concentration of 55.5 M.

Standard reduction potential ($\mathcal{E}^{\circ\prime}$). A measure of the tendency of a substance to gain electrons (to be reduced) under standard conditions.

Steady state. A set of conditions under which the formation and degradation of individual components are balanced such that the system does not change over time.

Stereocilia. Microfilament-stiffened cell processes on the surface of cells in the inner ear, which are deflected in response to sound waves.

Sticky ends. Single-stranded extensions of DNA that are complementary, often because they have been generated by the action of the same restriction endonuclease.

Stroma. The gel-like solution of enzymes and small molecules in the interior of the chloroplast; the site of carbohydrate synthesis.

Substrate. A reactant in an enzymatic reaction.

Substrate catalysis. The rate enhancement resulting from transition state stabilization by the substrate rather than the enzyme.

Subunit. One of several polypeptide chains that make up a protein.

Sugar. See carbohydrate.

Sugar–phosphate backbone. The chain of (deoxy)ribose groups linked by phosphodiester bonds in a polynucleotide chain.

Suicide substrate. A molecule that chemically inactivates an enzyme only after undergoing part of the normal catalytic reaction.

Sulfhydryl group. A portion of a molecule with the formula —SH.

Supercoiling. A topological state of DNA in which the helix is underwound or overwound so that the molecule tends to writhe or coil up on itself.

Symport. A transporter that allows the simultaneous transmembrane movement of two molecules in the same direction.

Synaptic vesicle. A vesicle loaded with neurotransmitters to be released from the end of an axon.

Synexpression group. A set of eukaryotic genes whose expression is coordinated.

Synonymous codon. A codon that specifies the same amino acid as another.

T state. One of two conformations of an allosteric protein; the other is the R state.

T → R transition. The shift in conformation of an allosteric protein induced by ligand binding.

TATA box. A eukaryotic promoter element with the consensus sequence TATA located 10 to 27 nucleotides upstream from the transcription start site.

Tautomer. One of a set of isomers that differ only in the positions of their hydrogen atoms.

Telomerase. An enzyme that uses an RNA template to polymerize deoxynucleotides and thereby extend the 3′-ending strand of a eukaryotic chromosome.

Telomere. The end of a linear eukaryotic chromosome, which consists of tandem repeats of a short G-rich sequence on the 3′-ending strand and its complementary sequence on the 5′-ending strand.

Tertiary structure. The entire three-dimensional structure of a single-chain polymer, including the conformations of its side chains.

Tetramer. An assembly consisting of four monomeric units.

Thermogenesis. The process of generating heat by muscular contraction or by metabolic reactions.

Thermophile. An organism that thrives at high temperatures.

Thick filament. A muscle cell structural element that is composed of several hundred myosin molecules.

Thin filament. A muscle cell structural element that consists primarily of an actin filament.

Thioester. A compound containing an ester linkage to a sulfur rather than an oxygen atom.

Thylakoid. The membranous structure in the interior of a chloroplast that is the site of the light reactions of photosynthesis.

T_m. See melting temperature.

Topoisomerase. An enzyme that alters DNA supercoiling by breaking and resealing one or both strands.

Trace element. An element that is present in small quantities in a living organism.

Transamination. The transfer of an amino group from an amino acid to an α-keto acid to yield a new α-keto acid and a new amino acid.

Transcription. The process by which RNA is synthesized using a DNA template, thereby transferring genetic information from the DNA to the RNA.

Transcription factor. A protein that promotes the transcription of a gene by binding to DNA sequences at or near the gene or by interacting with other proteins that do so.

Transcriptome. The set of all the RNA molecules produced by a cell.

Transcriptomics. The study of the genes that are transcribed in a certain cell type or at a certain time.

Transfer RNA (tRNA). The small L-shaped RNAs that deliver specific amino acids to ribosomes according to the sequence of a bound mRNA.

Transferase. An enzyme that catalyzes the transfer of a functional group from one molecule to another.

Transformation. (1) The process by which a normal cell becomes a cancer cell. (2) The permanent alteration of a bacterial cell's genetic message through the introduction of foreign DNA.

Transgenic organism. An organism that stably expresses a foreign gene.

Transition state. The point of highest free energy, or the structure that corresponds to that point, in the reaction coordinate diagram of a chemical reaction.

Transition state analog. A stable substance that geometrically and electronically resembles the transition state of a reaction and that therefore may inhibit an enzyme that catalyzes the reaction.

Translation. The process of transforming the information contained in the nucleotide sequence of an RNA to the corresponding amino acid sequence of a polypeptide as specified by the genetic code.

Translocase. An enzyme that catalyzes the movement of a lipid from one bilayer leaflet to another. Also called a flippase.

Translocation. The movement of tRNA and mRNA, relative to the ribosome, that occurs following formation of a peptide bond and that allows the next mRNA codon to be translated.

Translocon. The complex of membrane proteins that mediates the transmembrane movement of a polypeptide.

Transpeptidation. The ribosomal process in which the peptidyl group attached to a tRNA is transferred to the aminoacyl group of another tRNA, forming a new peptide bond and lengthening the polypeptide by one residue at its C-terminus.

Transposable element. A segment of DNA, sometimes including genes, that can move (be copied) from one position to another in a genome.

Transverse diffusion. The movement of a lipid from one leaflet of a bilayer to the other. Also called flip-flop.

Treadmilling. The addition of monomeric units to one end of a polymer and their removal from the opposite end such that the length of the polymer remains unchanged.

Triacylglycerol. A lipid in which three fatty acids are esterified to a glycerol backbone. Also called a triglyceride.

Triglyceride. See triacylglycerol.

Trimer. An assembly consisting of three monomeric units.

Triple helix. The right-handed helical structure formed by three left-handed helical polypeptide chains in collagen.

tRNA. See transfer RNA.

Tumor. A mass of cells resulting from the uncontrolled proliferation of cancer cells.

Tumor suppressor gene. A gene whose loss or mutation may lead to cancer.

Turnover number. See catalytic constant.

Uncoupler. A substance that allows the proton gradient across a membrane to dissipate without ATP synthesis so that electron transport proceeds without oxidative phosphorylation.

Unimolecular reaction. A reaction involving one molecule.

Uniport. A transporter that allows transmembrane movement of a single molecule.

Unsaturated fatty acid. A fatty acid that contains at least one double bond in its hydrocarbon chain.

Urea cycle. A cyclic metabolic pathway in which amino groups are converted to urea for disposal.

Usher syndrome. A genetic disease characterized by profound deafness and retinitis pigmentosa that leads to blindness, caused in some cases by a defective myosin protein.

van der Waals radius. The distance from an atom's nucleus to its effective electronic surface.

van der Waals interaction. A weak noncovalent association between molecules that arises from the attractive forces between polar groups (dipole–dipole interactions) or between nonpolar groups whose fluctuating electron distribution gives rise to temporary dipoles (London dispersion forces).

Variable residue. A position in a polypeptide that is occupied by different residues in evolutionarily related proteins; its substitution has little or no effect on protein function.

Vesicle. A fluid-filled sac enclosed by a lipid-bilayer membrane.

Vitamin. A metabolically required substance that cannot be synthesized by an animal and must therefore be obtained from the diet.

V_{max}. Maximal velocity of an enzymatic reaction.

Voltage-gated channel. A membrane transport channel that opens and closes in response to a change in membrane potential.

v_0. Initial velocity of an enzymatic reaction.

Wobble hypothesis. An explanation for the nonstandard base pairing between tRNA and mRNA at the third codon position, which allows a tRNA to recognize more than one codon.

X-Ray crystallography. A method for determining three-dimensional molecular structures from the diffraction pattern produced by exposing a crystal of a molecule to a beam of X-rays.

$\Delta\psi$. See membrane potential.

Y. See fractional saturation.

Z. The net charge of an ion.

Zinc finger. A protein structural motif consisting of 20–60 residues, including Cys and His residues to which one or two Zn^{2+} ions are tetrahedrally coordinated.

Z-scheme. A Z-shaped diagram indicating the electron carriers and their reduction potentials in the photosynthetic electron-transport system of plants and cyanobacteria.

Zymogen. The inactive precursor (proenzyme) of a proteolytic enzyme.

SOLUTIONS

Chapter 1 Solutions

1.

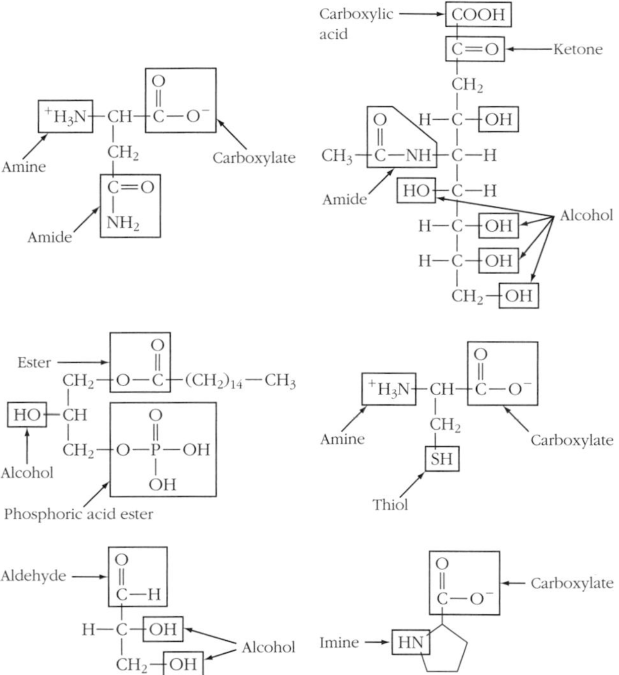

2. Amino acids, carbohydrates, nucleotides, and lipids are the four types of biological small molecules. Amino acids, monosaccharides, and nucleotides can form polymers of proteins, polysaccharides, and nucleic acids, respectively.

3. (a) *N*-acetylglucosamine is a monosaccharide derivative.
(b) Cholesteryl ester is a lipid derivative.
(c) Homocysteine is an amino acid derivative.
(d) rCMP is a nucleotide.

4. (a) All amino acids have carboxylate groups. All have primary amino groups except for proline, which has a secondary amino group.
(b) The carbon marked with an asterisk is chiral. This means that alanine has two possible enantiomers, or mirror-image isomers.

$$^{+}H_3N-\overset{\displaystyle CH_3}{\underset{}{C^*H}}-\overset{O}{\overset{\|}{C}}-O^-$$

5.

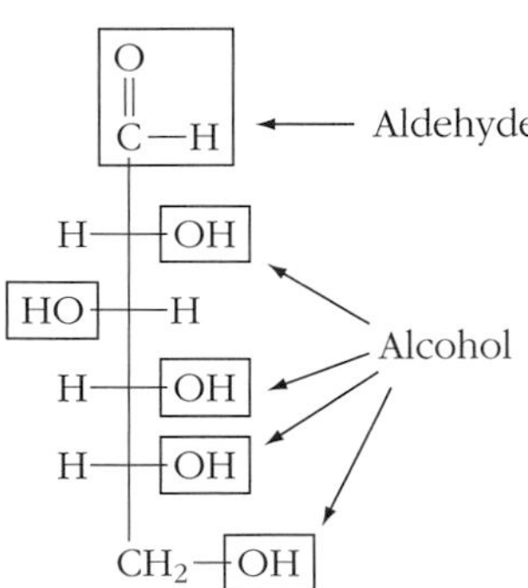

6. Nucleotides consist of a five-carbon sugar, a nitrogenous ring, and one or more phosphate groups linked covalently together.

7. As described in the text, palmitate and cholesterol are highly nonpolar and are therefore insoluble in water. Both are highly aliphatic. Alanine is water-soluble because its amino group and carboxylate group are ionized, which render the molecule "salt-like."

8. RNA forms a more regular structure because RNA consists of only four different nucleotides, whereas proteins are made up of as many as 20 different amino acids. This results in a more regular structure for RNA. The cellular role of RNA relies on the *sequence* of the nucleotides that make up the RNA and not on the overall shape of the RNA molecule itself. Proteins, on the other hand, fold into unique shapes, as illustrated by endothelin in Figure 1-4. The ability of proteins to fold into a wide variety of shapes means that proteins can also perform a wide variety of biochemical roles in the cell. According to Table 1-2, the major roles of proteins in the cell are to carry out metabolic reactions and to support cellular structures.

9. Polysaccharides serve as fuel storage molecules and as structural support for the cell.

10. A positive change in entropy indicates that the system has become more disordered; a negative change in entropy indicates that the system has become more ordered.
(a) negative
(b) positive
(c) positive
(d) positive
(e) negative
(f) positive

11. The polymeric molecule is more ordered and thus has less entropy. A mixture of constituent monomers has a large number of different arrangements (like the balls scattered on the pool table) and thus has greater entropy.

12. (a) The conversion of glucose to glucose-6-phosphate is not favorable because the $\Delta G^{\circ\prime}$ value for the reaction is positive.
(b) If the two reactions are coupled, the overall reaction would be the sum of the two individual reactions. The $\Delta G^{\circ\prime}$ value would be the sum of the $\Delta G^{\circ\prime}$ values for the two individual reactions.

$$\text{ATP} + \text{glucose} \rightleftharpoons \text{ADP} + \text{glucose-6-phosphate} \quad \Delta G^{\circ\prime} = -16.7 \text{ kJ} \cdot \text{mol}^{-1}$$

Coupling the conversion of glucose to glucose-6-phosphate with the hydrolysis of ATP has converted an unfavorable reaction to a favorable reaction. The $\Delta G^{\circ\prime}$ value of the coupled reaction is negative, which indicates that the reaction as written is favorable.

13. First, calculate ΔH and ΔS, as described in Sample Calculation 1-1:

$\Delta H = H_B - H_A$
$\Delta H = 60 \text{ kJ} \cdot \text{mol}^{-1} - 54 \text{ kJ} \cdot \text{mol}^{-1}$
$\Delta H = 6 \text{ kJ} \cdot \text{mol}^{-1}$
$\Delta S = S_B - S_A$
$\Delta S = 43 \text{ J} \cdot \text{K}^{-1} \cdot \text{mol}^{-1} - 22 \text{ J} \cdot \text{K}^{-1} \cdot \text{mol}^{-1}$
$\Delta S = 21 \text{ J} \cdot \text{K}^{-1} \cdot \text{mol}^{-1}$

(a) $\Delta G = (6000 \text{ J} \cdot \text{mol}^{-1}) - (4 + 273\text{K})(21 \text{ J} \cdot \text{K}^{-1} \cdot \text{mol}^{-1})$
$\Delta G = +183 \text{ J} \cdot \text{mol}^{-1}$
The reaction is not favorable at 4°C.

(b) $\Delta G = (6000 \text{ J} \cdot \text{mol}^{-1}) - (37 + 273\text{K})(21 \text{ J} \cdot \text{K}^{-1} \cdot \text{mol}^{-1})$
$\Delta G = -510 \text{ J} \cdot \text{mol}^{-1}$
The reaction is favorable at 37°C.

14. $0 > 15{,}000 \text{ J} \cdot \text{mol}^{-1} - (T)(51 \text{ J} \cdot \text{K}^{-1} \cdot \text{mol}^{-1})$
$-15{,}000 > -(T)(51 \text{ K}^{-1})$
$15{,}000 < (T)(51 \text{ K}^{-1})$
$T > 294\text{K}$

The reaction is favorable at temperatures higher than 21°C.

15. (a) decrease
(b) increase

16. Process d is never spontaneous.

17. (a) The reaction is exothermic because the value of ΔH is negative.
(b) $\Delta G = \Delta H - T\Delta S$
$\Delta G = -8190 \text{ J} \cdot \text{mol}^{-1} - (310\text{K})(2.2 \text{ J} \cdot \text{K}^{-1} \cdot \text{mol}^{-1})$
$\Delta G = -8872 \text{ J} \cdot \text{mol}^{-1}$

The negative value of ΔG indicates that the reaction is spontaneous.
(c) The ΔH term makes a greater contribution to the ΔG value. This indicates that the reaction is spontaneous largely because the reaction is exothermic. The reaction proceeds without much change in entropy.

18. (a) reduction
(b) oxidation

19. (a) oxidized
(b) oxidized
(c) oxidized
(d) reduced

20. The first biological molecules would have had to polymerize and they would have had to find some way to make copies of themselves.

21. (a) Palmitate's carbon atoms, which have the formula —CH_2—, are more reduced than CO_2, so their reoxidation to CO_2 can release free energy.
(b) Because the —CH_2— groups of palmitate are more reduced than those of glucose (—HCOH—), their conversion to the fully oxidized CO_2 would be even more thermodynamically favorable (have a larger negative value of ΔG) than the conversion of glucose carbons to CO_2. Therefore, palmitate carbons could provide more free energy than glucose carbons.

22. Morphological differences, which are useful for classifying large organisms, are not useful for bacteria, which often look alike. Furthermore, microscopic organisms do not leave an easily interpreted imprint in the fossil record, as vertebrates do. Thus, molecular information is often the only means for tracing the evolutionary history of bacteria.

23. It is difficult to envision how a single engulfment event could have given rise to a stable and heritable association of the eukaryotic host and the bacterial dependent within a single generation. It is much more likely that natural selection gradually promoted the interdependence of the cells. Over many generations, genetic information supporting the association would have become widespread.

Chapter 2 Solutions

1. The water molecule is not perfectly tetrahedral because only two of oxygen's four electron orbitals form covalent bonds. The orbitals containing unshared electron pairs "push away" the bonding orbitals so that they form an angle slightly less than 109°.

2. Because the partial negative charges are arranged symmetrically (and the shape of the molecule is linear), the molecule as a whole is not polar.

$$\overset{\delta^-}{O}=\overset{\delta^+}{C}=\overset{\delta^-}{O}$$

3. Ammonia is polar because it has one unshared electron pair. Its shape is trigonal pyramidal, and the molecule is not symmetrical. Nitrogen is more electronegative than hydrogen, so partial negative charges reside on the nitrogen and partial positive charges on the hydrogens.

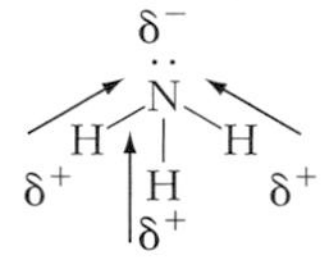

4. Water has the highest melting point because each water molecule forms hydrogen bonds with four neighboring water molecules, and hydrogen bonds are among the strongest intermolecular forces. Ammonia is also capable of forming hydrogen bonds, but they are not as strong (due to the electronegativity difference between hydrogen and nitrogen). Methane cannot form hydrogen bonds; the molecules are attracted to their neighbors only via weak London dispersion forces.

5.

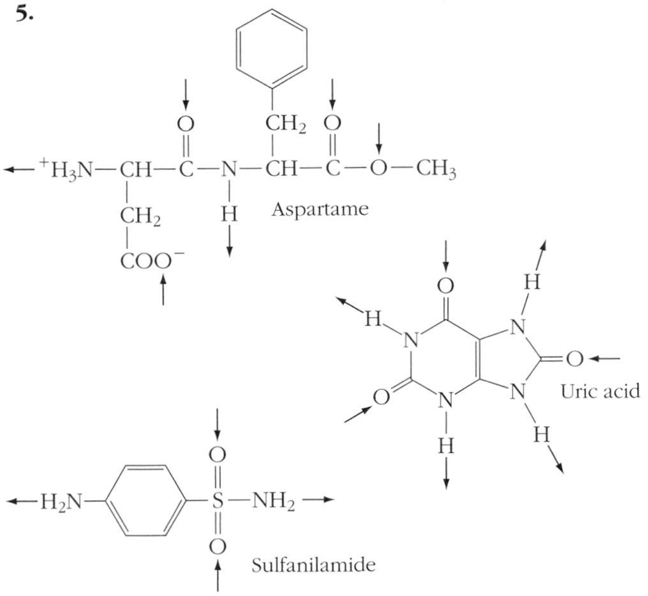

Aspartame

Uric acid

Sulfanilamide

6. Compound A does not form hydrogen bonds (the molecule has a hydrogen bond acceptor but no hydrogen bond donor). Compounds B and C form hydrogen bonds as shown because each molecule contains at least one hydrogen bond donor and a hydrogen bond acceptor. The molecules in D do not form hydrogen bonds with each other, whereas the molecules in E do because ammonia has a hydrogen bond donor and diethyl ether has a hydrogen bond acceptor.

(B)

(C) $H_3C-CH_2-O-H \cdots H_3C-CH_2-O-H$

(E)

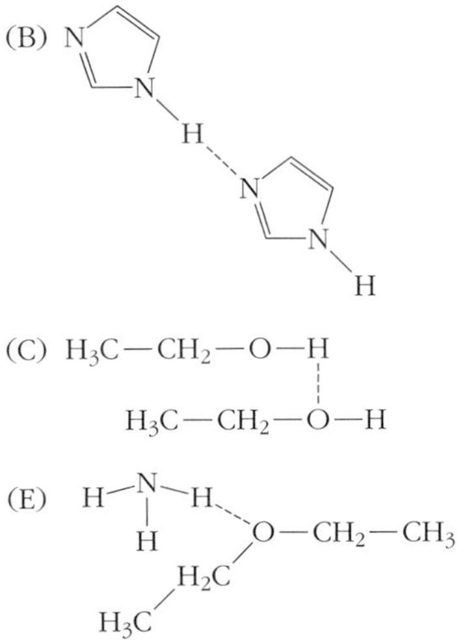

7. (a) Van der Waals forces (dipole–dipole interactions)
(b) Hydrogen bonding
(c) Van der Waals forces (London dispersion forces)
(d) Ionic interactions

8. Aquatic organisms that live in the pond are able to survive the winter. Since the water at the bottom of the pond remains in the liquid form instead of freezing, the organisms are able to move around. The ice on top of the pond also serves as an insulating layer from the cold winter air.

9. From most soluble (most polar) to least soluble (least polar): C, B, E, A, D. Compound C has three functional groups that can serve as hydrogen donors and/or acceptors. Compound B has one less NH_2 group than Compound C and is therefore less polar. Compound E has one functional group that can serve as a hydrogen bond acceptor, but has no donors. Compound A also has one functional group that can serve as a hydrogen bond acceptor (and has no donors) but has more hydrocarbon groups than compound E. Compound D is nonpolar and does not have the capability of forming hydrogen bonds and is the least water-soluble compound in the group.

10.

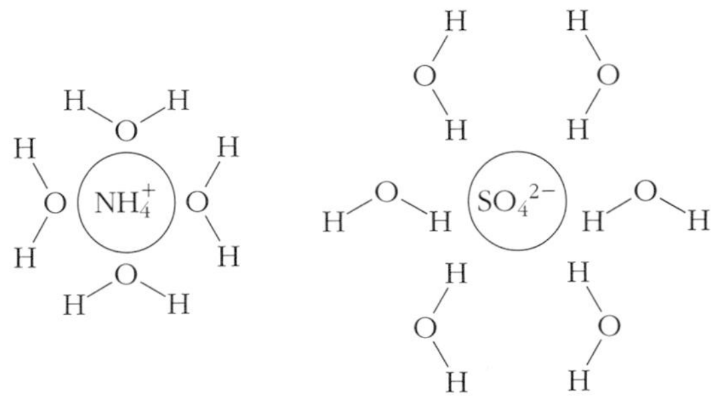

NH_4^+ SO_4^{2-}

The positively charged ammonium ion is surrounded by a shell of water molecules that are oriented so that their partially negatively charged oxygen atoms interact with the positive charge on the ammonium ion. Similarly, the negatively charged sulfate ion is hydrated with water molecules oriented so that the partially positively charged hydrogen atoms interact with the negative charge on the sulfate anion. (Not shown in the diagram is the fact that the ammonium ions outnumber the sulfate ions by a 2:1 ratio. Also note that the exact number of water molecules shown is unimportant.)

11. Structure A depicts a polar compound while structure B depicts an ionic compound similar to a salt like sodium chloride. This is more consistent with glycine's physical properties as a white crystalline solid with a high melting point. While structure A could be water soluble because of its ability to form hydrogen bonds, the high solubility of glycine in water is more consistent with an ionic compound whose positively and negatively charged groups are hydrated in aqueous solution by water molecules.

12. (a) First, calculate the number of moles of protein using Avogadro's number:

$$1000 \text{ molecules} \times \frac{1 \text{ mole}}{6.02 \times 10^{23} \text{ molecules}} = 1.66 \times 10^{-21} \text{ moles}$$

Next, calculate the volume of the cell, expressing r in centimeters:

$$\text{volume} = \frac{4\pi r^3}{3} = \frac{4\pi(5 \times 10^{-5}\text{cm})^3}{3} = 5.24 \times 10^{-13} \text{ cm}^3$$

Since 1 cm^3 = 1 mL, the volume is 5.24×10^{-13} mL or 5.24×10^{-16} L. Therefore, the concentration of the protein is

$$\frac{1.66 \times 10^{-21} \text{ moles}}{5.24 \times 10^{-16} \text{ L}} = 3.17 \times 10^{-6} \text{ M or } 3.17 \ \mu\text{M}$$

(b) $$\frac{5 \times 10^{-3} \text{ moles}}{\text{L}} \times \frac{6.02 \times 10^{23} \text{ molecules}}{\text{mole}} \times 5.24 \times 10^{-16} \text{ L} = 1.6 \times 10^{6} \text{ molecules}$$

13. (a) A 30X dilution is equivalent to multiplying one-tenth (10^{-1}) by itself 30 times: $(10^{-1})^{30} = 10^{-30}$. The concentration would be 10^{-30} M.

(b) Use Avogadro's number and multiply the concentration by the volume to show that there is much less than one molecule present in 1 mL:

$$(0.001 \text{ L})(10^{-30} \text{ moles/L})(6.022 \times 10^{23} \text{ molecules/mole}) = 6.022 \times 10^{-10} \text{ molecules}$$

(c) The ability of water molecules to form a hydrogen-bonded coating or cage around a solute molecule, particularly a hydrophobic one, might support the idea of water's memory. However, water molecules are constantly in motion, so a group of water molecules that have been in contact with a solute do not retain an imprint of it.

14. Compound A is amphiphilic and has a polar "head" and a nonpolar "tail" as indicated and can form a micelle (see Fig. 2-10). Compound B is nonpolar and cannot form a micelle or a bilayer. Compound C is polar (ionic) and cannot form a micelle or a bilayer. Compound D is amphiphilic and has a polar "head" and two nonpolar "tails" as indicated and can form a bilayer (see Fig. 2-11). Compound E is polar and cannot form a micelle or a bilayer.

(A) [$H_3C{-}(CH_2)_{11}$] (Nonpolar tail) — [$N^+(CH_3)_2{-}CH_2COO^-$] ← Polar head

(D) Polar head: [$CH_2{-}O{-}C(=O)$ / $HC{-}O{-}C(=O)$ / $HO{-}CH_2$] — Nonpolar tails: [$(CH_2)_{11}{-}CH_3$], [$(CH_2)_{11}{-}CH_3$]

15. (a) In the nonpolar solvent, AOT's polar head group faces the interior of the micelle, and its nonpolar tails face the solvent.

(b) The protein, which contains numerous polar groups, interacts with the polar AOT groups in the micelle interior.

Nonpolar tails: [$H_3C{-}(CH_2)_3{-}CH(CH_2CH_3){-}CH_2$], [$H_3C{-}(CH_2)_3{-}CH(CH_2CH_3){-}CH_2$] — Polar head: [$O{-}C(=O){-}CH_2$; $O{-}C(=O){-}CH{-}S(=O)_2{-}O^-$]

AOT

Isooctane

Water

Protein

16. (a) Nonpolar tail → [$H_3C{-}(CH_2)_{11}$] [$O{-}S(=O)_2{-}O^-Na^+$] ← Polar head

(b) Micelle of sodium dodecyl sulfate: eight $H_3C{-}(CH_2)_{11}O{-}SO_2{-}O^-Na^+$ molecules arranged with CH_3 tails pointing toward the center and O^-Na^+ sulfate heads on the outside.

(c) The hydrophobic grease can move into the hydrophobic core of the water-soluble soap micelle. The "dissolved" grease can then be washed away with the micelle.

17. (a) It is doubtful that the contents of the ball could influence the behavior of external water molecules separated by layers of rubber and plastic.

(b) Even if water clusters were disrupted (which they are not), the removal of dirt requires more than one individual water molecule. In order for a dirt molecule to be washed away, it must be surrounded (solubilized) by many water molecules.

(c) Hot water, because of the higher energy of its water molecules, has intrinsically better dirt-solubilizing power than cold water, regardless of the presence or absence of detergent. In the absence of detergent, hot water has significant cleaning power on its own, which could be attributed to the presence of a laundry ball.

18. The waxed car is a hydrophobic surface. To minimize its interaction with the hydrophobic molecules (wax), each water drop minimizes its surface area by becoming a sphere (the geometrical shape with the lowest possible ratio of surface to volume). Water does not bead on glass, because the glass presents a hydrophilic surface with which the water molecules can interact. This allows the water to spread out.

19. (a) The nonpolar core of the lipid bilayer helps prevent the passage of water since the polar water molecules cannot easily penetrate the hydrophobic core of the bilayer.

(b) Most human cells are surrounded by a fluid containing about 150 mM Na^+ and slightly less Cl^- (see Fig. 2-13). A solution containing 150 mM NaCl mimics the extracellular fluid and therefore helps maintain the isolated cells in near-normal conditions. If the cells were placed in pure water, water would tend to enter the cells by osmosis; this might cause the cells to burst.

20. Substances present at high concentration move to an area of low concentration spontaneously, or "down" a concentration gradient in a process that increases their entropy. The export of Na^+ ions out of the cell requires that the sodium ions be transported from an area of low concentration to an area of high concentration. The same is true for potassium transport. Thus, these processes are not spontaneous, and an input of cellular energy is required to accomplish the transport.

21. CO_2 is nonpolar and would be able to cross a bilayer. Glucose is polar and would not be able to pass through a bilayer because the presence of the hydroxyl groups means glucose is highly hydrated and would not be able to pass through the nonpolar "tails" of the molecules forming the bilayer. DNP is nonpolar and would be able to cross a bilayer. Calcium ions are charged and are, like glucose, highly hydrated and would not be able to cross a lipid bilayer.

22. The amount of Na^+ (atomic weight 23 g/mol) lost in 15 minutes, assuming a fluid loss rate of 2 L per hour and a sweat Na^+ concentration of 50 mM, is

$$0.25\ \text{h} \times \frac{2\ \text{L}}{\text{h}} \times \frac{0.05\ \text{mol}}{\text{L}} \times \frac{23\ \text{g}}{\text{mol}} = 0.575\ \text{g or } 575\ \text{mg}$$

It would take 2.9 ounces of potato chips (about a handful) to replace the lost sodium ions.

23. (a) In a high-solute medium, the cytoplasm loses water; therefore its volume decreases. In a low-solute medium, the cytoplasm gains water and therefore its volume increases.

(b) *E. coli* accumulates water when grown in low-osmolarity medium. However, regulation of only water content would cause a large increase in cytoplasmic volume. To avoid this large increase in volume, *E. coli* also exports K^+ ions. The opposite occurs when *E. coli* is grown in high-osmolarity medium—the cytoplasmic water content is decreased, but cytoplasmic osmolarity increases as *E. coli* imports K^+ ions.

24. Since the molecular mass of H_2O is 18.0 g · mol^{-1}, a given volume (for example, 1 L or 1000 g) has a molar concentration of 1000 g · L^{-1} ÷ 18.0 g · mol^{-1} = 55.5 M. By definition, a liter of water at pH 7.0 has a hydrogen ion concentration of 1.0×10^{-7} M. Therefore the ratio of $[H_2O]$ to $[H^+]$ is 55.5 M/(1.0×10^{-7} M) = 5.55×10^8.

25. In aqueous solution, where virtually all biochemical reactions take place, an extremely strong acid such as HCl dissociates completely, so that all its protons are donated to water: $HCl + H_2O \rightarrow H_3O^+ + Cl^-$. This leaves H_3O^+ as the only acidic species remaining.

26. Since pH = $-\log[H^+]$, $[H^+] = 10^{-pH}$.

For saliva, $[H^+] = 10^{-6.6} = 2.5 \times 10^{-7}$ M

For urine, $[H^+] = 10^{-4.5} = 3.2 \times 10^{-5}$ M

27.

	Acid, base, or neutral?	pH	$[H^+]$(M)	$[OH^-]$(M)
A	acid	5.60	2.5×10^{-6}	4.0×10^{-9}
B	base	7.65	2.2×10^{-8}	4.5×10^{-7}
C	neutral	7.00	1.0×10^{-7}	1.0×10^{-7}
D	acid	2.68	2.1×10^{-3}	4.8×10^{-12}

28. (a) The final concentration of HNO_3 is

$$\frac{(0.020\ \text{L})(1.0\ \text{M})}{0.520\ \text{L}} = 0.038\ \text{M}$$

Since HNO_3 is a strong acid and dissociates completely, the added $[H^+]$ is equal to $[HNO_3]$. (The existing hydrogen ion concentration in the water itself, 1.0×10^{-7} M, can be ignored because it is much smaller than the hydrogen ion concentration contributed by the nitric acid.)

$$\text{pH} = -\log[H^+]$$
$$\text{pH} = -\log(0.038)$$
$$\text{pH} = 1.4$$

(b) The final concentration of KOH is

$$\frac{(0.015\ \text{L})(1.0\ \text{M})}{0.515\ \text{L}} = 0.029\ \text{M}$$

Since KOH dissociates completely, the added $[OH^-]$ is equal to the [KOH]. (The existing hydroxide ion concentration in the water itself, 1.0×10^{-7} M, can be ignored because it is much smaller than the hydroxide ion concentration contributed by the KOH.)

$$K_w = 1.0 \times 10^{-14} = [H^+][OH^-]$$
$$[H^+] = \frac{1.0 \times 10^{-14}}{[OH^-]}$$
$$[H^+] = \frac{1.0 \times 10^{-14}}{(0.029\ \text{M})}$$
$$[H^+] = 3.4 \times 10^{-13}\ \text{M}$$
$$\text{pH} = -\log[H^+]$$
$$\text{pH} = -\log(3.4 \times 10^{-13})$$
$$\text{pH} = 12.5$$

29. (a) The final concentration of HCl is

$$\frac{(0.0015\ \text{L})(3.0\ \text{M})}{1.0015\ \text{L}} = 0.0045\ \text{M}$$

Since HCl is a strong acid and dissociates completely, the added $[H^+]$ is equal to [HCl]. (The existing hydrogen ion concentration in the water itself, 1.0×10^{-7} M, can be ignored because it is much smaller than the hydrogen ion concentration contributed by the hydrochloric acid.)

$$pH = -\log[H^+]$$
$$pH = -\log(0.0045)$$
$$pH = 2.35$$

(b) The final concentration of NaOH is

$$\frac{(0.0015\ L)(3.0\ M)}{1.0015\ L} = 0.0045\ M$$

Since NaOH dissociates completely, the added $[OH^-]$ is equal to the [NaOH]. (The existing hydroxide ion concentration in the water itself, 1.0×10^{-7} M, can be ignored because it is much smaller than the hydroxide ion concentration contributed by the NaOH.)

$$K_w = 1.0 \times 10^{-14} = [H^+][OH^-]$$
$$[H^+] = \frac{1.0 \times 10^{-14}}{[OH^-]}$$
$$[H^+] = \frac{1.0 \times 10^{-14}}{(0.0045\ M)}$$
$$[H^+] = 2.2 \times 10^{-12}\ M$$
$$pH = -\log[H^+]$$
$$pH = -\log(2.2 \times 10^{-12})$$
$$pH = 11.6$$

30. The HCl is a strong acid and dissociates completely. This means that the concentration of hydrogen ions contributed by the HCl is 1.0×10^{-9} M. But the concentration of the hydrogen ions contributed by the dissociation of water is 100-fold greater than this: 1.0×10^{-7} M. The concentration of the hydrogen ions contributed by the HCl is negligible in comparison. Therefore, the pH of the solution is equal to 7.0.

31. (a) $C_2O_4^{2-}$
(b) SO_3^{2-}
(c) HPO_4^{2-}
(d) CO_3^{2-}
(e) AsO_4^{3-}
(f) PO_4^{3-}
(g) O_2^{2-}

32. (a) $H_2C_2O_4$
(b) H_2SO_3
(c) H_3PO_4
(d) H_2CO_3
(e) $H_2AsO_4^-$
(f) $H_2PO_4^-$
(g) H_2O_2

33. (a) $H_3PO_4 + H_2O \rightleftharpoons H_2PO_4^- + H_3O^+$
$H_2PO_4^- + H_2O \rightleftharpoons HPO_4^{2-} + H_3O^+$
$HPO_4^{2-} + H_2O \rightleftharpoons PO_4^{3-} + H_3O^+$

(b) The pK values are read as midpoints on the titration curve. The value for pK_1 is estimated at 2.1, pK_2 = 7, and pK_3 = 12.5. (The actual literature values are 2.15, 6.82, and 12.38, respectively.) The K_a values (using the estimated pK values) are 7.9×10^{-3}, 1.0×10^{-7}, and 3.2×10^{-13} (recall that p$K = -\log K_a$, so $K_a = 10^{-pK}$).

(c)

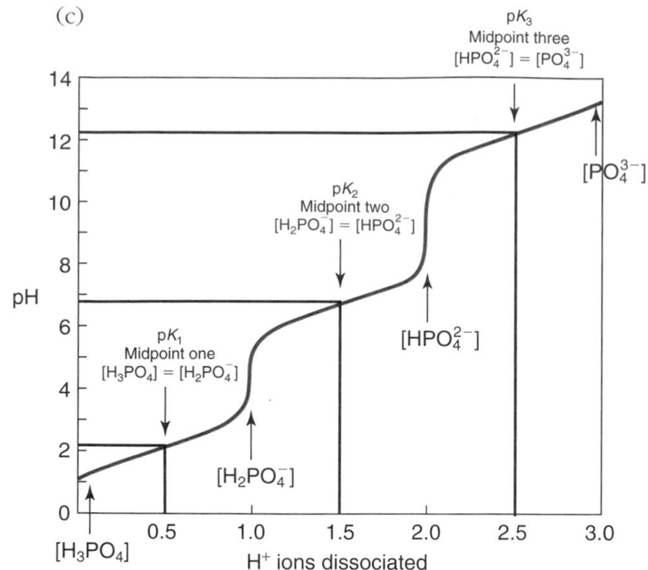

(d) The dissociation of the second proton has a pK of 6.82 and is the closest to physiological pH. Therefore, a buffer solution at pH = 7 would consist of the weak acid $H_2PO_4^-$ and its conjugate base, HPO_4^{2-} (each supplied as the sodium salts NaH_2PO_4 and Na_2HPO_4).

34.

CH_2-COOH
$HO-C-COOH$
CH_2-COOH

Citric acid

Piperidine

$^+H_2N-CH(-C(=O)-OH)-CH_2-CH_2-CH_2-CH_2-NH_2^+$

Lysine

$HO-C(=O)-C(=O)-OH$

Oxalic acid

Barbiturate

$O(C_4H_8)NH^+-CH_2-CH_2-SO_2-O^-$

4-Morphine ethanesulfonic acid (MES)

35. Convert all the data to either K_a or pK values to evaluate (p$K = -\log K_a$). The greater the K_a value, the stronger the acid, that is, the greater the tendency for the proton to be donated. (The lower the pK value, the stronger the acid.) From strongest to weakest acid: E, D, B, A, C. Note that the stronger the acid, the weaker its conjugate base. For example, citric acid is a stronger acid than citrate, and succinic acid is a stronger acid than succinate.

	Acid	K_a	pK
A	citrate	1.74×10^{-5}	4.76
B	succinic acid	6.17×10^{-5}	4.21
C	succinate	2.29×10^{-6}	5.64
D	formic acid	1.78×10^{-4}	3.75
E	citric acid	7.41×10^{-4}	3.13

36. pH 2 $\quad {}^{+}H_3N—CH_2—COOH$
pH 7 $\quad {}^{+}H_3N—CH_2—COO^-$
pH 10 $\quad H_2N—CH_2—COO^-$

The carboxylic acid group has a pK of 2.35, and the amino group has a pK of 9.78. The Henderson–Hasselbalch equation can be used to calculate the exact percentage of protonated/unprotonated forms of each functional group, but that really isn't necessary. Instead, the pK values for each group should be compared to the pH. At pH = 2, the pH is below both pK values, so both functional groups are mostly protonated. At pH = 7, the pH is well above the pK for the carboxylic acid group but below the pK for the amino group. Therefore the carboxylic acid group is unprotonated and the amino group is protonated. At pH = 10, the pH is above the pK values of both functional groups. Thus, both groups are mostly unprotonated.

37. The stomach contents have a low pH due to the contribution of gastric juice (pH 1.5). When the partially digested material enters the small intestine, the addition of pancreatic juice (pH 8.0) neutralizes the acid and increases the pH.

38. The final concentrations of the weak acid ($H_2PO_4^-$) and conjugate base (HPO_4^{2-}) are the following (note that K^+ is a spectator ion):

$$[H_2PO_4^-] = \frac{(0.025\ L)(2.0\ M)}{0.200\ L} = 0.25\ M$$

$$[HPO_4^{2-}] = \frac{(0.050\ L)(2.0\ M)}{0.200\ L} = 0.50\ M$$

Next, substitute these values into the Henderson–Hasselbalch equation using the pK value for the weak acid given in Sample Calculation 2-5:

$$\text{pH} = \text{p}K + \log\frac{[A^-]}{[HA]}$$
$$\text{pH} = 6.82 + \log(0.50\ M)/(0.25\ M)$$
$$\text{pH} = 6.82 + 0.30$$
$$\text{pH} = 7.12$$

39. Use the rearranged Henderson–Hasselbalch equation as shown in Sample Calculation 2-4 to isolate the [HA] term:

$$\log[HA] = \log[A^-] - \text{pH} + \text{p}K$$

The concentration of acetate is 0.20 M, since sodium acetate is a salt that dissociates completely in solution and the mole ratio of sodium acetate to acetate is 1:1. Substitute in the known quantities: the concentration of acetate (A^-), the desired pH, and the pK of acetic acid (from Table 2-1):

$$\log[HA] = \log(0.20\ M) - 5.0 + 4.76$$
$$\log[HA] = -0.70 - 5.0 + 4.76$$
$$\log[HA] = -0.94$$
$$[HA] = 10^{-0.94} = 0.115\ M$$

Next, calculate the volume of stock 6.0 M acetic acid that would need to be added to the solution to achieve a final concentration of 0.115 M:

$$\frac{(0.115\ M)(0.500\ L)}{(6.0\ M)} = 0.0096\ L = 9.6\ mL$$

40. Adding NaOH to the acetic acid will convert some of the acetic acid to acetate (note that spectator ions are not shown):

$$OH^- + CH_3COOH \rightleftharpoons CH_3COO^- + H_2O$$

The ratio of OH^- to CH_3COO^- is 1:1, so the number of moles of OH^- added to the solution will result in an equivalent number of moles of acetate produced ($[A^-] = [OH^-]$).

Use a similar strategy outlined in Sample Calculation 2-4 to rearrange the Henderson–Hasselbalch equation, this time isolating the $[A^-]$ term:

$$\text{pH} = \text{p}K + \log\frac{[A^-]}{[HA]}$$
$$\log\frac{[A^-]}{[HA]} = \text{pH} - \text{p}K$$
$$\log[A^-] - \log[HA] = \text{pH} - \text{p}K$$
$$\log[A^-] = \log[HA] + \text{pH} - \text{p}K$$

Substitute in the known quantities: the concentration of acetic acid (HA), the desired pH, and the pK of acetic acid (from Table 2-1):

$$\log[A^-] = \log(0.20\ M) + 5.0 - 4.76$$
$$\log[A^-] = -0.70 + 5.0 - 4.76$$
$$\log[A^-] = -0.46$$
$$[A^-] = 10^{-0.46} = 0.35\ M$$

Next, calculate the volume of stock 6.0 M sodium hydroxide that would need to be added to the solution to achieve a final concentration of 0.35 M:

$$\frac{(0.35\ M)(0.500\ L)}{(6.0\ M)} = 0.029\ L = 29\ mL$$

41. (a) 10 mM glycinamide buffer, because its pK is closer to the desired pH.
(b) 20 mM Tris buffer, because the higher the concentration of the buffering species, the more acid or base it can neutralize.
(c) Neither; each solution will contain an equilibrium mixture of the boric acid and its conjugate base (borate).

42. (a)

$$HO—(H_2C)_2—NH^+ \quad N—(CH_2)_2—SO_3^- + H_2O \rightleftharpoons$$

Weak acid (HA)

$$HO—(H_2C)_2—N \quad N—(CH_2)_2—SO_3^- + H_3O^+$$

Conjugate base (A)

(b) The pK for HEPES is 7.55; therefore its effective buffering range is 6.55–8.55.

(c) $1.0 \text{ L} \times \frac{0.10 \text{ mol}}{\text{L}} \times \frac{260.3 \text{ g}}{\text{mol}} = 26 \text{ g}$

Weigh 26 g of the HEPES salt and add to a beaker. Dissolve in slightly less than 1.0 liter of water (leave "room" for the NaOH solution that will be added in the next step).

(d) Adding NaOH to the HEPES will convert some of the weak acid to the conjugate base form of HEPES. The ratio of OH^- to conjugate base is 1:1, so the number of moles of OH^- added to the solution will result in an equivalent number of moles of conjugate base produced ($[A^-] = [OH^-]$).

Use a similar strategy outlined in Sample Calculation 2-4 to rearrange the Henderson–Hasselbalch equation, this time isolating the $[A^-]$ term:

$$\text{pH} = \text{p}K + \log\frac{[A^-]}{[HA]}$$

$$\log\frac{[A^-]}{[HA]} = \text{pH} - \text{p}K$$

$$\log[A^-] - \log[HA] = \text{pH} - \text{p}K$$

$$\log[A^-] = \log[HA] + \text{pH} - \text{p}K$$

Substitute in the known quantities: the concentration of HEPES (HA), the desired pH, and the pK of HEPES (from Table 2-2):

$$\log[A^-] = \log(0.10 \text{ M}) + 8.0 - 7.55$$

$$\log[A^-] = -1.0 + 8.0 - 7.55$$

$$\log[A^-] = -0.55$$

$$[A^-] = 10^{-0.55} = 0.28 \text{ M}$$

Next, calculate the volume of stock 6.0 M sodium hydroxide that would need to be added to the solution to achieve a final concentration of 0.28 M:

$$\frac{(0.28 \text{ M})(1.0 \text{ L})}{(6.0 \text{ M})} = 0.047 \text{ L} = 47 \text{ mL}$$

Add 47 mL to the 1.0 liter of 0.1 M HEPES prepared. Verify the pH with a pH meter and adjust if necessary. Dilute to a final volume of 1.0 liter.

43. (a)

Weak acid	Conjugate base
$HOH_2C{-}C(CH_2OH)_2{-}NH_3^+$	$HOH_2C{-}C(CH_2OH)_2{-}NH_2$

(b) The pK of Tris is 8.30; therefore its effective buffering range is 7.30–9.30.

(c) Use the Henderson-Hasselbalch equation to determine the ratio of conjugate base to weak acid:

$$\text{pH} = \text{p}K + \log\frac{[A^-]}{[HA]}$$

$$8.20 = 8.30 + \log\frac{[A^-]}{[HA]}$$

$$-0.1 = \log\frac{[A^-]}{[HA]}$$

$$\frac{0.79}{1.0} = \frac{[A^-]}{[HA]}$$

$0.79/1.79 = 44\%$ of the buffer or $0.44(0.1 \text{ M}) = 0.044 \text{ M} = [A^-]$

$1.0/1.79 = 56\%$ of the buffer or $0.56(0.1 \text{ M}) = 0.056 \text{ M} = [HA]$

(d) First, determine the $[H^+]$ contributed by the HCl (which dissociates completely):

$$[H^+] = [HCl] = \frac{(0.0015 \text{ L})(3.0 \text{ M})}{(1.0015 \text{ L})} = 0.0045 \text{ M}$$

The acid will convert some of the conjugate base to weak acid. Therefore the new concentrations are:

$$[A^-] = 0.044 \text{ M} - 0.0045 \text{ M} = 0.0395 \text{ M}$$

$$[HA] = 0.056 \text{ M} + 0.0045 \text{ M} = 0.0605 \text{ M}$$

The new pH is determined by substituting the new concentrations of H^- and HA into the Henderson–Hasselbalch equation:

$$\text{pH} = \text{p}K + \log\frac{[A^-]}{[HA]}$$

$$\text{pH} = 8.3 + \log\frac{(0.0395)}{(0.0605)}$$

$$\text{pH} = 8.3 + (-0.185)$$

$$\text{pH} = 8.1$$

The buffer has been effective: The pH has declined about 0.1 unit (from pH = 8.2 to pH = 8.1) with the addition of the strong acid. In comparison, the addition of the same amount of acid to water, which is not buffered, resulted in a pH change from approximately 7.0 to 2.35 (see Problem 29a).

(e) First, determine the $[OH^-]$ contributed by the NaOH (which dissociates completely):

$$[OH^-] = [NaOH] = \frac{(0.0015 \text{ L})(3.0 \text{ M})}{(1.0015 \text{ L})} = 0.0045 \text{ M}$$

The base will convert some of the weak acid to conjugate base. Therefore the new concentrations are:

$$[A^-] = 0.044 \text{ M} + 0.0045 \text{ M} = 0.0485 \text{ M}$$

$$[HA] = 0.056 \text{ M} - 0.0045 \text{ M} = 0.0515 \text{ M}$$

The new pH is determined by substituting the new concentrations of H^- and HA into the Henderson–Hasselbalch equation:

$$\text{pH} = \text{p}K + \log\frac{[A^-]}{[HA]}$$

$$\text{pH} = 8.3 + \log\frac{(0.0485)}{(0.0515)}$$

$$\text{pH} = 8.3 + (-0.026)$$

$$\text{pH} = 8.27$$

The buffer has been effective: The pH has increased only 0.07 unit (from pH = 8.2 to pH = 8.27) with the addition of the strong base. In comparison, the addition of the same amount of base to water, which is not buffered, resulted in a pH change from approximately 7.0 to 11.6 (see Problem 29b).

44. (a) Use the Henderson–Hasselbalch equation to determine the ratio of conjugate base to weak acid:

$$\text{pH} = \text{p}K + \log\frac{[A^-]}{[HA]}$$

$$2.0 = 8.30 + \log\frac{[A^-]}{[HA]}$$

$$-6.3 = \log\frac{[A^-]}{[HA]}$$

$$\frac{5.0 \times 10^{-7}}{1.0} = \frac{[A^-]}{[HA]}$$

Virtually all of the Tris is in the weak acid form. Therefore, the concentration of the weak acid, HA, is 0.1 M and the concentration of the conjugate base, A^-, is 5.0×10^{-8} M.

(b) First, determine the $[H^+]$ contributed by the HCl (which dissociates completely):

$$[H^+] = [HCl] = \frac{(0.0015\ L)(3.0\ M)}{(1.0015\ L)} = 0.0045\ M$$

In an effective buffer, the acid would convert some of the conjugate base to weak acid. But the concentration of conjugate base is already negligible. Therefore, the concentration of the additional acid added, 0.0045 M, should be added to the concentration of hydrogen ions already present (1.0×10^{-2} M), for a total concentration of 0.0145 M.

$$pH = -\log[H^+] = \log(0.0145\ M) = 1.84$$

The buffer has not functioned effectively. There was not enough conjugate base to react with the additional hydrogen ions added. The result is a decrease in pH from 2.0 to 1.84.

(c) First, determine the $[OH^-]$ contributed by the NaOH (which dissociates completely):

$$[OH^-] = [NaOH] = \frac{(0.0015\ L)(3.0\ M)}{(1.0015\ L)} = 0.0045\ M$$

The base will convert some of the weak acid to conjugate base. Therefore the new concentrations are:

$$[A^-] = 5.0 \times 10^{-8}\ M + 0.0045\ M = 0.0045$$
$$[HA] = 0.10\ M - 0.0045\ M = 0.0955$$

The new pH is determined by substituting the new concentrations of H^- and HA into the Henderson–Hasselbalch equation:

$$pH = pK + \log\frac{[A^-]}{[HA]}$$
$$pH = 8.3 + \log\frac{(0.0045)}{(0.0955)}$$
$$pH = 8.3 + (-1.33)$$
$$pH = 6.97$$

Tris is not an effective buffer at pH = 2.0, a pH more than 6 units lower than its pK value. Virtually all of the Tris is in the weak acid form at this pH. If acid is added, there is not enough base to absorb the excess added hydrogen ions, and the pH decreases. If base is added, some of the weak acid is converted to the conjugate base and the pH approaches the value of the pK.

45. Because it is small and nonpolar (see Solution 2-2), CO_2 can quickly diffuse across cell membranes to exit the tissues and enter red blood cells.

46. $H^+(aq) + HCO_3^-(aq) \rightleftharpoons H_2CO_3(aq) \rightleftharpoons H_2O(l) + CO_2(aq)$

Failure to eliminate CO_2 in the lungs would cause a buildup of $CO_2(aq)$. This would shift the equilibrium of the above equations to the left. The increase in $CO_2(aq)$ would lead to the increased production of carbonic acid, which would in turn dissociate to form additional hydrogen ions, causing acidosis.

47. $H^+(aq) + HCO_3^-(aq) \rightleftharpoons H_2CO_3(aq) \rightleftharpoons H_2O(l) + CO_2(aq)$

Loss of CO_2 through the shell would shift the above equations to the right. Carbonic acid in the egg would produce more water and CO_2 to make up for the loss of CO_2. This in turn would cause additional hydrogen ions and bicarbonate ions to form more carbonic acid. The loss of hydrogen ions would result in an increased pH of the contents of the egg.

48. (a) $H^+(aq) + HCO_3^-(aq) \rightleftharpoons H_2CO_3(aq) \rightleftharpoons H_2O(l) + CO_2(aq)$

Mechanical hyperventilation removes CO_2 from the patient's lungs. This would cause the above equations to shift to the right. Carbonic acid in the blood would produce more water and CO_2 to make up for the loss of CO_2. This in turn would cause additional hydrogen ions and bicarbonate ions to form more carbonic acid. The loss of hydrogen ions would result in an increased pH, bringing the patient's pH back to normal.

(b) $H^+(aq) + HCO_3^-(aq) \rightleftharpoons H_2CO_3(aq) \rightleftharpoons H_2O(l) + CO_2(aq)$

Adding HCO_3^- would cause the equations above to shift to the right. The additional bicarbonate would combine with hydrogen ions to form carbonic acid. The additional carbonic acid would dissociate to form water and carbon dioxide. This helps alleviate the acidosis because the bicarbonate combines with excess hydrogen ions, thus decreasing the hydrogen ion concentration and increasing the pH. However, it is not acceptable for use in patients with ALI, because of the increased production of aqueous CO_2 in the blood. The CO_2 produced would need to be exhaled in the lungs, which would be difficult in patients with ALI.

(c) Tris becomes protonated to form its conjugate base form. This removes H^+ from circulation and increases the pH back to normal. The protonated form of Tris is excreted in the urine. This method of acidosis treatment does not involve exhalation of CO_2 and is therefore an acceptable treatment for patients with ALI.

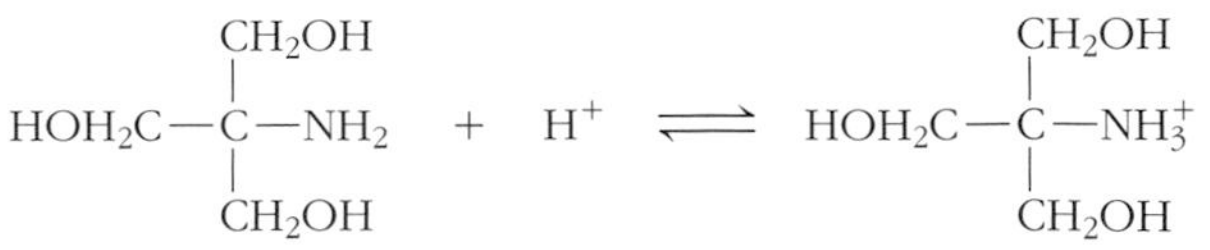

$$HOH_2C{-}C(CH_2OH)_2{-}NH_2 + H^+ \rightleftharpoons HOH_2C{-}C(CH_2OH)_2{-}NH_3^+$$

49. (a)

Acetylsalicylic acid (aspirin) + H_2O $\xrightarrow{\text{esterase}}$ Salicylic acid + $CH_3{-}COOH$ (Acetic acid)

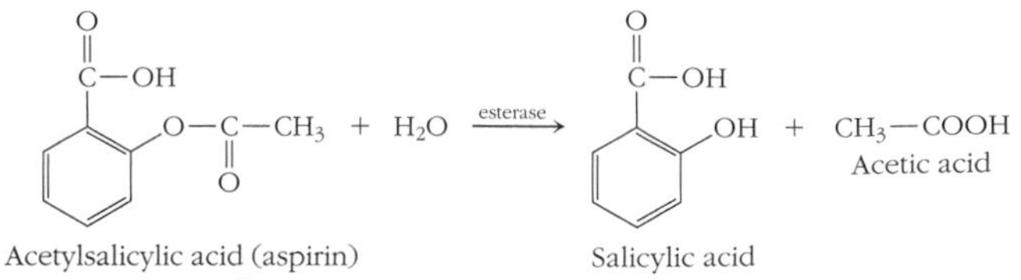

(b) These questions can be answered by using the Henderson–Hasselbalch equation:

$$\text{pH} = \text{p}K + \log\frac{[\text{salicylate}]}{[\text{salicylic acid}]}$$

$$2.0 = 3.0 + \log\frac{[\text{salicylate}]}{[\text{salicylic acid}]}$$

$$0.1 = \frac{[\text{salicylate}]}{[\text{salicylic acid}]}$$

At pH = 2, the percentage of salicylate (unprotonated) is 9% (0.1/1.1) and the percentage of salicylic acid (protonated) is 91% (1/1.1).

(c) $$\text{pH} = \text{p}K + \log\frac{[\text{salicylate}]}{[\text{salicylic acid}]}$$

$$8.5 = 3.0 + \log\frac{[\text{salicylate}]}{[\text{salicylic acid}]}$$

$$10^{5.5} = \frac{[\text{salicylate}]}{[\text{salicylic acid}]}$$

At pH = 8.5, virtually 100% of the salicylate is in the unprotonated form. The gastric lavage increases the solubility of the drug, because at this pH, the salicylate is unprotonated and negatively charged. At the pH of the stomach, pH = 2, about 90% of the salicylic acid is in the protonated, or uncharged, form. Negatively charged species are generally more water soluble than neutral species, so the gastric lavage increased the solubility of the drug. This facilitates the removal of the aspirin from the stomach.

(d) The patient experiences salicylate-induced hyperventilation, which means that carbon dioxide is being rapidly removed from the lungs. The patient's laboratory values show a low $p\text{CO}_2$ value and a high $p\text{O}_2$ value, indicating that oxygen is being taken in and carbon dioxide is being exhaled at a greater rate than normal.

(e) The carbonic acid/bicarbonate buffering system relies on these equilibria:

$$\text{H}^+(aq) + \text{HCO}_3^-(aq) \rightleftharpoons \text{H}_2\text{CO}_3(aq) \rightleftharpoons \text{H}_2\text{O}(l) + \text{CO}_2(aq)$$

The removal of CO_2 from the lungs shifts the equilibrium to the right. This causes $CO_2(aq)$ to be depleted, so the equilibrium shifts so that more carbonic acid dissociates into water and carbon dioxide, in an attempt to replenish the carbon dioxide. This depletes the carbonic acid, so the equilibrium shifts right to produce more carbonic acid. The result is that hydrogen ions are depleted and the blood becomes more basic. This is verified by looking at the laboratory values, which show that the patient's blood pH after 10 hours of aspirin ingestion is 7.55. This also causes bicarbonate ions to be depleted. This is why the bicarbonate concentration in the patient is lower than normal.

(f) The ratio of bicarbonate to carbonic acid in the patient's blood:

$$\text{pH} = \text{p}K + \log\frac{[\text{bicarbonate}]}{[\text{carbonic acid}]}$$

$$7.55 = 6.4 + \log\frac{[\text{bicarbonate}]}{[\text{carbonic acid}]}$$

$$14 = \frac{[\text{bicarbonate}]}{[\text{carbonic acid}]}$$

The ratio of bicarbonate to carbonic acid in normal blood:

$$\text{pH} = \text{p}K + \log\frac{[\text{bicarbonate}]}{[\text{carbonic acid}]}$$

$$7.4 = 6.4 + \log\frac{[\text{bicarbonate}]}{[\text{carbonic acid}]}$$

$$10 = \frac{[\text{bicarbonate}]}{[\text{carbonic acid}]}$$

In order to serve as an effective buffer, the concentration of the conjugate base to weak acid should range from 0.1/1 to 10/1 to ensure that there is some of each species present. Ratios lying outside of this range have an abundance of either the conjugate base or the weak acid alone. In the patient, the ratio of conjugate base to weak acid does not lie within the effective buffering range. The bicarbonate concentration is too high relative to the carbonic acid concentration. But it is not just the ratio of conjugate base to weak acid that is important—the *absolute concentration* of each is also important. The concentration of HCO_3^- can be calculated from the concentration of H_2CO_3 using the Henderson–Hasselbalch equation:

$$\text{pH} = \text{p}K + \log\frac{[\text{bicarbonate}]}{[\text{carbonic acid}]}$$

$$7.55 = 6.4 + \log\frac{2.1 \times 10^{-3}}{[\text{carbonic acid}]}$$

$$[\text{H}_2\text{CO}_3] = 1.49 \text{ mM}$$

The carbonic acid concentration in the patient is 1.49 mM. We can infer that in a normal patient, the carbonic acid concentration is 2.2–2.6 mM, since there is a 10:1 ratio between bicarbonate and carbonic acid. Thus both the carbonic acid and bicarbonate concentrations are lower than normal in our patient. If the concentration of buffering species is low, then the ability of the buffer to work effectively is compromised.

(g) As the salicylate is removed, the stimulus for salicylate-induced hyperventilation will decrease as a result. In the basic blood, OH^- reacts with the H^+ to form water. This depletes the H^+ so that carbonic acid dissociates to form more hydrogen and bicarbonate ions. This in turn depletes H_2CO_3, so water and carbon dioxide react to form more carbonic acid. This depletes $CO_2(aq)$, which results in shallow breathing. The shallow breathing results in a greater concentration of aqueous carbon dioxide in the blood, which can ultimately produce more hydrogen ions, which will bring the blood pH back to normal.

Chapter 3 Solutions

1. The heat treatment destroys the polysaccharide capsule of the wild-type *Pneumococcus,* but the DNA survives the heat treatment. The DNA then "invades" the mutant *Pneumococcus* and supplies the genes encoding the enzymes needed for capsule synthesis, which the mutant lacks. The mutant is now able to synthesize a capsule

and has the capacity to cause disease, which results in the death of the mice and the appearance of encapsulated *Pneumococcus* in the mouse tissue.

2. These experiments showed that the "transforming factor" was neither a protein nor RNA.

3. Some of the "parent" DNA label appears in the progeny, but none of the protein label appears in the progeny. This indicates that the DNA is involved in the production of progeny bacteriophages, but protein is not required.

4. The total amount of purines (A + G) in DNA must equal the total amount of pyrimidines (C + T) because each base pair in the double-stranded DNA molecule consists of a purine and a pyrimidine. This is not true for RNA, which is single-stranded.

5. The organism must also contain 19% C (since [A] = [T] according to Chargaff's rules) and 62% C + G (or 31% C and 31% G, since [C] = [G]). Each cell is a diploid, containing 60,000 kb or 6×10^7 bases. Therefore,

$$[A] = [T] = (0.19)(6 \times 10^7 \text{ bases}) = 1.14 \times 10^7 \text{ bases}$$
$$[C] = [G] = (0.31)(6 \times 10^7 \text{ bases}) = 1.86 \times 10^7 \text{ bases}$$

6. It is a T:A base pair.

T A

7.

Cytosine Inosine

Adenine Inosine

Uracil Inosine

8. The high pH eliminates hydrogen bonds between bases, making it easier to separate the strands of DNA.

9. The statement is false because the greater stability of GC-rich DNA is due to the stronger stacking interactions involving G:C base pairs and does not depend on the number of hydrogen bonds in the base pairs.

10. It is certainly the case that hydrogen bonds hold A:T and G:C base pairs together and that these interactions are very favorable. But upon denaturation of the DNA, each nitrogenous base has the opportunity to form equally favorable hydrogen bonds with water. Therefore, forces other than hydrogen bonds must contribute to the overall stability of the DNA molecule.

11.

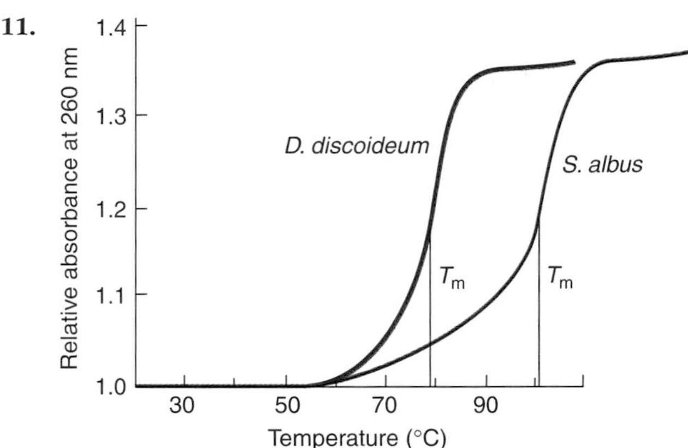

12. The DNA from the organisms that thrive in hot environments would contain more G and C than DNA from species living in a more temperate environment. The higher GC content increases the stability of DNA at high temperatures.

13. The positively charged sodium ions can form ion pairs with the negatively charged phosphate groups on the DNA backbone and "shield" the negative charges from one another. This increases the overall stability of DNA and makes it more difficult to melt.

14. You should increase the temperature to melt out imperfect matches between the probe and the DNA.

15. (a) An inherited characteristic could be determined by more than one gene.
(b) Some sequences of DNA encode RNA molecules that are not translated into protein (for example, rRNA and tRNA).
(c) Some genes are not transcribed during a cell's lifetime. This can occur if the gene is expressed only under certain environmental conditions or in certain specialized cells in a multicellular organism.

16. The DNA isolated after one generation is a homogeneous sample of DNA with a density intermediate between DNA containing all ^{14}N and all ^{15}N. The DNA isolated after the second generation is heterogeneous. Half of the DNA has the same density as the first generation; half of the DNA consists of all ^{14}N DNA and has a lower density.

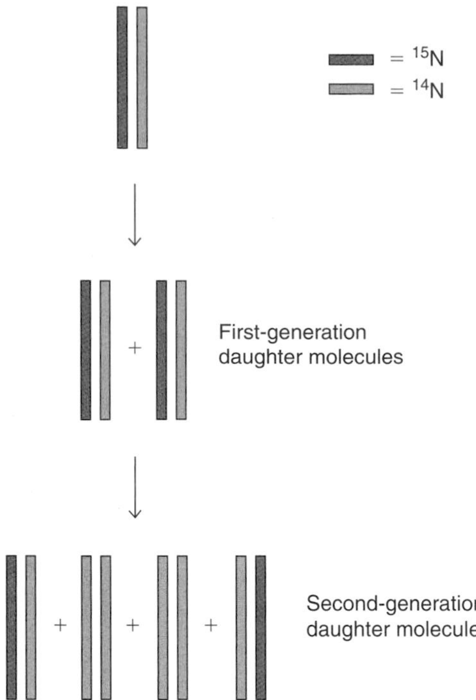

17. A restriction endonuclease is often used to prepare fragments of DNA for insertion into a cloning vector. Since the cloned DNA contains the recognition site (whose sequence is known), this sequence can be used as a starting point to sequence the unknown DNA segment.

18. The polymerization reaction must be carried out at high temperatures (hence the need for a heat-stable DNA polymerase) in order to ensure that the template DNA remains an unknotted single strand. The high temperature is necessary to melt GC-rich DNA, which is more stable than AT-rich DNA.

19. The number of possible sequences of four different nucleotides is 4^n where n is the number of nucleotides in the sequence. Therefore: (a) $4^1 = 4$, (b) $4^2 = 16$, (c) $4^3 = 64$, and (d) $4^4 = 256$.

20. First, identify the translation start site, the Met residue whose codon is AUG in the mRNA (see Table 3-3) or ATG in the DNA. Translation stops at the DNA sequence TAA, which corresponds to the stop codon UAA in the mRNA. Use Table 3-3 to decode the intervening codons, substituting U for T.

```
GTAATTCAAA ATG CCT TAC GCC CCT GGA GAC
           Met Pro Tyr Ala Pro Gly Asp

GAA AAG AAG GGT GCT ATT ACG TAT TTG AAG
Glu Lys Lys Gly Ala Ile Thr Tyr Leu Lys

AAG GCC ACC TCT GAG TAA ATGTGAC
Lys Ala Thr Ser Glu STOP
```

21. The genetic code (shown in Table 3-3) is redundant. Since there are 64 different possibilities for 3-base codons and only 20 amino acids, some amino acids have more than one codon. If a mutation just happens to occur in the third position (3′end), the mutation might not alter the protein sequence. For example, GUU, GUC, GUA, and GUG all code for valine. A mutation in the third position of a valine codon would still result in the selection of valine and would have no effect on the amino acid sequence of the protein.

22. *Msp*I, *Asu*I, *Eco*RI, *Pst*I, *Sau*I, and *Not*I generate sticky ends. *Alu*I and *Eco*RV generate blunt ends.

23. The restriction enzyme with the longer recognition sequence would be a rare cutter, because it is likely to encounter this sequence less often and therefore will cleave the DNA less frequently than a restriction enzyme with a shorter recognition sequence.

24.

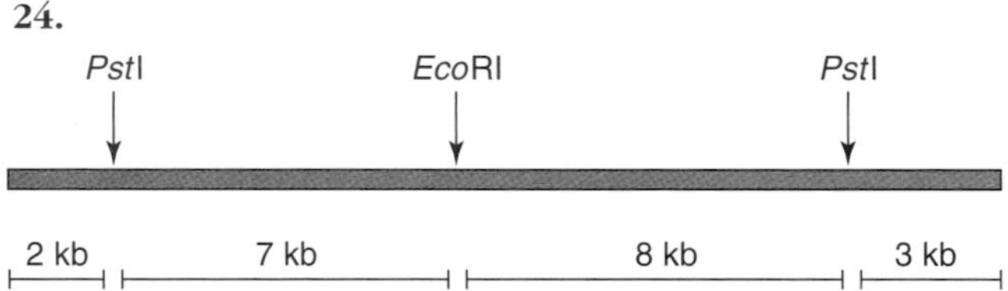

25. You can use a DNA polymerase that is not heat stable. You would have to cool the reaction mixture to a temperature at which the polymerase works best and you would have to add the enzyme at each reaction cycle because it would be lost every time the temperature was raised to melt the double-stranded DNA.

26. If you were to use plasmids, you would need at least 150,000 clones to accommodate the 3×10^6 kb genome in fragments of 20 kb each (see Table 3-5). This is an almost unmanageable number. However, if you were to use yeast artificial chromosomes, you would need only a minimum of 3000 clones to accommodate the 3×10^6 kb genome in fragments of 1000 kb.

27. To design an oligonucleotide probe for a gene, the researcher must apply the genetic code in reverse, that is, select codons that correspond to the amino acids in the protein. Most amino acids can be encoded by more than one codon, so the researcher would have to choose one of them and hope that it matched the DNA well enough for the probe to successfully hybridize with the DNA. Met and Trp, however, are encoded by only one codon each, so by using these codons as part of the probe, the researcher can be assured of a perfect match with the DNA, at least for these three nucleotides.

28. To amplify the protein-coding DNA sequence, the primers should correspond to the first three and last three residues of the protein (each amino acid represents three nucleotides, so the primers would each be nine bases long). Use Table 3-3 to find the codons that correspond to the first three residues:

Met	Pro	Tyr
AUG	CCU	UAU
	CCC	UAC
	CCA	
	CCG	

Using just the topmost set of codons, a possible DNA primer would therefore have the sequence 5′-ATGCCT-TAT-3′. This primer could base-pair with the gene's noncoding strand, and its extension from its 3′ end would yield a copy of the coding strand of the gene (see Fig. 3-19). The other primer must correspond to the last three amino acids of the protein:

Thr	Ser	Glu
ACU	UCU	GAA
ACC	UCC	GAG
ACA	UCA	
ACG	UCG	
	AGU	
	AGC	

Again, considering just the topmost set of codons, a probable DNA coding sequence would be 5′-ACTTCTGAA-3′. This sequence cannot be used as a primer. However, a suitable primer would be the complementary sequence 5′-TTCAGAAGT-3′, which can then be extended from its 3′ end to yield a copy of the noncoding strand of the gene. The number of possible primer pairs is quite large, because all but one of the amino acids has more than one codon. For the first primer, there are $1 \times 4 \times 2 = 8$ possibilities, and for the second, $4 \times 6 \times 2 = 48$ possibilities. There are $8 \times 48 = 384$ different pairs of primers that could be used to amplify the gene by PCR.

29. The same segment of DNA can encode two different proteins if each strand is a coding strand.

30. The presence of an Alu sequence between two cleavage sites for a particular restriction endonuclease would result in a restriction fragment that was increased in size by the length of the Alu sequence. Individuals with the Alu sequence would generate a larger restriction fragment than individuals without that Alu sequence.

Chapter 4 Solutions

1. This combination cannot occur at any pH. An unprotonated amino group cannot exist with a protonated carboxylate group because the amino group's pK is much greater than the carboxyl group's pK (therefore the carboxyl group ionizes at a lower pH than the amino group).

2. (a) His, Phe, Pro, Tyr, Trp
(b) His, Phe, Tyr, Trp
(c) His, Cys, Ser, Thr, Tyr
(d) Gly
(e) Arg, Lys
(f) Asp, Glu
(g) Cys, Met

3. Histones contain an abundance of the positively charged amino acids lysine and arginine. The positive charges of these amino acid side chains interact electrostatically to form ion pairs with the negatively charged phosphate groups on the backbone of the DNA molecule and to minimize charge–charge repulsion of the negatively charged phosphate groups.

4. The polypeptide would be even less soluble than free Tyr, because the amino and carboxylate groups that interact with water and make Tyr soluble are lost in forming the peptide bonds in poly(Tyr).

5. In a free amino acid, the charged amino and carboxylate groups, which are separated only by the alpha carbon, electronically influence each other. When the amino acid forms a peptide bond, one of these groups is neutralized, thereby altering the electronic properties of the remaining group.

6. (a) The chiral carbon is marked with an asterisk. D-Alanine, the mirror image isomer, is shown.

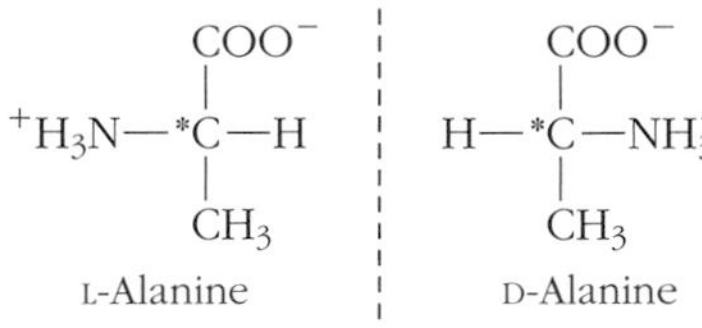

COO^- | ^+H_3N—*C—H | CH_3 (L-Alanine)

COO^- | H—*C—NH_3^+ | CH_3 (D-Alanine)

(b) Since the majority of proteins contain L-amino acids, the presence of D-amino acids in the bacterial cell wall renders the cell wall less susceptible to digestion by proteases (enzymes produced by certain organisms to destroy bacteria).

7. From least soluble to most soluble: Phe, Ile, Trp, Ser, His, Asp. You can use Table 4-4 as a guide, but you should also be able to do this type of problem without using the table.

8. (a) The three amino acids are Ser, Tyr, and Gly.
(b) Cyclization of the polypeptide backbone occurs between the carbonyl carbon of Ser and the amide nitrogen of Gly.
(c) Oxidation results in a double bond in the Tyr side chain between Cα and Cβ (the second carbon of the side chain).

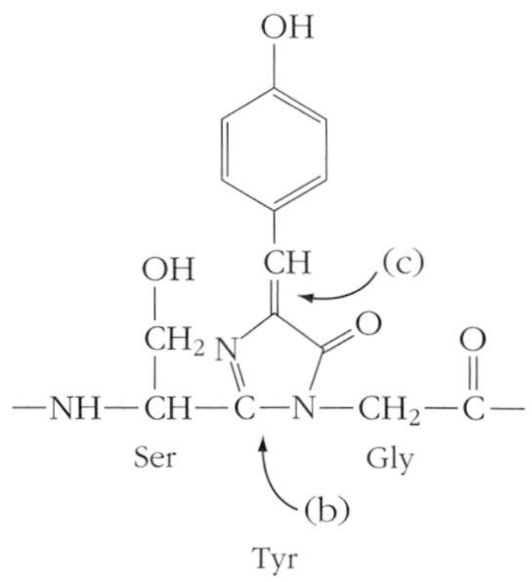

OH
CH (c)
OH
CH_2 N O O
—NH—CH—C—N—CH_2—C—
Ser Gly
(b)
Tyr

9. A polypeptide is a single polymer of amino acids. A protein may consist of one or more polypeptide chains.

10.

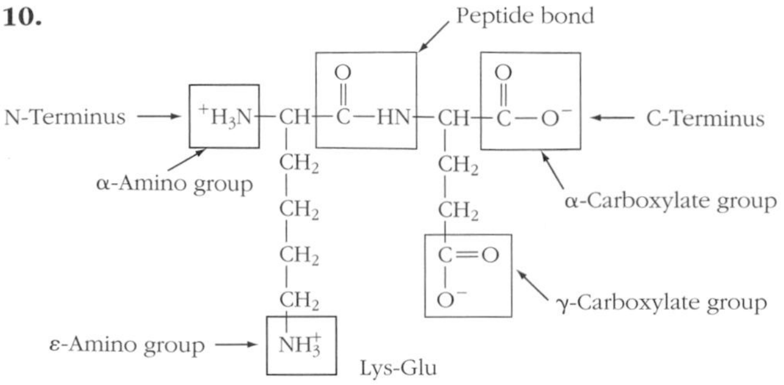

Peptide bond

N-Terminus → $^{+}H_3N$–CH–C(=O)–HN–CH–C(=O)–O^- ← C-Terminus

α-Amino group

α-Carboxylate group

γ-Carboxylate group

ε-Amino group → NH_3^+

Lys-Glu

11.

$^{+}H_3N$–CH(CH_2–C(=O)–O^-)–C(=O)–NH–CH(CH_2–C_6H_5)–C(=O)–OCH_3

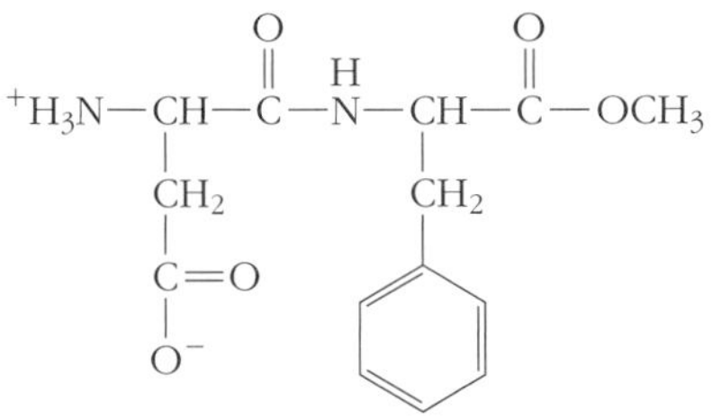

Aspartame (Asp–Phe–OMe)

12.

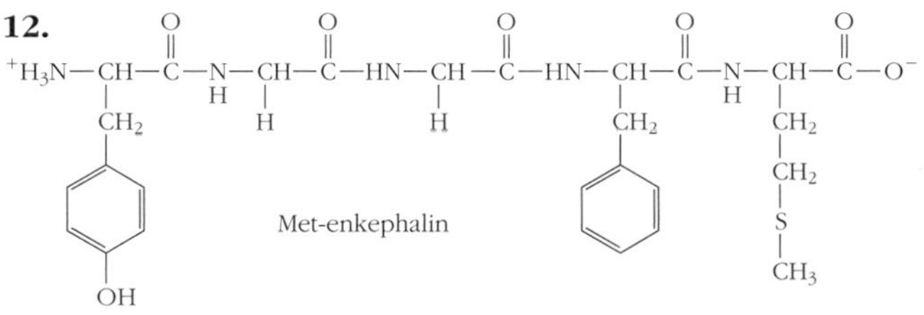

Met-enkephalin

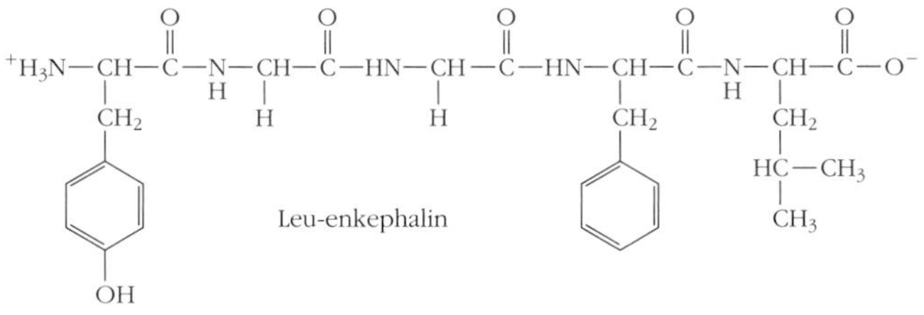

Leu-enkephalin

13. (a)

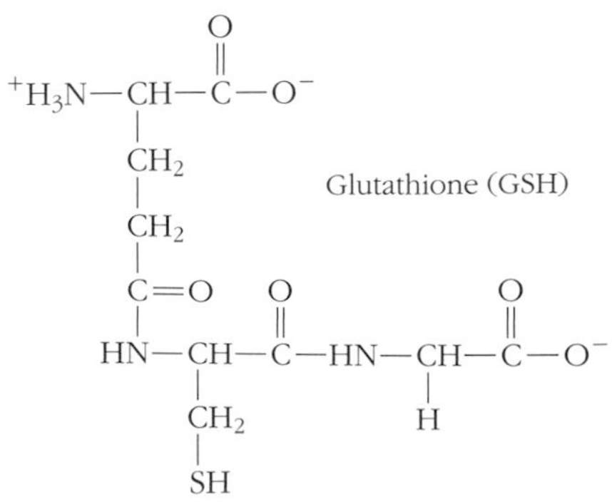

Glutathione (GSH)

(b)

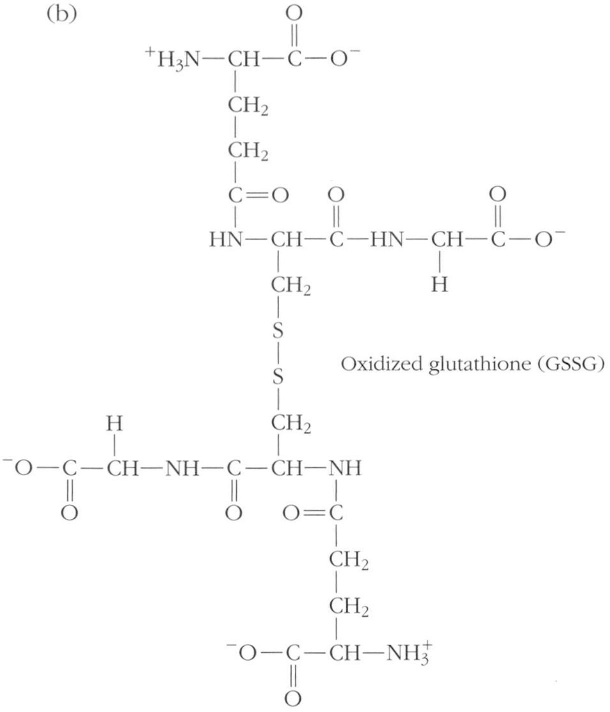

Oxidized glutathione (GSSG)

14. At pH 7.0, the peptide likely has a net positive charge since Arg (R) and Lys (K) outnumber Asp (D) and Glu (E). Therefore, the peptide is likely to bind to CM groups but not to DEAE groups.

15. The amino terminal residue is Ala. Chymotrypsin cleaves after Tyr, Phe, and Trp. Fragment II contains the Ala, so it appears in the sequence first, and Tyr must be the cleavage site. Trypsin cleaves after Lys and Arg. Since Fragment III contains Ala, it appears in the sequence before Fragment IV, and Arg must be the cleavage site (since Fragment I terminates with Arg). The positions of several of the amino acids can be identified by overlap. Elastase cleaves after Gly and Val (not after Ala since Ala is followed by Pro), so Fragment V must be the C-terminal fragment.

Elastase cleavage ↓ (Gly–Asp)   Trypsin cleavage ↓ (Arg–Ile)   Elastase cleavage ↓ (Val–His)

Ala—Pro—Gly—Asp—Arg—Ile—Tyr—Val—His—Pro—Phe

Chymotrypsin cleavage ↑ (Tyr–Val)

16. The cleavage site for each fragment is highlighted. A fragment not ending in Phe, Tyr, or Trp (for chymotrypsin cleavage) or Lys or Arg (for trypsin cleavage) must be the carboxyl terminal fragment. Use "overlap" to work backward from the C-terminus to determine the sequence of the polypeptide.

Chymotrypsin fragments

HSEGT**F**
SND**Y**
SK**Y**
LEDRKAQD**F**
VR**W**
LMNNKRSGAAE

Trypsin fragments
HSEGTFSNDYS**K**
YLEDR**K**
AQDFVRWLMNN**K**
RSGAAE

HSEGTFSNDYSKYLEDRKAQDFVRWLMNNKRSGAAE

17. Edman degradation of a polypeptide with a disulfide cross-link would not work properly when the first Cys became exposed at the N-terminus of the polypeptide (the Cys would not be released as a PTH derivative since it still would be covalently linked to a Cys residue further along the polypeptide chain). Reduction before sequencing breaks the disulfide bond, and alkylation of the two free Cys groups prevents re-formation of the bond.

18. Thermolysin would yield the most fragments (9) and chymotrypsin would yield the fewest (3).

19. (a)

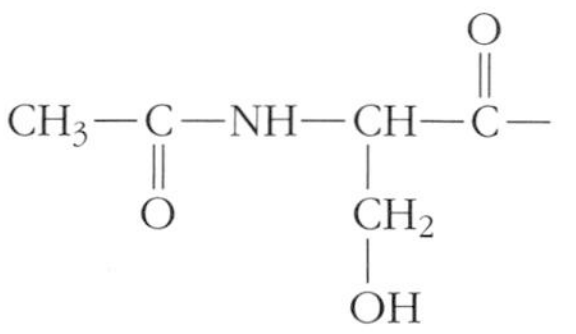

$$CH_3-\underset{\underset{O}{\|}}{C}-NH-\underset{\underset{\underset{OH}{|}}{\underset{CH_2}{|}}}{CH}-\overset{\overset{O}{\|}}{C}-$$

(b) The Edman reagent requires a primary amino group to react with. If an acetyl group is blocking the α-amino group on the N-terminus of the protein, the initial reaction with the Edman reagent cannot take place, and the fragment cannot be sequenced using this method.

(c) If only one proteolytic cleavage is carried out, the sequences of the fragments could be determined but it would not be possible to place the fragments in the proper order. If different enzymes are used to generate two sets of fragments, overlapping peptides would allow ordering of all the sequences.

(d) Trypsin cleaves after Lys and Arg residues. The fragments resulting from digestion of the heavy and light chains are shown below, identified by residue number.

Light chain	Heavy chain
6	1–7
7–9	8–18
10–11	19–31
12	32–35
13–20	36–38
21–28	39–42
29–38	43–57
	58–61
	62–71
	72–77
	78–85
	86–91

(e) Chymotrypsin would be a good choice for the second enzyme because it cleaves after Phe, Tyr, and Trp. The following fragments would be obtained:

Light chain	Heavy chain
6–13	1–63
14–25	64–83
26–38	84–91

20. The amino group of Pro is linked to its side chain (see Fig. 4-2), which limits the conformational flexibility of a peptide bond involving the amino group. The geometry of this peptide bond is incompatible with the bond angles required for a polypeptide to form an α helix.

21. Because Gly lacks a side chain, it often occurs in a protein conformation where no other residue can fit. Gly can take the place of a larger residue more easily than a larger residue, such as Val, can take its place.

22. (a) His
(b) Ser
(c) Tyr
(d) Cys
(e) Asn

23. (a) Phe. Ala and Phe are both hydrophobic, but Phe is much larger and might not fit as well in Val's place.
(b) Asp. Replacing a positively charged Lys residue with an oppositely charged Asp residue would be more disruptive.
(c) Glu. The amide-containing Asn would be a better substitute for Gln than the acidic Glu.
(d) His. The geometry of a Pro residue constrains the conformation of the polypeptide. Gly, which lacks a side chain, can adopt the same backbone conformation, but a residue with a bulkier side chain cannot.

24. There are many possible answers for this question. An example is shown for each.

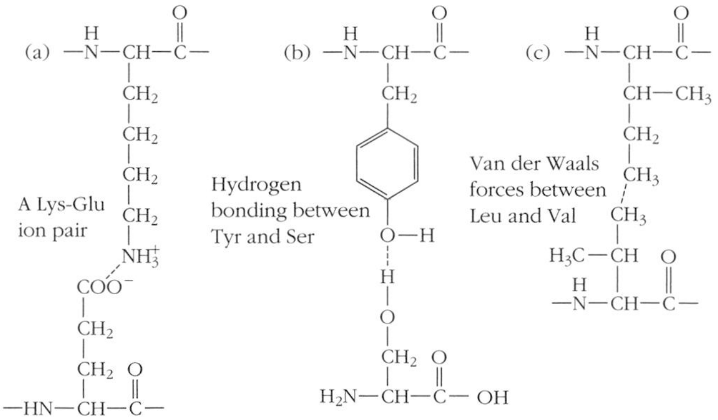

(a) $-NH-CH(CH_2CH_2CH_2CH_2NH_3^+)-CO-$ ⋯ $^-OOC-CH_2-CH_2-CH(-HN-)-CO-$
A Lys-Glu ion pair

(b) $-NH-CH(CH_2-C_6H_4-O-H)-CO-$ ⋯ $H-O-CH_2-CH(H_2N-)-C(=O)-OH$
Hydrogen bonding between Tyr and Ser

(c) $-NH-CH(CH(CH_3)-CH_2-CH_3)-CO-$ ⋯ $CH_3-CH(H_3C)-CH(-NH-)-CO-$
Van der Waals forces between Leu and Val

25. Substitution of a histidine for an arginine evidently causes a change in the three-dimensional structure of the protein, which adversely affects its function. This could occur for a variety of reasons. Histidine's side chain is composed of a five-membered ring, whereas arginine's side chain is a straight-chain structure. The change in shape of the side chain could lead to an overall change in the shape of the protein. Because of the difference in their structures (although both amino acids can form hydrogen bonds and ion pairs), the substituted histidine might not form an interaction that is crucial to the proper functioning of the protein. A "permanent" positive charge might also be necessary. Arginine has a pK of 12.5 and thus is always protonated at physiological pH. Even though the pK value of an amino acid in a protein is not necessarily the same as the pK value of the free amino acid, the great difference in the pK values indicates that the arginine is far more likely to be protonated than the histidine. If a full strong charge at this site is necessary for protein function, its replacement could result in a defective protein.

26. A polypeptide synthesized in a living cell has a sequence that has been optimized by natural selection so that it folds properly (with hydrophobic residues on the inside and polar residues on the outside). The random sequence of the synthetic peptide cannot direct a coherent folding process, so hydrophobic side chains on different molecules aggregate, causing the polypeptide to precipitate from solution.

27. (a) Heating causes a protein to "melt," or unfold, because heating increases the vibrational and rotational energy of atoms in the protein, which disrupts the weak interactions that keep the protein in its properly folded state.

(b) pH changes alter the ionization states of amino acid side chains. This affects the ability of side chains to form ion pairs. Hydrogen bonds may also be broken as protonation or deprotonation renders amino acid side chains unable to serve as hydrogen bond donors or acceptors.

(c) Detergents have a nonpolar domain that allows them to penetrate into the interior of the protein, thus interfering with the hydrophobic interactions responsible for the protein's tertiary structure.

(d) Reducing agents, such as 2-mercaptoethanol, break disulfide bonds (converting them to the —SH form) and destabilize proteins that require disulfide bonds in order to assume the correct conformation.

28. Anfinsen's ribonuclease experiment demonstrated that a protein's primary structure dictates its three-dimensional structure. Although some proteins, like ribonuclease, can renature spontaneously *in vitro,* most proteins require the assistance of molecular chaperones to fold properly *in vivo.*

29. The loss of the C chain means the loss of information that is essential to the proper folding of insulin. Removal of the C chain leaves two separate chains (A and B), and it is much more difficult for two chains to resume their native conformation than for one chain to do so. Proinsulin has no difficulty resuming its native conformation in a denaturation/renaturation experiment because proinsulin consists of only one peptide chain. When proinsulin is converted to insulin, two peptide bonds are cleaved to produce the correctly cross-linked A and B chains.

30. When the temperature increases, the vibrational and rotational energy of the atoms making up the protein molecules also increases, which increases the chance that the proteins will denature. Increasing the synthesis of chaperones under theses conditions allows the cell to renature, or refold, proteins that have been denatured by heat.

31. In a protein crystal, the residues at the end of a polypeptide chain may experience fewer intramolecular contacts and therefore tend to be less ordered (more mobile in the crystal). If their disorder prevents them from generating a coherent diffraction pattern, it may be impossible to map their electron density.

32. Protein loops are often at the protein surface, whereas regular secondary structure predominates in the protein core. The loops are better able to accommodate changes in size and amino acid composition than the core segments, where a change in size or composition may disrupt an α helix or β sheet that is an essential part of the protein's structure.

33. If the proteins were homodimers, they would be more likely to have two identical sites to interact with their palindromic recognition sites in the DNA. Heterodimeric proteins would likely lack the necessary symmetry. (In fact, these enzymes are homodimeric.)

34. Proteins and their component amino acids (and many other organic compounds) have inherent "handedness" and therefore cannot be interconverted through mirroring. Consider a right-handed α helix; its mirror image would be a left-handed helix, which is not found in nature.

35. (a) Position 6 (Gly) and Position 9 (Val) appear to be invariant.

(b) Conservative substitutions occur at Position 1 (Asp and Lys, both charged), Position 10 (Ile and Leu, similar in structure and hydrophobicity), and Position 2 (all uncharged bulky side chains). Positions 5 and 8 appear to tolerate some substitution.

(c) The most variable positions are 3, 4, and 7, where a variety of residues appear.

36. (a) Pig insulin is a good choice to replace human insulin. There is only one amino acid variation, at B30. Cow insulin differs at A8, A10, and B30, but these changes are conservative. Pig and cow would be the best sources because these animal products are readily available from slaughterhouses, whereas horse, dog, and rabbit are not. Chicken and duck have seven and six amino acid differences compared to human insulin and would not be good choices. The sequence of amino acids influences the tertiary structure of a protein, so it is possible that a change in the protein's sequence could alter its tertiary structure. A hormone molecule with a different shape might have difficulty interacting with insulin receptors.

(b) The pI values for all the mammalian insulins would be similar to that of human insulin, because the amino acid changes all involve neutral amino acids. Chicken insulin would be more basic (because His replaces Thr), whereas duck insulin would be more acidic (because Glu replaces Thr).

(c) The immune system produces antibodies to foreign proteins, in part because these proteins have shapes that differ from the individual's own proteins. As discussed above, a change in primary structure leads to a change in tertiary structure. Some people develop allergies and others do not because individual immune systems vary in sensitivity.

37. (a) Since each codon corresponds to an amino acid, the error rate is

$$\frac{5 \times 10^{-4} \text{ error}}{\text{residue}} \times 500 \text{ residues} = 0.25$$

About one-quarter of the polypeptides would contain a substitution.

(b) Virtually all the 2000-residue polypeptides would contain a substitution:

$$\frac{5 \times 10^{-4} \text{ error}}{\text{residue}} \times 2000 \text{ residues} = 1.0$$

38. In the arteries, nearly all the hemoglobin is oxygenated and therefore takes on the color of the Fe(II) in which the sixth coordination site is occupied by O_2. Blood that has passed through the capillaries and given up some of its oxygen contains a mixture of oxy and deoxyhemoglobin. Deoxyhemoglobin, in which the Fe(II) has only five ligands, imparts a bluish tinge to venous blood.

39. Globin lacks an oxygen-binding group and therefore cannot bind O_2. Heme alone is oxidized and therefore cannot bind O_2. The bound heme gives a protein such as myoglobin the ability to bind O_2. In turn, the protein helps prevent oxidation of the heme Fe atom.

40. His F8 and His E7 are invariant. The nitrogen in the imidazole ring of His F8 serves as one of the ligands to the iron in the heme group. His E7 forms a hydrogen bond with the bound oxygen. Both of these residues play a critical role in the structures of myoglobin and hemoglobin and are essential for the proper binding of oxygen; therefore, they are invariant.

41. Use Equation 4-4 to calculate the fractional saturation (Y) for hyperbolic binding, letting $K = 26$ torr:

$$Y = \frac{pO_2}{K + pO_2}$$

At 30 torr, $$Y = \frac{30 \text{ torr}}{26 \text{ torr} + 30 \text{ torr}} = 0.54$$

At 100 torr, $$Y = \frac{100 \text{ torr}}{26 \text{ torr} + 100 \text{ torr}} = 0.79$$

Therefore, if hemoglobin exhibited hyperbolic oxygen-binding behavior, it would be only 79% saturated in the lungs (where $pO_2 \approx 100$ torr) and would exhibit a loss of saturation of only 25% (79% − 54%) in the tissues (where $pO_2 \approx 30$ torr). Hemoglobin's sigmoidal binding behavior allows it to bind more oxygen in the lungs so that it can deliver relatively more oxygen to the tissues (for an overall change in saturation of about 40%; see Fig. 4-26).

42. The increased O_2 release is the result of the Bohr effect. The increase in $[H^+]$ promotes the shift from the oxy to the deoxy conformation of hemoglobin. The decrease in oxygen affinity improves oxygen delivery to the muscle, where it is needed.

43. (a) Asp 94 and His 146, which is protonated, form an ion pair in the deoxy form of hemoglobin. The protons are produced by cellular respiration. As described by the Bohr effect, an increase in hydrogen ion concentration favors the deoxy form of hemoglobin so that oxygen can be delivered to the tissues.

(b) The presence of the negatively charged Asp 94 increases the pK value of the imidazole ring of His and promotes the formation of the ion pair between the two side chains. The increased pK value means that the imidazole ring's affinity for protons is increased.

44. At high altitude, less oxygen is available to saturate hemoglobin in the lungs. Llama hemoglobin has lower oxygen affinity than human hemoglobin. This helps the molecule more easily give up its limited oxygen supply to the tissues.

45. (a) Oxygen transfer from the maternal circulation to the fetal circulation can be accomplished more efficiently if the fetal hemoglobin has a higher oxygen affinity than maternal hemoglobin.

(b) Curve A represents fetal hemoglobin, which has a higher oxygen affinity than adult hemoglobin (curve B). Curve A shows greater fractional saturation than curve B at any given oxygen concentration.

(c) Since the γ chain has fewer positive charges, the hemoglobin F central cavity has fewer positive charges, and BPG cannot bind to hemoglobin F as effectively as it binds to hemoglobin A (since BPG is negatively charged and forms ion pairs with the positively charged amino acid side chains in the cavity of hemoglobin A). Since BPG plays a role in decreasing oxygen binding affinity, the inability of BPG to bind to hemoglobin F means that hemoglobin F has a higher oxygen-binding affinity.

46. (a) Oxygen carrier A has a high affinity for oxygen in the lungs, and its affinity is similarly high in the tissues. This means that oxygen would not be effectively delivered to the tissues.

(b) Oxygen carrier B has a low affinity for oxygen. It would bind only a small amount in the lungs and would not have much available to deliver to the tissues.

(c) The curve drawn in between curves A and B represents a more effective oxygen carrier because it has a high oxygen affinity at the partial pressure of the lungs and a low affinity at the partial pressure of the tissues. In this manner, oxygen can be effectively delivered to the cells.

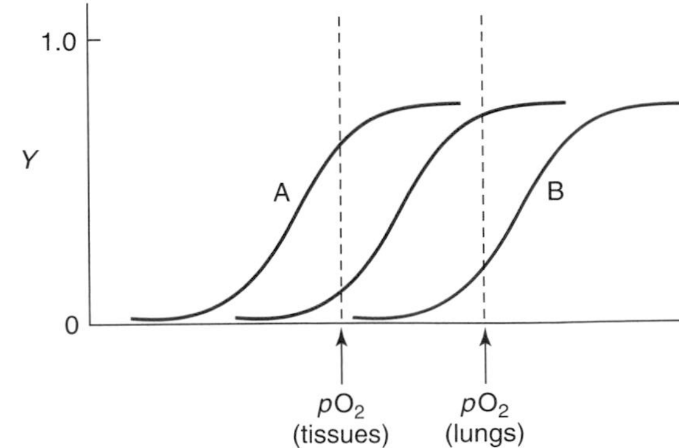

in the absence of bicarbonate, and 44 torr in the presence of bicarbonate. The higher p_{50} value in the presence of bicarbonate means that the hemoglobin has a lower affinity for oxygen.

(d) The lysine side chain is positively charged and can interact with the negative charge of the bicarbonate. The phenolate oxygens on the two tyrosine side chains can act as hydrogen bond acceptors for the bicarbonate hydrogen.

(e) The amino acids that don't directly interact with bicarbonate might cause a change in the three-dimensional conformation in such a way that binding of bicarbonate by hemoglobin is enhanced and conversion from the oxy form to the deoxy form occurs.

(f) All of the allosteric effectors are negatively charged. They bind to hemoglobin by forming ion pairs with positively charged amino acid side chains (such as Lys or Arg) or with the positively charged α-amino groups at the amino termini of the four polypeptide chains.

Chapter 5 Solutions

1. Both types of molecules are proteins and consist of polymers of amino acids. Both contain elements of secondary structure. But globular proteins are water soluble and nearly spherical in shape. Examples include proteins such as hemoglobin and myoglobin, and enzymes. Their cellular role involves participating in the chemical reactions of the cell in some way. In contrast, fibrous proteins are water insoluble and have an elongated shape. Their cellular role is structural—they form the fibers and filaments of the cytoskeleton of the cell and the connective tissue matrix between cells.

2. (a) Keratin and collagen; (b) myosin and kinesin; (c) actin and tubulin; (d) actin, myosin, tubulin, and kinesin.

3. Microfilaments and microtubules consist entirely of subunits that are assembled in a head-to-tail fashion so that the polarity of the subunits (actin monomers in microfilaments and tubulin dimers in microtubules) is preserved in the fully assembled fiber. In intermediate filament assembly, only the initial step (dimerization of parallel helices) maintains polarity. In subsequent steps, subunits align in an antiparallel fashion, so that in a fully assembled intermediate filament, each end contains heads and tails.

4. (a) Adding another protein was necessary because G-actin in solution tends to polymerize rather than form crystals. The added protein bound to actin and prevented its polymerization.
(b) The crystal structure of the protein alone was needed, so that it could be "subtracted" from the crystal structure of the actin–protein complex.

5. Myosin is both fibrous and globular. Its two heads are globular, with several layers of secondary structure. Its tail, however, consists of a single fibrous coiled coil.

6.

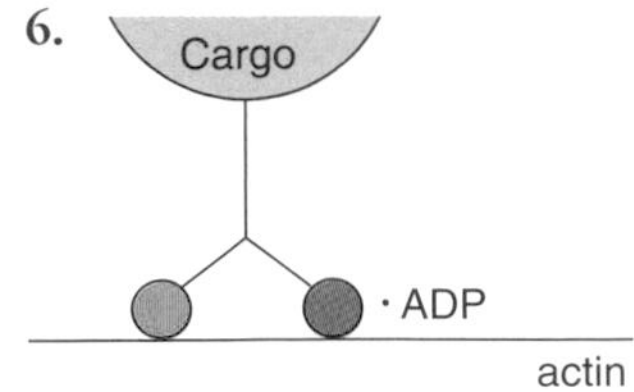

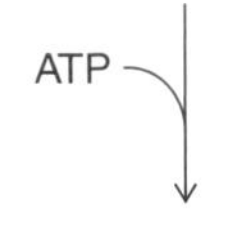

ATP binding to the trailing head induces a conformational change that causes it to release actin.

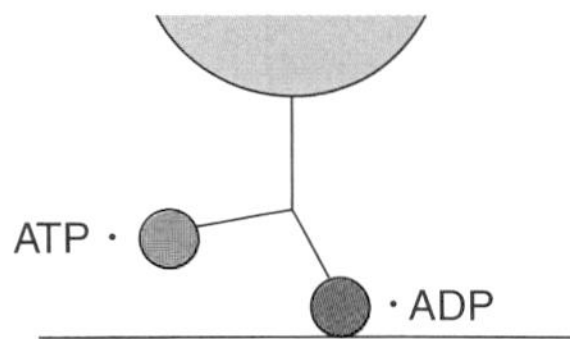

Hydrolysis of ATP to ADP + P_i triggers a conformational change that swings the trailing head forward. This also increases the affinity of the head for actin.

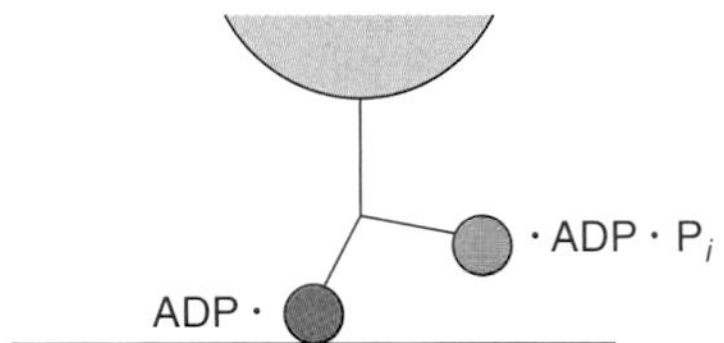

ADP

The new leading head binds to the actin filament. This causes the release of the ADP from the new trailing head.

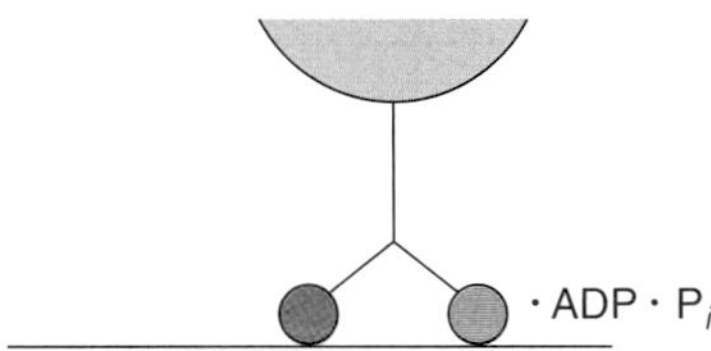

P_i

P_i dissociates from the leading head, preparing the myosin for another reaction cycle.

7. During rapid microtubule growth, β-tubulin subunits containing GTP accumulate at the (+) end because GTP hydrolysis occurs following subunit incorporation into the microtubule. In a slowly growing microtubule, the (+) end will contain relatively more GTP that has already been hydrolyzed to GDP. A protein that preferentially binds to (+) ends that contain GTP rather than GDP could thereby distinguish fast- and slow-growing microtubules.

8. Because each protofilament contains many tubulin dimers, the separation of protofilaments during fraying represents a greater loss of tubulin subunits than if the tubulin subunits left the microtubule one dimer at a time.

9. Polymers composed of β-tubulin molecules allowed to polymerize in the presence of a nonhydrolyzable analog

of GTP have increased stability. When the β-tubulin subunits are exposed to GTP in solution, the GTP binds to the β-tubulin and then is hydrolyzed to GDP, which remains bound to the β-tubulin. Additional αβ heterodimers are then added. The microtubule ends with GDP bound to the β-tubulin are less stable than those bound to GTP because protofilaments with GDP bound are curved rather than straight and tend to fray. If a nonhydrolyzable analog is bound, this will resemble GTP and the protofilament will be straight rather than curved. It is less likely to fray and the resulting protofilament is more stable as a result.

10. Although the drugs have opposite effects on microtubule dynamics, they both interfere with the normal formation of the mitotic spindle, which is required for cell division.

11. Microtubules form the mitotic spindle during cell division. Because cancer cells are rapidly dividing cells, and hence undergo mitosis at a rate more rapid than most other body cells, drugs that target tubulin and thus interfere with the formation of the mitotic spindle in some way will slow the growth of cancerous tumors.

12. Paclitaxel increases the stability of the microtubule if it is able to override the effects of GTP hydrolysis, since it is GTP hydrolysis to GDP that results in a curved protofilament that causes the microtubule to have a "frayed" appearance and increases the likelihood that the tubulin dimers in the protofilament will dissociate.

13. Colchicine, which promotes microtubule depolymerization, inhibits the mobility of the PMNs because cell mobility results from polymerization and depolymerization of microtubules.

14. Microtubules in the cell spindle are dynamic structures since the mitotic spindle must form during mitosis and degrade after cell division has occurred. Therefore these microtubule structures are less stable than those that make up the structures of the axons of nerve cells. Axonal microtubules are more stable because of their structural role.

15. (a) Diffusion is a random process. It tends to be slow (especially for large substances and over long distances). Because it is random, it operates in three dimensions (not linearly) and has no directionality.

(b) An intracellular transport system must have some sort of track (for linear movement of cargo) and an engine that moves cargo along the track by converting chemical energy to mechanical energy. The engine must operate irreversibly to promote rapid movement in one direction. Finally, some sort of addressing system is needed to direct cargo from its source to a certain destination.

16. Myosin's two heads act independently, so only one binds to the actin filament at a time. Consequently, when myosin advances from one actin subunit to another, the protein dissociates completely from its track, much like a one-legged hop. In contrast, kinesin's two heads work together so that one head remains bound to the microtubule track when the other is released. This mechanism is similar to a two-legged walk.

17. (a) The first and fourth side chains are buried in the coiled coil, but the remaining side chains are exposed to the solvent and therefore tend to be polar or charged.

(b) Although the residues at positions 1 and 4 in both sequences are hydrophobic, Trp and Tyr are much larger than Ile and Val and would therefore not fit as well in the area of contact between the two polypeptides in a coiled coil (see Fig. 5-22).

18. A fibrous protein such as keratin does not have a discrete globular core. Most of the residues in its coiled coil structure are exposed to the solvent. The exception is the strip of nonpolar side chains at the interface of the two coils.

19. The reducing conditions promote cleavage of the disulfide bonds that cross-link keratin molecules. This helps the larvae digest the wool clothing that they eat.

20. (a) Actin's primary structure is its amino acid sequence. Its secondary structure includes its α helices, β sheets, and other conformations of the polypeptide backbone. Its tertiary structure is the arrangement of its backbone and all its side chains in a globular structure. Monomeric actin by definition has no quaternary structure. However, when actin monomers associate to form a microfilament, the arrangement of subunits becomes the filament's quaternary structure. Thus, actin is an example of a protein that has quaternary structure under certain conditions.

(b) Collagen's primary structure is its amino acid sequence. Its secondary structure is the left-handed helical conformation characteristic of the Gly–Pro–Hyp repeating sequence. Its tertiary structure is essentially the same as its secondary structure, since most of the protein consists of one type of secondary structure. Collagen's quaternary structure is the arrangement of its three chains in a triple helix. It is also possible to view the triple helix as a form of tertiary structure, with quaternary structure referring to the association of collagen molecules.

21. The enzyme degrades collagen (which has a repetitive Gly–X–Y sequence with X often being Pro). Since collagen is the major protein involved in connective tissue, degradation of this tissue would facilitate the invasion of the bacterium into the host. The bacterium itself is not affected because bacteria do not contain collagen.

22. The enzyme is a collagenase. Cleaving the bond indicated will disrupt the triple helical structure of the collagen molecule and cause it to be denatured, which in effect destroys the collagen. This enzyme is necessary in the developing tadpole because the collagen that forms the structure of the tadpole's tail will need to be degraded when the tadpole develops into a frog (which has no tail).

23. (a) The Antarctic ice fish is collagen B, the shark skin is collagen C, and the chick skin is collagen A.

(b) The stability of each of these collagens is correlated with their imino acid content. The higher the percentage of hydroxyproline (imino acid content includes both proline and hydroxyproline), the more regular the structure and the more difficult it is to melt, resulting in more stable collagen. The chick has the most stable collagen of the three. The ice fish, which lives in freezing waters, has the least stable collagen of the three. But the melting temperatures of each collagen molecule are higher than the temperature at which each organism lives. Thus, each organism has stable collagen at the temperature of its environment.

Source of collagen	**Imino acid content per 1000 residues**	T_m**(°C)**
chick skin	212	41
Antarctic ice fish	143	6
shark skin	191	29

24. (a) The melting temperature, or T_m, can be used to rank the stability of the triple helices formed by the three peptides. The higher the melting temperature, the more stable the triple helix. Therefore, Peptide 1 has the highest stability, followed by Peptide 2 and then Peptide 3. The stability is correlated with imino acid content. Proline and hydroxyproline are responsible for the formation of the correct geometry of the left-handed helices of the three chains.

(b) Peptide 3 contains a great number of charged residues, both acidic and basic. Its maximum T_m is at pH = 7, when both the acidic and basic residues are charged. (At the low pH, the acidic residues are protonated and neutral, whereas at the high pH the basic residues are deprotonated and neutral.) The negatively charged acidic residues may form ion pairs with the positively charged basic residues, enhancing the stability of the triple helix at pH = 7.

(c) The T_m values of Peptide 1 do not vary as much when the pH changes because the only ionizable groups are the amino and carboxyl termini.

(d) The peptide with the greatest overall T_m values (and therefore the greatest stability) is Peptide 1, which does not have the ability to form ion pairs to as great an extent as Peptide 3. Therefore, imino acid content plays a greater role in stability than the ability of the amino acid side chains to form ion pairs.

25. (a) $(\text{Pro–Pro–Gly})_{10}$ has a melting temperature of 41°C while $(\text{Pro–Hyp–Gly})_{10}$ has a melting temperature of 60°C. $(\text{Pro–Hyp–Gly})_{10}$ and $(\text{Pro–Pro–Gly})_{10}$ both have an imino acid content of 67%, but $(\text{Pro–Hyp–Gly})_{10}$ contains hydroxyproline, whereas $(\text{Pro–Pro–Gly})_{10}$ does not. Hydroxyproline therefore has a stabilizing effect relative to proline.

(b) $(\text{Pro–Pro–Gly})_{10}$ and $(\text{Gly–Pro–Thr(Gal)})_{10}$ have the same melting point, indicating that they have equal stabilities. This is interesting because $(\text{Pro–Pro–Gly})_{10}$ has an imino acid content of 67% whereas $(\text{Gly–Pro–Thr(Gal)})_{10}$ has an imino acid content of only 33%. The glycosylated threonine must have an effect similar to that of proline. It is possible that the galactose, which contains many hydroxyl groups, provides additional sites for hydrogen bonding and would thus contribute to the stability of the triple helix.

(c) The inclusion of $(\text{Gly–Pro–Thr})_{10}$ is important because the results show that this molecule doesn't form a triple helix. This molecule is included as a control to show that the increased stability of the $(\text{Gly–Pro–Thr(Gal)})_{10}$ is due to the galactose and not to the threonine residue itself.

26. (a) The investigators chose fluorine because they were interested in looking at electronic effects on the pyrrolidine ring and fluorine is the most electronegative element that is not capable of forming hydrogen bonds.

(b) $(\text{Pro–Flp–Gly})_{10}$ is the most stable collagen, even though it doesn't contain hydroxyproline. However, it does contain a modified proline with a strongly electronegative element attached to it. This has a large inductive effect on the ring. The investigators have concluded that hydrogen bonds with bridging water molecules are not responsible for the stabilization of the collagen molecule. Instead, it is the inductive effect of the hydroxyl group. (They have suggested that the inductive effect favors the *trans* configuration of the hydroxyproline peptide bond, and that this *trans* conformation is responsible for conferring stability on the collagen molecule.)

27. Because collagen has such an unusual amino acid composition (almost two-thirds consists of Gly and Pro or Pro derivatives), it contains relatively fewer of the other amino acids and is therefore not as good a source of amino acids as proteins containing a greater variety of amino acids.

28. (a) The patient's eating habits are the cause of scurvy, since a diet lacking fresh fruits and vegetables is probably deficient in vitamin C.

(b) Ascorbic acid is necessary for the formation of hydroxyproline and hydroxylysine residues in newly synthesized collagen chains. Underhydroxylated collagen is less stable, so tissues containing the defective collagen are less sound, leading to bleeding gums, scaly skin, and fragile capillaries.

(c) In addition to minimizing stress and quitting smoking, your patient should immediately begin to include sources of vitamin C in his diet.

29. The individual collagen chains are synthesized on the ribosome from the constituent amino acids. Proline is one of the 20 amino acids for which a codon exists on the DNA. Some of the prolines are modified post-translationally (after the protein is synthesized) to hydroxyproline. Individual hydroxyproline amino acids are not incorporated into the protein during synthesis (there is no codon for hydroxyproline). So $[^{14}\text{C}]$-hydroxyproline is not used in the synthesis of collagen and no radioactivity appears in the collagen product.

30. (a) Minoxidil inhibits the lysyl hydroxylase enzyme. In the presence of minoxidil, less Lys is hydroxylated as demonstrated by the decrease of $[^{3}\text{H}]$-lysine incorporated into collagen.

(b) Since minoxidil inhibits lysyl hydroxylase, procollagen chains would be underhydroxylated. Lysines lacking hydroxyl groups cannot serve as attachment sites for sugars, which decreases the stability of the collagen and increases the likelihood that the collagen will be degraded once it is secreted from the fibroblast cell. This would be effective in reducing cellular collagen concentrations in patients with fibrosis.

(c) A similar explanation indicates that long-term minoxidil use in patients without fibrosis has the potential to compromise collagen synthesis in fibroblasts of the skin. The underhydroxylated collagen synthesized in the presence of minoxidil will be less stable, and skin structure might be affected as a result. The medical literature reports only scalp irritation, dryness, scaling, itching, and redness as a side effect in some men who received topical minoxidil treatments for nearly 2 years, however.

31. Lysyl oxidase oxidizes the side chains of lysine residues so that they can form cross-links. In the absence of this enzyme, fewer cross-links will form in collagen. As a result, the collagen produced is less stable, and produces abnormalities of the bones, joints, and blood vessels because these structures contain collagen as their main structural protein.

32. (a) The open reading frame is shown below. Collagen has the structure (Gly–X–Y)$_n$ where X is often proline and Y is often hydroxyproline. The codon for Gly is GGX and the codon for Pro is CCX (where X represents any nucleotide). In the mutant polypeptide, a Gly residue has been replaced by Asp.

Normal α2(I) procollagen gene

. . .CTGGTGCTGTTGGCCCAAGAGGTCCTAGTGGCCCAC. . .

. . .Gly Ala Val Gly Pro Arg **GLY** Pro Ser Gly Pro. . .

Mutant α2(I) procollagen gene

. . .CTGGTGCTGTTGGCCCAAGAGATCCTAGTGGCCCAC. . .

. . .Gly Ala Val Gly Pro Arg **Asp** Pro Ser Gly Pro. . .

(b) The T_m value for the mutant collagen would be lower than the T_m for normal collagen. The substitution of an Asp for a Gly would disrupt the triple helix in this region of the tropocollagen. In order for the three chains to pack together to form the triple helix, every third residue must be a Gly. There are three amino acids per turn; therefore the side chain of the Gly ends up on the interior of the triple helix. There is not sufficient room to accommodate a larger side chain of an amino acid such as Asp. The actual T_m values for normal collagen and the mutant collagen presented here are 41°C and 39°C, respectively.

(c) The patient suffered from osteogenesis imperfecta, a disease characterized by amino acid changes in Type I collagen, which is found mainly in bones and tendons. The patient's bones and tendons could not form properly due to the defects in the structure of collagen and the patient died as a result.

33. (a) The collagenase hydrolyzes the peptide bonds of collagen, thereby breaking down the meat to make it easier to chew and digest.

(b) The collagenase in the fresh fruit will degrade the gelatin (collagen) so that it cannot gel. Cooked fruit contains no collagenase activity because the heat of cooking destroys the collagenase.

34. For many proteins, bacterial expression systems offer a convenient source of protein for structural studies. However, even if collagen genes were successfully introduced into bacterial cells, the cells would not be able to produce mature collagen molecules because collagen is processed after it is synthesized. Bacteria are unable to undertake some of the processing steps, such as cleavage by extracellular proteases and covalent modification of Pro and Lys residues.

Chapter 6 Solutions

1. A globular protein can bind substrates in a sheltered active site and can support an arrangement of functional groups that facilitates the reaction and stabilizes the transition state. Most fibrous proteins are rigid and extended and therefore cannot surround the substrate to sequester it or promote its chemical transformation.

2. Myosin and kinesin are enzymes because they catalyze the hydrolysis of ATP. For myosin, the reaction is

$$\text{ATP} + H_2O \rightarrow \text{ADP} + P_i$$

For kinesin, the reaction is

$$2\ \text{ATP} + 2\ H_2O \rightarrow 2\ \text{ADP} + 2\ P_i$$

Two cycles of ATP hydrolysis are required to restore the two-headed kinesin motor to its original position.

3. (a) Pyruvate decarboxylase is a lyase. During the elimination of the carboxylate group ($—COO^-$) of pyruvate, a double bond is formed in CO_2 (O=C=O).

(b) Alanine aminotransferase is a transferase. The amino group is transferred from alanine to α-ketoglutarate.

(c) Alcohol dehydrogenase is an oxidoreductase. Acetaldehyde is reduced to ethanol or ethanol is oxidized to acetaldehyde.

(d) Hexokinase is a transferase. The phosphate group is transferred from ATP to glucose to form glucose-6-phosphate.

(e) Chymotrypsin is a hydrolase. Chymotrypsin catalyzes the hydrolysis of peptide bonds.

4. (a) Isomerase
(b) Lyase
(c) Oxidoreductase
(d) Hydrolase

5. Succinate dehydrogenase is an oxidoreductase.

$$\underset{\text{Succinate}}{{}^-OOC—CH_2—CH_2—COO^-} \xrightarrow{\text{succinate dehydrogenase}} \underset{\text{Fumarate}}{{}^-OOC—CH{=}CH—COO^-}$$

6. Malate dehydrogenase is an oxidoreductase.

$$\underset{\text{Malate}}{{}^-OOC—CH(OH)—CH_2—COO^-} \xrightarrow{\text{malate dehydrogenase}} \underset{\text{Oxaloacetate}}{{}^-OOC—C({=}O)—CH_2—COO^-}$$

7. A kinase transfers a phosphate group from ATP to a substrate.

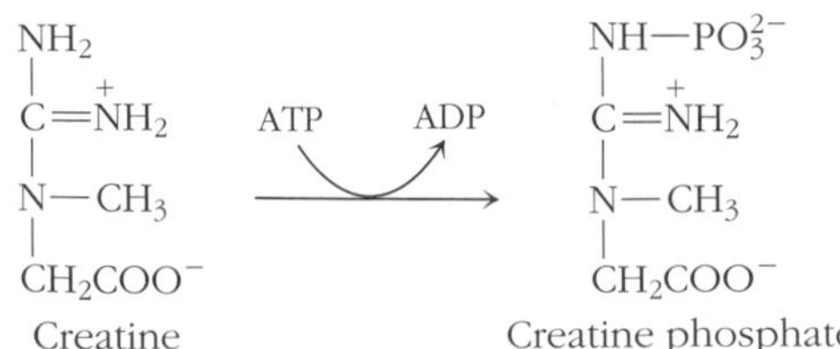

8. (a) Glucose-6-phosphate dehydrogenase
   (b) Isocitrate lyase
   (c) Phosphoglycerate kinase
   (d) Pyruvate carboxylase

9. As shown in Table 6-1, the only relationship between the rate of catalyzed and uncatalyzed reactions is that the catalyzed reaction is faster than the uncatalyzed reaction. The absolute rate of an uncatalyzed reaction does not correlate with the degree to which it is accelerated by an enzyme.

10. Every tenfold increase in rate corresponds to a decrease of about 5.7 kJ · mol^{-1} in $\Delta G^{\ddagger}$. For the nuclease, with a rate enhancement on the order of 10^{14}, $\Delta G^{\ddagger}$ is lowered about 14 × 5.7 kJ · mol^{-1}, or about 80 kJ · mol^{-1}.

11. At first, the temperature increases the reaction rate because heat increases the proportion of reacting groups that can achieve the transition state in a given time. However, when the temperature rises above a certain point, the heat causes the enzyme, which is a protein, to denature. Because most proteins are only marginally stable (see Section 4-3), denaturation occurs readily, accounting for the steep drop in enzymatic activity.

12. (a)

Fast G

Slow G

(b)

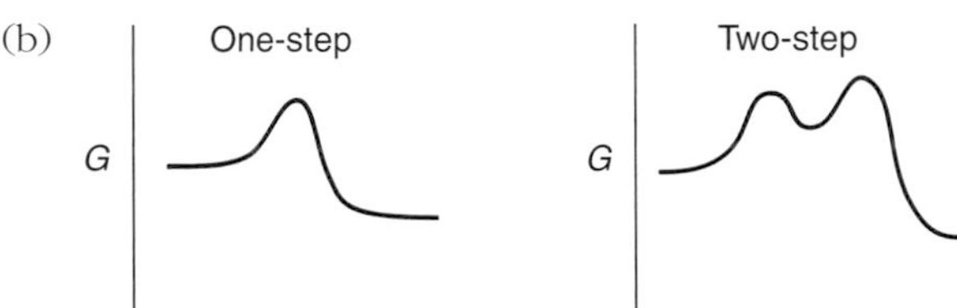

(c)

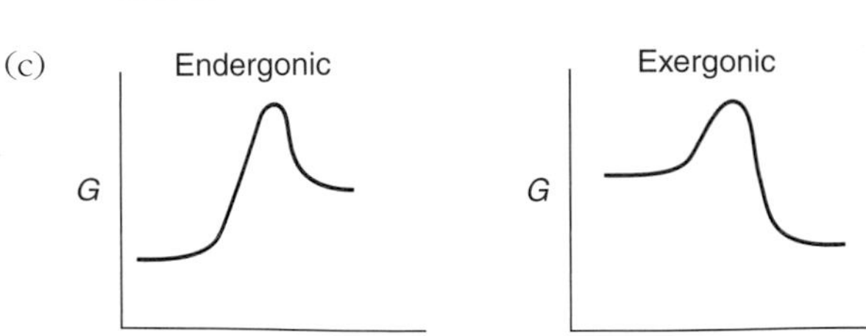

13. The ability of an enzyme to accelerate a reaction depends on the free energy difference between the enzyme-bound substrate and the enzyme-bound transition state. As long as this free energy difference is less than the free energy difference between the unbound substrate and the uncatalyzed transition state, the enzyme-mediated reaction proceeds more quickly.

14. The enzyme's conformation must be flexible enough to allow substrates access to the active site, to stabilize the changing electronic structure of the transition state, and to accommodate the reaction products.

15. Yes. An enzyme decreases the activation energy barrier for both the forward and the reverse directions of a reaction.

16. The nucleophilicity of a side chain is inversely related to its acidity. The less acidic (the more basic or electron-rich) a group is, the more nucleophilic it is and the more likely it is to react with an electrophilic group.

17. The transition state structure is likely tetrahedral at position 6 on the purine ring, since adenosine is planar whereas 1,6-dihydropurine is tetrahedral at this position. Enzymes bind the transition state much more tightly than the substrate.

18. (a) Gly, Ala, and Val have side chains that lack the functional groups required for acid–base or covalent catalysis.
   (b) Mutating one of these residues may alter the conformation at the active site enough to disrupt the arrangement of other groups that are involved in catalysis.

19. A mutation can increase or decrease an enzyme's catalytic activity, depending on how it affects the structure and activity of groups in the active site.

20. (a) In order for any molecule to act as an enzyme, it must be able to recognize and bind a substrate specifically, it must have the appropriate functional groups to effect a chemical reaction, and it must be able to position those groups for reaction.
   (b) DNA, as a double-stranded molecule, has limited conformational freedom. RNA, which is single-stranded, is able to assume a greater range of conformations. This flexibility allows it to bind to substrates and carry out chemical transformations.

21.

	Cation binding	**Anion binding**	**Proton transfer**
Asp and Glu	✓		✓
His		✓	✓
Lys and Arg		✓	
Cys and Ser			
Tyr			

	Forms covalent bonds with acyl groups	Possible hydrogen bond donor	Possible hydrogen bond acceptor
Asp and Glu			✓
His		✓	✓
Lys and Arg		✓	
Cys and Ser	✓	✓	✓
Tyr		✓	

22. Tosyl-lysine chloromethylketone (shown) or tosyl-arginine chloromethylketone could serve as an affinity label for trypsin. Tosyl-alanine chloromethylketone (shown) or tosyl-glycine chloromethylketone could serve as an affinity label for elastase.

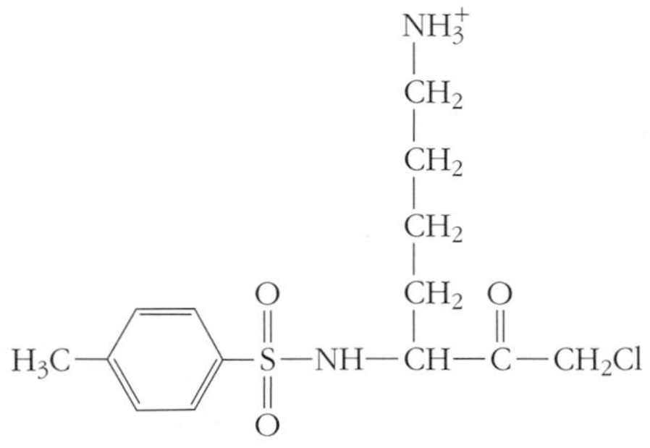

Tosyl-lysine chloromethylketone

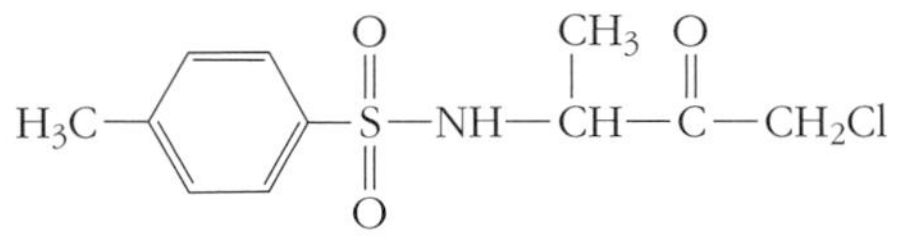

Tosyl-alanine chloromethylketone

23. His 57 ... HCl ... TPCK

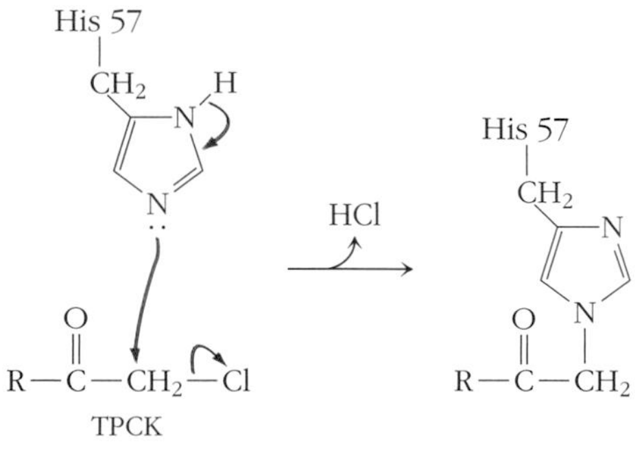

24. His 57 abstracts a proton from Ser 195, thus rendering the serine oxygen a better nucleophile. When Ser 195 is modified by formation of a covalent bond with DIP, the proton is no longer available and Ser 195 is unable to function as a nucleophile. Modification of His 57 with TPCK alkylates the nitrogen of the imidazole ring so that it can no longer abstract protons.

25. Cys 278 is highly exposed and unusually reactive compared to other cysteines in creatine kinase. Cys 278, because of its high reactivity, is probably one of the catalytic residues in the enzyme. The other cysteine residues are not as reactive because they are not directly involved in catalysis and/or because they are shielded in some way which prevents them from reacting with NEM.

26. Acetylcholinesterase (CH_2OH) + Sarin → HF + Modified enzyme

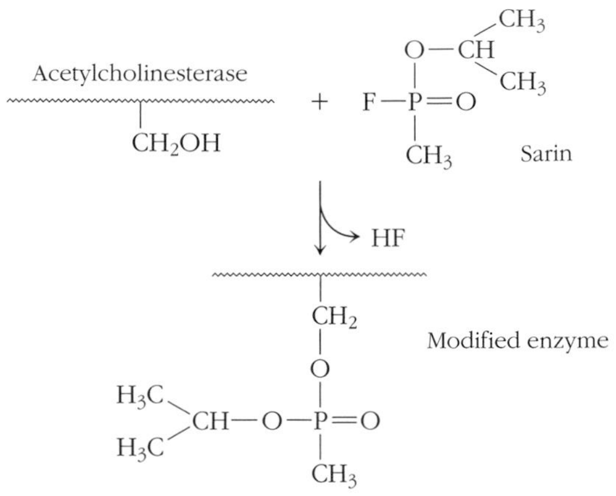

27. The enzyme would be inactive because Cys often functions as a nucleophile and would be unable to do so if a carboxymethyl group were covalently attached to the sulfur.

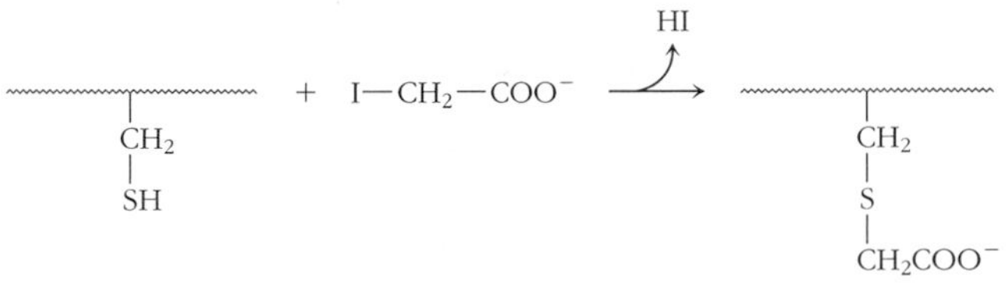

$-CH_2-SH + I-CH_2-COO^- \xrightarrow{HI} -CH_2-S-CH_2COO^-$

28. His residues are often involved in proton transfer. A carboxymethylated His would be unable to donate or accept protons.

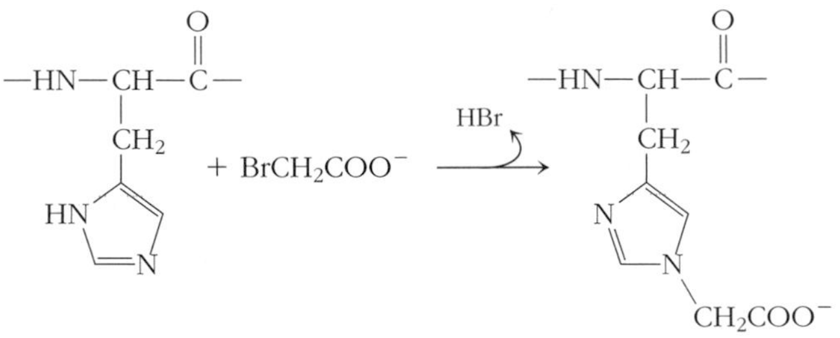

$+ BrCH_2COO^- \xrightarrow{HBr}$ ... CH_2COO^-

29. The experiment indicates that a Lys residue may be part of the active site. It is also possible that a modified Lys elsewhere on the enzyme affects catalytic activity by altering the conformation of the enzyme.

30. His ... Lys $(CH_2)_4$ ... Ser CH_2

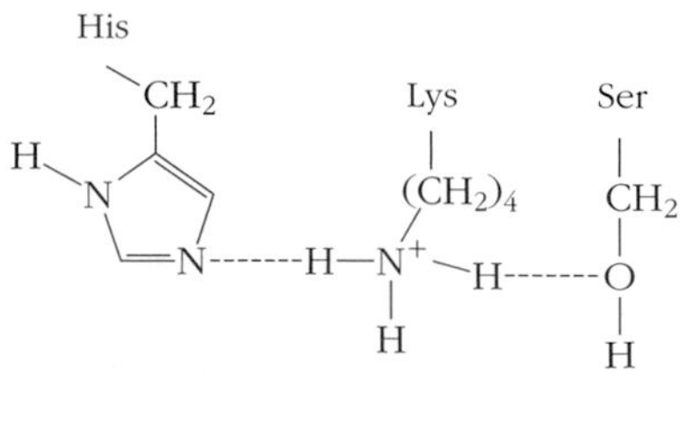

31. (a) In the first part of the reaction, the ester bond is cleaved and the chymotrypsin is acetylated. The *p*-nitrophenolate ion is quickly released, which accounts for the rapid increase in absorbance seen at 410 nm. The enzyme must be regenerated before a second round of catalysis can begin, which requires a deacetylation step. This step is much slower than the first step. Once the acetate is released, the enzyme is regenerated and another molecule of substrate can bind and react. Thus a "steady state" is reached and the absorbance increases at a uniform rate until the substrate is depleted.

Chymotrypsin–CH_2OH + $H_3C—C(=O)—O—C_6H_4—NO_2$
p-Nitrophenylacetate

fast ↓ → $^{-}O—C_6H_4—NO_2 + H^+$
p-Nitrophenolate (yellow)

Chymotrypsin–$CH_2—O—C(=O)—CH_3$

slow ↓ ← H_2O

Chymotrypsin–CH_2OH + $CH_3COO^- + H^+$

(b) The reaction coordinate diagram will look like the one in Figure 6-6, since this is a two-step reaction. Each step has a characteristic activation energy. The acetylated chymotrypsin is the intermediate.

(c) Yes, chymotrypsin and trypsin use the same catalytic mechanism, so trypsin can also act as an esterase as well as a protease.

32. Chymotrypsin can degrade neighboring molecules by catalyzing the hydrolysis of peptide bonds on the carboxyl side of Phe, Tyr, and Trp. If the chymotrypsin is stored in a solution of weak acid, His 57 would be protonated and would be unable to accept a proton from Ser 195 to begin hydrolysis.

33. A water molecule is able to enter the active site after the first product, the C-terminal portion of the substrate, diffuses away.

34.

Asp 25: $C(=O)—O^-$; $H—O—H$; $R_2—C(=O)—NHR_1$; $HO—C(=O)$: Asp 25′

↓

Asp 25: $C(=O)—OH$; $HO—C(R_2)(—O—H)—:NHR_1$; $^{-}O—C(=O)$: Asp 25′

↓

Asp 25: $C(=O)—O^-$; $HO—C(R_2)=O$ + $H—NHR_1$; $HO—C(=O)$: Asp 25′

35. (a)

Enz — Nucleophile (X:) , Acid catalyst (HB^+) ; Peptide $R_1—C(=O)—N(H)—R_2$ → [First tetrahedral intermediate: Enz—X—C(^{-}O)(R_1)—N(H)—R_2 ; Acid catalyst H—B^+]

→ Amine product $H—N(H)—R_2$

Acyl-enzyme adduct: Enz—X—C(=O)—R_1 ; Base catalyst B: ; Water H—O—H

↓

[Second tetrahedral intermediate: Enz—X—C(^{-}O)(R_1)—O—H ; Acid catalyst HB^+] → Enz — Nucleophile X: , Acid catalyst HB^+ + Carboxylate product $R_1—C(=O)—O^- + H^+$

(b) The mechanism employs acid–base catalysis as well as covalent catalysis.

(c) The reaction coordinate diagram would look like Figure 6-6, since it is a two-step reaction.

(d) Papain acts as a meat tenderizer because it is a protease and can hydrolyze the peptide bonds in the structural proteins of the meat. This makes the meat easier to chew and digest.

(e) Cys has a p*K* of 4.2 since it must be unprotonated to act as a nucleophile, and His has a p*K* of 8.2 since it must be protonated to be active; see part (a).

36. (a) His 12 has a pK value of 5.4 and His 119 has a pK value of 6.4. His 12 is unprotonated and has the lower pK while His 119 is protonated and has the higher pK.
(b) The pH optimum is 6 because at this pH His 12 is unprotonated (the pH is greater than the pK) and His 119 is protonated (the pH is less than the pK). At a pH less than 6, His 12 would be protonated and would not serve as a nucleophile to abstract the 2′-hydroxyl hydrogen. At pH greater than 6, His 119 would be unprotonated and would not be able to donate a hydrogen to the scissile bond.
(c) Ribonuclease does not hydrolyze DNA because DNA contains deoxyribose and is missing the 2′-hydroxyl group, which serves as the attacking nucleophile for the phosphorous (once the His 12 has removed the hydrogen).

37. (a)

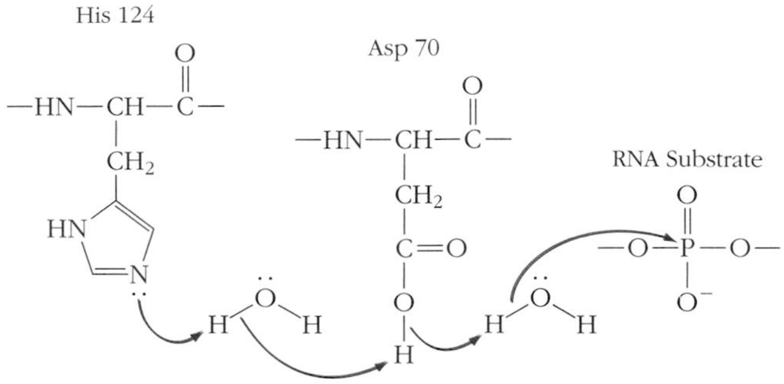

His 124 — Asp 70 — RNA Substrate

(b) His 124 most likely has a pK of less than 5.0 because the His shown in the relay is unprotonated. It must be unprotonated in order to accept a proton from the water molecule. If His 124 had a higher pK value, it would be more likely to be partially protonated at physiological pH and less able to accept a proton.
(c) Alanine has an aliphatic side chain and is unable to accept a proton in the relay as shown above.

38.

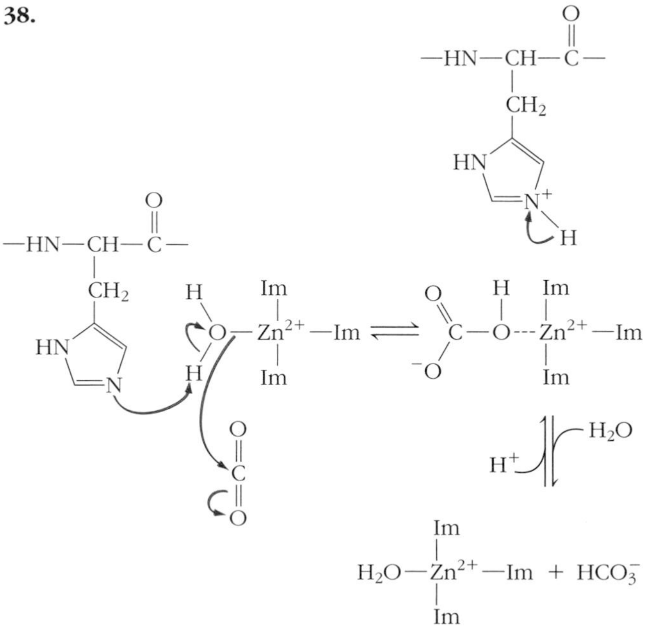

$H_2O{-}Zn^{2+}(Im)_3 + CO_2 \rightleftharpoons HCO_3^-{\cdots}Zn^{2+}(Im)_3$; $+ H_2O$, $- H^+$ → $H_2O{-}Zn^{2+}(Im)_3 + HCO_3^-$

39. (a) Trypsin cleaves peptide bonds on the carboxyl side of Lys and Arg residues, which are positively charged at physiological pH. These residues fit into the specificity pocket and interact electrostatically with Asp 189.
(b) A mutant trypsin with a positively charged Lys residue in its specificity pocket would no longer prefer basic side chains because the like charges would repel one another. The mutant trypsin might instead prefer to cleave peptide bonds on the carboxyl side of negatively charged residues such as Glu and Asp, whose side chains could interact electrostatically with the positively charged Lys residue.
(c) If the substrate specificity pocket does not include a positively charged Lys residue, then there would be no reason to expect the mutant enzyme to prefer substrates with acidic side chains. Instead, the mutant enzyme is more likely to prefer substrates with nonpolar side chains such as Leu or Ile.

40. (a) The pocket that holds the P1′ side chain is nonpolar, fairly large, and able to hold side chains of varying sizes. The Asp–Ala–Phe–Leu peptide is hydrolyzed fastest, followed by Asp–Phe–Ala–Leu. Thus, the pocket can accommodate small aliphatic side chains such as Ala as well as larger aromatic side chains such as Phe. A positively charged amino acid does not fit well into this pocket, since Asp–Lys–Ala–Leu is hydrolyzed more slowly than most of the other artificial peptides. The P2′ pocket most likely accommodates a large hydrophobic side chain, since the Asp–Ala–Phe–Leu peptide was hydrolyzed faster than the peptides with Ala at the P2′ position. A substrate with a charged residue such as Lys or Asp is hydrolyzed relatively slowly.
(b) Aspartame is a good candidate for hydrolysis by aspartyl aminopeptidase, since the enzyme catalyzes hydrolysis of peptides on the carboxyl side of Asp residues (that is, Asp is in the P1 position). The amino acid in the P1′ position would be Phe, a large hydrophobic residue that should fit in the P1′ pocket, as suggested by the studies of the artificial substrates.
(c)

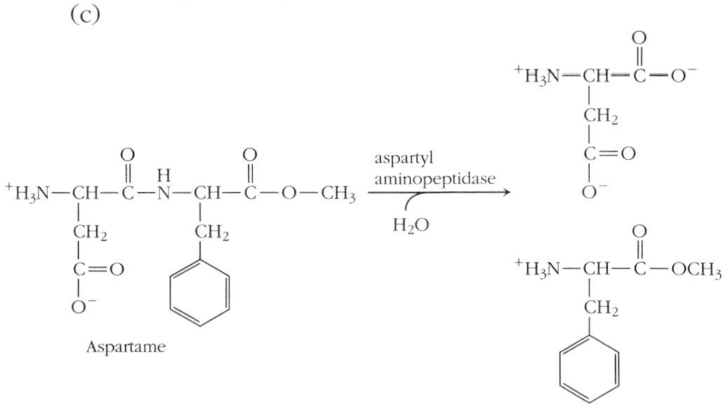

Aspartame + H_2O —(aspartyl aminopeptidase)→ Asp + Phe methyl ester

(d) The enzyme likely consists of eight identical subunits with a subunit molecular mass of 55 kD. The subunits are associated with one another noncovalently so that the entire enzyme complex has a molecular mass of 440 kD.

41. (a) The deamidation reaction for asparagine is shown. The deamidation reaction for glutamine is similar.

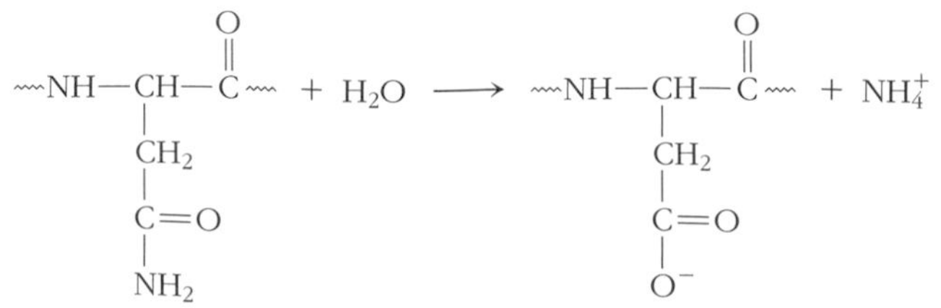

Asparagine $\quad$ Aspartate

(b) Asparagine and glutamine have neutral side chains that cannot form ion pairs. Deamidation produces negatively charged Asp and Glu side chains which could form ion pairs with nearby positively charged amino acid side chains. The conversion from an amide to a carboxylic acid functional group also changes the hydrogen-bonding capability of the side chains. An amide side chain can serve as both a hydrogen-bond donor and acceptor, whereas a carboxylic acid group can serve only as a hydrogen-bond acceptor. Changes in both ion pairing and hydrogen-bonding patterns could alter the three-dimensional structure of the protein, which might make it more susceptible to hydrolysis by proteases.

(c) Deamidated Asn residues are most likely to be preceded by Ser, Thr, or Lys and to be followed by Gly, Ser, or Thr.

(d)

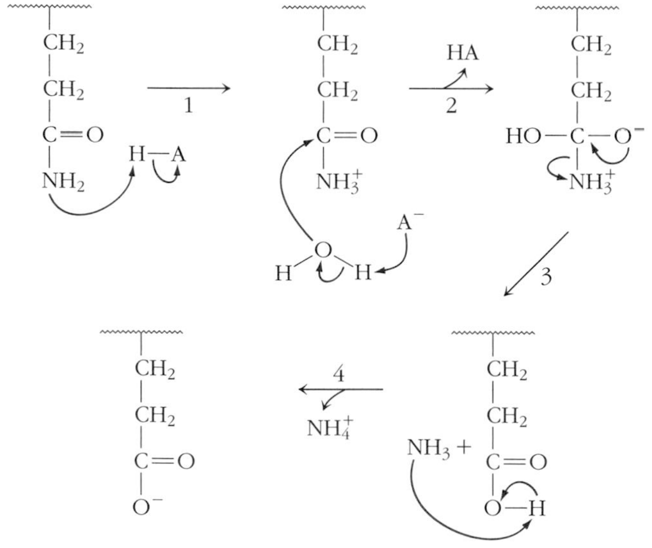

The product of Step 2 could be stabilized by ionic interactions. For example, the $—NH_3^+$ group could interact with Asp or Glu side chains on the protein. The $—O^-$ could interact with Lys or Arg residues. The intermediate could also be stabilized by hydrogen bonding. For example, the $—NH_3^+$ group could serve as a hydrogen bond donor, and the $—O^-$ could serve as a hydrogen bond acceptor. The product of Step 3 could similarly be stabilized by hydrogen bonding.

(e) Ser and Thr residues could stabilize the transition state. They could also serve as bases (if unprotonated) and accept a proton from water to form a hydroxide ion that would act as the attacking nucleophile. Ser and Thr (in the unprotonated form) could also act as attacking nucleophiles themselves.

An important point to keep in mind is that there may be amino acids close to the deamidated Asn or Gln in the three-dimensional structure of the protein that are not close by in primary sequence. For example, Asp, Glu, His, Lys, and Arg residues are not commonly found before and after the labile Asn according to the bar graph, but such residues could be close by when the three-dimensional conformation of the protein is considered. Consequently, Asp, Glu, and His residues (in the unprotonated form) might be able to function as nucleophiles even though these residues do not normally precede the deamidated amino acid. In the same manner, Lys and Arg, which are positively charged when protonated, could stabilize the negatively charged oxygen in the tetrahedral intermediate, even though these residues do not commonly follow the deamidated amino acid.

(f) The mechanism for the deamidation of an amino terminal Gln residue is shown. Amino terminal Asn residues are not deamidated because a four-membered ring would result, which is unstable.

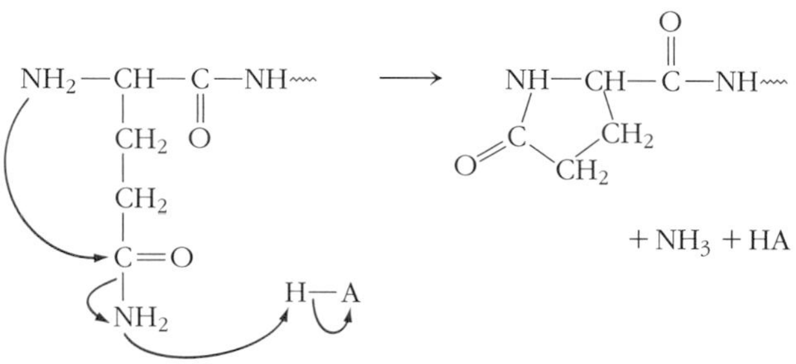

(g) Water is a substrate in the reaction. The Asn and Gln residues on the surface of the protein have much greater access to water molecules than interior Asn and Gln residues.

42. (a)

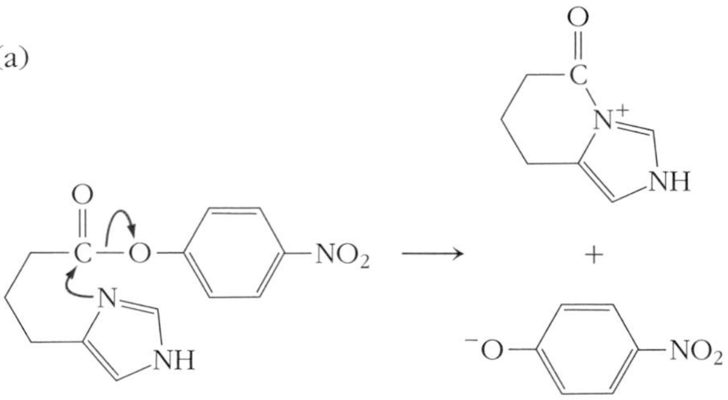

p-Nitrophenolate

(b) The imidazole ring is already tethered to the nitrophenyl group, so the reaction is unimolecular rather than bimolecular. In the bimolecular reaction, the reactants must first encounter each other via diffusion in solution. The unimolecular reaction proceeds faster because the reacting groups are already in close proximity.

(c) Enzymes speed up reaction rates in part by proximity and orientation effects. By binding to the enzymes, the reactants are in close proximity to one another. The enzyme also assists in binding the reactants in the proper orientation so that the reaction can occur with less added free energy, which results in a more rapid reaction.

43. During chymotrypsin activation, chymotrypsin cleaves other chymotrypsin molecules at a Leu, a Tyr, and an Asn residue. Only one of these (Tyr) fits the standard description of chymotrypsin's specificity. Clearly, chymotrypsin has wider substrate specificity, probably determined in part by the identities of residues near the scissile bond.

44. Chymotrypsin activation is a cascade mechanism, since chymotrypsinogen is activated by trypsin, which is in turn activated by enteropeptidase.

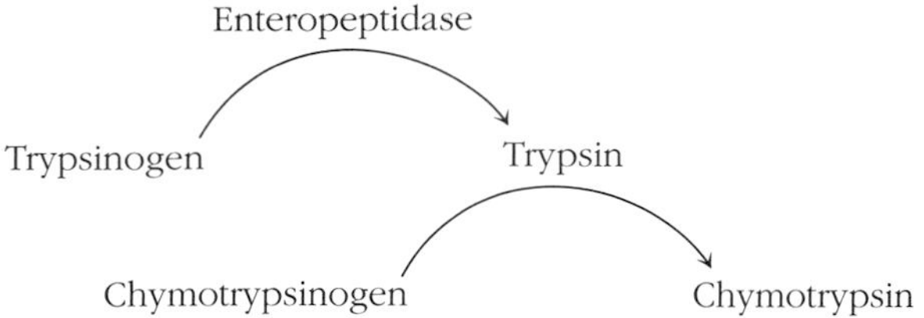

45. Activated factor IXa leads, via several steps, to the activation of the final coagulation protease, thrombin. The absence of factor IX therefore slows the production of thrombin, delaying clot formation, and causing the bleeding of hemophilia. Although activated factor XIa also leads to thrombin production, factor XI plays no role until it is activated by thrombin itself. By this point, coagulation is already well underway, so a deficiency of factor XI does not significantly delay coagulation.

46. As a digestive enzyme, chymotrypsin's function is to indiscriminately degrade a wide variety of ingested proteins, so that their component amino acids can be recovered. Broad substrate specificity would be dangerous for a protease that functions outside of the digestive system, since it might degrade proteins other than its intended target.

47. Since trypsin is at the "top" of the cascade (see Problem 6-44), it makes sense to inactivate it by using a trypsin inhibitor. Trypsin can activate chymotrypsinogen and proelastase, so if trypsin is inhibited, chymotrypsin and elastase will not be produced.

48. A trypsin inhibitor inactivates any trypsin that may have been activated prematurely. This prevents activation of other pancreatic zymogens, since trypsin is at the "top" of the activation cascade. Premature activation of pancreatic zymogens could destroy pancreatic tissue.

49. Serine protease inhibitors would interfere with the digestion of proteins in the small intestine. Undigested proteins could not be absorbed by the cells of the small intestine and thus their nutritional value would be lost. Gastric upset is also a possibility.

50. A protease with extremely narrow substrate specificity (that is, a protease with a single target) would pose no threat to nearby proteins, because these proteins would not be recognized as substrates for hydrolysis.

Chapter 7 Solutions

1.

Reaction	Molecularity	Rate equation
A → B + C	unimolecular	rate = k[A]
A + B → C	bimolecular	rate = k[A][B]
2 A → B	bimolecular	rate = k[A]^2
2 A → B + C	bimolecular	rate = k[A]^2

Units of k	Reaction velocity proportional to...	Order
s^{-1}	[A]	first
$M^{-1} \cdot s^{-1}$	[A] and [B]	second
$M^{-1} \cdot s^{-1}$	[A] squared	second
$M^{-1} \cdot s^{-1}$	[A] squared	second

2. The hyperbolic shape of the velocity vs. substrate curve suggests that the enzyme and substrate physically combine so that the enzyme becomes saturated at high concentrations of substrate. The lock-and-key model describes the interaction between an enzyme and its substrate in terms of a highly specific physical association between the enzyme (lock) and the substrate (key).

3.

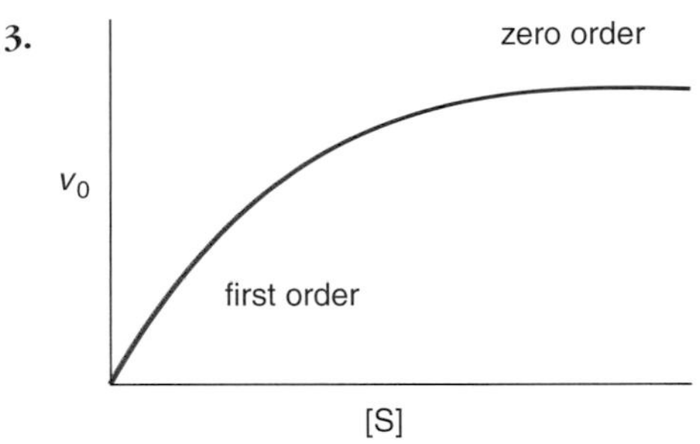

4. Enzyme activity is measured as an initial reaction velocity, which is the velocity before much substrate has been depleted and before much product has been generated. It is easier to measure the appearance of a small amount of product from a baseline of zero product than to measure the disappearance of a small amount of substrate against a background of a high concentration of substrate.

5.

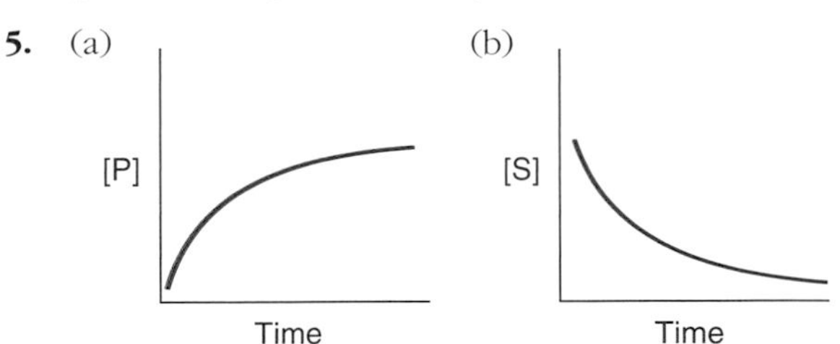

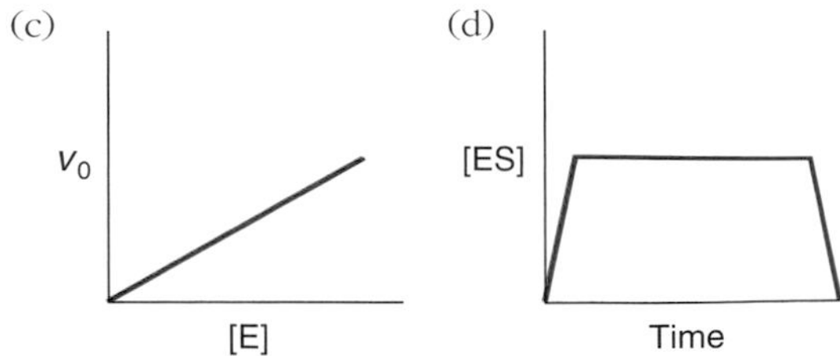

(e)

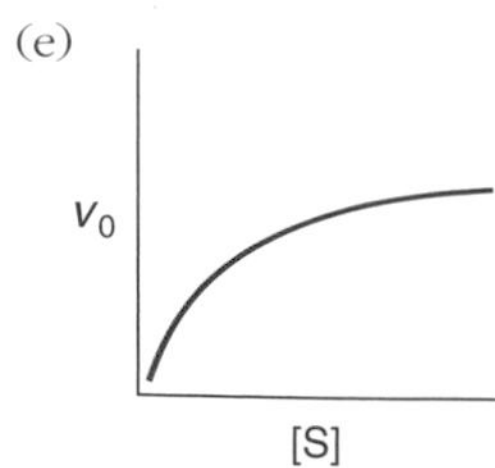

6. The apparent K_M would be greater than the true K_M because the experimental substrate concentration would be less than expected if some of the substrate has precipitated out of solution during the reaction.

7. The enzyme concentration is comparable to the lowest substrate concentration and therefore does not meet the requirement that [E] $\ll$ [S]. You could fix this problem by decreasing the amount of enzyme used for each measurement.

8. Velocity measurements can be made using any convenient units. K_M is by definition a substrate concentration, so its value does not reflect how the velocity is measured.

9. The V_{max} is approximately 30 μM/s and the K_M is approximately 5 μM.

10. The K_M of enzyme A is about 2 μM and the K_M for enzyme B is about 5 μM.
   (a) Enzyme A would generate product more rapidly when [S] = 1 μM.
   (b) Enzyme B would generate product more rapidly when [S] = 10 μM.

11. (a) Not necessary—need only to know variables of [S] and v_0 so that a Lineweaver–Burk plot can be constructed.
   (b) Not necessary—same as (a).
   (c) The value of $[E]_T$ is required to calculate k_{cat} since $k_{cat} = V_{max}/[E]_T$, according to Equation 7-23.

12. $$k_{cat} = \frac{V_{max}}{[E]_T}$$

$$k_{cat} = \frac{4.77 \times 10^{-3}\ M}{9.0 \times 10^{-6}\ M \cdot s^{-1}}$$

$$k_{cat} = 530\ s^{-1}$$

The k_{cat} is the turnover number, which is the number of catalytic cycles per unit time. Each molecule of the enzyme therefore undergoes 530 catalytic cycles per second.

13. (a) The reaction catalyzed by E_2 has the highest K_M value since it has the lowest $-1/K_M$ value of the three reactions.
   (b) The reaction catalyzed by E_3 has the highest V_{max}, since it has the lowest $1/V_{max}$ of the three reactions.

14. (a) *N*-acetyltyrosine ethyl ester, with its lower K_M value, has a higher affinity for the chymotrypsin enzyme. The aromatic tyrosine residue more easily fits into the nonpolar "pocket" on the enzyme than does the smaller aliphatic valine residue.
   (b) The value of V_{max} is not related to the value of K_M, so no conclusion can be drawn.

15. The higher K_M indicates that hexokinase has a lower affinity for fructose than for glucose. But once the substrate binds to the enzyme, the fructose is converted to product more rapidly than glucose.

16. Product P appears first because enzyme A has a much lower K_M for the substrate than enzyme B. Because V_{max} is approximately the same for the two enzymes, the relative efficiency of the enzymes depends almost entirely on their K_M values.

17. The maximum rate that two molecules can collide with one another in solution is $10^8 - 10^9\ M^{-1} \cdot s^{-1}$. Thus enzymes with k_{cat}/K_M values in this range can be considered to be "diffusion controlled," which means the reaction is catalyzed as rapidly as the two reactants can encounter each other in solution. Thus, enzymes B and C are diffusion controlled but enzyme A is not.

Enzyme	K_M	k_{cat}	k_{cat}/K_M
A	0.3 mM	5000 s^{-1}	$1.7 \times 10^7\ M^{-1} \cdot s^{-1}$
B	1 nM	2 s^{-1}	$2 \times 10^9\ M^{-1} \cdot s^{-1}$
C	2 μM	850 s^{-1}	$4.2 \times 10^8\ M^{-1} \cdot s^{-1}$

18. The simultaneous collision of three molecules (E, A, and B) is an unlikely event. It is much more likely that the enzyme binds first one and then the other substrate. For example, the first bimolecular reaction might be $E + A \rightleftharpoons EA$, and the second would be $EA + B \rightleftharpoons EAB$.

19. (a) $v_0 = 0.75\ V_{max}$, so substitute in:

$$0.75\ V_{max} = \frac{V_{max}\ [S]}{[S] + K_M}$$

V_{max} cancels out on both sides.

$$0.75 = \frac{[S]}{[S] + K_M}$$
$$0.75([S] + K_M) = [S]$$
$$0.75\ K_M = 0.25\ [S]$$
$$3\ K_M = [S]$$

Thus, the substrate concentration is three times as high as the K_M.

(b) $v_0 = 0.9\ V_{max}$, so substitute in:

$$0.9\ V_{max} = \frac{V_{max}\ [S]}{[S] + K_M}$$

V_{max} cancels out on both sides.

$$0.9 = \frac{[S]}{[S] + K_M}$$
$$0.9([S] + K_M) = [S]$$
$$0.9\ K_M = 0.1\ [S]$$
$$9\ K_M = [S]$$

Thus, the substrate concentration is nine times as high as the K_M.

20. $$v_0 = \frac{V_{max} \times 5\ K_M}{5\ K_M + K_M}$$
$$v_0 = \frac{V_{max} \times 5\ K_M}{6\ K_M}$$
$$v_0 = 0.83\ V_{max}$$

When [S] = 5 K_M, the velocity is 83% of the maximum velocity.

$$v_0 = \frac{V_{max} \times 20\ K_M}{20\ K_M + K_M}$$

$$v_0 = \frac{V_{max} \times 20\ K_M}{21\ K_M}$$

$$v_0 = 0.95\ V_{max}$$

When [S] = 20 K_M, the velocity is 95% of the maximum velocity. Therefore, a fourfold increase in substrate concentration causes a smaller proportional increase in velocity (from 83% to 95%). Estimating V_{max} from a plot of v_0 vs. [S] is difficult because the substrate concentration must be quite high in order to achieve a maximal velocity of close to 100%. (And it is possible that these points on the hyperbolic curve cannot be experimentally measured since at high concentration the substrate may not be soluble in the reaction medium.) It is better to obtain V_{max} by fitting experimental data to the equation of a hyperbola, or alternatively, from a Lineweaver–Burk plot.

21. (a) Indole is a competitive inhibitor of chymotrypsin because its structure resembles the side chain of tryptophan, which fits into the specificity pocket of chymotrypsin. Thus indole and tryptophan side chains compete with each other for binding to the active site.
(b) The V_{max} is the same in the presence and absence of the inhibitor since inhibition can be overcome at high substrate concentrations. The K_M increases because a higher concentration of substrate is needed to achieve half-maximal activity in the presence of an inhibitor.

22. (a) The reaction is a trisubstrate reaction and therefore does not obey Michaelis–Menten kinetics.
(b) The K_M value for one substrate is obtained by varying its concentration while holding the concentrations of the other two substrates constant at saturating levels.
(c) V_{max} is achieved by saturating the enzyme with each substrate. Therefore, the concentration of each substrate must be much greater than its K_M value.

23. By irreversibly reacting with chymotrypsin's active site, DIPF would decrease $[E]_T$. The apparent V_{max} would decrease since $V_{max} = k_{cat}[E]_T$ (Equation 7-23). K_M would not be affected since the unmodified enzyme would bind substrate normally.

24. (a) Since the structures are similar (both have choline groups), the inhibitor is competitive. Competitive inhibitors compete with the substrate for binding to the active site, so the structures must be similar.
(b) Yes, the inhibition can be overcome. If large amounts of substrate are added, the substrate will be able to effectively compete with the inhibitor such that very little inhibitor will be bound to the active site. The substrate "wins" the competition when it is in excess.
(c) Like all competitive inhibitors, the inhibitor binds reversibly.

25. It is difficult to envision how an inhibitor that interferes with the catalytic function (represented by k_{cat} or V_{max}) of amino acid side chains at the active site would not also interfere with the binding (represented by K_M) of a substrate to a site at or near those same amino acid side chains.

26. (a) If an irreversible inhibitor is present, the enzyme's activity would be exactly 100 times lower when the sample is diluted 100-fold. Dilution would not change the degree of inhibition.
(b) If a reversible inhibitor is present, dilution would lower the concentrations of both the enzyme and the inhibitor enough so that some inhibitor would dissociate from the enzyme. The enzyme's activity would therefore not be exactly 100 times less than the diluted sample, but would be slightly greater because the proportion of uninhibited enzyme would be greater at the lower concentration.

27. (a) NADPH is the most effective inhibitor because it has the lowest K_I value.
(b) Inorganic phosphate is likely to be completely ineffective because its K_I value is so much greater than the K_M values of either of the two substrates. This indicates that the enzyme has a much greater affinity for its substrates, and thus will bind the substrates rather than the inhibitor. Thus the inhibitor will be ineffective.

28. The inhibitor is a mixed noncompetitive inhibitor. The V_{max} is decreased and the K_M value is increased in the presence of the inhibitor.

29. The formation of a disulfide bond under oxidizing conditions, or its cleavage under reducing conditions, could act as an allosteric signal by altering the conformation of the enzyme in a way that affects the groups at the active site.

30. The compound is a transition state analog (it mimics the planar transition state of the reaction) and therefore acts as a competitive inhibitor.

31. Compound A has a K_I value of 1.2×10^{-12} M. The lower K_I value indicates that it is a more effective inhibitor. Compound A is more effective because its structure most closely resembles the structure of the transition state. Compound B lacks the large polar OH group that mimics the transition state of the reaction.

32. (a) ATCase is an allosteric enzyme because its activity vs. [S] curve is sigmoidally shaped.
(b) CTP is a negative effector, or inhibitor, because when CTP is added, the K_M increases and thus the affinity of the enzyme for the substrate decreases. CTP is the eventual product of the pyrimidine biosynthesis pathway; thus when the concentration of CTP is sufficient for the needs of the cell, CTP inhibits an early enzyme in the synthetic pathway, ATCase, by feedback inhibition.
(c) ATP is a positive effector, or activator, because when ATP is added, the K_M decreases and thus the affinity of the enzyme for its substrate increases. ATP is a reactant in the reaction sequence, so it serves as an activator. ATP is also a purine nucleotide whereas CTP is a pyrimidine nucleotide. Stimulation of ATCase by

A plot of α vs. inhibitor concentration, shown below, gives a slope of 2.26 μM^{-1}. The value of K_I is the reciprocal of the slope, or 0.44 μM.

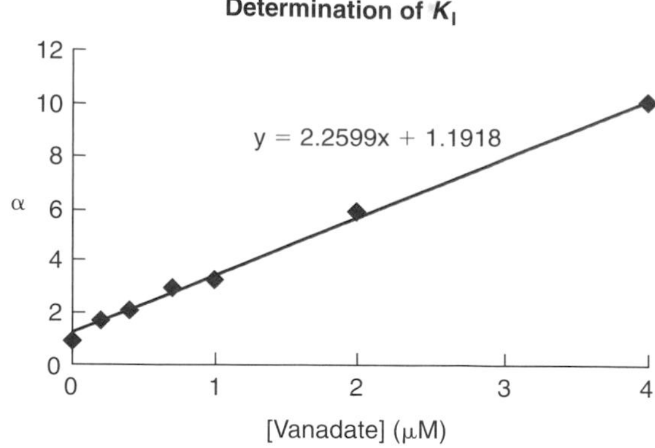

Chapter 8 Solutions

1.

$H_3C{-}(CH_2)_{12}{-}\overset{O}{\overset{\|}{C}}{-}O^-$

Myristate (14:0)

$H_3C{-}(CH_2)_5{-}CH{=}CH{-}(CH_2)_7{-}\overset{O}{\overset{\|}{C}}{-}O^-$

Palmitoleate (16:1 Δ^9)

$H_3C{-}CH_2{-}(CH{=}CH{-}CH_2)_3{-}(CH_2)_6{-}\overset{O}{\overset{\|}{C}}{-}O^-$

Linolenate (18:3 $\Delta^{9,12,15}$)

2.

$CH_2{-}O{-}\overset{O}{\overset{\|}{C}}{-}(CH_2)_{14}{-}CH_3$

$H_3C{-}(H_2C)_{14}{-}\overset{O}{\overset{\|}{C}}{-}O{-}CH$

$CH_2{-}O{-}\overset{O}{\overset{\|}{C}}{-}(CH_2)_{14}{-}CH_3$

Tripalmitin

3.

R_1 and R_2 each attached to $O{=}C{-}O{-}$ on $H_2C{-}CH{-}CH_2{-}O{-}P({=}O)(O^-){-}O{-}$ inositol ring (HO, H, HO, OH, H, OH, H, H, H, OH)

4.

$$O{=}P(O^-){-}O{-}(CH_2)_2{-}N^+(CH_3)_3$$

$$H_2C(O{-}C({=}O){-}(CH_2)_{14}{-}CH_3){-}CH(O{-}C({=}O){-}(CH_2)_{14}{-}CH_3){-}CH_2{-}O{-}P$$

DPPC

5. (a) A hydrocarbon chain is attached to the glycerol backbone at position 1 by a vinyl ether linkage. In a glycerophospholipid, an acyl group is attached by an ester linkage.

(b) The presence of this plasmalogen would not have a great effect since it has the same head group and same overall shape as phosphatidylcholine.

6.

$$H_2C(OH){-}CH(NH_3^+){-}CH(OH){-}CH{=}CH{-}(CH_2)_2{-}CH{=}C(CH_3){-}CH{=}CH{-}(CH_2)_6{-}CH_3$$

7. All except phosphatidylcholine have hydrogen-bonding head groups.

8. Both DNA and phospholipids have exposed phosphate groups that are recognized by the antibodies.

9. A glycerophospholipid with two saturated acyl chains has a cylindrical shape, whereas a glycerophospholipid with two unsaturated, kinked acyl chains would be more cone-shaped:

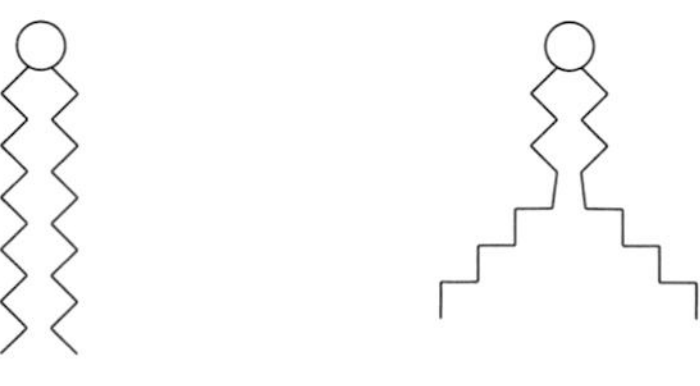

Saturated acyl chains Unsaturated acyl chains

10. Lipids that form bilayers are amphiphilic, whereas triacylglycerols are nonpolar. Amphiphilic molecules orient themselves so that their polar head groups face the aqueous medium on the inside and outside of the cell. Also, triacylglycerols are cone-shaped rather than cylindrical and thus would not fit well in a bilayer structure, as shown in Figure 8-3.

11. Vitamins A, D, and K are isoprenoids and are nonpolar. These dietary vitamins are soluble in the synthetic lipid Olestra® and pass out of the intestinal tract along with the Olestra® without being absorbed. Adding these vitamins to the product helps saturate the synthetic lipid with vitamins so that dietary vitamins are not excreted.

12.

(a)

$$\underset{\substack{|\\OH}}{H_2C}-\underset{\substack{|\\O\\|\\C=O\\|\\R}}{CH}-CH_2-O-\overset{\substack{O^-\\|}}{P}(=O)-O-CH_2-\underset{\substack{|\\NH_3^+}}{CH}-COO^- \quad + \quad R-\overset{\substack{O\\\|}}{C}-O^- \quad + \quad H^+$$

(b)

$$\underset{\substack{|\\O\\|\\O=C\\|\\R}}{H_2C}-\underset{\substack{|\\O\\|\\C=O\\|\\R}}{CH}-\overset{\substack{OH\\|}}{CH_2} \quad + \quad O=\overset{\substack{O^-\\|}}{\underset{\substack{|\\O^-}}{P}}-O-(CH_2)_2-\overset{\substack{CH_3\\|}}{\underset{\substack{|\\CH_3}}{N^+}}-CH_3 \quad + \quad H^+$$

(c)

$$\underset{\substack{|\\O\\|\\O=C\\|\\R}}{H_2C}-\underset{\substack{|\\O\\|\\C=O\\|\\R}}{CH}-CH_2-O-\overset{\substack{O^-\\|}}{P}(=O)-O^- \quad + \quad HO-CH_2-\underset{\substack{|\\OH}}{CH}-CH_2-OH \quad + \quad H^+$$

13. Two factors that influence the melting point of a fatty acid are the number of carbons and the number of double bonds. Double bonds are a more important factor than number of carbons, since a significant change in structure (a "kink") occurs when a double bond is introduced. An increase in the number of carbons increases the melting point, but the change is not nearly as dramatic. For example, the melting point of palmitate (16:0) is 63.1°C, whereas the melting point of stearate (18:0) is only slightly higher at 69.1°C. However, the melting point of oleate (18:1) is 13.2°C, a dramatic decrease with the introduction of a double bond.

14. B > A > C. *Trans*-oleate has a melting point similar to that of stearate (18:0) because the trans double bond does not produce a kink in the molecule. Its geometry more closely resembles that of a single bond.

15. In general, animal triacylglycerols must contain longer and/or more saturated acyl chains than plant triacylglycerols, since these chains have higher melting points and are more likely to be in the crystalline phase at room temperature. The plant triacylglycerols must contain shorter and/or less saturated acyl chains in order to remain fluid at room temperature.

16. Peanut oil has a higher melting point because the fatty acids that compose the monounsaturated triacylglycerols have a higher melting point than the more highly unsaturated fatty acids of the vegetable oil. Each double bond introduces a "kink" in the molecule, which means that the fatty acids don't pack together as well. The greater the number of double bonds, the greater the number of "kinks." Fatty acids that do not pack together well have fewer London dispersion forces among the chains, and less heat energy is required to disrupt the forces and melt the solid. Therefore, the vegetable oil, with a higher percentage of polyunsaturated triacylglycerols, has a lower melting point and does not freeze.

17. Unsaturated fatty acids closer to the hoof have a lower melting point due to the presence of double bonds that prevent them from packing together tightly. These lipids therefore help membranes remain fluid even at low temperatures as the reindeer walks through the snow. Saturated fatty acids, which pack together more efficiently and have higher melting points, would decrease membrane fluidity at low temperatures. By increasing the proportion of unsaturated fatty acids from the top of the leg (warm) to the hoof (cold), the reindeer can maintain constant membrane fluidity.

18. When phytanic acid is incorporated into membrane phospholipids, the resulting membrane is more fluid. The presence of the methyl groups on the phytanic acid result in an acyl chain that has a decreased ability to interact with neighboring acyl chains. This decreases the number of van der Waals interactions and decreases the melting point, which increases membrane fluidity.

19. The cyclopropane ring in lactobacillic acid produces a bend in the aliphatic chain, and thus the melting point of lactobacillic acid should be closer to the melting point of oleate, which also has a bend due to the double bond. The presence of bends decreases the opportunity for van der Waals forces to act among neighboring molecules, resulting in a melting point that is lower than that of a saturated fatty acid with the same number of carbons.

20. Increasing the temperature would make the membrane more fluid. To maintain constant fluidity, the bacteria synthesize fatty acids with more carbons and with fewer double bonds.

21. Cholesterol's planar ring system interferes with the movement of acyl chains and thus tends to decrease membrane fluidity. At the same time, cholesterol prevents close packing of the acyl chains, which tends to prevent their crystallization. The net result is that cholesterol helps the membrane resist melting at high temperatures and resist crystallization at low temperatures. Therefore, in a membrane containing cholesterol, the shift from the crystalline form to the fluid form is more gradual than it would be if cholesterol were absent.

22. No. Higher temperatures increase fatty acid fluidity. To counter the effect of temperature, the plants make relatively more fatty acids with higher melting points. Dienoic acids have higher melting points than trienoic acids because they are more saturated. Therefore, the plants convert fewer dienoic acids into trienoic acids.

23. In a bilayer, one end of each lipid acyl chain is fixed by its attachment to a head group. The methylene groups closer to the head group have the least conformational freedom, whereas the methylene groups farthest from the head group, near the bilayer center, have the greatest freedom.

24. (a) A or D
(b) A or C
(c) D
(d) D
(e) B
(f) B or E

25. Cytochrome *c* is a peripheral membrane protein that is only loosely associated with the membrane and can therefore be removed by gentle means such as a salt solution. Cytochrome oxidase is an integral membrane protein that completely spans the bilayer and has large hydrophobic portions where the protein spans the bilayer. Thus it is difficult to remove unless nonpolar organic solvents or amphipathic detergents are used to dissociate it from the membrane.

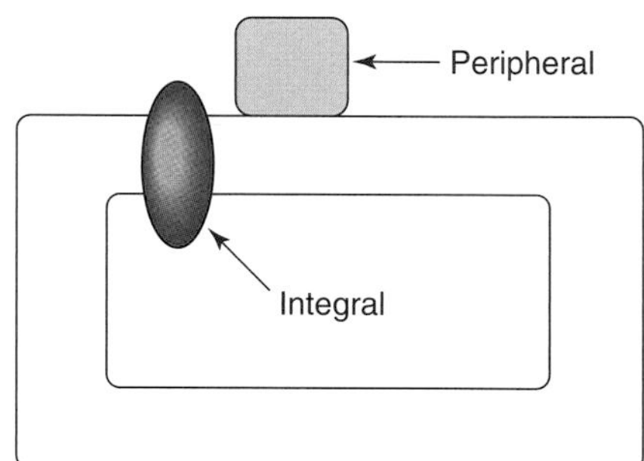

26. The membrane-spanning segment is a stretch of 19 residues that are all uncharged and mostly hydrophobic.

LSTTEVAMHTTTSSSVSKSYISSQTNDTHKRDTYAATPRA
HEVSEISVRTVYPPEEETGERVQLAHHFSEPEITLIIFGV
MAGVIGTILLISYGIRRLIKKSPSDVKPLPSPDTDVPLSS
VEIENPETSDQ

27. (a) A fully hydrogen-bonded β barrel can form only if the number of strands is even. A β sheet with an odd number of strands could not close up on itself to form a barrel.
(b) The strands are antiparallel because adjacent strands can be easily linked by loops on the solvent-exposed portions of the protein.
(c) A β barrel could contain some parallel β strands, but these could not be consecutive. A β barrel with consecutive parallel strands could occur only if the strands were linked by additional membrane-spanning segments (such as a transmembrane α helix or a structure that passed through the center of the barrel).

28. A steroid is a hydrophobic lipid that can easily cross a membrane to enter the cell. It does not require a cell-surface receptor, as does a polar molecule such as a peptide.

29. After fusion, the green and red markers were segregated because they represent cell-surface proteins derived from two different sets of cells. Over time, the cell-surface proteins that could diffuse in the lipid bilayer became distributed randomly over the surface of the hybrid cell, so the green and red markers were intermingled. At 15°C, the lipid bilayer was in a gel-like rather than a fluid state, which prevented membrane protein diffusion. Edidin's experiment supported the fluid mosaic model by demonstrating the ability of proteins to diffuse through a fluid membrane.

30. A. Fatty acyl–anchored protein (the acyl group is myristate)
B. Prenyl-anchored protein
C. Glycosylphosphatidylinositol (GPI)-anchored protein

31. The bleached area "recovers" its fluorescence as fluorophore-labeled molecules diffuse out of the small area and unbleached fluorophore-labeled molecules diffuse in. These experiments are useful for measuring diffusion rates of the target molecules.

32. CO_2 is small and nonpolar and therefore experiences no thermodynamic barrier to diffusing across the lipid bilayer.

33. The less polar a substance, the faster it can diffuse through the lipid bilayer. From slowest to fastest: D, C, A, B.

34. (a) A transport protein, like an enzyme, carries out a chemical reaction (in this case, the transmembrane movement of glucose) but is not permanently altered in the process. Because the transport protein binds glucose, its rate does not increase in direct proportion to increases in glucose concentration, and it becomes saturated at high glucose concentrations.
(b) The transport protein has a maximum rate at which it can operate (corresponding to V_{max}, the upper limit of the curve). It also binds glucose with a characteristic affinity (corresponding to K_M, the glucose concentration at half-maximal velocity).

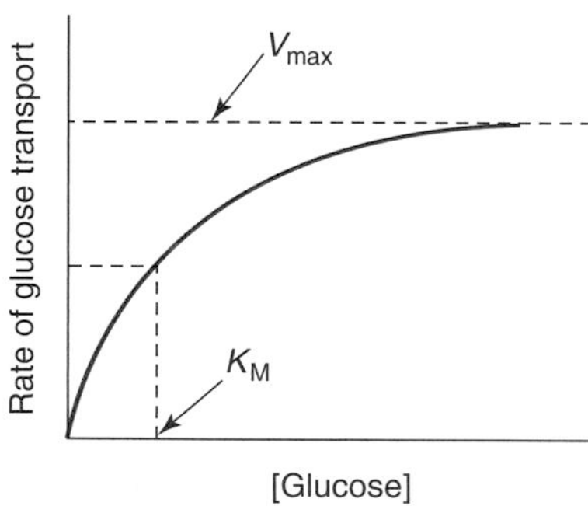

35. Use Equation 8-4 and let $Z = 2$ and $T = 310$K:

(a) $$\Delta G = RT \ln \frac{[Ca^{2+}_{in}]}{[Ca^{2+}_{out}]} + Z\mathcal{F}\Delta\psi$$

$$\Delta G = (8.3145 \text{ J} \cdot \text{K}^{-1} \cdot \text{mol}^{-1})(310\text{K}) \ln \frac{(10^{-7})}{(10^{-3})} + (2)(96{,}485 \text{ J} \cdot \text{V}^{-1} \cdot \text{mol}^{-1})(-0.05\text{V})$$

$$\Delta G = -23{,}700 \text{ J} \cdot \text{mol}^{-1} - 9600 \text{ J} \cdot \text{mol}^{-1}$$

$$\Delta G = -33{,}300 \text{ J} \cdot \text{mol}^{-1}$$

The negative value of ΔG indicates a thermodynamically favorable process.

(b) $\Delta G = RT \ln \frac{[Ca^{2+}_{in}]}{[Ca^{2+}_{out}]} + Z\mathcal{F}\Delta\psi$

$$\Delta G = (8.3145\ J \cdot K^{-1} \cdot mol^{-1})(310K) \ln \frac{(10^{-7})}{(10^{-3})} + (2)(96{,}485\ J \cdot V^{-1} \cdot mol^{-1})(+0.05V)$$

$$\Delta G = -23{,}700\ J \cdot mol^{-1} + 9600\ J \cdot mol^{-1}$$

$$\Delta G = 14{,}100\ J \cdot mol^{-1}$$

The positive value of ΔG indicates a thermodynamically unfavorable process.

36. (a) A transporter similar to a porin would be inadequate since even a large β barrel would be far too small to accommodate the massive ribosome. Likewise, a transport protein with alternating conformations would not be up to the task due to its small size relative to the ribosome. In addition, neither type of protein would be suited for transporting a particle across two membranes. In fact, ribosomes and other large particles move between the nucleus and cytoplasm via nuclear pores, which are constructed from many different proteins and form a structure, even larger than the ribosome, that spans both nuclear membranes.

(b) Initially, one might expect ribosomal transport to be a thermodynamically favorable process, since the concentration of ribosomes is greater in the nucleus, where they are synthesized. However, free energy would ultimately be required to establish a pore (which would span two membrane thicknesses) for the ribosome to pass through. In fact, the nucleocytoplasmic transport of all but very small substances requires the activity of GTPases that escort particles through the nuclear pore assembly and help ensure that transport proceeds in one direction.

37. (a) Since all the terms on the right side of Equation 8-1 are constant, except for T, the following proportion for the two temperatures (310K and 313K) applies:

$$\frac{-70\ \text{mV}}{310\text{K}} = \frac{\Delta\psi}{313\text{K}}$$

$$\Delta\psi = -70.7\ \text{mV}$$

The difference in membrane potential at the higher temperature would not significantly affect the neuron's activity.

(b) It is more likely that an increased temperature would increase the fluidity of cell membranes. This in turn might alter the activity of membrane proteins, including ion channels and pumps, which would have a more dramatic effect on membrane potential than temperature alone.

38. (a) Acetylcholine binding triggers the opening of the channel, an example of a ligand-gated transport protein.

(b) Na^{+} ions flow into the muscle cell, where their concentration is low.

(c) The influx of positive charges causes the membrane potential to increase.

39. Botulinum toxin destroys SNAREs, which are required for the fusion of synaptic vesicles with the neuronal plasma membrane. This interrupts communication between facial nerves and muscles. The result is paralysis of the muscles whose contraction accentuates wrinkles.

40. By adding a phosphate group, the kinase increases the size and negative charge of the lipid head group, which then occupies a larger volume and more strongly repels neighboring negatively charged lipid head groups. The phosphatidylinositol would become more cone-shaped, thereby increasing bilayer curvature, which is a necessary step in the formation of a new vesicle by budding.

41. (a) According to the results shown, DLPS is preferentially translocated. The DLPS is incorporated into red blood cells until there is an excess of phospholipid in the outer leaflet and the cells become spiky. In less than an hour, the DLPS is flipped to the inner leaflet, rendering the cells stomatocytic. DLPE is also transported by the flippase, but more slowly. DLPC is not transported at all. The DLPC becomes incorporated in the outer leaflet of the membrane, where it remains, and the cells remain echinocytic.

Because PS is negatively charged, whereas PE and PC are neutral, and because PS resides on the inner leaflet of the membrane, the membrane is more negatively charged on the inside and less negatively charged on the outside.

(b) The flippase requires a fatty acid in the 2 position on the phospholipid. If this fatty acid is missing, translocation is much less efficient.

(c) The ATP-depleted cells become less echinocytic but never completely regain their biconcave disk shape during the measurement period. In contrast, cells with normal concentrations of ATP were able to become stomatocytic. This indicates that the flippase can translocate PS from the outer leaflet to the inner leaflet of the membrane only if ATP is present. The flippase enzyme is therefore classified as an ATPase. The free energy of ATP is required to drive the thermodynamically unfavorable transfer of the polar lipid head group from one side of the bilayer to the other.

(d) The red blood cells treated with diamide remain echinocytic, which means that the flippase has not been able to translocate the lipid to the inner leaflet. Since diamide modifies sulfhydryl groups, and since the amino acid cysteine contains a sulfhydryl group, the data presented here indicate that a sulfhydryl group on a cysteine residue must be in the reduced form in order for translocation to take place.

(e) When deprived of ions, the red blood cells remain echinocytic, indicating that translocation cannot take place. But when magnesium ions are replaced, the cells rapidly become stomatocytic, indicating that magnesium ions are necessary for translocation activity.

(f) The flippase is an ATPase that depends on magnesium ions for full activity. It contains a cysteine side chain that is essential for translocation. The flippase preferentially translocates PS and PE to the inner leaflet. An acyl chain must be present at the 2 position of the phospholipid. The flippase maintains membrane asymmetry, since the PS and PE are found in the inner leaflet and PC is found in the outer leaflet.

42. (a) Glucose uptake by pericytes nearly doubles when sodium is present. In contrast, in endothelial cells, the uptake of glucose is constant, regardless of whether sodium is present or absent.

(b) Glucose uptake increases as sodium concentration increases in pericytes. In endothelial cells, glucose uptake is constant regardless of sodium ion concentration. The shape of the curve for the pericytes indicates that a protein transporter is involved: Glucose uptake increases linearly as sodium ion concentration increases, then reaches a plateau at high sodium ion concentrations, indicating that the transporter is saturated and is operating at its maximal capacity.

(c) With sodium ions

$$K_M = -1/(-0.25\ \text{mM}^{-1}) = 4\ \text{mM}$$
$$V_{max} = 1/(0.028\ \text{mg} \cdot \text{h}^{-1} \cdot \text{nmol}^{-1}) = 36\ \text{nmol} \cdot \text{h} \cdot \text{mg}^{-1}$$

Without sodium ions

$$K_M = -1/(-0.25\ \text{mM}^{-1}) = 4\ \text{mM}$$
$$V_{max} = 1/(0.063\ \text{mg} \cdot \text{h}^{-1} \cdot \text{nmol}^{-1}) = 16\ \text{nmol} \cdot \text{h} \cdot \text{mg}^{-1}$$

The K_M values for glucose transport in the presence and absence of sodium ions are the same, indicating that the transporter has equal affinity for glucose in the presence and absence of sodium ions. However, the maximal velocity is much greater in the presence of sodium ions than in their absence. Because the K_M values are the same, sodium ions probably do not increase the transporter's affinity for glucose. Instead, it is more likely that the sodium ions cause some kind of conformational change which allows the transporter to transport glucose more effectively.

The investigators have convincingly demonstrated that an SGLT exists in pericytes. Glucose transport increases as sodium ion concentration increases. The transport curve is hyperbolic, indicating that pericytes have a protein transporter with specific binding sites for glucose. There is probably not an SGLT in endothelial cells, since glucose transport is independent of sodium ion concentration.

(d) Galactose and 2-deoxyglucose inhibit glucose transport only slightly in the presence and absence of NaCl. Phlorizin inhibits glucose transport in the presence of sodium ions but does not inhibit glucose transport in the absence of sodium ions.

(e) When the glucose concentration increases, cell density increases, which is due to increased collagen synthesis in pericytes. Thus, increased glucose transport leads to increased collagen synthesis.

(f) In the absence or presence of phlorizin, glucose transport increases with increasing concentration of glucose. But in the presence of phlorizin, glucose transport is less. When the glucose concentration increases, the cell density increases in the absence or presence of phlorizin. In the presence of phlorizin, both Type IV and Type VI collagen synthesis decreases. Thus, a link has been established between glucose transport and collagen synthesis. When glucose concentration increases, transport into the cell increases, which stimulates collagen production.

(g) In the untreated diabetic, plasma glucose concentrations are high because insulin is either not present or fails to stimulate the uptake of glucose into the cells. Pericytes, however, do not rely on insulin for glucose uptake; instead glucose uptake is stimulated by sodium ions. The experimental data show that when the glucose concentration is increased, glucose uptake via the SGLT transporter also increases. The results also show that increased glucose uptake leads to increased collagen synthesis, which damages the pericytes and leads to retinopathy. In order to prevent this from occurring, diabetics should carefully control their plasma glucose concentration through exogenous insulin and diet. In addition, drugs that act in the same manner as phlorizin might be helpful, since the results show that in the presence of phlorizin, glucose uptake by the SGLT decreases, even in the presence of high glucose concentrations.

43. (a)

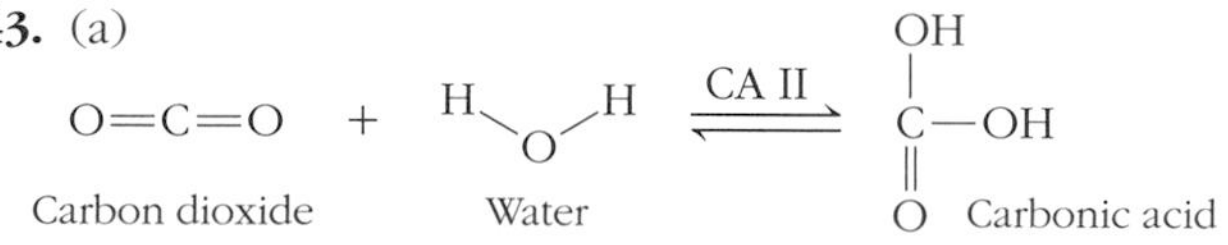

$$\underset{\text{Carbon dioxide}}{O{=}C{=}O} + \underset{\text{Water}}{H{-}O{-}H} \underset{}{\overset{\text{CA II}}{\rightleftharpoons}} \underset{\text{Carbonic acid}}{HO{-}C({=}O){-}OH}$$

$$H_2CO_3 \rightleftharpoons \underset{\text{Bicarbonate}}{HCO_3^-} + H^+$$

(b) The imidazole side chain of the histidine must be important in maintaining the proper three-dimensional conformation of the enzyme. The tyrosine side chain consists of a phenol ring and has different hydrogen-bonding capabilities than the imidazole side chain. The imidazole side chain can also be protonated, which would give it a positive charge so that it would form ion pairs with negatively charged amino acid side chains. Tyrosine does not have this capability.

(c)

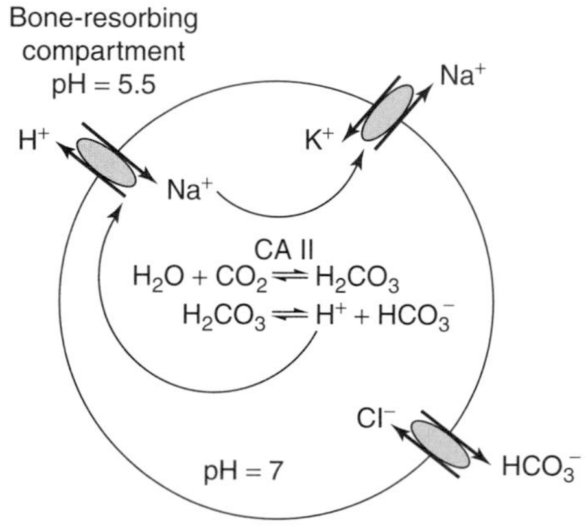

Hydrogen ions are produced from carbonic acid, which is produced from water and carbon dioxide via the carbonic anhydrase II reaction. The hydrogen ions exit the cell via the Na^+/H^+ exchanger. These hydrogen ions acidify the bone-resorbing compartment as sodium ions enter the cell. The sodium ions leave the cell via the Na^+,K^+ ATPase, while potassium ions enter the cell. Bicarbonate, the other product of the dissociation of carbonic acid, leaves the cell via the Cl^-/HCO_3^- exchanger as chloride enters the cell.

44. (a) The actual K_M and V_{max} values measured by the experimenters are 42 μM and 8.80 pmol · mg^{-1} · min^{-1}, respectively.

(b) The choline transporter responds to increasing concentration of choline by increasing its rate of transport so that efficient uptake occurs at a range of [choline] near the K_M. At low concentrations of choline (10 μM), the transporter operates at 20% of its maximal velocity, whereas at high concentrations of choline (80 μM), the transporter operates at nearly 100% of its maximal velocity. The K_M value of 42 μM is between the low and high physiological concentrations of choline.

(c) TEA is structurally similar to choline and acts as a competitive inhibitor. TEA binds to the choline transporter, preventing choline from binding. In this manner, TEA is brought into the cell instead of choline. Therefore, transport of choline from the portal circulation to the inside of the liver cell is inhibited.

(d) It is possible that the choline transporter cotransports hydrogen ions and choline. A hydrogen ion might be exported when choline is imported. This is an example of antiport transport.

45. There appears to be a link between HDL fluidity and LCAT activity (as indicated by V_{max}), with LCAT activity increasing with increasing fluidity. The PC with the 20:5 acyl chain is the most fluid, and the LCAT enzyme has the greatest activity with this phospholipid. Activity of LCAT decreases with decreasing fluidity. The affinity of LCAT for its substrate (measured as K_M) does not appear to be affected by fluidity, however, since there is no strong correlation between fluidity and affinity (a low K_M indicates a high affinity; a high K_M indicates low affinity for substrate). The increased LCAT activity might be due to the enzyme's increased ability to penetrate the more fluid phospholipid surface of the HDL to reach the nonpolar core where the cholesterol substrate is found.

Chapter 9 Solutions

1. (a)
$$K_{eq} = e^{-\Delta G^{\circ\prime}/RT}$$
$$K_{eq} = e^{-10\ \text{kJ}\cdot\text{mol}^{-1}/(8.3145\times10^{-3}\ \text{kJ}\cdot\text{K}^{-1}\cdot\text{mol}^{-1})(298\text{K})}$$
$$K_{eq} = e^{-4.04}$$
$$K_{eq} = 0.018$$
$$K_{eq} = e^{-\Delta G^{\circ\prime}/RT}$$
$$K_{eq} = e^{-20\ \text{kJ}\cdot\text{mol}^{-1}/(8.3145\times10^{-3}\ \text{kJ}\cdot\text{K}^{-1}\cdot\text{mol}^{-1})(298\text{K})}$$
$$K_{eq} = e^{-8.07}$$
$$K_{eq} = 0.000313$$

Small changes in $\Delta G^{\circ\prime}$ result in large changes in K_{eq}. Doubling the $\Delta G^{\circ\prime}$ value (a positive, unfavorable value) leads to a 60-fold decrease in K_{eq}.

(b)
$$K_{eq} = e^{-\Delta G^{\circ\prime}/RT}$$
$$K_{eq} = e^{-(-10\ \text{kJ}\cdot\text{mol}^{-1})/(8.3145\times10^{-3}\ \text{kJ}\cdot\text{K}^{-1}\cdot\text{mol}^{-1})(298\text{K})}$$
$$K_{eq} = e^{4.04}$$
$$K_{eq} = 56.6$$
$$K_{eq} = e^{-\Delta G^{\circ\prime}/RT}$$
$$K_{eq} = e^{-(-20\ \text{kJ}\cdot\text{mol}^{-1})/(8.3145\times10^{-3}\ \text{kJ}\cdot\text{K}^{-1}\cdot\text{mol}^{-1})(298\text{K})}$$
$$K_{eq} = e^{8.07}$$
$$K_{eq} = 3200$$

The same conclusion can be made: Small changes in $\Delta G^{\circ\prime}$ lead to large changes in K_{eq}. Doubling a (favorable) $\Delta G^{\circ\prime}$ results in a K_{eq} value that is nearly 60 times as large.

2. (a) The phosphate groups on the ATP molecule would be less negative at a lower pH. Therefore, there would be less charge–charge repulsion and therefore less energy released upon hydrolysis. The ΔG° would be less negative at a lower pH.

(b) Magnesium ions are positively charged and form ion pairs with the negatively charged phosphate groups. Thus magnesium ions serve to decrease the charge–charge repulsion associated with the phosphate groups. In the absence of magnesium ions, the charge–charge repulsion is greater; thus more free energy is released upon the removal of one of the phosphate groups. This results in a $\Delta G^{\circ\prime}$ value that is more negative.

3. The hydrolysis of pyrophosphate releases 33.5 kJ · mol^{-1} (see Table 9-2). These two reactions are coupled, and the overall ΔG value for the coupled reactions is negative; thus UDP–glucose formation occurs spontaneously.

4. (a) The synthesis of ATP from ADP requires 30.5 kJ · mol^{-1} of energy:

$$\text{ADP} + \text{P}_i \rightarrow \text{ATP} + \text{H}_2\text{O} \qquad \Delta G^{\circ\prime} = +30.5\ \text{kJ}\cdot\text{mol}^{-1}$$
$$\frac{2850\ \text{kJ}\cdot\text{mol}^{-1}}{30.5\ \text{kJ}\cdot\text{mol}^{-1}} \times 0.40 = 37\ \text{ATP}$$

(b)
$$\frac{9781\ \text{kJ}\cdot\text{mol}^{-1}}{30.5\ \text{kJ}\cdot\text{mol}^{-1}} \times 0.40 = 128\ \text{ATP}$$

(c) For glucose, 37 ATP/6 carbons = 6.2 ATP/carbon. For palmitate, 128 ATP/16 carbons = 8.0 ATP/carbon. Most of the carbon atoms of fatty acids are fully reduced $—CH_2—$ groups. Most of the carbon atoms of glucose have hydroxyl groups attached to them (—CHOH—) and are therefore already partly oxidized. Consequently, more free energy is available from a carbon in a triacylglycerol than from a carbon in a glycogen molecule.

5. (a) The polar glycogen molecule is fully hydrated, so its weight reflects a large number of closely associated water molecules. Fat is stored in anhydrous form. Therefore, a given weight of fat stores more free energy than the same weight of glycogen.

(b) Because it must be hydrated, a glycogen molecule occupies a large effective volume of the cytoplasm, which it shares with other glycogen molecules, enzymes, organelles, and so on. Because hydrophobic fat molecules are sequestered from the bulk of the cytoplasm, they do not have the same potential for interfering with other cellular constituents, so their collective volume is virtually unlimited.

6. (a)
$$\text{ADP} + \text{P}_i \rightarrow \text{ATP} + \text{H}_2\text{O} \qquad \Delta G^{\circ\prime} = +30.5\ \text{kJ}\cdot\text{mol}^{-1}$$
$$\frac{2000\ \text{kcal}}{\text{day}} \times 1\ \text{day} \times \frac{1\ \text{mol}}{30.5\ \text{kJ}} \times \frac{4.184\ \text{kJ}}{1\ \text{kcal}} \times 0.41 = 112\ \text{moles ATP}$$

(b)
$$112\ \text{moles ATP} \times \frac{505\ \text{g}}{\text{mol}} \times \frac{1\ \text{lb}}{2200\ \text{g}} = 25.7\ \text{lbs}$$

(c) ATP does not accumulate but instead is constantly recycled. As ATP is used, its hydrolysis products, ADP and P_i, serve as reactants for ATP synthesis in the process of oxidative phosphorylation.

7. (a) $K_{eq} = e^{-\Delta G^{\circ\prime}/RT}$

$K_{eq} = e^{-5\ kJ \cdot mol^{-1}/(8.3145 \times 10^{-3}\ kJ \cdot K^{-1} \cdot mol^{-1})(298K)}$

$K_{eq} = e^{-2.02}$

$K_{eq} = 0.133$

(b) Since $K_{eq} = \dfrac{[\text{isocitrate}]}{[\text{citrate}]} = 0.133$

$[\text{isocitrate}] = 0.133\ [\text{citrate}]$

The total concentration of isocitrate and citrate is 2 M, so

$[\text{isocitrate}] = 2\text{ M} - [\text{citrate}]$

Combining the two equations gives

$0.133\ [\text{citrate}] = 2\text{ M} - [\text{citrate}]$

$1.133\ [\text{citrate}] = 2\text{ M}$

$[\text{citrate}] = 1.77\text{ M}$

$[\text{isocitrate}] = 2\text{ M} - 1.77\text{ M} = 0.23\text{ M}$

(c) The preferred direction under standard conditions is toward the formation of citrate.

(d) The reaction occurs in the direction of isocitrate synthesis because standard conditions do not exist in the cell. Also, the reaction is the second step of an eight-step pathway, so isocitrate is removed as soon as it is produced in order to serve as the reactant for the next step of the pathway.

8. (a) $K_{eq} = \dfrac{[\text{arginine}][\text{ATP}]}{[\text{phosphoarginine}][\text{ADP}]}$

$K_{eq} = \dfrac{(4.78 \times 10^{-3})(3.87 \times 10^{-3})}{(0.737 \times 10^{-3})(0.750 \times 10^{-3})}$

$K_{eq} = 33.5$

(b) $\Delta G^{\circ\prime} = -RT \ln K_{eq}$

$\Delta G^{\circ\prime} = -(8.3145 \times 10^{-3}\ kJ \cdot K^{-1} \cdot mol^{-1})(298K) \ln 33.5$

$\Delta G^{\circ\prime} = -8.7\ kJ \cdot mol^{-1}$

The reaction is spontaneous under standard conditions.

9. The complete reaction is $ATP + H_2O \rightarrow ADP + P_i$. Use Equation 9-3 and the value of $\Delta G^{\circ\prime}$ from Table 9-2. The concentration of water is assumed to be equal to 1.

$$\Delta G = \Delta G^{\circ\prime} + RT \ln \frac{[ADP][P_i]}{[ATP]}$$

$$\Delta G = -30.5\ kJ \cdot mol^{-1} + (8.3145 \times 10^{-3}\ kJ \cdot K^{-1} \cdot mol^{-1})(310K) \ln \frac{(0.001)(0.005)}{(0.003)}$$

$$\Delta G = -30.5\ kJ \cdot mol^{-1} - 16.5\ kJ \cdot mol^{-1}$$

$$\Delta G = -47\ kJ \cdot mol^{-1}$$

10. (a) See Sample Calculation 9-1. The equilibrium constant can be derived by rearranging Equation 9-2:

$K_{eq} = e^{-\Delta G^{\circ\prime}/RT}$

$K_{eq} = e^{-7.9\ kJ \cdot mol^{-1}/(8.3145 \times 10^{-3}\ kJ \cdot K^{-1} \cdot mol^{-1})(298K)}$

$K_{eq} = e^{-3.19}$

$K_{eq} = 0.041$

(b) At 37°C, $T = 310K$:

$$\Delta G = \Delta G^{\circ\prime} + RT \ln \frac{[\text{dihydroxyacetone phosphate}]}{[\text{glyceraldehyde-3-phosphate}]}$$

$$\Delta G = 7.9\ kJ \cdot mol^{-1} + (8.3145 \times 10^{-3}\ kJ \cdot K^{-1} \cdot mol^{-1})(310K) \ln \frac{(5 \times 10^{-4})}{(1 \times 10^{-4})}$$

$$\Delta G = 7.9\ kJ \cdot mol^{-1} + 4.1\ kJ \cdot mol^{-1}$$

$$\Delta G = 12\ kJ \cdot mol^{-1}$$

(c) The reaction is not spontaneous as written. The reverse reaction, in which $\Delta G = -12\ kJ \cdot mol^{-1}$, would be spontaneous.

11. (a) $\Delta G^{\circ\prime} = -RT \ln K_{eq}$

$\Delta G^{\circ\prime} = -(8.3145 \times 10^{-3}\ kJ \cdot K^{-1} \cdot mol^{-1})(298K) \ln 0.41$

$\Delta G^{\circ\prime} = 2.2\ kJ \cdot mol^{-1}$

The reaction will proceed in the opposite direction as written.

(b)

$$\Delta G = \Delta G^{\circ\prime} + RT \ln \frac{[\text{fructose-6-phosphate}]}{[\text{glucose-6-phosphate}]}$$

$$\Delta G = 2.2\ kJ \cdot mol^{-1} + (8.3145 \times 10^{-3}\ kJ \cdot K^{-1} \cdot mol^{-1})(310K) \ln \frac{(5 \times 10^{-4})}{(2.0 \times 10^{-3})}$$

$$\Delta G = 2.2\ kJ \cdot mol^{-1} - 3.57\ kJ \cdot mol^{-1}$$

$$\Delta G = -1.37\ kJ \cdot mol^{-1}$$

Under these conditions, the reaction will proceed as written.

12. Mechanism 1: $\text{Glutamate} + NH_3 \rightleftharpoons \text{glutamine}$

$ATP + H_2O \rightleftharpoons ADP + P_i$

$\text{Glutamate} + NH_3 + ATP + H_2O \rightleftharpoons \text{glutamine} + ADP + P_i$

Mechanism 2: $\text{Glutamate} + ATP \rightleftharpoons \gamma\text{-glutamylphosphate} + ADP$

$\gamma\text{-glutamylphosphate} + H_2O + NH_3 \rightleftharpoons \text{glutamine} + P_i$

$\text{Glutamate} + NH_3 + ATP + H_2O \rightleftharpoons \text{glutamine} + ADP + P_i$

Mechanism 2 is the more likely mechanism because it proceeds through a phosphorylated intermediate that "captures" the energy of the phosphoanhydride bond of ATP. In the first mechanism, the two reactions are not linked to a common intermediate. ATP is hydrolyzed, but this energy is not "harnessed" in any way and is simply dissipated as heat.

13. (a) The equilibrium constant can be determined by rearranging Equation 9-2 (see Sample Calculation 9-1):

$K_{eq} = e^{-\Delta G^{\circ\prime}/RT}$

$K_{eq} = e^{-13.8\ kJ \cdot mol^{-1}/(8.3145 \times 10^{-3}\ kJ \cdot K^{-1} \cdot mol^{-1})(298K)}$

$K_{eq} = e^{-5.57}$

$K_{eq} = 0.0038$

(b) $K_{eq} = \dfrac{[\text{glucose-6-phosphate}]}{[\text{glucose}][P_i]}$

$0.0038 = \dfrac{[\text{glucose-6-phosphate}]}{(5.0 \times 10^{-3})(5.0 \times 10^{-3})}$

$0.0038 = \dfrac{[\text{glucose-6-phosphate}]}{(2.5 \times 10^{-6})}$

$[\text{glucose-6-phosphate}] = 9.5 \times 10^{-8}\text{ M}$

Under the given conditions, the reaction would produce only 9.5×10^{-8} M glucose-6-phosphate and thus is not a feasible route to the production of this compound for the glycolytic pathway.

(c) $K_{eq} = \dfrac{[\text{glucose-6-phosphate}]}{[\text{glucose}][P_i]}$

$0.0038 = \dfrac{(250 \times 10^{-6})}{[\text{glucose}](5.0 \times 10^{-3})}$

[glucose] = 13 M

Driving the reaction to the right using this method is not feasible because it is impossible to achieve a concentration of 13 M glucose inside the cell.

(d) Glucose + $P_i \rightleftharpoons$ glucose-6-phosphate + H_2O
$\Delta G^{\circ\prime} = 13.8$ kJ · mol^{-1}

ATP + $H_2O \rightleftharpoons$ ADP + P_i
$\Delta G^{\circ\prime} = -30.5$ kJ · mol^{-1}

Glucose + ATP $\rightleftharpoons$ glucose-6-phosphate + ADP
$\Delta G^{\circ\prime} = -16.7$ kJ · mol^{-1}

$K_{eq} = e^{-\Delta G^{\circ\prime}/RT}$
$K_{eq} = e^{-(-16.7 \text{ kJ} \cdot \text{mol}^{-1})/(8.3145 \times 10^{-3} \text{ kJ} \cdot \text{K}^{-1} \cdot \text{mol}^{-1})(298\text{K})}$
$K_{eq} = e^{6.74}$
$K_{eq} = 850$

(e) $K_{eq} = \dfrac{[\text{glucose-6-phosphate}][\text{ADP}]}{[\text{glucose}][\text{ATP}]}$

$850 = \dfrac{(250 \times 10^{-6})(1.25 \times 10^{-3})}{[\text{glucose}](5.0 \times 10^{-3})}$

[glucose] = 7.4×10^{-8} M

(f) The reaction can be accomplished at a much lower glucose concentration when the phosphorylation of glucose is coupled to ATP hydrolysis (7.4×10^{-8} M instead of 13 M). This can be done because the second reaction couples the phosphorylation of glucose with the exergonic hydrolysis of ATP. Thus an unfavorable reaction is converted to a favorable reaction.

14. (a) $K_{eq} = e^{-\Delta G^{\circ\prime}/RT}$
$K_{eq} = e^{-47.7 \text{ kJ} \cdot \text{mol}^{-1}/(8.3145 \times 10^{-3} \text{ kJ} \cdot \text{K}^{-1} \cdot \text{mol}^{-1})(298\text{K})}$
$K_{eq} = e^{-19.2}$
$K_{eq} = 4.4 \times 10^{-9}$

$K_{eq} = \dfrac{[\text{fructose-1,6-bisphosphate}]}{[\text{fructose-6-phosphate}][P_i]}$

$4.4 \times 10^{-9} = \dfrac{[\text{fructose-1,6-bisphosphate}]}{[\text{fructose-6-phosphate}](5.0 \times 10^{-3})}$

$\dfrac{[\text{fructose-1,6-bisphosphate}]}{[\text{fructose-6-phosphate}]} = 2.2 \times 10^{-11}$

(b)
Fructose-6-phosphate + $P_i \rightleftharpoons$ fructose-1,6-bisphosphate
$\Delta G^{\circ\prime} = 47.7$ kJ · mol^{-1}

ATP + $H_2O \rightleftharpoons$ ADP + P_i $\quad \Delta G^{\circ\prime} = -30.5$ kJ · mol^{-1}

Fructose-6-phosphate + ATP + $H_2O \rightleftharpoons$
fructose-1,6-bisphosphate + ADP $\quad \Delta G^{\circ\prime} = 17.2$ kJ · mol^{-1}

(c) $K_{eq} = e^{-\Delta G^{\circ\prime}/RT}$
$K_{eq} = e^{-17.2 \text{ kJ} \cdot \text{mol}^{-1}/(8.3145 \times 10^{-3} \text{ kJ} \cdot \text{K}^{-1} \cdot \text{mol}^{-1})(298\text{K})}$
$K_{eq} = e^{-6.94}$
$K_{eq} = 1.0 \times 10^{-3}$

$K_{eq} = \dfrac{[\text{fructose-1,6-bisphosphate}][\text{ADP}]}{[\text{fructose-6-phosphate}][\text{ATP}]}$

$1.0 \times 10^{-3} = \dfrac{[\text{fructose-1,6-bisphosphate}][\text{ADP}]}{[\text{fructose-6-phosphate}][\text{ATP}]}$

$\dfrac{(1.0 \times 10^{-3})[\text{ATP}]}{[\text{ADP}]} = \dfrac{[\text{fructose-1,6-bisphosphate}]}{[\text{fructose-6-phosphate}]}$

$\dfrac{(1.0 \times 10^{-3})(3.0 \times 10^{-3})}{(1.0 \times 10^{-3})} = \dfrac{[\text{fructose-1,6-bisphosphate}]}{[\text{fructose-6-phosphate}]}$

$(3.0 \times 10^{-3}) = \dfrac{[\text{fructose-1,6-bisphosphate}]}{[\text{fructose-6-phosphate}]}$

(d) The conversion of fructose-6-phosphate to fructose-1,6-bisphosphate is unfavorable. The ratio of products to reactants at equilibrium is 2.2×10^{-11} under standard conditions. But if the conversion of fructose-6-phosphate to fructose-1,6-bisphosphate is coupled with the hydrolysis of ATP, the reaction becomes more favorable and the ratio of fructose-1,6-bisphosphate to fructose-6-phosphate increases to 3×10^{-3}, a change of eight orders of magnitude.

(e) The second mechanism is biochemically feasible because the two steps are coupled via a common phosphorylated intermediate which "captures" the energy of ATP. In the first mechanism, the two steps are not coupled. The ATP is hydrolyzed and the energy is lost as heat instead of being used to assist the conversion of fructose-6-phosphate to fructose-1,6-bisphosphate.

15. I. GAP $\rightleftharpoons$ 1,3BPG $\quad \Delta G^{\circ\prime} = 6.7$ kJ · mol^{-1}
1,3BPG + $H_2O \rightleftharpoons$ 3PG + P_i $\quad \Delta G^{\circ\prime} = -49.3$ kJ · mol^{-1}

GAP + $H_2O \rightleftharpoons$ 3PG + P_i $\quad \Delta G^{\circ\prime} = -42.6$ kJ · mol^{-1}

II. GAP $\rightleftharpoons$ 1,3BPG $\quad \Delta G^{\circ\prime} = 6.7$ kJ · mol^{-1}
1,3BPG + ADP $\rightleftharpoons$ 3PG + ATP $\quad \Delta G^{\circ\prime} = -18.8$ kJ · mol^{-1}

GAP + ADP $\rightleftharpoons$ 3PG + ATP $\quad \Delta G^{\circ\prime} = -12.1$ kJ · mol^{-1}

The second scenario is more likely. The first coupled reaction is more exergonic, but the second coupled reaction "captures" some of this free energy in the form of ATP, which the cell can use.

16. (a) $K_{eq} = e^{-\Delta G^{\circ\prime}/RT}$
$K_{eq} = e^{-31.5 \text{ kJ} \cdot \text{mol}^{-1}/(8.3145 \times 10^{-3} \text{ kJ} \cdot \text{K}^{-1} \cdot \text{mol}^{-1})(298\text{K})}$
$K_{eq} = e^{-12.7}$
$K_{eq} = 3.0 \times 10^{-6}$

$K_{eq} = \dfrac{[\text{palmitoyl-CoA}]}{[\text{palmitate}][\text{CoA}]}$

$3.0 \times 10^{-6} = \dfrac{[\text{palmitoyl-CoA}]}{[\text{palmitate}][\text{CoA}]}$

Therefore the ratio of products to reactants is 3.0×10^{-6}:1. The reaction is not favorable.

(b)
$H_3C{-}(CH_2)_{14}{-}C({=}O){-}O^- + \text{CoA} \rightleftharpoons$
$H_3C{-}(CH_2)_{14}{-}C({=}O){-}S{-}\text{CoA} + H_2O \quad \Delta G^{\circ\prime} = 31.5$ kJ·mol^{-1}

ATP + $H_2O \rightleftharpoons$ ADP + P_i $\quad \Delta G^{\circ\prime} = 30.5$ kJ·mol^{-1}

$H_3C{-}(CH_2)_{14}{-}C({=}O){-}O^- + \text{CoA} + \text{ATP} \rightleftharpoons$
$H_3C{-}(CH_2)_{14}{-}C({=}O){-}S{-}\text{CoA} + \text{ADP} + P_i \quad \Delta G^{\circ\prime} = 1.0$ kJ·mol^{-1}

$$K_{eq} = e^{-\Delta G^{\circ\prime}/RT}$$
$$K_{eq} = e^{-1.0\ kJ \cdot mol^{-1}/(8.3145 \times 10^{-3}\ kJ \cdot K^{-1} \cdot mol^{-1})(298K)}$$
$$K_{eq} = e^{-0.40}$$
$$K_{eq} = 0.67$$
$$K_{eq} = \frac{[\text{palmitoyl-CoA}][\text{ADP}][P_i]}{[\text{palmitate}][\text{CoA}][\text{ATP}]}$$
$$0.67 = \frac{[\text{palmitoyl-CoA}][\text{ADP}][P_i]}{[\text{palmitate}][\text{CoA}][\text{ATP}]}$$

Coupling the synthesis of palmitoyl-CoA with the hydrolysis of ATP to ADP has improved the [product]/[reactant] ratio considerably, but the formation of products is still not favored.

(c)

$$H_3C{-}(CH_2)_{14}{-}\overset{O}{\overset{\|}{C}}{-}O^- + CoA \rightleftharpoons H_3C{-}(CH_2)_{14}{-}\overset{O}{\overset{\|}{C}}{-}S{-}CoA + H_2O \quad \Delta G^{\circ\prime} = 31.5\ kJ{\cdot}mol^{-1}$$
$$ATP + H_2O \rightleftharpoons AMP + PP_i \quad \Delta G^{\circ\prime} = -32.2\ kJ{\cdot}mol^{-1}$$

$$H_3C{-}(CH_2)_{14}{-}\overset{O}{\overset{\|}{C}}{-}O^- + CoA + ATP \rightleftharpoons H_3C{-}(CH_2)_{14}{-}\overset{O}{\overset{\|}{C}}{-}S{-}CoA + AMP + PP_i \quad \Delta G^{\circ\prime} = -0.7\ kJ{\cdot}mol^{-1}$$

$$K_{eq} = e^{-\Delta G^{\circ\prime}/RT}$$
$$K_{eq} = e^{-(-0.7\ kJ \cdot mol^{-1})/(8.3145 \times 10^{-3}\ kJ \cdot K^{-1} \cdot mol^{-1})(298K)}$$
$$K_{eq} = e^{0.28}$$
$$K_{eq} = 1.33$$
$$K_{eq} = \frac{[\text{palmitoyl-CoA}][\text{AMP}][PP_i]}{[\text{palmitate}][\text{CoA}][\text{ATP}]}$$
$$1.33 = \frac{[\text{palmitoyl-CoA}][\text{AMP}][PP_i]}{[\text{palmitate}][\text{CoA}][\text{ATP}]}$$

Coupling the synthesis of palmitoyl-CoA with the hydrolysis of ATP to AMP has improved the [product]/[reactant] ratio. The formation of products is now slightly favored.

(d)

$$H_3C{-}(CH_2)_{14}{-}\overset{O}{\overset{\|}{C}}{-}O^- + CoA + ATP \rightleftharpoons H_3C{-}(CH_2)_{14}{-}\overset{O}{\overset{\|}{C}}{-}S{-}CoA + AMP + PP_i \quad \Delta G^{\circ\prime} = -0.7\ kJ{\cdot}mol^{-1}$$
$$PP_i + H_2O \rightleftharpoons 2\ P_i \quad \Delta G^{\circ\prime} = -33.5\ kJ{\cdot}mol^{-1}$$

$$H_3C{-}(CH_2)_{14}{-}\overset{O}{\overset{\|}{C}}{-}O^- + CoA + ATP + H_2O \rightleftharpoons H_3C{-}(CH_2)_{14}{-}\overset{O}{\overset{\|}{C}}{-}S{-}CoA + AMP + 2\ P_i \quad \Delta G^{\circ\prime} = -34.2\ kJ{\cdot}mol^{-1}$$

$$K_{eq} = e^{-\Delta G^{\circ\prime}/RT}$$
$$K_{eq} = e^{-(-34.2\ kJ \cdot mol^{-1})/(8.3145 \times 10^{-3}\ kJ \cdot K^{-1} \cdot mol^{-1})(298K)}$$
$$K_{eq} = e^{13.8}$$
$$K_{eq} = 9.9 \times 10^5$$
$$K_{eq} = \frac{[\text{palmitoyl-CoA}][\text{AMP}][P_i]^2}{[\text{palmitate}][\text{CoA}][\text{ATP}]}$$
$$9.9 \times 10^5 = \frac{[\text{palmitoyl-CoA}][\text{AMP}][P_i]^2}{[\text{palmitate}][\text{CoA}][\text{ATP}]}$$

Coupling the activation of palmitate to palmitoyl-CoA with the hydrolysis of ATP to AMP, with subsequent hydrolysis of pyrophosphate, is a thermodynamically effective means of accomplishing the reaction. Coupling the reaction with hydrolysis of ATP to ADP is not effective.

17. (a) The value of C is determined from the slope of the plot. The slope is equal to $1.9 \times 10^{-6}\ mM^{-1}$ or $1.9 \times 10^3\ M^{-1}$.

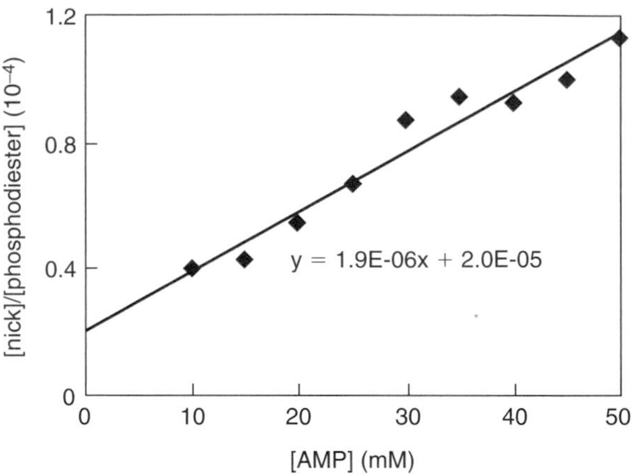

(b)
$$C = \frac{[PP_i]}{K_{eq}[\text{ATP}]}$$
$$K_{eq} = \frac{[PP_i]}{C[\text{ATP}]}$$
$$K_{eq} = \frac{(1.0 \times 10^{-3})}{(1.9 \times 10^{-3})(14 \times 10^{-6})}$$
$$K_{eq} = 3.8 \times 10^4$$

(c)
$$\Delta G^{\circ\prime} = -RT \ln K_{eq}$$
$$\Delta G^{\circ\prime} = -(8.3145 \times 10^{-3}\ kJ \cdot K^{-1} \cdot mol^{-1})(298K) \ln 3.8 \times 10^4$$
$$\Delta G^{\circ\prime} = -26\ kJ \cdot mol^{-1}$$

(d)

Nick ⇌ phosphodiester bond $\Delta G^{\circ\prime} = ?$
ATP ⇌ AMP + PP_i $\Delta G^{\circ\prime} = -48.5\ kJ \cdot mol^{-1}$

ATP + nick ⇌ AMP + PP_i + phosphodiester bond $\Delta G^{\circ\prime} = -26\ kJ \cdot mol^{-1}$

The $\Delta G^{\circ\prime}$ for the formation of the phosphodiester bond is $22.5\ kJ \cdot mol^{-1}$.

(e) The $\Delta G^{\circ\prime}$ value for the hydrolysis of a phosphodiester bond in DNA is $-22.5\ kJ \cdot mol^{-1}$, whereas the $\Delta G^{\circ\prime}$ value for the hydrolysis of a typical phosphomonoester bond is $-13.8\ kJ \cdot mol^{-1}$. Therefore, the phosphodiester bond in DNA is less stable than the typical phosphomonoester bond.

Chapter 10 Solutions

1. The β anomer is more stable because most of the bulky hydroxyl substituents are in the equatorial position.

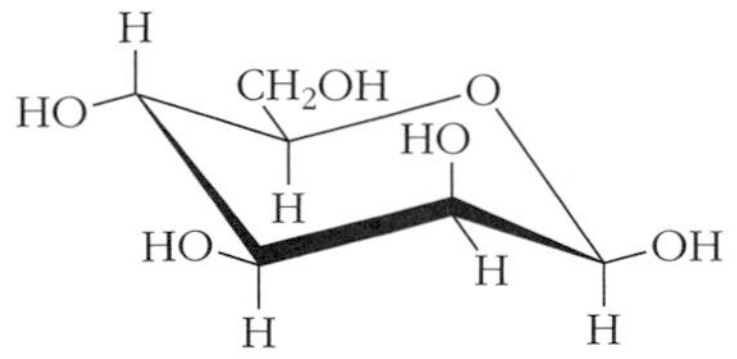

β-Mannose

2. The monosaccharide is glucose. The hydroxyl group at the C2 position has been replaced with an amino group ($—NH_2$), which is acetylated.

3. (a) Because the brain relies on glucose from the blood, it does not store glucose in the form of glycogen. Therefore, glucose rather than phosphorylated glucose is the substrate that enters the glycolytic pathway. The first step of glucose catabolism in the brain is catalyzed by hexokinase, so this step is the rate-determining step of the pathway. In other tissues that break down glycogen for glycolysis, the hexokinase step is bypassed.
(b) The low K_M means that the enzyme will be saturated with glucose and will therefore operate at maximum velocity. Even if the concentration of glucose were to fluctuate slightly, the brain's ability to catabolize glucose would not be affected.

4. (a) Hydroxyl groups on the glucose and fructose can supply both the —AH and —B functional groups, as shown below.

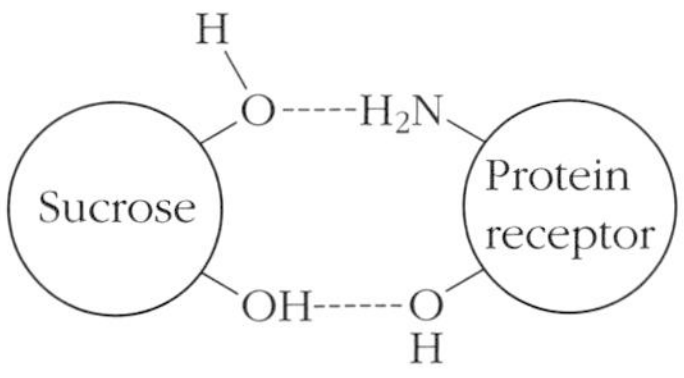

(b) The hydroxyl groups can supply the necessary —B and —AH groups while the chlorinated carbons can serve as hydrophobic —X groups.

5. In the presence of iodoacetate, fructose-l,6-bisphosphate accumulates, which suggests that iodoacetate inactivates the enzyme that uses fructose-l,6-bisphosphate as a substrate. Because iodoacetate reacts with Cys residues, the inactivation of the enzyme by the reagent suggests that a Cys residue is in the active site.

6. (a) Reactions 1, 3, 7, and 10; (b) Reactions 2, 5, and 8; (c) Reaction 6; (d) Reaction 9; (e) Reaction 4.

7. (a) The glucose–lactate pathway releases 196 $kJ \cdot mol^{-1}$ of free energy, enough to theoretically drive the synthesis of 196/30.5 or about 6 ATP.
(b) The complete oxidation of glucose releases 2850 $kJ \cdot mol^{-1}$ of free energy, enough for 2850/30.5 or about 93 ATP.

8. (a)

$$\Delta G^{\circ\prime} = -RT \ln \frac{[\text{glucose-6-phosphate}][\text{ADP}]}{[\text{glucose}][\text{ATP}]}$$

$$-16.7\ kJ \cdot mol^{-1} = -(8.3145 \times 10^{-3}\ kJ \cdot K^{-1} \cdot mol^{-1})(298K) \ln \frac{[\text{glucose-6-phosphate}][\text{ADP}]}{[\text{glucose}][\text{ATP}]}$$

$$16.7\ kJ \cdot mol^{-1} = 2.48\ kJ \cdot mol^{-1} \ln \frac{[\text{glucose-6-phosphate}][\text{ADP}]}{[\text{glucose}][\text{ATP}]}$$

$$6.73 = \ln \frac{[\text{glucose-6-phosphate}][\text{ADP}]}{[\text{glucose}][\text{ATP}]}$$

$$e^{6.73} = \frac{[\text{glucose-6-phosphate}][\text{ADP}]}{[\text{glucose}][\text{ATP}]}$$

$$840 = \frac{[\text{glucose-6-phosphate}][\text{ADP}]}{[\text{glucose}][\text{ATP}]}$$

$$840 = \frac{[\text{glucose-6-phosphate}](1)}{[\text{glucose}](10)}$$

$$8.4 \times 10^3 = \frac{[\text{glucose-6-phosphate}]}{[\text{glucose}]}$$

(b)

$$\Delta G^{\circ\prime} = -RT \ln \frac{[\text{glucose}][\text{ATP}]}{[\text{glucose-6-phosphate}][\text{ADP}]}$$

$$16.7\ kJ \cdot mol^{-1} = -(8.3145 \times 10^{-3}\ kJ \cdot K^{-1} \cdot mol^{-1})(298K) \ln \frac{[\text{glucose}][\text{ATP}]}{[\text{glucose-6-phosphate}][\text{ADP}]}$$

$$-16.7\ kJ \cdot mol^{-1} = 2.45\ kJ \cdot mol^{-1} \ln \frac{[\text{glucose}][\text{ATP}]}{[\text{glucose-6-phosphate}][\text{ADP}]}$$

$$-6.73 = \ln \frac{[\text{glucose}][\text{ATP}]}{[\text{glucose-6-phosphate}][\text{ADP}]}$$

$$e^{-6.73} = \frac{[\text{glucose}][\text{ATP}]}{[\text{glucose-6-phosphate}][\text{ADP}]}$$

$$1.2 \times 10^{-3} = \frac{[\text{glucose}][\text{ATP}]}{[\text{glucose-6-phosphate}][\text{ADP}]}$$

$$1.2 \times 10^{-3} = \frac{[\text{glucose}](10)}{[\text{glucose-6-phosphate}](1)}$$

$$1.2 \times 10^{-4} = \frac{[\text{glucose}]}{[\text{glucose-6-phosphate}]}$$

In order to reverse the reaction, the ratio of glucose-6-phosphate to glucose would have to be $8.3 \times 10^3:1$.

9. (a)

$$\Delta G^{\circ\prime} = -RT \ln \frac{[\text{fructose-6-phosphate}]}{[\text{glucose-6-phosphate}]}$$

$$2.2\ kJ \cdot mol^{-1} = -(8.3145 \times 10^{-3}\ kJ \cdot K^{-1} \cdot mol^{-1})(298K) \ln \frac{[\text{fructose-6-phosphate}]}{[\text{glucose-6-phosphate}]}$$

$$2.2\ kJ \cdot mol^{-1} = -2.47\ kJ \cdot mol^{-1} \ln \frac{[\text{fructose-6-phosphate}]}{[\text{glucose-6-phosphate}]}$$

$$-0.88 = \ln \frac{[\text{fructose-6-phosphate}]}{[\text{glucose-6-phosphate}]}$$

$$e^{-0.88} = \frac{[\text{fructose-6-phosphate}]}{[\text{glucose-6-phosphate}]}$$

$$0.41 = \frac{[\text{fructose-6-phosphate}]}{[\text{glucose-6-phosphate}]}$$

(b)
$$\Delta G = \Delta G^{\circ\prime} + RT \ln \frac{[\text{fructose-6-phosphate}]}{[\text{glucose-6-phosphate}]}$$

$$-1.4 \text{ kJ} \cdot \text{mol}^{-1} = 2.2 \text{ kJ} \cdot \text{mol}^{-1} + (8.3145 \times 10^{-3} \text{ kJ} \cdot \text{K}^{-1} \cdot \text{mol}^{-1}) (310\text{K}) \ln \frac{[\text{fructose-6-phosphate}]}{[\text{glucose-6-phosphate}]}$$

$$-3.6 \text{ kJ} \cdot \text{mol}^{-1} = 2.58 \text{ kJ} \cdot \text{mol}^{-1} \ln \frac{[\text{fructose-6-phosphate}]}{[\text{glucose-6-phosphate}]}$$

$$-1.4 = \ln \frac{[\text{fructose-6-phosphate}]}{[\text{glucose-6-phosphate}]}$$

$$e^{-1.4} = \frac{[\text{fructose-6-phosphate}]}{[\text{glucose-6-phosphate}]}$$

$$0.25 = \frac{[\text{fructose-6-phosphate}]}{[\text{glucose-6-phosphate}]}$$

The reaction will proceed in the direction of fructose-6-phosphate synthesis since $\Delta G < 0$.

10.
$$\Delta G = \Delta G^{\circ\prime} + RT \ln \frac{[\text{GAP}]}{[\text{DHAP}]}$$

$$4.4 \text{ kJ} \cdot \text{mol}^{-1} = 7.9 \text{ kJ} \cdot \text{mol}^{-1} + (8.3145 \times 10^{-3} \text{ kJ} \cdot \text{K}^{-1} \cdot \text{mol}^{-1})(310\text{K}) \ln \frac{[\text{GAP}]}{[\text{DHAP}]}$$

$$-3.5 \text{ kJ} \cdot \text{mol}^{-1} = 2.58 \text{ kJ} \cdot \text{mol}^{-1} \ln \frac{[\text{GAP}]}{[\text{DHAP}]}$$

$$-1.36 = \ln \frac{[\text{GAP}]}{[\text{DHAP}]}$$

$$e^{-1.36} = \frac{[\text{GAP}]}{[\text{DHAP}]}$$

$$0.26 = \frac{[\text{GAP}]}{[\text{DHAP}]}$$

The ratio of [GAP] to [DHAP] is 0.26:1, which seems to indicate that the formation of DHAP, and not the formation of GAP, is favored. However, GAP, the product of the triose phosphate isomerase reaction, is the substrate for the GAPDH reaction. The continuous removal of the product GAP by the action of GAPDH shifts the equilibrium toward formation of GAP from DHAP.

11. This would not be beneficial to the patient. In order to enter the glycolytic pathway, the glucose-6-phosphate would first have to enter cells. Glucose transporters recognize glucose and not glucose-6-phosphate; thus glucose-6-phosphate would be unable to enter cells for oxidation through the glycolytic pathway.

12. (a) When glucose is converted to pyruvate, NAD^+ is reduced to NADH in the glyceraldehyde-3-phosphate dehydrogenase reaction. Without a mechanism for regenerating NAD^+, glycolysis could not continue.
(b) If the trypanosome reduced pyruvate to lactate, this step would regenerate NAD^+ and no other pathway for regenerating NAD^+ would be needed.
(c) A drug that interfered with one of the enzymes in the trypanosome's NADH-oxidizing pathway would likely have no effect on the host since mammals lack the enzymes that would be inactivated by the drug.

13. Hexokinase-deficient erythrocytes have low levels of all glycolytic intermediates, since hexokinase catalyzes the first step of glycolysis. Therefore, the concentration of 2,3-BPG in the erythrocyte is decreased as well, favoring the oxygenated form of hemoglobin and decreasing its p_{50} value. Pyruvate kinase-deficient erythrocytes have high levels of 2,3-BPG since pyruvate kinase catalyzes the last step of glycolysis. This blockade at the last step causes the concentrations of all of the intermediates "ahead" of the block to be increased. Thus the oxygen affinity of hemoglobin decreases with increased 2,3-BPG concentration, and the p_{50} value increases as a result.

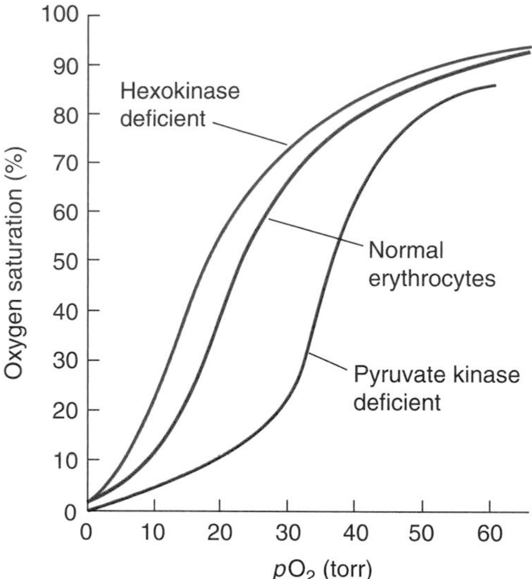

14. Arsenate is a metabolic poison, and the cells eventually die. In the presence of arsenate, 1,3-BPG is not formed; instead this step is essentially skipped. Two moles of ATP per mole of glucose are normally generated at this step. If ATP is not generated at this step, the net ATP yield for the glycolytic pathway is zero, and the cells die because they are unable to meet their energy requirements.

15. Phosphate levels increase because phosphate is a reactant for the GAPDH enzyme. ATP levels decrease, since ATP is a product of the GAPDH reaction. Levels of 2,3-BPG decrease as well, since levels of 1,3-BPG decrease as a result of GAPDH inhibition.

16. (a) The cancer cells may express the GAPDH protein at higher levels (i.e., transcription of the GAPDH gene and translation of its mRNA may occur at a higher rate).
(b) The structure of GAPDH in cancer cells is probably different from the structure of GAPDH in normal cells. The structure of the active site in GAPDH from cancer cells might be altered in such a way that the binding of methylglyoxal is permitted, which then precludes the binding of the substrate. Or the altered GAPDH might have a binding site for methylglyoxal elsewhere on the protein, which causes a conformational change in the protein that alters the substrate binding site so that the substrate can no longer bind.

17. (a) In hepatocytes, the phospho-His on phosphoglycerate mutase transfers its phosphate to the C2 position of 3-PG to form 2,3-BPG. The [^{32}P]-labeled phosphate at the C3 position is transferred back to the enzyme to form the 2-PG product, so initially the enzyme becomes labeled. In the next round of catalysis, the labeled phosphate on the enzyme is transferred to the C2 position of the next molecule of 3-PG substrate, so 2-PG becomes labeled. Eventually, this phosphate is transferred to ADP to form ATP, so ATP is labeled.
(b) In the plant, the labeled phosphate is transferred to C2 to form 2-PG, so 2-PG is labeled and then eventually ATP. The plant enzyme is not labeled.

18. Fluoride inhibits the enzyme enolase. If the enzyme is inactive, its substrate, 2-phosphoglycerate, will accumulate. The previous reaction is at equilibrium, so 3-phosphoglycerate will accumulate as 2-phosphoglycerate accumulates.

19. The alcohol dehydrogenase enzyme catalyzes the reduction of acetaldehyde to ethanol. Concomitantly, NADH is oxidized to NAD^+. The NADH reactant is produced by glycolysis in the GAPDH reaction. The NAD^+ produced in the alcohol dehydrogenase reaction can serve as a reactant for the glycolytic GAPDH reaction, allowing glycolysis to continue.

20. If hexokinase cannot be inhibited, all glucose that enters the yeast cell is phosphorylated at the expense of ATP to produce glucose-6-phosphate. Glucose-6-phosphate enters glycolysis and is isomerized to fructose-6-phosphate, which is then phosphorylated to fructose-1,6-bisphosphate, again at the expense of ATP. If concentrations of glucose are high, and if there is no mechanism to inhibit hexokinase, then cellular ATP will be depleted in these early steps. The early reactions of glycolysis proceed at a rate greater than the rate of the later ATP-generating steps of glycolysis. ATP is therefore used faster than it is regenerated, and the yeast mutants die as a result.

21. (a) One mole of ATP is invested when KDG is converted to KDPG. One mole of ATP is produced when 1,3-BPG is converted to 3-PG. One mole of ATP is produced when PEP is converted to pyruvate. Therefore, the net yield of this pathway (per mole of glucose) is one mole of ATP.
(b) In order to keep the pathway going, subsequent reactions would need to reoxidize NADPH produced when glucose is converted to gluconate and NADH produced by the GAPDH reaction.

22. This observation revealed that the pathways for glycogen degradation and synthesis must be different, since a defect in the degradative pathway has no effect on the synthetic pathway.

23. (a) In a liver cell, glucose-6-phosphate has four possible fates: It can be used to synthesize glycogen, it can be catabolized via glycolysis, it can be catabolized via the pentose phosphate pathway, and it can be converted to glucose and released from the cell.
(b) In a muscle cell, only the first three processes occur. Muscle cells lack glucose-6-phosphatase and therefore cannot release glucose from the cell.

24. (a) The first committed step of the pentose phosphate pathway is the first reaction, which is catalyzed by glucose-6-phosphate dehydrogenase and is irreversible. Once glucose-6-phosphate has passed this point, it has no other fate than conversion to a pentose phosphate.
(b) The hexokinase reaction does not commit glucose to the glycolytic pathway, since the product of the reaction, glucose-6-phosphate, can also enter the pentose phosphate pathway.

25. Full-grown cats, unlike kittens, are lactose intolerant and do not have the enzyme lactase to hydrolyze the milk sugar lactose into its component disaccharides, glucose and galactose. (Diarrhea results because lactose remains in the small intestine, increasing the osmotic pressure and drawing water out of the intestinal cells.)

26. An increase in the activity of glycogen phosphorylase in the fat body results in increased degradation of glycogen to glucose. Because fructose-2,6-bisphosphate concentrations are low, glycolysis is not stimulated (F26BP is a potent activator of the glycolytic enzyme PFK). Instead, glucose can be used to synthesize trehalose, which leaves the fat body and enters the hemolymph. In this way, the fat body produces sugars for use by other tissues in the fasting insect.

27. Production of glucose-1-phosphate requires only an isomerization reaction catalyzed by phosphoglucomutase to convert it to glucose-6-phosphate, which can enter glycolysis. This skips the hexokinase step and saves a molecule of ATP. Hydrolysis, which produces glucose, would require expenditure of an ATP to phosphorylate glucose to glucose-6-phosphate.

28. The pentose phosphate pathway in the red blood cell generates NADPH, which is used to regenerate oxidized glutathione. Glucose-6-phosphate dehydrogenase is the enzyme that catalyzes the first step of the oxidative branch of the pathway. Its deficiency results in a decreased output of NADPH from the pathway. As a result, glutathione remains in the oxidized form and cannot fulfill its roles of decreasing the concentrations of organic peroxides, maintaining red blood cell shape, and keeping the iron ion of hemoglobin in the +2 form. Hemolytic anemia is the likely result.

29. Fructose-1,6-bisphosphatase catalyzes the conversion of fructose-1,6-bisphosphate to fructose-6-phosphate in gluconeogenesis. When the enzyme is deficient, the gluconeogenic pathway is severely impaired and glucose synthesis from amino acids, glycerol, and lactate precursors occurs at a low level. At the beginning of a fast, blood glucose levels are maintained at normal levels because the source of blood glucose is glycogenolysis in the liver, which occurs normally. Once liver glycogen has been depleted, a normal individual would rely on gluconeogenesis for endogenous glucose production, but this pathway occurs at a low level in this patient because of the deficient enzyme; therefore blood glucose levels decrease. Pyruvate and alanine, both gluconeogenic precursors, are elevated because they cannot be converted to glucose via gluconeogenesis. The deficiency of fructose-1,6-bisphosphatase results in the buildup of the substrate, fructose-1,6-bisphosphate. This would promote the formation of GAP and DHAP, since the aldolase reaction is reversible. Because glycolysis does not occur during a fast, GAP is not consumed by the GAPDH reaction, so the ratio of GAP to DHAP, which is normally low, increases.

30. Glucose-6-phosphatase catalyzes the last reaction in gluconeogenesis (and glycogenolysis) in the liver. Glucose-6-phosphate is converted to glucose, and the glucose transporters export glucose to the circulation to make it available to other body tissues that do not carry out gluconeogenesis or store glycogen. In the absence of this enzyme, glucose-6-phosphate cannot be converted to glucose. Glucose-6-phosphate accumulates in the liver and is converted to glucose-1-phosphate, which is used for glycogen synthesis. Glycogen synthesis is therefore elevated in the livers of patients with this disease. The accumulation of glycogen enlarges the liver and causes the abdomen to protrude.

31.

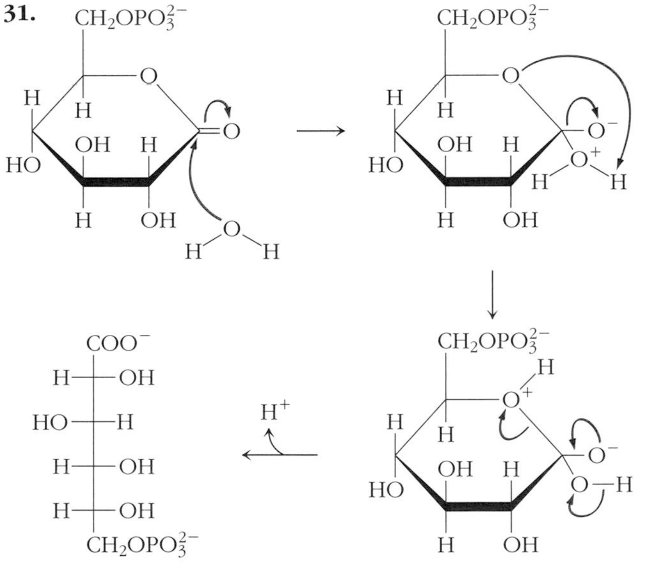

32. (a) All three allosteric effectors increase the catalytic activity of the enzyme by increasing the maximal velocity at saturating substrate concentrations and by increasing substrate affinity (measured as a decrease in K_M).

(b) The PFK-1C isozyme is less sensitive to the phosphate, AMP, and fructose-2,6-bisphosphate activators than isozymes A and B, as a higher concentration of these activators is required to achieve 50% of the maximal velocity. PFK-1C is intermediate to isozymes A and B in terms of its sensitivity to the inhibitor citrate.

The investigators have shown that PFK-1C differs from isozymes A and B because of its differential behavior toward allosteric activators and inhibitors. It's possible that the PFK-1C enzyme's different sensitivity to activators and inhibitors allows glycolysis to be regulated in a different manner in different tissues.

33. (a) Hexokinase, phosphofructokinase, and pyruvate kinase.

(b)

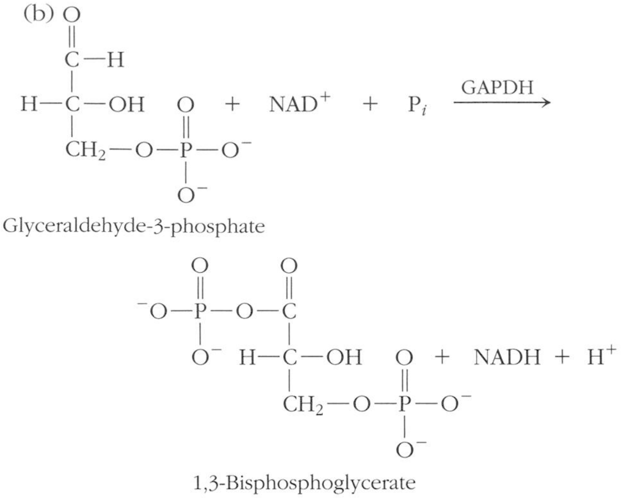

$$\text{Glyceraldehyde-3-phosphate} + NAD^+ + P_i \xrightarrow{\text{GAPDH}} \text{1,3-Bisphosphoglycerate} + NADH + H^+$$

(c) In the GAPDH reaction, NAD^+ is reduced to NADH and glyceraldehyde-3-phosphate is phosphorylated to 1,3-bisphosphoglycerate. Thus this reaction couples oxidation and phosphorylation. In the next step, ATP is generated when 1,3-bisphosphoglycerate donates one of its phosphoryl groups to ADP. Subsequent reactions (production of lactate, ethanol, or oxidative phosphorylation) must reoxidize the NADH to NAD^+ so that the glycolytic pathway can continue.

(d) The product, 1,3-bisphosphoglycerate, is not a substrate for the GAPDH enzyme. This indicates that the reaction takes place only in the forward direction and not in the reverse direction.

(e) $NADP^+$, NADH, and ATP are inhibitors of GAPDH because K_M values are greater than the control, indicating that GAPDH has a low affinity for substrate in the presence of these compounds. Glucose-1-phosphate, AMP, and ADP are activators of GAPDH because the K_M values are less than the control, indicating a greater affinity for substrate in the presence of these compounds.

NADH and ATP inhibit GAPDH because NADH is a product of the reaction and ATP is a product of the glycolytic pathway as a whole. Increased concentrations of NADH and ATP indicate a high energy charge of the cell. Glycolysis is inhibited when NADH and ATP are not needed. In the same manner, AMP and ADP activate GAPDH and thereby stimulate glycolysis, because AMP and ADP indicate that the energy charge of the cell is low and ATP is needed. Glucose-1-phosphate is an activator because it is the product of glycogenolysis. Glucose-1-phosphate is converted to glucose-6-phosphate, which enters glycolysis. Thus, glucose-1-phosphate is a feedforward activator that stimulates the activity of a "downstream" enzyme so that additional substrate can be accommodated in the glycolytic pathway. The inhibition by $NADP^+$ might be explained by noting that the other GAPDH isozyme is dependent on $NADP^+$. Inhibition of the NAD^+-dependent isozyme may ensure that both enzymes are not simultaneously active.

(f) The NAD^+-dependent GAPDH is a homotetramer and so may have multiple binding sites for its substrates. The plot shows that the $NADP^+$ inhibition generates a sigmoidal curve, indicating cooperative binding of the $NADP^+$ inhibitor.

Chapter 11 Solutions

1. The decarboxylation step is metabolically irreversible since the CO_2 product diffuses away from the enzyme. The other four reactions are transfer reactions or oxidation–reduction reactions (transfer of electrons) that are more easily reversed.

2. The purpose of Steps 4 and 5 is to regenerate the enzyme. In Step 3, the product acetyl-CoA is released, but the lipoamide prosthetic group of E2 is reduced. In Step 4, the FAD prosthetic group E3 reoxidizes the lipoamide group by accepting the protons and electrons from the reduced lipoamide. In Step 5, the $FADH_2$ is reoxidized by NAD^+. The product NADH then diffuses away.

3. TPP is a cofactor in two of the enzymes associated with the citric acid cycle—the pyruvate dehydrogenase complex and the α-ketoglutarate dehydrogenase complex. If thiamine was deficient, TPP would be deficient as well and the activities of both of these enzymes would decrease. As a result, the substrates of these two reactions, pyruvate and α-ketoglutarate, would accumulate.

4. Arsenite reacts with the reduced lipoamide group on E2 of the pyruvate dehydrogenase complex to form a compound with the structure shown in the figure. The enzyme cannot be regenerated and can no longer catalyze the conversion of acetyl-CoA to pyruvate. The α-ketoglutarate dehydrogenase complex also has a lipoamide group on its E2 subunits and is inhibited as well. The entire citric acid cycle cannot function, glucose cannot be oxidized aerobically, and respiration comes to a halt, which explains why these compounds are so toxic.

5.

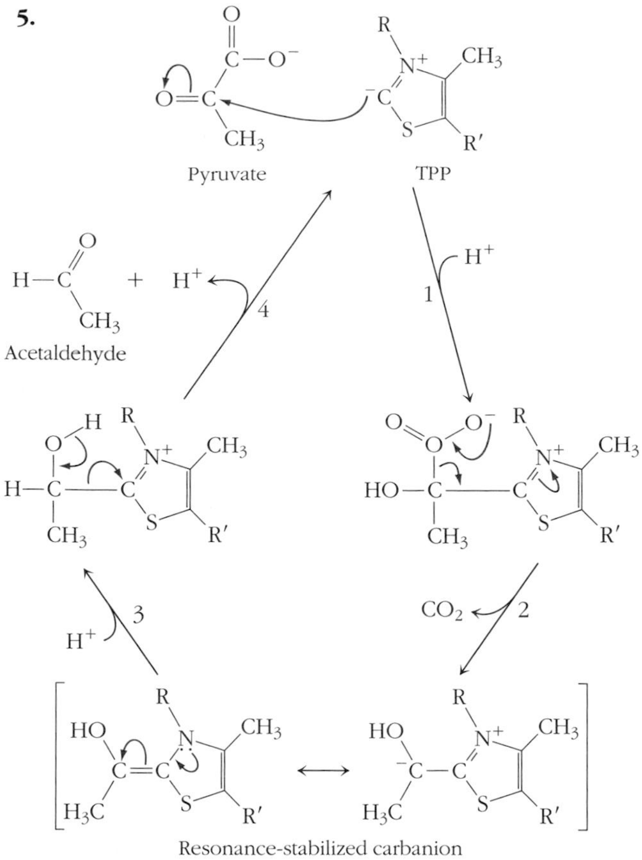

Pyruvate + TPP →(1, H^+) → (2, CO_2) → Resonance-stabilized carbanion →(3, H^+) → →(4) Acetaldehyde + H^+ + TPP

Pyruvate

TPP

Acetaldehyde

Resonance-stabilized carbanion

6. The phosphofructokinase reaction is the major rate-control point for the pathway of glycolysis. Inhibiting phosphofructokinase slows the entire pathway, so the production of acetyl-CoA by glycolysis followed by the pyruvate dehydrogenase complex can be decreased when the citric acid cycle is operating at maximum capacity and the citrate concentration is high.

7. Fluoroacetyl-CoA, an acetyl-CoA analog, is acted upon by citrate synthase to produce fluorocitrate from fluoroacetyl-CoA and oxaloacetate (OAA). The resulting fluorocitrate then serves as an inhibitor of aconitase. This leads to an accumulation of citrate, since the citric acid cycle can go no further if the aconitase reaction is inhibited.

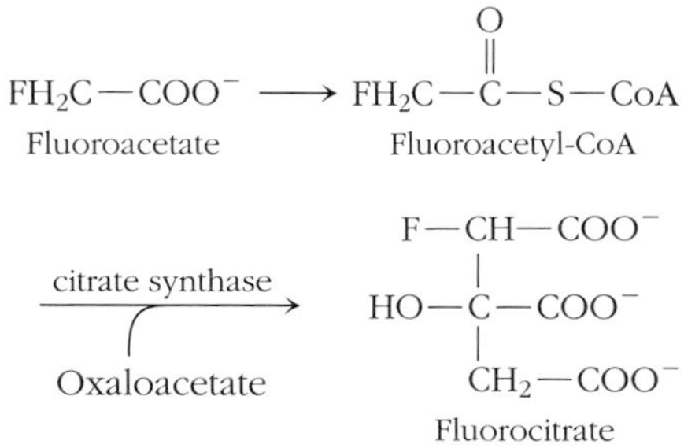

$$\underset{\text{Fluoroacetate}}{FH_2C{-}COO^-} \longrightarrow \underset{\text{Fluoroacetyl-CoA}}{FH_2C{-}\overset{O}{\overset{\|}{C}}{-}S{-}CoA}$$

$$\xrightarrow[\text{Oxaloacetate}]{\text{citrate synthase}} \text{F–CH(COO}^-\text{)–C(OH)(COO}^-\text{)–CH}_2\text{–COO}^-$$

Fluorocitrate

8. In the first step of the citrate synthase mechanism, a proton is removed from the acetyl group to form an enol. The histidine then acts as an acid to protonate the enol oxygen. The alanine side chain cannot donate a proton, thus the reaction cannot continue.

9. (a)

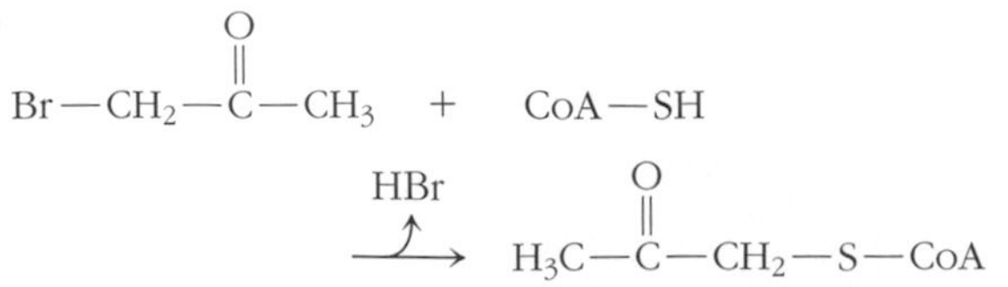

$$Br-CH_2-\overset{O}{\overset{\|}{C}}-CH_3 + CoA-SH \xrightarrow{HBr} H_3C-\overset{O}{\overset{\|}{C}}-CH_2-S-CoA$$

S-Acetonyl-CoA

(b) *S*-acetonyl-CoA is a competitive inhibitor. The V_{max} of the citrate synthase reaction is the same in the absence and in the presence of the inhibitor. The K_M has increased in the presence of the inhibitor, indicating that the enzyme's affinity for its substrate has decreased in the presence of the inhibitor. The inhibitor competes with acetyl-CoA for binding to the citrate synthase active site. *S*-acetonyl-CoA can do this because its structure resembles that of acetyl-CoA.

(c) Acetyl-CoA binds to pyruvate carboxylase and acts as an activator. A thioester functional group must be required for binding, since *S*-acetonyl-CoA, which lacks a thioester functional group, cannot bind to this binding site.

10. (a) Aconitase is the enzyme that catalyzes the reversible isomerization of citrate to isocitrate. Because this reaction is followed by and preceded by irreversible reactions, the inhibition of aconitase leads to an accumulation of citrate. The concentrations of other citric acid cycle intermediates will be decreased.

(b) If the citric acid cycle and mitochondrial respiration are not functioning, the cell turns to glycolysis to produce the ATP required for its energy needs. Consequently, flux through glycolysis increases. The increase in the rate of the pentose phosphate pathway is required to meet the increased demand for reducing equivalents during hyperoxia.

11. Step 1. In the first step, α-ketoglutarate is decarboxylated, a process which requires TPP. The carbonyl carbon becomes a carbanion, which forms a bond with TPP.

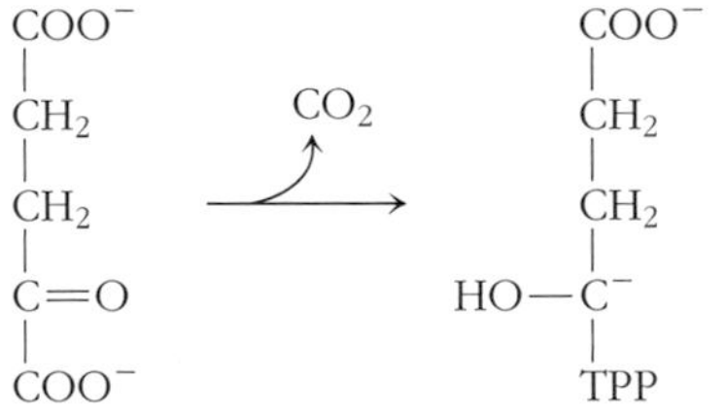

COO^-–CH_2–CH_2–C=O–COO^- (α-Ketoglutarate) $\xrightarrow{CO_2}$ COO^-–CH_2–CH_2–HO–C^-–TPP

α-Ketoglutarate

Step 2. The succinyl group is then transferred to the lipoamide prosthetic group of E2 of the α-ketoglutarate dehydrogenase complex.

COO^-–CH_2–CH_2–HO–C^-–TPP + Lipoamide (S–S, R) $\xrightarrow{TPP}$ COO^-–CH_2–CH_2–C=O–S–(lipoamide, HS, R)

Lipoamide

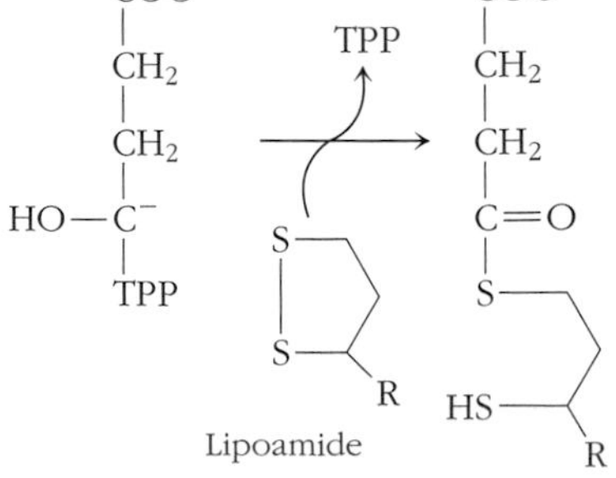

Step 3. The succinyl group is transferred to coenzyme A, and the lipoamide group is reduced.

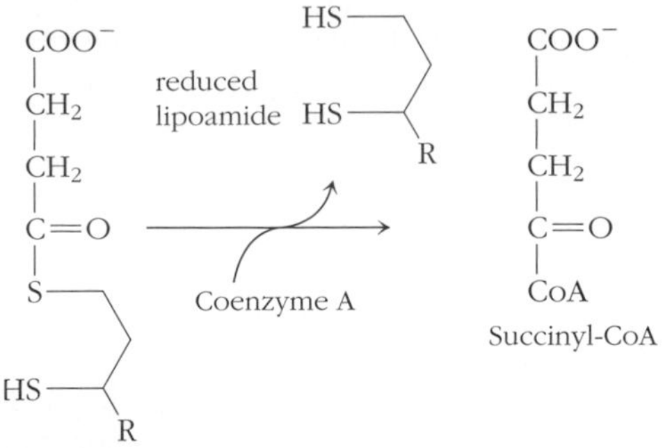

COO^-–CH_2–CH_2–C=O–S–(lipoamide, HS, R) + Coenzyme A $\rightarrow$ reduced lipoamide (HS, HS, R) + COO^-–CH_2–CH_2–C=O–CoA

Succinyl-CoA

Step 4. and 5. The last two steps are the same as for the pyruvate dehydrogenase complex. The FAD prosthetic group of E3 reoxidizes the lipoamide by accepting two protons and two electrons. $FADH_2$ is reoxidized by NAD^+. The NADH and H^+ products diffuse away.

12. When operating in reverse, succinyl-CoA synthetase catalyzes a kinase-type reaction, the transfer of a phosphoryl group from a nucleotide triphosphate (GTP or ATP).

13. Succinate accumulates because it cannot be converted to fumarate. Succinyl-CoA also accumulates because the succinyl-CoA synthetase reaction is reversible. However, the succinyl-CoA ties up some of the cell's CoA supply, so the α-ketoglutarate dehydrogenase reaction, which requires CoA, slows. As a result, α-ketoglutarate accumulates.

14. (a)

Malate + NAD^+ ⇌ oxaloacetate + NADH + H^+ $\quad \Delta G°' = 29.7\ kJ \cdot mol^{-1}$

Oxaloacetate + acetyl-CoA ⇌ citrate + coenzyme A $\quad \Delta G°' = -31.5\ kJ \cdot mol^{-1}$

Malate + NAD^+ + acetyl-CoA ⇌ citrate + NADH + H^+ + CoA $\quad \Delta G°' = -1.8\ kJ \cdot mol^{-1}$

(b) The equilibrium constant for the coupled reaction is 3.4×10^5 times greater than the equilibrium constant for the uncoupled reaction.

Reactions 1 and 8

$$K_{eq} = e^{-\Delta G°'/RT}$$
$$K_{eq} = e^{-(-1.8\ kJ \cdot mol^{-1})/(8.3145 \times 10^{-3}\ kJ \cdot K^{-1} \cdot mol^{-1})(298K)}$$
$$K_{eq} = e^{0.73}$$
$$K_{eq} = 2.1$$

Uncoupled Reaction 8

$$K_{eq} = e^{-\Delta G°'/RT}$$
$$K_{eq} = e^{-(29.7\ kJ \cdot mol^{-1})/(8.3145 \times 10^{-3}\ kJ \cdot K^{-1} \cdot mol^{-1})(298K)}$$
$$K_{eq} = e^{-12.0}$$
$$K_{eq} = 6.2 \times 10^{-6}$$

15. (a) The isotopic label on C4 of oxaloacetate is released as $^{14}CO_2$ in the α-ketoglutarate dehydrogenase reaction.

(b) The isotopic label on C1 of acetyl-CoA is scrambled at the succinyl-CoA synthetase step. Because succinate is symmetrical, C1 and C4 are chemically equivalent, so that in a population of molecules, both C1 and C4 would appear to be labeled (half the label would appear to be at C1 and half at C4). Consequently, one round of the citric acid cycle would yield oxaloacetate with half the labeled carbon at C1 and half at C4. Both of these labeled carbons would be lost as $^{14}CO_2$ in a second round of the citric acid cycle.

16. $\Delta G = \Delta G^{\circ\prime} + RT \ln \dfrac{[\text{malate}]}{[\text{fumarate}]}$

$$0 = -3.4 \text{ kJ} \cdot \text{mol}^{-1} + (8.3145 \times 10^{-3} \text{ kJ} \cdot \text{K}^{-1} \cdot \text{mol}^{-1})(310\text{K}) \ln \frac{[\text{malate}]}{[\text{fumarate}]}$$

$$-3.4 \text{ kJ} \cdot \text{mol}^{-1} = 2.58 \text{ kJ} \cdot \text{mol}^{-1} \ln \frac{[\text{malate}]}{[\text{fumarate}]}$$

$$1.32 = \ln \frac{[\text{malate}]}{[\text{fumarate}]}$$

$$e^{1.32} = \frac{[\text{malate}]}{[\text{fumarate}]}$$

$$3.7 = \frac{[\text{malate}]}{[\text{fumarate}]}$$

The ratio of malate to fumarate is 3.7 to 1, indicating that the reaction proceeds in the direction of formation of malate. This is not a control point for the citric acid cycle because the ΔG is close to zero, indicating it is a near-equilibrium reaction.

17. The volume increase that occurs when bread rises is due to the release of CO_2, which results from both aerobic and anaerobic oxidation of glucose (probably both processes occur, with aerobic respiration occurring on the surface of the dough and anaerobic fermentation occurring in the interior of the loaf).

In alcoholic fermentation (anaerobic), sucrose is hydrolyzed to glucose and fructose by enzymes in the yeast. Then the fructose and glucose both enter glycolysis. The end product is pyruvate. Pyruvate is then converted to acetaldehyde. In this step, CO_2 is released. The acetaldehyde is then reduced to ethanol, which evaporates when the bread is baked.

In aerobic respiration, the glucose and fructose enter glycolysis and are converted to pyruvate. Then pyruvate is converted to acetyl-CoA. This step releases CO_2. Acetyl-CoA then enters the citric acid cycle and is oxidized. For every acetyl-CoA that enters the citric acid cycle, two CO_2 molecules are released.

18. (a) Substrate availability: Acetyl-CoA and oxaloacetate levels regulate citrate synthase activity.
(b) Product inhibition: Citrate inhibits citrate synthase; NADH inhibits isocitrate dehydrogenase and α-ketoglutarate dehydrogenase; and succinyl-CoA inhibits α-ketoglutarate dehydrogenase.
(c) Feedback inhibition: NADH and succinyl-CoA inhibit citrate synthase.

19. The alternate pathway bypasses the succinyl-CoA synthetase reaction of the standard citric acid cycle, a step that is accompanied by the phosphorylation of a nucleoside diphosphate. The alternate pathway therefore generates one less nucleoside triphosphate than the standard citric acid cycle. There is no difference in the number of reduced cofactors generated.

20. Any metabolite that can be converted to oxaloacetate can enter gluconeogenesis and serve as a precursor for glucose. Biological molecules that are degraded to acetyl-CoA cannot be used as glucose precursors because acetyl-CoA enters the citric acid cycle and its two carbons are eventually oxidized to carbon dioxide. Thus, glyceraldehyde-3-phosphate, tryptophan, and phenylalanine can serve as gluconeogenic substrates because at least one of their breakdown products can be converted to oxaloacetate. Palmitate and leucine are not glucogenic because their breakdown products are acetyl-CoA or one of its derivatives. Acetyl-CoA cannot be converted to pyruvate in mammals.

21. An increase in the concentration of glucose increases flux through glycolysis, which produces more pyruvate. The pyruvate dehydrogenase complex activity increases in response to the increase in pyruvate so that the conversion of pyruvate to acetyl-CoA can increase proportionally. At the same time, more oxaloacetate is required, since equimolar amounts of acetyl-CoA and oxaloacetate are required for the first step of the citric acid cycle. Therefore, the activity of pyruvate carboxylase, the enzyme that catalyzes the conversion of pyruvate to oxaloacetate, also increases.

22. Succinyl-CoA is converted to oxaloacetate in the citric acid cycle. Oxaloacetate can be converted to glucose by gluconeogenesis.

23. The exercising muscle required greater concentrations of ATP to power it, so the rates of glycolysis and the citric acid cycle increased. Phosphoenolpyruvate was converted to pyruvate more rapidly, so the concentration of phosphoenolpyruvate decreased. Some of the pyruvate was converted to acetyl-CoA via the pyruvate dehydrogenase reaction. Since equimolar amounts of acetyl-CoA and oxaloacetate are required for the first step of the citric acid cycle, some of the pyruvate was converted to oxaloacetate via the pyruvate carboxylase reaction. This explains why oxaloacetate concentrations increased. The concentration of pyruvate did not increase because a steady state was reached: The rate of production of pyruvate from phosphoenolpyruvate was equal to the rate of consumption of pyruvate.

24. (a) Isocitrate is a branch point between the citric acid cycle and the glyoxylate cycle in organisms such as *E. coli,* which have the glyoxylate cycle. When acetate is the only food source, isocitrate dehydrogenase is inactive and acetate enters the glyoxylate cycle (as acetyl-CoA) instead. The acetate is converted to glucose by the glyoxylate cycle. A number of important intermediates are formed along the way, which can be used as precursors in other biosynthetic reactions.

Citrate ⟶ Isocitrate —(isocitrate lyase)⟶ Succinate + Glyoxylate

Glyoxylate + Acetyl-CoA

Acetate ⟶ Acetyl-CoA —(malate synthase)⟶ Malate —(malate dehydrogenase)⟶ Oxaloacetate

(b) When glucose is the substrate for the cultured *E. coli*, the glyoxylate cycle, which produces glucose, is no longer necessary. Isocitrate dehydrogenase then becomes active and glucose enters glycolysis and then the citric acid cycle. The oxidation of glucose provides the necessary ATP.

25. If the fatty acid is broken down, the resulting acetyl-CoA units can enter the citric acid cycle for further oxidation. Because the labeled carbons are not lost as CO_2 in the first round of the cycle (see Fig. 11-7), some of the label will appear in oxaloacetate. If some of this oxaloacetate is siphoned from the citric acid cycle for gluconeogenesis, some of the label will appear in glucose. This sequence of reactions does not violate the statement that fatty acids cannot be converted to carbohydrates, because two carbons have already been lost as CO_2. Thus, there is no *net* conversion of fatty acid carbons to glucose carbons.

26. (a) The citric acid cycle is a multistep catalyst. Degrading an amino acid to a citric acid cycle intermediate boosts the catalytic activity of the cycle but does not alter the stoichiometry of the overall reaction (acetyl-CoA $\rightarrow$ 2 CO_2).

(b) Pyruvate derived from the degradation of an amino acid can be converted to acetyl-CoA by the pyruvate dehydrogenase complex; these amino acid carbons can then be completely oxidized by the citric acid cycle.

27. In the reversible reaction shown, the amino acid aspartate and the glycolytic product pyruvate can undergo a transamination in which aspartate's amino group is transferred, leaving oxaloacetate, which is a citric acid cycle intermediate.

28. (a) Hydroxycitrate differs by the addition of a hydroxyl group.

(b) Based on its structural similarity to the substrate citrate, hydroxycitrate is most likely to act as a competitive inhibitor.

(c) Acetyl-CoA synthesized in the mitochondria from pyruvate, the product of carbohydrate catabolism, is made available in the cytosol by the action of ATP-citrate lyase (see Fig. 11-17). Inhibiting this enzyme might decrease the amount of acetyl-CoA available for fatty acid synthesis.

(d) Since cholesterol synthesis also begins with acetyl-CoA, hydroxycitrate might reduce the production of cholesterol and the structurally related steroid hormones. (Although hydroxycitrate inhibits ATP-citrate lyase *in vitro*, controlled clinical trials have shown that the compound has no significant effect on weight loss in humans.)

29. (a) NADH, citrate, and succinyl-CoA inhibit citrate synthase in mammals but do not inhibit citrate synthase in *H. pylori*. Isocitrate dehydrogenase is $NADP^+$ dependent rather than NAD^+ dependent and is regulated differently (by higher concentrations of its substrates $NADP^+$ and isocitrate instead of NADH). *H. pylori* lacks α-ketoglutarate dehydrogenase and instead has α-ketoglutarate oxidase. The enzyme succinyl-CoA synthetase is missing in *H. pylori*. This enzyme catalyzes the only substrate-level phosphorylation reaction in the citric acid cycle, therefore no GTP is produced in the citric acid cycle of this organism. Succinate dehydrogenase is missing. Fumarate reductase is present. Mammals do not have the glyoxylate cycle, but most bacteria do. *H. pylori* has only one step, catalyzed by malate synthase (isocitrate lyase is not present).

(b) High K_M values indicate a low affinity of the enzyme for the substrate. The high values indicate that substrate concentrations must be relatively high for the enzyme to attain half-maximal velocity. This indicates that the pathway does not operate unless the concentrations of the cycle intermediates are relatively high, which occurs when *H. pylori* is in a nutrient-rich environment with plentiful resources (which would occur when the human "host" ingests a meal). Since the primary purpose of the citric acid cycle in *H. pylori* is to provide biosynthetic intermediates, it makes sense that the pathway only operates when metabolic resources are plentiful.

(c) *H. pylori* citrate synthase is inhibited by ATP but is not affected by NADH or any of the other citric acid cycle intermediates. Since the citric acid cycle in this organism is not used to produce metabolic energy in the form of reducing equivalents for electron transport, it makes sense that NADH would not serve as an inhibitor.

(d) Citrate synthase, isocitrate dehydrogenase, and α-ketoglutarate oxidase may serve as regulatory control points since these enzymes catalyze irreversible reactions and are subject to activation and inhibition by allosteric modulators.

(e) Enzymes unique to *H. pylori* would be good therapeutic targets: α-ketoglutarate oxidase, fumarate reductase, and malate synthase.

30. (a)

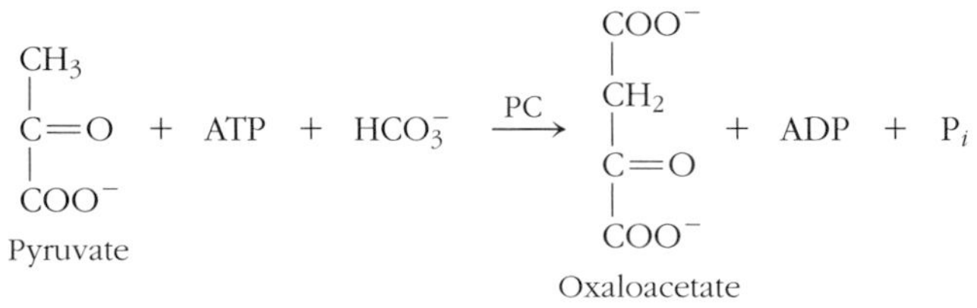

$$\underset{\text{Pyruvate}}{CH_3{-}C({=}O){-}COO^-} + ATP + HCO_3^- \xrightarrow{PC} \underset{\text{Oxaloacetate}}{{}^-OOC{-}CH_2{-}C({=}O){-}COO^-} + ADP + P_i$$

(b)

$$\underset{\text{Phosphoenolpyruvate}}{CH_2{=}C(O{-}P_i){-}COO^-} + HCO_3^- \xrightarrow{PPC} \underset{\text{Oxaloacetate}}{{}^-OOC{-}CH_2{-}C({=}O){-}COO^-} + P_i$$

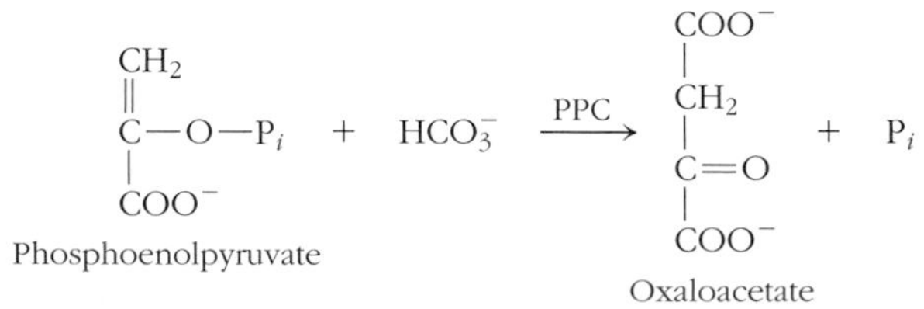

(c) In the first step of the citric acid cycle, acetyl-CoA condenses with oxaloacetate to form citrate. The net result at the end of the cycle is that the two acetyl-CoA carbons are oxidized to CO_2, reducing equivalents in the form of NADH and QH_2 are generated, and the oxaloacetate is regenerated so that the cycle can begin again. Without oxaloacetate, the citric acid cycle cannot operate. The cycle provides the last step for the oxidation of fuel molecules and also produces intermediates for biosynthetic reactions, so the proper functioning of the cycle is important to the health of the organism.

(d) The enzyme requires ATP for full activity. No other nucleotide can substitute. If ATP is present, additional nucleotides have no effect on the activity with the exception of ADP and ITP, which are mildly inhibitory.

(e) Acetyl-CoA and α-ketoglutarate are mildly inhibitory. Aspartate and glutamate have very little effect on the activity of the enzyme.

(f) Mg^{2+} ions are required for full activity. Co^{2+} and Mn^{2+} can substitute, but the activity of the enzyme is decreased. Zn^{2+} cannot substitute for Mg^{2+}.

Chapter 12 Solutions

1. Consult Table 12-1 for the relevant half-reactions involving pyruvate and NADH. Reverse the NADH half-reaction and the sign of its $\mathcal{E}°'$ value to indicate oxidation, then combine the half-reactions and their $\mathcal{E}°'$ values.

Pyruvate + $2H^+$ + $2e^-$ → lactate $\mathcal{E}°' = -0.185$ V
NADH → NAD^+ + H^+ + $2e^-$ $\mathcal{E}°' = 0.315$ V

NADH + pyruvate + H^+ → NAD^+ + lactate $\Delta\mathcal{E}°' = 0.130$ V

Use Equation 12-3 to calculate $\Delta G°'$ for this reaction:

$$\Delta G°' = -n\mathcal{F}\Delta\mathcal{E}°'$$
$$\Delta G°' = -(2)(96{,}485\ \text{J}\cdot\text{V}^{-1}\cdot\text{mol}^{-1})(0.130\ \text{V})$$
$$\Delta G°' = -25.1\ \text{kJ}\cdot\text{mol}^{-1}$$

The reduction of pyruvate by NADH (see Section 10-2) is spontaneous under standard conditions.

2. Consult Table 12-1 for the relevant half-reactions involving cytochrome a_3 and oxygen. Reverse the cytochrome a_3 half-reaction and the sign of its $\mathcal{E}°'$ value to indicate oxidation, multiply the coefficients by 2 so that the number of electrons transferred will be equal, then combine the half-reactions and their $\mathcal{E}°'$ values.

2 Cytochrome a_3 (Fe^{2+}) → 2 cytochrome a_3 (Fe^{3+}) + $2e^-$ $\mathcal{E}°' = -0.385$ V
1/2 O_2 + $2H^+$ + 2 e^- → H_2O $\mathcal{E}°' = 0.815$ V

2 Cyto a_3 (Fe^{2+}) + 1/2 O_2 + $2H^+$ → 2 cyto a_3 (Fe^{3+}) + H_2O $\Delta\mathcal{E}°' = 0.430$ V

Use Equation 12-3 to calculate $\Delta G°'$ for this reaction:

$$\Delta G°' = -n\mathcal{F}\Delta\mathcal{E}°'$$
$$\Delta G°' = -(2)(96{,}485\ \text{J}\cdot\text{V}^{-1}\cdot\text{mol}^{-1})(0.430\ \text{V})$$
$$\Delta G°' = -83\ \text{kJ}\cdot\text{mol}^{-1}$$

The reaction is spontaneous under standard conditions.

3. Consult Table 12-1 for the relevant half-reactions involving ubiquinol and cytochrome c_1. Reverse the ubiquinol half-reaction and the sign of its $\mathcal{E}°'$ value to indicate oxidation, multiply the coefficients in the cytochrome c_1 equation by 2 so that the number of electrons transferred will be equal, then combine the half-reactions and their $\mathcal{E}°'$ values.

Ubiquinol → ubiquinone + $2H^+$ + $2e^-$ $\mathcal{E}°' = -0.045$ V
2 Cytochrome c_1 (Fe^{3+}) + $2e^-$ → 2 cytochrome c_1 (Fe^{2+}) $\mathcal{E}°' = 0.220$ V

Ubiquinol + 2 cyto c_1 (Fe^{3+}) → ubiquinone + 2 cyto c_1 (Fe^{2+}) $\mathcal{E}°' = 0.205$ V

Use Equation 12-3 to calculate $\Delta G°'$ for this reaction:

$$\Delta G°' = -n\mathcal{F}\Delta\mathcal{E}°'$$
$$\Delta G°' = -(2)(96{,}485\ \text{J}\cdot\text{V}^{-1}\cdot\text{mol}^{-1})(0.205\ \text{V})$$
$$\Delta G°' = -39.6\ \text{kJ}\cdot\text{mol}^{-1}$$

The reaction is spontaneous under standard conditions.

4. Consult Table 12-1 for the relevant half-reactions involving acetoacetate and NADH. Reverse the NADH half-reaction and the sign of its $\mathcal{E}°'$ value to indicate oxidation, then combine the half-reactions and their $\mathcal{E}°'$ values.

Acetoacetate + $2H^+$ + $2e^-$ → 3-hydroxybutyrate $\mathcal{E}°' = -0.346$ V
NADH → NAD^+ + H^+ + $2e^-$ $\mathcal{E}°' = 0.315$ V

Acetoacetate + NADH + H^+ → NAD^+ + 3-hydroxybutyrate $\Delta\mathcal{E}°' = -0.031$ V

Use Equation 12-3 to calculate $\Delta G°'$ for this reaction:

$$\Delta G°' = -n\mathcal{F}\Delta\mathcal{E}°'$$
$$\Delta G°' = -(2)(96{,}485\ \text{J}\cdot\text{V}^{-1}\cdot\text{mol}^{-1})(-0.031\ \text{V})$$
$$\Delta G°' = 6.0\ \text{kJ}\cdot\text{mol}^{-1}$$

Consult Table 12-1 for the relevant half-reactions involving acetoacetate and $FADH_2$. Reverse the $FADH_2$ half-reaction and the sign of its $\mathcal{E}°'$ value to indicate oxidation, then combine the half-reactions and their $\mathcal{E}°'$ values.

Acetoacetate + $2H^+$ + $2e^-$ → 3-hydroxybutyrate $\mathcal{E}°' = -0.346$ V
$FADH_2$ → FAD + $2H^+$ + $2e^-$ $\mathcal{E}°' = 0.0$ V

Acetoacetate + $FADH_2$ → FAD + 3-hydroxybutyrate $\Delta\mathcal{E}°' = -0.346$ V

Use Equation 12-3 to calculate $\Delta G°'$ for this reaction:

$$\Delta G°' = -n\mathcal{F}\Delta\mathcal{E}°'$$
$$\Delta G°' = -(2)(96{,}485\ \text{J}\cdot\text{V}^{-1}\cdot\text{mol}^{-1})(-0.346\ \text{V})$$
$$\Delta G°' = 66.8\ \text{kJ}\cdot\text{mol}^{-1}$$

The reduction of acetoacetate by NADH is more favorable than reduction by $FADH_2$, as shown by the $\Delta G°'$ values calculated above. Although neither reaction is spontaneous under standard conditions, reduction by $FADH_2$ is far more unfavorable. The reduction of acetoacetate by NADH is, in fact, a favorable process under cellular conditions, which differ from standard conditions.

5. Consult Table 12-1 for the relevant half-reactions involving succinate and NAD^+. Reverse the succinate half-reaction and the sign of its $\mathcal{E}°'$ value to indicate oxidation, then combine the half-reactions and their $\mathcal{E}°'$ values.

Succinate → fumarate + $2H^+$ + $2e^-$ $\mathcal{E}°' = -0.031$ V
NAD^+ + H^+ + $2e^-$ → NADH $\mathcal{E}°' = -0.315$ V

Succinate + NAD^+ → NADH + fumarate + H^+ $\Delta\mathcal{E}°' = -0.346$ V

Use Equation 12-3 to calculate $\Delta G°'$ for this reaction:

$$\Delta G°' = -n\mathcal{F}\Delta\mathcal{E}°'$$
$$\Delta G°' = -(2)(96{,}485\ \text{J}\cdot\text{V}^{-1}\cdot\text{mol}^{-1})(-0.346\ \text{V})$$
$$\Delta G°' = 66.8\ \text{kJ}\cdot\text{mol}^{-1}$$

Consult Table 12-1 for the relevant half-reactions involving succinate and FAD. Reverse the succinate half-reaction and the sign of its $\mathcal{E}°'$ value to indicate oxidation, then combine the half-reactions and their $\mathcal{E}°'$ values.

Succinate $\rightarrow$ fumarate + $2H^+$ + $2e^-$	$\mathcal{E}°' = -0.031$ V
FAD + $2H^+$ + $2e^-$ $\rightarrow$ $FADH_2$	$\mathcal{E}°' = 0.0$ V
Succinate + FAD $\rightarrow$ fumarate + $FADH_2$	$\Delta\mathcal{E}°' = -0.031$ V

Use Equation 12-3 to calculate $\Delta G°'$ for this reaction:

$$\Delta G°' = -n\mathcal{F}\Delta\mathcal{E}°'$$
$$\Delta G°' = -(2)(96{,}485 \text{ J} \cdot \text{V}^{-1} \cdot \text{mol}^{-1})(-0.031 \text{ V})$$
$$\Delta G°' = 6.0 \text{ kJ} \cdot \text{mol}^{-1}$$

The oxidation of succinate by FAD is more favorable than oxidation by NAD^+, as shown by the $\Delta G°'$ values calculated above. Although neither reaction is spontaneous under standard conditions, oxidation by NAD^+ is far more unfavorable. The oxidation of succinate by FAD is, in fact, a favorable process under cellular conditions (citric acid cycle; see Chapter 11), which differ from standard conditions.

6. (a) Use the Nernst equation to determine the $\mathcal{E}$ values for these two half-reactions.

$QH_2 \rightarrow Q + 2H^+ + 2e^-$ $\quad \mathcal{E}°' = -0.045$ V

$$\mathcal{E} = \mathcal{E}°' + \frac{0.026 \text{ V}}{n} \ln \frac{[Q]}{[QH_2]}$$
$$\mathcal{E} = -0.045 \text{ V} + \frac{0.026 \text{ V}}{2} \ln \frac{1}{10}$$
$$\mathcal{E} = 0.075 \text{ V}$$

2 Cytochrome c (Fe^{3+}) + $2e^-$ $\rightarrow$ 2 cytochrome c (Fe^{2+}) $\quad \mathcal{E}°' = 0.235$ V

$$\mathcal{E} = \mathcal{E}°' + \frac{0.026 \text{ V}}{n} \ln \frac{[\text{cyto } c(Fe^{3+})]}{[\text{cyto } c(Fe^{2+})]}$$
$$\mathcal{E} = 0.235 \text{ V} + \frac{0.026 \text{ V}}{2} \ln 5$$
$$\mathcal{E} = 0.256 \text{ V}$$

$QH_2 \rightarrow Q + 2H^+ + 2e^-$	$\mathcal{E} = -0.075$ V
2 Cytochrome c (Fe^{3+}) + $2e^-$ $\rightarrow$ 2 cytochrome c (Fe^{2+})	$\mathcal{E} = 0.256$ V
QH_2 + 2 cyto c (Fe^{3+}) $\rightarrow$ Q + $2H^+$ + 2 cyto c (Fe^{2+})	$\Delta\mathcal{E} = 0.181$ V

(b) Use Equation 12-3 to calculate ΔG for this reaction:

$$\Delta G = -n\mathcal{F}\Delta\mathcal{E}$$
$$\Delta G = -(2)(96{,}485 \text{ J} \cdot \text{V}^{-1} \cdot \text{mol}^{-1})(0.181 \text{ V})$$
$$\Delta G = -34.9 \text{ kJ} \cdot \text{mol}^{-1}$$

7. Reverse the half-reaction for the iron-sulfur protein to indicate that it is being oxidized. Add up the two half-reactions to obtain the $\Delta\mathcal{E}°'$ for the reaction.

FeS(red) $\rightarrow$ FeS(ox) + e^-	$\mathcal{E}°' = -0.280$ V
Cytochrome c_1 (Fe^{3+}) + e^- $\rightarrow$ cytochrome c_1 (Fe^{2+})	$\mathcal{E}°' = 0.215$ V
FeS(red) + cyto c_1 (Fe^{3+}) $\rightarrow$ FeS(ox) + cyto c_1 (Fe^{2+})	$\Delta\mathcal{E}°' = -0.065$ V

Use Equation 12-3 to calculate $\Delta G°'$ for this reaction:

$$\Delta G°' = -n\mathcal{F}\Delta\mathcal{E}°'$$
$$\Delta G°' = -(1)(96{,}485 \text{ J} \cdot \text{V}^{-1} \cdot \text{mol}^{-1})(-0.065 \text{ V})$$
$$\Delta G°' = 6.27 \text{ kJ} \cdot \text{mol}^{-1}$$

The positive $\Delta G°'$ indicates that the electron transfer is unfavorable under standard conditions. However, cellular conditions are not necessarily standard conditions, and the ΔG for this reaction is likely to be negative. Also, since this reaction occurs as part of the electron transport chain, the electrons gained by cytochrome c_1 will be passed along to Complex IV, in effect coupling the two reactions, which would also tend to make the process more favorable than the $\Delta G°'$ indicates.

8. Consult Table 12-1 for the relevant half-reactions involving O_2 and NADH. Reverse the NADH half-reaction and the sign of its $\mathcal{E}°'$ value to indicate oxidation, then combine the half-reactions and their $\mathcal{E}°'$ values.

1/2 O_2 + 2 H^+ + $2e^-$ $\rightarrow$ H_2O	$\mathcal{E}°' = 0.815$ V
NADH $\rightarrow$ NAD^+ + H^+ + $2e^-$	$\mathcal{E}°' = 0.315$ V
NADH + 1/2 O_2 + H^+ $\rightarrow$ NAD^+ + H_2O	$\Delta\mathcal{E}°' = 1.13$ V

Use Equation 12-3 to calculate $\Delta G°'$ for this reaction:

$$\Delta G°' = -n\mathcal{F}\Delta\mathcal{E}°'$$
$$\Delta G°' = -(1)(96{,}485 \text{ J} \cdot \text{V}^{-1} \cdot \text{mol}^{-1})(1.13 \text{ V})$$
$$\Delta G°' = -218 \text{ kJ} \cdot \text{mol}^{-1}$$

The synthesis of 3 ATP requires a free energy investment of 3×30.5 kJ · mol^{-1}, or 91.5 kJ · mol^{-1}. The efficiency of oxidative phosphorylation is therefore 91.5/218 = 0.42, or 42%.

9. (a) Consult Table 12-1 for the relevant half-reactions involving NADH and coenzyme Q. Reverse the NADH half-reaction and the sign of its $\mathcal{E}°'$ value to indicate oxidation, then combine the half-reactions and their $\mathcal{E}°'$ values.

Ubiquinone + $2H^+$ + $2e^-$ $\rightarrow$ ubiquinol	$\mathcal{E}°' = 0.045$ V
NADH $\rightarrow$ NAD^+ + H^+ + $2e^-$	$\mathcal{E}°' = 0.315$ V
NADH + ubiquinone + H^+ $\rightarrow$ NAD^+ + ubiquinol	$\Delta\mathcal{E}°' = 0.360$ V

Use Equation 12-3 to calculate $\Delta G°'$ for this reaction:

$$\Delta G°' = -n\mathcal{F}\Delta\mathcal{E}°'$$
$$\Delta G°' = -(2)(96{,}485 \text{ J} \cdot \text{V}^{-1} \cdot \text{mol}^{-1})(0.360 \text{ V})$$
$$\Delta G°' = -69.5 \text{ kJ} \cdot \text{mol}^{-1}$$

The phosphorylation of ADP to ATP requires +30.5 kJ · mol^{-1} of free energy. Assuming that this reaction is 42% efficient, one mole of ATP can be synthesized under standard conditions.

$$\frac{69.5 \text{ kJ} \cdot \text{mol}^{-1}}{30.5 \text{ kJ} \cdot \text{mol}^{-1}}(0.42) = 0.95$$

(b) Consult Table 12-1 for the relevant half-reactions involving ubiquinol and cytochrome c. Reverse the ubiquinol half-reaction and the sign of its $\mathcal{E}°'$ value to indicate oxidation, multiply the coefficients in the cytochrome c equation by 2 so that the number of electrons transferred will be equal, then combine the half-reactions and their $\mathcal{E}°'$ values.

$QH_2 \rightarrow Q + 2H^+ + 2e^-$	$\mathcal{E}°' = -0.045$ V
2 Cytochrome c (Fe^{3+}) + $2e^-$ $\rightarrow$ 2 cytochrome c (Fe^{2+})	$\mathcal{E}°' = 0.235$ V
QH_2 + 2 cyto c (Fe^{3+}) $\rightarrow$ Q + $2H^+$ + cyto c (Fe^{2+})	$\Delta\mathcal{E}°' = 0.190$ V

Use Equation 12-3 to calculate $\Delta G°'$ for this reaction:

$$\Delta G°' = -n\mathcal{F}\Delta\mathcal{E}°'$$
$$\Delta G°' = -(2)(96{,}485 \text{ J} \cdot \text{V}^{-1} \cdot \text{mol}^{-1})(0.190 \text{ V})$$
$$\Delta G°' = -36.7 \text{ kJ} \cdot \text{mol}^{-1}$$

The phosphorylation of ADP to ATP requires +30.5 $kJ \cdot mol^{-1}$ of free energy. Assuming that this reaction is 42% efficient, 0.5 mole of ATP can be synthesized.

$$\frac{36.6 \text{ kJ} \cdot \text{mol}^{-1}}{30.5 \text{ kJ} \cdot \text{mol}^{-1}}(0.42) = 0.50$$

(c) Consult Table 12-1 for the relevant half-reactions involving cytochrome *c* and oxygen. Reverse the cytochrome *c* half-reaction and the sign of its $\mathcal{E}°'$ value to indicate oxidation, multiply the coefficients by 2 so that the number of electrons transferred will be equal, then combine the half-reactions and their $\mathcal{E}°'$ values.

$$2 \text{ Cytochrome } c\,(Fe^{2+}) + 2e^- \rightarrow 2 \text{ cytochrome } c\,(Fe^{3+}) \qquad \mathcal{E}°' = -0.235 \text{ V}$$
$$1/2\, O_2 + 2H^+ + 2e^- \rightarrow H_2O \qquad \mathcal{E}°' = 0.815 \text{ V}$$
$$2 \text{ Cyto } c\,(Fe^{2+}) + 1/2\, O_2 + 2H^+ \rightarrow 2 \text{ cyto } c\,(Fe^{3+}) + H_2O \qquad \Delta\mathcal{E}°' = 0.580 \text{ V}$$

Use Equation 12-3 to calculate $\Delta G°'$ for this reaction:

$$\Delta G°' = -n\mathcal{F}\Delta\mathcal{E}°'$$
$$\Delta G°' = -(2)(96{,}485 \text{ J} \cdot \text{V}^{-1} \cdot \text{mol}^{-1})(0.580 \text{ V})$$
$$\Delta G°' = -112 \text{ kJ} \cdot \text{mol}^{-1}$$

The phosphorylation of ADP to ATP requires +30.5 $kJ \cdot mol^{-1}$ of free energy. Assuming that this reaction is 42% efficient, 1.5 mole of ATP can be synthesized.

$$\frac{112 \text{ kJ} \cdot \text{mol}^{-1}}{30.5 \text{ kJ} \cdot \text{mol}^{-1}}(0.42) = 1.5$$

10. The relevant reactions and their $\mathcal{E}°'$ values are obtained from Table 12-1:

$$\text{Succinate} \rightarrow \text{fumarate} + 2H^+ + 2e^- \qquad \mathcal{E}°' = -0.031 \text{ V}$$
$$Q + 2H^+ + 2e^- \rightarrow QH_2 \qquad \mathcal{E}°' = 0.045 \text{ V}$$
$$\text{Succinate} + Q \rightarrow \text{fumarate} + QH_2 \qquad \Delta\mathcal{E}°' = 0.014 \text{ V}$$

Use Equation 12-3 to calculate $\Delta G°'$ for this reaction:

$$\Delta G°' = -n\mathcal{F}\Delta\mathcal{E}°'$$
$$\Delta G°' = -(2)(96{,}485 \text{ J} \cdot \text{V}^{-1} \cdot \text{mol}^{-1})(0.014 \text{ V})$$
$$\Delta G°' = -2.7 \text{ kJ} \cdot \text{mol}^{-1}$$

This is not enough free energy to drive ATP synthesis under standard conditions ($\Delta G°' = 30.5 \text{ kJ} \cdot \text{mol}^{-1}$).

11. A total of 38 ATP is obtained from the exergonic oxidation of glucose under aerobic conditions:

glycolysis	2 ATP	2 ATP
	2 NADH	2 × 3 = 6 ATP
2 pyruvate → 2 acetyl-CoA	2 NADH	2 × 3 = 6 ATP
citric acid cycle (2 rounds)	2 × 3 NADH 2 × 1 QH_2 2 × 1 GTP	6 × 3 = 18 ATP 2 × 2 = 4 ATP 2 × 1 = 2 ATP
Total		38 ATP

A total of 2 ATP per glucose is obtained when glucose is oxidized in the absence of oxygen by conversion to lactate or ethanol (Chapter 10).

12. (a) Aerobic respiration yields 38 ATP, each of which requires 30.5 $kJ \cdot mol^{-1}$ to synthesize:

$$\frac{38 \times 30.5 \text{ kJ} \cdot \text{mol}^{-1}}{2850 \text{ kJ} \cdot \text{mol}^{-1}} \times 100 = 41\%$$

Homolactic fermentation yields 2 ATP:

$$\frac{2 \times 30.5 \text{ kJ} \cdot \text{mol}^{-1}}{196 \text{ kJ} \cdot \text{mol}^{-1}} \times 100 = 31\%$$

Alcoholic fermentation yields 2 ATP:

$$\frac{2 \times 30.5 \text{ kJ} \cdot \text{mol}^{-1}}{235 \text{ kJ} \cdot \text{mol}^{-1}} \times 100 = 26\%$$

(b) Organisms which can oxidize glucose in the presence of oxygen have an advantage over anaerobic organisms because they can extract more energy per glucose. This may have been important in evolution.

13. (a) Aerobic oxidation of glucose yields 38 ATP per glucose, whereas alcoholic fermentation of glucose by the yeast yields only 2 ATP per glucose. Assuming that the energy needs of the yeast cell remain constant under both aerobic and anaerobic conditions described in the question, the catabolism of glucose by the yeast will be 19-fold greater in the absence of oxygen than in the presence of oxygen, in order to obtain the same amount of ATP. Thus, the rate of consumption of glucose decreases when the cells are exposed to oxygen, because fewer glucose molecules must be oxidized to yield the same amount of ATP.

(b) Both ratios will initially increase, as the citric acid cycle (which does not operate under anaerobic conditions) produces more NADH equivalents for electron transport. The [ATP]/[ADP] ratio will also increase, since aerobic oxidation of glucose produces more ATP per mole of glucose than anaerobic oxidation (as described in part a above). ATP and NADH will "reset" the equilibrium by inhibiting the regulatory enzymes of glycolysis and the citric acid cycle, slowing down these processes. Eventually, the $[NADH/NAD^+]$ and [ATP]/[ADP] ratios return to their "original" values.

14. (a) Since all of these inhibitors interfere with electron transfer somewhere in the electron transport chain, oxygen consumption will decrease if any of the inhibitors are added to a suspension of respiring mitochondria. Adding any of these inhibitors prevents electrons from being transferred to the oxygen, the final electron acceptor.

(b) In rotenone- or amytal-blocked mitochondria, NADH and Complex I redox centers are reduced while components from ubiquinone on are oxidized. In antimycin A-blocked mitochondria, NADH, Complex I redox centers, ubiquinol, and Complex III redox centers are reduced while cytochrome *c* and Complex IV redox centers are oxidized. In cyanide-blocked mitochondria, all of the electron transport components are reduced and only oxygen remains oxidized.

(c) Adding succinate to rotenone-blocked mitochondria effectively bypasses the block as succinate donates its electrons to ubiquinone and electron transport resumes. Adding succinate is not an effective bypass for antimycin A- or cyanide-blocked mitochondria because succinate donates its electrons upstream of the block.

15. The donation of a pair of electrons to Complex IV will result in the synthesis of about one ATP per atom of oxygen (1/2 O_2). Therefore, the P:O ratio of this compound is 1.

16. The donation of a pair of electrons to cytochrome *c* and then to Complex IV will result in the synthesis of about one ATP per atom of oxygen (1/2 O_2). Therefore, the P:O ratio of this compound is 1.

17. Adding tetramethyl-*p*-phenylenediamine to rotenone-blocked and antimycin A-blocked mitochondria effectively bypasses the block as the compound donates its electrons to Complex IV and electron transport resumes. Adding tetramethyl-*p*-phenylenediamine is not an effective bypass for cyanide-blocked mitochondria because cyanide inhibits electron transport in Complex IV. Similarly, ascorbate, which donates its electrons to cytochrome *c* and then to Complex IV, can act as an effective bypass for antimycin A-blocked mitochondria but not cyanide-blocked mitochondria.

18. Myxothiazol inhibits electron transfer in Complex III (specifically, it inhibits the transfer of electrons from reduced ubiquinol to both cytochrome *b* and the iron-sulfur protein redox centers in Complex III).

19. Cyanide binds to the Fe^{2+} in the Fe-Cu center of cytochrome a_3 (see Problem 14). When the iron in hemoglobin (Hb) is oxidized from Fe^{2+} to Fe^{3+}, cytochrome a_3 can donate an electron to reoxidize the hemoglobin to Fe^{2+}. This oxidizes the iron in cytochrome a_3 to Fe^{3+}. Cyanide does not bind to Fe^{3+} so it is released and the Complex IV can again function normally. The cyanide binds to the Fe^{2+} in hemoglobin, where it does not interfere with respiration (although it does interfere with oxygen delivery).

$HbO_2(Fe^{2+})$

↓

$HbO_2(Fe^{3+})$

cytochrome $a_3(Fe^{2+}—Cu)—CN^-$ → cytochrome $a_3(Fe^{3+}—Cu)$

↓

$HbO_2(Fe^{2+})$

20. (a) Pyruvate and malate are oxidized by pyruvate dehydrogenase and malate dehydrogenase, respectively, and one of the products is NADH, which can enter electron transport.

(b) Fluoxetine inhibits electron transport generally, since the rate of electron transport falls from an average of 163 to 77 when 0.15 mM fluoxetine is present. The inhibition is primarily at Complex I, since the rate of electron transport in the presence of fluoxetine is not decreased substantially when succinate (which donates its electrons to Complex II) and rotenone (which inhibits electron transfer in Complex I) are added. However, fluoxetine can also inhibit Complex IV somewhat, since electron transport in the presence of fluoxetine decreases in the presence of ascorbate (which donates its electrons to cytochrome *c*) and TMPD (which donates its electrons to Complex IV).

(c) By decreasing both the rate of electron transport and ATP synthesis, fluoxetine decreases the rate of ATP production in the brain. The brain relies on a constant source of ATP for proper function, so decreased ATP production could lead to an impairment of brain function.

21. Like the lipids that compose the membrane, coenzyme Q is amphiphilic, with a hydrophilic head and a hydrophobic tail. "Like dissolves like," and coenzyme Q literally dissolves in the membrane, which facilitates rapid diffusion.

22. Cytochrome *c* is a water-soluble, peripheral membrane protein and is easily dissociated from the membrane by adding salt solutions that would interfere with the ionic ineractions that tether it to the inner mitochondrial membrane. Cytochrome c_1 is an integral membrane protein and is largely water-insoluble due to the nonpolar amino acids that interact with the acyl chains of the membrane lipids. Detergents are required to dissociate cytochrome c_1 from the membrane because amphiphilic detergents can disrupt the membrane and coat membrane proteins, acting as substitute lipids in the solubilization process.

23. (a) The import of ADP (net charge −3) and the export of ATP (net charge −4) represents a loss of negative charge inside the mitochondria. This decreases the difference in electrical charge across the membrane, since the outside is positive due to the translocation of protons during electron transport. Consequently, the gradient is diminished by the activity of the adenine nucleotide translocase.

(b) The activity of the P_i–H^+ symport protein diminishes the proton gradient by allowing protons from the intermembrane space to reenter the matrix.

(c) Both transport systems are driven by the free energy of the electrochemical proton gradient.

24. If the translocase is unable to function, ATP will not be able to exit the mitochondrial matrix and ADP will not be transported inside. Without ADP, the substrate for ATP synthase, ATP synthesis will not occur. Since electron transport and oxidative phosphorylation are coupled, a decrease in the rate of oxidative phosphorylation will decrease the rate of electron transport.

25. (a) The pH of the intermembrane space is lower than the pH of the mitochondrial matrix because protons are pumped out of the matrix, across the inner membrane, and into the intermembrane space. The increase in concentration of protons in the intermembrane space decreases the pH; the deficit of protons in the matrix results in an increase in pH.

(b) Detergents disrupt membranes. An intact inner mitochondrial membrane is required for oxidative phosphorylation to take place. Without an intact membrane, an electrochemical gradient, which is the energy reservoir that drives ATP synthesis, cannot be established, and ATP synthesis does not occur.

(c) Uncouplers such as DNP ferry protons across the inner mitochondrial membrane and dissipate the proton gradient established by electron transport. In the presence of DNP, electron transport can still occur, but the free energy released by the process is dissipated as heat instead of being harnessed to synthesize ATP.

26. (a) The impermeability of the membrane to ions is an important feature of chemiosmosis, because the free movement of ions other than H^+ would dissipate the electrical component of the proton gradient and thereby compromise the ability of the gradient to supply free energy for ATP synthesis.

(b) If the membrane were permeable to other ions, the proton gradient would be a simple chemical gradient rather than an electrochemical gradient. However, it would still be able to serve as a source of free energy for ATP synthesis.

27. (a)

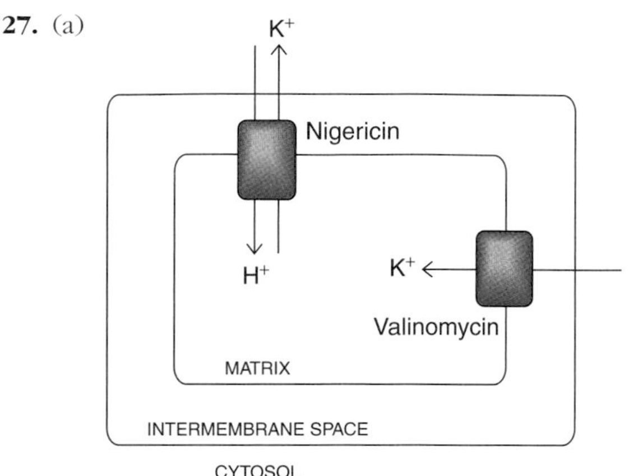

(b) Potassium ions enter the matrix with the assistance of valinomycin. These ions are then exported by nigericin in exchange for protons. Importing protons into the mitochondrial matrix dissipates the proton gradient. Since the proton gradient serves as the energy reservoir that drives ATP synthesis, no ATP is synthesized in the absence of the proton gradient.

28. Use the rearrangement of Equation 12-7 as shown in Sample Calculation 12-2.

$$\Delta G = 2.303\ RT(\text{pH}_{in} - \text{pH}_{out}) + Z\mathcal{F}\Delta\psi$$
$$\Delta G = 2.303(8.3145 \times 10^{-3}\ \text{kJ} \cdot \text{K}^{-1} \cdot \text{mol}^{-1})(310\text{K})$$
$$(7.6 - 7.2) + (1)(96{,}485\ \text{kJ}\cdot\text{V}^{-1}\cdot\text{mol}^{-1})(0.200\text{V})$$
$$\Delta G = 2.37\ \text{kJ} \cdot \text{mol}^{-1} + 19.3\ \text{kJ} = 21.7\ \text{kJ} \cdot \text{mol}^{-1}$$

29. Inactivation of one *c* subunit by DCCD blocks all proton translocation by F_0 since the movement of protons across the membrane requires continuous rotation of the *c* ring. Without this rotation, the γ subunit of F_1 cannot move, and therefore the β subunits cannot undergo the conformational changes necessary to synthesize or hydrolyze ATP by the binding change mechanism.

30. (a) ATP synthesis decreases dramatically in the presence of oligomycin, since proton transfer is required to stimulate rotation of the γ subunit of ATP synthase, which causes the sequential conformational change of the β subunits that catalyze the phosphorylation of ADP to ATP.

(b) Since oxidative phosphorylation and electron transport are coupled, a decrease in the rate of oxidative phosphorylation will also affect the rate of electron transport. If ATP synthesis is not occurring, the proton gradient is not "discharged" and the rate of electron transport decreases.

(c) A decrease in the rate of electron transport will also decrease the rate of oxygen consumption.

(d) If dinitrophenol is added, the proton gradient is dissipated, or "discharged," but not in a way that leads to ATP synthesis. Therefore, ATP synthesis still does not occur, but electron transport and oxygen consumption resume, with the free energy of the process released as heat.

31. DNP uncouples the electron transport chain from oxidative phosphorylation by dissipating the proton gradient. Electron transport still occurs, but the energy released by electron transport is dissipated as heat instead of being harnessed to synthesize ATP. One might think that DNP would be an effective diet aid because the sources of the electrons that go down the electron transport chain are dietary carbohydrates and fatty acids. If the energy of these compounds is dissipated as heat instead of used to synthesize ATP (which would then be used for, among other processes, the synthesis of fatty acids in adipocytes), weight gain from the ingestion of food could theoretically be prevented.

32. FCCP is a lipid-soluble compound that can pass through both the outer and inner mitochondrial membranes and into the matrix. FCCP has an acidic hydrogen on the central nitrogen that dissociates in the basic environment of the matrix. The FCCP passes back through the inner membrane to the intermembrane space where it is reprotonated and the entire cycle begins again. In this manner, the FCCP "shuttles" protons into the mitochondrial matrix and dissipates the proton gradient.

33. Organic compounds are oxidized by oxygen through the activity of the electron transport complexes of mitochondria. This activity generates the proton gradient that is dissipated during phosphorylation of ADP. When there is no more ADP, ATP synthase is inactive and unable to dissipate the proton gradient. As a result of the tight coupling between oxidative phosphorylation and electron transport, electron transport and oxygen consumption also come to a halt.

34. (a) The P:O ratio is decreased. For some reason, electron transport and oxidative phosphorylation have been uncoupled. NADH is oxidized in the electron transport chain, oxygen is reduced to water, but ATP may not be synthesized if ADP is not available. This decreases the P:O ratio because oxygen consumption occurs to a greater extent than ADP phosphorylation.

(b) If the energy released in electron transport is not used to synthesize ATP, it is released as heat, which accounts for the elevated body temperature. If sufficient ATP is not synthesized to meet energy needs, the rate of electron transport increases, which leads to an increased consumption of O_2 and an increase in the concentration of reduced coenzymes which enter electron transport, thus increasing metabolic rate.

(c) The patient will not be able to carry out strenuous exercise under aerobic conditions because her muscle cells are incapable of generating enough ATP to power the muscle.

35. F_0 acts as a proton channel as the *c* ring rotates, feeding protons through the α subunit (see Fig. 12-23). The addition of F_1 bocks proton movement because the γ shaft rotates along with the *c* ring. In this system, the γ subunit and the *c* ring can rotate only when the binding change mechanism is in operation, that is, when the β subunits are binding and releasing nucleotides. ATP or ADP + P_i must be added to the system in order for the γ subunit to move.

36. Since it has 9 *c* subunits, the bacterial enzyme can theoretically produce 3 ATP for every 9 protons translocated, or 1 ATP per 3 H^+. In the chloroplast, 3 ATP are synthesized for every 14 protons, or 1 ATP per 4.7 H^+. Thus, the bacterium is more efficient in its use of the proton gradient established during electron transport and has a higher ratio of ATP produced per oxygen consumed.

37. An increase in the amount of glucose led to an increase in glycolytic flux and an increase in the concentration of NADH, which was subsequently reoxidized during electron transport. Because this did not occur with the *pdc–* mutants, pyruvate decarboxylase must play a role in reoxidizing the NADH produced during glycolysis. In the absence of the enzyme, glycolytic flux could not increase due to the deficiency of NAD^+. The NAD^+ deficiency also prevented the oxidative decarboxylation of pyruvate by pyruvate dehydrogenase.

38. (a) When oxidation is uncoupled from phosphorylation, electron transport is still occurring. The energy that is released when electrons are passed down the chain from NADH/H^+ and QH_2 to oxygen is normally used to synthesize ATP, but if oxidation and phosphorylation are uncoupled, the energy is released as heat instead, and no ATP is synthesized.

(b) When UCP1 is stimulated in normal mice, oxidative phosphorylation is uncoupled. This means that the ATP yield per substrate molecule oxidized is decreased, since ATP can only be produced via substrate-level phosphorylation. The cell senses that its energy needs are not being met. This has the direct effect of increasing the rate of glycolysis and the citric acid cycle, and eventually electron transport, in a vain attempt to synthesize more ATP. Since oxygen is the final electron acceptor in electron transport, oxygen consumption increases in order to keep up with the increased rate of electron transport. In knockout mice, since there is no uncoupling protein, oxidative phosphorylation is not uncoupled. Therefore, ATP can be synthesized via oxidative phosphorylation. Since the energy needs of the cell are being met, the rate of electron transport does not dramatically increase. The slight increase in oxygen consumption may be due to the action of a second uncoupling protein. So perhaps there is some uncoupling of oxidative phosphorylation in the knockout mice, but not to the extent that is seen in normal mice.

(c) If the UCP1 is not present, then uncoupling of oxidative phosphorylation does not occur. Thus, ATP can be synthesized via oxidative phosphorylation, so less lipid needs to be broken down in order to meet the organism's energy needs.

(d) In the normal mice, the cold temperature activated the uncoupling protein. The result is that oxidative phosphorylation was uncoupled and the energy of electron transport was dissipated as heat rather than being used to synthesize ATP. This helped the mice maintain normal body temperature. But the UCP1-knockout mice lacked the uncoupling protein and were unable to uncouple oxidation from phosphorylation. Thus they were unable to generate "extra" heat and their body temperatures decreased as a result.

(e) When overfeeding occurs, adipose tissue content would increase in the absence of UCP2, especially if energy expenditure was less than the caloric intake. But if overfeeding induces synthesis of the uncoupling protein, then oxidative phosphorylation could be uncoupled. Thus, some of the energy content in the excess food is converted to heat rather than being stored as excess adipose tissue. The presence of UCP2 could explain why some people are able to eat all they want without gaining weight. Perhaps people who are obese have a deficiency of UCP2, but this has not yet been verified.

39. (a) In the root system of the skunk cabbage, starch is broken down enzymatically to yield glucose. Glucose is then broken down via aerobic respiration via glycolysis, the citric acid cycle, and the electron transport chain. The oxidation of glucose provides the NADH/H^+ and the QH_2 substrates required to keep electron transport going so that thermogenesis can occur.

(b) In the skunk cabbage, as temperature decreases, the need for thermogenesis increases. Thus the rate of aerobic oxidation of glucose increases to increase the rate of flux of NADH/H^+ and QH_2 through the electron transport chain. Since oxygen is the final electron acceptor in the electron transport chain, an increase in flux through electron transport will also increase oxygen consumption. In the lotus flower, when the ambient air temperature is warmer, the need for heat production is less, so there is less flux through the electron transport chain. Thus oxygen consumption decreases during the day.

(c) The synthesis of uncoupling protein increases with decreasing temperature, presumably by an increase in transcription of the mRNA that codes for the uncoupling protein. The Northern blot results indicate that the synthesis of mRNA increases when the temperature decreases. The increased amount of mRNA likely results in an increase in concentration of the uncoupling protein. The uncoupling protein would then dissipate the proton gradient, which would lead to the thermogenesis observed in plants at cold temperatures.

40. (a) Glutamate enters the mitochondrion through a specific transporter and is then acted upon by glutamate dehydrogenase in the mitochondrial matrix and is oxidized to α-ketoglutarate. Concomitantly, NAD^+ is reduced to NADH. NADH can then enter electron transport.

(b) Ceramide could inhibit the glutamate transporter, the glutamate dehydrogenase enzyme, any of the three complexes involved in the electron transport chain, ATP synthase, or the ATP translocase.

(c) Ceramide doesn't interfere with ATP synthase. If it did, respiration in the presence of the uncoupler would still occur, but no ATP would be synthesized.

(d) Ceramide does not inhibit the glutamate transporter or glutamate dehydrogenase. Ceramide must inhibit one of the three complexes involved in electron transport. (In fact, ceramide inhibits the activity of Complex III.)

Chapter 13 Solutions

1.

Proton translocation	C, M
Photophosphorylation	C
Photooxidation	C
Quinones	C, M
Oxygen reduction	M
Water oxidation	C
Electron transport	C, M
Oxidative phosphorylation	M
Carbon fixation	C
NADH oxidation	M
Mn cofactor	C
Heme groups	C, M
Binding change mechanism	C, M
Iron-sulfur clusters	C, M
$NADP^+$ reduction	C

2. (a) Use Planck's law multiplied by Avogadro's number to calculate the energy of the photons:

$$E = \frac{hc}{\lambda} \times N$$

$$E = \frac{(6.626 \times 10^{-34}\ \text{J} \cdot \text{s})(2.998 \times 10^{8}\ \text{m} \cdot \text{s}^{-1})}{4 \times 10^{-7}\ \text{m}} \times (6.022 \times 10^{23}\ \text{photons} \cdot \text{mol}^{-1})$$

$$E = 300\ \text{kJ} \cdot \text{mol}^{-1}$$

If the synthesis of each ATP requires 30.5 $\text{kJ} \cdot \text{mol}^{-1}$, then 9.8 ATP (300/30.5) could be synthesized.

(b) $$E = \frac{(6.626 \times 10^{-34}\ \text{J} \cdot \text{s})(2.998 \times 10^{8}\ \text{m} \cdot \text{s}^{-1})}{6 \times 10^{-7}\ \text{m}} \times (6.022 \times 10^{23}\ \text{photons} \cdot \text{mol}^{-1})$$

$$E = 200\ \text{kJ} \cdot \text{mol}^{-1}$$

About 6.6 ATP (200/30.5) could be synthesized.

3. Because the algae appear red, red light is transmitted rather than absorbed. Therefore, the photosynthetic pigments in the red algae do not absorb red light but absorb light of other wavelengths.

4. Photosynthetic bacteria have the ability to absorb light in the infrared and ultraviolet regions of the electromagnetic spectrum.

5. (a) The sequences would provide only limited information, since even highly homologous proteins such as myoglobin and hemoglobin (which have very similar structures) have limited sequence identity (see Section 4-5). The three-dimensional structures of the proteins would be more likely to indicate whether they function similarly.

(b) Low-resolution models that included the heme groups would be more useful, since the positions of the heme groups within the proteins might indicate whether they function similarly in accepting and donating electrons to their redox partners. It is possible that the heme groups could have similar orientations even if the apoproteins did not resemble each other in overall tertiary structure.

6. Consult Table 12-1 for the reduction potentials of the relevant half-reactions, reversing the sign for the water oxidation half-reaction.

$$H_2O \rightarrow 1/2\ O_2 + 2H^+ + 2e^- \qquad \mathcal{E}^{\circ\prime} = -0.815\ \text{V}$$
$$NADP^+ + H^+ + 2e^- \rightarrow NADPH \qquad \mathcal{E}^{\circ\prime} = -0.320\ \text{V}$$
$$H_2O + NADP^+ \rightarrow 1/2\ O_2 + NADPH + H^+ \qquad \mathcal{E}^{\circ\prime} = -1.135\ \text{V}$$

Use Equation 12-3 to calculate $\Delta G^{\circ\prime}$:

$$\Delta G^{\circ\prime} = -n\mathcal{F}\Delta\mathcal{E}^{\circ\prime}$$
$$\Delta G^{\circ\prime} = -(2)(96{,}485\ \text{J} \cdot \text{V}^{-1} \cdot \text{mol}^{-1})(-1.135\ \text{V})$$
$$\Delta G^{\circ\prime} = +219{,}000\ \text{J} \cdot \text{mol}^{-1}$$

Divide by Avogadro's number to obtain the free energy per molecule:

$$(219{,}000\ \text{J} \cdot \text{mol}^{-1}) \div (6.022 \times 10^{23}\ \text{molecules} \cdot \text{mol}^{-1}) = 3.6 \times 10^{-19}\ \text{J/molecule}$$

Next, use Equation 13-1 and multiply by 2 to calculate the energy of the photons:

$$E = \frac{hc}{\lambda} \times 2$$

$$E = \frac{(6.626 \times 10^{-34}\ \text{J} \cdot \text{s})(2.998 \times 10^{8}\ \text{m} \cdot \text{s}^{-1})(2)}{6 \times 10^{-7}\text{m}}$$

$$E = 6.6 \times 10^{-19}\ \text{J}$$

In theory, the two photons supply enough energy to drive the oxidation of one molecule of water by $NADP^+$.

7. (a) The buildup of the proton gradient indicates a high level of activity of the photosystems. A steep gradient could therefore trigger photoprotective activity to prevent further photooxidation when the proton-translocating machinery is operating at maximal capacity.

(b) Photooxidation would not be a good protective mechanism since it might interfere with the normal redox balance among the electron-carrying groups in the thylakoid membrane. Releasing the energy by exciton transfer or fluorescence (emitting light of a longer wavelength) could potentially funnel light energy back to the overactive photosystems. Dissipation of excess energy as heat would be the safest mechanism, since the photosystems do not have any way to harvest thermal energy to drive chemical reactions.

8. The order of action is water–plastoquinone oxidoreductase (Photosystem II), plastoquinone–plastocyanin oxidoreductase (cytochrome b_6f), then plastocyanin–ferredoxin oxidoreductase (Photosystem I).

9. If the photosystems were in close proximity, then Photosystem II might not undergo photooxidation and instead act as a light-harvesting complex for Photosystem I. Because the reaction center of Photosystem II absorbs light with a peak wavelength of 680 nm, it could pass energy to the reaction center of Photosystem I, which absorbs lower-energy (longer-wavelength) light at its P700 group.

10. Use Equation 12-7, as applied in Sample Calculation 12-2. The matrix and the stroma are both "in."

$$\Delta G = 2.303\ RT\,(\text{pH}_{in} - \text{pH}_{out}) + Z\mathcal{F}\Delta\psi$$

For the chloroplast,

$$\Delta G = 2.303(8.3145\ \text{J}\cdot\text{K}^{-1}\cdot\text{mol}^{-1})(298\text{K})(3.5) + (1)(96{,}485\ \text{J}\cdot\text{V}^{-1}\cdot\text{mol}^{-1})(-0.05\ \text{V})$$
$$\Delta G = 20{,}000\ \text{J}\cdot\text{mol}^{-1} - 4800\ \text{J}\cdot\text{mol}^{-1}$$
$$\Delta G = +15.2\ \text{kJ}\cdot\text{mol}^{-1}$$

For the mitochondrion,

$$\Delta G = 2.303(8.3145\ \text{J}\cdot\text{K}^{-1}\cdot\text{mol}^{-1})(298\text{K})(0.75) + (1)(96{,}485\ \text{J}\cdot\text{V}^{-1}\cdot\text{mol}^{-1})(0.20\ \text{V})$$
$$\Delta G = 4300\ \text{J}\cdot\text{mol}^{-1} + 19{,}300\ \text{J}\cdot\text{mol}^{-1}$$
$$\Delta G = +23.6\ \text{kJ}\cdot\text{mol}^{-1}$$

For both organelles, the translocation is endergonic. In the chloroplast, most of this energy is due to the concentration (pH) difference on the two sides of the membrane; in the mitochondrion, most of the energy is due to the membrane potential ($\Delta\psi$).

11. (a) An uncoupler dissipates the transmembrane proton gradient by providing a route for translocation other than ATP synthase. Therefore, chloroplast ATP production would decrease.
(b) The uncoupler would not affect NADP^+ reduction since light-driven electron transfer reactions would continue regardless of the state of the proton gradient.

12. Complex III of the mitochondrial electron transport chain is analogous to the cytochrome b_6f complex. In the presence of antimycin A, electron flow in the cytochrome b_6f complex would be inhibited. Electrons would not reach Photosystem I and NADPH would not be produced. Proton translocation from the stroma to the thylakoid lumen also would not occur, so CF_1 would not be stimulated and ATP synthesis would not occur.

13. If electrons cannot be transferred to Photosystem I, then Photosystem II remains reduced and cannot be reoxidized. The photosynthetic production of oxygen ceases. No proton gradient is generated, so ATP synthesis does not occur in the presence of DCMU.

14. Myxothiazol inhibits electron transfer in the cytochrome b_6f complex.

15. Like the lipids that compose the membrane, plastoquinone is amphiphilic, with a hydrophilic head and a hydrophobic tail. "Like dissolves like," so plastoquinone literally dissolves in the membrane, which facilitates rapid diffusion.

16. The net equation would be

$$\text{CO}_2 + 2\ \text{H}_2\text{S} \rightarrow (\text{CH}_2\text{O}) + \text{S}_2 + \text{H}_2\text{O}$$

17. (a) More c subunits means that more protons are required to rotate the ATP synthase through one ATP-synthesizing step. Therefore, more photons must be absorbed to drive the translocation of more protons, so the quantum yield decreases.
(b) Cyclic electron flow contributes to the proton gradient and therefore leads to ATP synthesis. However, carbon fixation by the Calvin cycle requires NADPH also, so the additional photons that drive cyclic flow do not lead to more carbon fixed. Consequently, the quantum yield decreases.

18. Because oligomycin inhibits the F_0 subunit of mitochondrial ATP synthase, protons from the intermembrane space cannot translocate to the mitochondrial matrix. F_1 is not stimulated, and ATP synthesis does not occur. The cytosolic ratio of ATP/ADP decreases because less ATP is exported to the cytosol from the matrix when ATP synthesis is inhibited. Because oligomycin does not inhibit CF_0, chloroplast ATP synthesis is not inhibited, so the chloroplast ATP/ADP ratio is not affected. In some cases the decreased cytosolic ATP/ADP ratio might stimulate ATP synthesis in the chloroplast, resulting in an increased chloroplastic ATP/ADP ratio.

19. Normally, plants must synthesize large quantities of rubisco, a protein whose constituent amino acids all contain nitrogen. If rubisco had greater catalytic activity, the plant might produce less of the enzyme, thereby decreasing its need for nitrogen.

20. (a) The compound resembles the transition state of the rubisco carboxylase reaction (see Fig. 13-24) and therefore inhibits the enzyme by binding in the active site.
(b) At night, the compound inhibits rubisco in order to prevent the Calvin cycle from consuming NADPH and ATP. During the day, when the light reactions are supplying NADPH and ATP, the inhibitor is broken down to reactivate rubisco so that it can fix carbon.

21.

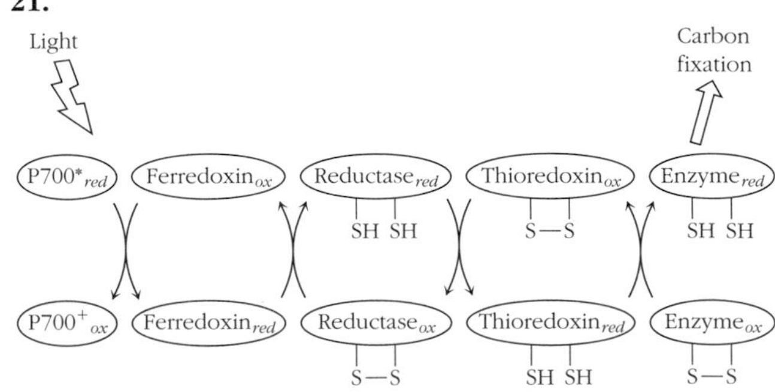

The activity of Photosystem I generates reduced ferredoxin, which is a substrate for the reductase. The product of the reductase reaction, reduced thioredoxin, can then reduce the disulfides of the Calvin cycle enzymes. Conformational changes in the enzymes upon exposure of free sulfhydryl groups could increase their activity.

22.

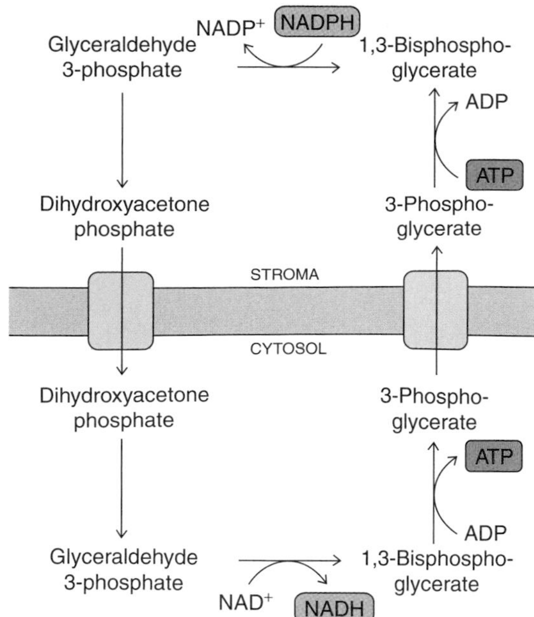

Operation of the transport cycle has the net effect of consuming stromal ATP and NADPH and generating ATP and NADH in the cytosol.

23. The unprotonated Lys side chain serves as a nucleophile when reacting with CO_2. At high pH, a higher percentage of ε-amino groups are in the unprotonated form.

24. $\Delta G = \Delta G^{\circ\prime} + RT \ln Q$

$$-41.0 \text{ kJ} \cdot \text{mol}^{-1} = -35.1 \text{ kJ} \cdot \text{mol}^{-1} + (8.3145 \times 10^{-3} \text{ kJ} \cdot \text{K}^{-1} \cdot \text{mol}^{-1})(298\text{K}) \ln Q$$

$$-5.9 \text{ kJ} \cdot \text{mol}^{-1} = (8.3145 \times 10^{-3} \text{ kJ} \cdot \text{K}^{-1} \cdot \text{mol}^{-1})(298\text{K}) \ln Q$$

$$-2.38 = \ln Q$$
$$e^{-2.38} = Q$$
$$0.092 = Q$$

25. $\Delta G = \Delta G^{\circ\prime} + RT \ln Q$

$$-29.7 \text{ kJ} \cdot \text{mol}^{-1} = -14.2 \text{ kJ} \cdot \text{mol}^{-1} + (8.3145 \times 10^{-3} \text{ kJ} \cdot \text{K}^{-1} \cdot \text{mol}^{-1})(298\text{K}) \ln Q$$

$$-15.5 \text{ kJ} \cdot \text{mol}^{-1} = (8.3145 \times 10^{-3} \text{ kJ} \cdot \text{K}^{-1} \cdot \text{mol}^{-1})(298\text{K}) \ln Q$$

$$-6.25 = \ln Q$$
$$e^{-6.25} = Q$$
$$0.0019 = Q$$

The enzyme is likely to be regulated because it catalyzes an irreversible step of the Calvin cycle (as evidenced by the large negative value of ΔG).

26. 3-Phosphoglycerate is the first stable radioactive intermediate that forms when algal cells are exposed to $^{14}CO_2$. The radioactive label is found on the carboxyl group of the compound.

27. The alanine residue at position 251 is essential for photosynthesis and photoautotrophic growth, and it also is part of the binding site for herbicides. If the alanine is mutated to a cysteine, which resembles alanine except that a sulfhydryl group has replaced a hydrogen, the mutated protein is similar to the wild type. If the alanine is changed to an amino acid that is similar in size but less hydrophobic (such as Gly, Pro, or Ser), the mutant is impaired in photosynthesis. Replacement with a larger, nonpolar amino acid results in a mutant that is further impaired, and replacement with a larger polar amino acid results in an organism that is not photosynthetically competent. Therefore, the residue at position 251 must be relatively small and nonpolar. Residues that are of similar size and polar, larger and nonpolar, and larger and polar result in an increasingly photosynthetically impaired organism.

Chapter 14 Solutions

1. (a) The hydrolysis of pyrophosphate releases -33.5 kJ · mol^{-1}.

$$\text{Fatty acid} + \text{CoA} + \text{ATP} \rightleftharpoons \text{acyl-CoA} + \text{AMP} + \text{PP}_i \qquad \Delta G^{\circ\prime} = 0$$

$$\text{PP}_i + \text{H}_2\text{O} \rightleftharpoons 2\ \text{P}_i \qquad \Delta G^{\circ\prime} = -33.5 \text{ kJ} \cdot \text{mol}^{-1}$$

$$\text{Fatty acid} + \text{CoA} + \text{ATP} + \text{H}_2\text{O} \rightleftharpoons \text{acyl-CoA} + \text{AMP} + 2\ \text{P}_i \qquad \Delta G^{\circ\prime} = -33.5 \text{ kJ} \cdot \text{mol}^{-1}$$

(b)

$$K_{eq} = e^{-\Delta G^{\circ\prime}/RT}$$
$$K_{eq} = e^{-(-33.5 \text{ kJ} \cdot \text{mol}^{-1})/(8.3145 \times 10^{-3} \text{ kJ} \cdot \text{K}^{-1} \cdot \text{mol}^{-1})(310\text{K})}$$
$$K_{eq} = e^{13}$$
$$K_{eq} = 4.4 \times 10^5$$

2.

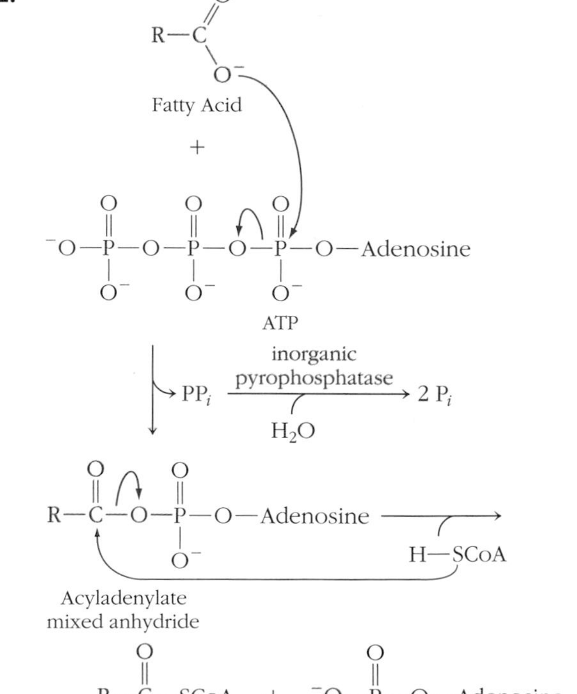

3. Oxidation of the methylene groups is accomplished by the action of a dehydrogenase, which removes electrons. The oxygen atom that becomes part of the carbonyl group is derived from a water molecule added by enoyl-CoA hydratase.

4. The conversion of fatty acyl-CoA to enoyl-CoA is similar to the conversion of succinate to fumarate because both reactions involve oxidation of the substrate and concomitant reduction of FAD to $FADH_2$ (see Section 11-2). The conversion of enoyl-CoA to hydroxyacyl-CoA is similar to the conversion of fumarate to malate because both reactions involve the addition of water across a trans double bond. The conversion of hydroxyacyl-CoA to ketoacyl-CoA is similar to the conversion of malate to oxaloacetate because both reactions involve the oxidation of an alcohol to a ketone with concomitant reduction of NAD^+ to NADH.

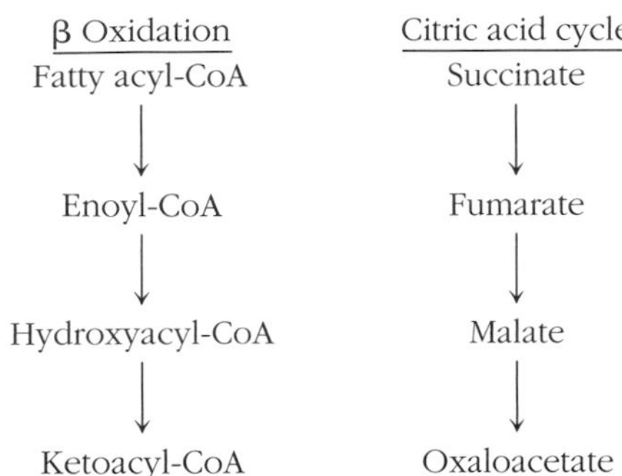

5. (a) Benzoate was produced when the dogs were fed phenylpropionate.
(b) Phenylacetate was produced when the dogs were fed phenylbutyrate.

$C_6H_5-CH_2-\!\!\wr\!\!-\overset{\beta}{C}H_2-\overset{\alpha}{C}OO^-$

Phenylpropionate

↓

$C_6H_5-COO^- + H_3C-\overset{O}{\overset{\|}{C}}-S-CoA$

Benzoate — Acetyl-CoA

$C_6H_5-CH_2-CH_2-\!\!\wr\!\!-\overset{\beta}{C}H_2-\overset{\alpha}{C}OO^-$

Phenylbutyrate

↓

$C_6H_5-CH_2-COO^- + H_3C-\overset{O}{\overset{\|}{C}}-S-CoA$

Phenylacetate — Acetyl-CoA

6. (a) If carnitine is deficient, fatty acid transport from the cytosol to the mitochondrial matrix (the site of β oxidation) is impaired. Fatty acid oxidation generates a great deal of ATP to power the muscle, so in the absence of fatty acid oxidation the muscle must rely on stored glycogen or uptake of circulating glucose to obtain the necessary ATP. Muscle cramping is exacerbated by fasting because the concentration of circulating glucose is decreased and glycogen stores are depleted. Exercise also increases muscle cramping because the demand for ATP by the muscle is greater.

(b) If medium-chain fatty acids are metabolized normally, then fatty acids of this size must be able to cross the inner mitochondrial membrane and enter the mitochondrial matrix without the aid of carnitine. Carnitine is required to transport fatty acids longer than 10 carbons.

7. Medium chain acyl-CoA would accumulate in individuals with MCAD deficiency, since the conversion of fatty acyl-CoA to enoyl-CoA is blocked. Acylcarnitine esters would also accumulate.

8. The products are three molecules of propionyl-CoA, three molecules of acetyl-CoA, and one molecule of 2-methylpropionyl-CoA.

Pristanate (β, α, C(=O)SCoA)

β oxidation → $H_3C-CH_2-\overset{O}{\overset{\|}{C}}-S-CoA$ Propionyl-CoA

(β, α, C(=O)SCoA)

β oxidation → $CH_3-\overset{O}{\overset{\|}{C}}-SCoA$ Acetyl-CoA

(β, α, SCoA)

β oxidation → Propionyl-CoA

β oxidation → Acetyl-CoA

β oxidation → Propionyl-CoA

β oxidation → Acetyl-CoA

$H_3C-CH(CH_3)-C(=O)-SCoA$

9. (a) Palmitate goes through seven cycles of β oxidation. The first six cycles produce 1 QH_2, 1 NADH, and 1 acetyl-CoA. The seventh cycle produces 1 QH_2, 1 NADH, and 2 acetyl-CoA. Each QH_2 generates 2 ATP in the electron transport chain, each NADH generates 3 ATP in the electron transport chain, and each acetyl-CoA generates a total of 12 ATP (1 $QH_2 \times 2 =$ 2 ATP; 3 NADH $\times$ 3 = 9 ATP; 1 GTP = 1 ATP for a total of 12 ATP per acetyl-CoA). The total is 131 ATP. Two ATP must be subtracted from this total to account for the ATP spent in activating palmitate to palmitoyl-CoA. This gives a total of 129 ATP.

(b) The same logic is used for stearate, except that stearate goes through eight cycles of β oxidation. The total is 146 ATP.

10. Adding up the ATP in the figure gives a total of 187 ATP.

$H_3C-(CH_2)_{20}-CH_2-CH_2-C(=O)-SCoA$

acyl-CoA oxidase ↓ FAD → $FADH_2$; O_2 → H_2O_2 → $2\,H_2O + O_2$

$H_3C-(CH_2)_{20}-C(H)=C(H)-C(=O)-SCoA$

enoyl-CoA

6 cycles of β oxidation (acyl-CoA dehydrogenase reaction is bypassed) ↓ 6 NADH, 6 acetyl-CoA = 6(3) + 6(12) ATP = 90 ATP

$H_3C-(CH_2)_8-CH_2-CH_2-C(=O)-SCoA$

PEROXISOME

MITOCHONDRION

5 cycles of β oxidation ↓ 5 QH_2, 5 NADH, 5 acetyl-CoA = 5(2) + 5(3) + 5(12) ATP = 85 ATP

acetyl-CoA (12 ATP)

11. (a) Adding up the ATP as shown in the figure yields a total of 146 ATP. Subtracting the 2 ATP required for activation brings the final total to 144 ATP (this is 2 fewer ATP than would be generated by the complete oxidation of stearate).

$H_3C-(CH_2)_7-CH=CH-(CH_2)_7-C(=O)-SCoA$

Oleoyl-CoA

3 cycles of β oxidation
- → 1 QH_2, 1 NADH, 1 acetyl-CoA = 2 + 3 + 12 ATP = 17 ATP
- → 1 QH_2, 1 NADH, 1 acetyl-CoA = 2 + 3 + 12 ATP = 17 ATP
- → 1 QH_2, 1 NADH, 1 acetyl-CoA = 2 + 3 + 12 ATP = 17 ATP

$H_3C-(CH_2)_7-C(H)=C(H)-CH_2-C(=O)-SCoA$

enoyl-CoA isomerase ↓

$H_3C-(CH_2)_8-C(H)=C(H)-C(=O)-SCoA$

1 cycle of β oxidation (acyl-CoA dehydrogenase reaction is bypassed) ↓ 1 NADH, 1 acetyl-CoA = 3 + 12 ATP = 15 ATP

$H_3C-(CH_2)_8-C(=O)-SCoA$

4 cycles of β oxidation
- → 1 QH_2, 1 NADH, 1 acetyl-CoA = 2 + 3 + 12 ATP = 17 ATP
- → 1 QH_2, 1 NADH, 1 acetyl-CoA = 2 + 3 + 12 ATP = 17 ATP
- → 1 QH_2, 1 NADH, 1 acetyl-CoA = 2 + 3 + 12 ATP = 17 ATP
- → 1 QH_2, 1 NADH, 1 acetyl-CoA = 2 + 3 + 12 ATP = 17 ATP

Acetyl-CoA (12 ATP)

(b) Adding up the ATP as shown in the figure yields a total of 143 ATP. Subtracting the 2 ATP required for activation brings the final total to 141 ATP (this is 5 fewer ATP than would be generated by the complete oxidation of stearate).

$H_3C-(CH_2)_4-CH=CH-CH_2-CH=CH-(CH_2)_7-C(=O)-SCoA$

Linoleyl-CoA

3 cycles of β oxidation
- → 1 QH_2, 1 NADH, 1 acetyl-CoA = 2 + 3 + 12 ATP = 17 ATP
- → 1 QH_2, 1 NADH, 1 acetyl-CoA = 2 + 3 + 12 ATP = 17 ATP
- → 1 QH_2, 1 NADH, 1 acetyl-CoA = 2 + 3 + 12 ATP = 17 ATP

$H_3C-(CH_2)_4-CH=CH-CH_2-C(H)=C(H)-CH_2-C(=O)-CoA$

enoyl-CoA isomerase ↓

$H_3C-(CH_2)_4-CH=CH-CH_2-CH_2-C(H)=C(H)-C(=O)-CoA$

1 cycle of β oxidation (acyl-CoA dehydrogenase reaction is bypassed) ↓ 1 NADH, 1 acetyl-CoA = 3 + 12 ATP = 15 ATP

$H_3C-(CH_2)_4-CH=CH-CH_2-CH_2-C(=O)-S-CoA$

acyl-CoA dehydrogenase ↓ 1 QH_2 = 2 ATP

$H_3C-(CH_2)_4-CH=CH-C(H)=C(H)-C(=O)-S-CoA$

2,4-dienoyl-CoA reductase ↓ NADPH → $NADP^+$ Costs 3 ATP

$H_3C-(CH_2)_4-CH_2-CH=CH-CH_2-C(=O)-S-CoA$

↓ enoyl-CoA-isomerase

$H_3C-(CH_2)_4-CH_2-CH_2-C(H)=C(H)-C(=O)-S-CoA$

1 cycle of β oxidation (acyl CoA dehydrogenase reaction is bypassed) → 1 NADH, 1 acetyl-CoA = 3 + 12 ATP = 15 ATP

→ 1 QH_2, 1 NADH, 1 acetyl-CoA = 2 + 3 + 12 ATP = 17 ATP

3 cycles of β oxidation → 1 QH_2, 1 NADH, 1 acetyl-CoA = 2 + 3 + 12 ATP = 17 ATP

→ 1 QH_2, 1 NADH, 1 acetyl-CoA = 2 + 3 + 12 ATP = 17 ATP

Acetyl-CoA (12 ATP)

12. A C_{17} fatty acid goes through seven cycles of β oxidation. The first six cycles produce 1 QH_2, 1 NADH, and 1 acetyl-CoA. The seventh cycle produces 1 QH_2, 1 NADH, 1 acetyl-CoA, and 1 propionyl-CoA. Each QH_2 generates 2 ATP in the electron transport chain, each NADH generates 3 ATP in the electron transport chain, and each acetyl-CoA generates a total of 12 ATP (1 $QH_2 \times 2 = 2$ ATP; 3 NADH $\times 3 = 9$ ATP; 1 GTP = 1 ATP for a total of 12 ATP per acetyl-CoA). The total is 119 ATP. Propionyl-CoA is metabolized to succinyl-CoA (at a cost of 1 ATP; see Fig. 14-7) and enters the citric acid cycle. Conversion of succinyl-CoA to succinate yields 1 GTP (which offsets the cost of the propionyl-CoA → succinyl-CoA conversion), and conversion of succinate to fumarate yields 1 QH_2 (equivalent to 2 ATP). Fumarate is converted to malate, then malate is converted to pyruvate, yielding 1 NADPH (equivalent to 3 ATP). The pyruvate dehydrogenase reaction converts pyruvate to acetyl-CoA (which is subsequently oxidized by the citric acid cycle to yield 12 ATP) and 1 NADH, which yields 3 ATP. Therefore, oxidation of propionyl-CoA yields an additional 20 ATP. The total is 119 ATP + 20 ATP = 139 ATP. Two ATP must be subtracted from this total to account for the ATP spent in activating the C_{17} fatty acid to fatty acyl-CoA. This gives a final total of 137 ATP. Note that a C_{17} fatty acid yields more ATP than palmitate (129 ATP) and less than oleate (146 ATP).

13. A fatty acid cannot be oxidized until it has been activated by its attachment to coenzyme A in an ATP-requiring step. The first phase of glycolysis also requires the investment of free energy in the form of ATP. Consequently, neither β oxidation nor glycolysis can produce any ATP unless some ATP is already available to initiate these catabolic pathways.

14. (a) The free energy cost of synthesizing ATP from ADP + P_i is 30.5 $kJ \cdot mol^{-1}$. Glucose oxidation could theoretically yield 2850/30.5, or 93.4 ATP. The ATP yield per carbon atom is 93.4/6 = 15.6. Palmitate oxidation could theoretically yield 9781/30.5, or 320.7 ATP. The ATP yield per carbon atom is 320.7/16 = 20.4 ATP.

(b) *In vivo,* the catabolism of glucose leads to the production of 38 ATP, which is 38/6, or 6.3 ATP per carbon atom. The catabolism of palmitate leads to the production of 129 ATP (see the solution to Problem 14-9a), which is 129/16, or 8.1 ATP per carbon atom.

(c) In theory, as well as *in vivo,* fatty acid oxidation yields more ATP per carbon atom than glucose oxidation (this is primarily because the carbons of carbohydrates are already partially oxidized, whereas fatty acid carbons are usually fully reduced). Both pathways are equally efficient from a thermodynamic point of view, recovering about 40% of the free energy available (for glucose 6.3/15.6, and for palmitate, 8.1/20.4).

15.

	Fatty acid degradation	**Fatty acid synthesis**
Cellular location	Mitochondrial matrix	Cytosol
Acyl group carrier	Coenzyme A	Acyl carrier protein
Electron carrier(s)	Ubiquinone and NAD^+ *accept* electrons to become ubiquinol and NADH	NADPH *donates* electrons and becomes oxidized to $NADP^+$
ATP requirement	One ATP (two high-energy phosphoanhydride bonds) required to activate the fatty acid	Consumes one ATP per two carbons incorporated into the growing fatty acyl chain
Unit product/unit donor	Two-carbon acetyl units (acetyl-CoA)	C_3 intermediate (malonyl-CoA)
Configuration of hydroxyacyl intermediate	L	D
Shortening/growth occurs at which end of the fatty acyl chain?	Thioester end	Thioester end

16.

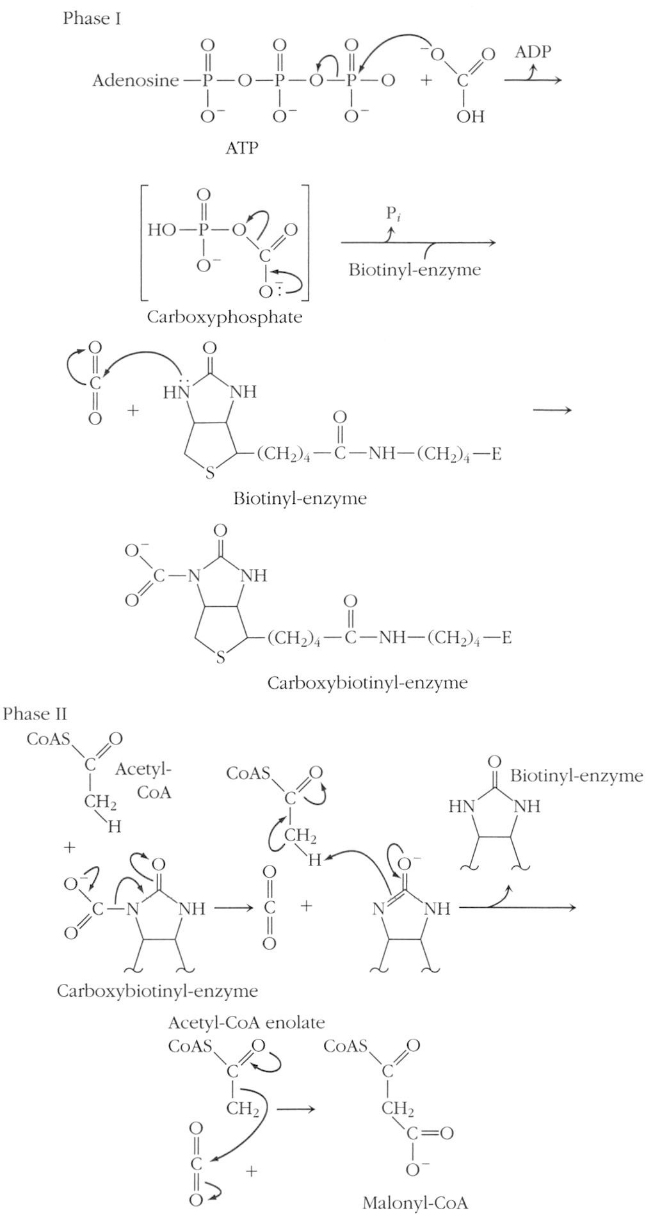

Phase I

Adenosine—P—O—P—O—P—O + C ADP

ATP

Carboxyphosphate

P_i

Biotinyl-enzyme

$(CH_2)_4$—C—NH—$(CH_2)_4$—E

Biotinyl-enzyme

Carboxybiotinyl-enzyme

Phase II

Acetyl-CoA

Biotinyl-enzyme

Carboxybiotinyl-enzyme

Acetyl-CoA enolate

Malonyl-CoA

17. As shown in the mechanism for malonyl-CoA formation (Problem 14-16), ATP is required to generate a carboxyphosphate intermediate. The result is that malonyl-CoA contains some of the free energy of ATP. This free energy is released when the acetyl group condenses with malonyl-ACP and CO_2 is released. The release of CO_2 is entropically favored and drives the reaction to completion.

18. β Oxidation of palmitate yields 129 ATP, as shown in Problem 14-9a. Palmitate synthesis costs 49 ATP, as described in the text (7 ATP required to convert each of 7 acetyl-CoA to malonyl-CoA, and 2 NADPH required for seven rounds of synthesis, which is equivalent to $2 \times 7 \times 3 = 42$ ATP for a total of 49 ATP).

19. In gluconeogenesis, the input of free energy is required to undo the exergonic pyruvate kinase reaction of glycolysis. Pyruvate is carboxylated to produce oxaloacetate, and then oxaloacetate is decarboxylated to produce phosphoenolpyruvate. Each of these reactions requires the cleavage of one phosphoanhydride bond (in ATP and GTP, respectively). In fatty acid synthesis, ATP is consumed in the acetyl-CoA carboxylase reaction, which produces malonyl-CoA. The decarboxylation reaction is accompanied by cleavage of a thioester bond, which has a similar change in free energy to cleaving a phosphoanhydride bond.

20. The label does not appear in palmitate because $^{14}CO_2$ is released in Reaction 3 of fatty acid synthesis (Fig. 14-13).

21. Fatty acids that cannot be synthesized from palmitate using the available cellular elongases and desaturases are termed essential fatty acids and must be obtained from the diet. Mammals do not have a desaturase enzyme that can introduce double bonds beyond C9. Oleate and palmitoleate, with a double bond at the 9,10 position, are not essential fatty acids. Linoleate has two double bonds, one at the 9,10 position and one at the 12,13 position, and therefore is an essential fatty acid. α-Linolenate has double bonds at positions 9,10, 12,13, and 15,16 and is also essential.

22. (a) Liver fatty acid synthase activity increases with consumption of a high-carbohydrate diet. Glucose in excess of what is required to meet immediate energy needs is oxidized to pyruvate by glycolysis, then converted to acetyl-CoA by pyruvate dehydrogenase. Excess acetyl-CoA is used to synthesize fatty acids, which are ultimately used to synthesize triacylglycerols for storage in adipose tissue.

(b) Liver fatty acid synthase activity decreases with consumption of a high-fat diet. Endogenous fatty acid synthesis is not required if fatty acids are obtained from the diet.

(c) Mammary gland fatty acid synthase activity increases in mid to late pregnancy to provide fatty acids for triacylglycerols in breast milk for the neonate.

23. (a) Acetoacetate is a ketone body. It is converted to acetyl-CoA, which can be oxidized by the citric acid cycle to supply free energy to the cell.

(b) Intermediates of the citric acid cycle are also substrates for other metabolic pathways. Unless they are replenished, the catalytic activity of the cycle is diminished. Ketone bodies are metabolic fuels, but they cannot be converted to citric acid cycle intermediates. A three-carbon glucose-derived compound such as pyruvate can be converted to oxaloacetate to increase the pool of citric acid cycle intermediates and keep the cycle operating at a high rate.

24. In the absence of intracellular glucose, the liver turns to other fuels to meet its energy needs. Fatty acid degradation produces acetyl-CoA, but in the absence of glucose, acetyl-CoA cannot enter the citric acid cycle because of insufficient oxaloacetate, a substrate for the first step. (When carbohydrate concentrations drop, any available oxaloacetate is diverted to gluconeogenesis to produce the necessary glucose for red blood cell and brain function.) If acetyl-CoA cannot enter the citric acid cycle (and acetyl-CoA cannot be converted to pyruvate in mammals), acetyl-CoA concentrations rise, and excess acetyl-CoA molecules are converted to ketone bodies. This explains why ketosis occurs in the untreated diabetic patient.

25. The synthesis of the ketone body acetoacetate does not require the input of free energy (the thioester bonds of 2 acetyl-CoA are cleaved; see Fig. 14-16). Conversion of acetoacetate to 3-hydroxybutyrate consumes NADH (which could otherwise generate 3 ATP by oxidative phosphorylation). However, the conversion of 3-hydroxybutyrate back to 2 acetyl-CoA regenerates the NADH (see Fig. 14-17). This pathway also requires a CoA group donated by succinyl-CoA. The conversion of succinyl-CoA to succinate by the citric acid cycle enzyme succinyl-CoA synthetase generates GTP from GDP + P_i, so the conversion of ketone bodies to acetyl-CoA has a free energy cost equivalent to this one phosphoanhydride bond.

26. (a) HDL remove excess cholesterol from tissues and transport it back to the liver. This helps prevent the accumulation of cholesterol in vessel walls that leads to atherosclerosis.

(b) HDL level alone does not indicate the risk of developing atherosclerosis, since the level of LDL, the activity of the LDL receptor, and other factors such as smoking or vessel wall injuries resulting from infection can all influence the likelihood of developing the disease.

27. (a) Because cholesterol is water-insoluble, it is found associated with other lipids, for example, in cell membranes. Only an integral membrane protein would be able to recognize cholesterol, which has a small OH head group and is mostly buried within the lipid bilayer.

(b) Proteolysis releases a soluble fragment of the SREBP that can travel from the cholesterol-sensing site to other areas of the cell, such as the nucleus.

(c) The DNA-binding portion of the protein might bind to a DNA sequence near the start of certain genes so as to mark them for transcription. In this way, the absence of cholesterol could stimulate the expression of proteins required to synthesize or take up cholesterol.

28. The soft margarine would be a better choice, since it is likely to contain more unsaturated fatty acids than the hard margarine (the melting point of a fatty acid decreases with increasing degree of unsaturation, because the double bonds induce kinks that make it more difficult for the acyl chains to pack together in a solid). The hard margarine is more likely to contain saturated fatty acids, and as a result of chemical hydrogenation, it may contain a higher proportion of trans double bonds.

29. Mammalian fatty acid synthase is structurally different from bacterial fatty acid synthase; thus Triclosan can act as an inhibitor of the bacterial enzyme but not the mammalian enzyme. The mammalian fatty acid synthase is a multifunctional enzyme made up of two identical polypeptides. In bacteria, the enzymes of the fatty acid synthetic pathway are separate proteins. Triclosan actually inhibits the bacterial enoyl-ACP reductase. The enzymes of the mammalian multifunctional enzyme must be arranged in such as way as to preclude the binding of Triclosan to the active site of the enoyl-ACP reductase.

30. (a) A portion of the inhibitor molecule mimics the structure of a fatty acid. The compound may act as a competitive inhibitor, binding to the enzyme active site to preclude binding of the substrate.

(b) The lower the ID_{50}, the lower the concentration of inhibitor required to kill the cells. It is desirable for the inhibitor to inhibit FAS in cancer cells but not normal cells. Thus an effective inhibitor would have a low ID_{50} in cancer cells but a high ID_{50} in normal cells. The ratios of ID_{50} values for normal cells and cancer cells were calculated to determine which of the inhibitors was most effective. As shown in the table below, compounds C81, C75, and C49 were the most effective inhibitors. These compounds had side chains of 11, 7, and 8 carbons, respectively. Of these three compounds, C75 has the shortest alkyl chain and is the most soluble of the three.

Compound	**Alkyl side chain (R)**	**SKBr3 cells ID_{50}/ normal cells ID_{50}**
C83	$—C_{13}H_{27}$	0.37
C81	$—C_{11}H_{23}$	0.17
C77	$—C_9H_{19}$	0.41
C75	$—C_8H_{17}$	0.23
C49	$—C_7H_{15}$	0.22
C73	$—C_6H_{13}$	0.68

(c) C75 is a slow inhibitor. It inhibits 50% of FAS activity in 5 minutes and 80% activity in 15 minutes.

(d) C75 inhibits incorporation of radiolabeled acetate into triglyceride significantly at a concentration of 2 μg/mL. At a concentration of 5 μg/mL, 80% of total acylglyeride synthesis is inhibited.

31. (a) Fumonisin inhibits ceramide synthase. Levels of the final product, ceramide, were decreased while other lipid synthetic pathways were not affected, indicating that fumonisin directly inhibits the ceramide synthetic pathway. The first enzyme in the pathway, serine palmitoyl transferase, was not the target of fumonisin B_1 because the enzyme's activity was not significantly decreased in the presence of fumonisin. The second enzyme in the pathway, 3-ketosphinganine reductase, was also not a target, because if it were, the substrate of this reaction, 3-ketosphinganine, would have accumulated in the presence of fumonisin. Accumulation of sphinganine indicates that ceramide synthase was inhibited. When ceramide synthase is inhibited, the reactant sphinganine cannot be converted to dihydroceramide and instead accumulates.

(b) Fumonisin likely acts as a competitive inhibitor. It is structurally similar to sphingosine and its derivatives and thus can bind to the active site and prevent substrate from binding. The fumonisin may form a covalent bond with the enzyme, or it may bind to the enzyme's active site noncovalently with high affinity. Alternatively, fumonisin may act as a substrate and be converted to a product that cannot be subsequently converted to ceramide.

Chapter 15 Solutions

1. The ATP-induced conformational change must decrease the $\mathcal{E}^{\circ\prime}$, from -0.29 V to about -0.40 V. The decrease in reduction potential allows the protein to donate electrons to N_2, since electrons flow spontaneously from a substance with a lower reduction potential to a substance with a higher reduction potential. Without the conformational change, nitrogenase could not reduce N_2.

2. (a) Method 1

$$\begin{aligned} &\text{Glu} + \text{ATP} + NH_4^+ \rightarrow \text{Gln} + P_i + \text{ADP}\\ &\text{Gln} + \alpha\text{-ketoglutarate} + \text{NADPH} \rightarrow 2\ \text{Glu} + \text{NADP}^+\\ &\underline{\text{Glu} + \alpha\text{-keto acid} \rightleftharpoons \alpha\text{-ketoglutarate} + \text{amino acid}}\\ &NH_4^+ + \text{ATP} + \text{NADPH} + \alpha\text{-keto acid} \rightarrow\\ &\qquad \text{ADP} + P_i + \text{NADP}^+ + \text{amino acid} \end{aligned}$$

(b) Method 2

$$\begin{aligned} &\alpha\text{-ketoglutarate} + NH_4^+ + \text{NAD(P)H} \rightleftharpoons\\ &\qquad \text{Glu} + H_2O + \text{NAD(P)}^+\\ &\underline{\text{Glu} + \alpha\text{-keto acid} \rightleftharpoons \alpha\text{-ketoglutarate} + \text{amino acid}}\\ &\alpha\text{-keto acid} + \text{NAD(P)H} + NH_4^+ \rightleftharpoons\\ &\qquad \text{NAD(P)}^+ + \text{amino acid} + H_2O \end{aligned}$$

(c) The low K_M of glutamine synthetase for ammonium ions ensures that Method 1 will be used when the concentration of available ammonia is low. Incorporation of one mole of ammonia into an amino acid costs one mole of ATP, whereas ATP is not required when the prokaryotic cell uses Method 2. The prokaryotic cell is at a disadvantage when the ammonia concentration in the growth medium is low because energy must be expended in order to synthesize amino acids under these conditions.

3. Leghemoglobin, like myoglobin, is an O_2-binding protein. Its presence decreases the concentration of free O_2 that would otherwise inactivate the bacterial nitrogenase.

4. Glutamate dehydrogenase, glutamine synthetase, and carbamoyl phosphate synthetase.

5. (a) The condensation of ammonia and α-ketoglutarate produces the amino acid glutamate.
(b) The reverse reaction, in which glutamate is deaminated to produce α-ketoglutarate, replenishes the citric acid cycle intermediate.

6. (a) $^+H_3N-CH_2-C(=O)-O^- \xrightarrow[\alpha\text{-Ketoglutarate}]{\text{Glu}} COO^--C(=O)-H$

(b) $^+H_3N-CH(CH_2-CH_2-CH_2-NH-C(=NH_2^+)-NH_2)-C(=O)-O^- \xrightarrow[\alpha\text{-Ketoglutarate}]{\text{Glu}} COO^--C(=O)-CH_2-CH_2-CH_2-NH-C(=NH_2^+)-NH_2$

(c) $^+H_3N-CH(CH_2-OH)-C(=O)-O^- \xrightarrow[\alpha\text{-Ketoglutarate}]{\text{Glu}} COO^--C(=O)-CH_2-OH$

(d) $^+H_3N-CH(CH_2-C_6H_5)-C(=O)-O^- \xrightarrow[\alpha\text{-Ketoglutarate}]{\text{Glu}} COO^--C(=O)-CH_2-C_6H_5$

7. (a) Asp: $^+H_3N-CH(CH_2-COO^-)-C(=O)-O^- \xrightarrow[\alpha\text{-Ketoglutarate}]{\text{Glu}} COO^--C(=O)-CH_2-COO^-$ Oxaloacetate

(b) Ala: $^+H_3N-CH(CH_3)-C(=O)-O^- \xrightarrow[\alpha\text{-Ketoglutarate}]{\text{Glu}} COO^--C(=O)-CH_3$ Pyruvate

(c) Glu: $^+H_3N-CH(CH_2-CH_2-COO^-)-C(=O)-O^- \xrightarrow[\text{Oxaloacetate}]{\text{Asp}} COO^--C(=O)-CH_2-CH_2-COO^-$ α-Ketoglutarate

The products are all intermediates of the citric acid cycle (oxaloacetate and α-ketoglutarate) or closely associated with the citric acid cycle (pyruvate).

8. (a) Leucine
(b) Methionine
(c) Tyrosine
(d) Valine

9.

$E{-}(CH_2)_4{-}NH_2$ + $H_3C{-}C_\beta H(OH){-}C_\alpha H(COO^-){=}N^+H{-}CH{-}$pyridoxal phosphate ($O_3^{2-}PO$, O^-, $N^+{-}H$) $\longrightarrow$

$E{-}(CH_2)_4{-}NH_3^+$ + $H_3C{-}C_\beta(=O){-}H$ + $H{-}C_\alpha(COO^-){=}N^+(H){-}CH{=}$pyridoxal phosphate ($O_3^{2-}PO$, O^-, $N{-}H$)

10. The SH group of cysteine is derived from methionine, which is an essential amino acid. Without an adequate supply of methionine, cysteine cannot be produced and must therefore be obtained in the diet.

11. (a) The sulfonamide is a structural analog of the *p*-aminobenzoate group of folate.

$H_2N{-}C_6H_4{-}C(=O){-}O^-$

p-Aminobenzoate

(b) Mammals obtain folate from their diet and lack the enzymes necessary to synthesize folate. They are therefore not affected by the drug.

12. (a) ADP and GDP both serve as allosteric inhibitors of ribose phosphate pyrophosphokinase.

(b) PRPP, the substrate of amidophosphoribosyltransferase, stimulates the enzyme by feedforward activation. AMP, ADP, ATP, GMP, GDP, and GTP are all products and inhibit the enzyme by feedback inhibition.

13. HPRT catalyzes a nucleotide salvage reaction in which the nucleotide IMP can be synthesized from the free base (hypoxanthine) plus a phosphorylated ribose derivative. Inhibiting HPRT would block production of IMP, which is a precursor of AMP and GMP. In order to be an effective drug target, HPRT must be essential for parasite growth; that is, the parasite cannot synthesize its own purine nucleotides from scratch but instead relies on salvage reactions using the host cell's hypoxanthine.

14. Phosphoribosylpyrophosphate (PRPP) is a reactant in the salvage reactions involving IMP and GMP, so if these reactions cannot occur, the PRPP that would normally be used in the salvage reactions would accumulate. PRPP stimulates amidophosphoribosyltransferase by feedforward activation (see Problem 12), which accelerates the synthesis of purine nucleotides and increases the concentration of their degradation product, uric acid.

15. The ketogenic amino acids are broken down to acetyl-CoA or acetoacetate, which can be converted to acetyl-CoA. For glucogenic amino acids, the carbon skeletons are broken down to either pyruvate (which is converted to acetyl-CoA by the pyruvate dehydrogenase complex) or a citric acid cycle intermediate (which can be converted to phosphoenolpyruvate by the reactions of gluconeogenesis; phosphoenolpyruvate is the precursor of pyruvate and therefore of acetyl-CoA).

16. The amino acids released by protein degradation are used as metabolic fuels. They can be completely catabolized to CO_2 by muscle cells to produce ATP via the citric acid cycle and oxidative phosphorylation, or they can be partially broken down and transported to the liver to be used in gluconeogenesis, which indirectly supplies the muscle cells with glucose.

17. Proteins, a polymeric form of amino acids, could be considered as a storage depot for amino acids, since the proteins can be degraded to release amino acids for use as metabolic fuels. However, proteins have functions other than fuel storage, which is not the case for glycogen and triacylglycerols (although triacylglycerols also function as thermal insulation in some species).

18. (a)

$CH_3{-}CH_2{-}CH(CH_3){-}CH(NH_3^+){-}COO^-$ (Isoleucine)

↓ 1 transaminase

$CH_3{-}CH_2{-}CH(CH_3){-}C(=O){-}COO^-$

↓ 2 branched-chain α-keto-acid dehydrogenase

$CH_3{-}CH_2{-}CH(CH_3){-}C(=O){-}SCoA$

↓ 3 acyl-CoA dehydrogenase

$CH_3{-}CH{=}C(CH_3){-}C(=O){-}SCoA$

↓ 4 enoyl-CoA hydratase

$CH_3{-}CH(OH){-}CH(CH_3){-}C(=O){-}SCoA$

↓ 5 hydroxyacyl-CoA dehydrogenase

$CH_3{-}C(=O){-}CH(CH_3){-}C(=O){-}SCoA$

↓ 6 thiolase

$CH_3{-}C(=O){-}SCoA$ (Acetyl-CoA) + $CH_3{-}CH_2{-}C(=O){-}SCoA$ (Propionyl-CoA)

(b) Reaction 2 is analogous to the pyruvate dehydrogenase reaction. The dehydrogenase catalyzes the release of carbon dioxide and the formation of a high energy bond with coenzyme A.
(c) Reaction 3 is analogous to the acyl-CoA dehydrogenase enzyme of fatty acid biosynthesis. Hydrogens are removed from the α and β carbons of the substrate.

19. The fate of propionyl-CoA produced upon degradation of isoleucine is identical to that of propionyl-CoA produced in the oxidation of odd-chain fatty acids (see Fig. 14-7). Propionyl-CoA is converted to (*S*)-methylmalonyl-CoA by propionyl-CoA carboxylase. A racemase converts the (*S*)-methylmalonyl-CoA to the (*R*) form. A mutase enzyme converts the (*R*)-methylmalonyl-CoA to succinyl-CoA, which enters the citric acid cycle.

20. (a) $C_6H_5{-}CH_2{-}C({=}O){-}COO^-$

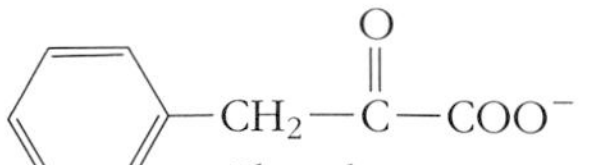

Phenylpyruvate

(b) A tetrahydrobiopterin deficiency prevents the conversion of phenylalanine to tyrosine in the phenylalanine catabolic pathway. Consequently, phenylalanine accumulates and undergoes transamination to phenylpyruvate, which is excreted.
(c) The growing child still requires some phenylalanine for growth. The low-phenylalanine diet should provide enough phenylalanine for growth but should not exceed amounts needed for growth since the excess cannot be metabolized due to the lack of the PAH enzyme.
(d) The artificial sweetener aspartame consists of a methylated Asp-Phe dipeptide. (The C-terminal carboxyl group is methylated.) Aspartame is broken down to Asp, Phe, and methanol. Since aspartame is a source of phenylalanine, patients with PKU should not use this product, and physicians should advise their patients to check labels carefully to avoid use of this product.
(e) Phenylalanine is converted to tyrosine in normal individuals. The tyrosine is then converted to melanin, which is the compound responsible for pigment in skin and hair.
(f) Phenylalanine is the precursor of tyrosine. If the phenylalanine intake is low, supplemental tyrosine may be needed.

21. If an essential amino acid is absent from the diet, then the rate of protein synthesis drops significantly, since most proteins contain an assortment of amino acids, including the deficient one. The other amino acids that would normally be used for protein synthesis are therefore broken down and their nitrogen excreted as urea. The decrease in protein synthesis coupled with the normal turnover of body proteins leads to the excretion of nitrogen in excess of the intake.

22. One nitrogen atom is derived from ammonia that is incorporated into carbamoyl phosphate for entrance into the urea cycle. The other nitrogen atom comes from aspartate, which serves as a substrate in the argininosuccinase reaction. Ultimately, both nitrogen atoms that appear in urea originate from excess dietary protein.

23. (a) A urea cycle enzyme deficiency decreases the rate at which nitrogen can be eliminated as urea. Since the sources of nitrogen for urea synthesis include free ammonia, low urea cycle activity may lead to high levels of ammonia in the body.
(b) A low-protein diet might reduce the amount of nitrogen to be excreted.
(c) Excretion of the nitrogen-containing amino acids glycine and glutamine (an important source of amino groups) helps reduce the amount of amino groups in the body. With Ucephan treatment, nitrogen is excreted as phenylacetylglutamine or hippuric acid.

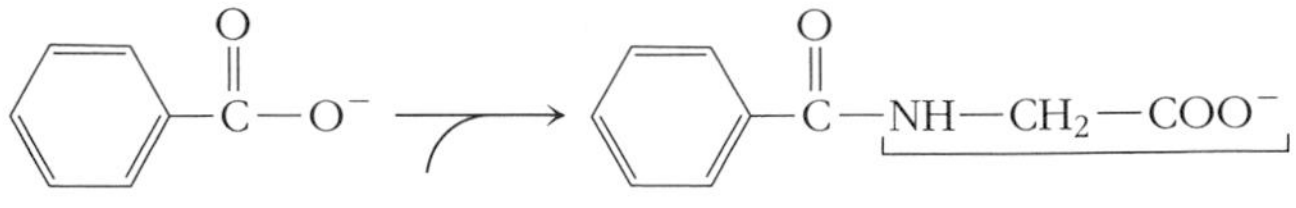

$C_6H_5{-}C({=}O){-}O^-$ (Benzoate) + Glycine $\longrightarrow$ $C_6H_5{-}C({=}O){-}NH{-}CH_2{-}COO^-$ (Hippuric acid; the $-NH-CH_2-COO^-$ portion is the Glycine residue)

$C_6H_5{-}CH_2{-}COO^-$ (Phenylacetate) + Glutamate $\longrightarrow$ $C_6H_5{-}CH_2{-}C({=}O){-}NH{-}CH(COO^-){-}(CH_2)_2{-}COO^-$ (Phenylacetylglutamine; the $-NH-CH(COO^-)-(CH_2)_2-COO^-$ portion is the Glutamate residue)

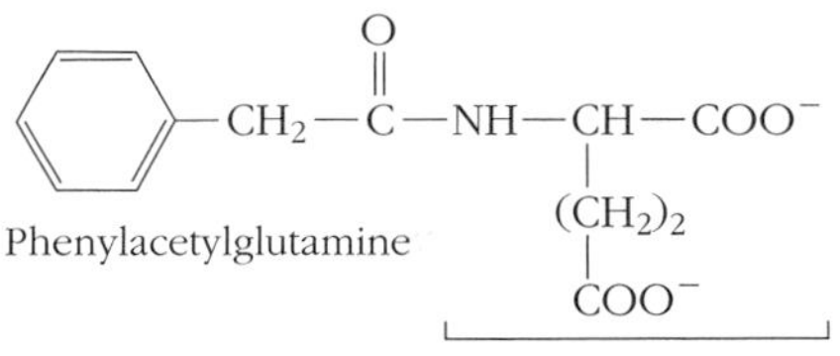

(d) Adding arginine, the product of the argininosuccinase reaction, would increase flux through the urea cycle.

24. Increased ammonium ion concentration would shift the equilibrium of the glutamate dehydrogenase reaction to form glutamate from NH_4^+ and α-ketoglutarate. This would eventually deplete cellular levels of α-ketoglutarate, a citric acid cycle intermediate. All citric acid cycle intermediates must be present in catalytic amounts for the cycle to operate efficiently. In the absence of α-ketoglutarate, the citric acid cycle, the major source of ATP synthesis in aerobic oxidation, ceases to function. Coma and death result from lack of sufficient ATP in the brain.

25. An individual consuming a high-protein diet uses amino acids as metabolic fuels. As the amino acid skeletons are converted to glucogenic or ketogenic compounds, the amino groups are disposed of as urea, leading to increased flux through the urea cycle. During starvation, proteins (primarily from muscle) are degraded to provide precursors for gluconeogenesis. Nitrogen from these protein-derived amino acids must be eliminated, which demands a high level of urea cycle activity.

26. [Structure: purine ring with O, HN, N, O⁻, N/H labels]

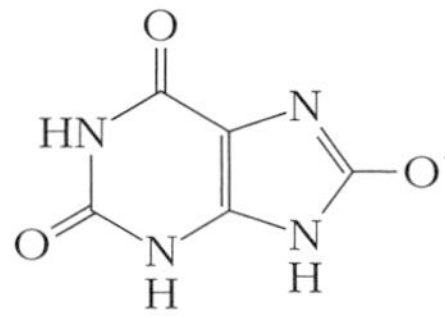

Urate

27. (a) *H. pylori* urease converts urea to NH_3 and CO_2. The ammonia has a p*K* of 9.25, so it combines with protons to produce NH_4^+. The resulting decrease in hydrogen ion concentration helps the bacteria maintain a high pH.
(b) Urease on the cell surface increases the pH of the fluid surrounding the cell, creating a more hospitable microenvironment for bacterial growth.

28. (a) UTase is stimulated by α-ketoglutarate and ATP, both reactants (either directly or indirectly) of the glutamine synthetase enzyme. UTase is inhibited by glutamine and inorganic phosphate, both products of the reaction. UR is stimulated by glutamine, a product of the glutamine synthetase reaction. High concentrations of glutamine decrease the activity of glutamine synthetase.
(b) Histidine, tryptophan, carbamoyl phosphate, glucosamine-6-phosphate, AMP, CTP, and NAD^+ are all end products of glutamine metabolic pathways. Alanine, serine, and glycine reflect the overall cellular nitrogen level. When nitrogen levels are adequate, glutamine synthetase activity is inhibited.
(c) The glutamine synthetase enzyme purified from the first batch was adenylylated and was in its less active form and more susceptible to inhibition by allosteric modulators. The enzyme was in the adenylylated form because the growth medium contained glutamate, and under these conditions, glutamine synthetase activity is inhibited. The enzyme from the second batch was not adenylylated and was in its fully active form, which is what would be expected when NH_4^+ is the sole nitrogen source. Because the second batch of enzyme was not adenylylated, it was not susceptible to inhibition by the allosteric modulators.

29. (a) The sigmoidal shape of the velocity vs. substrate concentration plot indicates that threonine deaminase binds its substrate in a positively cooperative manner. As the threonine concentration increases, threonine binds to the enzyme with increasing affinity.
(b) Isoleucine is an allosteric inhibitor of threonine deaminase and binds to the T form of the enzyme. Velocity decreases by about 15%, but the nearly 10-fold increase in K_M is more dramatic. The decrease in velocity and increase in K_M indicate that isoleucine, the end product of the pathway, acts as a negative allosteric inhibitor of the enzyme that catalyzes an early, committed step of its own synthesis. The velocity vs. substrate concentration curve obtained for threonine deaminase in the presence of isoleucine has greater sigmoidal character, which means that binding of threonine to the enzyme is even more cooperative in the presence of the inhibitor.
(c) Valine stimulates threonine deaminase by binding to the R form. The maximal velocity is somewhat increased, but the K_M is decreased, indicating that the threonine substrate has a higher affinity for the enzyme in the presence of valine. The cooperative binding of threonine to threonine deaminase is abolished in the presence of valine, however, as indicated by the hyperbolic shape of the curve.

Chapter 16 Solutions

1. Glycolysis produces two moles of ATP per mole of glucose. Synthesis of one mole of glucose via gluconeogenesis costs six moles of ATP. Therefore, the cost of running one round of the Cori cycle is 4 ATP.

2. The lactate dehydrogenase reaction which reduces pyruvate to lactate occurs with concomitant oxidation of NADH to NAD^+. NAD^+ serves as a reactant in the glyceraldehyde-3-phosphate dehydrogenase reaction in glycolysis. If pyruvate were the end product of glycolysis, all of the cellular NAD^+ would become reduced to NADH and the glycolytic pathway would grind to a halt for lack of NAD^+.

3. During starvation, muscle proteins are broken down to produce gluconeogenic precursors. The amino groups of the amino acids are transferred to pyruvate via transamination reactions. The resulting alanine travels to the liver, which can dispose of the nitrogen via the urea cycle and produce glucose from the alanine skeleton (pyruvate) and other amino acid skeletons. This glucose travels not just to the muscles but to all tissues that need it, so the metabolic pathway is not truly a cycle involving just the liver and muscles.

4. These two enzymes are part of the gluconeogenic pathway. Their concentrations increase when dietary fuels are not available, so that the liver can supply other tissues with newly synthesized glucose.

5. Because glucokinase is not saturated at physiological glucose concentrations, it can respond to changes in glucose availability with an increase or decrease in reaction velocity. Consequently, the entry of glucose into glycolysis and subsequent metabolic pathways depends on the glucose concentration. Because hexokinase is saturated at physiological glucose concentrations, its rate does not change with changes in glucose concentration.

6.

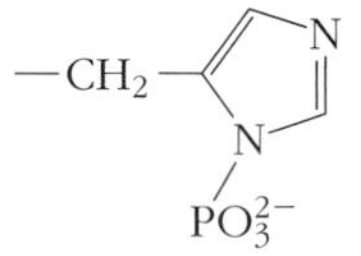

—CH_2— (imidazole ring: N, N)
PO_3^{2-}

7. Ingesting large amounts of glucose stimulates the β cells of the pancreas to release insulin, which causes liver and muscle cells to use the glucose to synthesize glycogen and causes adipose tissue to synthesize fatty acids. Insulin also inhibits the breakdown of metabolic fuels. The body is in a state of resting and digestion and is not prepared for running.

8. Insulin binding to its receptor stimulates the tyrosine kinase activity of the receptor. Proteins whose tyrosine residues are phosphorylated by the receptor tyrosine kinase can then interact with additional components of the signaling pathway. These interactions could not occur if a tyrosine phosphatase removed the phosphoryl groups attached to the Tyr residues.

9. Insulin stimulates uptake of glucose via GLUT4 receptors in adipocytes. The glucose is converted to glycerol-3-phosphate via the enzymes of the glycolytic pathway.

The glycerol-3-phosphate is required for the backbone of the triacylglycerol molecule.

10. The drugs can activate the intracellular tyrosine kinase domains of the insulin receptor, bypassing the need for insulin to bind to the receptor.

11. Insulin activation of cAMP phosphodiesterase destroys cAMP so that it can no longer activate protein kinase A, the enzyme that activates phosphorylase kinase as a step toward activating glycogen phosphorylase and inactivating glycogen synthase. The result is that insulin promotes glycogen synthesis.

12. The overall effect of the AMP-dependent kinase would be to inhibit anabolic pathways. Increased concentrations of AMP indicate that ATP levels are low (since ATP $\rightarrow$ ADP $\rightarrow$ AMP), so AMP would tend to promote catabolic processes that would replenish cellular ATP.

13. Phosphorolytic cleavage yields glucose-1-phosphate, which is negatively charged due to its phosphate group and will not diffuse out of the cell. In addition, glucose-1-phosphate can be isomerized to glucose-6-phosphate (and can enter glycolysis) without the expenditure of ATP. Hydrolytic cleavage yields neutral glucose, which could easily diffuse out of the cell. Free glucose would also need to be acted upon by hexokinase to convert it to glucose-6-phosphate (with the expenditure of ATP) before it could enter glycolysis.

14. GTPγS can bind to the G_α protein, but since it cannot be hydrolyzed, the G protein is in a persistently active state. If GTPγS binds to a stimulatory G protein, then adenylyl cyclase is continually active, which has the effect of increasing cellular cAMP concentrations. If GTPγS binds to an inhibitory G protein, adenylyl cyclase is continually inhibited, and cellular cAMP concentrations decrease.

15. Several days into a fast, muscle and liver glycogen have been depleted. In the absence of dietary glucose, the main source of endogenous glucose is gluconeogenesis. Citric acid cycle intermediates are used to synthesize oxaloacetate and then pyruvate, which enters the gluconeogenic pathway. Catalytic amounts of citric acid cycle intermediates are required for proper functioning of the cycle, so when these intermediates are diverted to gluconeogenesis, the citric acid cycle cannot function properly.

16. (a) In a normal person, glucagon binds to cell surface receptors on the liver, stimulating adenylyl cyclase to produce cAMP, and activating protein kinase A. Protein kinase A then activates glycogen phosphorylase via phosphorylation. Glycogen phosphorylase catalyzes the degradation of glycogen to glucose, which is released into the bloodstream. Blood glucose concentrations should rise shortly after an intravenous injection of glucagon. Glycogen degradation in Alex's liver thus appears to be normal.

(b) Glycogen metabolism in the liver appears to be normal, since glycogen content is normal and Alex's response to the glucagon test is normal. However, muscle glycogen is elevated, which indicates a defect in muscle glycogen metabolism. Most likely, glycogen synthesis is normal, since the biochemical structure of the glycogen is normal. It is possible that the defect is in the degradation of glycogen. Therefore, something could be wrong with the signaling mechanism or there could be a deficiency in one of the enzymes involved in glycogen degradation, such as muscle glycogen phosphorylase.

(c) Normally, muscle glycogen is degraded to glucose-6-phosphate, which enters glycolysis to be oxidized to yield ATP for muscle activity. In anaerobic conditions, pyruvate, the end product of glycolysis, is converted to lactate, which is released from the muscle into the blood and then enters the liver to be converted back to glucose via gluconeogenesis as part of the Cori cycle. Alex's muscles are unable to degrade glycogen to glucose-6-phosphate, so there is no glucose-6-phosphate to enter glycolysis, and lactate formation does not occur.

(d) During exercise, Alex's muscle cells become damaged and release myoglobin, which is excreted in the urine.

(e) Blood alanine concentrations normally increase as part of the glucose–alanine cycle. Glucose-6-phosphate, the product of glycogenolysis, enters glycolysis and produces pyruvate. Pyruvate undergoes a transamination reaction to form alanine, which is released from the muscle. Alanine then enters the liver, where the transamination reaction takes place in reverse to re-form pyruvate. Pyruvate can then enter gluconeogenesis, the resulting glucose product is released, and the cycle begins again. But Alex's tissues are not capable of the glucose–alanine cycle because his muscles are unable to produce glucose-6-phosphate. Instead, the muscles take up alanine and use it as a fuel. Thus, plasma alanine levels in the patient decrease rather than increase.

(f) Blood glucose concentrations are regulated by pancreatic hormones acting on the liver to stimulate glycogen synthesis or degradation, whatever is appropriate. Since Alex's liver enzymes appear to function normally, his blood glucose concentration is properly regulated and he is neither hypo- nor hyperglycemic.

(g) By ingesting glucose or fructose, blood sugar concentrations can be maintained at a high level. The muscles will thus be able to take up glucose or fructose and oxidize these sugars via glycolysis to yield the ATP required of active muscles. In this way, the muscles do not have to rely on stored glycogen as a fuel source.

17. (a) PAH is an oligomer with a subunit size of about 49,000 daltons. The subunits dissociate upon treatment with SDS, so one band at the low molecular weight is observed. The subunits are identical, since there is a single band in this lane. If PAH is subjected to electrophoresis under nondenaturing conditions, a single band with a molecular weight slightly less than 100,000 daltons is observed. Since this is double the molecular weight of the monomer, it can be concluded that the PAH consists of a dimer under nondenaturing conditions. Preincubation of PAH with phenylalanine yields a band that is four times the molecular weight of the monomer. Therefore, PAH is converted from a dimer to a tetramer in the presence of phenylalanine.

(b) We know from part (a) that PAH consists of a dimer that can be converted to a tetramer at high concentrations of phenylalanine. The sigmoidal curve indicates that the binding of the substrate is cooperative: The binding of one substrate molecule facilitates the binding of subsequent substrate molecules. Thus PAH is an allosteric enzyme and likely is affected by allosteric modulators that stimulate or inhibit the enzyme and thus control the degradation rate of phenylalanine.

(c) As discussed above, preincubation with phenylalanine converts PAH from the dimer to a tetrameric form. It is possible that the tetrameric form has a greater affinity for substrate than the dimeric form, which would explain the increase in velocity when the enzyme is preincubated with Phe. The dimer is the T form, which has a low affinity for the substrate, and the tetramer is the R form, which has a high affinity for the substrate.

(d) Insulin inhibits the activity of PAH, whereas glucagon stimulates the activity of the enzyme. Because PAH is phosphorylated in the presence of glucagon, it is likely that glucagon exerts its effects by activating a cAMP-dependent protein kinase, which in turn phosphorylates the enzyme. The effects of glucagon and Phe appear to be synergistic, since the enzyme is more active when glucagon and Phe are combined than when either stimulant is used alone. Since glucagon concentrations are high when blood sugar is low, it is possible that PAH activation occurs in these circumstances to stimulate the phenylalanine degradation pathway to produce fumarate, which can enter gluconeogenesis.

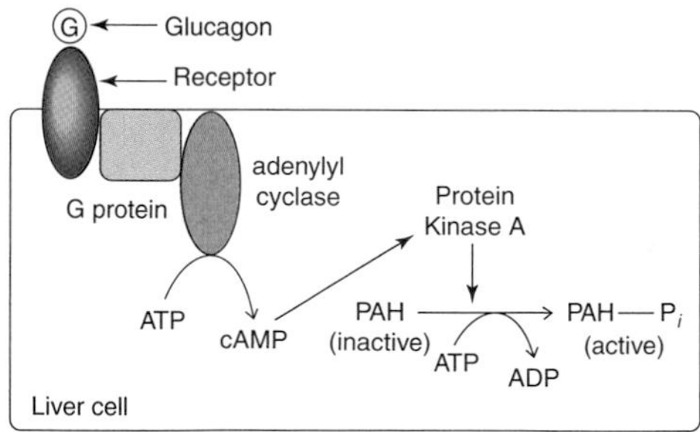

18. (a)

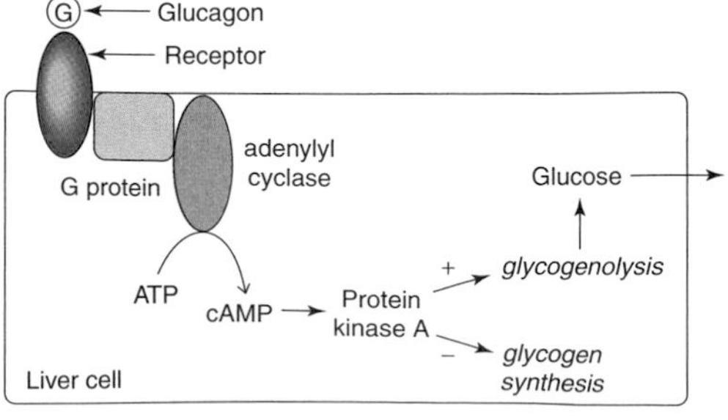

Glucagon binds to the extracellular domain of the receptor in the liver. Binding of glucagon activates a G protein, which in turn activates adenylyl cyclase, which converts intracellular ATP to the second messenger cAMP. cAMP activates protein kinase A, which phosphorylates several proteins, leading to the stimulation of glycogenolysis and the inhibition of glycogen synthesis. The end product of glycogenolysis, glucose-1-phosphate, is isomerized to glucose-6-phosphate. The phosphate is removed and glucose leaves the liver and enters the blood.

(b) These amino acids are positively charged. Since a negatively charged aspartate residue in the glucagon receptor has been shown to be essential for binding, it is possible that an ion pair forms between a positively charged amino acid side chain (His, Lys, or Arg) and the essential arginine. This hypothesis can be tested by modifying His 1, Lys 12, Arg 17, and Arg 18 to neutral or negatively charged side chains and assessing the resulting analogs' binding and signal-transducing capabilities.

(c) Eliminating the Asp at position 9 results in an analog with decreased affinity for the receptor but with little biological activity, indicating that the Asp plays a role in both binding and signal transduction. Substituting the Asp with a positively charged Lys reduces the binding affinity by about half but completely eliminates the biological activity. The Asp evidently plays an important role in binding, but conservation of the negative charge does not seem to be critical since a positive charge does not abolish binding. Hence some other aspect of the Asp side chain structure is important for binding. The Asp at position 9 does seem to be important in biological activity, since deletion or substitution of the Asp greatly decreases biological activity.

Abolishing the positive charge at position 12 decreases binding affinity by 50–90%. But once the analogs are bound, they are still capable of eliciting a biological response. The more nonpolar analogs (Ala and the acetylated Lys) bind more effectively than the Gly 12 analog, indicating that nonpolar interactions between the hormone and the receptor are important. The addition of a negative charge at position 12 virtually abolishes binding, so it is possible that the cationic group at position 12 forms an ion pair with a negatively charged amino acid on the glucagon receptor.

Leu 17 binds more effectively to the receptor than does Ala 17 and, once bound, has greater activity. This supports the hypothesis from part (a) that hydrophobic interactions between the hormone and the receptor are important, since leucine has a more hydrophobic side chain than alanine. Substitution with a Glu residue also decreases binding, but not as much as at position 12. These results confirm the hypothesis that hydrophobic interactions are important, since substitution by alanine decreases binding more than substitution with leucine. The positive charge is important, since replacement with the negatively charged glutamate abolishes more than 90% of the binding ability of the analog.

The des-His 1 glucagon has less binding ability and even less biological activity. This indicates that the histidine at position 1 is important for both binding and signal transduction but plays a greater role in signal transduction. This is also supported by the additional des-His 1 analogs. The des-His 1–des-Asp 9 analog does not bind well (only 7% of the control) and has no biological activity. The binding and biological activity of the des-His 1–des-Asp 9 analog are lower than the des-Asp 9 analog, indicating that histidine is important in both binding and signal transduction. Interestingly, the des-His 1–Lys 9 derivative binds well (70%) but has no biological activity. This indicates that the substitution of aspartate for lysine at position 9 preserves characteristics that are important for binding. However, once bound, the analog does not trigger signal transduction. The des-His 1 derivatives in which a negatively charged Glu replaces any of the positively charged residues at positions 12, 17, and 18 result in analogs that do not bind to the receptor.

(d) The negatively charged Asp at position 9 is important for glucagon binding to its receptor but appears to play a greater role in signal transduction. This also appears to be true for the His at position 1. Positively charged amino acid residues are important but not absolutely essential for receptor binding. It is possible that an ion pair forms between the positively charged amino acid residue and a negative charge on the receptor protein. However, ion pairs are not the only important interaction. Hydrophobic interactions between the hormone and its receptor also play an important role in binding. The positively charged amino acids at positions 12, 17, and 18 are important for binding, but not as important for biological activity. The analogs had difficulty binding to the receptor, but once bound, elicited a full (or nearly full) biological response.

(e) The des-His 1–Lys 9 is the best antagonist because it binds to the receptor with 70% of the affinity of the native hormone but has no biological activity. In this derivative, the two amino acids important in signal transduction have been modified while the positively charged residues at positions 12, 17, and 18, which are critical for binding, have been retained. It might be possible to synthesize an even better antagonist (one with greater binding affinity) by trying various amino acid replacements at position 9. The replacement amino acid must be chosen carefully so that the characteristics required for binding at this position are retained while the characteristics required for signal transduction are eliminated.

19. (a) The sugar in the grape is mainly fructose, or fruit sugar. Fructose can be converted to fructose-6-phosphate, which can then enter the glycolytic pathway to produce pyruvate. The enzyme pyruvate decarboxylase converts pyruvate to acetaldehyde. Acetaldehyde is then converted to ethanol by alcohol dehydrogenase. The acetaldehyde produced from pyruvate is converted to acetate (see Box 10-B).

(b) In the alcohol dehydrogenase reaction, acetaldehyde is reduced to ethanol with concomitant oxidation of NADH to NAD^+. The NADH used as a reactant by alcohol dehydrogenase is the product of the glyceraldehyde-3-phosphate dehydrogenase reaction in glycolysis. In order to keep glycolysis going, NADH must be reoxidized to NAD^+. If NAD^+ cannot be produced because of a lack of alcohol dehydrogenase, the oxidation must occur some other way, such as via the glycerol-3-phosphate shuttle (see Fig. 12-11b). Dihydroxyacetone phosphate is converted to glycerol-3-phosphate by a dehydrogenase that concomitantly oxidizes NADH to NAD^+. Phosphatases can then remove the phosphate from glycerol-3-phosphate to yield glycerol, which accumulates in the yeast cell.

(c) Ethanol is converted to acetaldehyde and then to acetate. Acetate is converted to acetyl-CoA, which then enters the glyoxylate cycle (see Box 11-B). The first step is the synthesis of citrate from acetyl-CoA and oxaloacetate. Citrate is isomerized to isocitrate. Isocitrate lyase splits isocitrate into succinate and glyoxylate. Succinate leaves the glyoxisome and enters the mitochondrion, where it can enter the citric acid cycle. The glyoxylate fuses with a second molecule of acetyl-CoA to yield malate. Malate leaves the glyoxisome and enters the cytosol, where malate dehydrogenase converts it to oxaloacetate. Oxaloacetate can then enter gluconeogenesis to form glucose.

(d) Glycerol can serve as an energy source because it can be converted to glyceraldehyde-3-phosphate, which can then enter the glycolytic pathway below the phosphofructokinase block. The mutants cannot grow on glucose because glucose enters the glycolytic pathway by first being converted to glucose-6-phosphate, then fructose-6-phosphate. The next step, conversion to fructose-1,6-bisphosphate, requires phosphofructokinase. Thus, glycerol is a suitable substrate for this mutant, but glucose is not.

(e) Yeast takes up the sucrose in the bread dough and hydrolyzes it to glucose and fructose, which enter glycolysis to yield pyruvate. Pyruvate decarboxylase then acts on the pyruvate to produce acetaldehyde and carbon dioxide. The acetaldehyde is converted to ethanol, which evaporates during the rising period or later when the bread is baked. The carbon dioxide causes the bread dough to rise.

(f) Enzymes of the glyoxylate pathway, particularly malate dehydrogenase and isocitrate lyase (which are unique to this pathway), are inactivated since the glyoxylate pathway produces glucose from noncarbohydrate sources. The glyoxylate pathway is not required when glucose is available. Enzymes involved in gluconeogenesis that are not involved in glycolysis would also be inactivated, mainly phosphoenolpyruvate carboxykinase and fructose-1,6-bisphosphatase.

(g) During anaerobic fermentation, only two ATP molecules can be obtained from each glucose molecule. Thus, in order to get enough ATP to satisfy the energy needs of the cell, glucose consumption by glycolysis is high, and a relatively small amount is left over for the pentose phosphate pathway. But during aerobic respiration, 36 to 38 ATP are generated per glucose molecule. Thus, less glucose is needed to synthesize a given amount of ATP, and a larger fraction of glucose can enter the pentose phosphate pathway.

20. (a) Acetyl-CoA could be used to synthesize glucose in the reductive branch of the pathway. Acetyl-CoA enters the glyoxylate pathway to produce malate, which can then be converted to oxaloacetate and used to synthesize glucose. Alternatively, acetyl-CoA can condense with oxaloacetate to form citrate, which is isomerized to isocitrate. Isocitrate is oxidized to α-ketoglutarate, which is then converted to succinate by α-ketoglutarate oxidase. In this branch of the pathway, acetyl-CoA provides intermediates for biosynthetic reactions.

(b) Aspartate undergoes transamination to form oxaloacetate, which then can be converted to malate to enter the reductive pathway. Or the oxaloacetate could condense with acetyl-CoA to form citrate and enter the oxidative pathway.

21. (a) Since pyruvate carboxylase catalyzes the carboxylation of pyruvate to oxaloacetate, a deficiency of the enzyme would result in increased pyruvate levels and decreased oxaloacetate levels. Some of the excess pyruvate would also be converted to alanine, so alanine levels would be elevated. Some of the excess pyruvate would also be converted to lactate, which explains why the patient suffers from lactic acidosis. Decreased oxaloacetate levels decrease the activity of the citrate synthase reaction, the first step of the citric acid cycle. This causes the accumulation of acetyl-CoA, which forms ketone bodies that accumulate in the blood to cause ketosis.

(b) A pyruvate carboxylase deficiency results in decreased oxaloacetate levels. Aspartate is formed by transamination of oxaloacetate, so aspartate levels are decreased as well. Low levels of oxaloacetate decrease the levels of all the citric acid cycle intermediates due to decreased activity of the citrate synthase reaction. This results in lower levels of α-ketoglutarate, which can be transaminated to glutamate. Therefore, glutamate levels are low, as are GABA levels, since GABA is produced from glutamate.

(c) Acetyl-CoA stimulates pyruvate carboxylase activity. Adding acetyl-CoA would allow the investigators to determine whether there was a small amount of pyruvate carboxylase activity that could be activated.

22. (a) Malonyl-CoA, the product of the acetyl-CoA carboxylase reaction (the committed reaction of fatty acid synthesis), inhibits simultaneous fatty acid oxidation by inhibiting carnitine acyltransferase, an enzyme required for shuttling fatty acids into the mitochondrial matrix for oxidation. This mechanism prevents the oxidation of newly synthesized fatty acids because they cannot enter the mitochondria.

(b) Endogenous fatty acid synthesis is required if dietary fatty acid intake is insufficient. Stimulation of acetyl-CoA carboxylase ensures that the cell will have a sufficient level of fatty acids in the absence of dietary lipids (although essential fatty acids will still be lacking under these circumstances). During starvation (and untreated diabetes, which is similar to the starved state), body tissues do not have the resources to synthesize fatty acids, so the acetyl-CoA carboxylase enzymes are inhibited. In the starved state, fatty acids are mobilized to provide fuel to body tissues.

(c) In the absence of ACC2, heart and muscle are unable to synthesize fatty acids. This increases the demand on the liver to provide fatty acids for heart and muscle. In the knockout mice, liver glycogen is degraded to glucose, then oxidized to pyruvate and acetyl-CoA in order to provide acetyl-CoA for fatty acid synthesis.

(d) Fatty acid levels decrease due to the lack of ACC2. Triacylglycerols are released by adipose tissue and travel to the muscle and heart for oxidation, since the muscle and heart cannot synthesize their own fatty acids. This accounts for the increased blood levels of triacylglycerols.

(e) Insulin stimulates the activity of ACC2 in the muscle cells of normal mice, which promotes fatty acid synthesis and inhibits fatty acid oxidation (due to increased malonyl-CoA levels). The muscle cells in the knockout mice lack ACC2 and are not subject to insulin-mediated control. Fatty acid synthesis does not occur, malonyl-CoA levels do not rise, and fatty acid oxidation proceeds normally, even in the presence of insulin.

(f) Knockout mice are leaner because their heart and muscle tissue cannot synthesize fatty acids, so triacylglycerols are mobilized to provide fatty acids for these tissues, as described in part (d). Knockout mice have a higher rate of fatty acid oxidation and a lower rate of synthesis, as described in part (e), which also accounts for their lower weight gain despite their increased caloric intake.

(g) Molecular modeling techniques could be used to design a drug that inhibits the enzyme activity of ACC2 but not ACC1. The drug would have to be designed in such a way that it would be delivered to the mitochondrial matrix, where ACC2 is located.

23. (a) Acetate is converted to acetyl-CoA, which serves as a substrate for acetyl-CoA carboxylase and is converted to malonyl-CoA. Fatty acid synthase normally acts on malonyl-CoA for incorporation into fatty acids, but in the presence of the inhibitor C75, fatty acid synthesis does not occur. Therefore, radioactive malonate accumulates, and virtually no label appears in fatty acids.

(b) C75 causes hypothalamic levels of NPY to decrease. This decreases appetite, which leads to weight loss. In this manner, C75 fools the body into thinking that it is in the fed state, when in fact it is being starved.

(c) Acetyl-CoA carboxylase (ACC) catalyzes the reaction that converts acetyl-CoA to malonyl-CoA. In the presence of the ACC inhibitor TOFA, malonyl-CoA cannot be produced and hepatic malonyl-CoA levels would decrease. Subsequent administration of C75 would have no effect on the TOFA-treated animals, because C75, a FAS inhibitor, acts downstream of ACC. If malonyl-CoA inhibits feeding, the TOFA- and C-75-treated mice would have lower levels of malonyl-CoA, and feeding would not be inhibited, that is, the mice would eat normally. The investigators carried out these experiments and this is in fact what occurred.

(d) When malonyl-CoA accumulates, long-chain fatty acyl-CoA molecules also accumulate in the cytosol. Malonyl-CoA inhibits carnitine acyltransferase, which prevents translocation of fatty acyl-CoAs into the mitochondrial matrix for β oxidation. It is possible that long-chain fatty acyl-CoAs serve as signaling molecules to begin the pathway that suppresses appetite.

24. (a) Leptin stimulates glucose uptake by skeletal muscle.
(b) Glycogenolysis is inhibited, probably by direct inhibition of glycogen phosphorylase, which catalyzes the committed step in glycogenolysis.
(c) Leptin increases the activity of cAMP phosphodiesterase; the result is that the cellular concentration of cAMP decreases. In this way, leptin acts as a glucagon antagonist in the same manner that insulin does; glucagon's signal-transduction pathway leads to an increase in cAMP concentration.

25. (a) Dihydroxyacetone phosphate is in equilibrium with glyceraldehyde-3-phosphate. If glucose-6-phosphate hydrolysis was stimulated, this would lead to an increase in the removal of phosphate from glucose-6-phosphate, which would explain why the glucose concentration increased and the glucose-6-phosphate concentration decreased. The phosphoenolpyruvate concentration increased because flux from dihydroxyacetone phosphate to phosphoenolpyruvate increased (the reactions involved are all near equilibrium).
(b) A process that removes the phosphate from glucose-6-phosphate enhances gluconeogenesis (and glycogenolysis), because this last step of both processes allows glucose to leave the liver. This is a necessary step in regulating blood glucose concentrations during periods of fasting or starvation, which is when glucagon is released. Glycolysis is inhibited at the same time because the very first step of glycolysis is the phosphorylation of glucose to yield glucose-6-phosphate. If phosphate groups are continually removed from C6, glucose-6-phosphate cannot enter glycolysis and thus glycolysis is effectively inhibited.

26. When AMP concentrations are high, the energy status of the cell is low, and the glycolytic pathway is needed to provide ATP for the cell. AMP deaminase will act on the increased concentrations of AMP to produce ammonium ions. These ions, along with AMP, stimulate the activities of phosphofructokinase and pyruvate kinase and increase the activity of the glycolytic pathway as a whole. Cellular metabolites that influence the activity of adenosine deaminase will thus also have an indirect effect on the rate of glycolysis.

Chapter 17 Solutions

1. Parental ^{15}N-labeled DNA strands are shown in black, and newly synthesized ^{14}N DNA strands are shown in gray. The original ^{15}N-labeled parental DNA strands persist throughout succeeding generations, but their proportion of the total DNA decreases as new DNA is synthesized.

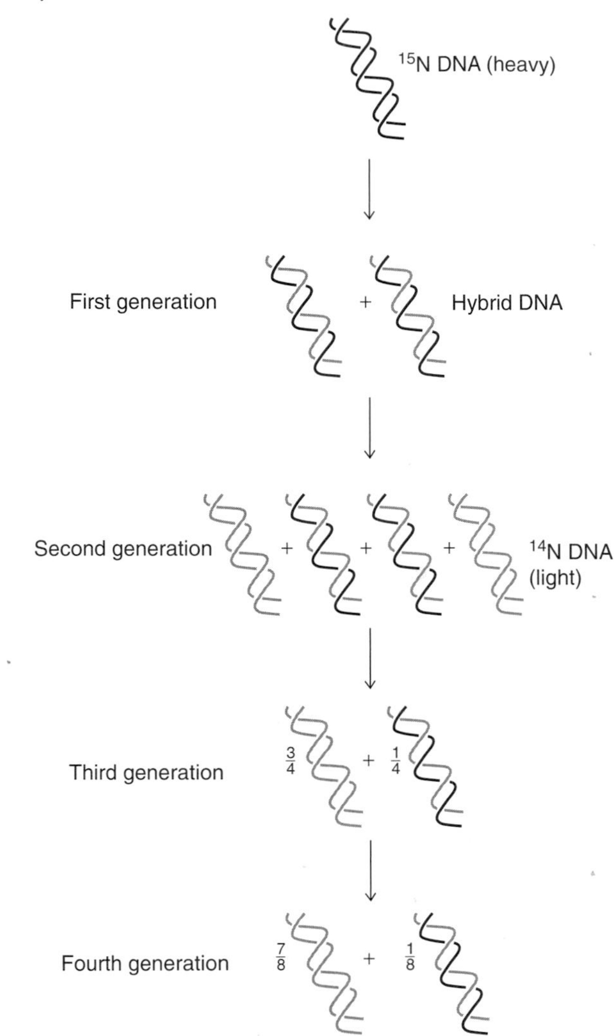

2. The origin is more likely to be richer in A:T base pairs, since these experience fewer stacking interactions and are more easily separated, which would allow easier access for the replication proteins.

3. The DNase can create a nick (a break in one strand). Then the exonuclease activity of DNA polymerase I can remove nucleotides on the 5′ side of the nick. At the same time, the polymerase active site can add radioactive deoxynucleotides to the 3′ side of the nick. The removal and replacement of nucleotides translates the nick in the 5′ → 3′ direction. DNA ligase can then seal the gap between the original DNA and the newly synthesized radioactive segment.

4. Novobiocin and ciprofloxacin are useful as antibiotics because they inhibit prokaryotic DNA gyrase but not eukaryotic topoisomerases. They can kill prokaryotic disease-causing microorganisms without harming host eukaryotic cells. Doxorubicin and etoposide inhibit eukaryotic topoisomerases and can be used as anticancer drugs. Although these drugs inhibit topoisomerases from both cancer cells and normal cells, cancer cells have a higher rate of DNA replication and are more susceptible to the effects of the inhibitors than normal cells.

5. The side chains of lysine and arginine residues have high p*K* values and are positively charged at physiological pH. The positively charged groups can form ion pairs with the negatively charged phosphate groups on the backbone of the DNA molecule.

6. (a) $-CH_2-CH_2-CH_2-CH_2-NH-C(=O)-CH_3$

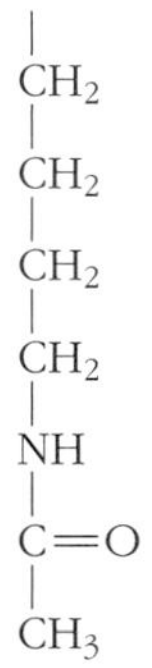

(b) $-CH_2-O-PO_3^{2-}$ ($^-O-P(=O)-O^-$)

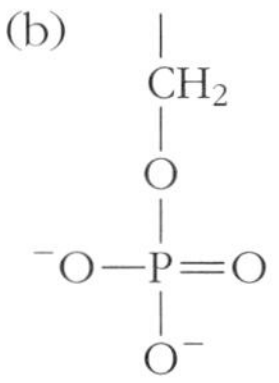

(c) $-CH_2-$(imidazole ring with N–phosphate: $^-O-P(=O)-O^-$)

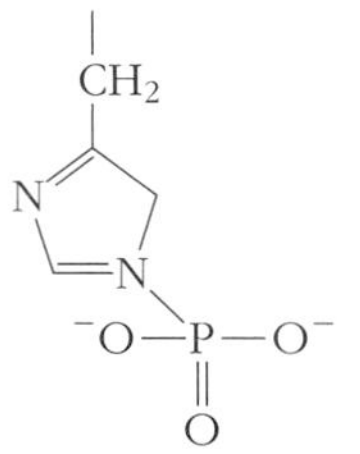

(d) $-CH_2-CH_2-CH_2-CH_2-NH-CH_3$

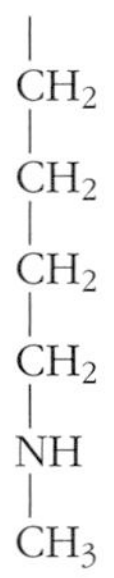

(e) $-CH_2-CH_2-CH_2-NH-C(=NH_2^+)-NH-CH_3$

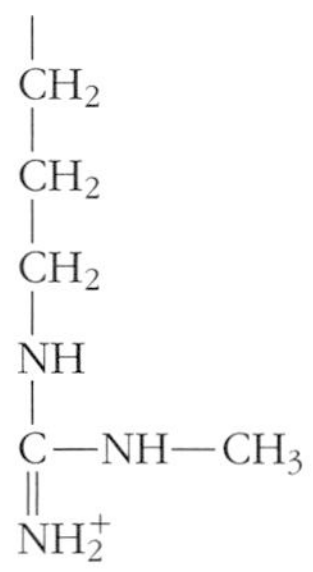

7. The ddNTP lacks the 3′ hydroxyl group that serves as the attacking nucleophile for the incoming dNTP.

8.

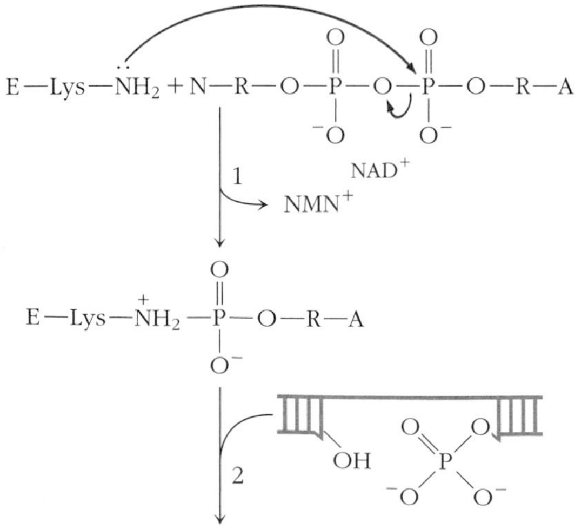

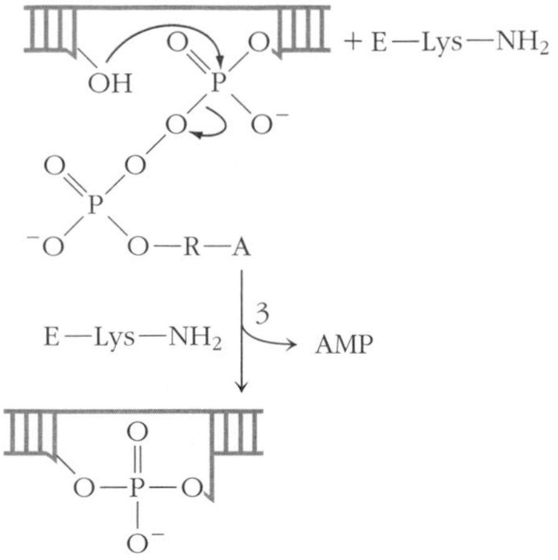

$E-Lys-\ddot{N}H_2 + N-R-O-P(=O)(O^-)-O-P(=O)(O^-)-O-R-A$

1: $NAD^+ \rightarrow NMN^+$

$E-Lys-\overset{+}{N}H_2-P(=O)(O^-)-O-R-A$

2: (nicked DNA with 3′ OH attacks the adenylylated 5′ phosphate)

$+\ E-Lys-NH_2$

3: $E-Lys-NH_2$, $\rightarrow AMP$

(sealed DNA: $O-P(=O)(O^-)-O$)

9. The drug would inhibit DNA synthesis because the polymerization reaction is accompanied by the release and hydrolysis of inorganic pyrophosphate (PP_i; see Fig. 17-9). Failure to hydrolyze the PP_i would remove the thermodynamic driving force for the overall process.

10. (a) SSB, which coats single-stranded DNA exposed at the replication fork, must have a relatively low affinity for DNA because it is displaced as the polymerase proceeds. The polymerase, along with its accessory proteins such as the clamp, has a relatively high affinity for the DNA.

(b) The cell most likely contains large amounts of SSB in order to coat all the single-stranded DNA and prevent it from forming secondary structure. Whereas the cell requires multiple copies of SSB for each replication fork, in theory, only two DNA polymerase enzymes are required (one for the leading strand and one for the lagging strand).

11. First, the cell contains roughly equal concentrations of the four deoxynucleotide substrates for DNA synthesis; this minimizes the chance for an overabundant dNTP to take the place of another or for the wrong dNTP to take the place of a scarce dNTP. Second, DNA polymerase requires accurate pairing between the template base and the incoming base. Third, the 3′ → 5′ exonuclease proofreads the newly formed base pair. Fourth, the removal of the RNA primer and some of the adjacent DNA helps minimize errors introduced by primase and by the DNA polymerase at the 5′ end of a new DNA segment. Finally, DNA repair mechanisms can excise mispaired or damaged nucleotides.

12. The enzyme most likely detects the geometry of the polynucleotide chain as it shifts from the wide and shallow A-form characteristic of an RNA helix to the narrower and steeper B-form of DNA.

13. The DNA molecule (chromosome) is much longer than an Okazaki fragment and must be condensed and packaged in some way to fit inside the nucleus (in a eukaryote) or cell (in a prokaryote). If the cell waited until the entire DNA molecule had been replicated, the newly synthesized lagging strand, in the form of many Okazaki fragments, might already be packaged and inaccessible to the endonuclease, polymerase, and ligase necessary to produce a continuous lagging strand.

14. (a) In all the variants, a neutral or negatively charged amino acid has been replaced with a positively charged amino acid (Lys or Arg). The +1 abbreviation means that one additional positive charge was introduced, a +2 indicates the introduction of two additional positive charges, etc. DNA is negatively charged because of its phosphodiester backbone. The protein engineers reasoned that an enzyme with an increased number of positive charges would bind more effectively to the negatively charged DNA.

(b) Intact, double-stranded DNA has a lower absorbance at 260 nm than single-stranded DNA. An increase in absorbance at 260 nm over time would be a useful measurement of the catalytic activity of DNase I, since the products of the reaction are short, single-stranded oligonucleotides.

(c) All of the variants have lower K_M values than the wild-type DNase I. This means that the DNase I variants bind more tightly to the DNA substrate than the wild-type enzyme. The tighter binding is no doubt due to the additional positive charges on the variants, which allows the formation of ion pairs between the variant enzymes and the DNA. There seems to be a rough correlation between the number of positively charged residues and the K_M value: An increase in the number of positively charged residues results in a lower K_M value, which indicates tighter binding. However, the V_{max} values must also be assessed. All of the variants (with the exception of a few variants with only one amino acid change) have a higher V_{max} value than the wild-type enzyme, indicating that the tighter binding leads to greater catalytic activity. Three amino acid changes seem to optimize K_M and V_{max} values. When four or five amino acids are replaced, the velocity is lower than when only three amino acids are replaced. Since the K_M value for the +4 and +5 variants is also generally smaller, it is possible that the enzyme and substrate bind to each other so tightly that catalytic efficiency is compromised.

(d) The pBR322 plasmid DNA normally exists in a supercoiled circle, as shown in the control lane. The wild-type DNase I can nick the DNA on one backbone to convert the plasmid to the relaxed circular DNA. The +3 mutant is the best of the three mutants in its ability to cleave supercoiled DNA. All three mutants are able to produce linear DNA, whereas the wild-type DNase I does not. This indicates that the variants can cut both strands, whereas the wild-type enzyme cuts only one strand.

(e) All of the mutants have a greater nicking activity than the wild type for low- and high-molecular-weight DNA at low concentrations.The +2 mutant is especially active under these conditions and would be a good choice to use for a lupus patient. The +1 and +2 mutants can degrade low-molecular-weight DNA at a high DNA concentration better than the wild-type enzyme, but The +3 and +4 mutants are not as active. Not all of the data are given, so it is difficult to make conclusions about the ability of the mutants to degrade high-molecular-weight DNA, but the +2 mutant is clearly more effective than the wild-type DNase I and would be a good choice to use for a CF patient.

(f) If three amino acids in DNase I are replaced with positively charged amino acids, the resulting variant is hyperactive: It catalyzes the reaction with a lower K_M and a greater V_{max}. The decreased K_M indicates that the enzyme has a high affinity for its substrate. The increased affinity is due to the replacement of the neutral or negatively charged amino acids with positively charged amino acids that can form ion pairs with the negatively charged phosphate groups on the DNA. Increased substrate affinity leads to an increase in the rate of hydrolysis of phosphodiester bonds. If the enzyme and the substrate bind more tightly, the enzyme might be able to make two cuts instead of one. For example, the +3, +4, and +6 mutants can convert supercoiled DNA to linear DNA. The +4, +5, and +6 mutants clearly have the ability to bind DNA, but their affinity for the substrate is perhaps too great, which causes the V_{max} values to decrease. The +2 mutants have a good ability to nick low-molecular-weight DNA even at low concentrations.

15. (a) Yes. By moving along a single DNA strand, the helicase can act as a wedge to push apart the double-stranded DNA ahead of it.

(b) The free energy of dTTP hydrolysis is similar to the free energy of ATP hydrolysis. Each hydrolysis reaction drives the helicase along two to three bases of DNA.

(c) The T7 helicase is probably a processive enzyme. Its hexameric ring structure is reminiscent of the clamp structure that promotes the processivity of DNA polymerase (see Fig. 17-12).

16. (a) The positively charged Mg^{2+} ion could decrease the affinity of the DNA's 3′ O atom for H, thereby increasing the nucleophilicity of the 3′ O atom (making it behave more like an O^- atom). The Mg^{2+} could also help neutralize the negative charge on the incoming nucleotide.

(b)

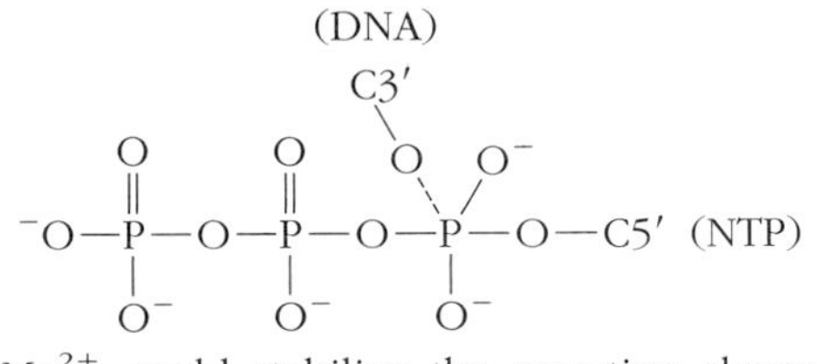

```
                    (DNA)
                     C3′
                       \
       O       O        O   O⁻
       ‖       ‖         \ /
⁻O—P—O—P—O—P—O—C5′ (NTP)
       |       |         |
       O⁻      O⁻        O⁻
```

The Mg^{2+} could stabilize the negative charge that develops on the pentacovalent transition state.

17. PP_i is the product of the polymerization reaction catalyzed by DNA polymerase. This reaction also requires a template DNA strand and a primer with a free 3′ end.

(a) There is no primer strand, so no PP_i is produced.
(b) There is no primer strand, so no PP_i is produced.
(c) PP_i is produced.
(d) No PP_i is produced because there is no 3′ end that can be extended.
(e) PP_i is produced.
(f) PP_i is produced.

18. (a) The polymerases that lack $3' \rightarrow 5'$ exonuclease activity have higher error rates than polymerases that contain the exonuclease activity.
(b) α, once every 6250 bases; β, once every 1493 bases; δ, once every 100,000 bases; ε, once every 100,000 bases; η, once every 29 bases.
(c) Polymerase η and HIV RT incorporate the correct base with approximately the same efficiency (for example, 420, 760, and 800 $\mu M \cdot min^{-1} \times 10^3$). However, polymerase η incorporates mispaired bases much more efficiently than HIV RT does (22, 1.6, and 8.7 vs. 0.07 $\mu M \cdot min^{-1} \times 10^3$). These results indicate that the high error rate of polymerase η results from its ability to incorporate the wrong base rather than its inability to incorporate the correct base.
(d) Overexpression of an error-prone DNA polymerase similar to polymerase η would increase the mutation rate.

19. (a)

(b)

20. (a) The nonstandard amino acid is homocysteine, which can accept a methyl group donated by methyl-tetrahydrofolate to regenerate methionine (see Section 15-3).

COO^-
$^+H_3N—C—H$
CH_2
CH_2
SH

(b) The other product is methanol, $HO—CH_3$.

Chapter 18 Solutions

1. If the kinase functions as a proto-oncogene, its aberrant activation could lead to cell transformation. The activity of the kinase is reversed by the activity of the phosphatase. The phosphatase is a tumor suppressor because its loss could lead to cell transformation.

2. Cancer results from damage to DNA, for example, from environmental agents (UV light or hydroxyl radicals) or viral infections. DNA damage most likely accumulates over time (more quickly if there is already an inherited genetic defect). Consequently, the cells of older individuals have more DNA mutations and are more likely than the cells of younger individuals to undergo transformation.

3. The triphosphatase removes nucleotides containing the modified base before they can be incorporated into DNA during replication.

4. A damaged unicellular organism does not harm others of its kind; it can go on to proliferate or die according to natural selection. In a multicellular organism, a runaway cell could threaten the well-being of the whole organism. Thus, apoptosis is necessary only in multicellular organisms.

5.

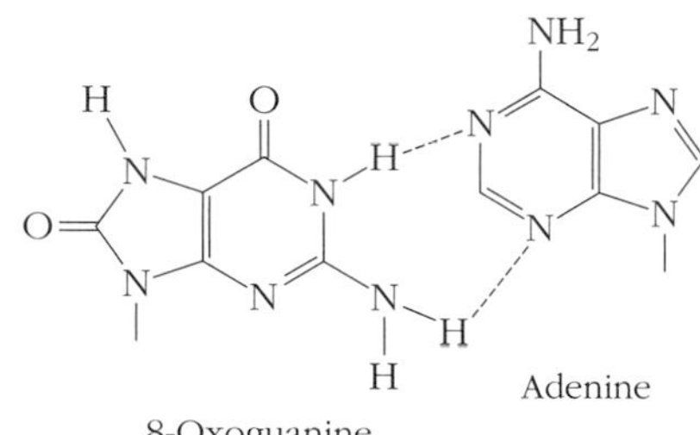

6. The O^6-methylguanine produced by the methylation of guanine produces a residue that can base-pair with either cytosine or thymine. If the O^6-methylguanine residue base-pairs with thymine, the G:C base pair will eventually be changed to an A:T base pair.

7. The structures of the intercalating agents resemble A:T and G:C base pairs, which explains why they are able to slip in between the stacked base pairs of DNA. This has the effect of creating what appears to the replication machinery as an "extra" base pair. An extra base incorporated into the newly synthesized DNA may eventually lead to a frameshift mutation (in which the additional nucleotide causes the translation apparatus to read a different set of successive three-nucleotide codons).

8. Bromouracil causes an A:T to G:C transition.

5-Bromouracil (keto tautomer)
5-Bromouracil (enol tautomer)
Guanine

9. The ultraviolet light is an additional precaution against contamination. Ultraviolet light causes the formation of

thymine dimers in bacteria. High levels of exposure to ultraviolet light would overwhelm the bacteria's ability to repair the dimers, which would result in the eventual death of the bacteria. Thus, the ultraviolet light helps keep the hood space free from bacteria that could contaminate the cultured cells.

10. (a)

Hypoxanthine

(b)

Hypoxanthine Cytosine

(c) An A:T base pair is converted to a C:G base pair (a transition; see Problem 8).

11. (a)

Xanthine

(b) Since cytosine also pairs with guanine, the deamination of guanine to xanthine does not induce a mutation.

12. All of these deaminations produce bases that are foreign to DNA; therefore they can be quickly spotted and repaired before DNA has replicated and the damage is passed on to the next generation.

13. (a)

5-Methylcytosine → Thymine

(b) Thymine

(c) A C:G to T:A transition occurs.

(d) The cell cannot repair the deaminated 5-methylcytosine since the resulting base is indistinguishable from thymine that occurs normally.

14. (a)

Cytosine 2-Aminopurine

(b)

15. (a)

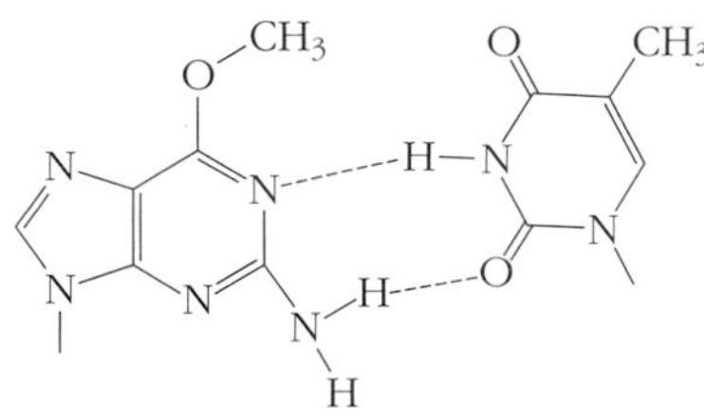

O^6-Methylguanine Thymine

(b) The methylation of guanine causes a G:C to A:T transition.

(c)

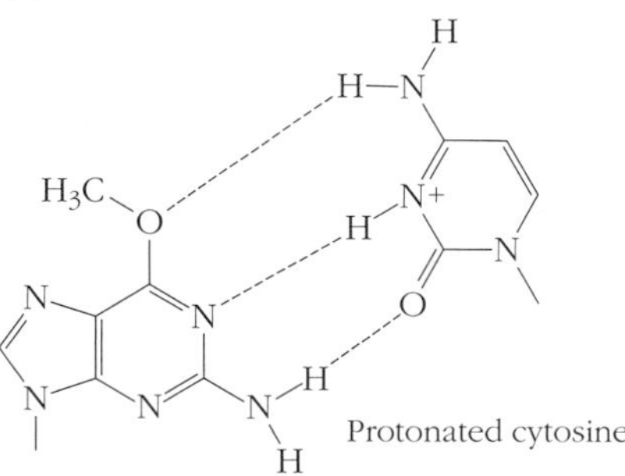

Protonated cytosine

O^6-Methylguanine

16. 5-Fluorouracil, like 5-bromouracil (see Problem 8), is a thymine analog and can be incorporated into the DNA of cancer cells. This causes a high rate of mutation, which eventually kills the cancer cells. Rapidly growing cancer cells carry out DNA replication at a higher rate than normal skin cells and are therefore more susceptible to the drug.

17. The mutant bacteria are unable to repair deaminated cytosine (uracil). In these cells, the rate of change of G:C base pairs to A:T base pairs is much greater than normal.

18. Most likely, the thymine–thymine dimer, since this lesion forms upon exposure of the DNA to ultraviolet light.

19. The amino acid code is degenerate so that several codons may specify the same amino acid. For example, the codons GCA, GCC, GCG, and GCU all code for alanine. So if a mutation occurred at the third position (the 3′ end) of the codon, alanine would still be incorporated into the protein.

20. DNA polymerase III replicates DNA until a thymine dimer is encountered. Polymerase III is accurate but cannot quickly bypass the damage. Polymerase V, which can more quickly proceed through the damaged site, does so, but at the cost of misincorporating G rather than A opposite T. Thus, replication can continue at a high rate. The tendency for DNA polymerase V to continue to introduce errors is minimized by its low processivity: Soon after passing the thymine dimer, it dissociates, and the more accurate polymerase III can continue replicating the DNA with high fidelity.

Chapter 19 Solutions

1. Organizing genes with related functions in an operon means that the genes can be turned on and off together. Furthermore, because the genes are transcribed as a unit, the products can be generated in roughly equivalent concentrations and in the same region of the cell (where the ribosome is located). Consequently, only one signal is needed to activate or deactivate the metabolic function carried out by the products of the operon's genes. A eukaryotic cell can coordinate the transcription of physically separated but functionally related genes if they are part of a synexpression group. Although the genes are not transcribed as a single mRNA molecule, as in an operon, they can be expressed at the same time and in roughly equivalent amounts through the use of shared regulatory mechanisms involving enhancers, silencers, activators, or repressors.

2. The accurate transmission of genetic information from one generation to the next requires a high degree of fidelity in DNA replication. A higher rate of error in RNA transcription is permitted because the cell's survival usually does not depend on accurately synthesized RNA. If translated, an RNA transcript containing an error may lead to a defective protein, which is likely to be destroyed by the cell before it can do much damage. The gene can be transcribed again and again to generate accurate transcripts.

3. RNA polymerase I transcribes rRNA genes. Although there are multiple copies of these genes, there are only a few types of rRNA molecules and the cell requires them in equal amounts. Therefore, rRNA gene transcription does not need to be selective, and one promoter suffices. In contrast, the protein-coding genes transcribed by RNA polymerase II must be expressed at different times and at different levels. The regulation of transcription of these genes is more elaborate, requiring different promoter sequences that can potentially interact with different sets of transcription factors.

4. Sequence-specific interactions require contact with the bases of DNA, which can participate in hydrogen bonding and van der Waals interactions with protein groups. Electrostatic interactions involve the ionic phosphate groups of the DNA backbone and are therefore sequence independent.

5. (a) The polymerase binds most tightly to the DNA segment with the largest bulge. This DNA mimics the transcription bubble, in which the DNA strands have already been separated.

(b) Since K_d is a dissociation constant, the apparent equilibrium constant for the binding reaction is $1/K_d$. Equation 9-2 gives the relationship between ΔG and K:

$$\Delta G = -RT \ln K \text{ or } \Delta G = -RT \ln(1/K_d)$$

(c) For double-stranded DNA,

$$\Delta G = -(8.3145 \text{ J} \cdot \text{K}^{-1} \cdot \text{mol}^{-1})(298\text{K}) \ln(1/315 \times 10^{-9}) = -37 \text{ kJ} \cdot \text{mol}^{-1}$$

For the eight-base bulge,

$$\Delta G = -(8.3145 \text{ J} \cdot \text{K}^{-1} \cdot \text{mol}^{-1})(298\text{K}) \ln(1/1.3 \times 10^{-12}) = -68 \text{ kJ} \cdot \text{mol}^{-1}$$

Polymerase binding to the melted DNA is more favorable than binding to double-stranded DNA.

(d) Melting open a DNA helix is thermodynamically unfavorable. Some of the favorable free energy of binding the polymerase to the DNA is spent in forming the transcription bubble. When the transcription bubble is preformed (for example, in the DNA with an eight-base bulge), this energy is not spent and is reflected in the apparent energy of polymerase binding. The difference in ΔG values for polymerase binding to double-stranded DNA and to the eight-base bulge is $-68 - (-37) = -31$ kJ · mol^{-1}. This value estimates the free energy cost (+31 kJ · mol^{-1}) of melting open eight base pairs of DNA.

(e) The AT-rich sequence is easier to melt open than the GC-rich sequence because GC-rich DNA experiences stronger stacking interactions.

6. Cordycepin, which resembles adenosine, can be phosphorylated and used as a substrate by RNA polymerase. However, it blocks further RNA polymerization because it lacks a 3′ OH group.

7. (a) $\text{mRNA}_{(n \text{ residues})} + \text{P}_i \rightarrow \text{NDP} + \text{mRNA}_{(n-1 \text{ residues})}$

(b) The reverse of the phosphorolysis reaction is an RNA polymerization reaction. PNPase uses an NDP substrate to extend the RNA by one nucleotide residue and releases P_i. RNA polymerase uses an NTP substrate and releases PP_i.

(c) High processivity would allow the exonuclease to rapidly degrade mRNA molecules. This would be important in cases where the gene product was no longer needed. An mRNA that was degraded more slowly could potentially continue to be translated.

8. The presence of the σ factor decreases the affinity of RNA polymerase for DNA. This allows the polymerase–σ factor complex to quickly scan long segments of DNA for promoter sequences. Once transcription has begun, the σ factor is no longer needed and dissociates from the enzyme. Now the RNA polymerase has a high affinity for DNA, which helps keep it associated with the template during transcription.

9. GUCCGAUCGAAUGCAUG

10. The protein must identify a gene to be expressed and must activate the transcription machinery so that the DNA can be transcribed to RNA. The gene is identified by the ability of the protein to bind specific DNA sequences associated with the gene. The resulting distortion in the DNA can then be recognized in a sequence-independent manner by transcription factors that assemble at the site in order to initiate RNA synthesis.

11. (a) The positively charged Lys residues most likely interact electrostatically with the negatively charged phosphate groups of the enzyme's substrate.
(b) The substrate RNA–DNA hybrid helix must have a conformation intermediate to that of double-stranded DNA (B helix) and double-stranded RNA (A helix).

12. (a) CAG codes for Gln. The resulting protein would contain a series of extra Gln residues. These polar residues would most likely be located on the protein surface but could interfere with protein folding, stability, interactions with other proteins, and catalytic activity.
(b) The longer transcripts could be due to transcription initiating upstream of the normal site or failing to stop at the usual termination point. Longer mRNA molecules could also result from the addition of an abnormally long poly(A) tail or the failure to undergo splicing.
(c)

```
       G C
     A     A
      C---G
      G---C
      A   A
      C---G
      G---C
      A   A
      C---G
 · · ·CAG---CAG· · ·
```

13. The capped mRNA has a 5′–5′ triphosphate linkage, which is not recognized by exonucleases (which normally cleave 5′–3′ phosphodiester bonds). Capping must occur as soon as the 5′ end of the mRNA emerges from the RNA polymerase so that the message will not be degraded by exonucleases.

14. (a) The active site of poly(A) polymerase is narrower because it does not need to accommodate a template strand.
(b) Poly(A) polymerase binds only ATP, whereas other RNA polymerases bind ATP, CTP, GTP, and UTP.

15. (a)

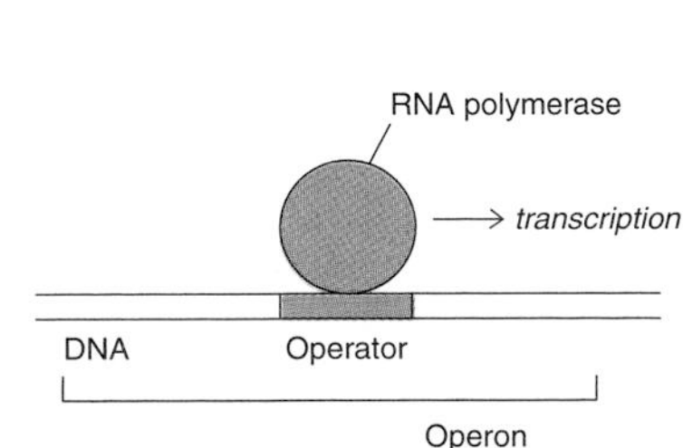

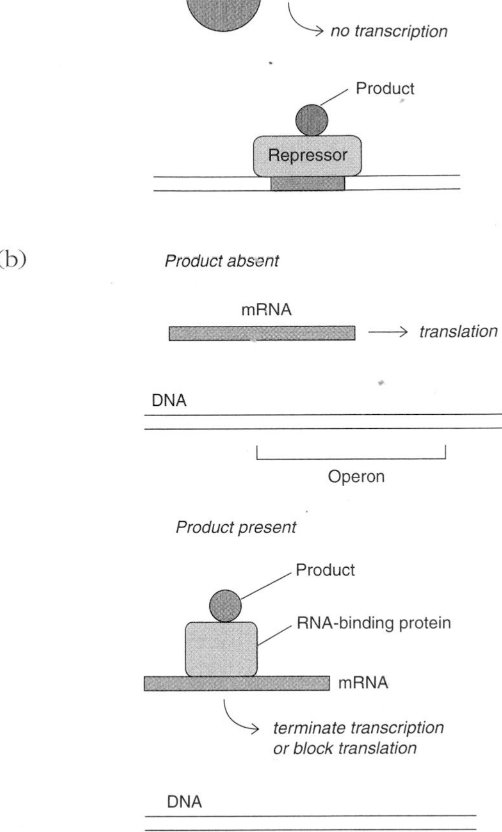

(b)

(c) If no protein were involved, the operon's product must interact directly with the mRNA.
(d)

```
       U A
     U     A
     U     G
      G---C
      A---U
      C---G
      U---A
 · · ·GAU---AGC
```

(e) Researchers could probe the conformation of an RNA molecule in the presence and absence of FAD by monitoring a conformation-dependent property of the RNA molecule, such as its susceptibility to an endonuclease.
(f) The FMN component of FAD is the most effective, since it has the lowest dissociation constant. The phosphate group is important for RNA binding, since riboflavin, which lacks a phosphate group, binds much less tightly.

16. The free energy change for the reaction must be close to zero. In excising the intron, two phosphodiester bonds are broken (one at each intron–exon junction). However, two new phosphodiester bonds are formed (one to join the exons and one 2′–5′ bond in the intron).

17. (a) The phosphate groups of the phosphodiester backbone of RNA will be labeled wherever α-[^{32}P]-ATP is used as a substrate by RNA polymerase.

(b) ^{32}P will appear only at the 5′ end of RNA molecules that have A as the first residue (this residue retains its α and β phosphates). In all other cases where β-[^{32}P]-ATP is used as a substrate for RNA synthesis, the β and γ phosphates are released as PP_i (see Fig. 17-9).

(c) No ^{32}P will appear in the RNA chain. During polymerization, the β and γ phosphates are released as PP_i. The terminal (γ) phosphate of an A residue at the 5′ end of an RNA molecule is removed during the capping process (see Section 19-3).

18. The splicing reactions are mediated by the spliceosome, a large RNA–protein complex. The intron must be large enough to include spliceosome binding site(s). In addition, the formation of a lariat-shaped intermediate (see Fig. 19-21) requires a segment of RNA long enough to curl back on itself without strain.

19. Although they catalyze their own excision from an RNA precursor, self-splicing introns are not truly catalytic because they emerge from the reaction in a different form (a true catalyst is fully regenerated) and they act only once (a true catalyst can promote multiple rounds of a reaction).

20. (a) Treat the ribonucleoprotein with proteases to destroy the protein component and test the remaining RNA component for endonuclease activity. Alternatively, use an *in vitro* transcription system to generate the RNA, purify it, and test it for activity.

(b) U69 is more important for catalysis, since deleting this residue or substituting it with another residue generally resulted in a dramatic decrease in the catalytic constant (k) but had a modest effect on K_d.

(c) Increasing the size of the bulge dramatically decreased substrate binding affinity, suggesting that although the presence of the bulge is necessary for catalysis, the geometry of the bulge is important for substrate binding.

21. The conformation of DNA is constrained by the requirement of forming a double-helical structure. Single-stranded RNA, in contrast, can fold into complex three-dimensional shapes that are better suited for catalytic activities such as binding substrate molecules, providing chemically reactive groups, and stabilizing transition states. In addition, the 2′ OH group of RNA (absent in DNA) participates in certain transesterification reactions, such as splicing (see Fig. 19-21).

Chapter 20 Solutions

1. (a) The RNA-binding site of the synthetase is necessary for binding tRNA for aminoacylation and for binding RNA during splicing.

(b) Because of their long evolutionary history, aminoacyl–tRNA synthetases have had many opportunities to diversify in structure and in function in order to assume additional roles in the cell.

2. The 5′ nucleotide is at the wobble position, which can participate in non–Watson-Crick base-pairings with the 3′ nucleotide of an mRNA codon. Because the first two codon positions are more important for specifying an amino acid (see Table 20-1), wobble at the third position may not affect translation.

3.

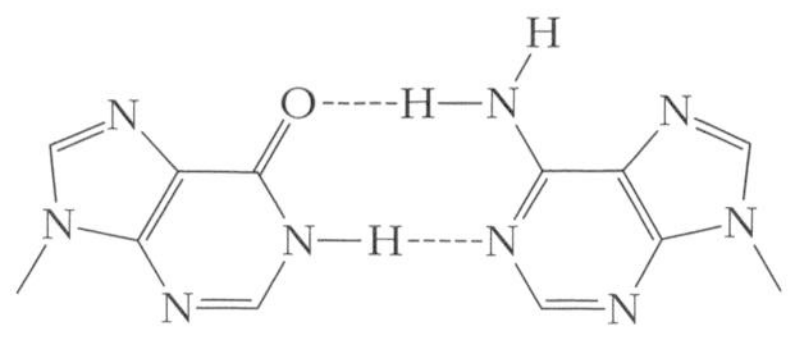

O----H—N (H)

N—H----N

I:A

4. If the adenosine of an mRNA codon were changed to inosine, then the codon could pair with a tRNA anticodon containing guanosine rather than uridine. This could result in the incorporation of a different amino acid at that codon position. A given gene could give rise to several different polypeptide products, depending on how many A residues were edited to I residues and how many of these changes resulted in amino acid substitutions.

5. The mRNA has the sequence

CGAUA**AUG**UCCGACCAAGCGAUCUCG**UAG**CA

The start codon and stop codon are highlighted. The encoded protein has the sequence

Met–Ser–Asp–Gln–Ala–Ile–Ser

6. The two Lys codons are AAA and AAG. Substitution with C would yield CAA and CAG, which code for Gln; substitution with G would yield GAA and GAG, which code for Glu, and substitution with U would yield AUU and UAG, which are stop codons. Replacing a Lys codon with a stop codon would terminate protein synthesis prematurely, most likely producing a nonfunctional protein. Replacing Lys with Glu or Gln could disrupt the protein's structure and therefore its function if the Lys residue were involved in a structurally essential interaction such as an ion pair in the protein interior. If the Lys residue were on the surface of the protein, replacing it with Glu or Gln, both of which are hydrophilic, might not have much impact on the protein's structure or function.

7. (a)

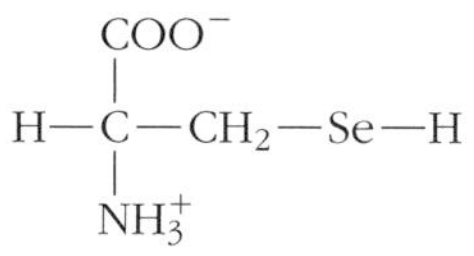

$$\begin{array}{c} COO^- \\ | \\ H—C—CH_2—Se—H \\ | \\ NH_3^+ \end{array}$$

Selenocysteine

(b) The mRNA codon is UGA.

(c) The UGA codon normally functions as a stop codon (see Table 20-1). In order for this codon to be read as a selenocysteine codon, the ribosome must recognize that it is not a stop codon, most likely through clues provided by the local mRNA sequence and/or structure.

8. Gly and Ala; Val and Leu; Ser and Thr; Asn and Gln; and Asp and Glu.

9. (a) The enzyme itself must act as a template to direct the addition of two C residues followed by one A residue.

(b) The enzyme must recognize only ATP and CTP as substrates, excluding GTP, UTP, and all dNTPs.

(c) In the two-domain enzyme, one polymerase adds the two C residues, and the other domain then adds the terminal A residue.

10. The correctly charged tRNAs (Ala–$tRNA^{Ala}$ and Gln–$tRNA^{Gln}$) bind to EF-Tu with approximately the same affinity, so they are delivered to the ribosomal A site with the same efficiency. The mischarged Gln–$tRNA^{Ala}$ binds to EF-Tu more loosely, indicating that it may dissociate from EF-Tu before it reaches the ribosome. The mischarged Ala–$tRNA^{Gln}$ binds to EF-Tu much more tightly, indicating that EF-Tu may not be able to dissociate from it at the ribosome. These results suggest that either a higher or a lower binding affinity could affect the ability of EF-Tu to carry out its function, which would decrease the rate at which mischarged aminoacyl–tRNAs bind to the ribosomal A site during translation.

11. The assembly of functional ribosomes requires equal amounts of the rRNA molecules. Therefore, it is advantageous for the cell to synthesize the rRNAs all at once.

12. In order for a cell to incorporate a nonstandard amino acid into a polypeptide, the amino acid must first be attached to a tRNA corresponding to one of the 20 standard amino acids. The aminoacyl–tRNA can then bind to the ribosome and its amino acid can be incorporated into the growing polypeptide at positions corresponding to the codon for the standard amino acid. The failure of cells to synthesize norleucine-containing peptides most likely reflects the inability of LeuRS to efficiently attach norleucine to $tRNA^{Leu}$. A mutant LeuRS, which presumably lacks the proofreading activity of the wild-type LeuRS, was able to produce norleucine–$tRNA^{Leu}$, and the cells' ribosomes used this aminoacylated tRNA to translate Leu codons.

13. The ribosome minimizes the chances of misreading the A-site codon by binding the A-site tRNA with lower affinity. If the tRNA bound with higher affinity, it would be less likely to dissociate as part of a kinetic proofreading mechanism.

14. In 16S rRNA, A1492 and A1493 act as a sensor to distinguish correctly and incorrectly paired codons and anticodons. tRNA binding triggers a conformational change in the rRNA that allows A1492 and A1493 to form hydrogen bonds with an mRNA that has correctly base-paired with a tRNA anticodon in the A site. Changing one of these two rRNA residues would inactivate the translational proofreading mechanism by eliminating the specific hydrogen bonding between the rRNA and the mRNA. As a result, incorrectly paired tRNAs could not be distinguished from correctly paired tRNAs, and the error rate of translation would increase.

15. In a living cell, EF-Tu and EF-G enhance the rate of protein synthesis by rendering various steps of translation irreversible. They also promote the accuracy of protein synthesis through kinetic proofreading. In the absence of the elongation factors, translation would be too slow and too inaccurate to support life. These constraints do not apply to an *in vitro* translation system, which can proceed in the absence of EF-Tu and EF-G. However, the resulting protein is likely to contain more misincorporated amino acids than a protein synthesized in a cell.

16. If EF-Tu formed a complex with fMet–$tRNA_f^{Met}$, the fMet–tRNA could be delivered to the ribosomal A site when a Met codon was positioned there. However, transpeptidation could not occur because the amino group of fMet is blocked by the formyl group. Polypeptide synthesis would be halted until the fMet–$tRNA_f^{Met}$ was replaced by Met–$tRNA^{Met}$ in the A site.

17. The number of phosphoanhydride bonds (about 30 kJ · mol^{-1} each) that are cleaved in order to synthesize a 20-residue polypeptide can be calculated as follows (the relevant ATP- or GTP-hydrolyzing proteins are indicated in parentheses):

Aminoacylation (AARS)	2 × 20 ATP
Translation initiation (IF2)	1 GTP
Positioning of each aminoacyl–tRNA (EF-Tu)	19 GTP
Translocation after each transpeptidation (EF-G)	19 GTP
Termination (RF3)	1 GTP
Total	80 ATP equivalents

Thus, approximately 80 × 30 kJ · mol^{-1}, or 2400 kJ, is required. In a cell, proofreading during aminoacylation and during translation requires the hydrolysis of additional phosphoanhydride bonds, making the cost of accurately synthesizing the 20-residue polypeptide greater than 2400 kJ · mol^{-1}.

18. In prokaryotes, both mRNA and protein synthesis take place in the cytosol, so a ribosome can assemble on the 5′ end of an mRNA even while RNA polymerase is synthesizing the 3′ end of the transcript. In eukaryotes, RNA is produced in the nucleus, but ribosomes are located in the cytosol. Because transcription and translation occur in separate compartments, they cannot occur simultaneously. A eukaryotic mRNA must be transported from the nucleus to the cytosol before it can be translated.

19. If a peptidyl–tRNA dissociates from the ribosome during translation, the hydrolase releases the peptide from the tRNA. Because peptide synthesis is prematurely terminated, the polypeptide is likely to be nonfunctional, and its amino acids must be recycled. Similarly, the tRNA, once released from the peptidyl group, can be reused. The essential nature of the peptidyl–tRNA hydrolase suggests that ribosomes that have initiated translation sometimes stop translating before reaching a stop codon.

20. (a) Transpeptidation involves the nucleophilic attack of the amino group of the aminoacyl–tRNA on the carbonyl carbon of the peptidyl–tRNA (see Fig. 20-16). The higher the pH, the more nucleophilic the amino group (the less likely it is to be protonated).

(b)

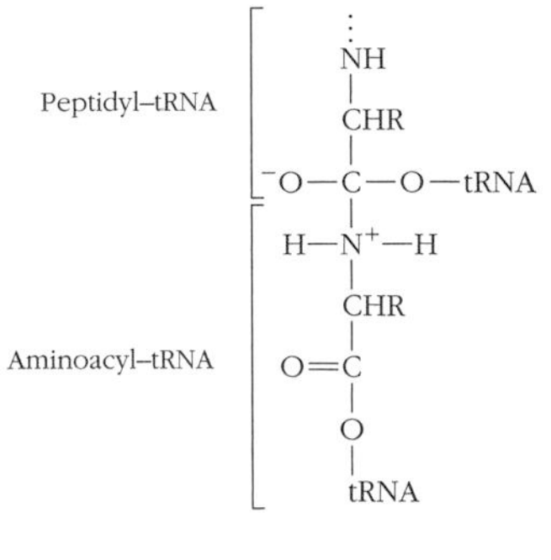

Peptidyl–tRNA:
⋮
NH
|
CHR
|
$^-$O—C—O—tRNA
|
Aminoacyl–tRNA:
H—N^+—H
|
CHR
|
O=C
|
O
|
tRNA

(c)

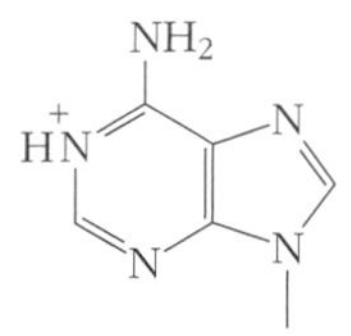

$$\text{protonated adenine: } \overset{+}{\text{HN}} \text{ (N1) with } NH_2 \text{ at C6, attached at N9}$$

The protonated A2451 residue, which has a positive charge, could stabilize the negatively charged oxyanion of the tetrahedral reaction intermediate. As the pH increases, A2451 would be less likely to be protonated, so this catalytic mechanism would be less effective at higher pH.

21. (a) The mutation would allow the aminoacylated tRNA rather than a release factor to enter the ribosome and pair with a stop codon. The result would be incorporation of an amino acid into a polypeptide rather than translation termination, so the ribosome would continue to read mRNA codons and produce elongated polypeptides. The inability of the mutated tRNA to recognize its amino acid–specifying codons would have a minor impact on protein synthesis, since the cell likely contains other isoacceptor tRNAs that can recognize the same codons.

(b) Not all proteins would be affected. Only proteins whose genes include the stop codon that is read by the mutated tRNA would be affected. Proteins whose genes include one of the other two stop codons would be synthesized normally.

(c) Aminoacyl–tRNA synthetases usually recognize both the anticodon and acceptor ends of their tRNA substrates. A mutation in the tRNA anticodon, such as a nonsense suppressor mutation, might interfere with tRNA recognition, so that the mutated tRNA molecule might not undergo aminoacylation. This would minimize the ability of the mutated tRNA to insert the amino acid at a position corresponding to a stop codon.

22. (a) Introduction of a stop codon would terminate protein synthesis prematurely, so no functional β-galactosidase would be synthesized. If the stop codon were located near the C-terminus of the protein, the polypeptide would be shorter than normal but might still retain activity, if its active site region were intact.

(b) These results indicate that the oxazolidinone increases the ability of ribosomes to overlook the stop codon and synthesize β-galactosidase polypeptides of normal size.

(c) β-Galactosidase activity would be extremely low because a nucleotide insertion or deletion changes the reading frame for translation. The resulting polypeptide would have a different amino acid sequence and would therefore be nonfunctional.

(d) These results indicate that the oxazolidinone promotes frameshifting in the ribosome so that despite the insertion or deletion, the correct reading frame is occasionally translated and some functional β-galactosidase is synthesized.

(e) There are two possible Glu codons: GAA and GAG. Changing a single base in these codons generates codons specifying four different amino acids. Substitutions at the second position yield codons for Ala (GCA and GCG), Gly (GGA and GGG), and Val (GUA and GUG). Substitution at the first position yields codons for Gln (CAA and CAG).

(f) The oxazolidinone does not promote codon misreading, that is, incorporation of an amino acid other than the one specified by a codon. If codon misreading were occurring, the encoded Ala, Gln, Gly, and Val codons would occasionally be read as Glu codons, and a functional enzyme would result.

(g) The ability of the oxazolidinone to cause the ribosome to read through stop codons would result in the synthesis of longer-than-normal polypeptides. These proteins might not fold properly or might not be catalytically active, which would interfere with normal cellular metabolism. The ability of the oxazolidinone to promote frameshifting would result in polypeptides with garbled amino acid sequences, which would be nonfunctional and potentially toxic to the cell.

(h) No; the oxazolidinone affects translational accuracy, which is primarily a function of the 30S ribosomal subunit, the location of the binding sites for mRNA and the tRNA anticodon loop. A binding site near the peptidyltransferase site would be more consistent with an effect on peptide bond formation. Presumably, communication between the 50S and 30S subunits allows binding at one site to influence events at another site.

23. (a) The α chains, when in excess, combine with all available β chains to form functional $\alpha_2\beta_2$ hemoglobin, thereby minimizing the formation of nonfunctional β_4 hemoglobin.

(b) The protein helps prevent the precipitation of the α chains that have not yet paired with β chains.

(c) When a deficiency of β chains is coupled with an excess of α chains, the α chains precipitate and destroy the red blood cells, worsening the anemia that results from the lack of β chains.

(d) The imbalance between the amounts of α and β chains is minimized when the synthesis of both globins is depressed due to mutations in both an α globin gene and a β globin gene.

(e) The process of initiating translation requires that eIF2 hydrolyze its bound GTP to GDP and P_i. In order for the protein to participate in subsequent translation initiation events, its GDP must be replaced with GTP. If this exchange does not occur, reinitiation is not possible, and protein synthesis comes to a halt.

(f) Heme prevents the phosphorylation of eIF2 so that translation initiation can proceed. This mechanism regulates the level of globin synthesis according to the availability of heme. Consequently, the cells can produce functional hemoglobin, which contains globin polypeptides as well as heme prosthetic groups.

INDEX

Page references followed by T indicate tables. Page references followed by F indicate figures.

AMINO ACID STRUCTURES AND ABBREVIATIONS

Hydrophobic amino acids

Alanine (Ala, A)

Valine (Val, V)

Phenylalanine (Phe, F)

Tryptophan (Trp, W)

Leucine (Leu, L)

Isoleucine (Ile, I)

Methionine (Met, M)

Proline (Pro, P)

Polar amino acids

Serine (Ser, S)

Threonine (Thr, T)

Tyrosine (Tyr, Y)

Cysteine (Cys, C)

Asparagine (Asn, N)

Glutamine (Gln, Q)

Histidine (His, H)

Glycine (Gly, G)

Charged amino acids

Aspartate (Asp, D)

Glutamate (Glu, E)

Lysine (Lys, K)

Arginine (Arg, R)

ESSENTIAL BIOCHEMISTRY

ESSENTIAL BIOCHEMISTRY

CHARLOTTE W. PRATT

Seattle Pacific University

KATHLEEN CORNELY

Providence College

John Wiley & Sons, Inc.

About the cover

Cytochrome *c*, a small protein whose structure, function, and evolution have been thoroughly studied, symbolizes the major role of proteins in biochemistry. In addition, cytochrome *c* and its bound heme group participate in a central pathway for energy transduction. To this background is added the small molecule ATP, the energy currency of all living cells and, as a nucleotide, a representative of another category of ancient and essential biological molecules.

Senior Acquisitions Editor	Patrick Fitzgerald
Senior Development Editor	Ellen Ford
Marketing Manager	Robert Smith
Production Editor	Sandra Dumas
Senior Designer	Kevin Murphy
Illustration Editor	Anna Melhorn
Production Management Services	Suzanne Ingrao/Ingrao Associates
Photo Editor	Hilary Newman
Photo Researcher	Teri Stratford

This book was typeset in 10/12 Garamond Light by TechBooks/GTS Companies, Inc. and printed and bound by Von Hoffmann Corporation, Inc. The cover was printed by Von Hoffmann Corporation.

The paper in this book was manufactured by a mill whose forest management programs include sustained yield harvesting of its timberlands. Sustained yield harvesting principles ensure that the number of trees cut each year does not exceed the amount of new growth.

This book is printed on acid-free paper. ♾

Pratt, Charlotte, W., Cornely, Kathleen
Essential Biochemistry

0-471-39387-8
0-471-45166-5 WIE ISBN

Printed in the United States of America.

10 9 8 7 6 5 4 3 2 1